TM 9-2320-272-24P-1
5 Ton M939 Series Truck
Direct and General Support
Maintenance Manual
Repair Parts and Special Tools List
Vol 1 of 2
February 1999

This manual contains parts and special tool information for the 5 ton M939 US Military Trucks. This is volume 1 of 2 in the Direct / General Support Manual Parts and Special Tools Series. M939 series trucks are a 5 ton heavy duty 6x6 truck. Cargo versions were designed to transport 10,000 pounds of cargo in all terrain and all weather conditions. Originally designed in the 1970's to replace the M39 and M809 series of vehicles. 44,590 units were produced. This manual is printed to help private owners in the maintenance of their vehicles.

Should you have suggestions or feedback on ways to improve this book please send email to Books@OcotilloPress.com

Edited 2021 Ocotillo Press
ISBN 978-1-954285-69-9

Ocotillo Press
Houston, TX 77017
Books@OcotilloPress.com

TECHNICAL MANUAL
UNIT, DIRECT SUPPORT, AND GENERAL SUPPORT MAINTENANCE
REPAIR PARTS AND SPECIAL TOOLS LIST
FOR
TRUCK, 5-TON, 6X6, M939, M939A1, M939A2
SERIES TRUCKS (DIESEL)

FEBRUARY 1999

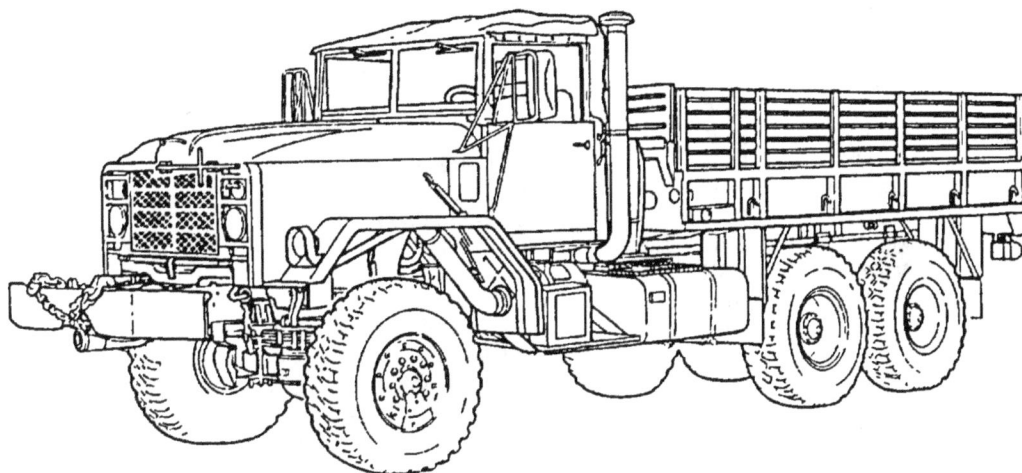

TRUCK, CARGO: 5-TON, 6X6, DROPSIDE,
M923 (2320-01-050-2084) (EIC: BRY); M923A1 (2320-01-206-4087) (EIC: BSS); M923A2 (2320-01-230-0307) (EIC: BS7);
M925 (2320-01-047-8769) (EIC: BRT); M925A1 (2320-01-206-4088) (EIC: BST); M925A2 (2320-01-230-0308) (EIC: BS8);

TRUCK, CARGO: 5-TON, 6X6 XLWB,
M927 (2320-01-047-8771) (EIC: BRV); M927A1 (2320-01-206-4089) (EIC: BSW); M927A2 (2320-01-230-0309) (EIC: BS9);
M928 (2320-01-047-8770) (EIC: BRU); M928A1 (2320-01-206-4090) (EIC: BSX); M928A2 (2320-01-230-0310) (EIC: BTM);

TRUCK, DUMP: 5-TON, 6X6,
M929 (2320-01-047-8756) (EIC: BTH); M929A1 (2320-01-206-4079) (EIC: BSY); M929A2 (2320-01-230-0305) (EIC: BTN);
M930 (2320-01-047-8755) (EIC: BTG); M930A1 (2320-01-206-4080) (EIC: BSZ); M930A2 (2320-01-230-0306) (EIC: BTO);

TRUCK, TRACTOR: 5-TON, 6X6,
M931 (2320-01-047-8753) (EIC: BTE); M931A1 (2320-01-206-4077) (EIC: BS2); M931A2 (2320-01-230-0302) (EIC: BTP);
M932 (2320-01-047-8752) (EIC: BTD); M932A1 (2320-01-205-2684) (EIC: BS55); M932A2 (2320-01-230-0303) (EIC: BTQ);

TRUCK, VAN, EXPANSIBLE: 5-TON, 6X6,
M934 (2320-01-047-8750) (EIC: BTB); M934A1 (2320-01-205-2682) (EIC: BS4); M934A2 (2320-01-230-0300) (EIC: BTR);

TRUCK, MEDIUM WRECKER: 5-TON, 6X6,
M936 (2320-01-047-8754) (EIC: BTF); M936A1 (2320-01-206-4078) (EIC: BS6); M936A2 (2320-01-230-0304) (EIC: BTT).

DEPARTMENTS OF THE ARMY AND THE AIR FORCE

TECHNICAL MANUAL
NO. 9-2320-272-24P-1

TECHNICAL ORDER
NO. 36A12-1C-1155-4-1

HEADQUARTERS
DEPARTMENTS OF THE ARMY AND THE AIR FORCE
Washington D.C., 1 February 1999

UNIT, DIRECT SUPPORT, AND GENERAL SUPPORT MAINTENANCE
REPAIR PARTS AND SPECIAL TOOLS LIST
FOR
TRUCK, 5-TON, 6X6, M939, M939A1, AND M939A2 SERIES TRUCKS (DIESEL)

TRUCK	MODEL	EIC	NSN WITHOUT WINCH	NSN WITH WINCH
Cargo, Dropside	M923	BRY	2320-01-050-2084	
Cargo, Dropside	M923A1	BSS	2320-01-206-4087	
Cargo, Dropside	M923A2	BS7	2320-01-230-0307	
Cargo, Dropside	M925	BRT		2320-01-047-8769
Cargo, Dropside	M925A1	BST		2320-01-206-4088
Cargo, Dropside	M925A2	BS8		2320-01-230-0308
Cargo	M927	BRV	2320-01-047-8771	
Cargo	M927A1	BSW	2320-01-206-4089	
Cargo	M927A2	BS9	2320-01-230-0309	
Cargo	M928	BRU		2320-01-047-8770
Cargo	M928A1	BSX		2320-01-206-4090
Cargo	M928A2	BTM		2320-01-230-0310
Dump	M929	BTH	2320-01-047-8756	
Dump	M929A1	BSY	2320-01-206-4079	
Dump	M929A2	BTN	2320-01-230-0305	
Dump	M930	BTG		2320-01-047-8755
Dump	M930A1	BSZ		2320-01-206-4080
Dump	M930A2	BTO		2320-01-230-0306
Tractor	M931	BTE	2320-01-047-8753	
Tractor	M931A1	BS2	2320-01-206-4077	
Tractor	M931A2	BTP	2320-01-230-0302	
Tractor	M932	BTD		2320-01-047-8752
Tractor	M932A1	BS5		2320-01-205-2684
Tractor	M932A2	BTQ		2320-01-230-0303
Van, Expansible	M934	BTB	2320-01-047-8750	
Van, Expansible	M934A1	BS4	2320-01-205-2682	
Van, Expansible	M934A2	BTR	2320-01-230-0300	
Medium Wrecker	M936	BTF		2320-01-047-8754
Medium Wrecker	M936A1	BS6		2320-01-206-4078
Medium Wrecker	M936A2	BTT		2320-01-230-0304

REPORTING OF ERRORS AND RECOMMENDING IMPROVEMENTS

You can help improve this manual. If you find any mistakes or if you know of a way to improve the procedures, please let us know. Mail your letter or DA Form 2028 (Recommended Changes to Publications and Blank Forms), or DA Form 2028-2 located in back of this manual, directly to: Director, Armament and Chemical Acquisition and Logistics Activity, ATTN: AMSTA-AC-NML, Rock Island, IL 61299-7630. A reply will be furnished to you. You may also provide DA Form 2028-2 information via datafax or e-mail:

- E-mail: amsta-ac-nml.@ria-emh2.army.mil
- Fax: DSN 783-0726 or commercial (309) 782-0726

TABLE OF CONTENTS

TABLE OF CONTENTS (Cont'd)

TABLE OF CONTENTS (Cont'd)

TABLE OF CONTENTS (Cont'd)

TABLE OF CONTENTS (Cont'd)

TABLE OF CONTENTS (Cont'd)

TABLE OF CONTENTS (Cont'd)

TABLE OF CONTENTS (Cont'd)

TABLE OF CONTENTS (Cont'd)

TABLE OF CONTENTS (Cont'd)

TABLE OF CONTENTS (Cont'd)

UNIT, DIRECT SUPPORT, AND GENERAL SUPPORT MAINTENANCE
REPAIR PARTS AND SPECIAL TOOLS LIST

Section I. INTRODUCTION

1. Scope.

This Repair Parts and Special Tools List (RPSTL) lists and authorizes spares and repair parts; special tools; Test, Measurement, and Diagnostic Equipment (TMDE); and other special support equipment required for performance of unit, direct support, and general support maintenance of the 5-ton, 6x6, M939, M939A1, and M939A2 series vehicles. It authorizes the requisitioning, issue, and disposition of spares, repair parts, and special tools as indicated by the Source, Maintenance, and Recoverability (SMR) codes.

2. General.

This RPSTL is divided into the following sections:

a. Section II. Repair Parts List. A list of spares and repair parts authorized for use in the performance of maintenance. The list also includes parts which must be removed for replacement of the authorized parts. Parts lists are composed of functional groups in ascending alphanumeric sequence, with the parts in each group listed in ascending figure and item number sequence. Bulk materials are listed in item name sequence. Repair parts kits are listed separately in their own functional group within section II. Repair parts for reparable special tools are also listed in this section. Items listed are shown on the associated illustration(s)/figure(s).

b. Section III. Special Tools List. A list of special tools, special TMDE, and other special support equipment authorized by this RPSTL (as indicated by Basis of Issue (BOI) information in DESCRIPTION AND USABLE ON CODE column) for the performance of maintenance.

c. Section IV. National Stock Number and Part Number Index. A list, in ascending National Item Identification Number (NIIN) (last nine numerals) sequence, of all National stock numbered items appearing in the listing, followed by a list, in alphanumeric sequence, of all part numbers appearing in the listings. National stock numbers and part numbers are cross-referenced to each illustration figure and item number appearance.

3. Explanation of Columns (Sections II and III).

a. Item No. (Column (1). Indicates the number used to identify items called out in the illustration.

b. SMR Code (Column (2). The Source, Maintenance, and Recoverability (SMR) code is a 5-position code containing supply/requisitioning information, maintenance category authorization criteria, and disposition instruction, as shown in the following breakout:

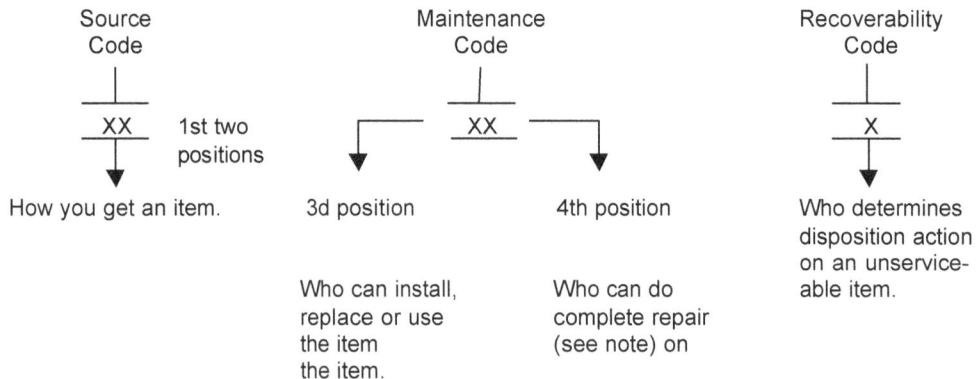

Source Code	Maintenance Code	Recoverability Code
XX 1st two positions	XX	X
How you get an item.	3d position 4th position	Who determines disposition action on an unservice- able item.
	Who can install, replace or use the item the item. Who can do complete repair (see note) on	

* Complete Repair: Maintenance capacity, capability, and authority to perform all corrective maintenance tasks of the "repair" function in a use/user environment in order to restore serviceability to a failed item.

(1) Source Code. The source code tells you how to get an item needed for maintenance, repair, or overhaul of an end item/equipment. Explanations of source codes follow:

Code   **Explanation**

PA
PB
PC**
PD
PE
PF
PG

Stocked items: use the applicable NSN to request/requisition items with these source codes. They are authorized to the category indicated by the code entered in the 3d position of the SMR code.

** NOTE: Items coded PC are subject to deterioration.

KD
KF
KB

Items with these codes are not to be requested/requisitioned individually. They are part of a kit which is authorized to the maintenance category indicated in the 3d position of the SMR code. The complete kit must be requisitioned and applied.

MO- (Made at Unit Level)
MF- (Made at DS Level)
MH- (Made at GS Level)
ML- (Made at Specialized
        Repair Activity (SRA))
MD- (Made at Depot)

Items with these codes are not to be requested/requisitioned individually. They must be made from bulk material which is identified by the part number in the DESCRIPTION AND USABLE ON CODE (UOC) column and listed in the bulk material group of the repair parts list in the RPSTL. If the item is authorized to you by the 3d position code of the SMR code, but the source code indicates it is made at a higher level, order the item from the higher level of maintenance.

AO- (Assembled by Unit Level)
AF- (Assembled by DS Level)
AH- (Assembled by GS Level)
AL- (Assembled by SRA)
AD- (Assembled by Depot)

Items with these codes are not to be requested/requisitioned individually. The parts that make up the assembled item must be requisitioned or fabricated and assembled at the level of maintenance indicated by the source code. If the 3d position code of the SMR code authorizes you to replace the item, but the source code indicates the item is assembled at a higher level, order the item from the higher level of maintenance.

XA- Do not requisition an XA-coded item. Order its next higher assembly. (Also, refer to the note below.)
XB- If an XB-coded item is not available from salvage, order it using the CAGEC and part number given.
XC- Installation drawing, diagram, instruction sheet, field service drawing, that is identified by manufacturer's part number.
XD- Item is not stocked. Order an XD-coded item through normal supply channels using the CAGEC and part number given if no NSN is available.

NOTE
Cannibalization or controlled exchange, when authorized, may be used as a source of supply for items with the above source codes, except for those source-coded XA.

(2) Maintenance Code. Maintenance codes tell you the level(s) of maintenance authorized to USE and REPAIR support items. The maintenance codes are entered in the 3d and 4th positions of the SMR code as follows:

(a) The maintenance code entered in the 3d position tells you the lowest maintenance level authorized to remove, replace, and use an item. The maintenance code entered in the 3d position will indicate authorization to one of the following levels of maintenance.

Code   **Application/Explanation**
C   - Crew or operator maintenance done within unit maintenance.
O   - Unit level can remove, replace, and use the item.
F   - Direct support level can remove, replace, and use the item.
H   - General support level can remove, replace, and use the item.
L   - Specialized repair activity can remove, replace, and use the item.
D   - Depot level can remove, replace, and use the item.

(b) The maintenance code entered in the 4th position tells you whether or not the item is to be repaired and identifies the lowest maintenance level with the capability to do complete repair (i.e., perform all authorized repair functions). (NOTE: Some limited repair may be done on the item at a lower level of maintenance if authorized by the Maintenance Allocation Chart (MIAC) and SMR codes.) This position will contain one of the following maintenance codes:

Code	Application/Explanation
O	- Unit is the lowest level that can do complete repair of the item.
F	- Direct support is the lowest level that can do complete repair of the item.
H	- General support is the lowest level that can do complete repair of the item.
L	- Specialized repair activity (designate the specialized repair activity) is the lowest level that can do complete repair of the item.
D	- Depot is the lowest level that can do complete repair of the item.
Z	- Nonreparable. No repair is authorized.
B	- No repair is authorized. (No parts or special tools are authorized for the maintenance of a B-coded item.) However, the item may be reconditioned by adjusting, lubricating, etc., at the user level.

(3) Recoverability Code. Recoverability codes are assigned to items to indicate the disposition action on unserviceable items. The recoverability code is entered in the 5th position of the SMR code as follows:

Recoverability Code	Application/Explanation
Z	- Nonreparable item. When unserviceable, condemn and dispose of the item at the level of maintenance shown in 3d position of SMR code.
O	- Reparable item. When uneconomically reparable, condemn and dispose of the item at unit level.
F	- Reparable item. When uneconomically reparable, condemn and dispose of the item at the direct support level.
H	- Reparable item. When uneconomically reparable, condemn and dispose of the item at the general support level.
D	- Reparable item. When beyond lower level repair capability, return to depot. Condemnation and disposal of item not authorized below depot level.
L	- Reparable item. Condemnation and disposal not authorized below SRA.
A	- Item requires special handling or condemnation procedures because of specific reasons (e g., precious metal content, high dollar value, critical material, or hazardous material). Refer to appropriate manuals/directives for specific instructions.

c. National Stock Number (Column (3). This column indicates the National Stock Number (NSN) assigned to the item. Use the NSN for requests/requisitions.

d. CAGE Code (Column (4). The Commercial and Government Entity Code (CAGEC) is a 5-digit numeric code which is used to identify the manufacturer, distributor, or Government agency, etc., that supplies the item.

e. Part Number (Column (5). Indicates the primary number used by the manufacturer (individual, company, firm, corporation, or Government activity), which controls the design and characteristics of the item by means of its engineering drawings, specifications standards, and inspection requirements to identify an item or range of items.

NOTE

When you use a NSN to requisition an item, the item you receive may have a different part number from the part ordered.

3

f. **Description and Usable on Codes (UOC) (Column (6).** This column includes the following information:

(1) The Federal item name and, when required, a minimum description to identify the item.

(2) Items that are included in kits and sets are listed below the name of the kit or set.

(3) Spare/repair parts that make up an assembled item are listed immediately following the assembled item line entry.

(4) Part numbers for bulk materials are referenced in this column in the line item entry for the item to be manufactured/fabricated.

(5) When the item is not used with all serial numbers of the same model, the effective serial numbers are shown on the last line(s) of the description (before UOC).

(6) The usable on code, when applicable (refer to paragraph 5, Special Information).

(7) In the Special Tools List section, the Basis of Issue (BOI) appears as the last line(s) in the entry for each special tool, special TMDE, and other special support equipment. When density of equipment supported exceeds density spread indicated in the basis of issue, the total authorization is increased proportionately.

(8) The statement END OF FIGURE appears just below the last item description in column 6 for a given figure in both section II and section III.

g. **QTY (Column (7).** The Quantity Incorporated in Unit (QTY INC IN UNIT) indicates the quantity of the item used in the breakout shown on the illustration figure, which is prepared for a functional group, subfunctional group, or an assembly. A "V" appearing in this column in place of a quantity indicates that no specific quantity is applicable (e.g., shims, spacers).

4. Explanation of Columns (Section IV).

a. **National Stock Number (NSN) Index.**

(1) **Stock number column.** This column lists the NSN by National Item Identification Number (NIIN)

$$\frac{NSN}{NIIN}$$

sequence. The NIIN consists of the last nine digits of the NSN (i.e., 5305-01-674-1467). When using this column to locate an item, ignore the first 4 digits of the NSN. However, the complete NSN should be used when ordering items by stock number.

(2) **Fig. column.** This column lists the number of the figure where the item is identified/located. The figures are in numerical order in section II and section III.

(3) **Item column.** The item number identifies the item associated with the figure listed in the adjacent FIG. column. This item is also identified by the NSN listed on the same line.

b. **Part number Index.** Part numbers in this index are listed by part number in ascending alphanumeric sequence (i.e., vertical arrangement of letter and number combination which places the first letter or digit of each group in order A through Z, followed by the numbers 0 through 9 and each following letter or digit in like order).

(1) **CAGEC column.** The Commercial and Government Entity Code (CAGEC) is a 5-digit numeric code used to identify the manufacturer, distributor, or Government agency, etc., that supplies the item.

(2) **Part number column.** Indicates the primary number used by the manufacturer (individual, company, firm, corporation, or Government activity), which controls the design and characteristics of the item by means of its engineering drawings, specifications standards, and inspection requirements to identify an item or range of items.

(3) **Stock number column.** This column lists the NSN for the associated part number and manufacturer identified in the PART NUMBER and CAGEC columns to the left.

4

(4) Fig. column. This column lists the number of the figure where the item is identified/located in section II and section III.

(5) Item column. The item number is that number assigned to the item as it appears in the figure referenced in the adjacent figure number column.

5. Special Information.

a. Usable on Code. The Usable on Code appears in the lower left corner of the DESCRIPTION column heading. Usable on Codes are shown as UOC: . . . . . . .in the DESCRIPTION column (justified left) on the first line of applicable item description/nomenclature. Uncoded items are applicable to all models and title description will indicate all M939, M939A1 or M939A2 models. Identification of the Usable on Codes used in the RPSTL are:

Code	Used On	Code	Used On	Code	Used On
DAA	M924A1 WO/W	V12	M926 W/W	V39	M945 W/W
DAB	M926A1 W/W	V13	M924 WO/W	ZAA	M923A2 WO/W
DAC	M927A1 WO/W	V14	M925 W/W	ZAB	M925A2 W/W
DAD	M928A1 W/W	V15	M923 WO/W	ZAC	M927A2 WO/W
DAE	M929A1 WO/W	V16	M928 W/W	ZAD	M928A2 W/W
DAF	M930A1 WA/	V17	M927 WO/W	ZAE	M929A2 W/W
DAG	M932A1 WO/W	V18	M936 W/W	ZAF	M930A2 W/W
DAH	M932A1 W/W	V19	M930 W/W	ZAG	M931A2 W/W
DAJ	M934A1 WO/W	V20	M929 WO/W	ZAH	M932A2 WAW
DAK	M935A1 WO/W	V21	M932 W/W	ZAJ	M934A2 WO/W
DAL	M936A1 WO/W	V22	M931 WO/W	ZAK	M935A2 WO/W
DAW	M923A1 WO/W	V24	M934 WO/W	ZAL	M936A2 W/W
DAX	M925A1 W/W	V25	M935 WO/W		

b. Fabrication Instructions. Bulk materials required to manufacture items are listed in the bulk material functional group of this RPSTL. Part numbers for bulk materials are also referenced in the description column of the line item entry for the item to be manufactured/fabricated. Detailed fabrication instructions for items source-coded to be manufactured or fabricated are found in TM 9-2320-272-24.

c. Assembly Instructions. Detailed assembly instructions for items source-coded to be assembled from component spare/repair parts are found in TM 9-2320-272-24. Items that make up the assembly are listed immediately following the assembly item entry, or reference is made to an applicable figure.

d. Kits. Line item entries for repair parts kits appear in LCN group kits in section II.

e. Index Numbers. Items which have the word BULK in the figure column will have an index number shown in the item number column. This index number is a cross-reference between the National Stock Number/Part Number Index and the bulk material list in section II.

f. Associated Publications. The publications listed below pertain to the 5-ton, 6x6, M939, M939A1, and M939A2 series vehicles and their components:

LO 9-2320-272-12	Lubrication Order
TM 9-2320-272-10	Operator's Manual
TM 9-2320-272-10HR	Hand Receipt, Operator's Manual
TM 9-2320-272-24	Unit, Direct Support, and General Support Maintenance Manual
DMWR 9-2520-508	Depot Maintenance Work Requirements, Front and Rear Axle Assemblies
DMWR 9-2520-522	Depot Maintenance Work Requirements, Automatic Transmission
DMWR 9-2520-530	Depot Maintenance Work Requirements, Transfer Assembly
DMAVR 9-3830-501	Depot Maintenance Work Requirements, Front and Rear Winch Assemblies

g. Illustrations. Item numbers on illustrations have been assigned in clockwise sequence, starting at the 11 o'clock position (upper left).

6. How to Locate Repair Parts.

    a. When National Stock Number or Part Number is Not Known.

        (1) First. Using the table of contents, determine the assembly group or subassembly group to which the item belongs. This is necessary since figures are prepared for assembly groups and subassembly groups, and listings are divided into the same groups.

        (2) Second. Find the figure covering the assembly group or subassembly group to which the item belongs.

        (3) Third. Identify the item on the figure and note the item number.

        (4) Fourth. Refer to the Repair Parts List for the figure to find the part number and NSN (if assigned) for the item number noted on the figure.

    b. When National Stock Number or Part Number is Known.

        (1) First. Using the Index of National Stock Numbers and Part Numbers, find the pertinent National Stock Number or Part Number. The NSN index is in National Item Identification Number (NIIN) sequence (see 4a(1)). The part numbers in the Part Number index are listed in ascending alphanumeric sequence (see 4b). Both indexes cross-reference you to the illustration figure and item number of the item you are looking for.

        (2) Second. After finding the figure and item number, verify that the item is the one you're looking for, then locate the item number in the repair parts list for the figure.

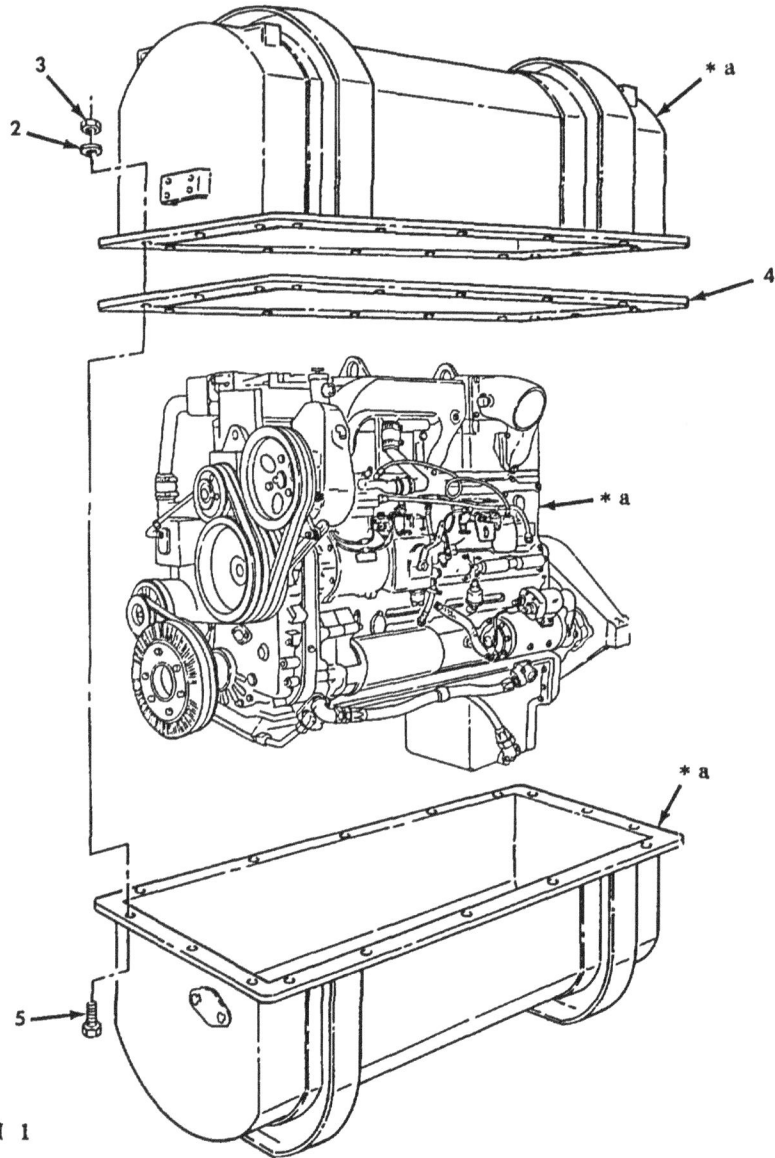

* a PART OF ITEM 1

Figure 1. Engine and Container Assembly (M939, M939A1).

(1) ITEM NO	(2) SMR CODE	(3) NSN	(4) CAGEC	(5) PART NUMBER	(6) DESCRIPTION AND USABLE ON CODES (UOC)	(7) QTY
					GROUP 01 ENGINE 0100 ENGINE ASSEMBLY	
					FIG. 1 ENGINE AND CONTAINER ASSEMBLY (M939,M939A1)	
1	PAFHH	2815011112262	19207	5704507	ENGINE, DIESEL .... UOC:DAA, DAB, DAC, DAD, DAE, DAF, DAG, DAH, DAJ, DAK, DAL, DAW, DAX, V12, V13, V14, V15, V16, V17, V18, V19, V20, V21, V22, V24, V25, V39	1
2	PAFZZ	5310008206653	96906	MS35338-50	.WASHER, LOCK .... UOC:DAA, DAB, DAC, DA D, DAE, DAF, DAG, DAH, DAJ, DAK, DAL, DAW, DAX, V12, V13, V14, V15, V16, V17, V18, V19, V20, V21, V22, V24, V25, V39	42
3	PAFZZ	5310007638920	96906	MS51967-20	.NUT, PLAIN, HEXAGON .... UOC:DAA, DAB, DAC, DAD, DAE, DAF, DAG, DAH, DAJ, DAK, DAL, DAW, DAX, V12, V13, V14, V15, V16, V17, V18, V19, V20, V21, V22, V24, V25, V39	42
4	PAFZZ	5330011377090	19207	11648657	.SEAL, NONMETALLIC RO .... UOC:DAA, DAB, DAC, DAD, DAE, DAF, DAG, DAH, DAJ, DAK, DAL, DAW, DAX, V12, V13, V14, V15, V16, V17, V18, V19, V20, V21, V22, V24, V25, V39	1
5	PAFZZ	5305007247222	80204	B1821BH063C200N	.SCREW, CAP, HEXAGON H .... UOC:DAA, DAB, DAC, DAD, DAE, DAF, DAG, DAH, DAJ, DAK, DAL, DAW, DAX, V12, V13, V14, V15, V16, V17, V18, V19, V20, V21, V22, V24, V25, V39	42

END OF FIGURE

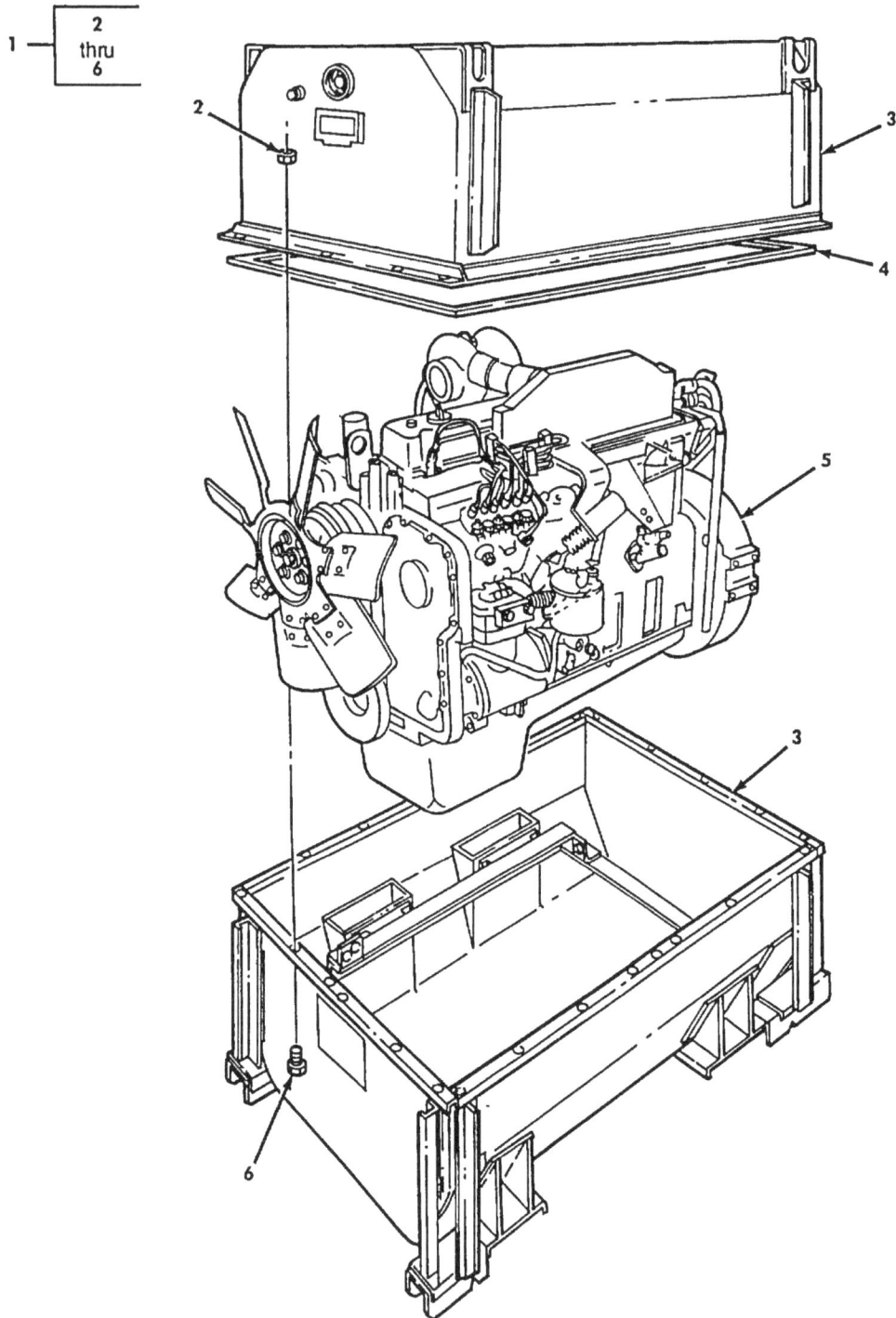

Figure 2. Engine and Container Assembly (M939A2).

(1) ITEM NO	(2) SMR CODE	(3) NSN	(4) CAGEC	(5) PART NUMBER	(6) DESCRIPTION AND USABLE ON CODES (UOC)	(7) QTY
					GROUP 0100 ENGINE ASSEMBLY	
					FIG. 2 ENGINE AND CONTAINER ASSEMBLY (M939A2)	
1	PAFHH	2815013747539	19207	57K3194	ENGINE,DIESEL ASSEMBLY W/CONTAINER......... UOC:ZAA,ZAB,ZAC,ZAD,ZAE,ZAF,ZAG,ZAH, ZAJ, ZAK, ZAL	1
2	PAFZZ	5310007680318	96906	MS51967-14	.NUT,PLAIN,HEXAGON .... UOC:ZAA,ZAB,ZAC,ZAD,ZAE,ZAF,ZAG,ZAH, ZAJ, ZAK, ZAL	24
3	XDFHH	B145013173498	19207	12363644	.SHIPPING AND STORAG .... UOC:ZAA,ZAB,ZAC,ZAD,ZAE,ZAF,ZAG,ZAH, ZAJ, ZAK, ZAL	1
4	PAFZZ	5330013506119	19207	12363654	.GASKET FOR STORAGE CONTAINER................ UOC:ZAA,ZAB,ZAC,ZAD,ZAE,ZAF,ZAG,ZAH, ZAJ,ZAK,ZAL	1
5	XAFHH		19207	12363270 XA	.ENGINE,DIESEL ASSEMBLY .... UOC:ZAA,ZAB,ZAC,ZAD,ZAE,ZAF,ZAG,ZAH, ZAJ,ZAK,ZAL	1
6	PAFZZ	5305000712067	80204	B1821BH050C125N	.SCREW,CAP,HEXAGON H................................... UOC:ZAA,ZAB,ZAC,ZAD,ZAE,ZAF,ZAG,ZAH, ZAJ,ZAK,ZAL	24

END OF FIGURE

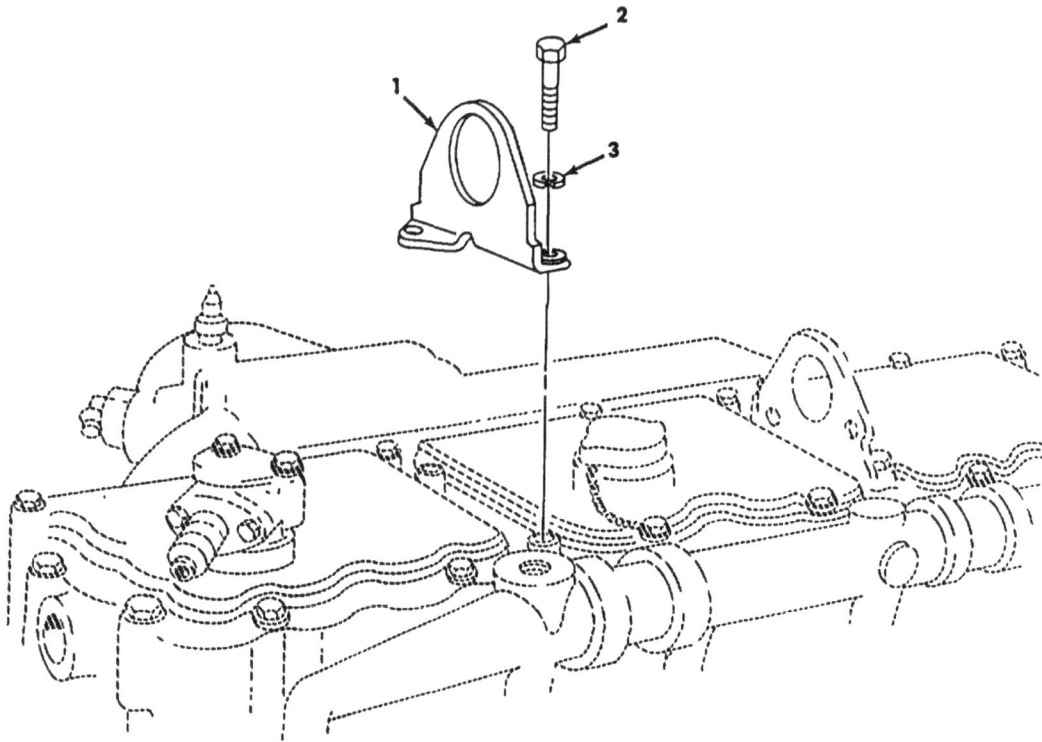

Figure 3. Engine Lifting Brackets (M939, M939A1).

(1) ITEM NO	(2) SMR CODE	(3) NSN	(4) CAGEC	(5) PART NUMBER	(6) DESCRIPTION AND USABLE ON CODES (UOC)	(7) QTY
					GROUP 0100 ENGINE ASSEMBLY	
					FIG. 3 ENGINE LIFTING BRACKETS(M939, M939A1)	
1	PAFZZ	5342004042946	15434	170226	BRACKET, ENGINE LIFT .... UOC: DAA, DAB, DAC, DAD, DAE, DAF, DAG, DAH, DAJ, DAK, DAL, DAW, DAX, V12, V13, V14, V15, V16, V17, V18, V19, V20, V21, V22, V24, V25, V39	2
2	PAFZZ	5305007959343	15434	S-176	SCREW,CAP,HEXAGON H .... UOC: DAA, DAB, DAC, DAD, DAE, DAF, DAG, DAH, DAJ, DAK, DAL, DAW, DAX, V12, V13, V14, V15, V16, V17, V18, V19, V20, V21, V22, V24, V25, V39	4
3	PAFZZ	5310012001318	15434	S-608	WASHER, LOCK .... UOC: DAA, DAB, DAC, DAD, DAE, DAF, DAG, DAH, DAJ, DAK, DAL, DAW, DAX, V12, V13, V14, V15, V16, V17, V18, V19, V20, V21, V22, V24, V25, V39	4

END OF FIGURE

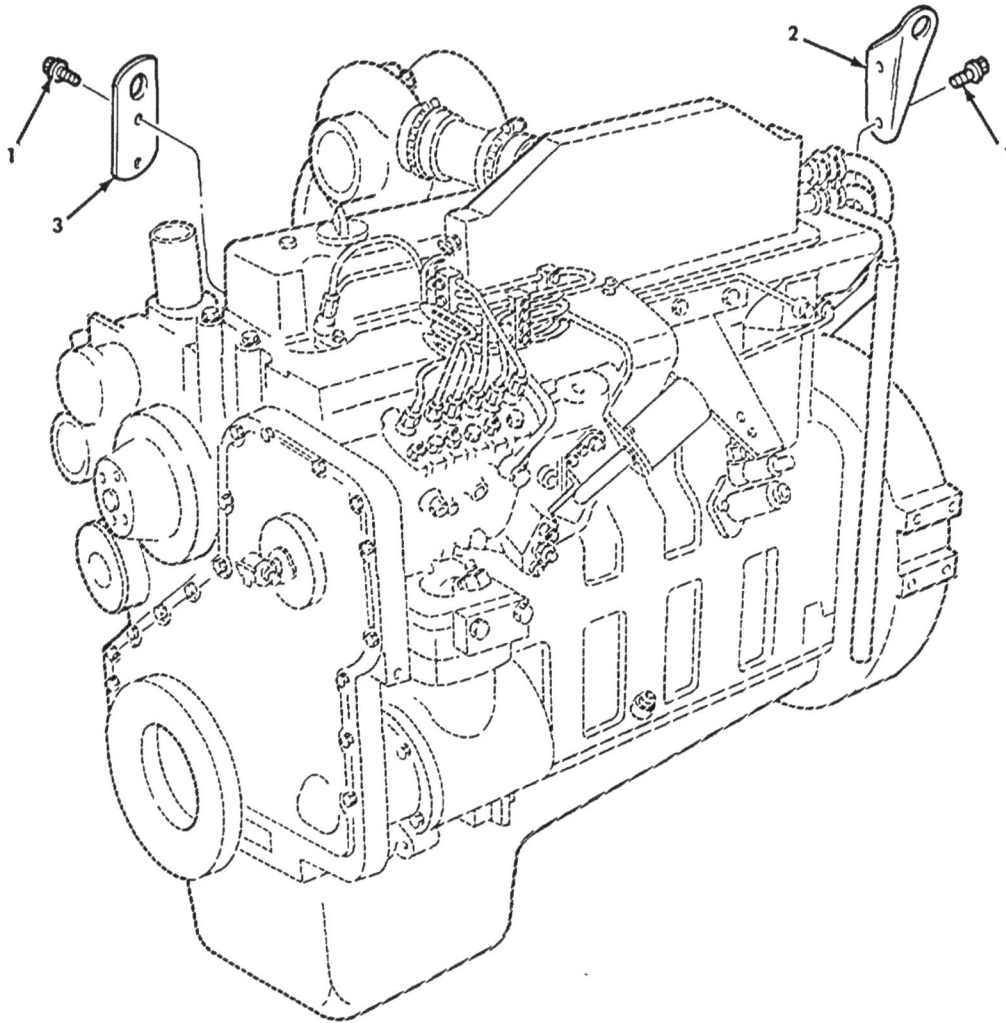

Figure 4. Engine Lifting Brackets (M939A2).

(1) ITEM NO	(2) SMR CODE	(3) NSN	(4) CAGEC	(5) PART NUMBER	(6) DESCRIPTION AND USABLE ON CODES (UOC)	(7) QTY
					GROUP 0100 ENGINE ASSEMBLY	
					FIG. 4 ENGINE LIFTING BRACKETS (M939A2)	
1	PAFZZ	5305012685558	15434	3903200	SCREW,CAP,HEXAGON H .... UOC: ZAA, ZAB, ZAC, ZAD, ZAE, ZAF, ZAG, ZAH, ZAJ, ZAK, ZAL	4
2	PAFZZ	5342012726724	15434	3906440	BRACKET, ENGINE ACCE .... UOC: ZAA, ZAB, ZAC, ZAD, ZAE, ZAF, ZAG, ZAH, ZAJ, ZAK, ZAL	1
3	PAFZZ	5342012726725	15434	3907861	BRACKET, ENGINE ACCE .... UOC: ZAA, ZAB, ZAC, ZAD, ZAE, ZAF, ZAG, ZAH, ZAJ, ZAK, ZAL	1

END OF FIGURE

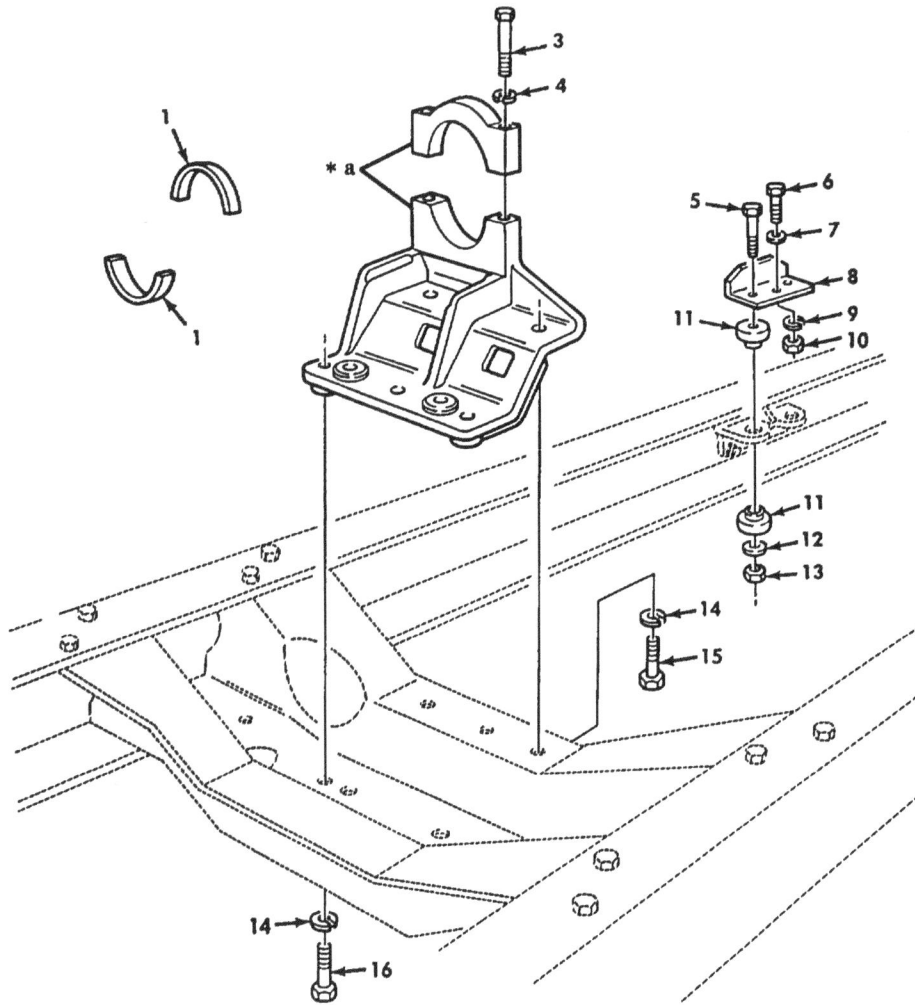

* a   PART OF ITEM 2

Figure 5. Engine Mounts (M939, M939A1).

(1) ITEM NO	(2) SMR CODE	(3) NSN	(4) CAGEC	(5) PART NUMBER	(6) DESCRIPTION AND USABLE ON CODES (UOC)	(7) QTY
					GROUP 0100 ENGINE ASSEMBLY	
					FIG. 5 ENGINE MOUNTS (M939, M939A1)	
1	PAFZZ	2815001344665	76005	J-9813-1	INSERT, MOUNT ENGINE .... UOC: DAA, DAB, DAC, DAD, DAE, DAF, DAG, DAH, DAJ, DAK, DAL, DAW, DAX, V12, V13, V14, V15, V16, V17, V18, V19, V20, V21, V22, V24, V25, V39	2
2	PAFZZ	2510001344663	19207	11664493	BRACKET, ENGINE MOUN .... UOC: DAA, DAB, DAC, DAD, DAE, DAF, DAG, DAH, DAJ, DAK, DAL, DAW, DAX, V12, V13, V14, V15, V16, V17, V18, V19, V20, V21, V22, V24, V25, V39	1
3	PAFZZ	5305007245939	96906	MS90725-173	.SCREW, CAP, HEXAGON H .... UOC: DAA, DAB, DAC, DAD, DAE, DAF, DAG, DAH, DAJ, DAK, DAL, DAW, DAX, V12, V13, V14, V15, V16, V17, V18, V19, V20, V21, V22, V24, V25, V39	2
4	PAFZZ	5310008206653	80045	23MS35338-50	.WASHER, LOCK .... UOC: DAA, DAB, DAC, DAD, DAE, DAF, DAG, DAH, DAJ, DAK, DAL, DAW, DAX, V12, V13, V14, V15, V16, V17, V18, V19, V20, V21, V22, V24, V25, V39	2
5	PAFZZ	5305007195270	96906	MS90727-123	SCREW, CAP, HEXAGON H .... UOC: DAA, DAB, DAC, DAD, DAE, DAF, DAG, DAH, DAJ, DAK, DAL, DAW, DAX, V12, V13, V14, V15, V16, V17, V18, V19, V20, V21, V22, V24, V25, V39	4
6	PAFZZ	5305007262561	80204	B1821BH063F450N	SCREW, CAP, HEXAGON H .... UOC: DAA, DAB, DAC, DAD, DAE, DAF, DAG, DAH, DAJ, DAK, DAL, DAW, DAX, V12, V13, V14, V15, V16, V17, V18, V19, V20, V21, V22, V24, V25, V39	2
7	PAFZZ	5310001499126	96906	MS20002-10	WASHER, FLAT .... UOC: DAA, DAB, DAC, DAD, DAE, DAF, DAG, DAH, DAJ, DAK, DAL, DAW, DAX, V12, V13, V14, V15, V16, V17, V18, V19, V20, V21, V22, V24, V25, V39	2
8	PAFZZ	5340000402314	19207	8330333	BRACKET, ANGLE .... UOC: DAA, DAB, DAC, DAD, DAE, DAF, DAG, DAH, DAJ, DAK, DAL, DAW, DAX, V12, V13, V14, V15, V16, V17, V18, V19, V20, V21, V22, V24, V25, V39	2
9	PAFZZ	5310008206653	80045	23MS35338-50	WASHER, LOCK .... UOC: DAA, DAB, DAC, DAD, DAE, DAF, DAG, DAH, DAJ, DAK, DAL, DAW, DAX, V12, V13, V14, V15, V16, V17, V18, V19, V20, V21, V22, V24, V25, V39	2
10	PAFZZ	5310002416664	96906	MS51943-44	NUT, SELF-LOCKING, HE .... UOC: DAA, DAB, DAC, DAD, DAE, DAF, DAG, DAH, DAJ, DAK, DAL, DAW, DAX, V12, V13, V14, V15,	2

(1) ITEM NO	(2) SMR CODE	(3) NSN	(4) CAGEC	(5) PART NUMBER	(6) DESCRIPTION AND USABLE ON CODES (UOC)	(7) QTY
					V16, V17, V18, V19, V20, V21, V22, V24, V25, V39	
11	PAFZZ	5342004848588	19207	11648643	MOUNT, RESILIENT .... UOC: DAA, DAB, DAC, DAD, DAE, DAF, DAG, DAH, DAJ, DAK, DAL, DAW, DAX, V12, V13, V14, V15, V16 V17, V18, V19, V20, V21, V22, V24, V25, V39	8
12	PAFZZ	5310004704271	19207	11664571	WASHER, FLAT... UOC: DAA, DAB, DAC, DAD, DAE, DAF, DAG, DAH, DAJ, DAK, DAL, DAW, DAX, V12, V13, V14, V15, V16, V17, V18, V19, V20, V21, V22, V24, V25, V39	4
13	PAFZZ	5310004883888	96906	MS51943-40	NUT, SELF-LOCKING, HE .... UOC: DAA, DAB, DAC, DAD, DAE, DAF, DAG, DAH, DAJ, DAK, DAL, DAW, DAX, V12, V13, V14, V15, V16, V17, V18, V19, V20, V21, V22, V24, V25, V39	4
14	PAFZZ	5310000034094	01276	210104-8S	WASHER, LOCK .... UOC: DAA, DAB, DAC, DAD, DAE, DAF, DAG, DAB, DAJ, DAK, DAL, DAW, DAX, V12, V13, V14, V15, V16, V17, V18, V19, V20, V21, V22, V24, V25, V39	5
15	PAFZZ	5305000712068	80204	B1821BH050C138N	SCREW, CAP, HEXAGON H .... UOC: DAA, DAB, DAC, DAD, DAE, DAF, DAG, DAH, DAJ, DAK, DAL, DAW, DAX, V12, V13, V14, V15, V16, V17, V18, V19, V20, V21, V22, V24, V25, V39	2
16	PAFZZ	5305000712069	80204	B1821BH050C150N	SCREW, CAP, HEXAGON H .... UOC: DAA, DAB, DAC, DAD, DAB, DAF, DAG, DAH, DAJ, DAK, DAL, DAW, DAX, V12, V13, V14, V15, V16, V17, V18, V19, V20, V21, V22, V24, V25, V39	2

END OF FIGURE

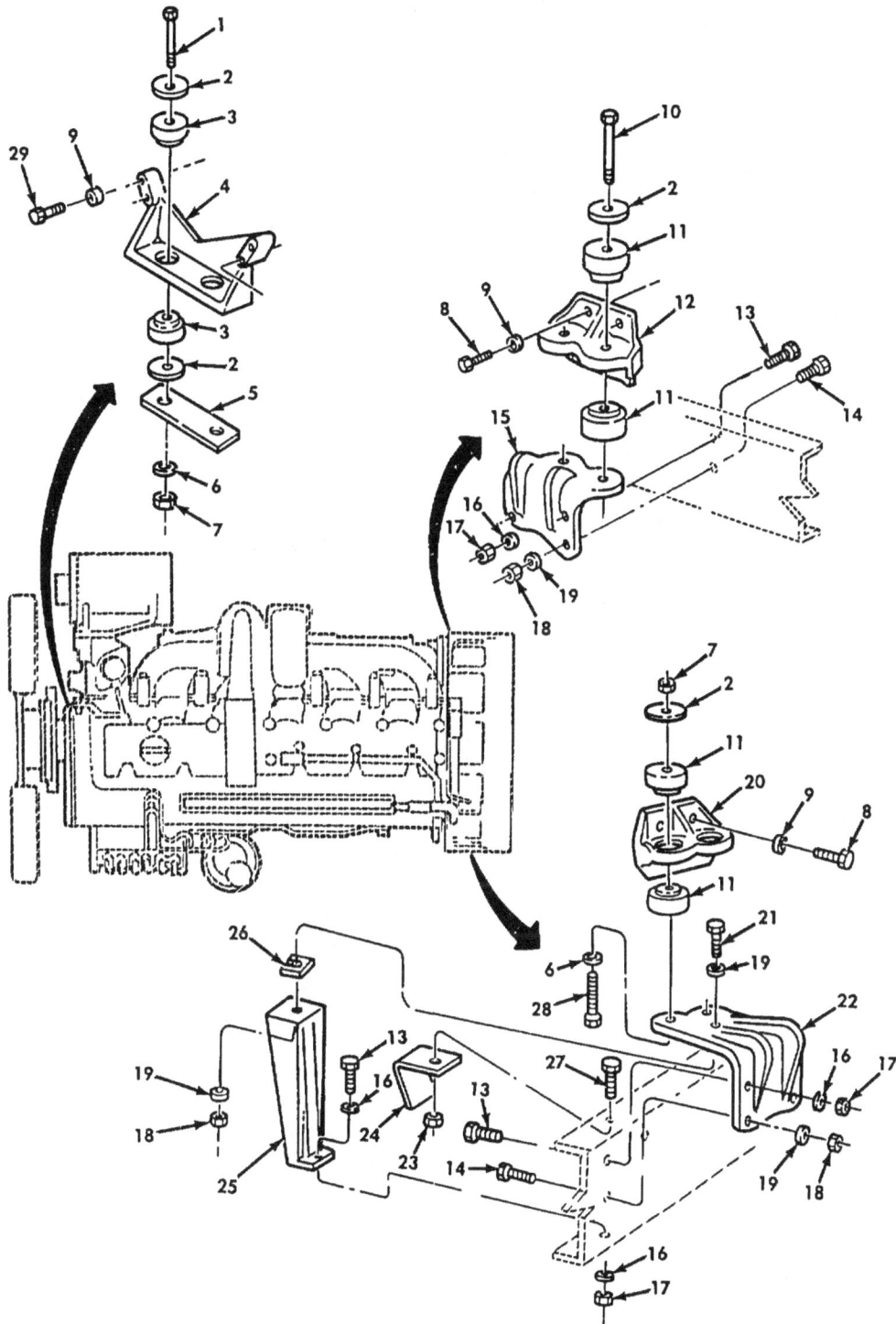

Figure 6. Engine Mounts (M939A2).

(1) ITEM NO	(2) SMR CODE	(3) NSN	(4) CAGEC	(5) PART NUMBER	(6) DESCRIPTION AND USABLE ON CODES (UOC)	(7) QTY

GROUP 0100 ENGINE ASSEMBLY

FIG. 6 ENGINE MOUNTS (M939A2)

1	PAFZZ	5305007247264	80204	B1821BH063C450N	SCREW, CAP, HEXAGON H .... UOC: ZAA, ZAB, ZAC, ZAD, ZAE, ZAF, ZAG, ZAH, ZAJ, ZAK, ZAL	2
2	PAFZZ	5310013098575	3T063	20510400	WASHER, FLAT .... UOC: ZAA, ZAB, ZAC, ZAD, ZAE, ZAF, ZAG, ZAH, ZAJ, ZAK, ZAL	8
3	PAFZZ	5342011853561	81860	22003-14	MOUNT, RESILIENT .... UOC: ZAA, ZAB, ZAC, ZAD, ZAE, ZAF, ZAG, ZAH, ZAJ, ZAK, ZAL	2
4	PAFZZ	2510012793089	47457	20510296	BRACKET, ENGINE MOUN .... UOC: ZAA, ZAB, ZAC, ZAD, ZAE, ZAF, ZAG, ZAH, ZAJ, ZAK, ZAL	1
5	PAFZZ	2510012863329	16567	587-12	BRACKET, ENGINE MOUN .... UOC: ZAA, ZAB, ZAC, ZAD, ZAE, ZAF, ZAG, ZAH, ZAJ, ZAK, ZAL	1
6	PAFZZ	5310012929504	3T063	20510370	WASHER, FLAT .... UOC: ZAA, ZAB, ZAC, ZAD, ZAE, ZAF, ZAG, ZAH, ZAJ, ZAK, ZAL	8
7	PAFZZ	5310011304274	24617	9422305	NUT, SELF-LOCKING, HE .... UOC: ZAA, ZAB, ZAC, ZAD, ZAE, ZAF, ZAG, ZAH, ZAJ, ZAK, ZAL	4
8	PAFZZ	5306012910136	73342	11502456	BOLT, MACHINE .... UOC: ZAA, ZAB, ZAC, ZAD, ZAE, ZAF, ZAG, ZAH, ZAJ, ZAK, ZAL	8
9	PAFZZ	5310012866075	24617	11511516	WASHER, FLAT .... UOC: ZAA, ZAB, ZAC, ZAD, ZAE, ZAF, ZAG, ZAH, ZAJ, ZAK, ZAL	12
10	PAFZZ	5305007247265	80204	B1821BH063C475N	SCREW, CAP, HEXAGON H .... UOC: ZAA, ZAB, ZAC, ZAD, ZAE, ZAF, ZAG, ZAH, ZAJ, ZAK, ZAL	2
11	PAFZZ	5342013317985	81860	25295-1	MOUNT, RESILIENT .... UOC: ZAA, ZAB, ZAC, ZAD, ZAE, ZAF, ZAG, ZAH, ZAJ, ZAK, ZAL	4
12	PAFZZ	2510012863328	47457	20510132	BRACKET, ENGINE MOUN R.H .... UOC: ZAA, ZAB, ZAC, ZAD, ZAE, ZAF, ZAG, ZAH, ZAJ, ZAK, ZAL	1
13	PAFZZ	5305010244775	72582	179827	SCREW, CAP, HEXAGON H .... UOC: ZAA, ZAB, ZAC, ZAD, ZAE, ZAF, ZAG, ZAH, ZAJ, ZAK, ZAL	5
14	PAFZZ	5305000712070	80204	B1821BH050C175N	SCREW, CAP, HEXAGON H .... UOC: ZAA, ZAB, ZAC, ZAD, ZAE, ZAF, ZAG, ZAH, ZAJ, ZAK, ZAL	8
15	PAFZZ	2510012804244	47457	20510134	BRACKET, ENGINE MOUN R.H .... UOC: ZAA, ZAB, ZAC, ZAD, ZAE, ZAF, ZAG, ZAH, ZAJ, ZAK, ZAL	1
16	PAFZZ	5310013017863	24617	9422849	WASHER, FLAT.. UOC: ZAA, ZAB, ZAC, ZAD, ZAE, ZAF, ZAG, ZAH, ZAJ, ZAK, ZAL	6

(1) ITEM NO	(2) SMR CODE	(3) NSN	(4) CAGEC	(5) PART NUMBER	(6) DESCRIPTION AND USABLE ON CODES (UOC)	(7) QTY
17	PAFZZ	5310000443342	19204	8712289-5	NUT, SELF-LOCKING, HE .... UOC: ZAA, ZAB, ZAC, ZAD, ZAE, ZAF, ZAG, ZAH, ZAJ, ZAK, ZAL	5
18	PAFZZ	5310011494407	24617	9422301	NUT, SELF-LOCKING, HE .... UOC: ZAA, ZAB, ZAC, ZAD, ZAE, ZAF, ZAG, ZAH, ZAJ, ZAK, ZAL	9
19	PAFZZ	5310010842362	63005	9411417	WASHER .... UOC: ZAA, ZAB, ZAC, ZAD, ZAE, ZAF, ZAG, ZAH, ZAJ, ZAK, ZAL	10
20	PAFZZ	2510012822526	47457	20510131	BRACKET, ENGINE MOUN L.H .... UOC: ZAA, ZAB, ZAC, ZAD, ZAE, ZAF, ZAG, ZAH, ZAJ, ZAK, ZAL	1
21	PAFZZ	5305000712074	80204	B1821BH050C275N	SCREW, CAP, HEXAGON H .... UOC: ZAA, ZAB, ZAC, ZAD, ZAE, ZAF, ZAG, ZAH, ZAJ, ZAK, ZAL	1
22	PAFZZ	2510012804246	47457	20510133	BRACKET, ENGINE MOUN L.H .... UOC: ZAA, ZAB, ZAC, ZAD, ZAE, ZAF, ZAG, ZAH, ZAJ, ZAK, ZAL	1
23	PAFZZ	5310008140672	96906	MS51943-36	NUT, SELF-LOCKING, HE .... UOC: ZAA, ZAB, ZAC, ZAD, ZAE, ZAF, ZAG, ZAH, ZAJ, ZAK, ZAL	1
24	PAFZZ	5340012880607	08627	3564	BRACKET, ANGLE .... UOC: ZAA, ZAB, ZAC, ZAD, ZAE, ZAF, ZAG, ZAH, ZAJ, ZAK, ZAL	1
25	PAFZZ	2510012804245	47457	20510006	BRACKET, ENGINE MOUN .... UOC: ZAA, ZAB, ZAC, ZAD, ZAE, ZAF, ZAG, ZAH, ZAJ, ZAK, ZAL	1
26	PAFZZ	5365013017871	47457	20511224-1	SPACER, PLATE ....V UOC: ZAA, ZAB, ZAC, ZAD, ZAE, ZAF, ZAG, ZAH, ZAJ, ZAK, ZAL	
26	PAFZZ	5365013017872	47457	20511224-2	SPACER, PLATE ....V UOC: ZAA, ZAB, ZAC, ZAD, ZAE, ZAF, ZAG, ZAH, ZAJ, ZAK, ZAL	
26	PAFZZ	5365013029485	47457	20511224-3	SPACER, PLATE ....V UOC: ZAA, ZAB, ZAC, ZAD, ZAE, ZAF, ZAG, ZAH, ZAJ, ZAK, ZAL	
26	PAFZZ	5365013040652	47457	20511225-1	SPACER, PLATE ....V UOC: ZAA, ZAB, ZAC, ZAD, ZAE, ZAF, ZAG, ZAH, ZAJ, ZAK, ZAL	
27	PAFZZ	5305002692804	96906	MS90726-61	SCREW, CAP, HEXAGON H .... UOC: ZAA, ZAB, ZAC, ZAD, ZAE, ZAF, ZAG, ZAH, ZAJ, ZAK, ZAL	1
28	PAFZZ	5305007247266	80204	B1821BH063C500N	SCREW, CAP, HEXAGON H .... UOC: ZAA, ZAB, ZAC, ZAD, ZAE, ZAF, ZAG, ZAH, ZAJ, ZAK, ZAL	2
29	PAFZZ	5305012910156	73342	11500899	SCREW, CAP, HEXAGON H .... UOC: ZAA, ZAB, ZAC, ZAD, ZAE, ZAF, ZAG, ZAH, ZAJ, ZAK, ZAL	4

END OF FIGURE

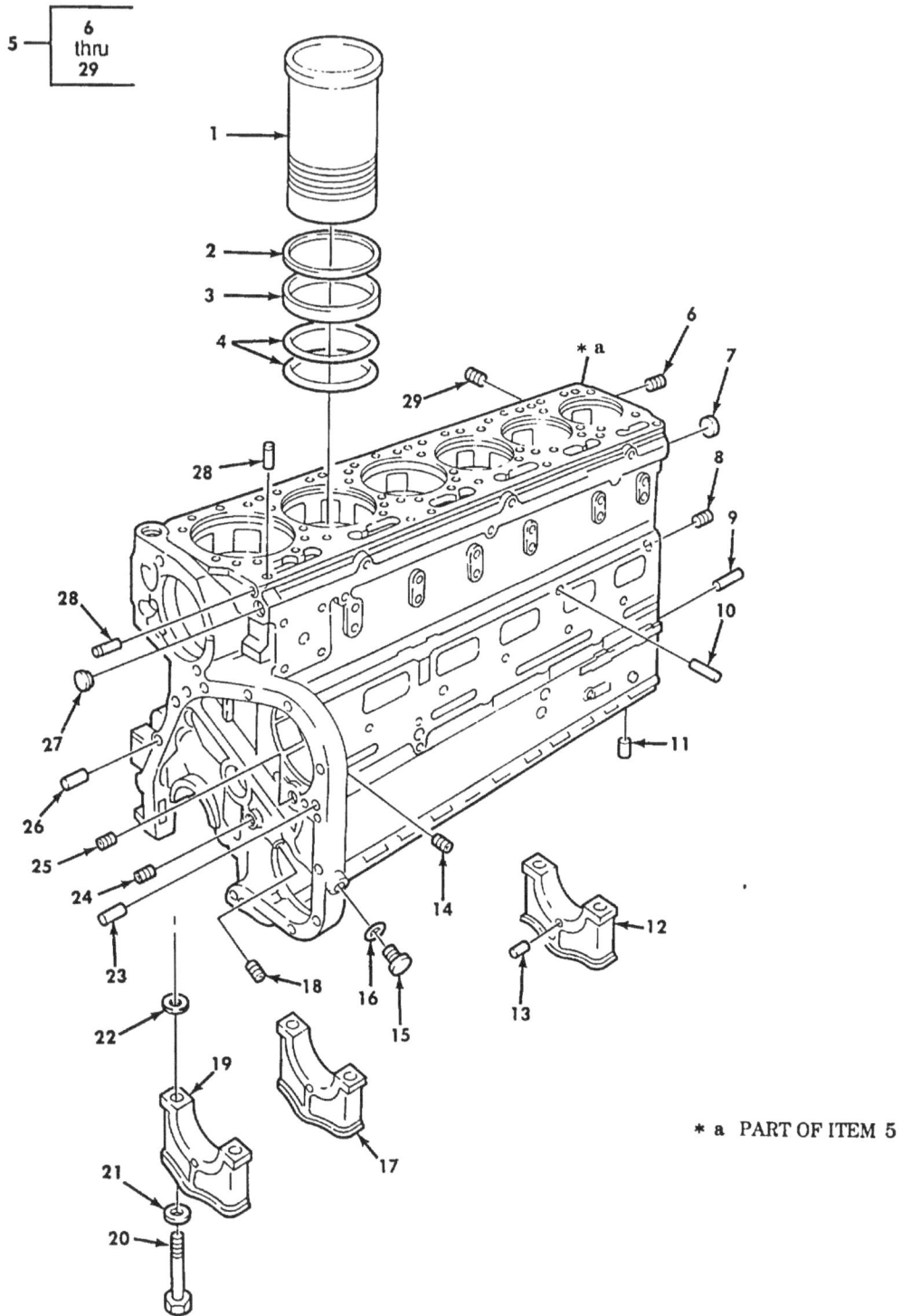

Figure 7. Cylinder Block Assembly (M939, M939A1).

(1) ITEM NO	(2) SMR CODE	(3) NSN	(4) CAGEC	(5) PART NUMBER	(6) DESCRIPTION AND USABLE ON CODES (UOC)	(7) QTY
					GROUP 0101 CRANKCASE, BLOCK, CYLINDER HEAD	
					FIG. 7 CYLINDER BLOCK ASSEMBLY (M939, M939A1)	
1	KFHZZ		15434	3042763	CYLINDER SLEEVE PART OF KIT P/N................... 3801826 PART OF KIT P/N 3804438 .... UOC: DAA, DAB, DAC, DAD, DAE, DAF, DAG, DAH, DAJ, DAK, DAL, DAW, DAX, V12, V13, V14, V15, V16, V17, V18, V19, V20, V21, V22, V24, V25, V39	6
2	PAHZZ	5365011479802	15434	3019955	SHIM .007, CYLINDER ....V UOC: DAA, DAB, DAC, DAD, DAE, DAF, DAG, DAH, DAJ, DAK, DAL, DAW, DAX, V12, V13, V14, V15, V16, V17, V18, V19, V20, V21, V22, V24, V25, V39	
2	PAHZZ	53650048890799	15434	3019956	SPACER, RING .008, CYLINDER ....V UOC: DAA, DAB, DAC, DAD, DAE, DAF, DAG, DAH, DAJ, DAK, DAL, DAW, DAX, V12, V13, V14, V15, V16, V17, V18, V19, V20, V21, V22, V24, V25, V39	
2	PAHZZ	5365011472496	15434	3019957	SHIM .009 CYLINDER ....V UOC: DAA, DAB, DAC, DAD, DAE, DAF, DAG, DAH, DAJ, DAK, DAL, DAW, DAX, V12, V13, V14, V15, V16, V17, V18, V19, V20, V21, V22, V24, V25, V39	
2	PAHZZ	5365011472497	15434	3019958	SPACER, RING .020 CYLINDER ....V UOC: DAA, DAB, DAC, DAD, DAE, DAF, DAG, DAH, DAJ, DAK, DAL, DAW, DAX, V12, V13, V14, V15, V16, V17, V18, V19, V20, V21, V22, V24, V25, V39	
2	PAHZZ	5365011488353	15434	3019959	SHIM .031 CYLINDER ....V UOC: DAA, DAB, DAC, DAD, DAE, DAF, DAG, DAH, DAJ, DAK, DAL, DAW, DAX, V12, V13, V14, V15, V16, V17, V18, V19, V20, V21, V22, V24, V25, V39	
2	PAHZZ	5365011472495	15434	3019960	SHIM .062, CYLINDER ....V UOC: DAA, DAB, DAC, DAD, DAE, DAF, DAG, DAH, DAJ, DAK, DAL, DAW, DAX, V12, V13, V14, V15, V16, V17, V18, V19, V20, V21, V22, V24, V25, V39	
3	KFHZZ	5330000644399	15434	215090	GASKET PART OF KIT P/N 3011472 PART............. OF KIT P/N 3804438 PART OF KIT P/N 3801826 .... UOC: DAA, DAB, DAC, DAD, DAE, DAF, DAG, DAH, DAJ, DAK, DAL, DAW, DAX, V12, V13, V14, V15, V16, V17, V18, V19, V20, V21, V22, V24, V25, V39	6
4	PAHZZ	5331012202389	15434	3032874	O-RING PART OF KIT P/N 3011472 PART............... OF KIT P/N 3804438... UOC: DAA, DAB, DAC, DAD, DAE, DAF, DAG, DAH,	12

(1) ITEM NO	(2) SMR CODE	(3) NSN	(4) CAGEC	(5) PART NUMBER	(6) DESCRIPTION AND USABLE ON CODES (UOC)	(7) QTY
					DAJ, DAK, DAL, DAW, DAX, V12, V13, V14, V15, V16, V17, V18, V19, V20, V21, V22, V24, V25, V39	
5	PAHHH	2815011599538	15434	3801314	BLOCK ASSEMBLY .... UOC: DAA, DAB, DAC, DAD, DAE, DAF, DAG, DAH, DAJ, DAK, DAL, DAW, DAX, V12, V13, V14, V15, V16, V17, V18, V19, V20, V21, V22, V24, V25, V39	1
6	PAFZZ	4730002030549	81348	WW-P-471ACBBUH	.PLUG, PIPE .... UOC: DAA, DAB, DAC, DAD, DAE, DAF, DAG, DAH, DAJ, DAK, DAL, DAW, DAX, V12, V13, V14, V15, V16, V17, V18, V19, V20, V21, V22, V24, V25, V39	1
7	PAFZZ	5340002765847	15434	S719	.PLUG, EXPANSION.. UOC: DAA, DAB, DAC, DAD, DAE, DAF, DAG, DAH, DAJ, DAK, DAL, DAW, DAX, V12, V13, V14, V15, V16, V17, V18, V19, V20, V21, V22, V24, V25, V39	1
8	PAFZZ	4730000819618	79470	C3159X2	.PLUG, PIPE.. UOC: DAA, DAB, DAC, DAD, DAE, DAF, DAG, DAH, DAJ, DAK, DAL, DAW, DAX, V12, V13, V14, V15, V16, V17, V18, V19, V20, V21, V22, V24, V25, V39	8
9	PAFZZ	5315000141284	15434	9226	.PIN, STRAIGHT, HEADLE............................... UOC: DAA, DAB, DAC, DAD, DAE, DAF, DAG, DAH, DAJ, DAK, DAL, DAW, DAX, V12, V13, V14, V15, V16, V17, V18, V19, V20, V21, V22, V24, V25, V39	2
10	PAHZZ	5315000141195	15434	68585	.PIN, STRAIGHT, HEADLE............................... UOC: DAA, DAB, DAC, DAD, DAE, DAF, DAG, DAH, DAJ, DAK, DAL, DAW, DAX, V12, V13, V14, V15, V16, V17, V18, V19, V20, V21, V22, V24, V25, V39	6
11	PAHZZ	5315005329388	23382	1AB2568	.PIN, STRAIGHT, HEADLE............................... UOC: DAA, DAB, DAC, DAD, DAE, DAF, DAG, DAH, DAJ, DAK, DAL, DAW, DAX, V12, V13, V14, V15, V16, V17, V18, V19, V20, V21, V22, V24, V25, V39	2
12	PAHZZ	3130011461228	15434	3008049	.CAP, PILLOW BLOCK .... UOC: DAA, DAB, DAC, DAD, DAE, DAF, DAG, DAH, DAJ, DAK, DAL, DAW, DAX, V12, V13, V14, V15, V16, V17, V18, V19, V20, V21, V22, V24, V25, V39	1
13	PAHZZ	5315010584551	15434	202903	.PIN STRAIGHT HEXAGO .... UOC: DAA, DAB, DAC, DAD, DAE, DAF, DAG, DAH, DAJ, DAK, DAL, DAW, DAX, V12, V13, V14, V15, V16, V17, V18, V19, V20, V21, V22, V24, V25, V39	2
14	PAFZZ	4730000189566	15434	S911B	.PLUG, PIPE .... UOC: DAA, DAB, DAC, DAD, DAE, DAF, DAG, DAH, DAJ, DAK, DAL, DAW, DAX, V12, V13, V14, V15, V16, V17, V18, V19, V20, V21, V22, V24, V25,	1

(1) ITEM NO	(2) SMR CODE	(3) NSN	(4) CAGEC	(5) PART NUMBER	(6) DESCRIPTION AND USABLE ON CODES (UOC)	(7) QTY
15	PAHZZ	5365011506257	15434	210884	V39 .PLUG, MACHINE THREAD .... UOC: DAA,DAB,DAC,DAD,DAE,DAF,DAG,DAH, DAJ,DAK,DAL,DAW,DAX,V12,V13,V14,V15, V16,V17,V18,V19,V20,V21,V22,V24,V25, V39	1
16	PAFZZ	5310001975304	15434	66292	.WASHER, FLAT .... UOC: DAA,DAB,DAC,DAD,DAE,DAF,DAG,DAH, DAJ,DAK,DAL,DAW,DAX,V12,V13,V14,V15, V16,V17,V18,V19,V20,V21,V22,V24,V25, V39	1
17	PAHZZ	3130011464504	15434	3008048	.CAP, BEARING .... UOC: DAA,DAB,DAC,DAD,DAE,DAF,DAG,DAH, DAJ,DAK,DAL,DAW,DAX,V12,V13,V14,V15, V16,V17,V18,V19,V20,V21,V22,V24,V25, V39	3
18	PAHZZ	4730011472223	15434	3008468	.PLUG, PIPE .... UOC: DAA,DAB,DAC,DAD,DAE,DAF,DAG,DAH, DAJ,DAK,DAL,DAW,DAX,V12,V13,V14,V15, V16,V17,V18,V19,V20,V21,V22,V24,V25, V39	1
19	PAHZZ	3130011461150	15434	3008047	.CAP, PILLOW BLOCK .... UOC: DAA,DAB,DAC,DAD,DAE,DAF,DAG,DAH, DAJ,DAK,DAL,DAW,DAX,V12,V13,V14,V15, V16,V17,V18,V19,V20,V21,V22,V24,V25, V39	3
20	PAHZZ	5305011792380	15434	208346	.SCREW, CAP, HEXAGON H 3/4 INCH, USE .... WITH BLOCK ASSY 3801314 ONLY.................. UOC: DAA,DAB,DAC,DAD,DAE,DAF,DAG,DAH, DAJ,DAK,DAL,DAW,DAX,V12,V13,V14,V15, V16,V17,V18,V19,V20,V21,V22,V24,V25, V39	14
20	PAHZZ	5306008042468	15434	105953	.BOLT, MACHINE 1 INCH, USE WITH..................... BLOCK ASSY AR-09883 ONLY........................... UOC: DAA,DAB,DAC,DAD,DAE,DAF,DAG,DAH, DAJ,DAK,DAL,DAW,DAX,V12,V13,V14,V15, V16,V17,V18,V19,V20,V21,V22,V24,V25, V39	14
21	PAHZZ	5310000821882	15434	140218	.WASHER, FLAT USE WITH BLOCK ASSY .... 3801314 ONLY... UOC: DAA,DAB,DAC,DAD,DAE,DAF,DAG,DAH, DAJ,DAK,DAL,DAW,DAX,V12,V13,V14,V15, V16,V17,V18,V19,V20,V21,V22,V24,V25, V39	14
21	PAHZZ	5310003561447	15434	3009213	.WASHER, LOCK USE WITH BLOCK ASSY .... AR-09883 ONLY .... UOC: DAA,DAB,DAC,DAD,DAE,DAF,DAG,DAH, DAJ,DAK,DAL,DAW,DAX,V12,V13,V14,V15, V16,V17,V18,V19,V20,V21,V22,V24,V25, V39	14
22	PAHZZ	5325012413888	15434	3037045	.RING, RETAINING USE WITH BLOCK................... ASSY 3801314 ONLY	7

(1) ITEM NO	(2) SMR CODE	(3) NSN	(4) CAGEC	(5) PART NUMBER	(6) DESCRIPTION AND USABLE ON CODES (UOC)	(7) QTY
					UOC:DAA,DAB,DAC,DAD,DAE,DAF,DAG,DAH, DAJ,DAK,DAL,DAW,DAX,V12,V13,V14,V15, V16,V17,V18,V19,V20,V21,V22,V24,V25, V39	
22	PAHZZ	5365004286201	15434	60575	.SPACER, RING USE WITH BLOCK ASSY .... AR-09883 ONLY UOC:V12,V13,V14,V15,V16,V17,V18,VI9, V20,V21,V22,V24,V25,V39	7
23	PAHZZ	2815007729434	15434	70653	.DOWEL, METALLIC .... UOC:DAA,DAB,DAC,DAD,DAE,DAF,DAG,DAH, DAJ,DAK,DAL,DAW,DAX,V12 ,V13,V14 ,V15, V16,V17,V18,V19,V20,V21,V22,V24,V25, V39	1
24	PAFZZ	4730011615115	15434	3013786	.PLUG, PIPE .... UOC:DAA,DAB,DAC,DAD,DAE,DAF, DAG,DAH, DAJ,DAK,DAL,DAW,DAX,V12,V13,V14,V15, V16,V17,V18,V19,V20,V21,V22,V24,V25, V39	5
25	PAHZZ	4730011060202	15434	3008469	.PLUG, PIPE .... UOC:DAA,DAB,DAC,DAD,DAE,DAF,DAG,DAH, DAJ,DAK,DAL,DAW,DAX,V12,V13,V14,V15, V16,V17,V18,V19,V20,V21,V22,V24,V25, V39	1
26	PAHZZ	5315002380882	15434	60408	.PIN, STRAIGHT, HEADLE .... UOC: DAA,DAB,DAC,DAD,DAE,DAF,DAG,DAH, DAJ,DAX,DAL,DAW,DAX,V12,V13,V14,V15, V16,V17,V18,V19,V20,V21,V22,V24,V25, V39	1
27	PAFZZ	5340000501600	96906	MS35648-8	.PLUG, EXPANSION .... UOC: DAA,DAB,DAC,DAD,DAE,DAF,DAG,DAH, DAJ,DAK,DAL,DAW,DAX,V12,V13,V14,V15, V16,V17,V18,V19,V20,V21,V22,V24,V25, V39	1
28	PAFZZ	5315002817610	15434	68445	.PIN, GROOVED, HEADLES................................. UOC:DAA,DAB,DAC,DAD,DAE,DAF,DAG,DAH, DAJ,DAK,DAL,DAW,DAX,V12,V13,V14,V15, V16, V17,V18 ,V19,V20,V21,V22,V24,V25, V39	6
29	PAFZZ	4730009541281	15434	3008466	.PLUG, PIPE .... UOC: DAA,DAB,DAC,DAD,DAE,DAF,DAG,DAH, DAJ,DAK,DAL,DAW,DAX,V12,V13,V14,V15, V16,V17,V18,V19,V20,V21,V22,V24,V25, V39	1

END OF FIGURE

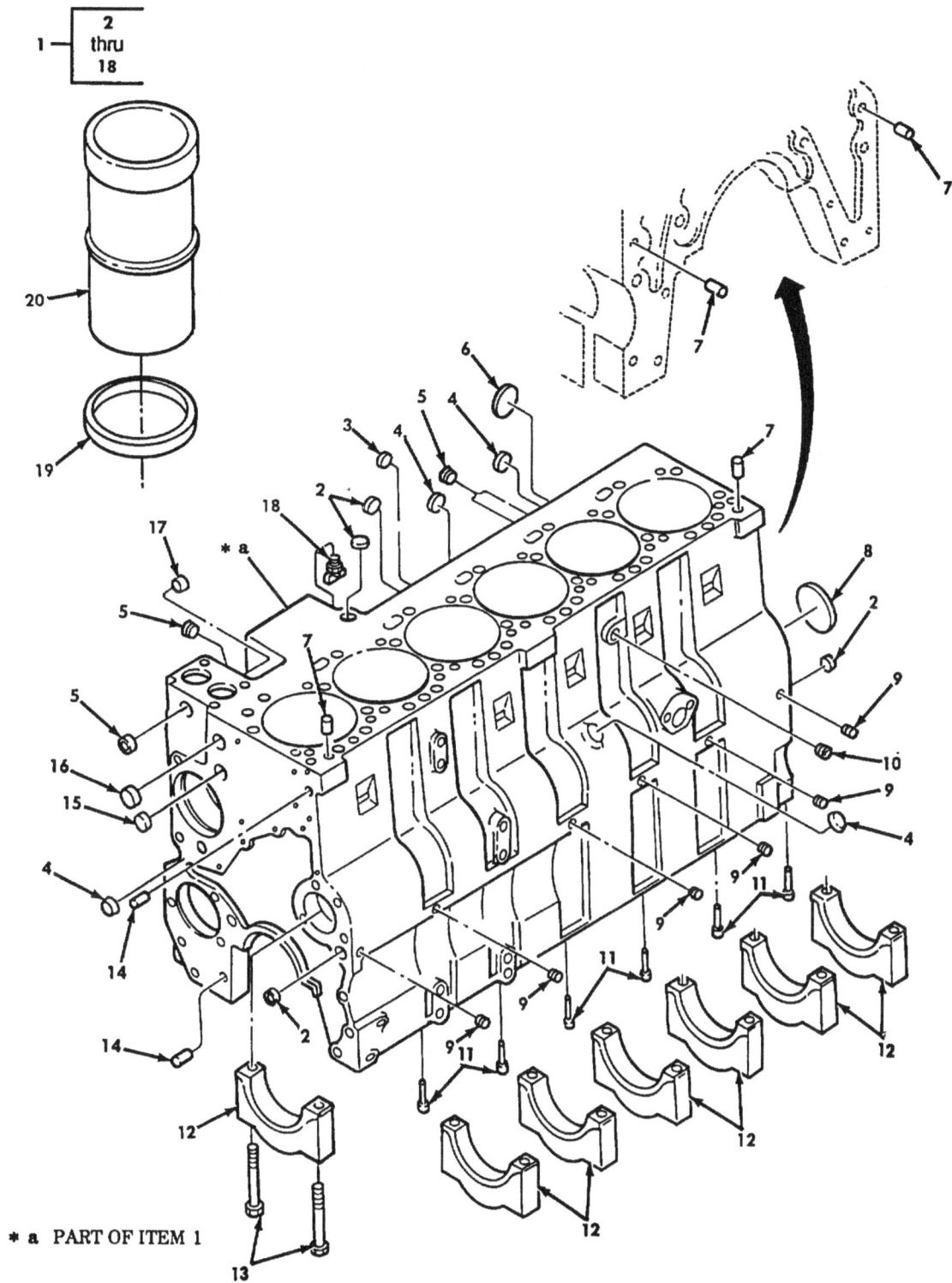

Figure 8. Cylinder Block Assembly (M939A2).

* a PART OF ITEM 1

(1) ITEM NO	(2) SMR CODE	(3) NSN	(4) CAGEC	(5) PART NUMBER	(6) DESCRIPTION AND USABLE ON CODES (UOC)	(7) QTY
					GROUP 0101 CRANKCASE, BLOCK, CYLINDER HEAD	
					FIG. 8 CYLINDER BLOCK ASSEMBLY (M939A2)	
1	PFHHH	2815013308069	15434	3802549	ENGINE BLOCK ASSEMB UP TO ENGINE S/N 44706125 UOC:ZAA,ZAB,ZAC,ZAD,ZAE,ZAF,ZAG,ZAH, ZAJ,ZAK,ZAL	1
1	PFHHH	2815013816486	15434	3802431	ENGINE BLOCK, DIESEL ENGINE S/N 44706126 AND ABOVE, THICK BLOCK UOC:ZAA,ZAB,ZAC,ZAD,ZAE,ZAF,ZAG,ZAH, ZAJ,ZAK,ZAL	1
2	PAFZZ	5340012712416	15434	3004258	.PLUG, EXPANSION. UOC:ZAA,ZAB,ZAC,ZAD,ZAE,ZAF,ZAG,ZAH, ZAJ,ZAK,ZAL	4
3	PAFZZ	5340012398606	15434	3027646	.PLUG, EXPANSION PART OF KIT P/N 3802367 PART OF KIT P/N 3909415 UOC:ZAA,ZAB,ZAC,ZAD,ZAE,ZAF,ZAG,ZAH, ZAJ,ZAK,ZAL	2
4	PAFZZ	5340012712417	15434	3905401	.PLUG, EXPANSION UOC:ZAA,ZAB,ZAC,ZAD,ZAE,ZAF,ZAG,ZAH, ZAJ,ZAK,ZAL	4
5	PAFZZ	4730011472223	15434	3008468	.PLUG, PIPE PART OF KIT P/N 3802367 PART OF KIT P/N 3909415 UOC:ZAA,ZAB,ZAC,ZAD,ZAE,ZAF,ZAG,ZAH, ZAJ,ZAK,ZAL	4
6	PAFZZ	5340011948936	15434	3900965	.PLUG, EXPANSION UOC:ZAA,ZAB,ZAC,ZAD,ZAE,ZAF,ZAG,ZAH, ZAJ,ZAK,ZAL	1
7	PAHZZ	5315012708284	15434	3901846	.PIN, STRAIGHT, HEADLE PART OF KIT P/N 3802367 PART OF KIT P/N 3909415 UOC:ZAA,ZAB,ZAC,ZAD,ZAE,ZAF,ZAG,ZAH, ZAJ,ZAK,ZAL	4
8	PAFZZ	5340012712418	15434	3901969	.PLUG, EXPANSION PART OF KIT P/N 3802367 PART OF KIT P/N 3909415 UOC:ZAA,ZAB,ZAC,ZAD,ZAE,ZAF,ZAG,ZAH, ZAJ,ZAK,ZAL	1
9	PAFZZ	4730012810812	15434	3906619	.PLUG, PIPE PART OF KIT P/N 3802367 PART OF KIT P/N 3909415 UOC:ZAA,ZAB,ZAC,ZAD,ZAE,ZAF,ZAG,ZAH, ZAJ,ZAK,ZAL	6
10	PAFZZ	4730011060202	15434	3008469	.PLUG, PIPE PART OF KIT P/N 3802367 PART OF KIT P/N 3909415 UOC:ZAA,ZAB,ZAC,ZAD,ZAE,ZAF,ZAG,ZAR, ZAJ,ZAK,ZAL	1
11	PAHZZ	4730012717181	15434	3905928	.RESTRICTOR, FLUID FL UP TO ENGINE S/N 44487829 PART OF KIT P/N 3802389 PART OF KIT P/N 3909415 UOC:ZAA,ZAB,ZAC,ZAD,ZAE,ZAF,ZAG,ZAH, ZAJ,ZAK,ZAL	6

(1) ITEM NO	(2) SMR CODE	(3) NSN	(4) CAGEC	(5) PART NUMBER	(6) DESCRIPTION AND USABLE ON CODES (UOC)	(7) QTY
11	PAHZZ	4730013312670	15434	3919003	.RESTRICTOR, FLUID FL ENGINE S/N.............. 4448730 AND ABOVE PART OF KIT P/N 3802389 PART OF KIT P/N 3802367 .... UOC:ZAA,ZAB,ZAC,ZAD,ZAE,ZAF,ZAG,ZAH, ZAJ,ZAK,ZAL	12
12	PAHZZ	3130012819164	15434	3917313	.CAP, PILLOW BLOCK UP TO ENGINE S/N............ 44487829.......................... UOC:ZAA,ZAB,ZAC,ZAD,ZAE,ZAF,ZAG,ZAH, ZAJ,ZAK,ZAL	7
13	PAHZZ	5305012715848	15434	3906655	.SCREW, CAP, HEXAGON H UP TO ENGINE .... S/N 44487829.......................... UOC:ZAA,ZAB,ZAC,ZAD,ZAE,ZAF,ZAG,ZAH, ZAJ,ZAK,ZAL	14
13	PAHZZ	5305013335382	15434	3916369	.SCREW, CAP, HEXAGON H ENGINE S/N.............. 44487830 AND ABOVE .... UOC: ZAA,ZAB,ZAC,ZAD,ZAE,ZAF,ZAG,ZAH, ZAJ,ZAK,ZAL	14
14	PAHZZ	5315011880761	15434	3900257	.PIN, STRAIGHT, HEADLE PART OF KIT P/N.......... 3802367 PART OF KIT P/N 3909415 .... UOC:ZAA,ZAB,ZAC,ZAD,ZAE,ZAF,ZAG,ZAH, ZAJ,ZAK,ZAL	2
15	PAFZZ	5340012712415	15434	143066	.PLUG, EXPANSION.......................... UOC:ZAA,ZAB,ZAC,ZAD,ZAE,ZAF,ZAG,ZAH, ZAJ,ZAK,ZAL	1
16	PAFZZ	5342011436045	15434	3008470	.PLUG PART OF KIT P/N 3802367 PART.............. OF KIT P/N 3909415 .... UOC:ZAA,ZAB,ZAC,ZAD,ZAE,ZAF,ZAG,ZAH, ZAJ,ZAK,ZAL	1
17	PAFZZ	5340012712419	15434	156075	.PLUG, EXPANSION PART OF KIT P/N.................. 3802367 PART OF KIT P/N 3909415 .... UOC:ZAA,ZAB,ZAC,ZAD,ZAE,ZAF,ZAG,ZAH, ZAJ,ZAK,ZAL	2
18	PAFZZ	4730002469217	24617	108686	.ELBOW, PIPE PART OF KIT P/N 3802367 .... PART OF KIT P/N 3909415 .... UOC:ZAA,ZAB,ZAC,ZAD,ZAE,ZAF,ZAG,ZAH, ZAJ,ZAK,ZAL	1
19	PAHZZ	5330012721282	15434	3917737	SEAL, NONMETALLIC RO PART OF KIT P/N........... 2DR671 PART OF KIT P/N 3802370 .... UOC:ZAA,ZAB,ZAC,ZAD,ZAE,ZAF,ZAG,ZAH, ZAJ,ZAK,ZAL	6
20	PFHZZ	2815012811125	15434	3907792	CYLINDER SLEEVE PART OF KIT P/N.................. 2DR671 PART OF KIT P/N 3802370 .... UOC:ZAA,ZAB,ZAC,ZAD,ZAE,ZAF,ZAG,ZAH, ZAJ,ZAK,ZAL	6
20	PFHZZ	2815014144120	15434	3802407	CYLINDER SLEEVE USE WITH THICK.................. BLOCK, P/N 3802431	6

END OF FIGURE

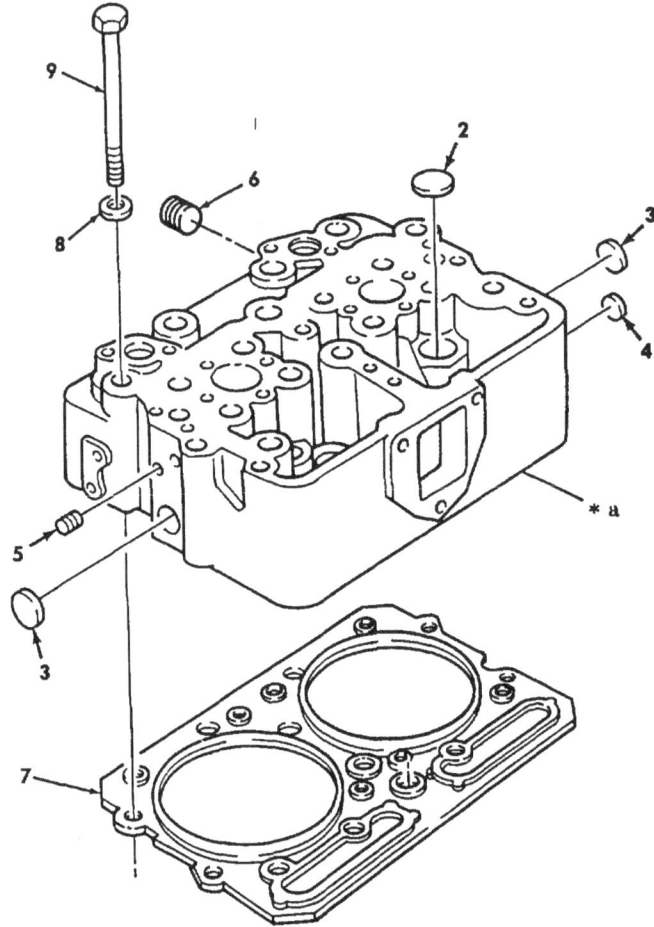

* a PART OF ITEM 1

Figure 9. Cylinder Head Assembly, Gasket, and Mounting Hardware (M939, M939A1).

(1) ITEM NO	(2) SMR CODE	(3) NSN	(4) CAGEC	(5) PART NUMBER	(6) DESCRIPTION AND USABLE ON CODES (UOC)	(7) QTY
					GROUP 0101 CRANKCASE, BLOCK, CYLINDER HEAD	
					FIG. 9 CYLINDER HEAD ASSEMBLY, GASKET, AND MOUNTING HARDWARE (M939, M939A1)	
1	PFFFF	2815011464182	15434	3007835	CYLINDER HEAD, DIESE .... UOC:DAA,DAB,DAC,DAD,DAE,DAF,DAG,DAH, DAJ,DAK,DAL,DAW,DAX,V12,V13,V14,V15, V16,V17,V18,V19,V20,V21,V22,V24,V25, V39	3
2	PAFZZ	5340010870681	15434	213395	.PLUG, EXPANSION .... UOC:DAA,DAB,DAC,DAD,DAE,DAF,DAG,DAH, DAJ,DAK,DAL,DAW,DAX,V12,V13,V14,V15, V16,V17,V18,V19,V20,V21,V22,V24,V25, V39	6
3	PAFZZ	5340010866193	15434	216524	.PLUG, EXPANSION .... UOC:DAA,DAB,DAC,DAD,DAE,DAF,DAG,DAH, DAJ,DAK,DAL,DAW,DAX,V12,V13,V14,V15, V16,V17,V18,V19,V20,V21,V22,V24,V25, V39	2
4	PAFZZ	5340010870682	15434	213394	.PLUG, EXPANSION .... UOC:DAA,DAB,DAC,DAD,DAE,DAF,DAG,DAH, DAJ,DAK,DAL,DAW,DAX,V12,V13,V14,V15, V16,V17,V18,V19,V20,V21,V22,V24,V25, V39	1
5	PAFZZ	4730000189566	24617	G1251	.PLUG, PIPE .... UOC:DAA,DAB,DAC,DAD,DAE,DAF,DAG,DAH, DAJ,DAK,DAL,DAW,DAX,V12,V13,V14,V15, V16,V17,V18,V19,V20,V21,V22,V24,V25, V39	10
6	PAFZZ	4730000444715	15434	S962	.PLUG, PIPE .... UOC:DAA,DAB,DAC,DAD,DAE,DAF,DAG,DAH, DAJ,DAK,DAL,DAW,DAX,V12,V13,V14,V15, V16,V17,V18,V19,V20,V21,V22,V24,V25, V39	4
7	PAFZZ	5330010805021	15434	3076189	GASKET .... UOC:DAA,DAB,DAC,DAD,DAE,DAF,DAG,DAH, DAJ,DAK,DAL,DAW,DAX,V12,V13,V14,V15, V16,V17,V18,V19,V20,V21,V22,V24,V25, V39	1
8	PAFZZ	5310002858833	17576	538174	WASHER, FLAT .... UOC:DAA,DAB,DAC,DAD,DAE,DAF,DAG,DAH, DAJ,DAK,DAL,DAW,DAX,V12,V13,V14,V15, V16,V17,V18,V19,V20,V21,V22,V24,V25, V39	36
9	PAFZZ	5305010724270	15434	177734	SCREW .... UOC:DAA,DAB,DAC,DAD,DAE,DAF,DAG,DAH, DAJ,DAK,DAL,DAW,DAX,V12,V13,V14,V15, V16,V17,V18,V19,V20,V21,V22,V24,V25, V39	36

END OF FIGURE

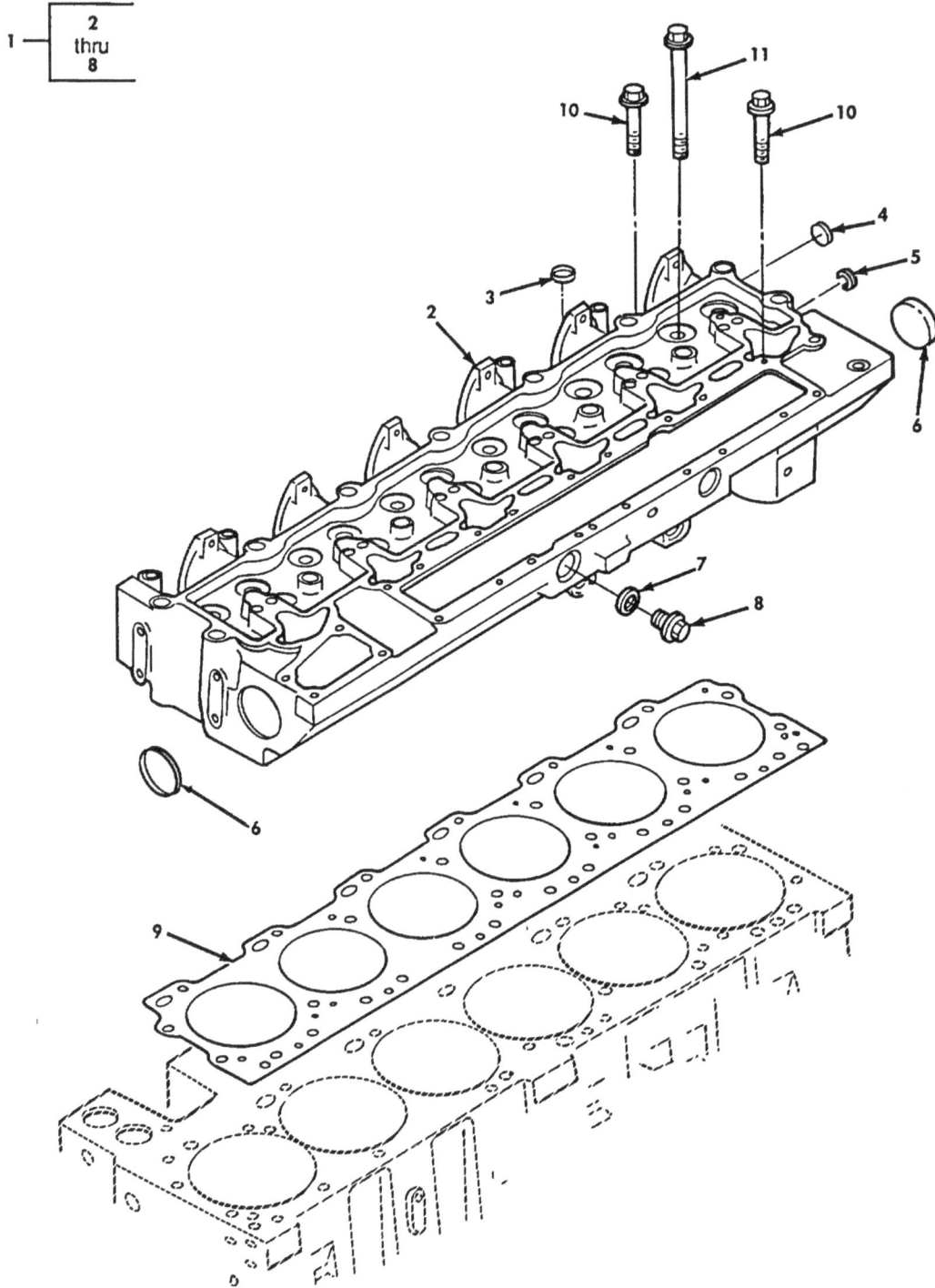

Figure 10. Cylinder Head Assembly, Gasket, and Mounting Hardware (M939A2).

(1) ITEM NO	(2) SMR CODE	(3) NSN	(4) CAGEC	(5) PART NUMBER	(6) DESCRIPTION AND USABLE ON CODES (UOC)	(7) QTY
					GROUP 0101 CRANKCASE, BLOCK, CYLINDER HEAD	
					FIG. 10 CYLINDER HEAD ASSEMBLY, GASKET, AND MOUNTING HARDWARE (M939A2)	
1	PAFHH	2815013303036	15434	3802467	CYLINDER HEAD, DIESE ENGINE S/N .... 44487830 AND ABOVE .... UOC:ZAA,ZAB,ZAC,ZAD,ZAE,ZAF,ZAG,ZAH, ZAJ,ZAK,ZAL	1
2	XAFZH	2815012688756	15434	3909416	.CYLINDER HEAD, DIESE UP TO ENGINE S........... /N 44487829 .... UOC:ZAA,ZAB,ZAC,ZAD,ZAE,ZAF,ZAG,ZAH, ZAJ,ZAK,ZAL	1
3	PAFZZ	5340012712420	15434	3032693	.PLUG, EXPANSION .... UOC:ZAA,ZAB,ZAC,ZAD,ZAE,ZAF,ZAG,ZAH, ZAJ,ZAK,ZAL	5
4	PAFZZ	5340012712419	15434	156075	.PLUG, EXPANSION PART OF KIT P/N................. 3909415 .... UOC:ZAA,ZAB,ZAC,ZAD,ZAE,ZAF,ZAG,ZAH, ZAJ,ZAK,ZAL	1
5	PAFZZ	5365011121524	15434	3008464	.PLUG, MACHINE THREAD .... UOC:ZAA,ZAB,ZAC,ZAD,ZAE,ZAF,ZAG,ZAH, ZAJ,ZAK,ZAL	1
6	PAFZZ	5340011948936	15434	3900965	.PLUG, EXPANSION .... UOC:ZAA,ZAB,ZAC,ZAD,ZAE,ZAF,ZAG,ZAH, ZAJ,ZAK,ZAL	2
7	PAFZZ	5310012090508	15434	3902425	.WASHER, FLAT PART OF KIT P/N 3802622 .... PART OF KIT P/N 3802623 PART OF KIT P/N 3802452.. UOC:ZAA,ZAB,ZAC,ZAD,ZAE,ZAF,ZAG,ZAH, ZAJ,ZAK,ZAL	2
8	PAFZZ	5365012171995	15434	3904386	.PLUG, MACHINE THREAD................................. UOC:ZAA,ZAB,ZAC,ZAD,ZAE,ZAF,ZAG,ZAH, ZAJ,ZAK,ZAL	2
9	PAFZZ	5330012714311	15434	3921850	GASKET (STANDARD) REFERENCE AGAINST....... MACHINE BLOCK PART OF KIT P/N 3802452 .... UOC:ZAA,ZAB,ZAC,ZAD,ZAE,ZAF,ZAG,ZAH, ZAJ,ZAK,ZAL	1
9	PAFZZ	5330012721144	15434	3921852	GASKET (.15MM O/S) REFERENCE .... AGAINST MACHINE BLOCK PART OF KIT P/N 3802622... UOC:ZAA,ZAB,ZAC,ZAD,ZAE,ZAF,ZAG,ZAH, ZAJ,ZAK,ZAL	1
9	PAFZZ	5330012721145	15434	3921853	GASKET (.50MM O/S) REFERENCE .... AGAINST MACHINE BLOCK PART OF KIT P/N 3802623... UOC:ZAA,ZAB,ZAC,ZAD,ZAE,ZAF,ZAG,ZAH, ZAJ,ZAK,ZAL	1
10	PAFZZ	5305012715852	15434	3907234	SCREW, CAP, HEXAGON H UP TO ENGINE S/....... N 44487829	12

(1) ITEM NO	(2) SMR CODE	(3) NSN	(4) CAGEC	(5) PART NUMBER	(6) DESCRIPTION AND USABLE ON CODES (UOC)	(7) QTY
10	PAFZZ	5305013319479	15434	3917729	UOC:ZAA,ZAB,ZAC,ZAD,ZAE,ZAF,ZAG,ZAH, ZAJ,ZAK,ZAL SCREW, CAP, HEXAGON H AFTER ENGINE S/ N 44487830	12
11	PAFZZ	5305012744407	15434	3907233	UOC:ZAA,ZAB,ZAC,ZAD,ZAE,ZAF,ZAG,ZAH, ZAJ, ZAK, ZAL SCREW, CAP, HEXAGON H UP TO ENGINE S/ N 44487829	14
11	PAFZZ	5305013319480	15434	3917728	UOC:ZAA,ZAB,ZAC,ZAD,ZAE,ZAF,ZAG,ZAH, ZAJ, ZA, ZAL SCREW, CAP, HEXAGON H AFTER ENGINE S/ N 44487830	14

UOC:ZAA,ZAB,ZAC,ZAD,ZAE,ZAF,ZAG,ZAH, ZAJ, ZAK, ZAL

END OF FIGURE

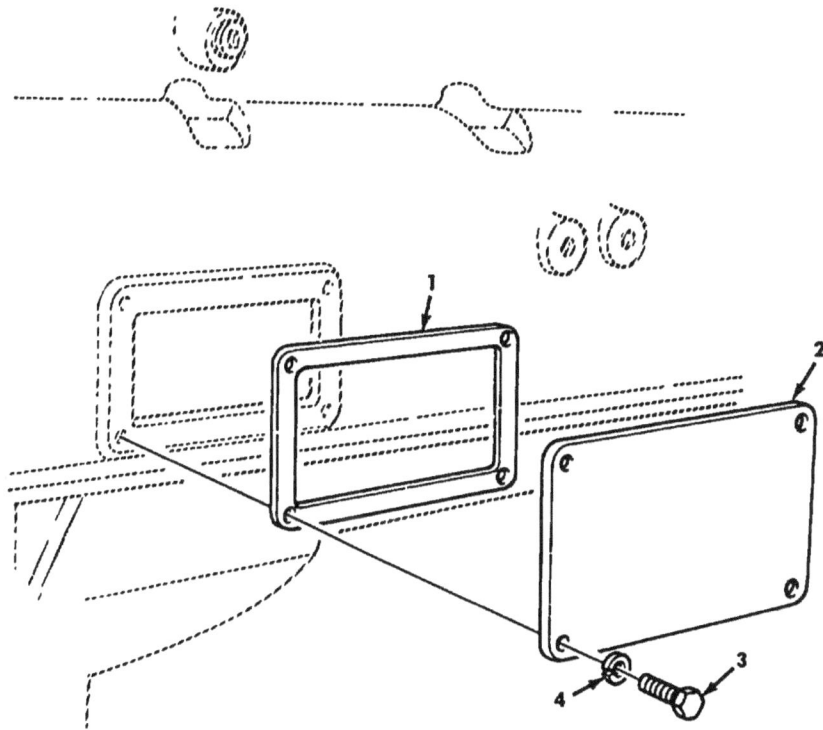

Figure 11. Engine Access Cover (M939, M939A1).

(1) ITEM NO	(2) SMR CODE	(3) NSN	(4) CAGEC	(5) PART NUMBER	(6) DESCRIPTION AND USABLE ON CODES (UOC)	(7) QTY
					GROUP 0101 CRANKCASE, BLOCK, CYLINDER HEAD	
					FIG. 11 ENGINE ACCESS COVER (M939, M939A1)	
1	PAOZZ	5330002460309	93568	00265-0014	GASKET PART OF KIT P/N 3011472 .... UOC:DAA,DAB,DAC,DAD,DAE,DAF,DAG,DAH, DAJ,DAK,DAL,DAW,DAX,V12,V13,V14,V15, V16,V17,V18,V19,V20,V21,V22,V24,V25, V39	1
2	PFOZZ	5340004042947	15434	158145	COVER, ACCESS .... UOC:DAA,DAB,DAC,DAD,DAE,DAF,DAG,DAH, DAJ,DAK,DAL,DAW,DAX,V12,V13,V14,V15, V16,V17,V18,V19,V20,V21,V22,V24,V25, V39	1
3	PAOZZ	5305010858197	15434	3010595	SCREW, MACHINE .... UOC:DAA,DAB,DAC,DAD,DAE,DAF,DAG,DAH, DAJ,DAK,DAL,DAW,DAX,V12,V13,V14,V15, V16,V17,V18,V19,V20,V21,V22,V24,V25, V39	4
4	PAOZZ	5310000806004	96906	MS27183-14	WASHER, FLAT .... UOC:DAA,DAB,DAC,DAD,DAE,DAF,DAG,DAH, DAJ,DAK,DAL,DAW,DAX,V12,V13,V14,V15, V16,V17,V18,V19,V20,V21,V22,V24,V25, V39	4

END OF FIGURE

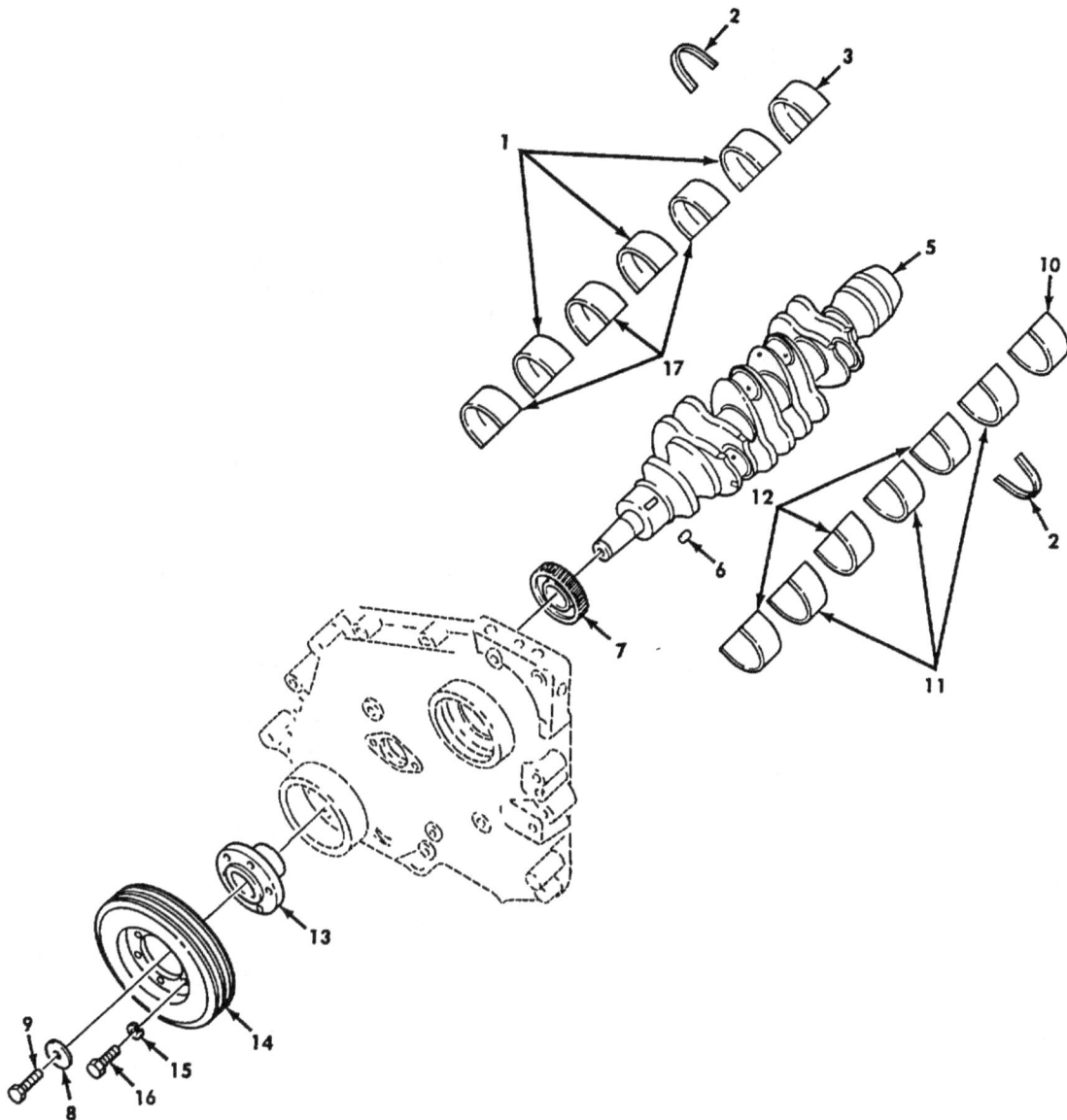

Figure 12. Crankshaft Assembly (M939, M939A1).

(1) ITEM NO	(2) SMR CODE	(3) NSN	(4) CAGEC	(5) PART NUMBER	(6) DESCRIPTION AND USABLE ON CODES (UOC)	(7) QTY
					GROUP 0102 CRANKSHAFT	

FIG. 12 CRANKSHAFT ASSEMBLY(M939, M939A1)

1	KFHZZ	15434	3019189	BEARING HALF .030 PART OF KIT P/N.................. 3801263	3
				UOC:DAA, DAB, DAC, DAD, DAE, DAF, DAG, DAH, DAJ, DAK, DAL, DAW, DAX, V12, V13, V14, V15, V16, V17, V18, V19, V20, V21, V22, V24, V25, V39	
1	KFHZZ	15434	3019190	BEARING HALF .040 PART OF KIT P/N.................. 3801264	3
				UOC:DAA, DAB, DAC, DAD, DAE, DAF, DAG, DAH, DAJ, DAK, DAL, DAW, DAX, V12, V13 , V13 4 , V15, V16, V17, V18, V19, V20, V21, V22, V24, V25, V39	
1	KFHZZ	15434	3019186	BEARING HALF, SLEEVE STD PART OF KIT........... P/N 3801260..............................	3
				UOC:DAA, DAB, DAC, DAD, DAE, DAF, DAG, DAH, DAJ, DAK, DAL, DAW, DAX, V12, V13, V14, V15, V16, V17, V18, V19, V20, V21, V22, V24, V25, V39	
1	KFHZZ	15434	3019187	BEARING HALF .010 PART OF KIT P/N.................. 3801261	3
				UOC:DAA, DAB, DAC, DAD, DAE, DAF, DAG, DAH, DAJ, DAK, DAL, DAW, DAX, V12, V13, V14, V15, V16 , V17, V18, V19, V20 , V21, V22, V24 , V25, V39	
1	KFHZZ	15434	3019188	BEARING HALF .020 PART OF KIT P/N.................. 3801262	3
				UOC:DAA, DAB, DAC, DAD, DAE, DAF, DAG, DAH, DAJ, DAK, DAL, DAW, DAX, V12, V13, V14, V15, V16, V17, V18, V19, V20, V21, V22, V24, V25, V39	
2	KFHZZ	15434	3019218	BEARING, WASHER, THRU PART OF KIT P/N........ 3801263 PART OF KIT P/N 3801264 PART OF KIT P/N 3801260 PART OF KIT P/N 3801261 PART OF KIT P/N 3801262	4
				UOC:DAA, DAB, DAC, DAD, DAE, DAF, DAG, DAH, DAJ, DAK, DAL, DAW, DAX, V12 , V13, V14, V15, V16, V17, V18, V19, V20, V21, V22, V24, V25, V39	
3	KFHZZ	15434	3019201	BEARING HALF .030 PART OF KIT P/N.................. 3801263	1
				UOC:DAA, DAB, DAC, DAD, DAE, DAF, DAG, DAH, DAJ, DAK, DAL, DAW, DAX, V12, V13, V14, V15, V16, V17, V18, V19, V20, V21, V22, V24, V25, V39	
3	KFHZZ	15434	3019202	BEARING HALF .040 PART OF KIT P/N.................. 3801264	1
				UOC:DAA, DAB, DAC, DAD, DAE, DAF, DAG, DAH,	

(1) ITEM NO	(2) SMR CODE	(3) NSN	(4) CAGEC	(5) PART NUMBER	(6) DESCRIPTION AND USABLE ON CODES (UOC)	(7) QTY
					DAJ,DAK,DAL,DAW,DAX,V12,V13,V14,V15, V16,V17,V1B,V19,V20,V21,V22,V24,V25, V39	
3	KFHZZ		15434	3019198	BEARING HALF,SLEEVE STD PART OF KIT............ P/N 3801260... UOC:DAA,DAB,DAC,DAD,DAE,DAF,DAG,DAH, DAJ,DAK,DAL,DAW,DAX,V12,V13,V14,V15, V16,V17,V18,V19,V20,V21,V22,V24,V25, V39	1
3	KFHZZ		15434	3019199	BEARING HALF .010 PART OF KIT P/N................. 3801261 .... UOC:DAA,DAB,DAC,DAD,DAE,DAF,DAG,DAH, DAJ,DAK,DAL,DAW,DAX,V12,V13,V14,V15, V16,V17,V18,V19,V20,V21,V22,V24,V25, V39	1
3	KFHZZ		15434	3019200	BEARING HALF .020 PART OF KIT P/N................. 3801262 .... UOC:DAA,DAB,DAC,DAD,DAE,DAF,DAG,DAH, DAJ,DAK,DAL,DAW,DAX,V12,V13,V14,V15, V16,V17,V18,V19,V20,V21,V22,V24,V25, V39	1
4	PBHHZ	2815004576311	15434	3801408	CRANKSHAFT ASSEMBLY .... UOC:DAA,DAB,DAC,DAD,DAE,DAF,DAG,DAH, DAJ,DAK,DAL,DAW,DAX,V12,V13,V14,V15, V16,V17,V18,V19,V20,V21,V22,V24,V25, V39	1
5	PBHHH	2815004538994	15434	208460	.CRANKSHAFT,ENGINE .... UOC:DAA,DAB,DAC,DAD,DAE,DAF,DAG,DAH, DAJ,DAK,DAL,DAW,DAX,V12,V13,V14,V15, V16,V17,V18,V19,V20,V21,V22,V24,V25, V39	1
6	PAHZZ	5315007812026	96906	MS20068-271	.KEY,MACHINE .... UOC:DAA,DAB,DAC,DAD,DAE,DAF,DAG,DAH, DAJ,DAK,DAL,DAW,DAX,V12,V13,V14,V15, V16,V17,V18,V19,V20,V21,V22,V24,V25, V39	1
7	KFHZZ		15434	142804	.GEAR,HELICAL PART OF KIT P/N 5704533 .... UOC:DAA,DAB,DAC,DAD,DAE,DAF,DAG,DAB, DAJ,DAK,DAL,DAW,DAX,V12,V13,V14,V15, V16,V17,V18,V19,V20,V21,V22,V24,V25, V39	1
8	PAFZZ	5310008295238	15434	140411	.WASHER,FLAT.. UOC: DAA,DAB, DAC, DAD, DAE, DAF, DAG, DAH, DAJ,DAK,DAL,DAW,DAX,V12,V13,V14,V15, V16,V17,V18,V19,V20,V21,V22,V24,V25, V39	1
9	PAFZZ	5306004201691	15434	140410	.BOLT,MACHINE.. UOC:DAA,DAB,DAC,DAD,DAE,DAF,DAG,DAH, DAJ,DAK,DAL,DAW,DAX,V12 ,V13,V14,V15, V16,V17,V18,V19,V20,V21,V22,V24,V25, V39	1

(1) ITEM NO	(2) SMR CODE	(3) NSN	(4) CAGEC	(5) PART NUMBER	(6) DESCRIPTION AND USABLE ON CODES (UOC)	(7) QTY
10	KFHZZ		15434	3019207	BEARING HALF .030 PART OF KIT P/N................ 3801263 .... UOC:DAA,DAB,DAC,DAD,DAE,DAF,DAG,DAH, DAJ,DAK,DAL,DAW,DAX,V12,V13,V14,V15, V16,V17,V18,V19,V20,V21,V22,V24,V25, V39	1
10	KFHZZ		15434	3019208	BEARING HALF .040 PART OF KIT P/N................ 3801264 .... UOC:DAA,DAB,DAC,DAD,DAE,DAF,DAG,DAH, DAJ,DAK,DAL,DAW,DAX,V12 ,V13 ,V14 ,V15, V16,V17,V18,V19,V20,V21,V22,V24,V25, V39	1
10	KFHZZ	3120005931507	15434	3019204	BEARING HALF,SLEEVE STD PART OF KIT............ P/N 3801260................................ UOC:DAA,DAB,DAC,DAD,DAE,DAF,DAG,DAH, DAJ,DAK,DAL,DAW,DAX,V12,V13,V14,V15, V16,V17 ,V18,V19 ,V20,V21 ,V22,V24,V25, V39	1
10	KFHZZ		15434	3019205	BEARING HALF .010 PART OF KIT P/N................ 3801261 .... UOC:DAA,DAB,DAC,DAD,DAE,DAF,DAG,DAB, DAJ,DAK,DAL,DAW,DAX,V12,V13,V14,V15, V16,V17,V18,V19,V20,V21,V22,V24,V25, V39	1
10	KFHZZ		15434	3019206	BEARING HALF .020 PART OF KIT P/N................ 3801262 .... UOC:DAA,DAB,DAC,DAD,DAE,DAF,DAG,DAH, DAJ,DAK,DAL,DAW,DAX,V12,V13,V14,V15, V16,V17 ,V1 ,V19 ,V20,V21,V22,V24,V25, V39	1
11	KFHZZ		15434	3019195	BEARING HALF .030 .030 PART OF KIT................ P/N 3801263................................ UOC:DAA,DAB,DAC,DAD,DAE,DAF,DAG,DAH, DAJ,DAK,DAL,DAW,DAX,V12,V13,V14,V15, V16,V17,V18,V19,V20,V21,V22,V24,V25, V39	3
11	KFHZZ		15434	3019196	BEARING HALF .040 .040 PART OF KIT................ P/N 3801264................................ UOC:DAA,DAB,DAC,DAD,DAE,DAF,DAG,DAH, DAJ,DAK,DAL,DAW,DAX,V12 ,V13,V14,V15, V16,V17,V18,V19,V20 1,V2,V22,V24,V25, V39	3
11	KFHZZ		15434	3019193	BEARING HALF .010 PART OF KIT P/N................ 3801261 .... UOC:DAA,DAB,DAC,DAD,DAE,DAF,DAG,DAH, DAJ,DAK,DAL,DAW,DAX,V12 ,V13,V14,V15, V16,V17,V18,V19,V20,V21,V22,V24,V25, V39	3
11	KFHZZ		15434	3019194	BEARING HALF .020 PART OF KIT P/N................ 3801262 .... UOC:DAA,DAB,DAC,DAD,DAE,DAF,DAG,DAB, DAJ,DAK,DAL,DAW,DAX,V12,V13,V14,V15,	3

(1) ITEM NO	(2) SMR CODE	(3) NSN	(4) CAGEC	(5) PART NUMBER	(6) DESCRIPTION AND USABLE ON CODES (UOC)	(7) QTY
					V16,V17,V18,V19,V20,V21,V22,V24,V25, V39	
11	KFHZZ		15434	3019192	BEARING HALF STD PART OF KIT P/N................. 3801260 .... UOC:DAA,DAB,DAC,DAD,DAE,DAF,DAG,DAH, DAJ,DAK,DAL,DAW,DAX,V12,V13,V14,V15, V16,V17,V18,V19,V20,V21,V22,V24,V25, V39	3
12	KFHZZ		15434	3019183	BEARING HALF .030 .030 PART OF KIT................. P/N 3801263.. UOC:DAA,DAB,DAC,DAD,DAE,DAF,DAG,DAH, DAJ,DAK,DAL,DAW,DAX,V12,V13,V14,V15, V16,V17,V18,V19,V20,V21,V22,V24,V25, V39	3
12	KFHZZ		15434	3019184	BEARING HALF .040 PART OF KIT P/N................. 3801264 .... UOC:DAA,DAB,DAC,DAD,DAE,DAF,DAG,DAH, DAJ,DAK,DAL,DAW,DAX,V12,V13,V14,V15, V16,V17,V18,V19,V20,V21,V22,V24,V25, V39	3
12	KFHZZ	3120006951232	15434	3019180	BEARING HALF,SLEEVE STD PART OF KIT............ P/N 3801260.. UOC:DAA,DAB,DAC,DAD,DAE,DAF,DAG,DAH, DAJ,DAK,DAL,DAW,DAX,V12,V13,V14,V15, V16,V17,V18,V19,V20,V21,V22,V24,V25, V39	3
12	KFHZZ		15434	3019181	BEARING HALF .010 PART OF KIT P/N................. 3801261 .... UOC:DAA,DAB,DAC,DAD,DAE,DAF,DAG,DAH, DAJ,DAK,DAL,DAW,DAX,V12,V13,V14,V15, V16,V17,V18,V19,V20,V21,V22,V24,V25, V39	3
12	KFHZZ		15434	3019182	BEARING,HALF 020.PART OF KIT P/N.................. 3801262 .... UOC:DAA,DAB,DAC,DAD,DAE,DAF,DAG,DAH, DAJ,DAK,DAL,DAW,DAX,V12,V13,V14,V15, V16,V17,V18,V19,V20,V21,V22,V24,V25, V39	3
13	PAFZZ	2815004722626	15434	202891	DAMPENER,VIBRATION, ... UOC:DAA,DAB,DAC,DAD,DAE,DAF,DAG,DAH, DAJ,DAK,DAL,DAW,DAX,V12,V13,V14,V15, V16,V17,V18,V19,V20,V21,V22,V24,V25, V39	1
14	PAFZZ	2815012029715	15434	3017946	DAMPENER,VIBRATION, ... UOC:DAA,DAB,DAC,DAD,DAE,DAF,DAG,DAH, DAJ,DAK,DAL,DAW,DAX,V12,V13,V14,V15, V16,V17,V18,V19,V20,V21,V22,V24,V25, V39	1
15	PAFZZ	5310000034094	01276	210104-8S	WASHER,LOCK .... UOC:DAA,DAB,DAC,DAD,DAE,DAF,DAG,DAH, DAJ,DAK,DAL,DAW,DAX,V12,V13,V14,V15, V16,V17,V18,V19,V20,V21,V22,V24,V25, V39	6

(1) ITEM NO	(2) SMR CODE	(3) NSN	(4) CAGEC	(5) PART NUMBER	(6) DESCRIPTION AND USABLE ON CODES (UOC)	(7) QTY
16	PAFZZ	5305011140895	15434	S-112-A	SCREW,CAP,HEXAGON H .... UOC:DAA,DAB,DAC,DAD,DAE,DAF,DAG,DAH, DAJ,DAK,DAL,DAW,DAX,V12,V13,V14,V15, V16,V17,V18,V19,V20,V21,V22,V24,V25, V39	6
17	KFHZZ		15434	3019177	BEARING HALF .030 .030 PART OF KIT.................. P/N 3801263.. UOC:DAA,DAB,DAC,DAD,DAE,DAF,DAG,DAH, DAJ,DAK,DAL,DAW,DAX,V12,V13,V14,V15, V16,V17,V18,V19,V20,V21,V22,V24,V25, V39	3
17	KFHZZ		15434	3019178	BEARING HALF .040 PART OF KIT P/N.................. 3801264 .... UOC:DAA,DAB,DAC,DAD,DAE,DAF,DAG,DAH, DAJ,DAK,DAL,DAW,DAX,V12,V13,V14,V15, V16,V17,V18,V19,V20,V21,V22,V24,V25, V39	3
17	KFHZZ	3120012414098	15434	3019174	BEARING HALF,SLEEVE STD PART OF KIT............ P/N 3801260.. UOC:DAA,DAB,DAC,DAD,DAE,DAF,DAG,DAH, DAJ,DAK,DAL,DAW,DAX,V12,V13,V14,V15, V16,V17,V18,V19,V20,V21,V22,V24,V25, V39	3
17	KFHZZ		15434	3019175	BEARING HALF 010 PART OF KIT P/N .... 3801261 .... UOC:DAA,DAB,DAC,DAD,DAE,DAF,DAG,DAH, DAJ,DAK,DAL,DAW,DAX,V12,V13,V14,V15, V16,V17,V18,V19,V20,V21,V22,V24,V25, V39	3
17	KFHZZ		15434	3019176	BEARING HALF 020 PART OF KIT P/N .... 3801262 .... UOC:DAA,DAB,DAC,DAD,DAE,DAF,DAG,DAH, DAJ,DAK,DAL,DAW,DAX,V12,V13,V14,V15, V16,V17,V18,V19,V20,V21,V22,V24,V25, V39	3

END OF FIGURE

Figure 13. Crankshaft Assembly (M939A2).

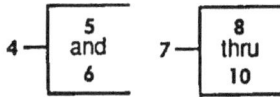

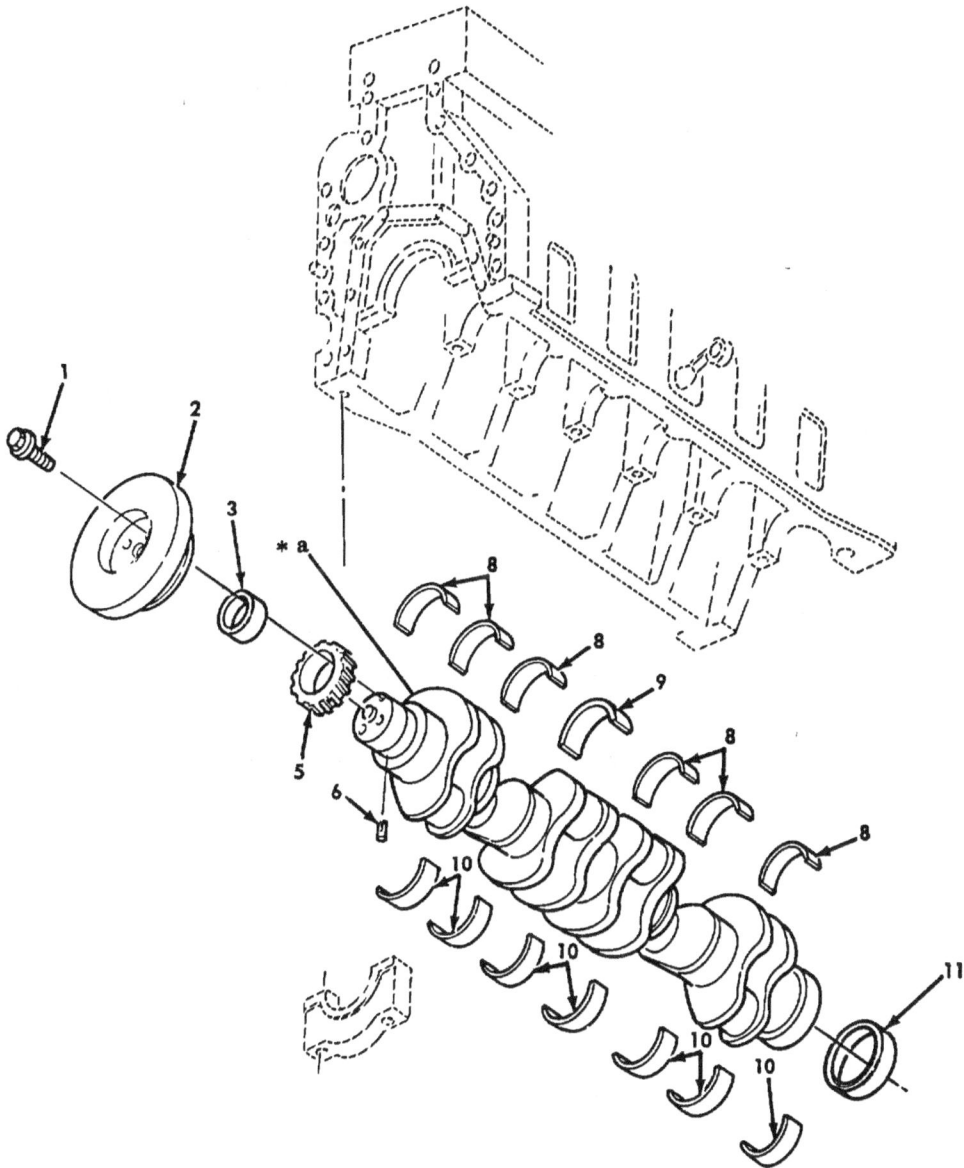

* a   PART OF ITEM 4

(1) ITEM NO	(2) SMR CODE	(3) NSN	(4) CAGEC	(5) PART NUMBER	(6) DESCRIPTION AND USABLE ON CODES (UOC)	(7) QTY
					GROUP 0102.CRANKSHAFT	
					FIG. 13 CRANKSHAFT ASSEMBLY(M939A2)	
1	PAFZZ	5305012745655	15434	3906733	SCREW,CAP,HEXAGON H .... UOC:ZAA,ZAB,ZAC,ZAD,ZAE,ZAF,ZAG,ZAH, ZAJ,ZAK,ZAL	4
2	PAFZZ	2815012717076	15434	3914456	DAMPENER,VIBRATION, ... UOC:ZAA,ZAB,ZAC,ZAD,ZAE,ZAF,ZAG,ZAH, ZAJ,ZAK,ZAL	1
3	PAHZZ	3120012661530	15434	3906081	BEARING,SLEEVE .... UOC:ZAA,ZAB,ZAC,ZAD,ZAE,ZAF,ZAG,ZAH, ZAJ,ZAK,ZAL	1
4	PBHHH	2815012715096	15434	3918986	CRANKSHAFT,ENGINE .... UOC:ZAA,ZAB,ZAC,ZAD,ZAE,ZAF,ZAG,ZAH, ZAJ,ZAK,ZAL	1
5	PAHZZ	3020012713812	15434	3918776	.GEAR,HELICAL .... UOC:ZAA,ZAB,ZAC,ZAD,ZAE,ZAF,ZAG,ZAH, ZAJ,ZAK,ZAL	1
6	PAHZZ	5315012708285	15434	3904483	.PIN,STRAIGHT,HEADLE .... UOC:ZAA,ZAB,ZAC,ZAD,ZAE,ZAF,ZAG,ZAH, ZAJ,ZAK,ZAL	1
7	PAHZZ	3120012734653	15434	3802210	BEARING SET,SLEEVE .... UOC:ZAA,ZAB,ZAC,ZAD,ZAE,ZAF,ZAG,ZAH, ZAJ,ZAK,ZAL	1
7	PAHZZ	3120012734654	15434	3802211	BEARING SET,SLEEVE UNDERSIZE, .25MM .... UOC:ZAA,ZAB,ZAC,ZAD,ZAE,ZAF,ZAG,ZAH, ZAJ,ZAK,ZAL	1
7	PAHZZ	3120012734655	15434	3802212	BEARING SET,SLEEVE UNDERSIZE, .50MM .... UOC:ZAA,ZAB,ZAC,ZAD,ZAE,ZAF,ZAG,ZAH, ZAJ,ZAK,ZAL	1
7	PAHZZ	3120012734656	15434	3802213	BEARING SET,SLEEVE UNDERSIZE, .75MM .... UOC:ZAA,ZAB,ZAC,ZAD,ZAE,ZAF,ZAG,ZAH, ZAJ,ZAK,ZAL	1
7	PAHZZ	3120012743377	15434	3802214	BEARING SET,SLEEVE UNDERSIZE, ... 1.00MM .... UOC:ZAA,ZAB,ZAC,ZAD,ZAE,ZAF,ZAG,ZAH, ZAJ,ZAK,ZAL	1
8	XAHZZ		15434	3916840	.BEARING, MAIN UPPER .... UOC:ZAA,ZAB,ZAC,ZAD,ZAE,ZAF,ZAG,ZAH, ZAJ,ZAK,ZAL	6
9	XAHZZ		15434	3916830	.BEARING, FLANGE THRUST .... UOC:ZAA,ZAB,ZAC,ZAD,ZAE,ZAF,ZAG,ZAH, ZAJ,ZAK,ZAL	1
10	XAHZZ		15434	3901590	.BEARING, MAIN LOWER .... UOC:ZAA,ZAB,ZAC,ZAD,ZAE,ZAF,ZAG,ZAH, ZAJ,ZAK,ZAL	7
11	PAHZZ	3120012806566	15434	3925626	BEARING,SLEEVE .... UOC:ZAA,ZAB,ZAC,ZAD,ZAE,ZAF,ZAG,ZAH, ZAJ,ZAK,ZAL	1

END OF FIGURE

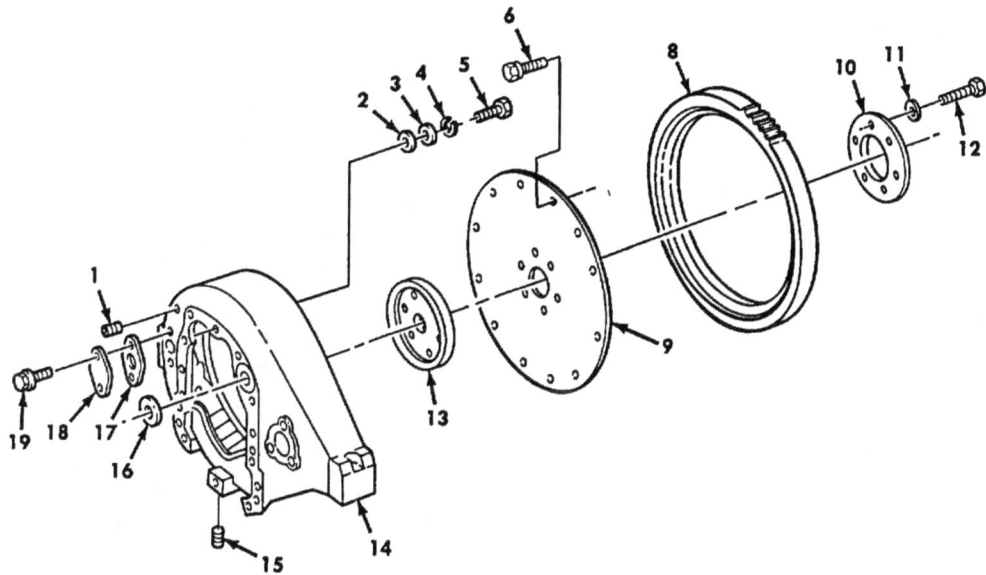

Figure 14. Flywheel Gear Assembly and Housing (M939, M939A1).

(1) ITEM NO	(2) SMR CODE	(3) NSN	(4) CAGEC	(5) PART NUMBER	(6) DESCRIPTION AND USABLE ON CODES (UOC)	(7) QTY
					GROUP 0103 FLYWHEEL ASSEMBLY	
					FIG. 14 FLYWHEEL GEAR ASSEMBLY AND HOUSING(M939,M939A1)	
1	PAFZZ	4730011659491	15434	3008465	PLUG,PIPE .... UOC:DAA,DAB,DAC,DAD,DAE,DAF,DAG,DAH, DAJ,DAK,DAL,DAW,DAX,V12,V13,V14,V15, V16,V17,V18,V19,V20,V21,V22,V24,V25, V39	1
2	PAFZZ	5330004042920	15434	172648	PACKING,PREFORMED PART OF KIT P/N .... 3011472 .... UOC:DAA,DAB,DAC,DAD,DAE,DAF,DAG,DAH, DAJ,DAK,DAL,DAW,DAX,V12,V13,V14,V15, V16,V17,V18,V19,V20,V21,V22,V24,V25, V39	9
3	PAFZZ	5310000818500	15434	127316	WASHER,RECESSED .... UOC:DAA,DAB,DAC,DAD,DAE,DAF,DAG,DAH, DAJ,DAK,DAL,DAW,DAX,V12,V13,V14,V15, V16,V17,V18,V19,V20,V21,V22,V24,V25, V39	9
4	PAFZZ	5310008206653	96906	MS35338-50	WASHER,LOCK .... UOC:DAA,DAB,DAC,DAD,DAE,DAF,DAG,DAH, DAJ,DAK,DAL,DAW,DAX,V12,V13,V14,V15, V16,V17,V18,V19,V20,V21,V22,V24,V25, V39	9
5	PAFZZ	5305012053407	15434	138042	SCREW,MACHINE .... UOC:DAA,DAB,DAC,DAD,DAE,DAF,DAG,DAH, DAJ,DAK,DAL,DAW,DAX,V12,V13,V14,V15, V16,V17,V18,V19,V20,V21,V22,V24,V25, V39	9
6	PAFZZ	5306012036299	15434	3011315	BOLT .... UOC:DAA,DAB,DAC,DAD,DAE,DAF,DAG,DAH, DAJ,DAK,DAL,DAW,DAX,V12,V13,V14,V15, V16,V17,V18,V19,V20,V21,V22,V24,V25, V39	12
7	PAFZZ	2815011147397	15434	3007279	FLYWHEEL ASSEMBLY .... UOC:DAA,DAB,DAC,DAD,DAE,DAF,DAG,DAH, DAJ,DAK,DAL,DAW,DAX,V12,V13,V14,V15, V16,V17,V18,V19,V20,V21,V22,V24,V25, V39	1
8	XAFZZ		15434	217381	.GEAR,STARTER RING .... UOC:DAA,DAB,DAC,DAD,DAE,DAF,DAG,DAH, DAJ,DAK,DAL,DAW,DAX,V12,V13,V14,V15, V16,V17,V18,V19,V20,V21,V22,V24,V25, V39	1
9	XAFZZ		15434	3019077	.FLEXPLATE... UOC:DAA,DAB,DAC,DAD,DAE,DAF,DAG,DAH, DAJ,DAK,DAL,DAW,DAX,V12,V13,V14,V15, V16,V17,V18,V19,V20,V21,V22,V24,V25, V39	1
10	PAFZZ	2520011276254	15434	217385	DISK,CLUTCH	1

(1) ITEM NO	(2) SMR CODE	(3) NSN	(4) CAGEC	(5) PART NUMBER	(6) DESCRIPTION AND USABLE ON CODES (UOC)	(7) QTY
					UOC:DAA,DAB,DAC,DAD,DAE,DAF,DAG,DAH, DAJ,DAK,DAL,DAW,DAX,V12,V13,V14,V15, V16,V17,V18,V19,V20,V21,V22,V24,V25, V39	
11	PAFZZ	5310001344171	15434	200861	WASHER,FLAT .... UOC:DAA,DAB,DAC,DAD,DAE,DAF,DAG,DAH, DAJ,DAK,DAL,DAW,DAX,V12,V13,V14,V15, V16,V17,V18,V19,V20,V21,V22,V24,V25, V39	6
12	PAFZZ	5305011129698	15434	180175	SCREW,CAP,HEXAGON H .... UOC:DAA,DAB,DAC,DAD,DAE,DAF,DAG,DAH, DAJ,DAK,DAL,DAW,DAX,V12,V13,V14,V15, V16,V17,V18,V19,V20,V21,V22,V24,V25, V39	6
13	PAFZZ	2815011267404	15434	3007148	ADAPTER,CRANKSHAFT .... UOC:DAA,DAB,DAC,DAD,DAE,DAF,DAG,DAH, DAJ,DAK,DAL,DAW,DAX,V12,V13,V14,V15, V16,V17,V18,V19,V20,V21,V22,V24,V25, V39	1
14	PFFZZ	2805004042917	15434	3014979	HOUSING,FLYWHEEL .... UOC:DAA,DAB,DAC,DAD,DAE,DAF,DAG,DAH, DAJ,DAK,DAL,DAW,DAX,V12,V13,V14,V15, V16,V17,V18,V19,V20,V21,V22,V24,V25, V39	1
15	PAFZZ	4730011060202	15434	3008469	PLUG,PIPE.. UOC:DAA,DAB,DAC,DAD,DAE,DAF,DAG,DAH, DAJ,DAK,DAL,DAW,DAX,V12,V13,V14,V15, V16,V17,V18,V19,V20,V21,V22,V24,V25, V39	1
16	PAFZZ	5330007294427	15434	9333-1	GASKET PART OF KIT P/N 3011472..... UOC:DAA,DAB,DAC,DAD,DAE,DAF,DAG,DAH, DAJ,DAK,DAL,DAW,DAX,V12,V13,V14,V15, V16,V17,V18,V19,V20,V21,V22,V24,V25, V39	1
17	PAFZZ	5330010826985	15434	3021735	GASKET .... UOC:DAA,DAB,DAC,DAD,DAE,DAF,DAG,DAH, DAJ,DAK,DAL,DAW,DAX,V12,V13,V14,V15, V16,V17,V18,V19,V20,V21,V22,V24,V25, V39	1
18	PFFZZ	5340011228002	15434	70657	COVER,ACCESS .... UOC:DAA,DAB,DAC,DAD,DAE,DAF,DAG,DAH, DAJ,DAK,DAL,DAW,DAX,V12,V13,V14,V15, V16,V17,V18,V19,V20,V21,V22,V24,V25, V39	1
19	PAFZZ	5305011355344	15434	3011342	SCREW .... UOC:DAA,DAB,DAC,DAD,DAE,DAF,DAG,DAH, DAJ,DAK,DAL,DAW,DAX,V12,V13,V14,V15, V16,V17,V18,V19,V20,V21,V22,V24,V25, V39	2

END OF FIGURE

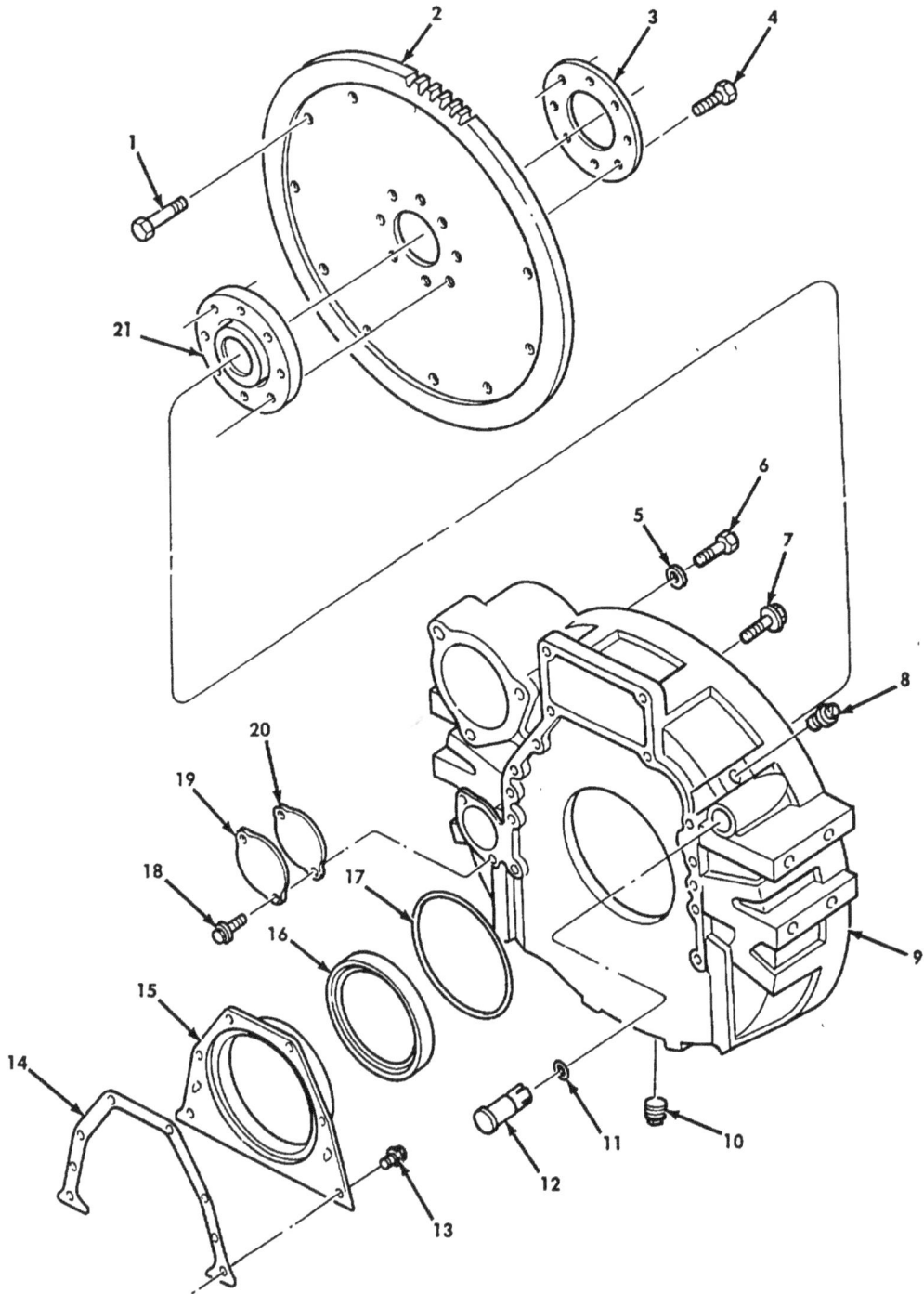

Figure 15. Flywheel Housing and Flexplate (M939A2).

(1) ITEM NO	(2) SMR CODE	(3) NSN	(4) CAGEC	(5) PART NUMBER	(6) DESCRIPTION AND USABLE ON CODES (UOC)	(7) QTY
					GROUP 0103 FLYWHEEL ASSEMBLY	
					FIG. 15 FLYWHEEL HOUSING AND FLEXPLATE(M939A2)	
1	PAFZZ	5306012813387	3T063	20510312	BOLT,SELF-LOCKING .... UOC:ZAA,ZAB,ZAC,ZAD,ZAE,ZAF,ZAG,ZAH, ZAJ,ZAK,ZAL	12
2	PAFZZ	3020012717114	15434	3911260	GEAR,SPUR................. UOC:ZAA,ZAB,ZAC,ZAD,ZAE,ZAF,ZAG,ZAH, ZAJ,ZAK,ZAL	1
3	PAFZZ	3040012717165	15434	3904361	PLATE,RETAINING,SHA .... UOC:ZAA,ZAB,ZAC,ZAD,ZAE,ZAF,ZAG,ZAH, ZAJ,ZAK,ZAL	1
4	PAFZZ	5305011925677	15434	3901395	SCREW,CAP,HEXAGON H .... UOC:ZAA,ZAB,ZAC,ZAD,ZAE,ZAF,ZAG,ZAH, ZAJ,ZAK,ZAL	8
5	PAFZZ	5310012822807	3T063	20510912	WASHER,FLAT .... UOC:ZAA,ZAB,ZAC,ZAD,ZAE,ZAF,ZAG,ZAH, ZAJ,ZAK,ZAL	12
6	PAFZZ	5306012834199	3T063	20510498	BOLT,SELF-LOCKING .... UOC:ZAA,ZAB,ZAC,ZAD,ZAE,ZAF,ZAG,ZAH, ZAJ,ZAK,ZAL	12
7	PAFZZ	5305013408395	15434	3920447	SCREW,CAP,HEXAGON H .... UOC:ZAA,ZAB,ZAC,ZAD,ZAE,ZAF,ZAG,ZAH, ZAJ,ZAK,ZAL	12
8	PAFZZ	4730012376950	15434	3904181	PLUG,PIPE................. UOC:ZAA,ZAB,ZAC,ZAD,ZAE,ZAF,ZAG,ZAH, ZAJ,ZAK,ZAL	1
9	PAFZZ	2815012726719	15434	3911604	HOUSING,FLYWHEEL .... UOC:ZAA,ZAB,ZAC,ZAD,ZAE,ZAF,ZAG,ZAH, ZAJ,ZAK,ZAL	1
10	PAFZZ	4730011060202	15434	3008469	PLUG,PIPE PART OF KIT P/N 3909415................. UOC:ZAA,ZAB,ZAC,ZAD,ZAE,ZAF,ZAG,ZAH, ZAJ, ZAK,ZAL	1
11	PAFZZ	5330012721124	15434	3910260	PACKING,PREFORMED .... UOC:ZAA,ZAB,ZAC,ZAD,ZAE,ZAF,ZAG,ZAH, ZAJ,ZAK,ZAL	1
12	PAFZZ	2815012730571	15434	3910248	PLUG,FLYWHEEL HOUSI .... UOC:ZAA,ZAB,ZAC,ZAD,ZAE,ZAF,ZAG,ZAH, ZAJ,ZAK,ZAL	1
13	PAFZZ	5305012715851	15434	3913638	SCREW,CAP,HEXAGON H .... UOC:ZAA,ZAB,ZAC,ZAD,ZAE,ZAF,ZAG,ZAH, ZAJ,ZAK,ZAL	8
14	PAFZZ	5330012714308	15434	3914301	GASKET PART OF KIT P/N 3802389 PART............. OF KIT P/N 3802389................. UOC:ZAA,ZAB,ZAC,ZAD,ZAE,ZAF,ZAG,ZAH, ZAJ,ZAK,ZAL	1
15	PAFZZ	5340012712497	15434	3907535	COVER,ACCESS................. UOC:ZAA,ZAB,ZAC,ZAD,ZAE,ZAF,ZAG,ZAH, ZAJ,ZAK,ZAL	1
16	PAFZZ	5330011922037	15434	3909410	SEAL,SPECIAL PART OF KIT P/N 3802389	1

(1) ITEM NO	(2) SMR CODE	(3) NSN	(4) CAGEC	(5) PART NUMBER	(6) DESCRIPTION AND USABLE ON CODES (UOC)	(7) QTY
					PART OF KIT P/N 3802389................................ UOC:ZAA,ZAB,ZAC,ZAD,ZAE,ZAF,ZAG,ZAH, ZAJ,ZAK,ZAL	
17	PAFZZ	5330012719375	15434	3911537	PACKING,PREFORMED PART OF KIT P/N............. 3802389 PART OF KIT P/N 3802389 .... UOC:ZAA,ZAB,ZAC,ZAD,ZAE,ZAF,ZAG,ZAR, ZAJ,ZAK,ZAL	1
18	PAFZZ	5305011922036	15434	3912072	SCREW .... UOC:ZAA,ZAB,ZAC,ZAD,ZAE,ZAF,ZAG,ZAH, ZAJ,ZAK,ZAL	2
19	PAFZZ	5340012712496	15434	3908095	COVER,ACCESS .... UOC:ZAA,ZAB,ZAC,ZAD,ZAE,ZAF,ZAG,ZAH, ZAJ,ZAK,ZAL	1
20	PAFZZ	5330012663294	15434	3908096	GASKET .... UOC:ZAA,ZAB,ZAC,ZAD,ZAE,ZAF,ZAG,ZAH, ZAJ,ZAK,ZAL	1
21	PAFZZ	2990012717086	15434	3911258	ADAPTER,FLYWHEEL HO .... UOC:ZAA,ZAB,ZAC,ZAD,ZAE,ZAF,ZAG,ZAH, ZAJ,ZAK,ZAL	1

END OF FIGURE

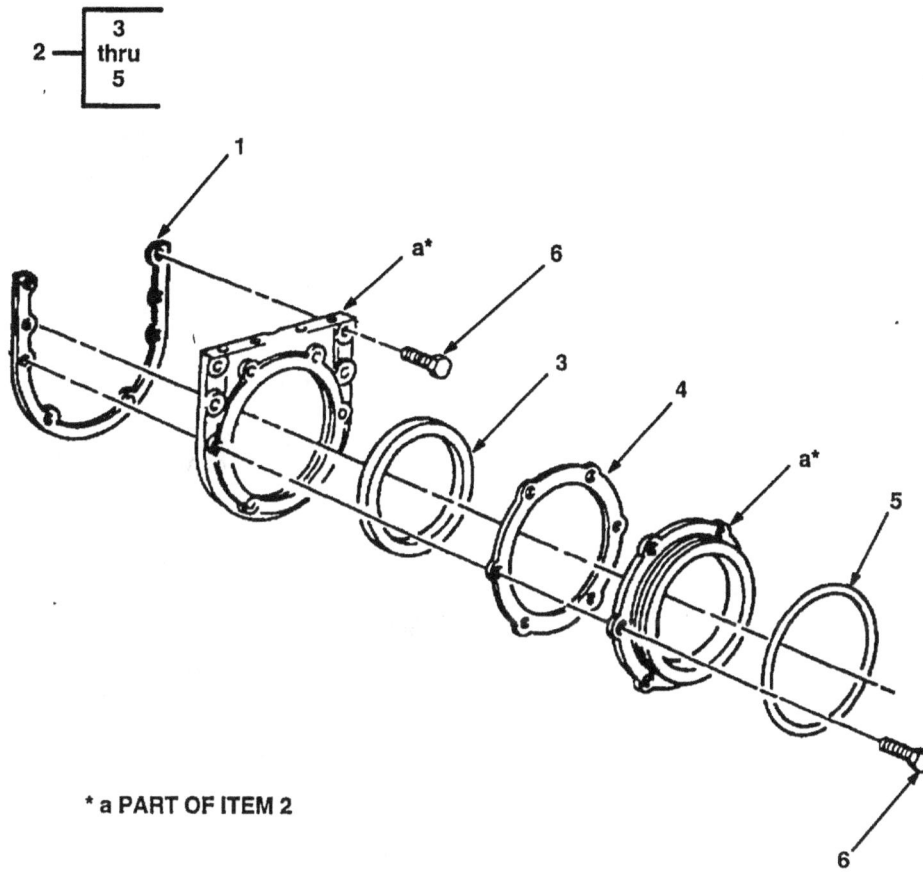

Figure 16. Crankshaft Rear Cover (M939, M939A1)

(1) ITEM NO	(2) SMR CODE	(3) NSN	(4) CAGEC	(5) PART NUMBER	(6) DESCRIPTION AND USABLE ON CODES (UOC)	(7) QTY
					GROUP 0103 FLYWHEEL ASSEMBLY	
					FIG. 16 CRANKSHAFT REAR COVER(M939, M939A1)	
1	PAFZZ	5330003612955	15434	3067616	GASKET.. UOC:DAA,DAB,DAC,DAD,DAE,DAF,DAG,DAH, DAJ,DAK,DAL,DAW,DAX,V12,V13,V14,V15, V16,V17,V18,V19,V20,V21,V22,V24,V25, V39	1
2	PAFZZ	2990011202883	15434	216165	COVER,REAR .... UOC:DAA,DAB,DAC,DAD,DAE,DAF,DAG,DAH, DAJ,DAK,DAL,DAW,DAX,V12,V13,V14,V15, V16,V17,V18,V19,V20,V21,V22,V24,V25, V39	1
3	PAFZZ	5330000050858	01212	M39807	.SEAL REAR COVER PART OF KIT P/N.................. 3011472 .... UOC:DAA,DAB,DAC,DAD,DAE,DAF,DAG,DAH, DAJ,DAK,DAL,DAW,DAX,V12,V13,V14,V15, V16,V17,V18,V19,V20,V21,V22,V24,V25, V39	1
4	PAFZZ	5330010821906	15434	151623	.GASKET PART OF KIT P/N 3011472 .... UOC:DAA,DAB,DAC,DAD,DAE,DAF,DAG,DAH, DAJ,DAK,DAL,DAW,DAX,V12,V13,V14,V15, V16,V17,V18,V19,V20,V21,V22,V24,V25, V39	1
5	PAFZZ	5330004209624	15434	137075	.O-RING PART OF KIT P/N 3011472 .... UOC:DAA,DAB,DAC,DAD,DAE,DAF,DAG,DAH, DAJ,DAK,DAL,DAW,DAX,V12,V13,V14,V15, V16,V17,V18,V19,V20,V21,V22,V24,V25, V39	1
6	PAFZZ	5305011306100	15434	3010594	SCREW,CAP,HEXAGON H .... UOC:DAA,DAB,DAC,DAD,DAE,DAF,DAG,DAH, DAJ,DAK,DAL,DAW,DAX,V12,V13,V14,V15, V16,V17,V18,V19,V20,V21,V22,V24,V25, V39	8

END OF FIGURE

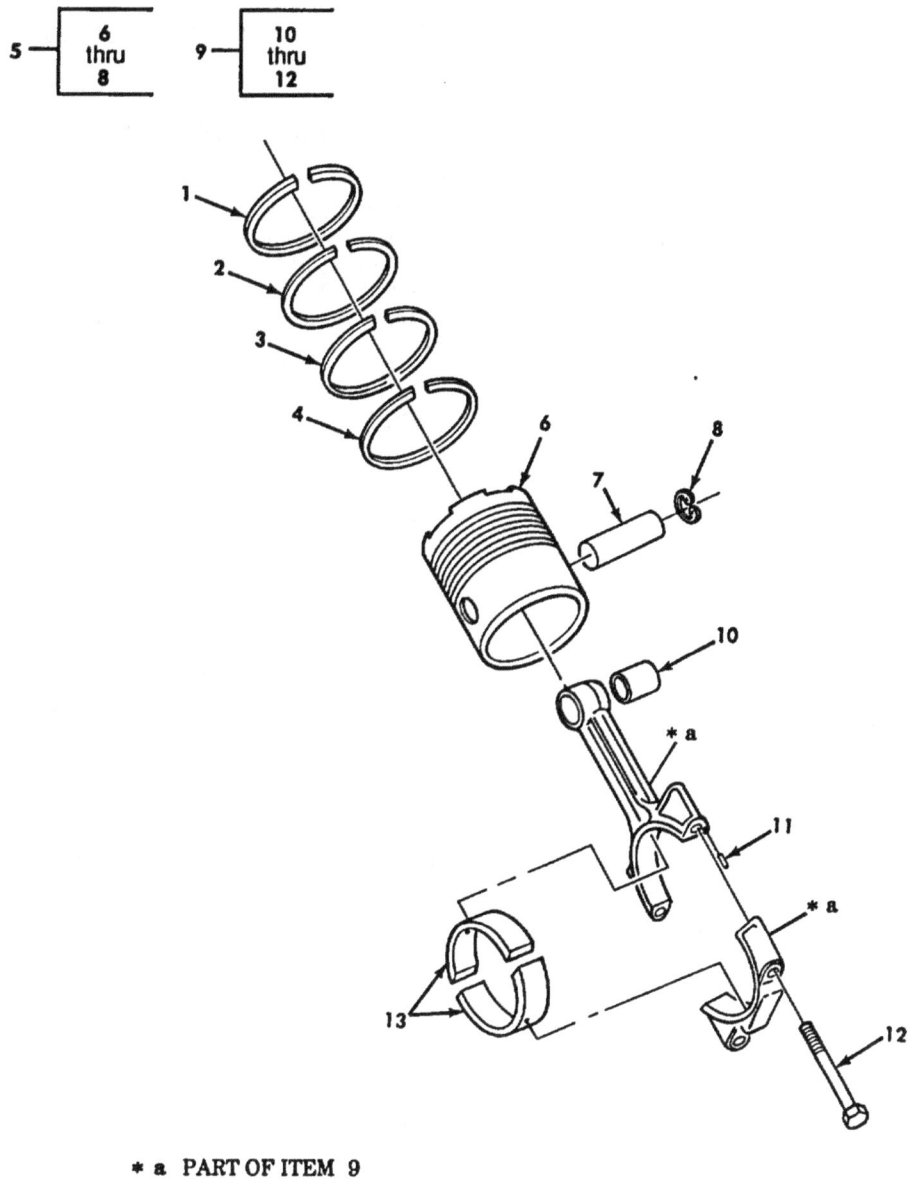

* a PART OF ITEM 9

Figure 17. Piston and Rod Assembly (M939, M939A1).

(1) ITEM NO	(2) SMR CODE	(3) NSN	(4) CAGEC	(5) PART NUMBER	(6) DESCRIPTION AND USABLE ON CODES (UOC)	(7) QTY
					GROUP 0104 PISTONS, CONNECTING RODS	
					FIG. 17 PISTON AND ROD ASSEMBLY(M939, M939A1)	
1	KFHZZ	2815011434140	15434	218025	RING,PISTON PART OF KIT P/N 3804438............... PART OF KIT P/N 3801056 .... UOC:DAA,DAB,DAC,DAD,DAE,DAF,DAG,DAH, DAJ,DAK,DAL,DAW,DAX,V12,V13,V14,V15, V16,V17,V18,V19,V20,V21,V22,V24,V25, V39	6
2	KFHZZ	2815011455547	15434	216983	RING,PISTON PART OF KIT P/N 3804438............... PART OF KIT P/N 3801056.................................. UOC:DAA,DAB,DAC,DAD,DAE,DAF,DAG,DAH, DAJ,DAK,DAL,DAW,DAX,V12,V13,V14,V15, V16,V17,V18,V19,V20,V21,V22,V24,V25, V39	6
3	KFHZZ	2815012416581	15434	214730	RING,PISTON PART OF KIT P/N 3801056............ 6 UOC:DAA,DAB,DAC,DAD,DAE,DAF,DAG,DAH, DAJ,DAK,DAL,DAW,DAX,V12,V13,V14,V15, V16,V17,V18,V19,V20,V21,V22,V24,V25, V39	
4	KFHZZ	2815012143802	15434	218732	RING,PISTON PART OF KIT P/N 3801056............... UOC:DAA,DAB,DAC,DAD,DAE,DAF,DAG,DAH, DAJ,DAK,DAL,DAW,DAX,V12,V13,V14,V15, V16,V17,V18,V19,V20,V21,V22,V24,V25, V39	6
5	PAHZZ	2815000081741	01212	2622PN	PISTON,INTERNAL COM PART OF KIT P/N........... 226-1747 .... UOC:DAA,DAB,DAC,DAD,DAE,DAF,DAG,DAH, DAJ,DAK,DAL,DAW,DAX,V12,V13,V14,V15, V16,V17,V18,V19,V20,V21,V22,V24,V25, V39	6
6	KFHZZ		15434	3042320	.PISTON,INTERNAL COM USE AFTER .... SERIAL NUMBER 11246663 PART OF KIT P/N 3804438.. UOC:DAA,DAB,DAC,DAD,DAE,DAF,DAG,DAH, DAJ,DAK,DAL,DAW,DAX,V12,V13,V14,V15, V16,V17,V18,V19,V20,V21,V22,V24,V25, V39	1
6	KFHZZ	2815012421455	15434	3025516	.PISTON,INTERNAL COM USE PRIOR TO .... SERIAL NUMBER 11246663 PART OF KIT P/N 3804438.. UOC:V12,V13,V14,V145,V15,V167,VS,17,V18,V9, V20,V21,V22,V24,V25,V39	1
7	KFHZZ	2815004804347	15434	191970	.PIN,PISTON PART OF KIT P/N 3804438............... UOC:DAA,DAB,DAC,DAD,DAE,DAF,DAG,DAH, DAJ,DAK,DAL,DAW,DAX,V12,V13,V14,V15, V16,V17,V18,V19,V20,V21,V22,V24,V25, V39	1
8	PAHZZ	5325008042784	96906	MS16625-1200	RING,RETAINING PART OF KIT P/N...................... 3804438...	2

(1) ITEM NO	(2) SMR CODE	(3) NSN	(4) CAGEC	(5) PART NUMBER	(6) DESCRIPTION AND USABLE ON CODES (UOC)	(7) QTY
					UOC:DAA,DAB,DAC,DAD,DAE,DAF,DAG,DAH, DAJ,DAK,DAL,DAW,DAX,V12,V13,V14,V15, V16,V17,V18,V19,V20,V21,V22,V24,V25, V39	
9	PAHZZ	2815010864508	15434	3013930	CONNECTING ROD,PIST	1
					UOC:DAA,DAB,DAC,DAD,DAE,DAF,DAG,DAH, DAJ,DAK,DAL,DAW,DAX,V12,V13,V14,V15, V16,V17,V18,V19,V20,V21,V22,V24,V25, V39	
10	PAHZZ	5365001320273	15434	187420	.BUSHING,PISTON PIN	1
					UOC:DAA,DAB,DAC,DAD,DAE,DAF,DAG,DAH, DAJ,DAK,DAL,DAW,DAX,V12,V13,V14,V15, V16,V17,V18,V19,V20,V21,V22,V24,V25, V39	
11	PAHZZ	2815011240232	15434	70550	.PIN,PISTON	2
					UOC:DAA,DAB,DAC,DAD,DAE,DAF,DAG,DAH, DAJ,DAK,DAL,DAW,DAX,V12,V13,V14,V15, V16,V17,V18,V19,V20,V21,V22,V24,V25, V39	
12	PAHZZ	5306010797027	15434	219153	.BOLT,MACHINE	2
					UOC:DAA,DAB,DAC,DAD,DAE,DAF,DAG,DAH, DAJ,DAK,DAL,DAW,DAX,V12,V13,V14,V15, V16,V17,V18,V19,V20,V21,V22,V24,V25, V39	
13	PAHZZ	3120010873004	15434	214950	BEARING	12
					UOC:DAA,DAB,DAC,DAD,DAE,DAF,DAG,DAH, DAJ,DAK,DAL,DAW,DAX,V12,V13,V14,V15, V16,V17,V18,V19,V20,V21,V22,V24,V25, V39	
13	PAHZZ	3120011554442	15434	214951	BEARING,SLEEVE 010 UNDERSIZE	12
					UOC:DAA,DAB,DAC,DAD,DAE,DAF,DAG,DAH, DAJ,DAK,DAL,DAW,DAX,V12,V13,V14,V15, V16,V17,V18,V19,V20,V21,V22,V24,V25, V39	
13	PAHZZ	3120011573316	15434	214952	BEARING,SLEEVE 020 UNDERSIZE	12
					UOC:DAA,DAB,DAC,DAD,DAE,DAF,DAG,DAH, DAJ,DAK,DAL,DAW,DAX,V12,V13,V14,V15, V16,V17,V18,V19,V20,V21,V22,V24,V25, V39	
13	PAHZZ	3120011558707	15434	214953	BEARING,SLEEVE 030 UNDERSIZE	12
					UOC:DAA,DAB,DAC,DAD,DAE,DAF,DAG,DAH, DAJ,DAK,DAL,DAW,DAX,V12,V13,V14,V15, V16,V17,V18,V19,V20,V21,V22,V24,V25, V39	

END OF FIGURE

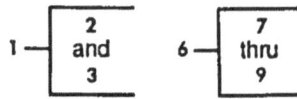

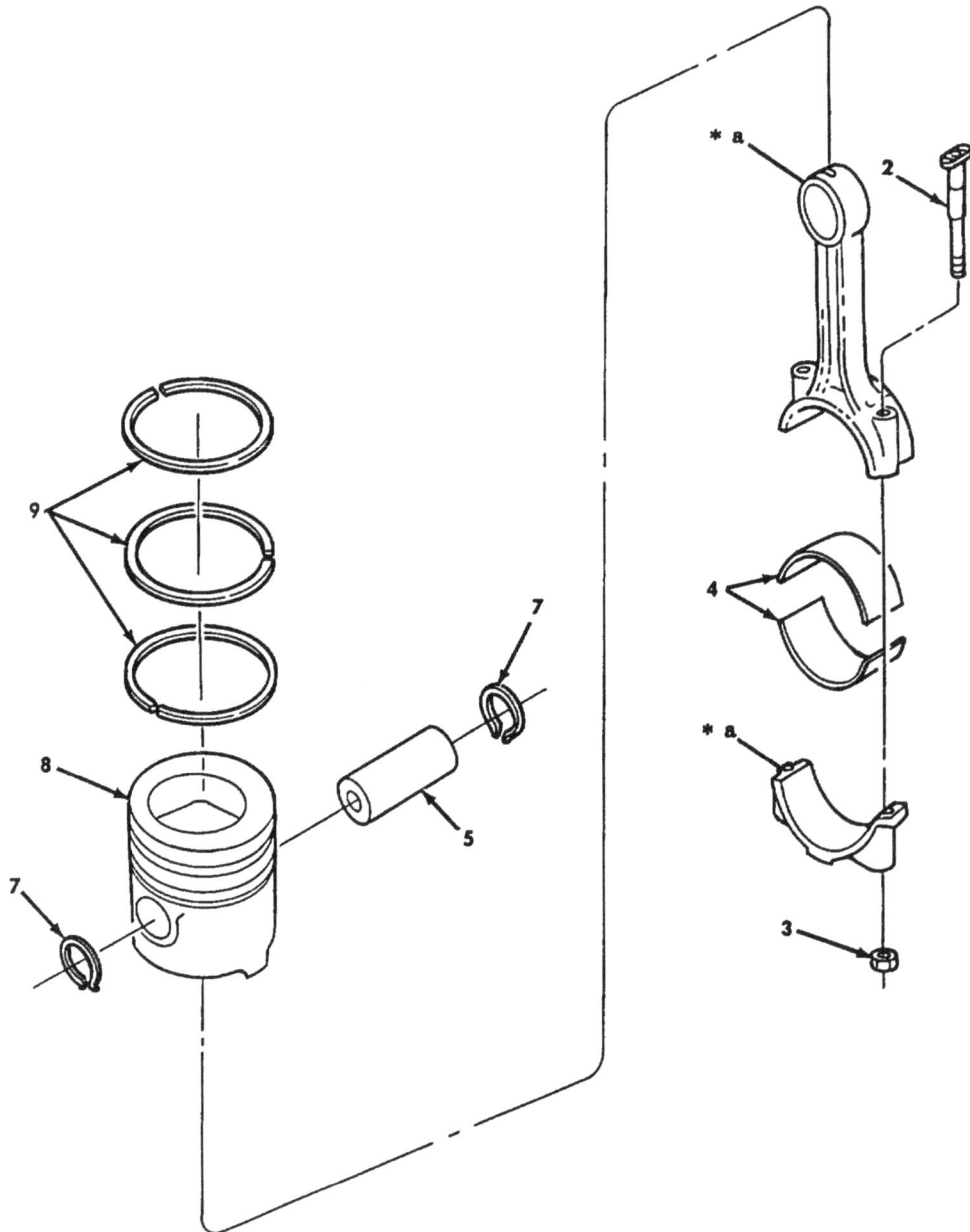

* a PART OF ITEM 1

Figure 18. Piston and Rod Assembly (M939A2).

(1) ITEM NO	(2) SMR CODE	(3) NSN	(4) CAGEC	(5) PART NUMBER	(6) DESCRIPTION AND USABLE ON CODES (UOC)	(7) QTY
					GROUP 0104 PISTONS, CONNECTING RODS	
					FIG. 18 PISTON AND ROD ASSEMBLY (M939A2)	
1	PAHZZ	2815012715119	15434	3901383	CONNECTING ROD,PIST .... UOC:ZAA,ZAB,ZAC,ZAD,ZAE,ZAF,ZAG,ZAH, ZAJ,ZAK,ZAL	6
2	PAHZZ	5306012716362	15434	3901380	.BOLT,MACHINE .... UOC:ZAA,ZAB,ZAC,ZAD,ZAE,ZAF,ZAG,ZAH, ZAJ,ZAK,ZAL	2
3	PAHZZ	5310012708246	15434	3901381	.NUT,PLAIN,HEXAGON .... UOC:ZAA,ZAB,ZAC,ZAD,ZAE,ZAF,ZAG,ZAH, ZAJ,ZAK,ZAL	2
4	PAHZZ	3120012743378	15434	3901430	BEARING,SLEEVE ROD .... UOC:ZAA,ZAB,ZAC,ZAD,ZAE,ZAF,ZAG,ZAH, ZAJ,ZAK,ZAL	12
4	PAHZZ	3120012723272	15434	3901431	BEARING,SLEEVE ROD, (.25MM O/S), ... USE AS REQUIRED .... UOC:ZAA,ZAB,ZAC,ZAD,ZAE,ZAF,ZAG,ZAH, ZAJ,ZAK,ZAL	12
4	PAHZZ	3120012723273	15434	3901432	BEARING,SLEEVE ROD, (.50MM O/S), ... USE AS REQUIRED .... UOC:ZAA,ZAB,ZAC,ZAD,ZAE,ZAF,ZAG,ZAH, ZAJ,ZAK,ZAL	12
4	PAHZZ	3120012757664	15434	3901433	BEARING,SLEEVE ROD, (.75MM O/S), ... USE AS REQUIRED .... UOC:ZAA,ZAB,ZAC,ZAD,ZAE,ZAF,ZAG,ZAH, ZAJ,ZAK,ZAL	12
4	PAHZZ	3120012757665	15434	3901434	BEARING,SLEEVE ROD, (1.00MM O/S), ... USE AS REQUIRED .... UOC:ZAA,ZAB,ZAC,ZAD,ZAE,ZAF,ZAG,ZAH, ZAJ,ZAK,ZAL	12
5	PFHZZ	2815012874502	15434	3901597	PIN,PISTON PART OF KIT P/N 2DR671 .... UOC:ZAA,ZAB,ZAC,ZAD,ZAE,ZAF,ZAG,ZAH, ZAJ,ZAK,ZAL	6
6	PAHZZ	2815012719802	45152	2DR672	PARTS KIT,PISTON AS PART OF KIT P/N....... 2DR671... UOC:ZAA,ZAB,ZAC,ZAD,ZAE,ZAF,ZAG,ZAH, ZAJ,ZAK,ZAL	6
7	PFHZZ	5325012805592	15434	3901996	.RING,RETAINING PART OF KIT P/N............. 2DR672.. UOC:ZAA,ZAB,ZAC,ZAD,ZAE,ZAF,ZAG,ZAH, ZAJ,ZAK,ZAL	2
8	KFHZZ		15434	3908750	.PISTON PART OF KIT P/N 2DR671............ UOC:ZAA,ZAB,ZAC, ZAD, ZAE, ZAF,ZAG, ZAH, ZAJ,ZAK,ZAL	1
9	PAHZZ	2815012719792	15434	3802110	.RING SET,PISTON PART OF KIT P/N............. 2DR671 PART OF KIT P/N 2DR672............. UOC:ZAA,ZAB,ZAC,ZAD,ZAE,ZAF,ZAG,ZAH, ZAJ,ZAK,ZAL	1

END OF FIGURE

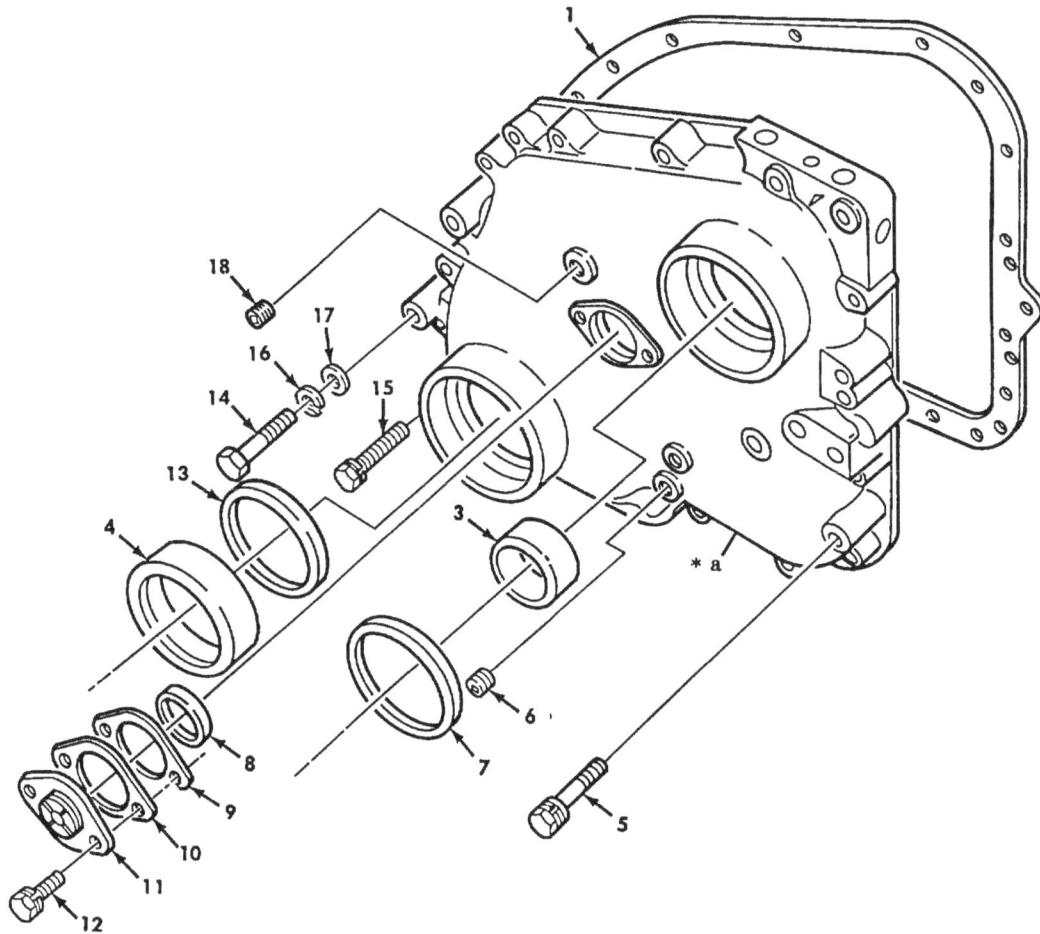

*a   PART OF ITEM 2

Figure 19. Front Gear Cover (M939, M939A1).

(1) ITEM NO	(2) SMR CODE	(3) NSN	(4) CAGEC	(5) PART NUMBER	(6) DESCRIPTION AND USABLE ON CODES (UOC)	(7) QTY
					GROUP 0105 VALVES, CAMSHAFTS, AND TIMING SYSTEM	
					FIG. 19 FRONT GEAR COVER(M939, M939A1)	
1	PAFZZ	5330001937652	15434	134276	GASKET PART OF KIT P/N 3011472 .... UOC:DAA,DAB,DAC,DAD,DAE,DAF,DAG,DAH, DAJ,DAK,DAL,DAW,DAX,V12,V13,V14,V15, V16,V17,V18,V19,V20,V21,V22,V24,V25, V39	1
2	PFFZZ	2815004042915	15434	3024416	COVER,TIMING GEAR,I FRONT .... UOC:DAA,DAB,DAC,DAD,DAE,DAF,DAG,DAH, DAJ,DAK,DAL,DAW,DAX,V12,Vl3,V134,V14,V15, V16,V17,V18,V19,V20,V21,V22,V24,V25, V39	1
3	PAFZZ	3120008772213	15434	3029852	.BEARING,SLEEVE .... UOC:DAA,DAB,DAC,DAD,DAE,DAF,DAG,DAH, DAJ,DAK,DAL,DAW,DAX,V12,V13,V14,V15, V16,V17,V18,V19,V20,V21,V22,V24,V25, V39	1
4	PAFZZ	3120008827960	15434	68226-1	.BUSHING,SLEEVE .... UOC:DAA,DAB,DAC,DAD,DAE,DAF,DAG,DAH, DAJ,DAK,DAL,DAW,DAX,V12,V13,V14,V15, V16,V17,V18,V19,V20,V21,V22,V24,V25, V39	1
5	PAFZZ	5305011474033	15434	3011711	SCREW,CAP,HEXAGON H .... UOC:DAA,DAB,DAC,DAD,DAE,DAF,DAG,DAH, DAJ,DAK,DAL,DAW,DAX,V12,V13,V14,V15, V16,V17,V18,V19,V20,V21,V22,V24,V25, V39	15
6	PAFZZ	4730011615115	15434	013786	PLUG,PIPE .... UOC:DAA,DAB,DAC,DAD,DAE,DAF,DAG,DAH, DAJ,DAK,DAL,DAW,DAX,V12,V13,V14,V15, V16,V17,V18,V19,V20,V21,V22,V24,V25, V39	1
7	PAFZZ	5330001356382	15434	211255	SEAL,PLAIN ENCASED PART OF KIT P/N .... 3011472 .... UOC:DAA,DAB,DAC,DAD,DAE,DAF,DAG,DAH, DAJ,DAK,DAL,DAW,DAX,V12,V13,V14,V15, V16,V17,V18,V19,V20,V21,V22,V24,V25, V39	1
8	PAFZZ	331001713879	5434	67270	O-RING PART OF KIT P/N 3011472 .... UOC:DAA,DAB,DAC,DAD,DAE,DAF,DAG,DAH, DAJ,DAK,DAL,DAW,DAX,V12,V13,V14,V15, V16,V17,V18,V19,V20,V21,V22,V24,V25, V39	1
9	KFFZZ	365003782885	15434	68192A	SHIM 0.010 PART OF KIT P/N BM56657 .... UOC:DAA,DAB,DAC,DAD,DAE,DAF,DAG,DAH, DAJ,DAK,DAL,DAW,DAX,V12,V13,V14,V15, V16,V17,V18,V19,V20,V21,V22,V24,V25, V39	1

(1) ITEM NO	(2) SMR CODE	(3) NSN	(4) CAGEC	(5) PART NUMBER	(6) DESCRIPTION AND USABLE ON CODES (UOC)	(7) QTY
9	KFFZZ	5365003782886	15434	68192B	SHIM 0.005 PART OF KIT P/N BM56657.................. UOC:DAA,DAB,DAC,DAD,DAE,DAF,DAG,DAH, DAJ,DAK,DAL,DAW,DAX,V12,V13,V14,V15, V16,V17,V18,V19,V20,V21,V22,V24,V25, V39	1
9	FFZZ	5365003782887	15434	68192C	SHIM 0.002 PART OF KIT P/N BM56657.................. UOC:DAA,DAB,DAC,DAD,DAE,DAF,DAG,DAH, DAJ,DAK,DAL,DAW,DAX,V12,V13,V14,V15, V16,V17,V18,V19,V20,V21,V22,V24,V25, V39	1
10	PAFZZ	5365011213068	15434	185574	SPACER,PLATE .... UOC:DAA,DAB,DAC,DAD,DAE,DAF,DAG,DAH, DAJ,DAK,DAL,DAW,DAX,V12,V13,V14,V15, V16,V17,V18,V19,V20,V21,V22,V24,V25, V39	1
11	PFFZZ	2815000339392	15434	138988	COVER PLATE,CAMSHAF .... UOC:DAA,DAB,DAC,DAD,DAE,DAF,DAG,DAH, DAJ,DAK,DAL,DAW,DAX,V12,V13,V14,V15, V16,V17,V18,V19,V20, V21,V22,V24,V25, V39	1
12	PAFZZ	5305011653892	15434	3011714	SCREW .... UOC:DAA,DAB,DAC,DAD,DAE,DAF,DAG,DAH, DAJ,DAK,DAL,DAW,DAX,V12,V13,V14,V15, V16,V17,V18,V19,V20,V21,V22,V24,V25, V39	2
13	PAFZZ	5330000062529	15434	208069	SEAL PART OF KIT P/N 3011472 .... UOC:DAA,DAB,DAC,DAD,DAE,DAF,DAG,DAH, DAJ,DAK,DAL,DAW,DAX,V12,V13,V14,V15, V16,V17,V18,V19,V20,V21,V22,V24,V25, V39	1
14	PAFZZ	5305009422196	80204	B1821BH038C100D	SCREW,CAP,HEXAGON H .... UOC:DAA,DAB,DAC,DAD,DAE,DAF,DAG,DAH, DAJ,DAK,DAL,DAW,DAX,V12,V13,V14,V15, V16,V17,V18,V19,V20,V21,V22,V24,V25, V39	2
15	PAFZZ	5305011458381	15434	3011713	SCREW,CAP,HEXAGON H .... UOC:DAA,DAB,DAC,DAD,DAE,DAF,DAG,DAH, DAJ,DAK,DAL,DAW,DAX,V12,V13,V14,V15, V16,V17,V18,V19,V20,V21,V22,V24,V25, V39	1
16	PAFZZ	5310006379541	96906	MS35338-46	WASHER,LOCK.................................. UOC:DAA,DAB,DAC,DAD,DAE,DAF,DAG,DAH, DAJ,DAK,DAL,DAW,DAX,V12,V13,V14,V15, V16,V17,V18,V19,V20,V21,V22,V24,V25, V39	2
17	PAFZZ	310000806004	96906	MS27183-14	WASHER,FLAT.................................. UOC:DAA,DAB,DAC,DAD,DAE,DAF,DAG,DAH, DAJ,DAK,DAL,DAW,DAX,V12,V13,V14,V15, V16,V17,V18,V19,V20,V21,V22,V24,V25, V39	2
18	PAFZZ	4730010304950	24617	272977	PLUG,PIPE .... UOC:DAA,DAB,DAC,DAD,DAE,DAF,DAG,DAH,	1

(1) ITEM NO	(2) SMR CODE	(3) NSN	(4) CAGEC	(5) PART NUMBER	(6) DESCRIPTION AND USABLE ON CODES (UOC)	(7) QTY

DAJ,DAK,DAL,DAW,DAX,V12,V13,V14,V15,
V16,V17,V18,V19,V20,V21,V22,V24,V25,
V39

END OF FIGURE

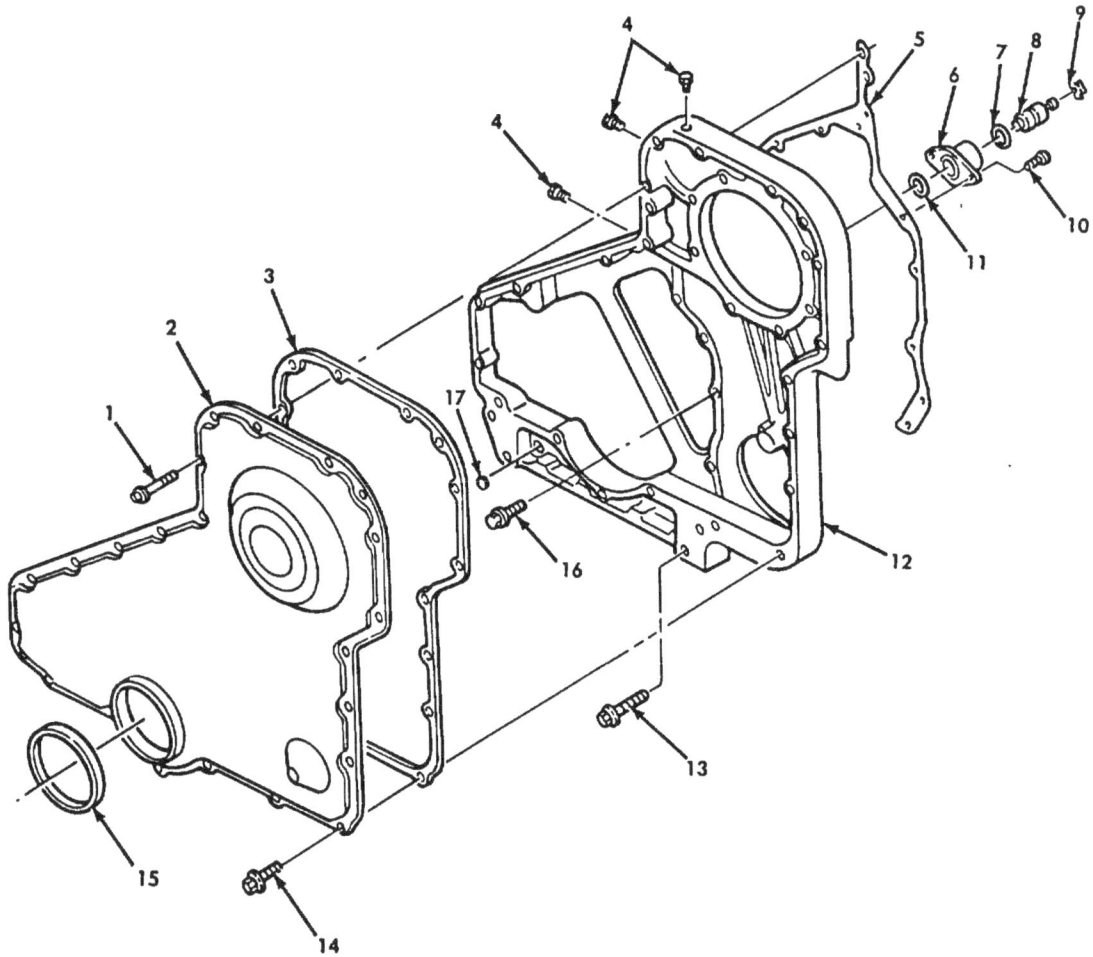

Figure 20. Front Gear Cover (M939A2).

(1) ITEM NO	(2) SMR CODE	(3) NSN	(4) CAGEC	(5) PART NUMBER	(6) DESCRIPTION AND USABLE ON CODES (UOC)	(7) QTY
					GROUP 0105 VALVES, CAMSHAFTS, AND TIMING SYSTEM	
					FIG. 20 FRONT GEAR COVER(M939A2)	
1	PAFZZ	5306012371166	15434	3900633	BOLT,MACHINE.. UOC:ZAA,ZAB,ZAC,ZAD,ZAE,ZAF, ZAG,ZAH, ZAJ,ZAK,ZAL	6
2	PAFZZ	3040013216365	15434	3916193	HOUSING,MECHANICAL .... UOC:ZAA,ZAB,ZAC,ZAD,ZAE,ZAF,ZAG,ZAH, ZAJ,ZAK,ZAL	1
3	PAFZZ	5330013212053	15434	3917780	GASKET PART OF KIT P/N 3802389 PART............. OF KIT P/N 3802389... UOC:ZAA,ZAB,ZAC,ZAD,ZAE,ZAF,ZAG,ZAH, ZAJ,ZAK,ZAL	1
4	PAFZZ	4730012810812	15434	3906619	PLUG,PIPE PART OF KIT P/N 3909415................. UOC:ZAA,ZAB,ZAC,ZAD,ZAE,ZAF,ZAG,ZAH, ZAJ,ZAK,ZAL	3
5	PAFZZ	330013173213	15434	3929253	GASKET PART OF KIT P/N 3802389 PART............. OF KIT P/N 3802389... UOC:ZAA,ZAB,ZAC,ZAD,ZAE,ZAF,ZAG,ZAH, ZAJ,ZAK,ZAL	1
6	PAFZZ	5342012750384	15434	3919683	HOUSNG,TIMING PIN .... UOC:ZAA,ZAB,ZAC,ZAD,ZAE,ZAF,ZAG,ZAH, ZAJ,ZAK,ZAL	1
7	PAFZZ	5330012916537	15434	3913994	PACKING,PREFORMED PART OF KIT P/N............. 3802389 PART OF KIT P/N 3802389 .... UOC:ZAA,ZAB,ZAC,ZAD,ZAE,ZAF,ZAG,ZAH, ZAJ,ZAK,ZAL	1
8	PAFZZ	3040011891760	15434	3903924	SHAFT,SHOULDERED .... UOC:ZAA,ZAB,ZAC,ZAD,ZAE,ZAF,ZAG,ZAH, ZAJ,ZAK,ZAL	1
9	PAFZZ	5365011880954	15434	3904849	RING,RETAINING .... UOC:ZAA,ZAB,ZAC,ZAD,ZAE,ZAF,ZAG,ZAH, ZAJ,ZAK,ZAL	1
10	PAFZZ	5305012632708	15434	3907998	SCREW,MACHINE .... UOC:ZAA,ZAB,ZAC,ZAD,ZAE,ZAF,ZAG,ZAH, ZAJ,ZAK,ZAL	2
11	PAFZZ	5330012636179	15434	3915772	GASKET PART OF KIT P/N 3802389 PART............. OF KIT P/N 3802389... UOC:ZAA,ZAB,ZAC,ZAD,ZAE,ZAF,ZAG,ZAH, ZAJ,ZAK,ZAL	1
12	PAFZZ	2835012712510	15434	3915074	HOUSING,GEARBOX,TUR .... UOC:ZAA,ZAB,ZAC,ZAD,ZAE,ZAF,ZAG,ZAH, ZAJ,ZAK,ZAL	1
13	PAFZZ	5305012744404	15434	3902116	SCREW,CAP,HEXAGON H .... UOC:ZAA,ZAB,ZAC,ZAD,ZAE,ZAF,ZAG,ZAH, ZAJ,ZAK,ZAL	4
14	PAFZZ	5305012374915	15434	3900629	SCREW,CAP,HEXAGON H .... UOC:ZAA,ZAB,ZAC,ZAD,ZAE,ZAF,ZAG,ZAH, ZAJ,ZAK,ZAL	17
15	AFZZ	5330012721108	5434	3353977	SEAL,PLAIN PART OF KIT P/N 3802389.................	1

(1) ITEM NO	(2) SMR CODE	(3) NSN	(4) CAGEC	(5) PART NUMBER	(6) DESCRIPTION AND USABLE ON CODES (UOC)	(7) QTY
					UOC:ZAA,ZAB,ZAC,ZAD,ZAE,ZAF,ZAG,ZAH, ZAJ,ZAK,ZAL	
16	PAFZZ	5305012343755	15434	3900630	SCREW,CAP,HEXAGON H	5
					UOC:ZAA,ZAB,ZAC,ZAD,ZAE,ZAF,ZAG,ZAH, ZAJ,ZAK,ZAL	
17	PAFZZ	5340011124280	15434	210036	PLUG,EXPANSION	1
					UOC:ZAA,ZAB,ZAC,ZAD,ZAE,ZAF,ZAG,ZAH, ZAJ,ZAK,ZAL	

END OF FIGURE

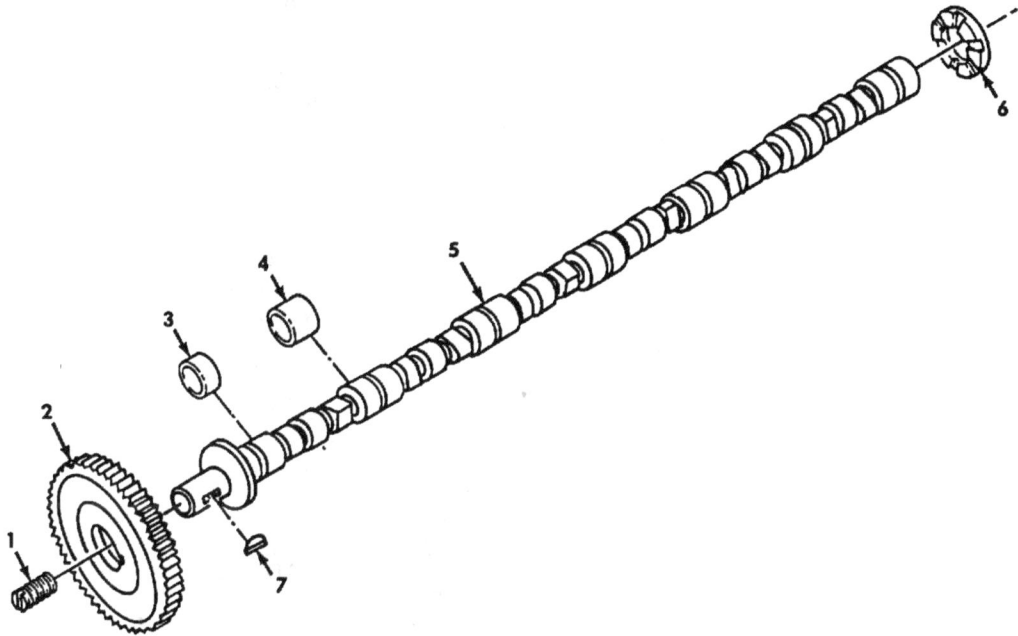

Figure 21. Camshaft, Gear, and Related Parts (M939, M939A1).

(1) ITEM NO	(2) SMR CODE	(3) NSN	(4) CAGEC	(5) PART NUMBER	(6) DESCRIPTION AND USABLE ON CODES (UOC)	(7) QTY
					GROUP 0105 VALVES, CAMSHAFTS, AND TIMING SYSTEM	
					FIG. 21 CAMSHAFT, GEAR, AND RELATED PARTS(M939,M939A1)	
1	PAHZZ	5340004042944	15434	68193	PLUG,VENT.. UOC: DAA, DAB, DAC, DAD, DAE,DAF, DAG, DAH, DAJ,DAK,DAL,DAW,DAX,V12,V13,V14,V15, V16,V17,V18,V19,V20,V21,V22,V24,V25, V39	1
2	KFHZZ		15434	3035194	GEAR,CAMSHAFT USED AFTER SERIAL.............. NUMBER 11247924 PART OF KIT P/N 5704533 .... UOC:DAA,DAB,DAC,DAD,DAE,DAF,DAG,DAH, DAJ,DAK,DAL,DAW,DAX,V12,V13,V14,V15, V16,V17,V18,V19,V20,V21,V22,V24,V25, V39	1
3	KFHZZ	3120005730391	15434	100670	BEARING,SLEEVE PART OF KIT P/N .... BM27253........................ UOC:DAA,DAB,DAC,DAD,DAE,DAF,DAG,DAH, DAJ,DAK,DAL,DAW,DAX,V12,V13,V14,V15, V16,V17,V18,V19,V20,V21,V22,V24,V25, V39	1
4	KFHZZ	3120009066657	15434	157870	BEARING,SLEEVE PART OF KIT P/N .... BM27253..... UOC:DAA,DAB,DAC,DAD,DAE,DAF,DAG,DAH, DAJ,DAK,DAL,DAW,DAX,V12 ,V13 ,V14 ,V15, V16,V17,V18,V19 ,V20,V21,V22,V24 ,V25, V39	6
5	PAHZZ	2815007911448	15434	3801030	CAMSHAFT,ENGINE .... UOC:DAA,DAB,DAC,DAD,DAE,DAF,DAG,DAH, DAJ,DAK,DAL,DAW,DAX,V12,V13,V14,V15, V16,V17,V18,V19,V20,V21,V22,V24,V25, V39	1
6	PAHZZ	3120003744342	15434	9235-1	BEARING,WASHER,THRU .... UOC:DAA,DAB,DAC,DAD,DAE,DAF,DA G,DAH, DAJ,DAK,DAL,DAW,DAX,V12,V13,V14,V15, V16,V17,V18,V19,V20,V21,V22,V24,V25, V39	1
7	PAHZZ	5315006165527	96906	MS35756-18	KEY,WOODRUFF .... UOC:DAA,DAB,DAC,DAD,DAE,DAF,DAG,DAH, DAJ,DAK,DAL,DAW,DAX,V12,V13,V14,V15, V16,V17,V18,V19,V20,V21,V22,V24,V25, V39	1

END OF FIGURE

2 — 3

1

* a

7

3

4

5

6

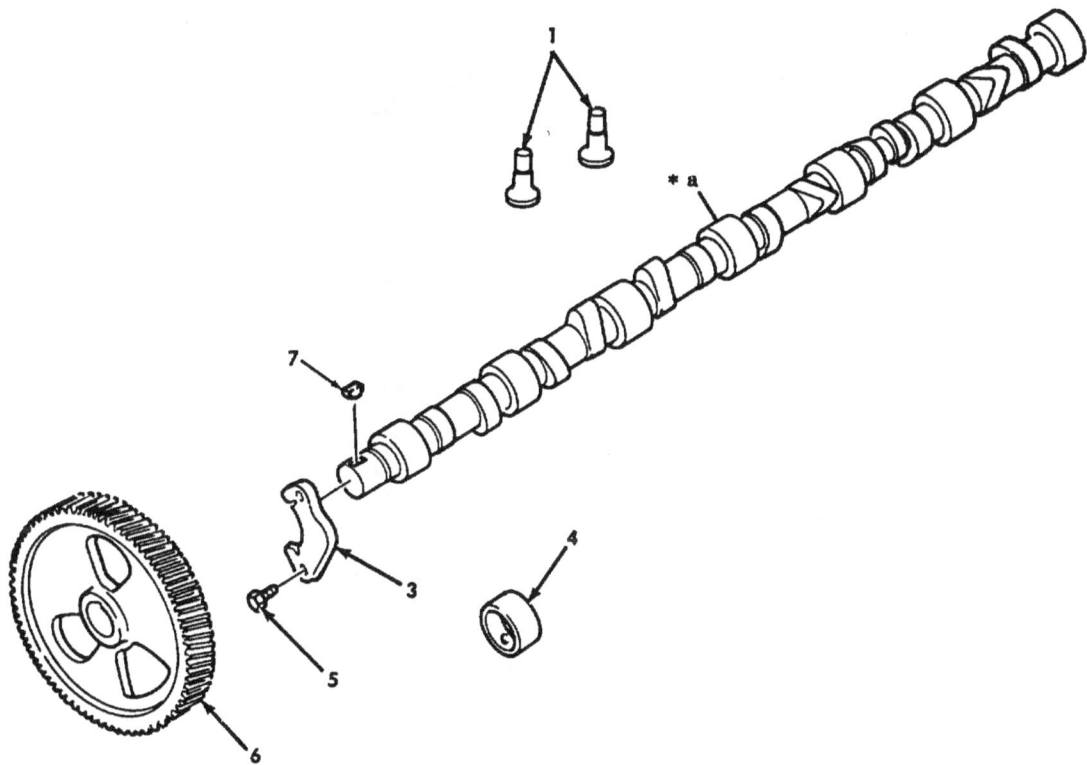

* a PART OF ITEM 2

Figure 22. Camshaft, Gear, and Related Parts (M939A2).

(1) ITEM NO	(2) SMR CODE	(3) NSN	(4) CAGEC	(5) PART NUMBER	(6) DESCRIPTION AND USABLE ON CODES (UOC)	(7) QTY
					GROUP 0105 VALVES, CAMSHAFTS, AND TIMING SYSTEM	
					FIG. 22 CAMSHAFT, GEAR, AND RELATED PARTS (M939A2)	
1	PAHZZ	2815013794920	15434	3925031	TAPPET,ENGINE POPPE .... UOC:ZAA,ZAB,ZAC,ZAD,ZAE,ZAF,ZAG,ZAH, ZAJ,ZAK, ZAL	12
2	PAHZZ	2815013888596	15434	3924471	CAMSHAFT,ENGINE.. UOC: ZAA, ZAB, ZAC, ZAD, ZAE, ZAF, ZAG, ZAH, ZAJ, ZAK, ZAL	1
3	PAHZZ	2815014244736	15434	3927155	.SUPPORT,CAMSHAFT TH .... UOC:ZAA,ZAB,ZAC,ZAD,ZAE,ZAF,ZAG,ZAH, ZAJ,ZAK,ZAL	1
4	PAHZZ	5365012711852	15434	3901685	BUSHING BLANK... UOC:ZAA,ZAB,ZAC,ZAD,ZAE,ZAF,ZAG,ZAH, ZAJ,ZAK,ZAL	7
5	PAHZZ	5305121689310	15434	3900620	SCREW,CAP,HEXAGON H .... UOC:ZAA,ZAB,ZAC,ZAD,ZAE,ZAF,ZAG,ZAH, ZAJ, ZAK, ZAL	2
6	PAHZZ	3020014148008	15434	3917328	GEAR,HELICAL .... UOC:ZAA,ZAB,ZAC,ZAD,ZAE,ZAF,ZAG,ZAH, ZAJ,ZAK,ZAL	1
7	PAHZZ	5315012354688	15434	3902332	KEY,WOODRUFF .... UOC:ZAA,ZAB,ZAC,ZAD,ZAE,ZAF,ZAG,ZAH, ZAJ,ZAK,ZAL	1

END OF FIGURE

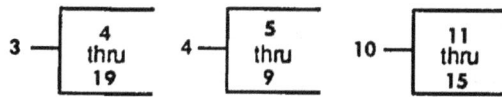

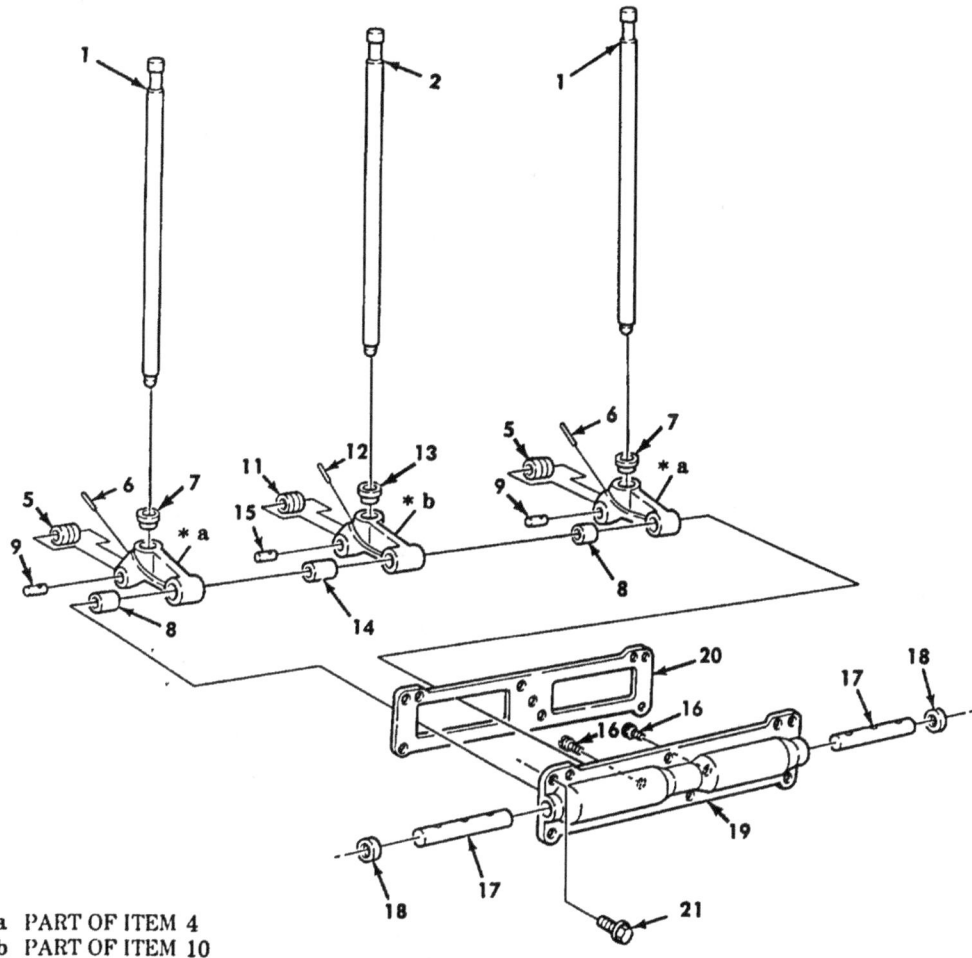

Figure 23. Cam Follower, Housing, and Push Rods (M939, M939A1).

(1) ITEM NO	(2) SMR CODE	(3) NSN	(4) CAGEC	(5) PART NUMBER	(6) DESCRIPTION AND USABLE ON CODES (UOC)	(7) QTY
					GROUP 0105 VALVES, CAMSHAFTS, AND TIMING SYSTEM	
					FIG. 23 CAM FOLLOWER, HOUSING, AND PUSH RODS(M939,M939A1)	
1	PAFZZ	2815011361987	15434	112700	PUSH ROD, ENGINE POP .... UOC:DAA,DAB,DAC,DAD,DAE,DAF,DAG,DAH, DAJ,DAK,DAL,DAW,DAX,V12,V13,V14,V15, V16,V17,V18,V19,V20,V21,V22,V24,V25, V39	12
2	PAFZZ	2815011361986	15434	3032682	PUSH ROD, ENGINE POP INTAKE .... UOC:DAA,DAB,DAC,DAD,DAE,DAF,DAG,DAH, DAJ,DAK,DAL,DAW,DAX,V12,V13,V14,V15, V16,V17,V18,V19,V20,V21,V22,V24,V25, V39	6
3	PAFZZ	2815011147398	15434	3018051	HOUSING ASSEMBLY,CA .... UOC:DAA,DAB,DAC,DAD,DAE,DAF,DAG,DAH, DAJ,DAK,DAL,DAW,DAX,V12,V13,V14,V15, V16,V17,V18,V19,V20,V21,V22,V24,V25, V39	3
4	PAFZZ	2815005055116	15434	BM-37634	.LEVER ASSEMBLY,CAM .... UOC:DAA,DAB,DAC,DAD,DAE,DAF,DAG,DAH, DAJ,DAK,DAL,DAW,DAX,V12,V13,V14,V15, V16,V17,V18,V19,V20,V21,V22,V24,V25, V39	4
5	PFFZZ	2815003112521	15434	9260-1	..ROLLER,VALVE CAM .... UOC:DAA,DAB,DAC,DAD,DAE,DAF,DAG,DAH, DAJ,DAK,DAL,DAW,DAX,V12,V13,V14,V15, V16,V17,V18,V19,V20,V21,V22,V24,V25, V39	2
6	PAFZZ	5315007773544	15434	118939	..PIN,STRAIGHT,HEADLE .... UOC:DAA,DAB,DAC,DAD,DAE,DAF,DAG,DAH, DAJ,DAK,DAL,DAW,DAX,V12,V13,V14,V15, V16,V17,V18,V19,V20,V21,V22,V24,V25, V39	2
7	PFFZZ	2815005055119	15434	107738	..SOCKET,CAM FOLLOWER .... UOC:DAA,DAB,DAC,DAD,DAE,DAF,DAG,DAH, DAJ,DAK,DAL,DAW,DAX,V12,V13,V14,V15, V16,V17,V18,V19,V20,V21,V22,V24,V25, V39	2
8	PFFZZ	3120006597808	15434	118378	..BEARING,SLEEVE .... UOC:DAA,DAB,DAC,DAD,DAE,DAF,DAG,DAH, DAJ,DAK,DAL,DAW,DAX,V12,V13,V14,V15, V16,V17,V18,V19,V20,V21,V22,V24,V25, V39	2
9	PFFZZ	5315000410916	15434	68513	..PIN,STRAIGHT,HEADLE .... UOC:DAA,DAB,DAC,DAD,DAE,DAF,DAG,DAH, DAJ,DAK,DAL,DAW,DAX,V12,V13,V14,V15, V16,V17,V18,V19,V20,V21,V22,V24,V25, V39	2
10	PAFZZ	2815007052851	15434	3018049	.LEVER ASSEMBLY,INJE	3

(1) ITEM NO	(2) SMR CODE	(3) NSN	(4) CAGEC	(5) PART NUMBER	(6) DESCRIPTION AND USABLE ON CODES (UOC)	(7) QTY
					UOC:DAA,DAB,DAC,DAD,DAE,DAF,DAG,DAH, DAJ,DAK,DAL,DAW,DAX,V12,V13,V14,V15, V16,V17,V18,V19,V20,V21,V22,V24,V25, V39	
11	PFFZZ	2815003621780	15434	7348-2	..ROLLER,INJECTOR CAM .... UOC:DAA,DAB,DAC,DAD,DAE,DAF,DAG,DAH, DAJ,DAK,DAL,DAW,DAX,V12,V13,V14,V15, V16,V17,V18,V19,V20,V21,V22,V24,V25, V39	1
12	PFFZZ	5315007773544	15434	118939	..PIN,STRAIGHT,HEADLE .... UOC:DAA,DAB,DAC,DAD,DAE,DAF,DAG,DABH, DAJ,DAK,DAL,DAW,DAX,V12,V13,V14,V15, V16,V17,V18,V19 ,V20,V21,V22,V24,V25, V39	1
13	PFFZZ	2815010486702	15434	213559	..SOCKET,CAM FOLLOWER .... UOC:DAA,DAB,DAC,DAD,DAE,DAF,DAG,DAH, DAJ,DAK,DAL,DAW,DAX,V12,V13,V14,V15, V16,V17,V18,V19,V20,V21,V22,V24,V25, V39	1
14	PFFZZ	3120007911440	15434	118377	..BEARING,SLEEVE .... UOC:DAA,DAB,DAC,DAD,DAE,DAF,DAG,DAH, DAJ,DAK,DAL,DAW,DAX,V12 ,V13 ,V14,V15, V16,V17,V18,V19,V20,V21,V22,V24,V25, V39	1
15	PFFZZ	5315011260601	15434	3017544	..PIN,STRAIGHT,HEADLE .... UOC:DAA,DAB,DAC,DAD,DAE,DAF,DAG,DAH, DAJ,DAK,DAL,DAW,DAX,V12,V13,V14,V15, V16,V17,V18,V19,V20,V21,V22,V24,V25, V39	1
16	PAFZZ	5305003391415	15434	69736	.SCREW,MACHINE .... UOC:DAA,DAB,DAC,DAD,DAE,DAF,DAG,DAH, DAJ,DAK,DAL,DAW,DAX,V12,V13,V14,V15, V16,V17,V18,V19,V20,V21,V22,V24,V25, V39	2
17	PAFZZ	3040003883126	15434	3065125	.SHAFT,STRAIGHT..................... UOC:DAA,DAB,DAC,DAD,DAE,DAF,DAG,DAH, DAJ,DAK,DAL,DAW,DAX,V12,V13,V14,V15, V16,V17,V18,V19,V20,V21,V22,V24,V25, V39	2
18	PAFZZ	5340004850945	15434	175831	.PLUG,EXPANSION..................... UOC:DAA,DAB,DAC,DAD,DAE,DAF,DAG,DAH, DAJ,DAK,DAL,DAW,DAX,V12,V13,V14,V15, V16,V17,V18,V19,V20,V21,V22,V24,V25, V39	2
19	XAFZZ		15434	44035	.HOUSING..................... UOC:DAA,DAB,DAC,DAD,DAE,DAF,DAG,DAH, DAJ,DAK,DAL,DAW,DAX,V12,V13,V14,V15, V16,V17,V18,V19,V20,V21,V22,V24,V25, V39	1
20	KFFZZ	5330007773545	15434	120819	GASKET .027-.033 PART OF KIT P/N..................V 3011472 PART OF KIT P/N AR51482 .... UOC:DAA,DAB,DAC,DAD, DAE ,DAF,DAG,DAH,	

(1) ITEM NO	(2) SMR CODE	(3) NSN	(4) CAGEC	(5) PART NUMBER	(6) DESCRIPTION AND USABLE ON CODES (UOC)	(7) QTY
					DAJ,DAK,DAL,DAW,DAX,V12 ,V13,V14,V15, V16,V17,V8,V199,V20,V21,V22,V24,V25, V39	
20	KFFZZ	5330003491219	15434	9266A	GASKET .007 PART OF KIT P/N........................V 3011472 PART OF KIT P/N AR51482 .... UOC:DAA,DAB,DAC,DAD, DAE,DAF,DAG,DAH, DAJ,DAK,DAL,DAW,DAX,V12,V13,V14,V15, V16,V17,V18,V19,V20,V21,V22,V24,V25, V39	
20	KFFZZ	5330001756585	15434	9266	GASKET .014-.020 PART OF KIT P/N..................V 3011472 .... UOC:DAA,DAB,DAC,DAD,DAE,DAF,DAG,DAH, DAJ,DAK,DAL,DAW,DAX,V12 ,V13,V14 ,V15, V16,V17,V18,V19,V20,V21,V22,V24,V25, V39	
20	KFFZZ		15434	3011272	GASKET .020-.024 PART OF KIT P/N..................V 3011472 .... UOC: DAA, DAB, DAC, DAD, DAE, DAF, DAG, DAH, DAJ,DAK,DAL,DAW,DAX,V12,V13,V14,V15, V16,V17,V18,V19,V20,V21,V22,V24,V25, V39	
20	KFFZZ		15434	3011273	GASKET .037-.041 PART OF KIT P/N..................V 3011472 .... UOC:DAA,DAB,DAC,DAD,DAE,DAF,DAG,DAH, DAJ,DAK,DAL,DAW,DAX,V12,V13,V14,V15, V16,V17,V18,V19,V20,V21,V22,V24,V25, V39	
21	PAFZZ	5305011973449	15434	3010593	SCREW,CAP,HEXAGON H .... UOC:DAA,DAB,DAC,DAD,DAE,DAF,DAG,DAH, DAJ,DAK,DAL,DAW,DAX,V12 ,V13 ,V14 ,V15, V16,V17,V18,V19,V20,V21,V22,V24,V25, V39	18

END OF FIGURE

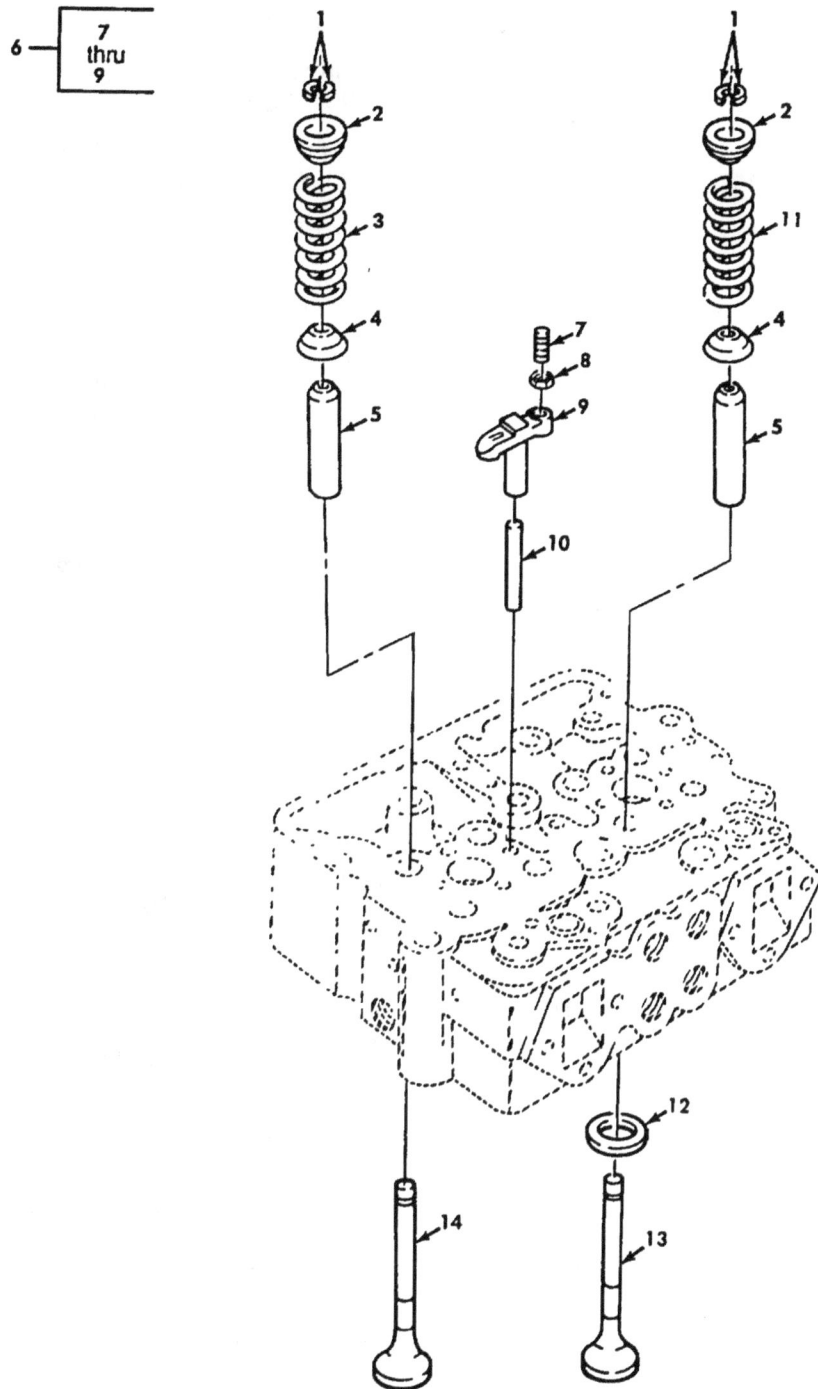

Figure 24. Valves, Springs, Guides, and Miscellaneous Parts (M939, M939A1).'

(1) ITEM NO	(2) SMR CODE	(3) NSN	(4) CAGEC	(5) PART NUMBER	(6) DESCRIPTION AND USABLE ON CODES (UOC)	(7) QTY
					GROUP 0105 VALVES, CAMSHAFTS, AND TIMING SYSTEM	
					FIG. 24 VALVES, SPRINGS, GUIDES, AND MISCELLANEOUS PARTS(M939,M939A 1)	
1	PAFZZ	5342011436048	15434	127554	HALF-COLLET .... UOC:DAA,DAB,DAC,DAD,DAE,DAF,DAG,DAH, DAJ,DAK,DAL,DAW,DAX,V12,V13,V14,V15, V16,V17,V18,V19,V20,V21,V22,V24,V25, V39	48
2	PAFZZ	5340009333009	15434	170296	SEAT,HELICAL COMPRE .... UOC:DAA,DAB,DAC,DAD,DAE,DAF,DAG,DAH, DAJ,DAK,DAL,DAW,DAX,V12,V13,V14,V15, V16,V17,V18,V19,V20,V21,V22,V24,V25, V39	24
3	PAFZZ	5360000099270	15434	211999	SPRING,HELICAL,COMP EXHAUST .... UOC:DAA,DAB,DAC,DAD,DAE,DAF,DAG,DAH, DAJ,DAK,DAL,DAW,DAX,V12,V13,V14,V15, V16,V17,V18,V19,V20,V21,V22,V24,V25, V39	12
4	PAFZZ	5340006326239	15434	172034	SEAT,HELICAL COMPRE SPRING,VALVE............ UOC:DAA,DAB,DAC,DAD,DAE,DAF,DAG,DAH, DAJ,DAK,DAL,DAW,DAX,V12,V13,V14,V15, V16,V17,V18,V19,V20,V21,V22,V24,V25, V39	24
5	PAHZZ	2815010852618	15434	3006456	GUIDE,VALVE STEM .... UOC:DAA,DAB,DAC,DAD,DAE,DAF,DAG,DAH, DAJ,DAK,DAL,DAW,DAX,V12,V13,V14,V15, V16,V17,V18,V19,V20,V21,V22,V24,V25, V39	24
6	PAFZZ	2815011591789	15434	3030038	VALVE,CROSSHEAD ASS .... UOC:DAA,DAB,DAC,DAD,DAE,DAF,DAG,DAH, DAJ,DAK,DAL,DAW,DAX,V12,V13,V14,V15, V16,V17,V18,V19,V20,V21,V22,V24,V25, V39	12
7	PAFZZ	5305000624378	15434	147389	.SETSCREW .... UOC:DAA,DAB,DAC,DAD,DAE,DAF,DAG,DAH, DAJ,DAK,DAL,DAW,DAX,V12,V13,V14,V15, V16,V17,V18,V19,V20,V21,V22,V24,V25, V39	1
8	PAFZZ	5310004263990	15434	203131	.NUT,CROSS HD .... UOC:DAA,DAB,DAC,DAD,DAE,DAF,DAG,DAH, DAJ,DAK,DAL,DAW,DAX,V12,V13,V14,V15, V16,V17,V18,V19,V20,V21,V22,V24,V25, V39	1
9	PAFZZ	2815011591789	15434	3030038	.VALVE,CROSSHEAD ASS .... UOC:DAA,DAB,DAC,DAD,DAE,DAF,DAG,DAH, DAJ,DAK,DAL,DAW,DAX,V12,V13,V14,V15, V16,V17,V18,V19,V20,V21,V22,V24,V25, V39	1
10	PAFZZ	5315008665015	15434	123558	PIN,STRAIGHT,HEADLE	12

(1) ITEM NO	(2) SMR CODE	(3) NSN	(4) CAGEC	(5) PART NUMBER	(6) DESCRIPTION AND USABLE ON CODES (UOC)	(7) QTY
					UOC:DAA,DAB,DAC,DAD,DAE,DAF,DAG,DAH, DAJ,DAX,DAL,DAW,DAX,V12,V13,V14,V15, V16,V17,V18,V19,V20,V21,V22,V24,V25, V39	
11	PAFZZ	5360000099270	15434	211999	SPRING,XELICAL,COMP INTAXE .... UOC:DAA,DAB,DAC,DAD,DAE,DAF,DAG,DAH, DAJ,DAK,DAL,DAW,DAX,V12,V13,V14,V15, V16,V17,V18,V19,V20,V21,V22,V24,V25, V39	12
12	PAFZZ	4820013002759	15434	127935	SEAT,VALVE .005 .... UOC:DAA,DAB,DAC,DAD,DAE,DAF,DAG,DAH, DAJ,DAK,DAL,DAW,DAX,V12,V13,V14,V15, V16,V17,V1B,V19,V20,V21,V22,V24,V25, V39	12
12	PAFZZ	2815001320240	15434	3014622	INSERT,ENGINE VALVE .010 .... UOC:DAA,DAB,DAC,DAD,DAE,DAF,DAG,DAH, DAJ,DAK,DAL,DAW,DAX,V12,V13,V14,V15, V16,V17,V18,V19,V20,V21,V22,V24,V25, V39	12
12	PAFZZ	2815011271060	15434	3014623	INSERT,VALVE SEAT .020 .... UOC:DAA,DAB,DAC,DAD,DAE,DAF,DAG,DAH, DAJ,DAK,DAL,DAW,DAX,V12,V13,V14,V15, V16,V17,V18,V19,V20,V21,V22,V24,V25, V39	12
12	PAFZZ	2815011273597	15434	3014624	INSERT,ENGINE VALVE .030 .... UOC:DAA,DAB,DAC,DAD,DAE,DAF,DAG,DAH, DAJ,DAK,DAL,DAW,DAX,V12,V13,V14,V15, V16,V17,V18,V19,V20,V21,V22,V24,V25, V39	12
12	PAFZZ	2815011273598	15434	3014625	INSERT,VALVE SEAT .040 .... UOC:DAA,DAB,DAC,DAD,DAE,DAF,DAG,DAH, DAJ,DAK,DAL,DAW,DAX,V12,V13 ,V114,V15, V16,V17,V18,V19,V20,V21,V22,V24,V25, V39	12
13	PAFZZ	2815013542702	15434	3803512	VALVE,POPPET,ENGINE .... UOC:DAA,DAB,DAC,DAD,DAE,DAF,DAG,DAH, DAJ,DAK,DAL,DAW,DAX,V12,V13,V14,V15, V16,V17,V18,V19,V20,V21,V22,V24,V25, V39	12
14	PAFZZ	2815009625623	01265	AX-1597	VALVE,POPPET,ENGINE EXHAUST .... UOC:DAA,DAB,DAC,DAD,DAE,DAF, DAG,DAH, DAJ,DAK,DAL,DAW,DAX,V12,V13,V14,V15, V16,V17,V18,V19 ,V20,V21,V22,V24,V25, V39	12

END OF FIGURE

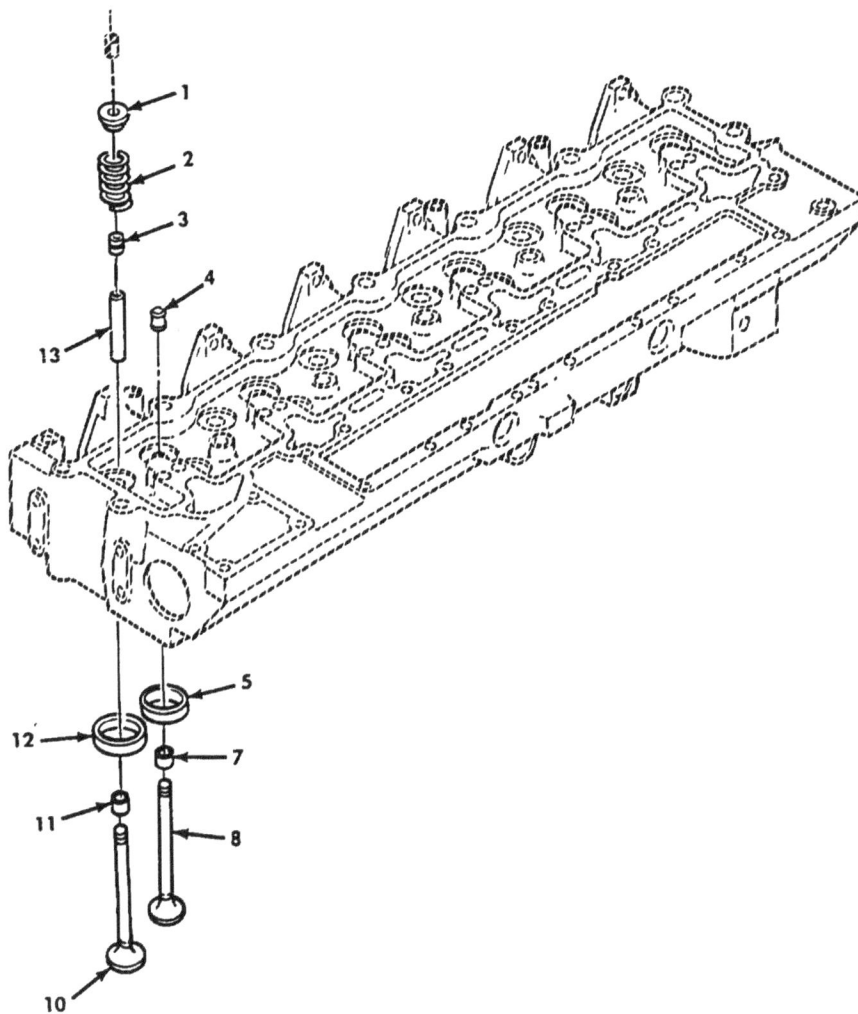

Figure 25. Valves, Springs, Guides, and Miscellaneous Parts (M939A2).

(1) ITEM NO	(2) SMR CODE	(3) NSN	(4) CAGEC	(5) PART NUMBER	(6) DESCRIPTION AND USABLE ON CODES (UOC)	(7) QTY
					GROUP 0105 VALVES, CAMSHAFTS, AND TIMING SYSTEM	
					FIG. 25 VALVES, SPRINGS, GUIDES, AND MISCELLANEOUS PARTS(M939A2)	
1	PAHZZ	5340012712485	15434	3912976	SEAT,HELICAL COMPRE .... UOC:ZAA,ZAB,ZAC,ZAD,ZAE,ZAF,ZAG,ZAH, ZAJ,ZAK,ZAL	12
2	PAHZZ	5360012718282	15434	3906412	SPRING,HELICAL,COMP .... UOC:ZAA,ZAB,ZAC,ZAD,ZAE,ZAF,ZAG,ZAH, ZAJ,ZAK,ZAL	12
3	PAHZZ	5340012817792	15434	3915707	RETAINER,HELICAL CO PART OF KIT P/N............. 3802622 PART OF KIT P/N 3802623 PART OF KIT P/N 3802452................................ UOC:ZAA,ZAB,ZAC,ZAD,ZAE,ZAF,ZAG,ZAH, ZAJ,ZAK,ZAL	12
4	PAHZZ	5330012825653	15434	3919038	SEAL,PLAIN EXHAUST PART OF KIT P/N............. 3802622 PART OF KIT P/N 3802623 PART OF KIT P/N 3802452 .... UOC:ZAA,ZAB,ZAC,ZAD,ZAE,ZAF,ZAG,ZAH, ZAJ,ZAK,ZAL	12
5	PAHZZ	2815012725538	15434	3929005	INSERT,ENGINE VALVE EXHAUST .... UOC:ZAA,ZAB,ZAC,ZAD,ZAE,ZAF,ZAG,ZAH, ZAJ,ZAK,ZAL	6
6	PAHZZ	2815012719794	15434	3802085	VALVE,POPPET,ENGINE EXHAUST..................... ASSEMBLY .... UOC:ZAA,ZAB,ZAC,ZAD,ZAE,ZAF,ZAG,ZAH, ZAJ,ZAK,ZAL	6
7	PAHZZ	2815012717171	15434	3901177	.LOCK,VALVE SPRING R .... UOC:ZAA,ZAB,ZAC,ZAD,ZAE,ZAF,ZAG,ZAH, ZAJ,ZAK,ZAL	2
8	PAHZZ	2815012808961	15434	3902254	.VALVE,POPPET,ENGINE EXHAUST .... UOC:ZAA,ZAB,ZAC,ZAD,ZAE,ZAF,ZAG,ZAH, ZAJ,ZAK,ZAL	1
9	PAHZZ	2815012726679	15434	3802275	VALVE,POPPET,ENGINE INTAKE ASSEMBLY .... UOC:ZAA,ZAB,ZAC,ZAD,ZAE,ZAF,ZAG,ZAH, ZAJ,ZAK,ZAL	6
10	PAHZZ	2815012815206	15434	3902253	.VALVE,POPPET,ENGINE INTAKE .... UOC:ZAA,ZAB,ZAC,ZAD,ZAE,ZAF,ZAG,ZAH, ZAJ,ZAK,ZAL	1
11	PAHZZ	2815012717171	15434	3901177	.LOCK,VALVE SPRING R .... UOC:ZAA,ZAB,ZAC,ZAD,ZAE,ZAF,ZAG,ZAH, ZAJ,ZAK,ZAL	2
12	PAHZZ	2815012725539	15434	3908830	INSERT,ENGINE VALVE INTAKE .... UOC:ZAA,ZAB,ZAC,ZAD,ZAE,ZAF,ZAG,ZAH, ZAJ,ZAK,ZAL	6
13	PFHZZ	2815012723980	45152	2DR636	GUIDE,VALVE STEM .... UOC:ZAA,ZAB,ZAC,ZAD,ZAE,ZAF,ZAG,ZAH, ZAJ,ZAK,ZAL	12

END OF FIGURE

Figure 26. Rocker Arms and Miscellaneous Hardware (M939, M939A1).

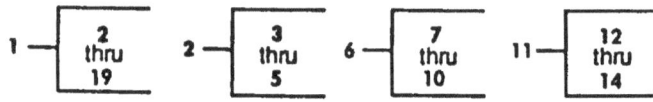

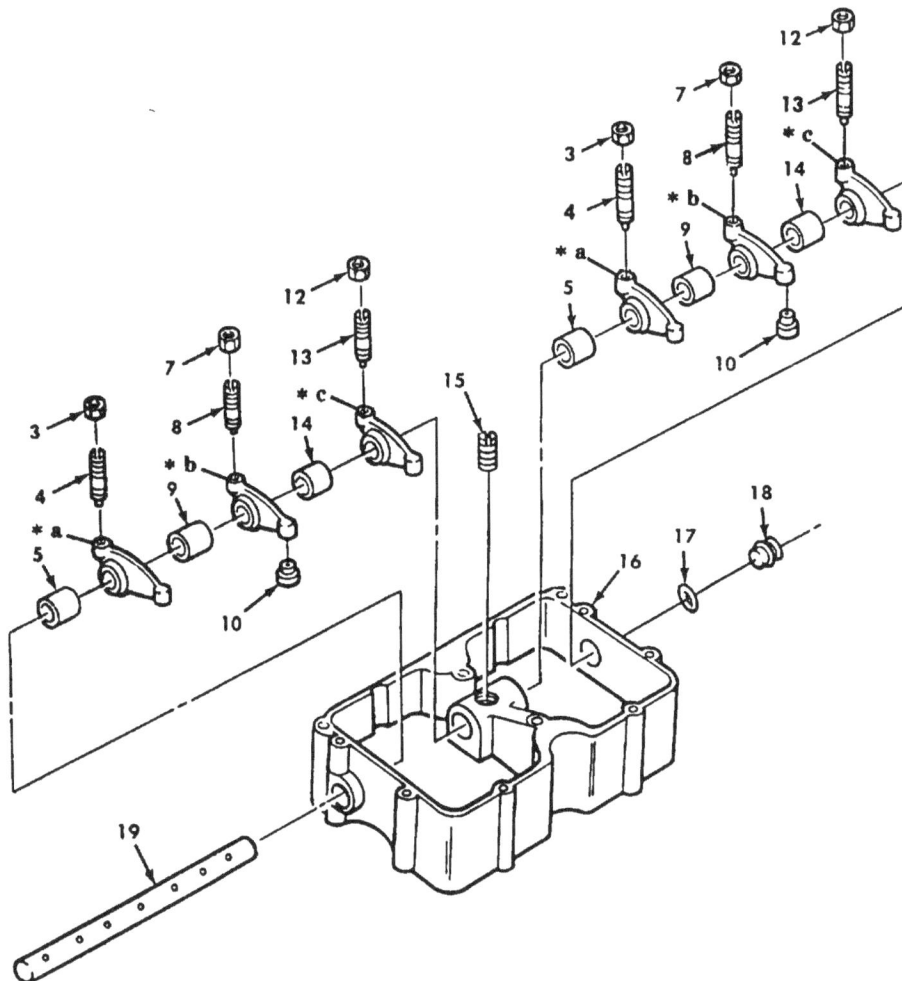

* a PART OF ITEM 2
* b PART OF ITEM 6
* c PART OF ITEM 11

(1) ITEM NO	(2) SMR CODE	(3) NSN	(4) CAGEC	(5) PART NUMBER	(6) DESCRIPTION AND USABLE ON CODES (UOC)	(7) QTY
					GROUP 0105 VALVES, CAMSHAFTS, AND TIMING SYSTEM	
					FIG. 26 ROCKER ARMS AND MISCELLANEOUS HARDWARE(M939,M939A1)	
1	PFFFF	2815010852569	15434	3035961	HOUSING AND ROCKER.................................... UOC:DAA,DAB,DAC,DAD,DAE,DAF,DAG,DAH, DAJ,DAK,DAL,DAW,DAX,V12,V13,V14,V15, V16,V17,V18,V19,V20,V21,V22,V24,V25, V39	3
2	PAFZZ	2815010969198	15434	BM95161	.ROCKER ARM,ENGINE P........................... UOC:DAA,DAB,DAC,DAD,DAE,DAF,DAG,DAH, DAJ,DAK,DAL,DAW,DAX,V12,V13,V14,V15, V16,V17,V18,V19,V20,V21,V22,V24,V25, V39	2
3	PAFZZ	5310007320560	96906	MS51968-14	..NUT,PLAIN,HEXAGON........................... UOC:DAA,DAB,DAC,DAD,DAE,DAF,DAG,DAH, DAJ,DAK,DAL,DAW,DAX,V12,V13,V14,V15, V16,V17,V18,V19,V20,V21,V22,V24,V25, V39	2
4	PAFZZ	5305009473437	15434	168306	..SETSCREW ROCKER ARM ADJUSTING.............. UOC:DAA,DAB,DAC,DAD,DAE,DAF,DAG,DAH, DAJ,DAK,DAL,DAW,DAX,V12,V13,V14,V15, V16,V17,V18,V19,V20,V21,V22,V24,V25, V39	2
5	PAFZZ	3120005893537	15434	140330	..BUSHING,SLEEVE................................ UOC:DAA,DAB,DAC,DAD,DAE,DAF,DAG,DAH, DAJ,DAK,DAL,DAW,DAX,V12 ,V13 ,V14 ,V15, V16,V17,V18,V19,V20,V21,V22,V24,V25, V39	2
6	PAFZZ	2815000057431	15434	AR-2308	.LEVER,INJECTOR,FUEL..................................... UOC:DAA,DAB,DAC,DAD,DAE,DAF,DAG,DAH, DAJ,DAK,DAL,DAW,DAX,V12,V13,V14,V15, V16,V17,V18,V19,V20,V21,V22,V24,V25, V39	2
7	PAFZZ	5310007320560	96906	MS51968-14	..NUT,PLAIN,HEXAGON........................... UOC:DAA,DAB,DAC,DAD,DAE,DAF,DAG,DAH, DAJ,DAK,DAL,DAW,DAX,V12,V13,V14,V15, V16,V17,V18,V19,V20,V21,V22,V24,V25, V39	2
8	PAFZZ	5305009473437	15434	168306	..SETSCREW................................ UOC:DAA,DAB,DAC,DAD,DAE,DAF,DAG,DAH, DAJ,DAK,DAL,DAW,DAX,V12,V13,V14,V15, V16,V17,V18,V19,V20,V21,V22,V24,V25, V39	2
9	PAFZZ	3120010795208	15434	218153	..BUSHING,SLEEVE................................ UOC:DAA,DAB,DAC,DAD,DAE,DAF,DAG,DAH, DAJ,DAK,DAL,DAW,DAX,V12,V13,V14,V15, V16,V17,V18,V19,V20,V21,V22,V24,V25, V39	2
10	PAFZZ	5340004042940	15434	194037	..SEAT,BALL SOCKET................................	2

(1) ITEM NO	(2) SMR CODE	(3) NSN	(4) CAGEC	(5) PART NUMBER	(6) DESCRIPTION AND USABLE ON CODES (UOC)	(7) QTY
					UOC:DAA,DAB,DAC,DAD,DAE,DAF,DAG,DAH, DAJ,DAK,DAL,DAW,DAX,V12,V13,V14,V15, V16,V17,V18,V19,V20,V21,V22,V24,V25, V39	
11	PAFZZ	2815008517637	15434	BM-95162	.ROCKER ARM,ENGINE P .... UOC:DAA,DAB,DAC,DAD,DAE,DAF,DAG,DAH, DAJ,DAK,DAL,DAW,DAX,V12,V13,V14,V15, V16,V17,V18,V19,V20,V21,V22,V24,V25, V39	2
12	PAFZZ	5310007320560	96906	MS51968-14	..NUT,PLAIN,HEXAGON .... UOC:DAA,DAB,DAC,DAD,DAE,DAF,DAG,DAH, DAJ,DAK,DAL,DAW,DAX,V12,V13,V14,V15, V16,V17,V18,V19,V20,V21,V22,V24,V25, V39	2
13	PAFZZ	5305009473437	15434	168306	..SETSCREW.. UOC:DAA,DAB,DAC,DAD,DAE,DAF,DAG,DAH, DAJ,DAK,DAL,DAW,DAX,V12,V13,V14,V15, V16,V17,V18,V19,V20,V21,V22,V24,V25, V39	2
14	PAFZZ	3120005893537	15434	140330	..BUSHING,SLEEVE .... UOC:DAA,DAB,DAC,DAD,DAE,DAF,DAG,DAH, DAJ,DAK,DAL,DAW,DAX,V12,V13,V14,V15, V16,V17,V18,V19,V20,V21,V22,V24,V25, V39	2
15	PAFZZ	5305002974022	15434	168319	.SETSCREW.. UOC:DAA,DAB,DAC,DAD,DAE,DAF,DAG,DAH, DAJ,DAK,DAL,DAW,DAX,V12,V13,V14,V15, V16,V17,V18,V19,V20,V21,V22,V24,V25, V39	1
16	PFFZZ	2815012106947	15434	3036285	.HOUSING,ROCKER ARM .... UOC:DAA,DAB,DAC,DAD,DAE,DAF,DAG,DAH, DAJ,DAK,DAL,DAW,DAX,V12,V13,V14,V15, V16,V17,V18,V19,V20,V21,V22,V24,V25, V39	1
17	PAFZZ	5331009843756	19207	7374401	.O-RING .... UOC:DAA,DAB,DAC,DAD,DAE,DAF,DAG,DAH, DAJ,DAK,DAL,DAW,DAX,V12,V13,V14,V15, V16,V17,V18,V19,V20,V21,V22,V24,V25, V39	1
18	PAFZZ	4730011240293	15434	218736	.PLUG,QUICK DISCONNE .... UOC:DAA,DAB,DAC,DAD,DAE,DAF,DAG,DAH, DAJ,DAK,DAL,DAW,DAX,V12,V13,V14,V15, V16,V17,V18,V19 V20,V221,V22,V24,V25, V39	1
19	PAFZZ	3040010791799	15434	3801433	.SHAFT,STRAIGHT .... UOC:DAA,DAB,DAC,DAD,DAE,DAF,DAG,DAH, DAJ,DAK,DAL,DAW,DAX,V12,V13,V14,V15, V16,V17,V18,V19,V20,V21,V22,V24,V25, V39	1

END OF FIGURE

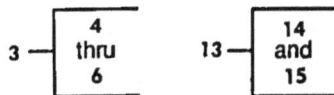

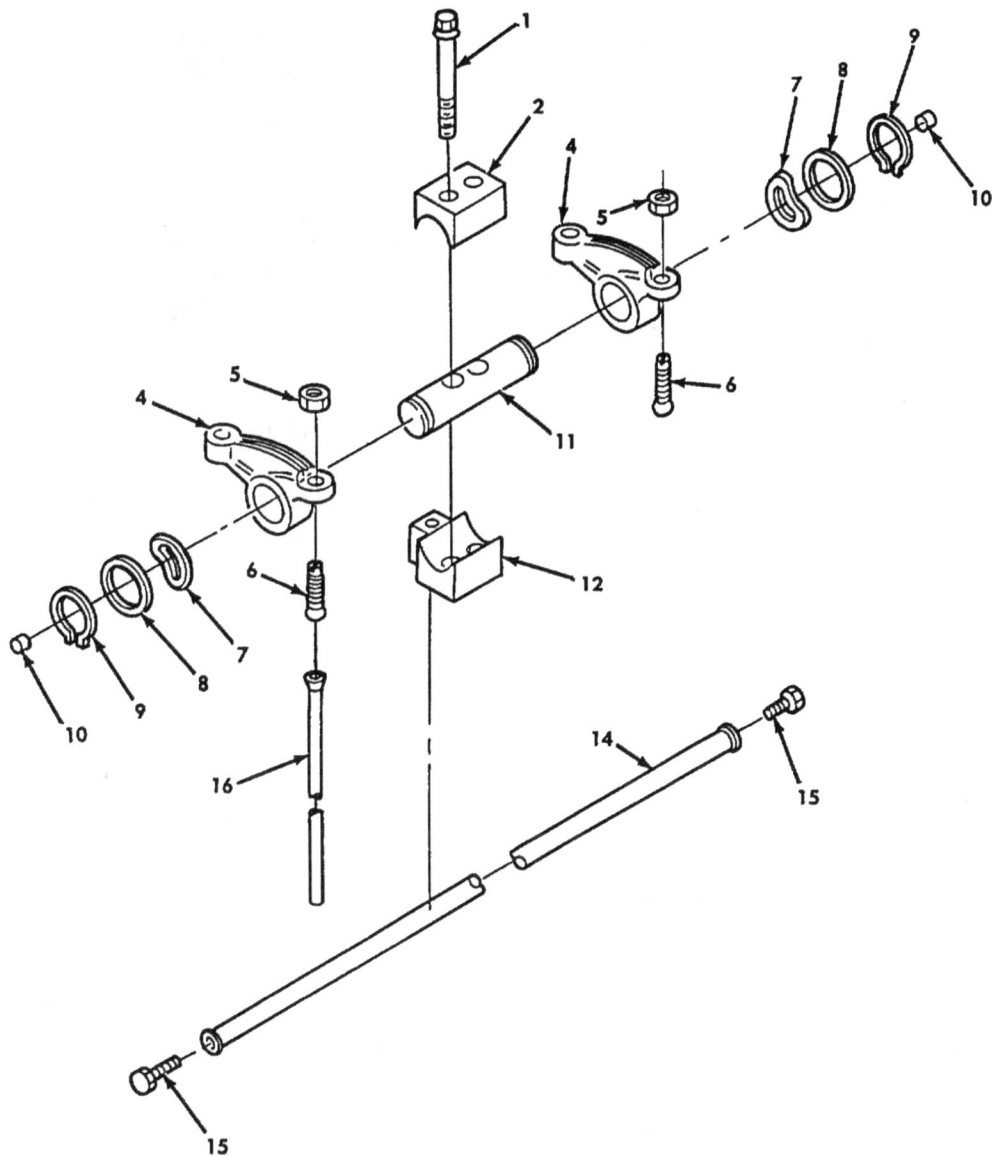

Figure 27. Rocker A and Miscellaneous Hardware (M939A2).

(1) ITEM NO	(2) SMR CODE	(3) NSN	(4) CAGEC	(5) PART NUMBER	(6) DESCRIPTION AND USABLE ON CODES (UOC)	(7) QTY
					GROUP 0105 VALVES, CAMSHAFTS, AND TIMING SYSTEM	
					FIG. 27 ROCKER ARMS AND MISCELLANEOUS HARDWARE(M939A2)	
1	PAFZZ	5305012714326	15434	3901617	SCREW,CAP,SOCKET HE .... UOC:ZAA,ZAB,ZAC,ZAD,ZAE,ZAF,ZAG,ZAB, ZAJ,ZAK,ZAL	12
2	PAFZZ	5342012715704	15434	3901693	CLAMP,RETAINING .... UOC:ZAA,ZAB,ZAC,ZAD,ZAE,ZAF,ZAG,ZAH, ZAJ,ZAK,ZAL	6
3	PAFZZ	2815012715098	15434	3901717	ROCKER ARM,ENGINE P .... UOC:ZAA,ZAB,ZAC,ZAD,ZAE,ZAF,ZAG,ZAH, ZAJ,ZAK,ZAL	12
4	XAFZZ		15434	3901717 XA	.ROCKER ARM,ENGINE P .... UOC:ZAA,ZAB,ZAC,ZAD,ZAE,ZAF,ZAG,ZAH, ZAJ,ZAK,ZAL	2
5	PAFZZ	5310011652184	15434	S-205	.NUT .... UOC:ZAA,ZAB,ZAC,ZAD,ZAE,ZAF,ZAG,ZAB, ZAJ,ZAK,ZAL	2
6	PAFZZ	5305011971663	15434	3900706	.SCREW,MACHINE .... UOC:ZAA,ZAB,ZAC,ZAD,ZAE,ZAF,ZAG,ZAH, ZAJ,ZAK,ZAL	2
7	PAFZZ	5310012708422	15434	3906100	WASHER,SPRING TENSI .... UOC:ZAA,ZAB,ZAC,ZAD,ZAE,ZAF,ZAG,ZAH, ZAJ,ZAK,ZAL	12
8	PAFZZ	5310012708386	15434	3907206	WASHER,FLAT .... UOC:ZAA,ZAB,ZAC,ZAD,ZAE,ZAF,ZAG,ZAH, ZAJ,ZAK,ZAL	12
9	PAFZZ	5325012708360	15434	3901764	RING,RETAINING .... UOC:ZAA,ZAB,ZAC,ZAD,ZAE,ZAF,ZAG,ZAH, ZAJ,ZAK,ZAL	12
10	PAFZZ	5340012397078	15434	203933	PLUG,EXPANSION .... UOC:ZAA,ZAB,ZAC,ZAD,ZAE,ZAF,ZAG,ZAH, ZAJ,ZAK,ZAL	12
11	PAFZZ	2815012713763	15434	3915416	SHAFT, ROCKER ARM, EN .... UOC:ZAA,ZAB,ZAC,ZAD,ZAE,ZAF,ZAG,ZAH, ZAJ,ZAK,ZAL	6
12	PAFZZ	2815012715074	15434	3905870	SUPPORT, ROCKER LEVE .... UOC:ZAA, ZAB, ZAC, ZAD, ZAE, ZAF, ZAG, ZAH, ZAJ, ZAK, ZAL	6
13	PAFZZ	4710012717921	15434	3910911	TUBE, METALLIC .... UOC:ZAA, ZAB, ZAC, ZAD, ZAE, ZAF, ZAG, ZAH, ZAJ, ZAK, ZAL	1
14	XAFZZ		15434	3910911 XA	.TUBE, METALLIC .... UOC:ZAA, ZAB, ZAC, ZAD, ZAE, ZAF, ZAG, ZAH, ZAJ, ZAK, ZAL	1
15	PAFZZ	5305012724809	15434	3910981	.SCREW, CAP, HEXAGON H .... UOC:ZAA, ZAB, ZAC, ZAD, ZAE, ZAF, ZAG, ZAH, ZAJ, ZAK, ZAL	2
16	PAFZZ	2815012715120	15434	3905194	PUSH ROD, ENGINE POP	12

(1) ITEM NO	(2) SMR CODE	(3) NSN	(4) CAGEC	(5) PART NUMBER	(6) DESCRIPTION AND USABLE ON CODES (UOC)	(7) QTY
					UOC:ZAA,ZAB,ZAC,ZAD,ZAE,ZAF,ZAG,ZAH, ZAJ,ZAK,ZAL	

END OF FIGURE

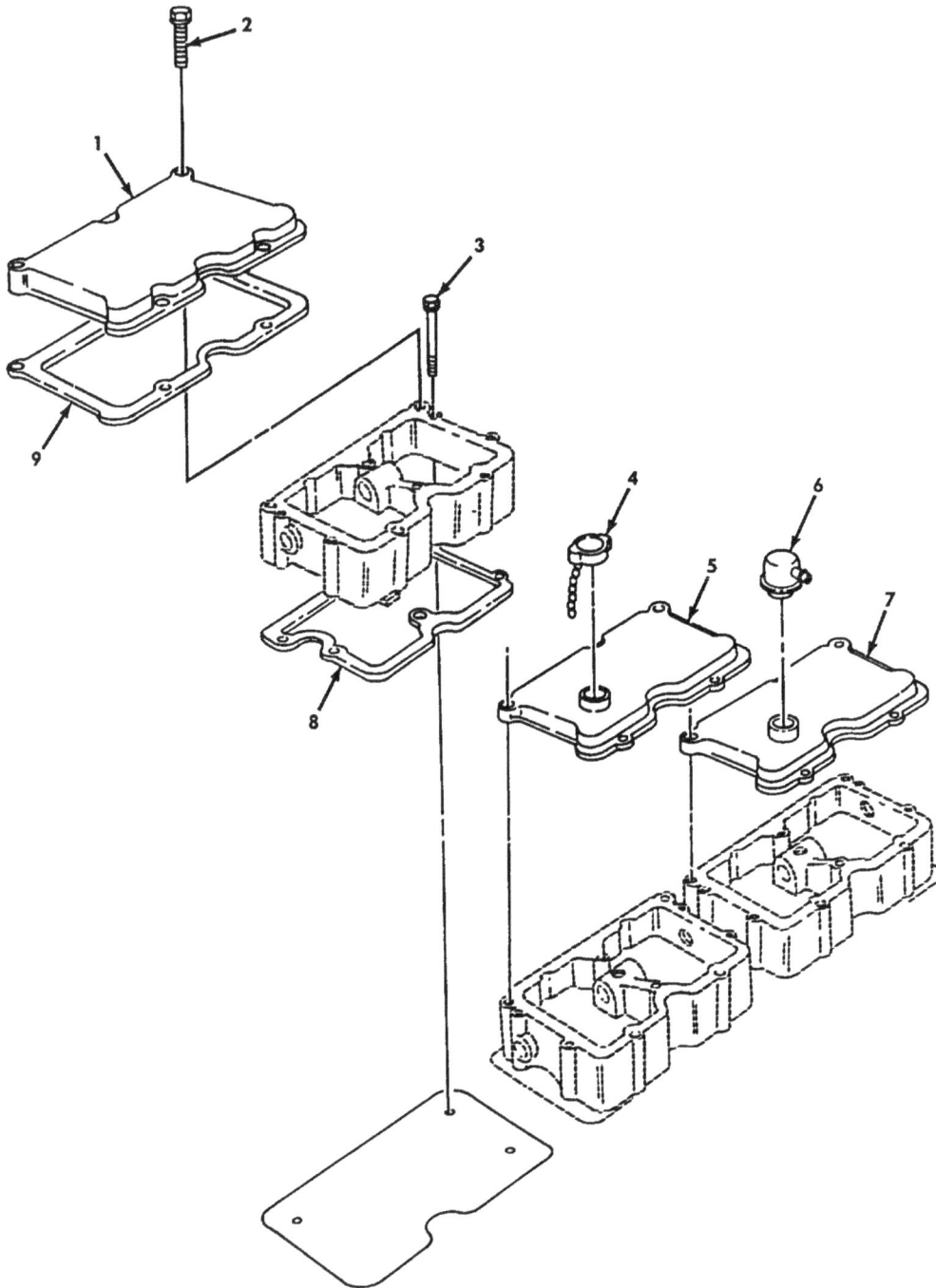

Figure 28. Value Cover and Gasket (M939, M939A1).

(1) ITEM NO	(2) SMR CODE	(3) NSN	(4) CAGEC	(5) PART NUMBER	(6) DESCRIPTION AND USABLE ON CODES (UOC)	(7) QTY
					GROUP 0105 VALVES, CAMSHAFTS, AND TIMING SYSTEM	
					FIG. 28 VALVE COVER AND GASKET(M939, M939A1)	
1	PFFZZ	2815011461024	15434	3006358	COVER,ENGINE POPPET .... UOC:DAA,DAB,DAC,DAD,DAE,DAF,DAG,DAH, DAJ,DAK,DAL,DAW,DAX,V12 ,V13,V14 ,V15, V16,V17,V18,V19,V20,V21,V22,V24,V25, V39	1
2	PAFZZ	5305011188826	15434	3006182	SCREW,CAP,HEXAGON H .... UOC:DAA,DAB,DAC,DAD,DAE,DAF,DAG,DAH, DAJ,DAK,DAL,DAW,DAX,V12 V13,V13 4 ,V14,V15, V16,V17,V18,V19,V20,V21,V22,V24,V25, V39	15
3	PAFZZ	5305011296901	15434	3010589	SCREW,ASSEMBLED WAS .... UOC:DAA,DAB,DAC,DAD,DAE,DAF,DAG,DAH, DAJ,DAK,DAL,DAW,DAX,V12 ,V13,V14 ,V15, V16,V17,V18,V19,V20,V21,V22,V24,V25, V39	18
4	PAOZZ	2590004219524	15434	208045	CAP,FILLER OPENING .... UOC:DAA,DAB,DAC,DAD,DAE,DAF,DAG,DAH, DAJ,DAK,DAL,DAW,DAX,V12,V13,V14,V15, V16,V17,V18,V19,V20,V21,V22,V24,V25, V39	1
5	PFFZZ	2815012151705	15434	3012297	COVER,ENGINE POPPET .... UOC:DAA,DAB,DAC,DAD,DAE, DAF,DAG,DAH, DAJ,DAK,DAL,DAW,DAX,V12,V13,V14,V15, V16,V17,V18,V19,V20,V21,V22,V24,V25, V39	1
6	PAOZZ	2815011461092	15434	3008595	BREATHER .... UOC:DAA,DAB,DAC,DAD,DAE,DAF,DAG,DAH, DAJ,DAK,DAL,DAW,DAX,V12 ,V13 ,V14,V15, V16,V17,V18,V19,V20,V21,V22,V24,V25, V39	1
7	PFFZZ	2815011421732	15434	3006183	COVER,ENGINE POPPET .... UOC:DAA,DAB,DAC,DAD,DAE,DAF, DAG,DAH, DAJ,DAK,DAL,DAW,DAX,V12,V13,V14,V15, V16,V17,V18,V19,V20,V21,V22,V24,V25, V39	1
8	PAFZZ	5330008618592	15434	3017750	GASKET PART OF KIT P/N 3011472 PART............. OF KIT P/N 3804272 .... UOC:DAA,DAB,DAC,DAD,DAE,DAF,DAG,DAH, DAJ,DAK,DAL,DAW,DAX,V12,V13,V14,V15, V16,V17,V18,V19,V20,V21,V22,V24,V25, V39	3
9	PAFZZ	5330012854827	15434	3054841	GASKET PART OF KIT P/N 3011472 PART............. OF KIT P/N 3804272 .... UOC:DAA,DAB,DAC,DAD,DAE,DAF,DAG,DAH, DAJ,DAK,DAL,DAW,DAX,V12,V13,V14,V15, V16,V17,V18,V19,V20,V21,V22,V24,V25,	3

(1) ITEM NO	(2) SMR CODE	(3) NSN	(4) CAGEC	(5) PART NUMBER	(6) DESCRIPTION AND USABLE ON CODES (UOC)	(7) QTY
					V39	

END OF FIGURE

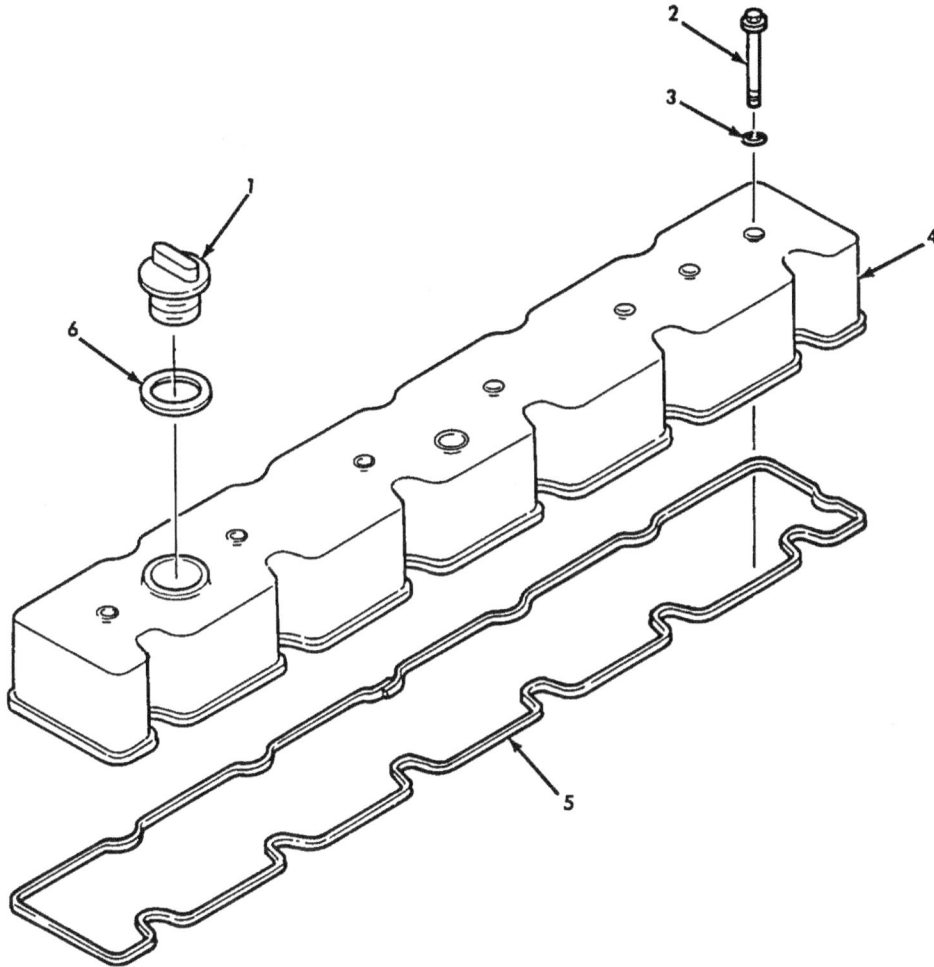

Figure 29. Valve Cover and Gasket (M939A2).

(1) ITEM NO	(2) SMR CODE	(3) NSN	(4) CAGEC	(5) PART NUMBER	(6) DESCRIPTION AND USABLE ON CODES (UOC)	(7) QTY
					GROUP 0105 VALVES, CAMSHAFTS, AND TIMING SYSTEM	
					FIG. 29 VALVE COVER AND GASKET (M939A2)	
1	PAOZZ	2590012733321	15434	3902468	CAP,FILLER OPENING .... UOC:ZAA,ZAB,ZAC,ZAD,ZAE,ZAF,ZAG,ZAH, ZAJ,ZAK,ZAL	1
2	PAOZZ	5305012731594	15434	3916585	SCREW,CAP,HEXAGON H .... UOC:ZAA,ZAB,ZAC,ZAD,ZAE,ZAF,ZAG,ZAH, ZAJ,ZAK,ZAL	6
3	PAOZZ	5331012818997	15434	3910824	O-RING PART OF KIT P/N 3802622 PART............... OF KIT P/N 3802623 PART OF KIT P/N 3802452 .... UOC:ZAA,ZAB,ZAC,ZAD,ZAE,ZAF,ZAG,ZAH, ZAJ,ZAK,ZAL	6
4	PAOZZ	2815012723954	15434	3907734	COVER,ENGINE POPPET .... UOC:ZAA,ZAB,ZAC,ZAD,ZAE,ZAF,ZAG,ZAH, ZAJ,ZAK,ZAL	1
5	PAOZZ	5330012718307	15434	3905449	GASKET PART OF KIT P/N 3802622 PART............. OF KIT P/N 3802623 PART OF KIT P/N 3802452 .... UOC:ZAA,ZAB,ZAC,ZAD,ZAE,ZAF,ZAG,ZAH, ZAJ,ZAK,ZAL	1
6	PAOZZ	5330012721246	15434	3902466	CAP,SEAL,NONMETALLI PART OF KIT P/N............ 3802622 PART OF KIT P/N 3802623 PART OF KIT P/N 3802452 .... UOC:ZAA,ZAB,ZAC,ZAD,ZAE,ZAF,ZAG,ZAH, ZAJ,ZAK,ZAL	1

END OF FIGURE

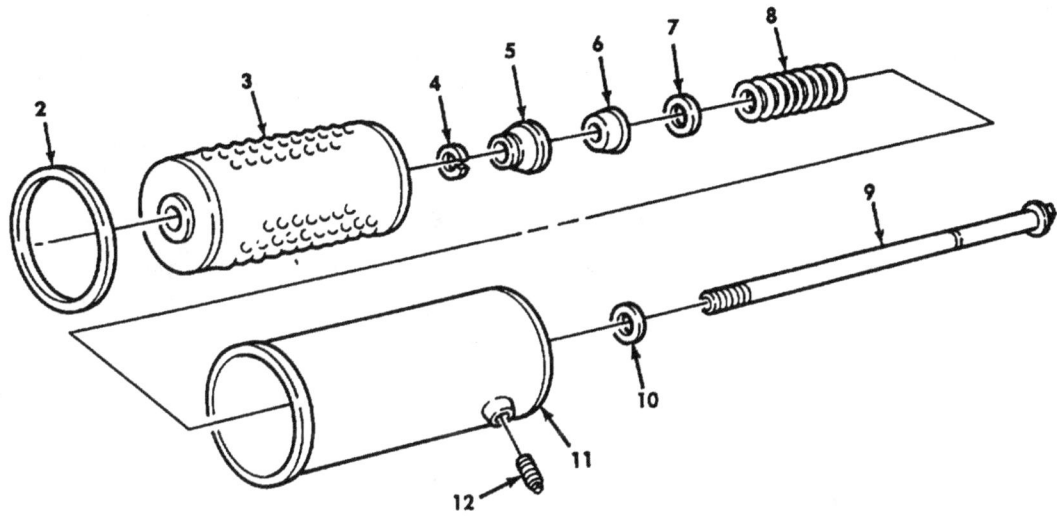

Figure 30. Engine Oil Filter (M939, M939A1).

(1) ITEM NO	(2) SMR CODE	(3) NSN	(4) CAGEC	(5) PART NUMBER	(6) DESCRIPTION AND USABLE ON CODES (UOC)	(7) QTY
					GROUP 0106 ENGINE LUBRICATION SYSTEM	
					FIG. 30 ENGINE OIL FILTER(M939, M939A1)	
1	PFOZZ	2940005523842	33457	252916	FILTER,FLUID .... UOC:DAA,DAB,DAC,DAD,DAE,DAF,DAG,DAH, DAJ,DAK,DAL,DAW,DAX,V12,V13,V14,V15, V16,V17,V18,V19,V20,V21,V22,V24,V25, V39	1
2	KFOZZ	5330009274373	15434	153528	.PACKING ASSEMBLY PART OF KIT P/N............... 3011472 PART OF KIT P/N AR51480 .... UOC:DAA,DAB,DAC,DAD,DAE,DAF,DAG,DAH, DAJ,DAK,DAL,DAW,DAX,V12,V13,V14,V15, V16,V17,V18,V19,V20,V21,V22,V24,V25, V39	1
3	KFOZZ	2940000733316	15434	158139	.FILTER ELEMENT,FLUI PART OF KIT P/N............. AR51480 .... UOC:DAA,DAB,DAC,DAD,DAE,DAF,DAG,DAH, DAJ,DAK,DAL,DAW,DAX,V12,V13,V14,V15, V16,V17,V18,V19,V20,V21,V22,V24,V25, V39	1
4	PAOZZ	5325012097625	33457	3301956	.RING,RETAINING .... UOC:DAA,DAB,DAC,DAD,DAE,DAF,DAG,DAH, DAJ,DAK,DAL,DAW,DAX,V12,V13,V14,V15, V16,V17,V18,V19,V20,V21,V22,V24,V25, V39	1
5	PAOZZ	1680011005608	15434	153521	.SUPPORT,CARTRIDGE .... UOC:DAA,DAB,DAC,DAD,DAE,DAF,DAG,DAH, DAJ,DAK,DAL,DAW,DAX,V12,V13,V14,V15, V16,V17,V18,V19,V20,V21,V22,V24,V25, V39	1
6	PAOZZ	5330010442096	33457	153518	.GASKET .... UOC:DAA,DAB,DAC,DAD,DAE,DAF,DAG,DAH, DAJ,DAK,DAL,DAW,DAX,V12,V13,V14,V15, V16,V17,V18,V19,V20,V21,V22,V24,V25, V39	1
7	PAOZZ	5310002496540	15434	153520	.WASHER,FLAT .... UOC:DAA,DAB,DAC,DAD,DAE,DAF,DAG,DAH, DAJ,DAK,DAL,DAW,DAX,V12,V13,V14,V15, V16,V17,V18,V19,V20,V21,V22,V24,V25, V39	1
8	PAOZZ	5360011185596	15434	153519-S	.SPRING,HELICAL,COMP .... UOC:DAA,DAB,DAC,DAD,DAE,DAF,DAG,DAH, DAJ,DAK,DAL,DAW,DAX,V12,V13,V14,V15, V16,V17,V18,V19,V20,V21,V22,V24,V25, V39	1
9	PAOZZ	5306011182300	15434	3301954	.BOLT,MACHINE .... UOC:DAA,DAB,DAC,DAD,DAE,DAF,DAG,DAH, DAJ,DAK,DAL,DAW,DAX,V12,V13,V14,V15, V16,V17,V18,V19,V20,V21,V22,V24,V25, V39	1

(1) ITEM NO	(2) SMR CODE	(3) NSN	(4) CAGEC	(5) PART NUMBER	(6) DESCRIPTION AND USABLE ON CODES (UOC)	(7) QTY
10	PAOZZ	5310002460221	15434	8265	.WASHER,FLAT .... UOC:DAA,DAB,DAC,DAD,DAE,DAF,DAG,DAH, DAJ,DAK,DAL,DAW,DAX,V12,V13,V14,V15, V16,V17,V18,V19,V20,V21,V22,V24,V25, V39	1
11	PAOZZ	4930011243523	15434	153516	.FILTER BODY,FLUID .... UOC:DAA,DAB,DAC,DAD,DAE,DAF,DAG,DAH, DAJ,DAK,DAL,DAW,DAX,V12,V13,V14,V15, V16,V17,V18,V19,V20,V21,V22,V24,V25, V39	1
12	PAOZZ	4730009247886	11083	5M6214	.PLUG,PIPE .... UOC:DAA,DAB,DAC,DAD,DAE,DAF,DAG,DAH, DAJ,DAK,DAL,DAW,DAX,V12,V13,V14,V15, V16,V17,V18,V19,V20,V21,V22,V24,V25, V39	1

END OF FIGURE

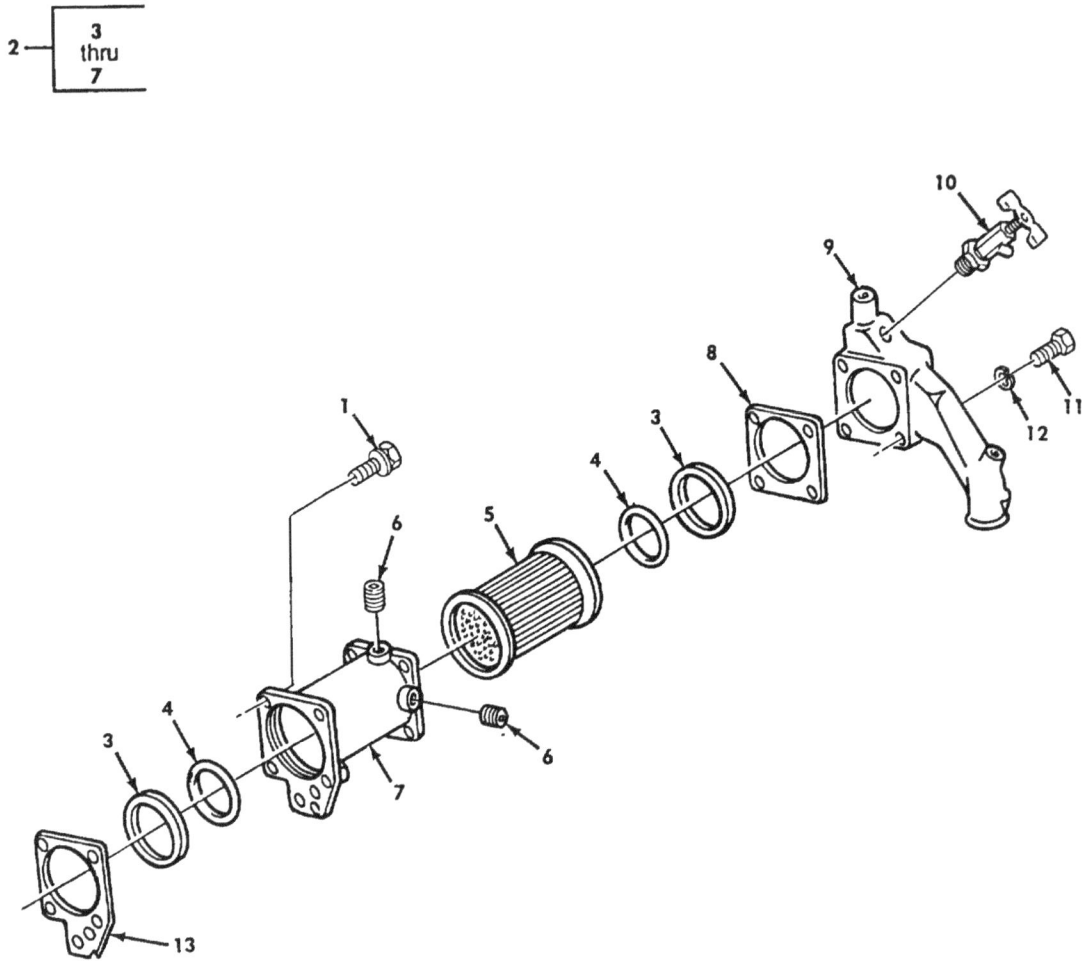

Figure 31. Engine Oil Cooler Assembly (M939, M939A1).

(1) ITEM NO	(2) SMR CODE	(3) NSN	(4) CAGEC	(5) PART NUMBER	(6) DESCRIPTION AND USABLE ON CODES (UOC)	(7) QTY
					GROUP 0106 ENGINE LUBRICATION SYSTEM	
					FIG. 31 ENGINE OIL COOLER ASSEMBLY (M939, M939A1)	
1	PAFZZ	5305010886019	15434	3010596	SCREW, ASSEMBLED WAS .... UOC: DAA, DAB, DAC, DAD, DAE, DAF, DAG, DAH, DAJ, DAK, DAL, DAW, DAX, V12, V13, V14, V15, V16, V17, V18, V19, V20, V21, V22, V24, V25, V39	5
2	PAFFF	2930011175238	15434	BM57837	COOLER, LUBRICATING .... UOC: DAA, DAB, DAC, DAD, DAE, DAF, DAG, DAH, DAJ, DAK, DAL, DAW, DAX, V12, V13, V14, V15, V16, V17, V18, V19, V20, V21, V22, V24, V25, V39	1
3	PAFZZ	5330007857894	15434	110827	.RETAINER, PACKING PART OF KIT P/N............... 3011472 .... UOC:DAA, DAB, DAC, DAD, DAE, DAF, DAG, DAH, DAJ, DAIK, DAL, DAW, DAX, V12, V13, V14, V15, V16, V17, V1B, V19, V20, V21, V22, V24, V25, V39	2
4	PAFZZ	5330011607458	15434	3019116	.O-RING PART OF KIT P/N 3011472 .... UOC:DAA, DAB, DAC, DAD, DAE, DAF, DAG, DAH, DAJ, DAK, DAL, DAW, DAX, V12 , V13 , V14 , V15, V16, V17, V18, V19, V20, V21, V22, V24, V25, V39	2
5	PAFZZ	2930007012091	15434	110848	.ELEMENT, OIL COOLER .... UOC:DAA, DAB, DAC, DAD, DAE, DAF, DAG, DAH, DAJ, DAK, DAL, DAW, DAX, V12, V13, V14, V15, V16, V17, V18, V19, V20, V21, V22, V24, V25, V39	1
6	PAOZZ	4730009541281	15434	3008466	.PLUG, PIPE................................ UOC:DAA, DAB, DAC, DAD, DAE, DAF, DAG, DAH, DAJ, DAK, DAL, DAW, DAX, V12, V13, V14, V15, V16, V17, V18, V19, V20, V21, V22, V24, V25, V39	2
7	PAOZZ	2815004302090	15434	127863	.HOUSING, LUBE OIL CO .... UOC:DAA, DAB, DAC, DAD, DAE, DAF, DAG, DAH, DAJ, DAK, DAL, DAW, DAX, V12, V13, V14, V15, V16, V17, V18, V19, V20, V21, V22, V24, V25, V39	1
8	PAFZZ	5330003288656	15434	68210	GASKET PART OF KIT P/N 3011472 .... UOC:DAA, DAB, DA C, DAD, DAE, DAF, DAG, DAH, DAJ, DAK, DAL, DAW, DAX, V12, V13, V14, V15, V16, V17, V18, V19, V20, V21, V22, V24, V25, V39	1
9	PAFZZ	5340011579898	15434	3030464	COVER, ACCESS .... UOC:DAA, DAB, DAC, DAD, DAE, DAF, DAG, DAH, DAJ, DAK, DAL, DAW, DAX, V12, V13, V14, V15, V16, V17, V18, V19, V20, V21, V22, V24, V25, V39	1
10	PAFZZ	4820008451096	96906	MS35783-2	COCK, DRAIN.AOAP OIL SAMPLER	1

(1) ITEM NO	(2) SMR CODE	(3) NSN	(4) CAGEC	(5) PART NUMBER	(6) DESCRIPTION AND USABLE ON CODES (UOC)	(7) QTY
					UOC:DAA, DAB, DAC, DAD, DAE, DAF, DAG, DAH, DAJ, DAK, DAL, DAW, DAX, V12, V13, V14, V15, V16, V17, V18, V19, V20, V21, V22, V24, V25, V39	
11	PAFZZ	5305001775552	15434	S126	SCREW, CAP, HEXAGON H .... UOC:DAA, DAB, DAC, DAD, DAE, DAF, DAG, DAH, DAJ, DAK, DAL, DAW, DAX, V12, V13, V14, V15, V16, V17, V18, V19, V20, V21, V22, V24, V25, V39	4
12	PAFZZ	5310006379541	96906	MS35338-46	WASHER, LOCK .... UOC:DAA, DAB, DAC, DAD, DAE, DAF, DAG, DAH, DAJ, DAK, DAL, DAW, DAX, V12, V13, V14, V15, V16, V17, V18, V19, V20, V21, V22, V24, V25, V39	4
13	PAFZZ	5330010796514	15434	3008017	GASKET PART OF KIT P/N 3011472 .... UOC:DAA, DAB, DAC, DAD, DAE, DAF, DAG, DAH, DAJ, DAK, DAL, DAW, DAX, V12, V13, V14, V15, V16, V17, V18, V19, V20, V21, V22, V24, V25, V39	1

END OF FIGURE

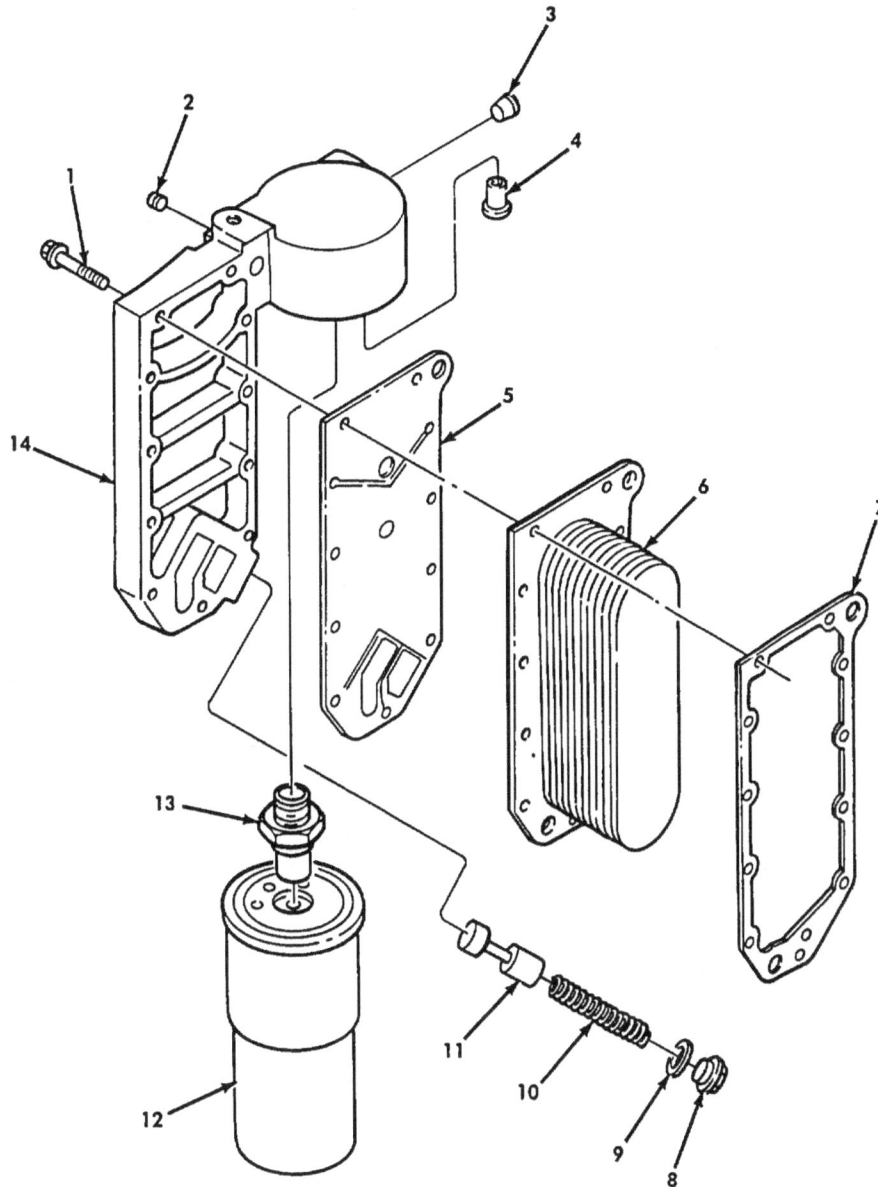

Figure 32. Engine Oil Cooler and Filter (M939A2).

(1) ITEM NO	(2) SMR CODE	(3) NSN	(4) CAGEC	(5) PART NUMBER	(6) DESCRIPTION AND USABLE ON CODES (UOC)	(7) QTY
					GROUP 0106 ENGINE LUBRICATION SYSTEM	
					FIG. 32 ENGINE OIL COOLER AND FILTER (M939A2)	
1	PAOZZ	5305012715850	15434	3901445	SCREW, CAP, HEXAGON H .... UOC:ZAA, ZAB, ZAC, ZAD, ZAE, ZAF, ZAG, ZAH, ZAJ, ZAK, ZAL	11
2	PAOZZ	4730011659491	15434	3008465	PLUG, PIPE .... UOC:ZAA, ZAB, ZAC, ZAD, ZAE, ZAF, ZAG, ZAH, ZAJ, ZAK, ZAL	1
3	PAOZZ	4730011060202	15434	3008469	PLUG, PIPE PART OF KIT P/N 3909415 .... UOC:ZAA, ZAB, ZAC, ZAD, ZAE, ZAF, ZAG, ZAH, ZAJ, ZAK, ZAL	1
4	PAOZZ	2815012115270	15434	3902338	VALVE PRESSURE RELI .... UOC:ZAA, ZAB, ZAC, ZAD, ZAE, ZAF, ZAG, ZAH, ZAJ, ZAK, ZAL	1
5	PAOZZ	5330012716404	15434	3914308	GASKET PART OF KIT P/N 3802389 PART............. OF KIT P/N 3802389.......................... UOC:ZAA, ZAB, ZAC, ZAD, ZAE, ZAF, ZAG, ZAH, ZAJ, ZAK, ZAL	1
6	PAOZZ	2930012715102	15434	3918175	CORE ASSEMBLY, FLUID .... UOC:ZAA, ZAB, ZAC, ZAD, ZAE, ZAF, ZAG, ZAH, ZAJ, ZAK, ZAL	1
7	PAOZZ	5330012715791	15434	3918174	GASKET PART OF KIT P/N 3802389 PART............. OF KIT P/N 3802389.......................... UOC:ZAA, ZAB, ZAC, ZAD, ZAE, ZAF, ZAG, ZAH, ZAJ, ZAK, ZAL	1
8	PAOZZ	5365012708290	15434	3906299	PLUG, MACHINE THREAD .... UOC:ZAA, ZAB, ZAC, ZAD, ZAE, ZAF, ZAG, ZAH, ZAJ, ZAK, ZAL	1
9	PAOZZ	5310012708388	15434	3901798	WASHER, FLAT PART OF KIT P/N 3802389 .... PART OF KIT P/N 3802389 .... UOC:ZAA, ZAB, ZAC, ZAD, ZAE, ZAF, ZAG, ZAH, ZAJ, ZAK, ZAL	1
10	PAOZZ	5360006645343	15434	68274	SPRING, HELICAL, COMP .... UOC:ZAA, ZAB, ZAC, ZAD, ZAE, ZAF, ZAG, ZAH, ZAJ, ZAK, ZAL	1
11	PAOZZ	5340013319625	15434	3918532	PLUNGER, DETENT............................ UOC:ZAA, ZAB, ZAC, ZAD, ZAE, ZAF, ZAG, ZAH, ZAJ, ZAK, ZAL	1
12	PAOZZ	2940011576309	15434	3313281	FILTER ELEMENT, FLUI .... UOC:ZAA, ZAB, ZAC, ZAD, ZAE, ZAF, ZAG, ZAH, ZAJ, ZAK, ZAL	1
13	PAOZZ	2940011459398	15434	3034578	ADAPTER, FILTER IEAD .... UOC:ZAA, ZAB, ZAC, ZAD, ZAE, ZAF, ZAG, ZAH, ZAJ, ZAK, ZAL	1
14	PAOZZ	2940012717203	15434	3918290	HEAD, FLUID FILTER .... UOC:ZAA, ZAB, ZAC, ZAD, ZAE, ZAF, ZAG, ZAR, ZAJ, ZAK, ZAL	1

END OF FIGURE

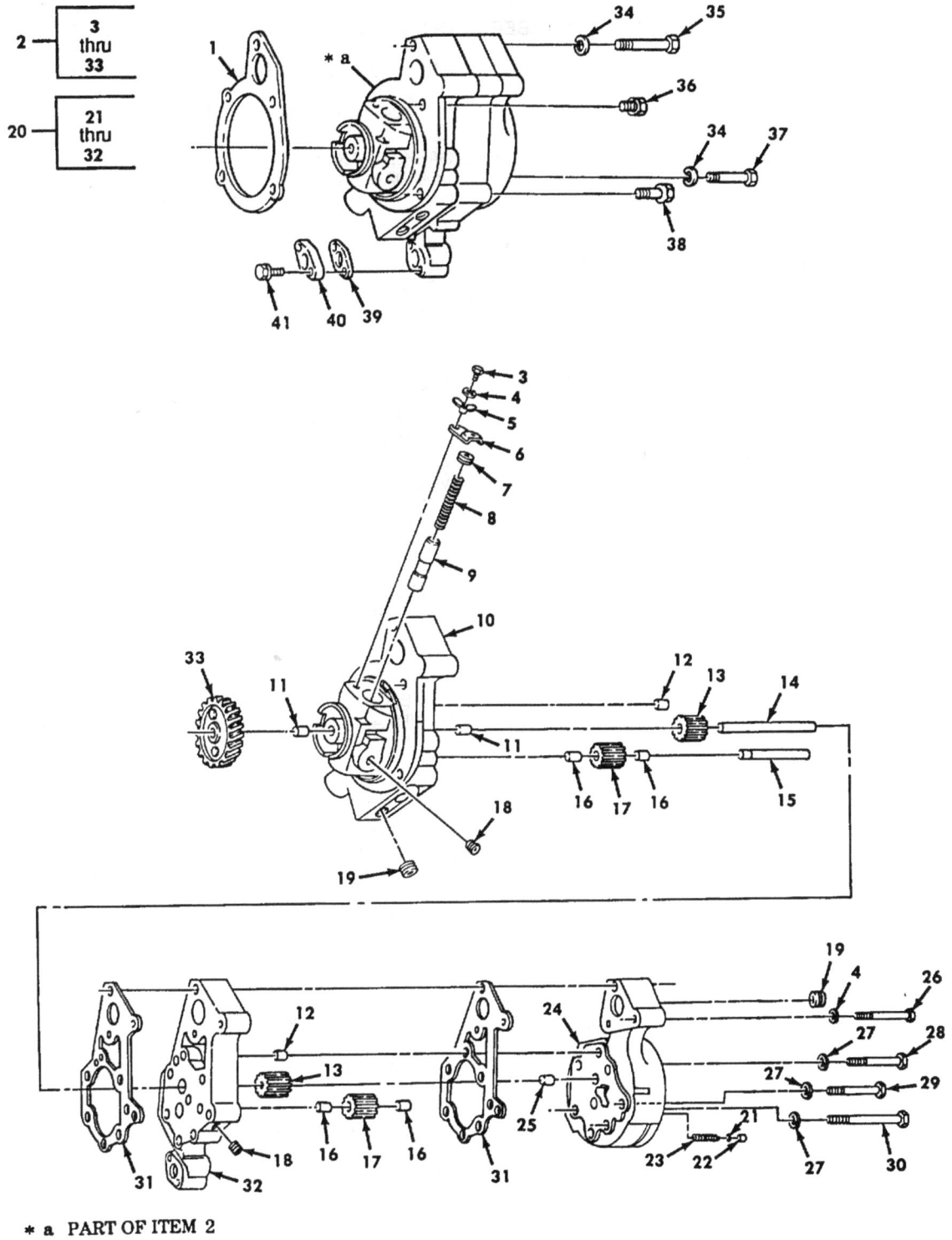

*a PART OF ITEM 2

Figure 33. Engine Oil Pump (M939, M939A1).

(1) ITEM NO	(2) SMR CODE	(3) NSN	(4) CAGEC	(5) PART NUMBER	(6) DESCRIPTION AND USABLE ON CODES (UOC)	(7) QTY
					GROUP 0106 ENGINE LUBRICATION SYSTEM	
					FIG. 33 ENGINE OIL PUMP(M939,M939A1)	
1	PAFZZ	5330011474071	15434	3067613	GASKET PART OF KIT P/N 3011472 .... UOC:DAA, DAB, DAC, DAD, DAE, DAF, DAG, DAH, DAJ, DAK, DAL, DAW, DAX, V12, V13, V14, V15, V16, V17, V18, V19, V20, V21, V22, V24, V25, V39	1
2	PAFHH	2815004042954	15434	BM94082	OIL PUMP ASSEMBLY, E .... UOC:DAA, DAB, DAC, DAD, DAE, DAF, DAG, DAH, DAJ, DAK, DAL, DAW, DAX, V12, V13, V14, V15, V16, V17, V18, V19, V20, V21, V22, V24, V25, V39	1
3	PAHZZ	5305005434372	80204	B1821BH038C075N	.SCREW, CAP, HEXAGON H .... UOC:DAA, DAB, DAC, DAD, DAE, DAF, DAG, DAH, DAJ, DAK, DAL, DAW, DAX, V12, V13, V14, V15, V16, V17, V18, V19, V20, V21, V22, V24, V25, V39	1
4	PAHZZ	5310004079566	96906	MS35338-45	.WASHER, LOCK .... UOC:DAA, DAB, DAC, DAD, DAE, DAF, DAG, DAH, DAJ, DAK, DAL, DAW, DAX, V12, V13, V14, V15, V16, V17, V18, V19, V20, V21, V22, V24, V25, V39	2
5	PAHZZ	2815004068936	15434	109319	.LOCK PLATE .... UOC:DAA, DAB, DAC, DAD, DAE, DAF, DAG, DAH, DAJ, DAK, DAL, DAW, DAX, V12, V13, V14, V15, V16, V17, V18, V19, V20, V21, V22, V24, V25, V39	1
6	PAHZZ	2815008287013	15434	126304	.YOKE, CAP RETAINING .... UOC:DAA, DAB, DAC, DAD, DAE, DAF, DAG, DAH, DAJ, DAK, DAL, DAW, DAX, V12, V13, V14, V15, V16, V17, V18, V19, V20, V21, V22, V24, V25, V39	1
7	PAHZZ	5340008337966	15434	134596	.PLUG, VENT .... UOC:DAA, DAB, DAC, DAD, DAE, DAF, DAG, DAH, DAJ, DAK, DAL, DAW, DAX, V12, V13, V14, V15, V16, V17, V18, V19, V20, V21, V22, V24, V25, V39	1
8	PAHZZ	5360006645343	15434	68274	.SPRING, HELICAL, COMP .... UOC:DAA, DAB, DAC, DAD, DAE, DAF, DAG, DAH, DAJ, DAK, DAL, DAW, DAX, V12, V13, V14, V15, V16, V17, V18, V19, V20, V21, V22, V24, V25, V39	1
9	PAHZZ	2815007911453	15434	127558	.PLUNGER, OIL PUMP RE .... UOC:DAA, DAB, DAC, DAD, DAE, DAF, DAG, DAH, DAJ, DAK, DAL, DAW, DAX, V12, V13, V14, V15, V16, V17, V18, V19, V20, V21, V22, V24, V25, V39	1
10	PAHZZ	2990011233454	15434	BM66076	.HOUSING, OIL PUMP AS .... UOC:DAA, DAB, DAC, DAD, DAE, DAF, DAG, DAH, DAJ, DAK, DAL, DAW, DAX, V12, V13, V14, V15,	1

(1) ITEM NO	(2) SMR CODE	(3) NSN	(4) CAGEC	(5) PART NUMBER	(6) DESCRIPTION AND USABLE ON CODES (UOC)	(7) QTY
					V16, V17, V18, V19, V20, V21, V22, V24, V25, V39	
11	PAHZZ	3120006616646	15434	68586	.BEARING, SLEEVE .... UOC:DAA, DAB, DAC, DAD, DAE, DAF, DAG, DAH, DAJ, DAK, DAL, DAW, DAX, V12, V13, V14, V15, V16, V17, V18, V19, V20, V21, V22, V24, V25, V39	2
12	PAHZZ	5315004752574	15434	69519	.PIN, STRAIGHT, HEADLE .... UOC:DAA, DAB, DAC, DAD, DAE, DAF, DAG, DAH, DAJ, DAK, DAL, DAW, DAX, V12, V13, V14, V15, V16, V17, V18, V19, V20, V21, V22, V24, V25, V39	2
13	PAHZZ	3020003539384	15434	68588	.GEAR, SPUR................................... UOC:DAA, DAB, DAC, DAD, DAE, DAF, DAG, DAH, DAJ, DAK, DAL, DAW, DAX, V12, V13, V14, V15, V16, V17, VV18, V19, V20, V21, V22, V24, V25, V39	2
14	PAHZZ	5315004250118	15434	164164	.PIN, STRAIGHT, HEADLE .... UOC:DAA, DAB, DAC, DAD, DAE, DAF, DAG, DAH, DAJ, DAK, DAL, DAW, DAX, V12, V13, V14, V15, V16, V17, V18, V19, V20, V21, V22, V24, V25, V39	1
15	PAHZZ	3040004859224	15434	133848	.SHAFT, SHOULDERED........................... UOC:DAA, DAB, DAC, DAD, DAE, DAF, DAG, DAH, DAJ, DAK, DAL, DAW, DAX, V12, V13, V14, V15, V16, V17, V18, V19, V20, V21, V22, V24, V25, V39	1
16	PAHZZ	3120006616646	15434	68586	.BEARING, SLEEVE............................... UOC:DAA, DAB, DAC, DAD, DAE, DAF, DAG, DAH, DAJ, DAK, DAL, DAW, DAX, V12, V13, V14, V15, V16, V17, V18, V19, V20, V21, V22, V24, V25, V39	2
17	PAHZZ	3020003317672	15434	3017065	.GEAR, SPUR .... UOC:DAA, DAB, DAC, DAD, DAE, DAF, DAG, DAH, DAJ, DAK, DAL, DAW, DAX, V12, V13, V14, V15, V16, V17, V18, V19, V20, V21, V22, V24, V25, V39	2
18	PAHZZ	4730000575555	29930	444697	.PLUG, PIPE .... UOC:DAA, DAB, DAC, DAD, DAE, DAF, DAG, DAH, DAJ, DAK, DAL, DAW, DAX, V12, V13, V14, V15, V16, V17, V18, V19, V20, V21, V22, V24, V25, V39	2
19	PAHZZ	4730011060202	15434	3008469	.PLUG, PIPE .... UOC:DAA, DAB, DAC, DAD, DAE, DAF, DAG, DAH, DAJ, DAK, DAL, DAW, DAX, V12, V13, V14, V15, V16, V17, V18, V19, V20, V21, V22, V24, V25, V39	2
20	PAHZZ	2940011234875	15434	BM94081	.HEAD, FILTER, OIL PUM .... UOC:DAA, DAB, DAC, DAD, DAE, DAF, DAG, DAH, DAJ, DAK, DAL, DAW, DAX, V12, V13, V14, V15, V16, V17, V18, V19, V20, V21, V22, V24, V25, V39	1

(1) ITEM NO	(2) SMR CODE	(3) NSN	(4) CAGEC	(5) PART NUMBER	(6) DESCRIPTION AND USABLE ON CODES (UOC)	(7) QTY
21	PAHZZ	2815001311700	15434	153526	..SEAT, LUBRICATION FI .... UOC:DAA, DAB, DAC, DAD, DAE, DAF, DAG, DAH, DAJ, DAK, DAL, DAW, DAX, V12, V13, V14, V15, V16, V17, V18, V19, V20, V21, V22, V24, V25, V39	1
22	PAHZZ	4820004005189	15434	200819	..DISK, VALVE .... UOC:DAA, DAB, DAC, DAD, DAE, DAF, DAG, DAH, DAJ, DAK, DAL, DAW, DAX, V12, V13, V14, V15, V16, V17, V18, V19, V20, V21, V22, V24, V25, V39	1
23	PAHZZ	5360009327452	15434	251152	..SPRING, HELICAL, COMP .... UOC:DAA, DAB, DAC, DAD, DAE, DAF, DAG, DAH, DAJ, DAK, DAL, DAW, DAX, V12, V13, V14, V15, V16, V17, V18, V19, V20, V21, V22, V24, V25, V39	1
24	XDHZZ		15434	BM94080	..HEAD, FLUID FILTER .... UOC:DAA, DAB, DAC, DAD, DAE, DAF, DAG, DAH, DAJ, DAK, DAL, DAW, DAX, V12, V13, V14, V15, V16, V17, V18, V19, V20, V21, V22, V24, V25, V39	1
25	PAHZZ	3120006616646	15434	68586	..BEARING, SLEEVE .... UOC:DAA, DAB, DAC, DAD, DAE, DAF, DAG, DAH, DAJ, DAK, DAL, DAW, DAX, V12, V13, V14, V15, V16, V17, V18, V19, V20, V21, V22, V24, V25, V39	1
26	PAHZZ	2815003276439	15434	101662	..SCREW .... UOC:DAA, DAB, DAC, DAD, DAE, DAF, DAG, DAH, DAJ, DAK, DAL, DAW, DAX, V12, V13, V14, V15, V16, V17, V18, V19, V20, V21, V22, V24, V25, V39	2
27	PAHZZ	5310006379541	96906	MS35338-46	..WASHER, LOCK .... UOC:DAA, DAB, DAC, DAD, DAE, DAF, DAG, DAH, DAJ, DAK, DAL, DAW, DAX, V12, V13, V14, V15, V16, V17, V18, V19, V20, V21, V22, V24, V25, V39	7
28	PAHZZ	5305009640565	96906	MS51095-374	..SCREW, CAP, HEXAGON H .... UOC:DAA, DAB, DAC, DAD, DAE, DAF, DAG, DAH, DAJ, DAK, DAL, DAW, DAX, V12, V13, V14, V15, V16, V17, V18, V19, V20, V21, V22, V24, V25, V39	2
29	PAHZZ	5305011268811	15434	S-199-A	..SCREW, CAP, HEXAGON H .... UOC:DAA, DAB, DAC, DAD, DAE, DAF, DAG, DAH, DAJ, DAK, DAL, DAW, DAX, V12, V13, V14, V15, V16, V17, V18, V19, V20, V21, V22, V24, V25, V39	1
30	PAHZZ	5305002064527	15434	S130A	..SCREW, CAP, HEXAGON H .... UOC:DAA, DAB, DAC, DA D, DAE, DAF, DAG, DAH, DAJ, DAK, DAL, DAW, DAX, V12, V13, V14, V15, V16, V17, V18, V19, V20, V21, V22, V24, V25, V39	3
31	PAHZZ	5330009616314	15434	151911	..GASKET PART OF KIT P/N 3011472 .... UOC:DAA, DAB, DAC, DAD, DAE, DAF, DAG, DAH,	2

(1) ITEM NO	(2) SMR CODE	(3) NSN	(4) CAGEC	(5) PART NUMBER	(6) DESCRIPTION AND USABLE ON CODES (UOC)	(7) QTY
					DAJ, DAK, DAL, DAW, DAX, V12, V13, V14, V15, V16, V17, V18, V19, V20, V21, V22, V24, V25, V39	
32	PAHZZ	2815011269367	15434	151917	..BODY, INNER LUBRICAT .... UOC:DAA, DAB, DAC, DAD, DAE, DAF, DAG, DAH, DAJ, DAK, DAL, DAW, DAX, V12, V13, V14, V15, V16, V17, V18, V19, V20, V21, V22, V24, V25, V39	1
33	PAHZZ	3020008207915	15434	3025198	.GEAR, HELICAL .... UOC:DAA, DAB, DAC, DAD, DAE, DAF, DAG, DAH, DAJ, DAX, DAL, DAW, DAX, V12, V13, V14, V15, V16, V17, V18, V19, V20, V21, V22, V24, V25, V39	1
34	PAFZZ	5310002090965	96906	MS35338-47	WASHER, LOCK .... UOC:DAA, DAB, DAC, DAD, DAE, DAF, DAG, DAH, DAJ, DAK, DAL, DAW, DAX, V12, V13, V14, V15, V16, V17, V18, V19, V20, V21, V22, V24, V25, V39	2
35	PAFZZ	5305007959341	15434	69960	SCREW, C AP .... UOC:DAA, DAB, DAC, DAD, DAE, DAF, DAG, DAH, DAJ, DAK, DAL, DAW, DAX, V12, V13, V14, V15, V16, V17, V18, V19, V20, V21, V22, V24, V25, V39	1
36	PAFZZ	5305007959352	15434	3012479	SCREW, CAP, HEXAGON H.................................. UOC:DAA, DAB, DAC, DAD, DAE, DAF, DAG, DAH, DAJ, DAK, DAL, DAW, DAX, V12, V13, V14, V15, V16, V17, V18, VI9, V20, V21, V22, V24, V25, V39	2
37	PAFZZ	5305007959308	15434	169657	SCREW, CAP .... UOC:DAA, DAB, DAC, DAD, DAE, DAF, DAG, DAH, DAJ, DAK, DAL, DAW, DAX, V12, V13, V14, V15, V16, V17, V18, V19, V20, V21, V22, V24, V25, V39	1
38	PAFZZ	5305012276249	15434	3012480	SCREW, ASSEMBLED WAS .... UOC:DAA, DAB, DAC, DAD, DAE, DAF, DAG, DAH, DAJ, DAK, DAL, DAW, DAX, V12, V13, V14, V15, V16, V17, V18, V19, V20, V21, V22, V24, V25, V39	1
39	PAFZZ	5330004655818	15434	134285	GASKET PART OF KIT P/N 3011472..................... UOC:DAA, DAB, DAC, DAD, DAE, DAF, DAG, DAH, DAJ, DAK, DAL, DAW, DAX, V12, V13, V14, V15, V16, V17, V18, V19, V20, V21, V22, V24, V25, V39	1
40	PAFZZ	2815011147399	15434	203502	FLANGE, BODY, PUMP .... UOC:DAA, DAB, DAC, DAD, DAE, DAF, DAG, DAH, DAJ, DAK, DAL, DAW, DAX, V12, V13, V14, V15, V16, V17, V18, V19, V20, V21, V22, V24, V25, V39	1
41	PAFZZ	5305010886019	15434	3010596	SCREW, ASSEMBLED WAS .... UOC:DAA, DAB, DAC, DAD, DAE, DAF, DAG, DAH, DAJ, DAK, DAL, DAW, DAX, V12, V13, V14, V15, V16, V17, V18, V19, V20, V21, V22, V24, V25,	2

(1) ITEM NO	(2) SMR CODE	(3) NSN	(4) CAGEC	(5) PART NUMBER	(6) DESCRIPTION AND USABLE ON CODES (UOC)	(7) QTY
					V39	

END OF FIGURE

Section II.

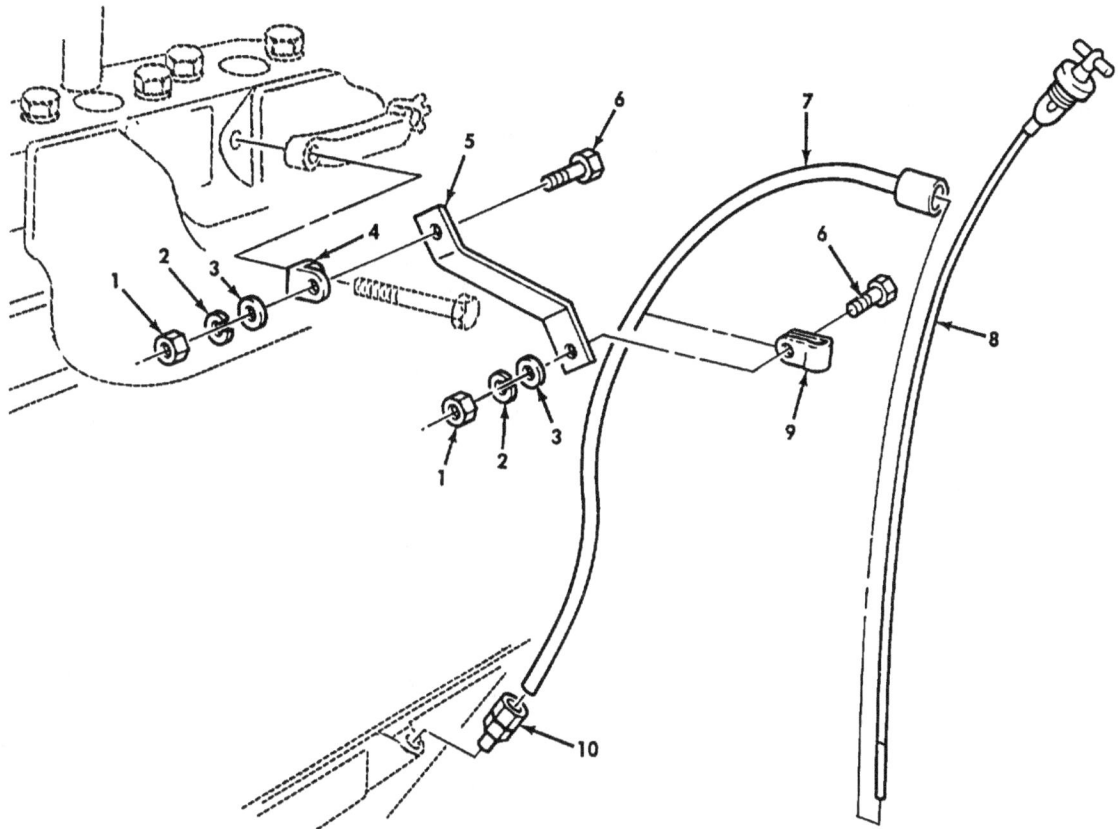

Figure 34. Engine Oil Dipstick and Tube (M939, M939A1).

(1) ITEM NO	(2) SMR CODE	(3) NSN	(4) CAGEC	(5) PART NUMBER	(6) DESCRIPTION AND USABLE ON CODES (UOC)	(7) QTY
					GROUP 0106 ENGINE LUBRICATION SYSTEM	
					FIG. 34 ENGINE OIL DIPSTICK AND TUBE (M939, M939A1)	
1	PAOZZ	5310005218595	15434	S223	NUT, HEXAGON .... UOC:DAA, DAB, DAC, DAD, DAE, DAF, DAG, DAH, DAJ, DAK, DAL, DAW, DAX, V12, V13, V14, V15, V16, V17, V18, V19, V20, V21, V22, V24, V25, V39	2
2	PAOZZ	5310006379541	96906	MS35338-46	WASHER, LOCK .... UOC:DAA, DAB, DAC, DAD, DAE, DAF, DAG, DAH, DAJ, DAK, DAL, DAW, DAX, V12, V13, V14, V15, V16, V17, V18, V19, V20, V21, V22, V24, V25, V39	2
3	PAOZZ	5310000806004	96906	MS27183-14	WASHER, FLAT .... UOC:DAA, DAB, DAC, DAD, DAE, DAF, DAG, DAH, DAJ, DAK, DAL, DAW, DAX, V12, V13, V14, V15, V16, V17, V18, V19, V20, V21, V22, V24, V25, V39	2
4	PFOZZ	5340002437145	34623	139395	BRACKET, ANGLE .... UOC:DAA, DAB, DAC, DAD, DAE, DAF, DAG, DAH, DAJ, DAK, DAL, DAW, DAX, V12, V13, V14, V15, V16, V17, V18, V19, V20, V21, V22, V24, V25, V39	1
5	PFOZZ	5340008212364	15434	204966	BRACKET, DOUBLE ANGL DIPSTICK TUBE .... UOC:DAA, DAB, DAC, DAD, DAE, DAF, DAG, DAH, DAJ, DAK, DAL, DAW, DAX, V12, V13, V14, V15, V16, V17, V18, V19, V20, V21, V22, V24, V25, V39	1
6	PAOZZ	5305002693236	80204	B1821BH038F100N	SCREW, CAP, HEXAGON H .... UOC:DAA, DAB, DAC, DAD, DAE, DAF, DAG, DAH, DAJ, DAK, DAL, DAW, DAX, V12, V13, V14, V15, V16, V17	2
7	PAOZZ	4710011336947	15434	204587	TUBE, DIPSTICK .... UOC:DAA, DAB, DAC, DAD, DAE, DAF, DAG, DAH, DAJ, DAK, DAL, DAW, DAX, V12, V13, V14, V15, V16, V17, V18, V19, V20, V21, V22, V24, V25, V39	1
8	PAOZZ	6680001591732	55783	851-204993	GAGE ROD, LIQUID LEV .... UOC:DAA, DAB, DAC, DAD, DAE, DAF, DAG, DAH, DAJ, DAK, DAL, DAW, DAX, V12, V13, V14, V15, V16, V17, V18, V19, V20, V21, V22, V24, V25, V39	1
9	PAOZZ	5340004175800	15434	200064	CLAMP, LOOP .... UOC:DAA, DAB, DAC, DAD, DAE, DAF, DAG, DAH, DAJ, DAK, DAL, DAW, DAX, V12, V13, V14, V15, V16, V17, V18, V19, V20, V21, V22, V24, V25, V39	1
10	PAOZZ	4730004926040	96906	MS51843-6P	ADAPTER, STRAIGHT, TU .... UOC:DAA, DAB, DAC, DAD, DAE, DAF, DAG, DAH, DAJ, DAK, DAL, DAW, DAX, V12, V13, V14, V15,	1

(1) ITEM NO	(2) SMR CODE	(3) NSN	(4) CAGEC	(5) PART NUMBER	(6) DESCRIPTION AND USABLE ON CODES (UOC)	(7) QTY
					V16, V17, V18, V19, V20, V21, V22, V24, V25, V39	

END OF FIGURE

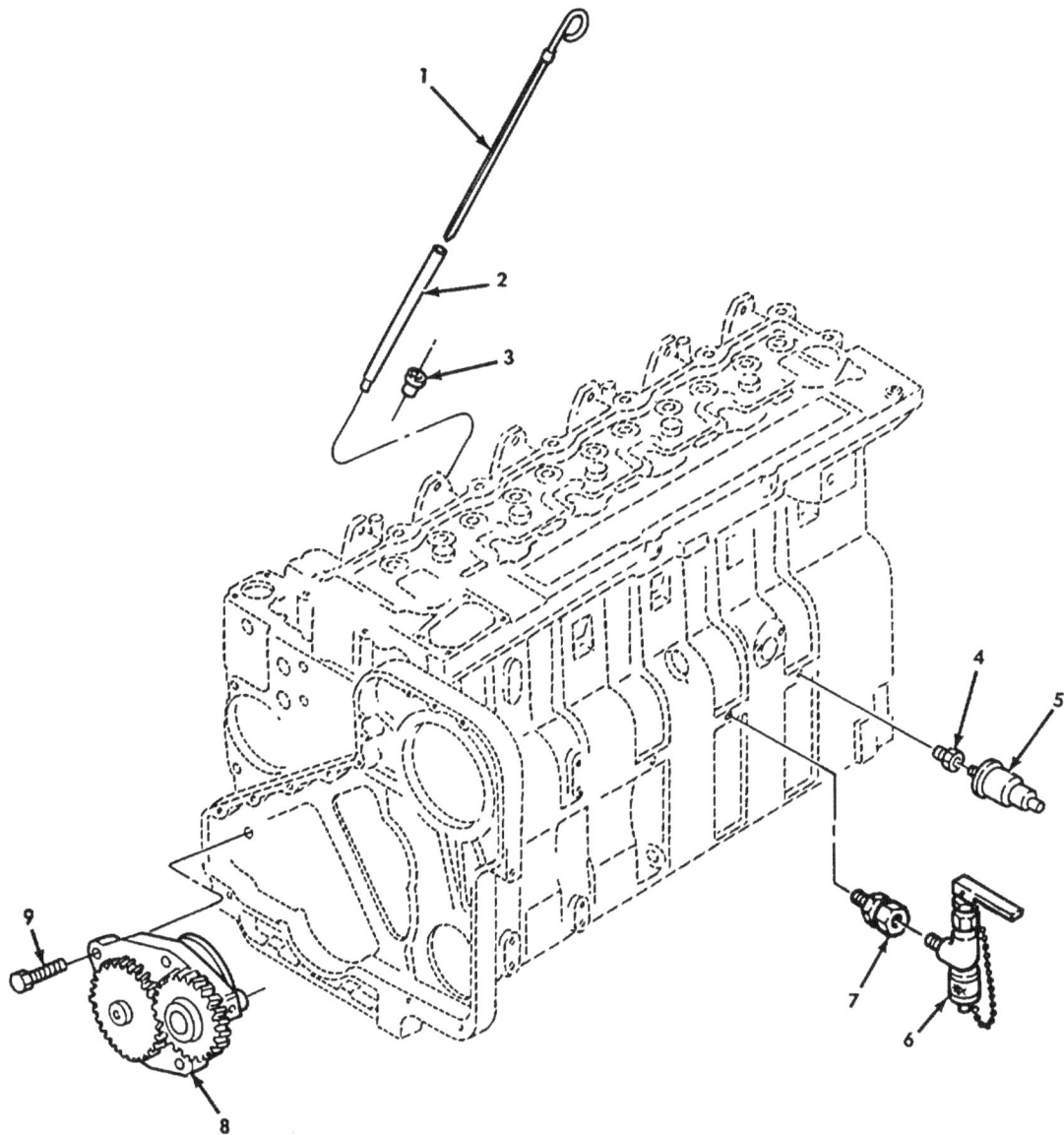

Figure 35. Engine Oil Pump and Dipstick (M939A2

(1) ITEM NO	(2) SMR CODE	(3) NSN	(4) CAGEC	(5) PART NUMBER	(6) DESCRIPTION AND USABLE ON CODES (UOC)	(7) QTY
					GROUP 0106 ENGINE LUBRICATION SYSTEM	
					FIG. 35 ENGINE OIL PUMP AND DIPSTICK (M939A2)	
1	PAOZZ	6680012721867	15434	3911630	GAGE ROD, LIQUID LEV .... UOC:ZAA, ZAB, ZAC, ZAD, ZAE, ZAF, ZAG, ZAH, ZAJ, ZAK, ZAL	1
2	PAOZZ	2590012819716	15434	3905779	FILLER NECK .... UOC:ZAA, ZAB, ZAC, ZAD, ZAE, ZAF, ZAG, ZAH, ZAJ, ZAK, ZAL	1
3	PAOZZ	5340012398607	15434	3900955	PLUG, EXPANSION .... UOC:ZAA, ZAB, ZAC, ZAD, ZAE, ZAF, ZAG, ZAH, ZAJ, ZAK, ZAL	1
4	PAOZZ	4730002268874	30554	71-4872	REDUCER, PIPE .... UOC:ZAA, ZAB, ZAC, ZAD, ZAE, ZAF, ZAG, ZAH, ZAJ, ZAK, ZAL	1
5	PAOZZ	6620009935546	02032	JHP85-430	TRANSMITTER, PRESSUR .... UOC:ZAA, ZAB, ZAC, ZAD, ZAE, ZAF, ZAG, ZAH, ZAJ, ZAK, ZAL	1
6	PAOZZ	4820010730080	19207	11669424	COCK, DRAIN .... UOC:ZAA, ZAB, ZAC, ZAD, ZAE, ZAF, ZAG, ZAH, ZAJ, ZAK, ZAL	1
7	PAOZA	4730006951133	19207	8378730	COUPLING, PIPE .... UOC:ZAA, ZAB, ZAC, ZAD, ZAE, ZAF, ZAG, ZAH, ZAJ, ZAK, ZAL	1
8	PAFZZ	2815012688753	15434	3802278	OIL PUMP ASSEMBLY, E .... UOC:ZAA, ZAB, ZAC, ZAD, ZAE, ZAF, ZAG, ZAH, ZAJ, ZAK, ZAL	1
9	PAFZZ	5305012077447	15434	3900677	SCREW, CAP, HEXAGON H .... UOC:ZAA, ZAB, ZAC, ZAD, ZAE, ZAF, ZAG, ZAH, ZAJ, ZAK, ZAL	4

END OF FIGURE

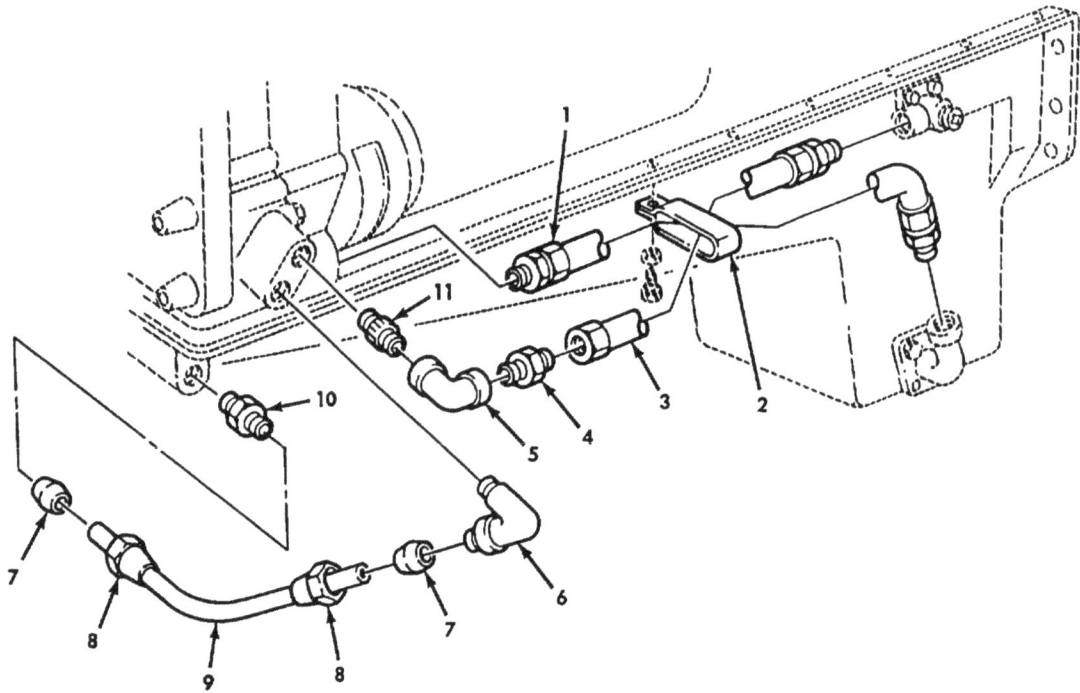

Figure 36. Engine Oil Pump Lines and Fittings (M939, M939A1).

(1) ITEM NO	(2) SMR CODE	(3) NSN	(4) CAGEC	(5) PART NUMBER	(6) DESCRIPTION AND USABLE ON CODES (UOC)	(7) QTY
					GROUP 0106 ENGINE LUBRICATION SYSTEM	
					FIG. 36 ENGINE OIL PUMP LINES AND FITTINGS(M939,M939A1)	
1	PAOZZ	4720011280331	15434	AS-1603506MS	HOSE ASSEMBLY, NONME .... UOC:DAA, DAB, DAC, DAD, DAE, DAF, DAG, DAH, DAJ, DAK, DAL, DAW, DAX, V12, V13, V14, V15, V16, V17, V18, V19, V20, V21, V22, V24, V25, V39	1
2	PAOZZ	5340004099978	15434	114421	CLAMP, LOOP.. UOC:DAA, DAB, DAC, DAD, DAE, DAF, DAG, DAH, DAJ, DAK, DAL, DAW, DAX, V12, V13, V14, V15, V16, V17, V18, V19, V20, V21, V22, V24, V25, V39	1
3	PAOZZ	4720006077645	15434	AS1602906MS	HOSE ASSEMBLY, NONME .... UOC:DAA, DAB, DAC, DAD, DAE, DAF, DAG, DAH, DAJ, DAK, DAL, DAW, DAX, V12, V13, V14, V15, V16, V17, V18, V19, V20, V21, V22, V24, V25, V39	1
4	PAOZZ	4730000974236	15434	67944	ADAPTER, STRAIGHT, PI................................. UOC:DAA, DAB, DAC, DAD, DAE, DAF, DAG, DAH, DAJ, DAK, DAL, DAW, DAX, V12, V13, V14, V15, V16, V17, V18, V19, V20, V21, V22, V24, V25, V39	1
5	PAOZZ	4730001937080	15434	S-987	ELBOW, PIPE... UOC:DAA, DAB, DAC, DAD, DAE, DAF, DAG, DAH, DAJ, DAK, DAL, DAW, DAX, V12, V13, V14, V15, V16, V17, V18, V19, V20, V21, V22, V24, V25, V39	1
6	PAOZZ	4730004285631	30327	69FL3-4	ELBOW, PIPE TO TUBE .... UOC:DAA, DAB, DAC, DAD, DAE, DAF, DAG, DAH, DAJ, DAK, DAL, DAW, DAX, V12, V13, V14, V15, V16, V17, VV18, V19, V20, V21, V22, V24, V25, V39	1
7	PAOZZ	5365003621880	15434	S1077	BUSHING, NONMETTALIC .... UOC:DAA, DAB, DAC, DAD, DAE, DAF, DAG, DAH, DAJ, DAK, DAL, DAW, DAX, V12, V13, V14, V15, V16, V17, V18, V19, V20, V21, V22, V24, V25, V39	2
8	PAOZZ	4730004304385	15434	S-1076	NUT, TUBE COUPLING .... UOC:DAA, DAB, DAC, DAD, DAE, DAF, DAG, DAH, DAJ, DAK, DAL, DAW, DAX, V12, V13, V14, V15, V16, V17, V18, V19, V20, V21, V22, V24, V25, V39	2
9	PAOZZ	4710004255921	15434	202603	TUBE ASSEMBLY, METAL .... UOC:DAA, DAB, DAC, DAD, DAE, DAF, DAG, DAH, DAJ, DAK, DAL, DAW, DAX, V12, V13, V14, V15, V16, V17, V18, V19, V20, V21, V22, V24, V25, V39	1
10	PAOZZ	4730011308766	30327	88-LB-12X12	ADAPTER, STRAIGHT PI .... UOC:DAA, DAB, DAC, DAD, DAE, DAF, DAG, DAH,	1

(1) ITEM NO	(2) SMR CODE	(3) NSN	(4) CAGEC	(5) PART NUMBER	(6) DESCRIPTION AND USABLE ON CODES (UOC)	(7) QTY
					DAJ, DAK, DAL, DAW, DAX, V12, V13, V14, V15, V16, V17, V18, V19, V20, V21, V22, V24, V25, V39	
11	PAOZZ	4730001961468	96906	MS51953-97	NIPPLE, PIPE	1
					UOC:DAA, DAB, DAC, DAD, DAE, DAF, DAG, DAH, DAJ, DAK, DAL, DAW, DAX, V12, V13, V14, V15, V16, V17, V18, V19, V20, V21, V22, V24, V25, V39	

END OF FIGURE

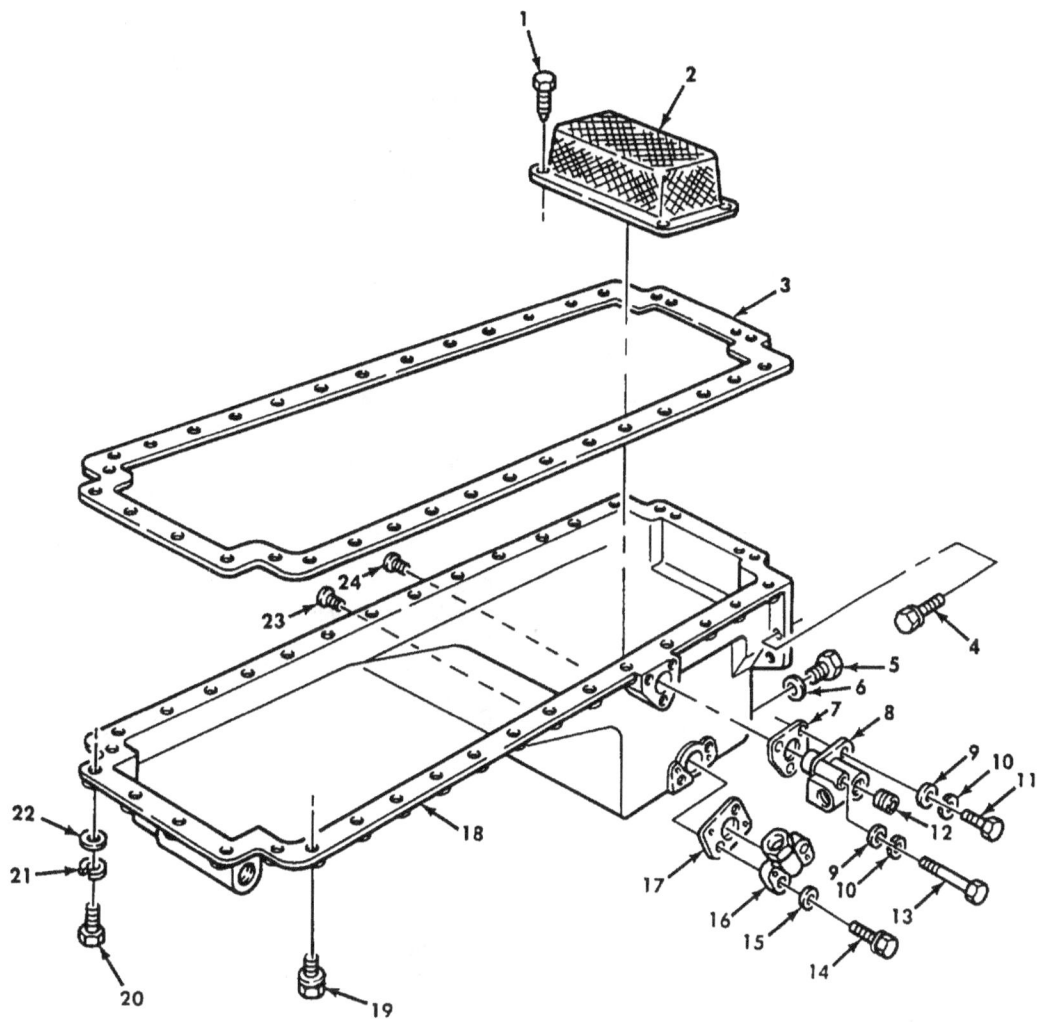

Figure 37. Engine Oil Pan and Related Parts (M939, M939A1).

(1) ITEM NO	(2) SMR CODE	(3) NSN	(4) CAGEC	(5) PART NUMBER	(6) DESCRIPTION AND USABLE ON CODES (UOC)	(7) QTY
					GROUP 0106 ENGINE LUBRICATION SYSTEM	
					FIG. 37 ENGINE OIL PAN AND RELATED PARTS (M939, M939A1)	
1	PAFZZ	5305000680523	96906	MS24621-42	SCREW, TAPPING .... UOC:DAA, DAB, DAC, DAD, DAE, DAF, DAG, DAH, DAJ, DAK, DAL, DAW, DAX, V12, V13, V14, V15, V16, V17, V18, V19, V20, V21, V22, V24, V25, V39	4
2	PAFZZ	4730003386839	34623	MA207-21139	STRAINER ELEMENT, SE .... UOC:DAA, DAB, DAC, DAD, DAE, DAF, DAG, DAH, DAJ, DAK, DAL, DAW, DAX, V12, V13, V14, V15, V16, V17, V18, V19, V20, V21, V22, V24, V25, V39	1
3	PAFZZ	5330011470748	15434	3032861	GASKET PART OF KIT P/N 3011472 .... UOC:DAA, DAB, DAC, DAD, DAE, DAF, DAG, DAH, DAJ, DAK, DAL, DAW, DAX, V12, V13, V14, V15, V16, V17, V18, V19, V20, V21, V22, V24, V25, V39	1
4	PAFZZ	5305004630429	15434	105574	SCREW, CAP, HEXAGON H .... UOC:DAA, DAB, DAC, DAD, DAE, DAF, DAG, DAH, DAJ, DAK, DAL, DAW, DAX, V12, V13, V14, V15, V16, V17, V18, V19, V20, V21, V22, V24, V25, V39	4
5	PAOZZ	5365006951247	15434	69962	PLUG, MACHINE THREAD .... UOC:DAA, DAB, DAC, DAD, DAE, DAF, DAG, DAH, DAJ, DAK, DAL, DAW, DAX, V12, V13, V14, V15, V16, V17, V18, V19, V20, V21, V22, V24, V25, V39	1
6	PAOZZ	5365001979327	15434	67946	SPACER, RING PART OF KIT P/N 3011472.......... UOC:DAA, DAB, DAC, DAD, DAE, DAF, DAG, DAH, DAJ, DAK, DAL, DAW, DAX, V12, V13, V14, V15, V16, V17, V18, V19, V20, V21, V22, V24, V25, V39	1
7	PAFZZ	5330001438371	15434	110453	GASKET PART OF KIT P/N 3011472 .... UOC:DAA, DAB, DAC, DAD, DAE, DAF, DAG, DAH, DAJ, DAK, DAL, DAW, DAX, V12, V13, V14, V15, V16, V17, V18, V19, V20, V21, V22, V24, V25, V39	1
8	PAFZZ	2815001388280	15434	204531	AERATOR, OIL PAN .... UOC:DAA, DAB, DAC, DAD, DAE, DAF, DAG, DAH, DAJ, DAK, DAL, DAW, DAX, V12, V13, V14, V15, V16, V17, V18, V19, V20, V21, V22, V24, V25, V39	1
9	PAFZZ	5310005626558	15434	S626	WASHER, FLAT .... UOC:DAA, DAB, DAC, DAD, DAE, DAF, DAG, DAH, DAJ, DAK, DAL, DAW, DAX, V12, V13, V14, V15, V16, V17, V18, V19, V20, V21, V22, V24, V25, V39	3
10	PAFZZ	5310004079566	19207	7410218	WASHER, LOCK .... UOC:DAA, DAB, DAC, DAD, DAE, DAF, DAG, DAH,	3

(1) ITEM NO	(2) SMR CODE	(3) NSN	(4) CAGEC	(5) PART NUMBER	(6) DESCRIPTION AND USABLE ON CODES (UOC)	(7) QTY
					DAJ, DAK, DAL, DAW, DAX, V12, V13, V14, V15, V16, V17, V18, V19, V20, V21, V22, V24, V25, V39	
11	PAFZZ	5306010758519	96906	MS90725-36	BOLT, MACHINE .... UOC:DAA, DAB, DAC, DAD, DAE, DAF, DAG, DAH, \K DAJ, DAK, DAL, DAW, DAX, V12, V13, V14, V15, V16, V17, V18, V19, V20, V21, V22, V24, V25, V39	2
12	PAFZZ	4730011060202	15434	3008469	PLUG, PIPE .... UOC:DAA, DAB, DAC, DAD, DAE, DAF, DAG, DAH, DAJ, DAK, DAL, DAW, DAX, V12, V13, V14, V15, V16, V17, V18, V19, V20, V21, V22, V24, V25, V39	2
13	PAFZZ	5306001369751	15434	S147B	BOLT, MACHINE... UOC:DAA, DAB, DAC, DAD, DAE, DAF, DAG, DAH, DAJ, DAK, DAL, DAW, DAX, V12, V13, V14, V15, V16, V17, V18, V19, V20, V21, V22, V24, V25, V39	1
14	PAFZZ	5305011376706	15434	3012473	SCREW .... UOC:DAA, DAB, DAC, DAD, DAE, DAF, DAG, DAH, DAJ, DAK, DAL, DAW, DAX, V12, V13, V14, V15, V16, V17, V18, V19, V20, V21, V22, V24, V25, V39	2
15	PAFZZ	5310000806004	96906	MS27183-14	WASHER, FLAT .... UOC:DAA, DAB, DAC, DAD, DAE, DAF, DAG, DAH, DAJ, DAK, DAL, DAW, DAX, V12, V13, V14, V15, V16, V17, V18, V19, V20, V21, V22, V24, V25, V39	2
16	PAFZZ	2815004042957	15434	179688	FLANGE, SUCTION .... UOC:DAA, DAB, DAC, DAD, DAE, DAF, DAG, DAH, DAJ, DAK, DAL, DAW, DAX, V12, V13, V14, V15, V16, V17, V18, V19, V20, V21, V22, V24, V25, V39	1
17	PAFZZ	5330001438376	15434	157551	GASKET .... UOC:DAA, DAB, DAC, DAD, DAE, DAF, DAG, DAH, DAJ, DAK, DAL, DAW, DAX, V12, V13, V14, V15, V16, V17, V18, V19, V20, V21, V22, V24, V25, V39	1
18	PAFZZ	2815004042956	15434	AR45247	OIL PAN .... UOC:DAA, DAB, DAC, DAD, DAE, DAF, DAG, DAH, DAJ, DAK, DAL, DAW, DAX, V12, V13, V14, V15, V16, V17, V18, V19, V20, V21, V22, V24, V25, V39	1
19	PAFZZ	5305012125210	15434	3008069	SCREW, CAP, HEXAGON H .... UOC:DAA, DAB, DAC, DAD, DAE, DAF, DAG, DAH, DAJ, DAK, DAL, DAW, DAX, V12, V13, V14, V15, V16, V17, V18, V19, V20, V21, V22, V24, V25, V39	32
20	PAFZZ	5305004630428	15434	185804	SCREW, CAP, HEXAGON H .... UOC:DAA, DAB, DAC, DAD, DAE, DAF, DAG, DAH, DAJ, DAK, DAL, DAW, DAX, V12, V13, V14, V15, V16, V17, V18, V19, V20, V21, V22, V24, V25,	4

(1) ITEM NO	(2) SMR CODE	(3) NSN	(4) CAGEC	(5) PART NUMBER	(6) DESCRIPTION AND USABLE ON CODES (UOC)	(7) QTY
21	PAFZZ	5310002090965	96906	MS35338-47	V39 WASHER, LOCK .... UOC:DAA, DAB, DAC, DAD, DAE, DAF, DAG, DAH, DAJ, DAK, DAL, DAW, DAX, V12, V13, V14, V15, V16, V17, V18, V19, V20, V21, V22, V24, V25, V39	4
22	PAFZZ	5310005626557	15434	S622	WASHER, FLAT .... UOC:DAA, DAB, DAC, DAD, DAE, DAF, DAG, DAH, DAJ, DAK, DAL, DAW, DAX, V12, V13, V14, V15, V16, V17, V18, V19, V20, V21, V22, V24, V25, V39	4
23	PAFZZ	4730011603579	15434	S-910-B	PLUG, PIPE .... UOC:DAA, DAB, DAC, DAD, DAE, DAF, DAG, DAH, DAJ, DAK, DAL, DAW, DAX, V12, V13, V14, V15, V16, V17, V18, V19, V20, V21, V22, V24, V25, V39	1
24	PAFZZ	4730008018186	15434	S915A	PLUG, PIPE .... UOC:DAA, DA B, DAC, DAD, DAE, DAF, DAG, DAH, DAJ, DAK, DAL, DAW, DAX, V12, V13, V14, V15, V16, V17, V18, V19, V20, V21, V22, V24, V25, V39	1

END OF FIGURE

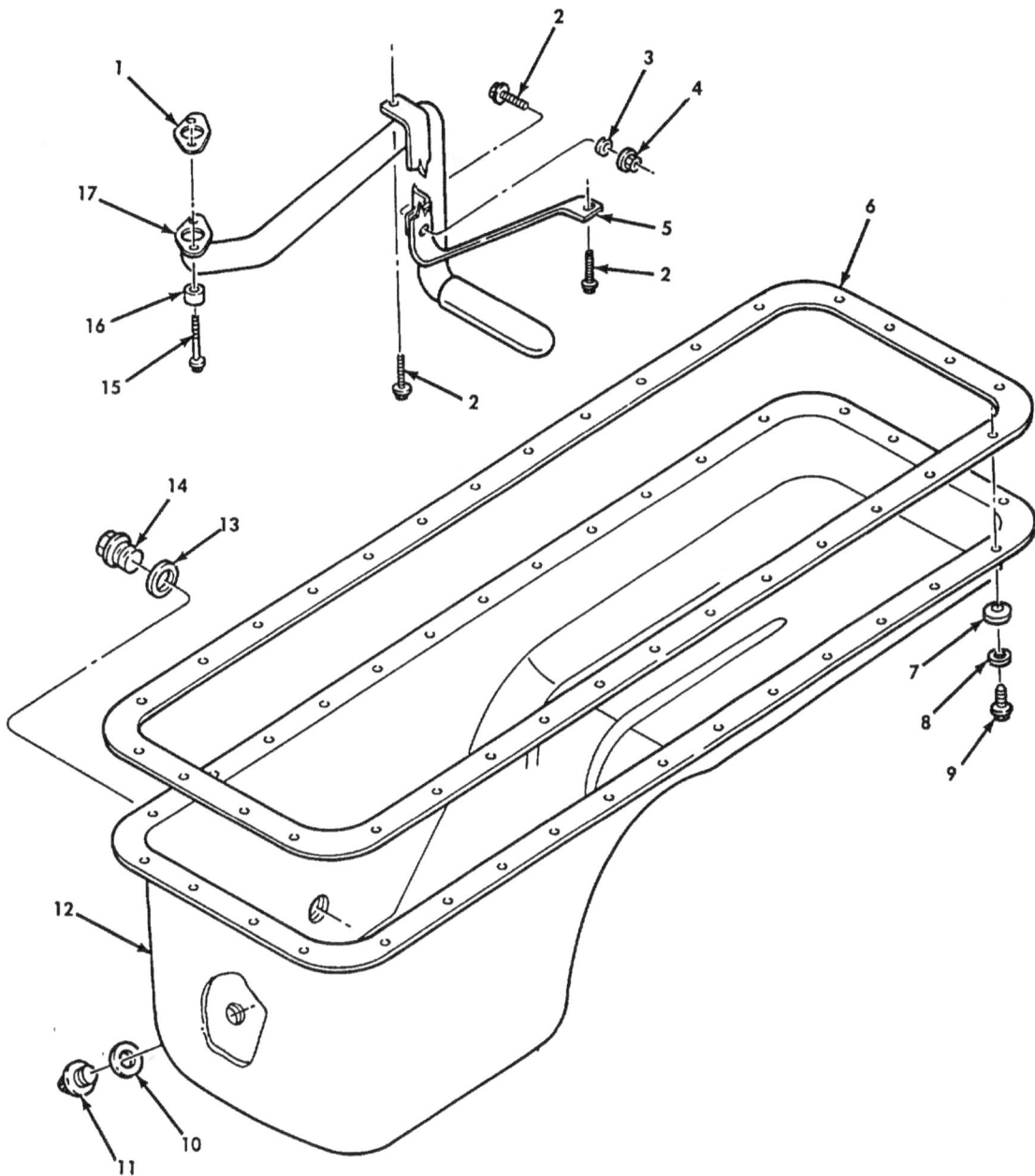

Figure 38. Engine Oil Pan and Related Parts (M939A2).

(1) ITEM NO	(2) SMR CODE	(3) NSN	(4) CAGEC	(5) PART NUMBER	(6) DESCRIPTION AND USABLE ON CODES (UOC)	(7) QTY
					GROUP 0106 ENGINE LUBRICATION SYSTEM	
					FIG. 38 ENGINE OIL PAN AND RELATED PARTS(M939A2)	
1	PAFZZ	5330012721143	15434	3914302	GASKET PART OF KIT P/N 3802389 .... UOC:ZAA, ZAB, ZAC, ZAD, ZAE, ZAF, ZAG, ZAH, ZAJ, ZAK, ZAL	1
2	PAFZZ	5305011936839	15434	3900627	SCREW, CAP, HEXAGON H .... UOC:ZAA, ZAB, ZAC, ZAD, ZAE, ZAF, ZAG, ZAH, ZAJ, ZAK, ZAL	3
3	PAFZZ	5310012708391	15434	3910960	WASHER, FLAT .... UOC:ZAA, ZAB, ZAC, ZAD, ZAE, ZAF, ZAG, ZAH, ZAJ, ZAK, ZAL	1
4	PAFZZ	5310012708251	15434	3906216	NUT, PLAIN, PLATE .... UOC:ZAA, ZAB, ZAC, ZAD, ZAE, ZAF, ZAG, ZAH, ZAJ, ZAK, ZAL	1
5	PAFZZ	5340012870751	7U263	3927611	STRAP, RETAINING .... UOC:ZAA, ZAB, ZAC, ZAD, ZAE, ZAF, ZAG, ZAH, ZAJ, ZAK, ZAL	1
6	PAFZZ	5330012893135	7U263	3914017	GASKET PART OF KIT P/N 3802389 .... UOC:ZAA, ZAB, ZAC, ZAD, ZAE, ZAF, ZAG, ZAH, ZAJ, ZAK, ZAL	1
7	PAFZZ	5310012708423	15434	3910266	WASHER, SPRING TENSI .... UOC:ZAA, ZAB, ZAC, ZAD, ZAE, ZAF, ZAG, ZAH, ZAJ, ZAK, ZAL	32
8	PAFZZ	5310012708390	15434	3908316	WASHER, FLAT .... UOC:ZAA, ZAB, ZAC, ZAD, ZAE, ZAF, ZAG, ZAH, ZAJ, ZAK, ZAL	32
9	PAFZZ	5305012668568	15434	3907860	SETSCREW .... UOC:ZAA, ZAB, ZAC, ZAD, ZAE, ZAF, ZAG, ZAH, ZAJ, ZAK, ZAL	32
10	PAOZZ	5310011880997	15434	3900216	WASHER, FLAT PART OF KIT P/N 3802389 .... UOC:ZAA, ZAB, ZAC, ZAD, ZAE, ZAF, ZAG, ZAH, ZAJ, ZAK, ZAL	1
11	PAOZZ	4730013312913	15434	3911638	PLUG, PIPE .... UOC:ZAA, ZAB, ZAC, ZAD, ZAE, ZAF, ZAG, ZAH, ZAJ, ZAK, ZAL	1
12	PAFZZ	2815012726714	15434	3914011	OIL PAN .... UOC:ZAA, ZAB, ZAC, ZAD, ZAE, ZAF, ZAG, ZAH, ZAJ, ZAK, ZAL	1
13	PAFZZ	5310012090508	15434	3902425	WASHER, FLAT PART OF KIT P/N 3802389 .... UOC:ZAA, ZAB, ZAC, ZAD, ZAE, ZAF, ZAG, ZAH, ZAJ, ZAK, ZAL	1
14	PAFZZ	5365012171995	15434	3904386	PLUG, MACHINE THREAD .... UOC:ZAA, ZAB, ZAC, ZAD, ZAE, ZAF, ZAG, ZAH, ZAJ, ZAK, ZAL	1
15	PAFZZ	5305012724812	15434	3900628	SCREW, CAP, HEXAGON H FLANGED .... UOC:ZAA, ZAB, ZAC, ZAD, ZAE, ZAF, ZAG, ZAH, ZAJ, ZAK, ZAL	2
16	PAFZZ	5365012708482	15434	3908738	SPACER, PLATE	2

(1) ITEM NO	(2) SMR CODE	(3) NSN	(4) CAGEC	(5) PART NUMBER	(6) DESCRIPTION AND USABLE ON CODES (UOC)	(7) QTY
17	PAFZZ	4710012717943	15434	3911617	UOC: ZAA, ZAB, ZAC, ZAD, ZAE, ZAF, ZAG, ZAH, ZAJ, ZAK, ZAL TUBE,BENT, METALLIC .... UOC: ZAA, ZAB, ZAC, ZAD, ZAE, ZAF, ZAG, ZAH, ZAJ, ZAK, ZAL	1

END OF FIGURE

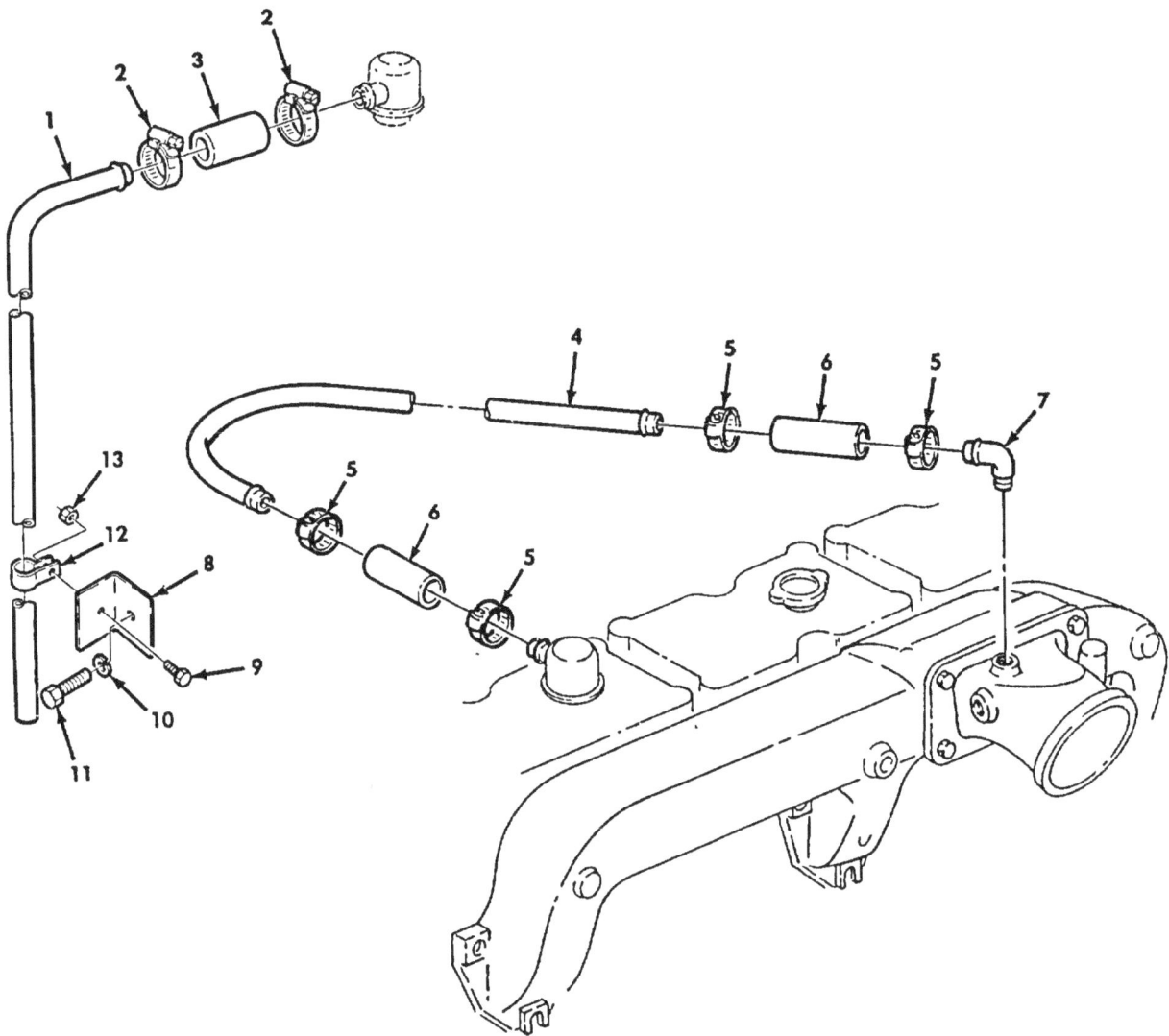

Figure 39. Engine Oil Draft Tube (M939, M939A1).

(1) ITEM NO	(2) SMR CODE	(3) NSN	(4) CAGEC	(5) PART NUMBER	(6) DESCRIPTION AND USABLE ON CODES (UOC)	(7) QTY
					GROUP 0106 ENGINE LUBRICATION SYSTEM	
					FIG. 39 ENGINE OIL DRAFT TUBE(M939, M939A1)	
1	PAOZZ	4710004938899	19207	11648495	TUBE, BENT, METALLIC ENGINE DRAFT .... UOC:DAA, DAB, DAC, DAD, DAE, DAF, DAG, DAH, DAJ, DAK, DAL, DAW, DAX, V12, V13, V14, V15, V16, V17, V18, V19, V20, V21, V22, V24, V25, V39	1
2	PAOZZ	4730009083194	96906	MS35842-11	CLAMP, HOSE USE PRIOR TO SERIAL .... NUMBER 11246663 .... UOC:V12, V13, V14, V15, V16, V17, V18, V19, V20, V21, V22, V24, V25, V39	2
3	MOOZZ		19207	11648497	HOSE USE PRIOR TO SERIAL NUMBER .... 11246663, MAKE FROM HOSE, P/N MS521303A203R .... UOC:DAA, V12, V13, V14, V15, V16, V17, V18, V19, V20, V21, V22, V24, V25, V39	1
4	PAOZZ	4710012572193	15434	3044873	TUBE, BENT, METALLIC USE AFTER .... SERIAL NUMBER 11246663 .... UOC:DAA, DAB, DAC, DAD, DAE, DAF, DAG, DAH, DAJ, DAK, DAL, DAW, DAX, V12, V13, V14, V15, V16, V17, V18, V19, V20, V21, V22, V24, V25, V39	1
5	PAOZZ	4730009083194	96906	MS35842-11	CLAMP, HOSE USE AFTER SERIAL NUMBER ...... 11246663 .... UOC:DAA, DAB, DAC, DAD, DAE, DAF, DAG, DAH, DAJ, DAK, DAL, DAW, DAX, V12, V13, V14, V15, V16, V17, V18, V19, V20, V21, V22, V24, V25, V39	4
6	MOOZZ		19207	11648497	HOSE MAKE FROM HOSE, P/N .... MS521303A203R .... UOC:DAA, DAB, DAC, DAD, DAE, DAF, DAG, DAH, DAJ, DAK, DAL, DAW, DAX, V12, V13, V14, V15, V16, V17, V18, V19, V20, V21, V22, V24, V25, V39	2
7	PAOZZ	4730012573328	15434	136521	ELBOW, PIPE TO HOSE USE AFTER .... SERIAL NUMBER 11246663 .... UOC:DAA, DAB, DAC, DAD, DAE, DAF, DAG, DAH, DAJ, DAK, DAL, DAW, DAX, V12, V13, V14, V15, V16, V17, V18, V19, V20, V21, V22, V24, V25, V39	1
8	PAOZZ	5340000079442	19207	11648729	BPACKET, ANGLE USE PRIOR TO SERIAL .... NUMBER 11246663 .... UOC:V12, V13, V14, V15, V16, V17, V18, V19, V20, V21, V22, V24, V25, V39	1
9	PAOZZ	5306000680513	60285	6893-2	BOLT, MACHINE USE PRIOR TO SERIAL .... NUMBER 11246663 .... UOC:V12, V13, V14, V15, V16, V17, V18, V19, V20, V21, V22, V24, V25, V39	1
10	PAOZZ	5310006379541	96906	MS35338-46	WASHER, LOCK USE PRIOR TO SERIAL	1

(1) ITEM NO	(2) SMR CODE	(3) NSN	(4) CAGEC	(5) PART NUMBER	(6) DESCRIPTION AND USABLE ON CODES (UOC)	(7) QTY
					NUMBER 11246663 .... UOC:V12,V13,V14,V15,V16,V17,V18,V19, V20,V21,V22,V24,V25,V39	
11	PAOZZ	5305005434372	80204	B1821BH038C075N	SCREW,CAP,HEXAGON H USE PRIOR TO .... SERIAL NUMBER 11246663................................ UOC:V12,V13,V14,V15,V16,V17,V18,V19, V20,V21,V22,V24,V25,V39	1
12	PAOZZ	5340002827521	96906	MS21333-40	CLAMP,LOOP USE PRIOR TO SERIAL .... NUMBER 11246663 .... UOC:V12,V13,V14,V15,V16,V17,V18,V19, V20,V21,V22,V24,V25,V39	1
13	PAOZZ	5310000617325	96906	MS21045-4	NUT,SELF-LOCKING,HE USE PRIOR TO .... SERIAL NUMBER 11246663 .... UOC:V12,V13,V14,V15,V16,V17,V18,V19, V20,V21,V22,V24,V25,V39	1

END OF FIGURE

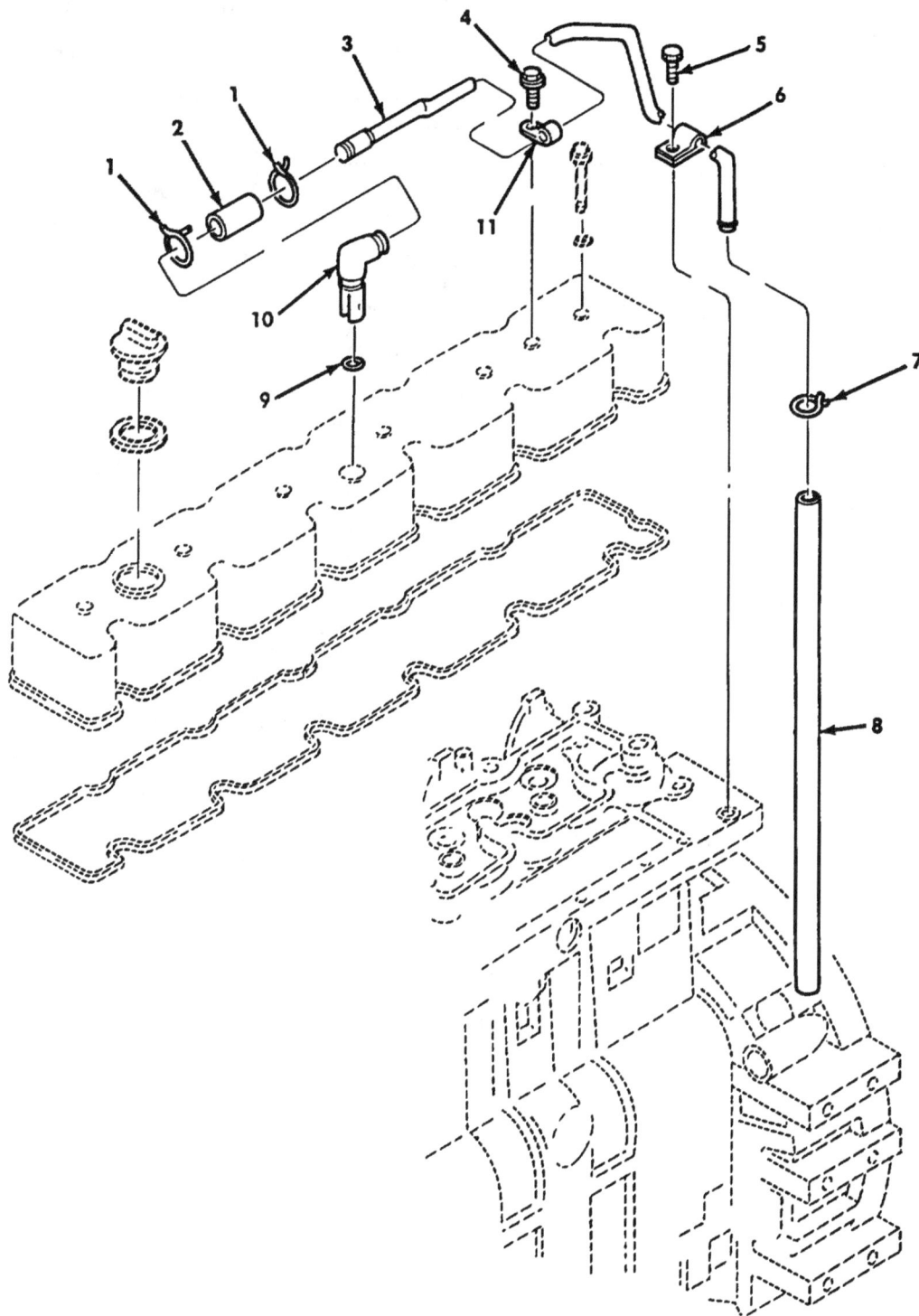

Figure 40. Crankshaft Breather Tube (M939A2).

(1) ITEM NO	(2) SMR CODE	(3) NSN	(4) CAGEC	(5) PART NUMBER	(6) DESCRIPTION AND USABLE ON CODES (UOC)	(7) QTY
					GROUP 0106 ENGINE LUBRICATION SYSTEM	
					FIG. 40 CRANKSHAFT BREATHER TUBE (M939A2)	
1	PAOZZ	4730012717903	15434	3918163	CLAMP, HOSE .... UOC:ZAA, ZAB, ZAC, ZAD, ZAE, ZAF, ZAG, ZAH, ZAJ, ZAK, ZAL	2
2	PAOZZ	4720013735652	15434	3918611	HOSE, NONMETALLIC .... UOC:ZAA, ZAB, ZAC, ZAD, ZAE, ZAF, ZAG, ZAH, ZAJ, ZAK, ZAL	1
3	PAOZZ	4710012717922	15434	3909669	TUBE, BENT, METALLIC .... UOC:ZAA, ZAB, ZAC, ZAD, ZAE, ZAF, ZAG, ZAH, ZAJ, ZAK, ZAL	1
4	PAOZZ	5305011922036	15434	3912072	SCREW .... UOC:ZAA, ZAB, ZAC, ZAD, ZAE, ZAF, ZAG, ZAH, ZAJ, ZARK, ZAL	1
5	PAOZZ	5305012453817	15434	3901249	SCREW, CAP, HEXAGON H .... UOC:ZAA, ZAB, ZAC, ZAD, ZAE, ZAF, ZAG, ZAH, ZAJ, ZAK, ZAL	1
6	PAOZZ	5340012712470	15434	70622	CLAMP, LOOP .... UOC:ZAA, ZAB, ZAC, ZAD, ZAE, ZAF, ZAG, ZAH, ZAJ, ZAK, ZAL	1
7	PAOZZ	4730012713739	15434	3907740	CLAMP, HOSE .... UOC:ZAA, ZAB, ZAC, ZAD, ZAE, ZAF, ZAG, ZAH, ZAJ, ZAK, ZAL	1
8	PAOZZ	4720012716950	15434	3918616	HOSE, NONMETALLIC .... UOC:ZAA, ZAB, ZAC, ZAD, ZAE, ZAF, ZAG, ZAH, ZAJ, ZAK, ZAL	1
9	PAOZZ	5331012721120	15434	3909397	O-RING PART OF KIT P/N 3802622 PART .... OF KIT P/N 3802623 PART OF KIT P/N 3802452 .... UOC:ZAA, ZAB, ZAC, ZAD, ZAE, ZAF, ZAG, ZAH, ZAJ, ZAK, ZAL	1
10	PAOZZ	4730012717955	15434	392049700	ELBOW, PIPE TO HOSE .... UOC:ZAA, ZAB, ZAC, ZAD, ZAE, ZAF, ZAG, ZAH, ZAJ, ZAK, ZAL	1
11	PAOZZ	5340012750487	15434	3907741	CLAMP, LOOP .... UOC:ZAA, ZAB, ZAC, ZAD, ZAE, ZAF, ZAG, ZAH, ZAJ, ZAK, ZAL	1

END OF FIGURE

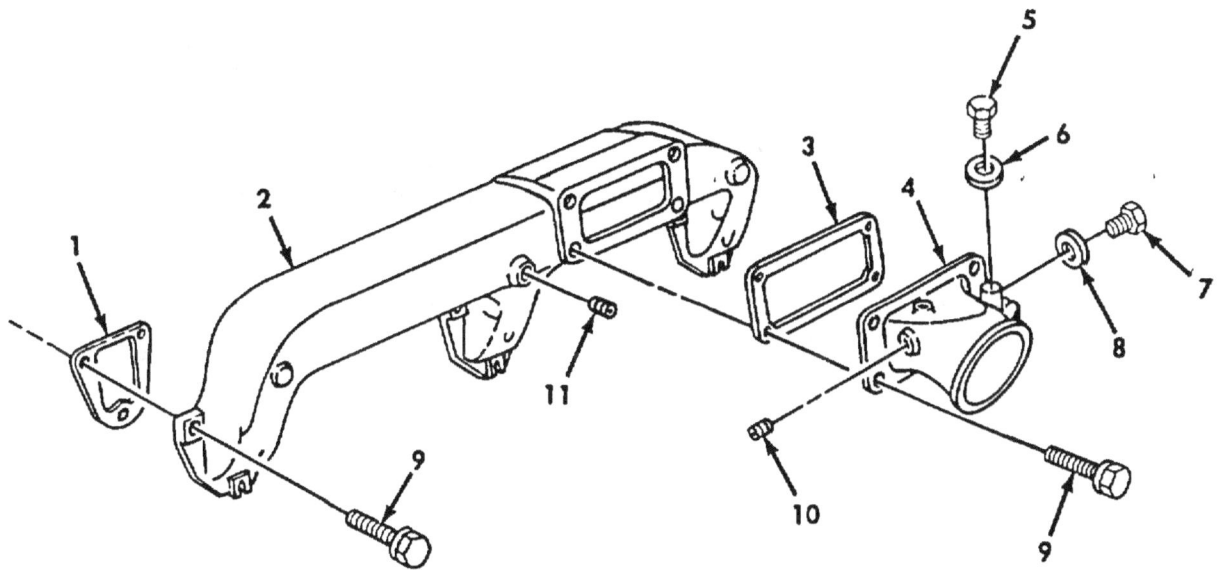

Figure 41. Air Intake Manifold (M939, M939A1).

(1) ITEM NO	(2) SMR CODE	(3) NSN	(4) CAGEC	(5) PART NUMBER	(6) DESCRIPTION AND USABLE ON CODES (UOC)	(7) QTY
					GROUP 0108 MANIFOLDS	
					FIG. 41 AIR INTAKE MANIFOLD(M939, M939A1)	
1	PAFZZ	5330010863523	15434	3008591	GASKET PART OF KIT P/N 3011472 PART .... OF KIT P/N 3804272 .... UOC:DAA, DAB, DAC, DAD, DAE, DAF, DAG, DAH, DAJ, DAK, DAL, DAW, DAX, V12, V13, V14, V15, V16, V17, V18, V19, V20, V21, V22, V24, V25, V39	3
2	PAFZZ	2815004042921	15434	202890	MANIFOLD, INTAKE .... UOC:DAA, DAB, DAC, DAD, DAE, DAF, DAG, DAH, DAJ, DAK, DAL, DAW, DAX, V12, V13, V14, V15, V16, V17, V18, V19, V20, V21, V22, V24, V25, V39	1
3	PAFZZ	5330011312967	15434	3012972	GASKET PART OF KIT P/N 3011472 PART .... OF KIT P/N 3804272 .... UOC:DAA, DAB, DAC, DAD, DAE, DAF, DAG, DAH, DAJ, DAK, DAL, DAW, DAX, V12, V13, V14, V15, V16, V17, V18, V19, V20, V21, V22, V24, V25, V39	1
4	PFFZZ	2815012570853	15434	3044876	CONNECTION, AIR INTA .... UOC:DAA, DAB, DAC, DAD, DAE, DAF, DAG, DAH, DAJ, DAK, DAL, DAW, DAX, V12, V13, V14, V15, V16, V17, V18, V19, V20, V21, V22, V24, V25, V39	1
5	PAFZZ	5365009749851	15434	138608	PLUG, MACHINE THREAD .... UOC:DAA, DAB, DAC, DAD, DAE, DAF, DAG, DAH, DAJ, DAK, DAL, DAW, DAX, V12, V13, V14, V15, V16, V17, V18, V19, V20, V21, V22, V24, V25, V39	1
6	PAFZZ	5330005994517	15434	138609	GASKET .... UOC:DAA, DAB, DAC, DAD, DAE, DAF, DAG, DAH, DAJ, DAK, DAL, DAW, DAX, V12, V13, V14, V15, V16, V17, V18, V19, V20, V21, V22, V24, V25, V39	1
7	PAFZZ	5365011973179	15434	108392	PLUG, MACHINE THREAD .... UOC:DAA, DAB, DAC, DAD, DAE, DAF, DAG, DAH, DAJ, DAK, DAL, DAW, DAX, V12, V13, V14, V15, V16, V17, V18, V19, V20, V21, V22, V24, V25, V39	1
8	PAFZZ	5365001979327	15434	67946	SPACER, RING .... UOC:DAA, DAB, DAC, DAD, DAE, DAF, DAG, DAH, DAJ, DAK, DAL, DAW, DAX, V12, V13, V14, V15, V16, V17, V18, V19, V20, V21, V22, V24, V25, V39	1
9	PAFZZ	5305010728816	15434	3011715	SCREW .... UOC:DAA, DAB, DAC, DAD, DAE, DAF, DAG, DAH, DAJ, DAK, DAL, DAW, DAX, V12, V13, V14, V15, V16, V17, V18, V19, V20, V21, V22, V24, V25, V39	13

(1) ITEM NO	(2) SMR CODE	(3) NSN	(4) CAGEC	(5) PART NUMBER	(6) DESCRIPTION AND USABLE ON CODES (UOC)	(7) QTY
10	PAFZZ	4730009541281	15434	3008466	PLUG, PIPE .... UOC:DAA, DAB, DAC, DAD, DAE, DAF, DAG, DAH, DAJ, DAK, DAL, DAW, DAX, V12, V13, V14, V15, V16, V17, V18, V19, V20, V21, V22, V24, V25, V39	2
11	PAFZZ	4730000189566	81348	WW-P-471ACABCA	PLUG, PIPE .... UOC:DAA, DAB, DAC, DAD, DAE, DAF, DAG, DAH, DAJ, DAK, DAL, DAW, DAX, V12, V13, V14, V15, V16, V17, V18, V19, V20, V21, V22, V24, V25, V39	1

END OF FIGURE

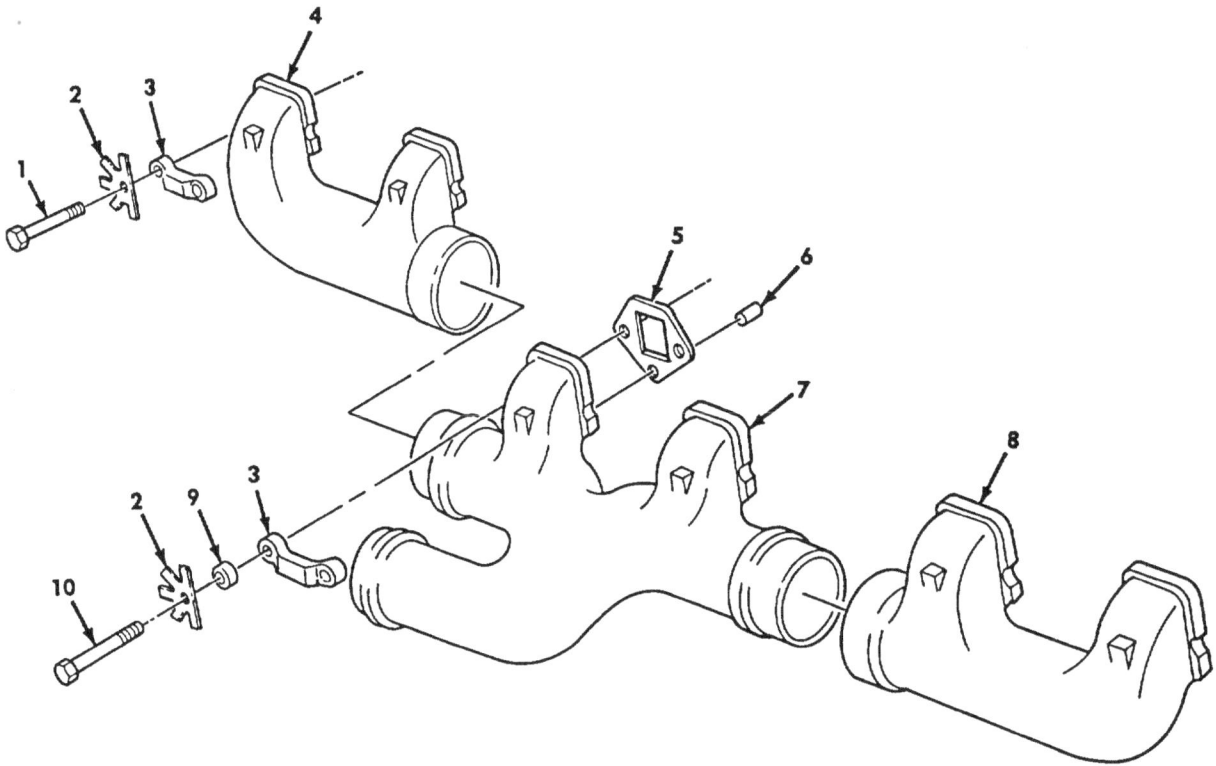

Figure 42. Exhaust Manifold (M939, M939A1).

(1) ITEM NO	(2) SMR CODE	(3) NSN	(4) CAGEC	(5) PART NUMBER	(6) DESCRIPTION AND USABLE ON CODES (UOC)	(7) QTY
					GROUP 0108 MANIFOLDS	
					FIG. 42 EXHAUST MANIFOLD(M939, M939A1)	
1	PAFZZ	5305000050666	15434	200908	SCREW, CAP, HEXAGON H .... UOC:DAA, DAB, DAC, DAD, DAE, DAF, DAG, DAH, DAJ, DAK, DAL, DAW, DAX, V12, V13, V14, V15, V16, V17, V18, V19, V20, V21, V22, V24, V25, V39	8
2	PAFZZ	5310008878325	15434	114638	WASHER, KEY PART OF KIT P/N 3011472 .... PART OF KIT P/N 3804272 .... UOC:DAA, DAB, DAC, DAD, DAE, DAF, DAG, DAH, DAJ, DAK, DAL, DAW, DAX, V12, V13, V14, V15, V16, V17, V18, V19, V20, V21, V22, V24, V25, V39	12
3	PAFZZ	5340001323203	15434	200919	STRAP, RETAINING .... UOC:DAA, DAB, DAC, DAD, DAE, DAF, DAG, DAH, DAJ, DAK, DAL, DAW, DAX, V12, V13, V14, V15, V16, V17, V18, V19, V20, V21, V22, V24, V25, V39	6
4	PFFZZ	2815004042926	15434	194921	MANIFOLD, EXHAUST .... UOC:DAA, DAB, DAC, DAD, DAE, DAF, DAG, DAH, DAJ, DAK, DAL, DAW, DAX, V12, V13, V14, V15, V16, V17, V18, V19, V20, V21, V22, V24, V25, V39	1
5	PAFZZ	5330006593178	15434	142234	GASKET PART OF KIT P/N 3011472 PART .... OF KIT P/N 3804272................................... UOC:DAA, DAB, DAC, DAD, DAE, DAF, DAG, DAH, DAJ, DAK, DAL, DAW, DAX, V12, V13, V14, V15, V16, V17, V18, V19, V20, V21, V22, V24, V25, V39	6
6	PAFZZ	2815008295227	15434	105199	DOWEL, MANIFOLD .... UOC:DAA, DAB, DAC, DAD, DAE, DAF, DAG, DAH, DAJ, DAK, DAL, DAW, DAX, V12, V13, V14, V15, V16, V17, V18, V19, V20, V21, V22, V24, V25, V39	6
7	PFFZZ	2815011379707	15434	210879	MANIFOLD, EXHAUST .... UOC:DAA, DAB, DAC, DAD, DAE, DAF, DAG, DAH, DAJ, DAK, DAL, DAW, DAX, V12, V13, V14, V15, V16, V17, V18, V19, V20, V21, V22, V24, V25, V39	1
8	PFFZZ	2815004042931	15434	194923	MANIFOLD, EXHAUST .... UOC:DAA, DAB, DAC, DAD, DAE, DAF, DAG, DAH, DAJ, DAK, DAL, DAW, DAX, V12, V13, V14, V15, V16, V17, V18, V19, V20, V21, V22, V24, V25, V39	1
9	PAFZZ	5310005626557	15434	S622	WASHER, FLAT .... UOC:DAA, DAB, DAC, DAD, DAE, DAF, DAG, DAH, DAJ, DAK, DAL, DAW, DAX, V12, V13, V14, V15, V16, V17, V18, V19, V20, V21, V22, V24, V25, V39	6

(1) ITEM NO	(2) SMR CODE	(3) NSN	(4) CAGEC	(5) PART NUMBER	(6) DESCRIPTION AND USABLE ON CODES (UOC)	(7) QTY
10	PAFZZ	5305003403492	15434	S152	SCREW, CAP, HEXAGON H .... UOC:DAA, DAB, DAC, DAD, DAE, DAF, DAG, DAH, DAJ, DAK, DAL, DAW, DAX, V12, V13, V14, V15, V16, V17, V18, V19, V20, V21, V22, V24, V25, V39	4

END OF FIGURE

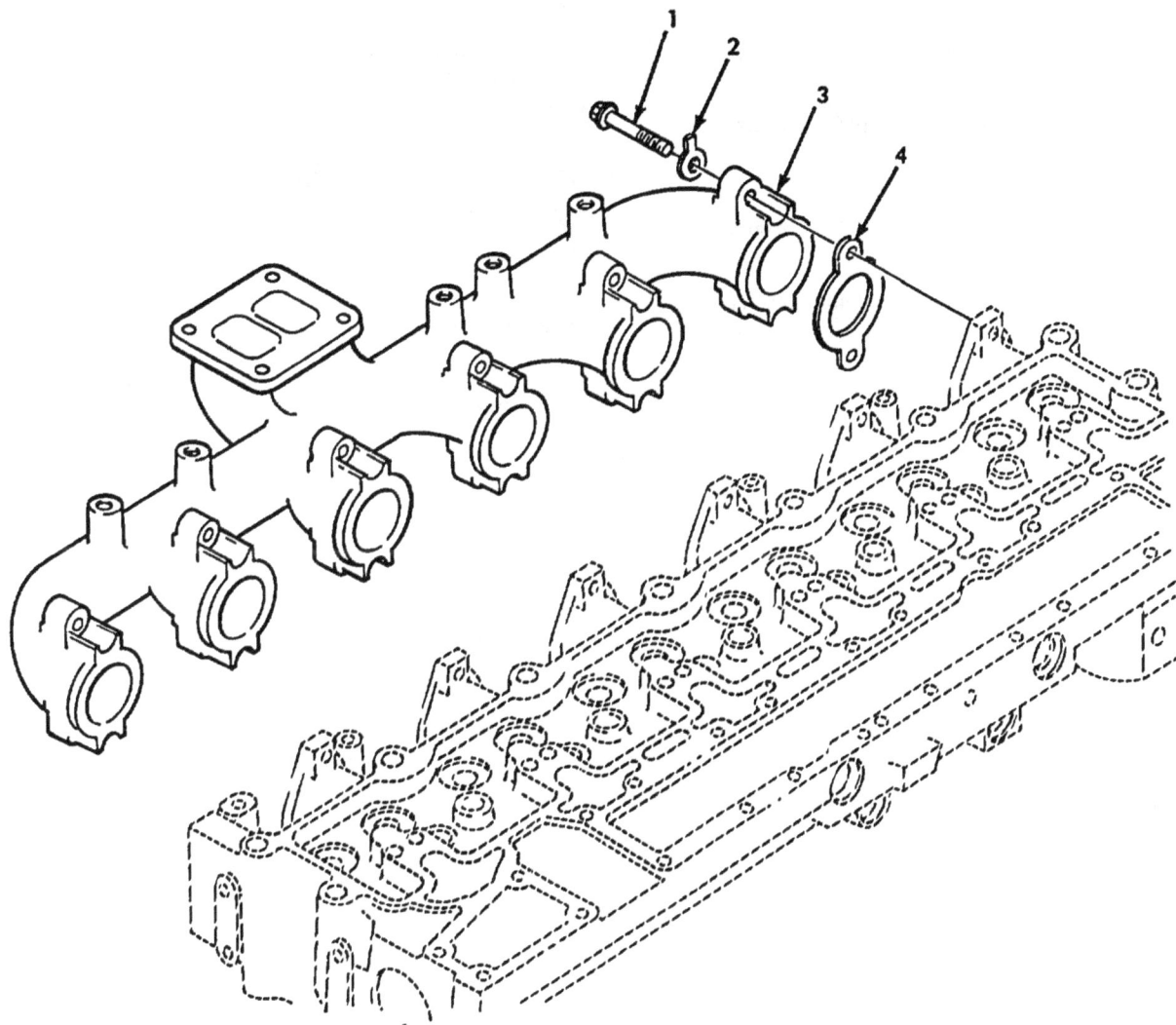

Figure 43. Exhaust Manifold (M939A2).

(1) ITEM NO	(2) SMR CODE	(3) NSN	(4) CAGEC	(5) PART NUMBER	(6) DESCRIPTION AND USABLE ON CODES (UOC)	(7) QTY
					GROUP 0108 MANIFOLDS	
					FIG. 43 EXHAUST MANIFOLD(M939A2)	
1	PAFZZ	5305012716448	15434	3081346	SCREW, CAP, HEXAGON H .... UOC:ZAA, ZAB, ZAC, ZAD, ZAE, ZAF, ZAG, ZAH, ZAJ, ZAK, ZAL	12
2	PAFZZ	5310013308313	15434	3914708	WASHER, KEY PART OF KIT P/N 3802623 .... PART OF KIT P/N 3802452 .... UOC:ZAA, ZAB, ZAC, ZAD, ZAE, ZAF, ZAG, ZAH, ZAJ, ZAK, ZAL	12
3	PAFZZ	2815012720547	15434	3906720	MANIFOLD, EXHAUST .... UOC:ZAA, ZAB, ZAC, ZAD, ZAE, ZAF, ZAG, ZAH, ZAJ, ZAK, ZAL	1
4	PAFZZ	5330012721139	15434	3921787	GASKET PART OF KIT P/N 3802622 PART .... OF KIT P/N 3802623 PART OF KIT P/N 3802452 .... UOC:ZAA, ZAB, ZAC, ZAD, ZAE, ZAF, ZAG, ZAH, ZAJ, ZAK, ZAL	6

END OF FIGURE

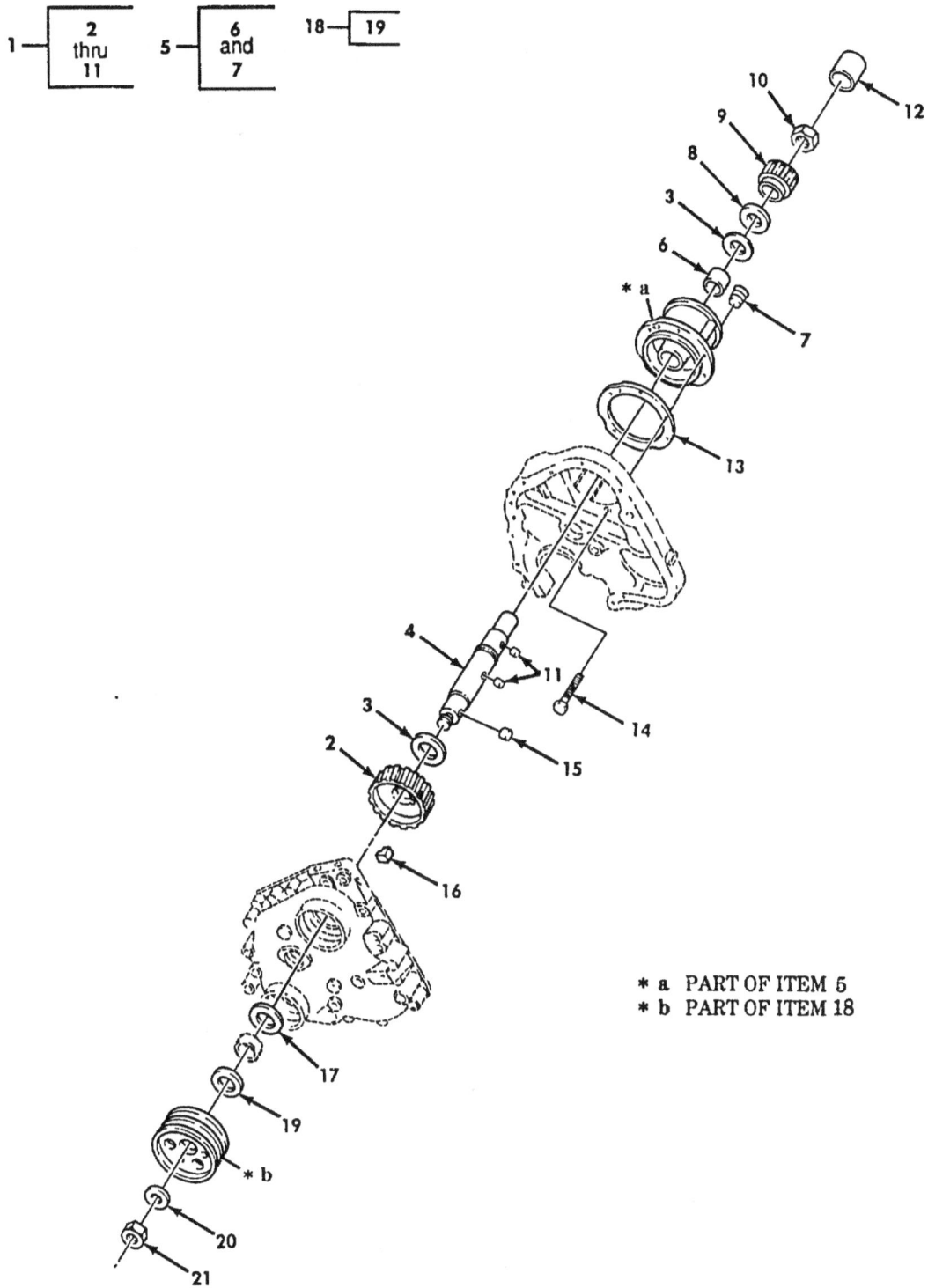

Figure 44. Accessory Drive, Pulley, and Related Parts (M939, M939A1).

* a PART OF ITEM 5
* b PART OF ITEM 18

(1) ITEM NO	(2) SMR CODE	(3) NSN	(4) CAGEC	(5) PART NUMBER	(6) DESCRIPTION AND USABLE ON CODES (UOC)	(7) QTY
					GROUP 0109 ACCESSORY DRIVING MECHANISMS	
					FIG. 44 ACCESSORY DRIVE, PULLEY, AND RELATED PARTS(M939,M939A1)	
1	PAFHH	4310010929815	15434	3005133	DRIVE ACCESSORY, COM .... UOC:DAA, DAB, DAC, DAD, DAE, DAF, DAG, DAH, DAJ, DAK, DAL, DAW, DAX, V12, V13, V14, V15, V16, V17, V18, V19, V20, V21, V22, V24, V25, V39	1
2	PAFZZ	3020008207914	15434	121933	.GEAR, HELICAL USE PRIOR TO SERIAL .... NUMBER 11246663 .... UOC:V12, V13, V14, V15, V16, V17, V18, V19, V20, V21, V22, V24, V25 , V39	1
2	KFHZZ	3020001609092	15434	142689	.GEAR, HELICAL USED AFTER SERIAL .... NUMBER 11246663 PART OF KIT P/N 5704533 .... UOC:DAA, DAB, DAC, DAD, DAE, DAF, DAG, DAH, DAJ, DAK, DAL, DAW, DAX, V12 , V13 , V14 , V15, V16, V17, V18 , V19 , V20, V21, V22, V24, V25, V39	1
3	PAFZZ	3120011447368	15434	3026557	.BEARING, WASHER, THRU .... UOC:DAA, DAB, DAC, DAD, DAE, DAF, DAG, DAH, DAJ, DAK, DAL, DAW, DAX, V12, V13, V14, V16, V17, V18, V19, V20, V21, V22, V24, V25, V39	2
4	PAFZZ	4310010929816	15434	3000171	.SHAFT, AIR COMPRESSO .... UOC:DAA, DAB, DAC, DAD, DAE, DAF, DAG, DAH, DAJ, DAK, DAL, DAW, DAX, V12, V13, V14, V15, V16, V17 , V18, V19, V20, V21, V22, V24, V25, V39	1
5	PAFZZ	3040011294302	15434	AR45724	.HOUSING, MECHANICAL .... UOC:DAA, DAB, DA C, DAD, DAE, DAF, DAG, DAH, DAJ, DAK, DAL, DAW, DAX, V12, V13, V14, V15, V16, V17, V18, V19, V20, V21, V22, V24, V25, V39	1
6	PAFZZ	3120007929834	15434	116391	..BEARING, SLEEVE .... UOC:DAA, DAB, DAC, DAD, DAE, DAF, DAG, DAH, DAJ, DAK, DAL, DAW, DAX, V12, V13, V14, V15, V16, V17, V18, V19, V20, V21, V22, V24, V25, V39	1
7	PAFZZ	4730000189566	15434	S911B	..PLUG, PIPE .... UOC:DAA, DAB, DAC, DAD, DAE, DAF, DAG, DAH, DAJ, DAK, DAL, DAW, DAX, V12, V13, V14, V15, V16, V17, V18, V19, V20, V21, V22, V24, V25, V39	2
8	PAFZZ	5310000819292	15434	3014103	.WASHER, FLAT .... UOC:DAA, DAB, DAC, DAD, DAE, DAF, DAG, DAH, DAJ, DAK, DAL, DAW, DAX, V12, V13, V14, V15, V16, V17, V18, V19, V20, V21, V22, V24, V25, V39	1

(1) ITEM NO	(2) SMR CODE	(3) NSN	(4) CAGEC	(5) PART NUMBER	(6) DESCRIPTION AND USABLE ON CODES (UOC)	(7) QTY
9	PAFZZ	3010010852732	15434	3000174	.COUPLING HALF, SHAFT .... UOC:DAA, DAB, DAC, DAD, DAE, DAF, DAG, DAH, DAJ, DAK, DAL, DAW, DAX, V12, V13, V14, V15, V16, V17, V1B, V19, V20, V21, V22, V24, V25, V39	1
10	PAFZZ	5310004426899	15434	191517	.NUT, SELF-LOCKING, HE .... UOC:DAA, DAB, DAC, DAD, DAE, DAF, DAG, DAH, DAJ, DAK, DAL, DAW, DAX, V12, V13, V14, V15, V16, V17, V18, V19, V20, V21, V22, V24, V25, V39	1
11	PAFZZ	5315011777507	15434	3034438	.PIN, STRAIGHT, HEADLE .... UOC:DAA, DAB, DAC, DAD, DAE, DAF, DAG, DAH, DAJ, DAK, DAL, DAW, DAX, V12, V13, V14, V15, V16, V17, V18, V19, V20, V21, V22, V24, V25, V39	2
12	PAFZZ	3010010793461	15434	199349	COUPLING, SHAFT, RIGI .... UOC:DAA, DAB, DAC, DAD, DAE, DAF, DAG, DAH, DAJ, DAK, DAL, DAW, DAX, V12, V13, V14, V15, V16, V17, V18, V19, V20, V21, V22, V24, V25, V39	1
13	PAFZZ	5330000262931	15434	3069101	GASKET PART OF KIT P/N 3011472 .... UOC:DAA, DAB, DAC, DAD, DAE, DAF, DAG, DAH, DAJ, DAK, DAL, DAW, DAX, V12, V13, V14, V15, V16, V17, V18, V19, V20, V21, V22, V24, V25, V39	1
14	PAFZZ	5305011198621	15434	3010590	SCREW .... UOC:DAA, DAB, DAC, DAD, DAE, DAF, DAG, DAH, DAJ, DAK, DAL, DAW, DAX, V12, V13, V14, V15, V16, V17, V18, V19, V20, V21, V22, V24, V25, V39	5
15	PAFZZ	2815011240232	15434	70550	PIN, PISTON .... UOC:DAA, DAB, DAC, DAD, DAE, DAF, DAG, DAH, DAJ, DAK, DAL, DAW, DAX, V12, V13, V14, V15, V16, V17, V18, V19, V20, V21, V22, V24, V25, V39	3
16	PAHZZ	5330011296541	15434	3008947	RUBBER STRIP PART OF KIT PIN 3011472 .... UOC:DAA, DAB, DAC, DAD, DAE, DAF, DAG, DAH, DAJ, DAK, DAL, DAW, DAX, V12, V13, V14, V15, V16, V17, V18, V19, V20, V21, V22, V24, V25, V39	1
17	PAFZZ	5330001356382	15434	211255	SEAL, PLAIN ENCASED .... UOC:DAA, DAB, DAC, DAD, DAE, DAF, DAG, DAH, DAJ, DAK, DAL, DAW, DAX, V12, V13, V14, V15, V16, V17, V18, V19, V20, V21, V22, V24, V25, V39	1
18	PAFZZ	3020011956990	15434	3030506	PULLEY, GROOVE .... UOC:DAA, DAB, DAC, DAD, DAE, DAF, DAG, DAH, DAJ, DAK, DAL, DAW, DAX, V12, V13, V14, V15, V16, V17, V18, V19, V20, V21, V22, V24, V25, V39	1
19	PAFZZ	2930004019531	15434	190397	.SLEEVE, PULLEY, PUMP .... UOC:DAA, DAB, DAC, DAD, DAE, DAF, DAG, DAH,	1

(1) ITEM NO	(2) SMR CODE	(3) NSN	(4) CAGEC	(5) PART NUMBER	(6) DESCRIPTION AND USABLE ON CODES (UOC)	(7) QTY
					DAJ, DAK, DAL, DAW, DAX, V12, V13, V14, V15, V16, V17, V18, V19, V20, V21, V22, V24, V25, V39	
20	PAFZZ	5310011246463	15434	193136	WASHER, FLAT	1
					UOC:DAA, DAB, DAC, DAD, DAE, DAF, DAG, DAH, DAJ, DAK, DAL, DAW, DAX, V12, V13, V14, V15, V16, V17, V17, V18, V19, V20, V21, V22, V24, V25, V39	
21	PAFZZ	5310011261045	15434	3012526	NUT, SELF-LOCKING, HE	1
					UOC:DAA, DAB, DAC, DAD, DAE, DAF, DAG, DAH, DAJ, DAK, DAL, DAW, DAX, V12, V13, V14, V15, V16, V17, V18, V19, V20, V21, V22, V24, V25, V39	

END OF FIGURE

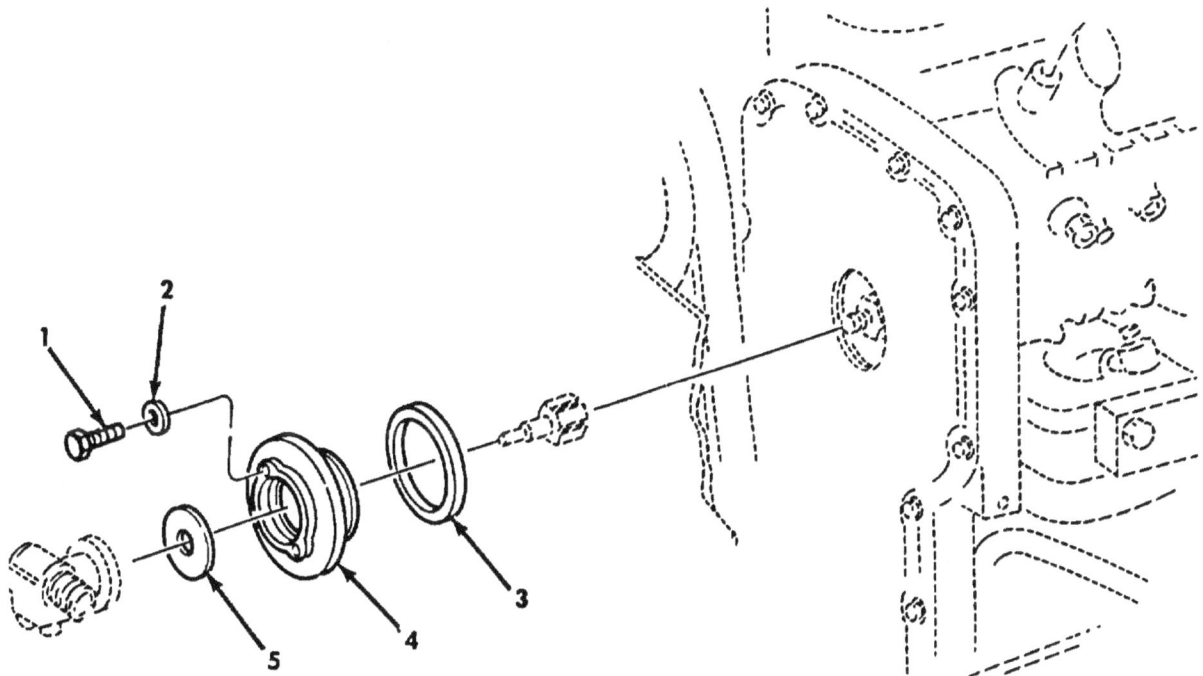

Figure 45.   Tachometer Access Cover (M939A2).

(1) ITEM NO	(2) SMR CODE	(3) NSN	(4) CAGEC	(5) PART NUMBER	(6) DESCRIPTION AND USABLE ON CODES (UOC)	(7) QTY
					GROUP 0109 ACCESSORY DRIVING MECHANISMS	
					FIG. 45 TACHOMETER ACCESS COVER (M939A2)	
1	PAOZZ	5305012632733	15434	3027685	SCREW, CAP, HEXAGON H .... UOC:ZAA, ZAB, ZAC, ZAD, ZAE, ZAF, ZAG, ZAH, ZAJ, ZAK, ZAL	2
2	PAOZZ	5310012708389	15434	62392	WASHER, FLAT .... UOC:ZAA, ZAB, ZAC, ZAD, ZAE, ZAF, ZAG, ZAH, ZAJ, ZAK, ZAL	2
3	PAOZZ	5330011918047	15434	3903475	GASKET .... UOC:ZAA, ZAB, ZAC, ZAD, ZAE, ZAF, ZAG, ZAH, ZAJ, ZAK, ZAL	1
4	PAOZZ	5340012663023	15434	3918215	COVER, ACCESS .... UOC:ZAA, ZAB, ZAC, ZAD, ZAE, ZAF, ZAG, ZAH, ZAJ, ZAK, ZAL	1
5	PAOZZ	5330012708144	15434	3915800	GASKET TACHOMETER DRIVE .... UOC:ZAA, ZAB, ZAC, ZAD, ZAE, ZAF, ZAG, ZAH, ZAJ, ZAK, ZAL	1

END OF FIGURE

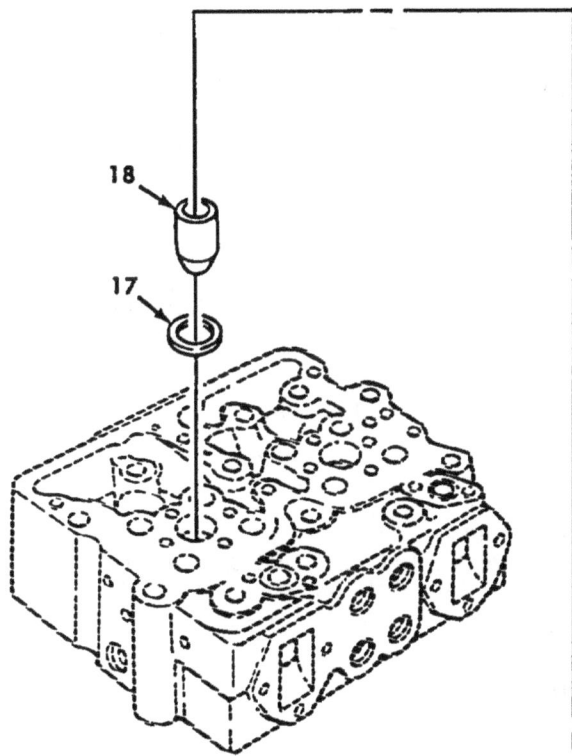

* a PART OF ITEM 6

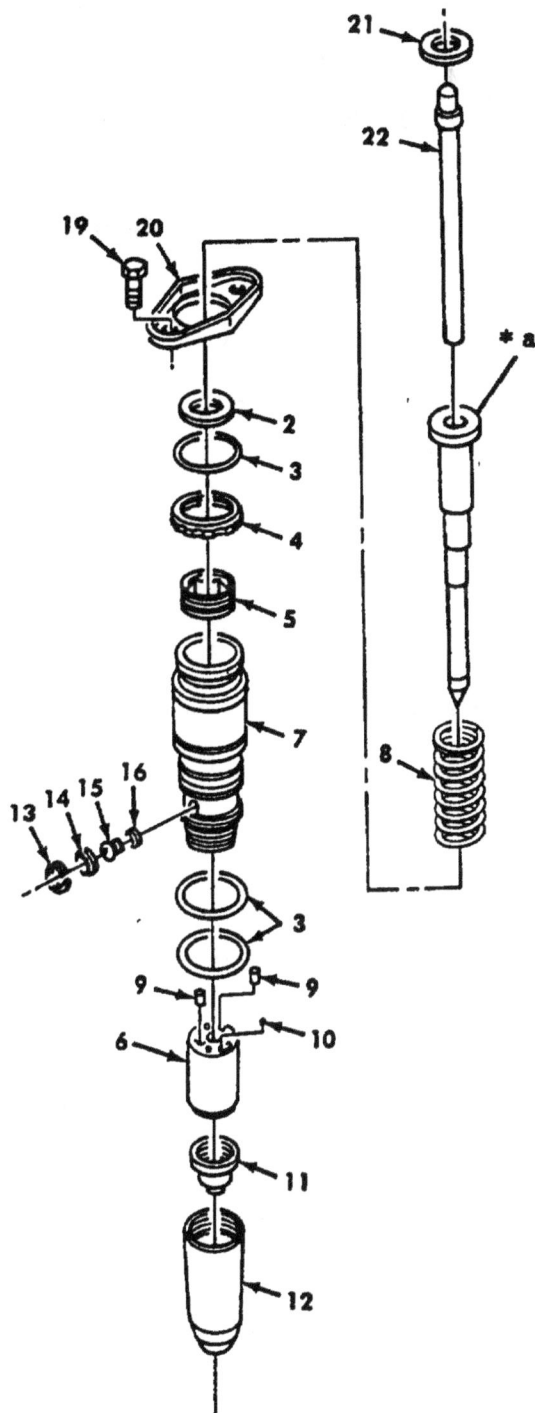

Figure 46. Fuel Injector Assembly and Related Parts (M939, M939A1).

(1) ITEM NO	(2) SMR CODE	(3) NSN	(4) CAGEC	(5) PART NUMBER	(6) DESCRIPTION AND USABLE ON CODES (UOC)	(7) QTY
					GROUP 03 FUEL SYSTEM 0301 CARBURETOR, FUEL INJECTOR FIG. 46 FUEL INJECTOR ASSEMBLY AND RELATED PARTS(M939,M939A1)	
1	PAFZZ	2910004043054	15434	3054250	INJECTOR ASSEMBLY, F USE PRIOR TO .... SERIAL NUMBER 11246663 .... UOC:DAA, V12, V13, V14, V15, V16, V17, V18, V19, V20, V21, V22, V24, V25, V39	6
1	PAFHH	2910012185155	15434	3046281	NOZZLE, FUEL INJECTI USE AFTER .... SERIAL NUMBER 11246663 .... UOC:DAA, DAB, DAC, DAD, DAE, DAF, DAG, DAH, DAJ, DAK, DAL, DAW, DAX, V12, V13, V14, V15, V16, V17, V18, V19, V20, V21, V22, V24, V25, V39	6
2	PAHZZ	5310011451114	15434	3015469	.WASHER USE AFTER SERIAL NUMBER .... 11246663... UOC:DAA, DAB, DAC, DAD, DAE, DAF, DAG, DAH, DAJ, DAK, DAL, DAW, DAX, V12, V13, V14, V15, V16, V17, V18, V19, V20, V21, V22, V24, V25, V39	1
3	KFHZZ	5330001320276	15434	193736	.GASKET PART OF KIT P/N 3011472 PART .... OF KIT P/N 3804272 PART OF KIT P/N AR-51483 .... UOC:DAA, DAB, DAC, DAD, DAE, DAF, DAG, DAH, DAJ, DAK, DAL, DAW, DAX, V12, V13, V14, V15, V16, V17, V18, V19, V20, V21, V22, V24, V25, V39	3
4	PAHZZ	5310010796708	15434	3000465	.NUT, PLAIN, DODECAGON .... UOC:DAA, DAB, DAC, DAD, DAE, DAF, DAG, DAH, DAJ, DAK, DAL, DAW, DAX, V12, V13, V14, V15, V16, V17, V18, V19, V20, V21, V22, V24, V25, V39	1
5	PAHZZ	5305010797028	15434	212954	.SCREW .... UOC:DAA, DAB, DAC, DAD, DAE, DAF, DAG, DAH, DAJ, DAK, DAL, DAW, DAX, V12, V13, V14, V15, V16, V17, V18, V19, V20, V21, V22, V24, V25, V39	1
6	PAHZZ	2910011918470	15434	3054532	.BARREL AND PLUNGER USE PRIOR TO .... SERIAL NUMBER 11246663 PART OF KIT P/N 3804272... UOC:DAA, V12, V13, V14, V15, V16, V17, V18, V19, V20, V21, V22, V24, V25, V39	1
6	PAHZZ	2910012192086	15434	3054533	.PARTS KIT, FUEL INJE USE AFTER................... SERIAL NUMBER 11246663 .... UOC:DAA, DAB, DAC, DAD, DAE, DAF, DAG, DAH, DAJ, DAK, DAL, DAW, DAX, V12, V13, V14, V15, V16, V17, V18, V19, V20, V21, V22, V24, V25, V39	1
7	PAHZZ	2910010768632	15434	3000464	.ADAPTER, INJECTOR USE AFTER SERIAL .... NUMBER 11246663	1

(1) ITEM NO	(2) SMR CODE	(3) NSN	(4) CAGEC	(5) PART NUMBER	(6) DESCRIPTION AND USABLE ON CODES (UOC)	(7) QTY
					UOC:DAA, DAB, DAC, DAD, DAE, DAF, DAG, DAH, DAJ, DAK, DAL, DAW, DAX, V12, V13, V14, V15, V16, V17, V18, V19, V20, V21, V22, V24, V25, V39	
7	PAHZZ	2910011056457	15434	185139	.ADAPTER, INJECTOR USE PRIOR TO .... SERIAL NUMBER 11246663 .... UOC:DAA, V12, V13, V14, V15, V16, V17, V18, V19, V20, V21, V22, V24, V25, V39	1
8	PAHZZ	5360001320245	15434	166009	.SPRING, HELICAL, COMP .... UOC:DAA, DAB, DAC, DAD, DAE, DAF, DAG, DAH, DAJ, DAK, DAL, DAW, DAX, V12, V13, V14, V15, V16, V17, V18, V19, V20, V21, V22, V24, V25, V39	1
9	PAHZZ	5315010796506	15434	203426	.PIN, SPRING .... UOC:DAA, DAB, DAC, DAD, DAE, DAF, DAG, DAH, DAJ, DAK, DAL, DAW, DAX, V12, V13, V14, V15, V16, V17, V18, V19, V20, V21, V22, V24, V25, V39	2
10	KFHZZ	4820010709710	15434	167157	.BALL, CHECK PART OF KIT P/N AR51522. .... UOC:DAA, DAB, DAC, DAD, DAE, DAF, DAG, DAH, DAJ, DAK, DAL, DAW, DAX, V12, V13, V14, V15, V16, V17, V18, V19, V20, V21, V22, V24, V25, V39	1
11	KFHZZ	2910011528531	15434	3012537	.CUP, INJECTOR PART OF KIT P/N .... 3804272 .... UOC:DAA, DAB, DAC, DAD, DAE, DAF, DAG, DAH, DAJ, DAK, DAL, DAW, DAX, V12, V13, V14, V15, V16, V17, V18, V19, V20, V21, V22, V24, V25, V39	1
12	PAHZZ	5342010794678	15434	185138	.RETAINER, CUP .... UOC:DAA, DAB, DAC, DAD, DAE, DAF, DAG, DAH, DAJ, DAK, DAL, DAW, DAX, V12, V13, V14, V15, V16, V17, V18, V19, V20, V21, V22, V24, V25, V39	1
13	KFHZZ	5365008151137	15434	174299	.RING, RETAINING PART OF KIT P/N .... AR51522 .... UOC:DAA, DAB, DAC, DAD, DAE, DAF, DAG, DAH, DAJ, DAK, DAL, DAW, DAX, V12, V13, V14, V15, V16, V17, V18, V19, V20, V21, V22, V24, V25, V39	1
14	KFHZZ	4730010772016	15434	3008706	.STRAINER ELEMENT, SE PART OF KIT P/N .... AR51522 .... UOC:DAA, DAB, DAC, DAD, DAE, DAF, DAG, DAH, DAJ, DAK, DAL, DAW, DAX, V12, V13, V14, V15, V16, V17, V18, V19, V20, V21, V22, V24, V25, V39	1
15	KFHZZ		15434	3044994	.PLUG, ORIFICE 019 PART OF KIT P/N .... AR51522 .... UOC:DAA, DAB, DAC, DAD, DAE, DAF, DAG, DAH, DAJ, DAK, DAL, DAW, DAX, V12, V13, V14, V15, V16, V17, V18, V19, V20, V21, V22, V24, V25, V39	1

(1) ITEM NO	(2) SMR CODE	(3) NSN	(4) CAGEC	(5) PART NUMBER	(6) DESCRIPTION AND USABLE ON CODES (UOC)	(7) QTY
15	KFHZZ		15434	3044993	.PLUG, ORIFICE 018 PART OF KIT P/N .... AR51522 .... UOC:DAA, DAB, DAC, DAD, DAE, DAF, DAG, DAH, DAJ, DAK, DAL, DAW, DAX, V12, V13, V14, V15, V16, V17, V18, V19, V20, V21, V22, V24, V25, V39	1
15	KFHZZ		15434	3044995	.PLUG, ORIFICE 020 PART OF KIT P/N .... AR51522 .... UOC:DAA, DAB, DAC, DAD, DAE, DAF, DAG, DAH, DAJ, DAK, DAL, DAW, DAX, V12, V13, V14, V15, V16, V17, V18, V19, V20, V21, V22, V24, V25, V39	1
15	KFHZZ		15434	3044992	.PLUG, ORIFICE 017 PART OF KIT P/N .... AR51522 .... UOC:DAA, DAB, DAC, DAD, DAE, DAF, DAG, DAH, DAJ, DAK, DAL, DAW, DAX, V12, V13, V14, V15, V16, V17, V18, V19, V20, V21, V22, V24, V25, V39	1
16	KFHZZ	5330001320247	15434	173086	.GASKET PART OF KIT P/N 3011472 PART .... OF KIT P/N 3804272 PART OF KIT P/N AR51522 .... UOC:DAA, DAB, DAC, DAD, DAE, DAF, DAG, DAH, DAJ, DAK, DAL, DAW, DAX, V12 , V13, V14 , V15, V16, V17, V18, V19, V20, V21, V22, V24, V25, V39	1
17	PAHZZ	5330010724436	15434	3007759	O-RING PART OF KIT P/N 3011472 PART .... OF KIT P/N 3804272.. UOC:DAA, DAB, DAC, DAD, DAE, DA-F, DAG, DAH, DAJ, DAK, DAL, DAW, DAX, V12 , V13 , V14, V15, V16, V17, V18, V19, V20, V21, V22, V24, V25, V39	6
18	PAHZZ	2910011460048	15434	3011934	SLEEVE, COOLING, FUEL .... UOC:DAA, DAB, DAC, DAD, DAE, DAF, DAG, DAH, DAJ, DAK, DAL, DAW, DAX, V12, V13, V14, V15, V16, V17, V18, V19, V20, V21, V22, V24, V25, V39	6
19	PAFZZ	5305011450777	15434	3028279	SCREW, CAP, HEXAGON H .... UOC:DAA, DAB, DAC, DAD, DAE, DAF, DAG, DAH, DAJ, DAK, DAL, DAW, DAX, V12, V13, V14, V15, V16, V17, V18, V19, V20, V21, V22, V24, V25, V39	12
20	PAFZZ	2910011504925	15434	3031137	CLAMP, INJECTOR MOUN. (NON TOP STOP) .... UOC:DAA, DAB, DAC, DAD, DAE, DAF, DAG, DAH, DAJ, DAK, DAL, DAW, DAX, V12, V13, V14, V15, V16, V17, V18, V19, V20, V21, V22, V24, V25, V39	6
20	PAFZZ	5342011450646	15434	2028171	CLAMP, INJECTOR MOUN. (TOP STOP) .... UOC:DAA, DAB, DAC, DAD, DAE, DAF, DAG, DAH, DAJ, DAK, DAL, DAW, DAX, V12 , V13, V14, V15, V16, V17, V18, V19, V20, V21, V22, V24, V25, V39	6

(1) ITEM NO	(2) SMR CODE	(3) NSN	(4) CAGEC	(5) PART NUMBER	(6) DESCRIPTION AND USABLE ON CODES (UOC)	(7) QTY
21	PAFZZ	5310010662942	15434	180626	WASHER, LOCK .... UOC:DAA, DAB, DAC, DAD, DAE, DA, DAFG, DAH, DAJ, DAK, DAL, DAW, DAX, V12, V13, V14, V15, V16, V17, V18, V19, V20, V21, V22, V24, V25, V39	1
22	PAFZZ	5340002385435	15434	191916	PLUNGER, DETENT .... UOC:DAA, DAB, DAC, DAD, DAE, DAF, DAG, DAH, DAJ, DAK, DAL, DAW, DAX, V12, V13, V14, V15, V16, V17, V18 , V19, V20 , V21, V22, V24, V25, V39	6

END OF FIGURE

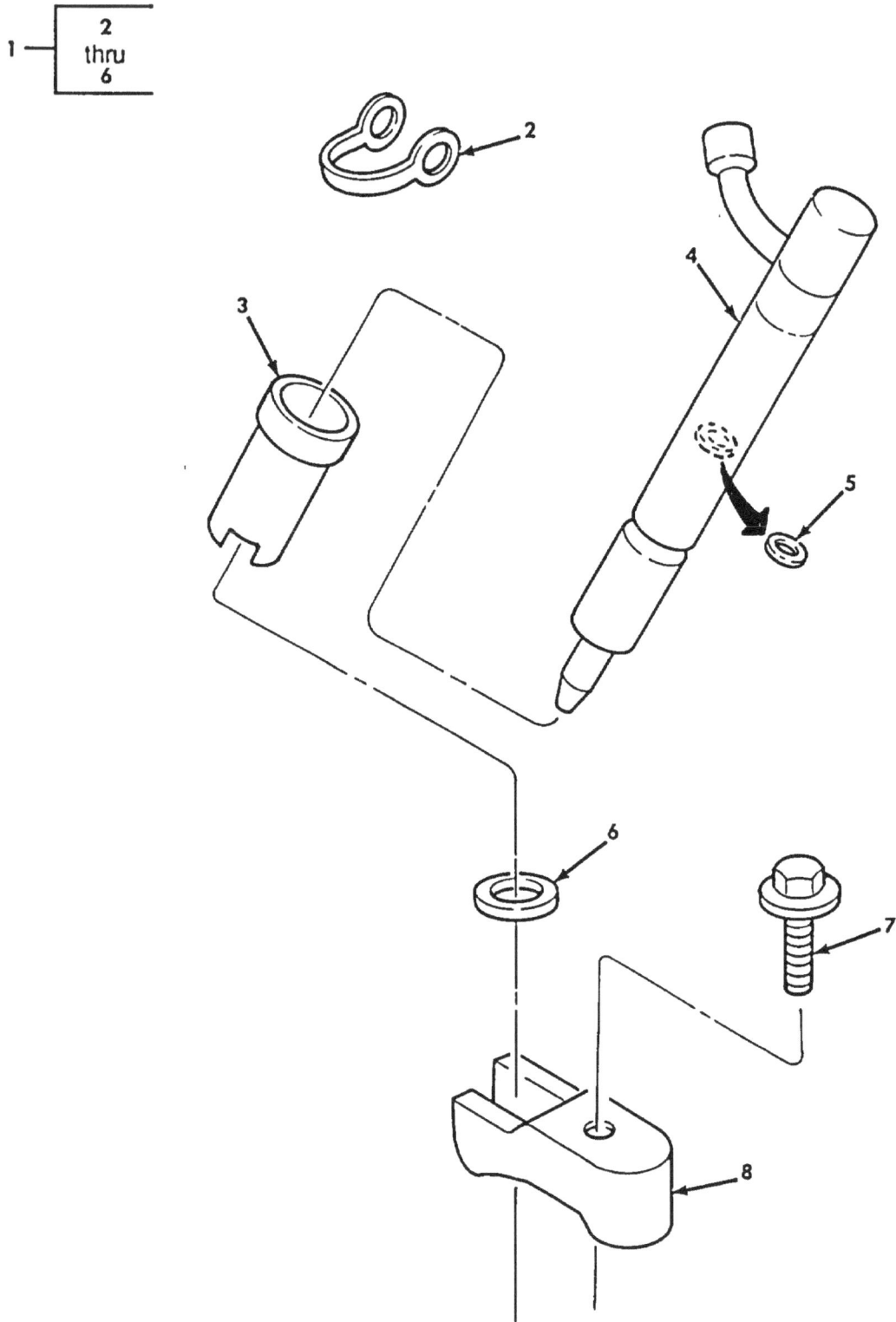

Figure 47. *Fuel Injector Assembly (M939A2).*

(1) ITEM NO	(2) SMR CODE	(3) NSN	(4) CAGEC	(5) PART NUMBER	(6) DESCRIPTION AND USABLE ON CODES (UOC)	(7) QTY
					GROUP 0301 CARBURETOR, FUEL INJECTOR	
					FIG. 47 FUEL INJECTOR ASSEMBLY (M939A2)	
1	PAFZZ	2910012925663	15434	3802091	PARTS KIT, FUEL INJE .... UOC:ZAA, ZAB, ZAC, ZAD, ZAE, ZAF, ZAG, ZAH, ZAJ, ZAK, ZAL	6
2	PAFZZ	5330011955268	15434	3903380	.SEAL, BANJO CONNECT PART OF KIT P/N .... 3802622 PART OF KIT P/N 3802623 PART OF KIT P/N 3802452 .... UOC:ZAA, ZAB , ZAC, ZAD, ZAE, ZAF, ZAG, ZAH, ZAJ, ZAK, ZAL	1
3	PAFZZ	2910012719826	15434	3909886	.SLEEVE, COOLING, FUEL .... UOC:ZAA, ZAB, ZAC, ZAD, ZAE, ZAF, ZAG, ZAH, ZAJ, ZAK, ZAL	1
4	XAHZZ		15434	3908513	.INJECTOR .... UOC:ZAA, ZAB, ZAC, ZAD, ZAE, ZAF, ZAG, ZAH, ZAJ, ZAK, ZAL	1
5	PAHZZ	5310011895413	15434	3905186	.WASHER, FLAT .... UOC:ZAA, ZAB, ZAC, ZAD, ZAE, ZAF, ZAG, ZAH, ZAJ, ZAX, ZAL	V
5	PAHZZ	5310011895412	15434	3905185	.WASHER, FLAT INJECTOR, 1.94MM .... UOC:ZAA, ZAB, ZAC, ZAD, ZAE, ZAF, ZAG, ZAH, ZAJ, ZAK, ZAL	V
5	PAHZZ	5310011895411	15434	3905184	.WASHER, FLAT INJECTOR, 1.90MM .... UOC:ZAA, ZAB, ZAC, ZAD, ZAE, ZAF, ZAG, ZAH, ZAJ, ZAK, ZAL	V
5	PAHZZ	5310011897499	15434	3905183	.WASHER, FLAT INJECTOR, 1.88MM .... UOC:ZAA, ZAB, ZAC, ZAD, ZAE, ZAF, ZAG, ZAH, ZAJ, ZAK, ZAL	V
5	PAHZZ	5310011895410	15434	3905182	.WASHER, FLAT INJECTOR, 1.84MM .... UOC:ZAA, ZAB, ZAC, ZAD, ZAE, ZAF, ZAG, ZAH, ZAJ, ZAK, ZAL	V
5	PAHZZ	5310011895409	15434	3905181	.WASHER, FLAT INJECTOR, 1.80MM .... UOC:ZAA, ZAB, ZAC, ZAD, ZAE, ZAF, ZAG, ZAH, ZAJ, ZAK, ZAL	V
5	PAHZZ	5365011912411	15434	3905180	.SHIM INJECTOR, 1.78MM .... UOC:ZAA, ZAB, ZAC, ZAD, ZAE, ZAF, ZAG, ZAH, ZAJ, ZAK, ZAL	V
5	PAHZZ	5365011913535	15434	3905178	.SPACER, PLATE INJECTOR, 1.74MM .... UOC:ZAA, ZAB, ZAC, ZAD, ZAE, ZAF, ZAG, ZAH, ZAJ, ZAK, ZAL	V
5	PAHZZ	5365011910776	15434	3905177	.SPACER, PLATE INJECTOR, 1.70MM .... UOC:ZAA, ZAB, ZAC, ZAD, ZAE, ZAF, ZAG, ZAH, ZAJ, ZAK, ZAL	V
5	PAHZZ	5310011895408	15434	3905176	.WASHER, FLAT INJECTOR, 1.68MM .... UOC:ZAA, ZAB, ZAC, ZAD, ZAE, ZAF, ZAG, ZAH, ZAJ, ZAK, ZAL	V
5	PAHZZ	5365011881056	15434	3905175	.SHIM INJECTOR, 1.64MM .... UOC:ZAA, ZAB, ZAC, ZAD, ZAE, ZAF, ZAG, ZAH, ZAJ, ZAK, ZAL	V

(1) ITEM NO	(2) SMR CODE	(3) NSN	(4) CAGEC	(5) PART NUMBER	(6) DESCRIPTION AND USABLE ON CODES (UOC)	(7) QTY
5	PAHZZ	5365011899030	15434	3905174	.SPACER, PLATE INJECTOR, 1.60MM .... UOC:ZAA, ZAB, ZAC, ZAD, ZAE, ZAF, ZAG, ZAH, ZAJ, ZAK, ZAL	V
5	PAHZZ	5365011899029	15434	3905173	.SPACER, PLATE INJECTOR, 1.58MM .... UOC:ZAA, ZAB, ZAC, ZAD, ZAE, ZAF, ZAG, ZAH, ZAJ, ZAK, ZAL	V
5	PAHZZ.	5365011899028	15434	3905172	.SPACER, PLATE INJECTOR, 1.54MM .... UOC:ZAA, ZAB, ZAC, ZAD, ZAE, ZAF, ZAG, ZAH, ZAJ, ZAK, ZAL	V
5	PAHZZ	5365011899027	15434	3905171	.SPACER, PLATE INJECTOR, 1.50MM .... UOC:ZAA, ZAB, ZAC, ZAD, ZAE, ZAF, ZAG, ZAH, ZAJ, ZAK, ZAL	V
5	PAHZZ	5310011916333	15434	3905170	.WASHER, FLAT INJECTOR, 1.48MM .... UOC:ZAA, ZAB, ZAC, ZAD, ZAE, ZAF, ZAG, ZAH, ZAJ, ZAK, ZAL	V
5	PAHZZ	5310011917512	15434	3905169	.WASHER, FLAT INJECTOR, 1.44MM .... UOC:ZAA, ZAB, ZAC, ZAD, ZAE, ZAF, ZAG, ZAH, ZAJ, ZAK, ZAL	V
5	PAHZZ	5365011910775	15434	3905168	.SPACER, PLATE INJECTOR, 1.40MM .... UOC:ZAA, ZAB, ZAC, ZAD, ZAE, ZAF, ZAG, ZAH, ZAJ, ZAK, ZAL	V
5	PAHZZ	5365011899026	15434	3905167	.SPACER, PLATE INJECTOR, 1.38MM .... UOC:ZAA, ZAB, ZAC, ZAD, ZAE, ZAF, ZAG, ZAH, ZAJ, ZAK, ZAL	V
5	PAHZZ	5310011912494	15434	3905166	.WASHER, FLAT INJECTOR, 1.34MM .... UOC:ZAA, ZAB, ZAC, ZAD, ZAE, ZAF, ZAG, ZAH, ZAJ, ZAK, ZAL	V
5	PAHZZ	5310011895407	15434	3905165	.WASHER, FLAT INJECTOR, 1.30MM .... UOC:ZAA, ZAB, ZAC, ZAD, ZAE, ZAF, ZAG, ZAH, ZAJ, ZAK, ZAL	V
5	PAHZZ	5365011913534	15434	3905164	.SHIM INJECTOR, 1.28MM .... UOC:ZAA, ZAB, ZAC, ZAD, ZAE, ZAF, ZAG, ZAH, ZAJ, ZAK, ZAL	V
5	PAHZZ	5365011897804	15434	3905163	.SPACER, PLATE INJECTOR, 1.24MM .... UOC:ZAA, ZAB, ZAC, ZAD, ZAE, ZAF, ZAG, ZAH, ZAJ, ZAK, ZAL	V
5	PAHZZ	5365011910774	15434	3905162	.SPACER, PLATE INJECTOR, 1.20MM .... UOC:ZAA, ZAB, ZAC, ZAD, ZAE, ZAF, ZAG, ZAH, ZAJ, ZAK, ZAL	V
5	PAHZZ	5365011913533	15434	3905161	.SHIM INJECTOR, 1.18MM .... UOC:ZAA, ZAB, ZAC, ZAD, ZAE, ZAF, ZAG, ZAH, ZAJ, ZAK, ZAL	V
5	PAHZZ	5365011912413	15434	3905160	.SHIM INJECTOR, 1.14MM .... UOC:ZAA, ZAB, ZAC, ZAD, ZAE, ZAF, ZAG, ZAH, ZAJ, ZAK, ZAL	V
5	PAHZZ	5365011913532	15434	3905159	.SHIM INJECTOR, 1.10MM .... UOC:ZAA, ZAB, ZAC, ZAD, ZAE, ZAF, ZAG, ZAH, ZAJ, ZAK, ZAL	V
5	PAHZZ	5365011881055	15434	3905158	.SHIM INJECTOR, 1.08MM .... UOC:ZAA, ZAB, ZAC, ZAD, ZAE, ZAF, ZAG, ZAH, ZAJ, ZAK, ZAL	V
5	PAHZZ	5365011902131	15434	3905157	.SHIM INJECTOR, 1.04MM	V

(1) ITEM NO	(2) SMR CODE	(3) NSN	(4) CAGEC	(5) PART NUMBER	(6) DESCRIPTION AND USABLE ON CODES (UOC)	(7) QTY
					UOC:ZAA, ZAB, ZAC, ZAD, ZAE, ZAF, ZAG, ZAH, ZAJ, ZAK, ZAL	
5	PAHZZ	5365011881054	15434	3905156	.SHIM INJECTOR, 1.00MM	V
					UOC:ZAA, ZAB, ZAC, ZAD, ZAE, ZAF , ZAG, ZAH, ZAJ, ZAX, ZAL	
6	PAFZZ	5310012708417	15434	3906659	.WASHER, SADDLE PART OF KIT P/N .... 3802622 PART OF KIT P/N 3802623 PART OF KIT P/N 3802452...	1
					UOC:ZAA, ZAB, ZAC, ZAD, ZAE, ZAF, ZAG, ZAH, ZAJ, ZAK, ZAL	
7	PAFZZ	5305012724811	15434	3902112	SCREW, CAP, HEXAGON H	6
					UOC:ZAA, ZAB, ZAC, ZAD, ZAE, ZAF, ZAG, ZAH, ZAJ, ZAK, ZAL	
8	PAFZZ	5340012712349	15434	3910279	CLAMP, INJECTOR	6
					UOC:ZAA, ZAB, ZAC, ZAD, ZAE, ZAF, ZAG, ZAH, ZAJ, ZAK, ZAL	

END OF FIGURE

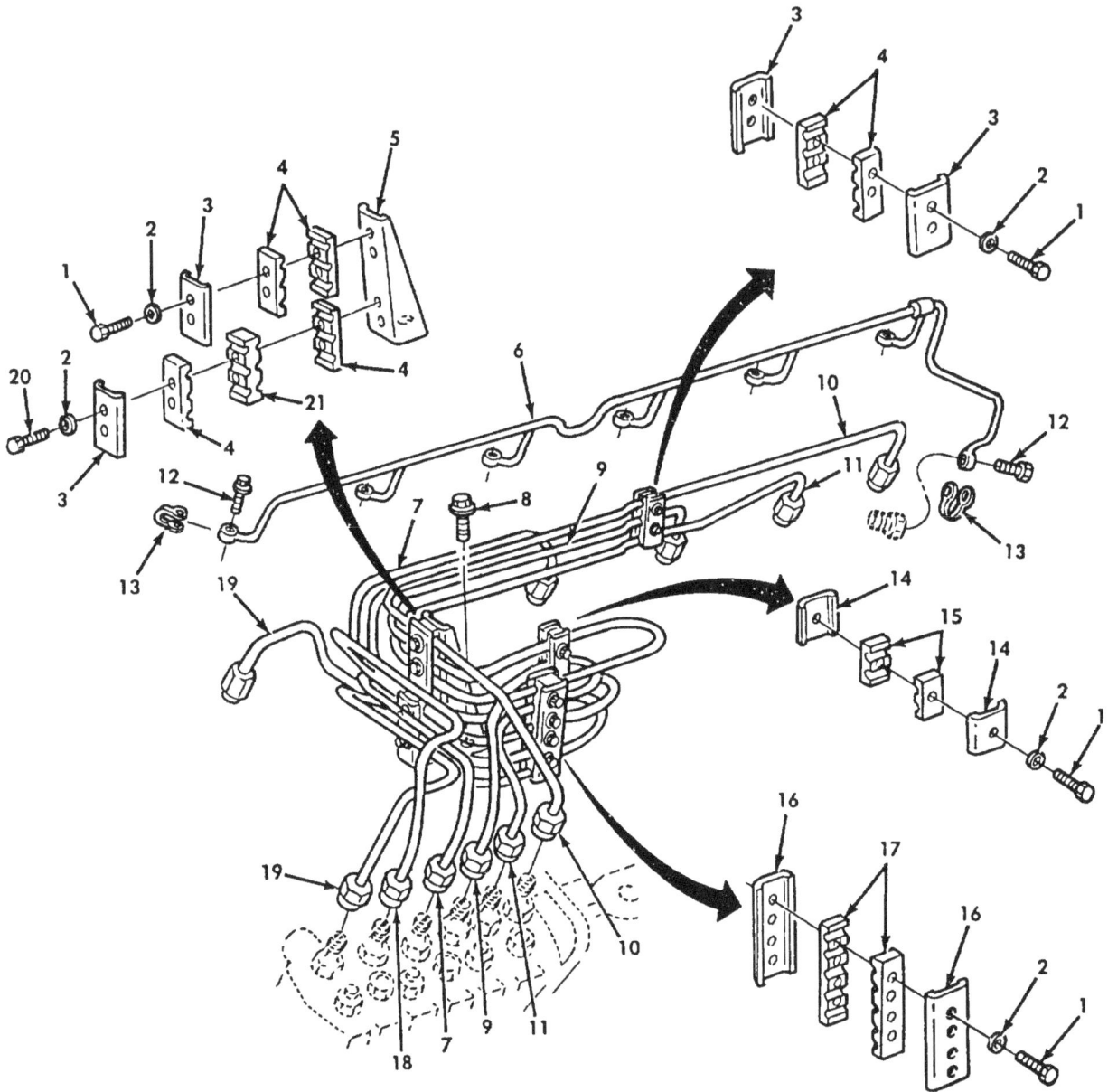

Figure 48. Fuel Injector Supply Tubes (M939A2).

(1) ITEM NO	(2) SMR CODE	(3) NSN	(4) CAGEC	(5) PART NUMBER	(6) DESCRIPTION AND USABLE ON CODES (UOC)	(7) QTY
					GROUP 0301 CARBURETOR, FUEL INJECTOR	
					FIG. 48 FUEL INJECTOR SUPPLY TUBES (M939A2)	
1	PAOZZ	5305012343756	15434	3903609	SCREW, CAP, HEXAGON H .... UOC: ZAA, ZAB, ZAC, ZAD, ZA, ZAF, ZAG, ZAH, ZAJ, ZAK, ZAL	9
2	PAOZZ	5310012341411	15434	3903723	WASHER, FLAT .... UOC:ZAA, ZAB, ZAC, ZAD, ZAE, ZAF, ZAG, ZAH, ZAJ, ZAK, ZAL	9
3	PAOZZ	5340011963680	15434	3904519	BRACKET, DOUBLE ANGL .... UOC:ZAA, ZAB, ZAC, ZAD, ZAE, ZAF, ZAG, ZAH, ZAJ, ZAK, ZAL	4
4	PAOZZ	5340013127652	15434	3914339	CLAMP, BLOCK, SECTION .... UOC:ZAA, ZAB, ZAC, ZAD, ZADE, ZAF, ZAG, ZAH, ZAJ, ZAK, ZAL	6
5	PAOZZ	5340012753403	15434	3910685	BRACKET, DOUBLE ANGL .... UOC:ZAA, ZAB, ZAC, ZAD, ZAE, ZAF, ZAG, ZAH, ZAJ, ZAK, ZAL	1
6	PAOZZ	4710013390584	15434	3922117	TUBE ASSEMBLY, METAL .... UOC:ZAA, ZAB, ZAC, ZAD, ZAE, ZAF, ZAG, ZAH, ZAJ, ZAK, ZAL	1
7	PAOZZ	4710012713838	15434	3910751	TUBE ASSEMBLY, METAL NO. 3 CYLINDER .... UOC:Z.AA, ZAB, ZAC, ZAD, ZAE, ZAF, ZAG, ZAH, ZAJ, ZAK, ZAL	1
8	PAOZZ	5305012453192	15434	3918109	SCREW, CAP, HEXAGON H .... UOC:ZAA, ZAB, ZAC, ZAD, ZAE, ZAF, ZAG, ZAH, ZAJ, ZAK, ZAL	1
9	PAOZZ	4710012713843	15434	3910752	TUBE ASSEMBLY, METAL NO. 4 CYLINDER .... UOC:ZAA, ZAB, ZAC, ZAD, ZAE, ZAF, ZAG, ZAH, ZAJ, ZAK, ZAL	1
10	PAOZZ	4710012713840	15434	3910754	TUBE ASSEMBLY, METAL NO. 6 CYLINDER .... UOC:ZAA, ZAB, ZAC, ZAD, ZAE, ZAF, ZAG, ZAH, ZAJ, ZAK, ZAL	1
11	PAOZZ	4710012713839	15434	3910753	TUBE ASSEMBLY, METAL NO. 5 CYLINDER .... UOC:ZAA, ZAB, ZAC, ZAD, ZAE, ZAF, ZAG, ZAH, ZAJ, ZAK, ZAL	1
12	PAOZZ	4730011950825	15434	3905307	BOLT, FLUID PASSAGE .... UOC:ZAA, ZAB, ZAC, ZAD, ZAE, ZAF, ZAG, ZAH, ZAJ, ZAK, ZAL	7
13	PAOZZ	5330011955268	15434	3903380	SEAL, BANJO CONNECT PART OF KIT P/N .... 3802622 PART OF KIT P/N 3802623 PART OF KIT P/N 3802452.......................... UOC: ZAA, ZAB, ZAC, ZAD, ZAE, ZAF, ZAG, ZAH, ZAJ, ZAK, ZAL	1
14	PAOZZ	5340012715705	15434	3907546	BRACKET, DOUBLE ANGL .... UOC:ZAA, ZAB, ZAC, ZAD, ZAE, ZAF, ZAG, ZAH, ZAJ, ZAK, ZAL	2
15	PAOZZ	5340012397140	15434	3914338	FAIRLEAD, BLOCK .... UOC:ZAA, ZAB, ZAC, ZAD, ZAE, ZAF, ZAG, ZAH, ZAJ, ZAK, ZAL	2

(1) ITEM NO	(2) SMR CODE	(3) NSN	(4) CAGEC	(5) PART NUMBER	(6) DESCRIPTION AND USABLE ON CODES (UOC)	(7) QTY
16	PAOZZ	2990012717081	15434	3910687	BRACKET, ENGINE ACCE .... UOC: ZAA , ZAB, ZAC, ZAD, ZAE, ZAF , ZAG, ZAH, ZAJ, ZAK, ZAL	2
17	PAOZZ	5340012806639	15434	3914341	MOUNT, RESILIENT .... UOC: ZAA, ZAB, ZAC, ZAD, ZAE, ZAF, ZAG, ZAH, ZAJ, ZAK, ZAL	1
18	PAOZZ	4710012717198	15434	3910750	TUBE ASSEMBLY, METAL NO. 2 CYLINDER .... UOC: ZAA, ZAB, ZAC, ZAD, ZAE, ZAF, ZAG, ZAH, ZAJ, ZAK, ZAL	1
19	PAOZZ	4710012713837	15434	3910749	TUBE ASSEMBLY, METAL NO. 1 CYLINDER .... UOC:ZAA, ZAB, ZAC, ZAD, ZAE, ZAF , ZAG, ZAH, ZAJ, ZAK, ZAL	1
20	PAOZZ	5305012721334	15434	3910037	SCREW, CAP, HEXAGON H .... UOC:ZAA, ZAB, ZAC, ZAD, ZAE, ZAF, ZAG, ZAH, ZAJ, ZAK, ZAL	2
21	PAOZZ	5340012715862	15434	3917748	FAIRLEAD, BLOCK .... UOC:ZAA, ZAB, ZAC, ZAD, ZAE, ZAFP, ZAG, ZAH, ZAJ, ZAK, ZAL	1

END OF FIGURE

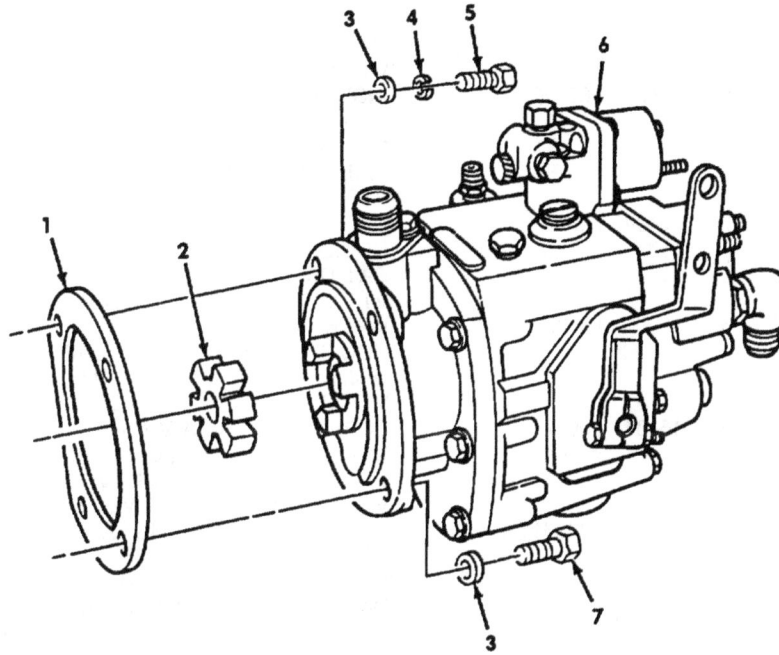

Figure 49. Fuel Pump Assembly, AFC (M939, M939AI).

(1) ITEM NO	(2) SMR CODE	(3) NSN	(4) CAGEC	(5) PART NUMBER	(6) DESCRIPTION AND USABLE ON CODES (UOC)	(7) QTY
					GROUP 0302 FUEL PUMPS	
					FIG. 49 FUEL PUMP ASSEMBLY,AFC (M939,M939A1)	
1	PAFZZ	5330013384829	15434	3069103	GASKET PART OF KIT P/N 3011472 PART OF KIT P/N 3010240 PART OF KIT P/N 5704519 .... UOC:DAA, DAB, DAC, DAD, DAE, DAF, DAG, DAH, DAJ, DAX, DAW , DAX, V12, V13, V14, V15, V16, V17, V19, V20, V21, V22, V24, V25, V39	1
2	PAFZZ	3010004479799	15434	162426	INSERT, FLEXIBLE COU PART OF KIT P/N 3011472 .... UOC:DAA, DAB, DAC, DAD, DAE, DAF, DAG, DAH, DAJ, DAK, DAW, DAX, V12, V13 , V14, V15, V16, V17, V19, V20, V21, V22, V24, V25, V39	1
3	PAFZZ	5310011124307	15434	69324	WASHER, FLAT .... UOC:DAA, DAB, DAC, DAD, DAE, DAF, DAG, DAH, DAJ, DAK, DAW, DAX, V12 , V13 , V14 , V15, V16, V17, V19, V20, V21, V22, V24, V25, V39	4
4	PAFZZ	5310002090965	96906	MS35338-47	WASHER, LOCK .... UOC:DAA, DAB, DAC, DAD, DAE, DAF, DAG, DAH, DAJ, DAK, DAW, DAX, V12 , V13 , V14 , V15 , V16, V17, V19, V20, V21, V22, V24, V25, V39	1
5	PAFZZ	5305011131179	15434	206326	SCREW, CAP, HEXAGON H .... UOC:DAA, DAB, DAC, DAD, DAE, DAF, DAG, DAH, DAJ, DAK, DAW, DAX, V12, V13, V14, V15, V16, V17, V19, V20, V21, V22, V24, V25, V39	1
6	PAFHH	2910012156721	15434	3060711-4144	PUMP, FUEL, METERING AFC, USE AFTER SERIAL NUMBER 1246663 .... UOC:DAA, DAB, DAC, DAD, DAE, DAF, DAG, DAH, DAJ, DAK, DAW, DAX, V12, V13, V14, V15, V16, V17, V19, V20, V21, V22, V24, V25, V39	1
7	PAFZZ	5305011294384	15434	3015282	SCREW, CAP, HEXAGON H .... UOC:DAA, DAB, DAC, DAD, DAE, DAF, DAG, DAH, DAJ, DAK, DAW, DAX, V12, V13, V14, V15, V16, V17, V19, V20, V21, V22, V24, V25, V39	3

END OF FIGURE

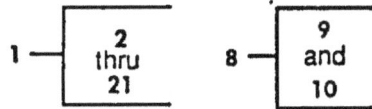

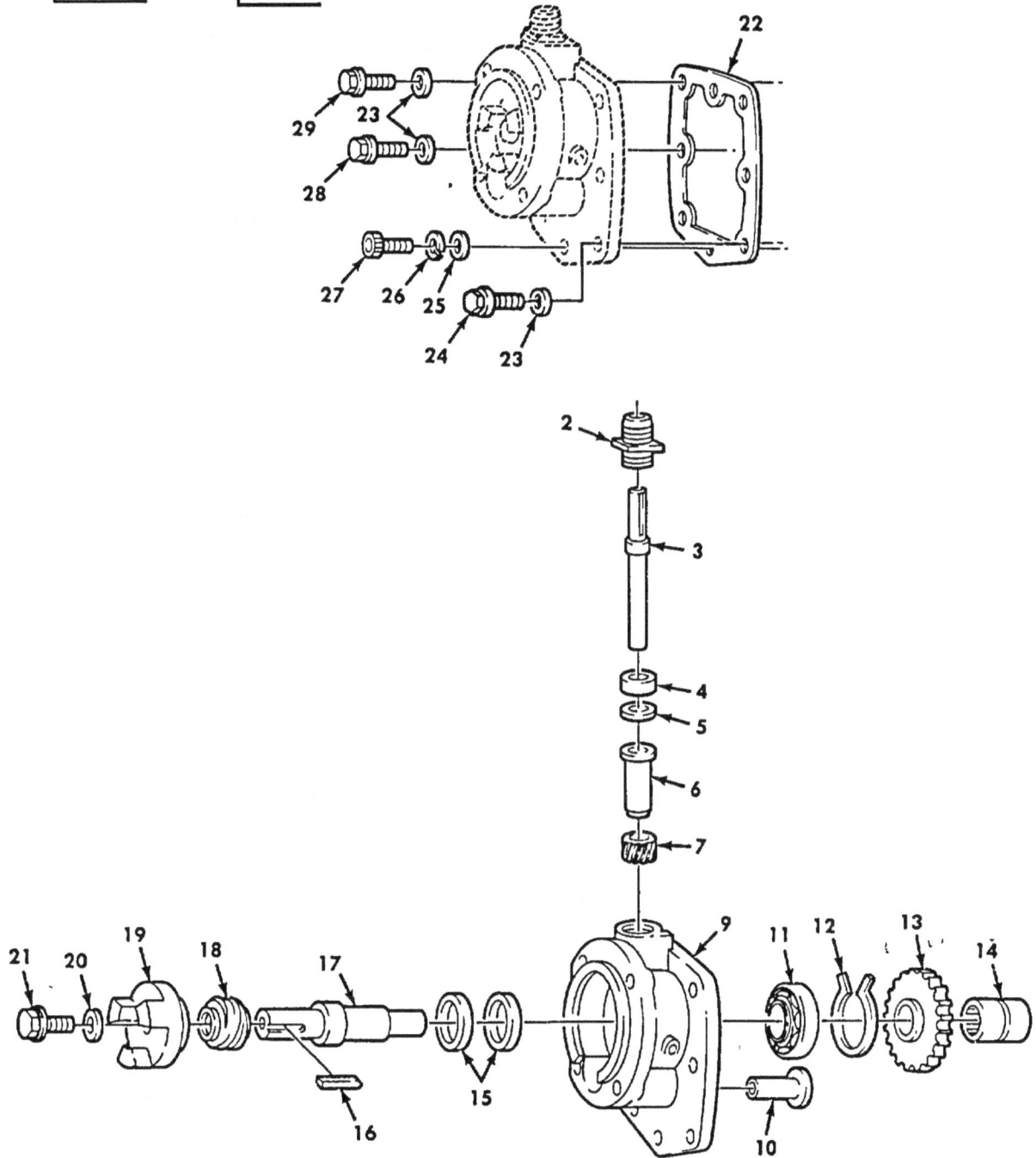

Figure 50. Fuel Pump Governor and Tachometer Drive Assembly, AFC (M939, M939A1).

(1) ITEM NO	(2) SMR CODE	(3) NSN	(4) CAGEC	(5) PART NUMBER	(6) DESCRIPTION AND USABLE ON CODES (UOC)	(7) QTY
					GROUP 0302 FUEL PUMPS	
					FIG. 50 FUEL PUMP GOVERNOR AND TACHOMETER DRIVE ASSEMBLY, AFC(M939, M939A1)	
1	PFHHH	2910011461999	15434	3030269	COVER ASSEMBLY, FRON .... UOC:DAA, DAB, DAC, DAD, DAE, DAF, DAG, DAH, DAJ, DAK, DAW, DAX, V12, V13, V14, V15, V16, V17, V19, V20, V21, V22, V24, V25, V39	1
2	PAHZZ	2910010867715	15434	3002110	.HOUSING, DRIVE ASSEM .... UOC:DAA, DAB, DAC, DAD, DAE, DAF, DAG, DAH, DAJ, DAK, DAW, DAX, V12, V13, V14, V15, V16, V17, V19, V20, V21, V22, V24, V25, V39	1
3	XAHZZ	3040011519348	15434	216908	.SHAFT, TACHOMETER DR .... UOC:DAA, DAB, DAC, DAD, DAE, DAF, DAG, DAH, DAJ, DAK, DAW, DAX, V12, V13, V14, V15, V16, V17, V19, V20, V21, V22, V24, V25, V39	1
4	PAHZZ	5330010728828	15434	212603	.SEAL .... UOC:DAA, DAB, DAC, DAD, DAE, DAF, DAG, DAH, DAJ, DAK, DAW, DAX, V12, V13, V14, V15, V16, V17, V19, V20, V21, V22, V24, V25, V39	1
5	PAHZZ	5365011263334	15434	3004724	.SPACER .... UOC:DAA, DAB, DAC, DAD, DAE, DAF, DAG, DAH, DAJ, DAK, DAW, DAX, V12, V13, V14, V15, V16, V17, V19, V20, V21, V22, V24, V25, V39	1
6	PAHZZ	3120010872539	15434	212609	.BEARING, SLEEVE .... UOC:DAA, DAB, DAC, DAD, DAE, DAF, DAG, DAH, DAJ, DAK, DAW, DAX, V12, V13, V14, V15, V16, V17, V19, V20, V21, V22, V24, V25, V39	1
7	PFHZZ	3020010868780	15434	212610	.GEAR, HELICAL .... UOC:DAA, DAB, DAC, DAD, DAE, DAF, DAG, DAH, DAJ, DAK, DAW, DAX, V12, V13, V1, V V15, V16, V17, V19, V20, V21, V22, V24, V25, V39	1
8	PAHZZ	2910011414337	15434	3803676	.HOUSING, FUEL PUMP .... UOC:DAA, DAB, DAC, DAD, DAE, DAF, DAG, DAH, DAJ, DAK, DAW, DAX, V12, V13, V14, V15, V16, V17, V19, V20, V21, V22, V24, V25, V39	1
9	XAHZZ	2990011474177	15434	3028302	..HOUSING, GOVERNOR AN .... UOC:DAA, DAB, DAC, DAD, DAE, DAF, DAG, DAH, DAJ, DAK, DAW, DAX, V12, V13 , V134, V1, V14, V15, V16, V17, V19, V20, V21, V22, V24, V25, V39	1
10	PAHZZ	3120009049595	15434	163944	..BUSHING, SLEEVE .... UOC:DAA, DAB, DAC, DAD, DAE, DAF, DAG, DAH, DAJ, DAK, DAW, DAX, V12, V13, V14 , V145 , V15, V16, V17, V19, V20, V21, V22, V24, V25, V39	1
11	PAHZZ	3110005165289	15434	S16052	.BEARING, BALL, ANNULA .... UOC:DAA, DAB, DAC, DAD, DAE, DAF, DAG, DAH, DAJ, DAK, DAW, DAX, V12, , V13, V14, V15, V16, V17, V19, V20, V21, V22, V24, V25, V39	1
12	PAHZZ	5325010810662	15434	212604	.RING, RETAINING .... UOC:DAA, DAB, DAC, DAD, DAE, DAF, DAG, DAH,	1

(1) ITEM NO	(2) SMR CODE	(3) NSN	(4) CAGEC	(5) PART NUMBER	(6) DESCRIPTION AND USABLE ON CODES (UOC)	(7) QTY
					DAJ, DAK, DAW, DAX, V12, V13, V14, V15, V16, V17, V19, V20, V21, V22, V24, V25, V39	
13	PAHZZ	3020010709003	15434	212605	.GEAR, SPUR	1
					UOC:DAA, DAB, DAC, DAD, DAE, DAF, DAG, DAH, DAJ, DAK, DAW, DAX, V12, V13, V14, V15, V16, V17, V19, V20, V21, V22, V24, V25, V39	
14	PBHZZ	3010010885727	15434	212639	.COUPLING, SHAFT, RIGI GEAR PUMP .... DRIVE..	1
					UOC:DAA, DAB, DAC, DAD, DAE, DA F, DAG, DAH, DAJ, DAK, DAW, DAX, V12, V13, V14, V15, V16, V17, V19, V20, V21, V22, V24, V25, V39	
15	PAHZZ	5330010728829	15434	3045173	.SEAL, OIL PART OF KIT P/N 3010240	2
					UOC:DAA, DAB, DAC, DAD, DAE, DAF, DAG, DAH, DAJ, DAK, DAW, DAX, V12, V13, V14, V15, V16, V17, V19, V20, V21, V22, V24, V25, V39	
16	PAHZZ	5315010870534	15434	212668	.KEY, MACHINE	1
					UOC:DAA, DAB, DAC, DAD, DAE, DAF, DAG, DAH, DAJ, DAK, DAW, DAX, V12, V13, V14, V15, V16, V17, V19, V20, V21, V22, V24, V25, V39	
17	PBHZZ	3040010709004	15434	212601	.SHAFT, SHOULDERED	1
					UOC:DAA, DAB, DAC, DAD, DAE, DAF, DAG, DAH, DAJ, DAK, DAW, DAX, V12, V13, V14, V15, V16, V17, V19, V20, V21, V22, V24, V25, V39	
18	PBHZZ	3020010864158	15434	212602	.GEAR, HELICAL	1
					UOC:DAA, DAB, DAC, DAD, DAE, DAF, DAG, DAH, DAJ, DAK, DAW, DAX, V12, V13, V14, V15, V16, V17, V19, V20, V21, V22, V24, V25, V39	
19	PBHZZ	3010010801529	15434	212613	.COUPLING HALF, SHAFT	1
					UOC:DAA, DAB, DAC, DAD, DAE, DAF, DAG, DAH, DAJ, DAK, DAW, DAX, V12, V13, V14, V15, V16, V17, V19, V20, V21, V22, V24, V25, V39	
20	PAHZZ	5310008093078	96906	MS27183-11	.WASHER, FLAT	1
					UOC:DAA, DAB, DAC, DAD, DAE, DAF, DAG, DAH, DAJ, DAK, DAW, DAX, V12, V13, V14, V15, V16, V17, V19, V20, V21, V22, V24, V25, V39	
21	PAHZZ	5306011198870	15434	3022589	.BOLT, MACHINE	1
					UOC:DAA, DAB, DAC, DAD, DAE, DAF, DAG, DAH, DAJ, DAK, DAW, DAX, V12, V13, V14, V15, V16, V17, V19, V20, V21, V22, V24, V25, V39	
22	PAHZZ	5330005064866	15434	100764	GASKET PART OF KIT P/N 5704519	1
					UOC:DAA, DAB, DAC, DAD, DAE, DAF, DAG, DAH, DAJ, DAK, DAW, DAX, V12, V13, V14, V15, V16, V17, V19, V20, V21, V22, V24, V25, V39	
23	PAHZZ	5310001411795	88044	AN960-416	WASHER, FLAT	6
					UOC:DAA, DAB, DAC, DAD, DAE, DAF, DAG, DAH, DAJ, DAK, DAW, DAX, V12, V13, V14, V15, V16, V17, V19, V20, V21, V22, V24, V25, V39	
24	PAHZZ	5305011129110	15434	3017051	SCREW, CAP, HEXAGON H	4
					UOC:DAA, DAB, DAC, DAD, DAE, DAF, DAG, DAH, DAJ, DAK, DAW, DAX, V12, V13, V14, V15, V16, V17, V19, V20, V21, V22, V24, V25, V39	
25	PAHZZ	5310000145850	96906	MS27183-42	WASHER, FLAT	11

(1) ITEM NO	(2) SMR CODE	(3) NSN	(4) CAGEC	(5) PART NUMBER	(6) DESCRIPTION AND USABLE ON CODES (UOC)	(7) QTY
					UOC:DAA, DAB, DAC, DAD, DAE, DAF, DAG, DAH, DAJ, DAK, DAW, DAX, V12, V13, V14, V15, V16, V17, V19, V20, V21, V22, V24, V25, V39	
26	PAHZZ	5310004841718	15434	181466	WASHER, LOCK	1
					UOC:DAA, DAB, DAC, DAD, DAE, DAF, DAG, DAH, DAJ, DAK, DAW, DAX, V12, V13, V14, V15, V16, V17, V19, V20, V21, V22, V24, V25, V39	
27	PAHZZ	5305001610902	15434	118226	SCREW	1
					UOC:DAA, DAB, DAC, DAD, DAE, DAF, DAG, DAH, DAJ, DAK, DAW, DAX, V12, V13, V14, V15, V16, V17, V19, V20, V21, V22, V24, V25, V39	
28	PAHZZ	5305011261128	15434	3017052	SCREW, CAP, HEXAGON H	1
					UOC:DAA, DAB, DAC, DAD, DAE, DAF, DAG, DAH, DAJ, DAK, DAW, DAX, V12, V13, V14, V15, V16, V17, V19, V20, V21, V22, V24, V25, V39	
29	PAHZZ	5305011345659	15434	3018682	SCREW, CAP, HEXAGON H	1
					UOC:DAA, DAB, DAC, DAD, DAE, DAF, DAG, DAH, DAJ, DAK, DAW, DAX, V12, V13, V14, V15, V16, V17, V19, V20, V21, V22, V24, V25, V39	

END OF FIGURE

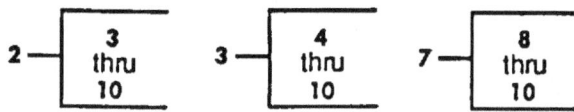

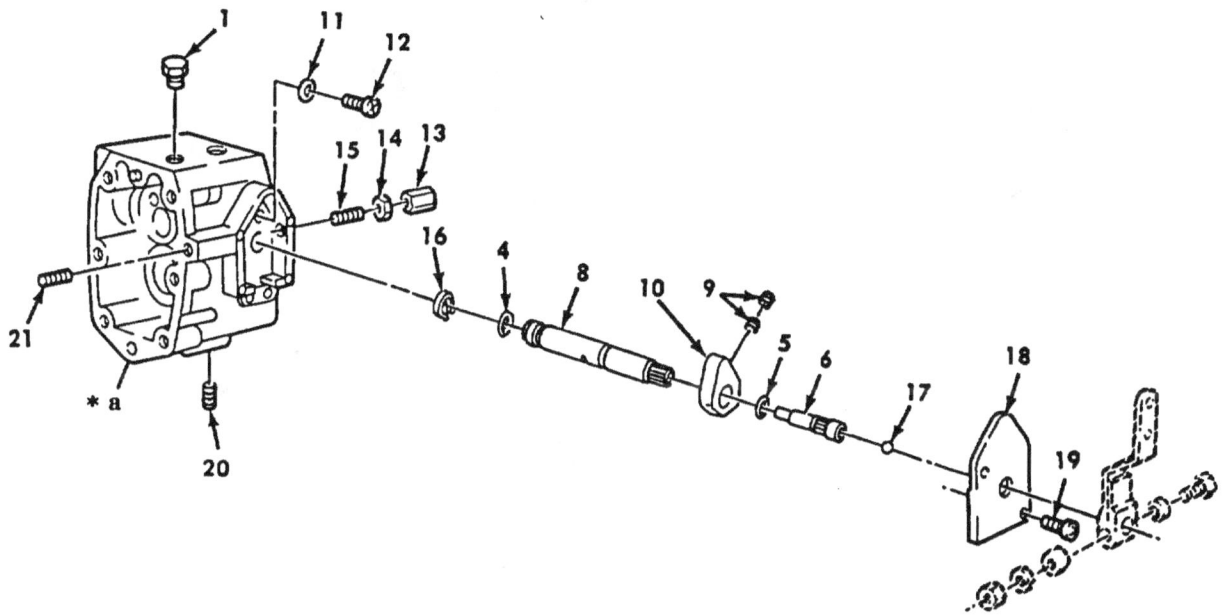

* a PART OF ITEM 2

Figure 51. Fuel Pump Housing, AFC (M939, M939AI).

(1) ITEM NO	(2) SMR CODE	(3) NSN	(4) CAGEC	(5) PART NUMBER	(6) DESCRIPTION AND USABLE ON CODES (UOC)	(7) QTY
					GROUP 0302 FUEL PUMPS	
					FIG. 51 FUEL PUMP HOUSING AFC(M939, M939A1)	
1	PAHZZ	5365011601832	15434	112076	PLUG, FUEL OUTLET .... UOC:DAA, DAB, DAC, DAD, DAE, DAF, DAG, DAH, DAJ, DAK, DAW, DAX, V12, V13, V14, V15, V16, V17, V19, V20, V21, V22, V24, V25, V39	1
2	PBHHH	2910013031195	15434	3043254	HOUSING, FUEL PUMP .... UOC:DAA, DAB, DAC, DAD, DAE, DAF, DAG, DAH, DAJ, DAK, DAW, DAX, V12, V13, V14, V15, V16, V17, V19, V20, V21, V22, V24, V25, V39	1
3	PBHZZ	2910010803149	15434	AR41010	.SHAFT ASSEIMBLY, THRO .... UOC:DAA, DAB, DAC, DAD, DAE, DAF, DAG, DAH, DAJ, DAK, DAW, DAX, V12, V13, V14, V15, V16, V17, V19, V20, V21, V22, V24, V25, V39	1
4	PAHZZ	5330000819289	15434	100478	..O-RING PART OF KIT P/N 3010240 .... UOC:DAA, DAB, DAC, DAD, DAE, DAF, DAG, DAH, DAJ, DAK, DAW, DAX, V12, V13, V14, V15, V16, V17, V19, V20, V21, V22, V24, V25, V39	1
5	PAHZZ	5330010728983	15434	213768	..O-RING .... UOC:DAA, DAB, DAC, DAD, DAE, DAF, DAG, DAH, DAJ, DAK, DAW, DAX, V12, V13, V14, V15, V16, V17, V19, V20, V21, V22, V24, V25, V39	1
6	PAHZZ	5305010728826	15434	3076040	..SCREW .... UOC:DAA, DAB, DAC, DAD, DAE, DAF, DAG, DAH, DAJ, DAK, DAW, DAX, V12, V13, V14, V15, V16, V17, V19, V20, V21, V22, V24, V25, V39	1
7	PBHZZ	2910011224015	15434	3006430	..THROTTLE ASSEMBLY .... UOC:DAA, DAB, DAC, DAD, DAE, DAF, DAG, DAH, DAJ, DAK, DAW, DAX, V12, V13, V14, V15, V16, V17, V19, V20, V21, V22, V24, V25, V39	1
8	PFHZZ	3040011504926	15434	3006350	...SHAFT, THROTTLE .... UOC:DAA, DAB, DAC, DAD, DAE, DAF, DAG, DAH, DAJ, DAK, DAW, DAX, V12, V13, V14, V15, V16, V17, V19, V20, V21, V22, V24, V25, V39	1
9	PAHZZ	5305011355446	15434	3006344	...SETSCREW .... UOC:DAA, DAB, DAC, DAD, DAE, DAF, DAG, DAH, DAJ, DAK, DAW, DAX, V12, V13, V1, VV15, V16, V17, V19, V20, V21, V22, V24, V25, V39	1
10	PBHZZ	3040010861449	15434	3006343	...COLLAR, SHAFT THROTTLE SHAFT STOP .... UOC:DAA, DAB, DAC, DAD, DAE, DAF, DAG, DAH, DAJ, DAK, DAW, DAX, V12, V13, V14, V15, V16, V17, V19, V20, V21, V22, V24, V25, V39	1
11	PAHZZ	5330009703461	15434	68061-A	O-RING PART OF KIT P/N 3011472 .... UOC:DAA, DAB, DAC, DAD, DAE, DAF, DAG, DAH, DAJ, DAK, DAW, DAX, V12, V13, V14, V15, V16, V17, V19, V20, V21, V22, V24, V25, V39	1
12	PAHZZ	4730010784703	15434	3004293	PLUG, TUBE FITTING, T .... UOC:DAA, DAB, DAC, DAD, DAE, DAF, DAG, DAH, DAJ, DAK, DAW, DAX, V12, V13, V14, V15, V16,	1

(1) ITEM NO	(2) SMR CODE	(3) NSN	(4) CAGEC	(5) PART NUMBER	(6) DESCRIPTION AND USABLE ON CODES (UOC)	(7) QTY
					V17, V19, V20, V21, V22, V24, V25, V39	
13	PAHZZ	5340007164975	15434	110058	POST, ELECTRICAL-MEC	1
					UOC:DAA, DAB, DAC, DAD, DAE, DAF, DAG, DAH, DAJ, DAK, DAW, DAX, V12, V13, V14, V15, V16, V17, V19, V20, V21, V22, V24, V25, V39	
14	PAHZZ	5310009717989	96906	MS35691-5	NUT, PLAIN, HEXAGON	2
					UOC:DAA, DAB, DAC, DAD, DAE, DAF, DAG, DAH, DAJ, DAK, DAW, DAX, V12, V13, V14, V15, V16, V17, V19, V20, V21, V22, V24, V25, V39	
15	PAHZZ	5305011099307	15434	195755	SCREW	1
					UOC:DAA, DAB, DAC, DAD, DAE, DAF, DAG, DAH, DAJ, DAK, DAW, DAX, V12, V13, V14, V15, V16, V17, V19, V20, V21, V22, V24, V25, V39	
16	PAHZZ	5325002562846	96906	MS16632-1050	RING, RETAINING	1
					UOC:DAA, DAB, DAC, DAD, DAE, DAF, DAG, DAH, DAJ, DAK, DAW, DAX, V12, V13, V14, V15, V16, V17, V19, V20, V21, V22, V24, V25, V39	
17	PAHZZ	3110010798190	15434	213769	BALL, BEARING PART OF KIT P/N 5704519	1
					UOC:DAA, DAB, DAC, DAD, DAE, DAF, DAG, DAH, DAJ, DAK, DAW, DAX, V12, V13, V14, V15, V16, V17, V19, V20, V21, V22, V24, V25, V39	
18	PAHZZ	4320010985115	15434	3000446	COVER, HYDRAULIC, PUM	1
					UOC:DAA, DAB, DAC, DAD, DAE, DAF, DAG, DAH, DAJ, DAK, DAW, DAX, V12, V13, V14, V15, V16, V17, V19, V20, V21, V22, V24, V25, V39	
19	PAHZZ	5305008046318	15434	S-2286	SCREW PART OF KIT P/N 5704519	2
					UOC:DAA, DAB, DAC, DAD, DAE, DAF, DAG, DAH, DAJ, DAK, DAW, DAX, V12, V13, V14, V15, V16, V17, V19, V20, V21, V22, V24, V25, V39	
20	PAHZZ	4730011243762	15434	3025460	PLUG, PIPE	3
					UOC:DAA, DAB, DAC, DAD, DAE, DAF, DAG, DAH, DAJ, DAK, DAW, DAX, V12, V13, V14, V15, V16, V17, V19, V20, V21, V22, V24, V25, V39	
21	PAHZZ	5305000635043	88044	AN565F428H24	SETSCREW	1
					UOC:DAA, DAB, DAC, DAD, DAE, DAF, DAG, DAH, DAJ, DAK, DAW, DAX, V12, V13, V14, V15, V16, V17, V19, V20, V21, V22, V24, V25, V39	

END OF FIGURE

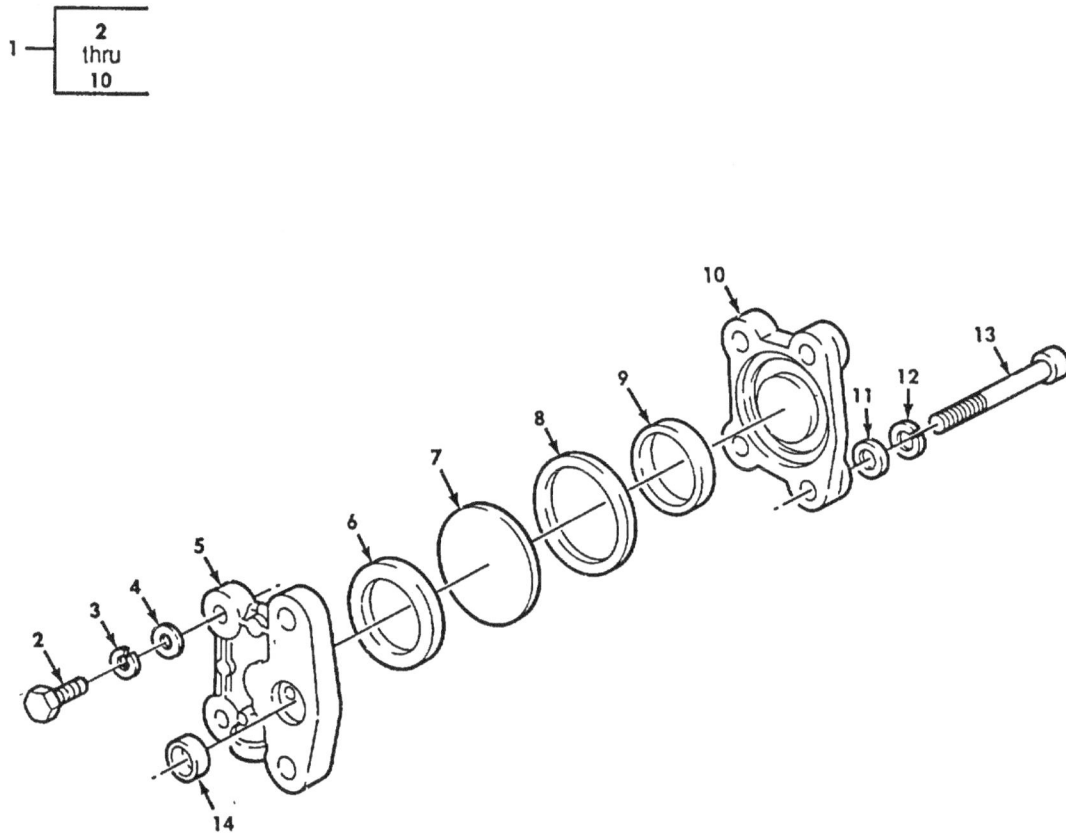

Figure 52. Fuel Pump Damper (M939, M939A1).

(1) ITEM NO	(2) SMR CODE	(3) NSN	(4) CAGEC	(5) PART NUMBER	(6) DESCRIPTION AND USABLE ON CODES (UOC)	(7) QTY
					GROUP 0302 FUEL PUMPS	
					FIG. 52 FUEL PUMP DAMPER(M939, M939A1)	
1	PBFZZ	6685008287126	15434	BM-76340	DAMPENER, FLUID PRES .... UOC:DAA, DAB, DAC, DAD, DAE, DAF, DAG, DAH, DAJ, DAK, DAW, DAX, V12, V13, V14, V15, V16, V17, V19, V20, V21, V22, V24, V25, V39	1
2	PAFZZ	5305000680509	80204	B1821BH025C125N	.SCREW, CAP, HEXAGON H .... UOC:DAA, DAB, DAC, DAD, DAE, DAF, DAG, DAH, DAJ, DAK, DAW, DAX, V12, V13, V14, V15, V16, V17, V19, V20, V21, V22, V24, V25, V39	2
3	PAFZZ	5310001596209	96906	MS122032	.WASHER, LOCK .... UOC:DAA, DAB, DAC, DAD, DAE, DAF, DAG, DAH, DAJ, DAK, DAW, DAX, V12, V13, V14, V15, V16, V17, V19, V20, V21, V22, V24, V25, V39	2
4	PAFZZ	5310001411795	88044	AN960-416	.WASHER, FLAT .... UOC:DAA, DAB, DAC, DAD, DAE, DAF, DAG, DAH, DAJ, DAK, DAW, DAX, V12, V13, V14, V15, V16, V17, V19, V20, V21, V22, V24, V25, V39	4
5	PFFZZ	2910008295616	15434	153336	.HOUSING, FUEL PUMP .... UOC:DAA, DAB, DAC, DAD, DAE, DAF, DAG, DAH, DAJ, DAK, DAW, DAX, V12, V13, V14, V15, V16, V17, V19, V20, V21, V22, V24, V25, V39	1
6	PAFZZ	5330008092667	15434	100099	.O-RING PART OF KIT P/N 3010240 .... UOC:DAA, DAB, DAC, DAD, DAE, DAF, DAG, DAH, DAJ, DAK, DAW, DAX, V12, V13, V14, V15, V16, V17, V19, V20, V21, V22, V24, V25, V39	1
7	PAFZZ	5340009513536	15434	202897	.DISK, SOLID, PLAIN .... UOC:DAA, DAB, DAC, DAD, DAE, DAF, DAG, DAH, DAJ, DAK, DAW, DAX, V12, V13, V14, V15, V16, V17, V19, V20, V21, V22, V24, V25, V39	1
8	PAFZZ	5330008093276	15434	139988	.O-RING PART OF KIT P/N 3010240 .... UOC:DAA, DAB, DAC, DAD, DAE, DAF, DAG, DAH, DAJ, DAK, DAW, DAX, V12, V13, V14, V15, V16, V17, V19, V20, V21, V22, V24, V25, V39	1
9	PAFZZ	5365009650870	15434	160514	.SPACER, RING PART OF KIT P/N 3010240 .... UOC:DAA, DAB, DAC, DAD, DAE, DAF, DAG, DAH, DAJ, DAK, DAW, DAX, V12, V13, V14, V15, V16, V17, V19, V20, V21, V22, V24, V25, V39	1
10	PFFZZ	5340008295617	15434	153338	.COVER, ACCESS .... UOC:DAA, DAB, DAC, DAD, DAE, DAF, DAG, DAH, DAJ, DAK, DAW, DAX, V12, V13, V14, V15, V16, V17, V19, V20, V21, V22, V24, V25, V39	1
11	PAFZZ	5310001411795	88044	AN960-416	WASHER , FLAT .... UOC:DAA, DAB, DAC, DAD, DAE, DAF, DAG, DAH, DAJ, DAK, DAW, DAX, V12, V13, V14, V15, V16, V17, V19, V20, V21, V22, V24, V25, V39	2
12	PAFZZ	5310004841718	15434	181466	WASHER, LOCK .... UOC:DAA, DAB, DAC, DAD, DAE, DAF, DAG, DAH, DAJ, DAK, DAW, DAX, V12, V13, V14, V15, V16,	2

(1) ITEM NO	(2) SMR CODE	(3) NSN	(4) CAGEC	(5) PART NUMBER	(6) DESCRIPTION AND USABLE ON CODES (UOC)	(7) QTY
13	PAFZZ	5305011332060	15434	133538	V17, V19, V20, V21, V22, V24, V25, V39 SCREW, CAP, SOCKET HE .... UOC:DAA, DAB, DAC, DAD, DAE, DAF, DAG, DAH, DAJ, DAK, DAW, DAX, V12, V13, V14, V15, V16, V17, V19, V20, V21, V22, V24, V25, V39	2
14	PAFZZ	5330002528888	9F512	691-10014	GASKET PART OF KIT P/N 3010240 .... UOC:DAA, DAB, DAC, DAD, DAE, DAF, DAG, DAH, DAJ, DAK, DAW, DAX, V12, V13, V14, V15, V16, V17, V19, V20, V21, V22, V24, V25, V39	1

END OF FIGURE

2 —[3]   4 —[5]

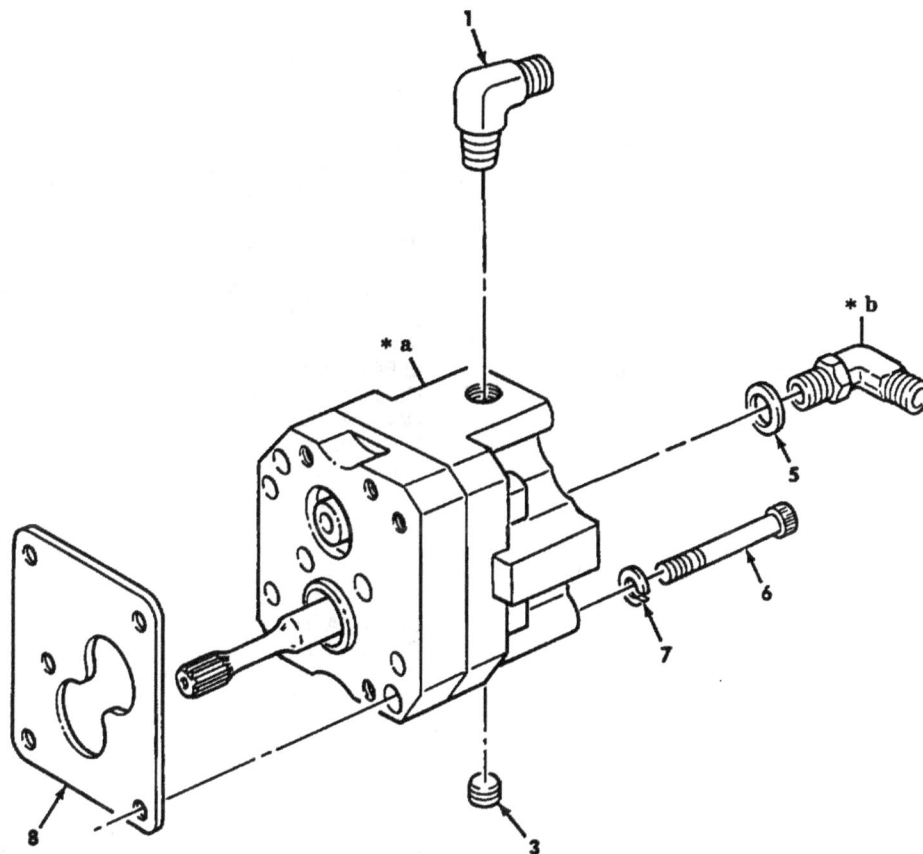

* a PART OF ITEM 2
* b PART OF ITEM 4

Figure 53. Fuel Gear Pump Assembly and Mounting Hardware (M939, M939A1).

(1) ITEM NO	(2) SMR CODE	(3) NSN	(4) CAGEC	(5) PART NUMBER	(6) DESCRIPTION AND USABLE ON CODES (UOC)	(7) QTY
					GROUP 0302 FUEL PUMPS	
					FIG. 53 FUEL GEAR PUMP ASSEMBLY AND MOUNTING HARDWARE(M939,M939A1)	
1	PAHZZ	4820011647002	15434	3033740	VALVE, CHECK .... UOC:DAA, DAB, DAC, DAD, DAE, DAF, DAG, DAH, DAJ, DAK, DAW, DAX, V12, V13, V14, V15, V16, V17, V19, V20, V21, V22, V24, V25, V39	1
2	PARHH	3020011665647	15434	3034217	GEAR, PUMP ASSEMBLY USE AFTER .... SERIAL NUMBER 11246663 .... UOC:DAA, DAB, DAC, DAD, DAE, DAF, DAG, DAH, DAJ, DAK, DAW, DAX, V12, V13, V14, V15, V16, V17, V19, V20, V21, V22, V24, V25, V39	1
3	PAHZZ	4730000500718	15434	3026624	.PLUG, PIPE .... UOC:DAA, DAB, DAC, DAD, DAE, DAF, DAG, DAH, DAJ, DAK, DAW, DAX, V12, V13, V14, V15, V16, V17, V19, V20, V21, V22, V24, V25, V39	1
4	PAHZZ	4730010789859	15434	203849	ELBOW, TUBE TO BOSS .... UOC:DAA, DAB, DAC, DAD, DAE, DAF, DAG, DAH, DAJ, DAK, DAW, DAX, V12, V13, V14, V15, V16, V17, V19, V20, V21, V22, V24, V25, V39	1
5	PAHZZ	5330010728984	15434	3046201	.O-RING .... UOC:DAA, DAB, DAC, DAD, DAE, DAF, DAG, DAH, DAJ, DAK, DAW, DAX, V12, V13, V14, VV1, V16, V17, V19, V20, V21, V22, V24, V25, V39	1
6	PAHZZ	5306004850790	15434	70790	BOLT, MACHINE USE AFTER SERIAL .... NUMBER 11246663.. UOC:DAA, DAB, DAC, DAD, DAE, DAF, DAG, DAH, DAJ, DAK, DAW, DAX, V12, V13, V14, V15, V16, V17, V19, V20, V21, V22, V24, V25, V39	4
7	PAHZZ	5310004841718	15434	181466	WASHER, LOCK .... UOC:DAA, DAB, DAC, DAD, DAE, DAF, DAG, DAH, DAJ, DAK, DAW, DAX, V12, V13, V14;V15, V16, V17, V19, V20, V21, V22, V24, V25, V39	4
8	PAHZZ	5330011368569	15434	3069017	GASKET PART OF KIT P/N 3010240 .... UOC:DAA, DAB, DAC, DAD, DAE, DAF, DAG, DAH, DAJ, DAK, DAW, DAX, V12, V13, V14, V15, V16, V17, V19, V20, V21, V22, V24, V25, V39	1

END OF FIGURE

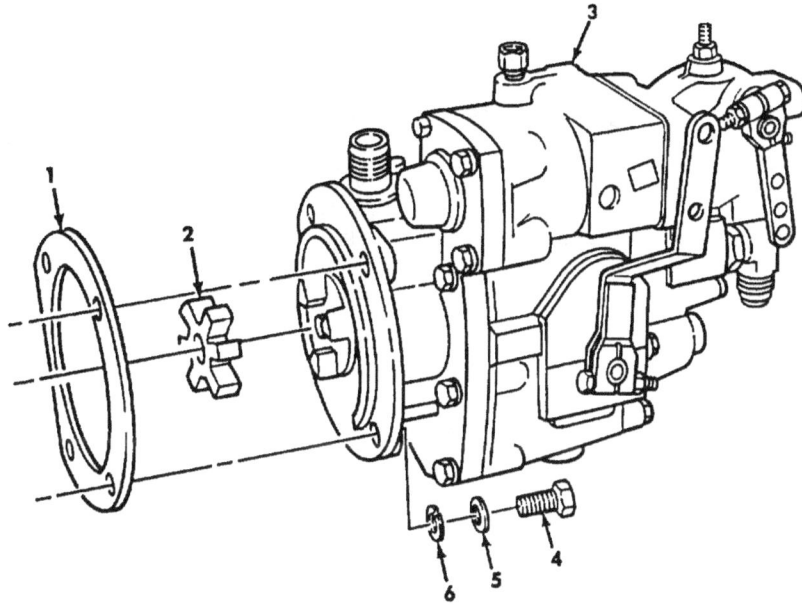

Figure 54. Fuel Pump Assembly with VS Governor (M939, M939AV).

(1) ITEM NO	(2) SMR CODE	(3) NSN	(4) CAGEC	(5) PART NUMBER	(6) DESCRIPTION AND USABLE ON CODES (UOC)	(7) QTY
					GROUP 0302 FUEL PUMPS	
					FIG. 54 FUEL PUMP ASSEMBLY WITH VS GOVERNOR(M939,M939A1)	
1	PAFZZ	5330011607460	15434	3035053	GASKET PART OF KIT P/N 3010240, ... 3011472 .... UOC:DAL, V18	1
2	PAFZZ	3010004479799	15434	162426	INSERT, FLEXIBLE COU PART OF KIT P/N .... 3011472 .... UOC:DAL, V18	1
3	PBFHH	2910012185158	15434	3045424-9254	PUMP, FUEL, METERING VS, USE AFTER .... SERIAL NUMBER 11246663 .... UOC:DAL, V18	1
4	PAFZZ	5305000712056	80204	B1821BH044C175N	SCREW, CAP, HEXAGON H .... UOC:DAL, V18	4
5	PAFZZ	5310005626557	15434	S-622	WASHER , FLAT .... UOC:DAL, V18	4
6	PAFZZ	5310002090965	96906	MS35338-47	WASHER, LOCK .... UOC:DAL, V18	1

END OF FIGURE

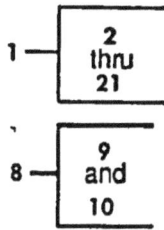

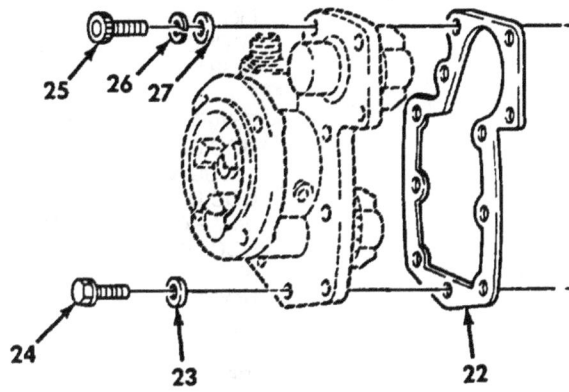

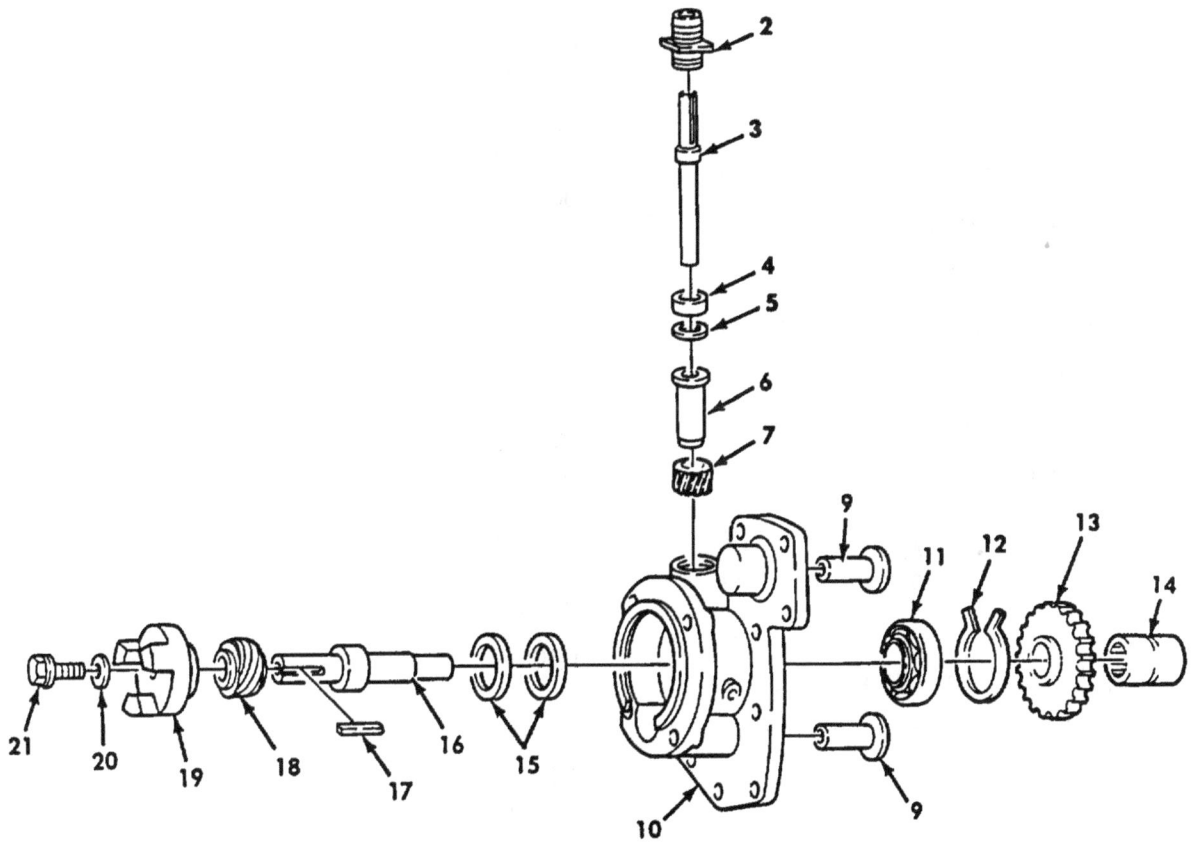

Figure 55. Fuel Pump Governor and Tachometer Drive Assembly, VS (M939, M939A1).

(1) ITEM NO	(2) SMR CODE	(3) NSN	(4) CAGEC	(5) PART NUMBER	(6) DESCRIPTION AND USABLE ON CODES (UOC)	(7) QTY
					GROUP 0302 FUEL PUMPS	
					FIG. 55 FUEL PUMP GOVERNOR AND TACHOMETER DRIVE ASSEMBLY, VS(M939, 1939A1)	
1	PFHHH	2910010909345	15434	3030276	SHAFT ASSEMBLY .... UOC:DAL, V18	1
2	PAHZZ	2910010867715	15434	3002110	.HOUSING, DRIVE ASSEM .... UOC:DAL, V18	1
3	XAHZZ	3040011519348	15434	216908	.SHAFT, TACHOMETER DR .... UOC:DAL, V18	1
4	PAHZZ	5330010728828	15434	212603	.SEAL .... UOC:DAL, V18	1
5	PAHZZ	5365011263334	15434	3004724	.SPACER .... UOC:DAL, V18	1
6	PAHZZ	3120010872539	15434	212609	.BEARING, SLEEVE .... UOC:DAL, V18	1
7	PFHZZ	3020010868780	15434	212610	.GEAR, HELICAL .... UOC:DAL, V18	1
8	PAHZZ	2815011797516	15434	30283770	.COVER AND BUSHING .... UOC:DAL, V18	1
9	PAHZZ	3120009049595	15434	163944	..BUSHING, SLEEVE .... UOC:DAL, V18	2
10	XAHZZ	2990011598738	15434	3023451	..COVER, MAINSHAFT, GOV .... UOC:DAL, V18	1
11	PAHZZ	3110001565453	15434	S16053	.BEARING, BALL, ANNULA .... UOC:DAL, V18	1
12	PAHZZ	5325010810662	15434	212604	.RING, RETAINING .... UOC:DAL, V18	1
13	PAHZZ	3020010709003	15434	212605	.GEAR, SPUR .... UOC:DAL, V18	1
14	PBHZZ	3010010885727	15434	212639	.COUPLING, SHAFT, RIGI .... UOC:DAL, V18	1
15	PAHZZ	5330010728829	15434	3045173	.SEAL, OIL PART OF KIT P/N 3010240 .... UOC:DAL, V18	2
16	PBHZZ	3040010709004	15434	212601	.SHAFT, SHOULDERED .... UOC:DAL, V18	1
17	PAHZZ	5315010870534	15434	212668	.KEY, MACHINE .... UOC:DAA, DAL, V18	1
18	PAHZZ	3020010864158	15434	212602	.GEAR, HELICAL .... UOC:DAL, V18	1
19	PBHZZ	3010010801529	15434	212613	.COUPLING HALF, SHAFT .... UOC:DAL, V18	1
20	PAHZZ	5310008093078	15434	146160	.WASHER, FLAT .... UOC:DAL, V18	1
21	PAHZZ	5306011198870	15434	3022589	.BOLT, MACHINE .... UOC:DAL, V18	1
22	PAHZZ	5330010511053	15434	200998	GASKET FUEL PUMP COVER .... UOC:DAL, V18	1
23	PAHZZ	5310001411795	88044	AN960-416	WASHER, FLAT .... UOC:DAL, V18	8

(1) ITEM NO	(2) SMR CODE	(3) NSN	(4) CAGEC	(5) PART NUMBER	(6) DESCRIPTION AND USABLE ON CODES (UOC)	(7) QTY
24	PAHZZ	5305011129110	15434	3017051	SCREW, CAP, HEXAGON H .... UOC:DAL, V18	8
25	PAHZZ	5305001610902	15434	118226	SCREW .... UOC:DAL, V18	2
26	PAHZZ	5310004106756	15434	S606	WASHER, LOCK .... UOC:DAL, V18	2
27	PAHZZ	5310000145850	96906	MS27183-42	WASHER, FLAT .... UOC:DAL, V18	2

END OF FIGURE

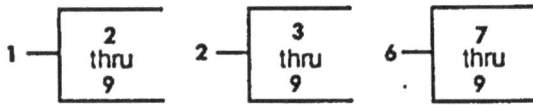

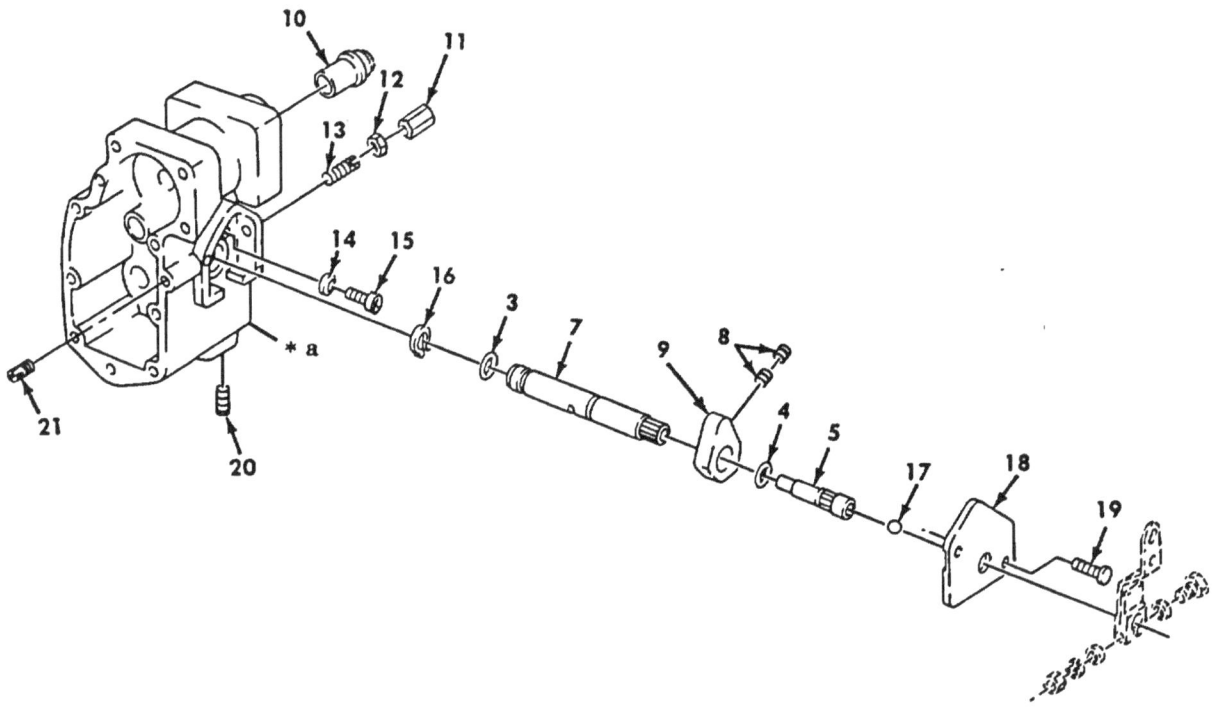

* a   PART OF ITEM 1

Figure 56. Fuel Pump Housing, VS (M939, M939A1).

(1) ITEM NO	(2) SMR CODE	(3) NSN	(4) CAGEC	(5) PART NUMBER	(6) DESCRIPTION AND USABLE ON CODES (UOC)	(7) QTY
					GROUP 0302 FUEL PUMPS	
					FIG. 56 FUEL PUMP HOUSING, VS(M939, M939A1)	
1	PBHZZ	3040012195697	5434	3038668	HOUSING,MECHANICAL USE AFTER .... NUMBER 11246663 .... UOC:DAL,V18	1
2	PBHZZ	2910010803149	15434	AR41010	.SHAFT ASSEMBLY,THRO.................... UOC:DAL,V18	1
3	PAHZZ	5330000819289	5434	100478	..O-RING PART OF KIT P/N 3010240 .... UOC:DAL,V18	1
4	PAHZZ	330010728983	15434	213768	..O-RING UOC:DAL,V18	1
5	PAHZZ	5305010728826	15434	3076040	..SCREW UOC:DAL,V18	1
6	PFHZZ	2910011224015	15434	3006430	..THROTTLE ASSEMBLY UOC:DAL,V18	1
7	PBHZZ	3040011504926	15434	3006350	...SHAFT,THROTTLE UOC:DAL,V18	1
8	PAHZZ	5305011355446	15434	3006344	...SETSCREW UOC:DAL,V18	2
9	PBHZZ	3040010861449	15434	3006343	...COLLAR,SHAFT.................. UOC:DAL,V18	1
10	PAHZZ	4820011061760	15434	201806	VALVE,DUAL BALL UOC:DAL,V18	1
11	PAHZZ	5340007164975	15434	110058	POST,ELECTRICAL-MEC UOC:DAL,V18	1
12	PAHZZ	5305011099307	15434	195755	SCREW.......................... UOC:DAL,V18	1
13	PAHZZ	5310009717989	96906	MS35691-5	NUT,PLAIN,HEXAGON UOC:DAL,V18	2
14	PAHZZ	5330001716600	83259	600-001 1-4	PACKING WITH RETAIN PART OF KIT P/N........... 3010240,3011472..................... UOC:DAA,DAL,V18	1
15	PFHZZ	4730010784703	15434	3004293	PLUG,TUBE FITTING,T UOC:DAL,V18	1
16	PAHZZ	5325002562846	96906	MS16632-1050	RING,RETAINING UOC:DAL,V18	1
17	PAHZZ	3110010798190	15434	213769	BALL,BEARING UOC:DAL,V18	1
18	PAHZZ	4320010985115	15434	3000446	COVER,HYDRAULIC,PUM UOC:DAL,V18	1
19	PAHZZ	5305008046318	15434	S-2286	SCREW.......................... UOC:DAL,V18	2
20	PAHZZ	4730011243762	15434	3025460	PLUG,PIPE UOC:DAL,V18	1
21	PAHZZ	5305000635043	88044	AN565F428H24	SETSCREW UOC:DAL,V18	1

END OF FIGURE

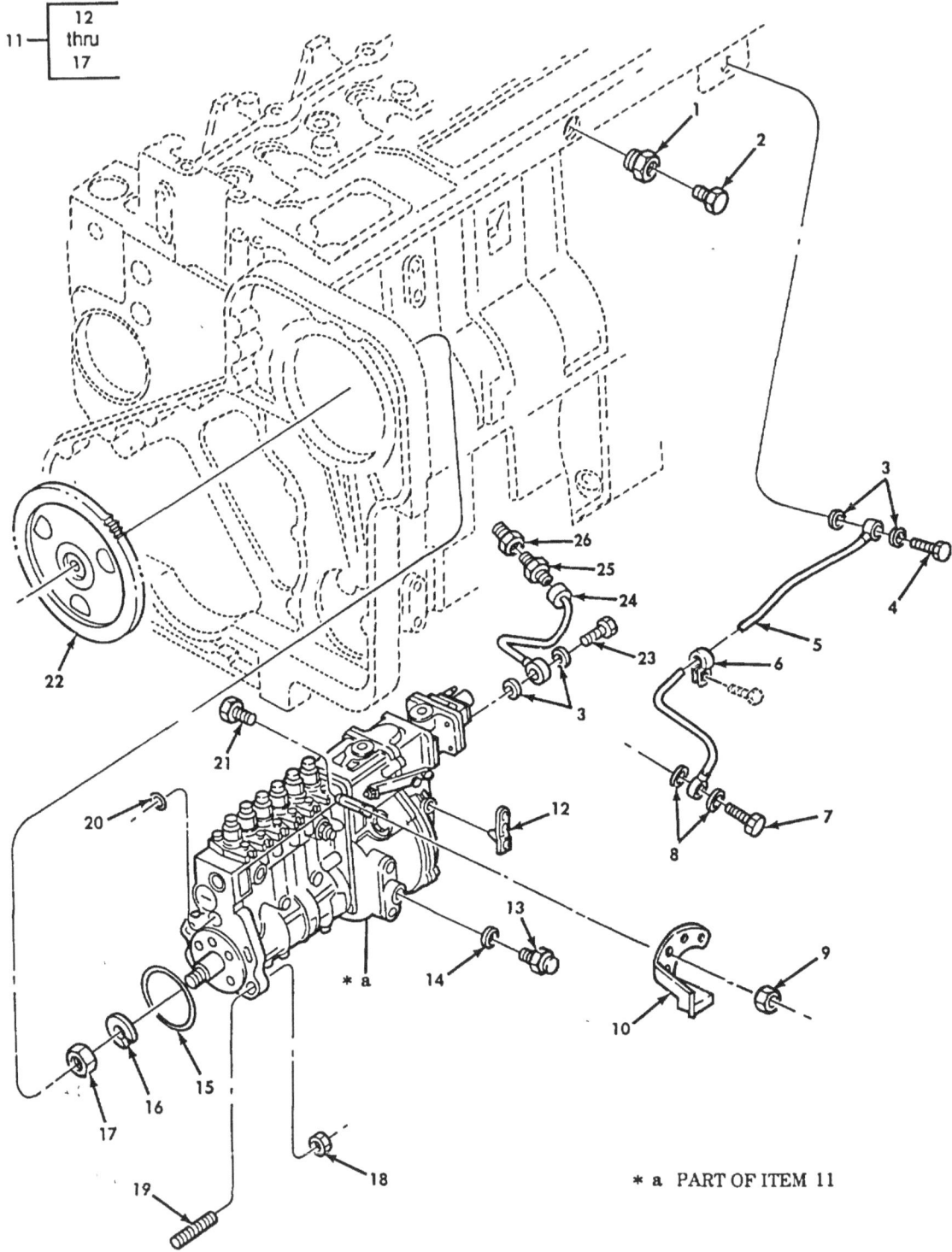

Figure 57. Fuel Injector Pump (M939A2).

* a   PART OF ITEM 11

(1) ITEM NO	(2) SMR CODE	(3) NSN	(4) CAGEC	(5) PART NUMBER	(6) DESCRIPTION AND USABLE ON CODES (UOC)	(7) QTY
					GROUP 0302 FUEL PUMPS	
					FIG. 57 FUEL INJECTOR PUMP(M939A2)	
1	PAOZZ	3120012723276	15434	3908110	BUSHING,SLEEVE................................ UOC:ZAA,ZAB,ZAC,ZAD,ZAE,ZAF,ZAG,ZAH, ZAJ,ZAK,ZAL	
2	PAOZZ	4730011659491	15434	3008465	PLUG,PIPE .... UOC:ZAA,ZAB,ZAC,ZAD,ZAE,ZAF,ZAG,ZAH, ZAJ,ZAK,ZAL	1
3	PAOZZ	5310013408469	15434	3918191	WASHER,SEAL PART OF KIT P/N 3802622............ PART OF KIT P/N 3802452 .... UOC:ZAA,ZAB,ZAC,ZAD,ZAE,ZAF,ZAG,ZAH, ZAJ,ZAK,ZAL	4
4	PAOZZ	5305012343714	15434	3905860	SCREW .... UOC:ZAA,ZAB,ZAC,ZAD,ZAE,ZAF,ZAG,ZAH, ZAJ,ZAK,ZAL	2
5	PAOZZ	4710012923701	15434	3912925	TUBE ASSEMBLY,METAL .... UOC:ZAA,ZAB,ZAC,ZAD,ZAE,ZAF,ZAG,ZAH, ZAJ,ZAK,ZAL	1
6	PAOZZ	5340008437825	96906	MS21333-68	CLAMP,LOOP .... UOC:ZAA,ZAB,ZAC,ZAD,ZAE,ZAF,ZAG,ZAH, ZAJ, ZAK, ZAL	1
7	PAOZZ	4730012810840	15434	3924725	BOLT,FLUID PASSAGE .... UOC:ZAA,ZAB,ZAC,ZAD,ZAE,ZAF,ZAG,ZAH, ZAJ,ZAK,ZAL	1
8	KFOZZ	5310013354861	15434	3918192	WASHER,FLAT PART OF KIT P/N 3802622............ PART OF KIT P/N 3802623 PART OF KIT P/N 3802452.. UOC:ZAA,ZAB,ZAC,ZAD,ZAE,ZAF,ZAG,ZAH, ZAJ,ZAK,ZAL	2
9	PAFZZ	5310012708341	15434	3910959	NUT,SELF-LOCKING,HE .... UOC:ZAA,ZAB,ZAC,ZAD,ZAE,ZAF,ZAG,ZAH, ZAJ,ZAK, ZAL	2
10	PAFZZ	3040012713821	15434	3912484	LEVER,REMOTE CONTRO .... UOC:ZAA,ZAB,ZAC,ZAD,ZAE,ZAF,ZAG,ZAH, ZAJ,ZAK,ZAL	1
11	PAFHH	2910012688757	53867	0 403 436 109	PUMP,FUEL,METERING .... UOC:ZAA,ZAB,ZAC,ZAD,ZAE,ZAF,ZAG,ZAH, ZAJ,ZAK, ZAL	1
12	PAHZZ	3040012713820	15434	3913326	.LEVER,REMOTE CONTRO .... UOC:ZAA,ZAB,ZAC,ZAD,ZAE,ZAF,ZAG,ZAH, ZAJ,ZAX,ZAL	1
13	PAHZZ	5342012712340	15434	3912888	.PLUG,TIMING.................................... UOC:ZAA,ZAB,ZAC,ZAD,ZAE,ZAF,ZAG,ZAH, ZAJ,ZAK,ZAL	1
14	PAHZZ	5310012715706	15434	3912889	.WASHER,SEALING .... UOC:ZAA,ZAB,ZAC,ZAD,ZAE,ZAF,ZAG,ZAH, ZAJ,ZAK,ZAL	1
15	PAFZZ	5331012721121	15434	3910503	.O-RING PART OF KIT P/N 3802389 PART.............. OF KIT P/N 3802389... UOC:ZAA,ZAB,ZAC,ZAD,ZAE,ZAF,ZAG,ZAH,	1

(1) ITEM NO	(2) SMR CODE	(3) NSN	(4) CAGEC	(5) PART NUMBER	(6) DESCRIPTION AND USABLE ON CODES (UOC)	(7) QTY
					ZAJ,ZAK,ZAL..K	
16	PAFZZ	5310012708405	15434	3912897	WASHER,LOCK .... UOC:ZAA,ZAB,ZAC,ZAD,ZAE,ZAF,ZAG,ZAH, ZAJ,ZAK,ZAL	1
17	PAFZZ	5310012708229	15434	3912896	NUT RETAINING .... UOC:ZAA,ZAB,ZAC,ZAD,ZAE,ZAF,ZAG,ZAH, ZAJ,ZAK,ZAL	1
18	PAFZZ	5310012708343	15434	3902662	NUT,SELF-LOCKING,HE .... UOC:ZAA,ZAB,ZAC,ZAD,ZAE,ZAF,ZAG,ZAH, ZAJ,ZAK,ZAL	4
19	PAFZZ	5307012719511	15434	3907978	STUD,PLAIN... UOC:ZAA,ZAB,ZAC,ZAD,ZAE,ZAF,ZAG,ZAH, ZAJ,ZAK,ZAL	4
20	PAFZZ	5331012721122	15434	3916284	O-RING PART OF KIT P/N 3802389 PART.............. OF KIT P/N 3802389............................... UOC:ZAA,ZAB,ZAC,ZAD,ZAE,ZAF,ZAG,ZAH, ZAJ,ZAK,ZAL	1
21	PAFZZ	5305012715851	15434	3913638	SCREW,CAP,HEXAGON H .... UOC:ZAA,ZAB,ZAC,ZAD,ZAE,ZAF,ZAG,ZAH, ZAJ,ZAK,ZAL	2
22	PAFZZ	3020012809024	15434	39317900	GEAR,SPUR.. UOC:ZAA,ZAB,ZAC,ZAD,ZAE,ZAF,ZAG,ZAH, ZAJ,ZAK,ZAL	1
23	PAOZZ	5305012077243	15434	3903035	SCREW.. UOC:ZAA,ZAB,ZAC,ZAD,ZAE,ZAF,ZAG,ZAH, ZAJ,ZAK,ZAL	1
24	PAOZZ	4710012713841	15434	3909556	TUBE ASSEMBLY,METAL .... UOC:ZAA,ZAB,ZAC,ZAD,ZAE,ZAF,ZAG,ZAH, ZAJ,ZAK,ZAL	2
25	PAOZZ	4730012722827	15434	3909552	ADAPTER,STRAIGHT,PI.................................. UOC:ZAA,ZAB,ZAC,ZAD, ZAE,ZAF,ZAG,ZAH, ZAJ,ZAK,ZAL	1
26	PAOZZ	4730012879100	7U263	3909557	REDUCER,PIPE.. UOC:ZAA,ZAB,ZAC,ZAD,ZAE,ZAF,ZAG,ZAH, ZAJ,ZAK,ZAL	1

END OF FIGURE

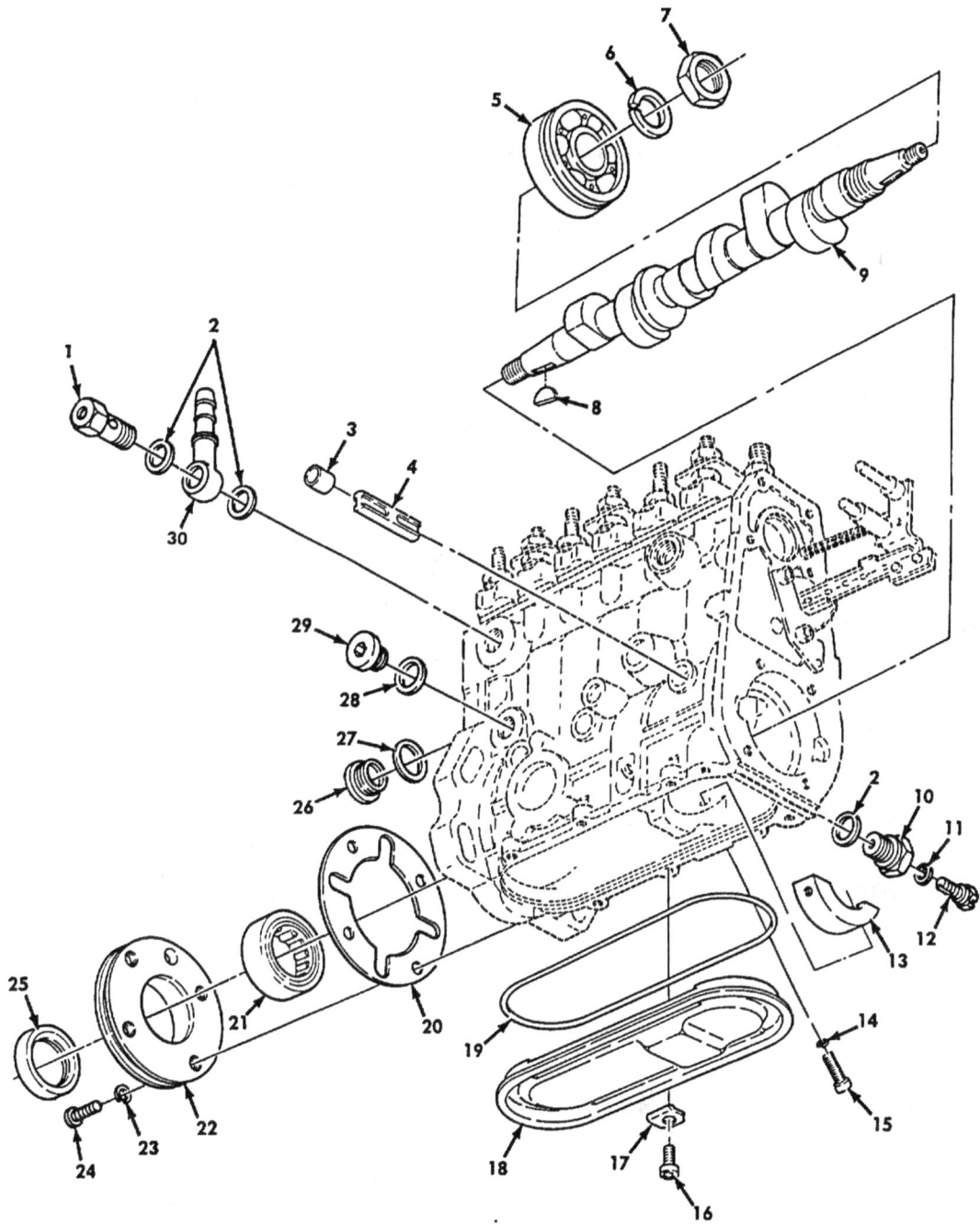

Figure 58. Fuel Injector Pump, Camshaft, and Related Parts (M939A2).

(1) ITEM NO	(2) SMR CODE	(3) NSN	(4) CAGEC	(5) PART NUMBER	(6) DESCRIPTION AND USABLE ON CODES (UOC)	(7) QTY
					GROUP 0302 FUEL PUMPS	
					FIG. 58 FUEL INJECTOR PUMP, CAMSHAFT, AND RELATED PARTS(M939A2)	
1	PFHZZ	4820013004051	53867	1 417 413 047	VALVE,CHECK .... UOC:ZAA,ZAB,ZAC,ZAD,ZAE,ZAF,ZAG,ZAH, ZAJ,ZAK,ZAL	1
2	PFHZZ	5330121564523	8046	007603014106	GASKET PART OF KIT P/N 1 417 010 008............... UOC:ZAA,ZAB,ZAC,ZAD,ZAE,ZAF,ZAG,ZAH, ZAJ,ZAK,ZAL	3
3	PFHZZ	5365013006885	538671	410 505 012	PLUG,MACHINE THREAD PART OF KIT P/N........... 1 417 010 008 .... UOC:ZAA,ZAB,ZAC,ZAD,ZAE,ZAF,ZAG,ZAH, ZAJ,ZAK,ZAL	3
4	PFHZZ	5315013013892	53867	1 413 106 002	PIN,STRAIGHT,HEADLE .... UOC:ZAA,ZAB,ZAC,ZAD,ZAE,ZAF,ZAG,ZAH, ZAJ,ZAK,ZAL	3
5	PFHZZ	3110013006796	53867	1 410 900 015	BEARING,BALL,ANNULA .... UOC:ZAA,ZAB,ZAC,ZAD,ZAE,ZAF,ZAG,ZAH, ZAJ,ZAK,ZAL	1
6	PFHZZ	5310013007037	3867	2 916 699 092	WASHER,LOCK...............22-FST PART OF KIT P/N 57K0144.. UOC:ZAA,ZAB,ZAC,ZAD,ZAE,ZAF,ZAG,ZAH, ZAJ,ZAK,ZAL	1
7	PFHZZ	5310013011798	3867	1 413 300 014	NUT,PLAIN,HEXAGON ....18 X 1.5 DIN 934- M-8... UOC:ZAA,ZAB,ZAC,ZAD,ZAE,ZAF,ZAG,ZAH, ZAJ,ZAK,ZAL	1
8	PFHZZ	5315013017912	53867	1 900 023 011	KEY,WOODRUFF.................5 X 7.5DIN68888 ST60- 2K.. UOC:ZAA,ZAB,ZAC,ZAD,ZAE,ZAF,ZAG,ZAH, ZAJ,ZAK,ZAL	1
9	XAHZZ	3040013374145	5T151	1 416 116 342	CAM,CONTROL .... UOC: ZAA,ZAB,ZAC,ZAD,ZAE,ZAF,ZAG,ZAH, ZAJ,ZAK,ZAL	1
10	PFHZZ	4730013008819	53867	1 413 457 001	BOLT,FLUID PASSAGE .... UOC:ZAA,ZAB,ZAC,ZAD,ZAE,ZAF,ZAG,ZAH, ZAJ,ZAK,ZAL	1
11	PFHZZ	5330012388316	53867	1 410 107 005	O-RING PART OF KIT P/N 57K0144 .... UOC:ZAA,ZAB,ZAC,ZAD,ZAE,ZAF,ZAG,ZAH, ZAJ,ZAK,ZAL	1
12	PFHZZ	5305012388438	53867	1413453025	SCREW,CAP,HEXAGON H .... UOC:ZAA,ZAB,ZAC,ZAD,ZAE,ZAF,ZAG,ZAH, ZAJ,ZAK,ZAL	1
13	PFHZZ	3120013007135	53867	1 415 800 014	BEARING HALF,SLEEVE .... UOC:ZAA,ZAB,ZAC,ZAD,ZAE,ZAF,ZAG,ZAH, ZAJ,ZAK,ZAL	1
14	PFHZZ	5331013006907	53867	2 410 113 004	O-RING PART OF KIT P/N 1 417 010 008............... UOC:ZAA,ZAB,ZAC,ZAD,ZAE,ZAF,ZAG,ZAH, ZAJ,ZAK,ZAL	2
15	PAHZZ	5305013017817	53867	2 914 552 158	SCREW,CAP,HEXAGON H	2

(1) ITEM NO	(2) SMR CODE	(3) NSN	(4) CAGEC	(5) PART NUMBER	(6) DESCRIPTION AND USABLE ON CODES (UOC)	(7) QTY
					UOC:ZAA,ZAB,ZAC,ZAD,ZAE,ZAF,ZAG,ZAH, ...ZAJ,ZAK,ZAL	
16	PFHZZ	5305013010533	53867	2 912 732 191	SCREW,CAP,HEXAGON H	6
					UOC:ZAA,ZAB,ZAC,ZAD,ZAE,ZAF,ZAG,ZAH, ZAJ,ZAK,ZAL	
17	PFHZZ	5340013007295	53867	1413335001	CLAMP,RIM CLENCHING	6
					UOC:ZAA,ZAB,ZAC,ZAD,ZAE,ZAF,ZAG,ZAH, ZAJ,ZAK,ZAL	
18	PFHZZ	5340013007203	53867	1 415 616 005	COVER,ACCESS	1
					UOC:ZAA,ZAB,ZAC,ZAD,ZAE,ZAF,ZAG,ZAH, ZAJ,ZAK,ZAL	
19	PFHZZ	5330013025738	53867	1 410 210 030	GASKET PART OF KIT P/N 1 417 010 008..............	1
					UOC:ZAA,ZAB,ZAC,ZAD,ZAE,ZAF,ZAG,ZAH, ZAJ,ZAK,ZAL	
20	PAHZZ	5330013015995	53867	1-410-137-021	GASKET PART OF KIT P/N 1 417 010 008..............	1
					UOC:ZAA,ZAB,ZAC,ZAD,ZAE,ZAF,ZAG,ZAH, ZAJ,ZAK,ZAL	
21	PFHZZ	3110013006804	53867	1 410 910 005	BEARING,ROLLER,NEED	1
					UOC:ZAA,ZAB,ZAC,ZAD,ZAE,ZAF,ZAG,ZAH, ZAJ,ZAK,ZAL	
22	PFHZZ	3110013007136	53867	1 415 511 047	PLATE,RETAINING,BEA................................	1
					UOC:ZAA,ZAB,ZAC,ZAD,ZAE,ZAF,ZAG,ZAH, ZAJ,ZAK,ZAL	
23	PAHZZ	5310011671964	56161	10505055	WASHER,LOCK	4
					UOC:ZAA,ZAB,ZAC,ZAD,ZAE,ZAF,ZAG,ZAH, ZAJ,ZAK,ZAL	
24	PAHZZ	5305013017818	53867	2 912 742 196	SCREW,CAP,HEXAGON H	4
					UOC:ZAA,ZAB,ZAC,ZAD,ZAE,ZAF,ZAG,ZAH, ZAJ,ZAK,ZAL	
25	PFHZZ	5330013006898	53867	1 410 283 005	SEAL,PLAIN PART OF KIT P/N 1 417 010.............. 008	1
					UOC:ZAA,ZAB,ZAC,ZAD,ZAE,ZAF,ZAG,ZAH, ZAJ,ZAK,ZAL	
26	PFHZZ	5365013365187	5T151	1 423 463 007	PLUG,MACHINE THREAD 3.00MM	1
					UOC:ZAA,ZAB,ZAC,ZAD,ZAE,ZAF,ZAG,ZAH, ZAJ,ZAK,ZAL	
27	PAHZZ	5330013011763	53867	2 916 710 619	SEAL RING DIM 7603-A26 X 31-E-CU	1
					UOC:ZAA,ZAB,ZAC,ZAD,ZAE,ZAF,ZAG,ZAH, ZAJ,ZAK,ZAL	
28	PFHZZ	5325012768488	53867	2916710613	RING,RETAINING PART OF KIT P/N 1 417.............. 010 008................................	1
					UOC:ZAA,ZAB,ZAC,ZAD,ZAE,ZAF,ZAG,ZAH, ZAJ,ZAK,ZAL	
29	PFHZZ	5365123050652	D8015	1413462898	PLUG,MACHINE THREAD	1
					UOC:ZAA,ZAB,ZAC,ZAD,ZAE,ZAF,ZAG,ZAH, ZAJ,ZAK,ZAL	
30	PFHZZ	4730013492436	8441	34982-14-6	CONNECTOR,MULTIPLE, ...	1
					UOC:ZAA,ZAB,ZAC,ZAD,ZAE,ZAF,ZAG,ZAH, ZAJ, ZAK,ZAL	

END OF FIGURE

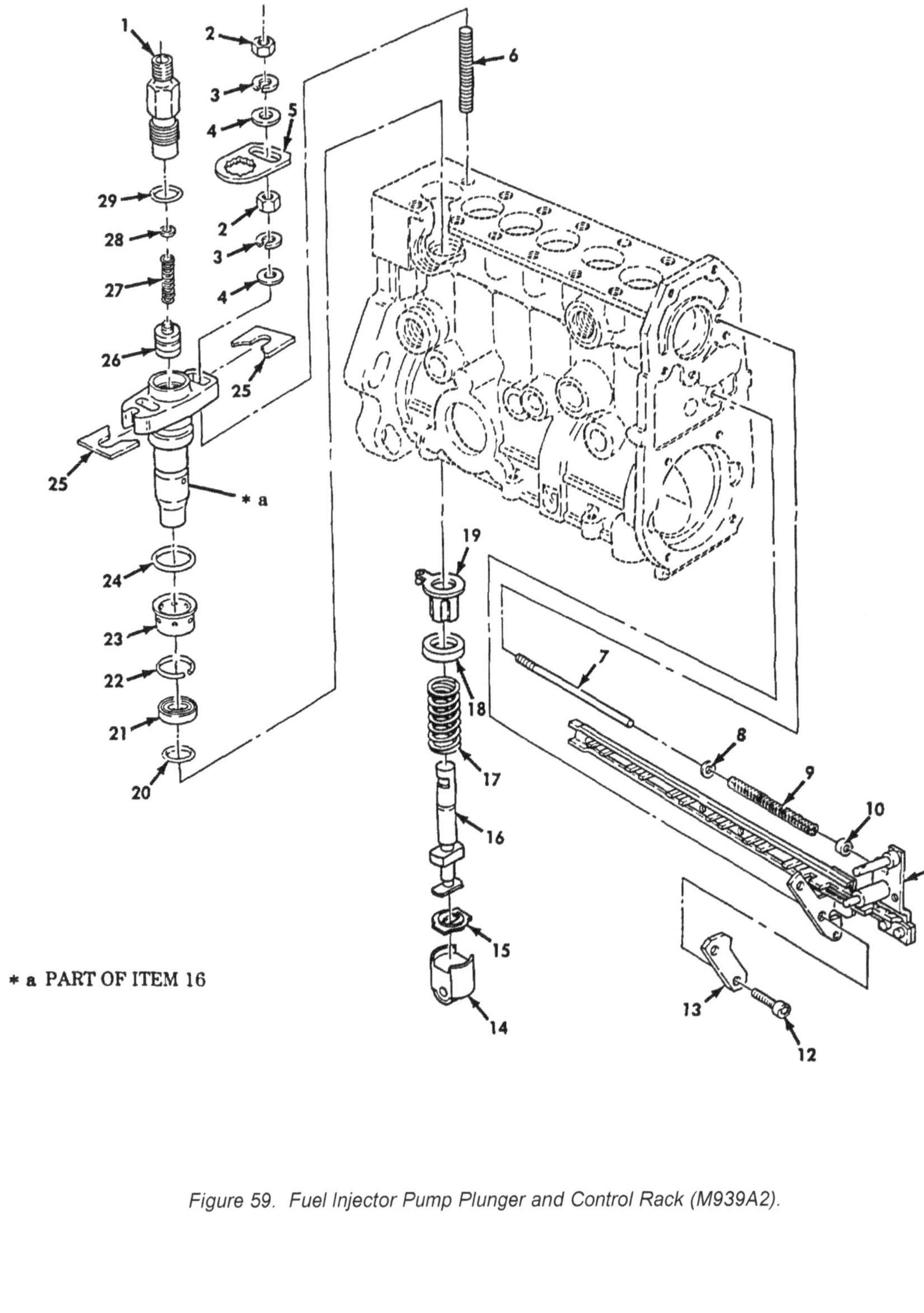

Figure 59. Fuel Injector Pump Plunger and Control Rack (M939A2).

* a PART OF ITEM 16

(1) ITEM NO	(2) SMR CODE	(3) NSN	(4) CAGEC	(5) PART NUMBER	(6) DESCRIPTION AND USABLE ON CODES (UOC)	(7) QTY
					GROUP 0302 FUEL PUMPS	
					FIG. 59 FUEL INJECTOR PUMP PLUNGER AND CONTROL RACK(M939A2)	
1	PFHZZ	4820013004257	53867	1 413 356 040	RETAINER,DISK,VALVE .... UOC: ZAA, ZAB, ZAC, ZAD,ZAE, ZAF,ZAG, ZAH, ZAJ,ZAK,ZAL	6
2	PFHZZ	5310013368721	5T151	1 413 300 023	NUT,PLAIN,HEXAGON PART OF KIT P/N.............. 57K0144.. UOC:ZAA,ZAB,ZAC,ZAD,ZAE,ZAF,ZAG,ZAH, ZAJ,ZAK,ZAL	12
3	PFHZZ	5310013368865	5T151	1 410 151 002	WASHER,LOCK PART OF KIT P/N 57K0144........... UOC:ZAA,ZAB,ZAC,ZAD,ZAE,ZAF,ZAG,ZAH, ZAJ, ZAK, ZAL	12
4	PFHZZ	5365013031612	3867	1 410 200 019	SPACER,RING PART OF KIT P/N 57K0144 .... UOC:ZAA,ZAB,ZAC,ZAD,ZAE,ZAF,ZAG,ZAH, ZAJ,ZAK,ZAL	12
5	PFHZZ	5310013011802	53867	1 410 149 001	WASHER,KEY.. UOC:ZAA,ZAB,ZAC,ZAD,ZAE,ZAF,ZAG,ZAH, ZAJ,ZAK,ZAL	6
6	PFHZZ	5307013017815	53867	1 413 500 006	STUD,PLAIN .... UOC:ZAA,ZAB,ZAC,ZAD,ZAE,ZAF,ZAG,ZAH, ZAJ,ZAK,ZAL	6
7	PFHZZ	5315013359941	53867	1 413 105 008	PIN,GROOVED,HEADLES .... UOC:ZAA,ZAB,ZAC,ZAD,ZAE,ZAF,ZAG,ZAH, ZAJ,ZAK,ZAL	1
8	PFHZZ	5310013366748	5T151	2 916 020 010	WASHER,FLAT... UOC:ZAA,ZAB,ZAC,ZAD,ZAE,ZAF,ZAG,ZAH, ZAJ,ZAK,ZAL	1
9	PFHZZ	5360013359947	5T151	1 424 610 053	SPRING,HELICAL,COMP .... UOC:ZAA,ZAB,ZAC,ZAD,ZAE,ZAF,ZAG,ZAH, ZAJ,ZAK,ZAL	1
10	PFHZZ	5340013390839	5T151	1 420 505 062	SEAT,HELICAL COMPRE .... UOC:ZAA,ZAB,ZAC,ZAD,ZAE,ZAF,ZAG,ZAH, ZAJ,ZAK,ZAL	1
11	PFHZZ	5340013363889	5T151	1 416 016 013	BRACKET,LEVER .... UOC:ZAA,ZAB,ZAC,ZAD,ZAE,ZAF,ZAG,ZAH, ZAJ,ZAK,ZAL	1
12	PAHZZ	5305013017817	53867	2 914 552 158	SCREW,CAP,HEXAGON H .... UOC:ZAA,ZAB,ZAC,ZAD,ZAE,ZAF,ZAG,ZAH, ZAJ,ZAK,ZAL	2
13	XBHZZ		5T1511	411 032 004	STOP PLATE .... UOC:ZAA,ZAB,ZAC,ZAD,ZAE,ZAF,ZAG,ZAH, ZAJ,ZAK,ZAL	1
14	PFHZZ	2910013019936	53867	1418-710-019	TAPPET,ROLLER,FUEL .... UOC:ZAA,ZAB,ZAC,ZAD,ZAE, ZAF, ZAG,ZAH, ZAJ,ZAK,ZAL	6
15	PFHZZ	5340013007154	53867	1 410 520 007	SEAT,HELICAL COMPRE .... UOC:ZAA,ZAB,ZAC,ZAD,ZAE,ZAF,ZAG,ZAH, ZAJ,ZAK,ZAL	6
16	PFHZZ	910013382335	5T151	1 418 415 082	PLUNGER ASSEMBLY,FU	6

(1) ITEM NO	(2) SMR CODE	(3) NSN	(4) CAGEC	(5) PART NUMBER	(6) DESCRIPTION AND USABLE ON CODES (UOC)	(7) QTY
					UOC:ZAA,ZAB,ZAC,ZAD,ZAE,ZAF,ZAG,ZAH, ZAJ,ZAK,ZAL	
17	PFHZZ	5360013006888	53867	1 414 618 030	SPRING,HELICAL,COMP	6
					UOC:ZAA,ZAB,ZAC,ZAD,ZAE,ZAF,ZAG,ZAH, ZAJ,ZAK,ZAL	
18	PFHZZ	5340013007153	53867	1 410 505 015	SEAT,HELICAL COMPRE	6
					UOC:ZAA,ZAB,ZAC,ZAD,ZAE,ZAF,ZAG,ZAH, ZAJ,ZAK,ZAL	
19	PFHZZ	2910013004271	53867	1 410 422 031	SLEEVE,GOVERNOR,FUE	6
					UOC:ZAA,ZAB,ZAC,ZAD,ZAE,ZAF,ZAG,ZAH, ZAJ,ZAK,ZAL	
20	PFHZZ	5331013015992	53867	1410210503	O-RING 16 X 3MM PART OF KIT P/N 1 .... 417 010 008..................	6
					UOC:ZAA,ZAB,ZAC,ZAD,ZAE,ZAF,ZAG,ZAH, ZAJ,ZAK,ZAL	
21	PFHZZ	5365013007149	53867	1 410 290 005	SPACER,RING PART OF KIT P/N 1 417 .... 010 008..................	6
					UOC:ZAA,ZAB,ZAC,ZAD,ZAE,ZAF,ZAG,ZAH,- ZAJ,ZAK,ZAL	
22	KFHZ		5T151	1 414 601 004	RETAINER RING PART OF KIT P/N 1 417 .... 010 008..................	6
					UOC:ZAA,ZAB,ZAC,ZAD,ZAE,ZAF,ZAG,ZAH, ZAJ,ZAK,ZAL	
23	PFHZZ	5365013007021	3867	1 410 505 023	SPACER,SLEEVE	6
					UOC:ZAA,ZAB,ZAC,ZAD,ZAE,ZAF,ZAG,ZAH, ZAJ,ZAK,ZAL	
24	PFHZZ	5331013017867	53867	1 410 210 501	O-RING 19 X 2MM PART OF. KIT P/N 1................. 417 010 008..................	6
					UOC:ZAA,ZAB,ZAC,ZAD,ZAE,ZAF,ZAG,ZAH, ZAJ,ZAK,ZAL K)	
25	PFHZZ	5365013007158	53867	1 411 030 134	SPACER,PLATE 1.00 MM THICK V	
					UOC:ZAA,ZAB,ZAC,ZAD,ZAE,ZAF,ZAG,ZAH, ZAJ,ZAK,ZAL	
25	PFHZZ	5365013007159	53867	1 411 030 135	SPACER,PLATE 1.05MM THICK V	
					UOC:ZAA,ZAB,ZAC,ZAD,ZAE,ZAF,ZAG,ZAH, ZAJ,ZAK,ZAL	
25	PFHZZ	5365013007160	53867	1 411 030 136	SPACER,PLATE 1.10MM THICK V	
					UOC:ZAA,ZAB,ZAC,ZAD,ZAE,ZAF,ZAG,ZAH, ZAJ,ZAK,ZAL	
25	PFHZZ	5365013010554	53867	1 411 030 137	SPACER,PLATE 1.15MM THICK V	
					UOC:ZAA,ZAB,ZAC,ZAD,ZAE,ZAF,ZAG,ZAH, ZAJ,ZAK,ZAL	
25	PFHZZ	5365013007161	53867	1 411 030 138	SPACER,PLATE 1.20MM THICK V	
					UOC:ZAA,ZAB,ZAC,ZAD,ZAE,ZAF,ZAG,ZAH, ZAJ,ZAK,ZAL	
25	PFHZZ	5365013007162	53867	1 411 030 139	SPACER,PLATE 1.25MM THICK V	
					UOC:ZAA,ZAB,ZAC,ZAD,ZAE,ZAF,ZAG,ZAH, ZAJ,ZAK,ZAL	
25	PFHZZ	5365013007163	53867	1 411 030 140	SPACER,PLATE 1.30MM THICK V	
					UOC:ZAA,ZAB,ZAC,ZAD,ZAE,ZAF,ZAG,ZAH, ZAJ,ZAK,ZAL	
25	PFHZZ	5365013007164	53867	1 411 030 141	SPACER,PLATE 1.35MM THICK V	

(1) ITEM NO	(2) SMR CODE	(3) NSN	(4) CAGEC	(5) PART NUMBER	(6) DESCRIPTION AND USABLE ON CODES (UOC)	(7) QTY
					UOC:ZAA,ZAB,ZAC,ZAD,ZAE,ZAF,ZAG,ZAH, ZAJ,ZAK,ZAL	
25	PFHZZ	5365013007165	53867	1 411 030 142	SPACER,PLATE 1.40MM THICK ....V UOC:ZAA,ZAB,ZAC,ZAD,ZAE,ZAF,ZAG,ZAH, ZAJ,ZAK,ZAL	
25	PFHZZ	5365013007166	53867	1 411 030 143	SPACER,PLATE 1.45MM THICK ....V UOC:ZAA,ZAB,ZAC,ZAD,ZAE,ZAF,ZAG,ZAH, ZAJ, ZAK, ZAL	
25	PFHZZ	5365013007167	53867	1 411 030 144	SPACER,PLATE 1.50MM THICK ....V UOC:ZAA,ZAB,ZAC,ZAD,ZAE,ZAF,ZAG,ZAH, ZAJ,ZAK, ZAL	
25	PFHZZ	5365013007168	53867	1 411 030 145	SPACER,PLATE 1.55MM THICK ....V UOC:ZAA,ZAB,ZAC,ZAD,ZAE,ZAF,ZAG,ZAH, ZAJ, ZAK,ZAL	
25	PFHZZ	5365013007169	53867	1 411 030 146	SPACER,PLATE 1.60MM THICK ....V UOC:ZAA,ZAB,ZAC,ZAD,ZAE,ZAF,ZAG,ZAH, ZAJ,ZAK, ZAL	
25	PFHZZ	5365013030937	58367	1 411 030 147	SPACER,PLATE 1.65MM THICK ....V UOC:ZAA,ZAB,ZAC,ZAD,ZAE,ZAF,ZAG,ZAH, ZAJ,ZAK,ZAL	
25	PFHZZ	5365013049530	53867	1 411 030 148	SPACER,PLATE 1.70MM THICK ....V UOC:ZAA,ZAB,ZAC,ZAD,ZAE,ZAF,ZAG,ZAH, ZAJ, ZAK, ZAL	
25	PFHZZ	5365013029953	53867	1 411 030 149	SPACER,PLATE 1.75MM THICK ....V UOC:ZAA,ZAB,ZAC,ZAD,ZAE,ZAF,ZAG,ZAH, ZAJ,ZAK,ZAL	
25	PFHZZ	5365013030938	53867	1 411 030 150	SPACER,PLATE 1.80MM THICK ....V UOC:ZAA,ZAB,ZAC,ZAD,ZAE,ZAF,ZAG,ZAH, ZAJ,ZAK,ZAL	
25	PFHZZ	5365013041802	53867	1 411 030 151	SPACER,PLATE 1.85MM THICK ....V UOC:ZAA,ZAB,ZAC,ZAD,ZAE,ZAF,ZAG,ZAH, ZAJ,ZAK,ZAL	
25	PFHZZ	5365013007170	53867	1 411 030 152	SPACER,PLATE 1.90MM THICK ....V UOC:ZAA,ZAB,ZAC,ZAD,ZAE,ZAF,ZAG,ZAH, ZAJ,ZAK,ZAL	
25	PFHZZ	5365013007171	53867	1 411 030 153	SPACER,PLATE 1.95MM THICK ....V UOC:ZAA,ZAB,ZAC,ZAD,ZAE,ZAF,ZAG,ZAH, ZAJ,ZAK,ZAL	
25	PFHZZ	5365013007172	53867	1 411 030 154	SPACER,PLATE 2.00MM THICK ....V UOC:ZAA,ZAB,ZAC,ZAD,ZAE,ZAF,ZAG,ZAH, ZAJ,ZAK,ZAL	
25	PFHZZ	5365013007173	53867	1 411 030 155	SPACER,PLATE 2.05MM THICK ....V UOC:ZAA,ZAB,ZAC,ZAD,ZAE,ZAF,ZAG,ZAH, ZAJ,ZAK,ZAL	
25	PFHZZ	5365013007174	53867	1 411 030 156	SPACER,PLATE 2.10MM THICK ....V UOC:ZAA,ZAB,ZAC,ZAD,ZAE,ZAF,ZAG,ZAH, ZAJ,ZAK,ZAL	
25	PFHZZ	5365013007175	53867	1 411 030 157	SPACER,PLATE 2.15MM THICK ....V UOC:ZAA,ZAB,ZAC,ZAD,ZAE,ZAF,ZAG,ZAH, ZAJ,ZAK,ZAL	
25	PFHZZ	5365013007176	53867	1 411 030 158	SPACER,PLATE 2.20MM THICK ....V UOC:ZAA,ZAB,ZAC,ZAD,ZAE,ZAP,ZAG,ZAH,	

(1) ITEM NO	(2) SMR CODE	(3) NSN	(4) CAGEC	(5) PART NUMBER	(6) DESCRIPTION AND USABLE ON CODES (UOC)	(7) QTY
25	PFHZZ	5365013007177	53867	1 411 030 159	SPACER,PLATE 2.25MM.THICK ....V UOC:ZAA,ZAB,ZAC,ZAD,ZAE,ZAF,ZAG,ZAH, ZAJ,ZAK,ZAL	
25	PFHZZ	5365013007178	53867	1 411 030 160	SPACER,PLATE 2.30MM THICK ....V UOC:ZAA,ZAB,ZAC,ZAD,ZAE,ZAF,ZAG,ZAH, ZAJ,ZAK,ZAL	
25	PFHZZ	5365013007179	53867	1 411 030 161	SPACER,PLATE 2.35MM THICK ....V UOC:ZAA,ZAB,ZAC,ZAD,ZAE,ZAF,ZAG,ZAH, ZAJ,ZAK,ZAL	
25	PFHZZ	5365013007180	53867	1 411 030 162	SPACER,PLATE 2.40MM THICK ....V UOC:ZAA,ZAB,ZAC,ZAD,ZAE,ZAF,ZAG,ZAH, ZAJ,ZAK,ZAL	
25	PFHZZ	5365013007181	53867	1 411 030 163	SPACER,PLATE 2.45MM THICK ....V UOC:ZAA,ZAB,ZAC,ZAD,ZAE,ZAF,ZAG,ZAH, ZAJ,ZAK,ZAL	
25	PFHZZ	5365013007182	53867	1 411 030 164	SPACER,PLATE 2.50MM THICK ....V UOC:ZAA,ZAB,ZAC,ZAD,ZAE,ZAF,ZAG,ZAH, ZAJ,ZAK,ZAL	
25	PFHZZ	5365013025848	53867	1 411 030 165	SPACER,PLATE 2.55MM THICK.........................V UOC:ZAA,ZAB,ZAC,ZAD,ZAD ,ZAF,ZAG,ZAH, ZAJ,ZAK,ZAL	
25	PFHZZ	5365013007183	53867	1 411 030 166	SPACER,PLATE 2.60MM THICK ....V UOC:ZAA,ZAB,ZAC,ZAD,ZAE,ZAF,ZAG,ZAH, ZAJ,ZAK,ZAL	
25	PFHZZ	5365013007184	53867	1 411 030 167	SPACER,PLATE 2.65MM THICK ....V UOC: ZAA, ZAB, ZAC, ZAD, ZAE, ZAF, ZAG, ZAH, ZAJ,ZAK,ZAL	
25	PFHZZ	5365013007185	53867	1 411 030 168	SPACER,PLATE 2.70MM THICK ....V UOC:ZAA,ZAB,ZAC,ZAD,ZAE,ZAF,ZAG,ZAH, ZAJ,ZAK,ZAL	
25	PFHZZ	5365013007186	53867	1 411 030 169	SPACER,PLATE 2.75MM THICK ....V UOC:ZAA,ZAB,ZAC,ZAD,ZAE,ZAF,ZAG,ZAH, ZAJ,ZAK,ZAL	
25	PFHZZ	5365013007187	53867	1 411 030 170	SPACER,PLATE 2.80MM THICK ....V UOC:ZAA,ZAB,ZAC,ZAD,ZAE,ZAF,ZAG,ZAH, ZAJ,ZAK,ZAL	
25	PFHZZ	5365013014006	53867	1 411 030 171	SPACER,PLATE 2.85MM THICK ....V UOC:ZAA,ZAB,ZAC,ZAD,ZAE,ZAF,ZAG,ZAH, ZAJ,ZAK,ZAL	
25	PFHZZ	5365013014007	53867	1 411 030 173	SPACER,PLATE 2.95MM THICK ....V UOC:ZAA,ZAB,ZAC,ZAD,ZAE,ZAF,ZAG,ZAH, ZAJ,ZAK,ZAL	
25	PFHZZ	5365013007188	53867	1 411 030 172	SPACER,PLATE 3.00MM THICK ....V UOC:ZAA,ZAB,ZAC,ZAD,ZAE,ZAF,ZAG,ZAH, ZAJ,ZAK,ZAL	
26	PFHZZ	2910013398598	5T151	1 418 512 225	VALVE,FUEL SYSTEM .... UOC:ZAA,ZAB,ZAC,ZAD,ZAE,ZAF,ZAG,ZAH, ZAJ,ZAK,ZAL	6
27	PFHZZ	5360013006889	53867	1 414 613 002	SPRING,HELICAL,COMP PART OF KIT P/N............ 57K0144 .... UOC:ZAA,ZAB,ZAC,ZAD,ZAE,ZAF,ZAG,ZAH,	6

(1) ITEM NO	(2) SMR CODE	(3) NSN	(4) CAGEC	(5) PART NUMBER	(6) DESCRIPTION AND USABLE ON CODES (UOC)	(7) QTY
					ZAJ,ZAK,ZAL	
28	PFHZZ	5310013017807	53867	1 410 100 002	WASHER,FLAT PART OF KIT P/N 57K0144............ UOC:ZAA,ZAB,ZAC,ZAD,ZAE,ZAF,ZAG,ZAH, ZAJ, ZAK, ZAL	6
29	PFHZZ	5331013031635	53867	1 410 210 041	O-RING 13 X 2.5MM PART OF KIT P/N 1................. 417 010 008.. UOC:ZAA,ZAB,ZAC,ZAD,ZAE,ZAF,ZAG,ZAH, ZAJ,ZAK,ZAL	6

END OF FIGURE

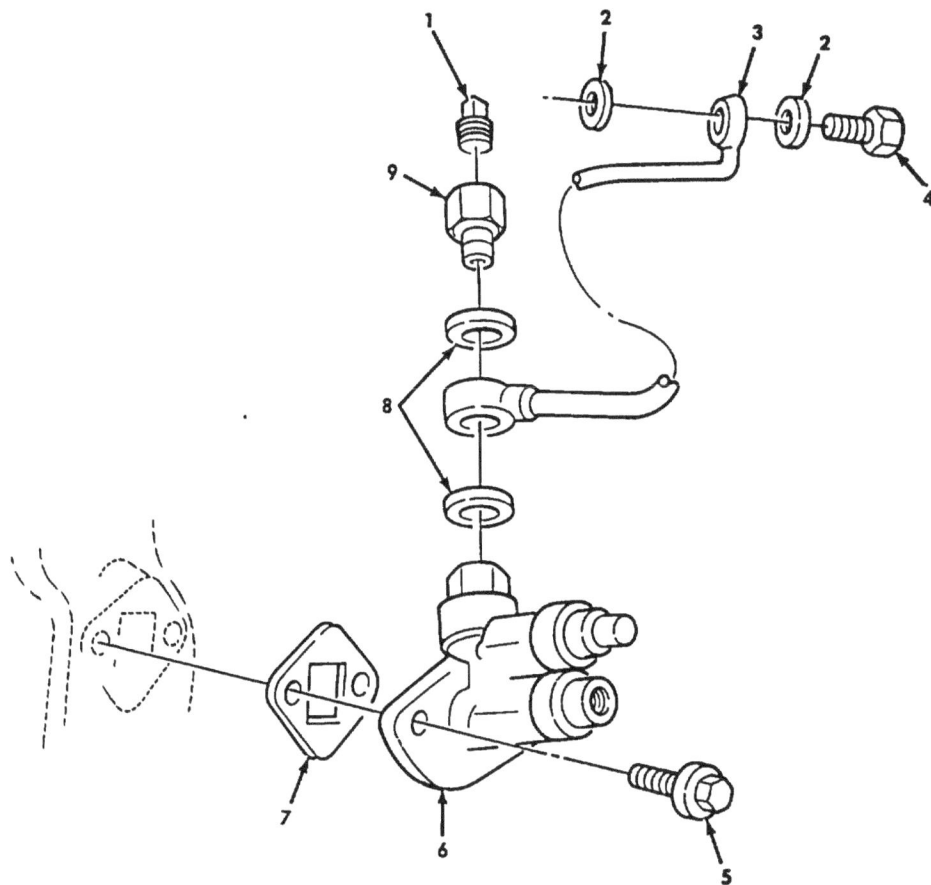

Figure 60. Fuel Transfer Pump (M939A2).

(1) ITEM NO	(2) SMR CODE	(3) NSN	(4) CAGEC	(5) PART NUMBER	(6) DESCRIPTION AND USABLE ON CODES (UOC)	(7) QTY
					GROUP 0302 FUEL PUMPS	
					FIG. 60 FUEL TRANSFER PUMP (M939A2)	
1	PAOZZ	4730000112578	96906	MS14314-2X	PLUG,PIPE.. UOC:ZAA,ZAB,ZAC,ZAD,ZAE,ZAF,ZAG,ZAH, ZAJ, ZAK, ZAL	1
2	PAOZZ	5310013408469	15434	3918191	WASHER,SEAL.. UOC:ZAA,ZAB,ZAC,ZAD,ZAE,ZAF,ZAG,ZAH, ZAJ,ZAK,ZAL	2
3	PAOZZ	4710012717199	15434	3914036	TUBE ASSEMBLY,METAL .... UOC:ZAA,ZAB,ZAC,ZAD,ZAE,ZAF,ZAG,ZAH, ZAJ, ZAK, ZAL	1
4	PAOZZ	5305012343714	15434	3905860	SCREW.. UOC:ZAA,ZAB,ZAC,ZAD,ZAE,ZAF,ZAG,ZAH, ZAJ,ZAK,ZAL	1
5	PAOZZ	5305012343755	15434	3900630	SCREW,CAP,HEXAGON H .... UOC:ZAA,ZAB,ZAC,ZAD,ZAE,ZAF,ZAG,ZAH, ZAJ,ZAK,ZAL	2
6	PAOZZ	2910012688736	15434	3917999	PUMP,FUEL TRANSFER.................................... UOC:ZAA,ZAB,ZAC,ZAD,ZAE,ZAF,ZAG,ZAH, ZAJ,ZAK,ZAL	1
7	PAOZZ	5330011909555	15434	3923054	GASKET PUMP PART OF KIT P/N 3802389............. UOC:ZAA,ZAB,ZAC,ZAD,ZAE,ZAF,ZAG,ZAH, ZAJ,ZAK,ZAL	1
8	PAOZZ	5310013319411	15434	3914896	WASHER,FLAT .... UOC:ZAA,ZAB,ZAC,ZAD,ZAE,ZAF,ZAG,ZAH, ZAJ,ZAK,ZAL	2
9	PAOZZ	4730013014286	15434	3914037	BOLT,FLUID PASSAGE .... UOC:ZAA,ZAB,ZAC,ZAD,ZAE,ZAF,ZAG,ZAH, ZAJ,ZAK,ZAL	1

END OF FIGURE

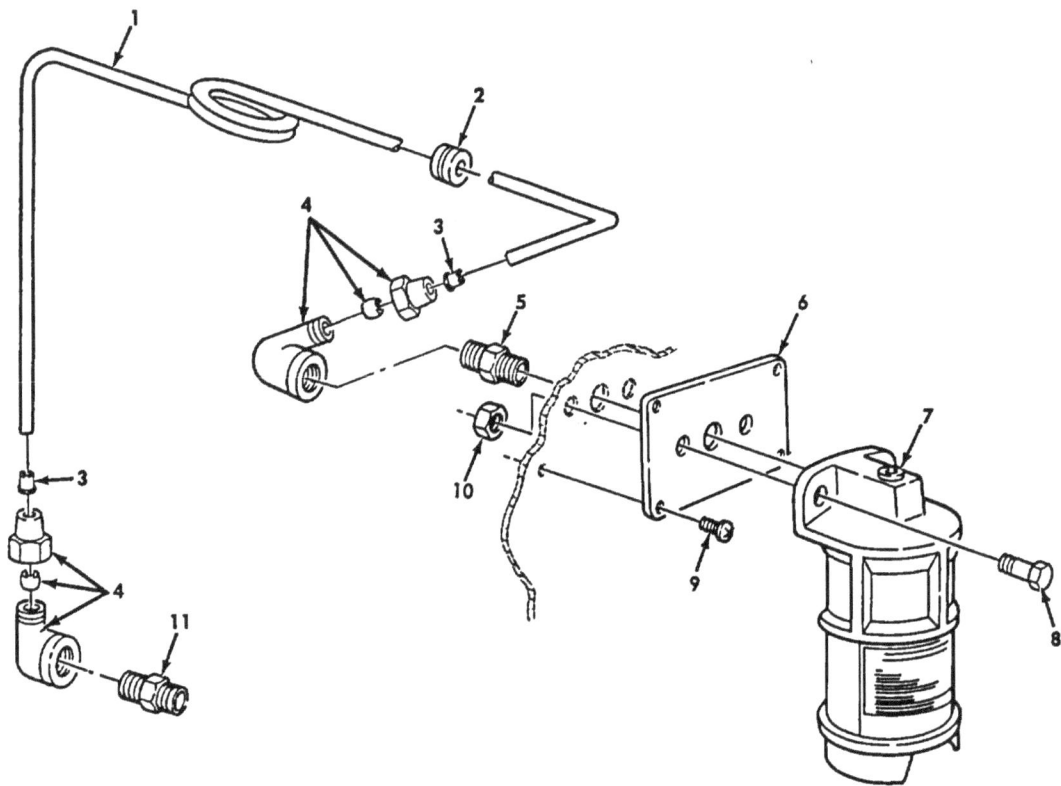

Figure 61. Air Cleaner Indicator and Related Parts.

(1) ITEM NO	(2) SMR CODE	(3) NSN	(4) CAGEC	(5) PART NUMBER	(6) DESCRIPTION AND USABLE ON CODES (UOC)	(7) QTY
					GROUP 0304 AIR CLEANER	
					FIG. 61 AIR CLEANER INDICATOR AND RELATED PARTS	
1	MOOZZ		19207	8675779-32	TUBE,METALLIC MAKE FROM TUBE, P/N 8675779, 32 INCHES LONG	1
2	PAOZZ	5325001642087	19207	11677607	GROMMET,NONMETALLIC	1
3	PAOZZ	4730001324588	79146	HO-159-4	INSERT,TUBE FITTING	2
4	PAOZZ	4730009129114	81343	4-2 120203BA	ELBOW,PIPE TO TUBE	2
5	PAOZZ	4730001961481	96906	MS51953-9	NIPPLE,PIPE	1
6	PBOZZ	4330011085109	19207	12277150	BRACKET,FILTER MOUN	1
7	PAOZZ	2940000712653	96906	MS53063-3	INDICATOR,FILTER WA	1
8	PAOZZ	5305002678953	80204	B1821BH025F063N	SCREW,CAP,HEXAGON H	2
9	PAOZZ	5305009897435	96906	MS35207-264	SCREW,MACHINE	4
10	PAOZZ	5310009359022	96906	MS51943-32	NUT,SELF-LOCKING,HE	2
11	PAOZZ	2910009115635	19207	8712058	FILTER,FLUID	1

END OF FIGURE

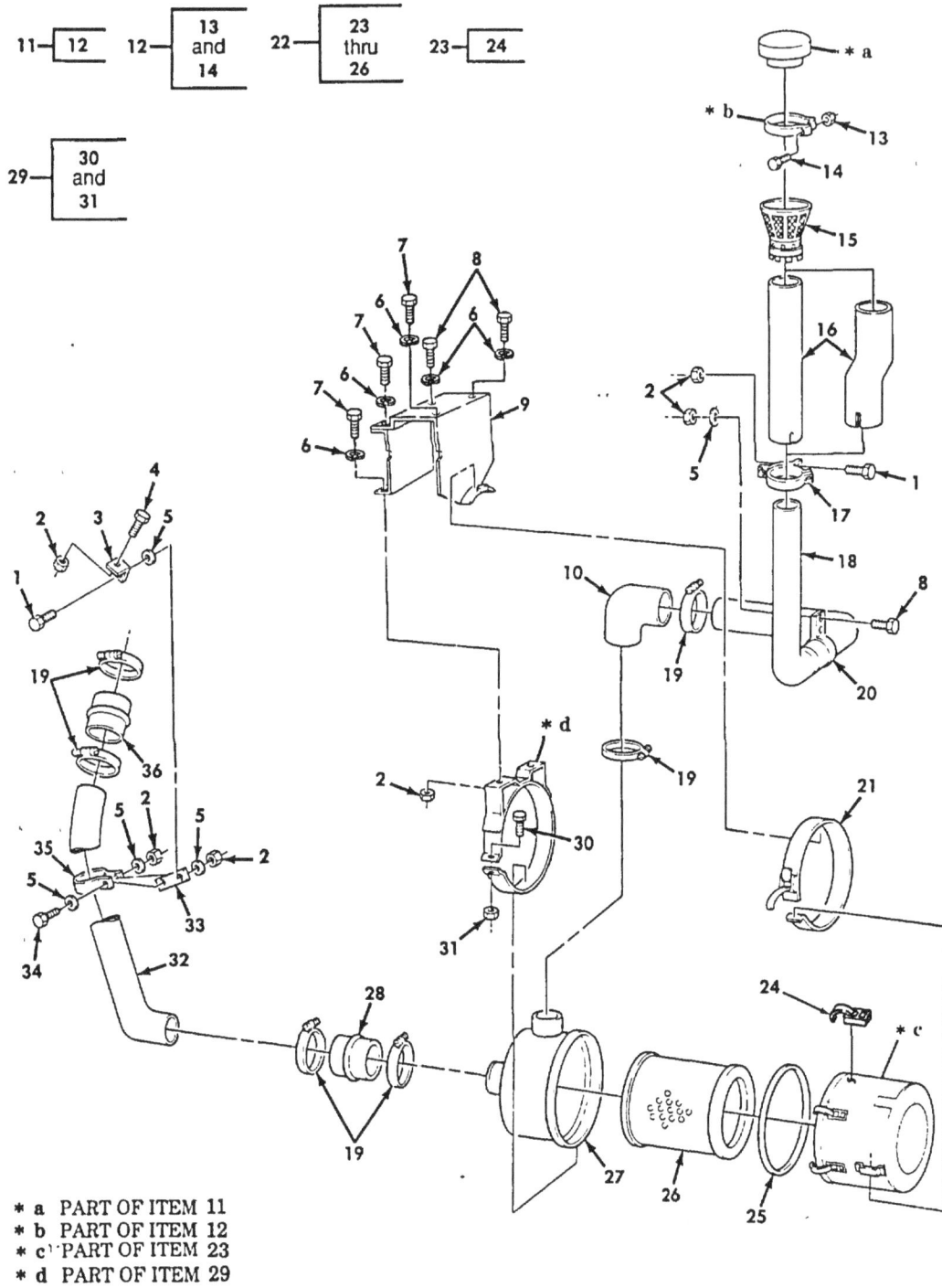

Figure 62. Air Intake System (M939, M939A1).

* a PART OF ITEM 11
* b PART OF ITEM 12
* c PART OF ITEM 23
* d PART OF ITEM 29

(1) ITEM NO	(2) SMR CODE	(3) NSN	(4) CAGEC	(5) PART NUMBER	(6) DESCRIPTION AND USABLE ON CODES (UOC)	(7) QTY
					GROUP 0304 AIR CLEANER	
					FIG. 62 AIR INTAKE SYSTEM(M939, M939A1)	
1	PAOZZ	5305002692804	96906	MS90726-61	SCREW,CAP,HEXAGON H .... UOC:DAA,DAB,DAC,DAD,DAE,DAF,DAG,DAH, DAJ,DAK,DAL,DAW,DAX,V12 ,V13 ,V14,V15, V16,V17,V18,V19,V20,V21,V22,V24,V25, V39	1
2	PAOZZ	5310009591488	96906	MS51922-21	NUT,SELF-LOCKING,HE .... UOC:DAA,DAB,DAC,DAD,DAE,DAF,DAG,DAH, DAJ,DAK,DAL,DAW,DAX,V12 ,V13,V14,V15, V16,V17,V18,V19,V20,V21,V22,V24,V25, V39	10
3	PFOZZ	5340010907627	19207	12256040	BRACKET,ANGLE . ... UOC:DAA,DAB,DAC,DAD,DAE,DAF,DAG,DAH, DAJ,DAK,DAL,DAW,DAX,V12,V13,V14,V15, V16,V17,V18,V19,V20,V21,V22,V24,V25, V39	1
4	PAOZZ	5305009125113	96906	MS51096-359	SCREW,CAP,HEXAGON H .... UOC:DAA,DAB,DAC,DAD,DAE,DAF,DAG,DAH, DAJ,DAK,DAL,DAW,DAX,V12,V13,V14,V15, V16,V17,V18,V19,V20,V21,V22,V24,V25, V39	1
5	PAOZZ	5310000877493	96906	MS27183-13	WASHER,FLAT .... UOC:DAA,DAB,DAC,DAD,DAE,DAF,DAG,DAH, DAJ,DAK,DAL,DAW,DAX,V12,V13,V14,V15, V16,V17,V18,V19,V20,V21,V22,V24,V25, V39	2
6	PAOZZ	5310006379541	96906	MS35338-46	WASHER,LOCK .... UOC:DAA,DAB,DAC,DAD,DAE,DAF,DAG,DAH, DAJ,DAK,DAL,DAW,DAX,V12,V13,V14,V15, V16,V17,V18,V19, V20,V21,V22,V24,V25, V39	5
7	PAOZZ	5305002692803	96906	MS90726-60	SCREW,CAP,HEXAGON H .... UOC:DAA,DAB,DAC,DAD,DAE,DAF,DAG,DAH, DAJ,DAK,DAL,DAW,DAX,V12,V13,V14,V15, V16,V17,V18,V19,V20,V21,V22,V24,V25, V39	11
8	PAOZZ	5305002693238	80204	B1821BH038F125N	SCREW,CAP,HEXAGON H .... UOC:DAA,DAB,DAC,DAD,DAE,DAF,DAG,DAH, DAJ,DAK,DAL,DAW,DAX,V12,V13,V14,V15, V16,V17,V18,V19,V20,V21,V22,V24,V25, V39	2
9	PFOZZ	5340010907628	19207	12256307	BRACKET,MO UNTING . ... UOC:DAA,DAB,DAC,DAD,DAE,DAF,DAG,DAH, DAJ,DAK,DAL,DAW,DAX,V12,V13,V14,V15, V16,V17,V18,V19,V20,V21,V22,V24,V25, V39	1
10	PAOZZ	4720001344655	19207	11664545	HOSE,PREFORMED .... UOC:DAA,DAB,DAC,DAD,DAE,DAF,DAG,DAH,	1

(1) ITEM NO	(2) SMR CODE	(3) NSN	(4) CAGEC	(5) PART NUMBER	(6) DESCRIPTION AND USABLE ON CODES (UOC)	(7) QTY
					DAJ,DAK,DAL,DAW,DAX,V12,V13,V14,V15, ... V16,V17,V18,V19,V20,V21,V22,V24,V25, V39	
11	PAOZZ	2940011290261	19207	11648582	CAP,AIR CLEANER INT .... UOC:DAA,DAB,DAC,DAD,DAE,DAF,DAG,DAH, DAJ,DAK,DAL,DAW,DAX,V12,V13,V14,V15, V16,V17,V18,V19,V20,V21,V22,V24,V25, V39	1
12	PAOZZ	5340011079688	19207	11663037	.CLAMP,LOOP UOC:DAG,DAH,V21,V22	1
13	PAOZZ	5310008299981	96906	MS35649-2312	..NUT,PLAIN,HEXAGON...................... UOC:DAG,DAH,V21,V22	1
14	PAOZZ	5306013706947	96906	MS16207-26	..BOLT,SQUARE NECK UOC:DAG,DAH,V21,V22	1
15	PAOZZ	2510011122170	19207	12277111	GRILLE,METAL UOC:DAG,DAH,V21,V22	1
16	PAOZZ	4710010904487	19207	12256314	TUBE,METALLIC BENT UOC:DAE,DAF,DAJ,DAK,V19,V20,V24,V25	1
16	PAOZZ	4710010904584	19207	12256315	TUBE,METALLIC STRAIGHT UOC:DAA,DAB,DAC,DAD,DAL,DAW,DAX,V12, V13,V14,V15,V16,V17,V18,V39	1
17	PAOZZ	5340010907650	19207	12255981	CLAMP,LOOP .... UOC:DAA,DAB,DAC,DAD,DAE,DAF,DAG,DAH, DAJ,DAK,DAL,DAW,DAX,V12,V13,V14,V15, V16,V17,V18,V19,V20,V21,V22,V24,V25, V39	1
18	PAOZZ	4710010904486	19207	12256096	TUBE ASSEMBLY .... UOC:DAA,DAB,DAC,DAD,DAE,DAF,DAG,DAH, DAJ,DAK,DAL,DAW,DAX,V12,V13,V14,V15, V16,V17,V18,V19,V20,V21,V22,V24,V25, V39	1
19	PAOZZ	4730009086294	75160	C16054	CLAMP,HOSE .... UOC:DAA,DAB,DAC,DAD,DAE,DAF,DAG,DAH, DAJ,DAK',DAL,DAW,DAX,V12,V13,V14,V15, V16,V17,V18,V19,V20,V21,V22,V24,V25, V39	6
20	PAOZZ	5340010907622	19207	12255862	CLAMP,LOOP .... UOC:DAA,DAB,DAC,DAD,DAE,DAF,DAG,DAH, DAJ,DAK,DAL,DAW,DAX,V12,V13,V14,V15, V16,V17,V18,V19,V20,V21,V22,V24,V25, V39	1
21	PAOZZ	5340010907629	19207	12256312	BAND,RETAINING . ... UOC:DAA,DAB,DAC,DAD,DAE,DAF,DAG,DAH, DAJ,DAK,DAL,DAW,DAX,V12,V13,V14,V15, V16,V17,V18,V19,V20,V21,V22,V24,V25, V39	1
22	PFOZZ	2940001348326	19207	11648617	AIR CLEANER,INTAKE .... UOC:DAA,DAB,DAC,DAD,DAE,DAF,DAG,DAH, DAJ,DAK,DAL,DAW,DAX,V12,V13,V14,V15, V16,V17,V18,V19,V20,V21,V22,V24,V25, V39	1
23	PFOZZ	4730010907621	19207	12256310	.CUP ASSEMBLY	1

(1) ITEM NO	(2) SMR CODE	(3) NSN	(4) CAGEC	(5) PART NUMBER	(6) DESCRIPTION AND USABLE ON CODES (UOC)	(7) QTY
					UOC:DAA,DAB,DAC,DAD,DAE,DAF,DAG,DAH, DAJ,DAK,DAL,DAW,DAX,V12,V13,V14,V15, V16,V17,V18,V19,V20,V21,V22,V24,V25, V39	
24	PAOZZ	5342008331236	19207	11602023	..LATCH,INTAKE AIR CL .... UOC:DA-,DAB,DAC,DAD,DAE,DAF,DAG,DAH, DAJ,DAK,DAL,DAW,DAX,V12,V13,V14,V15, V16,V17,V18,V19,V20,V21,V22,V24,V25, V39	1
25	PAOZZ	5330004322142	19207	11604520-5	.PACKING,PREFORMED PART OF KIT P/N........... 5702838 .... UOC:DAA,DAB,DAC,DAD,DAE,DAF,DAG,DAH, DAJ,DAK,DAL,DAW,DAX,V12,V13,V14,V15, V16,V17,V18,V19,V20,V21,V22,V24,V25, V39	1
26	PAOZZ	940001344657	19207	11604545	.FILTER ELEMENT,INTA PART OF KIT P/N............ 5702838 .... UOC:DAA,DAB,DAC,DAD,DAE,DAF,DAG,DAH, DAJ,DAK,DAL,DAW,DAX,V12,V13,V14,V15, V16,V17,V18,V19,V20,V21,V22,V24,V25, V39	1
27	PFOZZ	2940011450350	19207	11664491	AIR CLEANER,INTAKE .... UOC:DAA,DAB,DAC,DAD,DAE,DAF,DAG,DAH, DAJ,DAK,DAL,DAW,DAX,V12,V13,V14,V15, V16,V17,V18,V1B, V20,V21,V22,V24,V25, V39	1
28	PAOZZ	4720004778933	26151	236-0317	HOSE,PREFORMED .... UOC:DAA,DAB,DAC,DAD,DAE,DAF,DAG,DAH, DAJ,DAK,DAL,DAW,DAX,V12,V13,V14,V15, V16,V17,V18,V19,V20,V21,V22,V24,V25, V39	1
29	PAOZZ	2940010904480	19207	11664234-3	MOUNTING BAND .... UOC:DAA,DAB,DAC,DAD,DAE,DAF,DAG,DAH, DAJ,DAK,DAL,DAW,DAX,V12,V13,V14,V15, V16,V17,V18,V19,V20,V21,V22,V24,V25, V39	1
30	PAOZZ	5305002259092	96906	MS90726-37	.SCREW,CAP,HEXAGON H............................ UOC:DAA,DAB,DAC,DAD,DAE,DAF,DAG,DAH, DAJ,DAK,DAL,DAW,DAX,V12,V13,V14,V15, V16,V17,V18,V19,V20,V21,V22,V24,V25, V39	1
31	PAOZZ	5310009824912	6906	MS21045-5	.NUT,SELF-LOCKING,HE............................ UOC:DAA,DAB,DAC,DAD,DAE,DAF,DAG,DAH, DAJ,DAK,DAL,DAW,DAX,V12,V13,V14,V15, V16,V17,V18,V19,V20,V21,V22,V24,V25, V39	1
32	PAOZZ	4710010907620	19207	12256045	PIPE,AIR CLEANER OU............................. UOC:DAA,DAB,DAC,DAD,DAE,DAF,DAG,DAH, DAJ,DAK,DAL,DAW,DAX,V12,V13,V14,V15, V16,V17,V18,V19,V20,V21,V22,V24,V25, V39	1
33	PAOZZ	5340011122292	19207	12256042	BAND,RETAINING	1

(1) ITEM NO	(2) SMR CODE	(3) NSN	(4) CAGEC	(5) PART NUMBER	(6) DESCRIPTION AND USABLE ON CODES (UOC)	(7) QTY
					UOC: DAA, DAB, DAC, DAD, DAE, DAF, DAG, DAH, j DAJ,DAK,DAL,DAW,DAX,V12,V13,V14,V15, V16,V17,V18,V19,V20,V21,V22,V24,V25, V39	
34	PAOZZ	5305002693239	80204	B1821BH038F138N	SCREW,CA P,HEXAGON H .... UOC:DAA,DAB,DAC,DAD,DAE,DAF,DAG,DAH, DAJ,DAK,DAL,DAW,DAX,V12 ,V13,V14 ,V15, V16,V17,V18,V19,V20,V21,V22,V24,V25, V39	1
35	PAOZZ	5340010907619	19207	12256031-2	CLAMP,LOOP.. UOC:DAA,DAB,DAC,DAD,DAE,DAF,DAG,DAH, DAJ,DAK,DAL,DAW,DAX,V12 ,V13 ,V14 ,V15, V16,V17,V18 ,V19 ,V20,V21,V22,V24,V25, V39	1
36	PAOZZ	4720010907617	19207	12256068	HOSE,PREFORMED... UOC:DAA,DAB,DAC,DAD,DAE,DAF,DAG,DAH, DAJ,DAK,DAL,DAW,DAX,V12,V13,V14,V15, V16,V17,V18,V19,V20,V21,V22,V24,V25, V39	1

END OF FIGURE

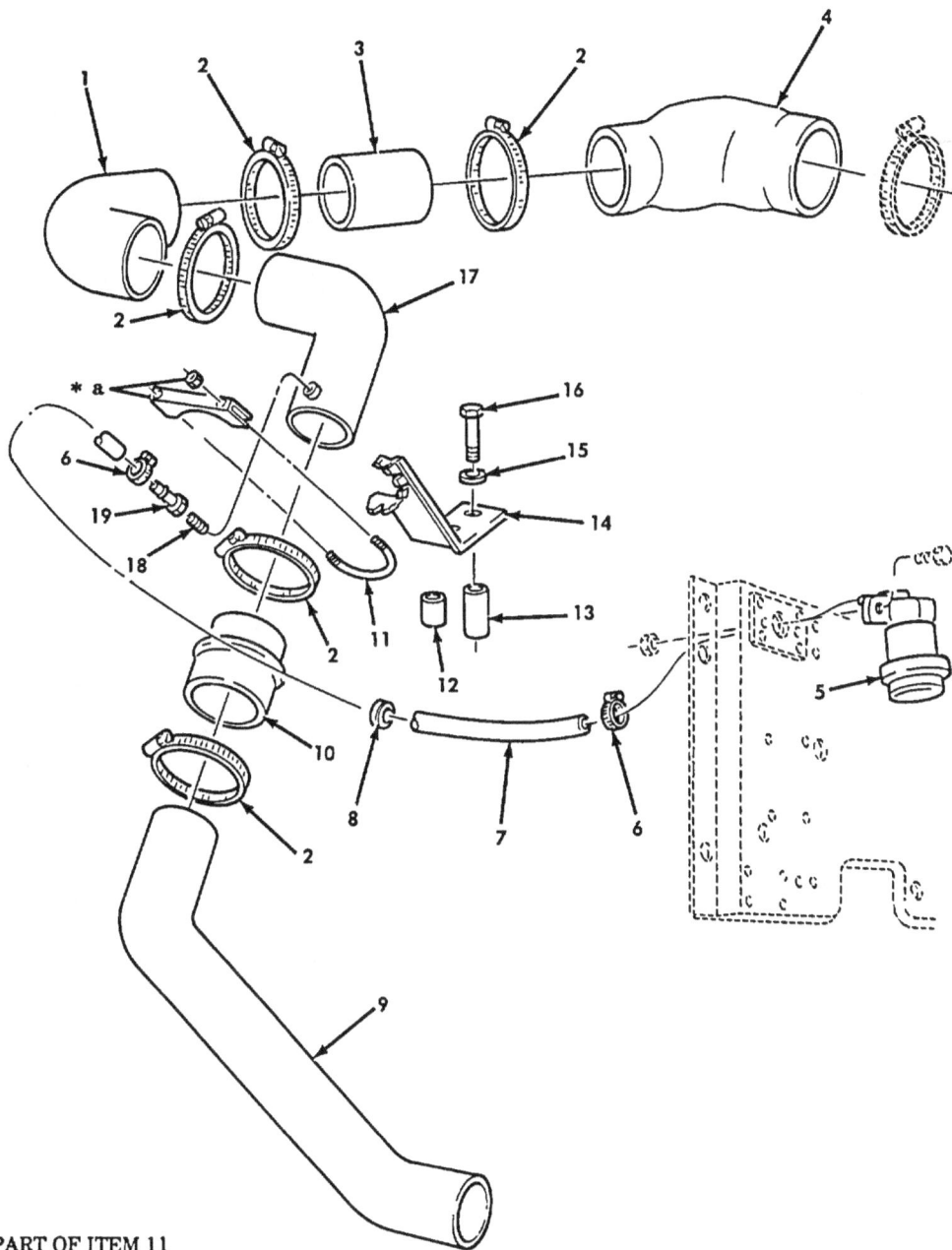

* a PART OF ITEM 11

Figure 63. Air Intake System (M939A2).

(1) ITEM NO	(2) SMR CODE	(3) NSN	(4) CAGEC	(5) PART NUMBER	(6) DESCRIPTION AND USABLE ON CODES (UOC)	(7) QTY
					GROUP 0304 AIR CLEANER	
					FIG. 63 AIR INTAKE SYSTEM(M939A2)	
1	PAOZZ	4720012793171	98441	38221-64-01RN	HOSE,PREFORMED .... UOC:ZAA,ZAB,ZAC,ZAD,ZAE,ZAF,ZAG,ZAH, ZAJ,ZAK,ZAL	1
2	PAOZZ	4730009086294	75160	C16054	CLAMP,HOSE .... UOC:ZAA,ZAB,ZAC,ZAD,ZAE,ZAF,ZAG,ZAH, ZAJ,ZAK,ZAL	5
3	PAOZZ	4710012791490	1HT81	Q46545	TUBE,METALLIC .... UOC:ZAA,ZAB,ZAC,ZAD,ZAE,ZAF,ZAG,ZAH, ZAJ,ZAK,ZAL	1
4	PAOZZ	4720012856312	15434	3037625	HOSE,PREFO RMED .... UOC:ZAA,ZAB,ZAC,ZAD,ZAE,ZAF,ZAG,ZAH, ZAJ,ZAK,ZAL	1
5	PAOZZ	5220013171436	59150	3875-425	MINDER,FILTER .... UOC:ZAA,ZAB,ZAC,ZAD,ZAE,ZAF,ZAG,ZAH, ZAJ,ZAK,ZAL	1
6	PAOZZ	4730003742045	70494	A6S	CLAMP,HOSE .... UOC:ZAA,ZAB,ZAC,ZAD,ZAE,ZAF,ZAG,ZAH, ZAJ,ZAK,ZAL	2
7	MOOZZ		19207	12277246 X 40 IN	TUBING MAKE FROM TUBING, P/N 211- .... 0109-300, 40 INCHES LONG............................. UOC:ZAA,ZAB,ZAC,ZAD,ZAE,ZAF,ZAG,ZAH, ZAJ,ZAK,ZAL	1
8	PAOZO	5325001745315	96906	MS35489-7	GROMMET,NONMETALLIC .... UOC:ZAA,ZAB,ZAC,ZAD,ZAE,ZAF,ZAG,ZAH, ZAJ,ZAK,ZAL	1
9	PAOZZ	2805012794630	1HT81	Q45405	PIPE,INTAKE... UOC:ZAA,ZAB,ZAC,ZAD,ZAE,ZAF,ZAG,ZAH, ZAJ,ZAK,ZAL	1
10	PAOZZ	4720013052440	47457	20510087	HOSE,PREFORMED .... UOC:ZAA,ZAB,ZAC,ZAD,ZAE,ZAF,ZAG,ZAH, ZAJ,ZA, ZAL	1
11	PAOZZ	5340012721471	96906	MS52150-37HE	CLAMP,LOOP .... UOC: ZAA,ZAB,ZAC,ZAD,ZAE,ZAF,ZAG,ZAH, ZAJ,ZAK,ZAL	1
12	PAOZZ	5365012809432	47457	20510492-2	SPACER,SLEEVE .... UOC:ZAA,ZAB,ZAC,ZAD,ZAE,ZAF,ZAG,ZAH, ZAJ,ZAK,ZAL	1
13	PAOZZ	5365012803668	47457	20510492-1	SPACER,SLEEVE .... UOC:ZAA,ZAB,ZAC,ZAD,ZAE,ZAF,ZAG,ZAH, ZAJ,ZAK,ZAL	1
14	PAOZZ	5340012807098	47457	20510807	BRACKET,MOUNTING.. UOC:ZAA,ZAB,ZAC,ZAD,ZAE,ZAF,ZAG,ZAH, ZAJ,ZAK,ZAL	1
15	PAOZZ	5310012866075	24617	11511516	WASHER,FLAT... UOC:ZAA,ZAB,ZAC,ZAD,ZAE,ZAF,ZAG,ZAH, ZAJ,ZAK,ZAL	2
16	PAOZZ	5305012718381	24617	11501092	SCREW,CAP,HEXAGON HHEAD M12X130.0.......... MM. LG..	2

(1) ITEM NO	(2) SMR CODE	(3) NSN	(4) CAGEC	(5) PART NUMBER	(6) DESCRIPTION AND USABLE ON CODES (UOC)	(7) QTY
					UOC:ZAA,ZAB,ZAC,ZAD,ZAE,ZAF,ZAG,ZAH, ZAJ,ZAK,ZAL	
17	PAOZZ	4710012791493	1HT81	033545	TUBE,BENT,METALLIC .... UOC:ZAA,ZAB,ZAC,ZAD,ZAE,ZAF,ZAG,ZAH, ZAJ,ZAK,ZAL	1
18	PAOZZ	2910009115635	19207	8712058	FILTER,FLUID .... 1 UOC:ZAA,ZAB,ZAC,ZAD,ZAE,ZAF,ZAG,ZAH, ZAJ, ZAK, ZAL	
19	PAOZZ	4730014432888	19207	12363593	ADAPTER,STRAIGHT,PI.................................... UOC:ZAA,ZAB,ZAC,ZAD,ZAE,ZAF,ZAG,ZAH, ZAJ, ZAK, ZAL	1

END OF FIGURE

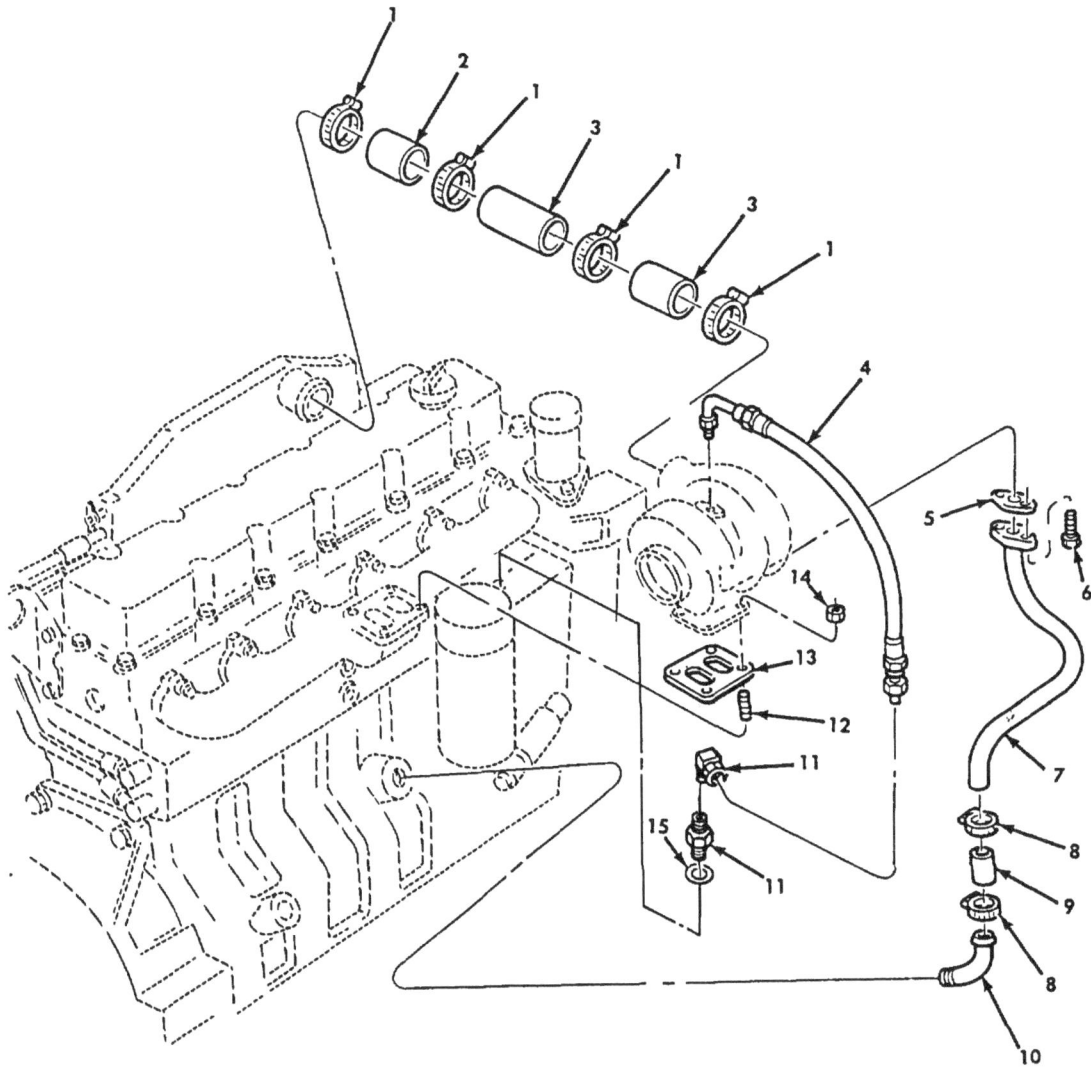

Figure 64. Turbocharger Hoses and Clamps (M939A2).

(1) ITEM NO	(2) SMR CODE	(3) NSN	(4) CAGEC	(5) PART NUMBER	(6) DESCRIPTION AND USABLE ON CODES (UOC)	(7) QTY
					GROUP 0305 TURBOCHARGER	
					FIG. 64 TURBOCHARGER HOSES AND CLAMPS (M939A2)	
1	PAOZZ	4730011100342	15434	3026396	CLAMP, HOSE .... UOC:ZAA, ZAB, ZAC, ZAD, ZAE, ZAF, ZAG, ZAH, ZAJ, ZAK, ZAL	4
2	PAOZZ	4710012717938	15434	3914943	TUBE, METALLIC .... UOC:ZAA, ZAB, ZAC, ZAD, ZAE, ZAF, ZAG, ZAH, ZAJ, ZAK, ZAL	1
3	PAOZZ	4720012716951	15434	3916165	HOSE, NONMETALLIC .... UOC:ZAA, ZAB, ZAC, ZAD, ZAE, ZAF, ZAG, ZAH, ZAJ, ZAK, ZAL	2
4	PAOZZ	4720012716945	15434	3909545	HOSE ASSEMBLY, NONME UP TO ENGINE S/...... N 44487829 .... UOC:ZAA, ZAB, ZAC, ZAD, ZAE, ZAF, ZAG, ZAH, ZAJ, ZAK, ZAL	1
4	PAOZZ	4720013318717	15434	3921530	HOSE ASSEMBLY, NONME AFTER ENGINE S/...... N 44487830 .... UOC:ZAA, ZAB, ZAC, ZAD, ZAE, ZAF, ZAG, ZAH, ZAJ, ZAK, ZAL	1
5	PAOZZ	5330011901905	15434	3914388	GASKET PART OF KIT P/N 3802622 PART........... OF KIT P/N 3802623 PART OF KIT P/N 3802452 .... UOC:ZAA, ZAB, ZAC, ZAD, ZAE, ZAF, ZAG, ZAH, ZAJ, ZAK, ZAL	1
6	PAOZZ	5305012374915	15434	3900629	SCREW, CAP, HEXAGON H .... UOC:ZAA, ZAB, ZAC, ZAD, ZAE, ZAF, ZAG, ZAH, ZAJ, ZAK, ZAL	2
7	PAOZZ	4710011937944	15434	3903744	TUBE, BENT, METALLIC .... UOC:ZAA, ZAB, ZAC, ZAD, ZAE, ZAF, ZAG, ZAH, ZAJ, ZAK, ZAL	1
8	PAOZZ	4730005558263	70403	A11	CLAMP, HOSE... UOC:ZAA, ZAB, ZAC, ZAD, ZAE, ZAF, ZAG, ZAH, ZAJ, ZAK, ZAL	2
9	PAOZZ	4720012716946	15434	3903745	HOSE, NONMETALLIC .... UOC:ZAA, ZAB, ZAC, ZAD, ZAE, ZAF, ZAG, ZAH, ZAJ, ZAK, ZAL	1
10	PAOZZ	4710012717941	15434	3917417	TUBE, BENT, METALLIC .... UOC:ZAA, ZAB, ZAC, ZAD, ZAE, ZAF, ZAG, ZAH, ZAJ, ZAK, ZAL	1
11	PAOZZ	4730002778274	15434	S-1014-1	ELBOW, PIPE TO TUBE UP TO ENGINE S/........... N 44487829, USE WITH HOSE P/N 3909545 UOC:ZAA, ZAB, ZAC, ZAD, ZAE, ZAF, ZAG, ZAH, ZAJ, ZAK, ZAL	1
11	PAOZZ	4730013330133	15434	3916857	ADAPTER, STRAIGHT, TU AFTER ENGINE S/........ N 44487830, USE WITH HOSE P/N 3918556 UOC:ZAA, ZAB, ZAC, ZAD, ZAE, ZAF, ZAG, ZAH, ZAJ, ZAK, ZAL	1
12	PAOZZ	5307012870854	7U263	3818823	STUD, PLAIN .... UOC:ZAA, ZAB, ZAC, ZAD, ZAE, ZAF, ZAG, ZAH,	4

(1) ITEM NO	(2) SMR CODE	(3) NSN	(4) CAGEC	(5) PART NUMBER	(6) DESCRIPTION AND USABLE ON CODES (UOC)	(7) QTY
					ZAJ, ZAK, ZAL	
13	PAOZZ	5330012721146	15434	3911941	GASKET PART OF KIT P/N 3802622 PART.............. OF KIT P/N 3802623 PART OF KIT P/N 3802452 PART OF KIT P/N 3802149 .... UOC:ZAA, ZAB, ZAC, ZAD, ZAE, ZAF, ZAG, ZAH, ZAJ, ZAK, ZAL	1
14	PAOZZ	5310012708244	15434	3818824	NUT, PLAIN, HEXAGON .... UOC: ZAA, ZAB, ZAC, ZAD, ZAE, ZAF, ZAG, ZAH, ZAJ, ZAK, ZAL	4
15	PAOZZ	5331013319293	15434	3037236	O-RING AFTER ENGINE S/N 44487830,................. USE WITH HOSE P/N 3918556............................ UOC: ZAA, ZAB, ZAC, ZAD, ZAE, ZAF, ZAG, ZAH, ZAJ, ZAK, ZAL	1

END OF FIGURE

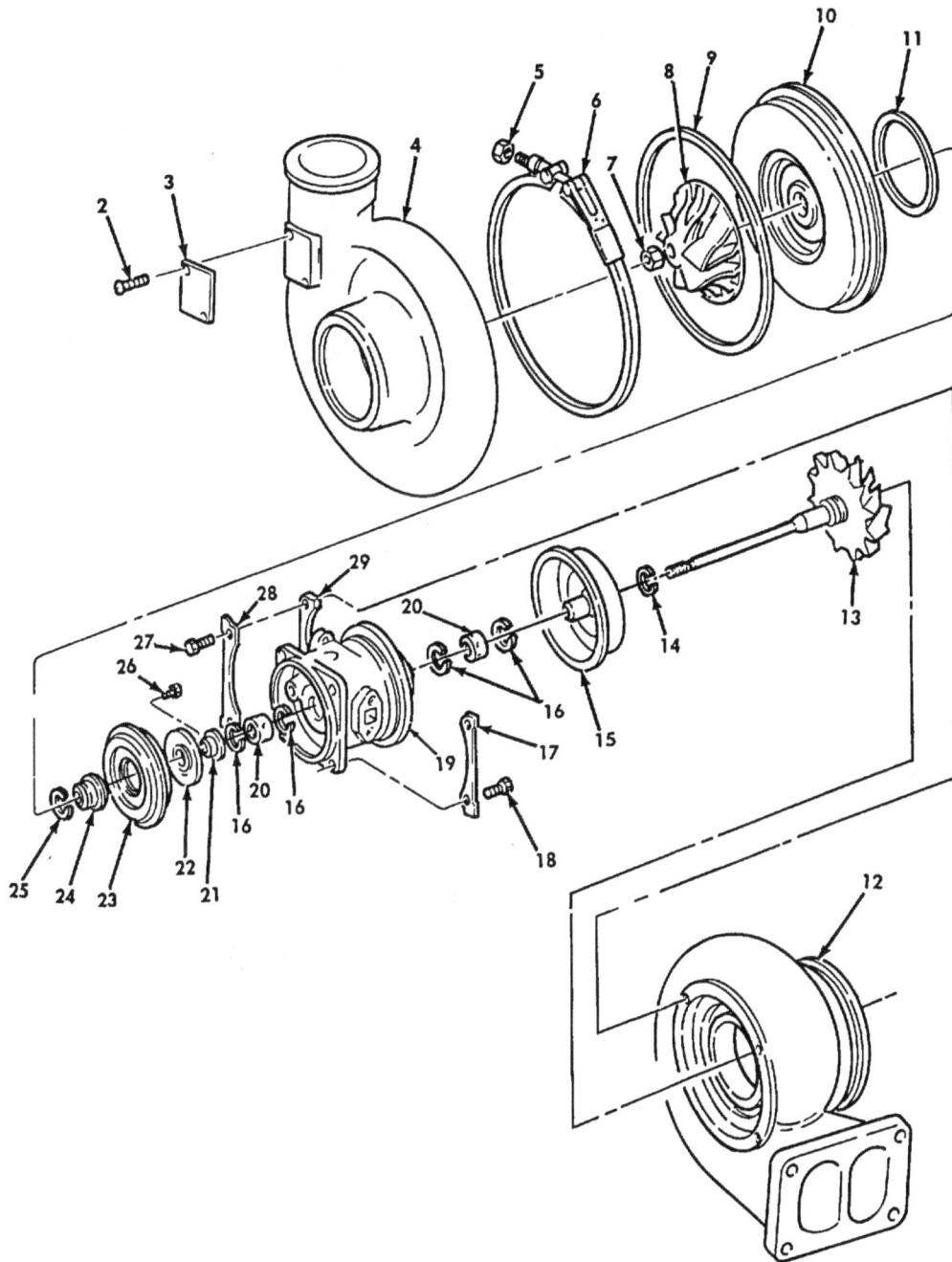

Figure 65. Turbocharger Assembly (M939A2).

(1) ITEM NO	(2) SMR CODE	(3) NSN	(4) CAGEC	(5) PART NUMBER	(6) DESCRIPTION AND USABLE ON CODES (UOC)	(7) QTY
					GROUP 0305 TURBOCHARGER	
					FIG. 65 TURBOCHARGER ASSEMBLY (M939A2)	
1	PAOFF	2990014465359	15434	3802257	TURBOSUPERCHARGER, N............................ UOC:ZAA, ZAB, ZAC, ZAD, ZAE, ZAF, ZAG, ZAH, ZAJ, ZAK, ZAL	1
2	PAFZZ	5305012718375	15434	3519163	.SCREW, DRIVE .... UOC:ZAA, ZAB, ZAC, ZAD, ZAE, ZAF , ZAG, ZAH, ZAJ, ZAK, ZAL	1
3	PAFZZ	9905012794690	15434	3519905	.PLATE, MARKING, BLANK .... UOC:ZAA, ZAB, ZAC, ZAD, ZAE, ZAF, ZAG, ZAH, ZAJ, ZAK, ZAL	1
4	PAFZZ	2950012712341	15434	3525359	.HOUSING, COMPRESSOR .... UOC:ZAA, ZAB, ZAC, ZAD, ZAE, ZAF, ZAG, ZAH, ZAJ, ZAK, ZAL	1
5	PAFZZ	5310012708245	15434	3529372	.NUT, PLAIN, HEXAGON .... UOC:ZAA, ZAB, ZAC, ZAD, ZAE, ZAF, ZAG, ZAH, ZAJ, ZAK, ZAL	1
6	PAFZZ	5340012712471	15434	3803431	.BAND, RETAINING .... UOC:ZAA, ZAB, ZAC, ZAD, ZAE, ZAF, ZAG, ZAH, ZAJ, ZAK, ZAL	1
7	PAFZZ	5310012885690	15434	3503562	.NUT, SELF-LOCKING, HE PART OF KIT P/N.......... 3545528 PART OF KIT P/N 3545745 PART OF KIT P/N 3802149 .... UOC:ZAA, ZAB, ZAC, ZAD, ZAE, ZAF, ZAG, ZAH, ZAJ, ZAK, ZAL	1
8	KFFZZ		15434	3525358	.IMPELLER, TURBO PART OF KIT P/N.................. 3545528 PART OF XIT P/N 3545745 .... UOC:ZAA, ZAB, ZAC, ZAD, ZAE, ZAF, ZAG, ZAH, ZAJ, ZAK, ZAL	1
9	PAFZZ	5330012721125	15434	3523958	.O-RING .... UOC:ZAA, ZAB, ZAC, ZAD, ZAE, ZAF, ZAG, ZAH, ZAJ, ZAK, ZAL	1
10	PAFZZ	2950012712342	15434	3522801	.DIFFUSER, SPECIAL .... UOC:ZAA, ZAB, ZAC, ZAD, ZAE, ZAF, ZAG, ZAH, ZAJ, ZAK, ZAL	1
11	PAFZZ	5330012721250	15434	3502449	.SEAL RING, METAL PART OF KIT P/N.................. 3802149 .... UOC:ZAA, ZAB, ZAC, ZAD, ZAE, ZAF, ZAG, ZAH, ZAJ, ZAK, ZAL	1
12	PAFZZ	2815013325462	15434	3527122	.HOUSING, SUPERCHARGE .... UOC:ZAA, ZAB, ZAC, ZAD, ZAE, ZAF, ZAG, ZAH, ZAJ, ZAK, ZAL	1

(1) ITEM NO	(2) SMR CODE	(3) NSN	(4) CAGEC	(5) PART NUMBER	(6) DESCRIPTION AND USABLE ON CODES (UOC)	(7) QTY
13	KFFZZ		15434	3522879	.SHAFT & WHEEL ASSY..................................... PART OF KIT P/N 3545745 UOC:ZAA, ZAB, ZAC, ZAD, ZAE, ZAF, ZAG, ZAH, ZAJ, ZAK, ZAL	1
14	PAFZZ	5330012033612	15434	3756754	.SEAL PART OF KIT P/N 3802149 .... UOC:ZAA, ZAB, ZAC, ZAD, ZAE, ZAF, ZAG, ZAH, ZAJ, ZAK, ZAL	
15	PAFZZ	2835012712511	15434	3519302	.SHIELD, HEAT, TURBINE .... UOC:ZAA, ZAB, ZAC, ZAD, ZAE, ZAF, ZAG, ZAH, ZAJ, ZAK, ZAL	1
16	PAFZZ	5325012708361	15434	3762259	.RING, RETAINING PART OF KIT P/N.................... 3802149 .... UOC:ZAA, ZAB, ZAC, ZAD, ZAE, ZAF, ZAG, ZAH, ZAJ, ZAK, ZAL	4
17	PAFZZ	5340012712504	15434	3502066	.PLATE, RETAINING, SEA PART OF KIT P/N........... 3802149 .... UOC:ZAA, ZAB, ZAC, ZAD, ZAE, ZAF, ZAG, ZAH, ZAJ, ZAK, ZAL	2
18	PAFZZ	5305012718379	15434	3522827	.SCREW, CAP, HEXAGON H................................ UOC:ZAA, ZAB, ZAC, ZAD, ZAE, ZAF, ZAG, ZAH, ZAJ, ZAK, ZAL	4
19	PAFZZ	3130013317182	15434	3530592	.HOUSING, BEARING UNI.................................. UOC:ZAA, ZAB, ZAC, ZAD, ZAE, ZAF, ZAG, ZAH, ZAJ, ZAK, ZAL	1
20	PAFZZ	3120012723270	15434	3503100	.BEARING, SLEEVE PART OF KIT P/N................... 3802149 .... UOC:ZAA, ZAB, ZAC, ZAD, ZAE, ZAF, ZAG, ZAH, ZAJ, ZAK, ZAL	2
21	KFFZZ		15434	3518980	.COLLAR, THRUST PART OF KIT P/N.................... 3545528 PART OF KIT P/N 3545745 .... UOC:ZAA, ZAB, ZAC, ZAD, ZAE, ZAF, ZAG, ZAH, ZAJ, ZAK, ZAL	1
22	PAFZZ	3120012723269	15434	3525739	.BEARING, SLEEVE .... UOC:ZAA, ZAB, ZAC, ZAD, ZAE, ZAF, ZAG, ZAH, ZAJ, ZAK, ZAL	1
23	PAFZZ	2950012110163	15434	3503668	.BAFFLE, OIL .... UOC:ZAA, ZAB, ZAC, ZAD, ZAE, ZAF, ZAG, ZAH, ZAJ, ZAK, ZAL	1
24	KFFZZ		15434	3503662	.SLINGER, OIL PART OF KIT P/N......................... 3545528 PART OF KIT P/N 3545745	1

(1) ITEM NO	(2) SMR CODE	(3) NSN	(4) CAGEC	(5) PART NUMBER	(6) DESCRIPTION AND USABLE ON CODES (UOC)	(7) QTY
					UOC:ZAA, ZAB, ZAC, ZAD, ZAE, ZAF, ZAG, ZAH, ZAJ, ZAK, ZAL	
25	PAFZZ	5330012131258	15434	3758848	.SEAL PART OF KIT P/N 3802149	1
					UOC:ZAA, ZAB, ZAC, ZAD, ZAE, ZAF, ZAG, ZAH, ZAJ, ZAK, ZAL	
26	PAFZZ	5305012923182	15434	3503347	.SCREW, MACHINE PART OF KIT P/N.................... 3802149	3
					UOC:ZAA, ZAB, ZAC, ZAD, ZAE, ZAF, ZAG, ZAH, ZAJ, ZAK, ZAL	
27	PAFZZ	5305012723305	15434	3759917	.SETSCREW PART OF KIT P/N 3802149................	4
					UOC:ZAA, ZAB, ZAC, ZAD, ZAE, ZAF, ZAG, ZAH, ZAJ, ZAK, ZAL	
28	PAFZZ	5340012712505	15434	3501102	.PLATE, RETAINING, SEA PART OF KIT P/N........... 3802149	2
					UOC:ZAA, ZAB, ZAC, ZAD, ZAE, ZAF, ZAG, ZAH, ZAJ, ZAK, ZAL	
29	PAFZZ	5342012712343	15434	3501188	.PLATE, CLAMPING	2
					UOC:ZAA, ZAB, ZAC, ZAD, ZAE, ZAF, ZAG, ZAH, ZAJ, ZAK, ZAL	

END OF FIGURE

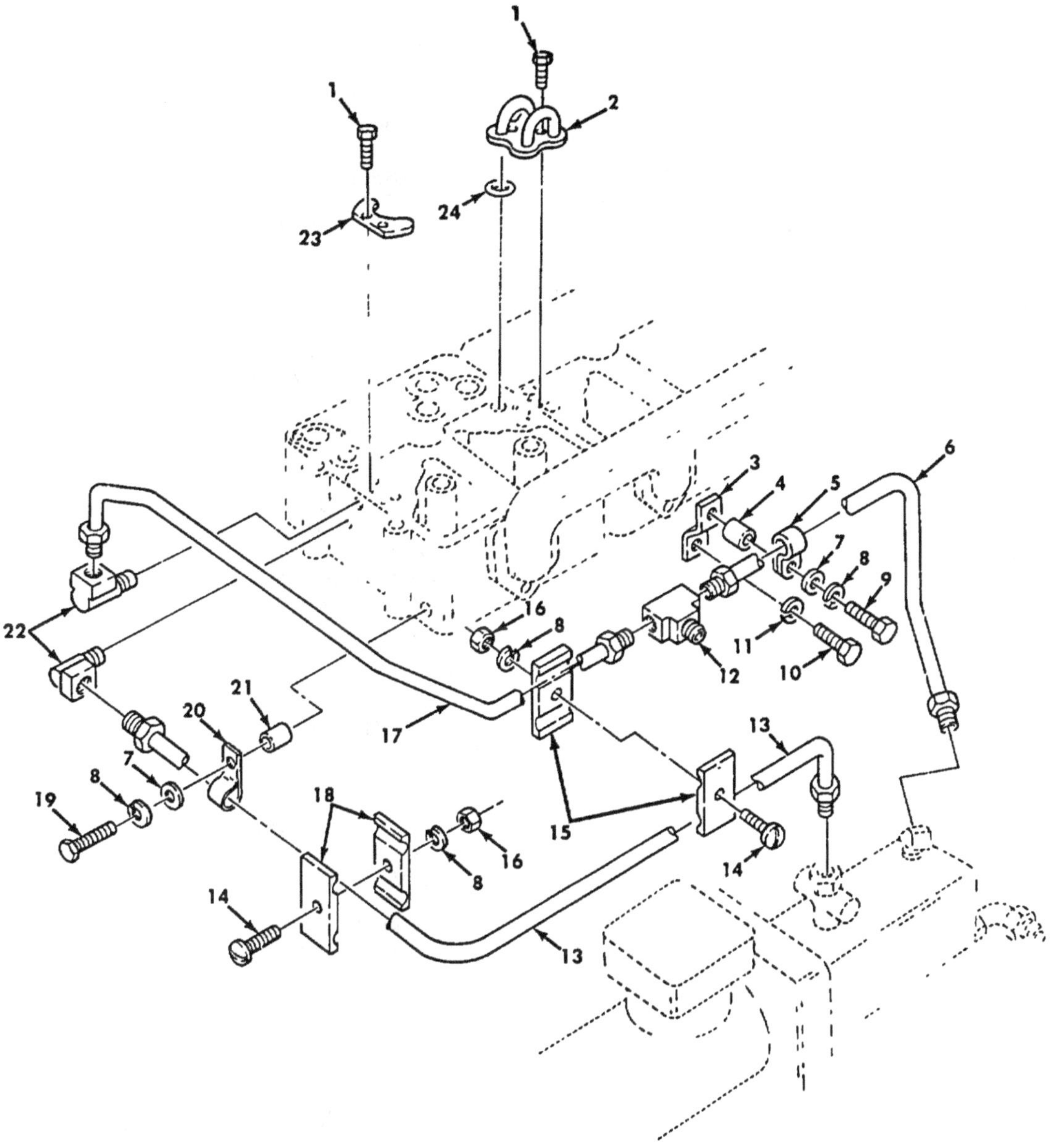

Figure 66. Fuel Crossover, Drain, and Bypass Tubing (M939, M939A1).

(1) ITEM NO	(2) SMR CODE	(3) NSN	(4) CAGEC	(5) PART NUMBER	(6) DESCRIPTION AND USABLE ON CODES (UOC)	(7) QTY
					GROUP 0306 TANKS, LINES, FITTINGS, HEADERS	
					FIG. 66 FUEL CROSSOVER, DRAIN, AND BYPASS TUBING(M939, M939A1)	
1	PAFZZ	5305004776769	15434	70772	SCREW, ASSEMBLED WAS .... UOC:DAA, DAB, DAC, DAD, DAE, DAF, DAG, DAH, DAJ, DAK, DAL, DAW, DAX, V12, V13, V14, V15, V16, V17, V18, V19, V20, V21, V22, V24, V25, V39	12
2	PAFZZ	2910009283505	15434	147100	CROSSOVER, FUEL .... UOC:DAA, DAB, DAC, DAD, DAE, DAF, DAG, DAH, DAJ, DAK, DAL, DAW, DAX, V12, V13, V14, V15, V16, V17, V18, V19, V20, V21, V22, V24, V25, V39	2
3	PAOZZ	5340012039261	15434	3033098	BRACKET, DOUBLE ANGL .... UOC:DAA, DAB, DAC, DAD, DAE, DAF, DAG, DAH, DAJ, DAK, DAL, DAW, DAX, V12, V13, V14, V15, V16, V17, V18, V19, V20, V21, V22, V24, V25, V39	1
4	PAOZZ	5365012218749	15434	3019301	SPACER, SLEEVE .... UOC:DAA, DAB, DAC, DAD, DAE, DAF, DAG, DAH, DAJ, DAK, DAL, DAW, DAX, V12, V13, V14, V15, V16, V17, V18, V19, V20, V21, V22, V24, V25, V39	1
5	PAOZZ	5340007194601	15434	180372	CLAMP, LOOP .... UOC:DAA, DAB, DAC, DAD, DAE, DAF, DAG, DAH, DAJ, DAK, DAL, DAW, DAX, V12, V13, V14, V15, V16, V17, V18, V19, V20, V21, V22, V24, V25, V39	1
6	PAOZZ	4710012186950	15434	3009846	TUBE ASSEMBLY, METAL .... UOC:DAA, DAB, DAC, DAD, DAE, DAF, DAG, DAH, DAJ, DAK, DAL, DAW, DAX, V12, V13, V14, V15, V16, V17, V18, V19, V20, V21, V22, V24, V25, V39	1
7	PAOZZ	5310005626560	15434	S-631	WASHER, FLAT .... UOC:DAA, DAB, DAC, DAD, DAE, DAF, DAG, DAH, DAJ, DAK, DAL, DAW, DAX, V12, V13, V14, V15, V16, V17, V18, V19, V20, V21, V22, V24, V25, V39	2
8	PAOZZ	5310001596209	96906	MS122032	WASHER, LOCK .... UOC:DAA, DAB, DAC, DAD, DAE, DAF, DAG, DAH, DAJ, DAK, DAL, DAW, DAX, V12, V13, V14, V15, V16, V17, V18, V19, V20, V21, V22, V24, V25, V39	4
9	PAOZZ	5305000680509	80204	B1821BH025C125N	SCREW, CAP, HEXAGON H .... UOC:DAA, DAB, DAC, DAD, DAE, DAF, DAG, DAH, DAJ, DAK, DAL, DAW, DAX, V12, V13, V14, V15, V16, V17, V18, V19, V20, V21, V22, V24, V25, V39	1
10	PAOZZ	5305007959345	15434	S-190-C	SCREW, CAP	1

(1) ITEM NO	(2) SMR CODE	(3) NSN	(4) CAGEC	(5) PART NUMBER	(6) DESCRIPTION AND USABLE ON CODES (UOC)	(7) QTY
					UOC:DAA, DAB, DAC, DAD, DAE, DAF, DAG, DAH, DAJ, DAK, DAL, DAW, DAX, V12 , V13 , V14 , V15, V16, V17, V18, V19, V20, V21, V22, V24, V25, V39	
11	PAOZZ	5310002617340	96906	MS35338-8	WASHER, LOCK .... UOC:DAA, DAB, DAC, DAD, DAE, DAF, DAG, DAH, DAJ, DAK, DAL, DAW, DAX, V12, V13, V14, V15, V16, V17, V18, V19, V20, V21, V22, V24, V25, V39	1
12	PAOZZ	4730010782731	15434	209726	TEE, TUBE .... UOC: DAA, DAB, DAC, DAD, DAE, DAF, DAG, DAH, DAJ, DAK, DAL, DAW, DAX, V12, V13, V14, V15, V16, V17, V18, V19, V20, V21, V22, V24, V25, V39	1
13	PAOZZ	4710012186951	15434	3035612	TUBE ASSEMBLY, METAL .... UOC:DAL, V18	1
13	PAOZZ	4710012186952	15434	3035616	TUBE ASSEMBLY, METAL .... UOC:DAA, DAB, DAC, DAD, DAE, DAF, DAG, DAH, DAJ, DAK, DAW, DAX, V12, V13, V14, V15, V16, V17, V19, V20, V21, V22, V24, V25, V39	1
14	PAOZZ	5305007576567	15434	S-1315	SCREW, MACHINE .... UOC:DAA, DAB, DAC, DAD, DAE, DAF, DAG, DAH, DAJ, DAK, DAL, DAW, DAX, V12, V13, V14, V15, V16, V17, V18, V19, V20, V21, V22, V24, V25, V39	2
15	PAOZZ	5340012466172	15434	3007025	STRAP, RETAINING .... UOC:DAA, DAB, DAC, DAD, DAE , DAF, DAG, DAH , DAJ, DAK, DAL, DAW, DAX, V12, V13, V14, V15, V16, V17, V18, V19, V20, V21, V22, V24, V25, V39	2
16	PAOZZ	5310009717989	96906	MS35691-5	NUT, PLAIN, HEXAGON .... UOC:DAA, DAB, DAC, DAD, DAE, DAF, DAG, DAH, DAJ, DAK, DAL, DAW, DAX, V12, V13, V14, V15, V16, V17, V18, V19, V20, V21, V22, V24, V25, V39	2
17	PAOZZ	4710012186949	15434	3035614	TUBE ASSEMBLY, METAL .... UOC:DAA, DAB, DAC, DAD, DAE, DAF, DAG, DAH, DAJ, DAK, DAL, DAW, DAX, V12, V13, V14, V15, V16, V17, V18, V19, V20, V21, V22, V24, V25, V39	1
18	PAOZZ	4730010706667	15434	3007024	CLAMP HALF, DOUBLE, H .... UOC:DAA, DAB, DAC, DAD, DAE, DAF, DAG, DAH, DAJ, DAK, DAL, DAW, DAX, V12, V13, V14, V15, V16, V17, V18, V19, V20, V21, V22, V24, V25, V39	2
19	PAOZZ	5305000712505	80204	B1821BH025C088N	SCREW, CAP, HEXAGON H .... UOC:DAA, DAB, DAC, DAD, DAE, DAF, DAG, DAH, DAJ, DAK, DAL, DAW, DAX, V12, V13 , V14 , V15, V16, V17, V18, V19, V20, V21, V22, V24, V25, V39	1
20	PAOZZ	5340010798097	15434	180371	CLAMP, LOOP .... UOC:DAA, DAB, DAC, DAD, DAE, DAF, DAG, DAH,	1

(1) ITEM NO	(2) SMR CODE	(3) NSN	(4) CAGEC	(5) PART NUMBER	(6) DESCRIPTION AND USABLE ON CODES (UOC)	(7) QTY
					DAJ, DAK, DAL, DAW, DAX, V12, V13, V14, V15, V16, V17, VV18, V19, V20, V21, V22, V24, V25, V39	
21	PAOZZ	5365002031281	15434	114850	SPACER, SPECIAL	1
					UOC:DAA, DAB, DAC, DAD, DAE, DAF, DAG, DAH, DAJ, DAK, DAL, DAW, DAX, V12, V13, V14, V15, V16, V17, V18, V19, V20, V21, V22, V24, V25, V39	
22	PAOZZ	4730004441710	15434	181213	ELBOW, PIPE TO TUBE	2
					UOC:DAA, DAB, DAC, DAD, DAE, DAF, DAG, DAH, DAJ, DAK, DAL, AWDAW, DAX, V12, V13, V14, V15, V16, V17, V18, V19, V20, V21, V22, V24, V25, V39	
23	PAOZZ	2815011464164	15434	135308	PLATE, FUEL CROSSOVE	2
					UOC:DAA, DAB, DAC, DAD, DAE, DAF, DAG, DAH, DAJ, DAK, DAL, DAW, DAX, V12, V13, V14, V15, V16, V17, V18, V19, V20, V21, V22, V24, V25, V39	
24	PAFZZ	5331001438485	15434	131026	O-RING PART OF KIT P/N 3011472	12
					UOC:DAA, DAB, DAC, DAD, DAE, DAF, DAG, DAH, DAJ, DAK, DAL, DAW, DAX, V12, V13, V14, V15, V16, V17, V18, V19, V20, V21, V22, V24, V25, V39	

END OF FIGURE

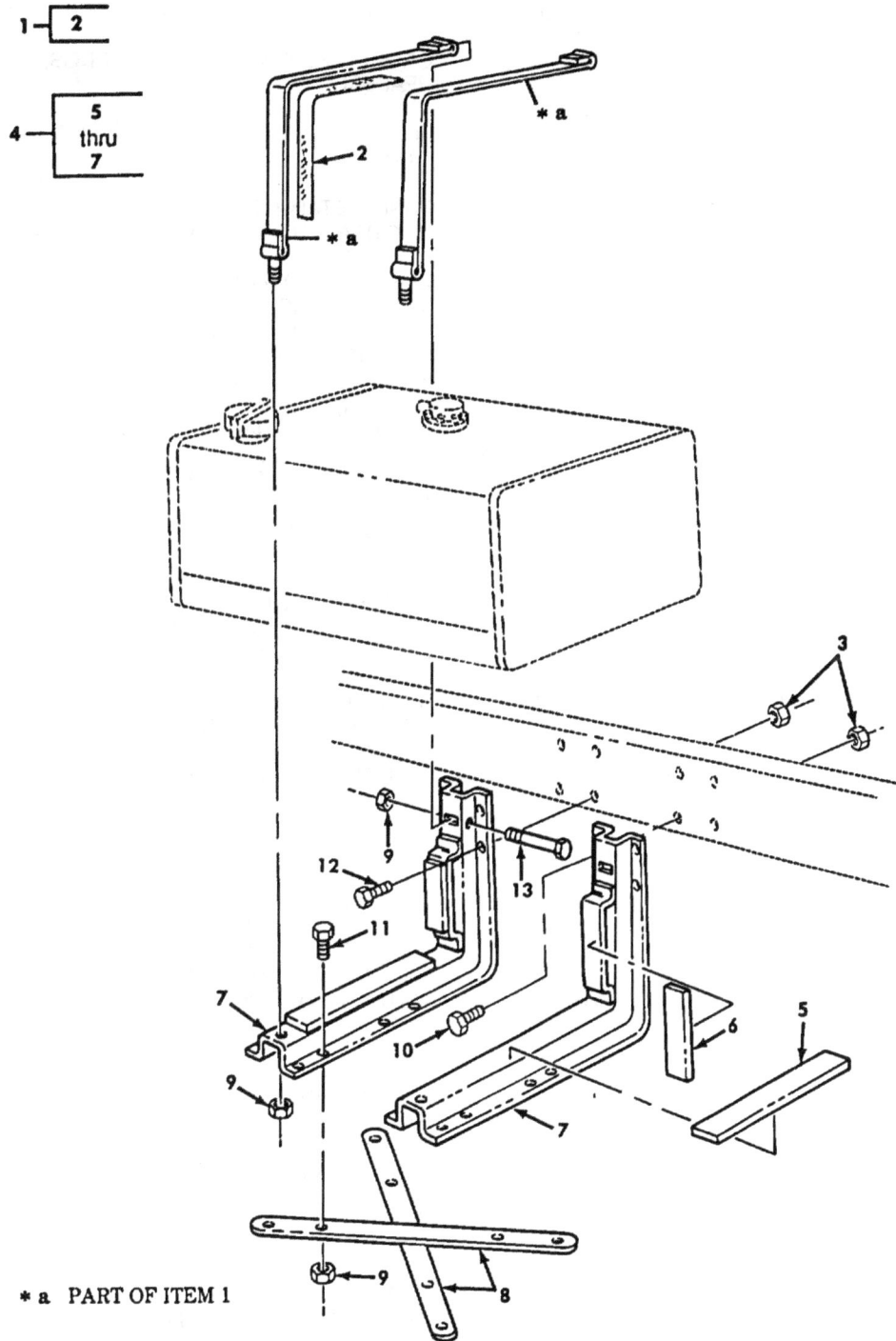

*a   PART OF ITEM 1

Figure 67. Fuel Tank Mounting, Single Tank, L.H.

(1) ITEM NO	(2) SMR CODE	(3) NSN	(4) CAGEC	(5) PART NUMBER	(6) DESCRIPTION AND USABLE ON CODES (UOC)	(7) QTY
					GROUP 0306 TANKS, LINES, FITTINGS, HEADERS	
					FIG. 67 FUEL TANK MOUNTING, SINGLE TANK, L.H.	
1	PAOZZ	5340010907657	19207	12276998-2	BAND, RETAINING............................ UOC:DAA, DAB, DAC, DAD, DAJ, DAK, DAW, DAX, V12, V13, V14, V15, V16, V17, V24, V25, ZAA, ZAB, ZAC, ZAD, ZAJ, ZAK	2
2	PAOZZ	9390011921610	19207	12277027-2	.NONMETALLIC CHANNEL .... UOC:DAA, DAB, DAC, DAD, DAJ, DAK, DAW, DAX, V12, V13, V14, V15, V16, V17, V24, V25, ZAA, ZAB, ZAC, ZAD, ZAJ, ZAK	1
3	PAOZZ	5310008775795	96906	MS21044-N8	NUT, SELF-LOCKING, HE .... UOC:DAA, DAB, DAC, DAD, DAJ, DAK, DAW, DAX, V12, V13, V14, V15, V16, V17, V24, V25, ZAA, ZAB, ZAC, ZAD, ZAJ, ZAK	8
4	PAOZZ	5340011122169	19207	12277030	BRACKET, ANGLE .... UOC:DAA, DAB, DAC, DAD, DAJ, DAK, DAW, DAX, V12, V13, V14, V15, V16, V17, V24, V25, ZAA, ZAB, ZAC, ZAD, ZAJ, ZAK	2
5	PAOZZ	5330011170608	19207	12277026-1	.RUBBER SHEET, SOLID .... UOC:DAA, DAB, DAC, DAD, DAJ, DAK, DAW, DAX, V12, V13, V14, V15, V16, V17, V24, V25, ZAA, ZAB, ZAC, ZAD, ZAJ, ZAK	1
6	PAOZZ	5330011170609	19207	12277026-2	.RUBBER SHEET, SOLID .... UOC:DAA, DAB, DAC, DAD, DAJ, DAK, DAW, DAX, V12, V13, V14, V15, V16, V17, V24, V25, ZAA, ZAB, ZAC, ZAD, ZAJ, ZAK	1
7	XAOZZ		19207	7017087-1	.HANGER............................ UOC:DAA, DAB, DAC, DAD, DAJ, DAK, DAW, DAX, V12, V13, V14, V15, V16, V17, V24, V25, ZAA, ZAB, ZAC, ZAD, ZAJ, ZAK	1
8	PAOZZ	5340010907649	19207	12256176-2	BRACKET, MOUNTIN G. .... UOC:DAA, DAB, DAC, DAD, DAJ, DAK, DAW, DAX, V12, V13, V14, V15, V16, V17, V24, V25, ZAA, ZAB, ZAC, ZAD, ZAJ, ZAK	2
9	PAOZZ	5310009591488	96906	MS51922-21	NUT, SELF-LOCKING, HE............................ UOC:DAA, DAB, DAC, DAD, DAJ, DAK, DAW, DAX, V12, V13, V14, V15, V16, V17, V24, V25	12
10	PAOZZ	5305007195235	80204	B1821BH050F175N	SCREW, CAP, HEXAGON H............................ UOC:DAC, DAD, DAJ, DAK, V16, V17, ZAC, ZAD, ZAJ, ZAK	8
10	PAOZZ	5305007195219	96906	MS90727-111	SCREW, CAP, HEXAGON H............................ UOC:DAA, DAB, DAW, DAX, V12, V13, V14, V15, V24, V25, ZAA, ZAB	8
11	PAOZZ	5305002692803	96906	MS90726-60	SCREW, CAP, HEXAGON H............................ UOC:DAA, DAB, DAC, DAD, DAJ, DAK, DAW, DAX, V12, V13, V14, V15, V16, V17, V24, V25, ZAA, ZAB, ZAC, ZAD, ZAJ, ZAK	8
12	PAOZZ	5305007254183	96906	MS90726-113	SCREW, CAP, HEXAGON H............................	4

(1) ITEM NO	(2) SMR CODE	(3) NSN	(4) CAGEC	(5) PART NUMBER	(6) DESCRIPTION AND USABLE ON CODES (UOC)	(7) QTY
13	PAOZZ	5305002693247	96906	MS90727-71	UOC:DAA, DAB, DAC, DADD, DAJ, DAK, DAW, DAX, V12, V13, V14, V15, V16, V17, V24, V25, ZAA, ZAB, ZAC, ZAD, ZAJ, ZAK SCREW, CAP, HEXAGON H .... UOC:DAA, DAB, DAC, DAD, DAJ, DAK, DAW, DAX, V12, V13, V14, V15, V16, V17, V24, V25, ZAA, ZAB, ZAC, ZAD, ZAJ, ZAK	2

END OF FIGURE

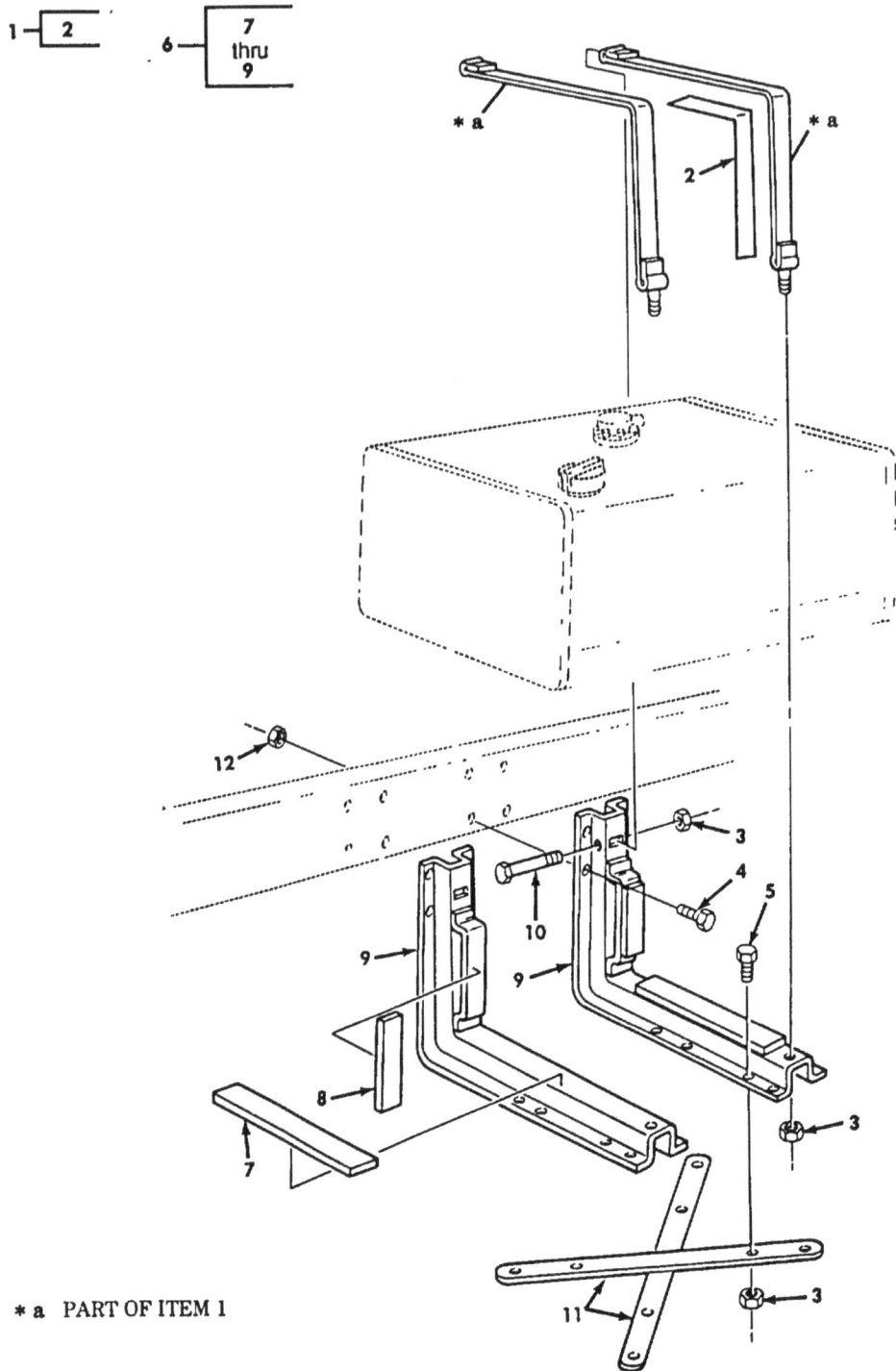

*a PART OF ITEM 1

Figure 68. Fuel Tank Mounting, Single Tank, R.H. (M945).

(1) ITEM NO	(2) SMR CODE	(3) NSN	(4) CAGEC	(5) PART NUMBER	(6) DESCRIPTION AND USABLE ON CODES (UOC)	(7) QTY
					GROUP 0306 TANKS, LINES, FITTINGS, HEADERS	
					FIG. 68 FUEL TANK MOUNTING, SINGLE TANK, R.H. (M945)	
1	PAOZZ	5340011170606	19207	12276998-1	STRAP, RETAINING UOC:V39	2
2	PAOZZ	9390012078125	19207	12277027-1	.NONMETALLIC CHANNEL UOC:V39	1
3	PAOZZ	5310009591488	96906	MS51922-21	NUT, SELF-LOCKING, HE UOC:V39	12
4	PAOZZ	5305007195235	80204	B1821BH050F175N	SCREW, CAP, HEXAGON H UOC:V39	5
5	PAOZZ	5305002692803	96906	MS90726-60	SCREW, CAP, HEXAGON H UOC:V39	8
6	PAOZZ	5340011122169	19207	12277030	BRACKET, ANGLE UOC:V39	2
7	PAOZZ	5330011170608	19207	12277026-1	.RUBBER SHEET, SOLID UOC:V39	1
8	PAOZZ	5330011170609	19207	12277026-2	.RUBBER SHEET, SOLID UOC:V39	1
9	XAOZZ		19207	7017087-1	.HANGER.. UOC:V39	1
10	PAOZZ	5305002693247	96906	MS90727-71	SCREW, CAP, HEXAGON H UOC:V39	2
11	PAOZZ	5340010907649	19207	12256176-2	BRACKET, MOUNTING UOC:V39	2
12	PAOZZ	5310008775795	96906	MS21044-N8	NUT, SELF-LOCKING, HE UOC:V39	5

END OF FIGURE

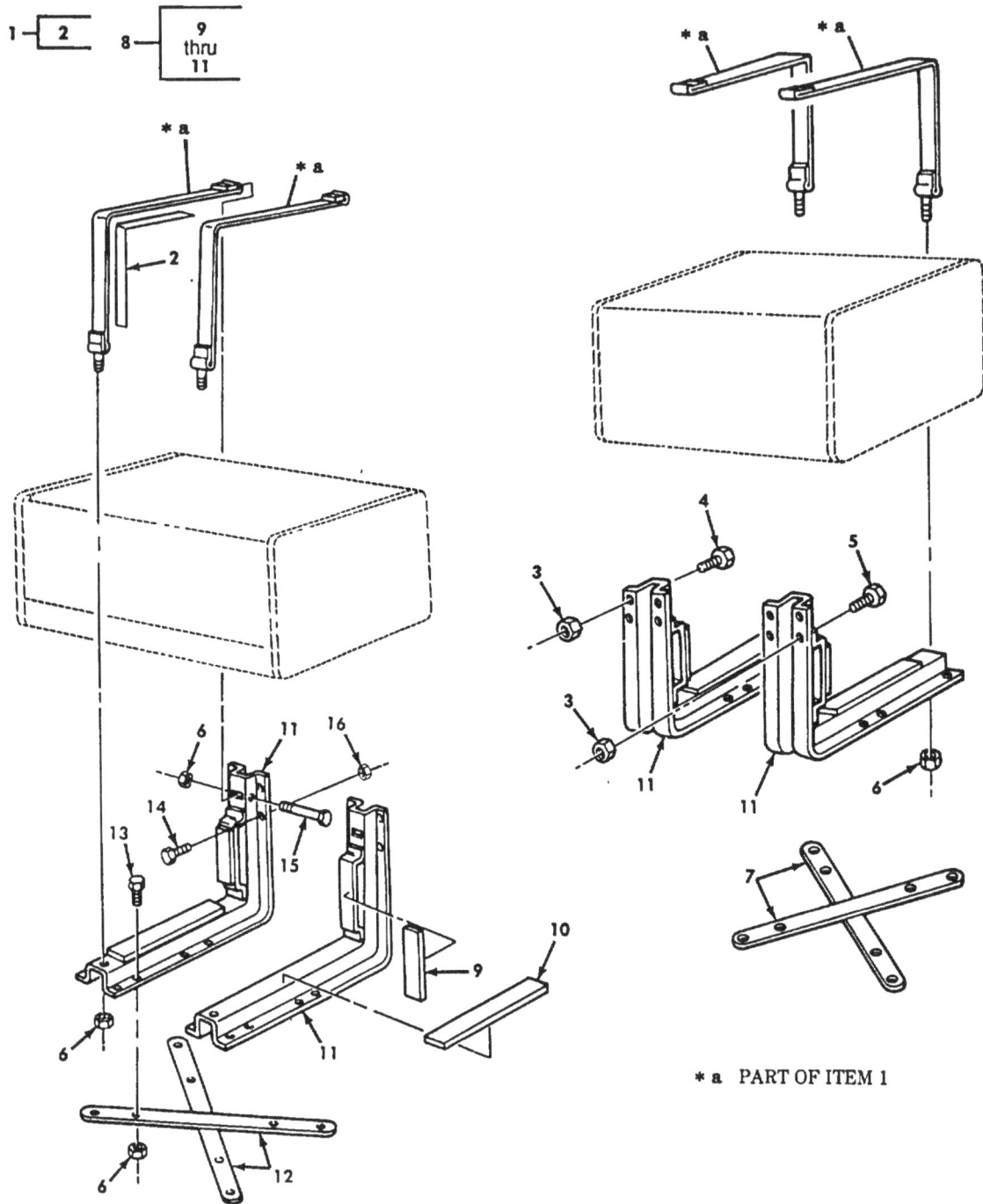

Figure 69. Fuel Tank Mounting, Dual Tank.

* a   PART OF ITEM 1

(1) ITEM NO	(2) SMR CODE	(3) NSN	(4) CAGEC	(5) PART NUMBER	(6) DESCRIPTION AND USABLE ON CODES (UOC)	(7) QTY
					GROUP 0306 TANKS, LINES, FITTINGS, HEADERS	
					FIG. 69 FUEL TANK MOUNTING, DUAL TANK	
1	PAOZZ	5340010907657	19207	12276998-2	BAND, RETAINING .... UOC:DAE, DAF, DAG, DAH, DAL, V18, V19, V20, V21, V22, ZAE, ZAF, ZAG, ZAH, ZAL	4
2	PAOZZ	9390011921610	19207	12277027-2	.NONMETALLIC CHANNEL .... UOC:DAE, DAF, DAG, DAH, DAL, V18, V19, V20, V21, V22, ZAE, ZAF, ZAG, ZAH, ZAL	1
3	PAOZZ	5310004883888	96906	MS51943-40	NUT, SELF-LOCKING, HE .... UOC:DAE, DAF, DAG, DAH, DAL, V18, V19, V20, V21, V22, ZAE, ZAF, ZAG, ZAH, ZAL	5
4	PAOZZ	5305007254183	96906	MS90726-113	SCREW, CAP, HEXAGON H .... UOC:DAE, DAF, DAG, DAH, DAL, V18, V19, V20, V21, V22, ZAE, ZAF, ZAG, ZAH, ZAL	1
5	PAOZZ	5305007195221	80204	B1821BH050F150N	SCREW, CAP, HEXAGON H .... UOC:DAE, DAF, DAG, DAH, DAL, V18, V19, V20, V21, V22, ZAE, ZAF, ZAG, ZAH, ZAL	4
6	PAOZZ	5310009591488	96906	MS51922-21	NUT, SELF-LOCKING, HE .... UOC:DAE, DAF, DAG, DAH, DAL, V18, V19, V20, V21, V22	24
7	PFOZZ	5340010907645	19207	12256176-1	PLATE, MENDING .... UOC:DAE, DAF, DAG, DAH, DAL, V18, V19, V20, V21, V22, ZAE, ZAF, ZAG, ZAH, ZAL	2
8	PAOZZ	5340011122169	19207	12277030	BRACKET, ANGLE .... UOC:DAE, DAF, DAG, DAH, DAL, V18, V19, V20, V21, V22, ZAE, ZAF, ZAG, ZAH, ZAL	4
9	PAOZZ	5330011170609	19207	12277026-2	.RUBBER SHEET, SOLID .... UOC:DAE, DAF, DAG, DAH, DAL, V18, V19, V20, V21, V22, ZAE, ZAF, ZAG, ZAH, ZAL	1
10	PAOZZ	5330011170608	19207	12277026-1	.RUBBER SHEET, SOLID .... UOC:DAE, DAF, DAG, DAH, DAL, V18, V19, V20, V21, V22, ZAE, ZAF, ZAG, ZAH, ZAL	1
11	XAOZZ		19207	7017087-1	.HANGER.. UOC:DAE, DAF, DAG, DAH, DAL, V18, V19, V20, V21, V22, ZAE, ZAF, ZAG, ZAH, ZAL	1
12	PAOZZ	5340010907649	19207	12256176-2	BRACKET, MOUNTING .... UOC:DAL, V18, ZAL	2
13	PAOZZ	5305002692803	96906	MS90726-60	SCREW, CAP, HEXAGON H .... UOC:DAE, DAF, DAG, DAH, DAL, V18, V19, V20, V21, V22, ZAE, ZAF, ZAG, ZAH, ZAL	16
14	PAOZZ	5305007195219	96906	MS90727-111	SCREW, CAP, HEXAGON H .... UOC:DAE, DAF, DAG, DAH, V19, V20, V21, V22, ZAE, ZAF, ZAG, ZAH	8
14	PAOZZ	5305007195235	80204	B1821BH050F175N	SCREW, CAP, HEXAGON H .... UOC:DAL, V18, ZAL	8
15	PAOZZ	5305002693247	96906	MS90727-71	SCREW, CAP, HEXAGON H .... UOC:DAE, DAF, DAG, DAH, V19, V20, V21, V22, ZAE, ZAF, ZAG, ZAH	4
16	PAOZZ	5310008775795	96906	MS21044-N8	NUT, SELF-LOCKING, HE	8

(1) ITEM NO	(2) SMR CODE	(3) NSN	(4) CAGEC	(5) PART NUMBER	(6) DESCRIPTION AND USABLE ON CODES (UOC)	(7) QTY

UOC:DAE, DAF, DAG, DAH, DAL, V18, V19, V20, V21, V22, ZAE, ZAF, ZAG, ZAH, ZAL

END OF FIGURE

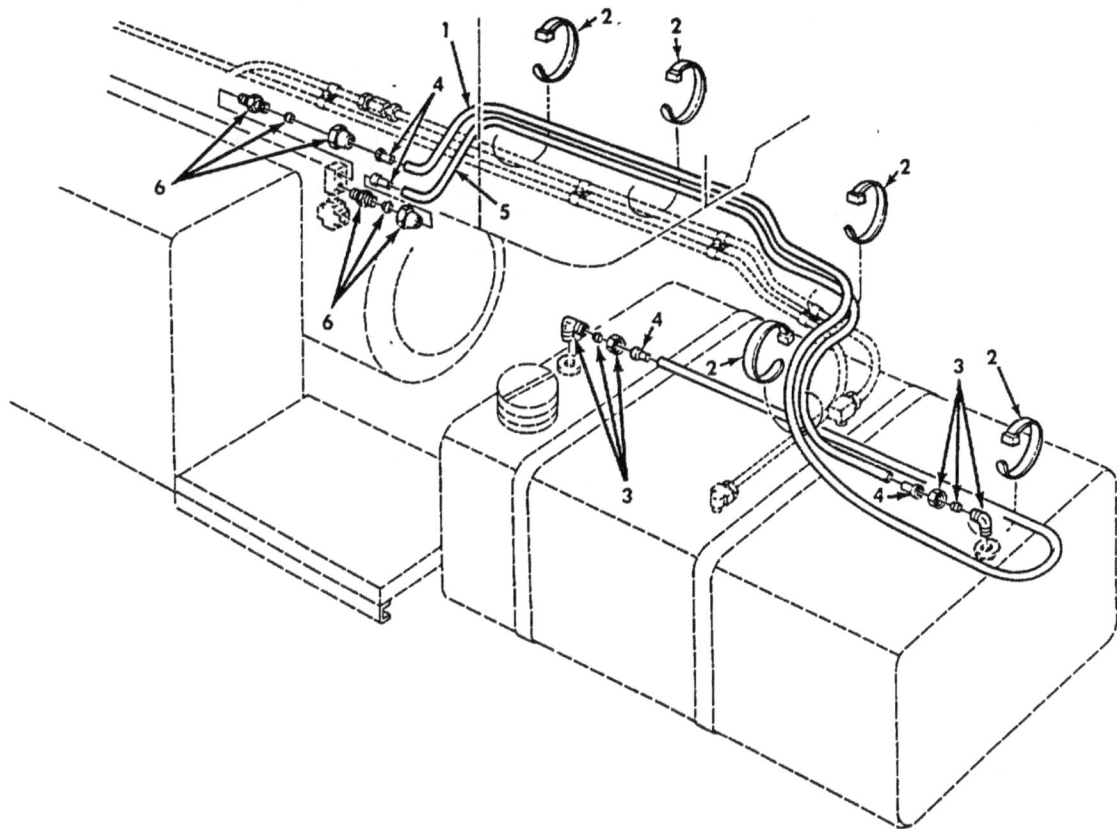

Figure 70. Fuel Tank Vent Lines, Single, L.H.

(1) ITEM NO	(2) SMR CODE	(3) NSN	(4) CAGEC	(5) PART NUMBER	(6) DESCRIPTION AND USABLE ON CODES (UOC)	(7) QTY
					GROUP 0306 TANKS, LINES, FITTINGS, HEADERS FIG. 70 FUEL TANK VENT LINES, SINGLE, L.H.	
1	MOOZZ		19207	CPR104420-1-144	TUBING, NYLON MAKE FROM TUBING, P/N.......... CPR104420-1, 144 INCHES LONG......................... UOC:DAA, DAB, DAC, DAD, DAJ, DAJ , DAK, DAW, DAX, V12, V13, V14, V15, V16, V17, V24, V25, ZAA, ZAB, ZAC, ZAD, ZAJ, ZAK	1
2	PAOZZ	5975009846582	96906	MS3367-1-0	STRAP, TIEDOWN, ELECT .... UOC:DAA, DAB, DAC, DAD, DAJ, DAK, DAW, DAX, V12, V13, V14, V15, V16, V17, V24, V25, ZAA, ZAB, ZAC, ZAD, ZAJ, ZAK	5
3	PAOZZ	4730009213240	81343	4-2 120202BA	ELBOW, PIPE TO TUBE .... UOC:DAA, DAB, DAC, DAD, DAJ, DAK, DAW, DAX, V12, V13, V14, V15, V16, V17, V24, V25, ZAA, ZAB, ZAC, ZAD, ZAJ, ZAK	2
4	PAOZZ	4730001324588	79470	2030X4A	INSERT, TUBE FITTING .... UOC:DAA, DAB, DAC, DAD, DAJ, DAK, DAW, DAX, V12, V13, V14, V15, V16, V17, V24, V25, ZAA, ZAB, ZAC, ZAD, ZAJ, ZAK	4
5	MOOZZ		19207	CPR104420-1-108	TUBING, NYLON MAKE FROM TUBING, P/N.......... CPR104420-1, 108 INCHES LONG....................... UOC:DAA, DAB, DAC, DAD, DAJ, DAK, DAW, DAX, V12, V13, V14, V15, V16, V17, V24, V25, ZAA, ZAB, ZAC, ZAD, ZAJ, ZAK	1
6	PAOZZ	4730002704580	96906	MS39179-2	ADAPTER, STRAIGHT, PI .... UOC:DAA, DAB, DAC, DAD, DAJ, DAK, DAW, DAX, V12, V13, V14, V15, V16, V17, V24, V25, ZAA, ZAB, ZAC, ZAD, ZAJ, ZAK	2

END OF FIGURE

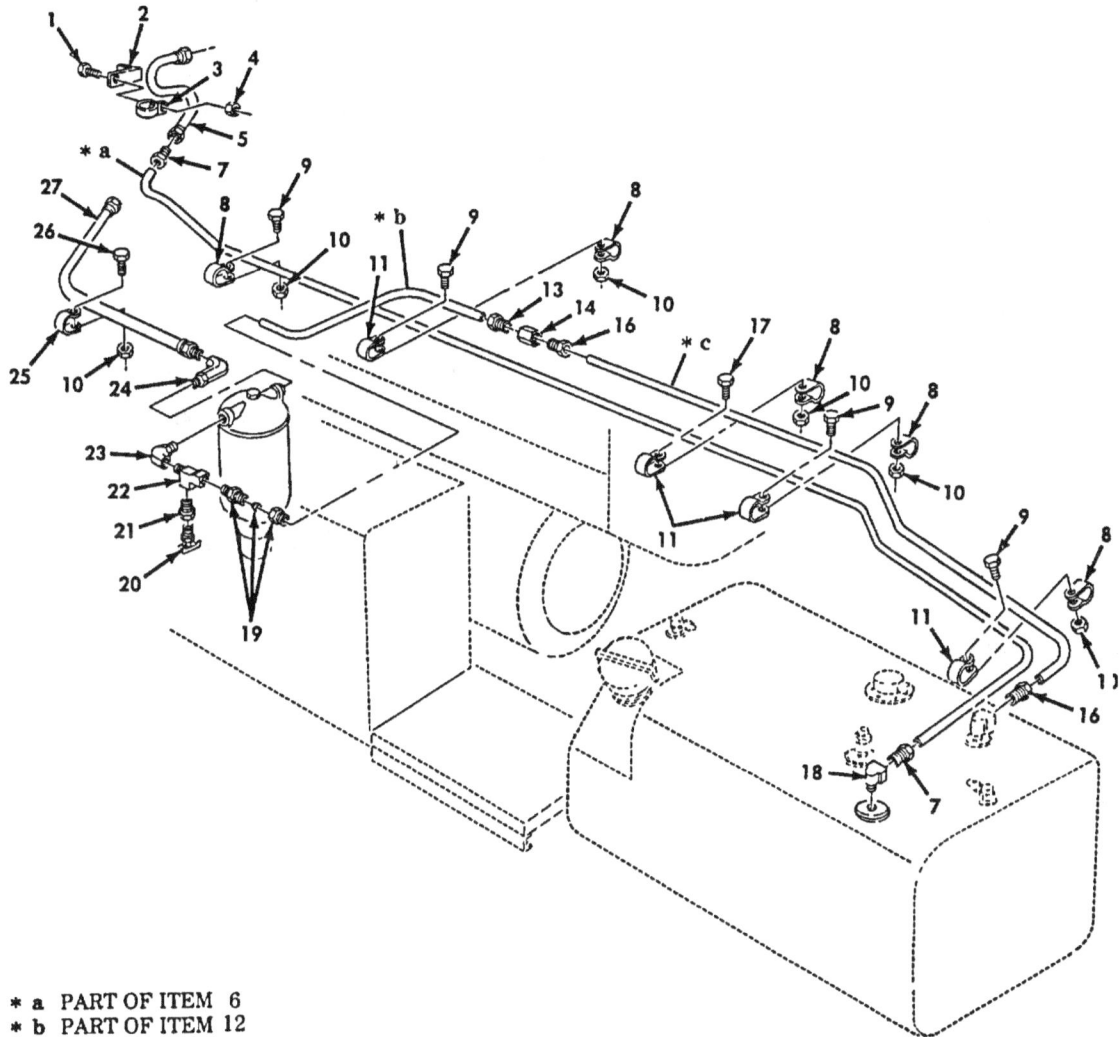

* a PART OF ITEM 6
* b PART OF ITEM 12
* c PART OF ITEM 15

Figure 71. Fuel Tank Supply and Return Lines, Single, L.H. (M939, M939A1).

SECTION II

(1) ITEM NO	(2) SMR CODE	(3) NSN	(4) CAGEC	(5) PART NUMBER	(6) DESCRIPTION AND USABLE ON CODES (UOC)	(7) QTY
					GROUP 0306 TANKS, LINES, FITTINGS, HEADERS	
					FIG. 71 FUEL TANK SUPPLY AND RETURN LINES, SINGLE, L.H.(M939, M939A1)	
1	PAOZZ	5305002678953	80204	B1821BH025F063N	SCREW, CAP, HEXAGON H .... UOC:DAA, DAB, DAC, DAD, DAJ, DAK, DAW, DAX, V12, V13, V14, V15, V16, V17, V24, V25	1
2	PFOZZ	5342010907658	19207	12256187	BRACKET, SUPPORT FUE .... UOC:DAA, DAB, DAC, DAD, DAJ, DAK, DAW, DAX, V12, V13, V14, V15, V16, V17, V24, V25	1
3	PAOZZ	5340000881254	96906	MS21333-104	CLAMP, LOOP .... UOC:DAA, DAB, DAC, DAD, DAJ, DAK, DAW, DAX, V12, V13, V14, V15, V16, V17, V24, V25	1
4	PAOZZ	5310001436102	96906	MS51922-6	NUT, SELF-LOCKING, HE .... UOC:DAA, DAB, DAC, DAD, DAJ, DAK, DAW, DAX, V12, V13, V14, V15, V16, V17, V24, V25	1
5	PAOZZ	4720010915169	19207	12256185	HOSE ASSEMBLY, METAL .... UOC:DAA, DAB, DAC, DAD, DAJ, DAK, DAW, DAX, V12, V13, V14, V15, V16, V17, V24, V25	1
6	MOOZZ		19207	12256160	TUBE ASSY FUEL BY-PASS, MAKE FROM .... TUBING, P/N 8689208 .... UOC:DAA, DAB, DAC, DAD, DAJ, DAK, DAW, DAX, V12, V13, V14, V15, V16, V17, V24, V25	1
7	PAOZZ	4730000142433	21450	142433	.INVERTED NUT, TUBE C .... UOC:DAA, DAB, DAC, DAD, DAJ, DAK, DAW, DAX, V12, V13, V14, V15, V16, V17, V24, V25	2
8	PAOZZ	5340009546014	96906	MS21333-121	CLAMP, LOOP .... UOC:DAA, DAB, DAC, DAD, DAJ, DAK, DAW, DAX, V12, V13, V14, V15, V16, V17, V24, V25	5
9	PAOZZ	5305002693233	80204	B1821BH038F063N	SCREW, CAP, HEXAGON H .... UOC:DAA, DAB, DAC, DAD, DAJ, DAK, DAW, DAX, V12, V13, V14, V15, V16, V17, V24, V25	5
10	PAOZZ	5310009591488	96906	MS51922-21	NUT, SELF-LOCKING, HE .... UOC:DAA, DAB, DAC, DAD, DAJ, DAK, DAW, DAX, V12, V13, V14, V15, V16, V17, V24, V25	6
11	PAOZZ	5340009891771	96906	MS21333-123	CLAMP, LOOP................................ UOC:DAA, DAB, DAC, DAD, DAJ, DAK, DAW, DAX, V12, V13, V14, V15, V16, V17, V24, V25	4
12	MOOZZ		19207	12256159	TUBE ASSY MAKE FROM TUBING, P/N................. 8689210 .... UOC:DAA, DAB, DAC, DAD, DAJ, DAK, DAW, DAX, V12, V13, V14, V15, V16, V17, V24, V25	1
13	PAOZZ	4730000142435	12204	142435	.INVERTED NUT, TUBE C .... UOC:DAA, DAB, DAC, DAD, DAJ, DAK, DAW, DAX, V12, V13, V14, V15, V16, V17, V24, V25	1
14	PAOZZ	4730011487397	24617	442340	COUPLING, TUBE................................ UOC:DAA, DAB, DAC, DAD, DAJ, DAK, DAW, DAX, V12, V13, V14, V15, V16, V17, V24, V25	1
15	MOOZZ		19207	12256158	TUBE ASSY MAKE FROM TUBING, P/N................. 8689208	1

(1) ITEM NO	(2) SMR CODE	(3) NSN	(4) CAGEC	(5) PART NUMBER	(6) DESCRIPTION AND USABLE ON CODES (UOC)	(7) QTY
					UOC:DAA, DAB, DAC, DAD, DAJ, DAK, DAW, DAX, V12, V13, V14, V15, V16, V17, V24, V25	
16	PAOZZ	4730000142435	12204	142435	.INVERTED NUT, TUBE C	2
					UOC:DAA, DAB, DAC, DAD, DAJ, DAK, DAW, DAX, V12, V13, V14, V15, V16, V17, V24, V25	
17	PAOZZ	53050002693235	80204	B1821BH038F088N	SCREW, CAP, HEXAGON H	1
					UOC:DAA, DAB, DAC, DAD, DAJ, DAK, DAW, DAX, V12, V13, V14, V15, V16, V17, V24, V25	
18	PAOZZ	4730002778274	21450	137423	ELBOW, PIPE TO TUBE	1
					UOC:DAA, DAB, DAC, DAD, DAJ, DAK, DAW, DAX, V12, V13, V14, V15, V16, V17, V24, V25	
19	PAOZZ	4730002029036	81343	8-8 120102BA	ADAPTER, STRAIGHT, PI	1
					UOC:DAA, DAB, DAC, DAD, DAJ, DAK, DAW, DAX, V12, V13, V14, V15, V16, V17, V24, V25	
20	PAOZZ	4820007529040	96906	MS35782-4	COCK, DRAIN	1
					UOC:DAA, DAB, DAC, DAD, DAJ, DAK, DAW, DAX, V12, V13, V14, V15, V16, V17, V24, V25	
21	PAOZZ	4730011336209	96906	MS14315-6X	BUSHING, PIPE	1
					UOC:DAA, DAB, DAC, DAD, DAJ, DAK, DAW, DAX, V12, V13, V14, V15, V16, V17, V24, V25	
22	PAOZZ	4730010952034	19207	8366166-1	TEE, PIPE	1
					UOC:DAA, DAB, DAC, DAD, DAJ, DAK, DAW, DAX, V12, V13, V14, V15, V16, V17, V24, V25	
23	PAOZZ	4730010906468	19207	11668054	ELBOW, PIPE TO TUBE	1
					UOC:DAA, DAB, DAC, DAD, DAJ, DAK, DAW, DAX, V12, V13, V14, V15, V16, V17, V24, V25	
24	PAOZZ	4730011044314	19207	11668054-1	ELBOW, PIPE TO TUBE	1
					UOC:DAA, DAB, DAC, DAD, DAJ, DAK, DAW, DAX, V12, V13, V14, V15, V16, V17, V24, V25	
25	PAOZZ	5340000538994	96906	MS21333-126	CLAMP, LOOP	1
					UOC:DAA, DAB, DAC, DAD, DAJ, DAK, DAW, DAX, V12, V13, V14, V15, V16, V17, V24, V25	
26	PAOZZ	5305002692804	96906	MS90726-61	SCREW, CAP, HEXAGON H	1
					UOC:DAA, DAB, DAC, DAD, DAJ, DAK, DAW, DAX, V12, V13, V14, V15, V16, V17, V24, V25	
27	PAOZZ	4720010907618	19207	12256186	HOSE ASSEMBLY, NONME	1
					UOC:DAA, DAB, DAC, DAD, DAJ, DAK, DAW, DAX, V12, V13, V14, V15, V16, V17, V24, V25	

END OF FIGURE

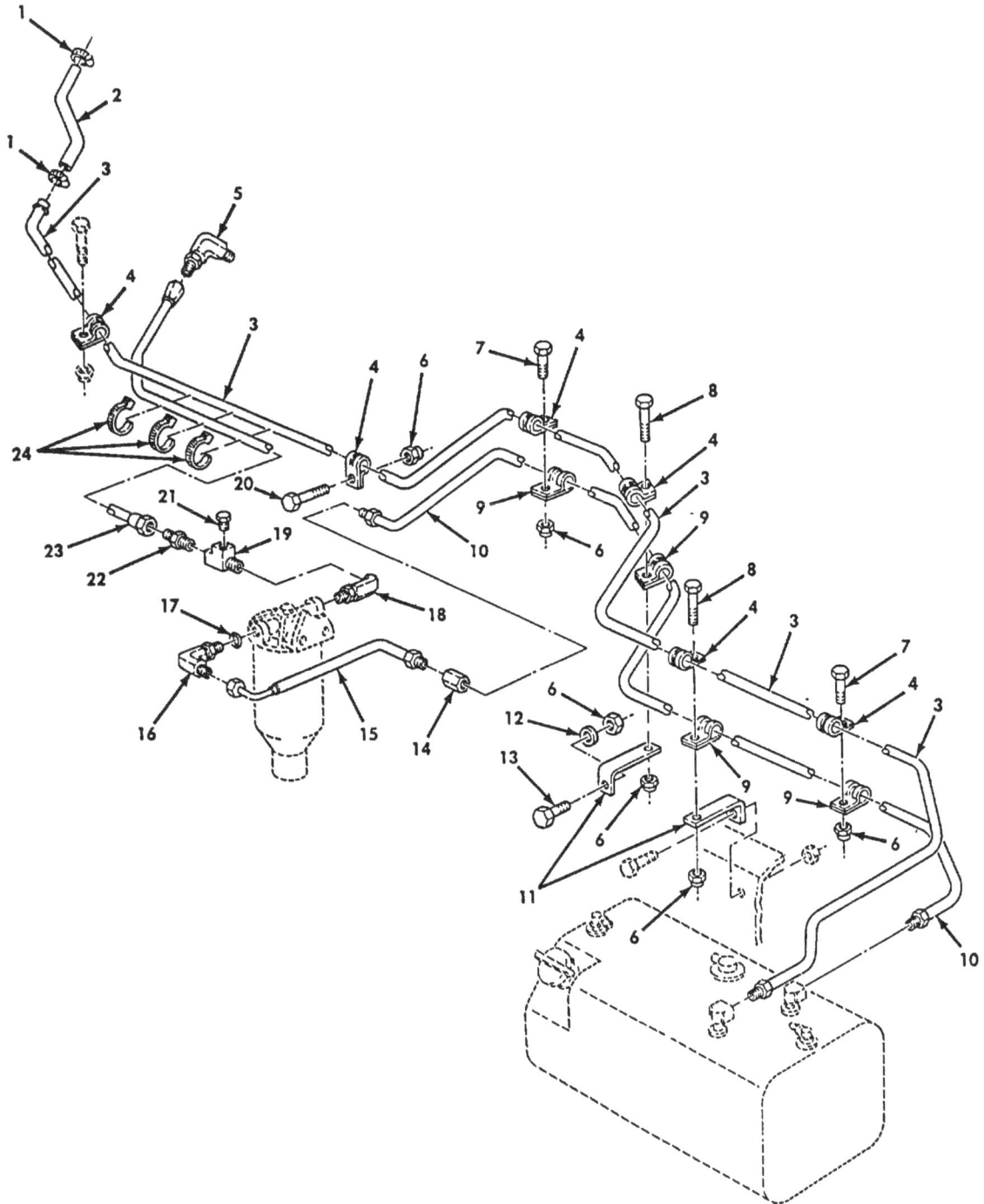

Figure 72. Fuel Tank Supply and Return Lines, Single, L.H. (M939A2).

(1) ITEM NO	(2) SMR CODE	(3) NSN	(4) CAGEC	(5) PART NUMBER	(6) DESCRIPTION AND USABLE ON CODES (UOC)	(7) QTY
					GROUP 0306 TANKS, LINES, FITTINGS, HEADERS	
					FIG. 72 FUEL TANK SUPPLY AND RETURN LINES, SINGLE, L.H.(M939A2)	
1	PAOZZ	4730011098001	19207	11662913	CLAMP, HOSE................................ UOC:ZAA, ZAB, ZAC, ZAD, ZAJ, ZAK	2
2	MOOZZ		98441	A2697 X 15 1/2IN	RETURN HOSE, FUEL MAKE FROM HOSE, 4720-00-274-1154, 15-1/2 IN LONG......................... UOC:ZAA, ZAB, ZAC, ZAD, ZAJ, ZAK	1
3	PAOZZ	4710012793163	4F744	20511273	TUBE ASSEMBLY, METAL .... UOC:ZAA, ZAB, ZAC, ZAD, ZAJ, ZAK	1
4	PAOZZ	5340009546014	96906	MS21333-121	CLAMP, LOOP .... UOC:ZAA, ZAB, ZAC, ZAD, ZAJ, ZAK	6
5	PAOZZ	4730002351483	33334	E1-8B	ELBOW, PIPE TO TUBE .... UOC:ZAA, ZAB, ZAC, ZAD, ZAJ, ZAK	1
6	PAOZZ	5310009591488	96906	MS51922-21	NUT, SELF-LOCKING, HE .... UOC:ZAA, ZAB, ZAC, ZAD, ZAJ, ZAK	6
7	PAOZZ	5305002693234	80204	B1821BH038F075N	SCREW, CAP, HEXAGON H .... UOC:ZAA, ZAB, ZAC, ZAD, ZAJ, ZAK	2
8	PAOZZ	5305002693235	80204	B1821BH038F088N	SCREW, CAP, HEXAGON H .... UOC:ZAA, ZAB, ZAC, ZAD, ZAJ, ZAK	2
9	PAOZZ	5340009891771	96906	MS21333-123	CLAMP, LOOP .... UOC:ZAA, ZAB, ZAC, ZAD, ZAJ, ZAK	4
10	PAOZZ	4710012793164	47457	20511274	TUBE ASSEMBLY, METAL FUEL .... UOC:ZAA, ZAB, ZAC, ZAD, ZAJ, ZAK	1
11	PFOZZ	5340012984861	47457	20510460	BRACKET, ANGLE STANDOFF .... UOC:ZAA, ZAB, ZAC, ZAD, ZAJ, ZAK	2
12	PAOZZ	5310000806004	96906	MS27183-14	WASHER, FLAT .... UOC:ZAA, ZAB, ZAC, ZAD, ZAJ, ZAK	1
13	PAOZZ	5305002692803	96906	MS90726-60	SCREW, CAP, HEXAGON H .... UOC:ZAA, ZAB, ZAC, ZAD, ZAJ, ZAK	1
14	PAOZZ	4730002033179	81343	8-8 010103B	ADAPTER, STRAIGHT, PI .... UOC:ZAA, ZAB, ZAC, ZAD, ZAJ, ZAK	1
15	PAOZZ	4720014422533	19207	12432371	HOSE ASSEMBLY, NONME .... UOC:ZAA, ZAB, ZAC, ZAD, ZAJ, ZAK	1
16	PAOZZ		81343	8-10 010220CA	ELBOW, PIPE .... UOC:ZAA, ZAB, ZAC, ZAD, ZAJ, ZAK	1
17	PAOZZ	5330012205498	56161	15557258	PACKING, PREFORMED .... UOC:ZAA, ZAB, ZAC, ZAD, ZAJ, ZAK	1
18	PAOZZ	4730011044314	92003	AE40GSS	ELBOW, PIPE TO TUBE .... UOC:ZAA, ZAB, ZAC, ZAD, ZAJ, ZAK	1
19	PAOZZ	4730010952034	81343	8-8-8 130424B	TEE, PIPE .... UOC:ZAA, ZAB, ZAC, ZAD, ZAJ, ZAK	1
20	PAOZZ	5305002693239	80204	B1821BH038F138N	SCREW, CAP, HEXAGON H .... UOC:ZAC, ZAD, ZAJ, ZAK	1
20	PAOZZ	5305002692803	96906	MS90726-60	SCREW, CAP, HEXAGON H .... UOC:ZAA, ZAB	1
21	PAOZZ	4730000144027	79470	3152X8	PLUG, PIPE .... UOC:ZAA, ZAB, ZAC, ZAD, ZAJ, ZAK	1
22	PAOZZ	4730008099427	81343	8-8 070102CA	ADAPTER, STRAIGHT, PI	1

(1) ITEM NO	(2) SMR CODE	(3) NSN	(4) CAGEC	(5) PART NUMBER	(6) DESCRIPTION AND USABLE ON CODES (UOC)	(7) QTY
					UOC:ZAA, ZAB, ZAC, ZAD, ZAJ, ZAK	
23	PAOZZ	4720014412922	01276	FG2722HHH0270	HOSE ASSEMBLY, NONME	1
					UOC: ZAA, ZAB, ZAC, ZAD, ZAJ, ZAK	
24	PAOZZ	5975008956630	96906	MS3367-3-0	STRAP, TIEDOWN, ELECT	3
					UOC:Z.AA, ZAB, ZAC, ZAD, ZAJ, ZAK	

END OF FIGURE

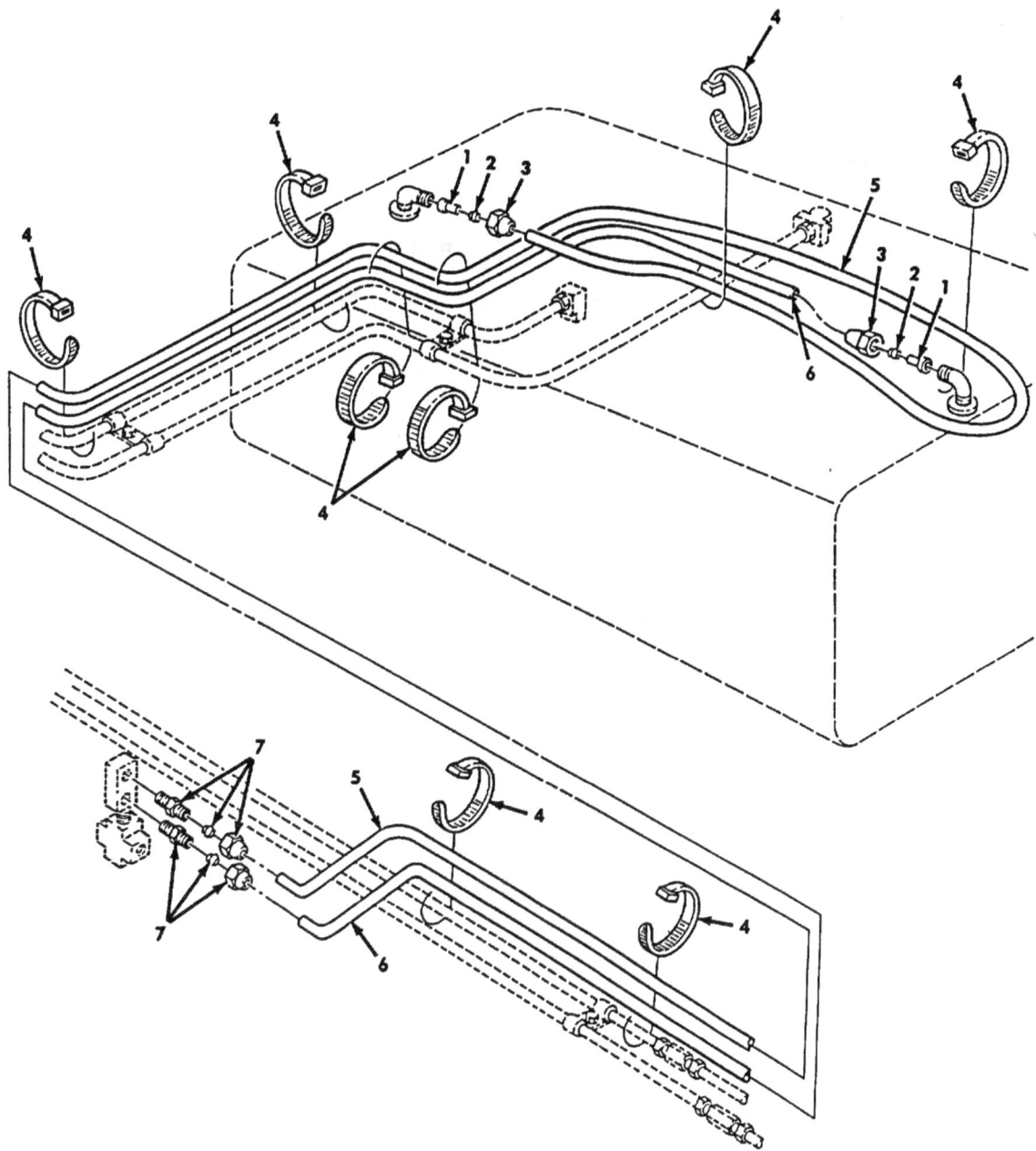

Figure 73. Fuel Tank Vent Lines, Single, R.H.

(1) ITEM NO	(2) SMR CODE	(3) NSN	(4) CAGEC	(5) PART NUMBER	(6) DESCRIPTION AND USABLE ON CODES (UOC)	(7) QTY
					GROUP 0306 TANKS, LINES, FITTINGS, HEADERS	
					FIG. 73 FUEL TANK VENT LINES, SINGLE, R.H.	
1	PAOZZ	4730001324588	79470	2030X4A	INSERT, TUBE FITTING UOC:V39	2
2	PAOZZ	4730001220477	16662	AC294	SLEEVE, COMPRESSION, ... UOC:DAA,DAB,DAC,DAD,DAL,DAW,DAX	1
4	PAOZZ	5975009846582	96906	MS3367-1-0	STRAP, TIEDOWN, ELECT UOC:V39	8
5	MOOZZ		19207	CPR104420-1-160	TUBING, NONMETALIC REAR LEFT HAND............. VENT, MAKE FROM TUBING, NYLON, P/N CPR104420-1.. UOC:V39	1
6	MOOZZ		19207	CPR104420-1-122	TUBING, NONMETALIC REAR LEFT HAND............. VENT, MAKE FROM TUBING, NYLON, P/N CPR104420-1.. UOC:V39	1
7	PAOZZ	4730002704580	81343	4-4 120101AB	ADAPTER, STRAIGHT UOC:V39	2

END OF FIGURE

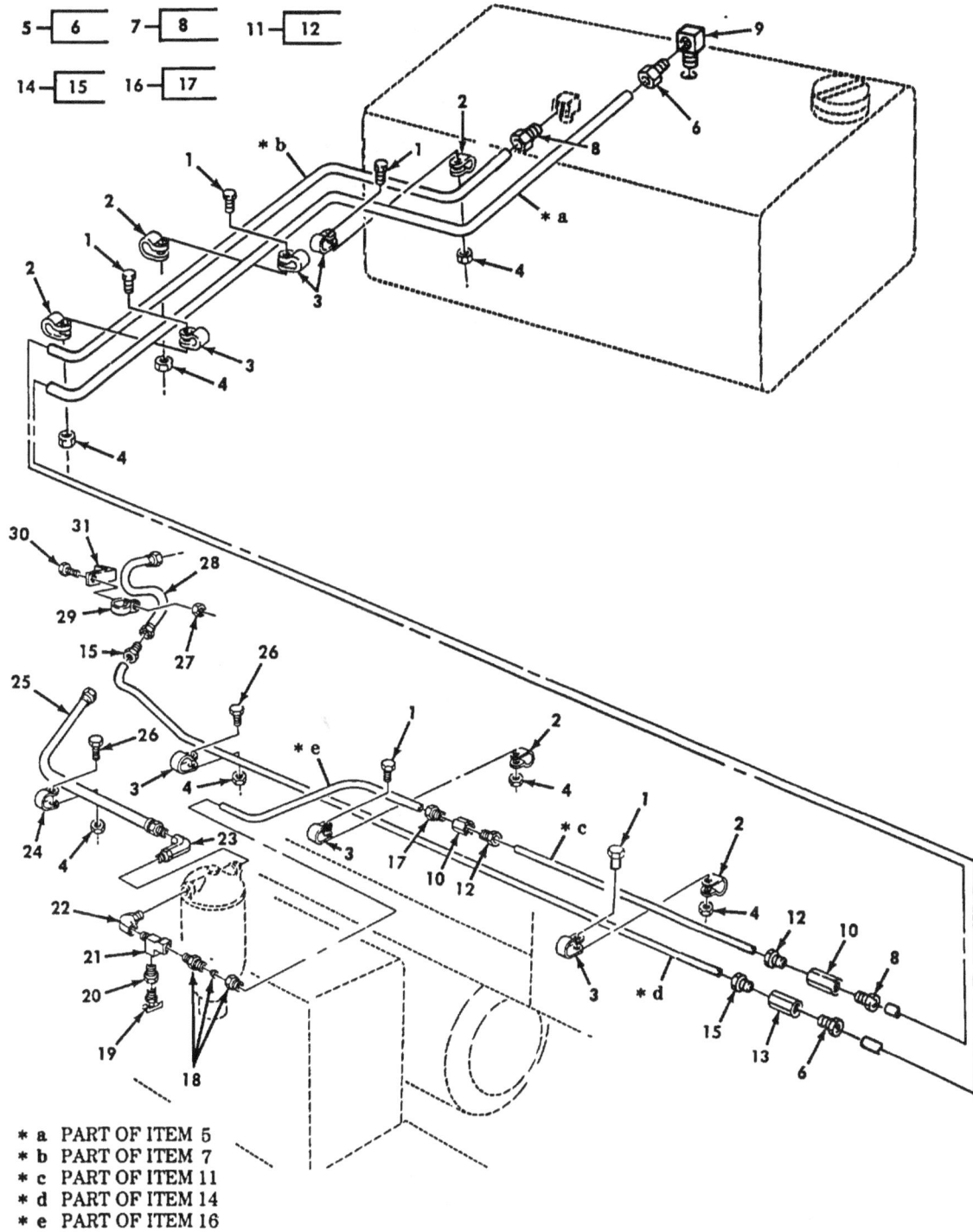

Figure 74. Fuel Tank Supply and Return Lines, Single, R.H.

* a PART OF ITEM 5
* b PART OF ITEM 7
* c PART OF ITEM 11
* d PART OF ITEM 14
* e PART OF ITEM 16

(1) ITEM NO	(2) SMR CODE	(3) NSN	(4) CAGEC	(5) PART NUMBER	(6) DESCRIPTION AND USABLE ON CODES (UOC)	(7) QTY
					GROUP 0306 TANKS, LINES, FITTINGS, HEADERS	
					FIG. 74 FUEL TANK SUPPLY AND RETURN LINES, SINGLE, R.H.	
1	PAOZZ	5305002693234	80204	B1821BH038F075N	SCREW, CAP, HEXAGON H UOC:V39	5
2	PAOZZ	5340009891771	96906	MS21333-123	CLAMP, LOOP UOC:V39	5
3	PAOZZ	5340009546014	96906	MS21333-121	CLAMP, LOOP UOC:V39	6
4	PAOZZ	5310009591488	96906	MS51922-21	NUT, SELF-LOCKING, HE UOC:V39	7
5	MOOZZ		19207	12257058	TUBE ASSY MAKE FORM TUBING P/N................ 8689208 .... UOC:V39	1
6	PAOZZ	4730000142433	21450	142433	.INVERTED NUT, TUBE C UOC:V39	2
7	MOOZZ		19207	12257057	TUBE ASSY FUEL RETURN, MAKE FROM............. TUBING, P/N 8689210 .... UOC:V39	1
8	PAOZZ	4730000142435	12204	142435	.INVERTED NUT, TUBE C UOC:V39	2
9	PAOZZ	4730002778274	21450	137423	ELBOW, PIPE TO TUBE UOC:V39	1
10	PAOZZ	4730002788889	21450	137417	COUPLING, TUBE................... UOC:V39	2
11	MOOZZ		19207	12257055	TUBE ASSY FUEL TANK SUPPLY LINE, ... MAKE FROM TUBING P/N 8689210 .... UOC:V39	1
12	PAOZZ	4730000142435	12204	142435	.INVERTED NUT, TUBE C UOC:V39	2
13	PAOZZ	4730002788888	21450	137415	COUPLING, TUBE UOC:V39	1
14	MOOZZ		19207	12257059	TUBE ASSY FUEL RETURN, MAKE FROM............. TUBING, P/N 8689208................... UOC:V39	1
15	PAOZZ	4730000142433	21450	142433	.INVERTED NUT, TUBE C UOC:V39	2
16	MOOZZ		19207	12256159	TUBE ASSEMBLY FUEL SUPPLY LINE, ... MAKE FROM TUBING, P/N 8689210................... UOC:V39	1
17	PAOZZ	4730000142435	12204	142435	.INVERTED NUT, TUBE C UOC:V39	1
18	PAOZZ	4730002029036	81343	8-8 120102BA	ADAPTER, STRAIGHT, PI................... UOC:V39	1
19	PAOZZ	4820007529040	96906	MS35782-4	COCK, DRAIN................... UOC:V39	1
20	PAOZZ	4730011336209	96906	MS14315-6X	BUSHING, PIPE................... UOC:V39	1
21	PAOZZ	4730010952034	19207	8366166-1	TEE, PIPE	1

(1) ITEM NO	(2) SMR CODE	(3) NSN	(4) CAGEC	(5) PART NUMBER	(6) DESCRIPTION AND USABLE ON CODES (UOC)	(7) QTY
					UOC:V39	
22	PAOZZ	4730010906468	19207	11668054	ELBOW, PIPE TO TUBE	1
					UOC:V39	
23	PAOZZ	4730011044314	19207	11668054-1	ELBOW, PIPE TO TUBE	1
					UOC:V39	
24	PAOZZ	5340000538994	96906	MS21333-126	CLAMP, LOOP	1
					UOC:V39	
25	PAOZZ	4720010907618	19207	12256186	HOSE ASSEMBLY, NONME	1
					UOC:V39	
26	PAOZZ	5305002692804	96906	MS90726-61	SCREW, CAP, HEXAGON H	2
					UOC:V39	
27	PAOZZ	5310001436102	96906	MS51922-6	NUT, SELF-LOCKING, HE	1
					UOC:V39	
28	PAOZZ	4720010915169	19207	12256185	HOSE ASSEMBLY, METAL	1
					UOC:V39	
29	PAOZZ	5340000881254	96906	MS21333-104	CLAMP, LOOP	1
					UOC:V39	
30	PAOZZ	5305002678953	80204	B1821BH025F063N	SCREW, CAP, HEXAGON H	1
					UOC:V39	
31	PFOZZ	5342010907658	19207	12256187	BRACKET, SUPPORT FUE	1
					UOC:V39	

END OF FIGURE

6 ─ | 7 | 12 ─ | 13 |

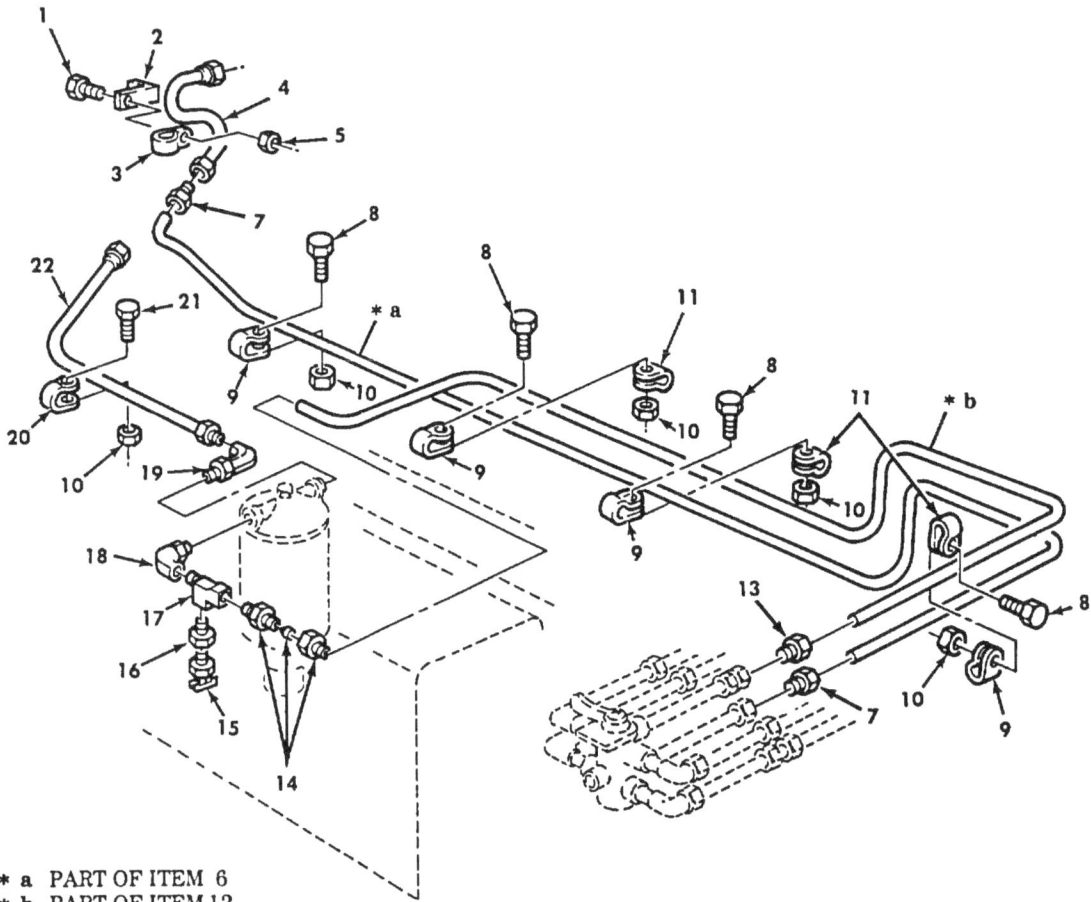

* a PART OF ITEM 6
* b PART OF ITEM 12

Figure 75. Fuel Tank Supply and Return Lines, Dual (M939, M939A1).

(1) ITEM NO	(2) SMR CODE	(3) NSN	(4) CAGEC	(5) PART NUMBER	(6) DESCRIPTION AND USABLE ON CODES (UOC)	(7) QTY
					GROUP 0306 TANKS, LINES, FITTINGS, HEADERS	
					FIG. 75 FUEL TANK SUPPLY AND RETURN LINES, DUAL (M939, M939A1)	
1	PAOZZ	5305002678953	80204	B1821BH025F063N	SCREW, CAP, HEXAGON H UOC:DAE,DAF,DAG,DAH,DAL,V18,V19,V20, V21,V22,ZAE,ZAF,ZAG,ZAH,ZAL	1
2	PFOZZ	5340010907658	19207	12256187	BRACKET, SUPPORT FUE UOC:DAE,DAF,DAG,DAH,DAL,V18,V19,V20, V21,V22,ZAE,ZAF,ZAG,ZAH,ZAL	1
3	PAOZZ	5340000881254	96906	MS21333-104	CLAMP, LOOP................................ UOC:DAE,DAF,DAG,DAH,DAL,V18,V19,V20, V21,V22,ZAE,ZAF,ZAG,ZAH,ZAL	1
4	PAOZZ	4720010915169	19207	12256185	HOSE ASSEMBLY, METAL UOC:DAE,DAF,DAG,DAH,DAL,V18,V19,V20, V21,V22,ZAE,ZAF,ZAG,ZAH,ZAL	1
5	PAOZZ	5310001436102	96906	MS51922-6	NUT, SELF-LOCKING, HE .... UOC:DAE,DAF,DAG,DAH,DAL,V18,V19,V20, V21,V22, ZAE,ZAF,ZAG,ZAH,ZAL	1
6	MOOZZ		19207	12256180	TUBE ASSEMBLY MAKE FROM TUBING, P/N......... 8689208 .... UOC:DAE,DAF,DAG,DAH,DAL,V18,V19,V20, V21,V22, ZAE,ZAF,ZAG, ZAH,ZAL	1
7	PAOZZ	4730000142433	21450	142433	.INVERTED NUT, TUBE C .... UOC:DAE,DAF,DAG,DAH,DAL,V18,V19,V20, V21,V22, ZAE,ZAF,ZAG,ZAH,ZAL	2
8	PAOZZ	5305002693233	80204	B1821BH038F063N	SCREW, CAP, HEXAGON H UOC:DAE,DAF,DAG,DAH,DAL,V18,V19,V20, V21,V22,ZAE,ZAF,ZAG,ZAH,ZAL	4
9	PAOZZ	5340009546014	96906	MS21333-121	CLAMP, LOOP UOC:DAE,DAF,DAG,DAH,DAL,V18,V19,V20, V21,V22,ZAE,ZAF,ZAG,ZAH,ZAL	4
10	PAOZZ	5310009591488	96906	MS51922-21	NUT, SELF-LOCKING .... UOC:DAE,DAF,DAG,DAH,DAL,V18,V19,V20, V21,V22, ZAE,ZAF,ZAG,ZAH,ZAL	5
11	PAOZZ	5340009891771	96906	MS21333-123	CLAMP, LOOP UOC:DAE,DAF,DAG,DAH,DAL,V18,V19,V20, V21,V22,ZAE,ZAF,ZAG,ZAH,ZAL	3
12	MOOZZ		19207	12256177	TUBE ASSEMBLY FUEL INLET, MAKE FROM......... TUBING, P/N 8689210 .... UOC:DAE,DAF,DAG,DAH,DAL,V18,V19,V20, V21,V22,ZAE,ZAF,ZAG,ZAH,ZAL	1
13	PAOZZ	4730000142435	12204	142435	.INVERTED NUT, TUBE C UOC:DAE,DAF,DAG,DAH,DAL,V18,V19,V20, V21,V22,ZAE,ZAF,ZAG,ZAH,ZAL	1
14	PAOZZ	4730002029036	81343	8-8 120102BA	ADAPTER, STRAIGHT, PI................................ UOC:DAE,DAF,DAG,DAH,DAL,V18,V19,V20, V21,V22, ZAE,ZAF,ZAG,ZAH,ZAL	1
15	PAOZZ	4820007529040	96906	MS35782-4	COCK, DRAIN................................ UOC:DAE,DAF,DAG,DAH,DAL,V18,V19,V20, V21,V22,ZAE,ZAF,ZAG,ZAH,ZAL	1

(1) ITEM NO	(2) SMR CODE	(3) NSN	(4) CAGEC	(5) PART NUMBER	(6) DESCRIPTION AND USABLE ON CODES (UOC)	(7) QTY
16	PAOZZ	4730011336209	96906	MS14315-6X	BUSHING, PIPE... UOC:DAE,DAF,DAG,DAH,DAL,V18,V19,V20, V21,V22,ZAE,ZAF,ZAG,ZAH,ZAL	1
17	PAOZZ	4730010952034	19207	8366166-1	TEE, PIPE.. UOC:DAE,DAF,DAG,DAH,DAL,V18,V19,V20, V21,V22, ZAE,ZAF,ZAG,ZAH,ZAL	1
18	PAOZZ	4730010906468	19207	11668054	ELBOW, PIPE TO TUBE .... UOC:DAE,DAF,DAG,DAH,DAL,V18,V19,V20, V21,V22,ZAE,ZAF,ZAG, ZAH,ZAL	1
19	PAOZZ	4730011044314	19207	11668054-1	ELBOW, PIPE TO TUBE UOC:DAE,DAF,DAG,DAH,DAL,V18,V19,V20, V21,V22,ZAE,ZAF,ZAG,ZAH,ZAL	1
20	PAOZZ	5340000538994	96906	MS21333-126	CLAMP, LOOP... UOC:DAE,DAF,DAG,DAH,DAL,V18,V19 ,V20, V21,V22,ZAE,ZAF,ZAG,ZAH, ZAL	1
21	PAOZZ	5305002692804	96906	MS90726-61	SCREW, CAP, HEXAGON H UOC:DAE,DAF,DAG,DAH,DAL,V18,V19,V20, V21,V22,ZAE,ZAF,ZAG,ZAH,ZAL	1
22	PAOZZ	4720010907618	19207	12256186	HOSE ASSEMBLY, NONME UOC:DAE,DAF,DAG,DAH,DAL,V18,V19,V20, V21,V22,ZAE,ZAF,ZAG,ZAH,ZAL	1

END OF FIGURE

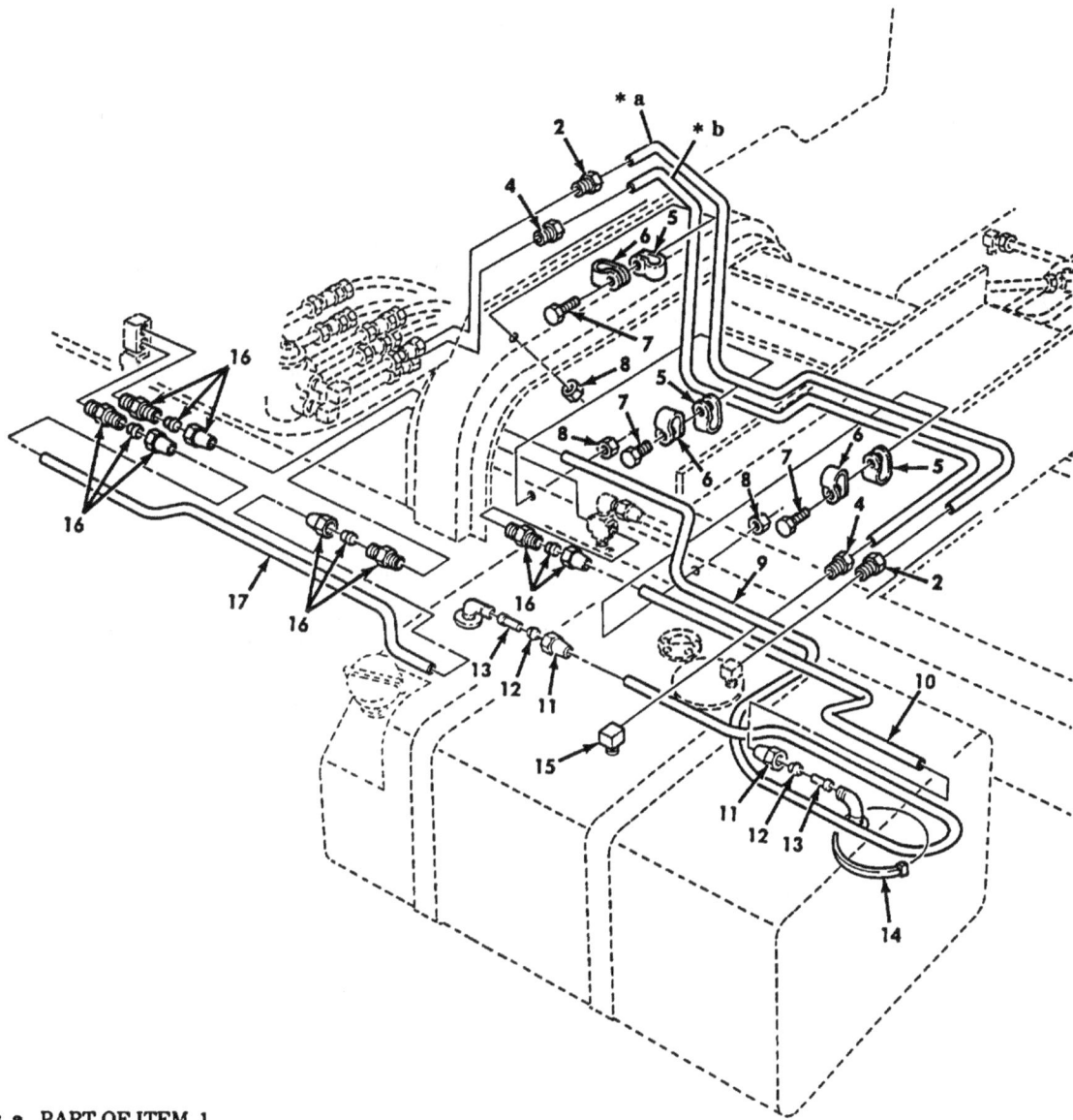

* a PART OF ITEM 1
* b PART OF ITEM 3

Figure 76. Fuel Tank Lines, Dual, L.H.

(1) ITEM NO	(2) SMR CODE	(3) NSN	(4) CAGEC	(5) PART NUMBER	(6) DESCRIPTION AND USABLE ON CODES (UOC)	(7) QTY
					GROUP 0306 TANKS, LINES, FITTINGS, HEADERS	
					FIG. 76 FUEL TANK LINES, DUAL, L.H.	
1	MOOZZ		19207	12256184	TUBE ASSY FUEL, L.H. INLET, MAKE.................... FROM TUBING, P/N 8689210 .... UOC:DAE,DAF,DAG,DAH,V19,V20,V21,V22, ZAE,ZAF,ZAG,ZAH	1
2	PAOZZ	4730000142435	12204	142435	.INVERTED NUT, TUBE C.................................. UOC:DAE,DAF,DAG,DAH,V19,V20,V21,V22, ZAE,ZAF,ZAG,ZAH	2
3	MOOZZ		19207	12256183	TUBE ASSY FUEL, L.H. RETURN, MAKE .... FROM TUBING, P/N 8689208 .... UOC:DAE,DAF,DAG,DAH,V19,V20,V21,V22, ZAE,ZAF,ZAG,ZAH	1
4	PAOZZ	4730000142433	21450	142433	. INVERTED NUT, TUBE C.................................. UOC:DAE,DAF,DAG,DAH,V19,V20,V21,V22, ZAE,ZAF,ZAG,ZAH	2
5	PAOZZ	5340009891771	96906	MS21333-123	CLAMP, LOOP.. UOC:DAE,DAF,DAG,DAH,V19,V20,V21,V22, ZAE,ZAF,ZAG,ZAH	3
6	PAOZZ	5340009546014	96906	MS21333-121	CLAMP, LOOP .... UOC:DAE,DAF,DAG,DAH,V19,V20,V21,V22, ZAE,ZAF,ZAG,ZAH	3
7	PAOZZ	5305002693233	80204	B1821BH038F063N	SCREW, CAP, HEXAGON H.............................. UOC:DAE,DAF,DAG,DAH,V19,V20,V21,V22, ZAE,ZAF,ZAG,ZAH	3
8	PAOZZ	5310009591488	96906	MS51922-21	NUT, SELF-LOCKING, HE................................. UOC:DAE,DAF,DAG,DAH,V19,V20,V21,V22, ZAE,ZAF,ZAG,ZAH	3
9	MOOZZ		19207	CPR104420-1-105	TUBING, NYLON MAKE FROM TUBING, P/N........... CPR104420-1, 105 INCHES LONG........................ UOC:DAE,DAF,DAG,DAH,V19,V20,V21,V22, ZAE,ZAF,ZAG,ZAH	1
10	MOOZZ		19207	CPR104420-1-37	TUBING, NYLON MAKE FROM TUBING, P/N........... CPR104420-1, 37 INCHES LONG .... UOC:DAE,DAF,DAG,DAH,V19,V20,V21,V22, ZAE,ZAF,ZAG,ZAH	1
11	PAOZZ	4730002788824	81343	4 120111B	NUT, TUBE COUPLING.................................... UOC:DAE,DAF,DAG,DAH,V19,V20,V21,V22, ZAE,ZAF,ZAG,ZAH	2
12	PAOZZ	4730001220477	81343	4 120115B	SLEEVE, COMPRESSION,................................. UOC:DAE,DAF,DAG,DAH,V19,V20,V21,V22, ZAE,ZAF,ZAG,ZAH	2
13	PAOZZ	4730001324588	79470	2030X4A	INSERT, TUBE FITTING.................................... UOC:DAE,DAF,DAG,DAH,V19,V20,V21,V22, ZAE,ZAF,ZAG,ZAH	4
14	PAOZZ	5975009846582	96906	MS3367-1-0	STRAP, TIEDOWN, ELECT .... UOC:DAE,DAF,DAG,DAH,V19,V20,V21,V22, ZAE,ZAF,ZAG,ZAH	3
15	PAOZZ	4730002778274	21450	137423	ELBOW, PIPE TO TUBE	1

(1) ITEM NO	(2) SMR CODE	(3) NSN	(4) CAGEC	(5) PART NUMBER	(6) DESCRIPTION AND USABLE ON CODES (UOC)	(7) QTY
					UOC:DAE,DAF,DAG,DAH,V19,V20,V21,V22, ZAE,ZAF,ZAG,ZAH	
16	PAOZZ	4730002704580	96906	MS39179-2	ADAPTER, STRAIGHT, PI......................................	4
					UOC:DAE,DAF,DAG,DAH,V19,V20,V21,V22, ZAE,ZAF,ZAG,ZAH	
17	MOOZZ		19207	CPR104420-1-36	TUBING, NYLON MAKE FROM TUBING, P/N........... CPR104420-1, 36 INCHES LONG .... UOC:DAE,DAF,DAG,DAH,V19,V20,V21,V22, ZAE, ZAF, ZAG, ZAH	1

END OF FIGURE

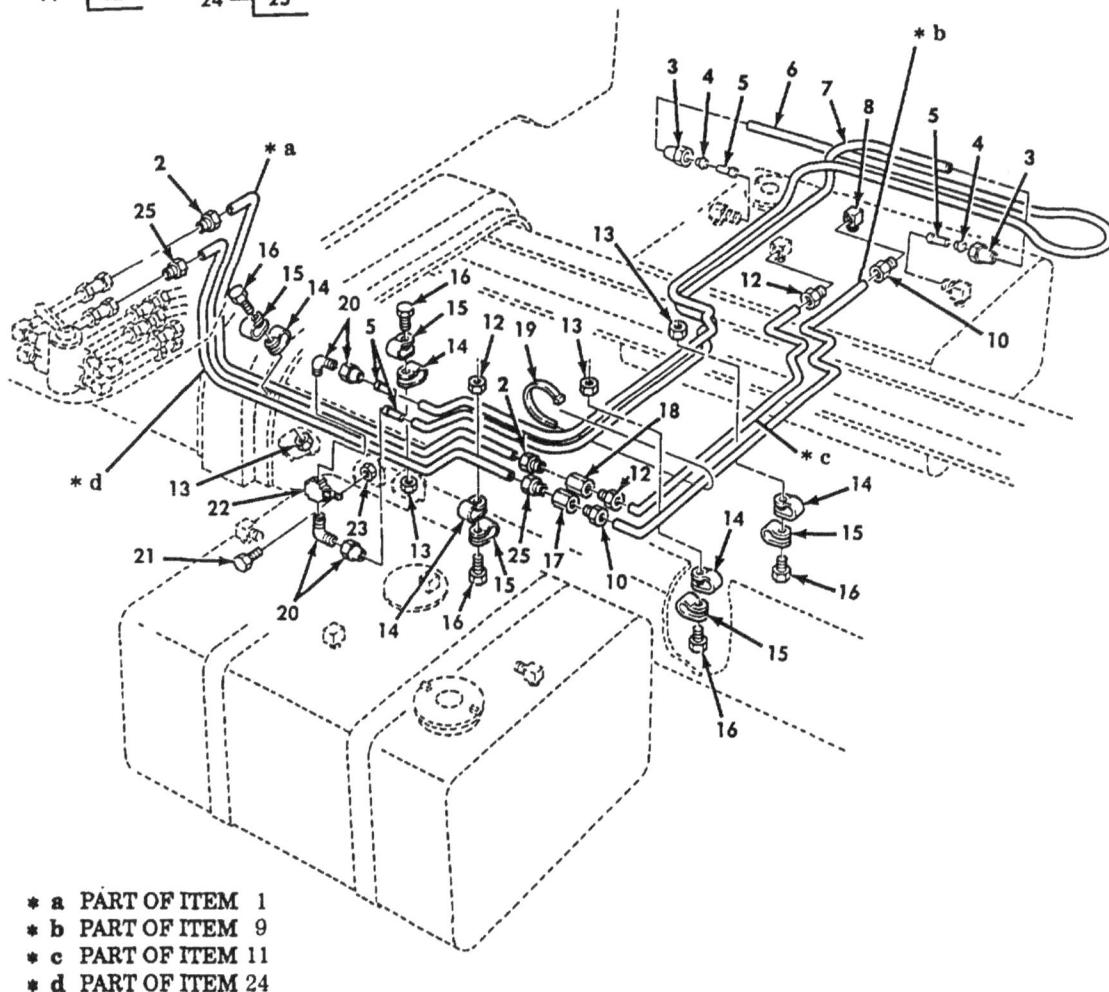

Figure 77. Fuel Tank Selector Valve to R.H. Fuel Tank Lines, Dual.

(1) ITEM NO	(2) SMR CODE	(3) NSN	(4) CAGEC	(5) PART NUMBER	(6) DESCRIPTION AND USABLE ON CODES (UOC)	(7) QTY
					GROUP 0306 TANKS, LINES, FITTINGS, HEADERS	
					FIG. 77 FUEL TANK SELECTOR VALVE TO R.H. FUEL TANK LINES, DUAL	
1	MOOZZ		19207	12256181	TUBE ASSY R.H. SUPPLY, FRONT, MAKE FROM TUBING, P/N 8689210 UOC:DAE,DAF,DAG,DAH,V19,V20,V21,V22, ZAE,ZAF,ZAG,ZAH	1
2	PAOZZ	4730000142435	12204	142435	.INVERTED NUT, TUBE C UOC:DAE,DAF,DAG,DAH,V19,V20,V21,V22, ZAE,ZAF,ZAG,ZAH	2
3	PAOZZ	4730001324588	79470	2030X4A	INSERT, TUBE FITTING UOC:DAE,DAF,DAG,DAH,V19,V20,V21,V22, ZAE,ZAF,ZAG,ZAH	2
4	PAOZZ	4730001220477	81343	4 120115B	SLEEVE, COMPRESSION UOC:DAE,DAF,DAG,DAH,V19,V20,V21,V22, ZAE,ZAF,ZAG,ZAH	2
5	PAOZZ	4730002788824	81343	4 120111B	NUT, TUBE COUPLING UOC:DAE,DAF,DAG,DAH,V19,V20,V21,V22, ZAE,ZAF,ZAG,ZAH	2
6	MOOZZ		19207	CPR104420-1-123	TUBING, NYLON VENT RIGHT REAR, MAKE FROM TUBING, P/N CPR104420-1, 123 INCHES LONG UOC:DAE,DAF,DAG,DAH,V19,V20,V21,V22, ZAE,ZAF,ZAG,ZAH	1
7	MOOZZ		19207	CPR104420-1-92	TUBING, NYLON VENT, R.H. REAR, MAKE FROM TUBING, P/N CPR104420-1, 92 INCHES LONG UOC:DAE,DAF,DAG,DAH,V19,V20,V21,V22, ZAE,ZAF,ZAG,ZAH	1
8	PAOZZ	4730002778274	21450	137423	ELBOW, PIPE TO TUBE UOC:DAE,DAF,DAG,DAH,V19,V20,V21,V22, ZAE,ZAF,ZAG,ZAH	1
9	MOOZZ		19207	12256179	TUBE ASSY FUEL R.H. RETURN, MAKE FROM TUBING, P/N 8689208 UOC:DAE,DAF,DAG,DAH,V19,V20,V21,V22, ZAE,ZAF,ZAG,ZAH	1
10	PAFZZ	4730000142433	21450	142433	.INVERTED NUT, TUBE C UOC:DAE,DAF,DAG,DAH,V19,V20,V21,V22, ZAE,ZAF,ZAG,ZAH	2
11	MOOZZ		19207	12256182	TUBE ASSY R.H. SUPPLY, REAR, MAKE FROM TUBING, P/N 8689210 UOC:DAE,DAF,DAG,DAH,V19,V20,V21,V22, ZAE,ZAF,ZAG,ZAH	1
12	PAOZZ	4730000142435	12204	142435	.INVERTED NUT, TUBE C UOC:DAE,DAF,DAG,DAH,V19,V20,V21,V22, ZAE,ZAF,ZAG,ZAH	2
13	PAOZZ	5310009591488	96906	MS51922-21	NUT, SELF-LOCKING, HE UOC:DAE,DAF,DAG,DAH,V19,V20,V21,V22, ZAE,ZAF,ZAG,ZAH	5

(1) ITEM NO	(2) SMR CODE	(3) NSN	(4) CAGEC	(5) PART NUMBER	(6) DESCRIPTION AND USABLE ON CODES (UOC)	(7) QTY
15	PAOZZ	5340009546014	96906	MS21333-121	CLAMP, LOOP .... UOC:DAE,DAF,DAG,DAH,V19,V20,V21,V22, ZAE,ZAF,ZAG,ZAH	5
16	PAOZZ	5305002693233	80204	B1821BH038F063N	SCREW, CAP, HEXAGON H .... UOC:DAE,DAF,DAG,DAH,V19,V20,V21,V22, ZAE,ZAF,ZAG,ZAH	5
17	PAOZZ	4730002788888	21450	137415	COUPLING, TUBE .... UOC:DAE,DAF,DAG,DAH,V19,V20,V21,V22, ZAE,ZAF,ZAG,ZAH	1
18	PAOZZ	4730011487397	24617	442340	COUPLING, TUBE .... UOC:DAE,DAF,DAG,DAH,V19,V20,V21,V22, ZAE,ZAF,ZAG,ZAH	1
19	PAOZZ	5975009846582	96906	MS3367-1-0	STRAP, TIEDOWN, ELECT .... UOC:DAE,DAF,DAG,DAH,V19,V20,V21,V22, ZAE,ZAF,ZAG,ZAH	10
20	PAOZZ	4730012085859	19207	8328777	ELBOW, PIPE TO TUBE .... UOC:DAE,DAF,DAG,DAH,V19,V20,V21,V22, ZAE,ZAF,ZAG,ZAH	2
21	PAOZZ	5306002259088	96906	MS90726-33	BOLT, MACHINE .... UOC:DAE,DAF,DAG,DAH,V19,V20,V21,V22, ZAE,ZAF,ZAG,ZAH	1
22	PAOZZ	4730011126561	19207	11669082	CROSS, PIPE .... UOC:DAE,DAF,DAG,DAH,V19,V20,V21,V22, ZAE,ZAF,ZAG,ZAH	1
23	PAOZZ	5310002416658	96906	MS51943-34	NUT, SELF-LOCKING, HE .... UOC:DAE,DAF,DAG,DAH,V19,V20,V21,V22, ZAE,ZAF,ZAG,ZAH	1
24	MOOZZ		19207	12256178	TUBE ASSY FUEL R.H. RETURN, MAKE............... FROM TUBING, P/N 8689802 .... UOC:DAE,DAF,DAG,DAH,V19,V20,V21,V22, ... ZAE,ZAF,ZAG,ZAH	1
25	PAOZZ	4730000142433	21450	142433	.INVERTED NUT, TUBE C .... UOC:DAE,DAF,DAG,DAH,V19,V20,V21,V22, ZAE,ZAF,ZAG,ZAH	2

END OF FIGURE

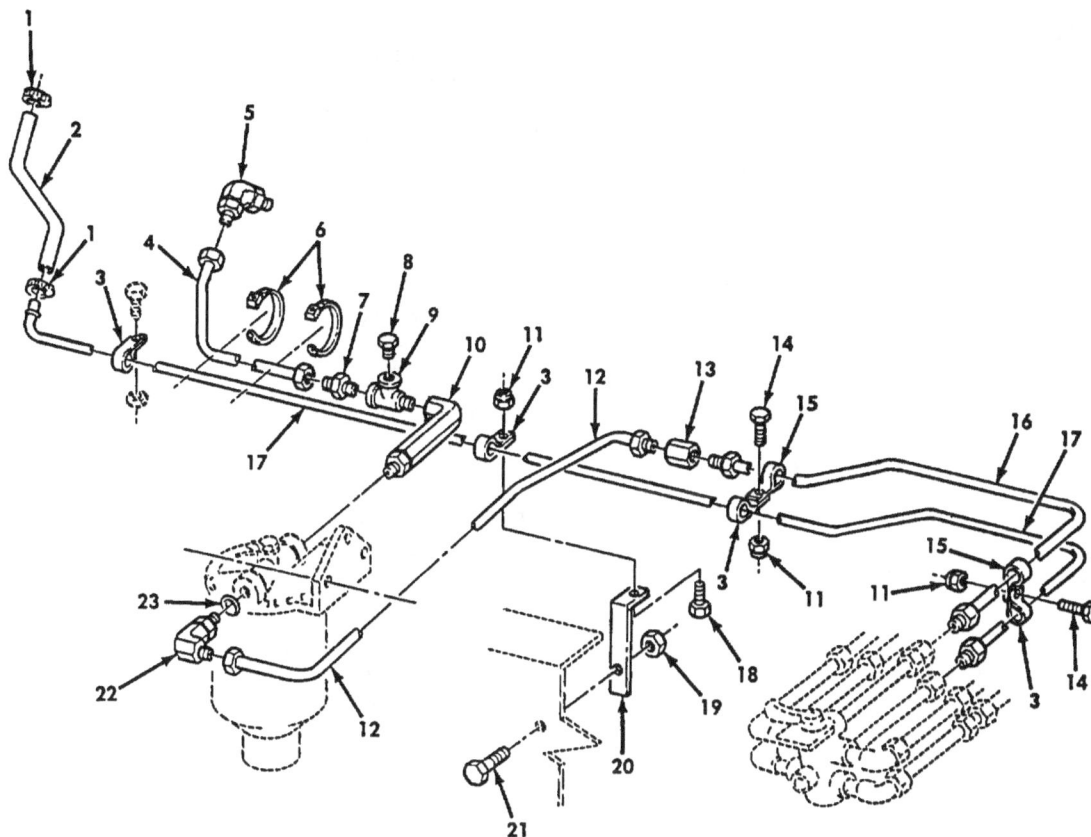

Figure 78. Dual Fuel Tank Supply and Return Lines (M939A2).

(1) ITEM NO	(2) SMR CODE	(3) NSN	(4) CAGEC	(5) PART NUMBER	(6) DESCRIPTION AND USABLE ON CODES (UOC)	(7) QTY
					GROUP 0306 TANKS, LINES, FITTINGS, HEADERS	
					FIG. 78 DUAL FUEL TANK SUPPLY AND RETURN LINES (M939A2)	
1	PAOZZ	4730011098001	19207	11662913	CLAMP, HOSE .... UOC:ZAE,ZAF,ZAG,ZAH,ZAL	2
2	MOOZZ		98441	A2697 X 15 1/2 IN	RETURN HOSE, FUEL MAKE FROM HOSE,........... 4720-00-274-1154, 15-1/2 IN LONG...................... UOC:ZAE,ZAF,ZAG,ZAH,ZAL	1
3	PAOZZ	5340009546014	96906	MS21333-121	CLAMP, LOOP .... UOC:ZAE,ZAF,ZAG,ZAH,ZAL	4
4	PAOZZ	4720014412922	01276	FG2722HHH0270	HOSE ASSEMBLY, NONME PART OF KIT P/N .... 57K0251... UOC:ZAA,ZAB,ZAC,ZAD,ZAE,ZAF,ZAG,ZAH, ZAJ,ZAK,ZAL	1
5	PAOZZ	4730002351483	33334	E1-8B	ELBOW, PIPE TO TUBE .... UOC:ZAE,ZAF,ZAG,ZAH,ZAL	1
6	PAOZZ	5975008956630	96906	MS3367-3-0	STRAP, TIEDOWN, ELECT .... UOC:ZAE,ZAF,ZAG,ZAH,ZAL	2
7	PAOZZ	4730008099427	81343	8-8 070102CA	ADAPTER, STRAIGHT, PI .... UOC:ZAE,ZAF,ZAG,ZAH,ZAL	1
8	PAOZZ	4730000144027	79470	3152X8	PLUG, PIPE .... UOC:ZAE,ZAF,ZAG,ZAH,ZAL	1
9	PAOZZ	4730010952034	81343	8-8-8 130424B	TEE .... UOC:ZAE,ZAF,ZAG,ZAH,ZAL	1
10	PAOZZ	4730011044314	92003	AE40GSS	ELBOW, PIPE TO TUBE .... UOC:ZAE,ZAF,ZAG,ZAH,ZAL	1
11	PAOZZ	5310009591488	96906	MS51922-21	NUT, SELF-LOCKING, HE .... UOC:ZAE,ZAF,ZAG,ZAH,ZAL	3
12	PAOZZ	4720014422533	19207	12432371	HOSE ASSEMBLY, NONME .... UOC:ZAE,ZAF,ZAG,ZAH,ZAL	1
13	PAOZZ	4730002033179	81343	8-8 010103B	ADAPTER, STRAIGHT, PI .... UOC:ZAE,ZAF,ZAG,ZAH,ZAL	1
14	PAOZZ	5305002693233	80204	B1821BH038F063N	SCREW, CAP, HEXAGON H .... UOC:ZAE,ZAF,ZAG,ZAH,ZAL	2
15	PAOZZ	4730012806402	08627	3545	CLAMP, HOSE, SPECIAL-............................... UOC:ZAE,ZAF,ZAG,ZAH,ZAL	2
16	PAOZZ	4710012822586	47457	12256177A	TUBE ASSEMBLY, METAL.............................. UOC:ZAE,ZAF,ZAG,ZAH,ZAL	1
17	PAOZZ	4710012822585	47457	12256180A	TUBE ASSEMBLY, METAL.............................. UOC:ZAE,ZAF,ZAG,ZAH,ZAL	1
18	PAOZZ	5305002693234	80204	B1821BH038F075N	SCREW, CAP, HEXAGON H .... UOC:ZAE,ZAF,ZAG,ZAH,ZAL	1
19	PAOZZ	5310002256408	96906	MS51922-53	NUT, SELF-LOCKING, HE .... UOC:ZAE,ZAF,ZAG,ZAH,ZAL	1
20	PFOZZ	5340012857762	47457	20510933	BRACKET, ANGLE.. UOC:ZAE,ZAF,ZAG,ZAH,ZAL	1
21	PAOZZ	5305007262543	96906	MS90727-160	SCREW, CAP, HEXAGON H .... UOC:ZAE,ZAF,ZAG,ZAH,ZAL	1
22	PAOZZ		81343	8-10 010220CA	ELBOW, PIPE... UOC:ZAE,ZAF,ZAG,ZAH,ZAL	1

(1) ITEM NO	(2) SMR CODE	(3) NSN	(4) CAGEC	(5) PART NUMBER	(6) DESCRIPTION AND USABLE ON CODES (UOC)	(7) QTY
23	PAOZZ	5330012205498	56161	15557258	PACKING, PREFORMED .... UOC:ZAE,ZAF,ZAG,ZAH,ZAL	1

END OF FIGURE

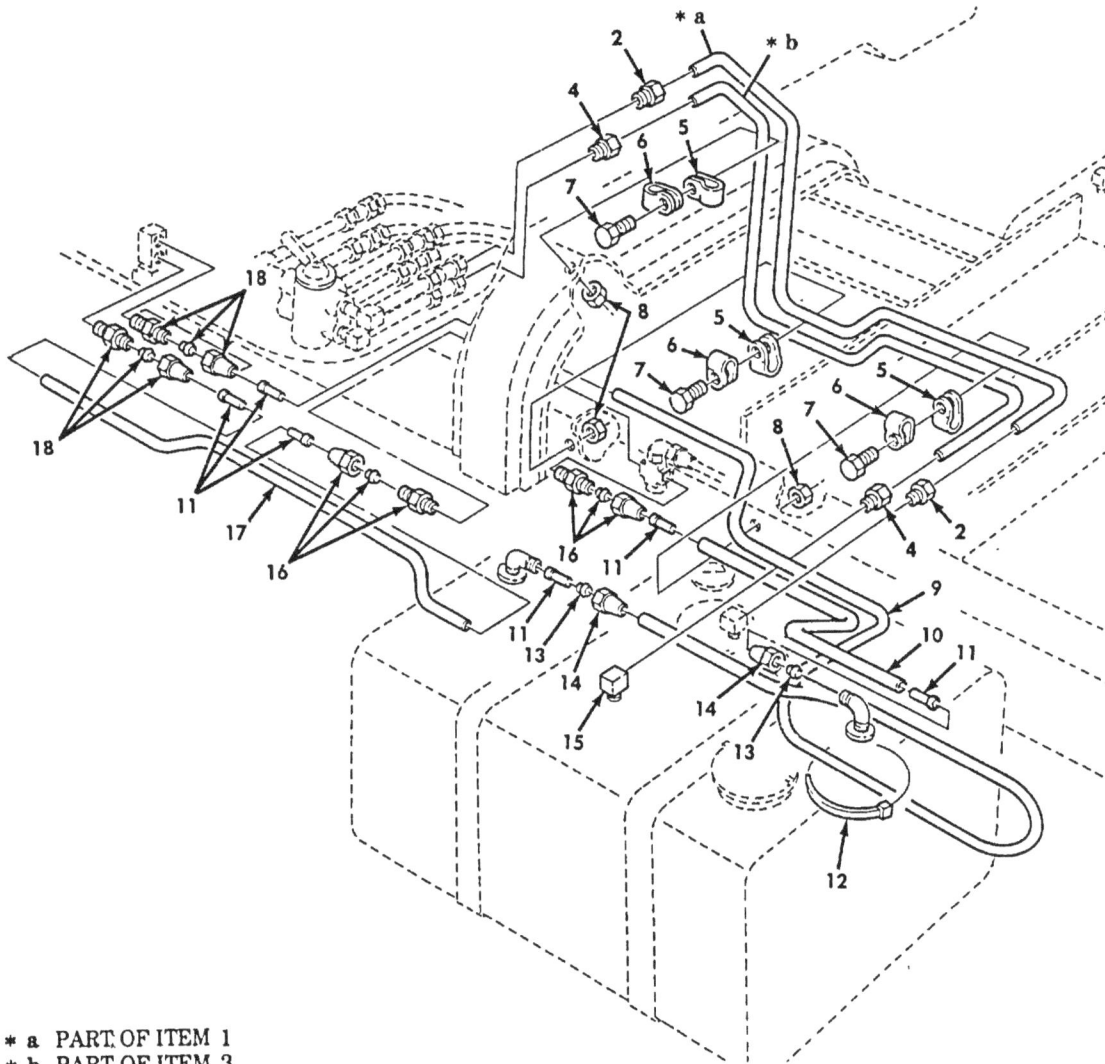

1 ─┤ 2 │ 3 ─┤ 4 │

* a   PART OF ITEM 1
* b   PART OF ITEM 3

Figure 79. Fuel Tank Fuel Lines to L.H. Tank, Wrecker.

(1) ITEM NO	(2) SMR CODE	(3) NSN	(4) CAGEC	(5) PART NUMBER	(6) DESCRIPTION AND USABLE ON CODES (UOC)	(7) QTY
					GROUP 0306 TANKS, LINES, FITTINGS, HEADERS	
					FIG. 79 FUEL TANK FUEL LINES TO L.H. TANK, WRECKER	
1	MOOZZ		19207	12256595	TUBE ASSEMBLY FUEL INLET, L.H., MAKE........... FROM TUBING, P/N 8689210 .... UOC:DAL,V18,ZAL	1
2	PAOZZ	4730000142435	12204	142435	.INVERTED NUT, TUBE C UOC:DAL,V18,ZAL	2
3	MOOZZ		19207	12256596	TUBE ASSEMBLY FUEL, RETURN, L.H................. MAKE FROM TUBING, P/N 8689208..................... UOC:DAL,V18,ZAL	1
4	PAOZZ	4730000142433	21450	142433	.INVERTED NUT, TUBE C UOC:DAL,V18,ZAL	2
5	PAOZZ	5340009891771	96906	MS21333-123	CLAMP, LOOP UOC:DAL,V18,ZAL	3
6	PAOZZ	5340009546014	96906	MS21333-121	CLAMP, LOOP UOC:DAL,V18,ZAL	3
7	PAOZZ	5305002693233	80204	B1821BH038F063N	SCREW, CAP, HEXAGON H................................. UOC:DAL,V18,ZAL	3
8	PAOZZ	5310009591488	96906	MS51922-21	NUT, SELF-LOCKING, HE................................. UOC:DAL,V18,ZAL	3
9	MOOZZ		19207	CPR104420-1-150	TUBING, NYLON L.H., MAKE FROM TUBING, ... P/N CPR104420-1,150 INCHES LONG UOC:DAL,V18,ZAL	1
10	MOOZZ		19207	CPR104420-1-60	TUBING, NYLON L.H., MAKE FROM TUBING, ... P/N CPR104420-1,60 INCHES LONG UOC:DAL,V18,ZAL	1
11	PAOZZ	4730001324588	79470	2030X4A	INSERT, TUBE FITTING UOC:DAL,V18,ZAL	4
12	PAOZZ	5975009846582	96906	MS3367-1-0	STRAP, TIEDOWN, ELECT................................. UOC:DAL,V18,ZAL	5
13	PAOZZ	4730001220477	81343	4 120115B	SLEEVE, COMPRESSION UOC:DAL,V18,ZAL	2
14	PAOZZ	4730002788824	81343	4 120111B	NUT, TUBE COUPLING UOC:DAL,V18,ZAL	2
15	PAOZZ	4730002778274	21450	137423	ELBOW, PIPE TO TUBE UOC:DAL,V18,ZAL	1
16	PAOZZ	4730002704580	96906	MS39179-2	ADAPTER, STRAIGHT, PI UOC:DAL,V18,ZAL	3
17	MOOZZ		19207	CPR104420-1-36	TUBING, NONMETALLIC MAKE FROM TUBING, ... P/N CPR104420-1,36 INCHES LONG..................... UOC:DAL,V18,ZAL	1
18	PAOZZ	4730012085859	19207	8328777	ELBOW, PIPE TO TUBE UOC:DAL,V18,ZAL	1

END OF FIGURE

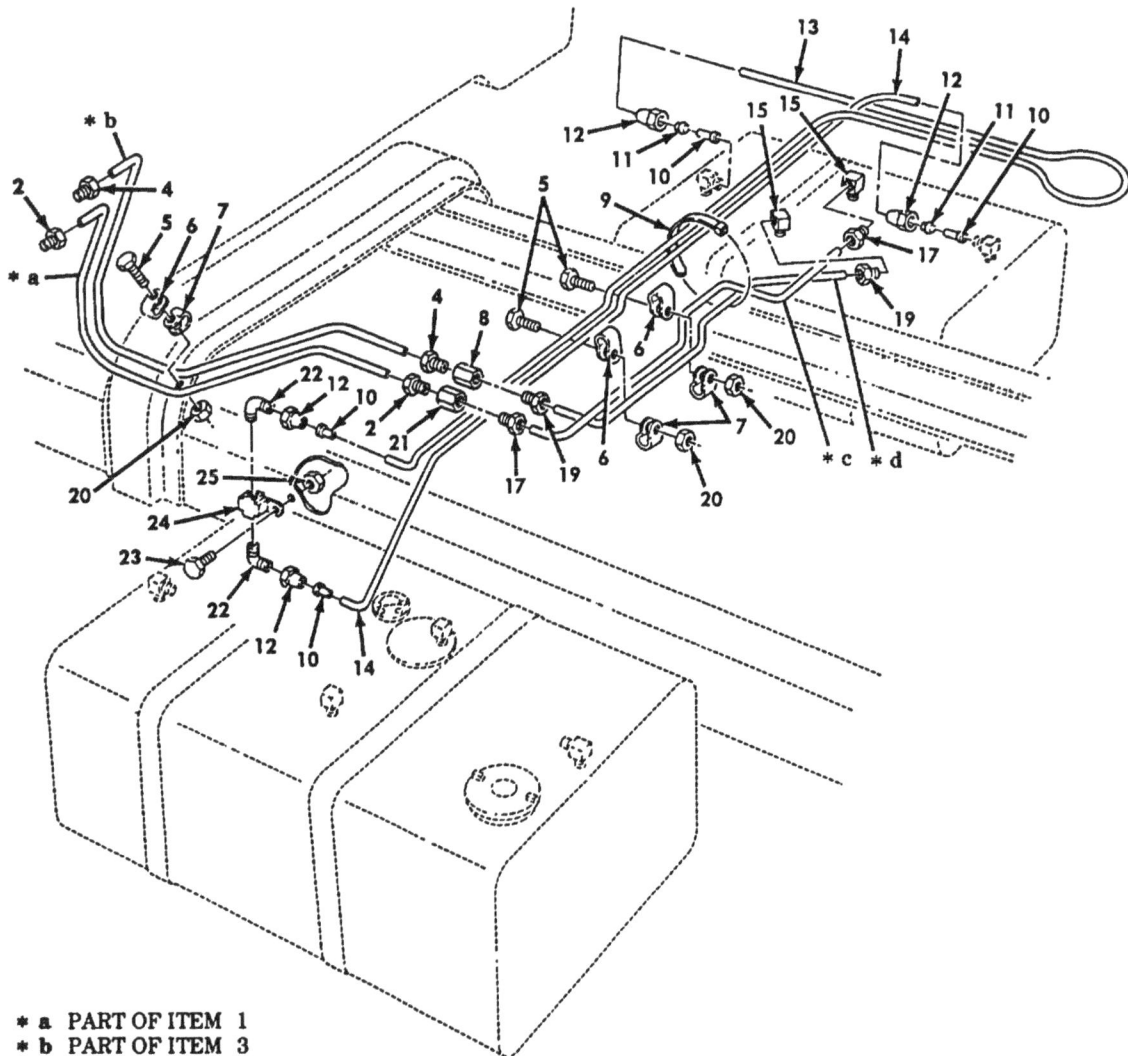

* a PART OF ITEM 1
* b PART OF ITEM 3
* c PART OF ITEM 16
* d PART OF ITEM 18

Figure 80. Dual Fuel Tank Fuel Lines to R.H. Tank, Wrecker.

(1) ITEM NO	(2) SMR CODE	(3) NSN	(4) CAGEC	(5) PART NUMBER	(6) DESCRIPTION AND USABLE ON CODES (UOC)	(7) QTY
					GROUP 0306 TANKS, LINES, FITTINGS, HEADERS	
					FIG. 80 DUAL FUEL TANK FUEL LINES TO R.H. TANK, WRECKER	
1	MOOZZ		19207	12256597	TUBE ASSEMBLY FUEL R.H. INLET REAR, ... MAKE FROM TUBING, P/N 8689208 .... UOC:DAL,V18,ZAL	1
2	PAOZZ	4730000142433	21450	142433	.INVERTED NUT, TUBE C................................... UOC:DAL,V18,ZAL	2
3	MOOZZ		19207	12256600	TUBE ASSEMBLY FUEL, R.H. INLET, MAKE .... FROM TUBING, P/N 8689210 .... UOC:DAL,V18,ZAL	1
4	PAOZZ	4730000142435	12204	142435	.INVERTED NUT, TUBE C................................... UOC:DAL,V18,ZAL	2
5	PAOZZ	5305002693233	80204	B1821BH038F063N	SCREW, CAP, HEXAGON H.............................. UOC:DAL,V18,ZAL	5
6	PAOZZ	5340009546014	96906	MS21333-121	CLAMP, LOOP... UOC:DAL,V18,ZAL	5
7	PAOZZ	5340009891771	96906	MS21333-123	CLAMP, LOOP... UOC:DAL,V18,ZAL	5
8	PAOZZ	4730002788889	21450	137417	COUPLING, TUBE UOC:DAL,V18,ZAL	1
9	PAOZZ	5975009846582	96906	MS3367-1-0	STRAP, TIEDOWN, ELECT UOC:DAL,V18,ZAL	3
10	PAOZZ	4730001324588	79470	2030X4A	INSERT, TUBE FITTING................................... UOC:DAL,V18,ZAL	2
11	PAOZZ	4730001220477	81343	4 120115B	SLEEVE, COMPRESSION................................. UOC:DAL,V18,ZAL	2
12	PAOZZ	4730002788824	81343	4 120111B	NUT, TUBE COUPLING UOC:DAL,V18,ZAL	2
13	MOOZZ		19207	CPR104420-1-142	TUBING, NYLON MAKE FROM TUBING, P/N.......... CPR104420-1,142 INCHES LONG UOC:DAL,V18,ZAL	1
14	MOOZZ		19207	CPR104420-1-128	TUBING, NYLON MAKE FROM TUBING, P/N.......... CPR104420-1,128 INCHES LONG UOC:DAL,V18,ZAL	1
15	PAOZZ	4730002778274	21450	137423	ELBOW, PIPE TO TUBE UOC:DAL,V18,ZAL	2
16	MOOZZ		19207	12256598	TUBE ASSEMBLY FUEL R.H. RETURN, ... MAKE FROM TUBING P/N 8689209 .... UOC:DAL,V18,ZAL	1
17	PAOZZ	4730000142433	21450	142433	.INVERTED NUT, TUBE C................................... UOC:DAL,V18,ZAL	2
18	MOOZZ		19207	12256599	TUBE ASSEMBLY FUEL R.H. INLET, REAR,.......... MAKE FROM TUBING, P/N 8689210 .... UOC:DAL,V18,ZAL	1
19	PAOZZ	4730000142435	12204	142435	.INVERTED NUT, TUBE C................................... UOC:DAL,V18,ZAL	2
20	PAOZZ	5310009591488	96906	MS51922-21	NUT, SELF-LOCKING, HE.................................. UOC:DAL,V18,ZAL	5

(1) ITEM NO	(2) SMR CODE	(3) NSN	(4) CAGEC	(5) PART NUMBER	(6) DESCRIPTION AND USABLE ON CODES (UOC)	(7) QTY
21	PAOZZ	4730002788888	21450	137415	COUPLING, TUBE.. UOC:DAL,V18,ZAL	1
22	PAOZZ	4730012085859	19207	8328777	ELBOW, PIPE TO TUBE UOC:DAL,V18,ZAL	2
23	PAOZZ	5306002259088	96906	MS90726-33	BOLT, MACHINE... UOC:DAL,V18,ZAL	1
24	PAOZZ	4730011126561	19207	11669082	CROSS, PIPE UOC:DAL,V18,ZAL	1
25	PAOZZ	5310002416658	96906	MS51943-34	NUT, SELF-LOCKING, HE..................................... UOC:DAL,V18,ZAL	1

END OF FIGURE

* a PART OF ITEM 2

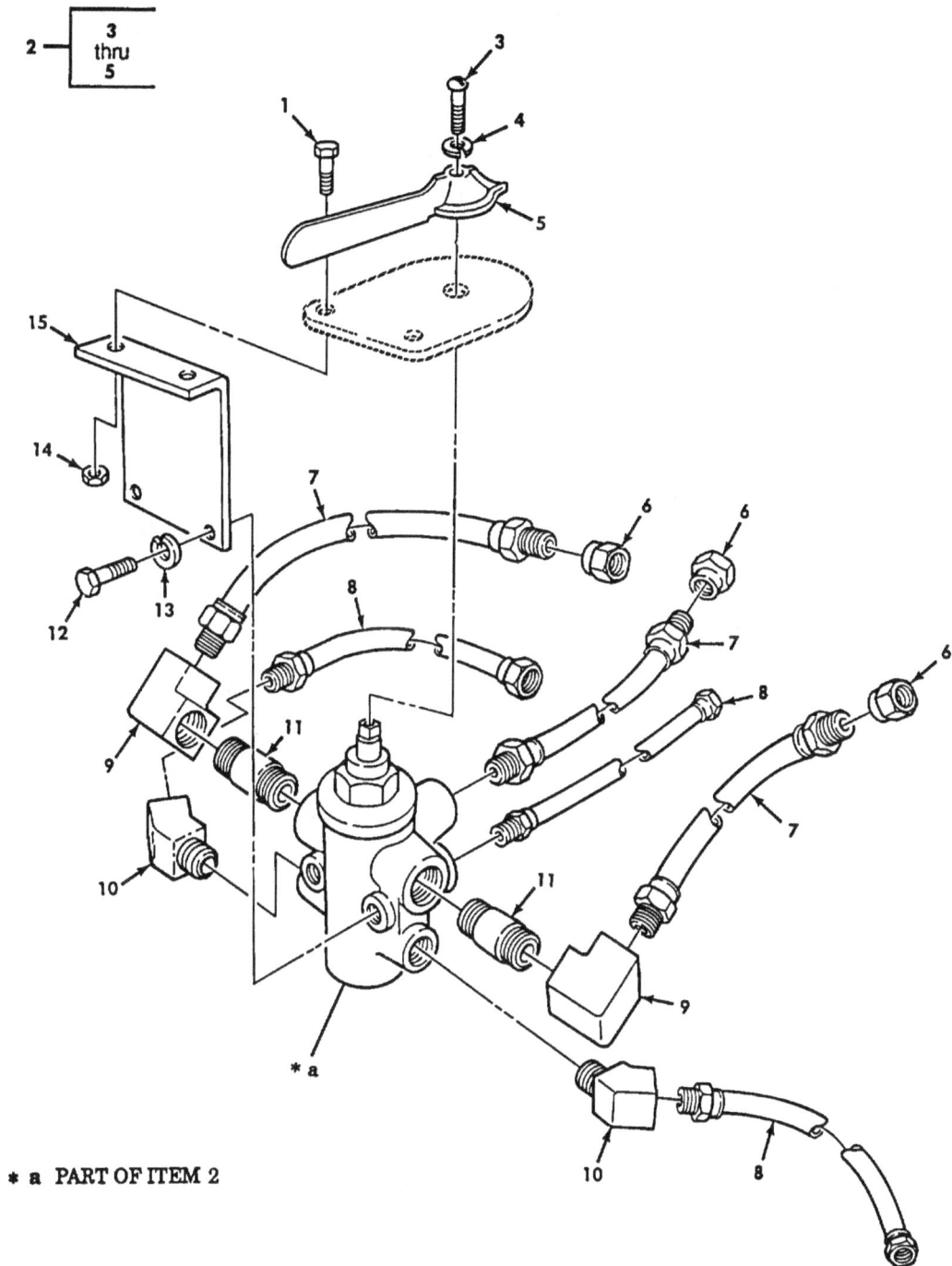

Figure 81. Fuel Tank Selector Valve Assembly and Related Parts.

(1) ITEM NO	(2) SMR CODE	(3) NSN	(4) CAGEC	(5) PART NUMBER	(6) DESCRIPTION AND USABLE ON CODES (UOC)	(7) QTY
					GROUP 0306 TANKS, LINES, FITTINGS, HEADERS	
					FIG. 81 FUEL TANK SELECTOR VALVE ASSEMBLY AND RELATED PARTS	
1	PAOZZ	5306000680514	80204	B1821BH025F088N	BOLT, MACHINE .... UOC:DAE, DAF, DAG, DAH, DAL, V18, V19, V20, V21, V22, ZAE, ZAF, ZAG, ZAH, ZAL	2
2	PAOZZ	4820011177949	70411	SP2483CM, REV C	VALVE, FUEL, 6 PORT .... UOC:DAE, DAF, DAG, DAH, DAL, V18, V19, V20, V21, V22, ZAE, ZAF, ZAG, ZAH, ZAL	1
3	PAOZZ	5305009846208	96906	MS35206-261	.SCREW, MACHINE .... UOC:DAE, DAF, DAG, DAH, DAL, V18, V19, V20, V21, V22, ZAE, ZAF, ZAG, ZAH, ZAL	1
4	PAOZZ	5310000453296	96906	MS35338-43	.WASHER, LOCK .... UOC:DAE, DAF, DAG, DAH, DAL, V18, V19, V20, V21, V22, ZAE, ZAF, ZAG, ZAH, ZAL	1
5	XAOZZ		70411	8X29	.HANDLE, LEVER .... UOC:DAE, DAF, DAG, DAH, DAL, V18, V19, V20, V21, V22, ZAE, ZAF, ZAG, ZAH, ZAL	1
6	PAOZZ	4730002784594	21450	191561	ADAPTER, STRAIGHT, PI .... UOC:DAE, DAF, DAG, DAH, DAL, V18, V19, V20, V21, V22, ZAE, ZAF, ZAG, ZAH, ZAL	3
7	PAOZZ	4720000017854	19207	11664614	HOSE ASSEMBLY, NONME .... UOC:DAE, DAF, DAG, DAH, DAL, V18, V19, V20, V21, V22, ZAE, ZAF, ZAG, ZAH, ZAL	3
8	PAOZZ	4720000969648	19207	11664615	HOSE ASSEMBLY, NONME .... UOC:DAE, DAF, DAG, DAH, DAL, V18, V19, V20, V21, V22, ZAE, ZAF, ZAG, ZAH, ZAL	3
9	PAOZZ	4730002535794	21450	444072	ELBOW, PIPE .... UOC:DAE, DAF, DAG, DAH, DAL, V18, V19, V20, V21, V22, ZAE, ZAF, ZAG, ZAH, ZAL	2
10	PAOZZ	4730002784822	19207	7364214	ELBOW, PIPE .... UOC:DAE, DAF, DAG, DAH, DAL, V18 , V18, V19, V20, V21, V22, ZAE, ZAF, ZAG, ZAH, ZAL	2
11	PAOZZ	4730001961504	24617	192075	NIPPLE, PIPE .... UOC:DAE, DAF, DAG, DAH, DAL, V18, V19, V20, V21, V22, ZAE, ZAF, ZAG, ZAH, ZAL	2
12	PAOZZ	5305007215492	80204	B1821BH038C063N	SCREW, CAP, HEXAGON H .... UOC:DAE, DAF, DAG, DAH, DAL, V18, V19, V20, V21, V22, ZAE, ZAF, ZAG, ZAH, ZAL	2
13	PAOZZ	5310006379541	96906	MS35338-46	WASHER, LOCK .... UOC:DAE, DAF, DAG, DAH, DAL, V18, V19, V20, V21, V22, ZAE, ZAF, ZAG, ZAH, ZAL	2
14	PAOZZ	5310009359022	96906	MS51943-32	NUT, SELF-LOCKING, HE .... UOC:DAE, DAF, DAG, DAH, DAL, V18, V19, V20, V21, V22, ZAE, ZAF, ZAG, ZAH, ZAL	2
15	PAOZZ	5340000360236	19207	10931962	BRACKET, ANGLE .... UOC:DAE, DAF, DAG, DAH, DAL, V18, V19, V20, V21, V22, ZAE, ZAF, ZAG, ZAH, ZAL	1

END OF FIGURE

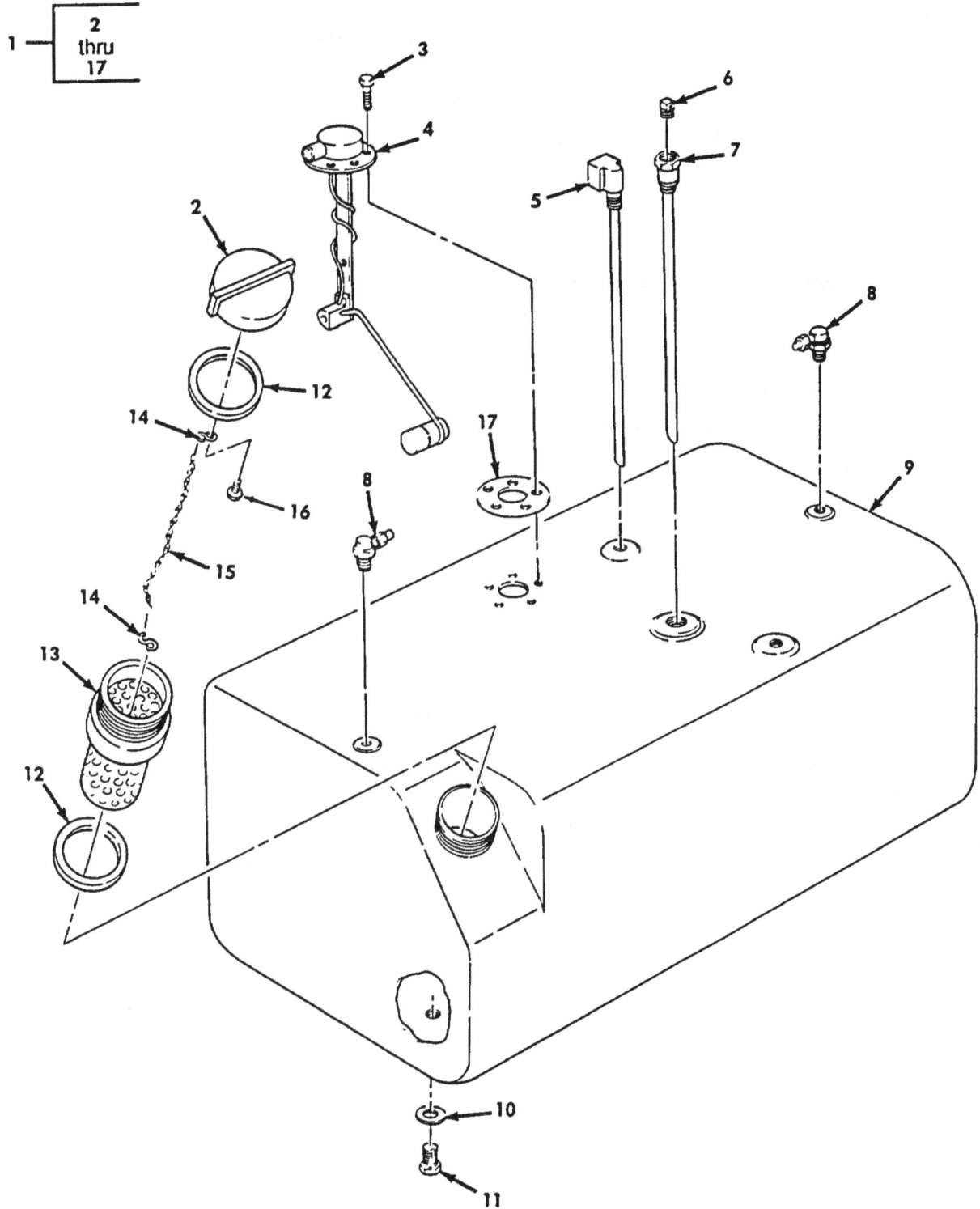

Figure 82. Fuel Tank Assembly.

(1) ITEM NO	(2) SMR CODE	(3) NSN	(4) CAGEC	(5) PART NUMBER	(6) DESCRIPTION AND USABLE ON CODES (UOC)	(7) QTY
					GROUP 0306 TANKS, LINES, FITTINGS, HEADERS	
					FIG. 82 FUEL TANK ASSEMBLY	
1	PAOFF	2910012879119	19207	12301297-5	TANK, FUEL, ENGINE 81 GALLONS .... UOC:DAA, DAB, DAC, DAD, DAE, DAF, DAG, DAH, DAL, DAW, DAX, V12, V13, V14, V15, V16, V17, V18, V19, V20, V21, V22, V39, ZAA, ZAB, ZAC, ZAD, ZAL	1
1	PAOFF	2910012859850	19207	12301297-6	TANK, FUEL, ENGINE 81 GALLONS .... UOC:ZAJ, ZAK, DAJ, DAK, V24, V25	1
1	PAOFF	2910012879120	19207	12301298-3	TANK, FUEL, ENGINE 58 GALLONS .... UOC:DAE, DAF, DAG, DAH, V19, V20, V21, V22, ZAE, ZAF, ZAG, ZAH	2
1	PAOFF	2910012879120	19207	12301298-3	TANK, FUEL, ENGINE 58 GALLONS .... UOC:DAL, V18, ZAL	1
2	PAOZZ	2590013101166	19207	12356774	.CAP, FILLER OPENING	1
3	PAOZZ	5305000977372	19207	11609358-2	.SCREW, SELF SEALING	5
4	PAOZZ	6680002264574	96906	MS500040-6	.TRANSMITTER, LIQUID	1
5	PAOZZ	4710011495078	19207	12302605	.TUBE ASSEMBLY, METAL	1
6	PAOZZ	4730002212137	96906	MS20913-2S	.PLUG, PIPE .... UOC:DAJ, DAK, V24, V25, ZAJ, ZAK	1
7	PAOZZ	4710011495079	19207	12302606	.TUBE ASSEMBLY, METAL .... UOC:DAJ, DAK, V24, V25, ZAJ, ZAK	1
8	PAOZZ	4820013093799	98168	261-3	.VALVE, CHECK ROLLOVER PROTECTION	2
9	XAOZZ		19207	12301298-1	.TANK, FUEL 58 GALLON .... UOC:DAE, DAF, DAG, DAH, DAL, V18, V19, V20, V21, V22, ZAE, ZAF, ZAG, ZAH, ZAL	1
9	XAOZZ		19207	12301297-4	.TANK, FUEL 81 GALLON .... UOC:DAA, DAB, DAC, DAD, DAE, DAF, DAG, DAH, DAJ, DAK, DAL, DAW, DAX, V12, V13, V14, V15, V16, V17, V18, V19, V20, V21, V22, V24, V25, V39, ZAA, ZAB, ZAC, ZAD, ZAJ, ZAK, ZAL	1
10	PAOZZ	5310011335847	72464	A5711	.WASHER, FLAT	1
11	PAOZZ	5365012984877	19207	12356772	.PLUG, MACHINE THREAD	1
12	PAOZZ	5330012996616	19207	12356775	.GASKET	2
13	PAOZZ	4730013145825	19207	12356776	.STRAINER ELEMENT, SE	1
14	PAOZZ	4030002705436	96906	MS87006-3	.HOOK, CHAIN, S	2
15	MOOZZ		16003	C43974 X 15 LG	.CHAIN, WELDLESS MAKE FROM CHAIN, ... P/N 42-C-15120-200, 8 INCHES LONG...................	1
16	PAOZZ	5305008550958	96906	MS24629-45	.SCREW, TAPPING	1
17	PAOZZ	5330007539072	19207	7539072	.GASKET	1

END OF FIGURE

(1) ITEM NO	(2) SMR CODE	(3) NSN	(4) CAGEC	(5) PART NUMBER	(6) DESCRIPTION AND USABLE ON CODES (UOC)	(7) QTY
					ZAJ, ZAK, ZAL	
15	MOOZZ		16003	C43974 X 15 LG	.CHAIN, WELDLESS MAKE FROM CHAIN, P/ .... N 42-C-15120-200, 8 INCHES LONG...................... UOC: ZAA, ZAB, ZAC, ZAD, ZAE, ZAF, ZAG, ZAH, ZAJ, ZAK, ZAL	1
16	PAOZZ	5305008550958	96906	MS24629-45	.SCREW, TAPPING .... UOC:ZAA, ZAB, ZAC, ZAD, ZAE, ZAF, ZAG, ZAH, ZAJ, ZAK, ZAL	1
17	PAOZZ	5330007539072	19207	7539072	.GASKET .... UOC:ZAA, ZAB, ZAC, ZAD, ZAE, ZAF, ZAG, ZAH, ZAJ, ZAK, ZAL	1

END OF FIGURE

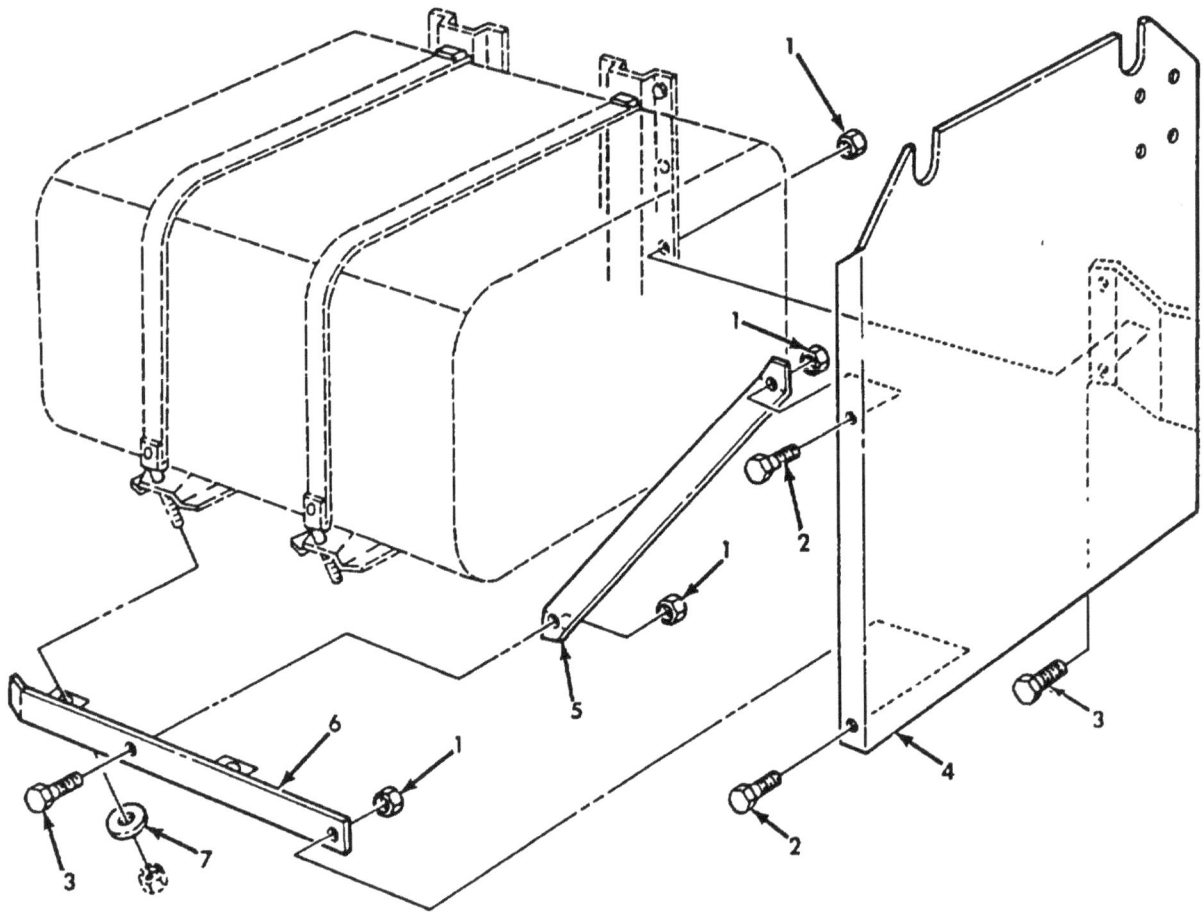

Figure 83. Fuel Tank Protector Plate and Mounting Hardware.

(1) ITEM NO	(2) SMR CODE	(3) NSN	(4) CAGEC	(5) PART NUMBER	(6) DESCRIPTION AND USABLE ON CODES (UOC)	(7) QTY
					GROUP 0306 TANKS, LINES, FITTINGS, HEADERS	
					FIG. 83 FUEL TANK PROTECTOR PLATE AND MOUNTING HARDWARE	
1	PAOZZ	5310008140672	96906	MS51943-36	NUT, SELF-LOCKING, HE .... UOC:DAE, DAF, V19, V20, ZAE, ZAF	8
2	PAOZZ	5305009125113	96906	MS51096-359	SCREW, CAP, HEXAGON H .... UOC:DAE, DAF, V19, V20, ZAE, ZAF	4
3	PAOZZ	5305002692803	96906	MS90726-60	SCREW, CAP, HEXAGON H .... UOC:DAE, DAF, V19, V20, ZAE, ZAF	4
4	PFOZZ	2510004076768	19207	10899256	PROTECTOR PLATE, FUE L.H. TANK .... UOC:DAE, DAF, V19, V20, ZAE, ZAF	1
4	PFOZZ	2510004172713	19207	10899372	PLATE, TANK, PROTECTO R.H. TANK .... UOC:DAE, DAF, V19, V20, ZAE, ZAF	1
5	PFOZZ	5340004076767	19207	10899255	PLATE, MENDING .... UOC:DAE, DAF, V19, V20, ZAE, ZAF	2
6	PFOZZ	2510004084631	19207	10899258	BRACKET, PROTECTOR L.H. PLATE .... UOC:DAE, DAF, V19, V20, ZAE, ZAF	1
6	PFOZZ	2510004084632	19207	10899259	BRACKET, PROTECTOR R.H. PLATE .... UOC:DAE, DAF, V19, V20, ZAE, ZAF	1
7	PAOZZ	5310000806004	96906	MS27183-14	WASHER, FLAT .... UOC:DAE, DAF, V19, V20, ZAE, ZAF	4

END OF FIGURE

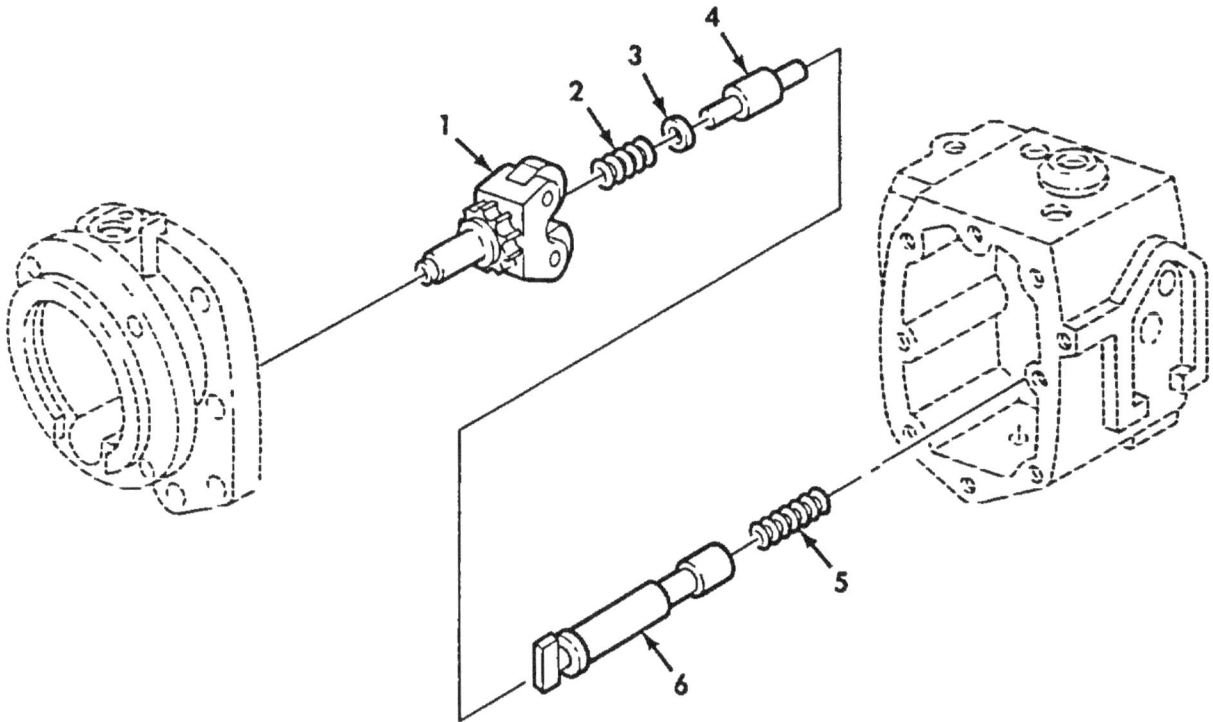

Figure 84. Fuel Pump Governor, AFC (M939, M939A1).

(1) ITEM NO	(2) SMR CODE	(3) NSN	(4) CAGEC	(5) PART NUMBER	(6) DESCRIPTION AND USABLE ON CODES (UOC)	(7) QTY
					GROUP 0308 ENGINE SPEED GOVERNOR AND CONTROLS	
					FIG. 84 FUEL PUMP GOVERNOR,AFC (M939, M939A1)	
1	PBHZZ	2990011435489	15434	3024989	WEIGHT AND CARRIER FUEL PUMP .... UOC:DAA, DAB, DAC, DAD, DAE, DAF, DAG, DAH, DAJ, DAK, DAW, DAX, V12, V13, V14, V15, V16, V17, V19, V20, V21, V22, V24, V25, V39	1
2	PAHZZ	5360010953661	15434	143848	SPRING, HELICAL, COMP .... UOC:DAA, DAB, DAC, DAD, DAE, DAF, DAG, DAH, DAJ, DAX, DAW, DAX, V12, V13, V14, V15, V16, V17, V19, V20, V21, V22, V24, V25, V39	1
3	PAHZZ	5310007278353	15434	144179	WASHER, FLAT GOVERNOR WEIGHT .... ASSEMBLY .... UOC:DAA, DAB, DAC, DAD, DAE, DAF, DAG, DAH, DAJ, DAK, DAW, DAX, V12, V13, V14, V15, V16, V17, V19, V20, V21, V22, V24, V25, V39	1
4	PAHZZ	5315000820448	15434	144178	PIN, SHOULDER, HEADLE .... UOC:DAA, DAB, DAC, DAD, DAE, DAF, DAG, DAH, DAJ, DAK, DAW, DAX, V12, V13, V14, V15, V16, V17, V19, V20, V21, V22, V24, V25, V39	1
5	PAHZZ	5360012155766	15434	138795	SPRING, HELICAL, COMP USE PRIOR TO .... SERIAL NUMBER 11246663 .... UOC:DAA, DAB, DAC, DAD, DAE, DAF, DAG, DAH, DAJ, DAK, DAW, DAX, V12, V13, V14, V15, V16, V17, V19, V20, V21, V22, V24, V25, V39	1
5	PAHZZ	5360011376707	15434	138999	SPRING USE AFTER SERIAL NUMBER .... 11246664... UOC:DAA, DAB, DAC, DAD, DAE, DAF, DAG, DAH, DAJ, DAK, DAW, DAX, V12, V13, V14, V15, V16, V17, V19, V20, V21, V22, V24, V25, V39	1
6	PBHZZ	5342011451549	15434	3010810	PLUNGER USE PRIOR TO SERIAL NUMBER .... 11246663... UOC:DAA, DAB, DAC, DAD, DAE, DAF, DAG, DAH, DAJ, DAK, DAW, DAX, V12, V13, V14, V15, V16, V17, V19, V20, V21, V22, V24, V25, V39	1
6	PBHZZ	2910012301919	15434	3040760	PLUNGER, GOVERNOR USE AFTER SERIAL ...... NUMBER 11246664 .... UOC:DAA, DAB, DAC, DAD, DAE, DAF, DAG, DAH, DAJ, DAK, DAW, DAX, V12, V13, V14, V15, V16, V17, V19, V20, V21, V22, V24, V25, V39	1

END OF FIGURE

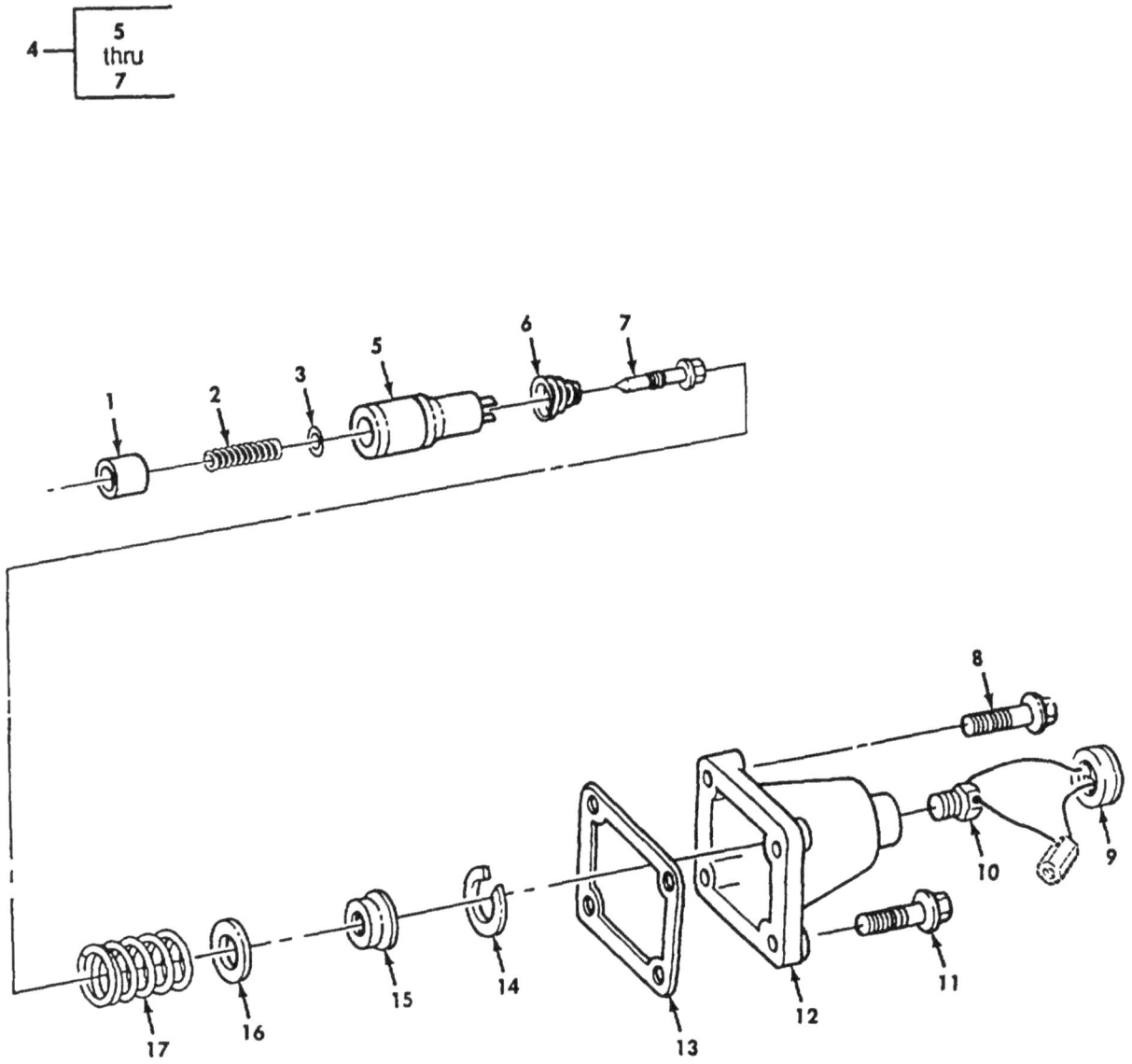

Figure 85. Fuel Pump Spring Pack, AFC (M939, M939A1).

(1) ITEM NO	(2) SMR CODE	(3) NSN	(4) CAGEC	(5) PART NUMBER	(6) DESCRIPTION AND USABLE ON CODES (UOC)	(7) QTY
					GROUP 0308 ENGINE SPEED GOVERNOR AND CONTROLS	
					FIG. 85 FUEL PUMP SPRING PACK, AFC (M939, M939A1)	
1	PAHZZ	2910002385434	15434	138862	PLUNGER, IDLE SPRING USE PRIOR TO .... SERIAL NUMBER 11246663 .... UOC:DAA, DAB, DAC, DAD, DAE, DAF, DAG, DAH, DAJ, DAK, DAW, DAX, V12, V13, V14, V15, V16, V17, V19, V20, V21, V22, V24, V25, V39	1
1	PAHZZ	2910011890901	15434	141634	PLUNGER, GOVERNOR USE AFTER SERIAL ...... NUMBER 11246663 UOC:DAA, DAB, DAC, DAD, DAE, DAF, DAG, DAH, DAJ, DAK, DAW, DAX, V12, V13, V14, V15, V16, V17, V19, V20, V21, V22, V24, V25, V39	1
2	PAHZZ	5360000820124	15434	144195	.SPRING, HELICAL, COMP .... UOC:DAA, DAB, DAC, DAD, DAE, DAF, DAG, DAH, DAJ, DAK, DAW, DAX, V12, V13, V14, V15, V16, V17, V19, V20, V21, V22, V24, V25, V39	1
3	PAHZZ	5310005073259	15434	70715	WASHER, FLAT .... UOC:DAA, DAB, DAC, DAD, DAE, DAF, DAG, DAH, DAJ, DAK, DAW, DAX, V12, V13, V14, V15, V16, V17, V19, V20, V21, V22, V24, V25, V39	1
4	PBHZZ	2910008032631	15434	3038215	SPRING GUIDE, ASSEMB .... UOC:DAA, DAB, DAC, DAD, DAE, DAF, DAG, DAH, DAJ, DAK, DAW, DAX V12, V13, V14, V15, V16, V17, V19, V20, V21, V22, V24, V25, V39	1
5	PAHZZ	2990012186935	15434	3036472	.RETAINER, HELICAL CO .... UOC:DAA, DAB, DAC, DAD, DAE, DAF, DAG, DAH, DAJ, DAK, DAW, DAX, V12, V13, V14, V15, V16, V17, V19, V20, V21, V22, V24, V25, V39	1
6	PAHZZ	5360012401626	15434	3032014	.SPRING, HELICAL, COMP .... UOC:DAA, DAB, DAC, DAD, DAE, DAF, DAG, DAH, DAJ, DAK, DAW, DAX, V12, V13, V14, V15, V16, V17, V19, V20, V21, V22, V24, V25, V39	1
7	PAHZZ	5305005065722	15434	70716	.SETSCREW .... UOC:DAA, DAB, DAC, DAD, DAE, DAF, DAG, DAH, DAJ, DAK, DAW, DAX, V12, V13, V14, V15, V16, V17, V19, V20, V21, V22, V24, V25, V39	1
8	PAHZZ	5305011129110	15434	3017051	SCREW, CAP, HEXAGON H .... UOC:DAA, DAB, DAC, DAD, DAE, DAF, DAG, DAH, DAJ, DAK, DAW, DAX, V12, V13, V14, V15, V16, V17, V19, V20, V21, V22, V24, V25, V39	3
9	PAHZZ	5330010728830	15434	3003156	SEAL, SPECIAL PART OF KIT P/N 5704519 .... UOC:DAA, DAB, DAC, DAD, DAE, DAF, DAG, DAH, DAJ, DAK, DAW, DAX, V12, V13, V14, V15 , V16, V17, V19, V20, V21, V22, V24, V25, V39	1
10	PAHZZ	5342011436046	15434	3025459	PLUG, COVER .... UOC:DAA, DAB, DAC, DAD, DAE, DAF, DAG, DAH, DAJ, DAK, DAW, DAX, V12, V13, V14, V15, V16, V17, V19, V20, V21, V22, V24, V25, V39	1

(1) ITEM NO	(2) SMR CODE	(3) NSN	(4) CAGEC	(5) PART NUMBER	(6) DESCRIPTION AND USABLE ON CODES (UOC)	(7) QTY
11	PAHZZ	5305011261128	15434	3017052	SCREW, CAP, HEXAGON H .... UOC:DAA, DAB, DAC, DAD, DAE, DAF, DAG, DAH, DAJ, DAK, DAW, DAX, V12, V13, V14, V15, V16, V17, V19, V20, V21, V22, V24, V25, V39	2
12	PBHZZ	2910008583522	15434	44678	COVER, SPRING PACK .... UOC:DAA, DAB, DAC, DAD, DAE, DAF, DAG, DAH, DAJ, DAK, DAW, DAX, V12, V13, V14, V15, V16, V17, V19, V20, V21, V22, V24, V25, V39	1
13	PAHZZ	5330005621176	15434	70705	GASKET PART OF KIT P/N 3010240, ... 5704519 .... UOC:DAA, DAB, DAC, DAD, DAE, DAF, DAG, DAH, DAJ, DAK, DAW, DAX, V12, V13, V14, V15, V16, V17, V19, V20, V21, V22, V24, V25, V39	1
14	PAHZZ	5365008072636	96906	MS16625-1100	RING, RETAINING .... UOC:DAA, DAB, DAC, DAD, DAE, DAF, DAG, DAH, DAJ, DAK, DAW, DAX, V12, V13, V14, V15, V16, V17, V19, V20, V21, V22, V24, V25, V39	1
15	PAHZZ	5365011194954	15434	203346	SPACER, SLEEVE .... UOC:DAA, DAB, DAC, DAD, DAE, DAF, DAG, DAH, DAJ, DAK, DAW, DAX, V12, V13, V14, V15, V16, V17, V19, V20, V21, V22, V24, V25, V39	1
16	PAHZZ	5365005073260	15434	70717	SPACER, RING PART OF KIT P/N 5704519 .... UOC:DAA, DAB, DAC, DAD, DAE, DAF, DAG, DAH, DAJ, DAK, DAW, DAX, V12, V13, V14, V15, V16, V17, V19, V20, V21, V22, V24, V25, V39	1
16	PAHZZ	5365005073261	15434	70717-A	SHIM PART OF KIT P/N 5704519 .... UOC:DAA, DAB, DAC, DAD, DAE, DAF, DAG, DAH, DAJ, DAK, DAW, DAX, V12, V13, V14, V15, V16, V17, V19, V20, V21, V22, V24, V25, V39	1
16	PAHZZ	5365005073262	15434	70717-B	SHIM PART OF KIT P/N 5704519 .... UOC:DAA, DAB, DAC, DAD, DAE, DAF, DAG, DAH, DAJ, DAK, DAW, DAX, V12, V13, V14, V15, V16, V17, V19, V20, V21, V22, V24, V25, V39	1
16	PAHZZ	5365004624504	15434	189800	SHIM PART OF KIT P/N 5704519 .... UOC:DAA, DAB, DAC, DAD, DAE, DAF, DAG, DAH, DAJ, DAK, DAW, DAX, V12, V13, V14, V15, V16, V17, V19, V20, V21, V22, V24, V25, V39	1
17	PAHZZ	5360012226879	15434	3000940	SPRING, HELICAL, COMP .... UOC:DAA, DAB, DAC, DAD, DAE, DAF, DAG, DAH, DAJ, DAK, DAW, DAX, V12, V13, V14, V15, V16, V17, V19, V20, V21, V22, V24, V25, V39	1

END OF FIGURE

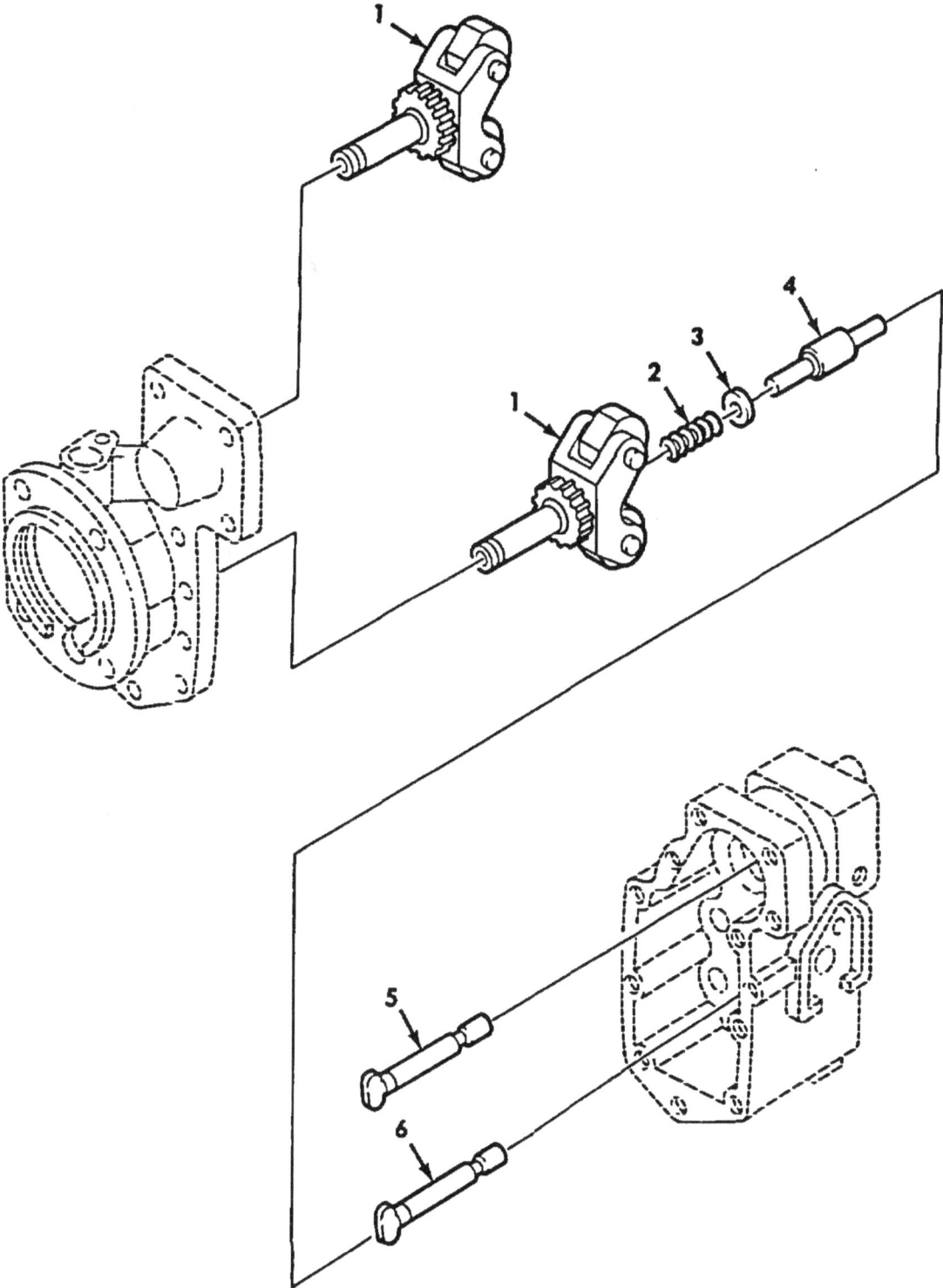

Figure 86. Fuel Pump Governor, VS (M939, M939A1).

(1) ITEM NO	(2) SMR CODE	(3) NSN	(4) CAGEC	(5) PART NUMBER	(6) DESCRIPTION AND USABLE ON CODES (UOC)	(7) QTY
					GROUP 0308 ENGINE SPEED GOVERNOR AND CONTROLS	
					FIG. 86 FUEL PUMP GOVERNOR, VS(M939, M939A1)	
1	PBHZZ	2990011435489	15434	3024989	WEIGHT AND CARRIER .... UOC:DAL, V18	2
2	PAHZZ	5360011912949	15434	143854	SPRING, HELICAL, COMP GOVERNOR .... WEIGHT ASSEMBLY .... UOC:DAL, V18, V24, V25, V39	1
3	PAHZZ	5310007278353	15434	144179	WASHER, FLAT....... LOWER CARRIER ASSEMBLY UOC:DAL, V18	3
4	PAHZZ	5315000820448	15434	144178	PIN, SHOULDER, HEADLE .... UOC:DAL, V18	1
5	PFHZZ	5342011607381	15434	AR-40230	PLUNGER .... UOC:DAL, V18	1
6	PBHZZ	5342011451549	15434	3010810	PLUNGER USE PRIOR TO SERIAL NUMBER .... 11246663... UOC:DAL, V18	1

END OF FIGURE

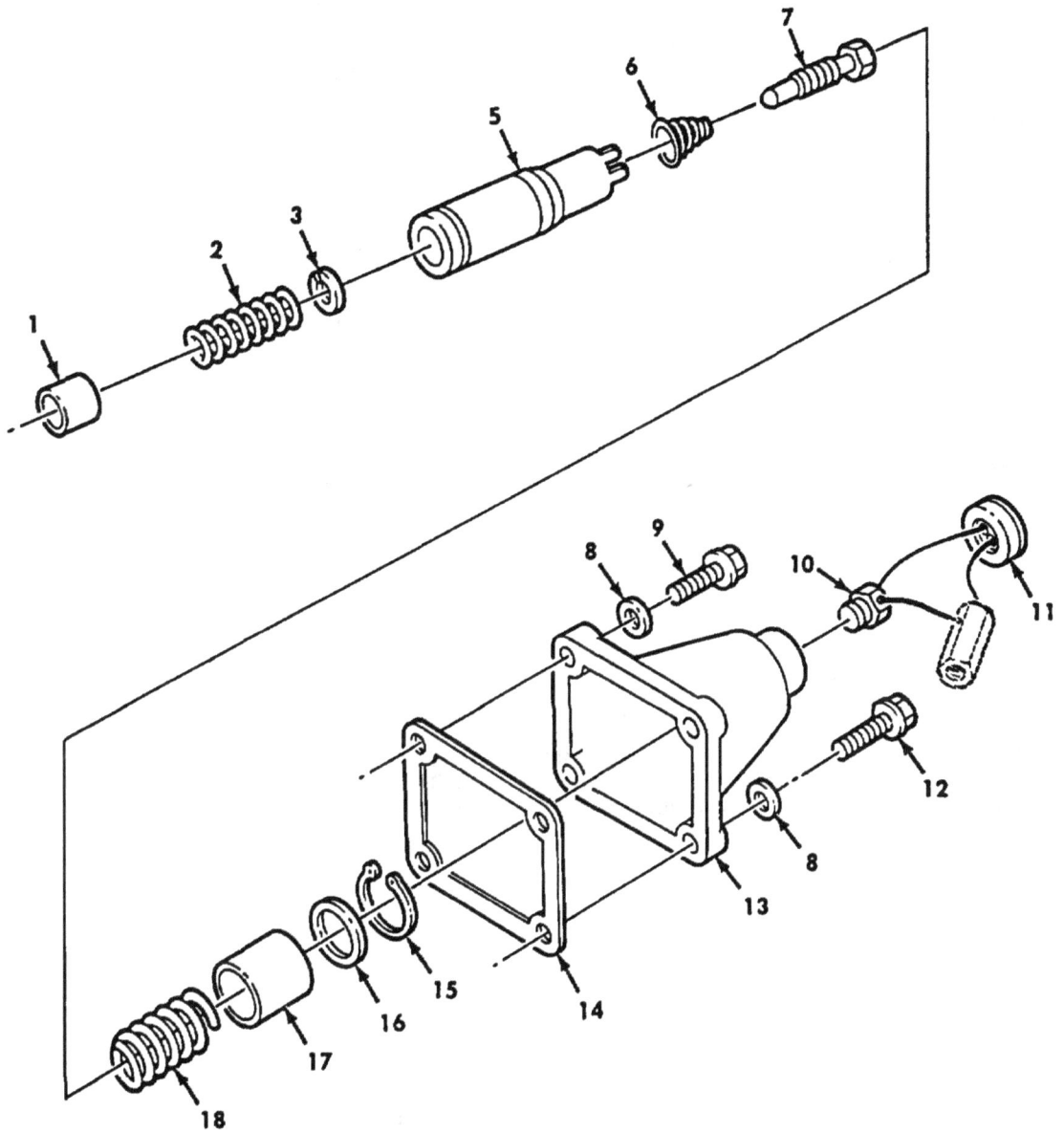

Figure 87. Fuel Pump Lower Spring Pack, VS (M939, M939A1).

(1) ITEM NO	(2) SMR CODE	(3) NSN	(4) CAGEC	(5) PART NUMBER	(6) DESCRIPTION AND USABLE ON CODES (UOC)	(7) QTY
					GROUP 0308 ENGINE SPEED GOVERNOR AND CONTROLS	
					FIG. 87 FUEL PUMP LOWER SPRING PACK, VS(M939,M939A1)	
1	PAHZZ	2910002385434	15434	138862	PLUNGER, IDLE SPRING USE PRIOR TO .... SERIAL NUMBER 11246663 .... UOC:DAL, V18	1
1	PAHZZ	2910011890901	15434	141634	PLUNGER, GOVERNOR USE AFTER SERIAL ...... NUMBER 11246663 .... UOC:DAL, V18	1
2	PAHZZ	5360000820124	15434	144195	.SPRING, HELICAL, COMP .... UOC:DAL, V18	1
3	PAHZZ	5310005073259	15434	70715	.WASHER, FLAT .... UOC:DAL, V18	1
4	PBHZZ	5360012280747	15434	3038216	SPRING ASSORTMENT .... UOC:DAL, V18	1
5	PAHZZ	5340012416939	15434	3036474	.SEAT, HELICAL COMPRE .... UOC:DAL, V18	1
6	PAHZZ	5360012401626	15434	3032014	.SPRING, HELICAL, COMP .... UOC:DAL, V18	1
7	PAHZZ	5305005065722	15434	70716	.SETSCREW .... UOC:DAL, V18	1
8	PAHZZ	5310001411795	88044	AN960-416	WASHER, FLAT .... UOC:DAL, V18	4
9	PAHZZ	5305011261128	15434	3017052	SCREW, CAP, HEXAGON H .... UOC:DAL, V18	1
10	PAHZZ	5340011436046	15434	3025459	PLUG, COVER .... UOC:DAL, V18	1
11	PAHZZ	5330010728830	15434	3003156	SEAL, SPECIAL .... UOC:DAL, V18	1
12	PAHZZ	5305011129110	15434	3017051	SCREW, CAP, HEXAGON H .... UOC:DAL, V18	2
13	PAHZZ	2910008583522	15434	44678	COVER, SPRING PACK .... UOC:DAL, V18	1
14	PAHZZ	5330005621176	15434	70705	GASKET PART OF KIT P/N 3010240 .... UOC:DAL, V18	1
15	PAHZZ	5365008072636	96906	MS16625-1100	RING, RETAINING .... UOC:DAL, V18	1
16	PAHZZ	5365005073260	15434	70717	SPACER, RING .... UOC:DAL, V18	1
16	PAHZZ	5365005073261	15434	70717-A	SHIM .... UOC:DAL, V18	1
16	PAHZZ	5365005073262	15434	70717-B	SHIM .... UOC:DAL, V18	1
16	PAHZZ	5365004624504	15434	189800	SHIM .... UOC:DAL, V18	1
17	PAHZZ	5365011194954	15434	203346	SPACER, SLEEVE .... UOC:DAL, V18	1
18	PAHZZ	5360011912950	15434	3001155	SPRING, HELICAL, COMP .... UOC:DAL, V18	1

END OF FIGURE

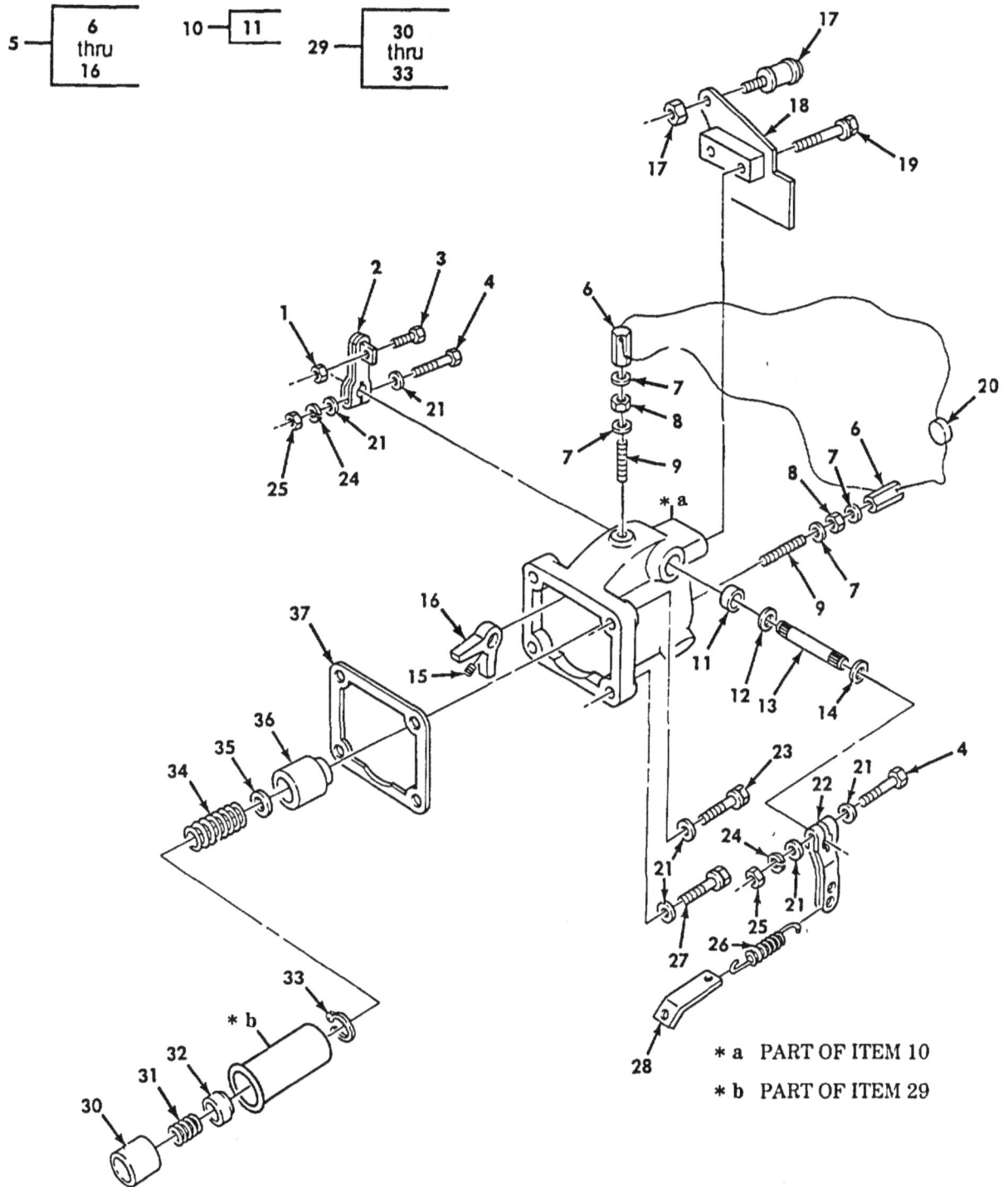

Figure 88. Fuel Pump Upper Spring Pack, VS (M939, M939A1).

(1) ITEM NO	(2) SMR CODE	(3) NSN	(4) CAGEC	(5) PART NUMBER	(6) DESCRIPTION AND USABLE ON CODES (UOC)	(7) QTY
					GROUP 0308 ENGINE SPEED GOVERNOR AND CONTROLS	
					FIG. 88 FUEL PUMP UPPER SPRING PACK, VS (M939,M939A1)	
1	PAHZZ	5310005218595	15434	S-223	NUT, HEXAGON .... UOC:DAL, V18	1
2	PAHZZ	2910010986741	15434	187126	LEVER, THROTTLE .... UOC:DAL, V18	1
3	PAHZZ	5306004616070	15434	187127	BOLT, CARRIAGE .... UOC:DAL, V18	1
4	PAHZZ	5305004933959	15434	S159B	SCREW, CAP, HEXAGON H .... UOC:DAL, V18	2
5	PBHZZ	5340011730133	15434	3020523	HOLDER, SPRING .... UOC:DAL, V18	1
6	PAHZZ	5340007164975	15434	110058	.POST, ELECTRICAL-MEC .... UOC:DAL, V18	2
7	PAHZZ	5310007357460	15434	112408	.WASHER, FLAT .... UOC:DAL, V18	4
8	PAHZZ	5310009717989	96906	MS35691-5	.NUT, PLAIN, HEXAGON .... UOC:DAL, V18	2
9	PAHZZ	5305011091292	15434	109917	.SETSCREW .... UOC:DAL, V18	2
10	PBHZZ	2910011778816	15434	3016021	.HOUSING, FUEL PUMP .... UOC:DAL, V18	1
11	PAHZZ	3120011601891	15434	3010941	..BUSHING, SLEEVE .... UOC:DAL, V18	2
12	PAHZZ	5331005144804	79500	1JH4350REVCPC	.O-RING .... UOC:DAL, V18	2
13	PAHZZ	3040011819509	15434	3010942	.SHAFT, STRAIGHT .... UOC:DAL, V18	1
14	PAHZZ	5330011610289	15434	3010945	.SEAL.. UOC:DAL, V18	2
15	PAHZZ	5305008998054	15434	70813	.SETSCREW .... UOC:DAL, V18	1
16	PAHZZ	2910007732108	15434	70834	.STOP, THROTTLE LEVER .... UOC:DAL, V18	1
17	PAHZZ	3040004848546	15434	185012	CYLINDER ASSEMBLY, A .... UOC:DAL, V18	1
18	PBHZZ	5342011101505	15434	211305	BRACKET .... UOC:DAL, V18	1
19	PAHZZ	5305011129021	15434	3013904	SCREW .... UOC:DAL, V18	2
20	PAHZZ	5330010728830	15434	3003156	SEAL, SPECIAL .... UOC:DAL, V18	1
21	PAHZZ	5310001411795	88044	AN960-416	WASHER, FLAT .... UOC:DAL, V18	8
22	PAHZZ	2910010985093	15434	3012497	LEVER, THROTTLE .... UOC:DAL, V18	1
23	PAHZZ	5305011261128	15434	3017052	SCREW, CAP, HEXAGON H .... UOC:DAL, V18	1

(1) ITEM NO	(2) SMR CODE	(3) NSN	(4) CAGEC	(5) PART NUMBER	(6) DESCRIPTION AND USABLE ON CODES (UOC)	(7) QTY
24	PAHZZ	5310001596209	96906	MS122032	WASHER, LOCK .... UOC:DAL, V18	2
25	PAOZZ	5310009717989	96906	MS35691-5	NUT, PLAIN, HEXAGON .... UOC:DAL, V18	2
26	PAHZZ	5360001402078	15434	203762	SPRING, HELICAL, EXTE .... UOC:DAL, V18	1
27	PAHZZ	5305011129110	15434	3017051	SCREW, CAP, HEXAGON H .... UOC:DAL, V18	3
28	PAHZZ	5342011094013	15434	3005543	LINK, SPRING .... UOC:DAL, V18	1
29	PBHHA	5360001945920	15434	BM30245	SPRING, GOVERNOR MAX .... UOC:DAL, V18	1
30	PAHZZ	2990005674367	15434	70820	.PLUNGER, GOVERNOR SP .... UOC:DAL, V18	1
31	PAHZZ	5360006987100	15434	70778	.SPRING, HELICAL, COMP .... UOC:DAL, V18	1
32	PAHZZ	5340005728042	15434	70798	.SEAT, HELICAL COMPRE .... UOC:DAL, V18	1
33	PAHZZ	5325008461637	96906	MS16627-1093	.RING, RETAINING .... UOC:DAL, V18	1
34	PAHZZ	5360012226880	15434	109686	SPRING, HELICAL, COMP .... UOC:DAL, V18	1
35	PAHZZ	5310005508124	15434	70811	WASHER, FLAT .020 .... UOC:DAL, V18	1
35	PAHZZ	5365005508125	15434	70811-A	SHIM .020 .... UOC:DAL, V18	1
35	PAHZZ	5365005508127	15434	70811-B	SHIM .005 .... UOC:DAL, V18	1
36	PAHZZ	5340007115372	15434	70836	PLUNGER, DETENT .... UOC:DAL, V18	1
37	PAHZZ	5330005621176	15434	70705	GASKET .... UOC:DAL, V1B	1

END OF FIGURE

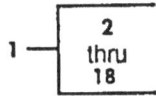

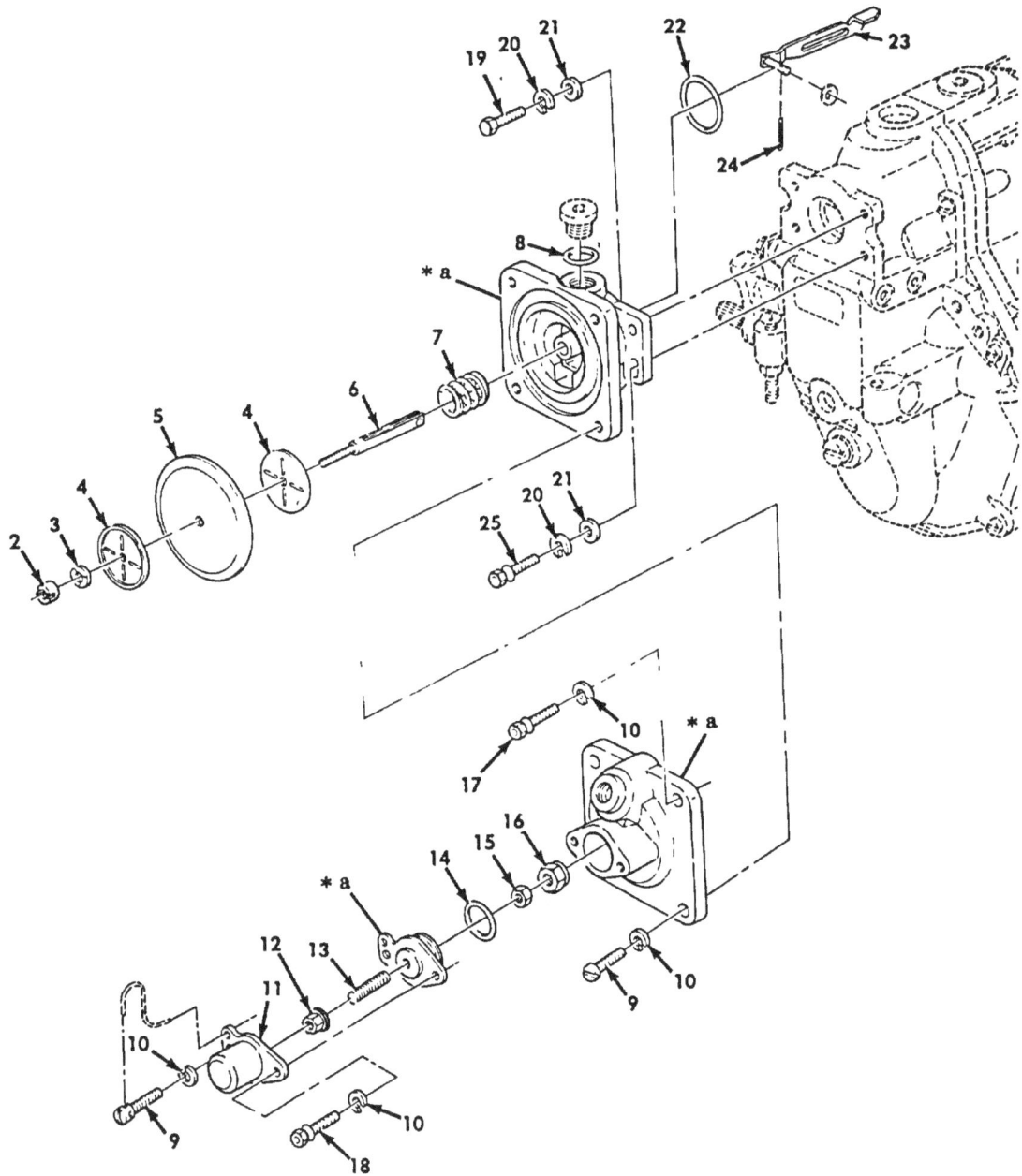

* a PART OF ITEM 1

Figure 89. Manifold Pressure Compensator Assembly and Related Parts (M939A2).

(1) ITEM NO	(2) SMR CODE	(3) NSN	(4) CAGEC	(5) PART NUMBER	(6) DESCRIPTION AND USABLE ON CODES (UOC)	(7) QTY
					GROUP 0308 ENGINE SPEED GOVERNOR AND CONTROLS	
					FIG. 89 MANIFOLD PRESSURE COMPENSATOR ASSEMBLY AND RELATED PARTS (M939A2)	
1	PFHHH	5365013368895	5T151	1 427 133 313	SHIM .... UOC:ZAA, ZAB, ZAC, ZAD, ZAE, ZAF, ZAG, ZAH, ZAJ, ZAK, ZAL	1
2	XAHZZ		5T151	2 420 360 004	.SLOTTED ROUND NUT .... UOC:ZAA, ZAB, ZAC, ZAD, ZAE, ZAF, ZAG, ZAH, ZAJ, ZAK, ZAL	1
3	PAHZZ	5310013418953	5T151	2 916 069 083	.WASHER, SPRING TENSI .... UOC:ZAA, ZAB, ZAC, ZAD, ZAE, ZAF, ZAG, ZAH, ZAJ, ZAK, ZAL	1
4	XAHZZ		5T151	2 420 551 005	.PLATE WASHER .... UOC:ZAA, ZAB, ZAC, ZAD, ZAE, ZAF, ZAG, ZAH, ZAJ, ZAK, ZAL	2
5	PFHZZ	4820013384642	5T151	2 420 503 013	.DIAPHRAGM, ACTUATOR PART OF KIT P/N .... 57K0143.. UOC:ZAA, ZAB, ZAC, ZAD, ZAE, ZAF, ZAG, ZAH, ZAJ, ZAK, ZAL	1
6	XAHZZ		5T151	2 423 050 056	.LIFTING ROD .... UOC:ZAA, ZAB, ZAC, ZAD, ZAE, ZAF, ZAG, ZAH, ZAJ, ZAK, ZAL	1
7	PFHZZ	5360013390684	5T151	1 424 619 179	.SPRING, HELICAL, COMP .... UOC:ZAA, ZAB, ZAC, ZAD, ZAE, ZAF, ZAG, ZAH, ZAJ, ZAK, ZAL	1
8	PFHZZ	5325012768488	53867	2916710613	.RING, RETAINING .... UOC:ZAA, ZAB, ZAC, ZAD, ZAE, ZAF, ZAG, ZAH, ZAJ, ZAK, ZAL	1
9	PAHZZ	5305013360006	5T151	2 423 450 005	.SCREW, CAP, HEXAGON H .... UOC:ZAA, ZAB, ZAC, ZAD, ZAE, ZAF, ZAG, ZAH, ZAJ, ZAK, ZAL	4
10	PFHZZ	5310013017811	53867	2 916 699 083	.WASHER, LOCK 6.00MM PART OF KIT P/N .... 57K0143.. UOC:ZAA, ZAB, ZAC, ZAD, ZAE, ZAF, ZAG, ZAH, ZAJ, ZAK, ZAL	6
11	PFHZZ	5340013359996	5T151	1 420 505 036	.PLUG, PROTECTIVE, DUS .... UOC:ZAA, ZAB, ZAC, ZAD, ZAE, ZAF, ZAG, ZAH, ZAJ, ZAK, ZAL	1
12	PAHZZ	5310013517542	5T151	1 423 314 003	.NUT, PLAIN, HEXAGON .... UOC:ZAA, ZAB, ZAC, ZAD, ZAE, ZAF, ZAG, ZAH, ZAJ, ZAK, ZAL	1
13	XAHZZ		5T151	2 423 400 034	.HEADLESS SETSCREW .... UOC:ZAA, ZAB, ZAC, ZAD, ZAE, ZAF, ZAG, ZAH, ZAJ, ZAK, ZAL	1
14	PFHZZ	5331013369559	5T151	1 410 210 014	.O-RING 12 X 2MM. PART OF KIT P/N .... 57K0143.. UOC:ZAA, ZAB, ZAC, ZAD, ZAE, ZAF, ZAG, ZAH, ZAJ, ZAK, ZAL	1

(1) ITEM NO	(2) SMR CODE	(3) NSN	(4) CAGEC	(5) PART NUMBER	(6) DESCRIPTION AND USABLE ON CODES (UOC)	(7) QTY
15	XAHZZ	5310013013885	53867	2 915 012 007	.NUT, PLAIN, HEXAGON DIM 934-M6-M8 .... UOC:ZAA, ZAB, ZAC, ZAD, ZAE, ZAF, ZAG, ZAH, ZAJ, ZAK, ZAL	1
16	XAHZZ		5T151	2 423 315 004	.ROUND-NECK NUT .... UOC:ZAA, ZAB, ZAC, ZAD, ZAE, ZAF, ZAG, ZAH, ZAJ, ZAK, ZAL	1
17	PFHZZ	5305013359965	5T151	2 910 172 197	.SCREW, MACHINE DIM 404-M6 X 16-5.8 .... UOC:ZAA, ZAB, ZAC, ZAD, ZAE, ZAF, ZAG, ZAH, ZAJ, ZAK, ZAL	1
18	PFHZZ	5305013425171	5T151	2 910 022 197	.SCREW, MACHINE PART OF KIT P/N .... 57K0143 UOC:ZAA, ZAB, ZAC, ZAD, ZAE, ZAF, ZAG, ZAH, ZAJ, ZAK, ZAL	1
19	PAHZZ	5305012645886	64678	000 824 08 71	SCREW, CAP, HEXAGON H DIM 84-AM6 X .... 16-8.8... UOC:ZAA, ZAB, ZAC, ZAD, ZAE, ZAF, ZAG, ZAH, ZAJ, ZAK, ZAL	3
20	PFHZZ	5310013017811	53867	2 916 699 083	WASHER, LOCK PART OF KIT P/N 57K0143 .... UOC:ZAA, ZAB, ZAC, ZAD, ZAE, ZAF, ZAG, ZAH, ZAJ, ZAK, ZAL	2
21	PAHZZ	5310011510113	18876	13142983-2/48	WASHER, LOCK .... UOC:ZAA, ZAB, ZAC, ZAD, ZAE, ZAF, ZAG, ZAH, ZAJ, ZAK, ZAL	4
22	KFHZZ		5T151	1 420 210 022	0-RING PART OF KIT P/N 57K0143 .... UOC:ZAA, ZAB, ZAC, ZAD, ZAE, ZAF, ZAG, ZAH, ZAJ, ZAK, ZAL	1
23	XAHZZ		5T151	2 421 335 025	STRAP .... UOC:ZAA, ZAB, ZAC, ZAD, ZAE, ZAF, ZAG, ZAH, ZAJ, ZAK, ZAL	1
24	KFHZZ		5T151	2 424 680 003	RETAINING SPRING PART OF KIT P/N .... 57K0143.. UOC:ZAA, ZAB, ZAC, ZAD, ZAE, ZAF, ZAG, ZAH, ZAJ, ZAK, ZAL	1
25	PAHZZ	5305013360007	53867	2 423 450 002	SCREW, CAP, HEXAGON H PART OF KIT P/N .... 57K0143.. UOC:ZAA, ZAB, ZAC, ZAD, ZAE, ZAF, ZAG, ZAH, ZAJ, ZAK, ZAL	1

END OF FIGURE

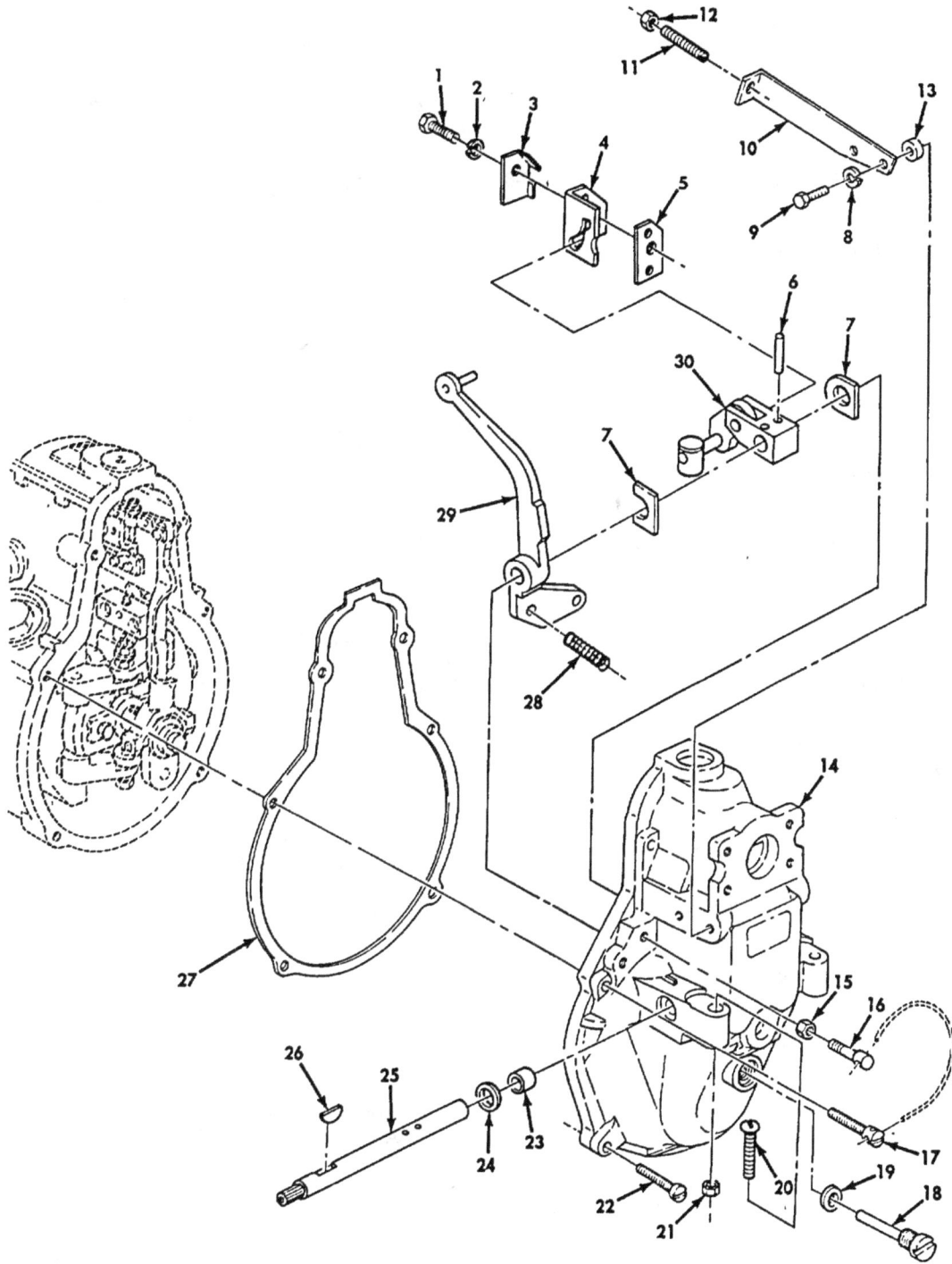

Figure 90. Governor Cover and Related Parts (M939A2).

(1) ITEM NO	(2) SMR CODE	(3) NSN	(4) CAGEC	(5) PART NUMBER	(6) DESCRIPTION AND USABLE ON CODES (UOC)	(7) QTY
					GROUP 0308 ENGINE SPEED GOVERNOR AND CONTROLS	
					FIG. 90 GOVERNOR COVER AND RELATED PARTS (M939A2)	
1	PFHZZ	5305121420188	D8286	M6X22DIN933-8-8A 2P	SCREW, CAP, HEXAGON H .... UOC:ZAA, ZAB, ZAC, ZAD, ZAE, ZAF, ZAG, ZAH, ZAJ, ZAK, ZAL	1
2	PFHZZ	5310013011885	53867	2 916 690 005	WASHER, SPRING TENSI PART OF KIT P/N .... 57K0142.. UOC:ZAA, ZAB, ZAC, ZAD, ZAE, ZAF, ZAG, ZAH, ZAJ, ZAK, ZAL	1
3	PFHZZ	5310013366747	5T151	2 421 321 009	WASHER, FLAT .... UOC:ZAA, ZAB, ZAC, ZAD, ZAE, ZAF, ZAG, ZAH, ZAJ, ZAK, ZAL	1
4	PFHZZ	3040013368168	53867	1 423 103 016	SHAFT, STRAIGHT .... UOC:ZAA, ZAB, ZAC, ZAD, ZAE, ZAF, ZAG, ZAH, ZAJ, ZAK, ZAL	1
5	PFHZZ	5365013368916	5T151	1 421 036 010	SPACER, PLATE 0.20MM .... UOC:ZAA, ZAB, ZAC, ZAD, ZAE, ZAF, ZAG, ZAH, ZAJ, ZAK, ZAL	1
5	PFHZZ	5340013390868	5T151	1 421 036 012	BRACKET, MOUNTING 0.80 MM .... UOC:ZAA, ZAB, ZAC, ZAD, ZAE, ZAF, ZAG, ZAH, ZAJ, ZAK, ZAL	1
5	PFHZZ	5340013416655	5T151	1 421 036 013	PLATE, MOUNTING 0.50 MM .... UOC:ZAA, ZAB, ZAC, ZAD, ZAE, ZAF, ZAG, ZAH, ZAJ, ZAK, ZAL	1
6	PFHZZ	5315013365185	5T151	2 423 202 001	PIN, STRAIGHT, HEADLE PART OF KIT P/N .... 57K0142.. UOC:ZAA, ZAB, ZAC, ZAD, ZAE, ZAF, ZAG, ZAH, ZAJ, ZAK, ZAL	2
7	PFHZZ	5365013365224	5T151	1 411 030 000	SPACER, PLATE .... UOC:ZAA, ZAB, ZAC, ZAD, ZAE, ZAF, ZAG, ZAH, ZAJ, ZAK, ZAL	2
8	PFHZZ	5310013017811	53867	2 916 699 083	WASHER, LOCK PART OF KIT P/N 57K0142 .... UOC:ZAA, ZAB, ZAC, ZAD, ZAE, ZAF, ZAG, ZAH, ZAJ, ZAK, ZAL	2
9	PFHZZ	5305012645886	64678	000 824 08 71	SCREW, CAP, HEXAGON H .... UOC:ZAA, ZAB, ZAC, ZAD, ZAE, ZAF, ZAG, ZAH, ZAJ, ZAK, ZAL	2
10	PFHZZ	5340013362292	5T151	2 421 332 022	BRACKET, ANGLE .... UOC:ZAA, ZAB, ZAC, ZAD, ZAE, ZAF, ZAG, ZAH, ZAJ, ZAK, ZAL	1
11	PFHZZ	5305013369380	5T151	2 423 400 031	SETSCREW .... UOC:ZAA, ZAB, ZAC, ZAD, ZAE, ZAF, ZAG, ZAH, ZAJ, ZAK, ZAL	1
12	PFHZZ	5310013013885	53867	2 915 012 007	NUT, PLAIN, HEXAGON .... UOC:ZAA, ZAB, ZAC, ZAD, ZAE, ZAF, ZAG, ZAH, ZAJ, ZAK, ZAL	1
13	PFHZZ	5365013368908	5T151	1 190 200 000	SPACER, RING .... UOC:ZAA, ZAB, ZAC, ZAD, ZAE, ZAF, ZAG, ZAH, ZAJ, ZAK, ZAL	2

(1) ITEM NO	(2) SMR CODE	(3) NSN	(4) CAGEC	(5) PART NUMBER	(6) DESCRIPTION AND USABLE ON CODES (UOC)	(7) QTY
14	PFHZZ	5340013373760	5T151	2 425 650 759	COVER, ACCESS .... UOC:ZAA, ZAB, ZAC, ZAD, ZAE, ZAF, ZAG, ZAH, ZAJ, ZAK, ZAL	1
15	PFHZZ	5310013366683	5T151	2 423 300 008	NUT, PLAIN, HEXAGON .... UOC:ZAA, ZAB, ZAC, ZAD, ZAE, ZAF, ZAG, ZAH, ZAJ, ZAK, ZAL	1
16	PFHZZ	5305013359966	5T151	2 423 410 026	SCREW, MACHINE .... UOC:ZAA, ZAB, ZAC, ZAD, ZAE, ZAF, ZAG, ZAH, ZAJ, ZAK, ZAL	1
17	PFHZZ	5305013359968	5T151	1 423 414 031	SCREW, MACHINE .... UOC:ZAA, ZAB, ZAC, ZAD, ZAE, ZAF, ZAG, ZAH, ZAJ, ZAK, ZAL	1
18	PFHZZ	5305013360026	5T151	1 423 412 025	SCREW, CLOSE TOLERAN .... UOC:ZAA, ZAB, ZAC, ZAD, ZAE, ZAF, ZAG, ZAH, ZAJ, ZAK, ZAL	1
19	PFHZZ	5330005035789	53867	2-916-710-605	GASKET PART OF KIT P/N 57K0142 .... UOC:ZAA, ZAB, ZAC, ZAD, ZAE, ZAF, ZAG, ZAH, ZAJ, ZAK, ZAL	1
20	PFHZZ	5305013359967	5T151	1 423 415 003	SCREW, MACHINE .... UOC:ZAA, ZAB, ZAC, ZAD, ZAE, ZAF, ZAG, ZAH, ZAJ, ZAK, ZAL	1
21	PFHZZ	5310011857191	53867	2-915-011-007	NUT, PLAIN, HEXAGON .... UOC:ZAA, ZAB, ZAC, ZAD, ZAE, ZAF, ZAG, ZAH, ZAJ, ZAK, ZAL	1
22	PFHZZ	5305013360015	5T151	2 912 732 203	SCREW, SHOULDER .... UOC:ZAA, ZAB, ZAC, ZAD, ZAE, ZAF, ZAG, ZAH, ZAJ, ZAK, ZAL	5
23	PFHZZ	3120121981758	5T151	1 420 301 002	BUSHING, SLEEVE .... UOC:ZAA, ZAB, ZAC, ZAD, ZAE, ZAF, ZAG, ZAH, ZAJ, ZAK, ZAL	2
24	PFHZZ	5340013366704	5T151	1 420 555 000	PLUG, EXPANSION .... UOC:ZAA, ZAB, ZAC, ZAD, ZAE, ZAF, ZAG, ZAH, ZAJ, ZAK, ZAL	1
25	PFHZZ	3040013398572	5T151	2 423 061 006	SHAFT, STRAIGHT .... UOC:ZAA, ZAB, ZAC, ZAD, ZAE, ZAF, ZAG, ZAH, ZAJ, ZAK, ZAL	1
26	PFHZZ	5315121564502	D8015	1900023005	KEY, WOODRUFF .... UOC:ZAA, ZAB, ZAC, ZAD, ZAE, ZAF, ZAG, ZAH, ZAJ, ZAK, ZAL	1
27	PFHZZ	5330013390141	5T151	2 421 015 066	GASKET PART OF KIT P/N 57K0142 .... UOC:ZAA, ZAB, ZAC, ZAD, ZAE, ZAF, ZAG, ZAH, ZAJ, ZAK, ZAL	1
28	PFHZZ	5360013359948	5T151	2 424 611 024	SPRING, HELICAL, COMP .... UOC:ZAA, ZAB, ZAC, ZAD, ZAE, ZAF, ZAG, ZAH, ZAJ, ZAK, ZAL	1
29	PFHZZ	2810013362199	5T151	2 422 060 021	ROCKER ARM, ENGINE P .... UOC:ZAA, ZAB, ZAC, ZAD, ZAE, ZAF, ZAG, ZAH, ZAJ, ZAK, ZAL	1
30	PFHZZ	3040013382545	53867	1 422 120 023	LEVER, REMOTE CONTRO .... UOC:ZAA, ZAB, ZAC, ZAD, ZAE, ZAF, ZAG, ZAH, ZAJ, ZAK, ZAL	1

END OF FIGURE

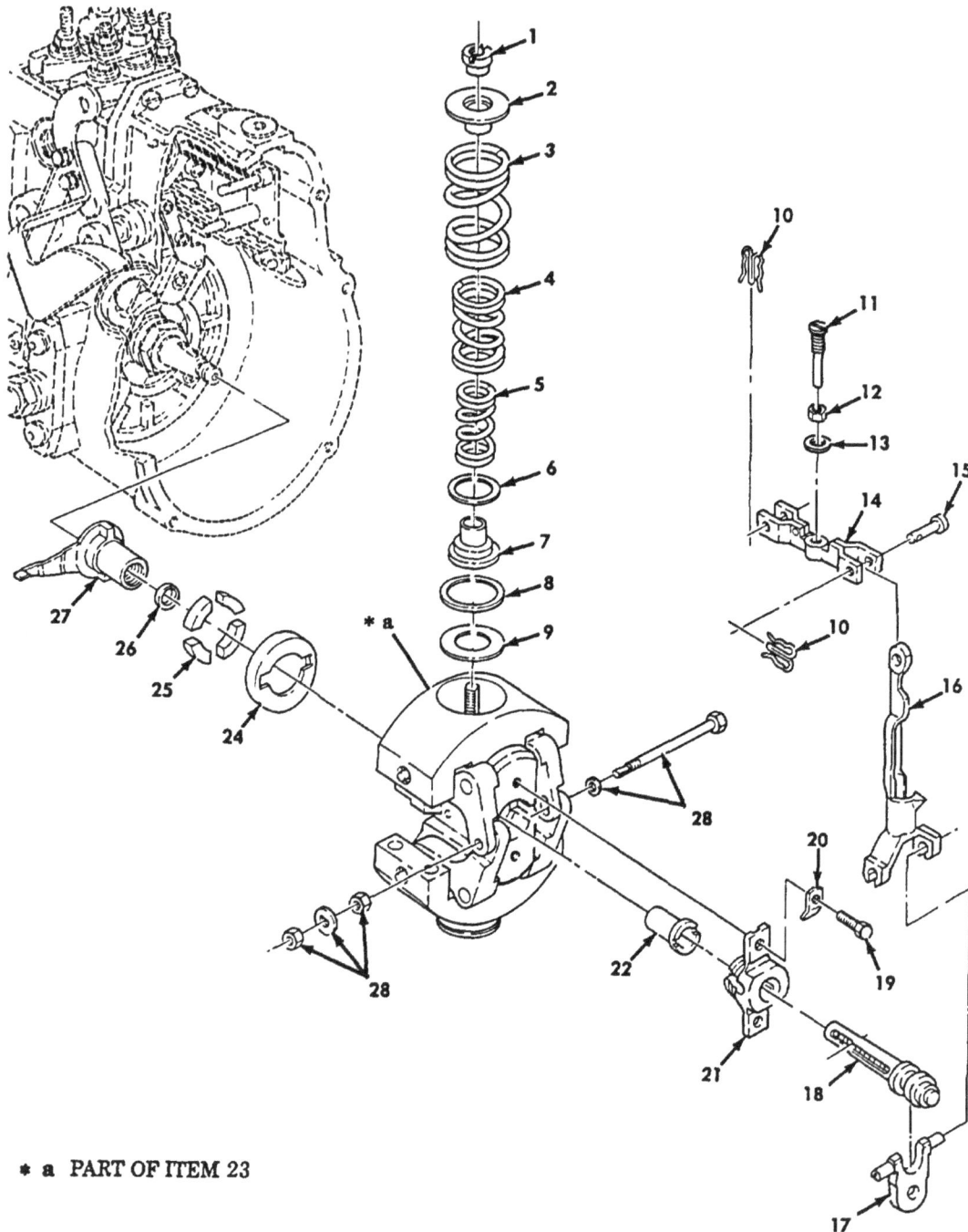

* a PART OF ITEM 23

Figure 91. Flyweight Assembly (M939A2).

(1) ITEM NO	(2) SMR CODE	(3) NSN	(4) CAGEC	(5) PART NUMBER	(6) DESCRIPTION AND USABLE ON CODES (UOC)	(7) QTY
					GROUP 0308 ENGINE SPEED GOVERNOR AND CONTROLS	
					FIG. 91 FLYWEIGHT ASSEMBLY(M939A2)	
1	PFHZZ	5310013368722	5T151	2 423 345 006	NUT,PLAIN,HEXAGON .... UOC:ZAA,ZAB,ZAC,ZAD,ZAE,ZAF,ZAG,ZAH, ZAJ,ZAK,ZAL	2
2	PFHZZ	5340013360021	5T151	2 420 520 005	SEAT,HELICAL COMPRE .... UOC:ZAA,ZAB,ZAC,ZAD,ZAE,ZAF,ZAG,ZAH, ZAJ,ZAK,ZAL	2
3	PFHZZ	5360013359951	5T151	1 424 618 035	SPRING,HELICAL,COMP .... UOC:ZAA,ZAB,ZAC,ZAD,ZAE,ZAF,ZAG,ZAH, ZAJ, ZAK, ZAL	2
4	PFHZZ	5360013359950	5T151	9 428 270 053	SPRING,HELICAL,COMP .... UOC:ZAA,ZAB,ZAC,ZAD,ZAE,ZAF,ZAG,ZAH, ZAJ,ZAK,ZAL	2
5	PFHZZ	5360013359949	5T151	2 424 619 172	SPRING,HELICAL,COMP .... UOC:ZAA,ZAB,ZAC,ZAD,ZAE,ZAF,ZAG,ZAH, ZAJ, ZAK, ZAL	2
6	PFHZZ	5365013365221	5T151	1 420 101 023	SPACER,RING 0.50MM .... UOC:ZAA,ZAB,ZAC,ZAD,ZAE,ZAF,ZAG,ZAH, ZAJ,ZAK,ZAL	2
7	PFHZZ	5340013360020	5T151	2 420 328 034	SEAT,HELICAL COMPRE .... UOC:ZAA,ZAB,ZAC,ZAD,ZAE,ZAF,ZAG,ZAH, ZAJ,ZAK,ZAL	2
8	PFHZZ	5365013365912	5T151	2 420 101 027	SPACER,RING 0.50MM .... UOC:ZAA,ZAB,ZAC,ZAD,ZAE,ZAF,ZAG,ZAH, ZAJ,ZAK,ZAL	2
9	PFHZZ	5365013365220	5T151	1 200 102 624	SPACER,RING 0.50MM .... UOC:ZAA,ZAB,ZAC,ZAD,ZAE,ZAF,ZAG,ZAH, ZAJ,ZAK,ZAL	2
10	KFHZZ		5T151	2 424 680 003	RETAINING SPRING PART OF KIT P/N.................. 57K0141.. UOC:ZAA,ZAB,ZAC,ZAD,ZAE,ZAF,ZAG,ZAH, ZAJ,ZAK,ZAL	2
11	PFHZZ	5305013359969	5T151	2 423 412 003	SCREW,MACHINE .... UOC:ZAA,ZAB,ZAC,ZAD,ZAE,ZAF,ZAG,ZAH, ZAJ,ZAK,ZAL	1
12	PFHZZ	5310013010455	53867	1 423 300 045	NUT,PLAIN,HEXAGON .... UOC:ZAA,ZAB,ZAC,ZAD,ZAE,ZAF,ZAG,ZAH, ZAJ,ZAK,ZAL	1
13	PAHZZ	5310013367368	5T151	2 916 069 007	WASHER,SPRING TENSI .... UOC:ZAA,ZAB,ZAC,ZAD,ZAE, ZAF,ZAG,ZAH, ZAJ,ZAK,ZAL	1
14	PFHZZ	5040013368293	53867	1 421 933 132	CONNECTING LINK,RIG .... UOC:ZAA,ZAB,ZAC,ZAD,ZAE,ZAF,ZAG,ZAH, ZAJ,ZAK,ZAL	1
15	PFHZZ	5315013528225	5T151	2 423 121 021	PIN,STRAIGHT,HEADLE .... UOC:ZAA,ZAB,ZAC,ZAD,ZAE,ZAF,ZAG,ZAH, ZAJ,ZAK,ZAL	1
16	PPHZZ	5340013360016	5T151	1 422 033 068	LEVER,LOCK-RELEASE	1

(1) ITEM NO	(2) SMR CODE	(3) NSN	(4) CAGEC	(5) PART NUMBER	(6) DESCRIPTION AND USABLE ON CODES (UOC)	(7) QTY
					UOC:ZAA,ZAB,ZAC,ZAD,ZAE,ZAF,ZAG,ZAH, ZAJ,ZAK,ZAL	
17	PFHZZ	3040013374368	53867	1 422 130 011	BRACKET,EYE,NONROTA	1
					UOC:ZAA,ZAB,ZAC,ZAD,ZAE,ZAF,ZAG,ZAH, ZAJ,ZAK,ZAL	
18	PFHZZ	5315013359940	5T151	1 423 521 014	PIN,STRAIGHT,THREAD	1
					UOC:ZAA,ZAB,ZAC,ZAD,ZAE,ZAF,ZAG,ZAH, ZAJ,ZAK,ZAL	
19	PFHZZ	5305011870531	53867	2911061196	SCREW,CAP,HEXAGON H	2
					UOC:ZAA,ZAB,ZAC,ZAD,ZAE,ZAF,ZAG,ZAH, ZAJ,ZAK,ZAL	
20	PFHZZ	5310013366687	5T151	1 421 331 009	WASHER,KEY	2
					UOC:ZAA,ZAB,ZAC,ZAD,ZAE,ZAF,ZAG,ZAH, ZAJ,ZAK,ZAL	
21	PFHZZ	3120013517779	5T151	2 425 703 005	BUSHING,SLEEVE...............................	1
					UOC:ZAA,ZAB,ZAC,ZAD,ZAE,ZAF,ZAG,ZAH, ZAJ,ZAK,ZAL	
22	PFHZZ	5310013366684	5T151	2 423 345 015	NUT,PLAIN,HEXAGON	1
					UOC:ZAA,ZAB,ZAC,ZAD,ZAE,ZAF,ZAG,ZAH, ZAJ,ZAK,ZAL	
23	PFHZZ	2910013372984	53867	1 428 199 009	FLYWEIGHT ASSEMBLY, ...	1
					UOC:ZAA,ZAB,ZAC,ZAD,ZAE,ZAF,ZAG,ZAH, ZAJ,ZAK,ZAL	
24	PFHZZ	3010013378937	53867	1 420 500 025	.INSERT,FLEXIBLE COU	1
					UOC:ZAA,ZAB,ZAC,ZAD,ZAE,ZAF,ZAG,ZAH, ZAJ,ZAK,ZAL	
25	KFHZZ		5T151	1 420 026 004	.RUBBER BUFFER PART OF KIT P/N..................... 57K0141...............................	4
					UOC:ZAA,ZAB,ZAC,ZAD,ZAE,ZAF,ZAG,ZAH, ZAJ,ZAK,ZAL	
26	PFHZZ	5365013393835	5T151	2 420 101 056	.SHIM 1.60MM THICK V	
					UOC:ZAA,ZAB,ZAC,ZAD,ZAE,ZAF,ZAG,ZAH, ZAJ,ZAK,ZAL	
26	PFHZZ	5365013393836	5T151	2 420 101 057	.SHIM 1.63MM THICK V	
					UOC:ZAA,ZAB,ZAC,ZAD,ZAE,ZAF,ZAG,ZAH, ZAJ,ZAK,ZAL	
26	PFHZZ	5340013366783	5T151	2 420 101 058	.RETAINER,HELICAL CO 1.66MM THICK V	
					UOC:ZAA,ZAB,ZAC,ZAD,ZAE,ZAF,ZAG,ZAH, ZAJ,ZAK,ZAL	
26	PFHZZ	5365013365212	5T151	2 420 101 059	.SPACER,RING 1.69MM THICK V	
					UOC:ZAA,ZAB,ZAC,ZAD,ZAE,ZAF,ZAG,ZAH, ZAJ,ZAK,ZAL	
26	PFHZZ	5365013365213	5T151	2 420 101 061	.SPACER,RING 1.75MM THICK V	
					UOC:ZAA,ZAB,ZAC,ZAD,ZAE,ZAF,ZAG,ZAH, ZAJ,ZAK,ZAL	
26	PFHZZ	5365013365910	5T151	2 420 101 062	.SPACER,RING 1.78MM THICK V	
					UOC:ZAA,ZAB,ZAC,ZAD,ZAE,ZAF,ZAG,ZAH, ZAJ,ZAK,ZAL	
26	PFHZZ	5365013365911	5T151	2 420 101 063	.SPACER,RING 1.81MM THICK V	
					UOC:ZAA,ZAB,ZAC,ZAD,ZAE,ZAF,ZAG,ZAH, ZAJ,ZAK,ZAL	
26	PFHZZ	5365013365214	5T151	2 420 101 064	.SPACER,RING 1.84MM THICK V	

(1) ITEM NO	(2) SMR CODE	(3) NSN	(4) CAGEC	(5) PART NUMBER	(6) DESCRIPTION AND USABLE ON CODES (UOC)	(7) QTY
					UOC:ZAA,ZAB,ZAC,ZAD,ZAE,ZAF,ZAG,ZAH, ZAJ,ZAK,ZAL	
26	PFHZZ	5365013393837	5T151	2 420 101 065	.SHIM 1.87MM THICK ....V UOC:ZAA,ZAB,ZAC,ZAD,ZAE,ZAF,ZAG,ZAH, ZAJ,ZAK,ZAL	
26	PFHZZ	5365013365215	5T151	2 420 101 066	.SPACER,RING 1.90MM THICK ....V UOC:ZAA,ZAB,ZAC,ZAD,ZAE,ZAF,ZAG,ZAH, ZAJ,ZAK,ZAL	
26	PFHZZ	5365013365216	5T151	2 420 101 067	.SPACER,RING 1.93MM THICK ....V UOC:ZAA,ZAB,ZAC,ZAD,ZAE,ZAF,ZAG,ZAH, ZAJ,ZAK,ZAL	
26	PFHZZ	5365013365217	5T151	2 420 101 068	.SPACER,RING 1.96MM THICK ....V UOC:ZAA,ZAB,ZAC,ZAD,ZAE,ZAF,ZAG,ZAH, ZAJ,ZAK,ZAL	
26	PFHZZ	5365013365218	5T151	2 420 101 069	.SPACER,RING 1.99MM THICK ....V UOC:ZAA,ZAB,ZAC,ZAD,ZAE,ZAF,ZAG,ZAH, ZAJ,ZAK,ZAL	
26	PFHZZ	5365013393838	5T151	2 420 101 070	.SHIM 2.02MM THICK ....V UOC:ZAA,ZAB,ZAC,ZAD,ZAE,ZAF,ZAG,ZAH, ZAJ,ZAK,ZAL	
26	PFHZZ	5365013368909	5T151	2 420 101 071	.SPACER,RING 2.05MM THICK ....V UOC:ZAA,ZAB,ZAC,ZAD,ZAE,ZAF,ZAG,ZAH, ZAJ,ZAK,ZAL	
26	PFHZZ	5365013368910	5T151	2 420 101 072	.SPACER,RING 2.08MM THICK ....V UOC:ZAA,ZAB,ZAC,ZAD,ZAE,ZAF,ZAG,ZAH, ZAJ,ZAK,ZAL	
26	PFHZZ	5365013368911	5T151	2 420 101 073	.SPACER,RING 2.11MM THICK ....V UOC:ZAA,ZAB,ZAC,ZAD,ZAE,ZAF,ZAG,ZAH, ZAJ,ZAK,ZAL	
26	PFHZZ	5365013365219	5T151	2 420 101 074	.SPACER,RING 2.14MM THICK ....V UOC:ZAA,ZAB,ZAC,ZAD,ZAE,ZAF,ZAG,ZAH, ZAJ,ZAK,ZAL	
26	PFHZZ	5365013366782	5T151	2 420 101 060	.SPACER,RING 1.72MM THICK ....V UOC:ZAA,ZAB,ZAC,ZAD,ZAE,ZAF,ZAG,ZAH, ZAJ,ZAK,ZAL	
27	PFHZZ	2910013372983	53867	2 426 449 018	.DRIVER,MECHANICAL G .... UOC:ZAA,ZAB,ZAC,ZAD,ZAE,ZAF,ZAG,ZAH, ZAJ,ZAK,ZAL	1
28	KFHZZ		5T151	1 423 450 900	PARTS SET, SCREW PART OF KIT P/N................. 57K0141... UOC:ZAA,ZAB,ZAC,ZAD,ZAE,ZAF,ZAG,ZAH, ZAJ,ZAK,ZAL	1

END OF FIGURE

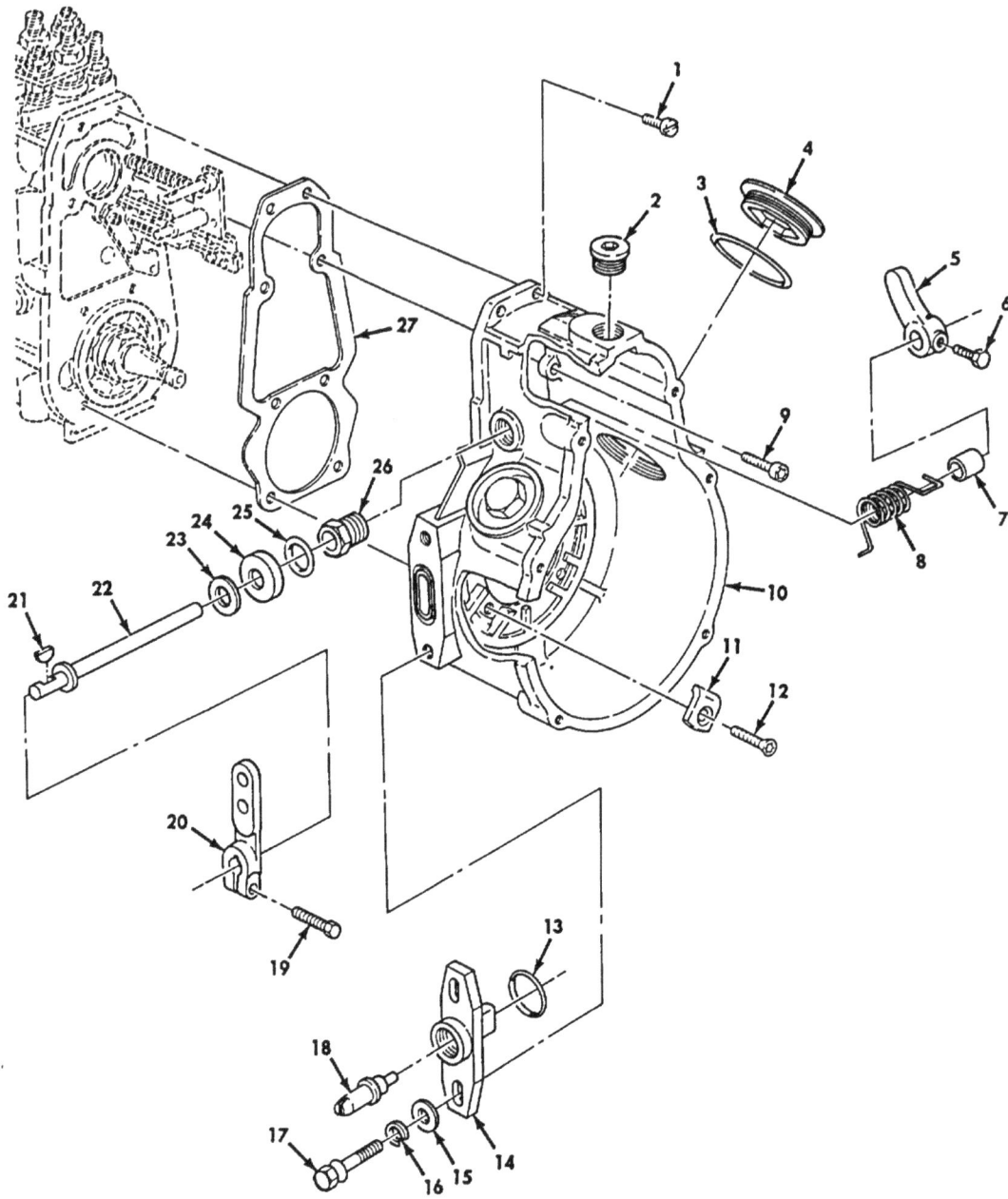

Figure 92. Governor Housing (M939A2).

(1) ITEM NO	(2) SMR CODE	(3) NSN	(4) CAGEC	(5) PART NUMBER	(6) DESCRIPTION AND USABLE ON CODES (UOC)	(7) QTY
					GROUP 0308 ENGINE SPEED GOVERNOR AND CONTROLS	
					FIG. 92 GOVERNOR HOUSING(M939A2)	
1	PFHZZ	5305013015034	53867	1 423 414 021	SCREW,MACHINE .... UOC:ZAA,ZAB,ZAC,ZAD,ZAE,ZAF,ZAG,ZAH, ZAJ, ZAK, ZAL	2
2	PFHZZ	5365011825468	53867	1-423-462-099	PLUG,MACHINE THREAD .... UOC:ZAA,ZAB,ZAC,ZAD,ZAE,ZAF,ZAG,ZAH, ZAJ,ZAK,ZAL	1
3	PFHZZ	5330013363150	5T151	1 420 113 004	GASKET.. UOC:ZAA,ZAB,ZAC,ZAD,ZAE,ZAF,ZAG,ZAH, ZAJ,ZAK,ZAL	2
4	PFHZZ	5340013362588	5T151	1 420 560 004	PLUG,PROTECTIVE,DUS .... UOC:ZAA,ZAB,ZAC,ZAD,ZAE,ZAF,ZAG,ZAH, ZAJ,ZAK,ZAL	2
5	PFHZZ	5340013360046	5T151	1-422-010-015	LEVER,MANUAL CONTRO .... UOC:ZAA,ZAB,ZAC,ZAD,ZAE,ZAF,ZAG,ZAH, ZAJ,ZAK,ZAL	1
6	PFHZZ	5305013360008	5T151	1 423 452 000	SCREW,CAP,HEXAGON H .... UOC:ZAA,ZAB,ZAC,ZAD,ZAE,ZAF,ZAG,ZAH, ZAJ,ZAK,ZAL	1
7	PFHZZ	5365013365913	5T151	2 420 200 009	SPACER,RING.................................. UOC:ZAA,ZAB,ZAC,ZAD,ZAE,ZAF,ZAG,ZAH, ZAJ,ZAK,ZAL	1
8	PFHZZ	5360013369366	5T151	2 424 651 025	SPRING,HELICAL,COMP .... UOC:ZAA,ZAB,ZAC,ZAD,ZAE,ZAF,ZAG,ZAH, ZAJ, ZAK, ZAL	1
9	PFHZZ	5305013017817	53867	2 914 552 158	SCREW,CAP,HEXAGON H .... UOC:ZAA,ZAB,ZAC,ZAD,ZAE,ZAF,ZAG,ZAH, ZAJ,ZAK,ZAL	2
10	PFHZZ	2910013374142	53867	1 425 100 323	HOUSING,GOVERNOR,DI .... UOC:ZAA,ZAB,ZAC,ZAD,ZAE,ZAF,ZAG,ZAH, ZAJ,ZAK,ZAL	1
11	PFHZZ	5340013366761	5T151	1 421 389 008	BRACKET,ANGLE .... UOC:ZAA,ZAB,ZAC,ZAD,ZAE,ZAF,ZAG,ZAH, ZAJ,ZAK,ZAL	4
12	KFHZZ		5T151	1 423 421 007	FLAT-HEAD SCREW PART OF KIT P/N.................. 57K0141... UOC:ZAA,ZAB,ZAC,ZAD,ZAE,ZAF,ZAG,ZAH, ZAJ,ZAK,ZAL	4
13	PFHZZ	5331013011825	53867	2420210034	O-RING PART OF KIT P/N 57K0141 .... UOC:ZAA,ZAB,ZAC,ZAD,ZAE,ZAF,ZAG,ZAH, ZAJ,ZAK,ZAL	1
14	PFHZZ	3040013005309	53867	1 425 703 013	PLATE,RETAINING,SHA 13.50MM .... UOC:ZAA,ZAB,ZAC,ZAD,ZAE,ZAF,ZAG,ZAH, ZAJ,ZAK,ZAL	1
15	PAHZZ	5310012375224	64678	000125 008413	WASHER,FLAT .... UOC:ZAA,ZAB,ZAC,ZAD,ZAE,ZAF,ZAG,ZAH, ZAJ,ZAK,ZAL	2
16	PAHZZ	5310013011875	53867	2 916 699 085	WASHER,LOCK	2

(1) ITEM NO	(2) SMR CODE	(3) NSN	(4) CAGEC	(5) PART NUMBER	(6) DESCRIPTION AND USABLE ON CODES (UOC)	(7) QTY
					UOC:ZAA,ZAB,ZAC,ZAD,ZAE,ZAF,ZAG,ZAH, ZAJ,ZAK,ZAL	
17	PAHZZ	5305013015112	53867	1 423 450 056	SCREW,CAP,HEXAGON H	2
					UOC:ZAA,ZAB,ZAC,ZAD,ZAE,ZAF,ZAG,ZAH, ZAJ,ZAK,ZAL	
18	PAHZZ	5306013014881	53867	1 423 124 108	BOLT,SPECIAL TIMING PIN	1
					UOC:ZAA,ZAB,ZAC,ZAD,ZAE,ZAF,ZAG,ZAH, ZAJ,ZAK,ZAL	
19	PFHZZ	5305012483222	D8286	MSX15DIN933-STL8 .8 CADMIUM PL	SCREW,CAP,HEXAGON H	1
					UOC:ZAA,ZAB,ZAC,ZAD,ZAE,ZAF,ZAG,ZAH, ZAJ,ZAK,ZAL	
20	PFHZZ	3040013382546	5T151	1 422 002 216	LEVER,REMOTE CONTRO	1
					UOC:ZAA,ZAB,ZAC,ZAD,ZAE,ZAF,ZAG,ZAH, ZAJ,ZAK,ZAL	
21	PFHZZ	5315013366689	5T151	1 900 023 002	KEY,WOODRUFF DIM 6888-2 X 3.7S160-............... 2A..	1
					UOC:ZAA,ZAB,ZAC,ZAD,ZAE,ZAF,ZAG,ZAH, ZAJ,ZAK,ZAL	
22	PAHZZ	3040013374368	53867	1 422 130 011	BRACKET,EYE,NONROTA	1
					UOC:ZAA,ZAB,ZAC,ZAD,ZAE,ZAF,ZAG,ZAH, ZAJ,ZAK,ZAL	
23	PFHZZ	5365013393839	5T151	1 420 100 602	SHIM 0.30MM THICK V	
					UOC:ZAA,ZAB,ZAC,ZAD,ZAE,ZAF,ZAG,ZAH, ZAJ,ZAK,ZAL	
24	PFHZZ	5340013366784	5T151	1420 505 057	RETAINER,HELICAL CO	1
					UOC:ZAA,ZAB,ZAC,ZAD,ZAE,ZAF,ZAG,ZAH, ZAJ,ZAK,ZAL	
25	PFHZZ	5331013366717	5T151	1 810 210 147	O-RING	1
					UOC:ZAA,ZAB,ZAC,ZAD,ZAE,ZAF,ZAG,ZAH, ZAJ,ZAK,ZAL	
26	PFHZZ	5325013366738	5T151	2 423 457 003	INSERT,SCREW THREAD	1
					UOC:ZAA,ZAB,ZAC,ZAD,ZAE,ZAF,ZAG,ZAH, ZAJ,ZAK,ZAL	
27	KFHZZ		5T151	1 421 015 082	GASKET PART OF KIT P/N 57K0141	1
					UOC:ZAA,ZAB,ZAC,ZAD,ZAE,ZAF,ZAG,ZAH, ZAJ,ZAK,ZAL	

END OF FIGURE

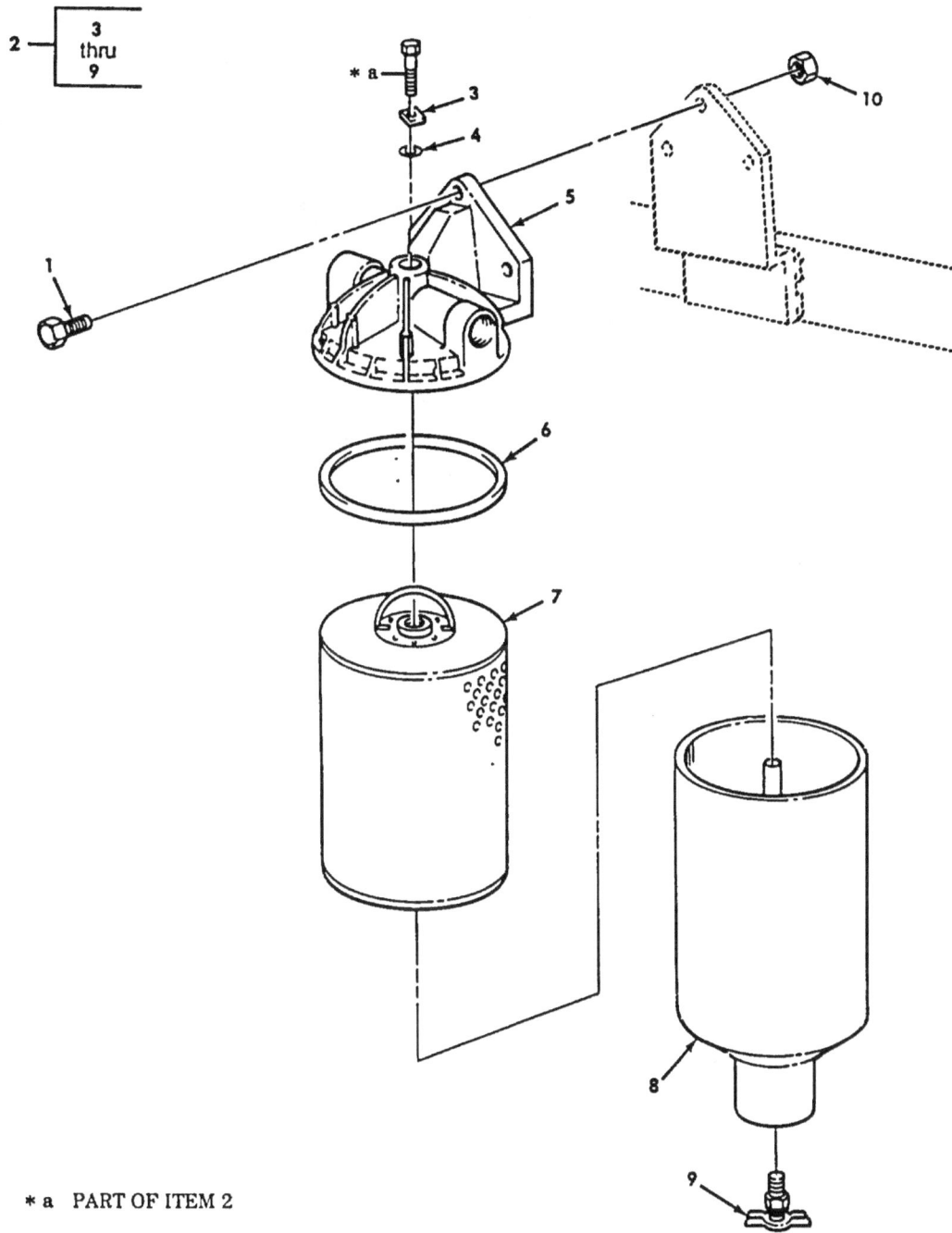

Figure 93. Fuel Filter, Water Separator, and Mounting Hardware.

* a   PART OF ITEM 2

(1) ITEM NO	(2) SMR CODE	(3) NSN	(4) CAGEC	(5) PART NUMBER	(6) DESCRIPTION AND USABLE ON CODES (UOC)	(7) QTY

GROUP 0309 FUEL FILTERS

FIG. 93 FUEL FILTER, WATER SEPARATOR, AND MOUNTING HARDWARE

(1) ITEM NO	(2) SMR CODE	(3) NSN	(4) CAGEC	(5) PART NUMBER	(6) DESCRIPTION AND USABLE ON CODES (UOC)	(7) QTY
1	PAOZZ	5305002693239	80204	B1821BH038F138N	SCREW,CAP,HEXAGON H	3
2	PAOZZ	4930004778276	15434	256172	SEPARATOR,WATER,LIQ................................	1
3	PAOZZ	5310004208044	15434	251081	.WASHER,FLAT..	1
4	KFOZZ	5331011607457	15434	3010937	.O-RING PART OF KIT P/N 256476......................	1
5	PAOZZ	2910010978591	15434	257548	.HEAD,FLUID FILTER..	1
6	KFOZZ		15434	251390	.GASKET PART OF KIT P/N 256476......................	1
7	KFOZZ		15434	256416	.ELEMENT ASSY PART OF KIT P/N 256476............	1
8	PAOZZ	4330011109054	33457	256424	.FILTER BODY,FLUID...	1
9	PAOZZ	4820012636410	79470	145	.COCK,DRAIN..	1
10	PAOZZ	5310009591488	96906	MS51922-21	NUT,SELF-LOCKING,HE	3

END OF FIGURE

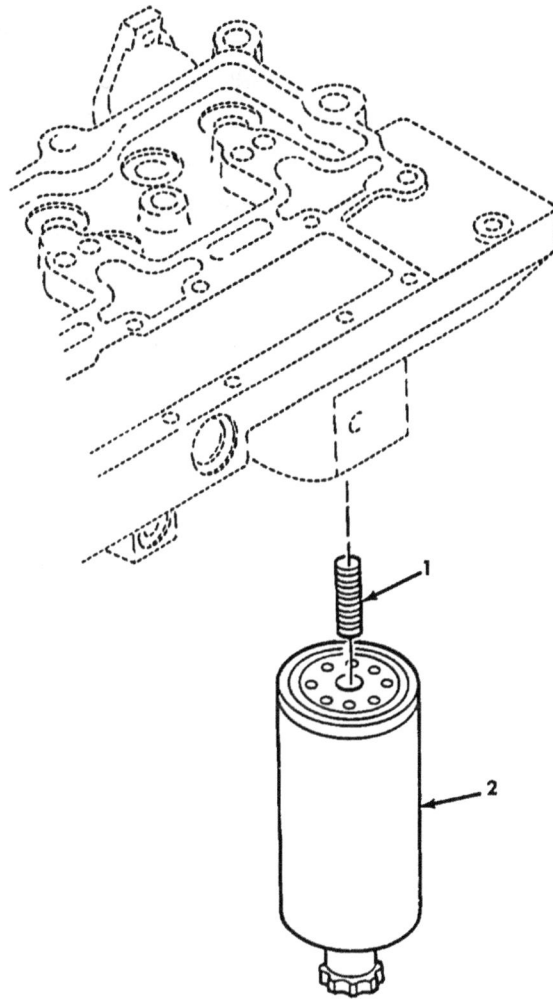

Figure 94. Fuel Filter/Water Separator (M939A2).

(1) ITEM NO	(2) SMR CODE	(3) NSN	(4) CAGEC	(5) PART NUMBER	(6) DESCRIPTION AND USABLE ON CODES (UOC)	(7) QTY
					GROUP 0309 FUEL FILTERS	
					FIG. 94 FUEL FILTER/WATER SEPARATOR (M939A2)	
1	PAOZZ	5307011964246	15434	3903845	STUD,CONTINUOUS THR...................................... UOC:ZAA,ZAB,ZAC,ZAD,ZAE,ZAF,ZAG,ZAH, ZAJ, ZAK, ZAL	1
2	PAOZZ	2910012017719	79396	33472	FILTER ELEMENT,FLUI.. UOC:ZAA,ZAB,ZAC,ZAD,ZAE,ZAF, ZAG,ZAH, ZAJ,ZAK,ZAL	1

END OF FIGURE

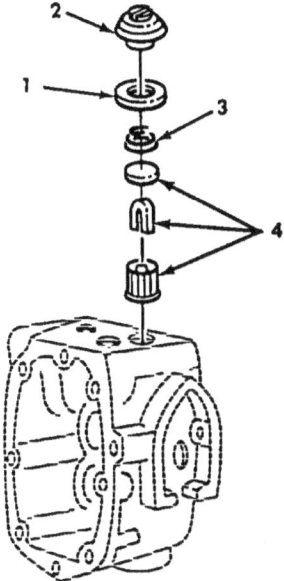

Figure 95. Fuel Pump Filter, AFC (M939, M939A1).

(1) ITEM NO	(2) SMR CODE	(3) NSN	(4) CAGEC	(5) PART NUMBER	(6) DESCRIPTION AND USABLE ON CODES (UOC)	(7) QTY
					GROUP 0309 FUEL FILTERS	
					FIG. 95 FUEL PUMP FILTER, AFC(M939, M939A1)	
1	PAOZZ	5330009619470	15434	154088	SEAL CAP PART OF KIT P/N 3010240, ... 5704519 .... UOC:DAA,DAB,DAC,DAD,DAE,DAF,DAG,DAH, DAJ,DAK,DAW,DAX,V12,V13,V14,V15,V16, V17,V19,V20,V21,V22,V24,V25,V39	1
2	PAOZZ	5365005073271	15434	157088	PLUG,MACHINE THREAD .... UOC:DAA,DAB,DAC,DAD,DAE,DAF,DAG,DAH, DAJ,DAK,DAW,DAX,V12,V13,V14,V15,V16, V17,V19,V20,V21,V22,V24,V25,V39	1
3	PAOZZ	5360005974570	15434	70700	SPRING,HELICAL,COMP .... UOC:DAA,DAB,DAC,DAD,DAE,DAF,DAG,DAH, DAJ,DAK,DAW,DAX,V12,V13,V14,V15,V16, V17,V19,V20,V21,V22,V24,V25,V39	1
4	PAOZZ	2910007908736	15434	146483	FILTER ELEMENT,FLUI .... UOC:DAA,DAB,DAC,DAD,DAE,DAF,DAG,DAH, DAJ,DAK,DAW,DAX,V12,V13,V14,V15,V16, V17,V19,V20,V21,V22,V24,V25,V39	1

END OF FIGURE

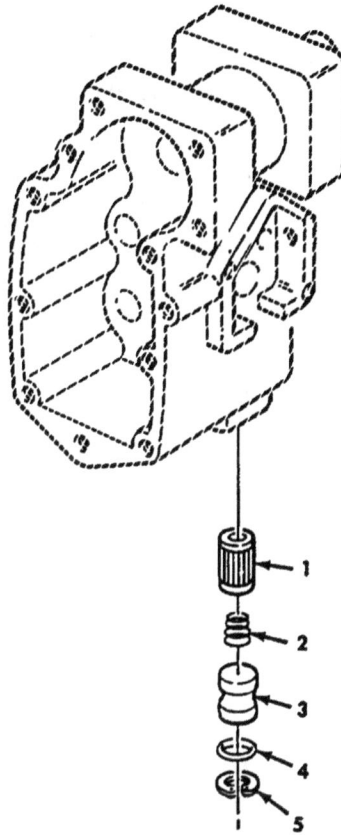

Figure 96. Fuel Pump Filter, VS (M939, M939A1).

(1) ITEM NO	(2) SMR CODE	(3) NSN	(4) CAGEC	(5) PART NUMBER	(6) DESCRIPTION AND USABLE ON CODES (UOC)	(7) QTY
					GROUP 0309 FUEL FILTERS	
					FIG. 96 FUEL PUMP FILTER, VS(M939, M939A1)	
1	PAOZZ	2910007908736	15434	146483	FILTER ELEMENT,FLUI....................................... UOC:DAL,V18	1
2	PAOZZ	5360005974570	15434	70700	SPRING,HELICAL,COMP...................................... UOC:DAL,V18	1
3	PAOZZ	2910010514292	15434	201007	COVER,FLUID FILTER.. UOC:DAL,V18	1
4	PAOZZ	5330010514243	15434	145504	PACKING,PREFORM ED..................................... UOC:DAL,V18	1
5	PAOZZ	5325008236002	15434	128807	RING,RETAINING UOC:DAL,V18	1

END OF FIGURE

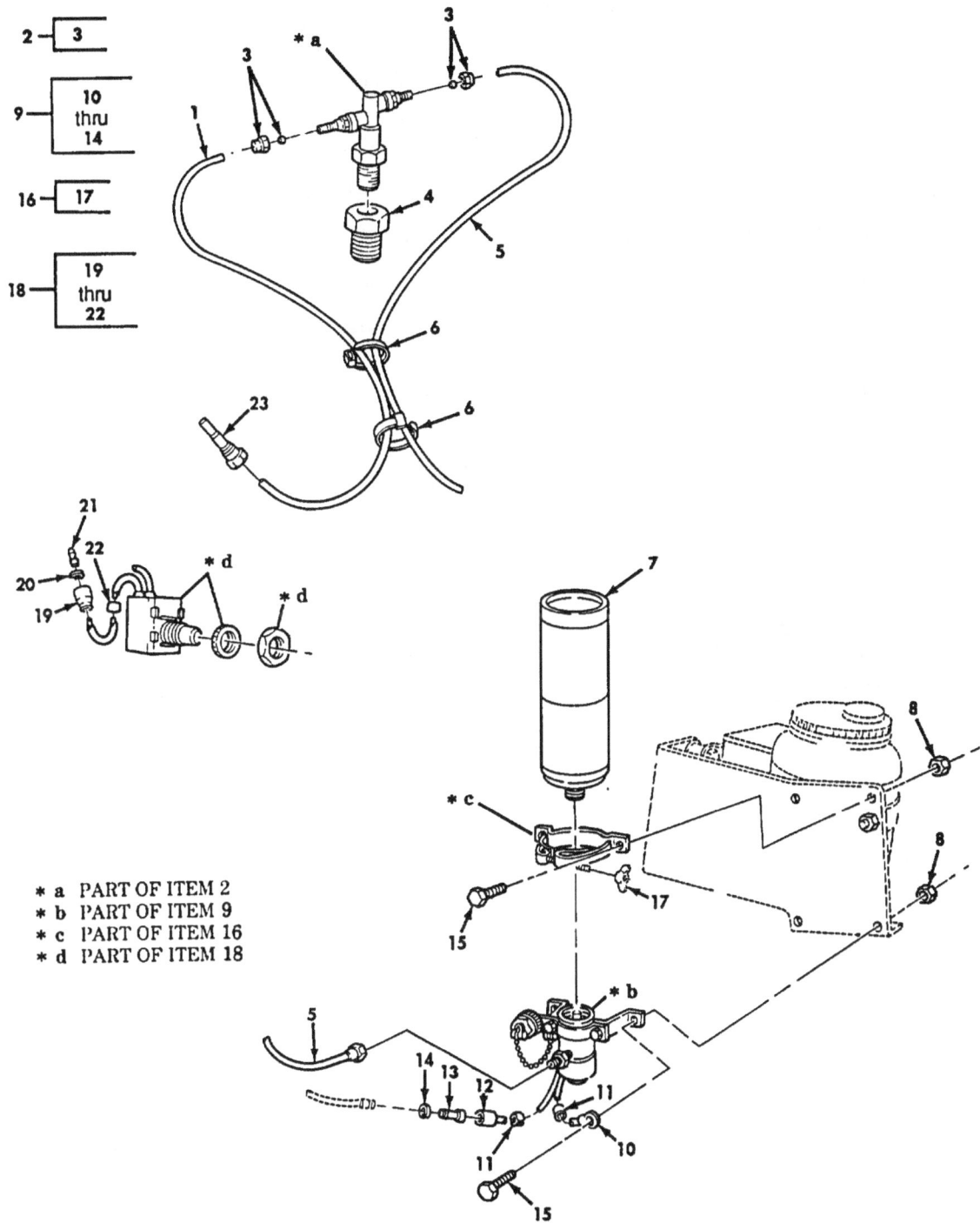

* a PART OF ITEM 2
* b PART OF ITEM 9
* c PART OF ITEM 16
* d PART OF ITEM 18

Figure 97. Ether Start Cylinder, Switch, and Related Lines and Fittings.

(1) ITEM NO	(2) SMR CODE	(3) NSN	(4) CAGEC	(5) PART NUMBER	(6) DESCRIPTION AND USABLE ON CODES (UOC)	(7) QTY
					GROUP 0311 STARTING AIDS	
					FIG. 97 ETHER START CYLINDER, SWITCH, AND RELATED LINES AND FITTINGS	
1	MOOZZ		19207	CPR104420-6-20	TUBING,NONMETALLIC MAKE FROM TUBE,......... P/N CPR104420-6	1
2	PAOZZ	5930010692776	61112	QS4577	SWITCH,THERMOSTATIC....................................	1
3	PAOZZ	4730010715740	61112	LP-5146-10	.SLEEVE,CLINCH,TUBE	2
4	PAOZZ	4730008176578	96906	MS14315-4	BUSHING,PIPE	1
5	MOOZZ		19207	CPR104420-6-30	TUBING,NONMETALLIC MAKE FROM TUBE,.......... P/N CPR104420-6	1
6	PAOZZ	5975009846582	96906	MS3367-1-0	STRAP,TIEDOWN ,ELECT................................	2
7	PAOZZ	2910011289537	19207	12277378	CYLINDER,ETHER,STAR	1
8	PAOZZ	5310002416658	96906	MS51943-34	NUT,SELF-LOCKING,HE	4
9	PAOZZ	4810011245056	19207	12277380	VALVE,SOLENOID	1
10	PAOZZ	5940002835281	96906	MS25036-109	.TERMINAL,LUG	1
11	PAOZZ	9905007524649	81349	M43436/1-1	.BAND,MAEKER	2
12	PAOZZ	5935006915591	19207	8724495	.SHELL,ELECTRICAL CO.................................	1
13	PAOZA	5999009263144	96906	MS27148-3	.CONTACT,ELECTRICAL.................................	1
14	PAOZZ	5310006560067	19207	8724497	.WASHER,SLOTTED	1
15	PAOZZ	5306000501238	96906	MS90727-32	BOLT,MACHINE..	4
16	PAOZZ	5340011289558	19207	12277381	CLAMP,LOOP	1
17	PAOZZ	5310011005199	96906	MS35426-27	.NUT,PLAIN,WING..	1
18	PAOZZ	5930011323247	19207	11669772	SWITCH,PUSH	1
19	PAOZZ	5935006915591	19207	8724495	.SHELL,ELECTRICAL CO.................................	2
20	PAOZZ	5310006560067	19207	8724497	.WASHER,SLOTTED	2
21	PAOOA	5999009263144	96906	MS27148-3	.CONTACT,ELECTRICAL.................................	2
22	PAOZZ	9905007524649	81349	M43436/1-1	.BAND,MARKE R	2
23	PAOZZ	2910011301535	19207	12277379	NOZZLE,FUEL INJECTI	1

END OF FIGURE

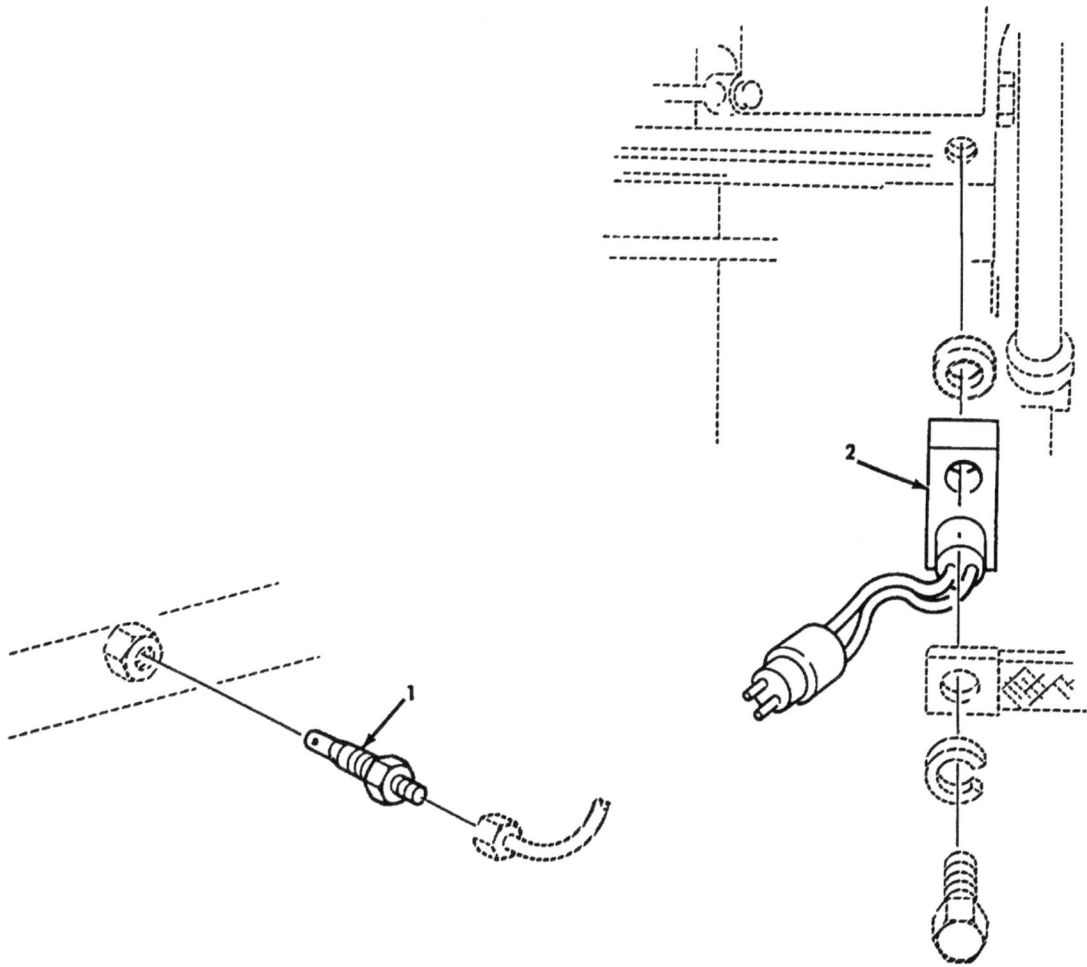

Figure 98. Atomizer Nozzle (M939A2).

(1) ITEM NO	(2) SMR CODE	(3) NSN	(4) CAGEC	(5) PART NUMBER	(6) DESCRIPTION AND USABLE ON CODES (UOC)	(7) QTY
					GROUP 0311 STARTING AIDS	
					FIG. 98 ATOMIZER NOZZLE(M939A2)	
1	PAOZZ	2835012853269	6Y402	8295204	NOZZLE,LUBRICATING, ... UOC:ZAA,ZAB,ZAC,ZAD,ZAE,ZAF,ZAG,ZAH, ZAJ,ZAK,ZAL	1
2	PAOZZ	5905012803388	6Y402	829-5220	SENSOR,TEMPERATURE ETHER START............... UOC:ZAA,ZAB,ZAC,ZAD,ZAE,ZAF,ZAG,ZAH, ZAJ, ZAK, ZAL	1

END OF FIGURE

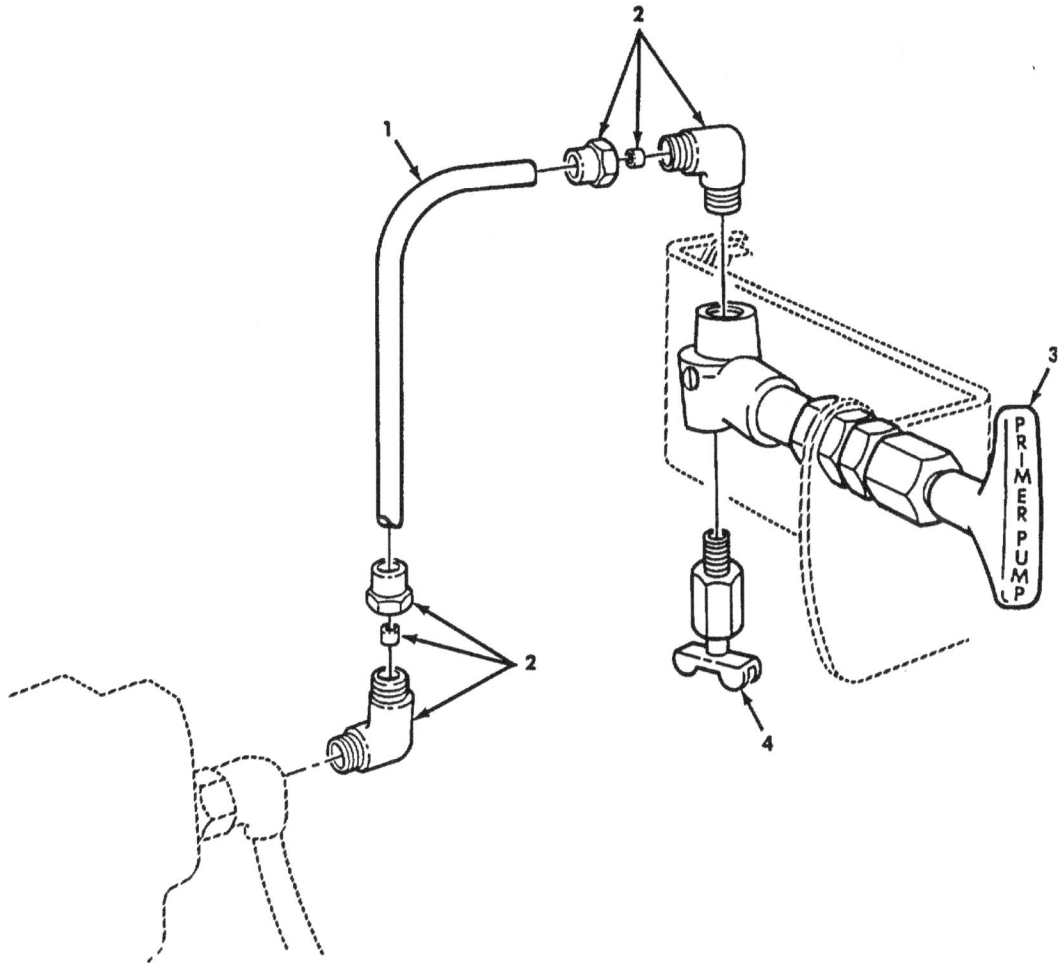

Figure 99. Fuel Primer Pump.

(1) ITEM NO	(2) SMR CODE	(3) NSN	(4) CAGEC	(5) PART NUMBER	(6) DESCRIPTION AND USABLE ON CODES (UOC)	(7) QTY
					GROUP 0311 STARTING AIDS	
					FIG. 99 FUEL PRIMER PUMP	
1	MOOZZ		19207	8675779-28	TUBING,NONMETALLIC MAKE FROM TUBE,.......... P/N 8675779...	1
2	PAOZZ	4730009213240	81343	4-2	120202BA ELBOW,PIPE TO TUBE........................	2
3	PAOZZ	2910004213967	19207	11664706	PUMP,ENGINE PRIMING....................................	1
4	PAOZZ	4820006840880	96906	MS35782-1	COCK,DRAIN..	1

END OF FIGURE

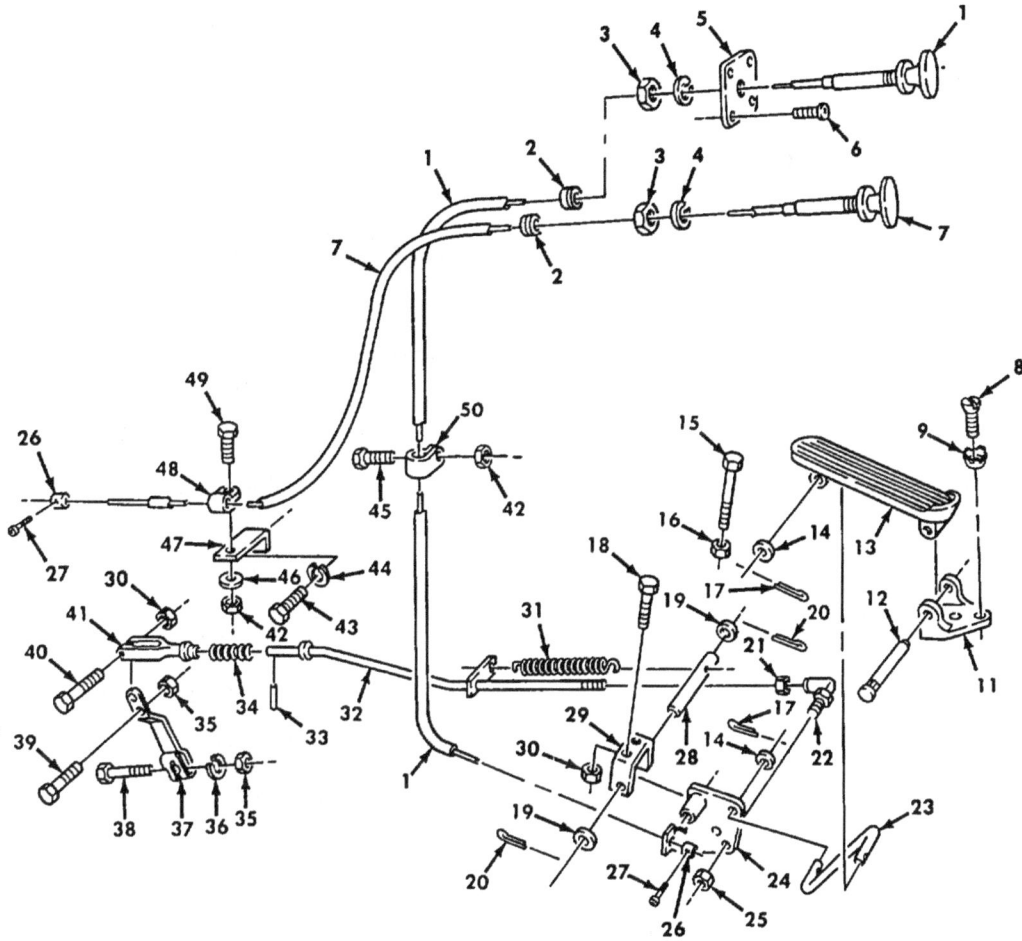

Figure 100. Accelerator Controls and Linkage.

(1) ITEM NO	(2) SMR CODE	(3) NSN	(4) CAGEC	(5) PART NUMBER	(6) DESCRIPTION AND USABLE ON CODES (UOC)	(7) QTY
					GROUP 0312 ACCELERATOR, THROTTLE OR CHOKE CONTROLS	
					FIG. 100 ACCELERATOR CONTROLS AND LINKAGE(M939,M939A1)	
1	PAOZZ	2590011079917	19207	11664388-3	CONTROL ASSEMBLY,PU............................ UOC:DAA,DAB,DAC,DAD,DAE,DAF,DAG,DAH, DAJ,DAK,DAL,DAW,DAX,V12,V13,V14,V15, V16,V17,V18,V19,V20,V21,V22,V24,V25, V39	1
2	PAOZZ	5325001745316	96906	MS35489-2	GROMMET,NONMETALLIC................................ UOC:DAA,DAB,DAC,DAD,DAE,DAF,DAG,DAH, DAJ,DAK,DAL,DAW,DAX,V12,V13 ,V14 ,V15, V16,V17,V18,V19 ,V20,V21,V22,V24,V25, V39	2
3	PAOZZ	5310000581626	96906	MS35650-3382	NUT,PLAIN,HEXAGON .... UOC:DAA,DAB,DAC,DAD,DAE,DAF,DAG,DAH, DAJ,DAK,DAL,DAW,DAX,V12,V13,V14,V15, V16,V17,V18,V19 ,V20,V21,V22,V24,V25, V39	2
4	PAOZZ	5310006379541	96906	MS35338-46	WASHER,LOCK............................ UOC:DAA,DAB,DAC,DAD,DAE,DAF,DAG,DAH, DAJ,DAK,DAL,DAW,DAX,V12,V13,V14,V15, V16, V17 ,V18,V19,V20,V21,V22,V24,V25, V39	2
5	PFOZZ	3040011307931	19207	12277149	PLATE,RETAINING,SHA............................ UOC:DAA,DAB,DAC,DAD,DAE,DAF,DAG,DAH, DAJ,DAK,DAL,DAW,DAX,V12 ,V13,V14 ,V15, V16,V17,V18,V19,V20,V21,V22,V24,V25, V39	1
6	PAOZZ	5305009897435	96906	MS35207-264	SCREW,MACHINE .... UOC:DAA,DAB,DAC,DAD,DAE,DAF,DAG,DAH, DAJ,DAK,DAL,DAW,DAX,V12,V13,V14,V15, V16,V17,V18,V19,V20,V21,V22,V24,V25, V39	4
7	PAOZZ	2590001971739	19207	11664388	CONTROL ASSEMBLY,PU............................ UOC:DAA,DAB,DAC,DAD,DAE,DAF,DAG,DAH, DAJ,DAK,DAL,DAW,DAX,V12,V13,V14,V15, V16,V17,V18,V19,V20,V21,V22,V24,V25, V39	1
8	PAOZZ	5305009847342	96906	MS35191-274	SCREW,MACHINE	2
9	PAOZZ	5310001949209	96906	MS35336-21	WASHER,LOCK	2
10	PFOZZ	2540001152565	19207	7397798	PEDAL,CONTROL THROTTLE PEDAL................	1
11	PAOZZ	3040007539130	19207	7539130	.BRACKET,EYE,ROTATI N............................	1
12	PAOZZ	5315003381621	19207	7539131	.PIN,STRAIGHT,HEADED	1
13	PAOZZ	2540007539114	19207	7539114	.PEDAL,CONTROL............................	1
14	PAOZZ	5310005157449	01496	AN960-C416L	WASHER,FLAT	2
15	PAOZZ	5305009146131	96906	MS18153-63	SCREW,CAP,HEXAGON H	1
16	PAOZZ	5310008911751	96906	MS35691-22	NUT,PLAIN,HEXAGON	1
17	PAOZZ	5315008392325	96152	A82-1	PIN,COTTER............................	2
18	PAOZZ	5306000501238	96906	M590727-32	BOLT,MACHINE............................	2

(1) ITEM NO	(2) SMR CODE	(3) NSN	(4) CAGEC	(5) PART NUMBER	(6) DESCRIPTION AND USABLE ON CODES (UOC)	(7) QTY
19	PAOZZ	5310000806004	96906	MS27183-14	WASHER,FLAT..	2
20	PAOZZ	5315008423044	96906	MS24665-283	PIN,COTTER..	2
21	PAOZZ	5310008807746	96906	MS51968-5	NUT,PLAIN,HEXAGON .... UOC:DAA,DAB,DAC,DAD,DAE,DAF,DAG,DAH, DAJ,DAK,DAL,DAW,DAX,V12 ,V13,V14,V15, V16,V17,V18,V19,V20,V21,V22,V24,V25, V39	1
22	PAOZZ	3040011079928	19207	11664437	BALL JOINT .... UOC:DAA,DAB,DAC,DAD,DAE,DAF,DAG,DAH, DAJ,DAK,DAL,DAW,DAX,V12,V13,V14,V15, V16,V17,V18,V19,V20,V21,V22,V24,V25, V39	1
23	PAOZZ	3040010909341	19207	12255972	CONNECTING LINK,RIG	1
24	PAOZZ	3040010904482	19207	12255965-1	BELL CRANK.. UOC:DAA,DAB,DAC,DAD,DAE,DAF,DAG,DAH, DAJ,DAK,DAL,DAW,DAX,V12,V13,V14,V15, V16,V17,V18,V19,V20,V21,V22,V24,V25, V39	1
25	PAOZZ	5310009843807	96906	MS51922-13	NUT,SELF-LOCKING,HE .... UOC:DAA,DAB,DAC,DAD,DAE,D,DA DAAG,DAH, DAJ,DAK,DAL,DAW,DAX,V12,V13,V14,V15, V16,V17,V18,V19,V20,V21,V22,V24,V25, V39	1
26	PAOZZ	2910007539184	56442	4B4H27	CONNECTOR,THROTTLE................................ UOC:DAA,DAB,DAC,DAD,DAE,DAF,DAG,DAH, DAJ,DAK,DAL,DAW,DAX,V12,V13,V14,V15, V16,V17,V18,V19,V20,V21,V22,V24,V25, V39	2
27	PAOZZ	5305008893002	96906	MS35206-242	SCREW,MACHINE .... UOC:DAA,DAB,DAC,DAD,DAE,DAF,DAG,DAH, DAJ,DAK,DAL,DAW,DAX,V12,V13,V14,V15, V16,V17,V18,V19,V20,V21,V22 ,V24,V25, V39	2
28	PAOZZ	3040010909342	19207	12256352	SHAFT,STRAIGHT	1
29	PAOZZ	2590004211595	19207	10896277	BRACKET,THROTTLE LI.................................	1
30	PAOZZ	5310002416658	96906	MS51943-34	NUT,SELF-LOCKING,HE	2
31	PAOZZ	5360009418684	19207	10938295	SPRING,HELICAL,EXTE .... UOC:DAA,DAB, DAC,DAD,DAE,DAF, DAG,DAH, DAJ,DAK,DAL,DAW,DAX,V12,V13,V14,V15, V16,V17,V18,V19,V20,V21,V22,V24,V25, V39	1
32	PAOZZ	5306010909344	19207	12256353	ROD,THREADED END..................................... UOC:DAA,DAB,DAC,DAD,DAE,DAF,DAG,DAH, DAJ,DAK,DAL,DAW,DAX,V12,V13,V14,V15, V16,V17,V18,V19,V20,V21,V22,V24,V25, V39	1
33	PAOZZ	5315008143530	96906	MS16562-35	PIN,SPRING .... UOC:DAA,DAB,DAC,DAD,DAE,DAF,DAG,DAH, DAJ,DAK,DAL,DAW,DAX,V12,V13,V14,V15, V16,V17,V18,V19,V20,V21,V22,V24,V25, V39	1

(1) ITEM NO	(2) SMR CODE	(3) NSN	(4) CAGEC	(5) PART NUMBER	(6) DESCRIPTION AND USABLE ON CODES (UOC)	(7) QTY
34	PAOZZ	5360010909333	19207	12256429	SPRING,HELICAL,EXTE .... UOC:DAA,DAB,DAC,DAD,DAE,DAF,DAG,DAH, DAJ,DAK,DAL,DAW,DAX,V12,V13,V14,V15, V16,V17,V18,V19,V20,V21,V22,V24,V25, V39	1
35	PAOZZ	5310008775796	96906	MS21044N4	NUT,SELF-LOCKING,HE .... UOC: DAA,DAB, DAC, DAD, DAE, DAF,DAG, DAH, DAJ,DAK,DAL,DAW,DAX,V12,V13,V14,V15, V16,V17,V18,V19,V20,V21,V22,V24,V25, V39	2
36	PAOZZ	5310005825965	96906	MS35338-44	WASHER,LOCK... UOC:DAA,DAB,DAC,DAD,DAE,DAF,DAG,DAH, DAJ,DAK,DAL,DAW,DAX,V12,V13,V14,V15, V16,V17,V18,V19,V20,V21,V22,V24,V25, V39	1
37	PAOZZ	3040011896406	19207	12302679-1	LEVER,REMOTE CONTRO .... UOC:DAA,DAB,DAC,DAD,DAE,DAF,DAG,DAH, DAJ,DAK,DAL,DAW,DAX,V12 ,V13 ,V14 ,V15, V16,V17,V18,V19,V20,V21,V22,V24,V25, V39	1
38	PAOZZ	5305002678955	80204	B1821BH025F138N	SCREW,CAP,HEXAGON H.................................. UOC:DAA,DAB,DAC,DAD,DAE,DAF,DAG,DAH, DAJ,DAK,DAL,DAW,DAX,V12,V13,V14,V15, V16,V17,V18,V19,V20,V21,V22,V24,V25, V39	1
39	PAOZZ	5305010907625	19207	12256366	SCREW,SHOULDER... UOC:DAA,DAB,DAC,DAD,DAE,DAF,DAG,DAH, DAJ,DAK,DAL,DAW,DAX,V12,V13,V14,V15, V16,V17,V18,V19,V20,V21,V22,V24,V25, V39	1
40	PAOZZ	5305002259092	96906	MS90726-37	SCREW,CAP,HEXAGON H.................................. UOC:DAA,DAB,DAC,DAD,DAE,DAF,DAG,DAH, DAJ,DAK,DAL,DAW,DAX,V12,V13,V14,V15, V16,V17,V18,V19,V20,V21,V22,V24,V25, V39	1
41	PAOZZ	5340010909332	19207	12256357	CLEVIS,ROD END.. UOC:DAA,DAB,DAC,DAD,DAE,DAF,DAG,DAH, DAJ,DAK,DAL,DAW,DAX,V12,V13,V14,V15, V16,V17,V18,V19,V20,V21,V22,V24,V25, V39	1
42	PAOZZ	5310009359022	96906	MS51943-32	NUT,SELF-LOCKING,HE .... UOC:DAA,DAB,DAC,DAD,DAE,DAF,DAG,DAH, DAJ,DAK,DAL,DAW,DAX,V12,V13,V14,V15, V16,V17,V18,V19,V20,V21,V22,V24,V25, V39	2
43	PAOZZ	5305012770423	96906	MS90727-132	SCREW,CAP,HEXAGON H.................................. UOC:DAA,DAB,DAC,DAD,DAE,DAF,DAG,DAH, DAJ,DAK,DAL,DAW,DAX,V12,V13,V14,V15, V16,V17,V18,V19,V20,V21,V22,V24,V25, V39	1

(1) ITEM NO	(2) SMR CODE	(3) NSN	(4) CAGEC	(5) PART NUMBER	(6) DESCRIPTION AND USABLE ON CODES (UOC)	(7) QTY
44	PAOZZ	5310001670680	96906	MS35338-49	WASHER,LOCK.. UOC:DAA,DAB,DAC,DAD,DAE,DAF,DAG,DAH, DAJ,DAK,DAL,DAW,DAX,V12,V13,V14,V15, V16,V17,V18,V19,V20,V21,V22,V24,V25, V39	1
45	PAOZZ	5305002678952	80204	B1821BH025F050N	SCREW,CAP,HEXAGON H............................... UOC:DAA,DAB,DAC,DAD,DAE,DAF,DAG,DAH, DAJ,DAK,DAL,DAW,DAX,V12,V13 ,V14,V15, V16,V17,V18,V19 ,V20,V21,V22,V24,V25, V39	1
46	PAOZZ		96906	MS27183-10	WASHER,FLAT... UOC:DAA,DAB,DAC,DAD,DAE,DAF,DAG,DAH, DAJ,DAK,DAL,DAW,DAX,V12,V13,V14,V15, V16,V17,V18,V19,V20,V21,V22,V24,V25, V39	1
47	PFOZZ	5340004721953	19207	11664613	BRACKET,ANGLE .... UOC:DAA,DAB,DAC,DAD,DAE,DAF,DAG,DAH, DAJ,DAK,DAL,DAW,DAX,V12,V13,V14,V15, V16,V17,V18,V19,V20,V21,V22,V24,V25, V39	1
48	PAOZZ	5340000881255	96906	MS21333-96	CLAMP,LOOP... UOC:DAA,DAB,DAC,DAD,DAE,DAF,DAG,DAH, DAJ,DAK,DAL,DAW,DAX,V12,V13,V14,V15, V16,V17,V18,V19,V20,V21,V22,V24,V25, V39	1
49	PAOZZ	5306000680513	60285	6893-2	BOLT,MACHINE... UOC:DAA,DAB,DAC,DAD,DAE,DAF,DAG,DAH, DAJ,DAK,DAL,DAW,DAX,V12,V13,V14,V15, V16,V17,V18,V19,V20,V21, V22,V24,V25, V39	1
50	PAOZZ	5340008278314	96906	MS21333-33	CLAMP,LOOP... UOC:DAA,DAB,DAC,DAD,DAE,DAF,DAG,DAH, DAJ,DAK,DAL,DAW,DAX,V12,V13,V14,V15, V16,V17,V18,V19,V20,V21,V22,V24,V25, V39	1

END OF FIGURE

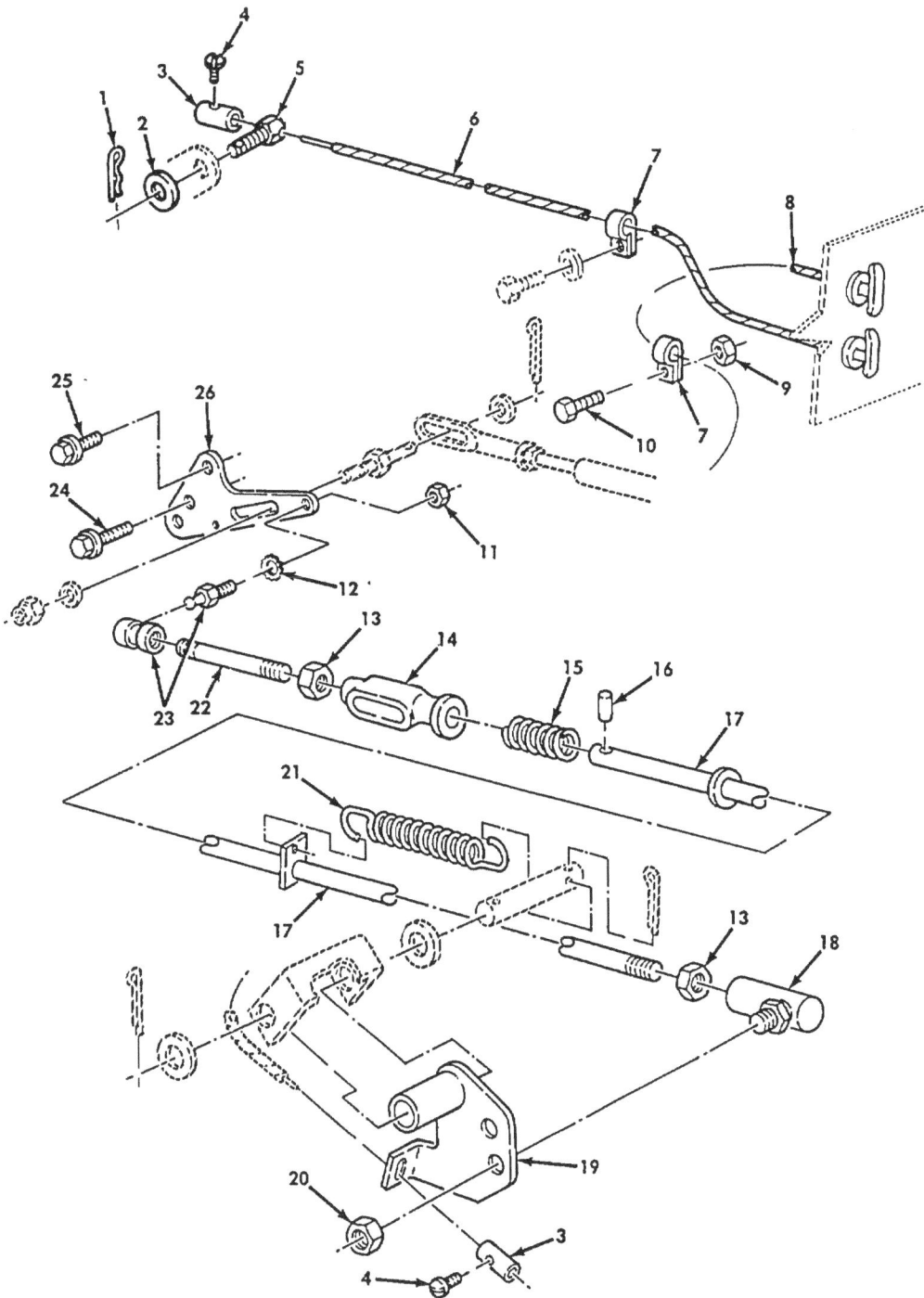

Figure 101. Accelerator and Throttle Linkage (M939A2).

(1) ITEM NO	(2) SMR CODE	(3) NSN	(4) CAGEC	(5) PART NUMBER	(6) DESCRIPTION AND USABLE ON CODES (UOC)	(7) QTY
					GROUP 0312 ACCELERATOR, THROTTLE OR CHOKE CONTROLS	
					FIG. 101 ACCELERATOR AND THROTTLE LINKAGE (M939A2)	
1	PAOZZ	5315012849812	60602	10166	PIN,LOCK.. UOC:ZAA,ZAB,ZAC,ZAD,ZAE,ZAF,ZAG,ZAH, ZAJ, ZAK, ZAL	1
2	PAOZZ	5310002363694	11083	8H9620	WASHER,FLAT .... UOC:ZAA,ZAB,ZAC,ZAD,ZAE,ZAF, ZAG,ZAH, ZAJ,ZAK,ZAL	1
3	PAOZZ	2910007539184	56442	4B4H27	CONNECTOR,THROTTLE.......................... UOC:ZAA,ZAB,ZAC,ZAD,ZAE,ZAF,ZAG,ZAH, ZAJ,ZAK,ZAL	2
4	PAOZZ	5305008893002	96906	MS35206-242	SCREW,MACHINE .... UOC:ZAA,ZAB,ZAC,ZAD,ZAE,ZAF,ZAG,ZAH, ZAJ,ZAK,ZAL	2
5	PAOZZ	5315012849583	60602	30223-3	PIN,SHOULDER,HEADLE .... UOC:ZAA,ZAB,ZAC,ZAD,ZAE,ZAF,ZAG,ZAH, ZAJ,ZAK,ZAL	1
6	PAOZZ	2590014438097	19207	11664388-4	CONTROL ASSEMBLY,PU EMERGENCY .... ENGINE STOP.. UOC:ZAA,ZAB,ZAC,ZAD,ZAE,ZAF,ZAG,ZAH, ZAJ,ZAK,ZAL	1
7	PAOZZ	5340008278314	96906	MS21333-33	CLAMP,LOOP .... UOC:ZAA,ZAB,ZAC,ZAD, ZAE,ZAF,ZAG,ZAH, ZAJ,ZAK,ZAL	2
8	PAOZZ	2590011079917	19207	11664388-3	CONTROL ASSEMBLY,PU THROTTLE	1
9	PAOZZ	5310009359022	96906	MS51943-32	NUT,SELF-LOCKING,HE .... UOC:ZAA,ZAB,ZAC,ZAD,ZAE,ZAF,ZAG,ZAH, ZAJ,ZAK,ZAL	1
10	PAOZZ	5305002678952	80204	B1821BH025F050N	SCREW,CAP,HEXAGON H.......................... UOC:ZAA,ZAB,ZAC,ZAD,ZAE,ZAF,ZAG,ZAH, ZAJ,ZAK,ZAL	1
11	PAOZZ	5310014378728	80204	B18244B06	NUT,PLAIN,EXTENDED .... UOC:ZAA,ZAB,ZAC,ZAD,ZAE,ZAF,ZAG,ZAH, ZAJ,ZAK,ZAL	1
12	PAOZZ	5310005273634	96906	MS35335-61	WASHER,LOCK...................................... UOC:ZAA,ZAB,ZAC,ZAD,ZAE,ZAF,ZAG,ZAH, AJ,ZAK,ZAL	1
13	PAOZZ	5310008807746	96906	MS51968-5	NUT,PLAIN,HEXAGON .... UOC:ZAA,ZAB,ZAC,ZAD,ZAE,ZAF,ZAG,ZAH, ZAJ,ZAK,ZAL	2
14	PAOZZ	3040012804153	08277	20510914	SLIP UNION,THROTTLE THROTTLE ROD............ UOC:ZAA,ZAB,ZAC,ZAD,ZAE,ZAF,ZAG,ZAH, ZAJ,ZAK,ZAL	1
15	PAOZZ	5360012856010	18028	161	SPRING,HELICAL,COMP .... UOC:ZAA,ZAB,ZAC,ZAD,ZAE,ZAF,ZAG,ZAH, ZAJ,ZAK,ZAL	1
16	PAOZZ	5315008263251	96906	MS16562-223	PIN,SPRING.. UOC:ZAA,ZAB,ZAC,ZAD,ZAE,ZAF,ZAG,ZAH,	1

(1) ITEM NO	(2) SMR CODE	(3) NSN	(4) CAGEC	(5) PART NUMBER	(6) DESCRIPTION AND USABLE ON CODES (UOC)	(7) QTY
					ZAJ,ZAK,ZAL	
17	PAOZZ	5340014441016	19207	12363294-2	ROD END,THREADED .... UOC:ZAA,ZAB,ZAC,ZAD,ZAE,ZAF,ZAG,ZAH, ZAJ,ZAK,ZAL	1
18	PAOZZ	3040011079928	19207	11664437	BALL JOINT... UOC:ZAA,ZAB,ZAC,ZAD,ZAE,ZAF,ZAG,ZAH, ZAJ,ZAK,ZAL	1
19	PAOZZ	3040012794658	33477	20510540	BELL CRANK.. UOC:ZAA,ZAB,ZAC,ZAD,ZAE,ZAF,ZAG,ZAH, ZAJ,ZAK,ZAL	1
20	PAOZZ	5310009843807	96906	MS51922-13	NUT,SELF-LOCKING,HE .... UOC:ZAA,ZAB,ZAC,ZAD,ZAE,ZAF,ZAG,ZAH, ZAJ,ZAK,ZAL	1
21	PAOZZ	5360009418684	19207	10938295	SPRING,HELICAL,EXTE .... UOC:ZAA,ZAB,ZAC,ZAD,ZAE,ZAF,ZAG,ZAH, ZAJ,ZAK,ZAL	1
22	PAOZZ	5306012816560	5Y952	670170	ROD,THREADED END... UOC:ZAA,ZAB,ZAC,ZAD,ZAE,ZAF,ZAG,ZAH, ZAJ,ZAK,ZAL	1
23	PAOZZ	3040012846230	60602	10718	BALL JOINT THROTTLE ROD .... UOC:ZAA,ZAB,ZAC,ZAD,ZAE,ZAF,ZAG,ZAH, ZAJ,ZAK,ZAL	1
24	PAOZZ	5305012871585	47457	20511190	SCREW,CAP,HEXAGON H .... UOC:ZAA,ZAB,ZAC,ZAD,ZAE,ZAF,ZAG,ZAH, ZAJ,ZAK,ZAL	1
25	PAOZZ	5306014353269	80204	B18234B06016N	BOLT,MACHINE.. UOC:ZAA,ZAB,ZAC,ZAD,ZAE,ZAF,ZAG,ZAH, ZAJ,ZAK,ZAL	1
26	PAOZZ	3040014447878	19207	12363633-1	LEVER,REMOTE CONTRO .... UOC:ZAA,ZAB,ZAC,ZAD,ZAE,ZAF,ZAG,ZAH, ZAJ,ZAK	1

END OF FIGURE

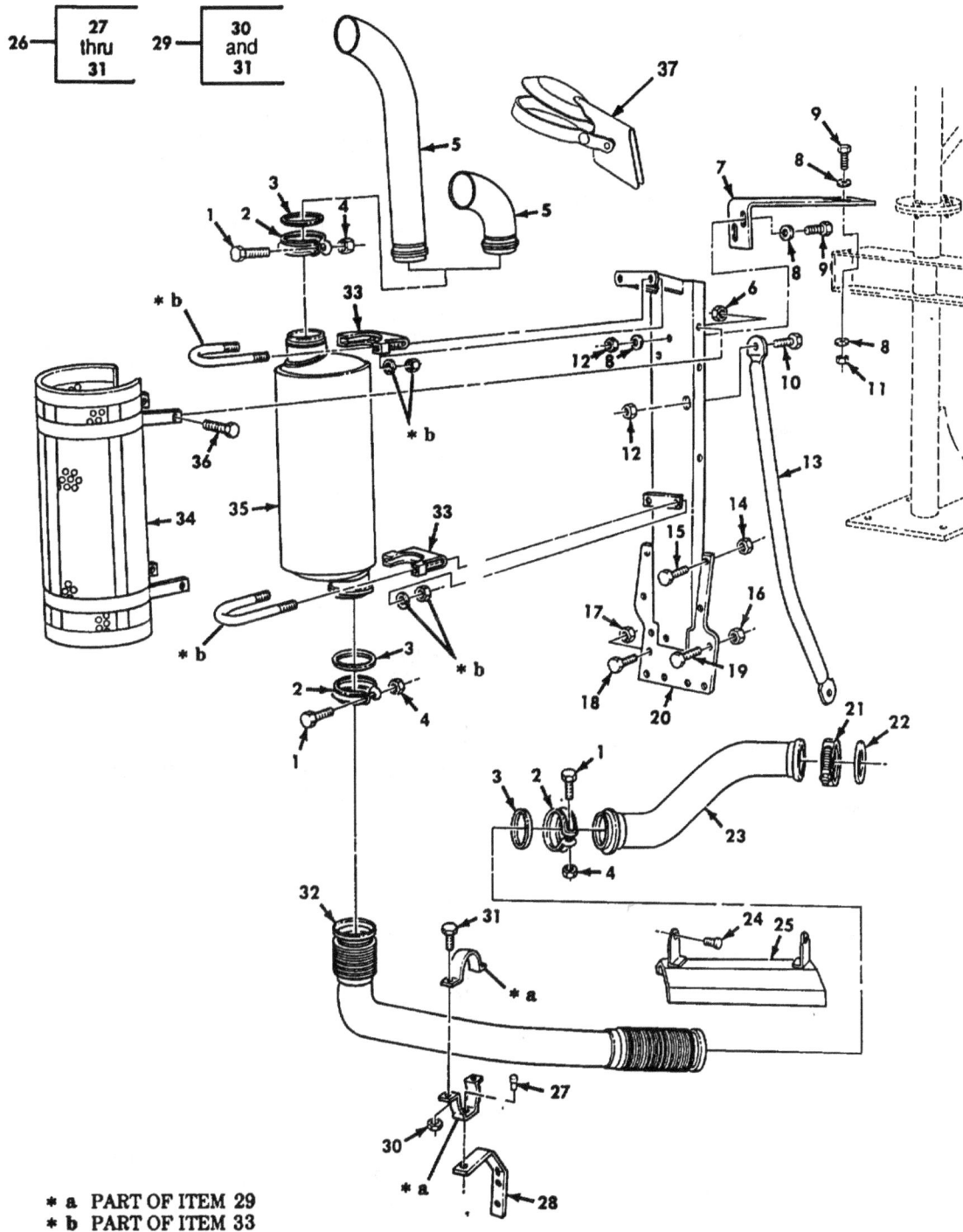

* a PART OF ITEM 29
* b PART OF ITEM 33

Figure 102. Muffler, Exhaust Pipes, and Related Parts.

(1) ITEM NO	(2) SMR CODE	(3) NSN	(4) CAGEC	(5) PART NUMBER	(6) DESCRIPTION AND USABLE ON CODES (UOC)	(7) QTY
					GROUP 04 EXHAUST SYSTEM 0401 MUFFLER AND PIPES	
					FIG. 102 MUFFLER, EXHAUST PIPES, AND RELATED PARTS	
1	PAOZZ	5306002259095	96906	MS90726-40	BOLT, MACHINE .... UOC:DAA, DAB, DAC, DAD, DAE, DAF, DAG, DAH, DAJ, DAK, DAL, DAW, DAX, V12, V13, V14, V15, V16, V17, V18, V19, V20, V21, V22, V24, V25, V39	3
2	PAOZZ	5342011065488	19207	11609348-11	COUPLING, CLAMP, GROO................................. UOC:DAA, DAB, DAC, DAD, DAE, DAF, DAG, DAH, DAJ, DAK, DAL, DAW, DAX, V12, V13, V14, V15, V16, V17, V18, V19, V20, V21, V22, V24, V25, V39	3
3	PAOZZ	5330013794345	19207	12255817 NON-ASB ESTOS	GASKET NON-ASBESTOS .... UOC:DAA, DAB, DAC, DAD, DAE, DAF, DAG, DAH, DAJ, DAK, DAL, DAW, DAX, V12, V13, V14, V15, V16, V17, V18, V19, V20, V21, V22, V24, V25, V39	3
4	PAOZZ	5310001766690	19207	11609727-2	NUT, SELF-LOCKING, HE .... UOC:DAA, DAB, DAC, DAD, DAE, DAF, DAG, DAH, DAJ, DAK, DAL, DAW, DAX, V12, V13, V14, V15, V16, V17, V18, V19, V20, V21, V22, V24, V25, V39	3
5	PAOZZ	2990010835715	19207	12255920	PIPE, EXHAUST .... UOC:DAA, DAB, DAC, DAD, DAG, DAH, DAL, DAW, DAX, V12, V13, V14, V15, V16, V17, V18, V21, V22, V39, ZAA, ZAB, ZAC, ZAD, ZAG, ZAH, ZAL	1
5	PAOZZ	2990010853833	19207	12256298	CONNECTOR, EXHAUST P .... UOC:DAE, DAF, DAJ, DAK, V19, V20, V24, V25, ZAE, ZAF, ZAJ, ZAK	1
6	PAOZZ	5310008140672	96906	MS51943-36	NUT, SELF-LOCKING, HE .... UOC:DAA, DAB, DAC, DAD, DAE, DAF, DAG, DAH, DAJ, DAK, DAL, DAW, DAX, V12, V13, V14, V15, V16, V17, V18, V19, V20, V21, V22, V24, V25, V39	4
7	PAOZZ	5340012102158	19207	12301092	BRACKET, ANGLE... UOC:DAA, DAB, DAC, DAD, DAJ, DAK, DAW, DAX, V12, V13, V14, V15, V16, V19, V24, V25, V39, ZAA, ZAB, ZAD, ZAF, ZAJ, ZAK	1
8	PAOZZ	5310000806004	96906	MS27183-14	WASHER, FLAT .... UOC:DAA, DAB, DAC, DAD, DAJ, DAK, DAW, DAX, V12, V13, V14, V15, V16, V17, V19, V24, V25, V39, ZAA, ZAB, ZAC, ZAD, ZAJ, ZAK	8
9	PAOZZ	5305002693238	80204	B1821BH038F125N	SCREW, CAP, HEXAGON H .... UOC:DAA, DAB, DAC, DAD, DAJ, DAK, DAW, DAX, V12, V13, V14, V15, V16, V17, V19, V24, V25, V39, ZAA, ZAB, ZAC, ZAD, ZAJ, ZAK	4
10	PAOZZ	5305007195219	96906	MS90727-111	SCREW, CAP, HEXAGON H .... UOC:DAE, DAF, DAG, DAH, V19, V20, V21, V22,	2

(1) ITEM NO	(2) SMR CODE	(3) NSN	(4) CAGEC	(5) PART NUMBER	(6) DESCRIPTION AND USABLE ON CODES (UOC)	(7) QTY
11	PAOZZ	5310008140672	96906	MS51943-36	V39, ZAE, ZAF, ZAG, ZAH NUT, SELF-LOCKING, HE .... UOC:DAA, DAB, DAC, DAD, DAJ, DAK, DAW, DAX, V12, V13, V14, V15, V16, V19, V24, V25, V39, ZAA, ZAB, ZAD, ZAF, ZAJ, ZAK	4
12	PAOZZ	5310004883888	96906	MS51943-40	NUT, SELF-LOCKING, HE .... UOC:DAE, DAF, DAG, DAH, V19, V20, V21, V22, V39, ZAE, ZAF, ZAG, ZAH	2
13	PAOZZ	5340010831117	19207	12256650	BRACKET, ANGLE . ... UOC:DAE, DAF, DAG, DAH, DAK, V19, V20, V21, V22, V25, V39, ZAE, ZAF, ZAG, ZAH, ZAK	1
14	PAOZZ	5310004883888	96906	MS51943-40	NUT, SELF-LOCKING, HE .... UOC:DAA, DAB, DAC, DAD, DAJ, DAK, DAL, DAW, DAX, V12, V13, V14, V15, V16, V17, V18, V24, V25, V39, ZAA, ZAB, ZAC, ZAD, ZAJ, ZAK, ZAL	4
15	PAOZZ	5305007195235	80204	B1821BH050F175N	SCREW, CAP, HEXAGON H .... UOC:DAE, DAF, DAG, DAH, V19, V20, V21, V22, ZAE, ZAF, ZAG, ZAH	4
16	PAOZZ	5310004883888	96906	MS51943-40	NUT, SELF-LOCKING, HE .... UOC:DAE, DAF, DAG, DAH, DAL, V18, V19, V20, V21, V22, ZAE, ZAF, ZAG, ZAH, ZAL	1
17	PAOZZ	5310004883888	96906	MS51943-40	NUT, SELF-LOCKING, HE .... UOC:DAA, DAB, DAC, DAD, DAE, DAF, DAG, DAH, DAJ;DAK, DAL, DAW, DAX, V12, V13, V14, V15, V16, V17, V18, V19, V20, V21, V22, V24, V25, V39	6
18	PAOZZ	5305007195238	80204	B1821BH050F200N	SCREW, CAP, HEXAGON H .... UOC:DAA, DAB, DAC, DAD, DAE, DAF, DAG, DAH, DAJ, DAK, DAL, DAW, DAX, V12, V13, V14, V15, V16, V17, V18, V19, V20, V21, V22, V24, V25, V39	6
19	PAOZZ	5305007195238	80204	B1821BH050F200N	SCREW, CAP, HEXAGON H .... UOC:DAE, DAF, DAG, DAH, V19, V20, V21, V22, ZAE, ZAF, ZAG, ZAH	1
20	PAOZZ	2990010855349	19207	12256632-1	HANGER, ENGINE EXHAU UOC:V12, V13, V14, V15, V16, V17, V24, V25, V39	1
20	PAOZZ	2990010855350	19207	12256632-2	HANGER, ENGINE EXHAU UOC:V18, V19, V20, V21, V22	1
20	PAOZZ	2990012104650	19207	12302949-1	BRACKET ASSEMBLY, MU .... UOC:DAA, DAB, DAC, DAD, DAJ, DAXK, DAW, DAX, ZAA, ZAB, ZAC, ZAD, ZAJ, ZAK	1
20	PAOZZ	2990012182100	19207	12302949-2	BRACKET ASSEMBLY, MU .... UOC:DAE, DAF, DAG, DAH, DAL, ZAE, ZAF, ZAG, ZAH, ZAL	1
21	PAOZZ	5340011516145	19207	10949040-4	COUPLING, CLAMP, GROO........................ UOC:DAA, DAB, DAC, DAD, DAE, DAF, DAG, DAH, DAJ, DAK, DAL, DAW, DAX, V12, V13, V14, V15, V16, V17, V18, V19, V20, V21, V22, V24, V25, V39	1
22	PAOZZ	5330011099410	19207	12256082-1	GASKET........................ UOC:DAA, DAB, DAC, DAD, DAE, DAF, DAG, DAH,	1

(1) ITEM NO	(2) SMR CODE	(3) NSN	(4) CAGEC	(5) PART NUMBER	(6) DESCRIPTION AND USABLE ON CODES (UOC)	(7) QTY
					DAJ, DAK, DAL, DAW, DAX, V12, V13, V14, V15, V16, V17, V18, V19, V20, V21, V22, V24, V25, V39	
23	PAOZZ	2990010829009	19207	12255964	PIPE, EXHAUST	1
					UOC:DAA, DAB, DAC, DAD, DAE, DAF, DAG, DAH, DAJ, DAK, DAL, DAW, DAX, V12, V13, V14, V15, V16, V17, V18, V19, V20, V21, V22, V24, V25, V39	
24	PAOZZ	5305002678953	80204	B1821BH025F063N	SCREW, CAP, HEXAGON H	2
25	PAOZZ	2990010822511	19207	12256292-1	GUARD, MUFFLER-EXHAU	1
26	PFOZZ	5342011037589	19207	12256802	CLAMP AND BRACKET A	1
27	PFOZZ	5320011453183	96906	MS35743-73	.RIVET, SOLID	1
28	PFOZZ	5340011307932	19207	12256801	.BRACKET, PIPE...	1
29	PFOZZ	5340011602299	19207	12256800	.CLAMP, LOOP	1
30	PAOZZ	5310008140672	96906	MS51943-36	..NUT, SELF-LOCKING, HE	2
31	PAOZZ	5305002692803	96906	MS90726-60	..SCREW, CAP, HEXAGON H.........................	2
32	PAOZZ	2990010831142	19207	12256075	PIPE, EXHAUST	1
33	PAOZZ	5340005329231	93742	B52-1111-2	CLAMP, LOOP	2
34	PAOZZ	2990010831123	19207	12255868	GUARD, MUFFLER-EXHAU	1
35	PAOZZ	2990010853786	19207	11669108	MUFFLER, EXHAUST	1
36	PAOZZ	5305002692803	96906	MS90726-60	SCREW, CAP, HEXAGON H	4
					UOC:DAA, DAB, DAC, DAD, DAE, DAF, DAG, DAH, DAJ, DAK, DAL, DAW, DAX, V12, V13, V14, V15, V16, V17, V18, V19, V20, V21, V22, V24, V25, V39	
37	PAOZZ	9390012041161	17284	FB0037	CAP ASSY, PROTECTOR	1
					UOC:DAA, DAB, DAC, DAD, DAG, DAH, DAL, DAW, DAX, V12, V13, V14, V15, V16, V17, V18, V21, V22, V39, ZAA, ZAB, ZAC, ZAD, ZAG, ZAH, ZAL	

END OF FIGURE

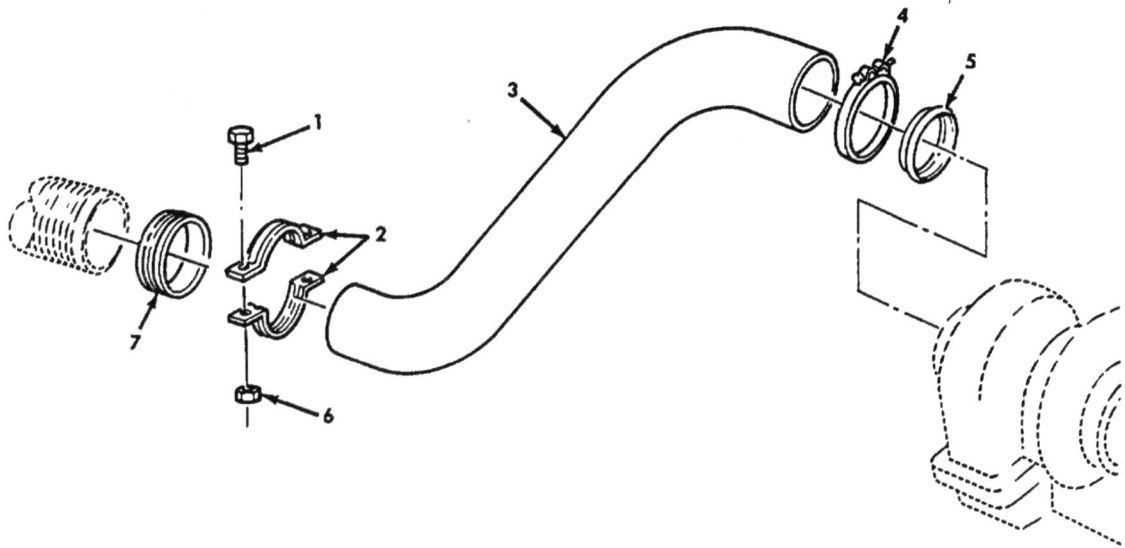

Figure 103. Exhaust System (M939A2).

(1) ITEM NO	(2) SMR CODE	(3) NSN	(4) CAGEC	(5) PART NUMBER	(6) DESCRIPTION AND USABLE ON CODES (UOC)	(7) QTY
					GROUP 0401 MUFFLER AND PIPES	
					FIG. 103 EXHAUST SYSTEM (M939A2)	
1	PAOZZ	5306002259095	96906	MS90726-40	BOLT, MACHINE.. UOC:ZAA, ZAB, ZAC, ZAD, ZAE, ZAF, ZAG, ZAH, ZAJ, ZAK, ZAL	2
2	PAOZZ	5342011065488	19207	11609348-11	COUPLING, CLAMP, GROO .... UOC:ZAA, ZAB, ZAC, ZAD, ZAE, ZAF, ZAG, ZAH, ZAJ, ZAK, ZAL	1
3	PAOZZ	2990012804284	1HT81	Q33562	PIPE, EXHAUST .... UOC:ZAA, ZAB, ZAC, ZAD, ZAE, ZAF, ZAG, ZAH, ZAJ, ZAK, ZAL	2
4	PAOZZ	5340012971187	15434	3903652	CLAMP, LOOP .... UOC:ZAA, ZAB, ZAC, ZAD, ZAE, ZAF, ZAG, ZAH, ZAJ, ZAK, ZAL	2
5	PAOZZ	5330013615600	OUBK6	20511420	GASKET .... UOC:ZAA, ZAB, ZAC, ZAD, ZAE, ZAF, ZAG, ZAH, ZAJ, ZAK, ZAL	1
6	PAOZZ	5310001766690	19207	11609727-2	NUT, SELF-LOCKING, HE .... UOC:ZAA, ZAB, ZAC, ZAD, ZAE, ZAF, ZAG, ZAH, ZAJ, ZAK, ZAL	2
7	PAOZZ	5330013794345	19207	12255817 NON-ASBESTOS	GASKET NON-ASBESTOS .... UOC:ZAA, ZAB, ZAC, ZAD, ZAE, ZAF, ZAG, ZAH, ZAJ, ZAK, ZAL	1

END OF FIGURE

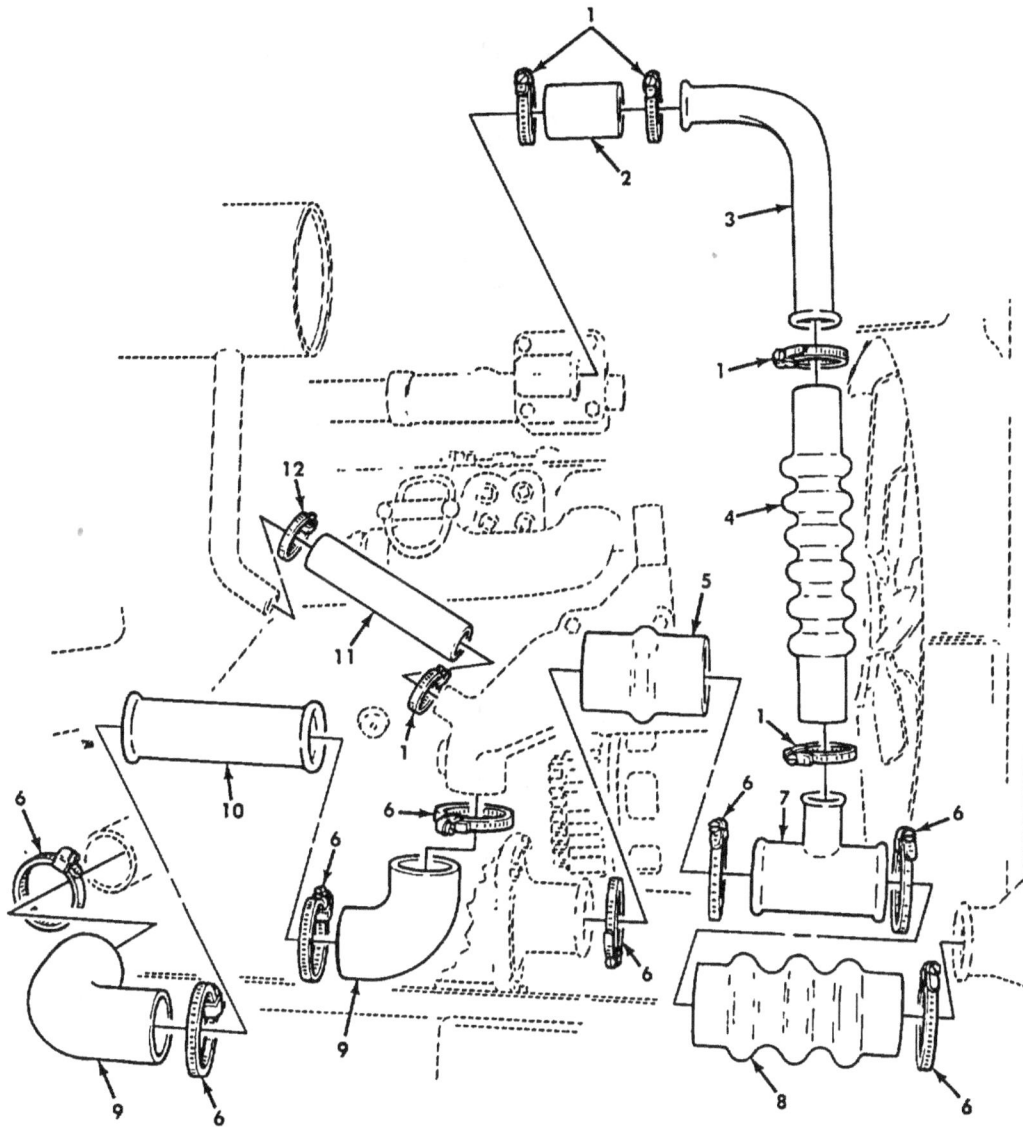

Figure 104. Radiator Hoses and Mounting Hardware (M939, M939A1).

(1) ITEM NO	(2) SMR CODE	(3) NSN	(4) CAGEC	(5) PART NUMBER	(6) DESCRIPTION AND USABLE ON CODES (UOC)	(7) QTY
					GROUP 05 COOLING SYSTEM 0501 RADIATOR, EVAPORATIVE COOLER OR HEAT EXCHANGER	
					FIG. 104 RADIATOR HOSES AND MOUNTING HARDWARE(M939,M939A1)	
1	PAOZZ	4730010894596	19207	11669079-1	CLAMP, HOSE.. UOC:DAA, DAB, DAC, DAD, DAE, DAF, DAG, DAH, DAJ, DAK, DAL, DAW, DAX, V12, V13, V14, V15, V16, V17, V18, V19, V20, V21, V22, V24, V25, V39	5
2	PAOZZ	4720011217749	19207	12277172-1	HOSE, NONMETALLIC .... UOC:DAA, DAB, DAC, DAD, DAE, DAF, DAG, DAH, DAJ, DAK, DAL, DAW, DAX, V12, V13, V14, V15, V16, V17, V178, V19, V20, V21, V22, V24, V25, V39	1
3	PAOZZ	4710011147759	19207	12277170	TUBE, BENT, METALLIC .... UOC:DAA, DAB, DAC, DAD, DAE, DAF, DAG, DAH, DAJ, DAK, DAL, DAW, DAX, V12, V13, V14, V15, V16, V17, V18, V19, V20, V21, V22, V24, V25, V39	1
4	PAOZZ	4720011085346	19207	12277172-3	HOSE, NONMETALLIC .... UOC:DAA, DAB, DAC, DAD, DAE, DAF, DAG, DAH, DAJ, DAK, DAL, DAW, DAX, V12, V13, V14, V15, V16, V17, V18, V19, V20, V21, V22, V24, V25, V39	1
5	PAOZZ	4720011085347	19207	12277173	HOSE, NONMETALLIC .... UOC:DAA, DAB, DAC, DAD, DAE, DAF, DAG, DAH, DAJ, DAK, DAL, DAW, DAX, V12, V13, V14, V15, V16, V17, V18, V19, V20, V21, V22, V24, V25, V39	1
6	PAOZZ	4730013316630	19207	11669079-5	CLAMP, HOSE.. UOC:DAA, DAB, DAC, DAD, DAE, DAF, DAG, DAH, DAJ, DAK, DAL, DAW, DAX, V12, V13, V14, V15, V16, V17, V18, V19, V20, V21, V22, V24, V25, V39	8
7	PAOZZ	4730011170602	19207	12277168	TEE, TUBE .... UOC:DAA, DAB, DAC, DAD, DAE, DAF, DAG, DAH, DAJ, DAK, DAL, DAW, DAX, V12, V13, V14, V15, V16, V17, V18, V19, V20, V21, V22, V24, V25, V39	1
8	PAOZZ	4720010899049	19207	12255978	HOSE, PREFORMED... UOC:DAA, DAB, DAC, DAD, DAE, DAF, DAG, DAH, DAJ, DAK, DAL, DAW, DAX, V12, V13, V14, V15, V16, V17, V18, V19, V20, V21, V22, V24, V25, V39	1
9	PAOZZ	4720011089118	19207	12255963	HOSE, PREFORMED... UOC:DAA, DAB, DAC, DAD, DAE, DAF, DAG, DAH, DAJ, DAK, DAL, DAW, DAX, V12, V13, V14, V15, V16, V17, V18, V19, V20, V21, V22, V24, V25, V39	2

(1) ITEM NO	(2) SMR CODE	(3) NSN	(4) CAGEC	(5) PART NUMBER	(6) DESCRIPTION AND USABLE ON CODES (UOC)	(7) QTY
10	PAOZZ	4710011217600	19207	12277171	TUBE, METALLIC .... UOC:DAA, DAB, DAC, DAD, DAE, DAF, DAG, DAH, DAJ, DAK, DAL, DAW, DAX, V12, V13, V14, V15, V16, V17, V18, V19, V20, V21, V22, V24, V25, V39	1
11	PAOZZ	4720010891108	19207	12256468	HOSE, PREFORMED .... UOC:DAA, DAB, DAC, DAD, DAE, DAF, DAG, DAH, DAJ, DAK, DAL, DAW, DAX, V12, V13, V14, V15, V16, V17, V18, V19, V20, V21, V22, V24, V25, V39	1
12	PAOZZ	4730012259003	19207	11669079-4	CLAMP, HOSE... UOC:DAA, DAB, DAC, DAD, DAE, DAF, DAG, DAH, DAJ, DAK, DAL, DAW, DAX, V12, V13, V14, V15, V16, V17, V18, V19, V20, V21, V22, V24, V25, V39	1

END OF FIGURE

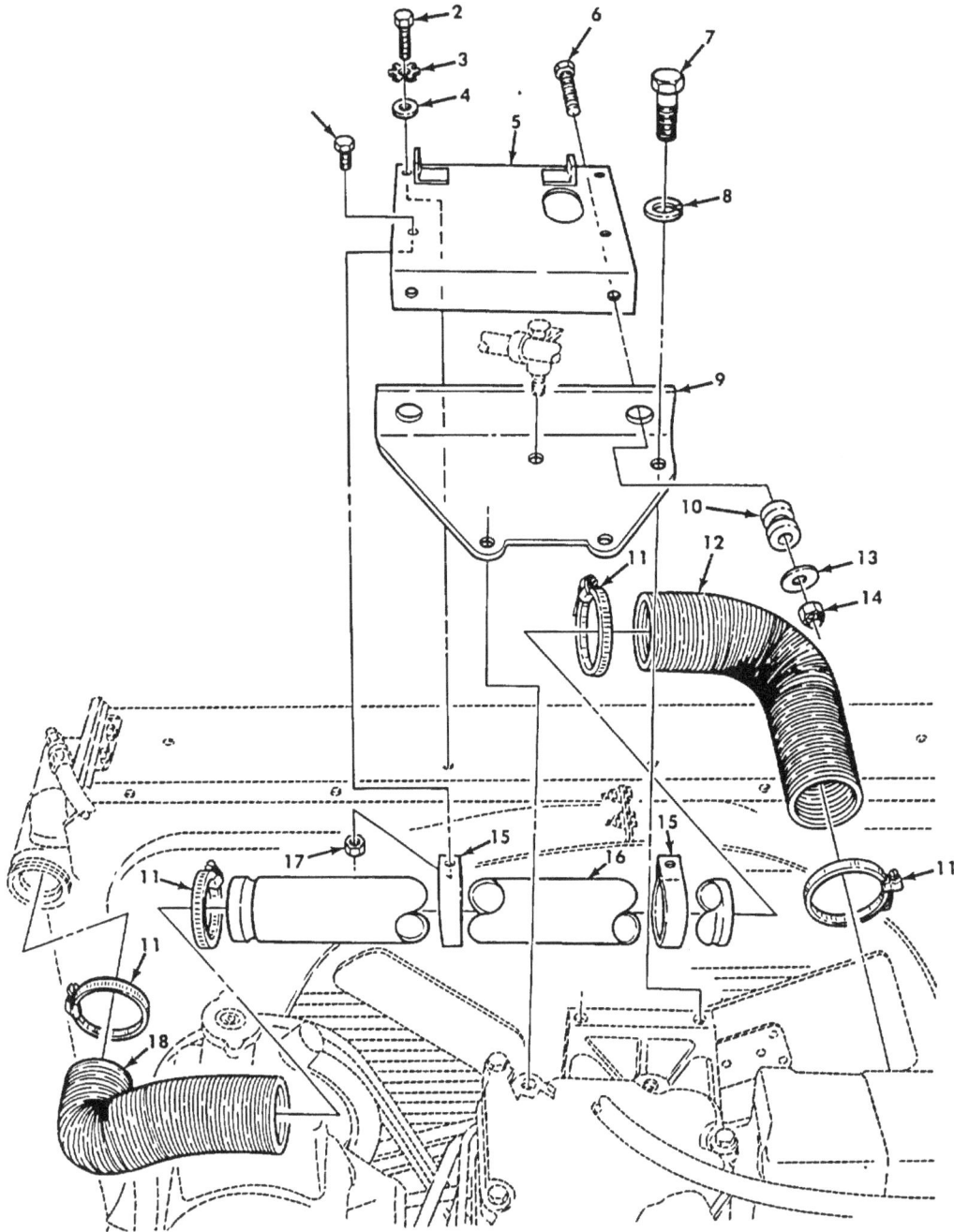

Figure 105. Upper Radiator Mounting Hardware and Hoses (M939, M939A1).

(1) ITEM NO	(2) SMR CODE	(3) NSN	(4) CAGEC	(5) PART NUMBER	(6) DESCRIPTION AND USABLE ON CODES (UOC)	(7) QTY
					GROUP 0501 RADIATOR, EVAPORATIVE COOLER, OR HEAT EXCHANGER	
					FIG. 105 UPPER RADIATOR MOUNTING HARDWARE AND HOSES (M939, M939A1)	
1	PAOZZ	5306000680514	80204	B1821BH025F088N	BOLT, MACHINE .... UOC:DAA, DAB, DAC, DAD, DAE, DAF, DAG, DAH, DAJ, DAK, DAL, DAW, DAX, V12, V13, V14, V15, V16, V17, V18, V19, V20, V21, V22, V24, V25, V39	2
2	PAOZZ	5305002692804	96906	MS90726-61	SCREW, CAP, HEXAGON H............................... UOC:DAA, DA B, DAC, DAD, DAE, DAF, DAG, DAH, DAJ, DAK, DAL, DAW, DAX, V12, V13, V14, V15, V16, V17, V18, V19, V20, V21, V22, V24, V25, V39	2
3	PAOZZ	5310006276128	96906	MS35335-35	WASHER, LOCK .... UOC:DAA, DAB, DAC, DAD, DAE, DAF, DAG, DAH, DAJ, DAK, DAL, DAW, DAX, V12, V13, V14, V15, V16, V17, V18, V19, V20, V21, V22, V24, V25, V39	2
4	PAOZZ	5310011208507	19207	12277033	WASHER, FLAT.. UOC:DAA, DAB, DAC, DAD, DAE, DAF, DAG, DAH, DAJ, DAK, DAL, DAW, DAX, V12, V13, V14, V15, V16, V17, V18, V19, V20, V21, V22, V24, V25, V39	2
5	PFOZZ	2930011307934	19207	12277035	RETAINER, RADIATOR .... UOC:DAA, DAB, DAC, DAD, DAE, DAF, DAG, DAH, DAJ, D, DAK, D, DAW, DAX, V12 , V13 , V14 , V15, V16, V17, V18, V19, V20, V21, V22, V24, V25, V39	1
6	PAOZZ	5305002693244	80204	B1821BH038F250N	SCREW, CAP, HEXAGON H................................. UOC:DAA, DAB, DAC, DAD, DAE, DAF, DAG, DAH, DAJ, DAK, DAL, DAW, DAX, V12 , V13 , V14 , V15, V16, V17, V18, V19, V20, V21, V22, V24, V25, V39	2
7	PAOZZ	5305000712068	80204	B1821BH050C138N	SCREW, CAP, HEXAGON H .... UOC:DAA, DAB, DAC, DAD, DAE, DAF, DAG, DAH, DAJ, DAK, DAL, DAW, DAX, V12, V13, V14, V15, V16, V17, V18, V19, V20, V21, V22, V24, V25, V39	1
8	PAOZZ	5310000034094	01276	210104-8S	WASHER, LOCK .... UOC:DAA, DAB, DAC, DAD, DAE, DAF, DAG, DAH, DAJ, DAK, DAL, DAW, DAX, V12, V13, V14, V15, V16, V17, V18, V19, V20, V21, V22, V24, V25, V39	1
9	PFOZZ	5340011307933	19207	12277034	BRACKET, ANGLE .... UOC:DAA, DAB, DAC, DAD, DAE, DAF, DAG, DAH, DAJ, DAK, DAL, DAW, DAX, V12, V13, V14, V15, V16, V17, V18, V19, V20, V21, V22, V24, V25, V39	1
10	PAOZZ	5342011010005	19207	11669109	MOUNT, RESILIENT	2

(1) ITEM NO	(2) SMR CODE	(3) NSN	(4) CAGEC	(5) PART NUMBER	(6) DESCRIPTION AND USABLE ON CODES (UOC)	(7) QTY
					UOC:DAA, DAB, DAC, DAD, DAE, DAF, DAG, DAH, DAJ, DAK, DAL, DAW, DAX, V12, V13, V14, V15, V16, V17, V18, V19, V20, V21, V22, V24, V25, V39	
11	PAOZZ	4730011173837	19207	11669079-2	CLAMP, HOSE...	4
					UOC:DAA, DAB, DAC, DAD, DAE, DAF, DAG, DAH, DAJ, DAK, DAL, DAW, DAX, V12, V13, V14, V15, V16, V17, V18, V19, V20, V21, V22, V24, V25, V39	
12	PAOZZ	4720010892061	19207	12256611	HOSE, PREFORMED...	1
					UOC:DAA, DAB, DAC, DAD, DAE, DAF, DAG, DAH; DAJ, DAK, DAL, DAW, DAX, V12, V13, V14, 15, V16, V17, V18, V19, V20, V21, V22, V24, V25, V39	
13	PAOZZ	5365011084811	19207	12256205	SHIM..	2
					UOC:DAA, DAB, DAC, DAD, DAE, DAF, DAG, DAH, DAJ, DAK, DAL, DAW, DAX, V12, V13, V14, V15, . V16, V17, V18, V19, V20, V21, V22, V24, V25; V39	
14	PAOZZ	5310008140672	96906	MS51943-36	NUT, SELF-LOCKING, HE......................................	2
					UOC:DAA, DAB, DAC, DAD, DAE, DAF, DAG, DAH, DAJ, DAK, DAL, DAW, DAX, V12, V13, V14, V15, V16, V17, V18, V19, V20, V21, V22, V24, V25, V39	
15	PAOZZ	5340011050993	19207	12276945	CLAMP, LOOP...	2
					UOC:DAA, DAB, DAC, DAD, DAE, DAF, DAG, DAH, DAJ, DAK, DAL, DAW, DAX, V12, V13, V14, V15, V16, V17, V18, V19, V20, V21, V22, V24, V25, V39	
16	PAOZZ	4710011303437	19207	12255959	TUBE, METALLIC...	1
					UOC:DAA, DAB, DAC, DAD, DAE, DAF, DAG, DAH, DAJ, DAK, DAL, DAW, DAX, V12, V13, V14, V15, V16, V17, V18, V19, V20, V21, V22, V24, V25, V39	
17	PAOZZ	5310009359022	96906	MS51943-32	NUT, SELF-LOCKING, HE	2
					UOC:DAA, DAB, DAC, DAD, DAE, DAF, DAG, DAH, DAJ, DAK, DAL, DAW, DAX, V12, V13, V14, V15, V16, V17, V18, V19, V20, V21, V22, V24, V25, V39	
18	PAOZZ	4720010894595	19207	12256446	HOSE, PREFORMED	1
					UOC:DAA, DAB, DAC, DAD, DAE, DAF, DAG, DAH, DAJ, DARK, DAL, DAW, DAX, V12, V13, V14, V15, V16, V17, V18, V19, V20, V21, V22, V24, V25, V39	

END OF FIGURE

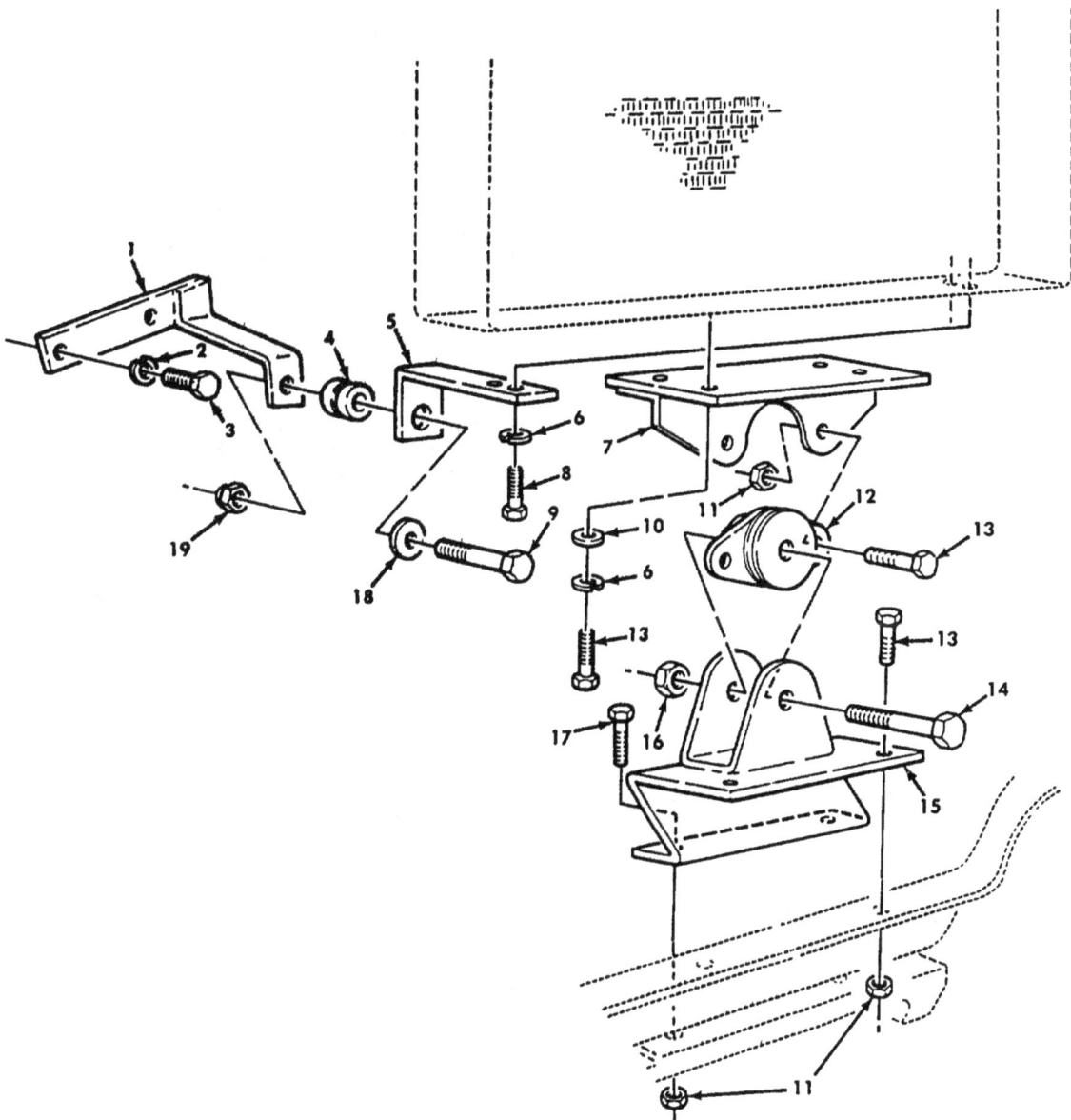

Figure 106. Lower Radiator Mounting Hardware (M939, M939A1).

(1) ITEM NO	(2) SMR CODE	(3) NSN	(4) CAGEC	(5) PART NUMBER	(6) DESCRIPTION AND USABLE ON CODES (UOC)	(7) QTY
					GROUP 0501 RADIATOR, EVAPORATIVE COOLER, OR HEAT EXCHANGER	
					FIG. 106 LOWER RADIATOR MOUNTING HARDWARE(M939, M939A1)	
1	PFOZZ	2930011310107	19207	12255852	RETAINER, RADIATOR .... UOC:DAA, DAB, DAC, DAD, DAE, DAF, DAG, DAH, DAJ, DAK, DAL, DAW, DAX, V12, V13, V14, V15, V16, V17, V18, V19, V20, V21, V22, V24, V25, V39	1
2	PAOZZ	5310002090965	96906	MS35338-47	WASHER, LOCK .... UOC:DAA, DAB, DAC, DAD, DAE, DAF, DAG, DAH, DAJ, DAK, DAL, DAW, DAX, V12, V13, V14, V15, V16, V17, V18, V19, V20, V21, V22, V24, V25, V39	2
3	PAOZZ	5305000711788	80204	B1821BH044C125N	SCREW, CAP, HEXAGON H .... UOC:DAA, DAB, DAC, DAD, DAE, DAF, DAG, DAH, DAJ, DAK, DAL, DAW, DAX, V12, V13, V14, V15, V16, V17, V18, V19, V20, V21, V22, V24, V25, V39	2
4	PAOZZ	5342011010005	19207	11669109	MOUNT, RESILIENT............................... UOC:DAA, DAB, DAC, DAD, DAE, DAF, DAG, DAH, DAJ, DAK, DAL, DAW, DAX, V12, V13, V14, V15, V16, V17, V18, V19, V20, V21, V22, V24, V25, V39	1
5	PFOZZ	5340011050992	19207	12255844	BRACKET, ANGLE................................. UOC:DAA, DAB, DAC, DAD, DAE, DAF, DAG, DAH, DAJ, DAK, DAL, DAW, DAX, V12, V13, V14, V15, V16, V17, V18, V19, V20, V21, V22, V24, V25, V39	1
6	PAOZZ	5310000034094	01276	210104-8S	WASHER, LOCK .... UOC:DAA, DAB, DAC, DAD, DAE, DAF, DAG, DAH, DAJ, DAK, DAL, DAW, DAX, V12, V13, V14, V15, V16, V17, V18, V19, V20, V21, V22, V24, V25, V39	6
7	PFOZZ	5340011048944	19207	12255837	BRACKET, MOUNTING............................ UOC:DAA, DAB, DAC, DAD, DAE, DAF, DAG, DAH, DAJ, DAK, DAL, DAW, DAX V12, V13, V14, V15, V16, V17, V18, V19, V20, V21, V22, V24, V25, V39	1
8	PAOZZ	5305002267767	96906	MS90726-109	SCREW, CAP, HEXAGON H .... UOC:DAA, DAB, DAC, DAD, DAE, DAF, DAG, DAH, DAJ, DAK, DAL, DAW, DAX, V12, V13, V14, V15, V16, V17, V18, V19, V20, V21, V22, V24, V25, V39	2
9	PAOZZ	5305002693244	80204	B1821BH038F250N	SCREW, CAP, HEXAGON H .... UOC:DAA, DAB, DAC, DAD, DAE, DAF, DAG, DAH, DAJ, DAK, DAL, DAW, DAX, V12, V13, V14, V15, V16, V17, V18, V19, V20, V21, V22, V24, V25, V39	1
10	PAOZZ	5310008095998	96906	MS27183-18	WASHER, FLAT	4

(1) ITEM NO	(2) SMR CODE	(3) NSN	(4) CAGEC	(5) PART NUMBER	(6) DESCRIPTION AND USABLE ON CODES (UOC)	(7) QTY
					UOC:DAA, DAB, DAC, DAD, DAE, DAF, DAG, DAH, DAJ, DAK, DAL, DAW, DAX, V12, V13, V14, V15, V16, V17, V18, V19, V20, V21, V22, V24, V25, V39	
11	PAOZZ	5310004883888	96906	MS51943-40	NUT, SELF-LOCKING, HE	6
					UOC:DAA, DA B, DAC, DAD, DAE, DAF, DAG, DAH, DAJ, DAK, DAL, DAW, DAX, V12, V13, V14, V15, V16, V17, V18, V19, V20, V21, V22, V24, V25, V39	
12	PAOZZ	2590011010075	81860	22696-5	MOUNT, RESILIENT	1
					UOC:DAA, DAB, DAC, DAD, DAE, DAF, DAG, DAH, DAJ, DAK, DAL, DAW, DAX, V12, V13, V14, V15, V16, V17, V18, V19, V20, V21, V22, V24, V25, V39	
13	PAOZZ	5305007195219	96906	MS90727-111	SCREW, CAP, HEXAGON H	8
					UOC:DAA, DAB, DAC, DAD, DAE, DAF, DAG, DAH, DAJ, DAK, DAL, DAW, DAX, V12, V13, V14, V15, V16, V17, V18, V19, V20, V21, V22, V24, V25, V39	
14	PAOZZ	5305009474360	80204	B1821BH075C450N	SCREW, CAP, HEXAGON H	1
					UOC:DAA, DAB, DAC, DAD, DAE, DAF, DAG, DAH, DAJ, DAK, DAL, DAW, DAX, V12, V13, V14, V15, V16, V17, V18, V19, V20, V21, V22, V24, V25, V39	
15	PFOZZ	5340010666086	19207	12255855	BRACKET, MOUNTING....................................	1
					UOC:DAA, DAB, DAC, DAD, DAE, DAF, DAG, DAH, DAJ, DAK, DAL, DAW, DAX, V12, V13, V14, V15, V16, V17, V18, V19, V20, V21, V22, V24, V25, V39	
16	PAOZZ	5310009353569	96906	MS51943-46	NUT, SELF-LOCKING, HE	1
					UOC:DAA, DAB, DAC, DAD, DAE, DAF, DAG, DAH, DAJ, DAK, DAL, DAW, DAX, V12, V13, V14, V15, V16, V17, V18, V19, V20, V21, V22, V24, V25, V39	
17	PAOZZ	5305007195235	80204	B1821BH050F175N	SCREW, CAP, HEXAGON H	2
					UOC:DAA, DAB, DAC, DAD, DAE, DAF, DAG, DAH, DAJ, DAK, DAL, DAW, DAX, V12, V13, V14, V15, V16, V17, V18 , V19, V20, V21, V22, V24, V25, V39	
18	PAOZZ	5365011084811	19207	12256205	SHIM	1
					UOC:DAA, DAB, DAC, DAD, DAE, DAF, DAG, DAH, DAJ, DAK, DAL, DAW, DAX, V12, V13, V14, V15, V16, V17, V18, V19, V20, V21, V22, V24, V25, V39	
19	PAOZZ	5310008140672	96906	MS51943-36	NUT, SELF-LOCKING, HE	1
					UOC:DAA, DAB, DAC, DAD, DAE, DAF, DAG, DAH, DAJ, DAK, DAL, DAW, DAX, V12, V13, V14, V15, V16, V17, V18, V19, V20, V21, V22, V24, V25, V39	

END OF FIGURE

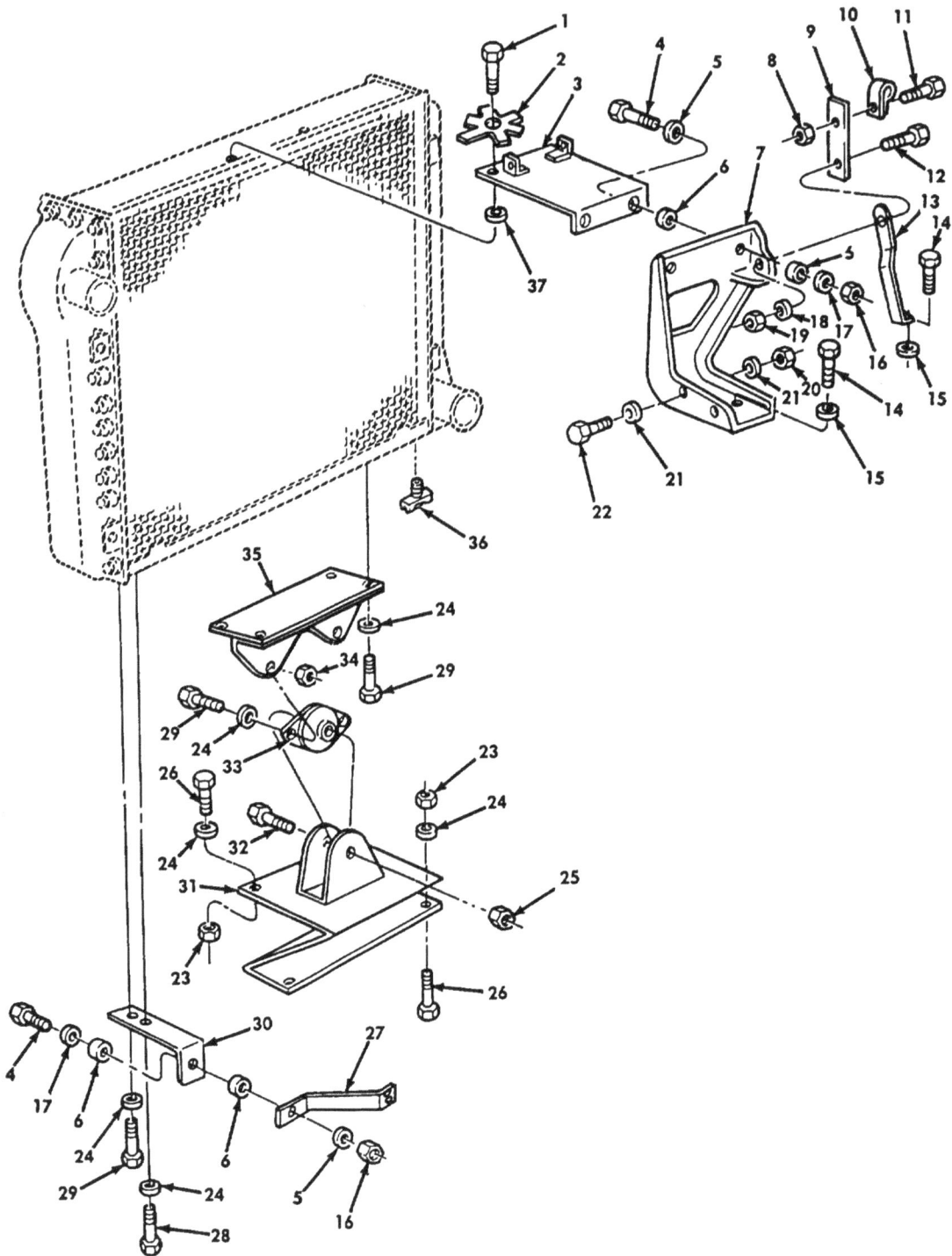

Figure 107. Radiator Assembly and Mounting Hardware (M939A2).

(1) ITEM NO	(2) SMR CODE	(3) NSN	(4) CAGEC	(5) PART NUMBER	(6) DESCRIPTION AND USABLE ON CODES (UOC)	(7) QTY
					GROUP 0501 RADIATOR, EVAPORATIVE COOLER, OR HEAT EXCHANGER	
					FIG. 107 RADIATOR ASSEMBLY MOUNTING HARDWARE(M939A2)	
1	PAOZZ	5305002693237	96906	MS90727-61	SCREW, CAP, HEXAGON H .... UOC:ZAA, ZAB, ZAC, ZAD, ZAE, ZAF, ZAG, ZAH, ZAJ, ZAK, ZAL	2
2	PAOZZ	5310006276128	96906	MS35335-35	WASHER, LOCK .... UOC:ZAA, ZAB, ZAC, ZAD, ZAE, ZAF, ZAG, ZAH, ZAJ, ZAK, ZAL	2
3	PAOZZ	5340013178144	47457	20510270	BRACKET, MOUNTING .... UOC:ZAA, ZAB, ZAC, ZAD, ZAE, ZAF, ZAG, ZAH, ZAJ, ZAK, ZAL	1
4	PAOZZ	5305006388920	80204	B1821BH038C225N	SCREW, CAP, HEXAGON H .... UOC:ZAA, ZAB, ZAC, ZAD, ZAE, ZAF, ZAG, ZAH, ZAJ, ZAK, ZAL	3
5	PAOZZ	5310002748041	90407	12084P11	WASHER, FLAT .... UOC:ZAA, ZAB, ZAC, ZAD, ZAE, ZAF, ZAG, ZAH, ZAJ, ZAK, ZAL	3
6	PAOZZ	5340011010005	19207	11669109	MOUNT, RESILIENT.................. UOC:ZAA, ZAB, ZAC, ZAD, ZAE, ZAF, ZAG, ZAH, ZAJ, ZAK, ZAL	3
7	PAOZZ	5340012807101	47457	20510480-1	BRACKET, MOUNTING.................. UOC:ZAA, ZAB, ZAC, ZAD, ZAE, ZAF, ZAG, ZAH, ZAJ, ZAK, ZAL	1
8	PAOZZ	5310009359021	96906	MS51943-35	NUT, SELF-LOCKING, HE.................. UOC:ZAA, ZAB, ZAC, ZAD, ZAE, ZAF, ZAG, ZAH, ZAJ, ZAK, ZAL	1
9	PAOZZ	5340012876660	16567	20511172	PLATE, MENDING .... UOC:ZAA, ZAB, ZAC, ZAD, ZAE, ZAF, ZAG, ZAH, ZAJ, ZAK, ZAL	1
10	PAOZZ	5340000538994	96906	MS21333-126	CLAMP, LOOP .... UOC:ZAA, ZAB, ZAC, ZAD, ZAE, ZAF, ZAG, ZAH, ZAJ, ZAK, ZAL	1
11	PAOZZ	5305007215492	80204	B1821BH038C063N	SCREW, CAP, HEXAGON H .... UOC:ZAA, ZAB, ZAC, ZAD, ZAE, ZAF, ZAG, ZAH, ZAJ, ZAK, ZAL	1
12	PAOZZ	5305000712055	80204	B1821BH044C150N	SCREW, CAP, HEXAGON H .... UOC:ZAA, ZAB, ZAC, ZAD, ZAE, ZAF, ZAG, ZAH, ZAJ, ZAK, ZAL	1
13	PAOZZ	5340012807100	47457	20510429	BRACKET, MOUNTING .... UOC:ZAA, ZAB, ZAC, ZAD, ZAE, ZAF, ZAG, ZAH, ZAJ, ZAK, ZAL	1
14	PAOZZ	5305012915138	73342	11500878	SCREW, CAP, HEXAGON H .... UOC:ZAA, ZAB, ZAC, ZAD, ZAE, ZAF, ZAG, ZAH, ZAJ, ZAK, ZAL	5
15	PAOZZ	5310012708392	24617	11511515	WASHER, FLAT.................. UOC:ZAA, ZAB, ZAC, ZAD, ZAE, ZAF, ZAG, ZAH, ZAJ, ZAK, ZAL	5
16	PAOZZ	5310011269404	34623	5590560	NUT, SELF-LOCKING, HE	3

(1) ITEM NO	(2) SMR CODE	(3) NSN	(4) CAGEC	(5) PART NUMBER	(6) DESCRIPTION AND USABLE ON CODES (UOC)	(7) QTY
					UOC:ZAA, ZAB, ZAC, ZAD, ZAE, ZAF, ZAG, ZAH, ZAJ, ZAK, ZAL	
17	PAOZZ	5310012927251	47457	20510333	WASHER, FLAT..............	3
					UOC:ZAA, ZAB, ZAC, ZAD, ZAE, ZAF, ZAG, ZAH, ZAJ, ZAK, ZAL	
18	PAOZZ	5310010925495	24617	9422848	WASHER, FLAT	1
					UOC:ZAA, ZAB, ZAC, ZAD, ZAE, ZAF, ZAG, ZAH, ZAJ, ZAK, ZAL	
19	PAOZZ	5310011504003	24617	9422299	NUT, SELF-LOCKING, HE............	1
					UOC:ZAA, ZAB, ZAC, ZAD, ZAE, ZAF, ZAG, ZAH, ZAJ, ZAK, ZAL	
20	PAOZZ	5310009843806	96906	MS51922-9	NUT, SELF-LOCKING, HE	2
					UOC:ZAA, ZAB, ZAC, ZAD, ZAE, ZAF, ZAG, ZAH, ZAJ, ZAK, ZAL	
21	PAOZZ	5310010841197	19207	8356625-2	WASHER, FL AT.........	4
					UOC:ZAA, ZAB, ZAC, ZAD, ZAE, ZAF, ZAG, ZAH, ZAJ, ZAK, ZAL	
22	PAOZZ	5306002264827	80204	B1B21BH031C100N	BOLT, MACHINE	2
					UOC:ZAA, ZAB, ZAC, ZAD, ZAE, ZAF, ZAG, ZAH, ZAJ, ZAK, ZAL	
23	PAOZZ	5310011494407	24617	9422301	NUT, SELF-LOCKING, HE...........	4
					UOC:ZAA, ZAB, ZAC, ZAD, ZAE, ZAF, ZAG, ZAH, ZAJ, ZAK, ZAL	
24	PAOZZ	5310010842362	63005	9411417	WASHER............	13
					UOC:ZAA, ZAB, ZAC, ZAD, ZAE, ZAF, ZAG, ZAH, ZAJ, ZAK, ZAL	
25	PAOZZ	5310009353569	96906	MS51943-46	NUT, SELF-LOCKING, HE	1
					UOC:ZAA, ZAB, ZAC, ZAD, ZAE, ZAF, ZAG, ZAH, ZAJ, ZAK, ZAL	
26	PAOZZ	5305000712069	80204	B1821BH050C150N	SCREW, CAP, HEXAGON H	4
					UOC:ZAA, ZAB, ZAC, ZAD, ZAE, ZAF, ZAG, ZAH, ZAJ, ZAK, ZAL	
27	PAOZZ	5340012853239	16567	20510277	BRACKET, DOUBLE ANGL STABILIZER............	1
					UOC:ZAA, ZAB, ZAC, ZAD, ZAE, ZAF, ZAG, ZAH, ZAJ, ZAK, ZAL	
28	PAOZZ	5305007195184	96906	MS90727-109	SCREW, CAP, HEXAGON H...............	2
					UOC:ZAA, ZAB, ZAC, ZAD, ZAE, ZAF, ZAG, ZAH, ZAJ, ZAK, ZAL	
29	PAOZZ	5305007195219	96906	MS90727-111	SCREW, CAP, HEXAGON H...............	7
					UOC:ZAA, ZAB, ZAC, ZAD, ZAE, ZAF, ZAG, ZAH, ZAJ, ZAK, ZAL	
30	PAOZZ	5340012817205	47457	20510341	BRACKET, ANGLE LOWER RADIATOR............	1
					UOC:ZAA, ZAB, ZAC, ZAD, ZAE, ZAF, ZAG, ZAH, ZAJ, ZAK, ZAL	
31	PAOZZ	5340011048944	19207	12255837	BRACKET, MOUNTING	1
					UOC:ZAA, ZAB, ZAC, ZAD, ZAE, ZAPF, ZAG, ZAH, ZAJ, ZAK, ZAL	
32	PAOZZ	5305009408069	80204	B1821BH075F450N	SCREW, CAP, HEXAGON H	1
					UOC:ZAA, ZAB, ZAC, ZAD, ZAE, ZAF, ZAG, ZAH, ZAJ, ZAK, ZAL	
33	PAOZZ	2590011010075	81860	22696-5	MOUNT, RESILIENT	1
					UOC:ZAA, ZAB, ZAC, ZAD, ZAE, ZAF, ZAG, ZAH, ZAJ, ZAK, ZAL	

(1) ITEM NO	(2) SMR CODE	(3) NSN	(4) CAGEC	(5) PART NUMBER	(6) DESCRIPTION AND USABLE ON CODES (UOC)	(7) QTY
					ZAJ, ZAK, ZAL	
34	PAOZZ	5310004883888	96906	MS51943-40	NUT, SELF-LOCKING, HE	3
					UOC:ZAA, ZAB, ZAC, ZAD, ZAE, ZAF, ZAG, ZAH, ZAJ, ZAK, ZAL	
35	PFOZZ	5340011048944	19207	12255837	BRACKET, MOUNTING..	1
					UOC: ZAA, ZAB, ZAC, ZAD, ZAE, ZAF, ZAG, ZAH, ZAJ, ZAK, ZAL	
36	PAOZZ	4820008491220	96906	MS35782-5	COCK, DRAIN	1
					UOC:ZAA, ZAB, ZAC, ZAD, ZAE, ZAF, ZAG, ZAH, ZAJ, ZAK, ZAL	
37	PAOZZ	5310012577590	96906	MS51412-7	WASHER, FLAT	1
					UOC:ZAA, ZAB, ZAC, ZAD, ZAE, ZAF, ZAG, ZAH, ZAJ, ZAK, ZAL	

END OF FIGURE

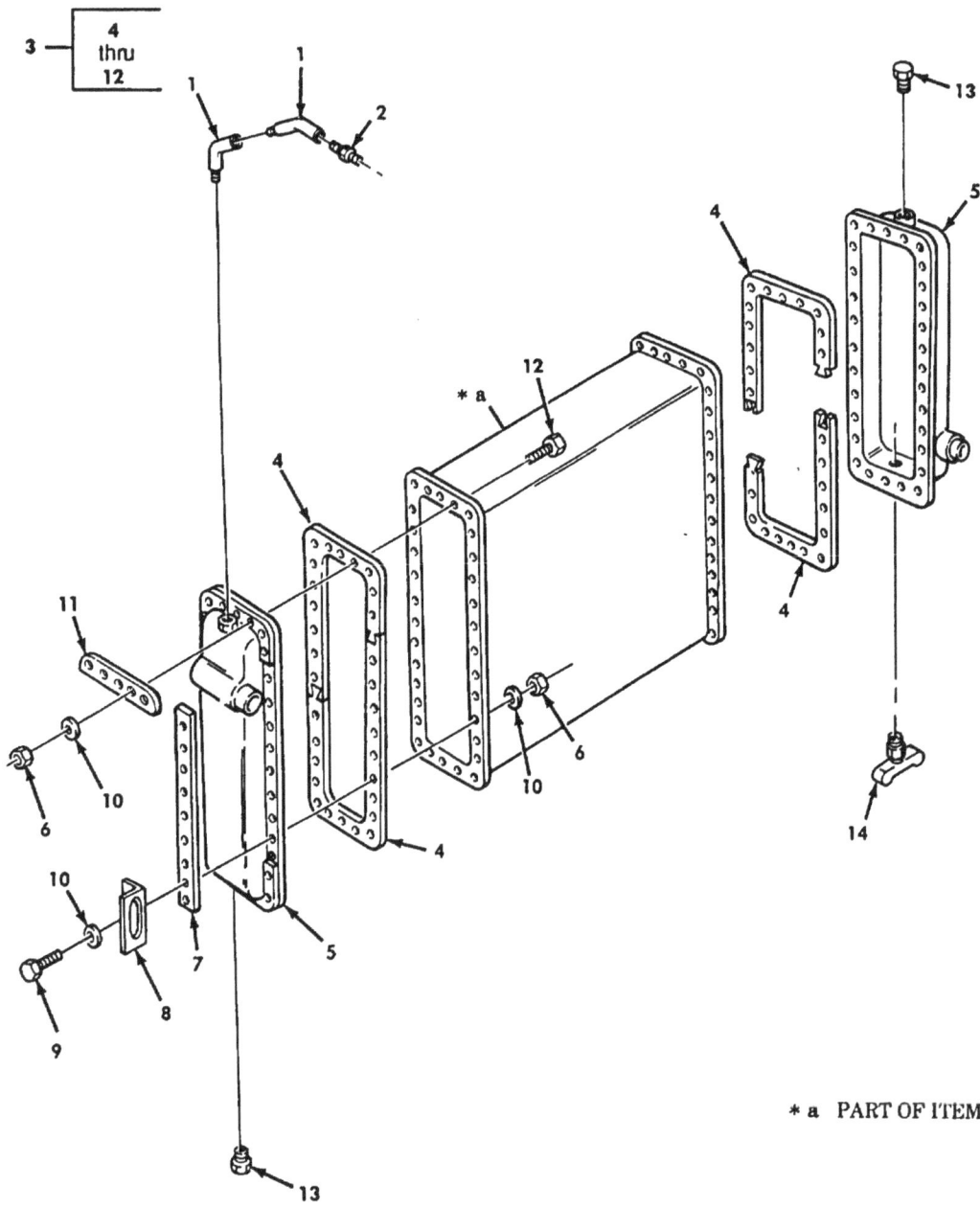

Figure 108. Radiator Assembly.

* a   PART OF ITEM 3

(1) ITEM NO	(2) SMR CODE	(3) NSN	(4) CAGEC	(5) PART NUMBER	(6) DESCRIPTION AND USABLE ON CODES (UOC)	(7) QTY
					GROUP 0501 RADIATOR, EVAPORATIVE COOLER, OR HEAT EXCHANGER	
					FIG. 108 RADIATOR ASSEMBLY	
1	PAOZZ	4730002775684	96906	MS14307-2	ELBOW, PIPE	2
2	PAOZZ	4730002660538	81343	6-4 010102B	ADAPTER, STRAIGHT, PI	1
3	PAOFF	2930011332143	19207	11669165	RADIATOR, ENGINE COO	1
4	PAFZZ	5330012321487	19207	12302621	.GASKET PART OF KIT P/N D-781964-A	4
5	PAFZZ	5430012355442	19207	12302618	.TANK SECTION, FLUID	2
6	PAOZZ		96906	MS51943-33	.NUT PART OF KIT P/N D-781964-A	80
7	PAFZZ	5340011317443	19207	12302624	.PLATE, MENDING PART OF KIT P/N D-781964-A	4
8	PAFZZ		19207	12301081	.BRACKET PART OF KIT P/N D-781964-A.	4
9	PAOZZ		96906	MS35356-33	.BOLT PART OF KIT P/N D-781964-A	8
10	PAOZZ		96906	MS27183-12	.WASHER PART OF KIT P/N D-781964-A	80
11	PAFZZ	5340011317444	19207	12302623	.PLATE, MENDING PART OF KIT P/N D-781964-A	4
12	PAOZZ		96906	MS90725-34	.BOLT PART OF KIT P/N D-781964-A	72
13	PAOZZ	4730000127951	21450	127951	PLUG, PIPE	2
14	PAOZZ	4820007204488	96906	MS35782-2	COCK, DRAIN	1

END OF FIGURE

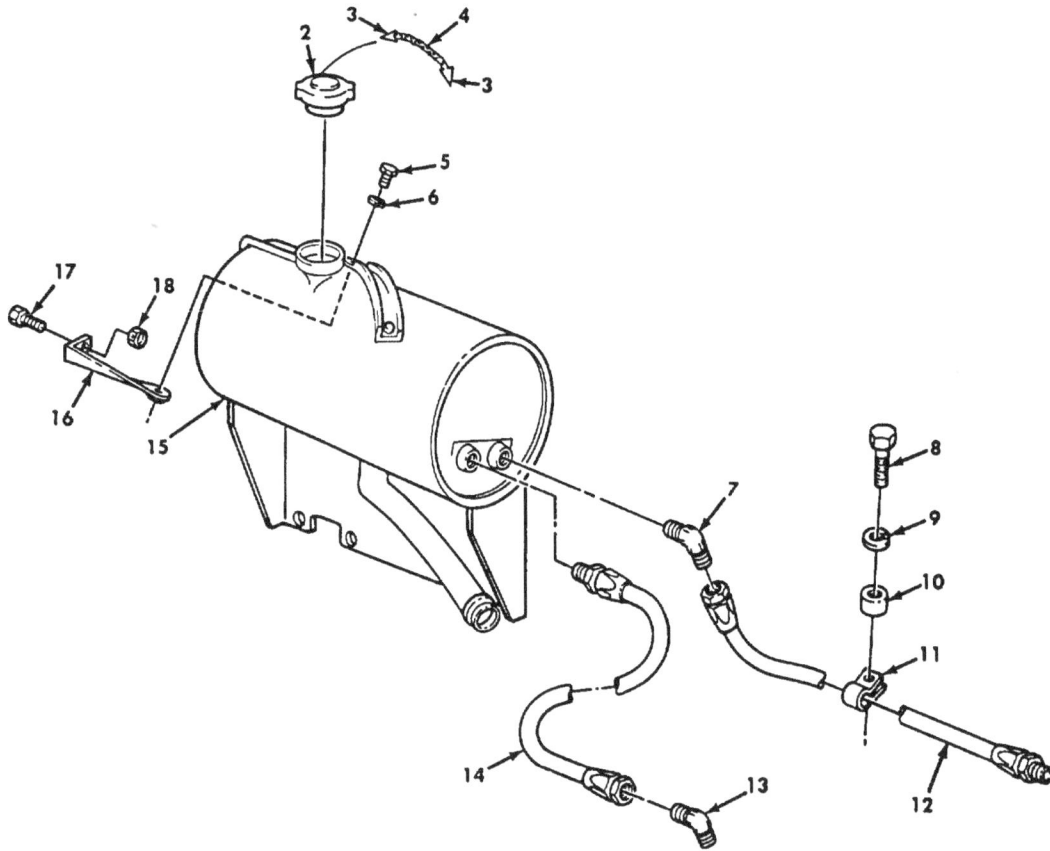

Figure 109. Surge Tank Assembly, Hoses, and Mounting Hardware (M939, M939A1).

(1) ITEM NO	(2) SMR CODE	(3) NSN	(4) CAGEC	(5) PART NUMBER	(6) DESCRIPTION AND USABLE ON CODES (UOC)	(7) QTY
					GROUP 0501 RADIATOR, EVAPORATIVE COOLER, OR HEAT EXCHANGER	
					FIG. 109 SURGE TANK ASSEMBLY, HOSES, AND MOUNTING HARDWARE(M939,M939A1)	
1	PAOZZ	2930001475202	19207	11648745	CAP, FILLER OPENING .... UOC:DAA, DAB, DAC, DAD, DAE, DAF, DAG, DAH, DAJ, DAK, DAL, DAW, DAX, V12, V13, V14, V15, V16, V17, V18, V19, V20, V21, V22, V24, V25, V39	1
2	XAOZZ		19207	11648745-1	.CAP, FILLER OPENING....................................... UOC:DAA, DAB, DAC, DAD, DAE, DAF, DAG, DAH, DAJ, DAK, DAL, DAW, DAX, V12, V13, V14, V15, V16, V17, V18, V19, V20, V21, V22, V24, V25, V39	1
3	PAOZZ	5315005142660	96906	MS29523-1	.PIN, RETAINING .... UOC:DAA, DAB, DAC, DAD, DAE, DAF, DAG, DAH, DAJ, DAK, DAL, DAW, DAX, V12, V13, V14, V15, V16, V17, V18, V19, V20, V21, V22, V24, V25, V39	2
4	MOOZZ		80244	42-C-16560-9	.CHAIN, SASH MAKE FROM CHAIN, P/N 42-.......... C-16560 .... UOC:DAA, DAB, DAC, DA D, DAE, DAF, DAG, DAH, DAJ, DAK, DAL, DAW, DAX, V12 , V13 , V14 , V15, V16, V17, V18, V19, V20, V21, V22, V24, V25, V39	1
5	PAOZZ	5306002258496	96906	MS90725-31	BOLT, MACHINE .... UOC:DAA, DAB, DAC, DAD, DAE , DAF, DAG, DAH, DAJ, DAK, DAL, DAW, DAX, V12, V13, V14, V15, V16, V17, V18, V19, V20, V21, V22, V24, V25, V39	1
6	PAOZZ	5310008093078	96906	MS27183-11	WASHER, FLAT .... UOC:DAA, DAB, DAC, DAD, DAE, DAF, DAG, DAH, DAJ, DAK, DAL, DAW, DAX, V12, V13, V14, V15, V16, V17, V18, V19, V20, V21, V22, V24, V25, V39	1
7	PAOZZ	4730002313906	88044	AN915-3	ELBOW, PIPE .... UOC:DAA, DAB, DAC, DAD, DAE, DAF, DAG, DAH, DAJ, DAK, DAL, DAW, DAX, V12, V13, V14, V15, V16, V17, V18, V19, V20, V21, V22, V24, V25, V39	1
8	PAOZZ	5305000712077	80204	B1821BH050C350N	SCREW, CAP, HEXAGON H .... UOC:DAA, DAB, DAC, DAD, DAE, DAF, DAG, DAH, DAJ, DAK, DAL, DAW, DAX, V12, V13, V14, V15, V16, V17, V18, V19, V20, V21, V22, V24, V25, V39	1
9	PAOZZ	5310000034094	01276	210104-8S	WASHER, LOCK .... UOC:DAA, DAB, DAC, DAD, DAE, DAF, DAG, DAH, DAJ, DAK, DAL, DAW, DAX, V12, V13, V14, V15, V16, V17, V18, V19, V20, V21, V22, V24, V25, V39	1

(1) ITEM NO	(2) SMR CODE	(3) NSN	(4) CAGEC	(5) PART NUMBER	(6) DESCRIPTION AND USABLE ON CODES (UOC)	(7) QTY
10	PAOZZ	5365011047846	19207	12256491	SPACER, SLEEVE .... UOC:DAA, DAB, DAC, DAD, DAE, DAF, DAG, DAH, DAJ, DAK, DAL, DAW, DAX, V12, V13, V14, V15, V16, V17, V18, V19, V20, V21, V22, V24, V25, V39	1
11	PAOZZ	5340004217254	19207	8352679	CLAMP, LOOP .... UOC:DAA, DAB, DAC, DAD, DAE, DAF, DAG, DAH, DAJ, DAK, DAL, DAW, DAX, V12, V13, V14, V15, V16, V17, V18, V19, V20, V21, V22, V24, V25, V39	1
12	PAOZZ	4720010889650	19207	11664472-1	HOSE ASSEMBLY, NONME .... UOC:DAA, DAB, DAC, DAD, DAE, DAF, DAG, DAH, DAJ, DAK, DAL, DAW, DAX, V12, V13, V14, V15, V16, V17, V18, V19, V20, V21, V22, V24, V25, V39	1
13	PAOZZ	4730002546211	99199	A335	ELBOW, PIPE TO TUBE .... UOC:DAA, DAB, DAC, DAD, DAE, DAF, DAG, DAH, DAJ, DAK, DAL, DAW, DAX, V12, V13, V14, V15, V16, V17, V18, V19, V20, V21, V22, V24, V25, V39	1
14	PAOZZ	4720001776184	19207	11664473	HOSE ASSEMBLY, NONME .... UOC:DAA, DAB, DAC, DAD, DAE, DAF, DAG, DAH, DAJ, DAK, DAL, DAW, DAX, V12, V13, V14, V15, V16, V17, V18, V19, V20, V21, V22, V24, V25, V39	1
15	PAOZZ	2930010831122	19207	12256019	TANK, RADIATOR, OVERF .... UOC:DAA, DAB, DAC, DAD, DAE, DAF, DAG, DAH, DAJ, DAK, DAL, DAW, DAX, V12, V13, V14, V15, V16, V17, V18, V19, V20, V21, V22, V24, V25, V39	1
16	PFOZZ	5340010822512	19207	12256284	BRACKET, MOUNTING .... UOC:DAA, DAB, DAC, DAD, DAE, DAF, DAG, DAH, DAJ, DAK, DAL, DAW, DAX, V12, V13, V14, V15, V16, V17, V18, V19, V20, V21, V22, V24, V25, V39	1
17	PAOZZ	5305002692803	96906	MS90726-60	SCREW, CAP, HEXAGON H .... UOC:DAA, DAB, DAC, DAD, DAE, DAF, DAG, DAH, DAJ, DAK, DAL, DAW, DAX, V12, V13, V14, V15, V16, V17, V18, V19, V20, V21, V22, V24, V25, V39	1
18	PAOZZ	5310009591488	96906	MS51922-21	NUT, SELF-LOCKING, HE .... UOC:DAA, DAB, DAC, DAD, DAE, DAF, DAG, DAH, DAJ, DAK, DAL, DAW, DAX, V12, V13, V14, V15, V16, V17, V18, V19, V20, V21, V22, V24, V25, V39	1

END OF FIGURE

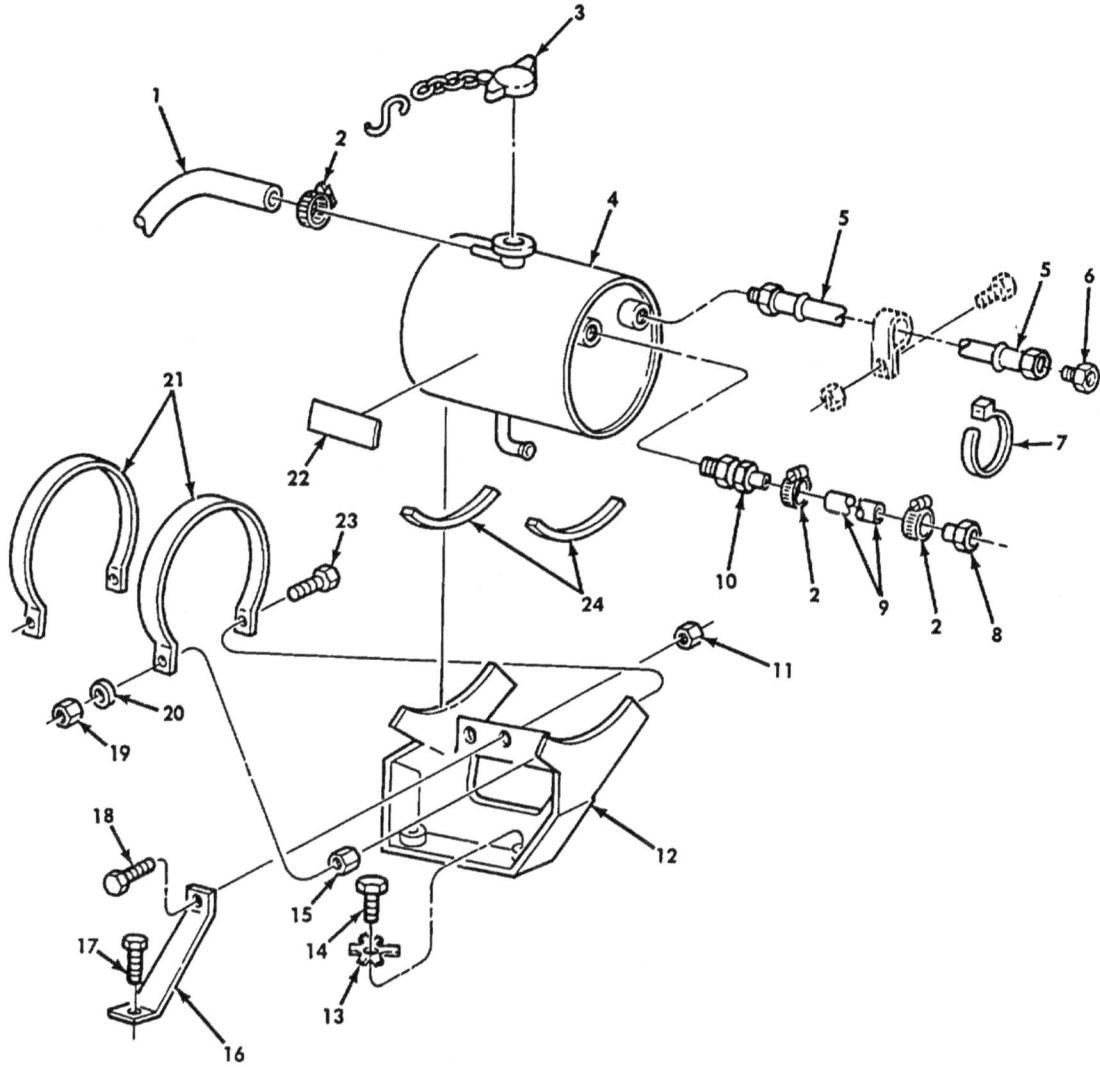

Figure 110. Surge Tank Assembly (M939A2).

(1) ITEM NO	(2) SMR CODE	(3) NSN	(4) CAGEC	(5) PART NUMBER	(6) DESCRIPTION AND USABLE ON CODES (UOC)	(7) QTY
					GROUP 0501 RADIATOR, EVAPORATIVE COOLER, OR HEAT EXCHANGER	
					FIG. 110 SURGE TANK ASSEMBLY(M939A2)	
1	MOOZZ		47457	20511258 X 25IN	HOSE,NONMETALLIC MAKE FROM HOSE, P/ .... N 483666, 25 INCHES LONG................... UOC:ZAA,ZAB,ZAC,ZAD,ZAE,ZAF,ZAG,ZAH, ZAJ,ZAK,ZAL	1
2	PAOZZ	4730011098001	19207	11662913	CLAMP,HOSE .... UOC:ZAA,ZAB,ZAC,ZAD,ZAE,ZAF,ZAG,ZAH, ZAJ,ZAK,ZAL	3
3	PAOZZ	2930001475202	19207	11648745	CAP,FILLER OPENING .... UOC:ZAA,ZAB,ZAC,ZAD,ZAE,Z,ZA ZAG,ZAH, ZAJ,ZAK,ZAL	1
4	PAOZZ	2590012914598	5U403	20510923	TANK,SURGE,TRUCK .... UOC:ZAA,ZAB,ZAC,ZAD,ZAE,ZAF,ZAG,ZAH, ZAJ,ZAK,ZAL	1
5	PAOZZ	4720010889650	19207	11664472-1	HOSE ASSEMBLY,NONME .... UOC:ZAA,ZAB,ZAC,ZAD,ZAE,ZAF,ZAG,ZAH, ZAJ,ZAK,ZAL	1
6	PAOZZ	4730012838148	98441	40483-4-6S25	ELBOW,PIPE TO TUBE .... UOC:ZAA,ZAB,ZAC,ZAD,ZAE,ZAF,ZAG,ZAH, ZAJ,ZAK,ZAL	1
7	PAOZZ	5975004515001	96906	MS3367-3-9	STRAP,TIEDOWN,ELECT .... UOC:ZAA,ZAB,ZAC,ZAD,ZAE,ZAF,ZAG,ZAH, ZAJ,ZAK,ZAL	2
8	PAOZZ	4730012849071	1GF04	31R82-2-4B	ADAPTER,STRAIGHT,PI...................... UOC:ZAA,ZAB,ZAC,ZAD,ZAE,ZAF,ZAG,ZAH, ZAJ,ZAK,ZAL	1
9	MOOZZ		47457	20511258 X 22IN	HOSE,NONMETALLIC .... MAKE FROM HOSE, P/ N 483666, 22 INCHES LONG................... UOC:ZAA,ZAB,ZAC,ZAD,ZAE,ZAF,ZAG,ZAH, ZAJ, ZAK, ZAL	1
10	PAOZZ	4730001941766	95879	2460-44	ADAPTER,STRAIGHT,PI...................... UOC:ZAA,ZAB,ZAC,ZAD,ZAE,ZAF,ZAG,ZAH, ZAJ,ZAK,ZAL	1
11	PAOZZ	5310011845784	11862	9422302	NUT,PLAIN,HEXAGON .... UOC:ZAA,ZAB,ZAC,ZAD,ZAE,ZAF,ZAG,ZAH, ZAJ,ZAK,ZAL	1
12	PAOZZ	5340012803716	47457	20511166	BRACKET,MOUNTI NG. .... UOC:ZAA,ZAB,ZAC,ZAD,ZAE,ZAF,ZAG,ZAH, ZAJ,ZAK,ZAL	1
13	PAOZZ	5310008245474	96906	MS15820-1	WASHER,KEY .... UOC:ZAA,ZAB,ZAC,ZAD,ZAE,ZAF,ZAG,ZAH, ZAJ,ZAK,ZAL	2
14	PAOZZ	5306012849663	24617	11503669	BOLT,MACHINE........................... UOC:ZAA,ZAB,ZAC,ZAD,ZAE,ZAF,ZAG,ZAH, ZAJ,ZAK,ZAL	2
15	PAOZZ	5310011073570	24617	271500	NUT,PLAIN,HEXAGON .... UOC: ZAA, ZAB, ZAC, ZAD, ZAE,ZAF, ZAG, ZAH, ZAJ,ZAK,ZAL	1

(1) ITEM NO	(2) SMR CODE	(3) NSN	(4) CAGEC	(5) PART NUMBER	(6) DESCRIPTION AND USABLE ON CODES (UOC)	(7) QTY
16	PAOZZ	5340012815150	47457	20511162	BRACKET,DOUBLE ANGL .... UOC:ZAA,ZAB,ZAC,ZAD,ZAE,ZAF,ZAG,ZAH, ZAJ,ZAK,ZAL	1
17	PAOZZ	5306011971513	24617	11502788	BOLT,SHOULDER .... UOC:ZAA,ZAB,ZAC,ZAD,ZAE,ZAF,ZAG,ZAH, ZAJ,ZAK,ZAL	1
18	PAOZZ	5305007195221	80204	B1821BH050F150N	SCREW,CAP,HEXAGON H .... UOC:ZAA,ZAB,ZAC,ZAD,ZAE,ZAF,ZAG,ZAH, ZAJ,ZAK,ZAL	1
19	PAOZZ	5310011936884	11862	9422279	NUT,SELF-LOCKING,HE .... UOC:ZAA,ZAB,ZAC,ZAD,ZAE,ZAF,ZAG,ZAH, ZAJ,ZAK,ZAL	2
20	PAOZZ	5310010925495	24617	9422848	WASHER,FLAT .... UOC:ZAA,ZAB,ZAC,ZAD,ZAE,ZAF,ZAG,ZAH, ZAJ,ZAK,ZAL	2
21	PAOZZ	5340012806996	47457	20510929	CLAMP,LOOP .... UOC:ZAA,ZAB,ZAC,ZAD,ZAE,ZAF,ZAG,ZAH, ZAJ,ZAK,ZAL	2
22	PAOZZ	7690011966355	19207	12302869	MARKER,IDENTIFICATI . ... UOC:ZAA,ZAB,ZAC,ZAD,ZAE,ZAF,ZAG,ZAH, ZAJ,ZAK,ZAL	1
23	PAOZZ	5305012809367	6N171	1039234	SCREW,CAP,HEXAGON H .... UOC:ZAA,ZAB,ZAC,ZAD,ZAE,ZAF,ZAG,ZAH, ZAJ,ZAK,ZAL	2
24	PAOZZ	9390012803408	57643	20511185	NONMETALLIC CHANNEL .... UOC:ZAA,ZAB,ZAC,ZAD,ZAE,ZAF,ZAG,ZAH, ZAJ,ZAK,ZAL	2

END OF FIGURE

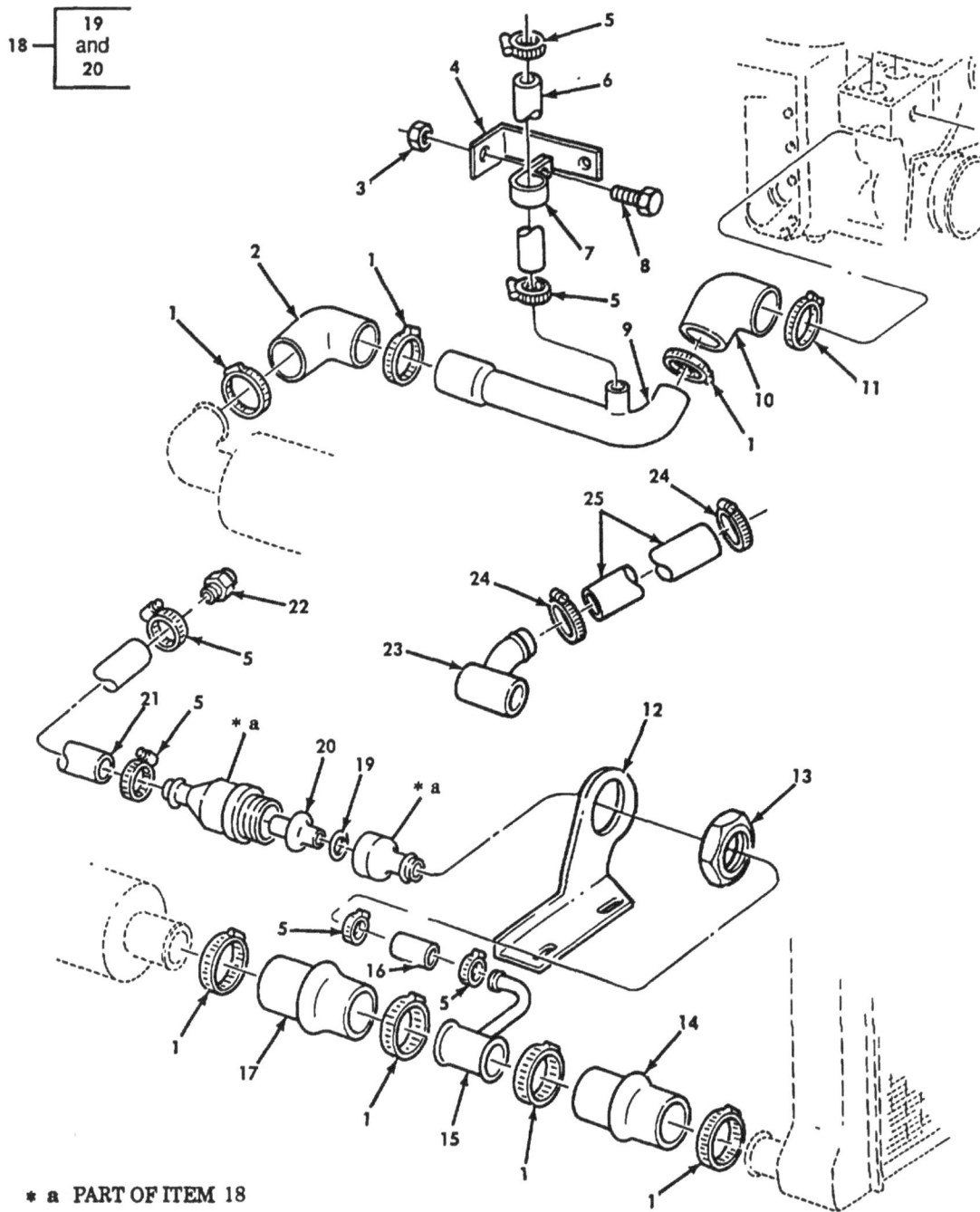

Figure 111. Surge Tank Lines and Clamps (M939A2).

* a PART OF ITEM 18

(1) ITEM NO	(2) SMR CODE	(3) NSN	(4) CAGEC	(5) PART NUMBER	(6) DESCRIPTION AND USABLE ON CODES (UOC)	(7) QTY
					GROUP 0501 RADIATOR, EVAPORATIVE COOLER, OR HEAT EXCHANGER	
					FIG. 111 SURGE TANK LINES AND CLAMPS (M939A2)	
1	PAOZZ	4730013316630	19207	11669079-5	CLAMP,HOSE.. UOC:ZAA,ZAB,ZAC,ZAD,ZAE,ZAF,ZAG,ZAH, ZAJ,ZAK,ZAL	7
2	PAOZZ	4720012793170	04827	20510329	HOSE,PREFORMED .... UOC:ZAA,ZAB,ZAC,ZAD,ZAE,ZAF,ZAG,ZAH, ZAJ,ZAK,ZAL	1
3	PAOZZ	5310002081918	88044	AN365-1024A	NUT,SELF-LOCKING,HE .... UOC:ZAA,ZAB,ZAC,ZAD,ZAE,ZAF,ZAG,ZAH, ZAJ,ZAK,ZAL	1
4	PAOZZ	5340012806875	08627	20510993	BRACKET,ANGLE . ... UOC:ZAA,ZAB,ZAC,ZAD,ZAE,ZAF,ZAG,ZAH, ZAJ,ZAK,ZAL	1
5	PAOZZ	4730012448434	76599	4820SS	CLAMP,HOSE .... UOC:ZAA,ZAB,ZAC,ZAD,ZAE,ZAF,ZAG,ZAH, ZAJ,ZAK,ZAL	6
6	MOOZZ		19207	8710557 X 44IN	HOSE,................MAKE FROM HOSE, P/N 8710557, 44 INCHES LONG.. UOC:ZAA,ZAB,ZAC,ZAD,ZAE,ZAF,ZAG,ZAH, ZAJ,ZAK,ZAL	1
7	PAOZZ	5340009226300	96906	MS21333-77	CLAMP,LOOP .... UOC:ZAA,ZAB,ZAC,ZAD,;ZAE,ZAF,ZAG,ZAH, ZAJ,ZAK,ZAL	1
8	PAOZZ	5306000440502	21450	440502	BOLT,MACHINE . ... UOC:ZAA,ZAB,ZAC,ZAD,ZAE,ZAF,ZAG,ZAH, ZAJ,ZAK,ZAL	1
9	PAOZZ	4710012849029	16251	20510409	TUBE,BENT,METALLIC .... UOC:ZAA,ZAB,ZAC,ZAD,ZAE,ZAF,ZAG,ZAH, ZAJ,ZAK,ZAL	1
10	PAOZZ	4720012717202	15434	3920762	HOSE,PREFORMED .... UOC:ZAA,ZAB,ZAC,ZAD,ZAE,ZAF,ZAG,ZAH, ZAJ,ZAK,ZAL	1
11	PAOZZ	4730013489510	7Z588	CT9444	CLAMP,HOSE... UOC:ZAA,ZAB,ZAC,ZAC,ZAD,ZAE,ZAF,ZAG,ZAH, ZAJ,ZAK,ZAL	1
12	PAOZZ	5340012806869	47457	20510980	BRACKET,ANGLE ....CANISTER UOC:ZAA,ZAB,ZAC,ZAD,ZAE,ZAF,ZAG,ZAH, ZAJ,ZAK,ZAL	1
13	PAOZZ	5310012878812	47457	20510996	NUT,PLAIN,ROUND .... UOC:ZAA,ZAB,ZAC,ZAD,ZAE,ZAF,ZAG,ZAH, ZAJ,ZAK,ZAL	1
14	PAOZZ	4720012793168	47457	12255978A	HOSE,PREFORMED . ... UOC:ZAA,ZAB,ZAC,ZAD,ZAE,ZAF,ZAG,ZAH, ZAJ,ZAK,ZAL	1
15	PAOZZ	4730012791510	4F744	20510924	TEE,HOSE .... USE WITH INTERNAL BYPASS SYSTEM ONLY .... UOC:ZAA,ZAB,ZAC,ZAD,ZAE,ZAF,ZAG,ZAH,	1

(1) ITEM NO	(2) SMR CODE	(3) NSN	(4) CAGEC	(5) PART NUMBER	(6) DESCRIPTION AND USABLE ON CODES (UOC)	(7) QTY
16	MOOZZ		19207	8710557 X 3 IN	ZAJ,ZAK,ZAL HOSE, BYPASS MAKE FROM HOSE, P,N................ 8710557, 3 INCHES LONG, USE WITH INTERNAL BYPASS SYSTEM ONLY .... UOC:ZAA,ZAB,ZAC,ZAD,ZAE,ZAF,ZAG,ZAH, ZAJ,ZAK,ZAL	1
17	PAOZZ	720012853008	04827	20510979	HOSE,PREFORMED USE WITH INTERNAL............ BYPASS SYSTEM ONLY .... UOC:ZAA,ZAB,ZAC,ZAD,ZAE,ZAF,ZAG,ZAH, ZAJ,ZAK,ZAL	1
18	PAOOO	2910012856253	47457	20510922	CANISTER,CARBON,EMI THERMOSTAT,USE .... WITH INTERNAL BYPASS SYSTEM ONLY .... UOC:ZAA,ZAB,ZAC,ZAD,ZAE,ZAF,ZAG,ZAH, ZAJ,ZAK,ZAL	1
19	PAOZZ	5330013592316	59342	CT-2900-2	.GASKET .... UOC:ZAA,ZAB,ZAC,ZAD,ZAE,ZAF,ZAG,ZAH, ZAJ,ZAK,ZAL	1
20	PAOZZ	6685013617552	47457	20510922-5	.THERMOSTAT,FLOW CON............................... UOC:ZAA,ZAB,ZAC,ZAD,ZAE,ZAF,ZAG,ZAH, ZAJ,ZAK,ZAL	1
21	MOOZZ		19207	8710557X 25IN	HOSE, CANISTER, MAKE FROM HOSE, P/ .... N 8710557, 25 INCHES LONG, USE WITH INTERNAL BYPASS SYSTEM ONLY .... UOC:ZAA,ZAB,ZAC,ZAD,ZAE,ZAF,ZAG,ZAH, ZAJ,ZAK,ZAL	1
22	PAOZZ	4730011059466	93061	68HB-10-8	ADAPTER,STRAIGHT,PI USE WITH...................... INTERNAL BYPASS SYSTEM ONLY...................... UOC:ZAA,ZAB,ZAC,ZAD,ZAE,ZAF,ZAG,ZAH, ZAJ,ZAK,ZAL	1
23	PAOZZ	4730013004112	47457	20511285	TEE,HOSE USE WITH EXTERNAL BYPASS............ SYSTEM ONLY .... UOC:ZAA,ZAB,ZAC,ZAD,ZAE,ZAF,ZAG,ZAH, ZAJ,ZAK,ZAL	1
24	PAOZZ	4730009098627	01276	FF9311-36	CLAMP,HOSE USE WITH EXTERNAL.................... BYPASS SYSTEM ONLY .... UOC:ZAA,ZAB,ZAC,ZAD,ZAE,ZAF,ZAG,ZAH, ZAJ,ZAK,ZAL	2
25	PAOZZ	4720013004146	98441	4244-24	HOSE,NONMETALLIC USE WITH EXTERNAL......... BYPASS SYSTEM ONLY .... UOC:ZAA,ZAB,ZAC,ZAD,ZAE,ZAF,ZAG,ZAH, ZAJ,ZAK,ZAL	1

END OF FIGURE

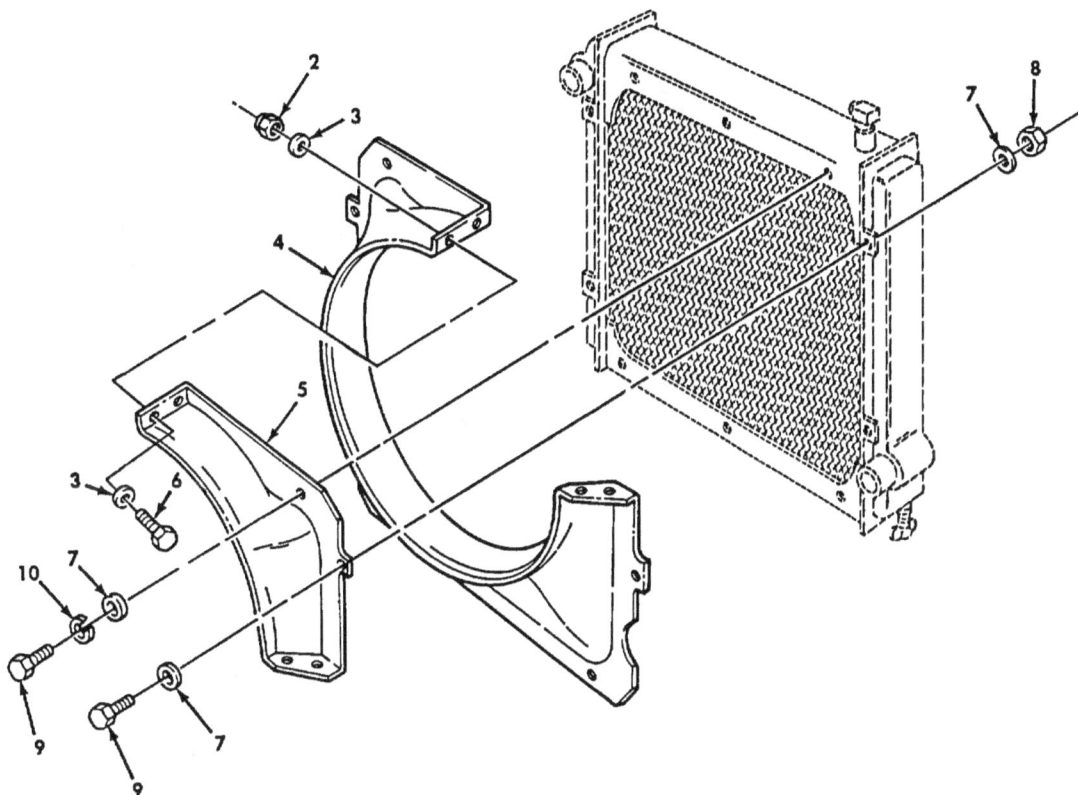

Figure 112. Shroud Assembly and Mounting Hardware (M939, M939A1).

(1) ITEM NO	(2) SMR CODE	(3) NSN	(4) CAGEC	(5) PART NUMBER	(6) DESCRIPTION AND USABLE ON CODES (UOC)	(7) QTY
					GROUP 0502 SHROUDS	
					FIG. 112 SHROUD ASSEMBLY AND MOUNTING HARDWARE(M939,M939A1)	
2	PAOZZ	5310009359022	96906	MS51943-32	.NUT,SELF-LOCKING,HE .... UOC:DAA,DAB,DAC,DAD,DAE,DAF,DAG,DAH, DAJ,DAK,DAL,DAW,DAX,V12 ,V13,V14,V15, V16,V17,V18,V19,V20,V21,V22,V24,V25, V39	4
3	PAOZZ	5310008238804	96906	MS27183-9	.WASHER,FLAT .... UOC:DAA,DAB,DAC,DAD,DAE,DAF,DAG,DAH, DAJ,DAK,DAL,DAW,DAX,V12,V13,V14,V15, V16,V17,V18,V19,V20,V21,V22,V24,V25, V39	8
4	PAOZZ	2930010823631	19207	12256280-1	.SHROUD,FAN,RADIATOR .... UOC:DAA,DAB,DAC,DAD,DAE,DAF,DAG,DAH, DAJ,DAK,DAL,DAW,DAX,V12,V13,V14,V15, V16,V17,V18,V19,V20,V21,V22,V24,V25, V39	1
5	PAOZZ	2930010822513	19207	12256280-2	.SHROUD,FAN,RADIATOR .... UOC:DAA,DAB,DAC,DAD,DAE,DAF,DAG,DAH, DAJ,DAK,DAL,DAW,DAX,V12,V13,V14,V15, V16,V17,V18,V19,V20,V21,V22,V24,V25, V39	1
6	PAOZZ	5305000680515	80204	B1821BH025F100N	.SCREW,CAP,HEXAGON H SHROUD.................... ASSEMBLY,RADIATOR .... UOC:DAA,DAB,DAC,DAD,DAE,DAF,DAG,DAH, DAJ,DAK,DAL,DAW,DAX,V12 ,V13 ,V14,V15, V16,V17,V18,V19,V20,V21,V22,V24,V25, V39	4
7	PAOZZ	5310011328275	24617	9418924	WASHER,FLAT .... UOC:DAA,DAB,DAC,DAD,DAE,DAF,DAG,DAH, DAJ,DAK,DAL,DAW,DAX,V12,V13,V14,V15, V16,V17,V18,V19,V20,V21,V22,V24,V25, V39	14
8	PAOZZ	5310002416658	96906	MS51943-34	NUT,SELF-LOCKING,HE .... UOC:DAA,DAB,DAC,DAD,DAE,DAF,DAG,DAH, DAJ,DAK,DAL,DAW,DAX,V12,V13,V14,V15, V16,V17,V18,V19,V20,V21,V22,V24,V25, V39	4
9	PAOZZ	5306000501238	96906	MS90727-32	BOLT,MACHINE .... UOC:DAA,DAB,DAC,DAD,DAE,DAF,DAG,DAH, DAJ,DAK,DAL,DAW,DAX,V12,V13,V14,V15, V16,V17,V18,V19,V20,V21,V22,V24,V25, V39	10
10	PAOZZ	5310004079566	96906	MS35338-45	WASHER,LOCK.. UOC:DAA,DAB,DAC,DAD,DAE,DAF,DAG,DAH, DAJ,DAK,DAL,DAW,DAX,V12,V13,V14,V15, V16,V17,V18,V19,V20,V21,V22,V24,V25, V39	6

END OF FIGURE

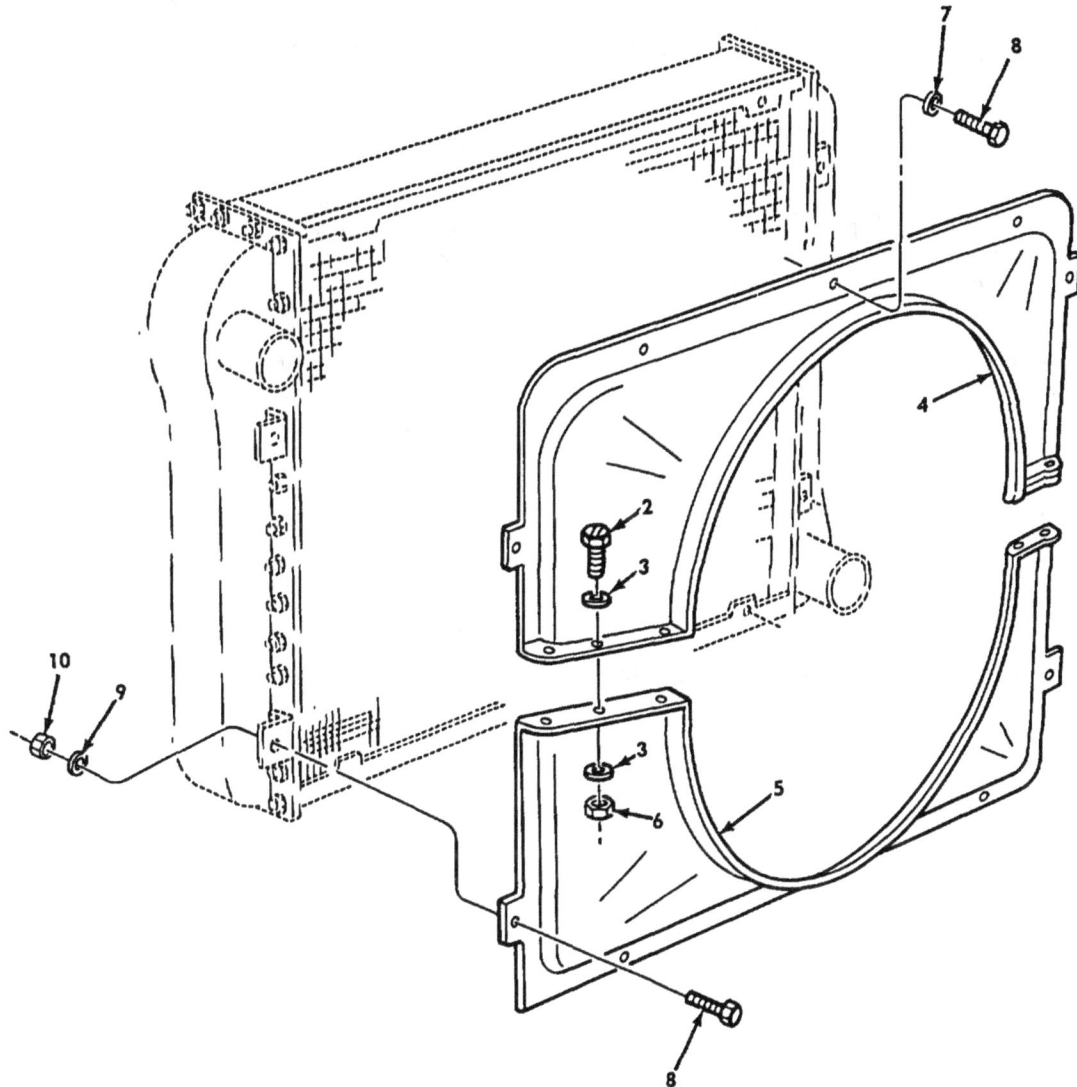

Figure 113. Radiator Fan Shroud (M939A2).

(1) ITEM NO	(2) SMR CODE	(3) NSN	(4) CAGEC	(5) PART NUMBER	(6) DESCRIPTION AND USABLE ON CODES (UOC)	(7) QTY
					GROUP 0502 SHROUDS	
					FIG. 113 RADIATOR FAN SHROUD(M939A2)	
1	AOOOZ	2930012874545	47457	20510266-2	SHROUD,FAN,RADIATOR .... UOC:ZAA,ZAB,ZAC,ZAD,ZAE,ZAF,ZAG,ZAH, ZAJ,ZAK,ZAL	1
2	PAOZZ	5305000680515	80204	B1821BH025F100N	.SCREW,CAP,HEXAGON H .... UOC:ZAA,ZAB,ZAC,ZAD,ZAE,ZAF,ZAG,ZAH, ZAJ, ZAK, ZAL	4
3	PAOZZ	5310008238804	96906	MS27183-9	.WASHER,FLAT . ... UOC:ZAA,ZAB,ZAC,ZAD,ZAE,ZAF,ZAG,ZAH, ZAJ,ZAK,ZAL	8
4	PAOZZ	2930012873180	47457	20510266-1	.SHROUD,FAN,RADIATOR .... UOC:ZAA,ZAB,ZAC,ZAD,ZAE,ZAF,ZAG,ZAH, ZAJ,ZAK,ZAL	1
5	PAOZZ	2930012874545	47457	20510266-2	.SHROUD,FAN,RADIATOR . ... UOC:ZAA,ZAB,ZAC,ZAD,ZAE,ZAF,ZAG,ZAH, ZAJ,ZAK,ZAL	1
6	PAOZZ	5310009359022	96906	MS51943-32	.NUT,SELF-LOCKING,HE .... UOC:ZAA,ZAB,ZAC,ZAD,ZAE,ZAF,ZAG,ZAH, ZAJ,ZAK,ZAL	4
7	PAOZZ	5310012590296	96906	MS51412-6	WASHER,FLAT .... UOC:ZAA,ZAB,ZAC,ZAD,ZAE,ZAF,ZAG,ZAH, ZAJ,ZAK,ZAL	1
8	PAOZZ	5306002259087	96906	MS90726-32	BOLT,MACHINE .... UOC:ZAA,ZAB,ZAC,ZAD,ZAE,ZAF ,ZAG,ZAH, ZAJ, ZAK, ZAL	10
9	PAOZZ	5310011328275	24617	9418924	WASHER,FLAT .... UOC:ZAA,ZAB,ZAC,ZAD,ZAE,ZAF,ZAG,ZAH, ZAJ,ZAK,ZAL	13
10	PAOZZ	5310002416658	96906	MS51943-34	NUT,SELF-LOCKING,HE .... UOC:ZAA,ZAB,ZAC,ZAD,ZAE,ZAF,ZAG,ZAH, ZAJ,ZAK,ZAL	4

END OF FIGURE

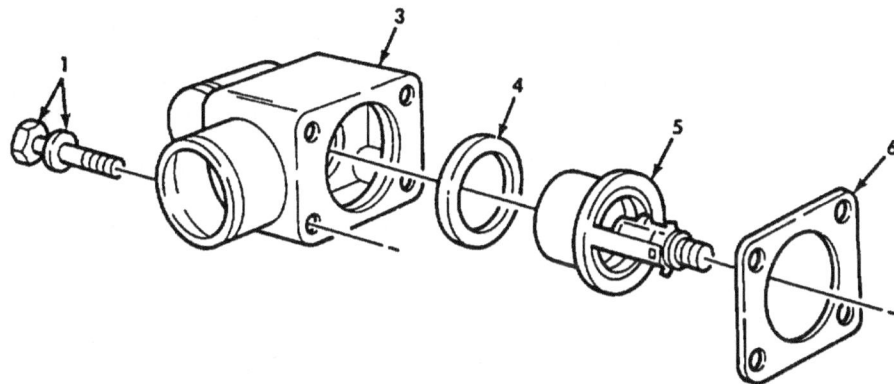

Figure 114. Thermostat, Housing, Seal, and Mounting Hardware (M939, M939A1).

(1) ITEM NO	(2) SMR CODE	(3) NSN	(4) CAGEC	(5) PART NUMBER	(6) DESCRIPTION AND USABLE ON CODES (UOC)	(7) QTY
					GROUP 0503 WATER MANIFOLD, HEADERS, THERMOSTATS, AND HOUSING GASKET	
					FIG. 114 THERMOSTAT, HOUSING, SEAL, AND MOUNTING HARDWARE(M939, M939A1)	
1	PAOZZ	5305010867036	15434	3010597	SCREW .... UOC:DAA,DAB,DAC,DAD,DAE,DAF,DAG,DAH, DAJ,DAK,DAL,DAW,DAX,V12 ,V13,V14,V15, V16,V17,V18,V19,V20,V21,V22,V24,V25, V39	4
2	PBOZZ	930004043053	15434	AR-51264	WATER OUTLET,ENGINE .... UOC:DAA,DAB,DAC,DAD,DAE,DAF,DAG,DAH, DAJ,DAK,DAL,DAW,DAX,V12,V13,V14,V15, V16,V17,V18,V19,V20,V21,V22,V24,V25, V39	1
3	PAOZZ	6620011583125	15434	3013607	.HOUSING,THERMOSTAT .... UOC:DAA,DAB,DAC,DAD,DAE,DAF,DAG,DAH, DAJ,DAK,DAL,DAW,DAX,V12 ,V13 ,V14 ,V15, V16,V17,V18,V19,V20,V21,V22,V24,V25, V39	1
4	PAOZZ	5330008645422	15434	186780	.SEAL,THERMO PART OF KIT P/N 3011472 .... PART OF KIT P/N 3804272................................ UOC:DAA,DAB,DAC,DAD,DAE,DAF,DAG,DAH, DAJ,DAK,DAL,DAW,DAX,V12,V13,V14,V15, V16,V17,V18,V19,V20,V21,V22,V24,V25, V39	1
5	PAOZZ	6685011410907	15434	201737	THERMOSTAT,FLOW CON .... UOC:DAA,DAB,DAC,DAD,DAE,DAF,DAG,DAH, DAJ,DAK,DAL,DAW,DAX,V12,V13,V14,V15, V16,V17,V18,V19,V20,V21,V22,V24,V25, V39	1
6	PAOZZ	5330005080411	15434	70441	GASKET THERMOSTAT HOUSING PART OF.......... KIT P/N 3011472 PART OF KIT P/N 3804272 .... UOC:DAA,DAB,DAC,DAD,DAE,DAF,DAG,DAH, DAJ,DAK,DAL,DAW,DAX,V12,V13,V14,V15, V16,V17,V18,V19,V20,V21,V22,V24,V25, V39	1

END OF FIGURE

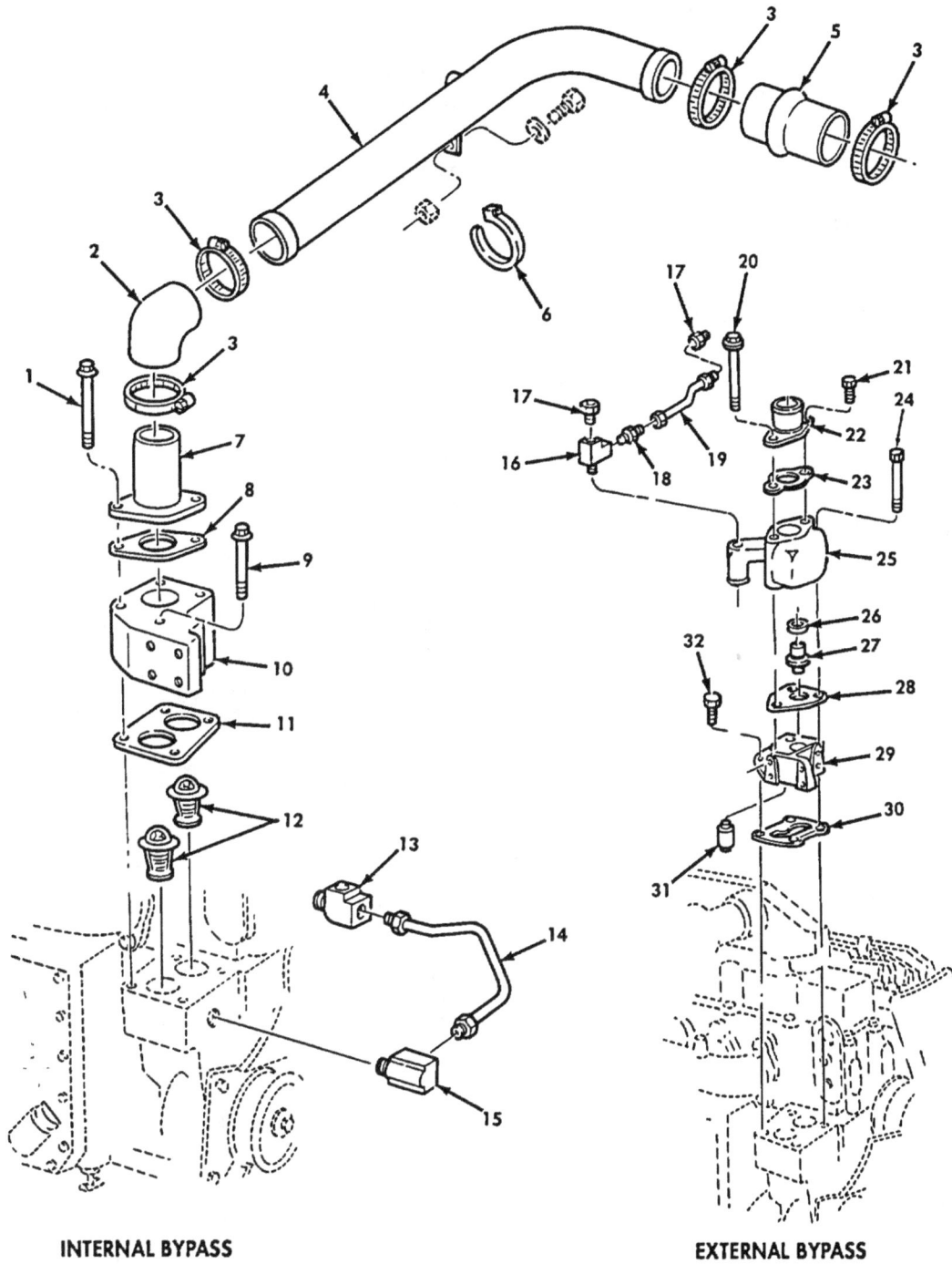

Figure 115. Thermostat and Housing Assembly (M939A2).

INTERNAL BYPASS

EXTERNAL BYPASS

(1) ITEM NO	(2) SMR CODE	(3) NSN	(4) CAGEC	(5) PART NUMBER	(6) DESCRIPTION AND USABLE ON CODES (UOC)	(7) QTY
					GROUP 0503 WATER MANIFOLD, HEADERS, THERMOSTATS, AND HOUSING GASKET	
					FIG. 115 THERMOSTAT AND HOUSING ASSEMBLY (M939A2)	
1	PAOZZ	5305012724810	15434	3906715	SCREW,CAP,HEXAGON H USED WITH .... INTERNAL BYPASS SYSTEM ONLY...................... UOC:ZAA,ZAB,ZAC,ZAD,ZAE,ZAF,ZAG,ZAH, ZAJ,ZAK,ZAL	2
2	PAOZZ	4720013026687	47457	20510473	HOSE,PREFORMED .... UOC:ZAA,ZAB,ZAC,ZAD,ZAE,ZAF,ZAG,ZAH, ZAJ,ZAK,ZAL	1
3	PAOZZ	4730013316630	19207	11669079-5	CLAMP,HOSE .... UOC:ZAA,ZAB,ZAC,ZAD,ZAE,ZAF,ZAG,ZAH, ZAJ,ZAK,ZAL	4
4	PAOZZ	4710012849032	47457	20510518	TUBE,BENT,METALLIC RADIATOR .... UOC:ZAA,ZAB,ZAC,ZAD,ZAE,ZAF,ZAG,ZAH, ZAJ,ZAK,ZAL	1
5	PAOZZ	4720012793169	04827	20510481	HOSE,PREFORMED RADIATOR .... UOC:ZAA,ZAB,ZAC,ZAD,ZAE,ZAF,ZAG,ZAH, ZAJ,ZAK,ZAL	1
6	PAOZZ	5975004515001	96906	MS3367-3-9	STRAP,TIEDOWN,ELECT .... UOC:ZAA,ZAB,ZAC,ZAD,ZAE,ZAF,ZAG,ZAH, ZAJ,ZAK,ZAL	1
7	PAOZZ	2930012726716	15434	3912800	WATER OUTLET,ENGINE OUTLET, USED.............. WITH INTERNAL BYPASS SYSTEM ONLY .... UOC:ZAA,ZAB,ZAC,ZAD,ZAE,ZAF,ZAG,ZAH, ZAJ,ZAK,ZAL	1
8	PAOZZ	5330012721142	15434	3060912	GASKET USED WITH INTERNAL BYPASS.............. SYSTEM ONLY PART OF KIT P/N 3802622 PART OF KIT P/N 3802623 PART OF KIT P/N 3802452…………………………………………… UOC:ZAA,ZAB,ZAC,ZAD,ZAE,ZAF,ZAG,ZAH, ZAJ,ZA, ZAL	1
9	PAOZZ	5305012724811	15434	3902112	SCREW,CAP,HEXAGON H .... UOC:ZAA,ZAB,ZAC,ZAD,ZAE,ZAF,ZAG,ZAH, ZAJ,ZAK,ZAL	2
10	PAOZZ	2930012915867	15434	3911493	WATER OUTLET,ENGINE USED WITH .... INTERNAL BYPASS SYSTEM ONLY...................... UOC:ZAA,ZAB,ZAC,ZAD,ZAE,ZAF,ZAG,ZAH, ZAJ,ZAK,ZAL	1
11	PAOZZ	5330012878656	7U263	3914310	GASKET USED WITH INTERNAL BYPASS.............. SYSTEM ONLY PART OF KIT P/N 3802622 PART OF KIT P/N 3802623 PART OF KIT P/N 3802452…………………………………………… UOC:ZAA,ZAB,ZAC,ZAD,ZAE,ZAF,ZAG,ZAH, ZAJ,ZAK,ZAL	1
12	PAOZZ	6620012721716	15434	3907242	THERMOSTAT,FLOW CON USED WITH .... INTERNAL BYPASS SYSTEM ONLY...................... UOC:ZAA,ZAB,ZAC,ZAD,ZAE,ZAF,ZAG,ZAH, ZAJ,ZAK,ZAL	2

(1) ITEM NO	(2) SMR CODE	(3) NSN	(4) CAGEC	(5) PART NUMBER	(6) DESCRIPTION AND USABLE ON CODES (UOC)	(7) QTY
13	PAOZZ	4730012717977	15434	3908763	TEE,TUBE USED WITH INTERNAL BYPASS .... SYSTEM ONLY .... UOC:ZAA,ZAB,ZAC,ZAD,ZAE,ZAF,ZAG,ZAH, ZAJ,ZAK,ZAL	1
14	PAOZZ	4710012718033	15434	3926545	TUBE ASSEMBLY,METAL USED WITH .... INTERNAL BYPASS SYSTEM ONLY...................... UOC:ZAA,ZAB,ZAC,ZAD,ZAE,ZAF,ZAG,ZAH, ZAJ,ZAK,ZAL	1
15	PAOZZ	4730011516316	15434	S1004-1	ELBOW,TUBE USED WITH INTERNAL................... BYPASS SYSTEM ONLY .... UOC:ZAA,ZAB,ZAC,ZAD,ZAE,ZAF,ZAG,ZAH, ZAJ,ZAK,ZAL	1
16	PAOZZ	4730013009030	15434	3023198	TEE,PIPE TO TUBE USED WITH .... EXTERNAL BYPASS SYSTEM ONLY .... UOC:ZAA,ZAB,ZAC,ZAD,ZAE,ZAF,ZAG,ZAH, ZAJ,ZAK,ZAL	1
17	PAOZZ	4730013009024	15434	3018888	ADAPTER,STRAIGHT,PI USED WITH .... EXTERNAL BYPASS SYSTEM ONLY...................... UOC:ZAA,ZAB,ZAC,ZAD,ZAE,ZAF,ZAG,ZAH, ZAJ,ZAK,ZAL	2
18	PAOZZ	4730011314884	15434	S-1097	ADAPTER,STRAIGHT,PI USED WITH .... EXTERNAL BYPASS SYSTEM ONLY...................... UOC:ZAA,ZAB,ZAC,ZAD,ZAE,ZAF,ZAG,ZAH, ZAJ,ZAK,ZAL	1
19	PAOZZ	4710013014358	15434	3913033	TUBE ASSEMBLY,METAL USED WITH .... EXTERNAL BYPASS SYSTEM ONLY .... UOC:ZAA,ZAB,ZAC,ZAD,ZAE,ZAF,ZAG,ZAH, ZAJ,ZAK,ZAL	1
20	PAOZZ	5305013019756	15434	3903118	SCREW,CAP,HEXAGON HUSED WITH .... EXTERNAL BYPASS SYSTEM ONLY...................... UOC:ZAA,ZAB,ZAC,ZAD,ZAE,ZAF,ZAG,ZAH, ZAJ,ZAK,ZAL	1
21	PAOZZ	5305012343755	15434	3900630	SCREW,CAP,HEXAGON H USED WITH .... EXTERNAL BYPASS SYSTEM ONLY .... UOC:ZAA,ZAB,ZAC,ZAD,ZAE,ZAF,ZAG,ZAH, ZAJ,ZAK,ZAL	1
22	PAOZZ	2930013053303	15434	3913030	WATER OUTLET,ENGINE USED WITH .... EXTERNAL BYPASS SYSTEM ONLY .... UOC:ZAA,ZAB,ZAC,ZAD,ZAE,ZAF,ZAG,ZAH, ZAJ,ZAK,ZAL	1
23	PAOZZ	5330013011828	15434	3913032	GASKET USED WITH EXTERNAL BYPASS............ SYSTEM ONLY .... UOC:ZAA,ZAB,ZAC,ZAD,ZAE,ZAF,ZAG,ZAH, ZAJ,ZAK,ZAL	1
24	PAOZZ	5305013019757	15434	3913034	SCREW,CAP,HEXAGON H USED WITH.................. EXTERNAL BYPASS SYSTEM ONLY...................... UOC:ZAA,ZAB,ZAC,ZAD,ZAE,ZAF,ZAG,ZAH, ZAJ,ZAK,ZAL	2
25	PAOZZ	2930013005943	15434	3913024	WATER OUTLET,ENGINE USED WITH .... EXTERNAL BYPASS SYSTEM ONLY...................... UOC:ZAA,ZAB,ZAC,ZAD,ZAE,ZAF,ZAG,ZAH, ZAJ,ZAK,ZAL	1

(1) ITEM NO	(2) SMR CODE	(3) NSN	(4) CAGEC	(5) PART NUMBER	(6) DESCRIPTION AND USABLE ON CODES (UOC)	(7) QTY
26	PAOZZ	5330013011761	15434	188318	SEAL,SPECIAL USED WITH EXTERNAL BYPASS SYSTEM ONLY UOC:ZAA,ZAB,ZAC,ZAD,ZAE,ZAF,ZAG,ZAH, ZAJ, ZAK,ZAL	1
27	PAOZZ	6620013020045	15434	3913028	THERMOSTAT,FLOW CON USED WITH EXTERNAL BYPASS SYSTEM ONLY UOC: ZAA, ZAB, ZAC, ZAD, ZAE, ZAF, ZAG, ZAH, ZAJ, ZAK, ZAL	1
28	PAOZZ	5330013020780	15434	3913025	GASKET USED WITH EXTERNAL BYPASS SYSTEM ONLY UOC:ZAA,ZAB,ZAC,ZAD,ZAE,ZAF,ZAG,ZAH, ZAJ,ZAK,ZAL	1
29	PAOZZ	2930013005944	15434	3926094	WATER OUTLET,ENGINE USED WITH EXTERNAL BYPASS SYSTEM ONLY UOC:ZAA,ZAB,ZAC,ZAD,ZAE,ZAF,ZAG,ZAH, ZAJ,ZAK,ZAL	1
30	PAOZZ	5330013011829	15434	3913027	GASKET USED WITH EXTERNAL BYPASS SYSTEM ONLY UOC:ZAA,ZAB,ZAC,ZAD,ZAE,ZAF,ZAG,ZAH, ZAJ,ZAK,ZAL	1
31	PAOZZ	5365013026555	15434	3913029	PLUG,MACHINE THREAD USED WITH EXTERNAL BYPASS SYSTEM ONLY UOC:ZAA,ZAB,ZAC,ZAD,ZAE,ZAF,ZAG,ZAH, ZAJ,ZAK,ZAL	1
32	PAOZZ	5305012696274	15434	3902114	SCREW,CAP,HEXAGON H USED WITH EXTERNAL BYPASS SYSTEM ONLY UOC:ZAA,ZAB,ZAC,ZAD,ZAE,ZAF,ZAG,ZAH, ZAJ,ZAK, ZAL	2

END OF FIGURE

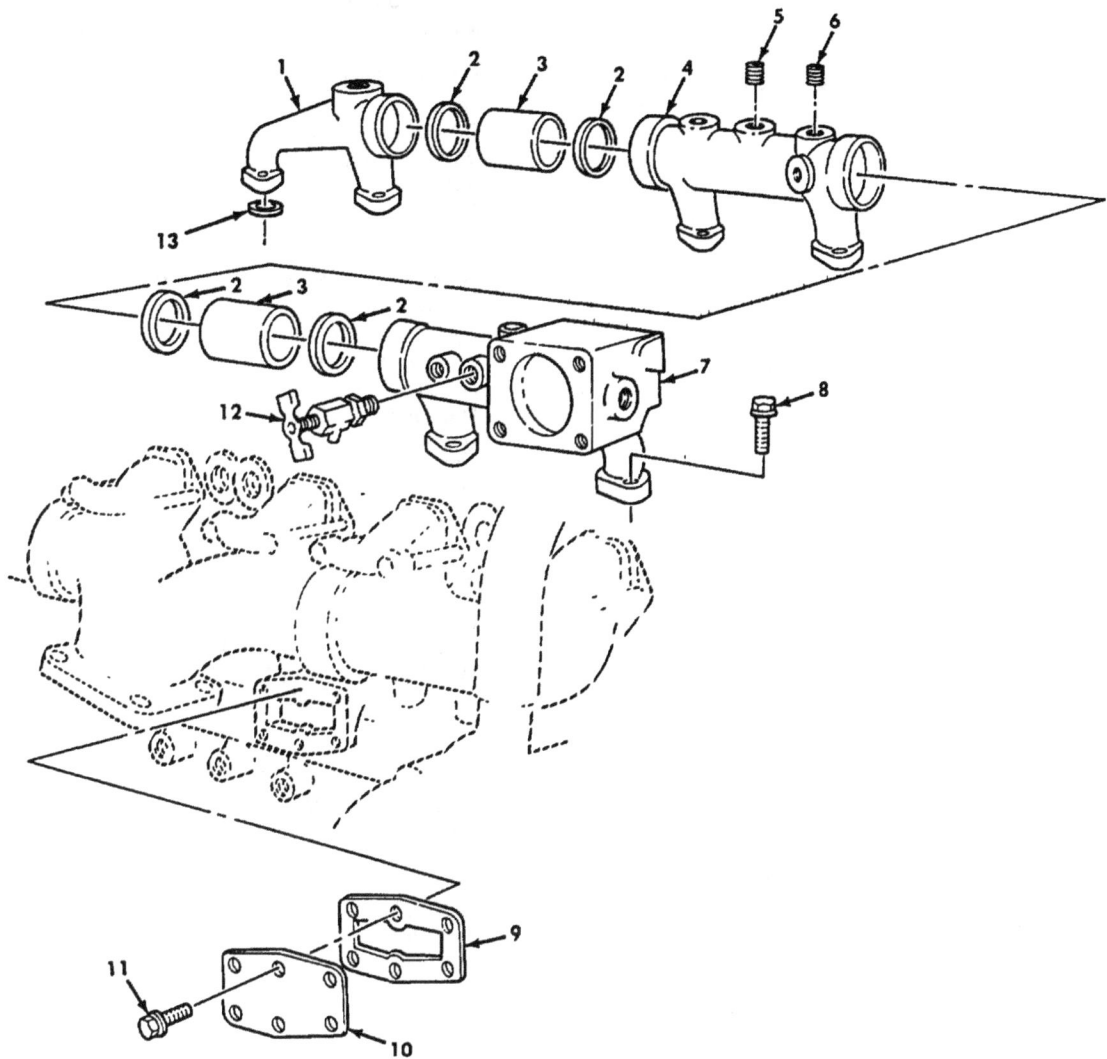

Figure 116. Water Manifold, Cover, and Mounting Hardware (M939, M939A1).

(1) ITEM NO	(2) SMR CODE	(3) NSN	(4) CAGEC	(5) PART NUMBER	(6) DESCRIPTION AND USABLE ON CODES (UOC)	(7) QTY
					GROUP 0503 WATER MANIFOLD, HEADERS, THERMOSTATS, AND HOUSING GASKET	
					FIG. 116 WATER MANIFOLD, COVER, AND MOUNTING HARDWARE(M939, M939A1)	
1	PAFZZ	2930011342238	15434	3013001	MANIFOLD,FLUID COOL REAR .... UOC:DAA,DAB,DAC,DAD,DAE,DAF,DAG,DAH, DAJ,DAK,DAL,DAW,DAX,V12,V13,V14,V15, V16,V17,V18,V19,V20,V21,V22,V24,V25, V39	1
2	PAFZZ	5331005064874	15434	70624	O-RING PART OF KIT P/N 3011472 PART.............. OF KIT P/N 3804272... UOC:DAA,DAB,DAC,DAD,DAE,DAF,DAG,DAH, DAJ,DAK,DAL,DAW,DAX,V12,V13,V14,V15, V16,V17,V18,V19,V20,V21,V22,V24,V25, V39	4
3	PAFZZ	4730004042906	15434	130394	COUPLING,MANIFOLD .... UOC:DAA,DAB,DAC,DAD,DAE,DAF,DAG,DAH, DAJ,DAK,DAL,DAW,DAX,V12,V13,V14,V15, V16,V17,V18,V19,V20,V21,V22,V24,V25, V39	2
4	PAFZZ	2930012044475	15434	3030866	MANIFOLD,FLUID COOL CENTER .... UOC:DAA,DAB,DAC,DAD,DAE,DAF,DA G,DAH, DAJ,DAK,DAL,DAW,DAX,V12,V13,V14,V15, V16,V17,V18,V19,V20,V21,V22,V24,V25, V39	1
5	PAOZZ	4730010857328	15434	217632	PLUG,PIPE .... UOC:DAA,DAB,DAC,DAD,DAE,DAF,DAG,DAH, DAJ,DAK,DAL,DAW,DAX,V12,V13,V14,V15, V16,V17,V18,V19,V20,V21,V22,V24,V25, V39	1
6	PAOZZ	4730011615115	15434	3013786	PLUG,PIPE .... UOC:DAA,DAB,DAC,DAD,DAE,DAF,DAG,DAH, DAJ,DAK,DAL,DAW,DAX,V12,V13,V1,VV15, V16,V17,V18,V19,V20,V21,V22,V24,V25, V39	1
7	PAFZZ	2930004043050	15434	202334	MANIFOLD,FLUID COOL .... UOC:DAA,DAB,DAC,DAD,DAE,DAF,DAG,DAH, DAJ,DAK,DAL,DAW,DAX,V12,V13,V14,V15, V16,V17,V18,V19,V20,V21,V22,V24,V25, V39	1
8	PAFZZ	5305010886019	15434	3010596	SCREW,ASSEMBLED WAS .... UOC:DAA,DAB,DAC,DAD,DAE,DAF,DAG,DAH, DAJ,DAK,DAL,DAW,DAX,V12,V13,V14,V15, V16,V17,V18,V19,V20,V21,V22,V24,V25, V39	12
9	PAOZZ	5330005372382	15434	70089-1	GASKET PART OF KIT P/N 3011472...................... UOC:DAA,DAB,DAC,DAD,DAE,DAF,DAG,DAH, DAJ,DAK,DAL,DAW,DAX,V12,V13,V14,V15, V16,V17,V18,V19,V20,V21,V22,V24,V25, V39	2

(1) ITEM NO	(2) SMR CODE	(3) NSN	(4) CAGEC	(5) PART NUMBER	(6) DESCRIPTION AND USABLE ON CODES (UOC)	(7) QTY
10	PBOZZ	5340007990843	15434	132019	COVER,ACCESS .... UOC:DAA,DAB,DAC,DAD,DAE,DAF,DAG,DAH, DAJ,DAK,DAL,DAW,DAX,V12,V13,V14,V15, V16,V17,V1B,V19,V20,V21,V22,V24,V25, V39	2
11	PAOZZ	5305011129021	15434	3013904	SCREW .... UOC:DAA,DAB,DAC,DAD,DAE,DAF,DAG,DAH, DAJ,DAK,DAL,DAW,DAX,V12,V13,V14,V15, V16,V17,V18,V19,V20,V21,V22,V24,V25, V39	12
12	PAOZZ	4820007529040	96906	MS35782-4	COCK,DRAIN .... UOC:DAA,DAB,DAC,DAD,DAE,DAF,DAG,DAH, DAJ,DAK,DAL,DAW,DAX,V12,V13,V14,V15, V16,V17,V18,V19,V20,V21,V22,V24,V25, V39	1
13	PAFZZ	5330011455381	15434	3024709	GASKET PART OF KIT P/N 3011472 PART.............. OF KIT P/N 3804272.. UOC:DAA,DAB,DAC,DAD,DAE,DAF,DAG,DAH, DAJ,DAK,DAL,DAW,DAX,V12,V13,V14,V15, V16,V17,V18,V19,V20,V21,V22,V24,V25, V39	6

END OF FIGURE

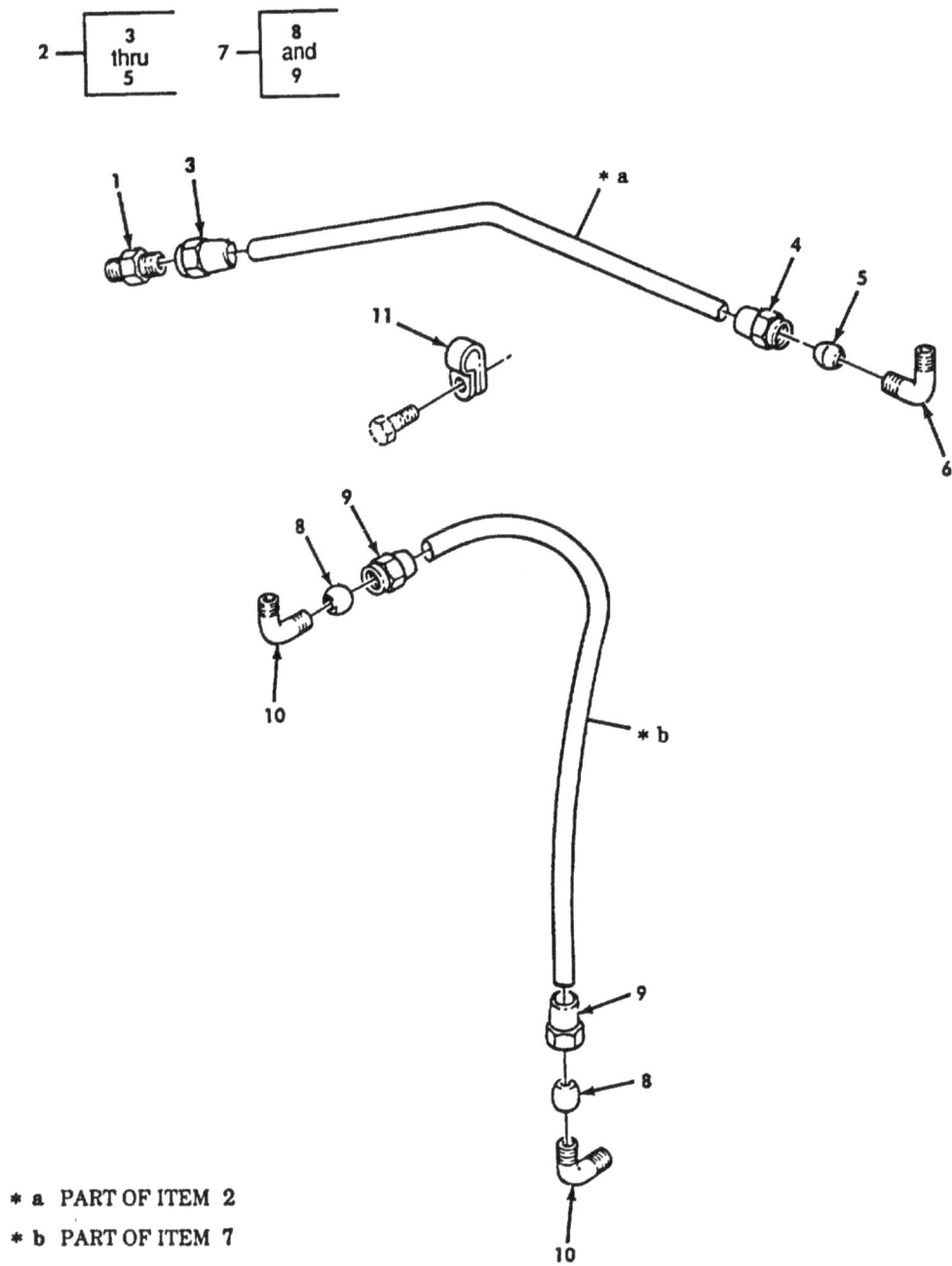

* a PART OF ITEM 2
* b PART OF ITEM 7

Figure 117. Air Compressor Coolant Lines and Related Parts (M939, M939A).

(1) ITEM NO	(2) SMR CODE	(3) NSN	(4) CAGEC	(5) PART NUMBER	(6) DESCRIPTION AND USABLE ON CODES (UOC)	(7) QTY
					GROUP 0503 WATER MANIFOLD, HEADERS, THERMOSTATS, AND HOUSING GASKET	
					FIG. 117 AIR COMPRESSOR COOLANT LINES AND RELATED PARTS(M939,M939A1)	
1	PAOZZ	4730003652690	15434	S1002A	ADAPTER,STRAIGHT,TU .... UOC:DAA,DAB,DAC,DAD,DAE,DAF,DAG,DAH, DAJ,DAK,DAL,DAW,DAX,V12,V13,V14,V15, V16,V17,V18 ,V19,V20,V21,V22,V24,V25, V39	1
2	MFOZZ	4720013510573	15434	204898	TUBING,NONMETALLIC MAKE FROM .... TUBING,P/N 8689210 .... UOC:DAA,DAB,DAC,DAD,DAE,DAF,DAG,DAH, DAJ,DAK,DAL,DAW,DAX,V12 ,V13 ,V14,V15, V16,V17,V18,V19,V20,V21,V22,V24,V25, V39	1
3	PAOZZ	4730000137401	72582	137401	.INVERTED NUT,TUBE C .... UOC:DAA,DAB,DAC,DAD,DAE,DAF,DAG,DAH, DAJ,DAK,DAL,DAW,DAX,V12,V13,V14,V15, V16,V17,V18,V19,V20,V21,V22,V24,V25, V39	1
4	PAOZZ	4730002895179	15434	S1004A	.NUT,TUBE COUPLING .... UOC:DAA,DAB,DAC,DAD,DAE,DAF,DAG,DAH, DAJ,DAK,DAL,DAW,DAX,V12,V13,V14,V15, V16,V17,V18,V19,V20,V21,V22,V24,V25, V39	1
5	PAOZZ	5365005985255	15434	S1003A	.BUSHING,NONMETALLIC .... UOC:DAA,DAB,DAC,DAD,DAE,DAF,DAG,DAH, DAJ,DAK,DAL,DAW,DAX,V12,V13,V14,V15, V16,V17,V18,V19,V20,V21,V22,V24,V25, V39	1
6	PAOZZ	4730012010717	15434	S1090	ELBOW,PIPE TO TUBE .... UOC:DAA,DAB,DAC,DAD,DAE,DAF,DAG,DAH, DAJ,DAK,DAL,DAW,DAX,V12,V13,V14,V15, V16,V17,V18,V19 ,V20,V21,V22,V24,V25, V39	1
7	MFOFF		15434	3031619	TUBE .... UOC:DAA,DAB,DAC,DAD,DAE,DAF,DAG,DAH, DAJ,DAK,DAL,DAW,DAX,V12 ,V13 ,V14 ,V15, V16,V17,V18,Vi9,V20,V21,V22,V24,V25, V39	1
8	PAOZZ	5365005985255	15434	S1003A	.BUSHING,NONMETALLIC AIR................... COMPRESSOR TUBE .... UOC:DAA,DAB,DAC,DAD,DAE,DAF,DAG,DAH, DAJ,DAK,DAL,DAW,DAX,V12,V13,V14,V15, V16,V17 ,V18 ,V19 ,V20,V21,V22,V24,V25, V39	2
9	PAOZZ	4730002895179	15434	S1004A	.NUT,TUBE COUPLING AIR COMPRESSOR .... TUBE .... UOC:DAA,DAB,DAC,DAD,DAE,DAF,DAG,DAH, DAJ,DAK,DAL,DAW,DAX,V12,V13,V14,V15,	2

(1) ITEM NO	(2) SMR CODE	(3) NSN	(4) CAGEC	(5) PART NUMBER	(6) DESCRIPTION AND USABLE ON CODES (UOC)	(7) QTY
					V16,V17,V18,V19,V20,V21,V22,V24,V25, V39	
10	PAOZZ	4730003744282	15434	S-1005-A	ELBOW,PIPE TO TUBE .... UOC:DAA,DAB,DAC,DAD,DAE,DAF,DAG,DAH, DAJ,DAK,DAL,DAW,DAX,V12 ,V13,V14,V15, V16,V17 ,18,V19,V20,V21,V22,V24,V25, V39	2
11	PAOZZ	5340002861868	15434	68152	CLAMP,LOOP .... UOC:DAA,DAB,DAC,DAD,DAE,DAF,DAG,DAH, DAJ,DAK,DAL,DAW,DAX,V12,V13,V14,V15, V16,V17,V18,V19,V20,V21,V22,V24,V25, V39	1

END OF FIGURE

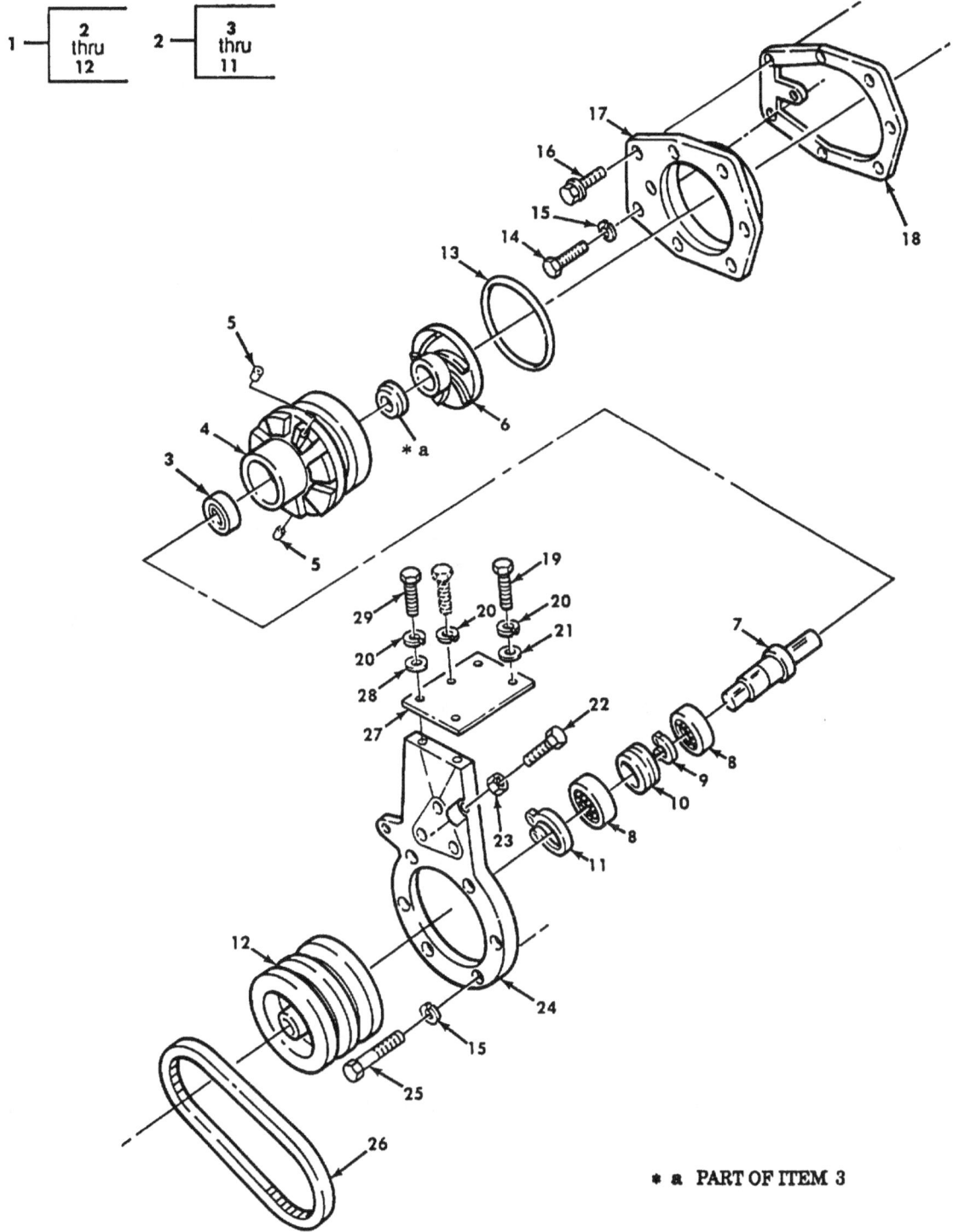

Figure 118. Water Pump Assembly, Seals, and Mounting Hardware (M939, M939A).

* a   PART OF ITEM 3

(1) ITEM NO	(2) SMR CODE	(3) NSN	(4) CAGEC	(5) PART NUMBER	(6) DESCRIPTION AND USABLE ON CODES (UOC)	(7) QTY
					GROUP 0504 WATER PUMP	
					FIG. 118 WATER PUMP ASSEMBLY, SEALS, AND MOUNTING HARDWARE(M939, M939A1)	
1	PAOFF	2930008902440	15434	AR4284	PUMP,COOLING SYSTEM .... UOC:DAA,DAB,DAC,DAD,DAE,DAF,DAG,DAH, DAJ,DAK,DAL,DAW,DAX,V12,V13,V14,V15, V16,V17,V18,V19,V20,V21,V22,V24,V25, V39	1
2	PAFZZ	2930000048420	98349	FP1411	.PUMP,WATER,ENGINE C .... UOC:DAA,DAB,DAC,DAD,DAE,DAF,DAG,DAH, DAJ,DAK,DAL,DAW,DAX,V12,V13,V14,V15, V16,V17,V18,V19,V20,V21,V22,V24,V25, V39	1
3	PAFZZ	5330000050407	15434	3071085	..PACKING WITH RETAIN .... UOC:DAA,DAB,DAC,DAD,DAE,DAF,DAG,DAH, DAJ,DAK,DAL,DAW,DAX,V12,V13,V14,V15, V16,V17,V18,V19,V20,V21,V22,V24,V25, V39	1
4	XAFZZ		15434	196845	..BODY.. UOC:DAA,DAB,DAC,DAD,DAE,DAF,DAG,DAH, DAJ,DAK,DAL,DAW,DAX,V12,V13,V14,V15, V16,V17,V18,V19,V20,V21,V22,V24,V25, V39	1
5	PAFZZ	4730000113175	15434	70295	..PLUG,PIPE... UOC:DAA,DAB,DAC,DAD,DAE,DAF,DAG,DAH, DAJ,DAK,DAL,DAW,DAX,V12,V13,V14,V15, V16,V17,V18,V19,V20,V21,V22,V24,V25, V39	2
6	PAFZZ	2930011342192	15434	3020479	..IMPELLER, WATER PUM .... UOC:DAA,DAB,DAC,DAD,DAE,DAF,DAG,DAH, DAJ,DAK,DAL,DAW,DAX,V12,V13,V14,V15, V16,V17,V18,V19,V20,V21,V22,V24,V25, V39	1
7	PAFZZ	3040011343480	15434	199410	..SHAFT,SHOULDERED.............................. UOC:DAA,DAB,DAC,DAD,DAE,DAF,DAG,DAH, DAJ,DAK,DAL,DAW,DAX,V12 ,V13 ,V14 ,V15, V16,V17,V18,V19,V20,V21,V22,V24,V25, V39	1
8	PAFZZ	3110001077564	15434	115519	..BEARING,BALL,ANNULA .... UOC:DAA,DAB,DAC,DAD,DAE,DAF,DAG,DAH, DAJ,DAK,DAL,DAW,DAX,V12 ,V13 ,V14 ,V15, V16,V17,V18,V19,V20,V21,V22,V24,V25, V39	2
9	PAFZZ	5325004209696	15434	112302	..RING,RETAINING .... UOC:DAA,DAB,DAC,DAD,DAE,DAF,DAG,DAH, DAJ,DAK,DAL,DAW,DAX,V12 ,V13 ,V14 ,V15, V16,V17,V18,V19,V20,V21,V22,V24,V25, V39	1
10	PAFZZ	365011321984	15434	196844	..SPACER.. UOC:DAA,DAB,DAC,DAD,DAE,DAF,DAG,DAH,	1

(1) ITEM NO	(2) SMR CODE	(3) NSN	(4) CAGEC	(5) PART NUMBER	(6) DESCRIPTION AND USABLE ON CODES (UOC)	(7) QTY
					DAJ,DAK,DAL,DAW,DAX,V12,V13,V14,V15, V16,V17,V18,V19,V20,V21,V22,V24,V25, V39	
11	PAFZZ	2815008150355	15434	S-16255	..RING,BEARING RETAIN .... UOC:DAA,DAB,DAC,DAD,DAE,DAF,DAG,DAH, DAJ,DAK,DAL,DAW,DAX,V12,V13,V14,V15, V16,V17,V18,V19,V20,V21,V22,V24,V25, V39	1
12	PAFZZ	3020009290713	15434	154966	.PULLEY,GROOVE .WATER PUMP .... UOC:DAA,DAB,DAC,DAD,DAE,DAF,DAG,DAH, DAJ,DAK,DAL,DAW,DAX,V12,V13,V14,V15, V16,V17,V18,V19,V20,V21,V22,V24,V25, V39	1
13	PAFZZ	5330001066369	15434	130240	GASKET PART OF KIT P/N 3011472 .... UOC:DAA,DAB,DAC,DAD,DAE,DAF,DAG,DAH, DAJ,DAK,DAL,DAW,DAX,V12,V13,V14 ,V15, V16,V17,V18,V19,V20,V21,V22,V24,V25, V39,	1
14	PAFZZ	5305012101608	15434	106549	SCREW,CAP,HEXAGON H .... UOC:DAA,DAB,DAC,DAD,DAE,DAF,DAG,DAH, DAJ,DAK,DAL,DAW,DAX,V12,V13,V14,V15, V16,V17,V18,V19,V20,V21,V22,V24,V25, V39	1
15	PAFZZ	5310002617340	15434	S-604	WASHER,LOCK .... UOC:DAA,DAB,DAC,DAD,DAE,DAF,DAG,DAH, DAJ,DAK,DAL,DAW,DAX,V12,V13,V14,V15, V16,V17,V18,V19,V20,V21,V22,V24,V25, V39	1
16	PAFZZ	5305011306100	5434	3010594	SCREW,CAP,HEXAGON H .... UOC:DAA,DAB,DAC,DAD,DAE,DAF,DAG,DAH, DAJ,DAK,DAL,DAW,DAX,V12,V13,V14,V15, V16,V17,V18,V19,V20,V21,V22,V24,V25, V39	1
17	PBFZZ	2930004089404	15434	130227	SUPPORT,PUMP,WATER...................... UOC:DAA,DAB,DAC,DAD, DAE,DAF,DAG,DAH, DAJ,DAK,DAL,DAW,DAX,V12,V13,V14,V15, V16,V17,V18,V19,V20,V21,V22,V24,V25, V39	1
18	PAFZZ	5330001066370	15434	130226	GASKET WATER PUMP SUPPORT PART OF .... KIT P/N 3011472........................... UOC: DAA,DAB,DAC,DAD,DAE,DAF,DAG,DAH, DAJ,DAK,DAL,DAW,DAX,V12,V13,V14,V15, V16,V17,V18,V19,V20,V21,V22,V24,V25, V39	1
19	PAFZZ	5305007959343	15434	S-176	SCREW,CAP,HEXAGON H .... UOC:DAA,DAB,DAC,DAD,DAE,DAF,DAG,DAH, DAJ,DAK,DAL,DAW,DAX,V12,V13,V14,V15, V16,V17,V18,V19,V20,V21,V22,V24,V25, V39	2
20	PAFZZ	5310000034094	01276	210104-8S	WASHER,LOCK .... UOC:DAA,DAB,DAC,DAD,DAE,DAF,DAG,DAH, DAJ,DAK,DAL,DAW,DAX,V12,V13,V14,V15,	7

(1) ITEM NO	(2) SMR CODE	(3) NSN	(4) CAGEC	(5) PART NUMBER	(6) DESCRIPTION AND USABLE ON CODES (UOC)	(7) QTY
					V16,V17,V18,V19,V20,V21,V22,V24,V25, V39	
21	PAFZZ	5310009307013	11083	4B4280	WASHER,FLAT.. UOC:DAA,DAB,DAC,DAD,DAE,DAF,DAG,DAH, DAJ,DAK,DAL,DAW,DAX,V12 ,V13 ,V14,V15, V16,V17,V18,V19,V20,V21,V22,V24,V25, V39	2
22	PAOZZ	5305010912498	15434	166777	SCREW.. UOC:DAA,DAB,DAC,DAD,DAE,DAF,DAG,DAH, DAJ,DAK,DAL,DAW,DAX,V12,V13,V14,V15, V16,V17,V18,V19,V20,V21,V22,V24,V25, V39	1
23	PAOZZ	5310004706154	15434	S-285	NUT,PLAIN,HEXAGON .... UOC:DAA,DAB,DAC,DAD,DAE,DAF,DAG,DAH, DAJ,DAX,DAL,DAW,DAX,V12,V13,V14,V15, V16,V17,V18,V19,V20,V21,V22,V24,V25, V39	1
24	BFZZ	5342011343860	15434	147588	BRACKET,FAN SUPPORT .... UOC:DAA,DAB,DAC,DAD , DAE,DAF,DAG,DAH, DAJ,DAK,DAL,DAW,DAX,V12,V13,V14,V15, V16,V17,V18,V19,V20,V21,V22,V24,V25, V39	1
25	PAFZZ	5305009146131	96906	MS18153-63	SCREW,CAP,HEXAGON H .... UOC:DAA,DAB,DAC,DAD,DAE,DAF,DAG,DAH, DAJ,DAK,DAL,DAW,DAX,V12,V13,V14,V15, V16,V17,V18,V19,V20,V21,V22,V24,V25, V39	6
26	PAOZZ	3030012006004	15434	3019061	BELT,V .... UOC:DAA,DAB,DAC,DAD,DAE,DAF,DAG,DAH, DAJ,DAK,DAL,DAW,DAX,V12,V13,V14,V15, V16,V17,V18,V19,V20,V21,V22,V24,V25, V39	1
27	PAFZZ	2930011967521	15434	131622	BRACE,FAN SUPPORT .... UOC:DAA,DAB,DAC,DAD,DAE,DAF,DAG,DAH, DAJ,DAK,DAL,DAW,DAX,V12,V13,V14,V15, V16,V17,V18,V19,V20,V21,V22,V24,V25, V39	1
28	PAFZZ	5310005956153	11083	1F7960	WASHER,FLAT .... UOC:DAA,DAB,DAC,DAD,DAE,DAF,DAG,DAH, DAJ,DAK,DAL,DAW,DAX,V12,V13,V14,V15, V16,V17,V18,V19,V20,V21,V22,V24,V25, V39	2
29	PAFZZ	5305012097068	15434	69952	SCREW,CAP,HEXAGON H .... UOC:DAA,DAB,DAC,DAD,DAE,DAF,DAG,DAH, DAJ,DAK,DAL,DAW,DAX,V12,V13,V14,V15, V16,V17,V18,V19,V20,V21,V22,V24,V25, V39	2

END OF FIGURE

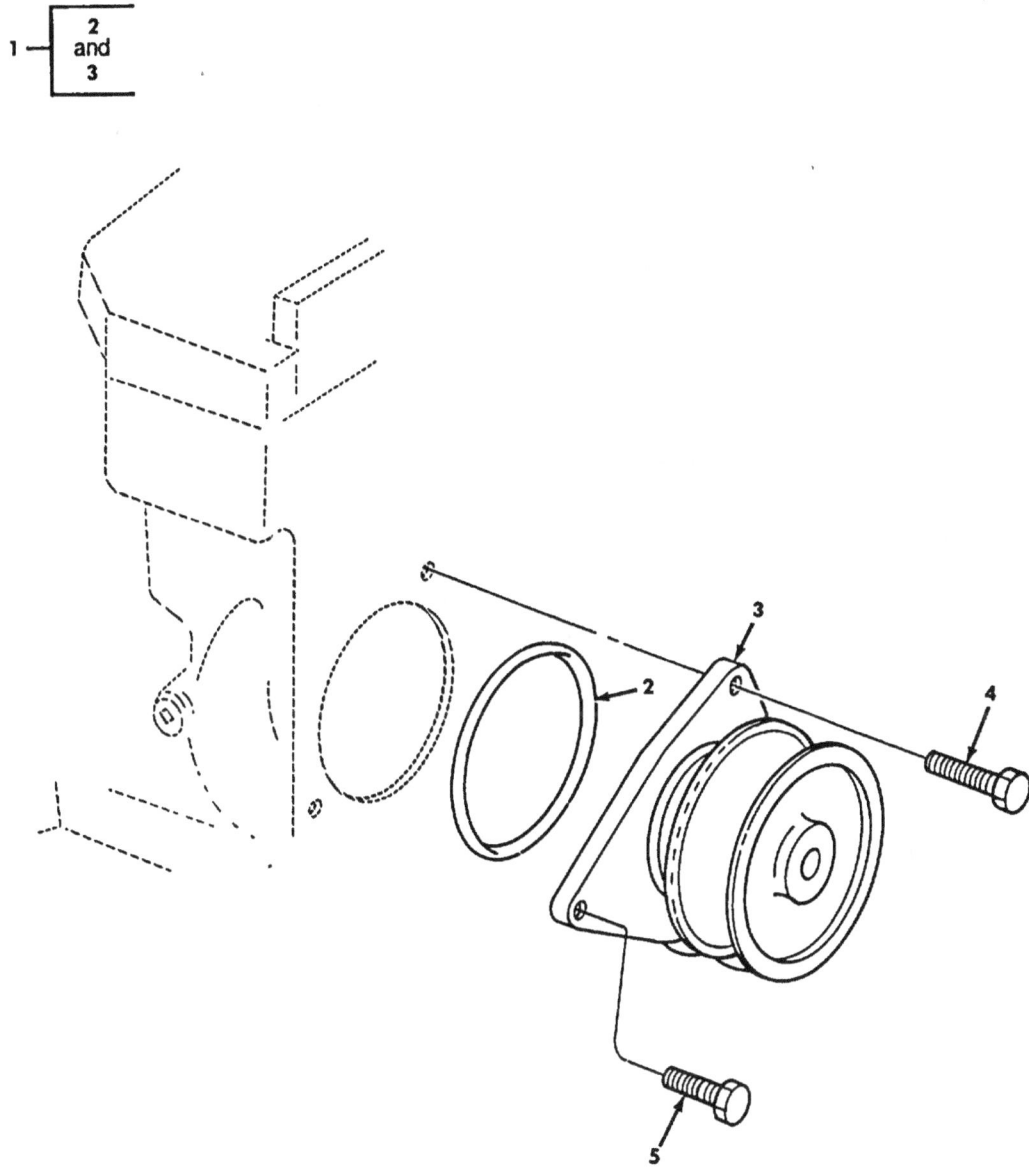

Figure 119. Water Pump (M939A2).

(1) ITEM NO	(2) SMR CODE	(3) NSN	(4) CAGEC	(5) PART NUMBER	(6) DESCRIPTION AND USABLE ON CODES (UOC)	(7) QTY
					GROUP 0504 WATER PUMP	
					FIG. 119 WATER PUMP (M939A2)	
1	PAOZZ	2930012688751	15434	3802081	PUMP,COOLING SYSTEM .... UOC:ZAA,ZAB,ZAC,ZAD,ZAE,ZAF,ZAG,ZAH, ZAJ,ZAKX,ZAL	1
2	PAOZZ	5331012721123	15434	3902089	. O-RING PART OF KIT P/N 3802389 PART.......... OF KIT P/N 3802389 UOC:ZAA,ZAB,ZAC,ZAD,ZAE,ZAF,ZAG,ZAH, ZAJ,ZAK,ZAL	1
3	XAOZZ	2930012688751	15434	3802081	. PUMP,COOLING SYSTEM .... UOC:ZAA,ZAB,ZAC,ZAD,ZAE,ZAF,ZAG,ZAH, ZAJ,ZAK,ZAL	1
4	PAOZZ	5305012077447	15434	3900677	SCREW,CAP,HEXAGON H................................. UOC:ZAA,ZAB,ZAC,ZAD,ZAE,ZAF,ZAG,ZAH, ZAJ, ZAK, ZAL	1
5	PAOZZ	5305012453193	15434	3900621	SCREW,CAP,HEXAGON H .... UOC:ZAA,ZAB,ZAC,ZAD,ZAE,ZAF,ZAG,ZAH, ZAJ,ZAK,ZAL	4

END OF FIGURE

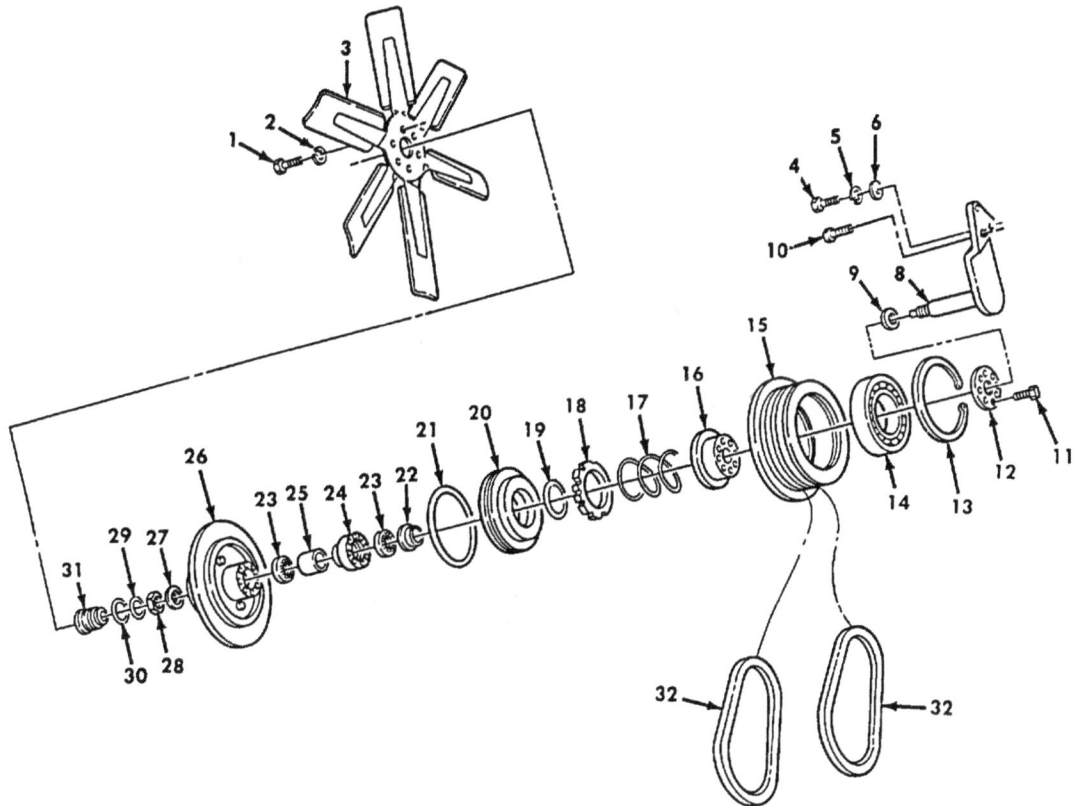

Figure 120. Fan Clutch Assembly, Fan, and Belts (M939, M939A1).

(1) ITEM NO	(2) SMR CODE	(3) NSN	(4) CAGEC	(5) PART NUMBER	(6) DESCRIPTION AND USABLE ON CODES (UOC)	(7) QTY
					GROUP 0505 FAN ASSEMBLY	
					FIG. 120 FAN CLUTCH ASSEMBLY, FAN, AND BELTS(M939,M939A1)	
1	PAOZZ	5305012104595	97403	13222E0109	SCREW,ROLLER ASSEMB .... UOC:DAA,DAB,DAC,DAD,DAE,DAF,DAG,DAH, DAJ,DAK,DAL,DAW,DAX,V12,V13,V14,V15, V16,V17,V18,V19,V20,V21,V22,V24,V25, V39	6
2	PAOZZ	5310006379541	96906	MS35338-46	WASHER,LOCK .... UOC:DAA,DAB,DAC,DAD,DAE,DAF,DAG,DAH, DAJ,DAK,DAL,DAW,DAX,V12,V13,V14,V15, V16,V17,V18,V19,V20,V21,V22,V24,V25, V39	6
3	PAOZZ	4140010893058	19207	11669185	IMPELLER,FAN,AXIAL RIGHT HAND .... SUCKER UOC:DAA,DAB,DAC,DAD,DAE,DAF,DAG,DAH, DAJ,DAK,DAL,DAW,DAX,V12,V13,V14,V15, V16,V17,V18,V19,V20,V21,V22,V24,V25, V39	1
4	PAOZZ	5305007195235	80204	B1821BH050F175N	SCREW,CAP,HEXAGON H .... UOC:DAA,DAB,DAC,DAD,DAE,DAF,DAG,DAH, DAJ,DAK,DAL,DAW,DAX,V12,V13,V14,V15, V16,V17,V18,V19,V20,V21,V22,V24,V25, V39	3
5	PAOZZ	5310000034094	01276	210104-8S	WASHER,LOCK .... UOC:DAA,DAB,DAC,DAD,DAE,DAF,DAG,DAH, DAJ,DAK,DAL,DAW,DAX,V12,V13,V14,V15, V16,V17,V18,V19,V20,V21,V22,V24,V25, V39	3
6	PAOZZ	5310011211703	06032	2310-0143-001	WASHER,FLAT .... UOC:DAA,DAB,DAC,DAD,DAE,DAF,DAG,DAH, DAJ,DAK,DAL,DAW,DAX,V12,V13,V14,V15, V16,V17,V18,V19,V20,V21,V22,V24,V25, V39	2
7	PAOFF	2520011152285	21102	FC212000	CLUTCH ASSEMBLY,FRI ENGINE COOLING. .... UOC:DAA,DAB,DAC,DAD,DAE,DAF,DAG,DAH, DAJ,DAK,DAL,DAW,DAX,V12,V13,V14,V15, V16,V17,V18,V19,V20,V21,V22,V24,V25, V39	1
8	PBFZZ	3020011246079	21102	FC200007	BRACKET AND SHAFT A .... UOC:DAA,DAB,DAC,DAD,DAE,DAF,DAG,DAH, DAJ,DAK,DAL,DAW,DAX,V12,V13,V14,V15, V16,V17,V18,V19,V20,V21,V22,V24,V25, V39	1
9	PAFZZ	5365011265192	21102	FC030002	. SPACER,RING................................. UOC:DAA,DAB,DAC,DAD,DAE,DAF,DAG,DAH, DAJ,DAK,DAL,DAW,DAX,V12,V13,V14,V15, V16,V17,V18,V19,V20,V21,V22,V24,V25, V39	1
10	PAFZZ	5305011616000	21102	FD613	. SCREW,CAP,HEXAGON H	2

(1) ITEM NO	(2) SMR CODE	(3) NSN	(4) CAGEC	(5) PART NUMBER	(6) DESCRIPTION AND USABLE ON CODES (UOC)	(7) QTY
					UOC:DAA,DAB,DAC,DAD,DAE,DAF,DAG,DAH, DAJ,DAK,DAL,DAW,DAX,V12,V13,V14,V15, V16,V17,V18,V19,V20,V2 1,V22,V24,V25, V39	
11	PAFZZ	5305011208438	21102	FC030011	. SCREW,CAP,SOCKET HE	8
					UOC:DAA,DAB,DAC,DAD,DAE,DAF,DAG,DAH, DAJ,DAK,DAL,DAW,DAX,V12,V13,V14,V15, V16,V17,V18,V19,V20,V21,V22,V24,V25, V39	
12	PAFZZ	2930011225982	21102	FC030010	.PLATE,FAN,THRUST CA	1
					UOC:DAA,DAB,DAC,DAD,DAE,DAF,DAG,DAH, DAJ,DAK,DAL,DAW,DAX,V12,V13,V14,V15, V16,V17,V18,V19,V20,V21,V22,V24,V25, V39	
13	PAFZZ	5325009145837	79136	N5002-500MD	. RING,RETAINING	1
					UOC:DAA,DAB,DAC,DAD,DAE,DAF,DAG,DAH, DAJ,DAK,DAL,DAW,DAX,V12,V13,V14,V15, V16,V17,V18,V19,V20,V21,V22,V24,V25, V39	
14	PAFZZ	3110011261287	52676	6017-2RS	. BEARING,BALL,ANNULA	1
					UOC:DAA,DAB,DAC,DAD,DAE,DAF,DAG,DAH, DAJ,DAK,DAL,DAW,DAX,V12,V13,V14,V15, V16,V17,V18,V19,V20,V21,V22,V24,V25, V39	
15	PAFZZ	3020011249417	21102	FC200006	. PULLEY,CONE	1
					UOC:DAA,DAB,DAC,DAD,DAE,DAF,DAG,DAH, DAJ,DAK,DAL,DAW,DAX,V12,V13,V14,V15, V16,V17,V18,V19,V20,V21,V22,V24,V25, V39	
16	PAFZZ	2930011336948	21102	FC030035	. CAP,THRUST,FAN CLUT	1
					UOC:DAA,DAB,DAC,DAD,DAE,DAF,DAG,DAH, DAJ,DAK,DAL,DAW,DAX,V12,V13,V14,V15, V16,V17,V18,V19,V20,V21,V22,V24,V25, V39	
17	PAFZZ	5360011243373	21102	FD019	. SPRING,HELICAL,COMP	1
					UOC:DAA,DAB,DAC,DAD,DAE,DAF,DAG,DAH, DAJ,DAK,DAL,DAW,DAX,V12,V13,V14,V15, V16,V17,V18,V19,V20,V21,V22,V24,V25, V39	
18	PAFZZ	2520010774009	21102	FC030009	. LINING,FRICTION	1
					UOC:DAA,DAB,DAC,DAD,DAE,DAF,DAG,DAH, DAJ,DAK,DAL,DAW,DAX,V12,V13,V14,V15, V16,V17,V18,V19,V20,V21,V22,V24,V25, V39	
19	PAFZZ	5331012719374	21102	FD145A	. O-RING	1
					UOC:DAA,DAB,DAC,DAD,DAE,DAF,DAG,DAH, DAJ,DAK,DAL,DAW,DAX,V12,V13,V14,V15, V16,V17,V18,V19,V20,V21,V22,V24,V25, V39	
20	PAFZZ	2520010888172	34623	MA207-22-742	. PISTON..	1
					UOC:DAA,DAB,DAC,DAD,DAE,DAF,DAG,DAH, DAJ,DAK,DAL,DAW,DAX,V12,V13,V14,V15,	

(1) ITEM NO	(2) SMR CODE	(3) NSN	(4) CAGEC	(5) PART NUMBER	(6) DESCRIPTION AND USABLE ON CODES (UOC)	(7) QTY
					V16,V17,V18,V19,V20,V21,V22,V24,V25, V39	
21	PAFZZ	5330010796513	21102	FD0077	. O-RING .... UOC:DAA,DAB,DAC,DAD,DAE,DAF,DAG,DAH, DAJ,DAK,DAL,DAW,DAX,V12,V13,V14,V15, V16,V17 ,V18 ,V19 ,V20 ,V21,V22,V24,V25, V39	1
22	PAFZZ	5365011231638	21102	FD0202	. SPACER,CLUTCH.................................... UOC: DAA,DAB, DAC ,DAD,DAE,DAF,DAG, DAH, DAJ,DAK,DAL,DAW,DAX,V12 ,V13 ,V14 ,V15, V16,V17,V18,V19,V20,V21,V22,V24,V25, V39	1
23	PAFZZ	3110001091179	19207	10951608	. BEARING,BALL,ANNULA .... UOC:DAA,DAB,DAC,DAD,DAE,DAF,DAG,DAH, DAJ,DAK,DAL,DAW,DAX,V12 ,V13 ,V14 ,V15, V16,V17,V18,V19,V20,V21,V22,V24,V25, V39	2
24	PAFZZ	5365011231638	21102	FD0202	. SPACER,CLUTCH .... UOC:DAA,DAB,DAC,DAD,DAE,DAF,DAG,DAH, DAJ,DAK,DAL,DAW,DAX,V12,V13,V14,V15, V16,V17,V18,V19,V20,V21,V22,V24,V25, V39	1
25	PAFZZ	5365011268689	21102	FC030029	. SPACER,SLEEVE .... UOC:DAA,DAB,DAC,DAD,DAE,DAF,DAG,DAH, DAJ,DAKX,DAL,DAW,DAX,V12 ,V13 ,V14,V15, V16,V17,V18,V19,V20,V21,V22,V24,V25, V39	1
26	PAFZZ	2520011238704	91929	FC200008	. HOUSING,CLUTCH .... UOC:DAA,DAB,DAC,DAD,DAE,DAF,DAG,DAH, DAJ,DAK,DAL,DAW,DAX,V12 ,V13 ,V14 ,V15, V16,V17,V18,V19,V20,V21,V22,V24,V25, V39	1
27	PAFZZ	5310010797036	21102	FD70A	. WASHER,RECESSED .... UOC: DAA,DAB, DAC,DAD, DAD , DAF, DAG ,DAH, DAJ,DAK,DAL,DAW,DAX,V12 ,V13 ,V14 ,V15, V16,V17,V18,V19,V20,V21,V22,V24,V25, V39	1
28	PAFZZ	5310012708342	21102	FC030005	. NUT,SELF-LOCKING,HE .... UOC:DAA,DAB,DAC,DAD,DAE,DAF,DAG,DAH, DAJ,DAK,DAL,DAW,DAX,V12 ,V13 ,V14 ,V15, V16,V17,V18,V19,V20,V21,V22,V24,V25, V39	1
29	PAFZZ	5331012719372	21102	FC030020	. PACKING,PREFORMED .... UOC:DAA,DAB,DAC,DAD,DAE,DAF,DAG,DAH, DAJ,DAK,DAL,DAW,DAX,V12,V13,V14,V15, V16,V17,V18,V19,V20,V21,V22,V24,V25, V39	1
30	PAFZZ	5330012719371	21102	FC030021	. O-RING .... UOC:DAA,DAB,DAC,DAD,DAE,DAF,DAG,DAH, DAJ,DAK,DAL,DAW,DAX V12 ,V13 ,V14 ,V15, V16,V17,V1B,V19,V20,V21,V22,V24,V25, V39	1

(1) ITEM NO	(2) SMR CODE	(3) NSN	(4) CAGEC	(5) PART NUMBER	(6) DESCRIPTION AND USABLE ON CODES (UOC)	(7) QTY
31	PAFZZ	5365011345534	21102	FC212143	. PLUG,MACHINE THREAD .... 1 UOC:DAA,DAB,DAC,DAD,DAE,DAF,DAG ,DAH, DAJ,DAK,DAL,DAW,DAX,V12,V13,V14,V15, V16,V17,V18,V19,V20,V21,V22,V24,V25, V39	
32	PAOZZ	3030011181318	96906	MS51069RC49-2	BELTS,V,MATCHED SET .... 1 UOC:DAA,DAB,DAC,DAD,DAE,DAF,DAG,DAH, DAJ,DAK,DAL,DAW,DAX,V12,V13,V14,V15, V16,V17,V18,V19,V20,V21,V22,V24,V25, V39	

END OF FIGURE

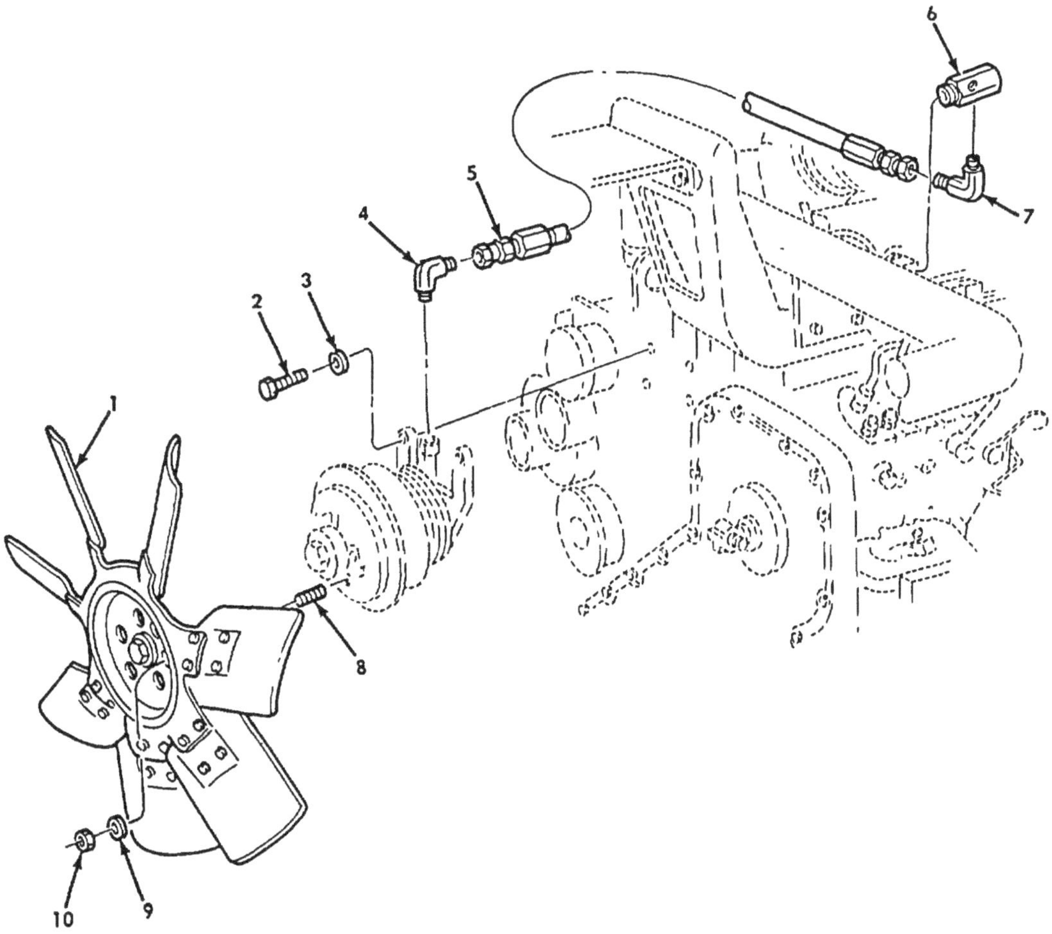

Figure 121. Fan Assembly (M939A2).

(1) ITEM NO	(2) SMR CODE	(3) NSN	(4) CAGEC	(5) PART NUMBER	(6) DESCRIPTION AND USABLE ON CODES (UOC)	(7) QTY
					GROUP 0505 FAN ASSEMBLY	
					FIG. 121 FAN ASSEMBLY (M939A2)	
1	PAOZZ	2930012868357	74080	219132	IMPELLER,FAN,AXIAL .... UOC:ZAA,ZAB,ZAC,ZAD,ZAE,ZAF,ZAG,ZAH, ZAJ,ZAK,ZAL	1
2	PAOZZ	5306012761601	24617	11500869	BOLT,MACHINE .... UOC:ZAA,ZAB,ZAC,ZAD,ZAE,ZAF,ZAG,ZAH, ZAJ,ZAK,ZAL	4
3	PAOZZ	5310012865452	24617	11511514	WASHER,FLAT.. UOC:ZAA,ZAB,ZAC,ZAD,ZAE,ZAF,ZAG,ZAH, ZAJ, ZAK,ZAL	4
4	PAOZZ	4730012838150	93061	159F-4-4	ELBOW,PIPE TO TUBE.. UOC:ZAA,ZAB,ZAC,ZAD,ZAE,ZAF,ZAG,ZAH, ZAJ,ZAK,ZAL	1
5	PAOZZ	4720012848184	98441	A3915-4	HOSE ASSEMBLY,NONME .... UOC:ZAA,ZAB,ZAC,ZAD,ZAE,ZAF,ZAG,ZAH, ZAJ,ZAK,ZAL	1
6	PAOZZ	2930012855027	21102	CA200011	ACTUATOR,FAN CLUTCH .... UOC:ZAA,ZAB,ZAC,ZAD,ZAE,ZAF,ZAG,ZAH, ZAJ,ZAK,ZAL	1
7	PAOZZ	4730013155596	1Z155	FRS-4-100	ELBOW,PIPE TO TUBE... UOC:ZAA,ZAB,ZAC,ZAD,ZAE,ZAF,ZAG,ZAH, ZAJ,ZAK,ZAL	1
8	PAOZZ	5305012838462	5Y952	670263	SETSCREW .... UOC:ZAA,ZAB,ZAC,ZAD,ZAE,ZAF,ZAG,ZAH, ZAJ,ZAK,ZAL	6
9	PAOZZ	5310002748041	27618	12084P11	WASHER,FLAT.. UOC:ZAA,ZAB,ZAC,ZAD,ZAE,ZAF,ZAG,ZAH, ZAJ,ZAK,ZAL	6
10	PAOZZ	5310011269404	24617	9422277	NUT,SELF-LOCKING,HE .... UOC:ZAA,ZAB,ZAC,ZAD,ZAE,ZAF,ZAG,ZAH, ZAJ,ZAK,ZAL	6

END OF FIGURE

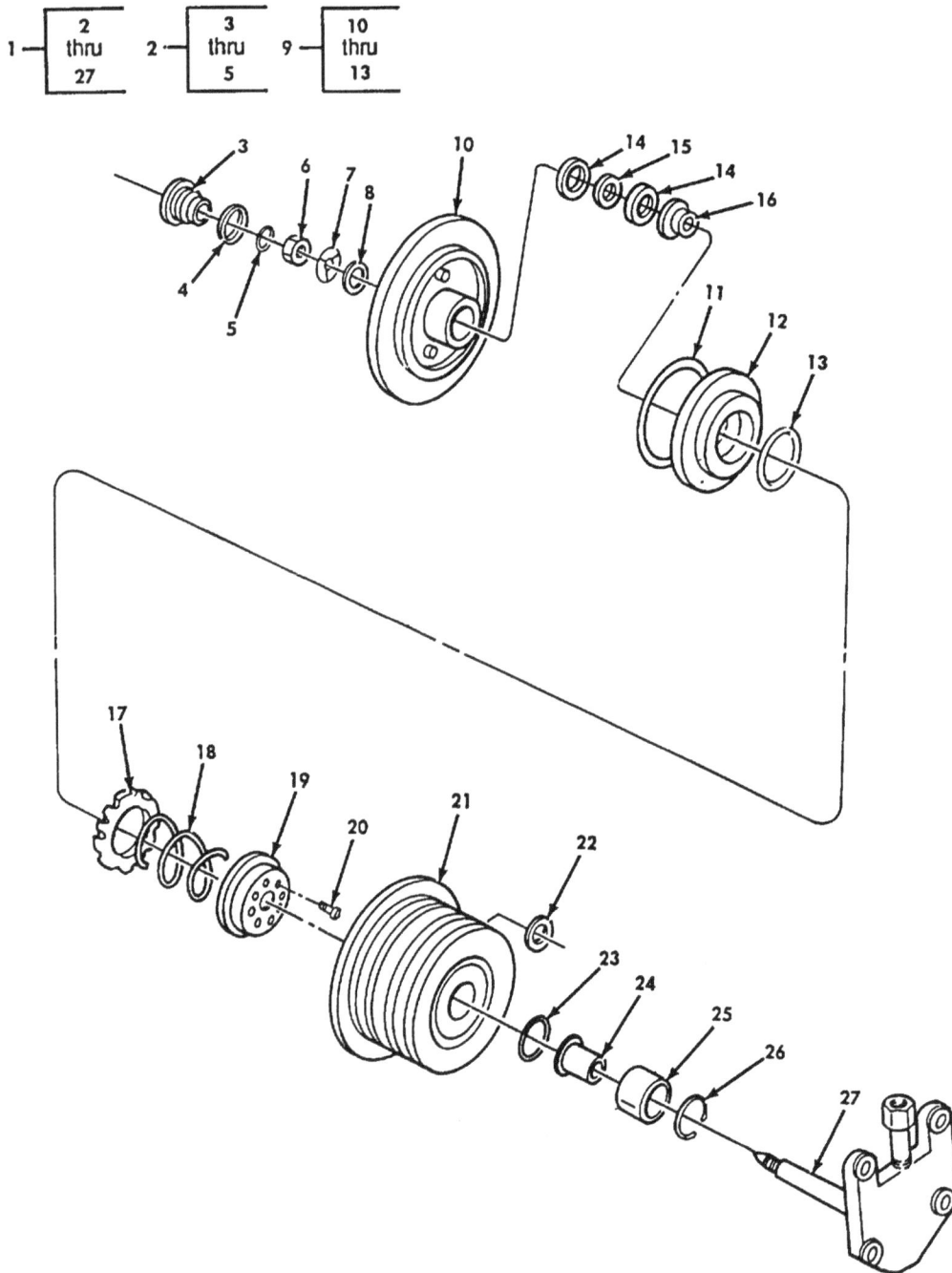

Figure 122. Evans Fan Clutch Assembly (M939A2)

(1) ITEM NO	(2) SMR CODE	(3) NSN	(4) CAGEC	(5) PART NUMBER	(6) DESCRIPTION AND USABLE ON CODES (UOC)	(7) QTY
					GROUP 0505 FAN ASSEMBLY	
					FIG. 122 FAN DRIVE CLUTCH ASSEMBLY (M939A2)	
1	PAOFF	2930012688752	62038	SD-3835300	CLUTCH,FAN,ENGINE(HORTON) .... UOC:ZAA,ZAB,ZAC,ZAD,ZAE,ZAF,ZAG,ZAH, ZAJ,ZAK,ZAL	1
1	PAOFF	2930012688752	21102	FC200158	CLUTCH,FAN,ENGINE(EVANS).EVANS PARTS ...... FOLLOW... UOC:ZAA,ZAB,ZAC,ZAD,ZAE,ZAF,ZAG,ZAH, ZAJ,ZAK,ZAL	1
2	PAFZZ	5340012721038	21102	FC200002	. CAP,SPECIAL PART OF KIT P/N .... FC212028... UOC:ZAA,ZAB,ZAC,ZAD,ZAE,ZAF,ZAG,ZAH, ZAJ,ZAK,ZAL	1
3	XAFZZ		21102	FC200002 XA	. .CAP .... UOC:ZAA,ZAB,ZAC,ZAD,ZAE,ZAF,ZAG,ZAH, ZAJ,ZAK,ZAL	1
4	PAFZZ	5330012719371	21102	FC030021	.. O-RING PART OF KIT P/N FC212028 .... UOC:ZAA,ZAB,ZAC,ZAD,ZAE,ZAF,ZAG,ZAH, ZAJ,ZAK,ZAL	1
5	PAFZZ	5330012719372	21102	FC030020	.. PACKING,PREFORMED PART OF KIT P/N.......... FC212028... UOC:ZAA,ZAB,ZAC,ZAD,ZAE,ZAF,ZAG,ZAH, ZAJ,ZAK,ZAL	1
6	PAFZZ	5310012708342	21102	FC030005	. NUT,SELF-LOCKING,HE PART OF KIT P/N.......... FC212028... UOC:ZAA,ZAB,ZAC,ZAD,ZAE,ZAF,ZAG,ZAH, ZAJ,ZAK,ZAL	1
7	PAFZZ	5310012711796	21102	FC030004	. WASHER,SPRING TENSI PART OF KIT P/N.......... FC212028... UOC:ZAA,ZAB,ZAC,ZAD,ZAE,ZAF,ZAG,ZAH, ZAJ,ZAK,ZAL	1
8	PAFZZ	5365012708472	21102	FC030003	. SPACER,RING PART OF KIT P/N........................ FC212028... UOC:ZAA,ZAB,ZAC,ZAD,ZAE,ZAF,ZAG,ZAH, ZAJ,ZAK,ZAL	1
9	PAFZZ	2520010888172	21102	FC200008	. PISTON .... UOC:ZAA,ZAB,ZAC,ZAD,ZAE,ZAF,ZAG,ZAH, ZAJ,ZAK,ZAL	1
10	XAFZZ	21102		FC200008 XA	.. PISTON .... UOC:ZAA,ZAB,ZAC,ZAD,ZAE,ZAF,ZAG,ZAH, ZAJ,ZAK,ZAL	1
11	PAFZZ	5330012719373	21102	FC030027	..O-RING PART OF KIT P/N FC212028 .... UOC:ZAA,ZAB,ZAC,ZAD,ZAE,ZAF,ZAG,ZAH, ZAJ,ZAK,ZAL	1
12	PAFZZ	2520010888172	21102	FC030025	.. PISTON .... UOC:ZAA,ZAB,ZAC,ZAD,ZAE,ZAF,ZAG,ZAH, ZAJ,ZAK,ZAL	1
13	PAFZZ	5330012719374	21102	FC030028	.. O-RING PART OF KIT P/N FC212028 .... ,UOC:ZAA,ZAB,ZAC,ZAD,ZAE,ZAF,ZAG,ZAH, ZAJ, ZAK, ZAL	1

(1) ITEM NO	(2) SMR CODE	(3) NSN	(4) CAGEC	(5) PART NUMBER	(6) DESCRIPTION AND USABLE ON CODES (UOC)	(7) QTY
14	PAFZZ	3110001002368	21102	FC030024	. BEARING,BALL,ANNULA PART OF KIT P/N FC212028 UOC:ZAA,ZAB,ZAC,ZAD,ZAE,ZAF,ZAG,ZAH, ZAJ,ZAK,ZAL	2
15	PAFZZ	5365012711854	21102	FC030043	.SPACER,RING PART OF KIT P/N FC212028 UOC:ZAA,ZAB,ZAC,ZAD,ZAE,ZAF,ZAG,ZAH, ZAJ,ZAK,ZAL	1
16	PAFZZ	5365012708459	21102	FC030060	.SPACER,STEPPED PART OF KIT P/N FC212028 UOC:ZAA,ZAB,ZAC,ZAD,ZAE,ZAF,ZAG,ZAH, ZAJ,ZAK,ZAL	1
17	PAFZZ	2520010774009	21102	FC030009	.LINING,FRICTION PART OF KIT P/N FC212028 UOC: ZAA,ZAB,ZAC,ZAD,ZAE,ZAF,ZAG,ZAH, ZAJ,ZAK,ZAL	1
18	PAFZZ	5360012715767	21102	FC030007	.SPRING,HELICAL,COMP UOC:ZAA,ZAB,ZAC,ZAD,ZAE,ZAF,ZAG,ZAH, ZAJ, ZAK,ZAL	1
19	PAFZZ	2520012717152	21102	FC030061	.DISK,CLUTCH UOC:ZAA,ZAB,ZAC,ZAD,ZAE,ZAF,ZAG,ZAH, ZAJ,ZAK,ZAL	1
20	PAFZZ	5305012563046	21102	FC030047	.SCREW,CAP,SOCKET HE UOC:ZAA,ZAB,ZAC,ZAD,ZAE,ZAF,ZAG,ZAH, ZAJ, ZAK, ZAL	8
21	PAFZZ	3020012715126	21102	FC200161	.PULLEY,GROOVE UOC:ZAA,ZAB,ZAC,ZAD,ZAE,ZAFP,ZAG, ZAH, ZAJ,ZAK,ZAL	1
22	PAFZZ	5365012708473	21102	FC030098	.SPACER,RING UOC:ZAA,ZAB,ZAC,ZAD,ZAE,ZAF,ZAG,ZAH, ZAJ,ZAK,ZAL	1
23	PAFZZ	5325012708363	21102	FC030064	.RING,RETAINING PART OF KIT P/N FC212028 UOC:ZAA,ZAB,ZAC,ZAD,ZAE,ZAF,ZAG,ZAH, ZAJ,ZAK,ZAL	1
24	PAFZZ	3120012771036	21102	FC030065	.BUSHING,SLEEVE PART OF KIT P/N FC212028 UOC:ZAA,ZAB,ZAC,ZAD,ZAE,ZAF,ZAG,ZAH, ZAJ,ZAK,ZAL	1
25	PAFZZ	3110012716353	21102	FC030066	.BEARING,ROLLER,CYLI PART OF KIT P/N FC212028 UOC:ZAA,ZAB,ZAC,ZAD,ZAE,ZAF,ZAG,ZAH, ZAJ, ZAK, ZAL	1
26	PAFZZ	5325012708362	21102	FC030067	.RING,RETAINING UOC:ZAA,ZAB,ZAC,ZAD,ZAE,ZAF,ZAG,ZAH, ZAJ, ZAK, ZAL	1
27	PAFZZ	3040012713597	21102	FC200162	.BRACKET AND SHAFT A UOC:ZAA,ZAB,ZAC,ZAD,ZAE,ZAF,ZAG,ZAH, ZAJ, ZAK, ZAL	1

END OF FIGURE

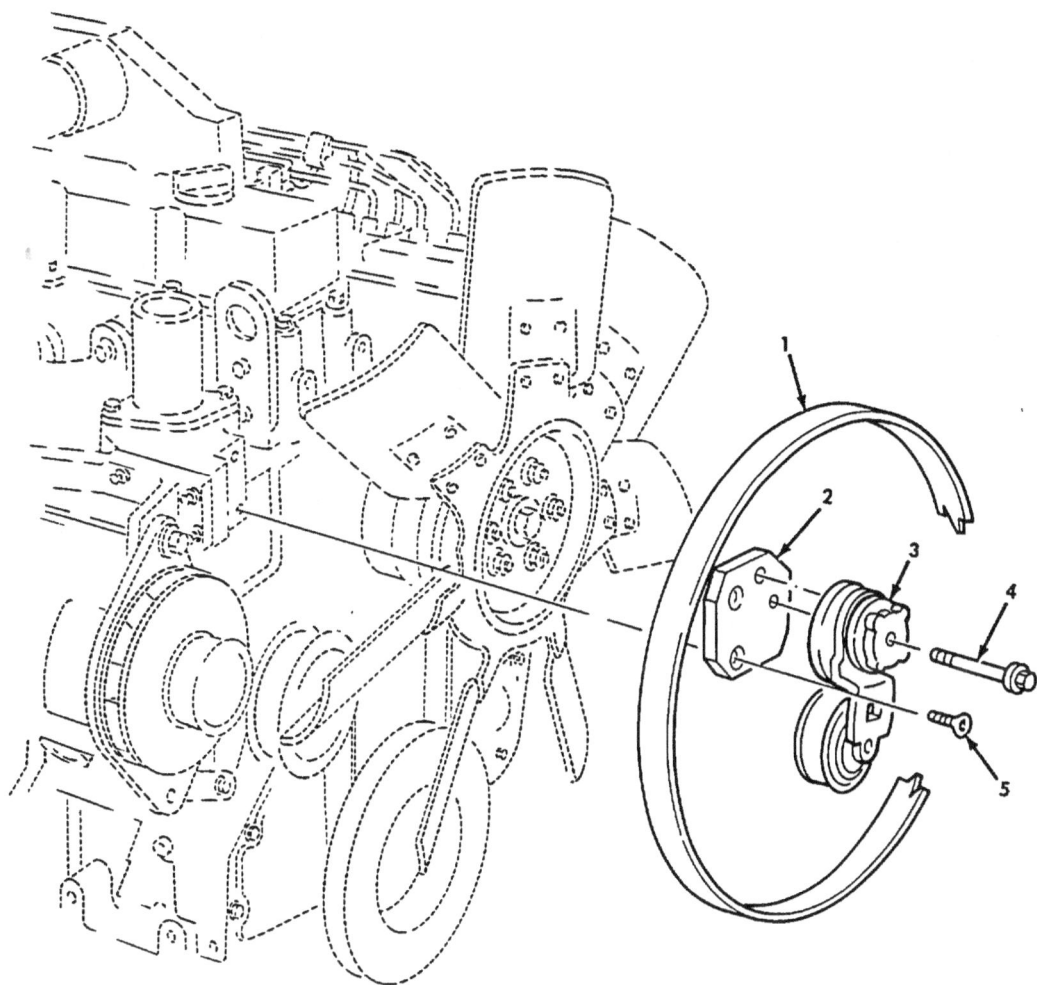

Figure 123. Belt Tensioner Pulley (M939A2).

(1) ITEM NO	(2) SMR CODE	(3) NSN	(4) CAGEC	(5) PART NUMBER	(6) DESCRIPTION AND USABLE ON CODES (UOC)	(7) QTY
					GROUP 0505 FAN ASSEMBLY	
					FIG. 123 BELT TENSIONER PULLEY (M939A2)	
1	PAOZZ	3030012713754	15434	3912004	BELT,V .... UOC:ZAA,ZAB,ZAC,ZAD,ZAE,ZAF,ZAG,ZAH, ZAJ,ZAK,ZAL	1
2	PAOZZ	5342012719818	15434	3909897	BRACKET,ENGINE ACCE .... UOC:ZAA,ZAB,ZAC,ZAD,ZAE,ZAF,ZAG,ZAH, ZAJ, ZAK, ZAL	1
3	PAOZZ	2990012719816	15434	3922901	ADJUSTING DEVICE,BE .... UOC:ZAA,ZAB,ZAC,ZAD,ZAE,ZAF,ZAG,ZAH, ZAJ,ZAX,ZAL	1
4	PAOZZ	5306012377531	15434	3901757	BOLT,MACHINE..................... UOC:ZAA,ZAB,ZAC,ZAD,ZAE,ZAF,ZAG,ZAH, ZAJ,ZAK,ZAL	1
5	PAOZZ	5305012366157	15434	3903095	SCREW,CAP,SOCKET HE .... UOC:ZAA,ZAB,ZAC,ZAD,ZAE,ZAF,ZAG,ZAH, ZAJ,ZAK,ZAL	2

END OF FIGURE

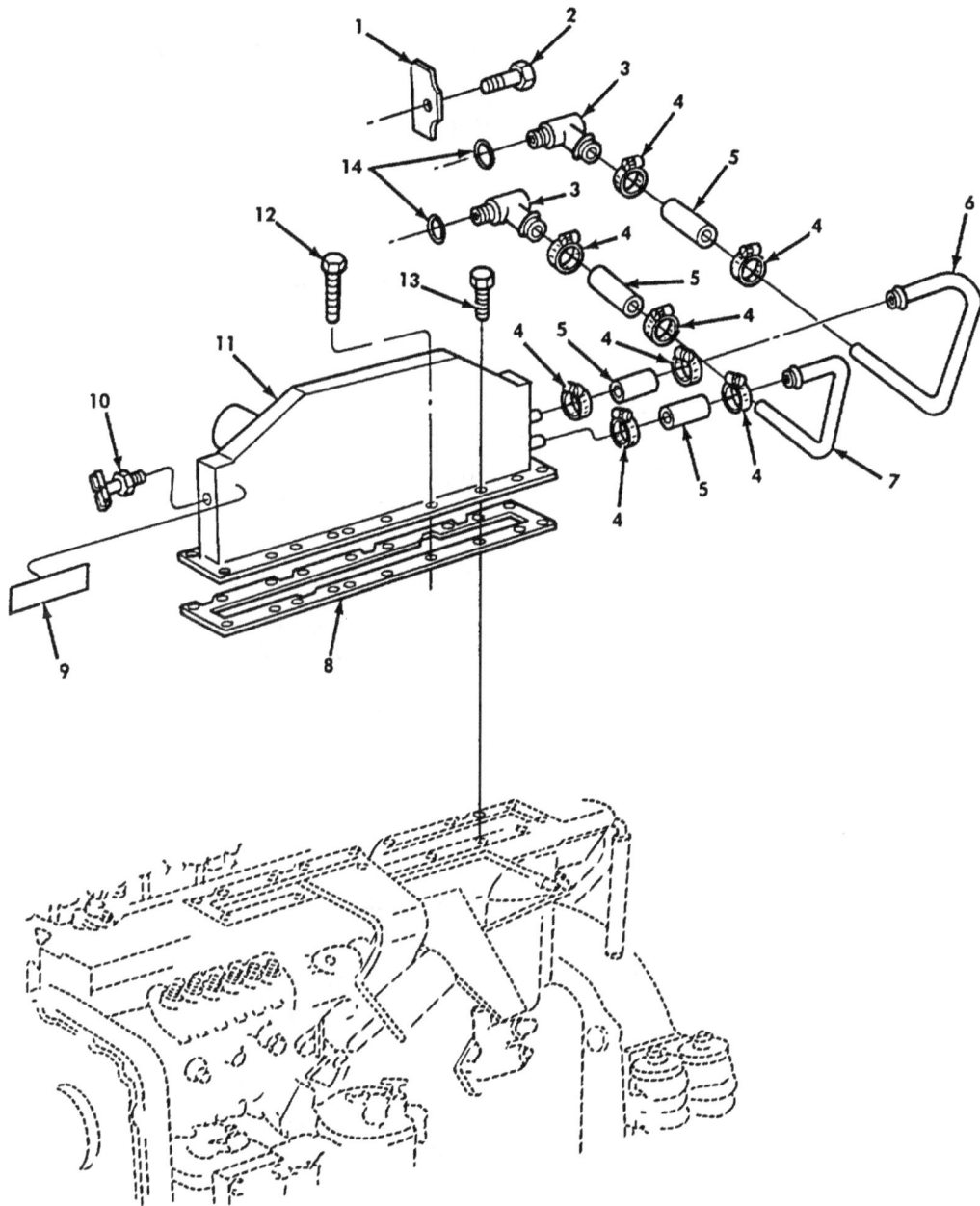

Figure 124. Engine Aftercooler (M939A2).

(1) ITEM NO	(2) SMR CODE	(3) NSN	(4) CAGEC	(5) PART NUMBER	(6) DESCRIPTION AND USABLE ON CODES (UOC)	(7) QTY
					GROUP 0507 AUXILIARY COOLING	
					FIG. 124 ENGINE AFTERCOOLER(M939A2)	
1	PAOZZ	5340012712441	15434	3906439	PLATE,MENDING .... UOC:ZAA,ZAB,ZAC,ZAD,ZAE,ZAF,ZAG,ZAH, ZAJ,ZAK,ZAL	1
2	PAOZZ	5305012374915	15434	3900629	SCREW,CAP,HEXAGON H .... UOC:ZAA,ZAB,ZAC,ZAD,ZAE,ZAF,ZAG,ZAH, ZAJ,ZAK,ZAL	1
3	PAOZZ	4730012717874	15434	3917394	ELBOW,PIPE TO HOSE .... UOC:ZAA,ZAB,ZAC,ZAD,ZAE,ZAF,ZAG,ZAH, ZAJ,ZAK,ZAL	2
4	PAOZZ	4730012717903	15434	3918163	CLAMP,HOSE............................. UOC:ZAA,ZAB,ZAC,ZAD,ZAE,ZAF,ZAG,ZAH, ZAJ,ZAK,ZAL	8
5	PAOZZ	4720013735652	15434	3918611	HOSE,NONMETALLIC .... UOC:ZAA,ZAB,ZAC,ZAD,ZAE,ZAF,ZAG,ZAH, ZAJ, ZAK, ZAL	4
6	PAOZZ	4710012717940	15434	3908402	TUBE,BENT,METALLIC .... UOC:ZAA,ZAB,ZAC,ZAD,ZAE,ZAF,ZAG,ZAH, ZAJ,ZAK,ZAL	1
7	PAOZZ	4710012722882	15434	3906436	TUBE,BENT,METALLIC .... UOC:ZAA,ZAB,ZAC,ZAD,ZAE,ZAF,ZAG,ZAH, ZAJ,ZAK,ZAL	1
8	PAOZZ	5330012721138	15434	3914311	GASKET PART OF KIT P/N 3802622 PART..... OF KIT P/N 3802623 PART OF KIT P/N 3802452 .... UOC:ZAA,ZAB,ZAC,ZAD,ZAE,ZAF,ZAG,ZAH, ZAJ,ZAK, ZAL	1
9	PAOZZ	7690012711926	15434	3910778	DECAL .... UOC:ZAA,ZAB,ZAC,ZAD,ZAE,ZAF,ZAG,ZAH, ZAJ,ZAK,ZAL	1
10	PAOZZ	4820007529040	96906	MS35782-4	COCK,DRAIN .... UOC:ZAA,ZAB,ZAC,ZAD,ZAE,ZAF,ZAG,ZAH, ZAJ,ZAK,ZAL	1
11	PAOZZ	2815012688737	15434	3914501	AFTERCOOLER,ENGINE .... UOC:ZAA,ZAB,ZAC,ZAD,ZAE,ZAF,ZAG,ZAH, ZAJ,ZAK,ZAL	1
12	PAOZZ	5305012397202	15434	3900632	SCREW,CAP,HEXAGON H .... UOC:ZAA,ZAB,ZAC,ZAD,ZAE,ZAF,ZAG,ZAH, ZAJ, ZAK, ZAL	2
13	PAOZZ	5305012453192	15434	3918109	SCREW,CAP,HEXAGON H .... UOC:ZAA,ZAB,ZAC,ZAD,ZAE,ZAF,ZAG,ZAH, ZAJ,ZAK,ZAL	16
14	PAOZZ	5331012013623	15434	145530	O-RING PART OF KIT P/N 3802622 PART...... OF KIT P/N 3802623 PART OF KIT P/N 3802452 .... UOC:ZAA,ZAB,ZAC,ZAD,ZAE,ZAPF,ZAG,ZAH, ZAJ,ZAK,ZAL	2

END OF FIGURE

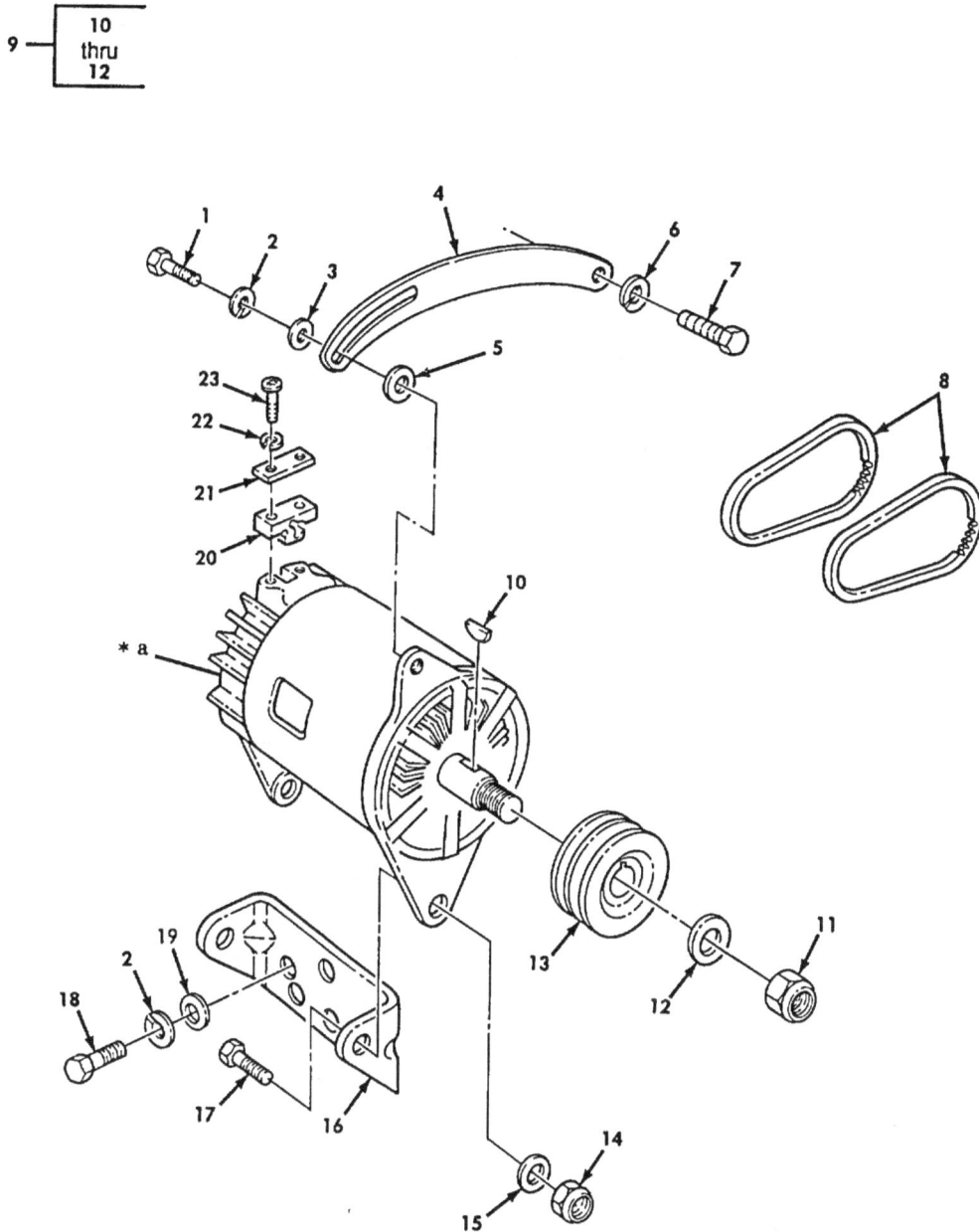

*a PART OF ITEM 9

Figure 125. Alternator Assembly, Mounting Hardware, Belts, and Pulley (M939, M939A1).

(1) ITEM NO	(2) SMR CODE	(3) NSN	(4) CAGEC	(5) PART NUMBER	(6) DESCRIPTION AND USABLE ON CODES (UOC)	(7) QTY
					GROUP 06 ELECTRICAL SYSTEM 0601 GENERATOR, ALTERNATOR	
					FIG. 125 ALTERNATOR ASSEMBLY, MOUNTING HARDWARE, BELTS, AND PULLEY (M939,M939A1)	
1	PAOZZ	5305000680511	80204	B1821BH03BC125N	SCREW,CAP,HEXAGON H .... UOC:DAA,DAB,DAC,DAD,DAE,DAF,DAG,DAH, DAJ,DAK,DAL,DAW,DAX,V12,V13,V14,V15, V16,V17,V18,V19,V20,V21,V22,V24,V25, V39	1
2	PAOZZ	5310006379541	96906	MS35338-46	WASHER,LOCK .... UOC:DAA,DAB,DAC,DAD,DAE,DAF,DAG,DAH, DAJ,DAK,DAL,DAW,DAX,V12,V13,V14,V15, V16,V17,V18,V19,V20,V21,V22,V24,V25, V39	5
3	PAOZZ	5310008094061	96906	MS27183-15	WASHER,FLAT .... UOC:DAA,DAB,DAC,DAD,DAE,DAF,DAG,DAH, DAJ,DAK,DAL,DAW,DAX,V12,V13,V14,V15, V16,V17,V18,V19,V20,V21,V22,V24,V25, V39	1
4	PAOZZ	2920010835408	19207	11669323	ARM,ADJUSTING,BELT .... UOC:DAA,DAB,DAC,DAD,DAE,DAF,DAG,DAH, DAJ,DAK,DAL,DAW,DAX,V12,V13,V14,V15, V16,V17,V18,V19,V20,V21,V22,V24,V25, V39	1
5	PAOZZ	5310008094085	96906	MS27183-16	WASHER,FLAT .... UOC:DAA,DAB,DAC,DAD,DAE,DAF,DAG,DAH, DAJ,DAK,DAL,DAW,DAX,V12 ,V13 ,V14 ,V15, V16,V17,V18,V19,V20,V20,V21,V22,V24,V25, V39	6
6	PAOZZ	5310002090965	96906	MS35338-47	WASHER,LOCK . ... UOC:DAA,DAB,DAC,DAD,DAE,DAF,DAG,DAH, DAJ,DAK,DAL,DAW,DAX,V12,V13,V14,V15, V16,V17,V18,V19,V20,V21,V22,V24,V25, V39	1
7	PAOZZ	5305007098482	96906	MS90727-96	SCREW,CAP,HEXAGON H .... UOC:DAA,DAB,DAC,DAD,DAE,DAF,DAG,DAH, DAJ,DAK,DAL,DAW,DAX,V12,V13,V14,V15, V16,V17,V18,V19,V20,V21,V22,V24,V25, V39	1
8	PAOZZ	3030009832873	96906	MS51065RP48-2	BELTS,V,MATCHED SET .... UOC:DAA,DAB,DAC,DAD,DAE,DAF,DAG,DAH, DAJ,DAK,DAL,DAW,DAX,V12,V13,V14,V15, V16,V17,V18,V19,V20,V21,V22,V24,V25, V39	1
9	PAOFF	2920009092483	19207	10929868	GENERATOR,ENGINE AC .... UOC:DAA,DAB,DAC,DAD,DAE,DAF,DAG,DAH, DAJ,DAK,DAL,DAW,DAX,V12,V13,V14,V15, V16,V17,V18,V19,V20,V21,V22,V24,V25, V39	1
10	PAOZZ	5315006165526	96906	MS35756-8	.KEY,WOODRUFF.............	1

(1) ITEM NO	(2) SMR CODE	(3) NSN	(4) CAGEC	(5) PART NUMBER	(6) DESCRIPTION AND USABLE ON CODES (UOC)	(7) QTY
					UOC:DAA,DAB,DAC,DAD,DAE,DAF,DAG,DAH, DAJ,DAK,DAL,DAW,DAX,V12,V13,V14,V15, V16,V17,V18,V19,V20,V21,V22,V24,V25, V39	
11	PAOZZ	5310004492381	96906	MS21245-L10	.NUT,SELF-LOCKING,HE .... UOC:DAA,DAB,DAC,DAD,DAE,DAF,DAG,DAH, DAJ,DAK,DAL,DAW,DAX,V12,V13,V14,V15, V16,V17,V18,V19,V20,V21,V22,V24,V25, V39	1
12	PAOZZ	5310006143505	96906	MS15795-820	WASHER,FLAT .... UOC:DAA,DAB,DAC,DAD,DAE,DAF,DAG,DAH, DAJ,DAK,DAL,DAW,DAX,V12,V13,V14,V15, V16,V17,V18,V19,V20,V21,V22,V24,V25, V39	1
13	PAOZZ	3020010906695	19207	11669322	PULLEY,GROOVE .... UOC:DAA,DAB,DAC,DAD,DAE,DAF,DAG,DAH, DAJ,DAK,DAL,DAW,DAX,V12,V13,V14,V15, V16,V17,V18,V19,V20,V21,V22,V24,V25, V39	1
14	PAOZZ	5310002749364	96906	MS21045-7	NUT,SELF-LOCKING,HE .... UOC:DAA,DAB,DAC,DAD,DAE,DAF,DAG,DAH, DAJ,DAK,DAL,DAW,DAX,V12,V13,V14,V15, V16,V17,V18,V19,V20,V21,V22,V24,V25, V39	2
15	PAOZZ	5310008095997	96906	MS27183-17	WASHER,FL AT .... UOC:DAA,DAB,DAC,DAD,DAE,DAF,DAG,DAH, DAJ,DAK,DAL,DAW,DAX,V12,V13,V14,V15, V16,V17,V18,V19,V20,V21,V22,V24,V25, V39	2
16	PFOZZ	5340011972183	19207	11669324-1	BRACKET .... UOC:DAA,DAB,DAC,DAD,DAE,DAF,DAG,DAH, DAJ,DAK,DAL,DAW,DAX,V12,V13,V14,V15, V16,V17,V1B,V19,V20,V21,V22,V24,V25, V39	1
17	PAOZZ	5305009146133	96906	MS18153-88	SCREW,CAP,HEXAGON H .... UOC:DAA,DAB,DAC,DAD,DAE,DAF,DAG,DAH, DAJ,DAK,DAL,DAW,DAX,V12,V13,V14,V15, V16,V17,V18,V19,V20,V21;V22,V24,V25, V39	2
18	PAOZZ	5305005434372	80204	B1821BH038C075N	SCREW,CAP,HEXAGON H .... UOC:DAA,DAB,DAC,DDAD,DAE,DAF,DAG,DAH, DAJ,DAK,DAL,DAW,DAX,V12,V13,V14,V15, V16,V17,V18,V19,V20,V21,V22,V24,V25, V39	4
19	PAOZZ	5310000806004	96906	MS27183-14	WASHER ,FLAT . ... UOC:DAA,DAB,DAC,DAD,DAE,DAF,DAG,DAH, DAJ,DAIC,DAL,DAW,DAX,V12,V13,V14,V15, V16,V17,V18,V19,V20,V21,V22,V24,V25, V39	4
20	PAOZZ	5365012894434	19207	12357116	SPACER,PLATE .... UOC:DAA,DAB,DAC,DAD,DAE,DAF,DAG,DAH, DAJ,DAK,DAL,DAW,DAX,V12,V13,V14,V15, V16,V17,V1B ,V19,V20,V21,V22,V24,V25, V39	1

(1) ITEM NO	(2) SMR CODE	(3) NSN	(4) CAGEC	(5) PART NUMBER	(6) DESCRIPTION AND USABLE ON CODES (UOC)	(7) QTY
21	PAOZZ	5365012897852	19207	12357126	SPACER,PLATE .... UOC:DAA,DAB,DAC,DAD,DAE,DAF,DAG,DAH, DAJ,DAK,DAL,DAW,DAX,V12, V13,V14 ,V15, V16,V17,V18 ,V19 ,V20,V21,V22,V24,V25, V39	1
22	PAOZZ	5310007217809	96906	MS35340-43	WASHER,LOCK .... UOC:DAA,DAB,DAC,DAD,DAE,DAF,DAG,DAH, DAJ,DAK,DAL,DAW,DAX,V12,V13,V14,V15, V16,V17,V18,V19,V20,V21,V22,V24,V25, V39	2
23	PAOZZ	5305009846212	96906	MS35206-265	SCREW,MACHINE .... UOC:DAA,DAB,DAC,DAD,DAE,DAF,DAG,DAH, DAJ,DAK,DAL,DAW,DAX,V12,V13,V14,V15, V16,V17,V18,V19,V20,V21,V22,V24,V25, V39	2

END OF FIGURE

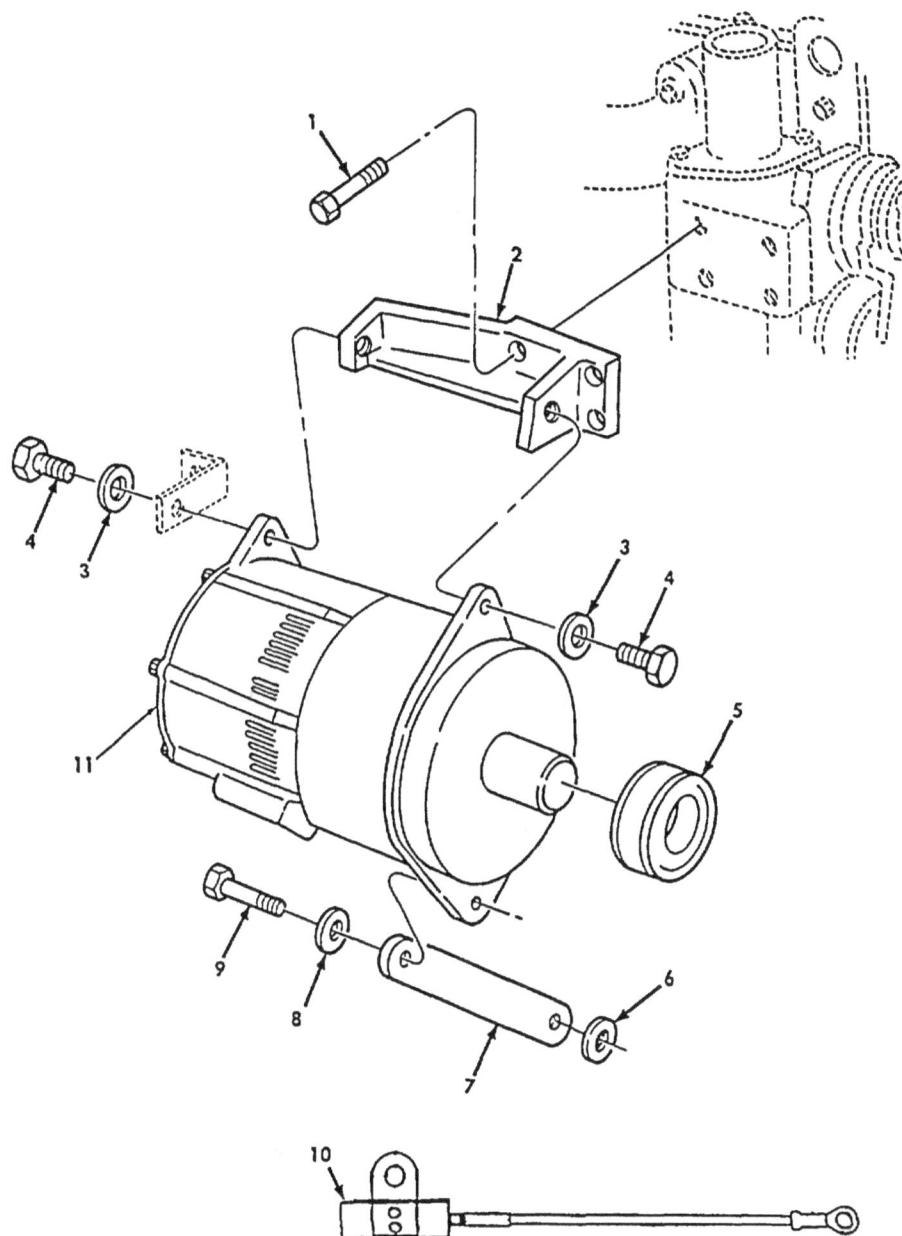

Figure 126. Alternator Mounting (M939A2).

(1) ITEM NO	(2) SMR CODE	(3) NSN	(4) CAGEC	(5) PART NUMBER	(6) DESCRIPTION AND USABLE ON CODES (UOC)	(7) QTY
					GROUP 0601 GENERATOR, ALTERNATOR	
					FIG. 126 ALTERNATOR MOUNTING(M939A2)	
1	PAOZZ	5305012453193	15434	3900621	SCREW,CAP,HEXAGON H .... UOC:ZAA,ZAB,ZAC,ZAD,ZAE,ZAF,ZAG,ZAH, ZAJ,ZAK,ZAL	4
2	PAOZZ	2990012715088	15434	3909898	BRACKET,ENGINE ACCE .... UOC:ZAA,ZAB,ZAC,ZAD,ZAE,ZAF,ZAG,ZAH, ZAJ,ZAK,ZAL	1
3	PAOZZ	5310010925495	24617	9422848	WASHER,FLAT .... UOC:ZAA,ZAB,ZAC,ZAD,ZAE,ZAF,ZAG,ZAH, ZAJ,ZAK,ZAL	2
4	PAOZZ	5305000711788	80204	B1821BH044C125N	SCREW,CAP,HEXAGON H .... UOC:ZAA,ZAB,ZAC,ZAD,ZAE,ZAF,ZAG,ZAH, ZAJ,ZAK,ZAL	2
5	PAOZZ	3020012713832	15434	3919390	PULLEY,GROOVE .... UOC:ZAA,ZAB,ZAC,ZAD,ZAE,ZAF,ZAG,ZAH, ZAJ,ZAK,ZAL	1
6	PAOZZ	5310012865452	24617	11511514	WASHER,FLAT .... UOC:ZAA,ZAB,ZAC,ZAD,ZAE,ZAF,ZAG,ZAH, ZAJ,ZAK,ZAL	1
7	PAOZZ	2990012715089	15434	3909899	BRACKET,ENGINE ACCE................................... UOC:ZAA,ZAB,ZAC,ZAD,ZAE,ZAF,ZAG,ZAH, ZAJ,ZAK,ZAL	1
8	PAOZZ	5310002748041	90407	12084P11	WASHER,FLAT.. UOC:ZAA,ZAB,ZAC,ZAD,ZAE,ZAF,ZAG,ZAH, ZAJ,ZAK,ZAL	1
9	PAOZZ	5305000680510	80204	B1821BH038C100N	SCREW,CAP,HEXAGON H .... UOC:ZAA,ZAB,ZAC,ZAD,ZAE,ZAF,ZAG,ZAH, ZAJ,ZAK,ZAL	1
10	PAOZZ	5910014590289	19207	12432245	CAPACITOR,FIXED,PLA...................................... UOC:ZAA,ZAB,ZAC,ZAD,ZAE,ZAF,ZAG,ZAH, ZAJ,ZAK,ZAL	1
11	PAOZZ	2920009092483	19207	10929868	GENERATOR,ENGINE AC.................................... UOC:ZAA,ZAB,ZAC,ZAD,ZAE,ZAF,ZAG,ZAH, ZAJ,ZAK,ZAL	1

END OF FIGURE

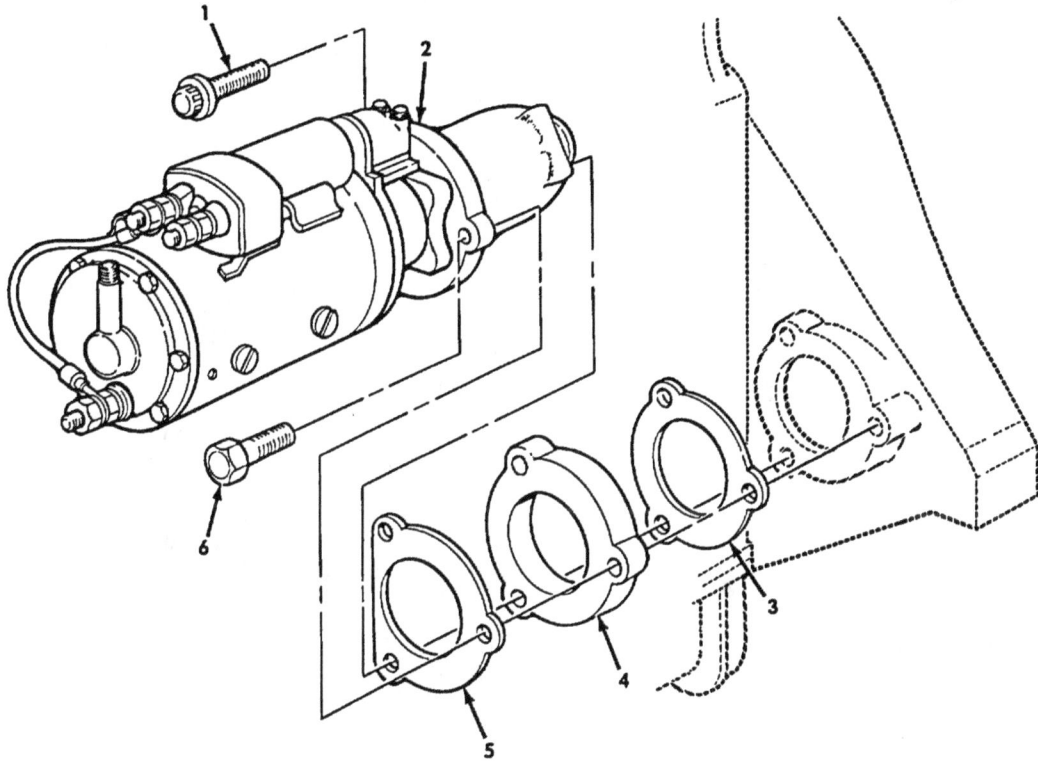

Figure 127. Starting Motor Assembly and Mounting Hardware (M939, M939A1).

(1) ITEM NO	(2) SMR CODE	(3) NSN	(4) CAGEC	(5) PART NUMBER	(6) DESCRIPTION AND USABLE ON CODES (UOC)	(7) QTY
					GROUP 0603 STARTING MOTOR	
					FIG. 127 STARTING MOTOR ASSEMBLY AND MOUNTING HARDWARE(M939,M939A1)	
1	PAOZZ	5306011063850	19207	11664479-1	BOLT,MACHINE.. UOC:DAA,DAB,DAC,DAD,DAE,DAF,DAG,DAH, DAJ,DAK,DAL,DAW,DAX,V12,V13,V14,V15, V16,V17,V18,V19,V20,V21,V22,V24,V25, V39	1
2	PAOFH	2920003043493	96906	MS53011-1	STARTER,ENGINE,ELEC .... UOC:DAA,DAB,DAC,DAD,DAE,DAF,DAG,DAH, DAJ,DAK,DAL,DAW,DAX,V12,V13,V14,V15, V16,V17,V18,V19,V20,V21,V22,V24,V25, V39	1
3	PAOZZ	5330001437737	19207	11664431	GASKET... UOC:DAA,DAB,DAC,DAD,DAE,DAF,DAG,DAH, DAJ,DAK,DAL,DAW,DAX,V12,V13,V14,V15, V16,V17,V18,V19 ,V20,V21,V22,V24,V25, V39	1
4	PAOZZ	5310011043804	19207	11664432-1	WASHER,FLAT... UOC:DAA,DAB,DAC,DAD,DAE,DAF,DAG,DAH, DAJ,DAK,DAL,DAW,DAX,V12,V13,V14,V15, V16,V17,V18,V19,V20,V21,V22,V24,V25, V39	1
5	PAOZZ	5330002523274	19207	11664480	GASKET PART OF KIT P/N 3011472 .... UOC:DAA,DAB,DAC,DAD,DAE,DAF,DAG,DAH, DAJ,DAK,DAL,DAW,DAX,V12,V13,V14,V15, V16,V17,V18,Vi9,V20,V21,V22,V24,V25, V39	1
6	PAOZZ	5305007247221	80204	B1821BH063C175N	SCREW,CAP,HEXAGON H .... UOC:DAA,DAB,DAC,DAD,DAE,DAF,DAG,DAH, DAJ,DAK,DAL,DAW,DAX,V12,V13,V14,V15, V16,V17,V1B,V19,V20,V21,V22,V24,V25, V39	2

END OF FIGURE

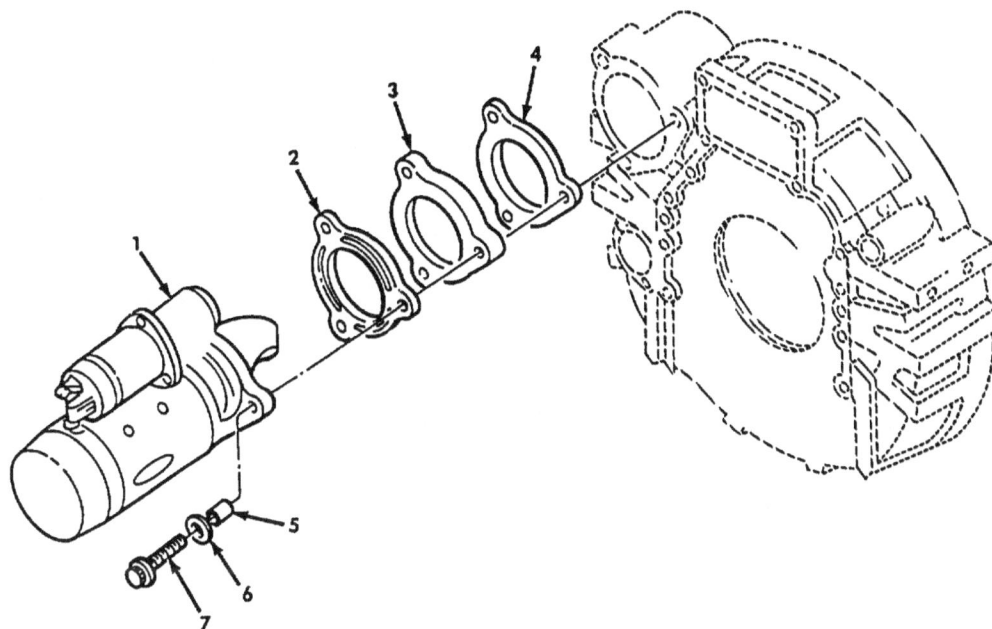

Figure 128. Starter Assembly (M939A2).

(1) ITEM NO	(2) SMR CODE	(3) NSN	(4) CAGEC	(5) PART NUMBER	(6) DESCRIPTION AND USABLE ON CODES (UOC)	(7) QTY
					GROUP 0603 STARTING MOTOR	
					FIG. 128 STARTER ASSEMBLY (M939A2)	
1	PAOFF	2920010319027	96906	MS53011-2	STARTER,ENGINE,ELEC .... UOC:ZAA,ZAB,ZAC,ZAD,ZAE,ZAF,ZAG,ZAH, ZAJ,ZAK,ZAL	1
2	PAOZZ	5330011374487	15434	33069826	GASKET................... UOC:ZAA,ZAB,ZAC,ZAD,ZAE,ZAF,ZAG,ZAH, ZAJ,ZAK,ZAL	1
3	PAOZZ	5365012708481	15434	3908328	SPACER,PLATE .... UOC:ZAA,ZAB,ZAC,ZAD,ZAE,ZAF,ZAG,ZAH, ZAJ,ZAK,ZAL	1
4	PAOZZ	5330011652314	15434	3031007	GASKET .... UOC:ZAA,ZAB,ZAC,ZAD,ZAE,ZAF,ZAG,ZAH, ZAJ,ZAK,ZAL	1
5	PAOZZ	5365012806924	33477	20510304	SPACER,SLEEVE .... UOC:ZAA,ZAB,ZAC,ZAD,ZAE,ZAF,ZAG,ZAH, ZAJ,ZAK,ZAL	3
6	PAOZZ	5310012823391	3T063	20510303	WASHER,FLAT................... UOC:ZAA,ZAB,ZAC,ZAD,ZAE,ZAF,ZAG,ZAH, ZAJ,ZAK,ZAL	3
7	PAOZZ	5305012744406	15434	3908321	SCREW,CAP,HEXAGON H .... UOC:ZAA,ZAB,ZAC,ZAD,ZAE,ZAF,ZAG,ZAH, ZAJ,ZAK,ZAL	3

END OF FIGURE

]

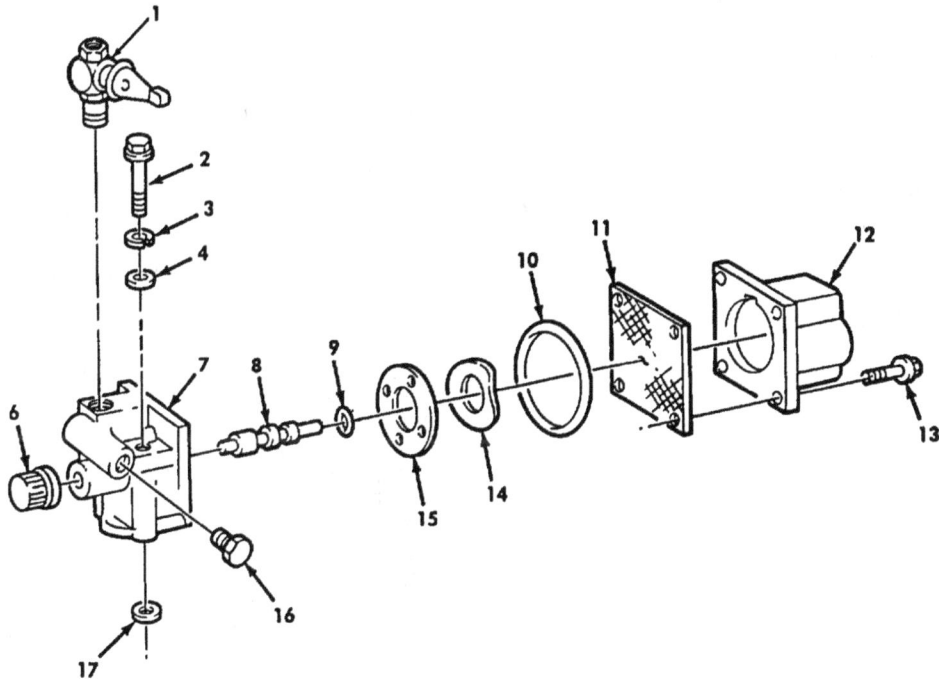

Figure 129. Shutdown Valve and Mounting Hardware (M939, M939A1).

(1) ITEM NO	(2) SMR CODE	(3) NSN	(4) CAGEC	(5) PART NUMBER	(6) DESCRIPTION AND USABLE ON CODES (UOC)	(7) QTY
					GROUP 0606 ENGINE SAFETY CONTROLS	
					FIG. 129 SHUTDOWN VALVE AND MOUNTING HARDWARE(M939,M939A1)	
1	PAOZZ	4820005888604	30327	51807	VALVE,PLUG .... UOC:DAA,DAB,DAC,DAD,DAE,DAF,DAG,DAH, DAJ,DAK,DAL,DAW,DAX,V12,V13,V14,V15, V16,V17,V18,V19,V20,V21,V22,V24,V25, V39	1
2	PAFZZ	5305005098106	15434	S189-C	SCREW,CAP,SOCKET HE .... UOC:DAA,DAB,DAC,DAD,DAE,DAF,DAG,DAH, DAJ,DAK,DAL,DAW,DAX,V12,V13,V14,V15, V16,V17,V18,V19,V20,V21,V22,V24,V25, V39	2
3	PAFZZ	5310004841718	15434	181466	WASHER,LOCK... UOC:DAA,DAB,DAC,DAD,DAE,DAF,DAG,DAH, DAJ,DAK,DAL,DAW,DAX,V12,V13,V14,V15, V16,V17,V18,V19,V20,V21,V22,V24,V25, V39	2
4	PAFZZ	5310002622986	15434	67684	. WASHER,FLAT .... UOC:DAA,DAB,DAC,DAD,DAE,DAF,DAG,DAH, DAJ,DAK,DAL,DAW,DAX,V12,V13,V14,V15, V16,V17,V18,V19,V20,V21,V22,V24,V25, V39	2
5	PAFZZ	481001187492515434		3035344	VALVE,SOLENOID .... UOC:DAA,DAB,DAC,DAD,DAE,DAF,DAG,DAH, DAJ,DAK,DAL,DAW,DAX,V12,V13,V14,V15, V16,V17,V18,V19,V20,V21,V22,V24,V25, V39	1
6	PAFZZ	5355000821189	15434	129838	. KNOB.. UOC:DAA,DAB,DAC,DAD,DAE,DAF,DAG,DAH, DAJ,DAK,DAL,DAW,DAX,V12,V13,V14,V15, V16,V17,V18,V19,V20,V21,V22,V24,V25, V39	1
7	PAFZZ	482000829560015434		129826	. BODY,VALVE .... UOC:DAA,DAB,DAC,DAD,DAE,DAF,DAG,DAH, DAJ,DAK,DAL,DAW,DAX,V12,V13,V14,V15, V16,V17,V18,V19,V20,V21,V22,V24,V25, V39	1
8	PAFZZ	3040010852616	15434	3000266	. SHAFT,SHOULDERED .... UOC:DAA,DAB,DAC,DAD,DAE,DAF,DAG,DAH, DAJ,DAK,DAL,DAW,DAX,V12,V13,V14,V15, V16,V17,V18,V19,V20,V21,V22,V24,V25, V39	1
9	PAFZZ	5330001320274	15434	190876	. O-RING PART OF KIT P/N 3010240 .... UOC:DAA,DAB,DAC,DAD,DAE,DAF,DAG,DAH, DAJ,DAK,DAL,DAW,DAX,V12,V13,V14,V15, V16,V17,V18,V19,V20,V21,V22,V24,V25, V39	1
10	PAFZZ	5330000819299	15434	129888	. O-RING PART OF KIT P/N 3010240 .... UOC:DAA,DAB,DAC,DAD,DAE,DAF,DAG,DAH,	1

(1) ITEM NO	(2) SMR CODE	(3) NSN	(4) CAGEC	(5) PART NUMBER	(6) DESCRIPTION AND USABLE ON CODES (UOC)	(7) QTY
					DAJ,DAK,DAL,DAW,DAX,V12,V13,V14,V15, V16,V17,V18,V19,V20,V21,V22,V24,V25, V39	
11	PAFZZ	5340000847787	15434	129839	.COVER,ACCESS .... UOC:DAA,DAB,DAC,DAD,DAE,DAF,DAG,DAH, DAJ,DAK,DAL,DAW,DAX,V12,V13,V14,V15, V16,V17,V18,V19,V20,V21,V22,V24,V25, V39	1
12	PAFZZ	2910000857436	15434	134074	.COIL ASSEMBLY,SHUT .... UOC:DAA,DAB,DAC,DAD,DAE,DAF,DAG,DAH, DAJ,DAK,DAL,DAW,DAX,V12,V13,V14,V15, V16,V17,V18,V19,V20,V21,V22,V24,V25, V39	1
13	PAFZZ	5305001389848	15434	187556	.SCREW,MACHINE .... UOC:DAA,DAB,DAC,DAD,DAE,DAF,DAG,DAH, DAJ,DAK,DAL,DAW,DAX,V12,V13,V14,V15, V16,V17,V18,V19,V20,V21,V22,V24,V25, V39	4
14	PAFZZ	5310000821888	15434	129768	.WASHER,SPRING TENSI .... UOC:DAA,DAB,DAC,DAD,DAE,DAF,DAG,DAH, DAJ,DAK,DAL,DAW,DAX,V12,V13,V14,V15, V16,V17,V18,V19,V20,V21,V22,V24,V25, V39	1
15	PAFZZ	4820011461048	15434	3030970	.DISK,VALVE .... UOC:DAA,DAB,DAC,DAD,DAE,DAF,DAG,DAH, DAJ,DAK,DAL,DAW,DAX,V12,V13,V14,V15, V16,V17,V18,V19,V20,V21,V22,V24,V25, V39	1
16	PAFZZ	4730000113175	15434	3025458	.PLUG,PIPE... UOC:DAA,DAB,DAC,DAD,DAE,DAF,DAG,DAH, DAJ,DAK,DAL,DAW,DAX,V12,V13,V14,V15, V16,V17,V18,V19,V20,V21,V22,V24,V25, V39	1
17	PAFZZ	5330009513538	91265	TS33-016 70 DURO BUNA N	GASKET PART OF KIT P/N 3010240 .... UOC:DAA,DAB,DAC,DAD,DAE,DAF,DAG,DAH, DAJ,DAK,DAL,DAW,DAX, V12,V13,V14,V15, V16,V17,V18,V19,V20,V21,V22,V24,V25, V39	1

END OF FIGURE

4 —▭ 5 ▭

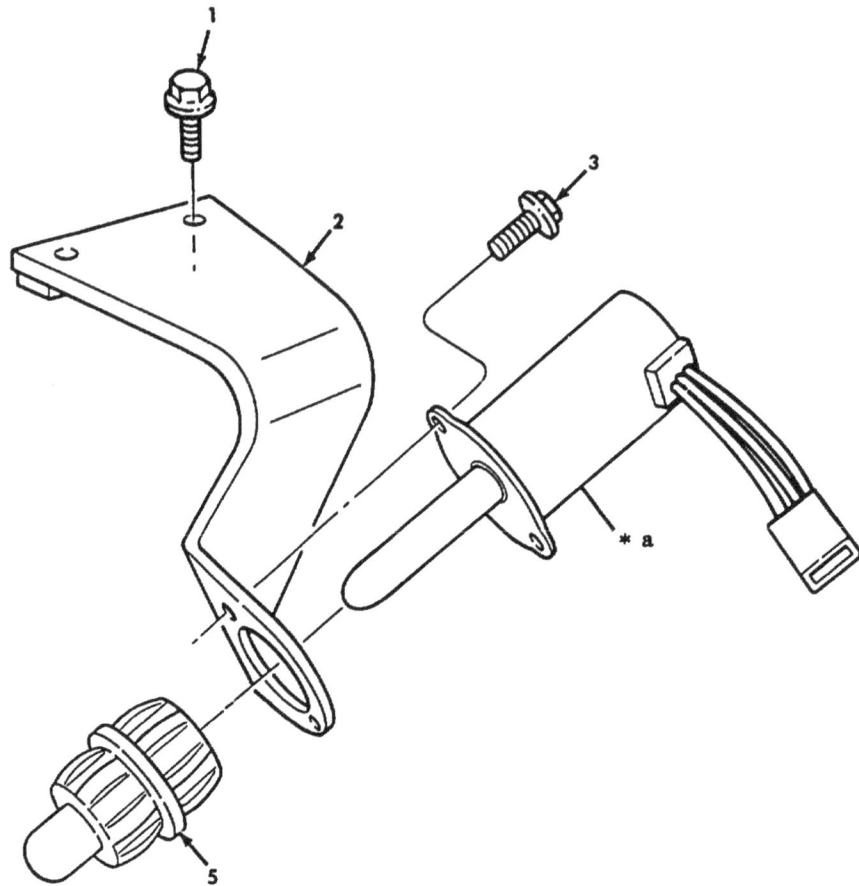

* a   PART OF ITEM 4

Figure 130. Throttle Control Solenoid (M939A2).

(1) ITEM NO	(2) SMR CODE	(3) NSN	(4) CAGEC	(5) PART NUMBER	(6) DESCRIPTION AND USABLE ON CODES (UOC)	(7) QTY
					GROUP 0606 ENGINE SAFETY CONTROLS	
					FIG. 130 THROTTLE CONTROL SOLENOID (M939A2)	
1	PAOZZ	5305012397202	15434	3900632	SCREW,CAP,HEXAGON H.................................. UOC:ZAA,ZAB,ZAC,ZAD,ZAE,ZAF,ZAG,ZAH, ZAJ,ZAK,ZAL	2
2	PAOZZ	5340012712475	15434	3912487	BRACKET,DOUBLE ANGL.................................... UOC:ZAA,ZAB,ZAC,ZAD,ZAE,ZAF,ZAG,ZAH, ZAJ,ZAK,ZAL	1
3	PAOZZ	5305011936839	15434	3900627	SCREW,CAP,HEXAGON H.................................. UOC:ZAA,ZAB,ZAC,ZAD,ZAE,ZAF,ZAG,ZAH, ZAJ,ZAK,ZAL	2
4	PAOZZ	5945012794802	15434	3921980	SOLENOID,ELECTRICAL...................................... UOC:ZAA,ZAB,ZAC,ZAD,ZAE,ZAF,ZAG,ZAH, ZAJ,ZAK,ZAL	1
5	PAOZZ	5340013319487	15434	3924912	.BOOT,DUST AND MOIST.................................... UOC:ZAA,ZAB,ZAC,ZAD,ZAE,ZAF,ZAG,ZAH, ZAJ,ZAK,ZAL	1

END OF FIGURE

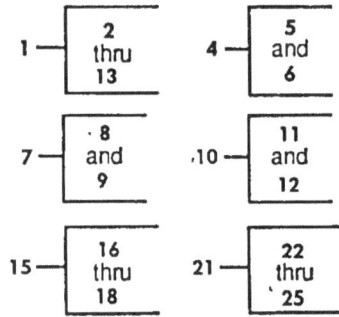

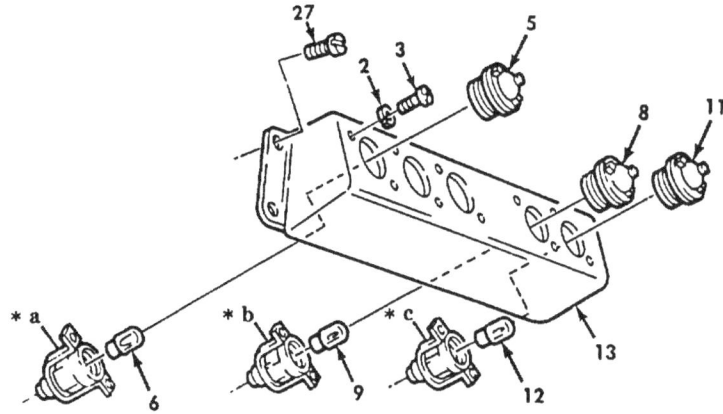

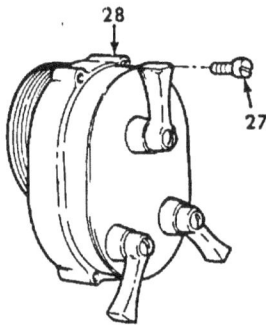

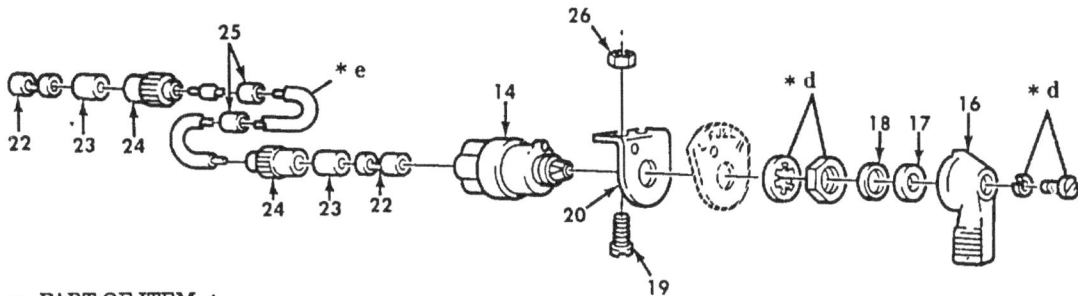

*a PART OF ITEM 4
*b PART OF ITEM 7
*c PART OF ITEM 10
*d PART OF ITEM 14
*e PART OF ITEM 21

Figure 131. Control Panel, Light Switch, and Fuel Tank Selector Switch.

(1) ITEM NO	(2) SMR CODE	(3) NSN	(4) CAGEC	(5) PART NUMBER	(6) DESCRIPTION AND USABLE ON CODES (UOC)	(7) QTY
					GROUP 0607 SWITCHES AND CIRCUIT BREAKERS	
					FIG. 131 CONTROL PANEL, LIGHT SWITCH, AND FUEL TANK SELECTOR SWITCH	
1	PAOZZ	6210011089125	19207	12277153-1	LIGHT SET,INDICATOR..	1
2	PAOZZ	5310000453299	96906	MS35338-42	.WASHER,LOCK	10
3	PAOZZ	5305009846189	96906	MS35206-241	.SCREW,MACHINE	10
4	PAOZZ	6210006355686	19207	7971111	.INDICATOR,LIGHT...	3
5	PAOZZ	6220014230209	19207	7358672-4	..LENS,LIGHT	3
6	PAOZZ	5980012856688	19207	12360890-1	..LAMP,INCANDESCENT	3
7	PAOZZ	6220011085108	19207	12277167	.INDICATOR,LIGHT...	1
8	PAOZZ	6220011295740	19207	12277166	..LENS,LIGHT	1
9	PAOZZ	6240002669940	96906	MS25231-1829	..LAMP,INCANDESCENT	1
10	PAOZZ	6210011249301	19207	8376499-1	.LIGHT,INDICATOR	1
11	PAOZZ	6220004518161	19204	7358622-1	..LENS,LIGHT	1
12	PAOZZ	6240002669940	96906	MS25231-1829	..LAMP,INCANDESCENT	1
13	XAOZZ		19207	12277152	.PANEL,WARNING LIGHT...	1
14	PAOZZ	5930008980500	81349	MIL-S-13623/1-1	SWITCH,ROTARY UOC:DAE,DAF,DAG,DAH,DAL,V18,V19,V20, V21,V22,ZAE,ZAF,ZAG,ZAH,ZAL	1
15	PAOZZ	5930001305349	19207	5381088	HANDLE,SWITCH... UOC:DAE,DAF,DAG,DAH,DAL,V18,V19,V20, V21,V22,ZAE,ZAF,ZAG,ZAH,ZAL	1
16	PAOZZ	5930011978661	19207	5381051	.LEVER,MANUAL CONTRO....................................... UOC:DAE,DAF,DAG,DAH,DAL,V18,V19,V20, V21,V22,ZAE,ZAF,ZAG,ZAH,ZAL	1
17	XDOZZ		19207	7392454	.FELT,MECHANICAL,PRE.. UOC:DAE,DAF,DAG,DAH,DAL,V18,V19,V20, V21,V22,ZAE,ZAF,ZAG,ZAH,ZAL	1
18	PAOZZ	5310011085236	19207	5381087	.WASHER,SHOULDERED UOC:DAE,DAF,DAG,DAH,DAL,V18,V19,V20, V21,V22,ZAE,ZAF,ZAG,ZAH,ZAL	1
19	PAOZZ	5305009906444	96906	MS35207-261	SCREW,MACHINE UOC:DAE,DAF,DAG,DAH,DAL,V18,V19,V20, V21,V22,ZAE,ZAF,ZAG,ZAH,ZAL	2
20	PFOZZ	5340001760868	19207	7059461	BRACKET,ANGLE UOC:DAE,DAF,DAG,DAH,DAL,V18,V19,V20, V21,V22,ZAE,ZAF,ZAG,ZAH,ZAL	1
21	PFOZZ	6150011450361	19207	12277231	LEAD,ELECTRICAL... UOC:DAE,DAF,DAG,DAH,DAL,V18,V19,V20, V21,V22,ZAE,ZAF,ZAG,ZAH,ZAL	1
22	PAOZZ	5940003996676	19207	8338564	.TERMINAL ASSEMBLY FUEL TANK...................... SWITCH LEAD... UOC:DAE,DAF,DAG,DAH,DAL,V18,V19,V20, V21,V22,ZAE,ZAF,ZAG,ZAH,ZAL	2
23	PAOZZ	5970008338562	19207	8338562	.INSULATOR,BUSHING FUEL TANK...................... SWITCH LEAD .... UOC:DAE,DAF,DAG,DAH,DAL,V18,V19,V20, V21,V22,ZAE,ZAF,ZAG,ZAH,ZAL -	2
24	PFOZZ	5935003996673	19207	7982401	.SHELL,ELECTRICAL CO FUEL TANK....................	2

(1) ITEM NO	(2) SMR CODE	(3) NSN	(4) CAGEC	(5) PART NUMBER	(6) DESCRIPTION AND USABLE ON CODES (UOC)	(7) QTY
					SWITCH LEAD .... UOC:DAE,DAF,DAG,DAH,DAL,V18,V19,V20, V21,V22, ZAE,ZAF,ZAG,ZAH,ZAL	
25	PAOZZ	9905007524649	81349	M43436/1-1	.BAND,MARKER FUEL TANK SWITCH LEAD........... UOC:DAE,DAF,DAG,DAH,DAL,V18,V19,V20, V21,V22, ZAE,ZAF,ZAG,ZAH,ZAL	2
26	PAOZZ	5310008775797	96906	MS21044-N3	NUT,SELF-LOCKING,HE UOC:DAE,DAF,DAG,DAH,DAL,V18,V19,V20, V21,V22,ZAE,ZAF,ZAG,ZAH,ZAL	2
27	PAOZZ	5305009897435	96906	MS35207-264	SCREW,MACHINE	8
28	PAOZZ	5930003078856	96906	MS51113-1	SWITCH ASSEMBLY	1

END OF FIGURE

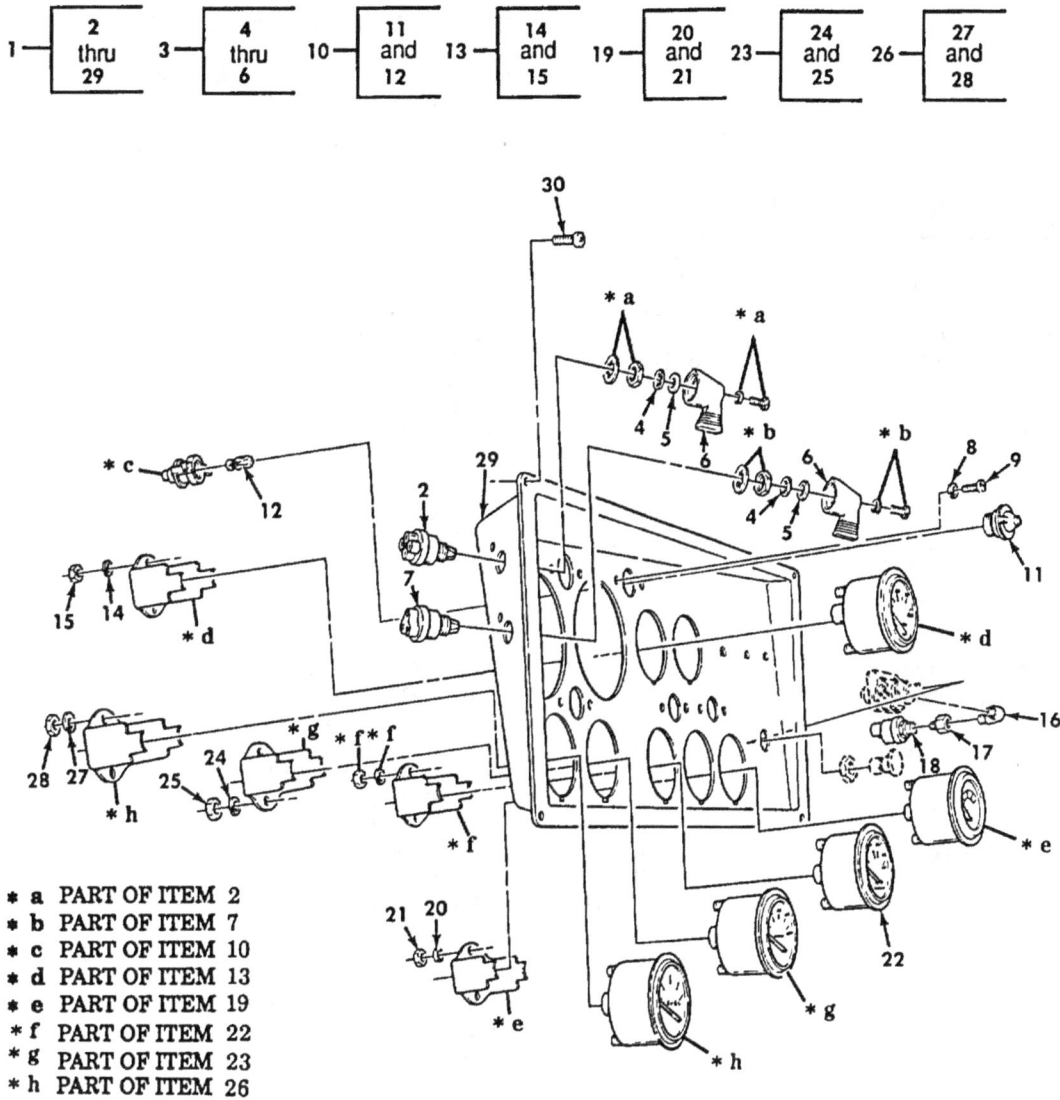

* a   PART OF ITEM 2
* b   PART OF ITEM 7
* c   PART OF ITEM 10
* d   PART OF ITEM 13
* e   PART OF ITEM 19
* f   PART OF ITEM 22
* g   PART OF ITEM 23
* h   PART OF ITEM 26

Figure 132. Instrument Cluster Assembly.

(1) ITEM NO	(2) SMR CODE	(3) NSN	(4) CAGEC	(5) PART NUMBER	(6) DESCRIPTION AND USABLE ON CODES (UOC)	(7) QTY
					GROUP 0607 SWITCHES AND CIRCUIT BREAKERS	
					FIG. 132 INSTRUMENT CLUSTER ASSEMBLY	
1	PBOZZ	6695011495830	19207	12277176-1	INDICATOR ASSEMBLY	1
2	PAOZZ	5930006999438	96906	MS39060-2	.SWITCH,ROTARY	1
3	PAOZZ	5930001305349	19207	5381088	.HANDLE,SWITCH...	2
4	PAOZZ	5310011085236	19207	5381087	..WASHER,SHOULDERED...............................	2
5	XDOZZ		19207	7392454	..FELT,MECHANICAL,PRE	2
6	PAOZZ	5930011978661	19207	5381051	..LEVER,MANUAL CONTRO............................	2
7	PAOZZ	5930001345036	19207	11614131	.SWITCH,ROTARY	1
8	PAOZZ	5310000453299	96906	MS35338-42	.WASHER,LOCK	10
9	PAOZZ	5305009846189	96906	MS35206-241	.SCREW,MACHINE	10
10	PAOZZ	6210006999458	83058	160403	.LIGHT,INDICATOR	5
11	PFOZZ	6210003377345	19207	7358621	..LENS,LIGHT	5
12	PAOZZ	6240002669940	96906	MS25231-1829	..LAMP,INCANDESCENT	5
13	PAOZZ	6625009362139	96906	MS24543-2	.METER,SPECIAL SCALE	1
14	PAOZZ	5310000453296	96906	MS35338-43	..WASHER,LOCK	2
15	PAOZZ	5310009349751	96906	MS35650-302	..NUT,PLAIN,HEXAGON.................................	2
16	PAOZZ	4730002871649	21450	127960	.ELBOW,PIPE	1
17	PAOZZ	4730002268874	30780	1-4X1-8FGS	.REDUCER,PIPE...	1
18	PAOZZ	5930011890494	19207	12375453	.SWITCH,PRESSURE	1
19	PAOZZ	6625010869580	96906	MS24532-2REVG	.METER,SPECIAL SCALE	1
20	PAOZZ	5310000453296	96906	MS35338-43	..WASHER,LOCK.	2
21	PAOZZ	5310009349751	96906	MS35650-302	..NUT,PLAIN,HEXAGON.................................	2
22	PAOZZ	6685010985110	19207	11669355	.GAGE,PRESSURE,DIAL..................................	1
23	PAOZZ	6620001159042	96906	MS24540-2	.INDICATOR,PRESSURE	1
24	PAOZZ	5310000453296	96906	MS35338-43	..WASHER,LOCK	2
25	PAOZZ	5310009349751	96906	MS35650-302	..NUT,PLAIN,HEXAGON.................................	2
26	PAOZZ	6680009333600	96906	MS24544-2	.INDICATOR,LIQUID QU	1
27	PAOZZ	5310000453296	96906	MS35338-43	..WASHER,LOCK	2
28	PAOZZ	5310009349751	96906	MS35650-302	..NUT,PLAIN,HEXAGON.................................	2
29	PFOZZ	2590011307935	19207	12277175	.PANEL,BLANK .... UOC:DAA,DAB,DAC,DAD,DAE,DAF,DAG,DAH, DAJ,DAK,DAL,DAW,DAX,V12,V13,V14,V15, V16,V17,V18,V19,V20,V21,V22,V24,V25, V39	1
30	PAOZZ	5305009897435	96906	MS35207-264	SCREW,MACHINE	8

END OF FIGURE

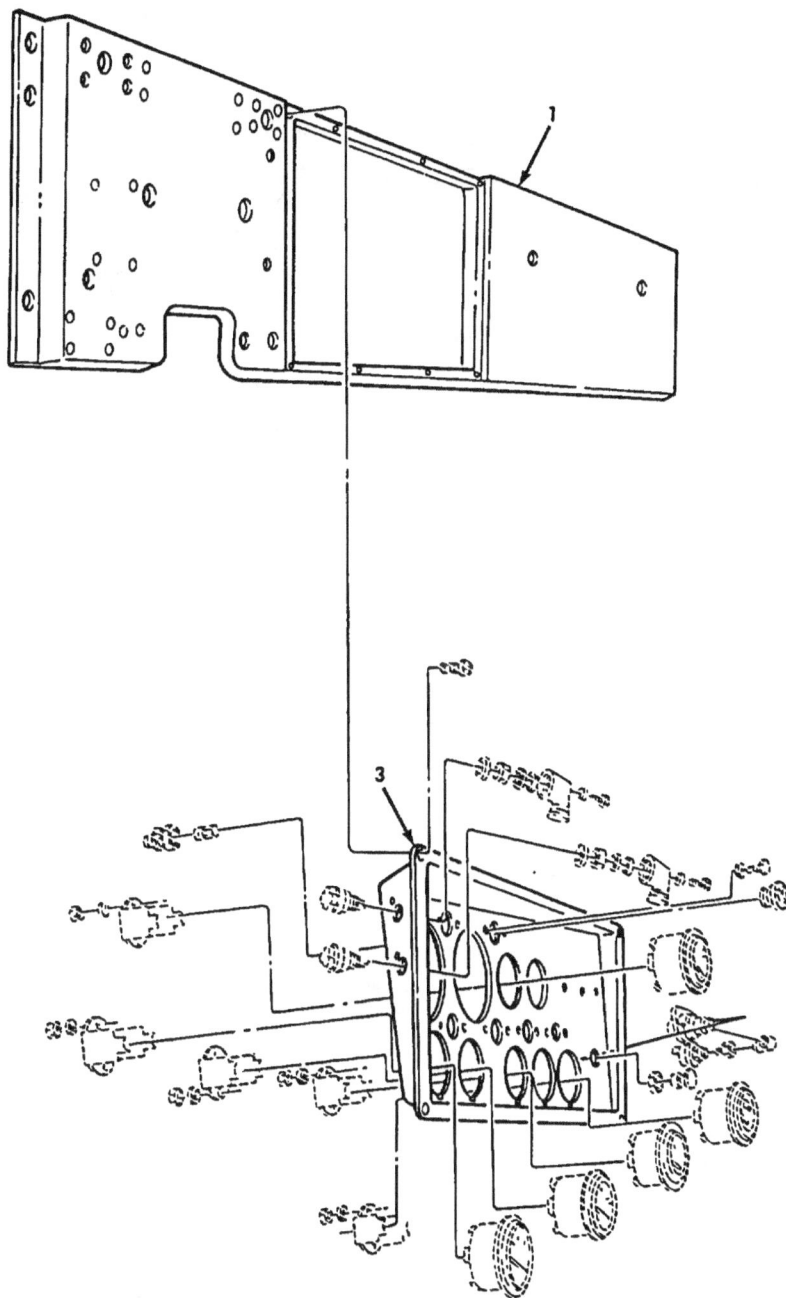

Figure 133. Instrument Cluster Assembly (M939A2).

(1) ITEM NO	(2) SMR CODE	(3) NSN	(4) CAGEC	(5) PART NUMBER	(6) DESCRIPTION AND USABLE ON CODES (UOC)	(7) QTY

SECTION II

GROUP 0607 SWITCHES AND CIRCUIT BREAKERS

FIG. 133 INSTRUMENT CLUSTER ASSEMBLY (M939A2)

(1) ITEM NO	(2) SMR CODE	(3) NSN	(4) CAGEC	(5) PART NUMBER	(6) DESCRIPTION AND USABLE ON CODES (UOC)	(7) QTY
1	PFOZZ	2510012804155	47457	12277175A	PANEL,VEHICULAR OPE .... UOC:ZAA,ZAB,ZAC,ZAD,ZAE,ZAF,ZAG,ZAH, ZAJ,ZAK,ZAL	1
1	PFOZZ	5340014181936	19207	12277175-1	PANEL,BLANK.. UOC:ZAA,ZAB,ZAC,ZAD,ZAE,ZAF,ZAG,ZAH, ZAJ,ZAK,ZAL	1
2	XDOZZ		47457	12277176A	PANEL,VEHICULAR.. UOC:ZAA,ZAB,ZAC,ZAD,ZAE,ZAF,ZAG,ZAH, ZAJ,ZAK,ZAL	1
2	PBOOO	6695014174453	19207	12277176-3	INDICATOR ASSEMBLY.. UOC:ZAA,ZAB,ZAC,ZAD,ZAE,ZAF,ZAG,ZAH, ZAJ,ZAK,ZAL	1
3	PFOZZ	2510012854592	47457	12277177A	.PANEL,VEHICULAR OPE.. UOC:ZAA,ZAB,ZAC,ZAD,ZAE,ZAF,ZAG,ZAH, ZAJ,ZAK,ZAL	1
3	PFOZZ	2510011896401	19207	12277177	.PANEL,VEHICULAR OPE.. UOC:ZAA,ZAB,ZAC,ZAD,ZAE,ZAF,ZAG,ZAH, ZAJ,ZAK,ZAL	1

END OF FIGURE

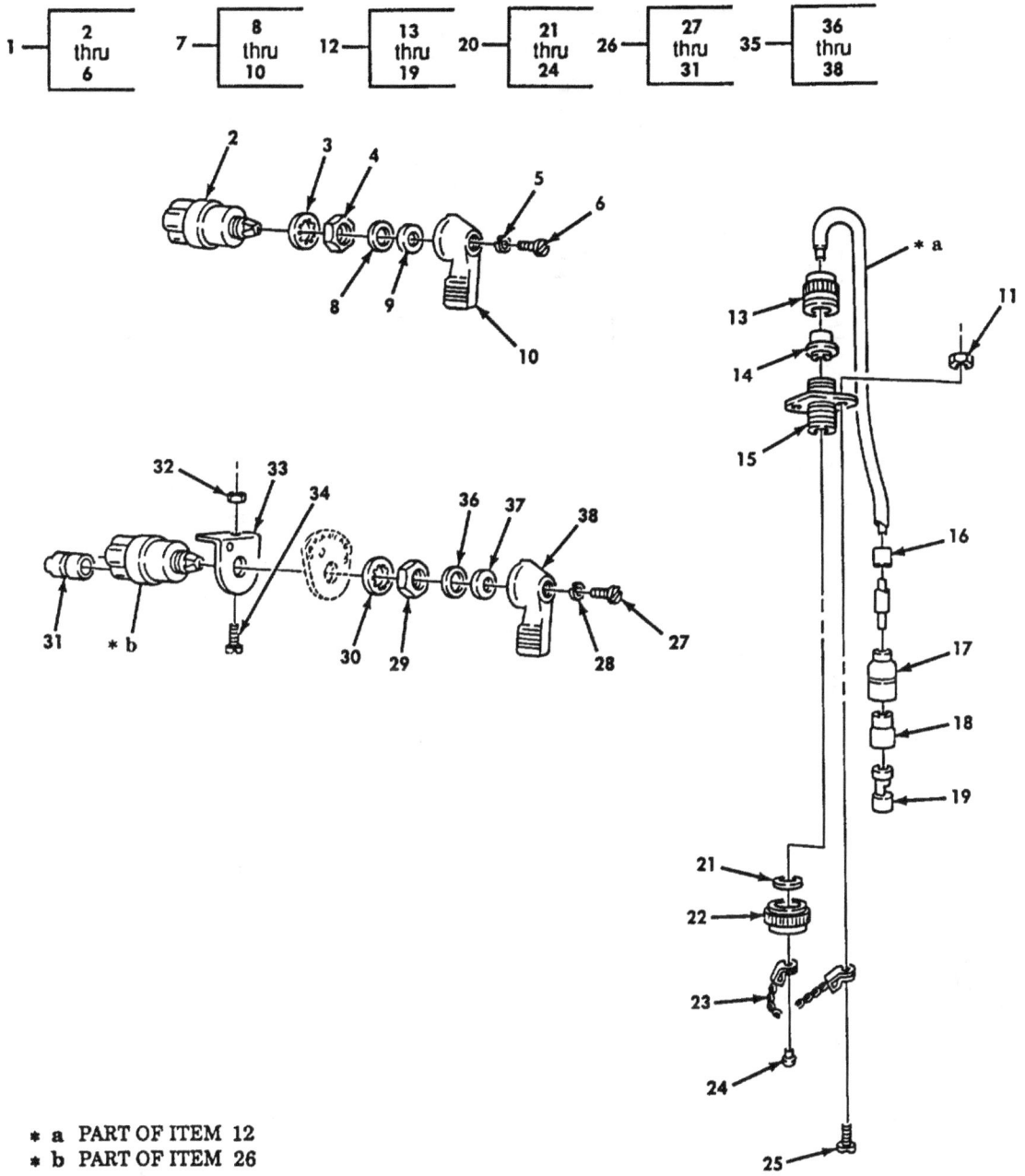

*a PART OF ITEM 12
*b PART OF ITEM 26

Figure 134. Instrument Panel, Floodlight, Heater, Warning Switches, Lamp Holder, and Auxiliary Receptacle
(Sheet 1 of 2).

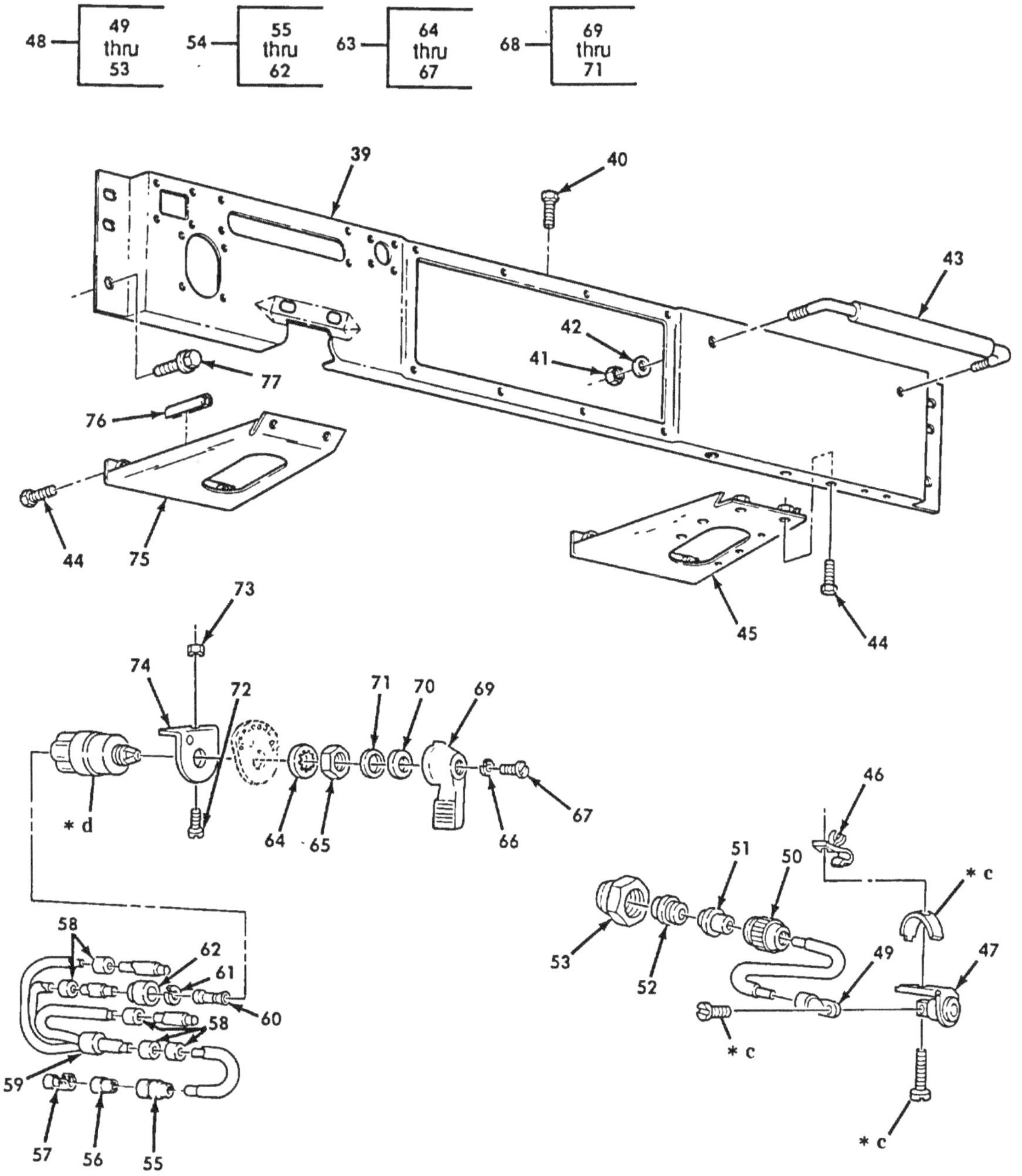

*c PART OF ITEM 47
*d PART OF ITEM 63

Figure 134. Instrument Panel, Floodlight, Heater, Warning Switches, Lamp Holder, and Auxiliary Receptacle
(Sheet 2 of 2).

(1) ITEM NO	(2) SMR CODE	(3) NSN	(4) CAGEC	(5) PART NUMBER	(6) DESCRIPTION AND USABLE ON CODES (UOC)	(7) QTY
					GROUP 0607 SWITCHES AND CIRCUIT BREAKERS	
					FIG. 134 INSTRUMENT PANEL, FLOODLIGHT, HEATER,WARNING SWITCHES, LAMP HOLDER, AND AUXILIARY RECEPTACLE	
1	PAOZZ	5930010986743	19207	11669214	SWITCH,ROTARY	1
2	XAOZZ		19207	12375495	.SWITCH,ROTARY	1
3	PAOZZ	5310001941483	96906	MS35333-44	.WASHER,LOCK	1
4	PAOZZ	5310008326852	19207	5381233	.NUT,PLAIN,HEXAGON	1
5	PAOZZ	5310000453299	96906	MS35338-42	.WASHER,LOCK	1
6	PAOZZ	5305006143423	96906	MS35265-43	.SCREW,MACHINE	1
7	PAOZZ	5930001305349	19207	5381088	HANDLE,SWITCH..	1
8	PAOZZ	5310011085236	19207	5381087	.WASHER,SHOULDERED..	1
9	XDOZZ		19207	7392454	.FELT,MECHANICAL,PRE..	1
10	PAOZZ	5930011978661	19207	5381051	.LEVER,MANUAL CONTRO.......................................	1
11	PAOZZ	5310009349755	96906	MS35650-362	NUT,PLAIN,HEXAGON UOC:DAL,V18,ZAL	4
12	PAOZZ	5995001778220	19207	7363002	LEAD,ELECTRICAL... UOC:DAL,V18,ZAL	1
13	PAOZZ	5935003333088	19207	7723306	.NUT,BUSHING RETAINE UOC:DAL,V18,ZAL	1
14	PAOZZ	5365007722343	19207	7722343	.BUSHING,NONMETALLIC.. UOC:DAL,V18,ZAL	1
15	PAOZZ	5935011303536	19207	7720497	.CONNECTOR BODY,RECE UOC:DAL,V18,ZAL	1
16	PAOZZ	9905007524649	81349	M43436/1-1	.BAND,MARKER UOC:DAL,V18,ZAL	1
17	PAOZZ	5935008338561	19207	8338561	.SHELL,ELECTRICALCO.. UOC:DAL,V18,ZAL	1
18	PAOZZ	5970008338562	19207	8338562	.INSULATOR,BUSHING UOC:DAL,V18,ZAL	1
19	PAOZZ	1015007982997	19207	7982997	.TERMINAL,SOLDEREDF UOC:DAL,V18,ZAL	1
20	PAOZZ	5340007261670	77820	60-37398-12	.CAP,PROTECTIVE,DUST... UOC:DAL,V18,ZAL	
21	PAOZZ	5330009905804	19207	8701226	..DISK,SOLID,PLAIN UOC:DAL,V18,ZAL	1
22	PAOZZ	5999011234557	19207	8701287	..CAP,ELECTRICAL.. UOC:DAL,V18,ZAL	1
23	PAOZZ	4010010394831	19207	8701233	..CHAIN,BEAD UOC:DAL,V18,ZAL	1
24	PFOZZ	5320010297722	19207	8701223	..RIVET,SOLID.: UOC:DAL,V18,ZAL	1
25	PAOZZ	5305011251181	21450	425544	SCREW,ASSEMBLED WAS UOC:DAL,V18,ZAL	4
26	PAOZZ	5930008980500	81349	MIL-S-13623/1-1	SWITCH,ROTARY UOC:DAL,V18,ZAL	1
27	PAOZZ	5305006143423	96906	MS35265-43	SCREW,MACHINE UOC:DAL,V18,ZAL	1

(1) ITEM NO	(2) SMR CODE	(3) NSN	(4) CAGEC	(5) PART NUMBER	(6) DESCRIPTION AND USABLE ON CODES (UOC)	(7) QTY
29	PAOZZ	5310008326852	19207	5381233	NUT,PLAIN,HEXAGON UOC:DAL,V18,ZAL	1
30	PAOZZ	5310001941483	96906	MS35333-44	WASHER,LOCK UOC:DAL,V18,ZAL	1
31	PAOZZ	5935008338561	19207	8338561	SHELL,ELECTRICAL CO UOC:DAL,V18,ZAL	1
32	PAOZZ	5310008775797	96906	MS21044-N3	NUT,SELF-LOCKING,HE UOC:DAL,V18,ZAL	4
33	PFOZZ	5340001760868	19207	7059461	BRACKET,ANGLE UOC:DAL,V18,ZAL	2
34	PAOZZ	5305009906444	96906	MS35207-261	SCREW,MACHINE UOC:DAL,V18,ZAL	4
35	PAOZZ	5930001305349	19207	5381088	HANDLE,SWITCH..................................... UOC:DAL,V18,ZAL	2
36	PAOZZ	5310011085236	19207	5381087	.WASHER,SHOULDERED............................. UOC:DAL,V18,ZAL	2
37	XDOZZ		19207	7392454	.FELT,MECHANICAL,PRE UOC:DAL,V18,ZAL	2
38	PAOZZ	5930011978661	19207	5381051	.LEVER,MANUAL CONTRO.......................... UOC:DAL,V18,ZAL	2
39	PGOZZ	2510011896401	19207	12277177	PANEL,VEHICULAR OPE	1
40	PAOZZ	5305002678953	80204	B1821BH025F063N	SCREW,CAP,HEXAGON H	5
41	PAOZZ	5310002416658	96906	MS51943-34	NUT,SELF-LOCKING,HE	2
42	PAOZZ	5310000814219	96906	MS27183-12	WASHER,FLAT...	2
43	PFOZZ	5340011310108	19207	12255921	HANDLE,MANUAL CONTR	1
44	PAOZZ	5305010907626	19207	7372083-1	SCREW,ASSEMBLED WAS	6
45	PFOZZ	5340010907644	19207	12255731-2	BRACKET,MOUNTING................................	1
46	PAOZZ	5340011076971	96906	MS51928-1	CLIP,SPRING TENSION UOC:DAL,V18,ZAL	1
47	PAOZZ	6250007418960	13445	EX2235	LAMPHOLDER... UOC:DAL,V18,ZAL	1
48	PFOZZ	5995000571642	19207	7418959	LEAD,ELECTRICAL.................................... UOC:DAL,V18,ZAL	1
49	PAOZZ	5940007056708	19207	7056708	.TERMINAL,LUG UOC:DAL,V18,ZAL	1
50	PAOZA	5935007720495	19207	8724198	.CONNECTOR,PLUG,ELEC UOC:DAL,V18,ZAL	1
51	PFOZZ	5365007722343	19207	7722343	.BUSHING,NONMETALLIC AUXILIARY .... OUTLET CABLE....................................... UOC:DAL,V18,ZAL	1
52	PAOZZ	5935003333088	19207	7723306	.NUT,BUSHING RETAINE UOC:DAL,V18,ZAL	1
53	PAOZZ	5975005227125	77820	10-40457-12S	.NUT,COUPLING,ELECTR UOC:DAL,V18,ZAL	1
54	PAOZZ	2590011576240	19207	12277230	LEAD ASSEMBLY,ELECT UOC:DAL,V18,ZAL	1
55	PAOZZ	5935008338561	19207	8338561	.SHELL,ELECTRICAL CO............................. UOC:DAL,V18,ZAL	3
56	PAOZZ	5970008338562	19207	8338562	.INSULATOR,BUSHING UOC:DAL,V18,ZAL	3
57	PAOZZ	5940003996676	19207	8338564	.TERMINAL ASSEMBLY FLOODLIGHT................... SWITCH	3

(1) ITEM NO	(2) SMR CODE	(3) NSN	(4) CAGEC	(5) PART NUMBER	(6) DESCRIPTION AND USABLE ON CODES (UOC)	(7) QTY
					UOC:DAL,V18 ,ZAL	
59	PAOZA	5999000572929	19204	572929	.CONTACT,ELECTRICAL.....................................	1
					UOC:DAL,V18,ZAL	
60	PAOZZ	5310008338567	19207	8338567	.WASHER,SLOTTED..	1
					UOC:DAL,V18,ZAL	
61	PAOZZ	5935005729180	19207	8338566	.SHELL,ELECTRICAL CO..	1
					UOC:DAL,V18,ZAL	
62	PAOZZ	9905007524649	81349	M43436/1-1	.BAND,MARKER FLOODLIGHT SWITCH LEAD........	5
					UOC:DAL,V18,ZAL	
63	PAOZZ	5930006999438	96906	MS39060-2	SWITCH,ROTARY	1
					UOC:DAL,V18,ZAL	
64	PAOZZ	5310001941483	96906	MS35333-44	WASHER,LOCK..	1
					UOC:DAL,V18,ZAL	
65	PAOZZ	5310008326852	19207	5381233	NUT,PLAIN,HEXAGON	1
					UOC:DAL,V18,ZAL	
66	PAOZZ	5310000453299	96906	MS35338-42	WASHER,LOCK..	1
					UOC:DAL,V18,ZAL	
67	PAOZZ	5305006143423	96906	MS35265-43	SCREW,MACHINE	1
					UOC:DAL,V18,ZAL	
68	PAOZZ	5930001305349	19207	5381088	HANDLE,SWITCH..	1
					UOC:DAL,V18,ZAL	
69	PAOZZ	5930011978661	19207	5381051	.LEVER,MANUAL CONTRO..................................	1
					UOC:DAL,V18,ZAL	
70	XDOZZ		19207	7392454	.FELT,MECHANICAL,PRE....................................	1
					UOC:DAL,V18,ZAL	
71	PAOZZ	5310011085236	19207	5381087	.WASHER,SHOULDERED....................................	1
					UOC:DAL,V18,ZAL	
72	PAOZZ	5305009906444	96906	MS35207-261	SCREW,MACHINE FUEL TANK GAGE.................... SELECTOR	1
					UOC:DAL,V18,ZAL	
73	PAOZZ	5310008775797	96906	MS21044-N3	NUT,SELF-LOCKING,HE	2
					UOC:DAL,V18,ZAL	
74	PFOZZ	5340001760868	19207	7059461	BRACKET,ANGLE	1
					UOC:DAL,V18,ZAL	
75	PAOZZ	5340010907643	19207	12255731-1	PLATE,MENDING..	1
76	XDOZZ	9330001656006	19207	11599001	TUBING,NONMETALLIC	1
77	PAOZZ	5306000425859	24617	425859	BOLT,ASSEMBLED WASH..................................	2

END OF FIGURE

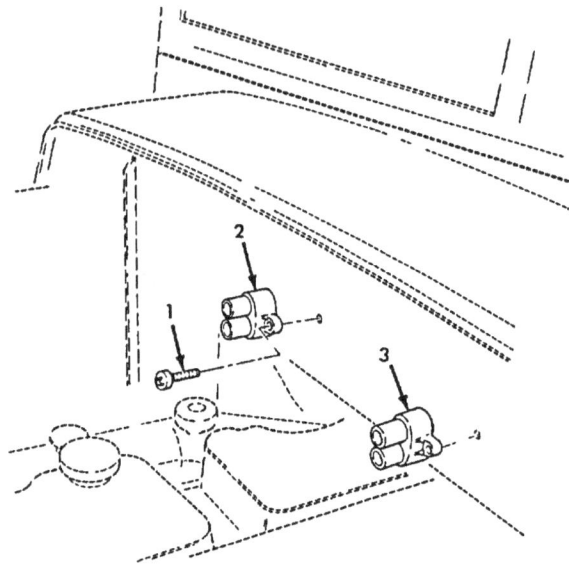

FRONT COWL

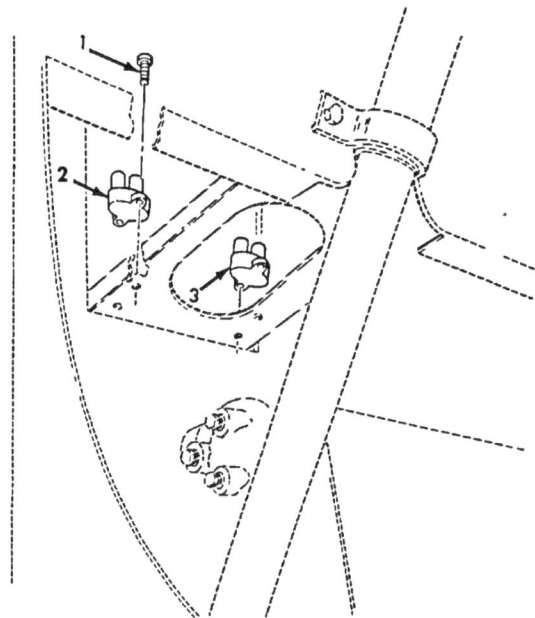

DASH UNDERSIDE

Figure 135. Circuit Breakers (M939, M939A1).

(1) ITEM NO	(2) SMR CODE	(3) NSN	(4) CAGEC	(5) PART NUMBER	(6) DESCRIPTION AND USABLE ON CODES (UOC)	(7) QTY
					GROUP 0607 SWITCHES AND CIRCUIT BREAKERS	
					FIG. 135 CIRCUIT BREAKERS(M939, M939A1)	
1	PAOZZ	5305008550965	96906	MS24629-38	SCREW,TAPPING...	8
2	PAOZZ	5925014302318	58536	AA55571/01-001	CIRCUIT BREAKER...	2
3	PAOZZ	5925003331584	58536	AA55571/01-004	CIRCUIT BREAKER...	2

END OF FIGURE

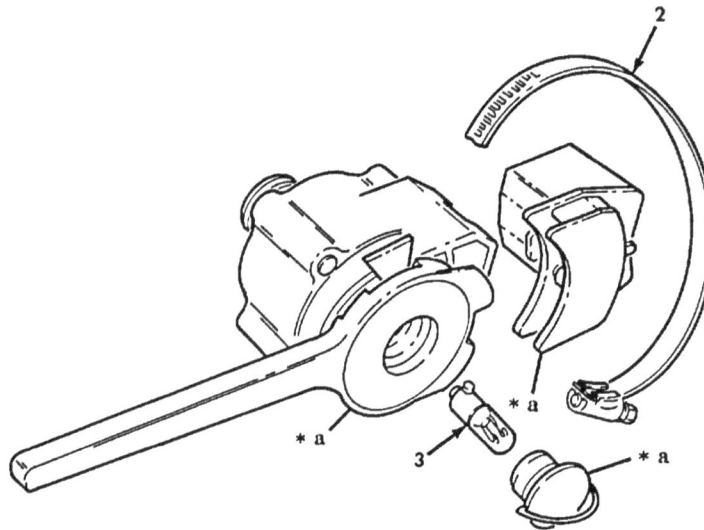

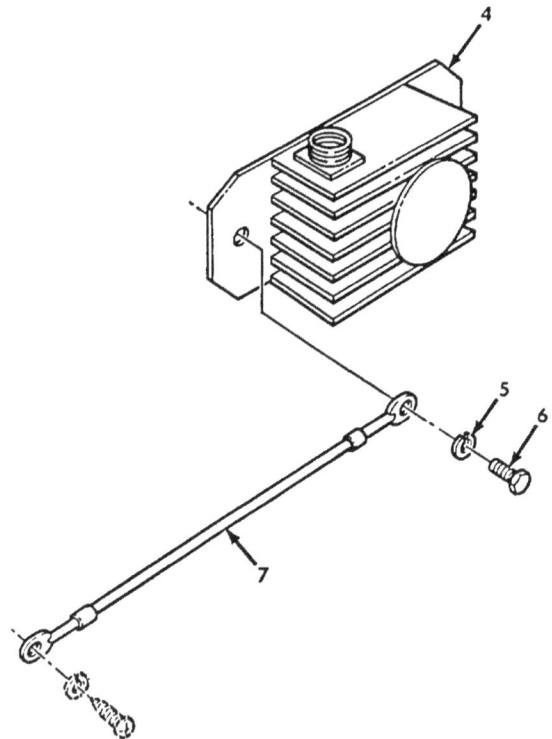

* a   PART OF ITEM 1

Figure 136. Control Unit and Directional Flasher.

(1) ITEM NO	(2) SMR CODE	(3) NSN	(4) CAGEC	(5) PART NUMBER	(6) DESCRIPTION AND USABLE ON CODES (UOC)	(7) QTY
					GROUP 0607 DIRECTIONAL TURN INDICATOR CONTROL	
					FIG. 136 CONTROL UNIT AND DIRECTIONAL FLASHER	
1	PAOZZ		19207	11613632-3	CONTROL,DIRECTIONAL..................................	1
2	PAOZZ		96906	MS35842-14	.CLAMP...	1
3	PAOZZ	6240004193185	96906	MS25231-1873	.LAMP,INCANDESCENT..............................	1
4	PAOZZ	5945007893706	19207	11613631	FLASHER,THERMAL..................................	1
5	PAOZZ		96906	MS35333-40	WASHER,LOCK.....................................	2
6	PAOZZ		96906	MS90727-9	SCREW,CAP,HEXAGON H............................	2
7	PFOZZ	5995011475423	19207	12256051	LEAD,ELECTRICAL..................................	1

END OF FIGURE

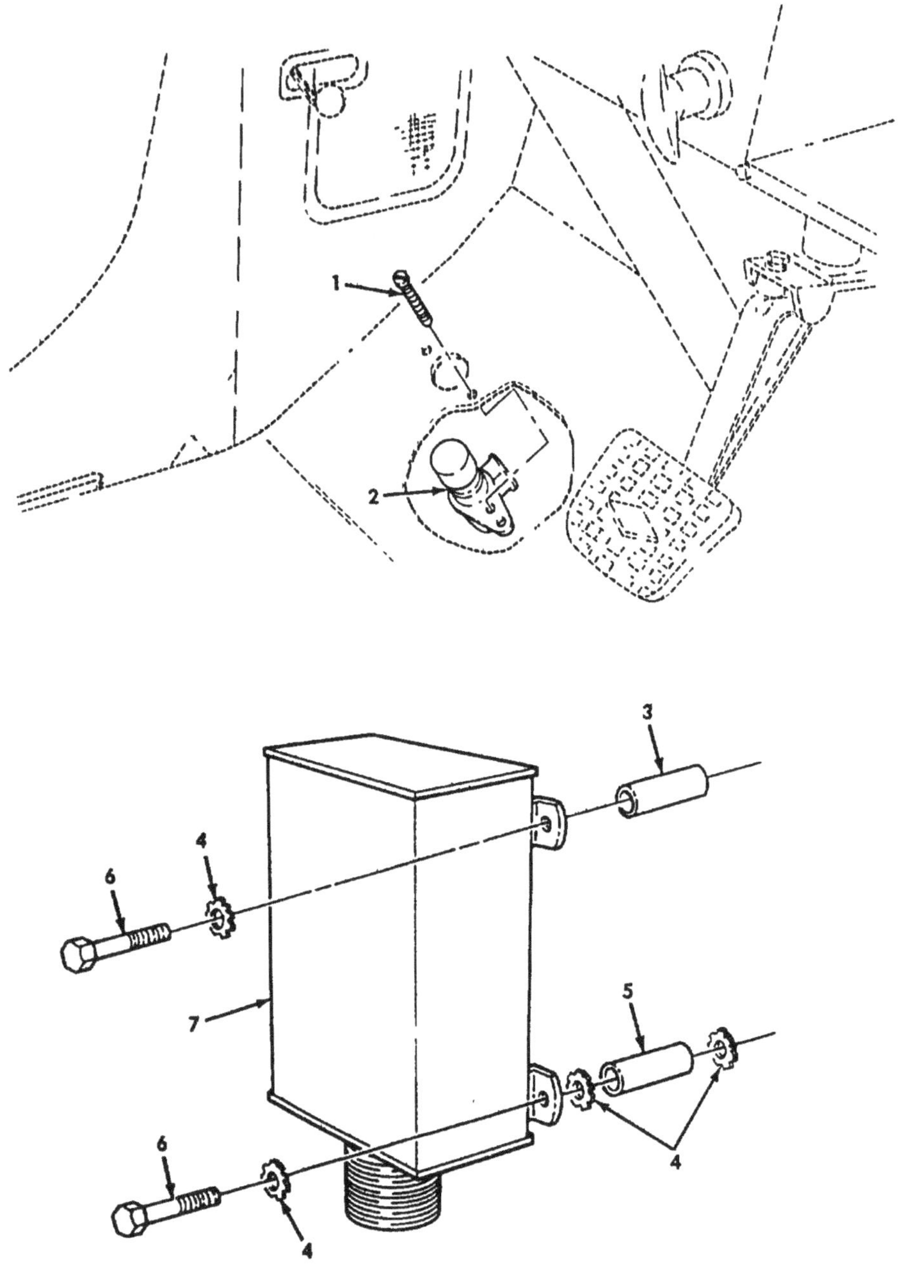

Figure 137. Protective Control Box, High Beam Switch, and Mounting Hardware.

(1) ITEM NO	(2) SMR CODE	(3) NSN	(4) CAGEC	(5) PART NUMBER	(6) DESCRIPTION AND USABLE ON CODES (UOC)	(7) QTY
					GROUP 0607 DIRECTIONAL TURN INDICATOR CONTROL	
					FIG. 137 PROTECTIVE CONTROL BOX, HIGH BEAM SWITCH, AND MOUNTING HARDWARE	
1	PAOZZ	5305009881724	96906	MS35206-280	SCREW,MACHINE	2
2	PAOZZ	2590008012355	96906	MS53000-1	SWITCH,BEAM SELECTI	1
3	PAOZZ	5365011066060	19207	12256063-1	SPACER,SLETVE	2
4	PAOZZ	5310007282044	96906	MS45904-73	WASHER,LOCK...	8
5	PAOZZ	5365011176655	19207	12256063-3	SPACER,SLEEVE	2
6	PAOZZ	5306000514084	96906	MS90727-42	BOLT,MACHINE..	4
7	PAOZZ	6110014442546	19207	12450333	CONTROL,LIGHT SOURC....................................	1

END OF FIGURE

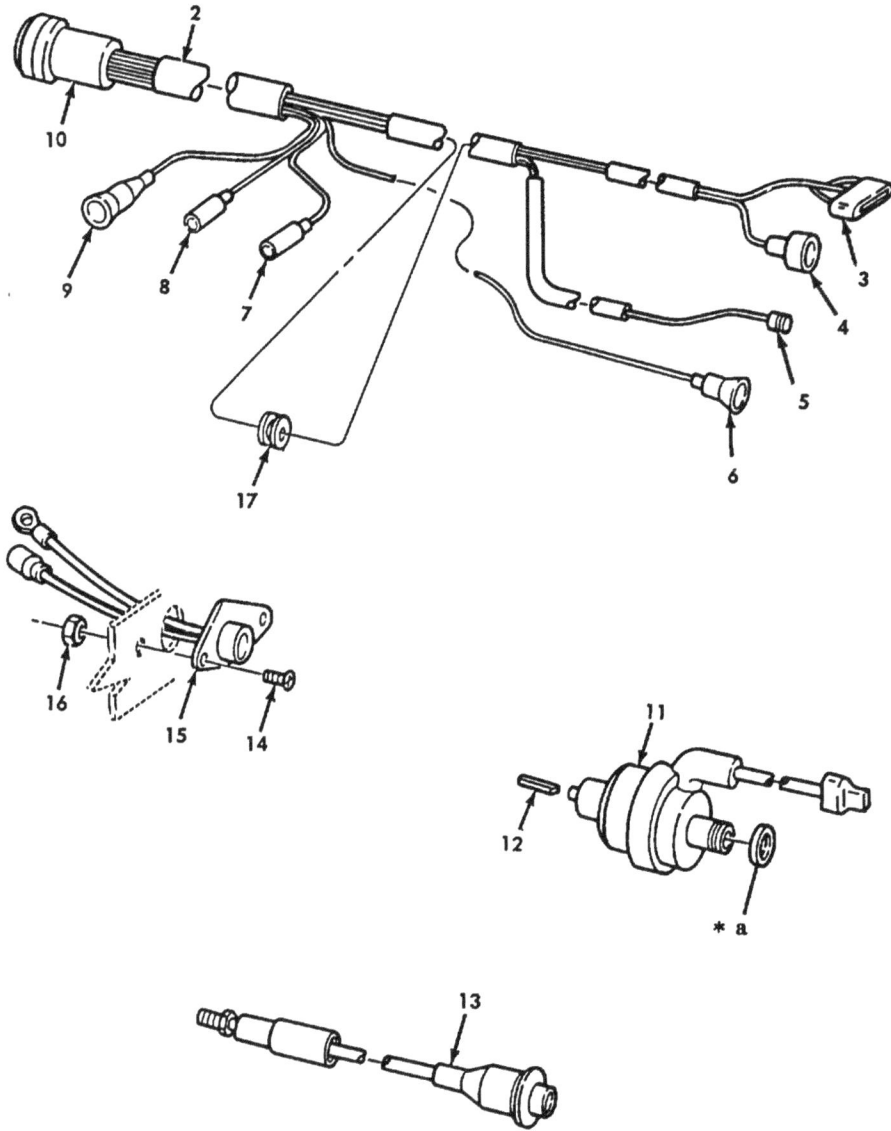

*a PART OF ITEM 11

Figure 138. CTIS Electrical Wiring Harness (M939A2).

(1) ITEM NO	(2) SMR CODE	(3) NSN	(4) CAGEC	(5) PART NUMBER	(6) DESCRIPTION AND USABLE ON CODES (UOC)	(7) QTY
					GROUP 0608 MISCELLANEOUS ITEMS	
					FIG. 138 CTIS ELECTRICAL WIRING HARNESS (M939A2)	
1	PAFFF	5995012901293	52304	599669	WIRING HARNESS .... UOC:ZAA,ZAB,ZAC,ZAD,ZAE,ZAF,ZAG,ZAH, ZAJ,ZAK	1
1	PAFFF	5995012901294	52304	599763	WIRING HARNESS .... UOC:ZAL	1
2	XAFZF	5995012901293	52304	599669	.WIRING HARNESS.................................. UOC:ZAA,ZAB,ZAC,ZAD,ZAE,ZAF,ZAG,ZAH, ZAJ,ZAK	1
3	XAOZZ	6150012821888	52304	599747	.LEAD,ELECTRICAL................................. UOC:ZAA,ZAB,ZAC,ZAD,ZAE,ZAF,ZAG,ZAH, ZAJ,ZAK	1
3	XBOZZ		52304	599980	.LEAD,ELECTRICAL................................. UOC:ZAL	1
4	XAOZZ	6150012822785	52304	599752	.CABLE ASSEMBLY,SPEC TRANSDUCER............. UOC:ZAA,ZAB,ZAC,ZAD,ZAE,ZAF,ZAG,ZAH, ZAJ,ZAK,ZAL	1
5	PAOZZ	5935012874286	52304	599753	.CONNECTOR,PLUG,ELEC COIL .... UOC:ZAA,ZAB,ZAC,ZAD,ZAE,ZAF,ZAG,ZAH, ZAJ,ZAK,ZAL	1
6	XAOZZ	6150012822786	52304	599754	.CABLE ASSEMBLY,SPEC SWITCH..................... UOC:ZAA,ZAB,ZAC,ZAD,ZAE,ZAF,ZAG,ZAH, ZAJ,ZAK,ZAL	1
7	XAOZZ	6150012821885	52304	599971	.CABLE ASSEMBLY,SPEC SENSOR .... UOC:ZAA,ZAB,ZAC,ZAD,ZAE,ZAF,ZAG,ZAH, ZAJ,ZAK,ZAL	1
8	XAOZZ		52304	599751	.LEAD,ELECTRICAL LIGHT................................ UOC:ZAA,ZAB,ZAC,ZAD,ZAE,ZAF,ZAG,ZAH, ZAJ,ZAK,ZAL	1
9	XAOZZ	6150012821890	52304	599982	.LEAD,ELECTRICAL................................. UOC:ZAA,ZAB,ZAC,ZAD,ZAE,ZAF,ZAG,ZAH, ZAJ,ZAK,ZAL	1
10	XAOZZ	5935012821881	65884	SIA-2017	.CONNECTOR,PLUG,ELEC AND GROUND............. UOC:ZAA,ZAB,ZAC,ZAD,ZAE,ZAF,ZAG,ZAH, ZAJ,ZAK,ZAL	1
11	PAOZZ	6625012892062	52304	599602	GENERATOR,SIGNAL SPEED .... UOC:ZAA,ZAB,ZAC,ZAD,ZAE,ZAF,ZAG,ZAH, ZAJ,ZAK,ZAL	1
12	PAOZZ	5315013842149	78388	SA-2677	PIN,SHOULDER,HEADLE .... UOC:ZAA,ZAB,ZAC,ZAD,ZAE,ZAF,ZAG,ZAH, ZAJ,ZAK,ZAL	1
13	PAOZZ	5930012873965	52304	599603	SWITCH,PRESSURE................................. UOC:ZAA,ZAB,ZAC,ZAD,ZAE,ZAF,ZAG,ZAH, ZAJ,ZAK,ZAL	1
14	PAOZZ	5305012718332	24617	159358	SCREW,MACHINE .... UOC:ZAA,ZAB,ZAC,ZAD,ZAE,ZAF,ZAG,ZAH, ZAJ,ZAK,ZAL	2
15	PAOZZ	6220012730177	96139	IRS-WP-1254AP	LIGHT,WARNING	1

(1) ITEM NO	(2) SMR CODE	(3) NSN	(4) CAGEC	(5) PART NUMBER	(6) DESCRIPTION AND USABLE ON CODES (UOC)	(7) QTY
					UOC:ZAA,ZAB,ZAC,ZAD,ZAE,ZAF,ZAG,ZAH, ZAJ,ZAK,ZAL	
16	PAOZZ	5310009824935	96906	MS27040-6	NUT,PLAIN,SQUARE...	2
					UOC:ZAA,ZAB,ZAC,ZAD,ZAE,ZAF,ZAG,ZAH, ZAJ,ZAK,ZAL	
17	PAOZZ	5325002636648	96906	MS35489-135	GROMMET,NONMETALLIC....................................	1
					UOC:ZAA,ZAB,ZAC,ZAD,ZAE,ZAF,ZAG,ZAH, ZAJ,ZAK,ZAL	

END OF FIGURE

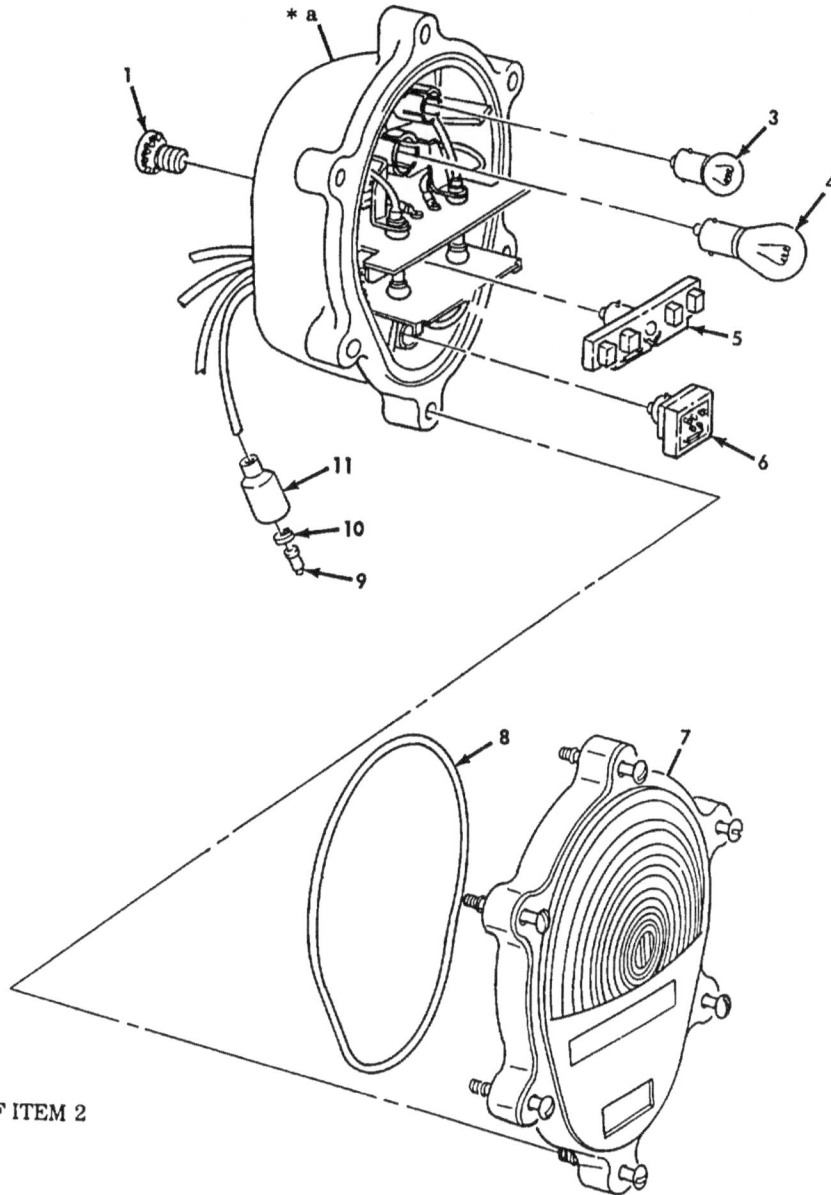

Figure 139. Taillight Assembly and Mounting Hardware.

* a PART OF ITEM 2

(1) ITEM NO	(2) SMR CODE	(3) NSN	(4) CAGEC	(5) PART NUMBER	(6) DESCRIPTION AND USABLE ON CODES (UOC)	(7) QTY
					GROUP 0609 LIGHTS	
					FIG. 139 TAILLIGHT ASSEMBLY AND MOUNTING HARDWARE	
1	PAOZZ	5305000423568	21450	423568	SCREW,ASSEMBLED WAS	4
2	PAOOO	6220013723883	19207	12375837	TAILLIGHT,VEHICULAR	2
3	PAOZZ	6240000193093	58536	A52463-1-09	.LAMP,INCANDESCENT	1
4	PAOZZ	6240000446914	08805	1683	.LAMP,INCANDESCENT	1
5	PAOZZ	6220012842709	19207	12360850-1	.LIGHT,MARKER,CLEARA	1
6	PAOZZ	6220012973217	19207	12360870-2	.STOP LIGHT,VEHICULA	1
7	PAOZZ	6220013592870	19207	12375841	.LENS,LIGHT USE WITH PLASTIC .... HOUSING ASSEMBLY ONLY	1
8	PAOZZ	5331004620907	19207	11639519-2	.O-RING	1
9	PAOZA	5999000572929	19204	572929	.CONTACT,ELECTRICAL	4
10	PAOZZ	5310008338567	19207	8338567	.WASHER,SLOTTED	4
11	PAOZZ	5935005729180	19207	8338566	.SHELL,ELECTRICAL CO	4

END OF FIGURE

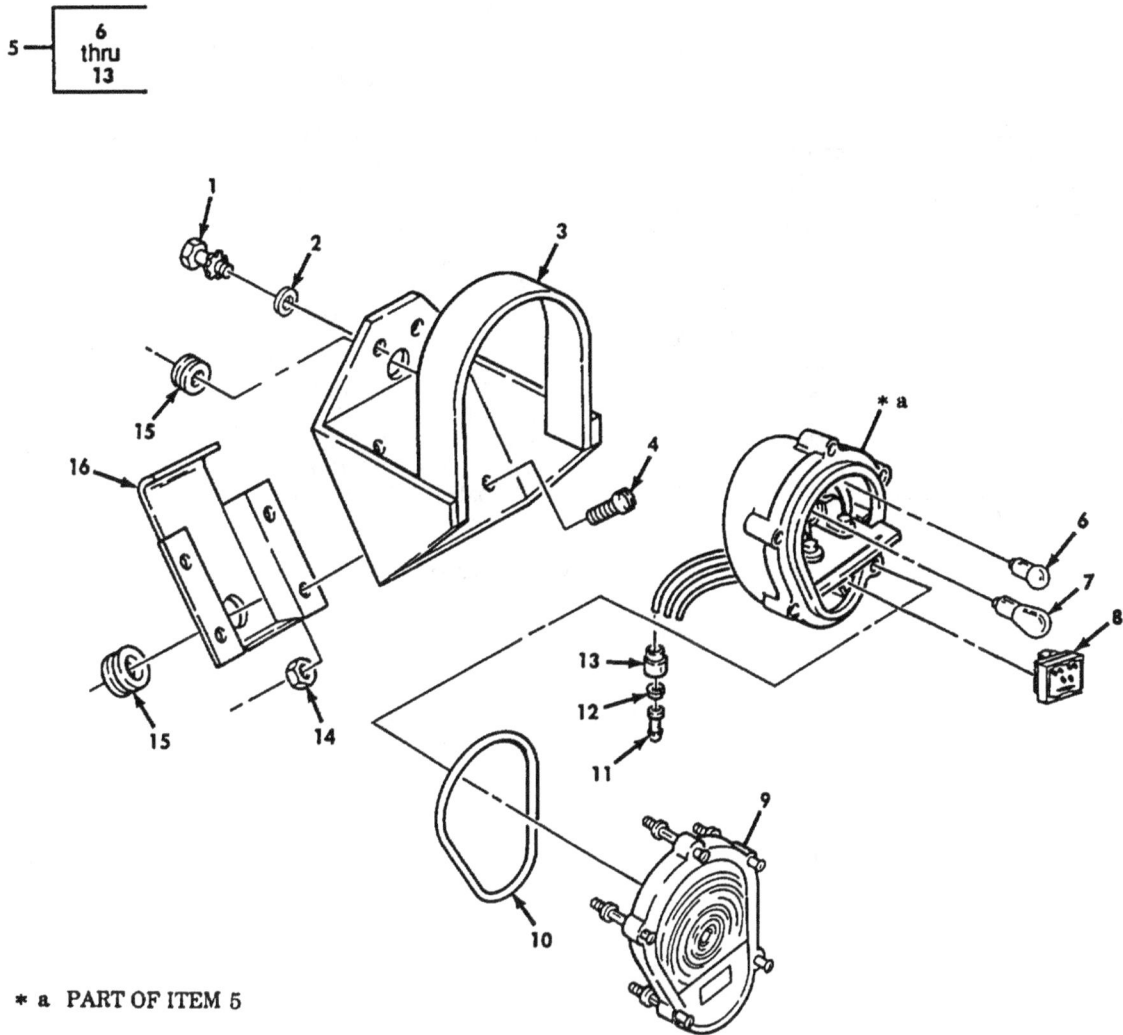

Figure 140. Front Composite Light and Mounting Hardware.

* a PART OF ITEM 5

(1) ITEM NO	(2) SMR CODE	(3) NSN	(4) CAGEC	(5) PART NUMBER	(6) DESCRIPTION AND USABLE ON CODES (UOC)	(7) QTY
					GROUP 0609 LIGHTS	
					FIG. 140 FRONT COMPOSITE LIGHT AND MOUNTING HARDWARE	
1	PAOZZ	5305000423568	21450	423568	SCREW, ASSEMBLED WAS	4
2	PAOZZ	5310005957237	96906	MS35333-42	WASHER, LOCK	4
3	PAOZZ	6220004285943	19207	11658687	LIGHT, HOOD, VEHICLE.................................	2
4	PAOZZ	5306000501238	96906	MS90727-32	BOLT, MACHINE	8
5	PAOOO	6220014438813	19207	12432437-1	LIGHT, PARKING	1
6	PAOZZ	6240000193093	58536	A52463-1-09	.LAMP, INCANDESCENT	1
7	PAOZZ	6240000446914	58536	A52463-2-10	.LAMP, INCANDESCENT	1
8	PAOZZ	5961013058848	19207	12360865	.SEMICONDUCTOR DEVIC	1
9	PAOZZ	6220014438805	19207	12432440	.LENS, LIGHT	1
10	PAOZZ	5331004630200	19207	11639519-1	.0-RING	1
11	PAOZA	5999000572929	19204	572929	.CONTACT, ELECTRICAL...............................	3
12	PAOZZ	5310008338567	19207	8338567	.WASHER, SLOTTED	3
13	PAOZZ	5935005729180	19207	8338566	.SHELL, ELECTRICAL CO	3
14	PAOZZ	5310002416658	96906	MS51943-34	NUT, SELF-LOCKING, HE	8
15	PAOZZ	5325002766091	96906	MS35489-19	GROMMET, NONMETALLIC	4
16	PAOZZ	5340001781036	19207	7397758	COVER, ACCESS	2

END OF FIGURE

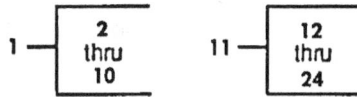

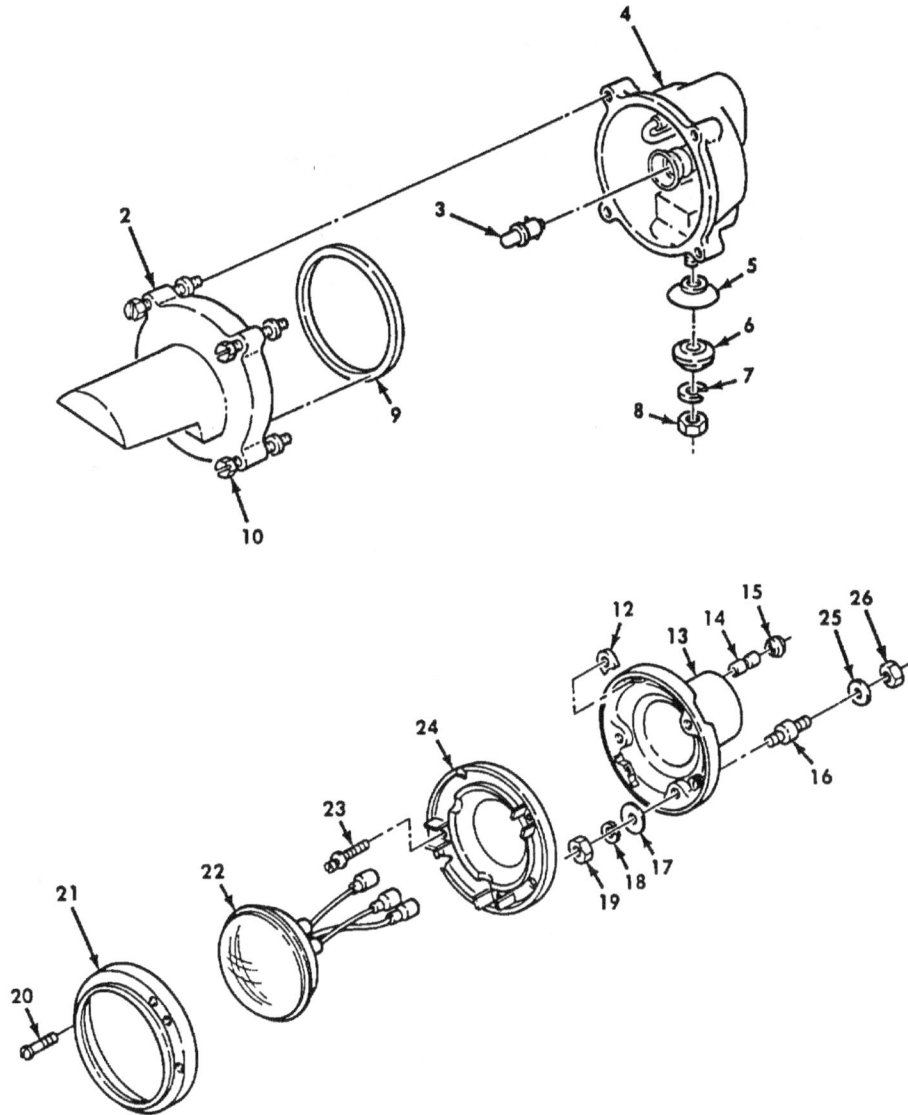

Figure 141. Blackout Lamp and Headlight.

(1) ITEM NO	(2) SMR CODE	(3) NSN	(4) CAGEC	(5) PART NUMBER	(6) DESCRIPTION AND USABLE ON CODES (UOC)	(7) QTY

GROUP 0609 LIGHTS

FIG. 141 BLACKOUT LAMP AND HEADLIGHT

(1) ITEM NO	(2) SMR CODE	(3) NSN	(4) CAGEC	(5) PART NUMBER	(6) DESCRIPTION AND USABLE ON CODES (UOC)	(7) QTY
1	PFOZZ	6220013003643	19207	12360910-1	HEADLIGHT...	1
2	PAOZZ	6220014113584	19207	12360912	.LENS, LIGHT	1
3	PAOZZ	6240012841925	19207	12360840-1	.LAMP, INCANDESCENT................................	1
4	XAOZZ		19207	12360911	.BODY ASSEMBLY	1
5	PAOZZ	5310011002067	19207	11668979	.WASHER, LOCK	1
6	PAOZZ	5310003502655	19207	5294507	.WASHER, FINISHING	1
7	PAOZZ	5310006379541	96906	MS35338-46	.WASHER, LOCK	1
8	PAOZZ	5310007320558	96906	MS51967-8	.NUT, PLAIN, HEXAGON...............................	1
9	PAOZZ	5331011896351	81349	M25988/1-246	.O-RING	1
10	PAOZZ	5306010835536	19207	12287561-1	.BOLT, EXTERNALLY REL............................	4
11	PAOZZ	6220011931970	19207	12338611	HEADLIGHT...	2
12	PAOZZ	5310002416921	19207	8741435	.NUT BLANK..	2
13	PAOZZ	6220004430589	19207	8741461	.HOUSING, LIGHT..	1
14	PAOZZ	5935008074109	19207	8741492	.ADAPTER, CONNECTOR.............................	3
15	PAOZZ	5325000886147	19207	8741442	.GROMMET, NONMETALLIC..........................	3
16	PAOZZ	5342001823726	19207	8741441	.MOUNT, RESILIENT....................................	3
17	PAOZZ	5310004630268	19207	5310615	.WASHER, FLAT	3
18	PAOZZ	5310005825965	96906	MS35338-44	.WASHER, LOCK	3
19	PAOZZ	5310007680319	96906	MS51968-2	.NUT, PLAIN, HEXAGON...............................	6
20	PAOZZ	5305008325743	19207	8741437	.SCREW, EXTERNALLY RE	3
21	PAOZZ	5325008325650	19207	8741446	.RING, RETAINING	1
22	PAOZZ	6240009663831	19207	8741491	.LAMP, INCANDESCENT...............................	1
23	PAOZZ	5305003530969	19207	7538146	.SCREW, EXTERNALLY RE	2
24	PAOZZ	6220009986142	19207	8741447	.RETAINER, LENS	1
25	PAOZZ	5310002090786	96906	MS35335-33	WASHER, LOCK	6
26	PAOZZ	5310007680319	96906	MS51968-2	NUT, PLAIN, HEXAGON................................	6

END OF FIGURE

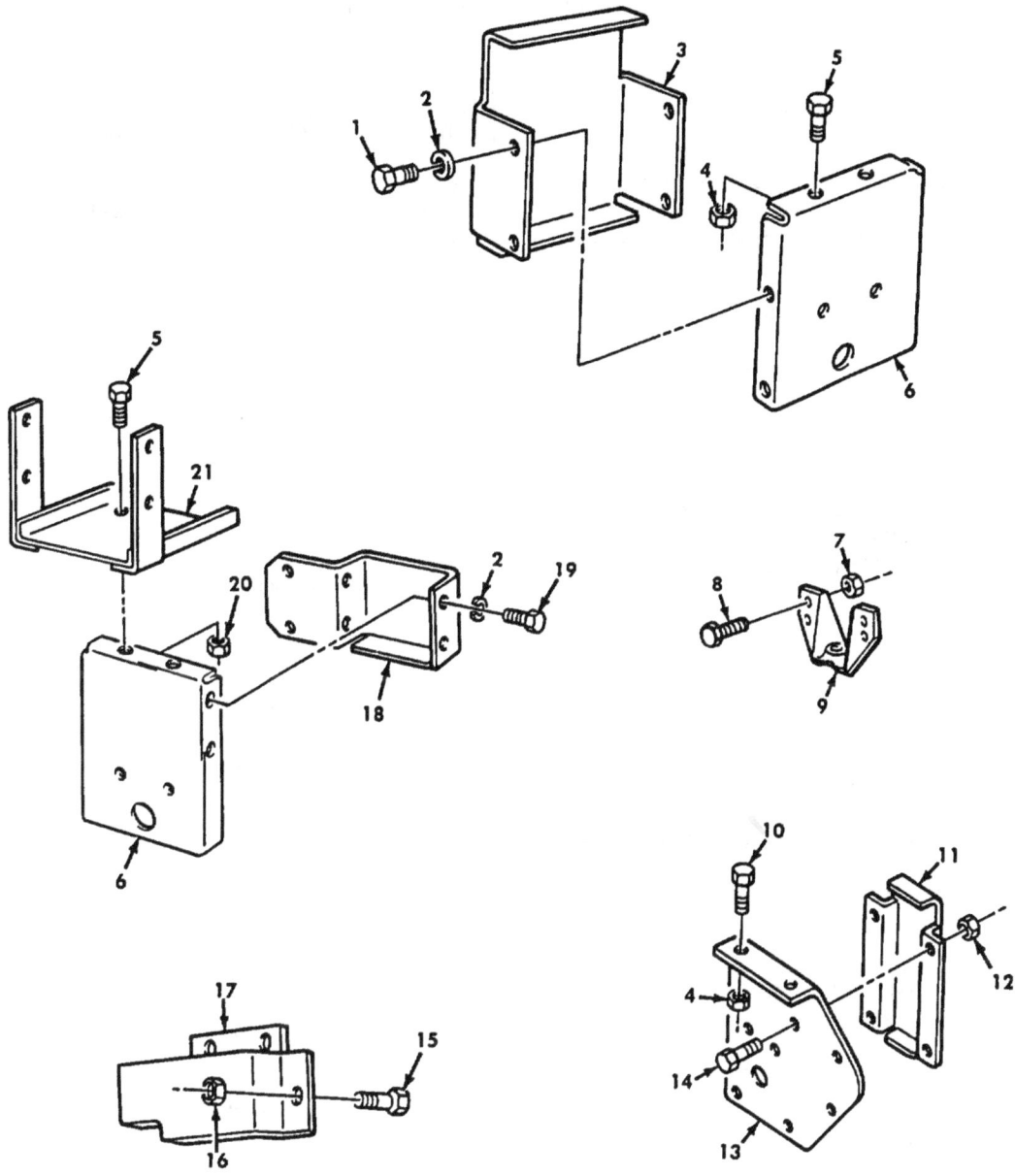

Figure 142. Taillight and Blackout Light Brackets and Mounting Hardware.

(1) ITEM NO	(2) SMR CODE	(3) NSN	(4) CAGEC	(5) PART NUMBER	(6) DESCRIPTION AND USABLE ON CODES (UOC)	(7) QTY
					GROUP 0609 LIGHTS	
					FIG. 142 TAILLIGHT AND BLACKOUT LIGHT BRACKETS AND MOUNTING HARDWARE	
1	PAOZZ	5305000526920	96906	MS24629-56	SCREW, TAPPING UOC:DAA,DAJ,DAK,V24,V25,V39,ZAJ,ZAK	8
2	PAOZZ	5310002090786	96906	MS35335-33	WASHER, LOCK .... UOC:DAA,DAJ,DAK,DAL,V18,V24,V25,V39, ZAJ,ZAK,ZAL	8
3	PAOZZ	5340004072612	19207	11658679	COVER, ACCESS UOC:DAA,DAJ,DAK,V24,V25,V39,ZAJ,ZAK	2
4	PAOZZ	5310009591488	96906	MS51922-21	NUT, SELF-LOCKING, HE	4
5	PAOZZ	5305009125113	96906	MS51096-359	SCREW, CAP, HEXAGON H .... UOC:DAA,DAJ,DAK,DAL,V18,V24,V25,V39, ZAJ,ZAK,ZAL	4
6	PFOZZ	2510004072617	19207	11658685	BRACKET, TAILLIGHT, V UOC:DAA,DAJ,DAK,V24,V25,V39,ZAJ,ZAK	2
7	PAOZZ	5310009359022	96906	MS51943-32	NUT, SELF-LOCKING, HE	4
8	PAOZZ	5305002678953	80204	B1821BH025F063N	SCREW, CAP, HEXAGON H	4
9	PFOZZ	2590007539545	19207	7539545	BRACKET, MOUNTING, HE	1
10	PAOZZ	5305002692803	96906	MS90726-60	SCREW, CAP, HEXAGON H UOC:DAA,DAG,DAH,V21,V22,ZAG,ZAH	4
11	PAOZZ	5340004075087	19207	10883106-1	COVER, ACCESS UOC:DAA,DAG,DAH,V21,V22,ZAG,ZAH	4
12	PAOZZ	5310009359022	96906	MS51943-32	NUT, SELF-LOCKING, HE UOC:DAA,DAG,DAH,V21,V22,ZAG,ZAH	8
13	PFOZZ	2510004075093	19207	11658683-1	BRACKET, TAIL LIGHT LEFT UOC:DAA,DAG,DAH,V21,V22,ZAG,ZAH	1
13	PFOZZ	2510004075095	19207	11658683-2	BRACKET, TAIL LIGHT RIGHT UOC:DAA,DAG,DAH,V21,V22,ZAG,ZAH	1
14	PAOZZ	5305002692803	96906	MS90726-60	SCREW, CAP, HEXAGON H UOC:DAA,DAG,DAH,V21,V22,ZAG,ZAH	8
15	PAOZZ	5306000680513	60285	6893-2	BOLT, MACHINE UOC:DAA,DAE,DAF,V19,V20,ZAE,ZAF	6
16	PAOZZ	5310009359022	96906	MS51943-32	NUT, SELF-LOCKING, HE UOC:DAA,DAE,DAF,V19,V20,ZAE,ZAF	6
17	PAOZZ	2510004075085	19207	11658682-1	COVER, PROTECTOR, TAI RIGHT HAND .... TAILLIGHT UOC:DAA,DAE,DAF,V19,V20,ZAE,ZAF	1
17	PAOZZ	2510004075086	19207	11658682-2	COVER, PROTECTOR, TAI LEFT HAND .... TAILLIGHT UOC:DAA,DAE,DAF,V19,V20,ZAE,ZAF	1
18	PAOZZ	2510002317437	19207	11658686-1	GUARD, TAILLIGHT BRA LEFT UOC:DAA,DAL,V18,ZAL	1
18	PAOZZ	2510002317438	19207	11658686-2	GUARD, TAILLIGHT BRA RIGHT UOC:DAA,DAL,V18,ZAL	1
19	PAOZZ	5305000526920	96906	MS24629-56	SCREW, TAPPING UOC:DAA,DAL,V18,ZAL	8
20	PAOZZ	5310008140672	96906	MS51943-36	NUT, SELF-LOCKING, HE UOC:DAA,DAL,V18,ZAL	4
21	PFOZZ	2510002228973	19207	10883450	BRACKET, TAILLIGHT UOC:DAA,DAL,V18,ZAL	2

END OF FIGURE

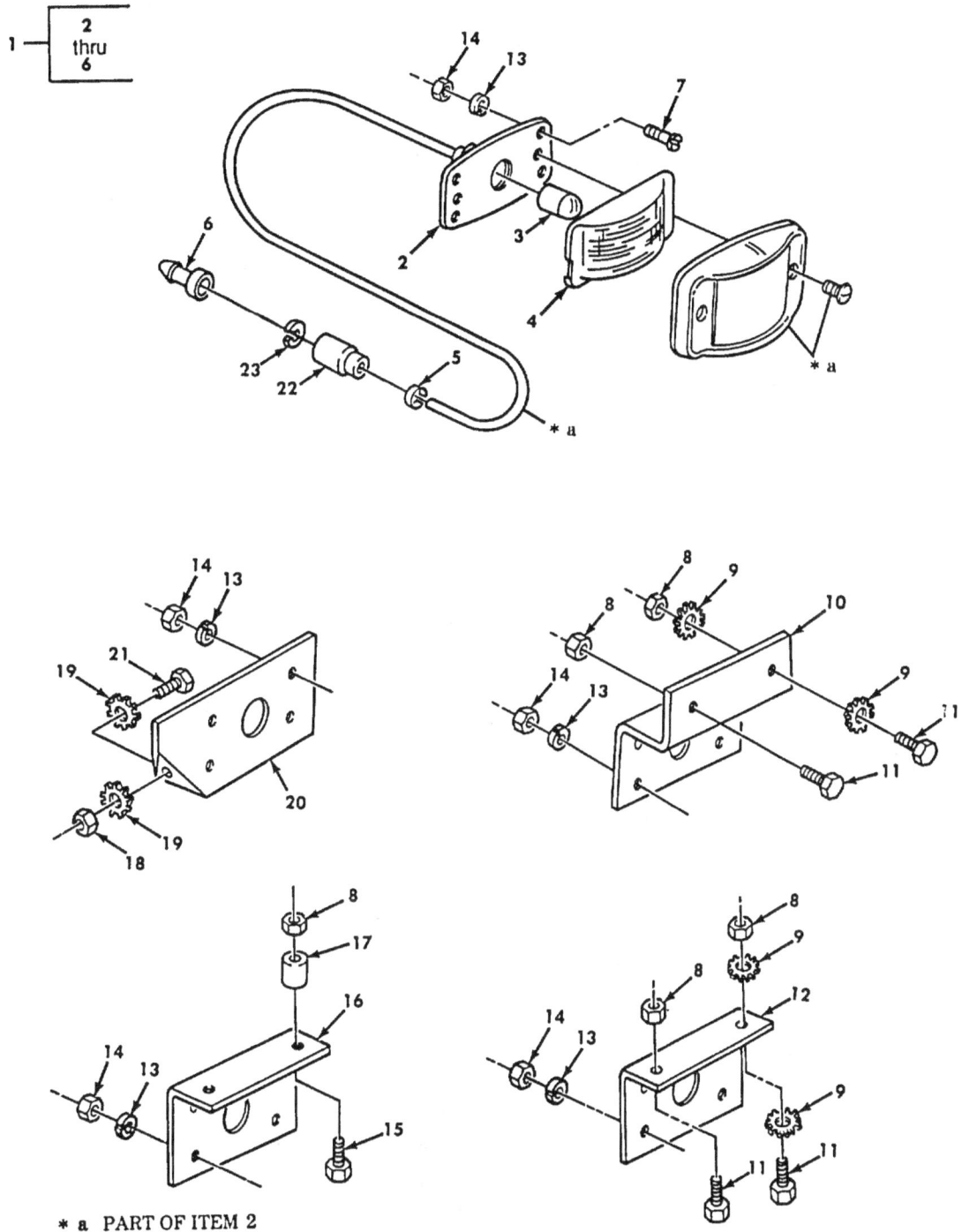

Figure 143. Side Marker Lights, Brackets, and Mounting Hardware.

SECTION II

(1) ITEM NO	(2) SMR CODE	(3) NSN	(4) CAGEC	(5) PART NUMBER	(6) DESCRIPTION AND USABLE ON CODES (UOC)	(7) QTY
					GROUP 0609 LIGHTS	
					FIG. 143 SIDE MARKER LIGHTS, BRACKETS ,AND MOUNTING HARDWARE	
1	PAOZZ	6220005773434	96906	MS35423-1	LIGHT, MARKER, CLEARA AMBER, BODY .... UOC:DAA,DAB,DAC,DAD,DAE,DAF,DAL,DAW, DAX,V12,V13,V14,V15,V16,V17,V18,V19, V20,ZAA,ZAB,ZAC, ZAD,ZAE,ZAF,ZAL	2
1	PAOZZ	6220007261916	96906	MS35423-2	LIGHT, MARKER, CLEARA RED, BODY .... UOC:DAA,DAB,DAC,DAD,DAE,DAF,DAL,DAW, DAX,V12,V13,V14,V15,V16,V17,V18,V19, V20,ZAA,ZAB,ZAC,ZAD,ZAE,ZAF,ZAL	2
1	PAOZZ	6220005773434	96906	MS35423-1	LIGHT, MARKER, CLEARA AMBER, BODY .... FRONT	2
2	PAOZZ	6220007299295	96906	MS35422-1	.LIGHT, MARKER, CLEARA	1
3	PAOZZ	6240000190877	58536	A52463-1-08	.LAMP, INCANDESCENT	1
4	PAOZZ	6220002997425	96906	MS35421-1	.LENS, LIGHT AMBER	1
4	PAOZZ	6220002997426	96906	MS35421-2	.LENS, LIGHT RED	1
5	PAOZZ	9905007524649	81349	M43436/1-1	.BAND, MARKER	1
6	PAOZA	5999000572929	19204	572929	.CONTACT, ELECTRICAL	1
7	PAOZZ	5305009846212	96906	MS35206-265	SCREW, MACHINE .... UOC:DAA,DAB,DAC,DAD,DAE,DAF,DAL,DAW, DAX,V12,V13,V14,V15,V16,V17,V18,V19, V20,ZAA,ZAB,ZAC,ZAD,ZAE,ZAF,ZAL	16
7	PAOZZ	5305009846212	96906	MS35206-265	SCREW, MACHINE	8
8	PAOZZ	5310009591488	96906	MS51922-21	NUT, SELF-LOCKING, HE	8
9	PAOZZ	5310005957237	96906	MS35333-42	WASHER, LOCK .... UOC:DAA,DAB,DAC,DAD,DAE,DAF,DAW,DAX, V12,V13,V14,V15,V16,V17,V19,V20,ZAA, ZAB,ZAC,ZAD,ZAE,ZAF	8
10	PFOZZ	5340011290474	19207	12277192	BRACKET, DOUBLE ANGL UOC:DAA,DAB,DAC,DAD,DAE,DAF,V12,V13, V16,V17,VI9,V20,ZAC,ZAD,ZAE,ZAF	4
11	PAOZZ	5305002693234	80204	B1821BH038F075N	SCREW, CAP, HEXAGON H .... UOC:DAA,DAB,DAC,DAD,DAE ,DAF,DAW,DAX, V12,V13,V14,V15,V16,V17,V18,V19,V20, ZAA,ZAB,ZAC,ZAD,ZAE,ZAF	8
12	PAOZZ	5340011089268	19207	12277002	BRACKET, ANGLE UOC:DAW,DAX,V14,V15,ZAA,ZAB	4
13	PAOZZ	5310005967691	96906	MS35335-32	WASHER, LOCK .... UOC:DAA,DAB,DAC,DAD,DAE,DAF,DAL,DAW, DAX,V12,V13,V14,V15,V16,V18,V19,V20, ZAA,ZAB,ZAC,ZAD,ZAE,ZAF,ZAL	16
13	PAOZZ	5310005967691	96906	MS35335-32	WASHER, LOCK	8
14	PAOZZ	5310009349758	96906	MS35649-202	NUT, PLAIN, HEXAGON .... UOC:DAA,DAB,DAC,DAD,DAE,DAF,DAL,DAW, DAX,V12,V13,V14,V15,V16,V17,V18,V19, V20,ZAA,ZAB,ZAC,ZAD,ZAE,ZAF,ZAL	16
14	PAOZZ	5310009349758	96906	MS35649-202	NUT, PLAIN, HEXAGON	8
15	PAOZZ	5305002692803	96906	MS90726-60	SCREW, CAP, HEXAGON H UOC:DAA,DAL,V18,ZAL	4

(1) ITEM NO	(2) SMR CODE	(3) NSN	(4) CAGEC	(5) PART NUMBER	(6) DESCRIPTION AND USABLE ON CODES (UOC)	(7) QTY
15	PAOZZ	5305002693242	80204	B1821BH038F200N	SCREW, CAP, HEXAGON H UOC:DAA,DAL,V18,ZAL	8
16	PBOZZ	5340012319291	19207	12277002-1	BRACKET, ANGLE UOC:DAA,DAL,V18,ZAL	4
17	PFOZZ	5365011089258	19207	12277047	SPACER, SLEEVE UOC:DAA,DAL,V18,ZAL	4
18	PAOZZ	5310009359022	96906	MS51943-32	NUT, SELF-LOCKING, HE SIDE MARKER.............. BRACKET...	8
19	PAOZZ	5310008892528	96906	MS45904-68	WASHER, LOCK	4
20	PAOZZ	5340011217601	19207	12277042-1	BRACKET, MOUNTING RIGHT	1
20	PAOZZ	5340011062068	19207	12277042-2	BRACKET, MOUNTING SIDE MARKER	1
21	PAOZZ	5306000680513	60285	6893-2	BOLT, MACHINE SIDE MARKER BRACKET...........	1
22	PAOZZ	5935005729180	19207	8338566	SHELL, ELECTRICAL CO .... UOC:DAA,DAB,DAC,DAD,DAE,DAF,DAL,DAW, DAX,V12,V13,V14,V15,V16,V17,V18,V19, V20,ZAA,ZAB,ZAC,ZAD,ZAE,ZAF,ZAL	4
23	PAOZZ	5310008338567	19207	8338567	WASHER, SLOTTED .... UOC:DAA,DAB,DAC,DAD,DAE,DAF,DAL,DAW, DAX,V12,V13,V14,V15,V16,V17,V18,V19, V20,ZAA,ZAB,ZAC,ZAD,ZAE,ZAF,ZAL	4

END OF FIGURE

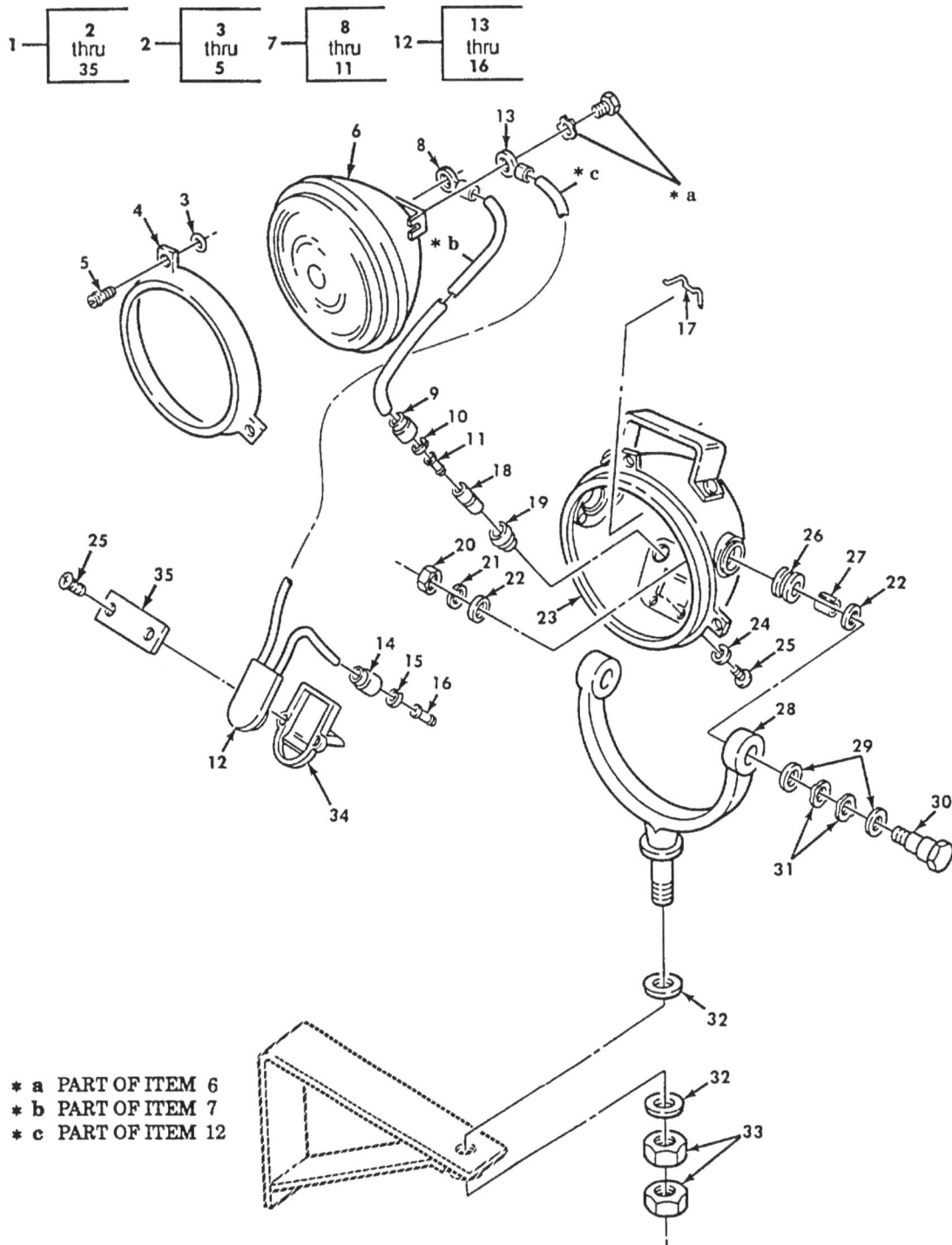

Figure 144. Floodlight Assembly.

* a PART OF ITEM 6
* b PART OF ITEM 7
* c PART OF ITEM 12

(1) ITEM NO	(2) SMR CODE	(3) NSN	(4) CAGEC	(5) PART NUMBER	(6) DESCRIPTION AND USABLE ON CODES (UOC)	(7) QTY
					GROUP 0609 LIGHTS	
					FIG. 144 FLOODLIGHT ASSEMBLY	
1	PAOZZ	6220008787301	19207	8739551	FLOODLIGHT, ELECTRIC UOC:DAL,V18,ZAL	3
2	PAOZZ	6220011552357	19207	7385905	.RETAINER, LENS UOC:DAL,V18,ZAL	1
3	PAOZZ	5340007320642	19207	7320642	..CLIP, RETAINING... UOC:DAL,V18,ZAL	3
4	XAOZZ		19207	7212871	..DOOR UOC:DAL,V18,ZAL	1
5	PAOZZ	5305007525938	19207	7525938	..SCREW, MACHINE... UOC:DAL,V18,ZAL	3
6	PAOZZ	6240002952421	96906	MS18006-4572	.LAMP, INCANDESCENT UOC:DAL,V18,ZAL	1
7	PFOZZ	5995009916707	19207	8739552	.LEAD, ELECTRICAL UOC:DAL,V18,ZAL	1
8	PAOZZ	5940000506207	21450	506207	..TERMINAL, LUG FLOODLIGHT LEAD.................. UOC:DAL,V18,ZAL	1
9	PAOZZ	5935006915591	19207	8724495	..SHELL, ELECTRICAL CO FLOODLIGHT.............. LEAD.. UOC:DAL,V18,ZAL	1
10	PAOZZ	5310006560067	19207	8724497	..WASHER, SLOTTED FLOODLIGHT LEAD............. UOC:DAL,V18,ZAL	1
11	PAOOA	5999009263144	96906	MS27148-3	..CONTACT, ELECTRICAL FLOODLIGHT............... LEAD UOC:DAL,V18,ZAL	1
12	PFOZZ	5930009916713	19207	8739573	.SWITCH, SENSITIVE....................................... UOC:DAL,V18,ZAL	1
13	PAOZZ	5940000506207	21450	506207	..TERMINAL, LUG FLOODLIGHT SWITCH UOC:DAL,V18,ZAL	1
14	PAOZZ	5935006915591	19207	8724495	..SHELL, ELECTRICAL CO FLOODLIGHT.............. SWITCH UOC:DAL,V18,ZAL	1
15	PAOZZ	5310006560067	19207	8724497	..WASHER, SLOTTED FLOODLIGHT SWITCH......... UOC:DAL,V18,ZAL	1
16	PAOOA	5999009263144	96906	MS27148-3	..CONTACT, ELECTRICAL FLOODLIGHT............... SWITCH UOC:DAL,V1B,ZAL	1
17	PAOZZ	5340007416869	19207	7416869	.CLIP, RETAINING UOC:DAL,V18,ZAL	4
18	PAOZZ	5935008074109	19207	8741492	.ADAPTER, CONNECTOR.................................. UOC:DAL,V18,ZAL	2
19	PAOZZ	5325000886147	19207	8741442	.GROMMET, NONMETALLIC UOC:DAL,V18,ZAL	2
20	PAOZZ	5310007320558	96906	MS51967-8	.NUT, PLAIN, HEXAGON................................... UOC:DAL,V18,ZAL	2
21	PAOZZ	5310006379541	96906	MS35338-46	.WASHER, LOCK UOC:DAL,V18,ZAL	2
22	PAOZZ	5310007416862	19207	7416862	.WASHER, FLAT.. UOC:DAL,V18,ZAL	4

(1) ITEM NO	(2) SMR CODE	(3) NSN	(4) CAGEC	(5) PART NUMBER	(6) DESCRIPTION AND USABLE ON CODES (UOC)	(7) QTY
23	PFOZZ	6220003973335	19207	8739559	.HOUSING, LIGHT................................ UOC:DAL,V1B,ZAL	1
24	PAOZZ	5310000453299	96906	MS35338-42	.WASHER, LOCK UOC:DAL,V18,ZAL	2
25	PAOZZ	5305008893002	96906	MS35206-242	.SCREW, MACHINE UOC:DAL,V18,ZAL	4
26	PAOZZ	5325007414180	19207	7414180	.GROMMET, NONMETALLIC UOC:DAL,V18,ZAL	2
27	PAOZZ	5365009234253	19207	7416861	.SPACER, SLEEVE UOC:DAL,V18,ZAL	2
28	PFOZZ	5340010086448	19207	7416866	.BRACKET, MOUNTING UOC:DAL,V18,ZAL	1
29	PAOZZ	5310008095998	96906	MS27183-18	.WASHER, FLAT UOC:DAL,V18,ZAL	4
30	PAOZZ	5306007416865	19207	7416865	.BOLT, SHOULDER UOC:DAL,V18,ZAL	2
31	PAOZZ	5310000799573	82679	WA132	.WASHER, SPRING TENSI UOC:DAL,V18,ZAL	4
32	PAOZZ	5310007416867	19207	7416867	.WASHER, FLAT UOC:DAL,V18,ZAL	2
33	PAOZZ	5310007638905	96906	MS51968-20	.NUT, PLAIN, HEXAGON UOC:DAL,V18,ZAL	2
34	PAOZZ	2510005670130	19207	7414388	.HOUSING, SWITCH, FLOO UOC:DAL,V18,ZAL	1
35	PAOZZ	5935009916708	19207	8739556	.RETAINER, ELECTRICAL UOC:DAL,V18,ZAL	1

END OF FIGURE

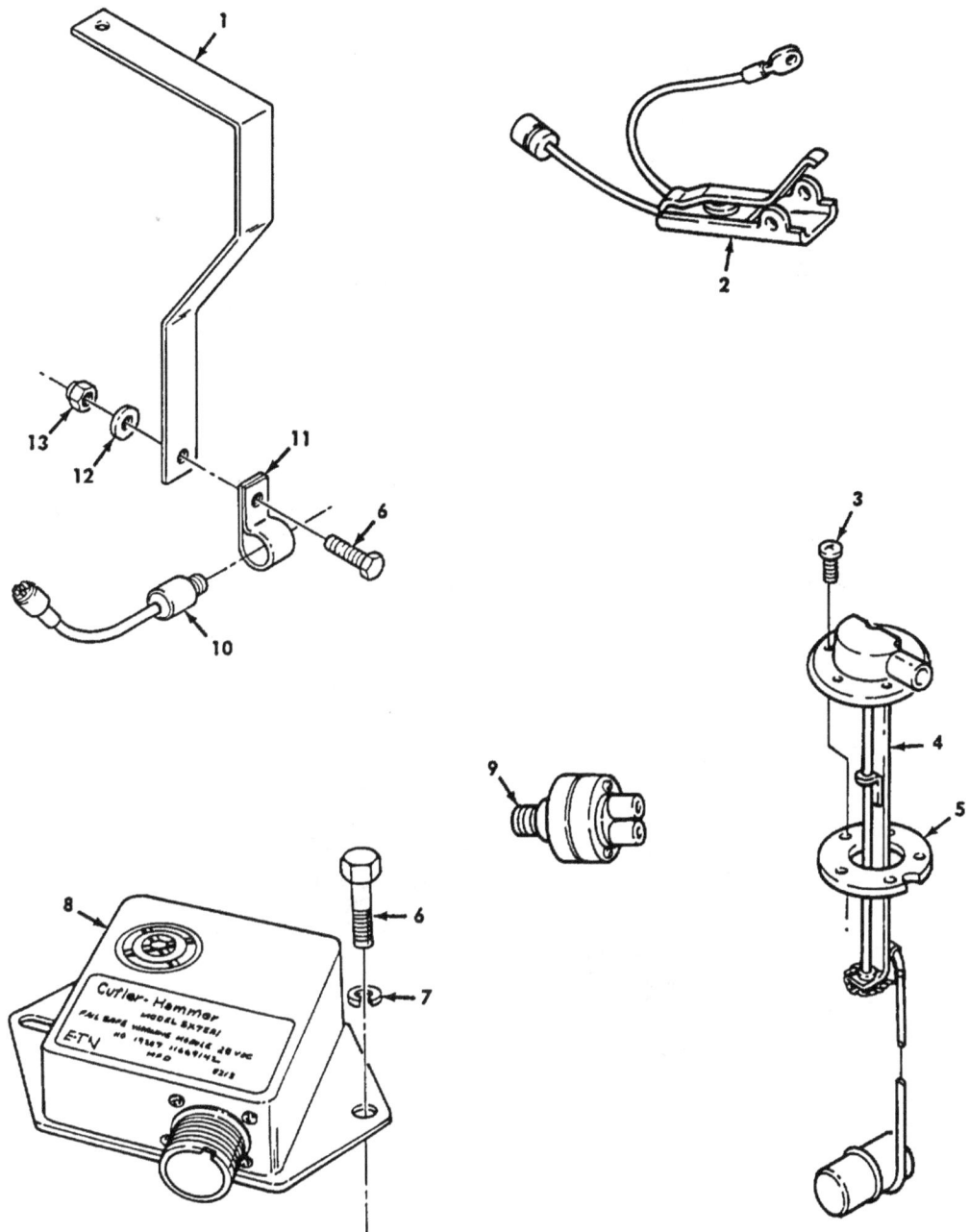

Figure 145. Fuel Pressure Transducer and Support, Fuel Level Transmitter, Parking Brake, and Warning Control Device.

(1) ITEM NO	(2) SMR CODE	(3) NSN	(4) CAGEC	(5) PART NUMBER	(6) DESCRIPTION AND USABLE ON CODES (UOC)	(7) QTY
					GROUP 0610 SENDING UNITS AND WARNING SWITCHES	
					FIG. 145 FUEL PRESSURE TRANSDUCER AND SUPPORT, FUEL LEVEL TRANSMITTER, PARKING BRAKE, AND WARNING CONTROL DEVICE	
1	PAOZZ	5340012059262	19207	12302859	BRACKET, ANGLE	1
2	PAOZZ	2530011208441	19207	11669094	SWITCH, BRAKE, PARKIN	1
3	PAOZZ	5305000977372	19207	11609358-2	SCREW, SELF SEALING .... UOC:DAA,DAB,DAC,DAD,DAJ,DAK,DAW,DAX, V12,V13,V14,V15,V16,V17,V24,V25,V39, ZAA,ZAB,ZAC,ZAD,ZAJ,ZAK	5
3	PAOZZ	5305000977372	19207	11609358-2	SCREW, SELF SEALING UOC:DAE,DAF,DAG,DAH,DAL,V18,V19,V20, V21,V22,ZAE,ZAF,ZAG,ZAH,ZAL	10
4	PAOZZ	6680002264574	96906	MS500040-6	TRANSMITTER, LIQUID .... UOC:DAA,DAB,DAC,DAD,DAJ,DAK,DAW,DAX, V12,V13,V14,V15,V16,V17,V24,V25,ZAA, ZAB,ZAC,ZAD,ZAJ,ZAK	1
4	PAOZZ	6680002264574	96906	MS500040-6	TRANSMITTER, LIQUID UOC:DAE,DAF,DAG,DAH,DAL,V18,V19,V20, V21,V22,ZAE,ZAF,ZAG,ZAH,ZAL	2
5	PAOZZ	5330007539072	19207	7539072	GASKET.. UOC:DAA,DAB,DAC,DAD,DAJ,DAK,DAW,DAX, V12,V13,V14,V15,V16,V17,V24,V25,ZAA, ZAB,ZAC,ZAD,ZAJ,ZAK	1
5	PAOZZ	5330007539072	19207	7539072	GASKET.. UOC:DAE,DAF,DAG,DAH,DAL,V18,V19,V20, V21,V22,ZAE,ZAF,ZAG,ZAH,ZAL	2
6	PAOZZ	5306000680513	60285	6893-2	BOLT, MACHINE...	3
7	PAOZZ	5310005501130	96906	MS35333-40	WASHER, LOCK	2
8	PAOZZ	6350010892987	19207	11669142	ALARM-MONITOR FAILSAFE	1
9	PAOZZ	5930011619580	19207	12302684	SWITCH, PRESSURE ETHER START	1
10	PAOZZ	6695011467132	19207	12258932-7	TRANSDUCER, MOTIONAL.................................	1
11	PAOZZ	5340008091500	96906	MS21333-107	CLAMP, LOOP	1
12	PAOZZ	5310008094058	96906	MS27183-10	WASHER, FLAT	1
13	PAOZZ	5310009359022	96906	MS51943-32	NUT, SELF-LOCKING, HE	1

END OF FIGURE

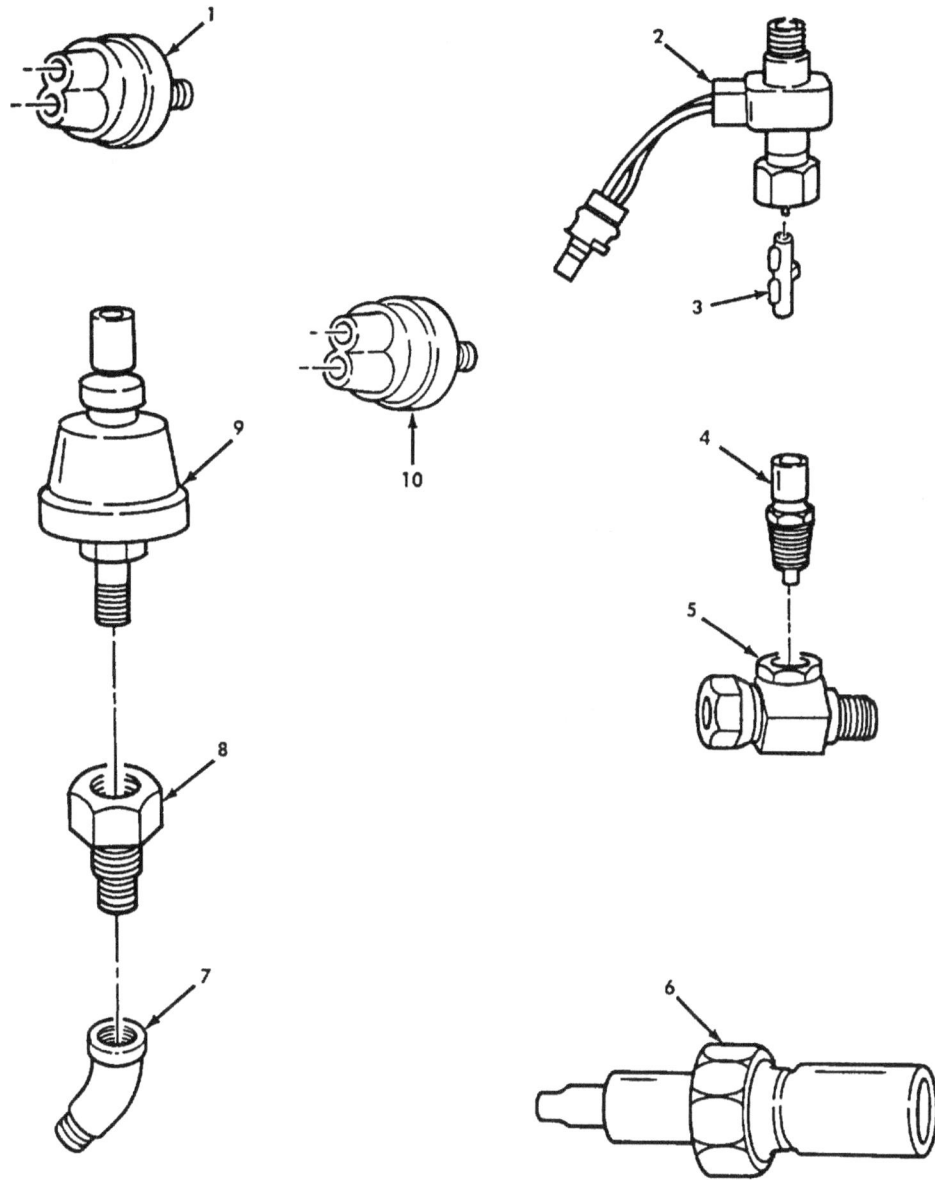

Figure 146. Low Air Warning Switch, Mechanical Tachometer, Transmission Temperature Transmitter, Coolant Temperature Sending Unit, Oil Pressure Transmitter, and Stoplight Switch.

(1) ITEM NO	(2) SMR CODE	(3) NSN	(4) CAGEC	(5) PART NUMBER	(6) DESCRIPTION AND USABLE ON CODES (UOC)	(7) QTY
					GROUP 0610 SENDING UNITS AND WARNING SWITCHES	
					FIG. 146 LOW AIR WARNING SWITCH, MECHANICAL TACHOMETER, TRANSMISSION TEMPERATURE TRANSMITTER, COOLANT TEMPERATURE SENDING UNIT, OIL PRESSURE TRANSMITTER, AND STOPLIGHT SWITCH	
1	PAOZZ	5930011890494	19207	12375453	SWITCH, PRESSURE	2
2	PAOZZ	6680013192354	7Z588	12258931-3	ADAPTER, SPEEDOMETER	1
3	PAOZZ	5340012901738	19207	12301264	PLUNGER, DETENT..	1
4	PAOZZ	6685008145271	96906	MS24537-1	TRANSMITTER, TEMPERA ENGINE COOLANT....... TEMPERATURE	2
5	PAOZZ	4730010910266	19207	12256531	TEE, PIPE TO TUBE..	1
6	PAOZZ	6685008145271	96906	MS24537	TRANSMITTER, TEMPERA	1
7	PAOZZ	4730002493932	96906	MS39231-1	ELBOW, PIPE..	1
8	PAOZZ	4730005291487	30327	120-B-04X02	REDUCER, PIPE...	1
9	PAOZZ	6620009935546	96906	MS24539	TRANSMITTER, PRESSUR	1
10	PAOZZ	2540007896192	19207	11602160	SWITCH, PRESSURE, STOPLIGHT	1

END OF FIGURE

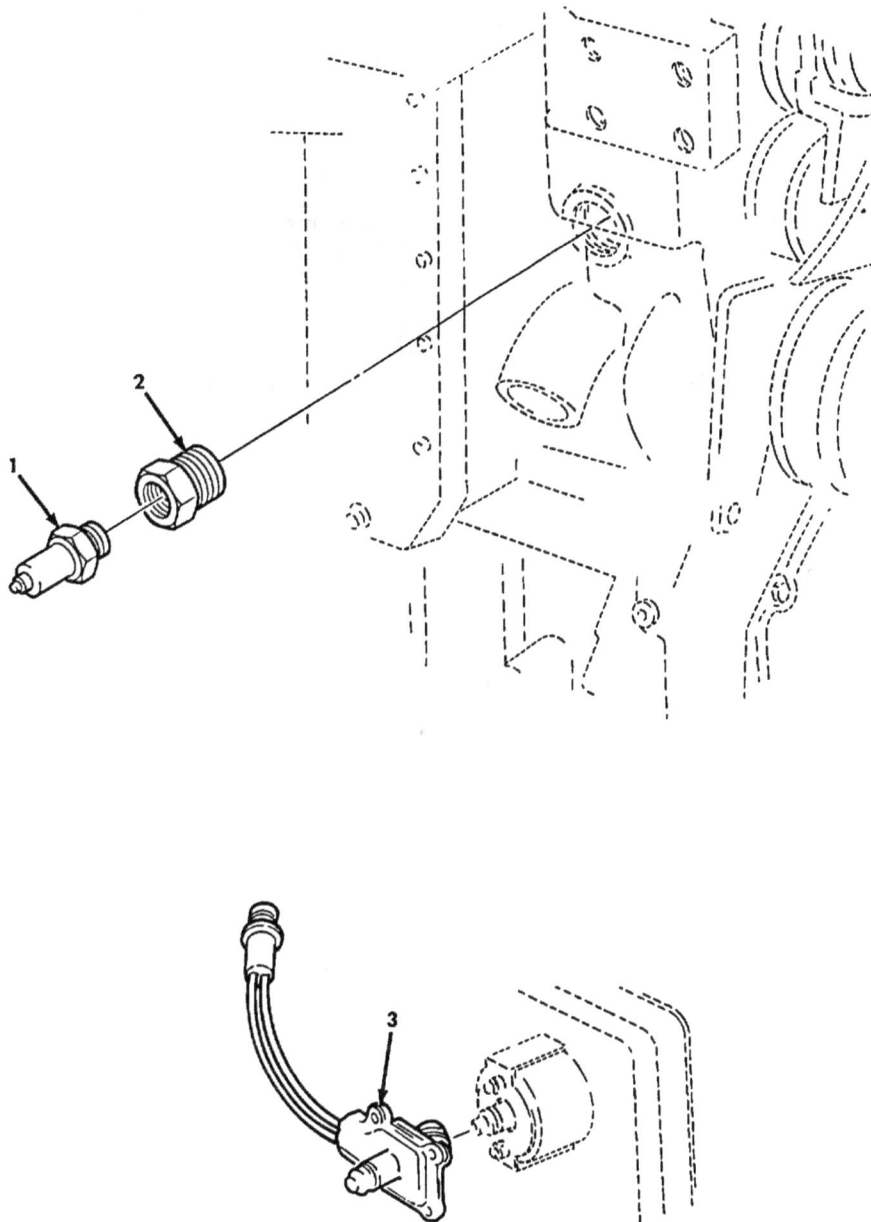

Figure 147. Temperature Transmitter and Tachometer Pulse Sensor (M939A2).

(1) ITEM NO	(2) SMR CODE	(3) NSN	(4) CAGEC	(5) PART NUMBER	(6) DESCRIPTION AND USABLE ON CODES (UOC)	(7) QTY
					GROUP 0610 SENDING UNITS AND WARNING SWITCHES	
					FIG. 147 TEMPERATURE TRANSMITTER AND TACHOMETER PULSE SENSOR(M939A2)	
1	PAOZZ	6685008145271	96906	MS24537-1	TRANSMITTER, TEMPERA .... UOC:ZAA,ZAB,ZAC,ZAD,ZAE,ZAF,ZAG,ZAH, ZAJ,ZAK,ZAL	1
2	PAOZZ	4730001960888	24617	14042	BUSHING, PIPE .... UOC:ZAA,ZAB,ZAC,ZAD,ZAE,ZAF,ZAG,ZAH, ZAJ,ZAK,ZAL	1
3	PAOZZ	6680012729204	47457	20511338	DRIVE UNIT, TACHOMET .... UOC:ZAA,ZAB,ZAC,ZAD,ZAE,ZAF,ZAG,ZAH, ZAJ,ZAK,ZAL	1

END OF FIGURE

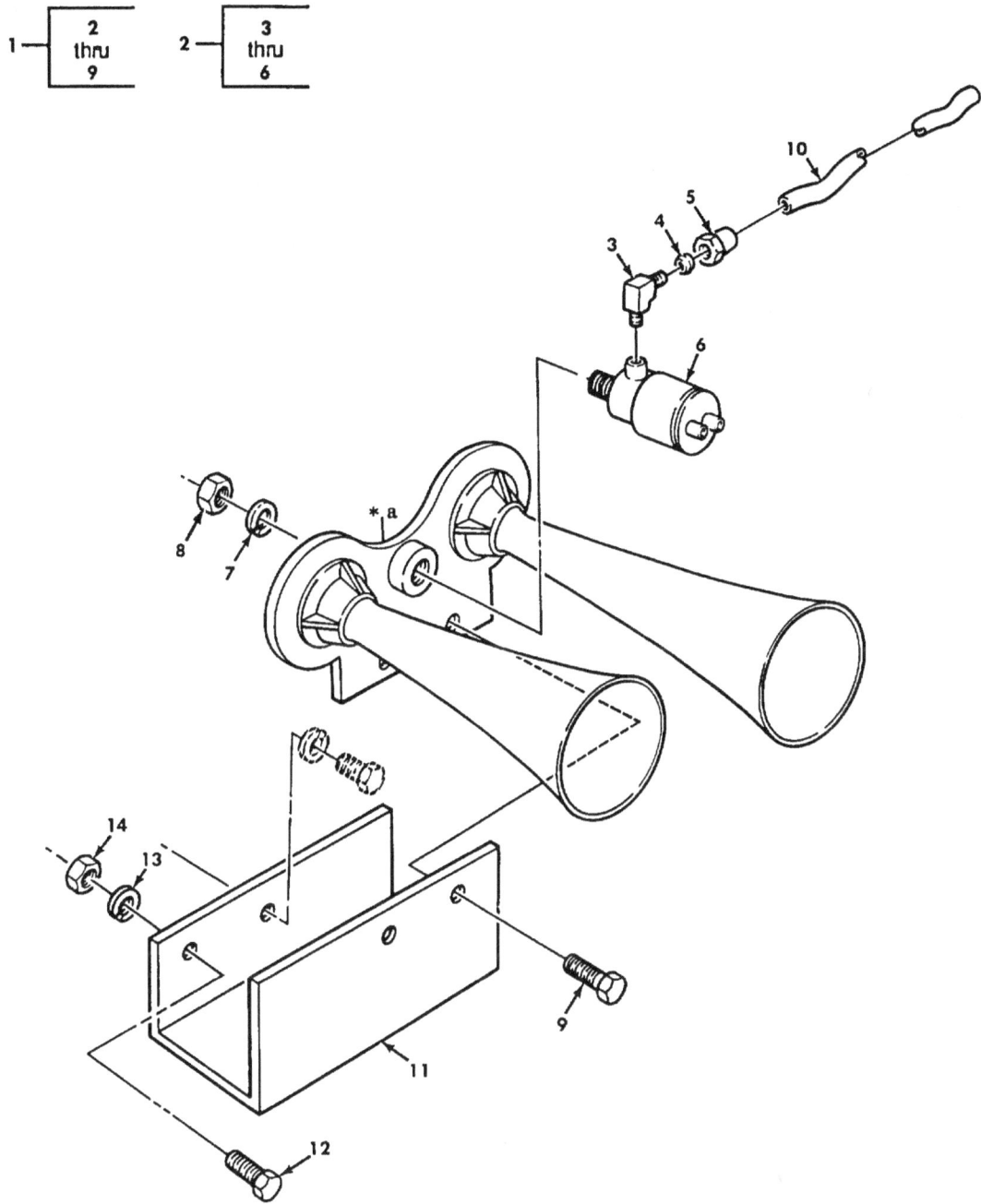

*a   PART OF ITEM 1

Figure 148. Horn Assembly, Related Tubing, and Mounting Bracket.

(1) ITEM NO	(2) SMR CODE	(3) NSN	(4) CAGEC	(5) PART NUMBER	(6) DESCRIPTION AND USABLE ON CODES (UOC)	(7) QTY
					GROUP 0611 HORN	
					FIG. 148 HORN ASSEMBLY, RELATED TUBING, AND MOUNTING BRACKET	
1	PAOZZ	2590009001640	96906	MS51301-1	HORN, FLUID OPERATED....................................	1
2	PAOZZ	5945002401684	96906	MS17981-1	.SOLENOID, ELECTRICAL	1
3	PAOZZ	4730002871604	81343	6-2 120202BA	..ELBOW, PIPE TO TUBE	1
4	PAOZZ	4730002937108	81343	6 120115B	..SLEEVE, COMPRESSION, ...	1
5	PAOZZ	4730002788824	81343	4 120111B	..NUT, TUBE COUPLING..................................	1
6	PAOZZ	5945011700553	55061	H813-1	..SOLENOID, ELECTRICAL..............................	1
7	PAOZZ	5310005501130	96906	MS35333-40	.WASHER, LOCK	2
8	PAOZZ	5310007680319	96906	MS51968-2	.NUT, PLAIN, HEXAGON	2
9	PAOZZ	5305002678954	80204	B1821BH025F125N	.SCREW, CAP, HEXAGON H	2
10	PAOZZ	4710011317529	19207	12256065	TUBE, BENT, METALLIC	1
11	PFOZZ	5340011048946	19207	12256092	BRACKET, MOUNTING	1
12	PAOZZ	5305009125113	96906	MS51096-359	SCREW, CAP, HEXAGON H	1
13	PAOZZ	5310005957237	96906	MS35333-42	WASHER, LOCK	1
14	PAOZZ	5310007320559	96906	MS51968-8	NUT, PLAIN, HEXAGON	1

END OF FIGURE

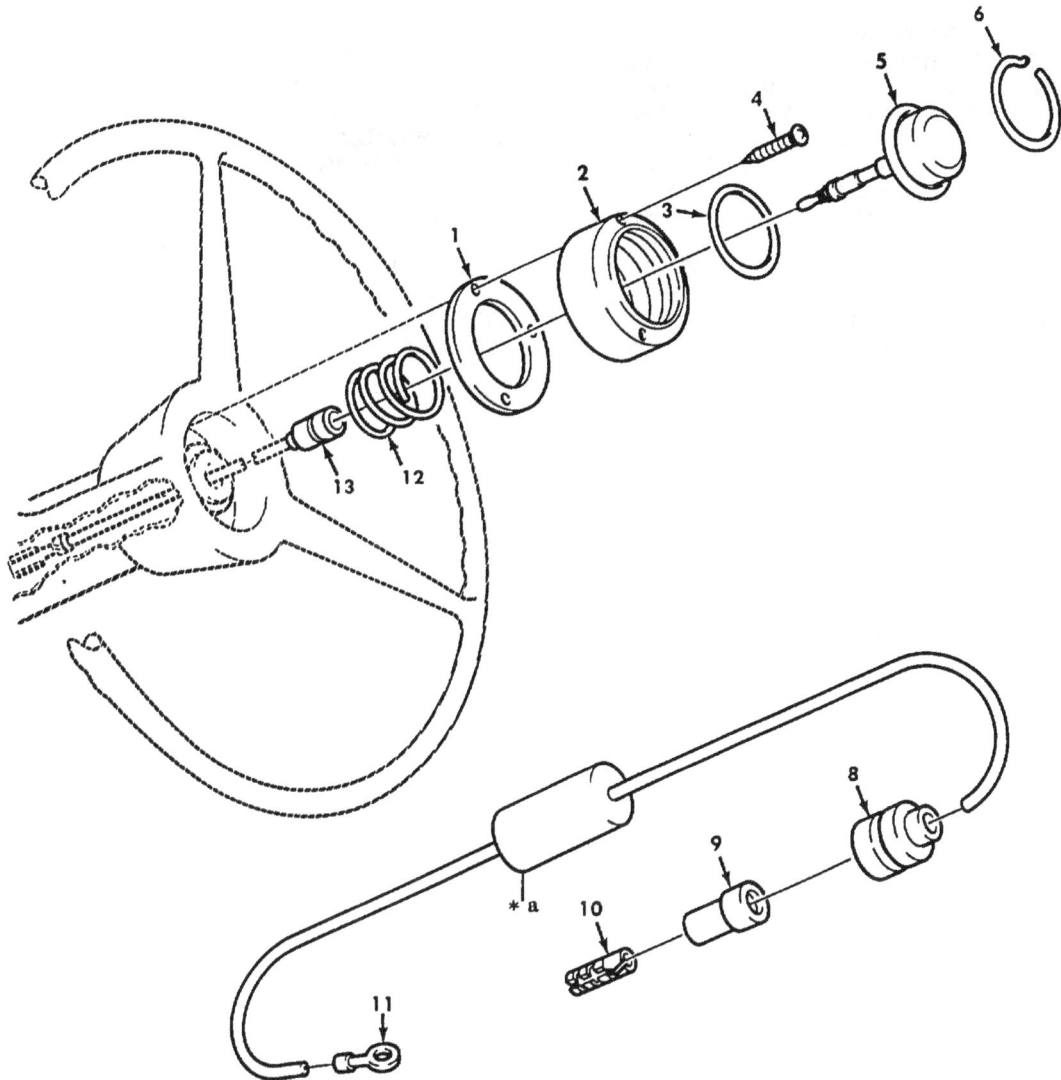

* a   PART OF ITEM 7

Figure 149. Horn Button and Diode Coupling Assembly.

(1) ITEM NO	(2) SMR CODE	(3) NSN	(4) CAGEC	(5) PART NUMBER	(6) DESCRIPTION AND USABLE ON CODES (UOC)	(7) QTY
					GROUP 0611 HORN	
					FIG. 149 HORN BUTTON AND DIODE COUPLING ASSEMBLY	
1	PAOZZ	5330010990554	19207	12255677	GASKET	1
2	PAOZZ	2590011115391	19207	11677306	ADAPTER, HORN BUTTON	1
3	PAOZZ	5330010497374	96906	MS28775-032	PACKING, PREFORMED	1
4	PAOZZ	5305004327956	96906	MS51861-40	SCREW, TAPPING	3
5	PAOZZ	2530011361098	19207	10921898-1	HORN BUTTON, VEHICLE	1
6	PAOZZ	5325011444871	19207	8754124	RING, RETAINING	1
7	PAOZZ	5961011805634	19207	12302643	SEMICONDUCTOR DEVIC	1
8	PAOZZ	5935008338561	19207	8338561	.SHELL, ELECTRICAL CO	1
9	PAOZZ	5970008338562	19207	8338562	.INSULATOR, BUSHING	1
10	PAOZZ	1015007982997	19207	7982997	.TERMINAL, SOLDERED F	1
11	PAOZZ	5940008923151	96906	MS35436-9	.TERMINAL, LUG	2
12	PAOZZ	5360011204610	19207	12255660	SPRING, HELICAL, COMP	1
13	PAOZZ	5999009256495	96906	MS27148-1	CONNECTOR	1

END OF FIGURE

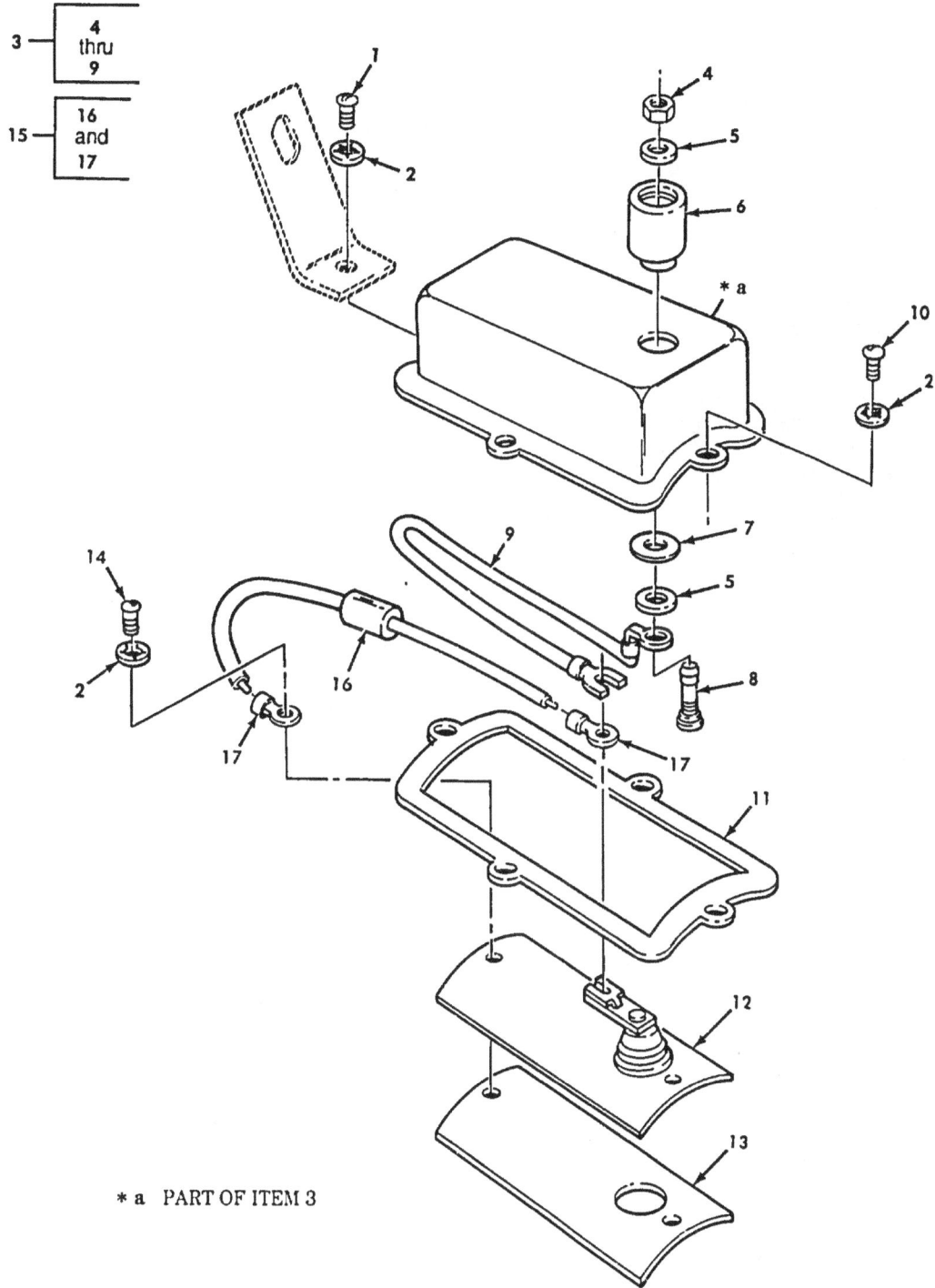

*a   PART OF ITEM 3

Figure 150. Horn Cover, Brush, and Capacitor Assembly.

(1) ITEM NO	(2) SMR CODE	(3) NSN	(4) CAGEC	(5) PART NUMBER	(6) DESCRIPTION AND USABLE ON CODES (UOC)	(7) QTY
					GROUP 0611 HORN	
					FIG. 150 HORN COVER, BRUSH, AND CAPACITOR ASSEMBLY	
1	PAOZZ	5305008893002	96906	MS35206-242	SCREW, MACHINE	1
2	PAOZZ	5310005590070	96906	MS35333-38	WASHER, LOCK	6
3	PAOZZ	4320009224933	1DD64	402235-A1	COVER ASSEMBLY, CONT	1
4	PAOZZ	5310000873946	96906	MS25082-12	.NUT, PLAIN, HEXAGON	1
5	PAOZZ	5310000455207	96906	MS15795-908	.WASHER, FLAT	2
6	PAOZZ	5970009060159	19207	7982399	.INSULATION SLEEVING	1
7	PAOZZ	5970011216587	19207	7060041	.INSULATOR, WASHER	1
8	PAOZZ	5940011214280	19207	7060039	.TERMINAL, STUD	1
9	PAOZZ	6150011143697	19207	7060040	.LEAD, ELECTRICAL	1
10	PAOZZ	5305009846189	96906	MS35206-241	SCREW, MACHINE	3
11	PAOZZ	5330007351272	19207	7351272	GASKET	1
12	PAOZZ	5977005786495	77640	032156	ELECTRICAL CONTACT	1
13	PAOZZ	2590005251352	19207	8728234	PAD, CUSHIONING	1
14	PAOZZ	5305009846226	96906	MS35206-240	.SCREW, MACHINE	2
15	PAOZZ	5910010965021	19207	12277055	CAPACITOR, FIXED, PLA	1
16	XAOZZ	5910010965021	19207	12277055-1	.CAPACITOR, FIXED, PLA	1
17	PAOZZ	5940001133138	96906	MS20659-102	.TERMINAL, LUG	2

END OF FIGURE

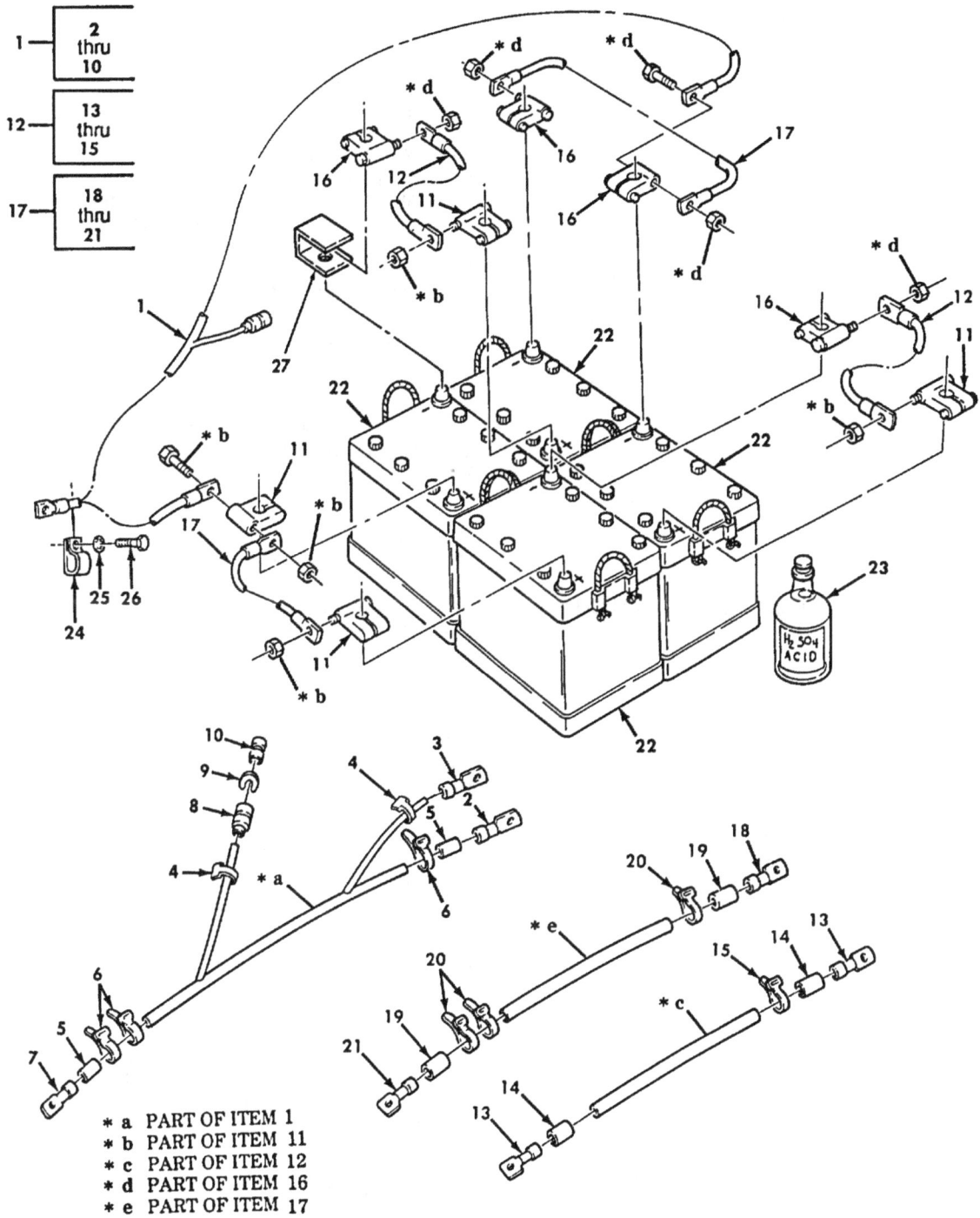

* a PART OF ITEM 1
* b PART OF ITEM 11
* c PART OF ITEM 12
* d PART OF ITEM 16
* e PART OF ITEM 17

Figure 151. Batteries, Cables, and Related Parts.

(1) ITEM NO	(2) SMR CODE	(3) NSN	(4) CAGEC	(5) PART NUMBER	(6) DESCRIPTION AND USABLE ON CODES (UOC)	(7) QTY
					GROUP 0612 BATTERIES, STORAGE	
					FIG. 151 BATTERIES, CABLES, AND RELATED PARTS(M939, M939A1)	
1	MOOZZ		19207	12277050	LEAD, ELECTRICAL .... UOC:DAA,DAB,DAC,DAD,DAE,DAF,DAG,DAH, DAJ,DAK,DAL,DAW,DAX,V12,V13 ,V14,V15, V16,V17,V18,V19,V20,V21,V22,V24,V25, V39	1
1	MOOZZ		19207	12277051	LEAD, ELECTRICAL .... UOC:DAA,DAB,DAC,DAD,DAE,DAF,DAG,DAH, DAJ,DAK,DAL,DAW,DAX,V12,V13,V14,V15, V16,V17,V18,V19,V20,V21,V22,V24,V25, V39	1
2	PAOZZ	5940001155006	96906	MS25036-133	.TERMINAL, LUG .... UOC:DAA,DAB,DAC,DAD,DAE,DAF,DAG,DAH, DAJ,DAK,DAL,DAW,DAX,V12 ,V13,V14 ,V15, V16,V17,V18,V19,V20,V21,V22,V24,V25, V39	1
3	PAOZZ	5940010354212	19207	7728780	.TERMINAL, LUG .... UOC:DAA,DAB,DAC,DAD,DAE,DAF,DAG,DAH, DAJ,DAK,DAL,DAW,DAX,V12,V13,V14,V15, V16,V17,V18,V19,V20,V21,V22,V24,V25, V39	1
4	PAOZZ	9905010138723	81349	M43436/3-1	.BAND, MARKER .... UOC:DAA,DAB,DAC,DAD,DAE,DAF,DAG,DAH, DAJ,DAK,DAL,DAW,DAX,V12,V13,V14,V15, V16,V17,V18,V19,V20,V21,V22,V24,V25, V39	2
5	PAOZZ	5970011749449	19207	7056640	.INSULATION SLEEVING .... UOC:DAA,DAB,DAC,DAD,DAE,DAF,DAG,DAH, DAJ,DAK,DAL,DAW,DAX,V12,V13,V14,V15, V16,V17,V18,V19,V20,V21,V22,V24,V25, V39	2
6	PAOZZ	9905008933570	81349	M43436/1-3	.BAND, MAKER .... UOC:DAA,DAB,DAC,DAD,DAE,DAF,DAG,DAH, DAJ,DAK,DAL,DAW,DAX,V12,V13,V14,V15, V16,V17,V18,V19,V20,V21,V22,V24,V25, V39	2
7	PAOZZ	5940007355520	19207	7355520	.TERMINAL, LUG..................................... UOC:DAA,DAB,DAC,DAD,DAE,DAF,DAG,DAH, DAJ,DAK,DAL,DAW,DAX,V12,V13,V14,V15, V16,V17,V18,V19,V20,V21,V22,V24,V25, V39	1
8	PAOZZ	5935006915591	19207	8724495	.SHELL, ELECTRICAL CO .... UOC:DAA,DAB,DAC,DAD,DAE,DAF,DAG,DAH, DAJ,DAK,DAL,DAW,DAX,V12,V13,V14,V15, V16,V17,V18,V19,V20,V21,V22,V24,V25, V39	1
9	PAOZZ	5310006560067	19207	8724497	.WASHER, SLOTTED.. UOC:DAA,DAB,DAC,DAD,DAE,DAF,DAG,DAH, DAJ,DAK,DAL,DAW,DAX,V12,V13,V14,V15, V16,V17,V18,V19,V20,V21,V22,V24,V25, V39	1

(1) ITEM NO	(2) SMR CODE	(3) NSN	(4) CAGEC	(5) PART NUMBER	(6) DESCRIPTION AND USABLE ON CODES (UOC)	(7) QTY
					DAJ,DAK,DAL,DAW,DAX,V12,V13,V14,V15, V16,V17,V18,V19,V20,V21,V22,V24,V25, V39	
10	PAOOA	5999009263144	96906	MS27148-3	.CONTACT, ELECTRICAL	1
					UOC:DAA,DAB,DAC,DAD,DAE,DAF,DAG,DAH, DAJ,DAK,DAL,DAW,DAX,V12,V13,V14,V15, V16,V17,V18,V19,V20,V21,V22,V24,V25, V39	
11	PAOZZ	5940005496583	96906	MS75004-2	TERMINAL, LUG	4
12	MOOZZ	6150007762738	19207	7762738	LEAD, ELECTRICAL	2
13	PAOZZ	5940001155006	96906	MS25036-133	.TERMINAL, LUG	2
14	PAOZZ		19207	7056640	.INSULATION SLEEVING	2
15	PAOZZ	9905008933570	81349	M43436/1-3	.BAND, MARKER	1
16	PAOZZ	5940005496581	96906	MS75004-1	TERMINAL, LUG	4
17	PAOZZ	6150010831154	19207	12256008	CABLE ASSEMBLY, SPEC	2
18	PAOZZ	5940001155006	96906	MS25036-133	.TERMINAL, LUG	2
19	PAOZZ		19207	7056640	.INSULATION SLEEVING	2
20	PAOZZ	9905008933570	81349	M43436/1-3	.BAND, MARKER	3
21	PAOZZ	5940007355520	19207	7355520	.TERMINAL, LUG	1
22	PAOFA	6140014311172	04055	6TLFP	BATTERY, STORAGE	4
23	PAOZZ	6810002499354	19207	10875529	SULFURIC ACID, ELECT	1
24	PAOZZ	5340011179876	19207	7397785	CLAMP, LOOP	2
25	PAOZZ	5310005501130	96906	MS35333-40	WASHER, LOCK	2
26	PAOZZ	5306000680513	60285	6893-2	BOLT, MACHINE	2
27	PAOZZ	5940007386272	19207	10942521	COVER, BATTERY TERMI	8

END OF FIGURE

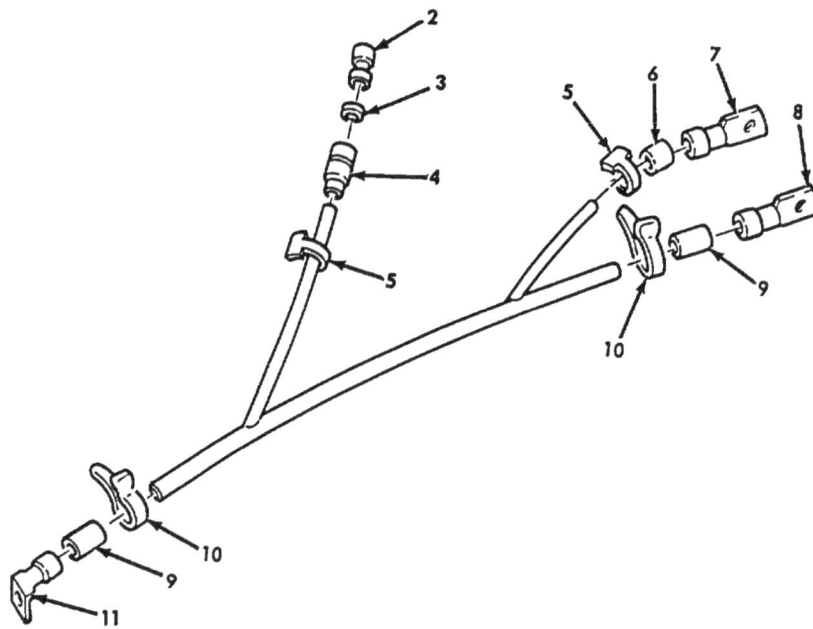

Figure 152. Battery Cables (M939A2).

(1) ITEM NO	(2) SMR CODE	(3) NSN	(4) CAGEC	(5) PART NUMBER	(6) DESCRIPTION AND USABLE ON CODES (UOC)	(7) QTY
					GROUP 0612 BATTERIES,STORAGE	
					FIG. 152 BATTERY CABLES(M939A2)	
1	MOOZZ		1U238	12277050A XX	CABLE ASSEMBLY, SPEC POSITIVE .... BATTERY, MAKE FROM WIRE, P/N M13486-1-3, 73 INCHES LONG, AND P/N M13486-1-14, 112.50 INCHES LONG .... UOC:ZAA, ZAB, ZAC, ZAD, ZAE, ZAF, ZAG, ZAH, ZAJ, ZAK, ZAL	1
1	MOOZZ		1U238	12277051A XX	CABLE ASSEMBLY, POSITIVE BATTERY, ... MAKE FROM WIRE, P/N M13486-1-3, 70 INCHES LONG, AND P/N M13486-1-14, 102.50 INCHES LONG .... UOC:ZAA, ZAB, ZAC, ZAD, ZAE, ZAF, ZAG, ZAH, ZAJ, ZAK, ZAL	1
2	PAOZA	5999009263144	96906	MS27148-3	.CONTACT, ELECTRICAL .... UOC:ZAA, ZAB, ZAC, ZAD, ZAE, ZAF, ZAG, ZAH, ZAJ, ZAK, ZAL	2
3	PAOZZ	5310006560067	19207	8724497	.WASHER, SLOTTED.............................. UOC:ZAA, ZAB, ZAC, ZAD, ZAE, ZAF, ZAG, ZAH, ZAJ, ZAK, ZAL	1
4	PAOZZ	5935006915591	19207	8724495	.SHELL, ELECTRICAL CO .... UOC:ZAA, ZAB, ZAC, ZAD, ZAE, ZAF, ZAG, ZAH, ZAJ, ZAK, ZAL	1
5	PAOZZ	9905010138723	81349	M43436/3-1	.BAND, MARKER .... UOC:ZAA, ZAB, ZAC, ZAD, ZAE, ZAF, ZAG, ZAH, ZAJ, ZAK, ZAL	2
6	MOOZZ		81349	M23053/5 X 3 IN	.INSULATION SLEEVING MAKE FROM .... INSULATION, P/N M23053/5-107-0, 3 INCHES LONG .... UOC:ZAA, ZAB, ZAC, ZAD, ZAE, ZAF, ZAG, ZAH, ZAJ, ZAK, ZAL	1
7	PAOZZ	5940010354212	19207	7728780	.TERMINAL, LUG .... UOC:ZAA, ZAB, ZAC, ZAD, ZAE, ZAF, ZAG, ZAH, ZAJ, ZAK, ZAL	1
8	PAOZZ	5940001155006	96906	MS25036-133	.TERMINAL, LUG .... UOC:ZAA, ZAB, ZAC, ZAD, ZAE, ZAF, ZAG, ZAH, ZAJ, ZAK, ZAL	1
9	PAOZZ	5970011749449	19207	7056640	.INSULATION SLEEVING .... UOC:ZAA, ZAB, ZAC, ZAD, ZAE, ZAF, ZAG, ZAH, ZAJ, ZAK, ZAL	2
10	PAOZZ	9905008933570	81349	M43436/1-3	.BAND, MARKER .... UOC:ZAA, ZAB, ZAC, ZAD, ZAE, ZAF, ZAG, ZAH, ZAJ, ZAK, ZAL	2
11	PAOZZ	5940007355520	19207	7355520	.TERMINAL, LUG .... UOC:ZAA, ZAB, ZAC, ZAD, ZAE, ZAF, ZAG, ZAH, ZAJ, ZAK, ZAL	1

END OF FIGURE

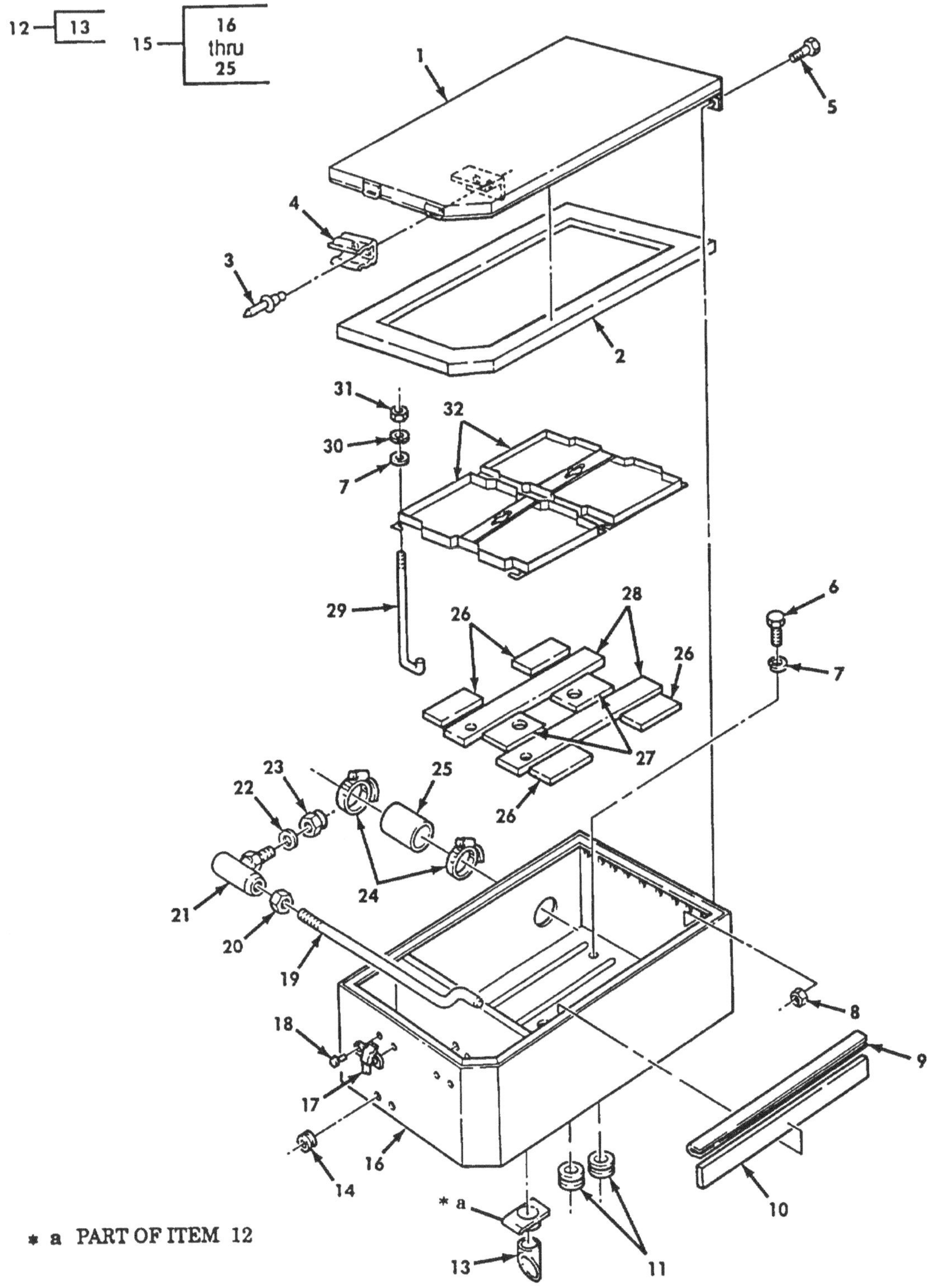

12 — 13   15 — 16 thru 25

* a PART OF ITEM 12

Figure 153. Battery Box Assembly, Cover, and Mounting Hardware.

(1) ITEM NO	(2) SMR CODE	(3) NSN	(4) CAGEC	(5) PART NUMBER	(6) DESCRIPTION AND USABLE ON CODES (UOC)	(7) QTY

GROUP 0612 BATTERIES, STORAGE

FIG. 153 BATTERY BOX ASSEMBLY, COVER, AND MOUNTING HARDWARE

(1) ITEM NO	(2) SMR CODE	(3) NSN	(4) CAGEC	(5) PART NUMBER	(6) DESCRIPTION AND USABLE ON CODES (UOC)	(7) QTY
1	PFOZZ	6160011308045	19207	12255880	COVER, BATTERY BOX	1
2	PAOZZ	5330011047702	19207	12256041	RUBBER STRIP	1
3	PAOZZ	5320010554452	81349	M24243/1-B304	RIVET, BLIND	2
4	PAOZZ	5340007256033	81349	M24066/2-321	CLIP, SPRING TENSION	1
5	PAOZZ	5305009897435	96906	MS35207-264	SCREW, MACHINE	6
6	PAOZZ	5305002692803	96906	MS90726-60	SCREW, CAP, HEXAGON H	4
7	PAOZZ	5310000806004	96906	MS27183-14	WASHER, FLAT	10
8	PAOZZ	5310008775797	96906	MS21044-N3	NUT, SELF-LOCKING, HE	6
9	PAOZZ	5970011444841	19207	12257069	INSULATION SLEEVING	1
10	PAOZZ	5970011345093	19207	12257070	INSULATOR, PLATE	1
11	PAOZZ	5325002024005	96906	MS35489-110	GROMMET, NONMETALLIC	2
12	PFOZZ	6140011566187	19207	12277135	VENT TUBE, BATTERY	1
13	PAOZZ	4720011223656	19207	12277138	.HOSE, NONMETALLIC	1
14	PAOZZ	5325012427083	96906	MS35489-17	GROMMET, NONMETALLIC	4
15	PAOZZ	6160010935836	19207	12255881	BATTERY BOX	1
16	XAOZZ		19207	12255882	.BATTERY BOX	1
17	PAOZZ	5340011280191	19207	7413565	.CATCH, CLAMPING	2
18	PFOZZ	5320010696365	96906	MS35743-16	.RIVET, SOLID	4
19	PAOZZ	6160010934256	19207	12302883	.RETAINER, BATTERY	1
20	PAOZZ	5310008911751	96906	MS35691-22	.NUT, PLAIN, HEXAGON	1
21	PAOZZ	3040005410995	81601	109417	.BALL JOINT	1
22	PAOZZ	5310000877493	96906	MS27183-13	.WASHER, FLAT	1
23	PAOZZ	5310008140672	96906	MS51943-36	.NUT, SELF-LOCKING, HE	1
24	PAOZZ	4730009098627	01276	FF9311-36	CLAMP, HOSE	2
25	PAOZZ	4720011324868	19207	12277340	TUBING, NONMETALLIC	1
26	PAOZZ	6140011256075	19207	12277132	.BLOCK, SUPPORT, BATTE	4
27	PAOZZ	6140011256074	19207	12277133	.BLOCK, SUPPORT, BATTE	2
28	PAOZZ	6140011256073	19207	12277134	.BLOCK, SUPPORT, BATTE	2
29	PAOZZ	5306007397754	19207	7397754	BOLT, HOOK	10
30	PAOZZ	5310006379541	96906	MS35338-46	WASHER, LOCK	10
31	PAOZZ	5310006559544	96906	MS35690-604	NUT, PLAIN, HEXAGON	10
32	PAOZZ	6160004042669	96906	MS53046-3	RETAINER, BATTERY	2

END OF FIGURE

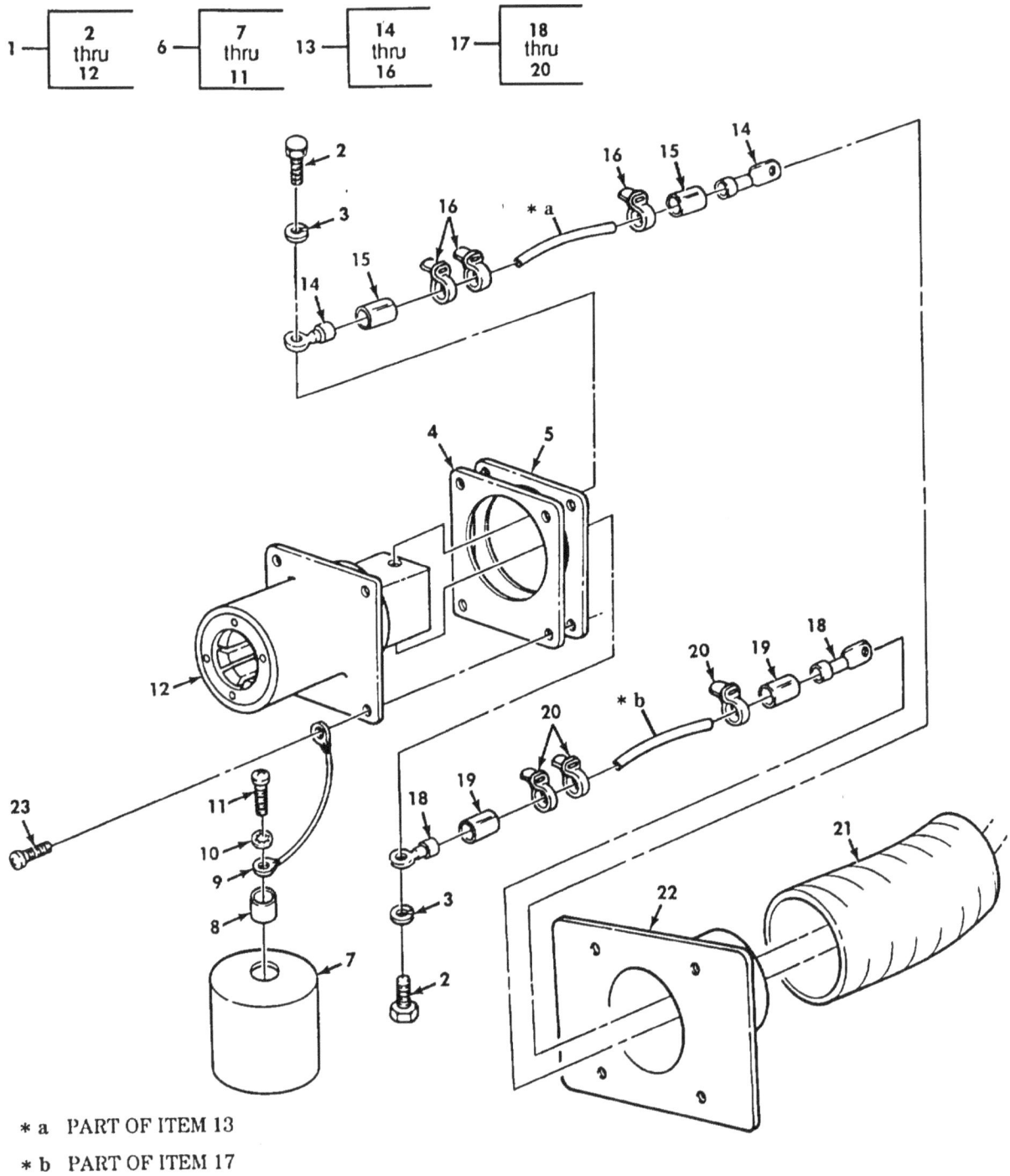

*a PART OF ITEM 13

*b PART OF ITEM 17

Figure 154. Slave Receptacle, Cable Assemblies, and Mounting Hardware.

(1) ITEM NO	(2) SMR CODE	(3) NSN	(4) CAGEC	(5) PART NUMBER	(6) DESCRIPTION AND USABLE ON CODES (UOC)	(7) QTY
					GROUP 0613 CHASSIS WIRING HARNESS	
					FIG. 154 SLAVE RECEPTACLE, CABLE ASSEMBLIES, AND MOUNTING HARDWARE	
1	PAOZZ	5935010979974	19207	11674728	CONNECTOR, RECEPTACL	1
2	PAOZZ	5305002693233	80204	B1821BH038F063N	.SCREW, CAP, HEXAGON H	2
3	PAOZZ	5310006379541	96906	MS35338-46	.WASHER, LOCK	2
4	PAOZZ	5970010448391	19207	11674730	.INSULATOR, PLATE	1
5	PAOZZ	5330010594286	19207	11674729	.GASKET	1
6	PAOZZ	5340010590114	19207	11675004	.CAP, PROTECTIVE, DUST	1
7	XAOZZ		19207	11675004-1	..BODY	1
8	XAOZZ	5325010062246	96906	MS35914-148	..INSERT, SCREW THREAD	1
9	XAOZZ	4020010920331	19207	12269868	..FIBER ROPE ASSEMBLY	1
10	XAOZZ	5310005795554	96906	MS35333-35	..WASHER, LOCK	1
11	XAOZZ	5305009584357	96906	MS35207-242	..SCREW, MACHINE	1
12	PAOZZ	5935010448382	19207	11682345	.CONNECTOR, RECEPTACL	1
13	MOOZZ		19207	12256699-2	LEAD ASSEMBLY, ELECT MAKE FROM WIRE	1
					P/N M13486/1-12 (81349)...................................	
14	PAOZZ	5940007056730	19207	7056730	.TERMINAL, LUG	2
15	PAOZZ	5970008110640	80244	17-I-1728-651	.INSULATION SLEEVING	2
16	PAOZZ	9905008933570	81349	M43436/1-3	.BAND, MARKER	3
17	MOOZZ		19207	12256699-1	LEAD ASSEMBLY, ELECT	1
18	PAOZZ	5940007056730	19207	7056730	.TERMINAL, LUG	2
19	PAOZZ	5970007056639	19207	7056639	.INSULATOR, BUSHING	2
20	PAOZZ	9905008933570	81349	M43436/1-3	.BAND, MARKER	3
21	PAOZZ	4720011957604	19207	12277340-1	HOSE, PREFORMED	1
22	PAOZZ	5340012122464	19207	12277338	PLATE, MOUNTING	1
23	PAOZZ	5305000813728	96906	MS9316-08	SCREW, MACHINE	4

END OF FIGURE

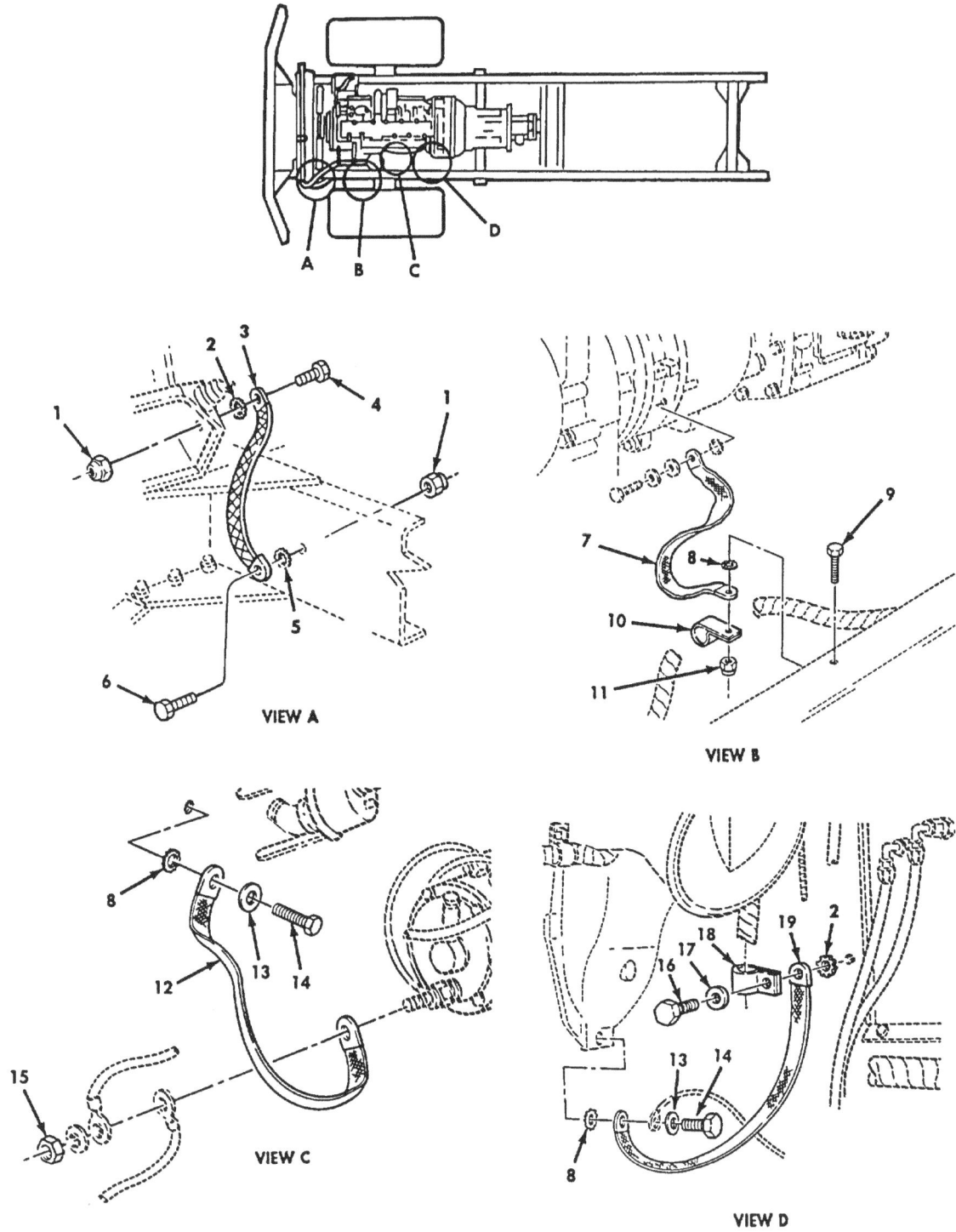

Figure 155. Electrical Ground Straps (M939, M939A1).

(1) ITEM NO	(2) SMR CODE	(3) NSN	(4) CAGEC	(5) PART NUMBER	(6) DESCRIPTION AND USABLE ON CODES (UOC)	(7) QTY

SECTION II

GROUP 0613 CHASSIS WIRING HARNESS

FIG. 155 ELECTRICAL GROUND STRAPS (M939,M939A1)

(1) ITEM NO	(2) SMR CODE	(3) NSN	(4) CAGEC	(5) PART NUMBER	(6) DESCRIPTION AND USABLE ON CODES (UOC)	(7) QTY
1	PAOZZ	5310004883888	96906	MS51943-40	NUT, SELF-LOCKING, HE .... UOC:DAA, DAB, DAC, DAD, DAE, DAF, DAG, DAH, DAJ, DAK, DAL, DAW, DAX, V12, V13, V14, V15, V16, V17, V18, V19, V20, V21, V22, V24, V25, V39	2
2	PAOZZ	5310009358984	96906	MS45904-84	WASHER, LOCK .... UOC:DAA, DAB, DAC, DAD, DAE, DAF, DAG, DAH, DAJ, DAK, DAL, DAW, DAX, V12, V13, V14, V15, V16, V17, V18, V19, V20, V21, V22, V24, V25, V39	2
3	PAOZZ	5999011282755	19207	12277229-2	STRIP, ELECTRICAL GR .... UOC:DAA, DAB, DAC, DAD, DAE, DAF, DAG, DAH, DAJ, DAK, DAL, DAW, DAX, V12, V13, V14, V15, V16, V17, V18, V19, V20, V21, V22, V24, V25, V39	1
4	PAOZZ	5305002267767	96906	MS90726-109	SCREW, CAP, HEXAGON H .... UOC:DAA, DAB, DAC, DAD, DAE, DAF, DAG, DAH, DAJ, DAK, DAL, DAW, DAX, V12, V13, V14, V15, V16, V17, V18, V19, V20, V21, V22, V24, V25, V39	1
5	PAOZZ	5310009011339	96906	MS45904-87	WASHER, LOCK .... UOC:DAA, DAB, DAC, DAD, DAE, DAF, DAG, DAB, DAJ, DAK, DAL, DAW, DAX V12, V13 , V13 4 , V14, V15, V16, V17, V18, V19, V20, V21, V22, V24, V25, V39	1
6	PAOZZ	5305007195219	96906	MS90727-111	SCREW, CAP, HEXAGON H .... UOC:DAA, DAB, DAC, DAD, DAE, DAF, DAG, DAB, DAJ, DAK, DAL, DAW, DAX, V12, V13, V14, V15, V16, V17, V18, V19, V20, V21, V22, V24, V25, V39	1
7	PAOZZ	2590011325195	19207	10896285-4	LEAD, ELECTRICAL .... UOC:DAA, DAB, DAC, DAD, DAE, DAF, DAG, DAH, DAJ, DAK, DAL, DAW, DAX, V12 , V13 , V14 , V15, V16, V17, V18, V19, V20, V21, V22, V24, V25, V39	1
8	PAOZZ	5310000611258	96906	MS45904-76	WASHER, LOCK .... UOC:DAA, DAB, DAC, DAD, DAE, DAF, DAG, DAH, DAJ, DAK, DAL, DAW, DAX, V12, V13, V14, V15, V16, V17, V18, V19, V20, V21, V22, V24, V25, V39	3
9	PAOZZ	5305002693238	80204	B1821BH038F125N	SCREW, CAP, HEXAGON H .... UOC:DAA, DAB, DAC, DAD, DAE, DAF, DAG, DAH, DAJ, DAK, DAL, DAW, DAX, V12, V13, V14, V15, V16, V17, V18, V19, V20, V21, V22, V24, V25, V39	1
10	PAOZZ	5340007022848	96906	MS21333-128	CLAMP, LOOP .... UOC:DAA, DAB, DAC, DAD, DAE, DAF, DAG, DAH, DAJ, DAK, DAL, DAW, DAX, V12, V13, V14, V15, V16, V17, V18, V19, V20, V21, V22, V24, V25, V39	1

(1) ITEM NO	(2) SMR CODE	(3) NSN	(4) CAGEC	(5) PART NUMBER	(6) DESCRIPTION AND USABLE ON CODES (UOC)	(7) QTY
11	PAOZZ	5310008140672	96906	MS51943-36	NUT, SELF-LOCKING .... UOC:DAA, DAB, DAC, DAD, DAE, DAF, DAG, DAH, DAJ, DAK, DAL, DAW, DAX, V12, V13, V14, V15, V16, V17, V18, V19, V20, V21, V22, V24, V25, V39	1
12	PAOZZ	6150012259535	19207	12302703	LEAD, ELECTRICAL .... UOC:DAA, DAB, DAC, DAD, DAE, DAF, DAG, DAH, DAJ, DAK, DAL, DAW, DAX, V12, V13, V14, V15, V16, V17, V18, V19, V20, V21, V22, V24, V25, V39	1
13	PAOZZ	5310000877493	96906	MS27183-13	WASHER, FLAT .... UOC:DAA, DAB, DAC, DAD, DAE, DAF, DAG, DAH, DAJ, DAK, DAL, DAW, DAX, V12, V13, V14, V15, V16, V17, V18, V19, V20, V21, V22, V24, V25, V39	2
14	PAOZZ	5305002693234	80204	B1821BH038F075N	SCREW, CAP, HEXAGON H .... UOC:DAA, DAB, DAC, DAD, DAE, DAF, DAG, DAH, DAJ, DAX, DAL, DAW, DAX, V12, V13, V14, V15, V16, V17, V18, V19, V20, V21, V22, V24, V25, V39	2
15	PAOZZ	5310009895945	96906	MS35691-35	NUT, PLAIN, HEXAGON .... UOC:DAA, DAB, DAC, DAD, DAE, DAF, DAG, DAH, DAJ, DAK, DAL, DAW, DAX, V12, V13, V14, V15, V16, V17, V18, V19, V20, V21, V22, V24, V25, V39	1
16	PAOZZ	5305000680515	80204	B1821BH025F100N	SCREW, CAP, HEXAGON H .... UOC:DAA, DAB, DAC, DAD, DAE, DAF, DAG, DAH, DAJ, DAK, DAL, DAW, DAX, V12, V13 , V14 , V15, V16, V17, V18, V19, V20, V21, V22, V24, V25, V39	1
17	PAOZZ	5310009391061	96906	MS35333-108	WASHER, LOCK .... UOC:DAA, DAB, DAC, DAD, DAE, DAF, DAG, DAH, DAJ, DAK, DAL, DAW, DAX, V12, V13, V14, V15, V16, V17, V18, V19, V20, V21, V22, V24, V25, V39	1
18	PAOZZ	5340000881254	96906	MS21333-104	CLAMP, LOOP .... UOC:DAA, DAB, DAC, DAD, DAE, DAF, DAG, DAH, DAJ, DAK, DAL, DAW, DAX, V12, V13, V14, V15, V16, V17, V18, V19, V20, V21, V22, V24, V25, V39	1
19	PAOZZ	5999013318578	19207	10896285-2	STRIP, ELECTRICAL .... UOC:DAA, DAB, DAC, DAD, DAE, DAF, DAG, DAH, DAJ, DAK, DAL, DAW, DAX, V12, V13, V14, V15, V16, V17, V18, V19, V20, V21, V22, V24, V25, V39	1

END OF FIGURE

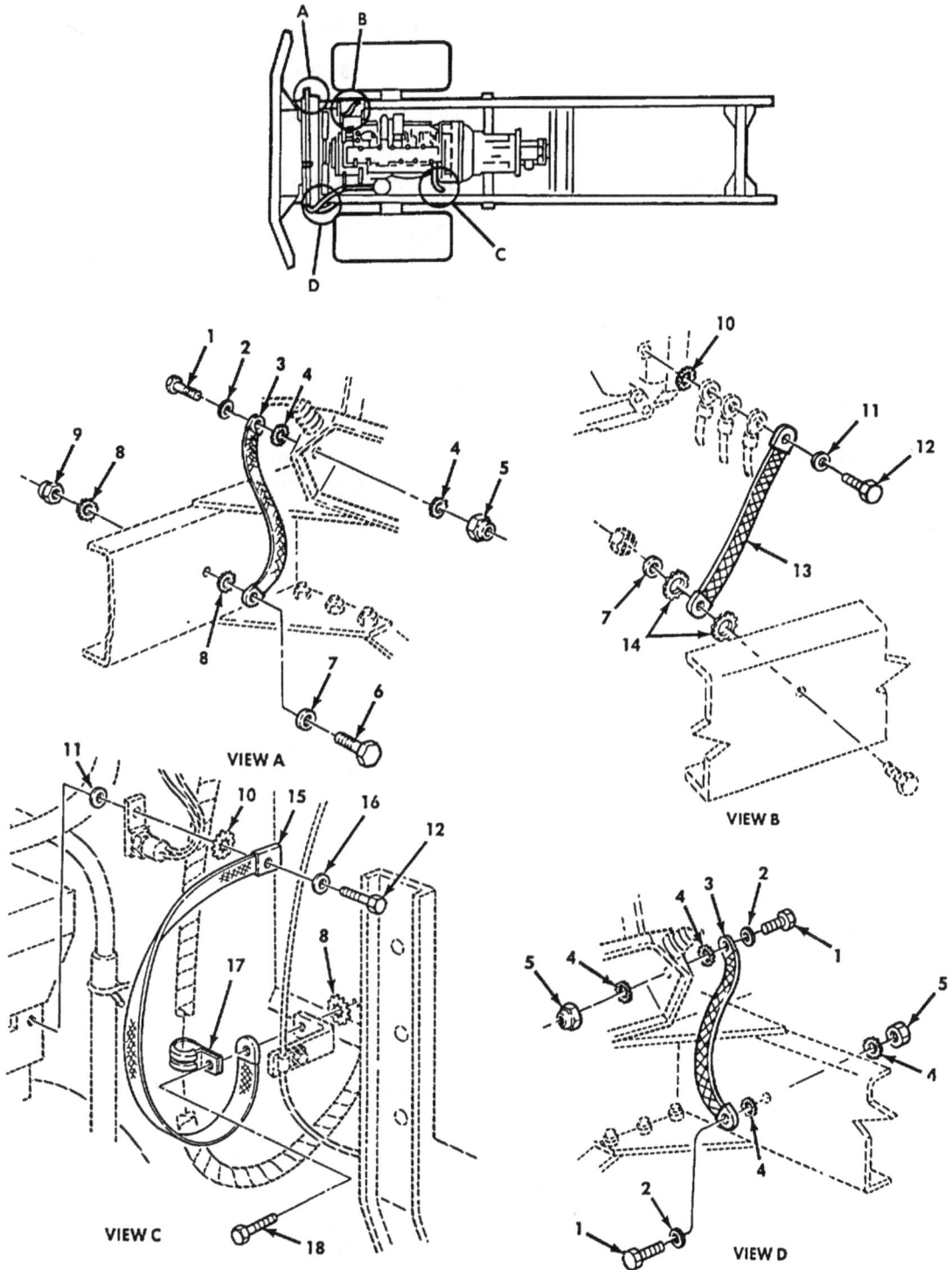

Figure 156. Electrical Ground Straps (M939A2).

(1) ITEM NO	(2) SMR CODE	(3) NSN	(4) CAGEC	(5) PART NUMBER	(6) DESCRIPTION AND USABLE ON CODES (UOC)	(7) QTY
					GROUP 0613 CHASSIS WIRING HARNESS	
					FIG. 156 ELECTRICAL GROUND STRAPS (M939A2)	
1	PAOZZ	5305007195219	96906	MS90727-111	SCREW, CAP, HEXAGON H .... UOC:ZAA, ZAB, ZAC, ZAD, ZAE, ZAF, ZAG, ZAH, ZAJ, ZAK, ZAL	3
2	PAHZZ	5310008095998	96906	MS27183-18	WASHER, FLAT .... UOC: ZAA, ZAB, ZAC, ZAD, ZAE, ZAF, ZAG, ZAH, ZAJ, ZAK, ZAL	3
3	PAOZZ	5999011282755	19207	12277229-2	STRIP, ELECTRICAL GR .... UOC:ZAA, ZAB, ZAC, ZAD, ZAE, ZAF, ZAG, ZAH, ZAJ, ZAK, ZAL	2
4	PAOZZ	5310009358984	96906	MS45904-84	WASHER, LOCK .... UOC:ZAA, ZAB, ZAC, ZAD, ZAE, ZAF, ZAG, ZAH, ZAJ, ZAK, ZAL	6
5	PAOZZ	5310004883888	96906	MS51943-40	NUT, SELF-LOCKING .... UOC:ZAA, ZAB, ZAC, ZAD, ZAE, ZAF, ZAG, ZAH, ZAJ, ZAK, ZAL	3
6	PAOZZ	5305002693238	80204	B1821BH038F125N	SCREW, CAP, HEXAGON .... UOC:ZAA, ZAB, ZAC, ZAD, ZAE, ZAF, ZAG, ZAH, ZAJ, ZAK, ZAL	1
7	PAOZZ	5310000806004	96906	MS27183-14	WASHER, FLAT .... UOC:ZAA, ZAB, ZAC, ZAD, ZAE, ZAF, ZAG, ZAH, ZAJ, ZAK, ZAL	2
8	PAOZZ	5310000611258	96906	MS45904-76	WASHER, LOCK .... UOC:ZAA, ZAB, ZAC, ZAD, ZAE, ZAF, ZAG, ZAH, ZAJ, ZAK, ZAL	3
9	PAOZZ	5310008140672	96906	MS51943-36	NUT, SELF-LOCKING .... UOC:ZAA, ZAB, ZAC, ZAD, ZAE, ZAF, ZAG, ZAH, ZAJ, ZAK, ZAL	1
10	PAOZZ		24617	138485	WASHE R, LOCK .... UOC:ZAA, ZAB, ZAC, ZAD, ZAE, ZAF, ZAG, ZAH, ZAJ, ZAK, ZAL	2
11	PAOZZ	5310014073451	24617	9422772	WASHER, FLAT .... UOC:ZAA, ZAB, ZAC, ZAD, ZAE, ZAF, ZAG, ZAH, ZAJ, ZAK, ZAL	2
12	PAOZZ	5306011971492	24617	11501872	BOLT, MACHINE .... UOC:ZAA, ZAB, ZAC, ZAD, ZAE, ZAF, ZAG, ZAH, ZAJ, ZAK, ZAL	2
13	PAOZZ	5999013280524	19207	10896285-1	STRIP, ELECTRICAL GR .... UOC:ZAA, ZAB, ZAC, ZAD, ZAE, ZAF, ZAG, ZAH, ZAJ, ZAK, ZAL	1
14	PAOZZ	5310006276128	24617	138489	WASHER, LOCK .... UOC:ZAA, ZAB, ZAC, ZAD, ZAE, ZAF, ZAG, ZAH, ZAJ, ZAK, ZAL	2
15	PAOZZ	5999013318578	19207	10896285-2	STRIP, ELECTRICAL GR .... UOC:ZAA, ZAB, ZAC, ZAD, ZAE, ZAF, ZAG, ZAH, ZAJ, ZAK, ZAL	1
16	PAOZZ		24617	446364	WASHER, FLAT .... UOC:ZAA, ZAB, ZAC, ZAD, ZAE, ZAF, ZAG, ZAH, ZAJ, ZAK, ZAL	1

(1) ITEM NO	(2) SMR CODE	(3) NSN	(4) CAGEC	(5) PART NUMBER	(6) DESCRIPTION AND USABLE ON CODES (UOC)	(7) QTY
17	PAOZZ	5340008091500	96906	MS21333-107	CLAMP, LOOP .... UOC: ZAA, ZAB, ZAC, ZAD, ZAE, ZAF, ZAG, ZAH, ZAJ, ZAK, ZAL	1
18	PAOZZ		24617	454849	BOLT, MACHINE .... UOC:ZAA, ZAB, ZAC, ZAD, ZAE, ZAF, ZAG, ZAH, ZAJ, ZAK, ZAL	1

END OF FIGURE

RIGHT HAND LIGHTS

B O HEAD LAMP

HEAD LAMP UPPER BEAM

LOWER BEAM

SIDE MARKER

PARKING LAMP

TURN SIGNAL LAMP

INTERMEDIATE TURN SIGNAL

B O. MARKER

UPPER BEAM   LOWER BEAM

HEAD LAMP

B O. MARKER

TURN SIGNAL LAMP

INTERMEDIATE TURN SIGNAL

SIDE MARKER

PARKING LAMP

LEFT HAND LIGHTS

VIEW A   VIEW B   VIEW C   VIEW D

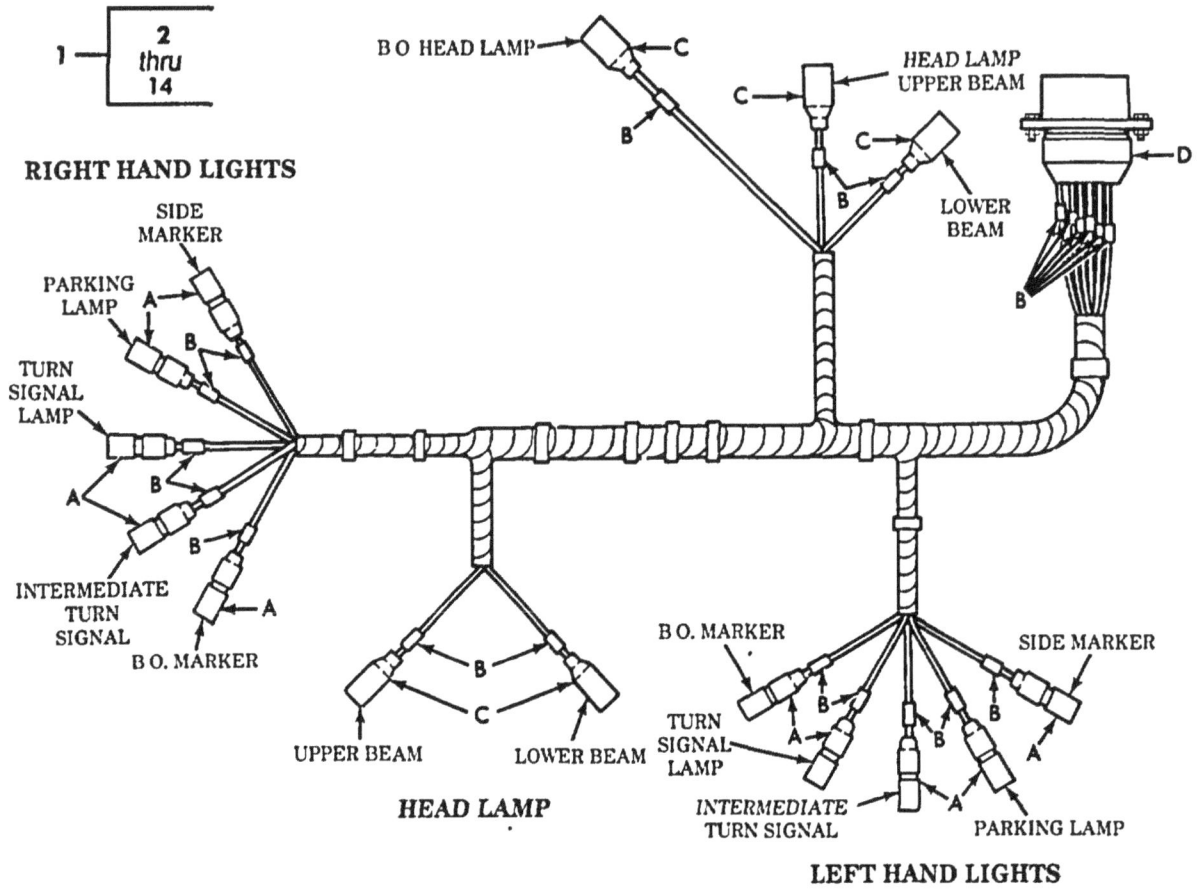

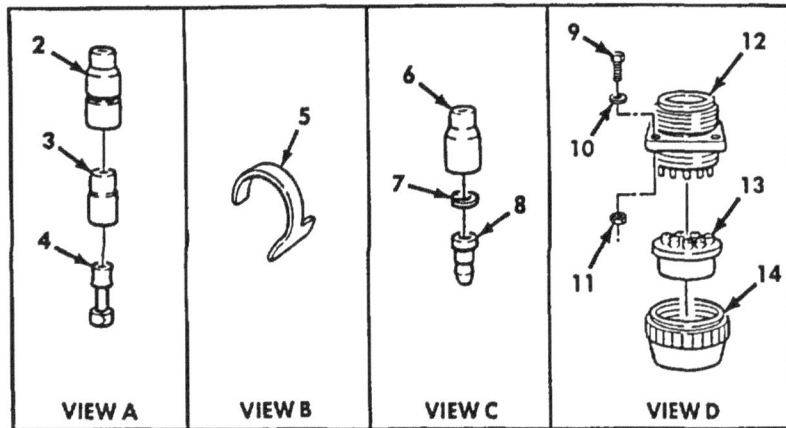

Figure 157. Harness Assembly, Front Vehicle Lights (Sheet 1 of 2).

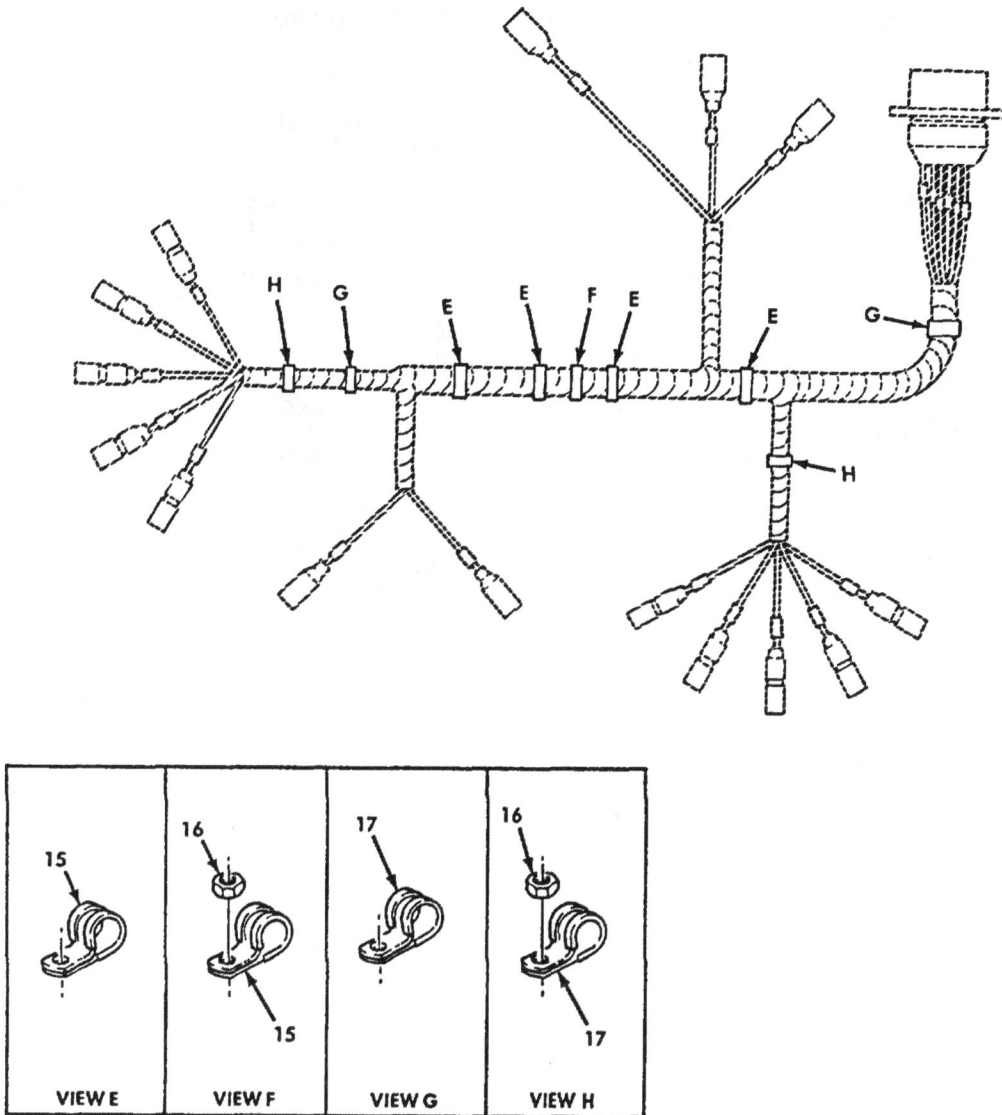

Figure 157. Harness Assembly, Front Vehicle Lights (Sheet 2 of 2).

(1) ITEM NO	(2) SMR CODE	(3) NSN	(4) CAGEC	(5) PART NUMBER	(6) DESCRIPTION AND USABLE ON CODES (UOC)	(7) QTY
					GROUP 0613 CHASSIS WIRING HARNESS	
					FIG. 157 HARNESS ASSEMBLY, FRONT VEHICLE LIGHTS	
1	PAFFF	2590011222419	19207	12277040	WIRING HARNESS, BRAN	1
2	PAOZZ	5935008338561	19207	8338561	.SHELL, ELECTRICAL CO	10
3	PAOZZ	5970008338562	19207	8338562	.INSULATOR, BUSHING	10
4	PAOZZ	5940003996676	19207	8338564	.TERMINAL ASSEMBLY	10
5	PAOZZ	9905007524649	81349	M43436/1-1	.BAND, MARKER	23
6	PAOZZ	5935005729180	19207	8338566	.SHELL, ELECTRICAL CO	5
7	PAOZZ	5310008338567	19207	8338567	.WASHER, SLOTTED	5
8	PAOZA	5999000572929	19204	572929	.CONTACT, ELECTRICAL	5
9	PAOZZ	5305009846192	96906	MS35206-244	.SCREW, MACHINE	4
10	PAOZZ	5310005967693	96906	MS35335-31	.WASHER, LOCK	4
11	PAOZZ	5310009349757	96906	MS35649-282	.NUT, PLAIN, HEXAGON	4
12	PAOZZ	5935007722353	19207	7722353	.CONNECTOR, RECEPTACL	1
13	PAOZZ	5365007722322	19207	7722322	.BUSHING, NONMETALLIC	1
14	PAOZZ	5935003339414	19207	7723308	.NUT, BUSHING RETAINE	1
15	PAOZZ	5340000538994	96906	MS21333-126	CLAMP, LOOP	5
16	PAOZZ	5310008140672	96906	MS51943-36	NUT, SELF-LOCKING, HE	3
17	PAOZZ	5340009645267	96906	MS21333-120	CLAMP, LOOP	4

END OF FIGURE

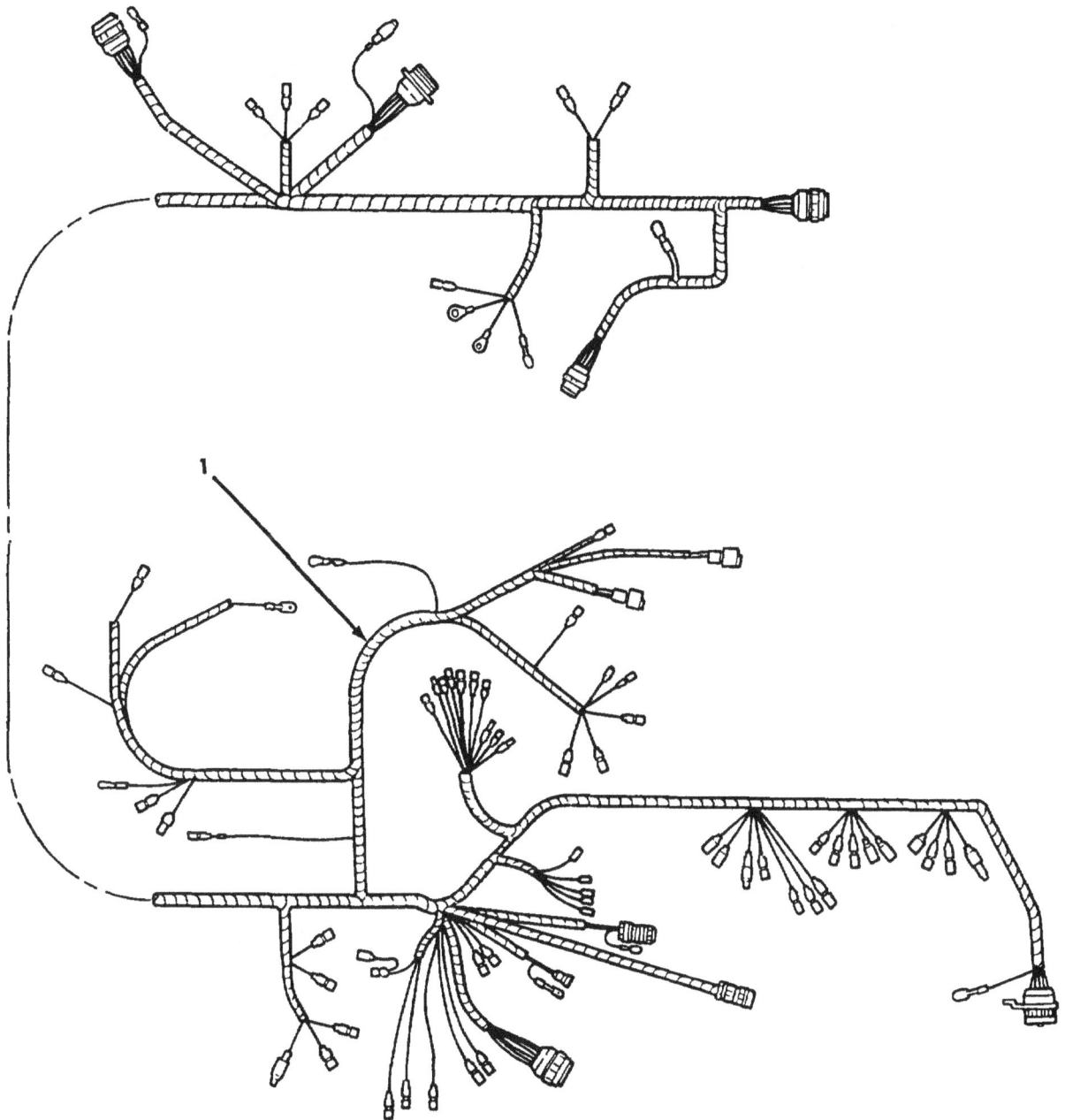

Figure 158. Front Wiring Harness Assembly.

(1) ITEM NO	(2) SMR CODE	(3) NSN	(4) CAGEC	(5) PART NUMBER	(6) DESCRIPTION AND USABLE ON CODES (UOC)	(7) QTY
					GROUP 0613 CHASSIS WIRING HARNESS	
					FIG. 158 FRONT WIRING HARNESS ASSEMBLY	
1	PAFFF	2590010835728	19207	12256006	WIRING HARNESS, BRAN .... UOC:DAA, DAB, DAC, DAD, DAE, DAF, DAG, DAH, DAJ, DAK, DAL, DAW, DAX, V12, V13, V14, V15, V16, V17, V18, V19, V20, V21, V22, V24, V25, V39	1
1	PBFFF	5995012873988	19207	12363340	WIRING HARNESS, BRAN .... UOC:ZAA, ZAB, ZAC, ZAD, ZAE, ZAF, ZAG, ZAH, ZAJ, ZAK, ZAL	1
					END OF FIGURE	

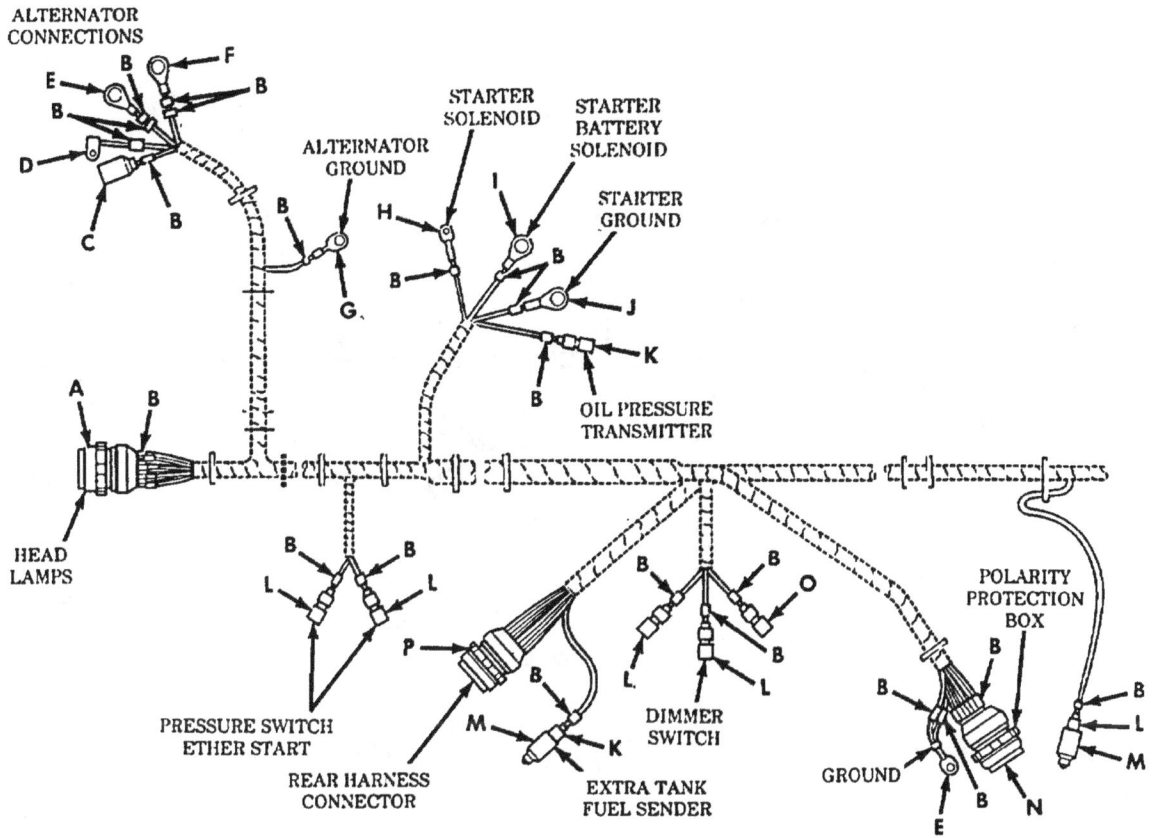

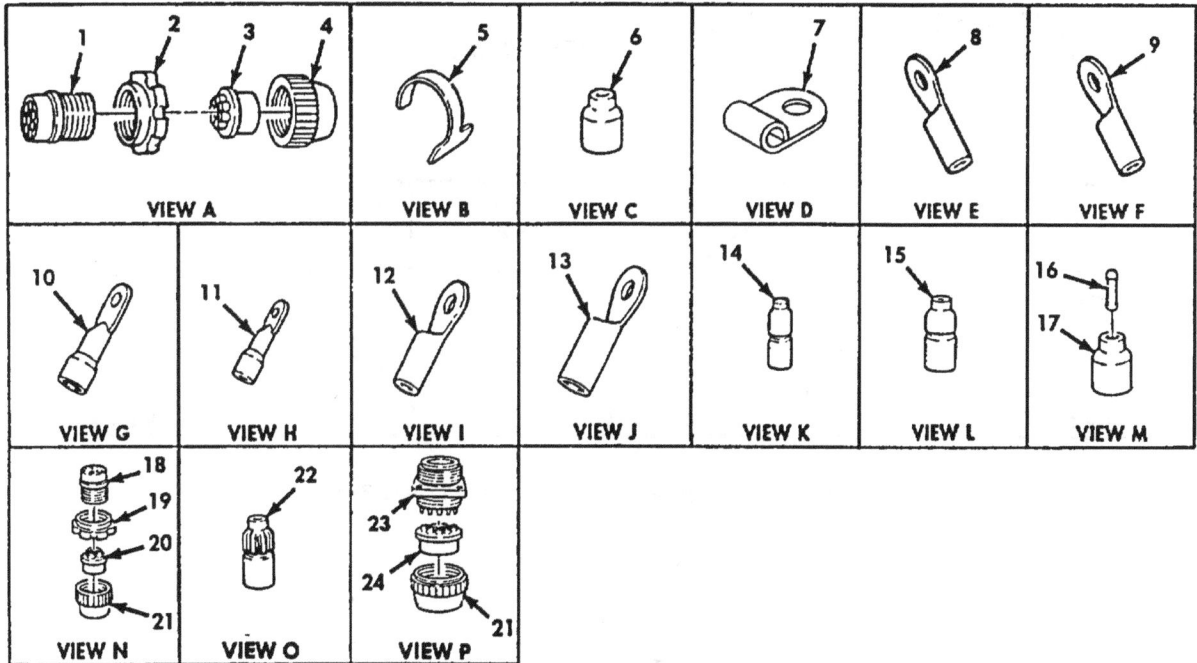

Figure 159. Front Wiring Harness, Partial View (Sheet 1 of 3).

FUEL
SOLENOID

T

PARKING
BRAKE
SWITCH

K

12V
SUPPLY

S

L

COOLANT TEMP
SENSOR

K

PULSE TACH
SENDER PLUG

CAPTACITOR
5th GEAR LOCK-UP
SWITCH

CAPACITOR
T/CASE
SWITCH

L

U

L

L

V

FUEL PRESSURE
TRANSDUCER
PLUG

M

ENGINE
GROUND

L

M

AUXILIARY
POWER

W

ETHER
TANK
VALVE

R

K

L

DIODE
COUPLING
ASSEMBLY

AA

L

HEATER

TO BATTERY
CABLES

K

L

HORN

SEE SHEET 1

SEE SHEET 3

25

VIEW R

26

VIEW S

27

VIEW T

28

29

30

VIEW U

31

29

30

VIEW V

32

VIEW W

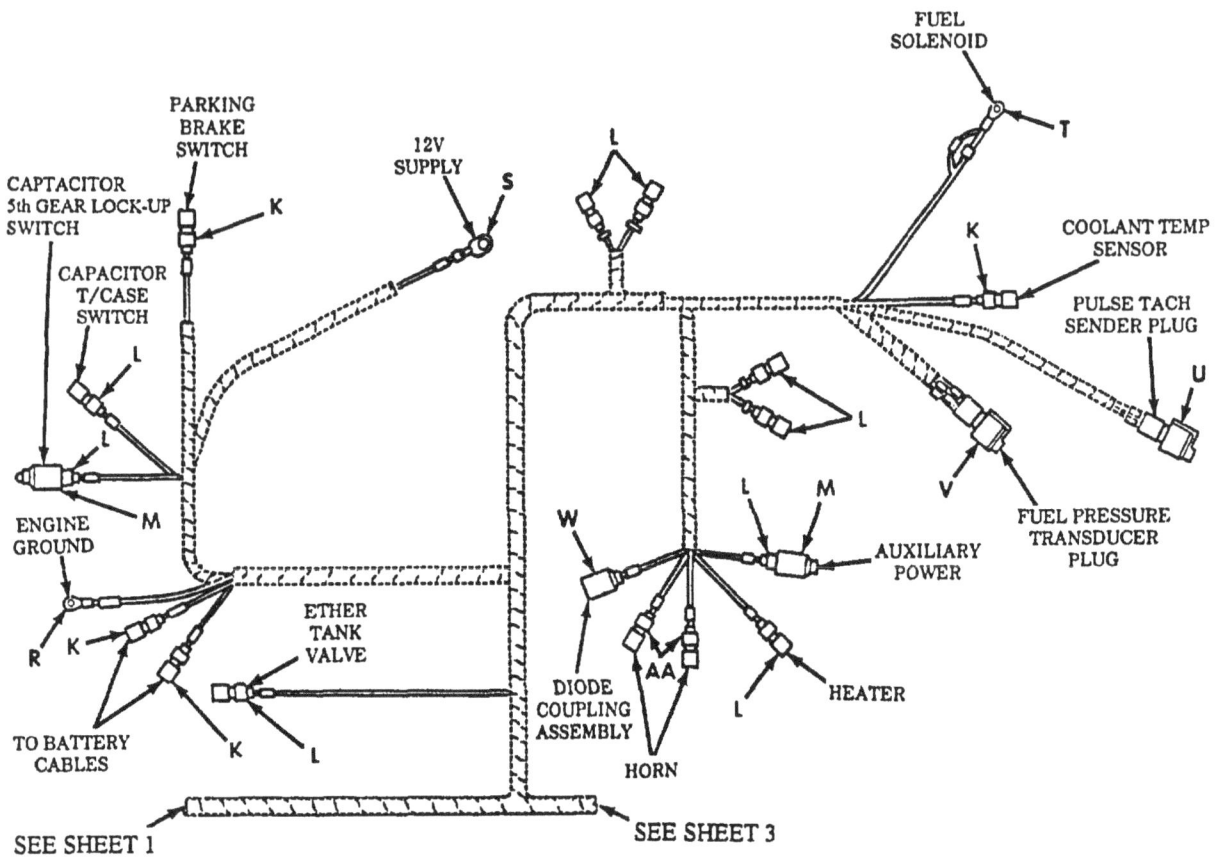

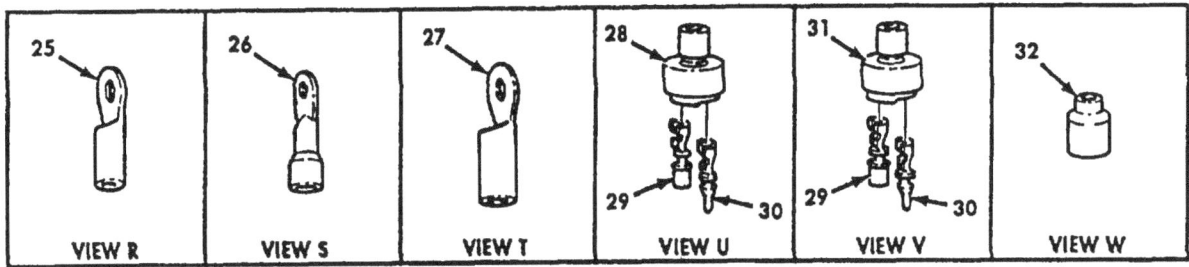

Figure 159. Front Wiring Harness, Partial View (Sheet 2 of 3).

(1) ITEM NO	(2) SMR CODE	(3) NSN	(4) CAGEC	(5) PART NUMBER	(6) DESCRIPTION AND USABLE ON CODES (UOC)	(7) QTY
					GROUP 0613 CHASSIS WIRING HARNESS	
					FIG. 159 FRONT WIRING HARNESS, PARTIAL VIEW	
1	PAOZZ	5935006143959	77820	10-42622-235	CONNECTOR, PLUG, ELEC	1
2	PAOZZ	5310006559860	19207	8701325	NUT, SLEEVE	1
3	PAOZZ	5365007722322	19207	7722322	BUSHING, NONMETALLIC	1
4	PAOZZ	5935003339414	19207	7723308	NUT, BUSHING RETAINE	1
5	PAOZZ	9905007524649	81349	M43436/1-1	BAND, MARKER	138
6	PAOZZ	5935001152306	96906	MS27142-3	CONNECTOR, PLUG, ELEC	5
7	PAOZA	5940010211874	19207	11669805	TERMINAL, LUG	1
8	PAOZZ	5940001152674	96906	MS20659-108	TERMINAL, LUG .... UOC:DAA, DAB, DAC, DAD, DAE, DAF, DAG, DAH, DAJ, DAK, DAL, DAW, DAX, V12, V13, V14, V15, V16, V17, V18, V19, V20, V21, V22, V24, V25, V39	2
9	PAOZZ	5940001139825	96906	MS20659-141	TERMINAL, LUG .... UOC:DAA, DAB, DAC, DAD, DAE, DAF, DAG, DAH, DAJ, DAK, DAL, DAW, DAX, V12, V13, V14, V15, V16, V17, V18, V19, V20, V21, V22, V24, V25, V39	1
10	PAOZZ	5940007056715	56501	TG15	TERMINAL, LUG	1
11	PAOZZ	5940007056702	10001	7056702	TERMINAL, LUG	1
12	PAOZZ	5940001141315	96906	MS20659-142	TERMINAL, LUG .... UOC:DAA, DAB, DAC, DAD, DAE, DAF, DAG, DAH, DAJ, DAK, DAL, DAW, DAX, V12, V13, V14, V15, V16, V17, V18, V19, V20, V21, V22, V24, V25, V39	1
13	PAOZZ	5940001139821	96906	MS20659-166	TERMINAL, LUG .... UOC:DAA, DAB, DAC, DAD, DAE, DAF, DAG, DAH, DAJ, DAK, DAL, DAW, DAX, V12, V13, V14, V15, V16, V17, V18, V19, V20, V21, V22, V24, V25, V39	1
14	PAOZA	5935001152307	96906	MS27144-2	CONNECTOR, PLUG, ELEC	2
15	PAOZA	5935001677775	96906	MS27144-1	CONNECTOR, PLUG, ELEC	28
16	PAOZZ	5935002140904	19207	7982907	DUMMY CONNECTOR, PLU	4
17	PAOZZ	5935005729180	19207	8338566	SHELL, ELECTRICAL CO	2
18	PAOZZ	5935000810401	19207	8724260	CONNECTOR, PLUG, ELEC	1
19	PAOZZ	5975007716634	19207	7716634	NUT, COUPLING, ELECTR	1
20	PAOZZ	5325006788435	19207	8724264	GROMMET, NONMETALLIC	1
21	PAOZZ	5310003936685	19207	7723309	NUT, PLAIN, KNURLED	3
22	PAOZA	5935001846707	96906	MS27144-3	CONNECTOR, PLUG, ELEC	1
23	PAOZZ	5935001786075	19207	7388359	CONNECTOR, RECEPTACL	1
24	PAOZZ	5365005078766	19207	7388366	BUSHING, NONMETALLIC	1
25	PAOZZ	5940001133148	96906	MS20659-164	TERMINAL, LUG	1
26	PAOZZ	5940001139826	96906	MS25036-114	TERMINAL, LUG	1
27	PAOZZ	5940001141300	96906	MS20659-105	TERMINAL, LUG	1
28	PAOZZ	5935011495165	11083	7N9738	CONNECTOR BODY, PLUG	1
29	PAOZZ	5999011508808	19207	12258939-2	CONTACT, ELECTRICAL	3
30	PAOZZ	5315011566314	19207	12258939-1	PIN, GROOVED, HEADLES	3
31	PAOZZ	5935011546233	19207	12258940-4	CONNECTOR, PLUG, ELEC	1
32	PAOZZ	5935004626603	96906	MS27142-2	CONNECTOR, PLUG, ELEC	8

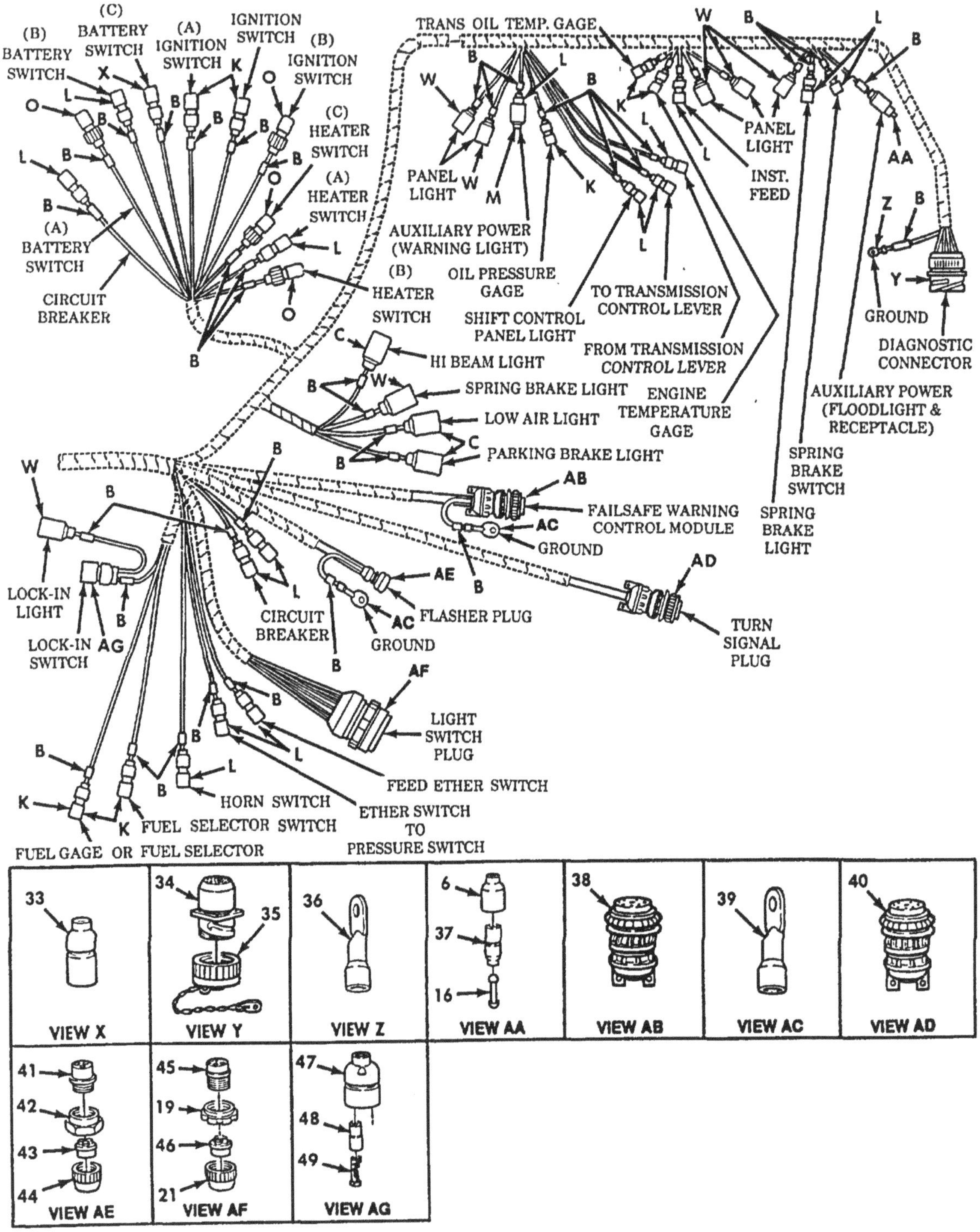

Figure 159. Front Wiring Harness, Partial View (Sheet 3 of 3).

(1) ITEM NO	(2) SMR CODE	(3) NSN	(4) CAGEC	(5) PART NUMBER	(6) DESCRIPTION AND USABLE ON CODES (UOC)	(7) QTY
33	PAOZZ	5935001140607	16528	MS27143-1	CONNECTOR, PLUG, ELEC	1
34	PAOZZ	5935011027124	19207	12258941	CONNECTOR, RECEPTACL	1
35	PAOZZ	5935011145354	80064	2601167	COVER, ELECTRICAL CO	1
36	PAOZZ	5940007056707	19207	7056707	TERMINAL, LUG	1
37	PAOZZ	5935008338561	19207	8338561	SHELL, ELECTRICAL CO	1
38	PAOZZ	5935006222830	96906	MS3456W18-1S	CONNECTOR, PLUG, ELEC	1
39	PAOZZ	5940007056709	19207	7056709	TERMINAL, LUG	2
40	PAOZZ	5935010772622	96906	MS3456W18-8S	CONNECTOR, PLUG, ELEC	1
41	PAOZZ	5935006059322	80064	1755683	CONNECTOR, PLUG, ELEC	1
42	PAOZZ	5975006977769	19207	7527645	NUT, COUPLING, ELECTR	1
43	PAOZZ	5935007722344	19207	7722344	INSERT, ELECTRICAL C	1
44	PAOZZ	5935007723307	19204	7723307	NUT, BUSHING RETAINE	1
45	PAOZZ	5935006862599	19207	8724258	CONNECTOR, PLUG, ELEC	1
46	PAOZZ	5365000905426	19207	7722333	BUSHING, NONMETALLIC	1
47	PAOZA	5935007677936	96906	MS27145-1	CONNECTOR, PLUG, ELEC	1
48	PAOZZ	5970008338562	19207	8338562	INSULATOR, BUSHING	2
49	PAOZZ	1015007982997	19207	7982997	TERMINAL, SOLDERED F	2

END OF FIGURE

Fuel Solenoid Feed

Air Dryer Heater

Starter Ground

Z Splice

Polarity Protection Box

AC Signal

Field

Starter Battery Solenoid

Starter Solenoid

Oil Pressure Transmitter

Starter Ground To Engine Block

Aux. Power

VIEW A	VIEW B	VIEW C	VIEW D	VIEW E
VIEW F	VIEW G	VIEW H	VIEW I	

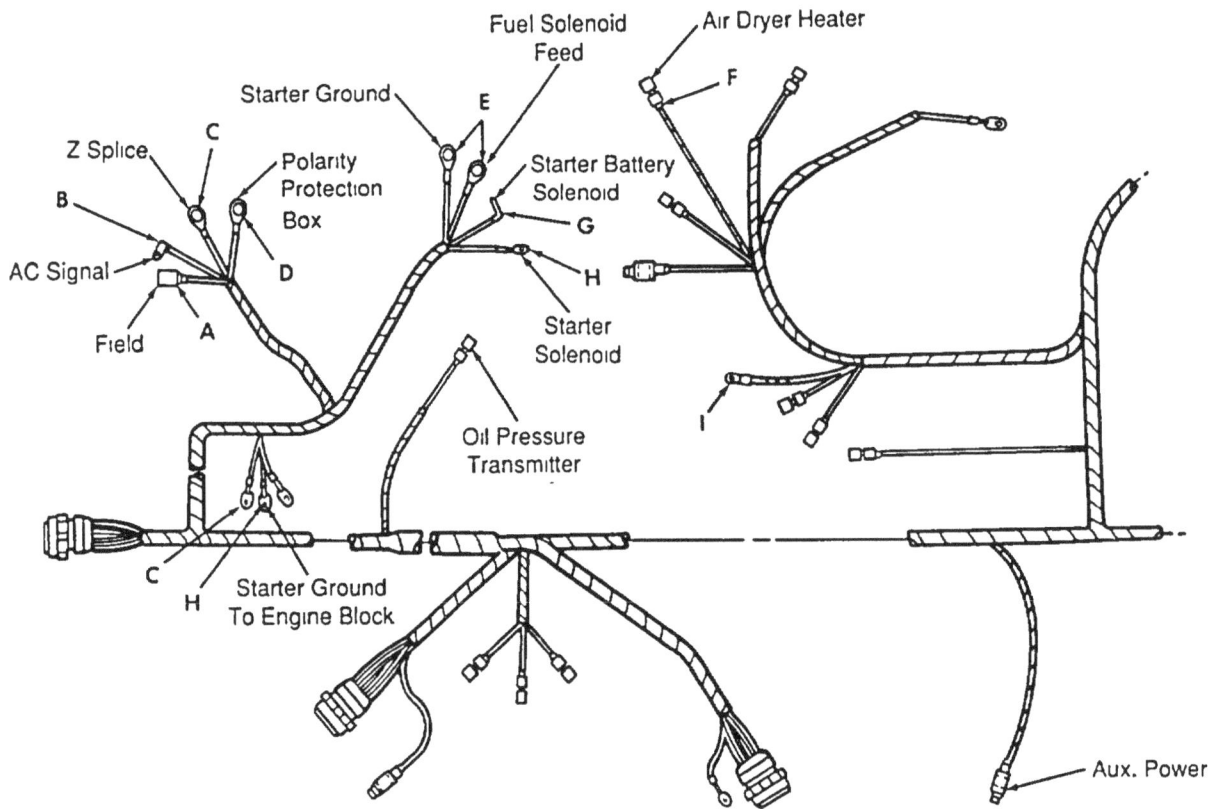

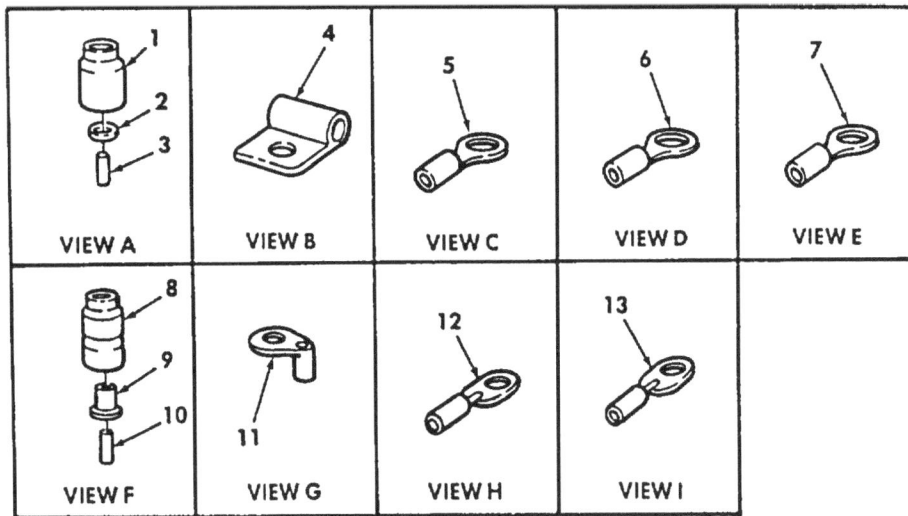

Figure 160. Front Wiring Harness, Partial View, (M939A2) (Sheet 1 of 2).

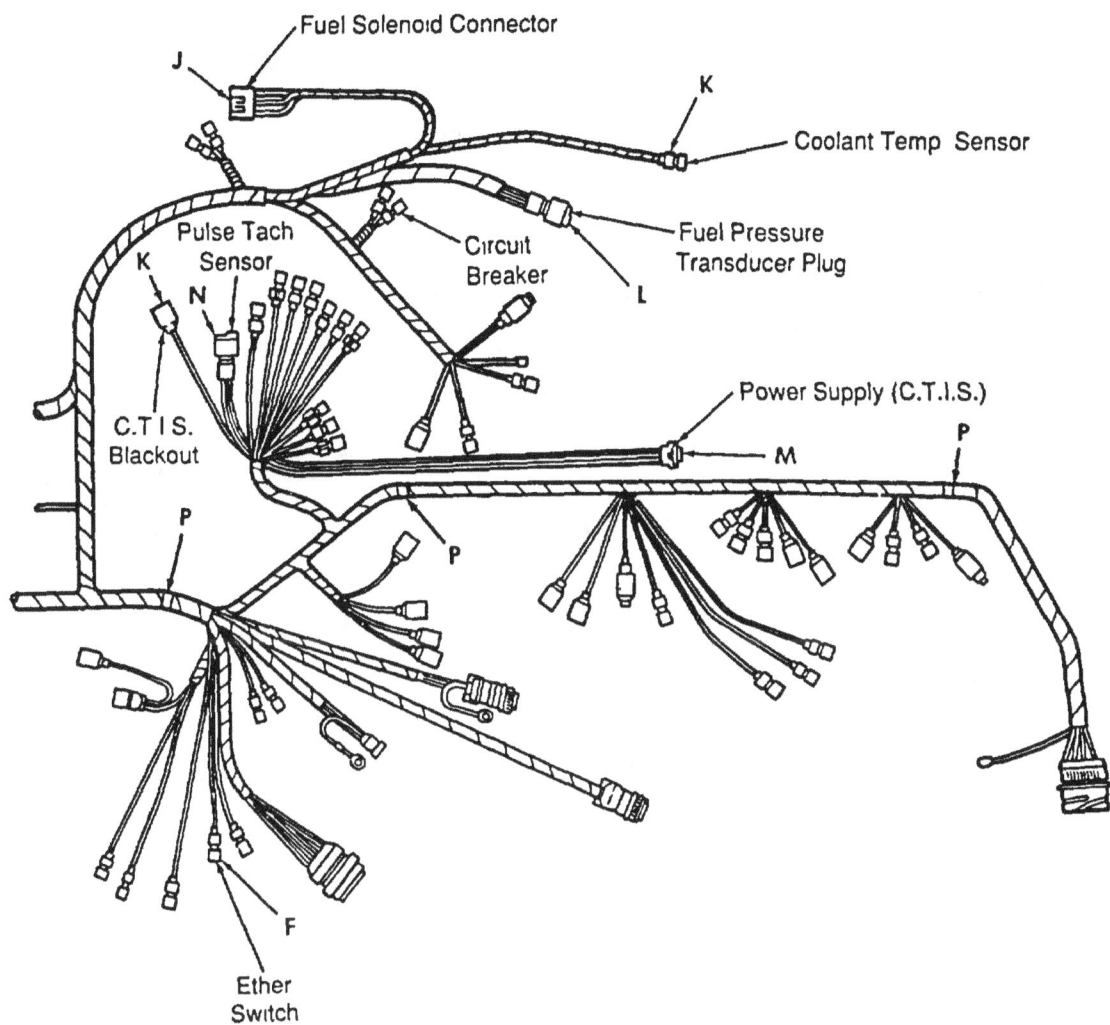

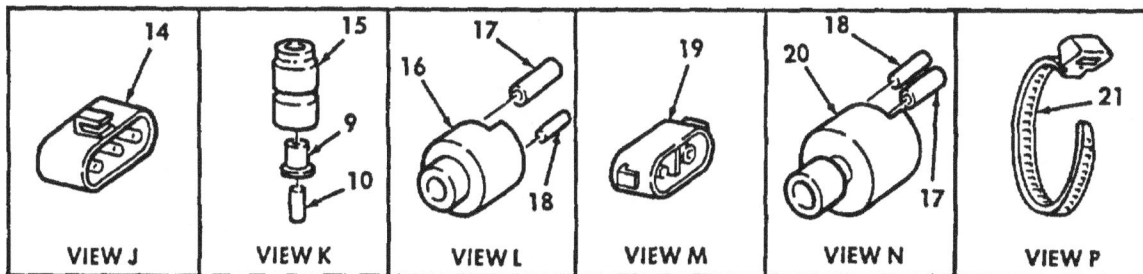

Figure 160. Front Wiring Harness, Partial View, (M939A2) (Sheet 2 of 2).

(1) ITEM NO	(2) SMR CODE	(3) NSN	(4) CAGEC	(5) PART NUMBER	(6) DESCRIPTION AND USABLE ON CODES (UOC)	(7) QTY
					GROUP 0613 CHASSIS WIRING HARNESS	
					FIG. 160 FRONT WIRING HARNESS, PARTIAL VIEW (M939A2)	
1	PAOZZ	5935006915591	19207	8724495	SHELL, ELECTRICAL CO .... UOC:ZAA, ZAB, ZAC, ZAD, ZAE, ZAF, ZAG, ZAH, ZAJ, ZAK, ZAL	1
2	PAOZZ	5310006560067	19207	8724497	WASHER, SLOTTED .... UOC:ZAA, ZAB, ZAC, ZAD, ZAE, ZAF, ZAG, ZAH, ZAJ, ZAK, ZAL	1
3	PAOZA	5999009263144	96906	MS27148-3	CONTACT, ELECTRICAL .... UOC:ZAA, ZAB, ZAC, ZAD, ZAE, ZAF, ZAG, ZAH, ZAJ, ZAK, ZAL	1
4	PAOZA	5940010211874	56501	AB53	TERMINAL, LUG .... UOC:ZAA, ZAB, ZAC, ZAD, ZAE, ZAF, ZAG, ZAH, ZAJ, ZAK, ZAL	1
5	PAOZZ	5940001152674	96906	MS20659-108	TERMINAL, LUG .... UOC:ZAA, ZAB, ZAC, ZAD, ZAE, ZAF, ZAG, ZAH, ZAJ, ZAK, ZAL	2
6	PAOZZ	5940001139825	96906	MS20659-141	TERMINAL, LUG .... UOC:ZAA, ZAB, ZAC, ZAD, ZAE, ZAF, ZAG, ZAH, ZAJ, ZAK, ZAL	1
7	PAOZZ	5940001139821	96906	MS20659-166	TERMINAL, LUG .... UOC:ZAA, ZAB, ZAC, ZAD, ZAE, ZAF, ZAG, ZAH, ZAJ, ZAK, ZAL	2
8	PAOZZ	5935008338561	19207	8338561	SHELL, ELECTRICAL CO .... 2 UOC:ZAA, ZAB, ZAC, ZAD, ZAE, ZAF, ZAG, ZAH, ZAJ, ZAK, ZAL	
9	PAOZZ	5970008338562	19207	8338562	INSULATOR, BUSHING .... UOC:ZAA, ZAB, ZAC, ZAD, ZAE, ZAF, ZAG, ZAH, ZAJ, ZAK, ZAL	4
10	PAOZZ	5940003996676	19207	8338564	TERMINAL ASSEMBLY .... UOC:ZAA, ZAB, ZAC, ZAD, ZAE, ZAF, ZAG, ZAH, ZAJ, ZAK, ZAL	4
11	PAOZZ	5940001141315	96906	MS20659-142	TERMINAL, LUG .... UOC:ZAA, ZAB, ZAC, ZAD, ZAE, ZAF, ZAG, ZAH, ZAJ, ZAK, ZAL	1
12	PAOZZ	5940007056702	56501	TG2	TERMINAL, LUG .... UOC:ZAA, ZAB, ZAC, ZAD, ZAE, ZAF, ZAG, ZAH, ZAJ, ZAK, ZAL	1
13	PAOZZ	5940007056715	19207	7056715	TERMINAL, LUG .... UOC:ZAA, ZAB, ZAC, ZAD, ZAE, ZAF, ZAG, ZAH, ZAJ, ZAK, ZAL	1
14	PAOZZ	5935012922336	45152	1618210	CONNECTOR, PLUG, ELEC .... UOC:ZAA, ZAB, ZAC, ZAD, ZAE, ZAF, ZAG, ZAH, ZAJ, ZAK, ZAL	1
15	PAOZZ	5975006605962	19207	8724494	CABLE NIPPLE, ELECTR .... UOC:ZAA, ZAB, ZAC, ZAD, ZAE, ZAF, ZAG, ZAH, ZAJ, ZAK, ZAL	2
16	PAOZZ	5935011546233	19207	12258940-4	CONNECTOR, PLUG, ELEC .... UOC:ZAA, ZAB, ZAC, ZAD, ZAE, ZAF, ZAG, ZAH, ZAJ, ZAK, ZAL	1

(1) ITEM NO	(2) SMR CODE	(3) NSN	(4) CAGEC	(5) PART NUMBER	(6) DESCRIPTION AND USABLE ON CODES (UOC)	(7) QTY
17	PAOZZ	5999011508808	19207	12258939-2	CONTACT, ELECTRICAL .... UOC:ZAA, ZAB, ZAC, ZAD, ZAE, ZAF, ZAG, ZAH, ZAJ, ZAK, ZAL	3
18	PAOZZ	5315011566314	19207	12258939-1	PIN, GROOVED, HEADLES .... UOC:ZAA, ZAB, ZAC, ZAD, ZAE, ZAF, ZAG, ZAH, ZAJ, ZAK, ZAL	3
19	PAOZZ	5935013087866	77060	15300027	CONNECTOR BODY, PLUG .... UOC:ZAA, ZAB, ZAC, ZAD, ZAE, ZAF, ZAG, ZAH, ZAJ, ZAK, ZAL	1
20	PAOZZ	5935011495165	19207	12258940-2	CONNECTOR BODY, PLUG .... UOC:ZAA, ZAB, ZAC, ZAD, ZAE, ZAF, ZAG, ZAH, ZAJ, ZAK, ZAL	1
21	PAOZZ	5975009846582	96906	MS3367-1-0	STRAP, TIEDOWN, ELECT .... UOC:ZAA, ZAB, ZAC, ZAD, ZAE, ZAF, ZAG, ZAH, ZAJ, ZAK, ZAL	1
21	PAOZZ	5975001563253	96906	MS3367-2-9	STRAP, TIEDOWN, ELECT .... UOC:ZAA, ZAB, ZAC, ZAD, ZAE, ZAF, ZAG, ZAH, ZAJ, ZAK, ZAL	1
21	PAOZZ	5975004515001	96906	MS3367-3-9	STRAP, TIEDOWN, ELECT .... UOC:ZAA, ZAB, ZAC, ZAD, ZAE, ZAF, ZAG, ZAH, ZAJ, ZAK, ZAL	1

END OF FIGURE

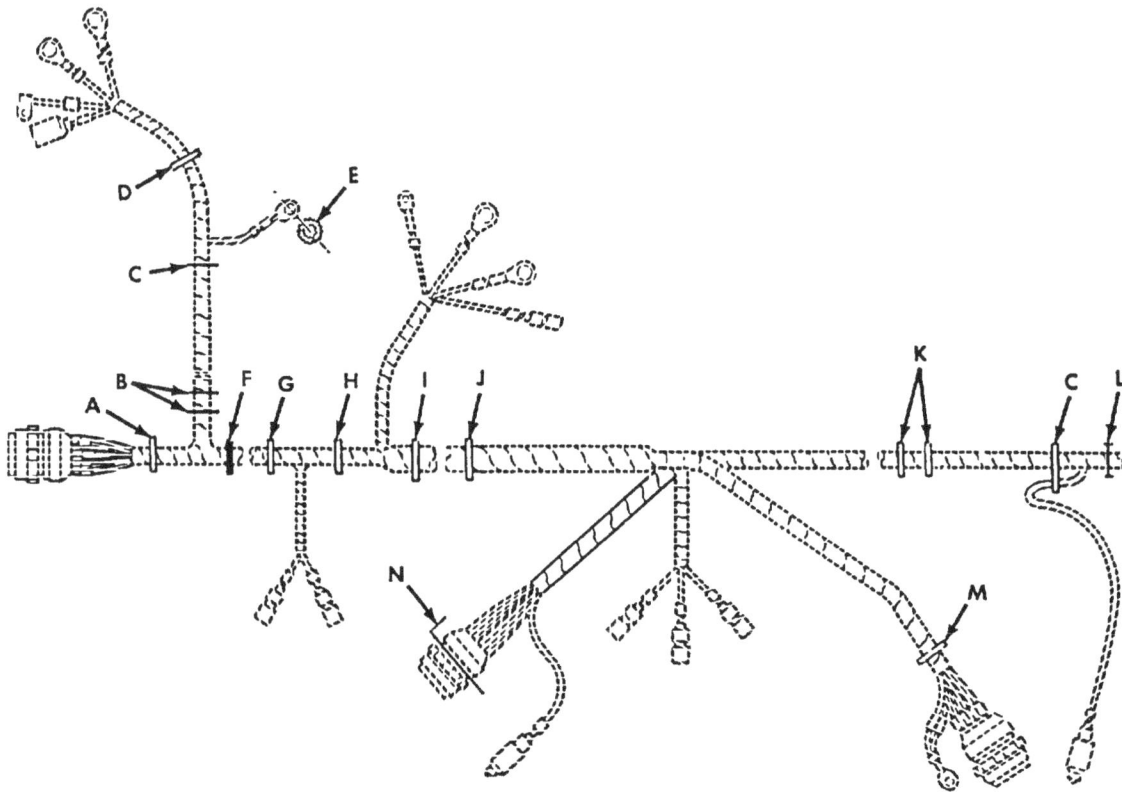

Figure 161. Front Wiring Harness Clamps and Hardware.

(1) ITEM NO	(2) SMR CODE	(3) NSN	(4) CAGEC	(5) PART NUMBER	(6) DESCRIPTION AND USABLE ON CODES (UOC)	(7) QTY
					GROUP 0613 CHASSIS WIRING HARNESS	
					FIG. 161 FRONT WIRING HARNESS CLAMPS AND HARDWARE	
1	PAOZZ	5340000573043	96906	MS21333-112	CLAMP,LOOP	1
2	PAOZZ	5340011885087	19207	12302768	CLIP,RETAINING	2
3	PAOZZ	5975001563253	96906	MS3367-2-9	STRAP,TIEDOWN,ELECT	2
4	PAOZZ	5340000538994	96906	MS21333-126	CLAMP,LOOP	1
5	PAOZZ	5305002693238	80204	B1821BH038F125N	SCREW,CAP,HEXAGON H	1
6	PAOZZ	5310008140672	96906	MS51943-36	NUT,SELF-LOCKING,HE	4
7	PAOZZ	5310000611258	96906	MS45904-76	WASHER,LOCK	3
8	PAOZZ	5325007397776	19207	7397776	EYELET,METALLIC	1
9	PAOZZ	5340007647052	96906	MS21333-116	CLAMP,LOOP	2
10	PAOZZ	5340007022848	96906	MS21333-128	CLAMP,LOOP	1
11	PAOZZ	5305009125113	96906	MS51096-359	SCREW,CAP,HEXAGON H	1
12	PAOZZ	5310001788631	96906	MS35333-75	WASHER,LOCK	1
13	PAOZZ	5306000501238	96906	MS90727-32	BOLT,MACHINE..	1
14	PAOZZ	5310000544892	96906	MS35650-3312	NUT,PLAIN,HEXAGON	1
15	PAOZZ	5340001519651	96906	MS21333-129	CLAMP,LOOP	1
16	PAOZZ	5340009040933	96906	MS21333-80	CLAMP,LOOP	2
17	PAOZZ	5306000425570	21450	425570	BOLT,ASSEMBLED WASH	2
18	PAOZZ	5310009010279	96906	MS45904-74	WASHER,LOCK	2
19	PAOZZ	5310008094058	96906	MS27183-10	WASHER,FLAT...	2
20	PAOZZ	5306000680513	60285	6893-2	BOLT,MACHINE..	2
21	PAOZZ	5325012315963	84324	7397744-1	GROMMET,METALLIC	2
22	PAOZZ	5975004515001	96906	MS3367-3-9	STRAP,TIEDOWN,ELECT	1
23	PAOZZ	5310009349757	96906	MS35649-282	NUT,PLAIN,HEXAGON	4
24	PAOZZ	5340001584078	19207	10899434	BRACKET,ANGLE	1
25	PAOZZ	5305000191675	24617	191675	SCREW,ASSEMBLED WAS	2
26	PAOZZ	5310004005503	96906	MS35650-3254	NUT,PLAIN,HEXAGON	2
27	PAOZZ	5305009846193	96906	MS35206-245	SCREW,MACHINE	4
28	PAOZZ	5310005967693	96906	MS35335-31	WASHER,LOCK..	4

END OF FIGURE

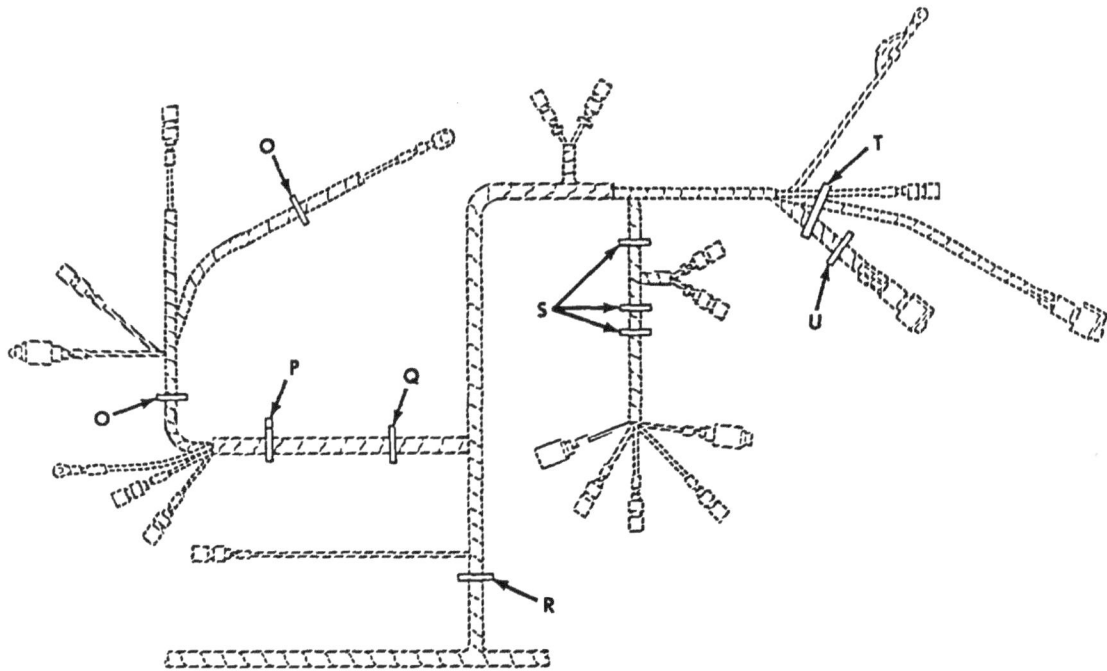

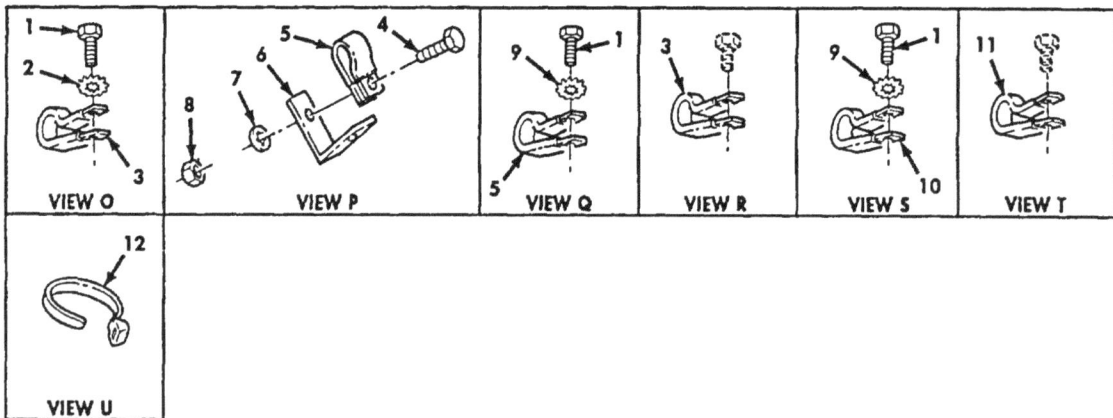

Figure 162. Front Wiring Harness Clamps and Hardware.

(1) ITEM NO	(2) SMR CODE	(3) NSN	(4) CAGEC	(5) PART NUMBER	(6) DESCRIPTION AND USABLE ON CODES (UOC)	(7) QTY
					GROUP 0613 CHASSIS WIRING HARNESS	
					FIG. 162 FRONT WIRING HARNESS CLAMPS AND HARDWARE	
1	PAOZZ	5306000680513	60285	6893-2	BOLT,MACHINE	5
2	PAOZZ	5310005501130	96906	MS35333-40	WASHER,LOCK	3
3	PAOZZ	5340011179876	19207	7397785	CLAMP,LOOP	4
4	PAOZZ	5305000680515	80204	B1821BH025F100N	SCREW,CAP,HEXAGON H	1
5	PAOZZ	5340000881254	96906	MS21333-104	CLAMP,LOOP	2
6	PFOZZ	5340011476759	19207	12256266	BRACKET,ANGLE	1
7	PAOZZ	5310000228834	96906	MS35333-108	WASHER,LOCK.	1
8	PAOZZ	5310009359022	96906	MS51943-32	NUT,SELF-LOCKING,HE	1
9	PAOZZ	5310009010279	96906	MS45904-74	WASHER,LOCK	4
10	PAOZZ	5340009936207	96906	MS21333-99	CLAMP,LOOP	3
11	PAOZZ	5340000673868	96906	MS21333-109	CLAMP,LOOP	1
12	PAOZZ	5975007275153	96906	MS3367-4-9	STRAP,TIEDOWN,ELECT	1

END OF FIGURE

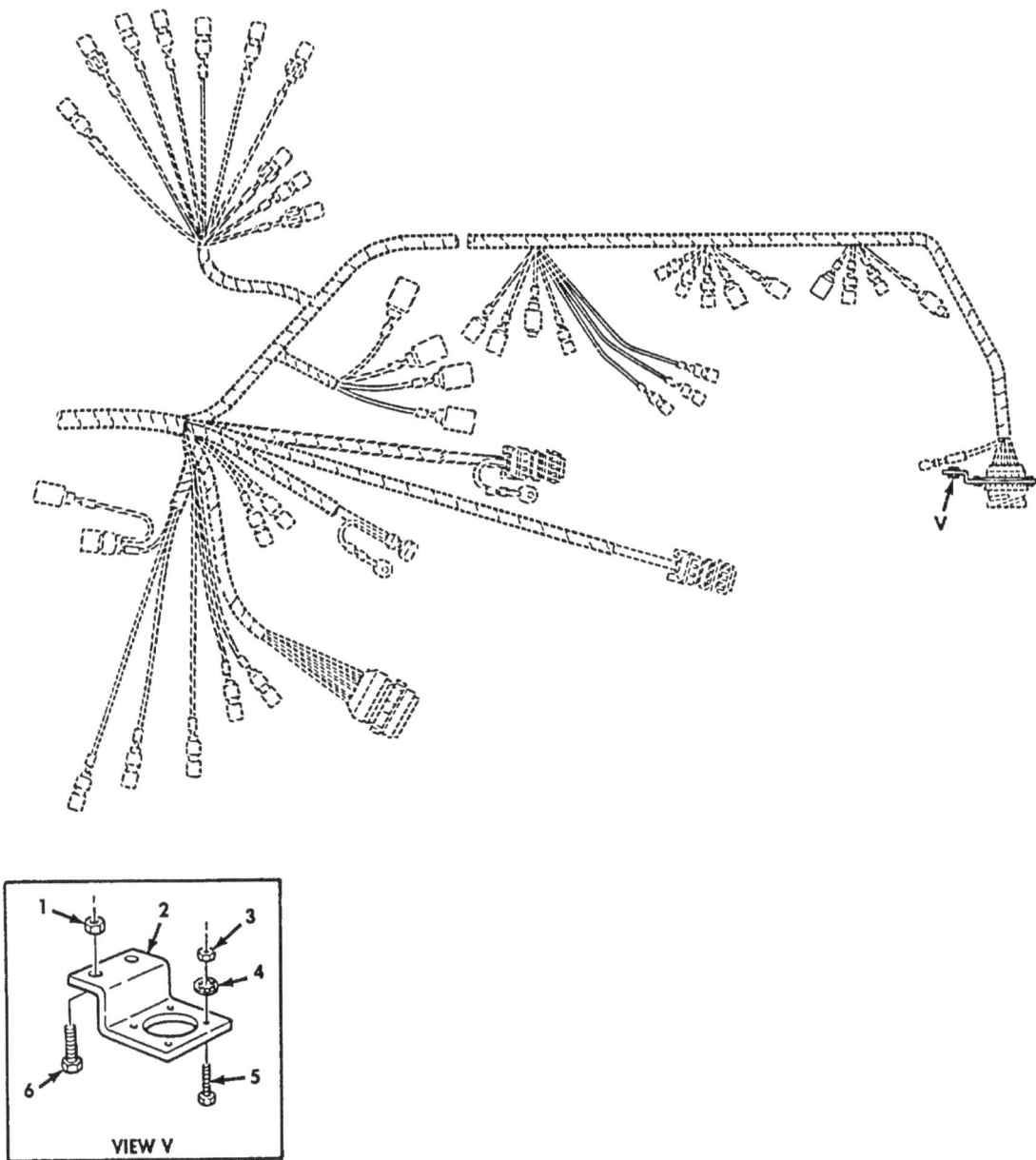

VIEW V

Figure 163. Front Wiring Harness Bracket and Hardware.

(1) ITEM NO	(2) SMR CODE	(3) NSN	(4) CAGEC	(5) PART NUMBER	(6) DESCRIPTION AND USABLE ON CODES (UOC)	(7) QTY
					GROUP 0613 CHASSIS WIRING HARNESS	
					FIG. 163 FRONT WIRING HARNESS BRACKET AND HARDWARE	
1	PAOZZ	5310008140672	96906	MS51943-36	NUT,SELF-LOCKING,HE	2
2	PAOZZ	5935011203744	19207	12277052	BRACKET,ELECTRICAL	1
3	PAOZZ	5310009349754	96906	MS35650-382	NUT,PLAIN,HEXAGON	4
4	PAOZZ	5310005501130	96906	MS35333-40	WASHER,LOCK	1
5	PAOZZ	5305009931848	96906	MS35207-265	SCREW,MACHINE	4
6	PAOZZ	5305002693234	80204	B1821BH038F075N	SCREW,CAP,HEXAGON H	2

END OF FIGURE

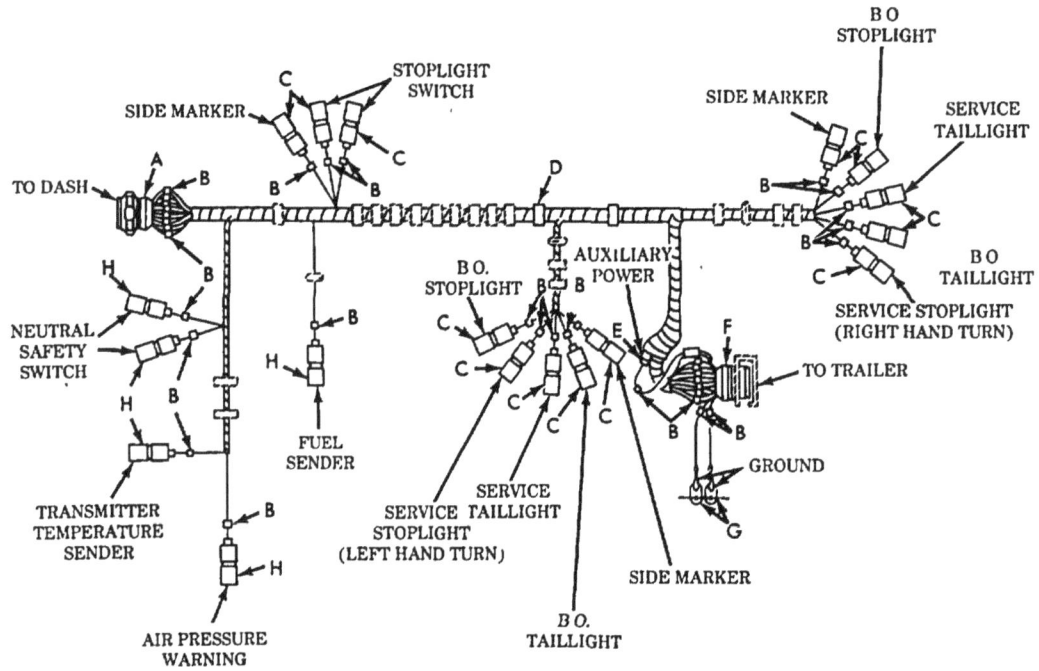

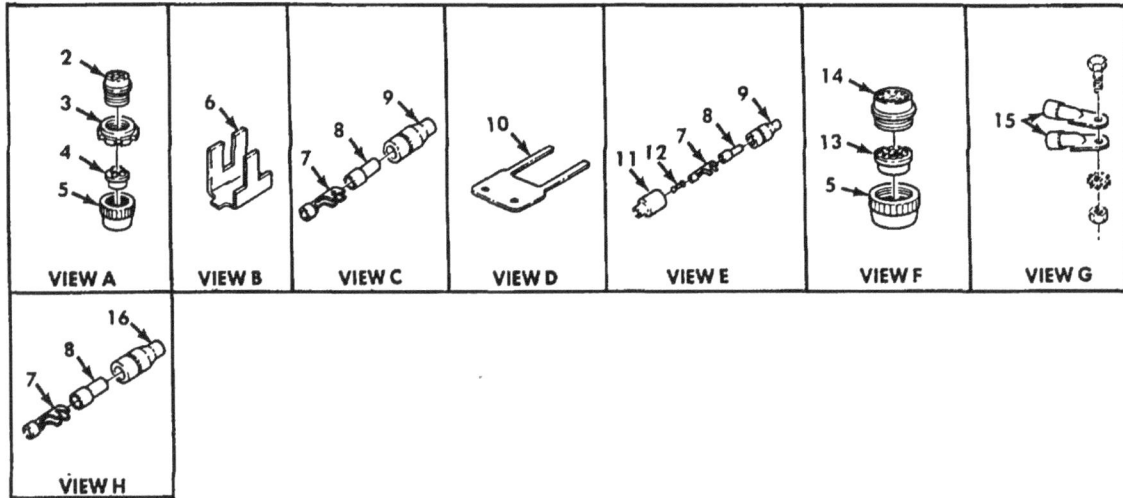

Figure 164. Rear Wiring Harness Assembly, Cargo (Sheet 1 of 2).

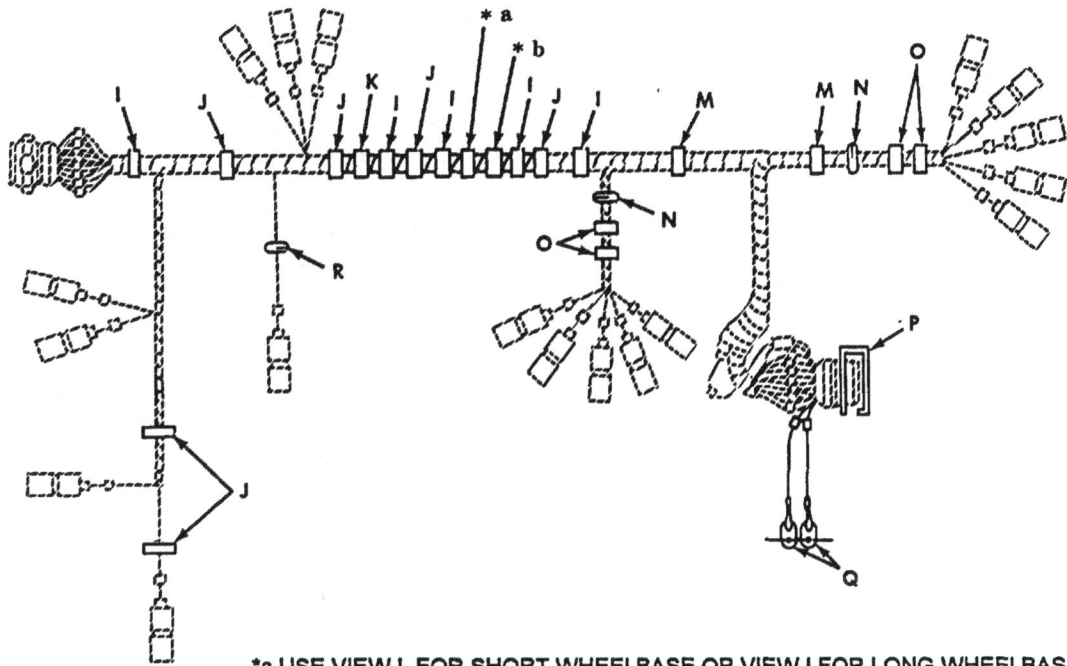

*a USE VIEW L FOR SHORT WHEELBASE OR VIEW I FOR LONG WHEELBASE
*b USE VIEW I FOR SHORT WHEELBASE OR VIEW L FOR LONG WHEELBASE

| VIEW I | VIEW J | VIEW K | VIEW L | VIEW M |
| VIEW N | VIEW O | VIEW P | VIEW Q | VIEW R |

Figure 164. Rear Wiring Harness Assembly, Cargo (Sheet 2 of 2).

(1) ITEM NO	(2) SMR CODE	(3) NSN	(4) CAGEC	(5) PART NUMBER	(6) DESCRIPTION AND USABLE ON CODES (UOC)	(7) QTY
					GROUP 0613 CHASSIS WIRING HARNESS	
					FIG. 164 REAR WIRING HARNESS ASSEMBLY, CARGO	
1	PBOOO	2590010853834	19207	12256002-2	WIRING HARNESS,BRAN UOC:DAC,DAD,V16,V17,ZAC,ZAD	1
1	PAOOO	2920010831187	19207	12256002-1	WIRING HARNESS,BRAN .... UOC:DAA,DAB,DAW,DAX,V12,V13,V14,V15, ZAA,ZAB	1
2	PAOZZ	5935000801020	19207	8724255	.CONNECTOR,PLUG,ELEC UOC:DAA,DAB,DAC,DAD,DAW,DAX,V12,V13, V14,V15,V16,V17,ZAA,ZAB,ZAC,ZAD	1
3	PAOZZ	5975007716634	19207	7716634	.NUT,COUPLING,ELECTR UOC:DAA,DAB,DAC,DAD,DAW,DAX,V12,V13, V14,V15,V16,V17,ZAA,ZAB,ZAC,ZAD	1
4	PAOZZ	5365005078766	19207	7388366	.BUSHING,NONMETALLIC UOC:DAA,DAB,DAC,DAD,DAW,DAX,V12,V13, V14,V15,V16,V17,ZAA,ZAB,ZAC,ZAD	1
5	PAOZZ	5310003936685	19207	7723309	.NUT,PLAIN,KNURLED .... UOC:DAA,DAB,DAC,DAD,DAW,DAX,V12,V13, V14,V15,V16,V17,ZAA,ZAB,ZAC, ZAD	2
6	PAOZZ	9905010138723	81349	M43436/3-1	.BAND,MARKER UOC:DAA,DAB,DAC,DAD,DAW,DAX,V12,V13, V14,V15,V16,V17,ZAA,ZAB,ZAC,ZAD	49
7	PAOZZ	1015007982997	19207	7982997	.TERMINAL,SOLDERED F UOC:DAA,DAB,DAC,DAD,DAW,DAX,V12,V13, V14,V15,V16,V17,ZAA,ZAB,ZAC,ZAD	19
8	PAOZZ	5970008338562	19207	8338562	.INSULATOR,BUSHING UOC:DAA,DAB,DAC,DAD,DAW,DAX,V12,V13, V14,V15,V16,V17,ZAA,ZAB,ZAC,ZAD	19
9	PAOZZ	5935008338561	19207	8338561	.SHELL,ELECTRICAL CO .... UOC:DAA,DAB,DA C,DAD,DAW,DAX,V12,V13, V14,V15,V16,V17,ZAA,ZAB,ZAC,ZAD	14
10	PAOZZ	9905001141334	97403	13207E6882-1	.BAND,MARKER UOC:DAA,DAB,DAC,DAD,DAW,DAX,V12,V13, V14,V15,V16,V17,ZAA,ZAB,ZAC,ZAD	1
11	PAOZZ	5935005729180	19207	8338566	.SHELL,ELECTRICAL CO UOC:DAA,DAB,DAC,DAD,DAW,DAX,V12,V13, V14,V15,V16,V17,ZAA,ZAB,ZAC,ZAD	1
12	PAOZZ	5935002140904	19207	7982907	.DUMMY CONNECTOR,PLU UOC:DAA,DAB,DAC,DAD,DAW,DAX,V12,V13, V14,V15,V16,V17,ZAA,ZAB,ZAC,ZAD	1
13	PAOZZ	5365000905426	19207	7722333	.BUSHING, NONMETALLIC UOC:DAA,DAB,DAC,DAD,DAW,DAX,V12,V13, V14,V15,V16,V17,ZAA,ZAB,ZAC,ZAD	1
14	PAOZA	5935008463884	96906	MS75021-2	.CONNECTOR,RECEPTACL UOC:DAA,DAB,DAC,DAD,DAW,DAX,V12,V13, V14,V15,V16,V17,ZAA,ZAB,ZAC,ZAD	1
15	PAOZZ	5940007056711	19207	7056711	.TERMINAL,LUG UOC:DAA,DAB,DAC,DAD,DAW,DAX,V12,V13, V14,V15,V16,V17,ZAA,ZAB,ZAC,ZAD	2

(1) ITEM NO	(2) SMR CODE	(3) NSN	(4) CAGEC	(5) PART NUMBER	(6) DESCRIPTION AND USABLE ON CODES (UOC)	(7) QTY
16	PAOZZ	5975006605962	19207	8724494	CABLE NIPPLE,ELECTR UOC:DAA,DAB,DAC,DAD,DAW,DAX,V12,V13, V14,V15,V16,V17,ZAA,ZAB,ZAC,ZAD	5
17	PAOZZ	5340000538994	96906	MS21333-126	CLAMP,LOOP .... UOC:DAA,DAB,B,DAW,DAX,V12,V13,V14,V15, ZAA,ZAB	8
18	PAOZZ	5975001563253	96906	MS3367-2-9	STRAP,TIEDOWN,ELECT .... UOC:DAA,DAB,DAW,DAX,V12,V13,V14,V15, ZAA,ZAB	7
19	PAOZZ	5305002693240	80204	B1821BH038F150N	SCREW,CAP,HEXAGON H .... UOC:DAA,DAB,DAW,DAX,V12,V13,V14,V15, ZAA,ZAB	1
20	PAOZZ	5310008140672	96906	MS51943-36	NUT,SELF-LOCKING,HE .... UOC:DAA,DAB,DAW,DAX,V12,V13,V14,V15, ZAA,ZAB	1
21	PAOZZ	5305002693238	80204	B1821BH038F125N	SCREW,CAP,HEXAGON H UOC:DAA,DAC,DAD,V16,V17,ZAC,ZAD	2
22	PAOZZ	5306000680513	60285	6893-2	BOLT,MACHINE UOC:DAA,DAB,DAC,DAD,DAW,DAX,V12,V13, V14,V15,V16,V17,ZAA,ZAB,ZAC,ZAD	2
23	PAOZZ	5340008091492	96906	MS21333-100	CLAMP,LOOP UOC:DAA,DAB,DAC,DAD,DAW,DAX,V12,V13, V14,V15,V16,V17,ZAA,ZAB,ZAC,ZAD	2
24	PAOZZ	5310009359022	96906	MS51943-32	NUT,SELF-LOCKING,HE UOC:DAA,DAB,DAC,DAD,DAW,DAX,V12,V13, V14,V15,V16,V17,ZAA,ZAB,ZAC,ZAD	2
25	PAOZZ	3120000733163	19207	7397775	BUSHING,SLEEVE UOC:DAA,DAB,DAC,DAD,DAW,DAX,V12,V13, V14,V15,V16,V17,ZAA,ZAB,ZAC,ZAD	2
26	PAOZZ	5305009125113	96906	MS51096-359	SCREW,CAP,HEXAGON H UOC:DAA,DAB,DAC,DAD,DAW,DAX,V12,V13, V14,V15,V16,V17,ZAA,ZAB,ZAC,ZAD	4
27	PAOZZ	5340009546014	96906	MS21333-121	CLAMP,LOOP UOC:DAA,DAB,DAC,DAD,DAW,DAX,V12,V13, V14,V15,V16,V17,ZAA,ZAB,ZAC,ZAD	4
28	PAOZZ	5310009591488	96906	MS51922-21	NUT,SELF-LOCKING,HE UOC:DAA,DAB,DAC,DAD,DAW,DAX,V12,V13, V14,V15,V16,V17,ZAA,ZAB,ZAC,ZAD	5
29	PAOZZ	5935007731428	19207	7731428	COVER,ELECTRICAL CO UOC:DAA,DAB,DAC,DAD,DAW,DAX,V12,V13, V14,V15,V16,V17,ZAA,ZAB,ZAC,ZAD	1
30	PAOZZ	5310008775796	96906	MS21044N4	NUT,SELF-LOCKING,HE UOC:DAA,DAB,DAC,DAD,DAW,DAX,V12,V13, V14,V15,V16,V17,ZAA,ZAB,ZAC,ZAD	4
31	PAOZZ	5305002678957	80204	B1821BH025F175N	SCREW,CAP,HEXAGON H UOC:DAA,DAB,DAC,DAD,DAW,DAX,V12,V13, V14,V15,V16,V17,ZAA,ZAB,ZAC,ZAD	4
32	PAOZZ	5305000425603	24617	425603	SCREW,ASSEMBLED WAS UOC:DAA,DAB,DAC,DAD,DAW,DAX,V12,V13, V14,V15,V16,V17,ZAA,ZAB,ZAC,ZAD	1
33	PAOZZ	5310006276128	96906	MS35335-35	WASHER,LOCK... UOC:DAA,DAB,DAC,DAD,DAW,DAX,V12,V13, V14,V15,V16,V17,ZAA,ZAB,ZAC,ZAD	1

(1) ITEM NO	(2) SMR CODE	(3) NSN	(4) CAGEC	(5) PART NUMBER	(6) DESCRIPTION AND USABLE ON CODES (UOC)	(7) QTY
34	PAOZZ	5325000531116	19207	7397790	GROMMET,REAR WIRING UOC:DAA,DAB,DAC,DAD,DAW,DAX,V12,V13, V14,V15,V16,V17,ZAA,ZAB,ZAC,ZAD	1

END OF FIGURE

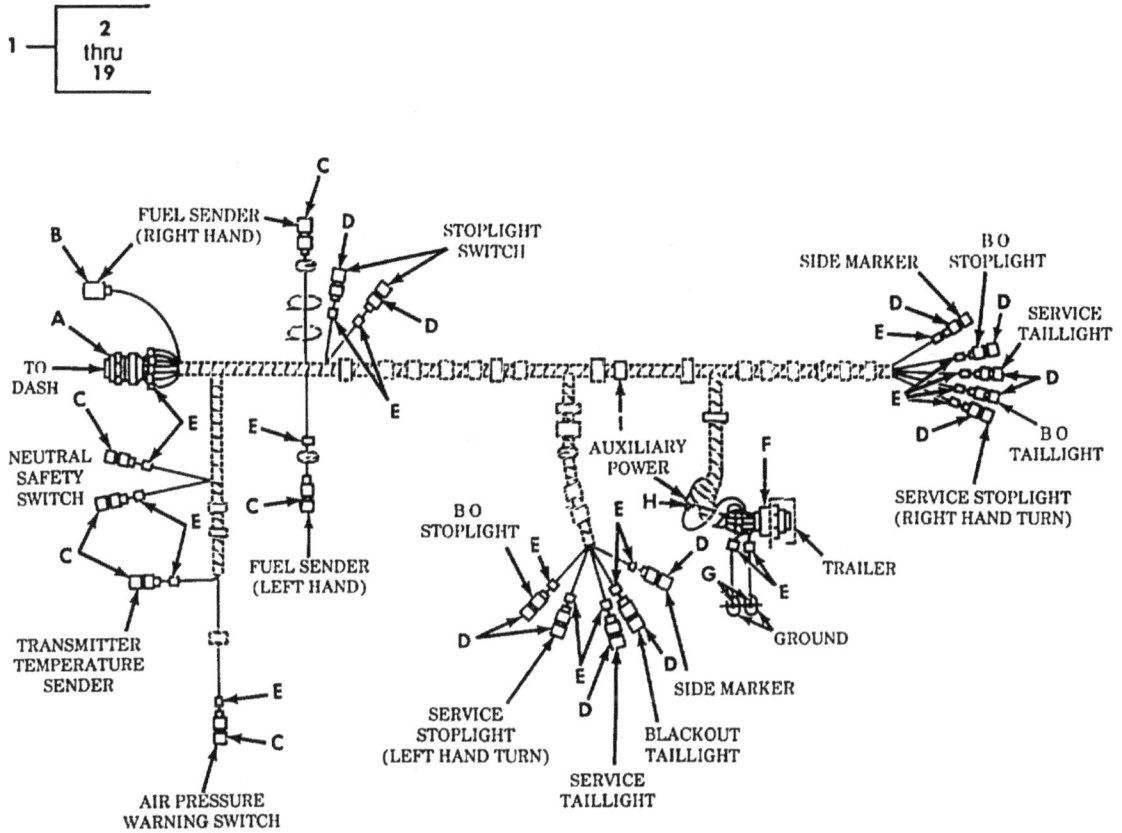

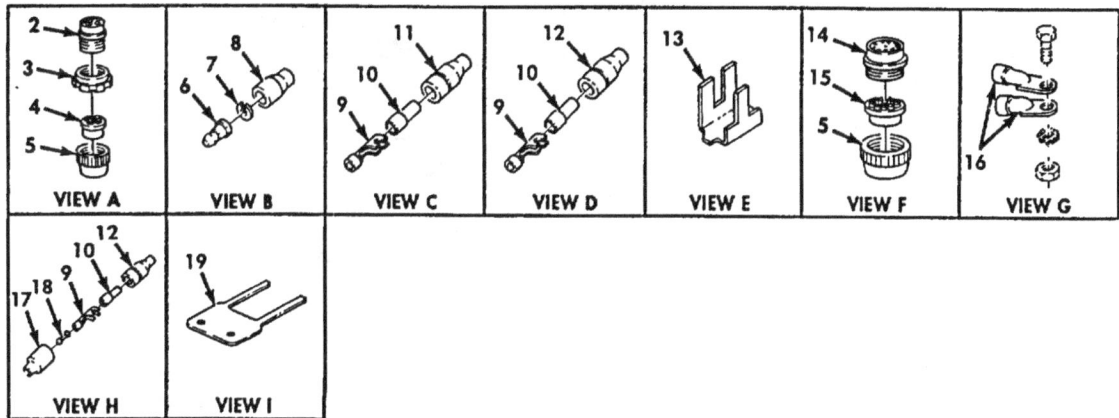

Figure 165. Rear Wiring Harness Assembly, Wrecker (Sheet 1 of 2).

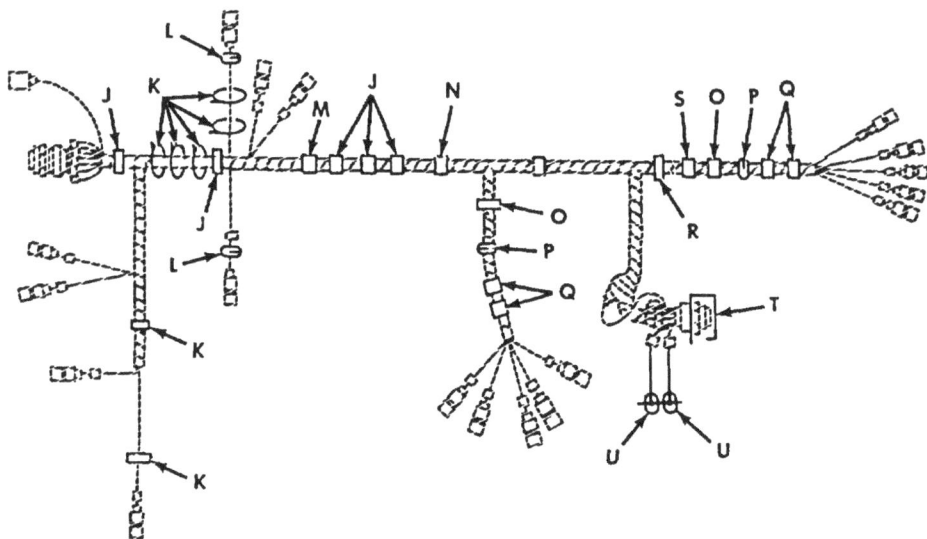

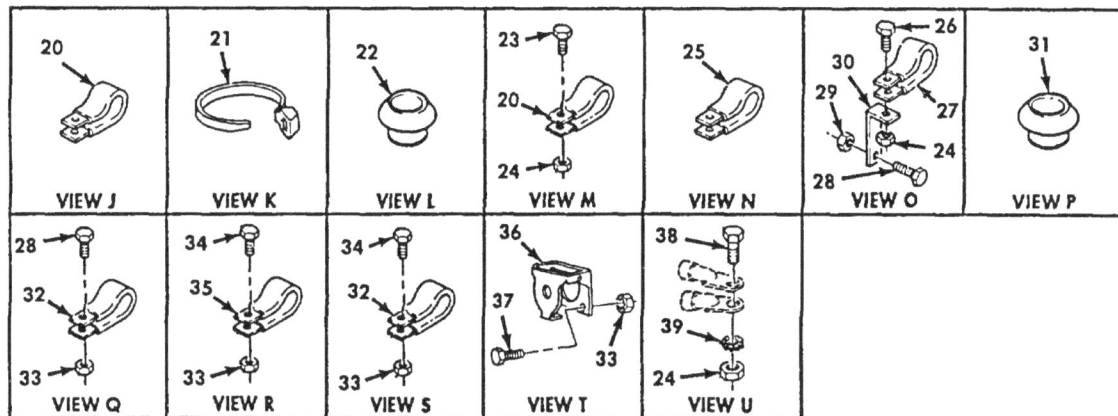

Figure 165. Rear Wiring Harness Assembly, Wrecker (Sheet 2 of 2).

(1) ITEM NO	(2) SMR CODE	(3) NSN	(4) CAGEC	(5) PART NUMBER	(6) DESCRIPTION AND USABLE ON CODES (UOC)	(7) QTY
					GROUP 0613 CHASSIS WIRING HARNESS	
					FIG. 165 REAR WIRING HARNESS ASSEMBLY, WRECKER	
1	PAFZZ	2590010823828	19207	12256102	WIRING HARNESS,BRAN UOC:DAL,V18,ZAL	1
2	PAOZZ	5935000801020	19207	8724255	.CONNECTOR,PLUG,ELEC UOC:DAL,V18,ZAL	1
3	PAOZZ	5975007716634	19207	7716634	.NUT,COUPLING,ELECTR UOC:DAL,V18,ZAL	1
4	PAOZZ	5365005078766	19207	7388366	.BUSHING,NONMETALLIC UOC:DAL,V18,ZAL	1
5	PAOZZ	5310003936685	19207	7723309	.NUT,PLAIN,KNURLED UOC:DAL,V18,ZAL	2
6	PAOOA	5999009263144	96906	MS27148-3	.CONTACT,ELECTRICAL UOC:DAL,V18,ZAL	1
7	PAOZZ	5310006560067	19207	8724497	.WASHER,SLOTTED UOC:DAL,V18,ZAL	1
8	PAOZZ	5935006915591	19207	8724495	.SHELL,ELECTRICAL CO UOC:DAL,V18,ZAL	1
9	PAOZZ	1015007982997	19207	7982997	.TERMINAL,SOLDERED F UOC:DAL,V18,ZAL	19
10	PAOZZ	5970008338562	19207	8338562	.INSULATOR,BUSHING UOC:DAL,V18,ZAL	19
11	PAOZZ	5975006605962	19207	8724494	.CABLE NIPPLE,ELECTR UOC:DAL,V18,ZAL	6
12	PAOZZ	5935008338561	19207	8338561	.SHELL,ELECTRICAL CO UOC:DAL,V18,ZAL	13
13	PAOZZ	9905010138723	81349	M43436/3-1	.BAND,MARKER WRECKER UOC:DAL,V18,ZAL	20
14	PAOZA	5935008463884	96906	MS75021-2	.CONNECTOR,RECEPTACL .... UOC:DAL,V18, ZAL	1
15	PAOZZ	5365000905426	19207	7722333	.BUSHING,NONM ETALLIC UOC:DAL,V18,ZAL	1
16	PAOZZ	5940007056711	19207	7056711	.TERMINAL,LUG UOC:DAL,V18,ZAL	2
17	PAOZZ	5935005729180	19207	8338566	.SHELL,ELECTRICAL CO UOC:DAL,V18,ZAL	1
18	PAOZZ	5935002140904	19207	7982907	.DUMMY CONNECTOR,PLU UOC:DAL,V18,ZAL	1
19	PAOZZ	9905001141334	97403	13207E6882-1	.BAND,MARKER UOC:DAL,V18,ZAL	1
20	PAOZZ	5340000538994	96906	MS21333-126	CLAMP,LOOP UOC:DAL,V18,ZAL	5
21	PAOZZ	5975001563253	96906	MS3367-2-9	STRAP,TIEDOWN,ELECT UOC:DAL,V18,ZAL	6
22	PAOZZ	5325000531116	19207	7397790	GROMMET,REAR WIRING UOC:DAL,V18,ZAL	2
23	PAOZZ	5305002693239	80204	B1821BHO38F138N	SCREW,CAP,HEXAGON H UOC:DAL,V18,ZAL	1
24	PAOZZ	5310008140672	96906	MS51943-36	NUT,SELF-LOCKING,HE	5

(1) ITEM NO	(2) SMR CODE	(3) NSN	(4) CAGEC	(5) PART NUMBER	(6) DESCRIPTION AND USABLE ON CODES (UOC)	(7) QTY
					UOC:DAL,V18,ZAL	
25	PAOZZ	5340004217254	19207	8352679	CLAMP,LOOP	1
					UOC:DAL,V18,ZAL	
26	PAOZZ	5305002693234	80204	B1821BH038F075N	SCREW,CAP,HEXAGON H	4
					UOC:DAL,V18,ZAL	
27	PAOZZ	5340009546014	96906	MS21333-121	CLAMP,LOOP	4
					UOC:DAL,V18,ZAL	
8	PAOZZ	5306000680513	60285	6893-2	BOLT,MACHINE..	6
					UOC:DAL,V18,ZAL	
29	PAOZZ	5310009359022	96906	MS51943-32	NUT,SELF-LOCKING,HE	2
					UOC:DAL,V18,ZAL	
30	PAOZZ	5340004199473	19207	7414917	BRACKET,ANGLE	2
					UOC:DAL,V18,ZAL	
31	PAOZZ	3120000733163	19207	7397775	BUSHING,SLEEVE..	2
					UOC:DAL,V18,ZAL	
32	PAOZZ	5340008091492	56232	310-8	CLAMP,LOOP	5
					UOC:DAL,V18,ZAL	
33	PAOZZ	5310008775796	96906	MS21044N4	NUT,SELF-LOCKING,HE	6
					UOC:DAL,V18,ZAL	
34	PAOZZ	5306000680514	80204	B1821BH025F088N	SCREW,CAP,HEXAGON H	6
					UOC:DAL,V18,ZAL	
35	PAOZZ	5340008091494	96906	MS21333-105	CLAMP,LOOP	1
					UOC:DAL,V18,ZAL	
36	PAOZZ	5935007731428	19207	7731428	COVER,ELECTRICAL CO	1
					UOC:DAL,V18,ZAL	
37	PAOZZ	5305002678957	80204	B1821BH025F175N	SCREW,CAP,HEXAGON H	4
					UOC:DAL,V18,ZAL	
38	PAOZZ	5305000425801	24617	425603	SCREW,ASSEMBLED	1
					UOC:DAL,V18,ZAL	
39	PAOZZ	5310006276128	96906	MS35335-35	WASHER,LOCK..	1
					UOC:DAL,V18,ZAL	

END OF FIGURE

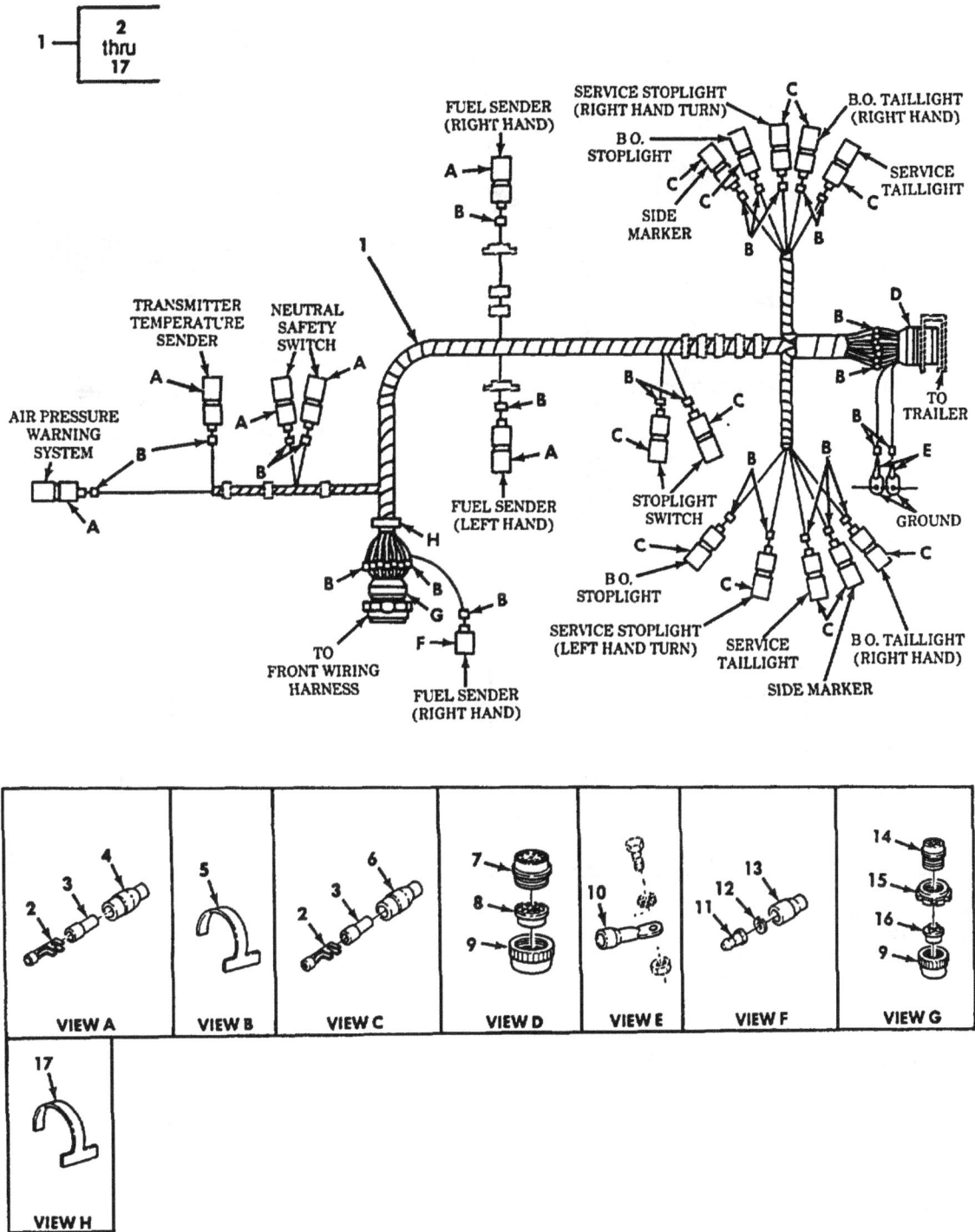

Figure 166. Rear Wiring Harness Assembly, Dump (Sheet 1 of 2).

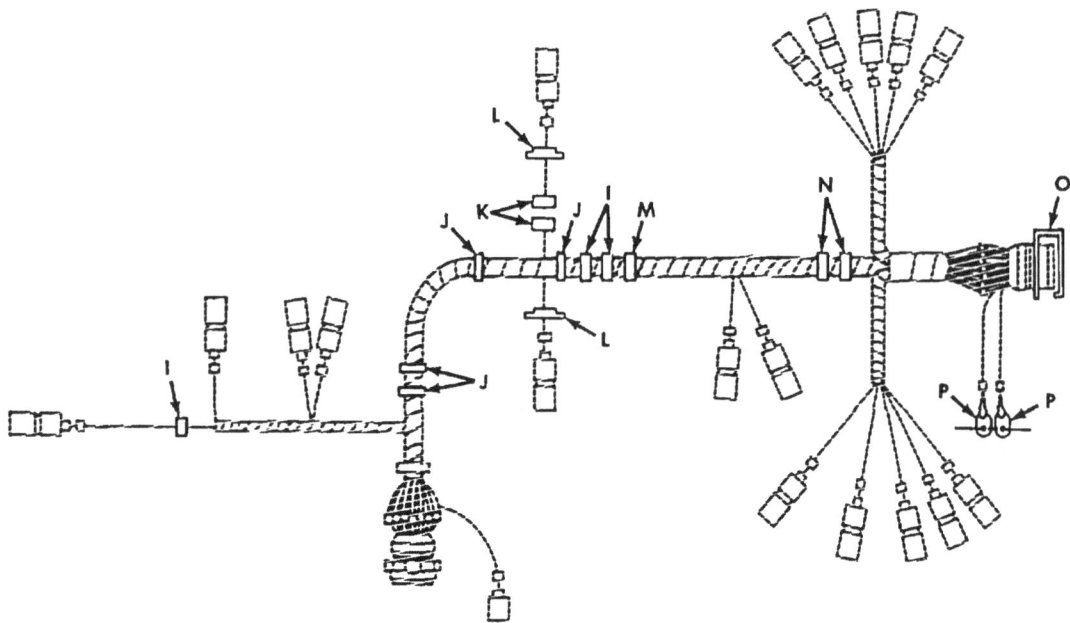

Figure 166. Rear Wiring Harness Assembly, Dump (Sheet 2 of 2).

GROUP 0613 CHASSIS WIRING HARNESS

FIG. 166 REAR WIRING HARNESS
ASSEMBLY, DUMP

(1) ITEM NO	(2) SMR CODE	(3) NSN	(4) CAGEC	(5) PART NUMBER	(6) DESCRIPTION AND USABLE ON CODES (UOC)	(7) QTY
1	PAFFF	2590010835729	19207	12256406	WIRING HARNESS,BRAN UOC:DAE,DAF,V19,V20,ZAE,ZAF	1
2	PAOZZ	1015007982997	19207	7982997	TERMINAL,SOLDERED F UOC:DAE,DAF,V19,V20,ZAE,ZAF	18
3	PAOZZ	5970008338562	19207	8338562	INSULATOR,BUSHING UOC:DAE,DAF,V19,V20,ZAE,ZAF	18
4	PAOZZ	5975006605962	19207	8724494	CABLE NIPPLE,ELECTR UOC:DAE,DAF,V19,V20,ZAE,ZAF	6
5	PAOZZ	9905010138723	81349	M43436/3-1	BAND,MARKER UOC:DAE,DAF,V19,V20,ZAE,ZAF	48
6	PAOZZ	5935008338561	19207	8338561	SHELL,ELECTRICAL CO UOC:DAE,DAF,V19,V20,ZAE,ZAF	12
7	PAOZA	5935008463884	96906	MS75021-2	CONNECTOR,RECEPTACL UOC:DAE,DAF,V19,V20,ZAE,ZAF	1
8	PAOZZ	5365000905426	19207	7722333	BUSHING,NONMETALLIC UOC:DAE,DAF,V19,V20,ZAE,ZAF	1
9	PAOZZ	5310003936685	19207	7723309	NUT,PLAIN,KNURLED UOC:DAE,DAF,V19,V20,ZAE,ZAF	2
10	PAOZZ	5940007056711	19207	7056711	TERMINAL,LUG UOC:DAE,DAF,V19,V20,ZAE,ZAF	2
11	PAOZA	5999000572929	19204	572929	CONTACT,ELECTRICAL UOC:DAE,DAF,V19,V20,ZAE,ZAF	1
12	PAOZZ	5310008338567	19207	8338567	WASHER,SLOTTED UOC:DAE,DAF,V19,V20,ZAE,ZAF	1
13	PAOZZ	5935005729180	19207	8338566	SHELL,ELECTRICAL CO UOC:DAE,DAF,V19,V20,ZAE,ZAF	1
14	PAOZZ	5935000801020	19207	8724255	CONNECTOR,PLUG,ELEC UOC:DAE,DAF,V19,V20,ZAE,ZAF	1
15	PAOZZ	5975007716634	19207	7716634	NUT,COUPLING,ELECTR UOC:DAE,DAF,V19,V20,ZAE,ZAF	1
16	PAOZZ	5365005078766	19207	7388366	BUSHING,NONMETALLIC UOC:DAE,DAF,V19,V20,ZAE,ZAF	1
17	PAOZZ	9905009357777	81349	M43436/4-2	BAND,MARKER UOC:DAE,DAF,V19,V20,ZAE,ZAF	1
18	PAOZZ	5340000881254	96906	MS21333-104	CLAMP,LOOP UOC:DAE,DAF,V19,V20,ZAE,ZAF	3
19	PAOZZ	5975001563253	96906	MS3367-2-9	STRAP,TIEDOWN,ELECT UOC:DAE,DAF,V19,V20,ZAE,ZAF	6
20	PAOZZ	5975009846582	96906	MS3367-1-0	STRAP,TIEDOWN,ELECT UOC:DAE,DAF,V19,V20,ZAE,ZAF	2
21	PAOZZ	5325000531116	19207	7397790	GROMMET,REAR WIRING UOC:DAE,DAF,V19,V20,ZAE,ZAF	2
22	PAOZZ	5340004217254	19207	8352679	CLAMP,LOOP UOC:DAE,DAF,V19,V20,ZAE,ZAF	1
23	PAOZZ	5306000680514	80204	B1821BH025F088N	BOLT,MACHINE... UOC:DAE,DAF,V19,V20,ZAE,ZAF	2
24	PAOZZ	5340008091492	96906	MS21333-100	CLAMP,LOOP	2

(1) ITEM NO	(2) SMR CODE	(3) NSN	(4) CAGEC	(5) PART NUMBER	(6) DESCRIPTION AND USABLE ON CODES (UOC)	(7) QTY
25	PAOZZ	5310009359022	96906	MS51943-32	UOC:DAE,DAF,V19,V20,ZAE,ZAF NUT,SELF-LOCKING,HE	6
26	PAOZZ	5305002678957	80204	B1821BH025F175N	UOC:DAE,DAF,V19,V20,ZAE,ZAF SCREW,CAP,HEXAGON H	4
27	PAOZZ	5935007731428	19207	7731428	UOC:DAE,DAF,V19,V20,ZAE,ZAF COVER,ELECTRICAL CO	1
28	PAOZZ	5310008775796	96906	MS21044N4	UOC:DAE,DAF,V19,V20,ZAE,ZAF NUT,SELF-LOCKING,HE	4
29	PAOZZ	5305000425603	24617	425603	UOC:DAE,DAF,V19,V20,ZAE,ZAF SCREW,ASSEMBLED WAS	1
30	PAOZZ	5310006276128	96906	MS35335-35	UOC:DAE,DAF,V19,V20,ZAE,ZAF WASHER,LOCK	2
31	PAOZZ	5310009591488	96906	MS51922-21	UOC:DAE,DAF,V19,V20,ZAE,ZAF NUT,SELF-LOCKING,HE UOC:DAE,DAF,V19,V20,ZAE,ZAF	2

END OF FIGURE

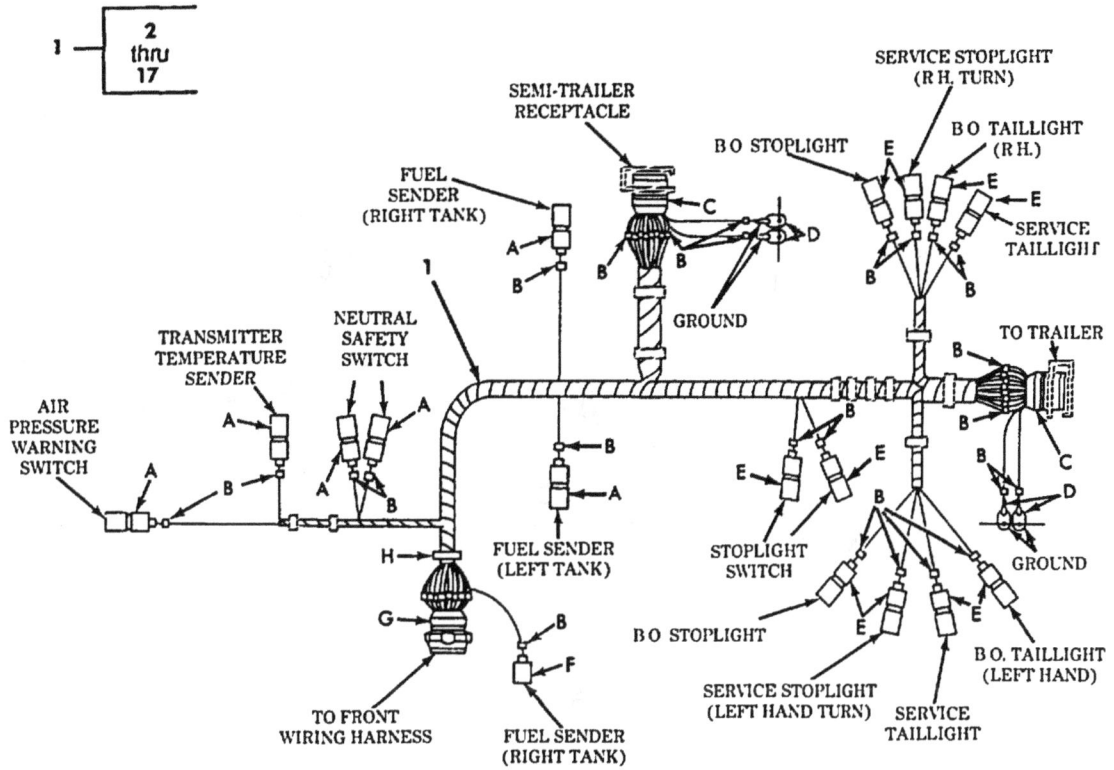

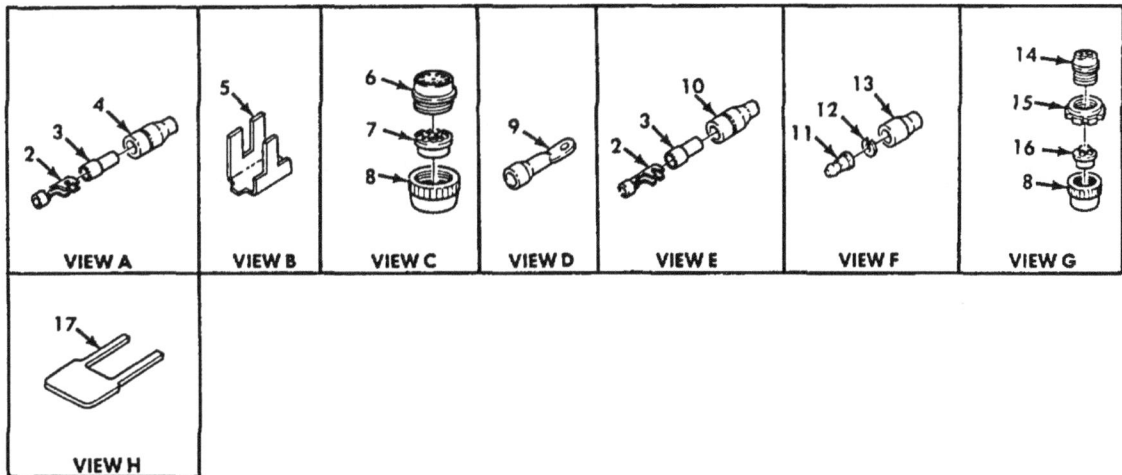

Figure 167. Rear Wiring Harness Assembly, Tractor (Sheet 1 of 2).

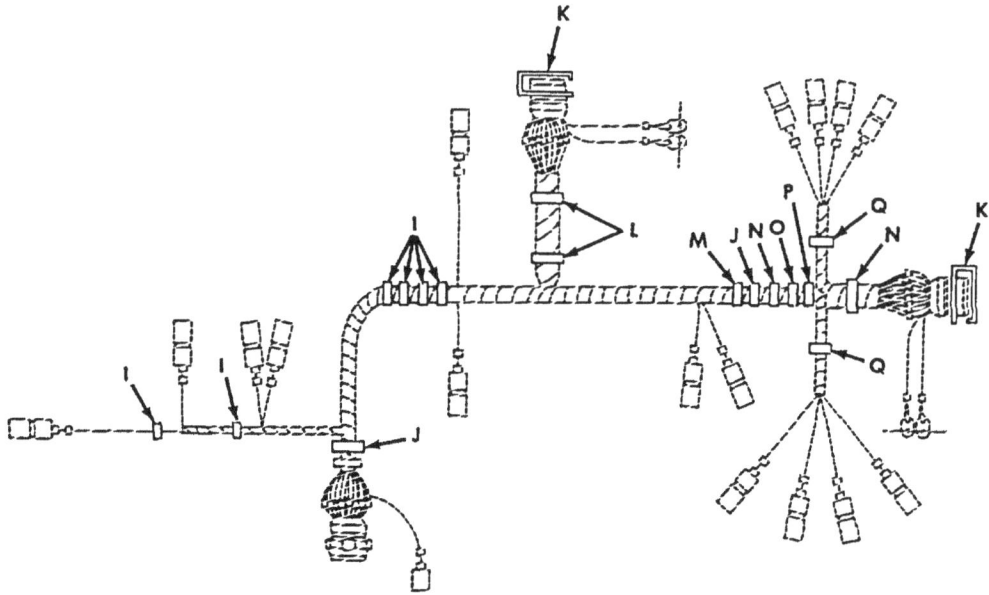

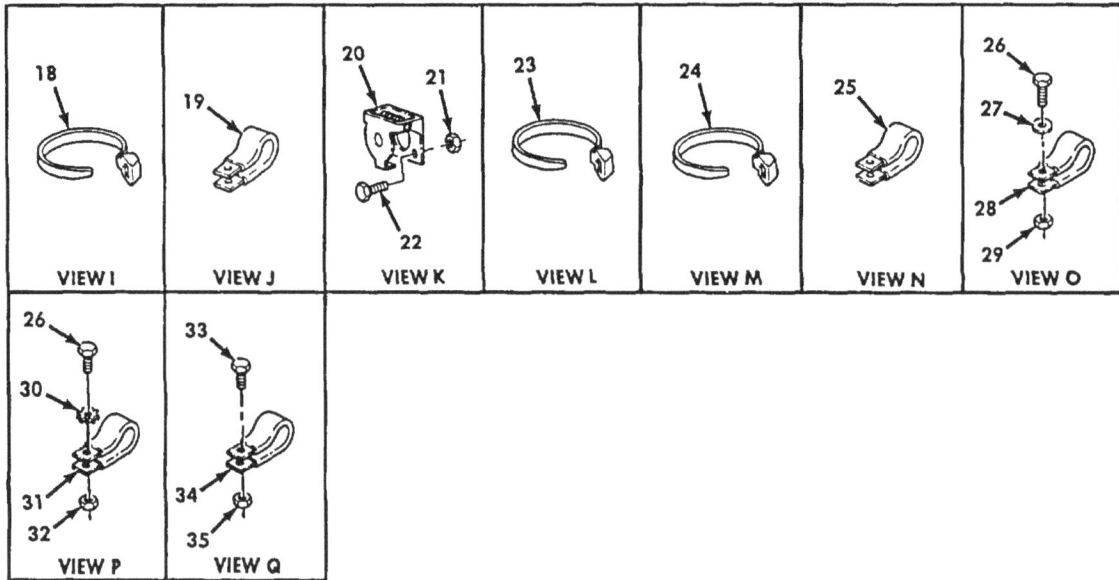

Figure 167. Rear Wiring Harness Assembly, Tractor (Sheet 2 of 2).

(1) ITEM NO	(2) SMR CODE	(3) NSN	(4) CAGEC	(5) PART NUMBER	(6) DESCRIPTION AND USABLE ON CODES (UOC)	(7) QTY

GROUP 0613 CHASSIS WIRING HARNESS

FIG. 167 REAR WIRING HARNESS
ASSEMBLY, TRACTOR

(1) ITEM NO	(2) SMR CODE	(3) NSN	(4) CAGEC	(5) PART NUMBER	(6) DESCRIPTION AND USABLE ON CODES (UOC)	(7) QTY
1	PBOOO	6150010831152	19207	12256097	WIRING HARNESS,BRAN UOC:DAG,DAH,V21,V22,ZAG,ZAH	1
2	PAOZZ	1015007982997	19207	7982997	.TERMINAL,SOLDERED F UOC:DAG,DAH,V21,V22,ZAG,ZAH	16
3	PAOZZ	5970008338562	19207	8338562	.INSULATOR,BUSHING UOC:DAG,DAH,V21,V22,ZAG,ZAH	16
4	PAOZZ	5975006605962	19207	8724494	.CABLE NIPPLE,ELECTR UOC:DAG,DAH,V21,V22,ZAG,ZAH	6
5	PAOZZ	9905010138723	81349	M43436/3-1	.BAND,MARKER UOC:DAG,DAH,V21,V22,ZAG,ZAH	57
6	PAOZA	5935008463884	96906	MS75021-2	.CONNECTOR,RECEPTACL UOC:DAG,DAH,V21,V22,ZAG,ZAH	2
7	PAOZZ	5365000905426	19207	7722333	.BUSHING,NONMETALLIC UOC:DAG,DAH,V21,V22,ZAG,ZAH	2
8	PAOZZ	5310003936685	19207	7723309	.NUT,PLAIN,KNURLED UOC:DAG,DAH,V21,V22,ZAG,ZAH	3
9	PAOZZ	5940007056711	19207	7056711	.TERMINAL,LUG UOC:DAG,DAH,V21,V22,ZAG,ZAH	4
10	PAOZZ	5935008338561	19207	8338561	.SHELL,ELECTRICAL CO UOC:DAG,DAH,V21,V22,ZAG,ZAH	10
11	PAOOA	5999009263144	96906	MS27148-3	.CONTACT,ELECTRICAL UOC:DAG,DAH,V21,V22,ZAG,ZAH	1
12	PAOZZ	5310006560067	19207	8724497	.WASHER,SLOTTED UOC:DAG,DAH,V21,V22,ZAG,ZAH	1
13	PAOZZ	5935006915591	19207	8724495	.SHELL,ELECTRICAL CO UOC:DAG,DAH,V21,V22,ZAG,ZAH	1
14	PAOZZ	5935000801020	19207	8724255	.CONNECTOR,PLUG,ELEC UOC:DAG,DAH,V21,V22,ZAG,ZAH	1
15	PAOZZ	5975007716634	19207	7716634	.NUT,COUPLING,ELECTR UOC:DAG,DAH,V21,V22,ZAG,ZAH	1
16	PAOZZ	5365005078766	19207	7388366	.BUSHING,NONMETALLIC UOC:DAG,DAH,V21,V22,ZAG,ZAH	1
17	PAOZZ	9905001141334	97403	13207E6882-1	.BAND,MARKER UOC:DAG,DAH,V21,V22,ZAG,ZAH	1
18	PAOZZ	5975001563253	96906	MS3367-2-9	.STRAP,TIEDOWN,ELECT UOC:DAG,DAH,V21,V22,ZAG,ZAH	6
19	PAOZZ	5340000538994	96906	MS21333-126	CLAMP,LOOP UOC:DAG,DAH,V21,V22,ZAG,ZAH	2
20	PAOZZ	5935007731428	19207	7731428	COVER,ELECTRICAL CO UOC:DAG,DAH,V21,V22,ZAG,ZAH	2
21	PAOZZ	5310008775796	96906	MS21044N4	NUT,SELF-LOCKING,HE UOC:DAG,DAH,V21,V22,ZAG,ZAH	8
22	PAOZZ	5305002678957	80204	B1821BH025F175N	SCREW,CAP,HEXAGON H UOC:DAG,DAH,V21,V22,ZAG,ZAH	8
23	PAOZZ	5975009846582	96906	MS3367-1-0	STRAP,TIEDOWN,ELECT UOC:DAG,DAH,V21,V22,ZAG,ZAH	2
24	PAOZZ	5975005709598	96906	MS3367-7-9	STRAP,TIEDOWN,ELECT	1

(1) ITEM NO	(2) SMR CODE	(3) NSN	(4) CAGEC	(5) PART NUMBER	(6) DESCRIPTION AND USABLE ON CODES (UOC)	(7) QTY
25	PAOZZ	5340004217254	19207	8352679	UOC:DAG,DAH,V21,V22,ZAG,ZAH CLAMP,LOOP	1
26	PAOZZ	5305000680515	80204	B1821BH025F100N	UOC:DAG,DAH,V21,V22,ZAG,ZAH SCREW,CAP,HEXAGON H	1
27	PAOZZ	5310008094058	96906	MS27183-10	UOC:DAG,DAH,V21,V22,ZAG,ZAR WASHER,FLAT	1
28	PAOZZ	5340008091494	96906	MS21333-105	UOC:DAG,DAH,V21,V22,ZAG,ZAH CLAMP,LOOP	1
29	PAOZZ	5305000425603	24617	425603	UOC:DAG,DAH,V21,V22,ZAG,ZAH SCREW,ASSEMBLED WAS	1
30	PAOZZ	5310006276128	96906	MS35335-35	UOC:DAG,DAH,V21,V22,ZAG,ZAH WASHER,LOCK	1
31	PAOZZ	5340007022848	96906	MS21333-128	UOC:DAG,DAH,V21,V22,ZAG,ZAH CLAMP,LOOP	1
32	PAOZZ	5310009591488	96906	MS51922-21	UOC:DAG,DAH,V21,V22,ZAG, ZAH NUT,SELF-LOCKING,HE	1
33	PAOZZ	5306000680513	60285	6893-2	UOC:DAG,DAH,V21,V22,ZAG,ZAH BOLT,MACHINE	2
34	PAOZZ	5340008091492	96906	MS21333-100	UOC:DAG,DAH,V21,V22,ZAG,ZAH CLAMP,LOOP	2
35	PAOZZ	5310009359022	96906	MS51943-32	UOC:DAG,DAH,V21,V22,ZAG,ZAH NUT,SELF-LOCKING,HE UOC:DAG,DAH,V21,V22,ZAG,ZAH	2

END OF FIGURE

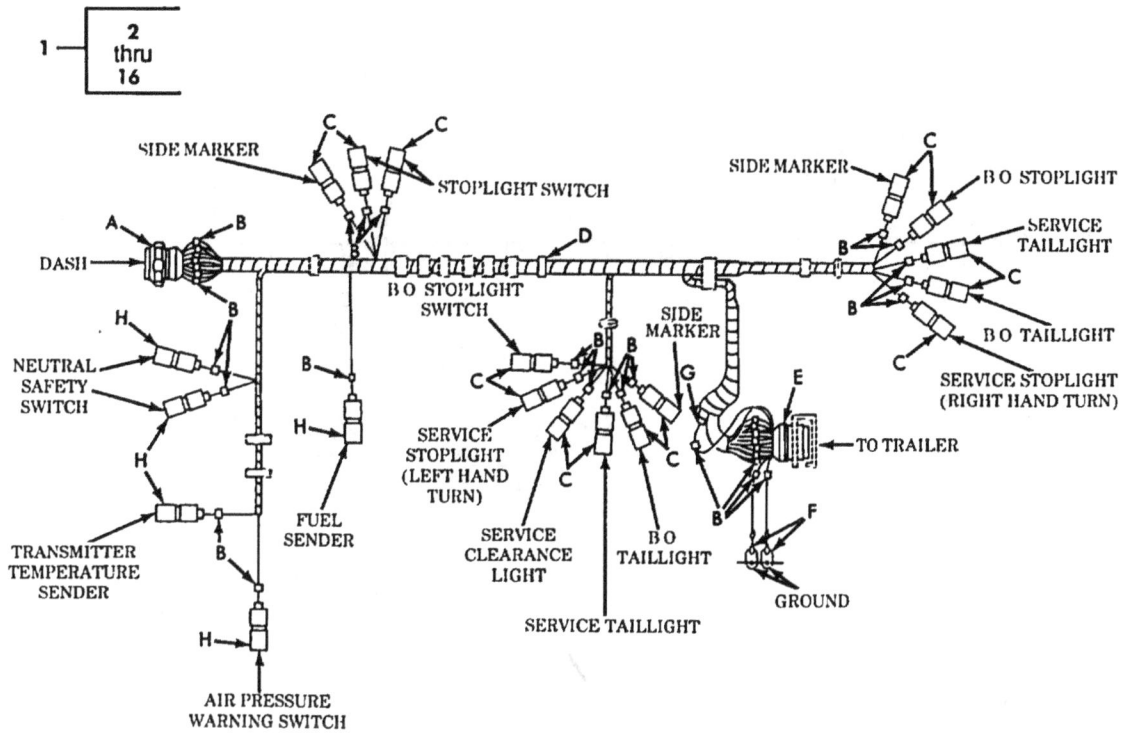

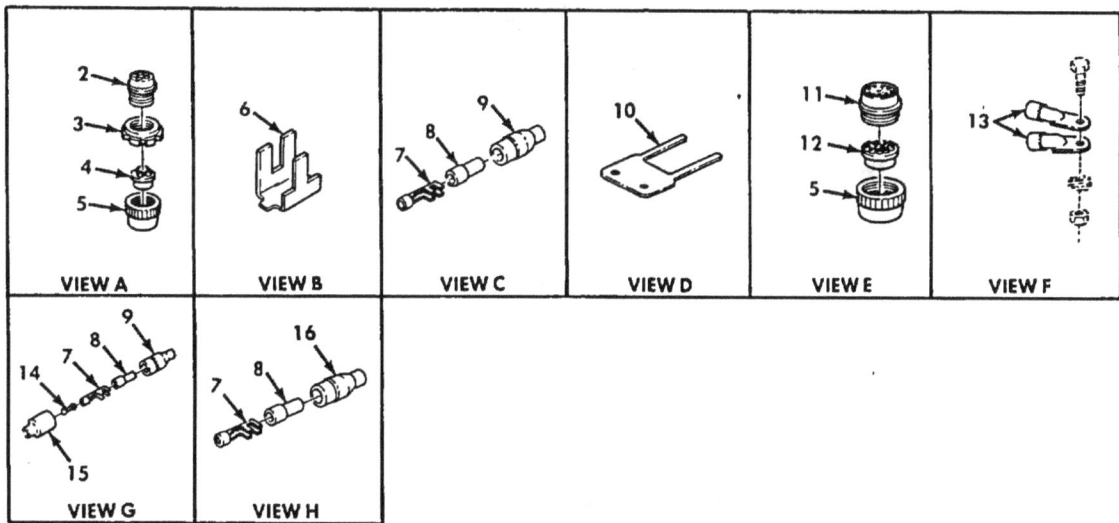

Figure 168. Rear Wiring Harness Assembly, Van (Sheet 1 of 2).

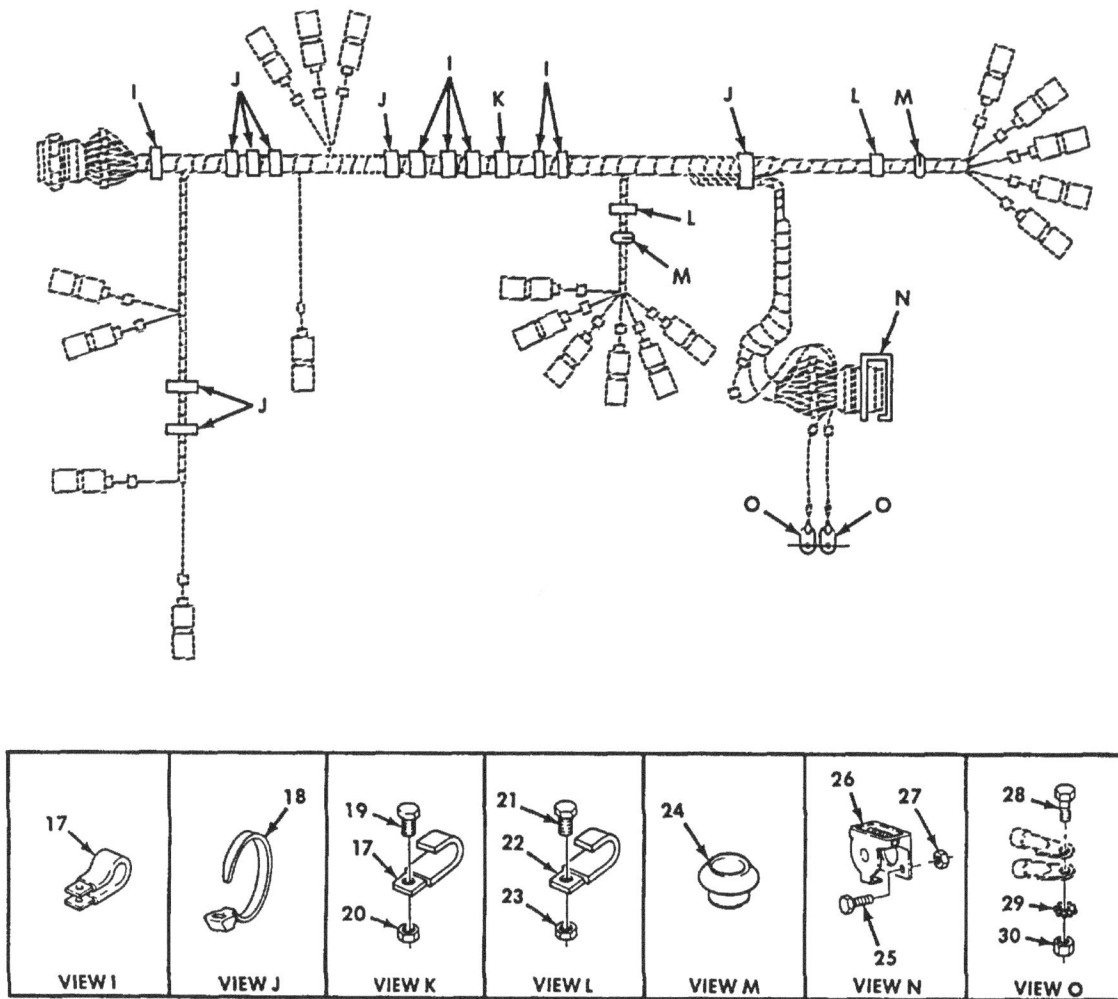

Figure 168. Rear Wiring Harness Assembly, Van (Sheet 2 of 2).

(1) ITEM NO	(2) SMR CODE	(3) NSN	(4) CAGEC	(5) PART NUMBER	(6) DESCRIPTION AND USABLE ON CODES (UOC)	(7) QTY
					GROUP 0613 CHASSIS WIRING HARNESS	
					FIG. 168 REAR WIRING HARNESS ASSEMBLY, VAN	
1	PAFFF	2590010849634	19207	12256002-3	WIRING HARNESS,BRAN UOC:DAJ,DAK,V24,V25,ZAJ,ZAK	1
2	PAOZZ	5935000801020	19207	8724255	.CONNECTOR,PLUG,ELEC UOC:DAJ,DAK,V24,V25,ZAJ,ZAK	1
3	PAOZZ	5975007716634	19207	7716634	.NUT,COUPLING,ELECTR UOC:DAJ,DAK,V24,V25,ZAJ,ZAK	1
4	PAOZZ	5365005078766	19207	7388366	.BUSHING,NONMETALLIC UOC:DAJ,DAK,V24,V25,ZAJ,ZAK	1
5	PAOZZ	5310003936685	19207	7723309	.NUT,PLAIN,KNURLED UOC:DAJ,DAK,V24,V25,ZAJ,ZAK	2
6	PAOZZ	9905010138723	81349	M43436/3-1	.BAND,MARKER UOC:DAJ,DAK,V24,V25,ZAJ,ZAK	49
7	PAOZZ	1015007982997	19207	7982997	.TERMINAL,SOLDERED F UOC:DAJ,DAK,V24,V25,ZAJ,ZAK	21
8	PAOZZ	5970008338562	19207	8338562	.INSULATOR,BUSHING UOC:DAJ,DAK,V24,V25,ZAJ,ZAK	21
9	PAOZZ	5935008338561	19207	8338561	.SHELL,ELECTRICAL CO UOC:DAJ,DAK,V24,V25,ZAJ,ZAK	16
10	PAOZZ	9905001141334	97403	13207E6882-1	.BAND,MAKER UOC:DAJ,DAK,V24,V25,ZAJ,ZAK	1
11	PAOZA	5935008463884	96906	MS75021-2	.CONNECTOR,RECEPTACL UOC:DAJ,DAX,V24,V25,ZAJ,ZAK	1
12	PAOZZ	5365000905426	19207	7722333	.BUSHING,NONMETALLIC UOC:DAJ,DAK,V24,V25,ZAJ,ZAK	1
13	PAOZZ	5940007056711	19207	7056711	.TERMINAL,LUG UOC:DAJ,DAK,V24,V25,ZAJ,ZAK	2
14	PAOZZ	5935002140904	19207	7982907	.DUMMY CONNECTOR,PLU UOC:DAJ,DAK,V24,V25,ZAJ,ZAK	4
15	PAOZZ	5935005729180	19207	8338566	.SHELL,ELECTRICAL CO UOC:DAJ,DAK,V24,V25,ZAJ,ZAK	4
16	PAOZZ	5975006605962	19207	8724494	.CABLE NIPPLE,ELECTR UOC:DAJ,DAK,V24,V25,ZAJ,ZAK	5
17	PAOZZ	5340000538994	96906	MS21333-126	CLAMP,LOOP UOC:DAJ,DAK,V24,V25,ZAJ,ZAK	7
18	PAOZZ	5975001563253	96906	MS3367-2-9	STRAP,TIEDOWN,ELECT UOC:DAJ,DAK,V24,V25,ZAJ,ZAK	7
19	PAOZZ	5305000680515	80204	B1821BH025F100N	SCREW,CAP,HEXAGON H UOC:DAJ,DAK,V24,V25,ZAJ,ZAK	1
20	PAOZZ	5310008140672	96906	MS51943-36	NUT,SELF-LOCKING,HE UOC:DAJ,DAK,V24,V25,ZAJ,ZAK	1
21	PAOZZ	5306000680513	60285	6893-2	BOLT,MACHINE UOC:DAJ,DAK,V24,V25,ZAJ,ZAK	2
22	PAOZZ	5340000881254	96906	MS21333-104	CLAMP,LOOP UOC:DAJ,DAK,V24,V25,ZAJ,ZAK	7
23	PAOZZ	5310009359022	96906	MS51943-32	NUT,SELF-LOCKING,HE UOC:DAJ,DAK,V24,V25,ZAJ,ZAK	2
24	PAOZZ	3120000733163	19207	7397775	BUSHING,SLEEVE	2

(1) ITEM NO	(2) SMR CODE	(3) NSN	(4) CAGEC	(5) PART NUMBER	(6) DESCRIPTION AND USABLE ON CODES (UOC)	(7) QTY
25	PAOZZ	5305002678957	80204	B1821BH025F175N	UOC:DAJ,DAK,V24,V25,ZAJ,ZAK SCREW,CAP,HEXAGON H	4
26	PAOZZ	5935007731428	19207	7731428	UOC:DAJ,DAK,V24,V25,ZAJ,ZAK COVER,ELECTRICAL CO	1
27	PAOZZ	5310008775796	96906	MS21044N4	UOC:DAJ,DAK,V24,V25,ZAJ,ZAK NUT,SELF-LOCKING,HE	4
28	PAOZZ	5305000425601	24617	425601	UOC:DAJ,DAK,V24,V25,ZAJ,ZAK SCREW,ASSEMBLED WAS	1
29	PAOZZ	5310006276128	96906	MS35335-35	UOC:DAJ,DAK,V24,V25,ZAJ,ZAK WASHER,LOCK	1
30	PAOZZ	5310009591488	96906	MS51922-21	UOC:DAJ,DAK,V24,V25,ZAJ,ZAK NUT,SELF-LOCKING,HE UOC:DAJ,DAK,V24,V25,ZAJ,ZAK	1

END OF FIGURE

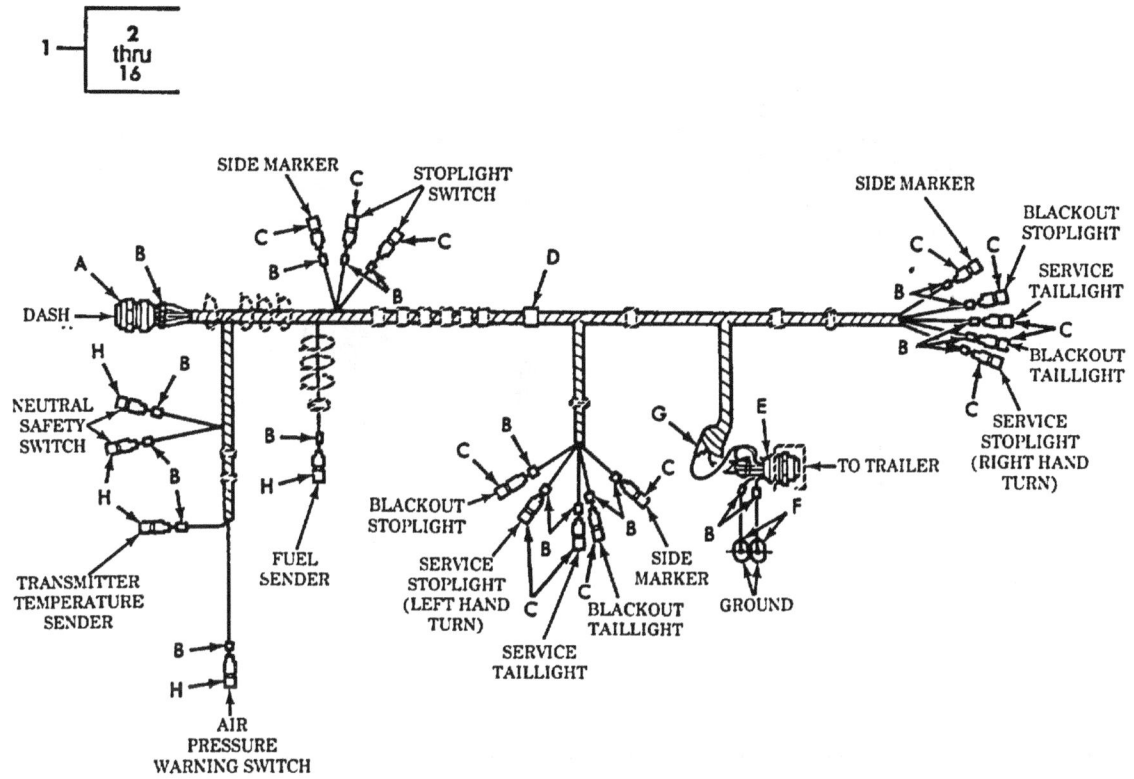

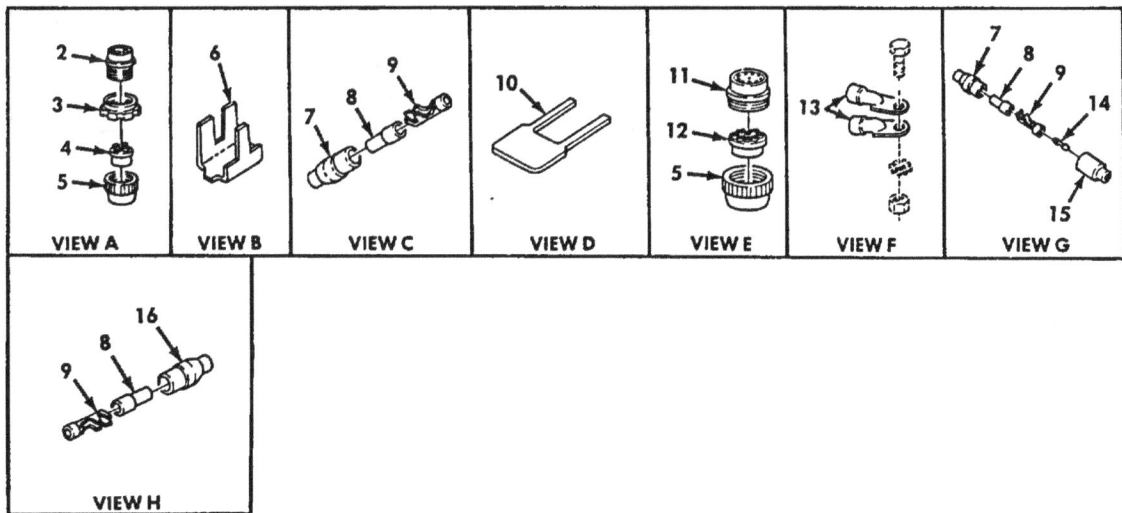

Figure 169. Rear Wiring Harness Assembly, Chassis (Sheet 1 of 2).

FLOODLIGHT
SWITCH

17 —| 18 thru 23 |

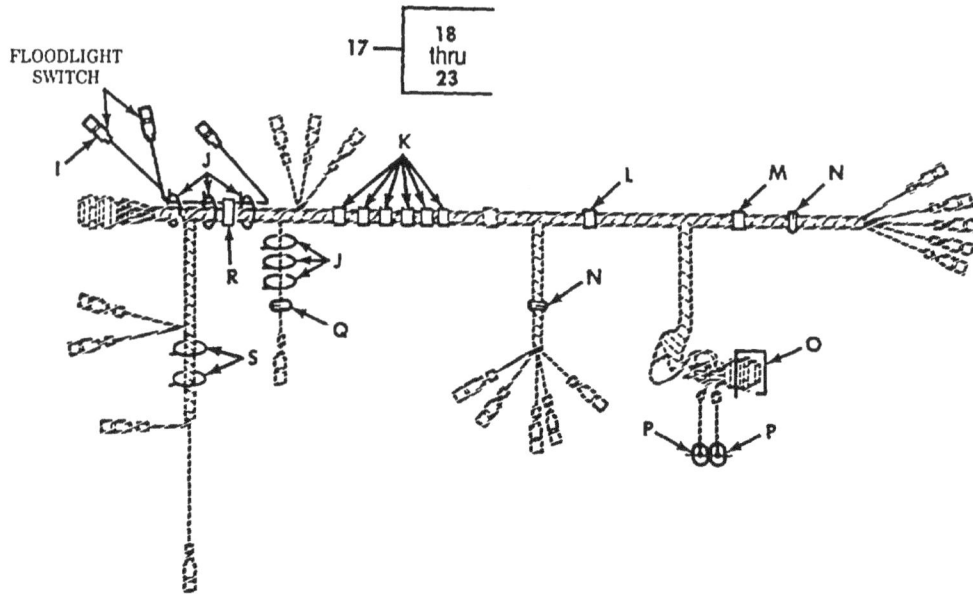

* a   PART OF ITEM 17

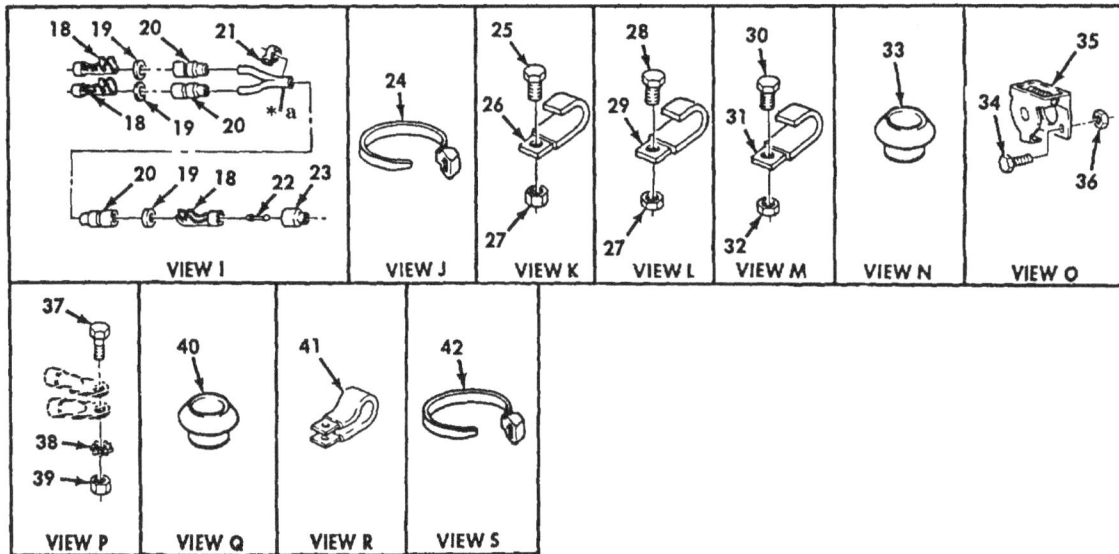

VIEW I

VIEW J

VIEW K

VIEW L

VIEW M

VIEW N

VIEW O

VIEW P

VIEW Q

VIEW R

VIEW S

Figure 169. Rear Wiring Harness Assembly, Chassis (Sheet 2 of 2).

(1) ITEM NO	(2) SMR CODE	(3) NSN	(4) CAGEC	(5) PART NUMBER	(6) DESCRIPTION AND USABLE ON CODES (UOC)	(7) QTY
					GROUP 0613 CHASSIS WIRING HARNESS	
					FIG. 169 REAR WIRING HARNESS ASSEMBLY, CHASSIS	
1	PAFFF	2920010831188	19207	12256002-4	WIRING HARNESS,BRAN UOC:V39	1
2	PAOZZ	5935000801020	19207	8724255	.CONNECTOR,PLUG,ELEC UOC:V39	1
3	PAOZZ	5975007716634	19207	7716634	.NUT,COUPLING,ELECTR UOC:V39	1
4	PAOZZ	5365005078766	19207	7388366	.BUSHING,NONMETALLIC UOC:V39	1
5	PAOZZ	5310003936685	19207	7723309	.NUT,PLAIN,KNURLED UOC:V39	2
6	PAOZZ	9905010138723	81349	M43436/3-1	.BAND,MARKER UOC:V39	49
7	PAOZZ	5935008338561	19207	8338561	.SHELL,ELECTRICAL CO UOC:V39	14
8	PAOZZ	5970008338562	19207	8338562	INSULATOR,BUSHING UOC:V39	19
9	PAOZZ	1015007982997	19207	7982997	TERMINAL,SOLDERED F UOC:V39	19
10	PAOZZ	9905001141334	97403	13207E6882-1	.BAND,MARXKER UOC:V39	1
11	PAOZA	5935008463884	96906	MS75021-2	.CONNECTOR,RECEPTACL UOC:V39	1
12	PAOZZ	5365000905426	19207	7722333	.BUSHING,NONMETALLIC UOC:V39	1
13	PAOZZ	5940007056711	19207	7056711	.TERMINAL,LUG UOC:V39	2
14	PAOZZ	5935002140904	19207	7982907	.DUMMY CONNECTOR,PLU UOC:V39	1
15	PAOZZ	5935005729180	19207	8338566	.SHELL,ELECTRICAL CO UOC:V39	1
16	PAOZZ	5975006605962	19207	8724494	.CABLE NIPPLE,ELECTR UOC:V39	5
17	PAOZZ	6150010823818	19207	7363004-1	.CABLE ASSEMBLY UOC:V39	1
18	PAOZZ	1015007982997	19207	7982997	.TERMINAL,SOLDERED F UOC:V39	3
19	PAOZZ	5310002988903	19207	8338570	.WASHER,FLAT FLOODLIGHT SWITCH CABLE UOC:V39	3
20	PBOZZ	2590006959076	19207	8338569	SHELL,HEAD LIGHT CI FLOODLIGHT SWITCH CABLE UOC:V39	3
21	PAOZZ	9905010697222	81349	M43436/2-1	.BAND,MARKER FLOODLIGHT SWITCH CABLE UOC:V39	3
22	PAOZZ	5935002140904	19207	7982907	.DUMMY CONNECTOR,PLU UOC:V39	1

(1) ITEM NO	(2) SMR CODE	(3) NSN	(4) CAGEC	(5) PART NUMBER	(6) DESCRIPTION AND USABLE ON CODES (UOC)	(7) QTY
23	PAOZZ	5975006605962	19207	8724494	.CABLE NIPPLE,ELECTR UOC:V39	1
24	PAOZZ	5975009846582	96906	MS3367-1-0	STRAP,TIEDOWN,ELECT UOC:V39	6
25	PAOZZ	5305000680515	80204	B1821BH025F100N	SCREW,CAP,HEXAGON H UOC:V39	6
26	PAOZZ	5340004454557	19207	8330341	STRAP,RETAINING UOC:V39	6
27	PAOZZ	5310009359022	96906	MS51943-32	NUT,SELF-LOCKING,HE UOC:V39	7
28	PAOZZ	5306000680513	60285	6893-2	BOLT,MACHINE.. UOC:V39	1
29	PAOZZ	5340004193077	19207	8330343	STRAP,RETAINING UOC:V39	1
30	PAOZZ	5305002259091	96906	MS90726-36	SCREW,CAP,HEXAGON H UOC:V39	1
31	PAOZZ	5340003442767	19207	8330338	STRAP,RETAINING UOC:V39	1
32	PAOZZ	5310002416658	96906	MS51943-34	NUT,SELF-LOCKING,HE UOC:V39	1
33	PAOZZ	3120000733163	19207	7397775	BUSHING,SLEEVE.. UOC:V39	2
34	PAOZZ	5305002678957	80204	B1821BH025F175N	SCREW,CAP,HEXA GON H UOC:V39	4
35	PAOZZ	5935007731428	19207	7731428	COVER,ELECTRICAL CO UOC:V39	1
36	PAOZZ	5310008775796	96906	MS21044N4	NUT,SELF-LOCKING,HE UOC:V39	4
37	PAOZZ	5305000425603	24617	425603	SCREW,ASSEMBLED UOC:V39	1
38	PAOZZ	5310006276128	96906	MS35335-35	WASHER,LOCK... UOC:V39	1
39	PAOZZ	5310009591488	96906	MS51922-21	NUT,SELF-LOCKING,HE UOC:V39	1
40	PAOZZ	5325000531116	19207	7397790	GROMMET,REAR WIRING.................................... UOC:V39	1
41	PAOZZ	5340007255280	96906	MS21333-125	CLAMP,LOOP REAR WIRING HARNESS UOC:V39	1
42	PAOZZ	5975001563253	96906	MS3367-2-9	STRAP,TIEDOWN,ELECT UOC:V39	2

END OF FIGURE

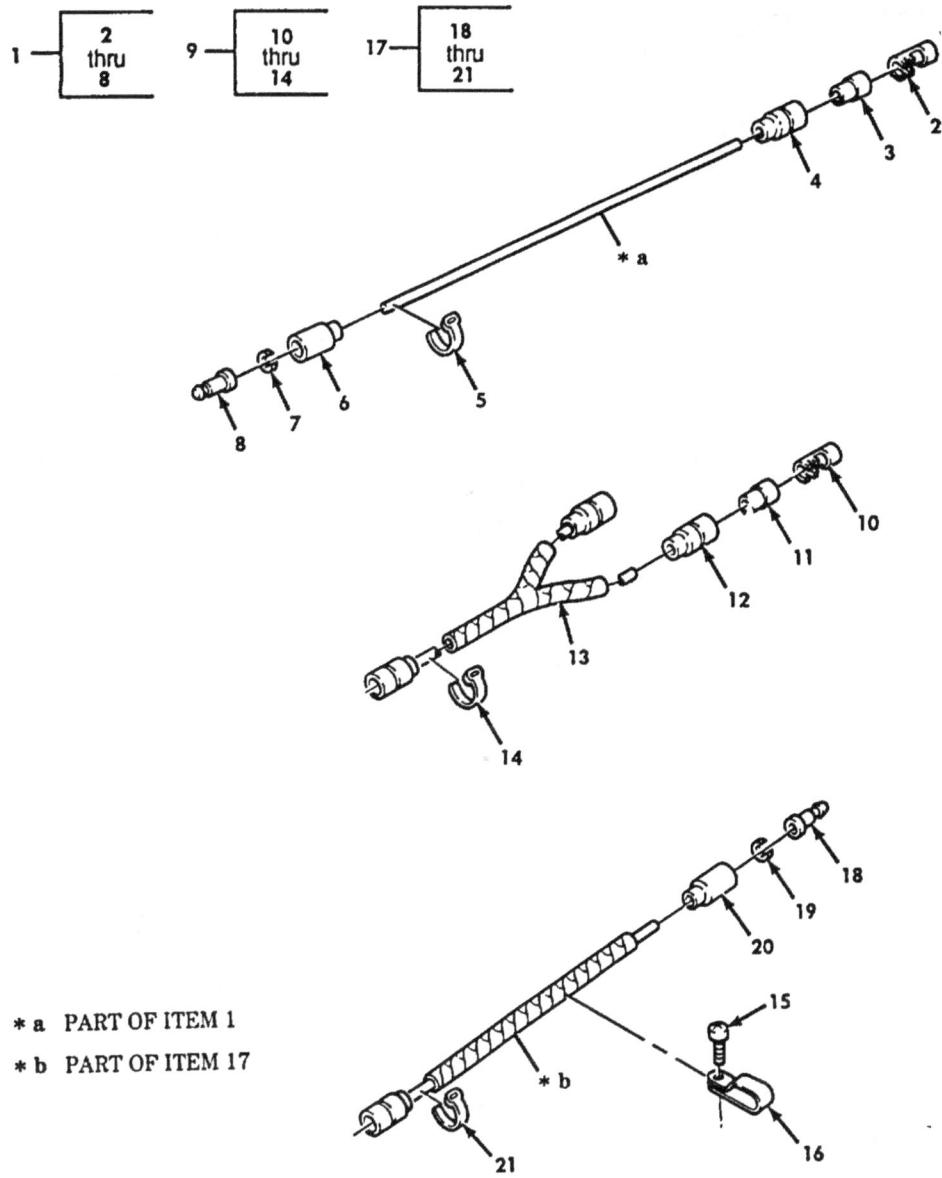

* a   PART OF ITEM 1
* b   PART OF ITEM 17

Figure 170. Crane Wiring Leads (Sheet 1 of 2).

Figure 170. Crane Wiring Leads (Sheet 2 of 2).

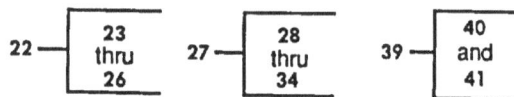

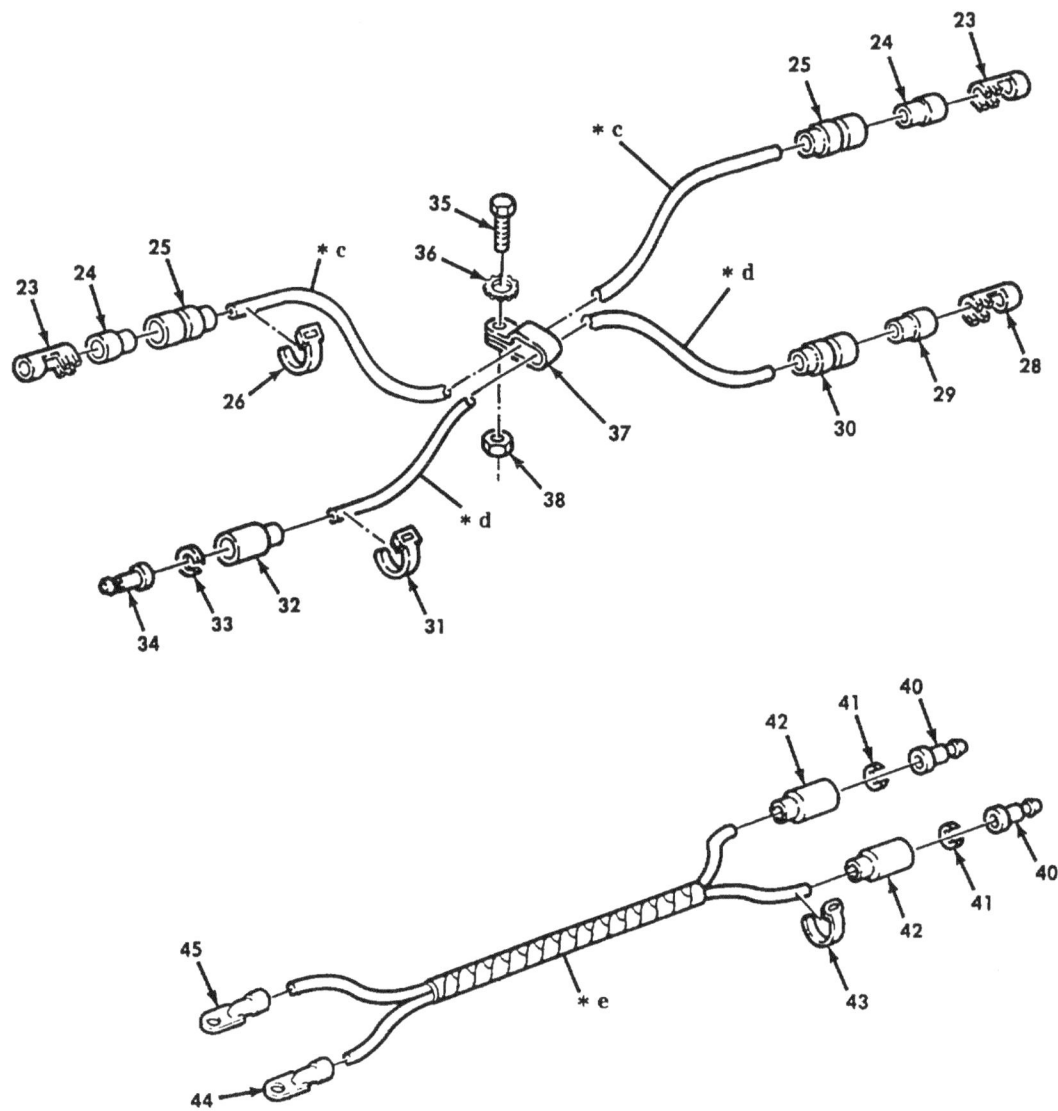

* c PART OF ITEM 22
* d PART OF ITEM 27
* e PART OF ITEM 39

(1) ITEM NO	(2) SMR CODE	(3) NSN	(4) CAGEC	(5) PART NUMBER	(6) DESCRIPTION AND USABLE ON CODES (UOC)	(7) QTY
					GROUP 0613 CHASSIS WIRING HARNESS	
					FIG. 170 CRANE WIRING LEADS	
1	MOOZZ		19207	12277326-3	LEAD,ELECTRICAL MAKE FROM CABLE P/N........ M13486/1-5, LENGTH AS REQUIRED.................... UOC:DAL,V18,ZAL	1
2	PAOZZ	5940003996676	19207	8338564	.TERMINAL ASSEMBLY CRANE SWIVEL.............. VALVE LEAD... UOC:DAL,V18,ZAL	1
3	PAOZZ	5970008338562	19207	8338562	.INSULATOR,BUSHING CRANE SWIVEL............... VALVE LEAD... UOC:DAL,V18,ZAL	1
4	PAOZZ	5935008339561	19207	8338561	.SHELL,ELECTRICAL CO CRANE SWIVEL............. VALVE LEAD... UOC:DAL,V18,ZAL	1
5	PAOZZ	9905007524649	81349	M43436/1-1	.BAND,MARKER CRANE SWIVEL VALVE .... LEAD... UOC:DAL,V18,ZAL	2
6	PAOZZ	5935005729180	19207	8338566	.SHELL,ELECTRICAL CO CRANE SWIVEL............. VALVE LEAD... UOC:DAL,V18,ZAL	1
7	PAOZZ	5310008338567	19207	8338567	.WASHER,SLOTTED CRANE SWIVEL VALVE......... LEAD... UOC:DAL,V18,ZAL	1
8	PAOZA	5999000572929	19204	572929	.CONTACT,ELECTRICAL CRANE SWIVEL............. VALVE LEAD... UOC:DAL,V18,ZAL	1
9	MOOZZ		19207	12277334 XB	LEAD AND CONDUIT AS MAKE FROM .... CABLE,P/N M13486/1-5,AS REQUIRED................. UOC:DAL,V18,ZAL	1
10	PAOZZ	5940003996676	19207	8338564	.TERMINAL ASSEMBLY FLOODLIGHT LEAD. UOC:DAL,V18,ZAL	3
11	PAOZZ	5970008338562	19207	8338562	.INSULATOR,BUSHING FLOODLIGHT LEAD........... UOC:DAL,V18,ZAL	3
12	PAOZZ	5935008338561	19207	8338561	.SHELL,ELECTRICAL CO FLOODLIGHT................ LEAD D.. UOC:DAL,V18,ZAL	3
13	MOOZZ		80244	17-C-18035-50-11 9	.CONDUIT,NONMETALLIC MAKE FROM............... CONDUIT,P/N 17-C-18035-50............................ UOC:DAL,V18,ZAL	1
14	PAOZZ	9905007524649	81349	M43436/1-1	.BAND,MARKER FLOODLIGHT LEAD................... UOC:DAL,V18,ZAL	3
15	PAOZZ	5305008550957	96906	MS24629-46	SCREW,TAPPING.. UOC:DAL,V18,ZAL	4
16	PAOZZ	5340009301754	96906	MS21334-2	CLAMP,LOOP... UOC:DAL,V18,ZAL	4
17	MOOOZ	5995002317422	19207	10876527	LEAD,ELECTRICAL RIGHT,MAKE FROM............... CABLE P/N M13486/1-5,LENGTH AS REQUIRED... UOC:DAL,V18,ZAL	1
17	MOOOZ	5995002317423	19207	10876543	LEAD,ELECTRICAL BOOM FLOODLIGHT.............. LEFT,MAKE FROM CABLE P/N M13486/1-5, LENGTH AS REQUIRED................................... UOC:DAL,V18,ZAL	1

(1) ITEM NO	(2) SMR CODE	(3) NSN	(4) CAGEC	(5) PART NUMBER	(6) DESCRIPTION AND USABLE ON CODES (UOC)	(7) QTY
					MAKE FROM CABLE,P/N M13486/1-5 .... UOC:DAL,V18,ZAL	
18	PAOZA	5999000572929	19204	572929	.CONTACT,ELECTRICAL BOOM .... FLOODLIGHT LEAD... UOC:DAL,V18,ZAL	2
19	PAOZZ	5310008338567	19207	8338567	.WASHER,SLOTTED BOOM FLOODLIGHT............... LEAD UOC:DAL,V18,ZAL	2
20	PAOZZ	5935005729180	19207	8338566	.SHELL,ELECTRICAL CO BOOM .... FLOODLIGHT LEAD... UQC:DAL,V18,ZAL	2
21	PAOZZ	9905007524649	81349	M43436/1-1	.BAND,MARKER BOOM FLOODLIGHT LEAD UOC:DAL,V18,ZAL	2
22	MOOZZ	5995011662235	19207	12277328	LEAD,ELECTRICAL MAKE FROM WIRE,P/N .... M13486/1-3... UOC:DAL,V18,ZAL	1
23	PAOZZ	5940003996676	19207	8338564	.TERMINAL ASSEMBLY WARNING LIGHT............... LEAD UOC:DAL,V18,ZAL	2
24	PAOZZ	5970008338562	19207	8338562	.INSULATOR,BUSHING WARNING LIGHT............... LEAD UOC:DAL,V18,ZAL	2
25	PAOZZ	5975006605962	19207	8724494	.CABLE NIPPLE,ELECTR WARNING LIGHT............. LEAD UOC:DAL,V18,ZAL	2
26	PAOZZ	9905007524649	81349	M43436/1-1	.BAND,MARXER WARNING LIGHT LEAD................ UOC:DAL,V18,ZAL	2
27	MOOZZ	6150011643358	19207	12277327-1	LEAD,ELECTRICAL MAKE FROM WIRE,P/N .... M13486/1-3... UOC:DAL,V18,ZAL	1
28	PAOZZ	5940003996676	19207	8338564	.TERMINAL ASSEMBLY WARNING LIGHT............... LEAD... UOC:DAL,V18,ZAL	1
29	PAOZZ	5970008338562	19207	8338562	.INSULATOR,BUSHING WARNING LIGHT............... LEAD UOC:DAL,V18,ZAL	1
30	PAOZZ	5975006605962	19207	8724494	.CABLE NIPPLE,ELECTR WARNING LIGHT............. LEAD UOC:DAL,V18,ZAL	1
31	PAOZZ	9905007524649	81349	M43436/1-1	.BAND,MARKER WARNING LIGHT LEAD................ UOC:DAL,V18,ZAL	2
32	PAOZZ	5935006915591	19207	8724495	.SHELL,ELECTRICAL CO WARNING LIGHT.......... LEAD UOC:DAL,V18,ZAL	1
33	PAOZZ	5310006560067	19207	8724497	.WASHER,SLOTTED WARNING LIGHT LEAD.......... UOC:DAL,V18,ZAL	1
34	PAOOA	5999009263144	96906	MS27148-3	.CONTACT,ELECTRICAL WARNING LIGHT............ LEAD UOC:DAL,V18,ZAL	1
35	PAOZZ	5306000680514	96906	MS90727-7	SCREW,CAP,HEXAGON H UOC:DAL,V18,ZAL	1
36	PAOZZ	5310008892528	96906	MS45904-68	WASHER,LOCK UOC:DAL,V18,ZAL	1

(1) ITEM NO	(2) SMR CODE	(3) NSN	(4) CAGEC	(5) PART NUMBER	(6) DESCRIPTION AND USABLE ON CODES (UOC)	(7) QTY
37	PAOZZ	5340002827544	96906	MS21333-35	CLAMP,LOOP UOC:DAL,V18,ZAL	1
38	PAOZZ	5310009359022	96906	MS51943-32	NUT,SELF-LOCKING,HE UOC:DAL,V18,ZAL	1
39	MOOZZ	5995011641900	19207	12277329	WIRING HARNESS,BRAN MAKE FROM WIRE,........ P/N M13486/1-3 .... UOC:DAL,V18,ZAL	1
40	PAOOA	5999009263144	96906	MS27148-3	.CONTACT,ELECTRICAL WARNING LIGHT............ LEAD.. UOC:DAL,V18,ZAL	2
41	PBOZZ	5310006560067	19207	8724497	.WASHER,SLOTTED WARNING LIGHT LEAD.......... UOC:DAL,V18,ZAL	2
42	PAOZZ	5935006915591	19207	8724495	SHELL,ELECTRICAL CO WARNING LIGHT............ LEAD.. UOC:DAL,V18,ZAL	2
43	PAOZZ	9905007524649	81349	M43436/1-1	BAND,MARKER WARNING LIGHT LEAD................ UOC:DAL,V18,ZAL	4
44	PAOZZ	5940001434780	96906	MS25036-108	TERMINAL,LUG WARNING LIGHT LEAD UOC:DAL,V18,ZAL	1
45	PBOZZ	5940001434774	96906	MS25036-153	TERMINAL,LUG WARNING LIGHT LEAD UOC:DAL,V18,ZAL	1

END OF FIGURE

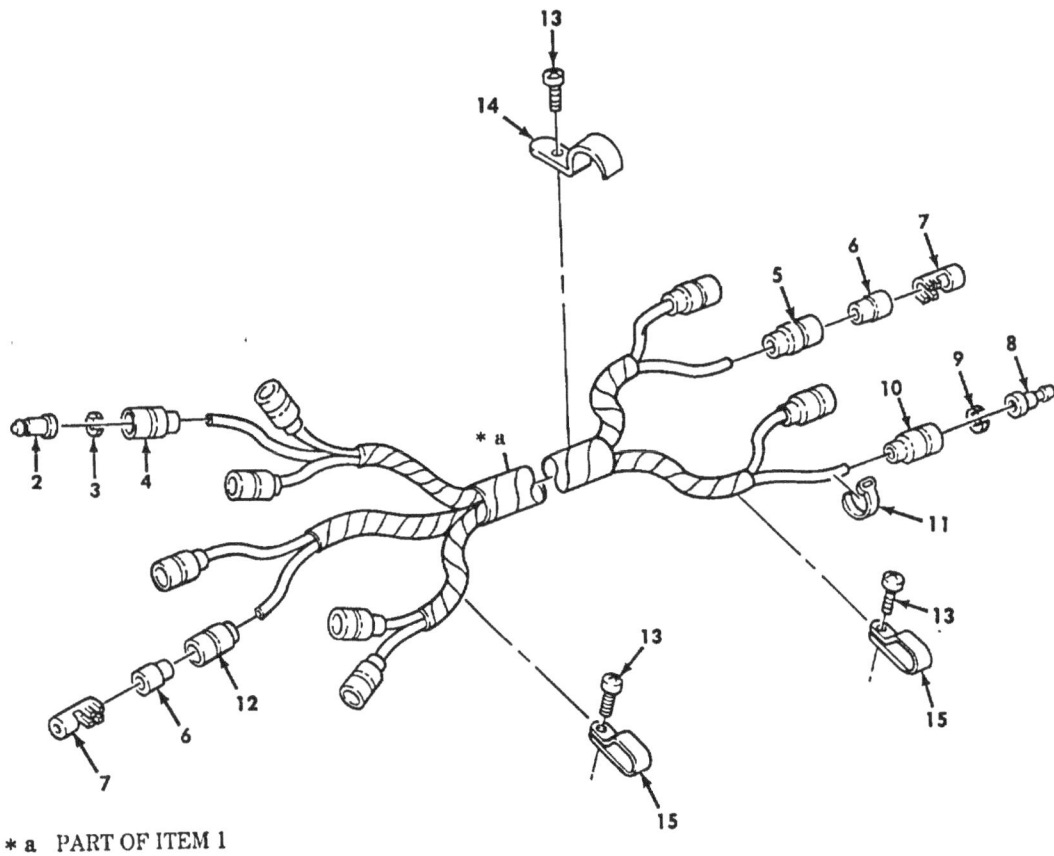

*a PART OF ITEM 1

Figure 171. Wrecker Crane Boom Wiring Harness

(1) ITEM NO	(2) SMR CODE	(3) NSN	(4) CAGEC	(5) PART NUMBER	(6) DESCRIPTION AND USABLE ON CODES (UOC)	(7) QTY
					GROUP 0613 CHASSIS WIRING HARNESS	
					FIG. 171 WRECKER CRANE BOOM WIRING HARNESS	
1	MOOZZ	2590011643440	19207	12277332	WIRING HARNESS,BRAN................................ UOC:DAL,V18,ZAL	1
2	PAOOA	5999009263144	96906	MS27148-3	.CONTACT,ELECTRICAL BOOM WIRING............... HARNESS UOC:DAL,V18,ZAL	1
3	PBOZZ	5310006560067	19207	8724497	.WASHER,SLOTTED BOOM WIRING HARNESS...... UOC:DAL,V18,ZAL	1
4	PAOZZ	5935006915591	19207	8724495	.SHELL,ELECTRICAL CO BOOM WIRING .... HARNESS UOC:DAL,V18,ZAL	1
5	PAOZZ	5935008338561	19207	8338561	.SHELL,ELECTRICAL CO CRANE WIRING............. HARNESS UOC:DAL,V18,ZAL	4
6	PAOZZ	5970008338562	19207	8338562	.INSULATOR,BUSHING CRANE WIRING................ HARNESS UOC:DAL,V18,ZAL	6
7	PAOZZ	5940003996676	19207	8338564	.TERMINAL ASSEMBLY BOOM WIRING................. HARNESS UOC:DAL,V18,ZAL	6
8	PAOZA	5999000572929	19204	572929	.CONTACT,ELECTRICAL HARNESS..................... ASSEMBLY,CRANE TO GONDOLA FLOODLIGHT UOC:DAL,V18,ZAL	4
9	PAOZZ	5310008338567	19207	8338567	.WASHER,SLOTTED UOC:DAL,V18,ZAL	4
10	PAOZZ	5935005729180	19207	8338566	.SHELL,ELECTRICAL CO.................................... UOC:DAL,V18,ZAL	4
11	PAOZZ	9905007524649	81349	M43436/1-1	.BAND,MARKER BOOM WIRING HARNESS............ UOC:DAL,V18,ZAL	11
12	PAOZZ	5975006605962	19207	8724494	.CABLE NIPPLE,ELECTR BOOM WIRING............... HARNESS UOC:DAL,V18,ZAL	2
13	PAOZZ	5305008550957	96906	MS24629-46	SCREW,TAPPING.. UOC:DAL,V18,ZAL	17
14	PAOZZ	5340002827793	96906	MS35140-6	STRAP,RETAINING UOC:DAL,V18,ZAL	5
15	PAOZZ	5340009301754	96906	MS21334-2	CLAMP,LOOP.. UOC:DAL,V18,ZAL	18

END OF FIGURE

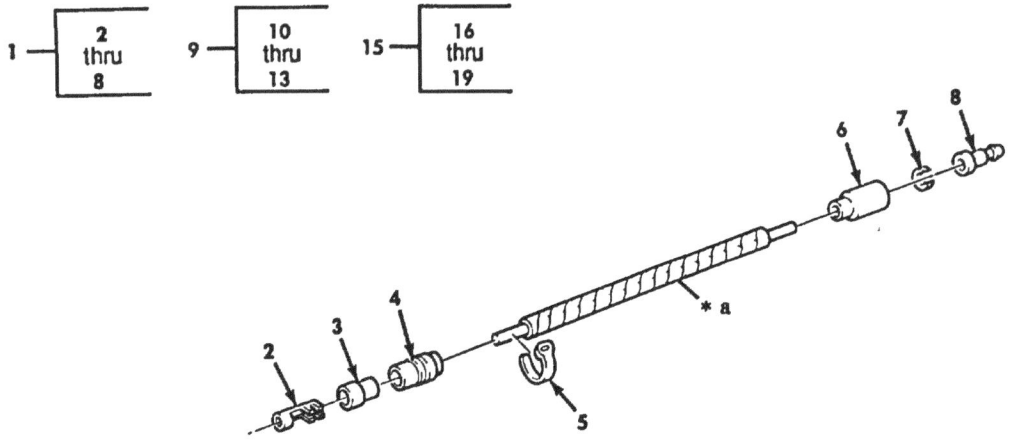

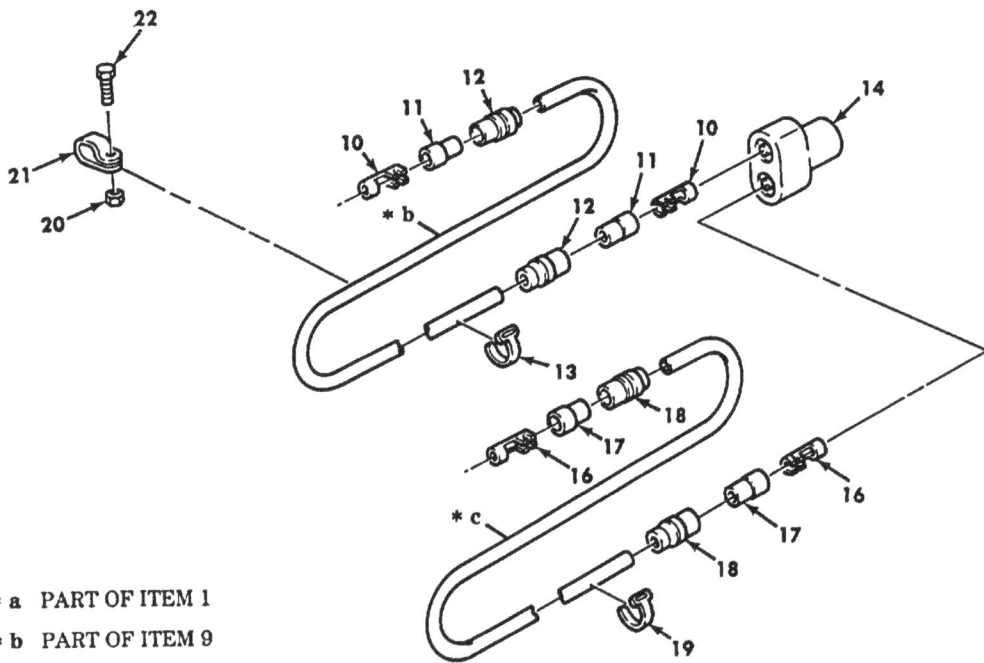

*a   PART OF ITEM 1
*b   PART OF ITEM 9
*c   PART OF ITEM 15

Figure 172. Side Marker Light Leads.

(1) ITEM NO	(2) SMR CODE	(3) NSN	(4) CAGEC	(5) PART NUMBER	(6) DESCRIPTION AND USABLE ON CODES (UOC)	(7) QTY
					GROUP 0613 CHASSIS WIRING HARNESS	
					FIG. 172 SIDE MARKER LIGHT LEADS	
1	PAOZZ	6150011308042	19207	12277005-2	LEAD,ELECTRICAL............................ UOC:DAA,DAB,DAC,DAD,DAW,DAX,V12,V13, V14,V15,V16,V17,ZAA,ZAB,ZAC,ZAD	2
2	PAOZZ	1015007982997	19207	7982997	.TERMINAL,SOLDERED F SIDE MARKER.............. TO TAILLIGHT LEAD.................... UOC:DAA,DAB,DAC,DAD,DAW,DAX,V12,V13, V14,V15,V16,V17,ZAA,ZAB,ZAC,ZAD	1
3	PAOZZ	5970008338562	19207	8338562	.INSULATOR,BUSHING SIDE MARKER TO .... TAILLIGHT LEAD UOC:DAA,DAB,DAC,DAD,DAW,DAX,V12,V13, V14,V15,V16,V17,ZAA,ZAB,ZAC,ZAD	1
4	PAOZZ	5935008338561	19207	8338561	.SHELL,ELECTRICAL CO SIDE MARKER............... TO TAILLIGHT LEAD.................... UOC:DAA,DAB,DAC,DAD,DAW,DAX,V12,V13, V14,V15,V16,V17,ZAA,ZAB,ZAC,ZAD	1
5	PAOZZ	9905007524649	81349	M43436/1-1	.BAND,MARKER SIDE MARKER TO .... TAILLIGHT LEAD UOC:DAA,DAB,DAC,DAD,DAW,DAX,V12,V13, V14,V15,V16,V17,ZAA,ZAB,ZAC,ZAD	2
6	PAOZZ	5935005729180	19207	8338566	.SHELL,ELECTRICAL CO SIDE MARKER............... TO TAILLIGHT LEAD.................... UOC:DAA,DAB,DAC,DAD,DAW,DAX,V12,V13, V14,V15,V16,V17,ZAA,ZAB,ZAC,ZAD	1
7	PAOZZ	5310008338567	19207	8338567	.WASHER,SLOTTED SIDE MARKER TO .... TAILLIGHT LEAD UOC:DAA,DAB,DAC,DAD,DAW,DAX,V12,V13, V14,V15,V16,V17,ZAA,ZAB,ZAC,ZAD	1
8	PAOZA	5999000572929	19204	572929	.CONTACT,ELECTRICAL SIDE MARKER TO .... TAILLIGHT LEAD UOC:DAA,DAB,DAC,DAD,DAW,DAX,V12,V13, V14,V15,V16,V17,ZAA,ZAB,ZAC,ZAD	1
9	PFOZZ	6150011310147	19207	12277004-1	LEAD ASSEMBLY,ELECT UOC:DAA,DAB,DAC,DAD,DAW,DAX,V12,V13, V14,V15,V16,V17,ZAA,ZAB,ZAC,ZAD	1
10	PAOZZ	1015007982997	19207	7982997	.TERMINAL,SOLDERED F LEFT SIDE .... MARKER LIGHT LEAD UOC:DAA,DAB,DAC,DAD,DAW,DAX,V12,V13, V14,V15,V16,V17,ZAA,ZAB,ZAC,ZAD	2
11	PAOZZ	5970008338562	19207	8338562	.INSULATOR,BUSHING LEFT SIDE .... MARKER LIGHT LEAD UOC:DAA,DAB,DAC,DAD,DAW,DAX,V12,V13, V14,V15,V16,V17,ZAA,ZAB,ZAC,ZAD	2
12	PAOZZ	5935008338561	19207	8338561	.SHELL,ELECTRICAL CO LEFT SIDE.................... MARKER LIGHT LEAD UOC:DAA,DAB,DAC,DAD,DAW,DAX,V12,V13, V14,V15,V16,V17,ZAA,ZAB,ZAC,ZAD	2
13	PAOZZ	9905007524649	81349	M43436/1-1	.BAND,MARKER LEFT SIDE MARKER................... LIGHT LEAD...................	2

(1) ITEM NO	(2) SMR CODE	(3) NSN	(4) CAGEC	(5) PART NUMBER	(6) DESCRIPTION AND USABLE ON CODES (UOC)	(7) QTY
14	PAOZA	5935009006281	96906	MS27147-1	UOC:DAA,DAB,DAC,DAD,DAW,DAX,V12,V13, V14,V15,V16,V17,ZAA,ZAB,ZAC,ZAD ADAPTER,CONNECTOR	1
15	PFOZZ	6150011310148	19207	12277004-2	UOC:DAA,DAB,DAC,DAD,DAW,DAX,V12,V13, V14,V15,V16,V17,ZAA,ZAB,ZAC,ZAD LEAD ASSEMBLY,ELECT	1
16	PAOZZ	1015007982997	19207	7982997	UOC:DAA,DAB,DAC,DAD,DAW,DAX,V12,V13, V14,V15,V16,V17,ZAA,ZAB,ZAC,ZAD .TERMINAL,SOLDERED F RIGHT SIDE .... MARKER LIGHT	2
17	PAOZZ	5970008338562	19207	8338562	UOC:DAA,DAB,DAC,DAD,DAW,DAX,V12,V13, V14,V15,V16,V17,ZAA,ZAB,ZAC,ZAD .INSULATOR,BUSHING RIGHT SIDE .... MARKER LIGHT LEAD	2
18	PAOZZ	5935008338561	19207	8338561	UOC:DAA,DAB,DAC,DAD,DAW,DAX,V12,V13, V14,V15,V16,V17,ZAA,ZAB,ZAC,ZAD .SHELL,ELECTRICAL CO RIGHT SIDE................... MARKER LIGHT LEAD	2
19	PAOZZ	9905007524649	81349	M43436/1-1	UOC:DAA,DAB,DAC,DAD,DAW,DAX,V12,V13, V14,V15,V16,V17,ZAA,ZAB,ZAC,ZAD .BAND,MARKER RIGHT SIDE MARKER.................. LIGHT LEAD	2
20	PAOZZ	5310009591488	96906	MS51922-21	UOC:DAA,DAB,DAC,DAD,DAW,DAX,V12,V13, V14,V15,V16,V17,ZAA,ZAB,ZAC,ZAD NUT,SELF-LOCKING,HE	2
21	PAOZZ	5340011391002	19207	7539108	UOC:DAA,DAB,DAC,DAD,DAW,DAX,V12,V13, V14,V15,V16,V17,ZAA,ZAB,ZAC,ZAD CLAMP,LOOP UOC:DAA,DAB,DAC,DAD,DAW,DAX,V12,V13, V14,V15,V16,V17,ZAA,ZAB,ZAC,ZAD	22

END OF FIGURE

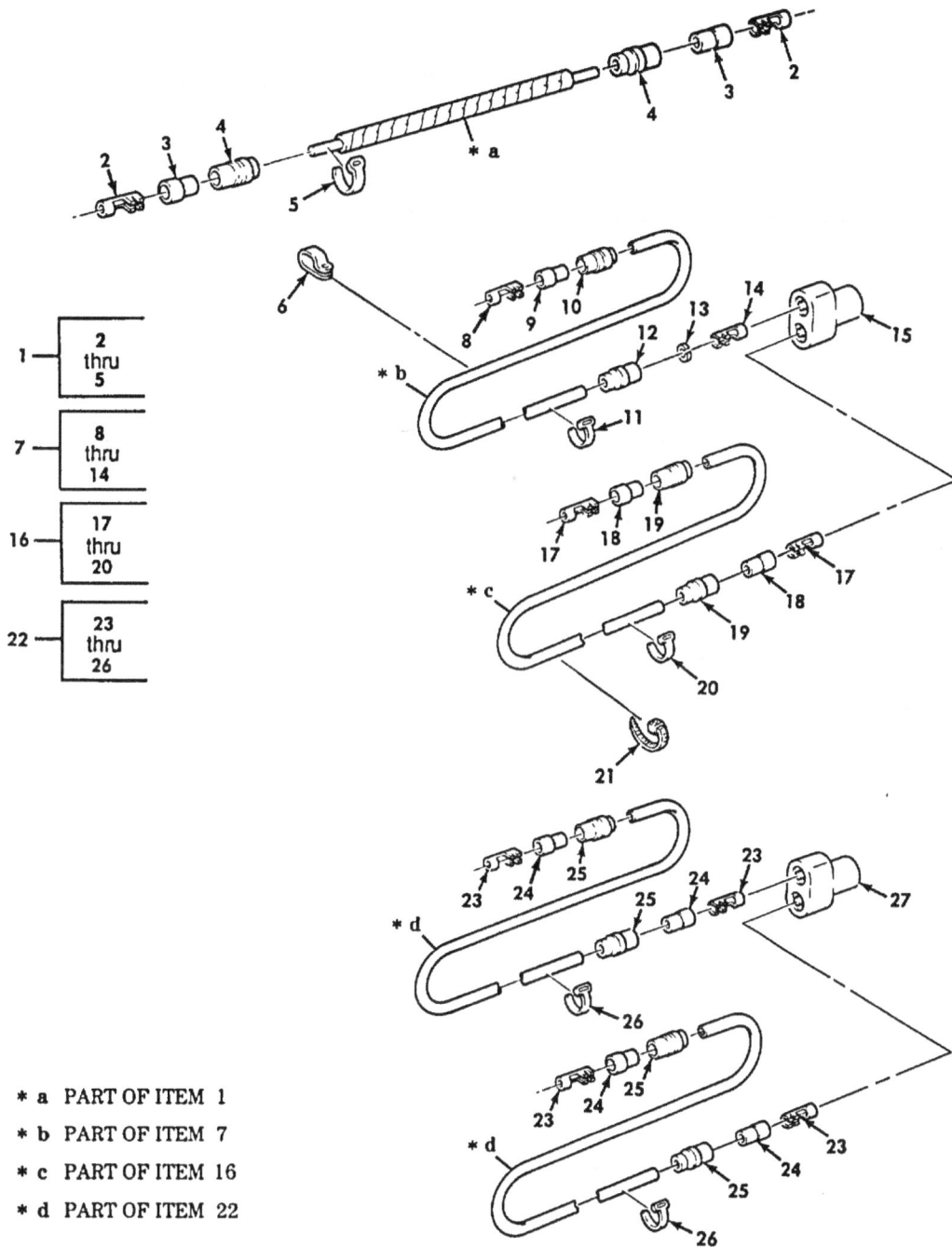

Figure 173. Side Marker Light Leads.

* a PART OF ITEM 1
* b PART OF ITEM 7
* c PART OF ITEM 16
* d PART OF ITEM 22

(1) ITEM NO	(2) SMR CODE	(3) NSN	(4) CAGEC	(5) PART NUMBER	(6) DESCRIPTION AND USABLE ON CODES (UOC)	(7) QTY
					GROUP 0613 CHASSIS WIRING HARNESS	
					FIG. 173 SIDE MARKER LIGHT LEADS	
1	PAOZZ	6150010958309	19207	12277004-5	CABLE ASSEMBLY,SPEC................................ UOC:DAL,V18,ZAL	2
2	PAOZZ	1015007982997	19207	7982997	.TERMINAL,SOLDERED F SIDE MARKER.............. LIGHT CABLE.. UOC:DAL,V18,ZAL	2
3	PAOZZ	5970008338562	19207	8338562	.INSULATOR,BUSHING SIDE MARKER................. LIGHT CABLE.. UOC:DAL,V18,ZAL	2
4	PAOZZ	5935008338561	19207	8338561	.SHELL,ELECTRICAL CO SIDE MARKER............... LIGHT CABLE.. UOC:DAL,V18,ZAL	2
5	PAOZZ	9905007524649	81349	M43436/1-1	.BAND,MARKER SIDE MARKER LIGHT.................. CABLE.. UOC:DAL,V18,ZAL	2
6	PAOZZ	5340000673868	96906	MS21333-109	CLAMP,LOOP.. UOC:DAL,V18,ZAL	2
7	PAOZZ	6150011308042	19207	12277005-2	LEAD,ELECTRICAL... UOC:DAL,V18,ZAL	2
8	PAOZZ	1015007982997	19207	7982997	.TERMINAL,SOLDERED F UOC:DAL,V18,ZAL	1
9	PAOZZ	5970008338562	19207	8338562	.INSULATOR,BUSHING UOC:DAL,V18,ZAL	1
10	PAOZZ	5935008338561	19207	8338561	.SHELL,ELECTRICAL CO................................... UOC:DAL,V18,ZAL	1
11	PAOZZ	9905007524649	81349	M43436/1-1	.BAND,MARKER UOC:DAL,V18,ZAL	2
12	PAOZZ	5935005729180	19207	8338566	.SHELL,ELECTRICAL CO................................... UOC:DAL,V18,ZAL	1
13	PAOZZ	5310008338567	19207	8338567	.WASHER,SLOTTED UOC:DAL,V18,ZAL	1
14	PAOZA	5999000572929	19204	572929	.CONTACT,ELECTRICAL................................... UOC:DAL,V18,ZAL	1
15	PAOZA	5935009006281	96906	MS27147-1	ADAPTER,CONNECTOR UOC:DAL,V18,ZAL	2
16	PFOZZ	6150010958310	19207	12277004-6	LEAD,ELECTRICAL... UOC:DAL,V18,ZAL	2
17	PAOZZ	1015007982997	19207	7982997	.TERMINAL,SOLDERED F UOC:DAL,V18,ZAL	2
18	PAOZZ	5970008338562	19207	8338562	.INSULATOR,BUSHING SIDE MARKER................. LIGHT CABLE.. UOC:DAL,V18,ZAL	2
19	PAOZZ	5935008338561	19207	8338561	.SHELL,ELECTRICAL CO SIDE MARKER............... LIGHT CABLE.. UOC:DAL,V18,ZAL	2
20	PAOZZ	9905007524649	81349	M43436/1-1	.BAND,MARKER SIDE MARKER LIGHT.................. CABLE.. UOC:DAL,V18,ZAL	2
21	PAOZZ	5975001563253	96906	MS3367-2-9	STRAP,TIEDOWN,ELECT...................................	2

(1) ITEM NO	(2) SMR CODE	(3) NSN	(4) CAGEC	(5) PART NUMBER	(6) DESCRIPTION AND USABLE ON CODES (UOC)	(7) QTY
					UOC:DAL,V18,ZAL	
22	PAOZZ	6150011310150	19207	12277004-4	LEAD ASSEMBLY,ELECT	4
					UOC:DAE,DAF,V19,V20,ZAE,ZAF	
23	PAOZZ	1015007982997	19207	7982997	.TERMINAL,SOLDERED F SIDE MARKER...............	2
					LIGHT CABLE..	
					UOC:DAE,DAF,V19,V20,ZAE,ZAF	
24	PAOZZ	5970008338562	19207	8338562	.INSULATOR,BUSHING SIDE MARKER	2
					LIGHT CABLE..	
					UOC:DAE,DAF,V19,V20,ZAE,ZAF	
25	PAOZZ	5935008338561	19207	8338561	.SHELL,ELECTRICAL CO SIDE MARKER...............	2
					LIGHT CABLE..	
					UOC:DAE,DAF,V19,V20,ZAE,ZAF	
26	PAOZZ	9905007524649	81349	M43436/1-1	.BAND,MARKER SIDE MARKER LIGHT..................	2
					CABLE	
					UOC:DAE,DAF,V19,V20,ZAE,ZAF	
27	PAOZA	5935009006281	96906	MS27147-1	ADAPTER,CONNECTOR	2
					UOC:DAE,DAF,V19,V20,ZAE,ZAF	

END OF FIGURE

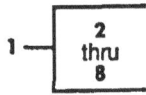

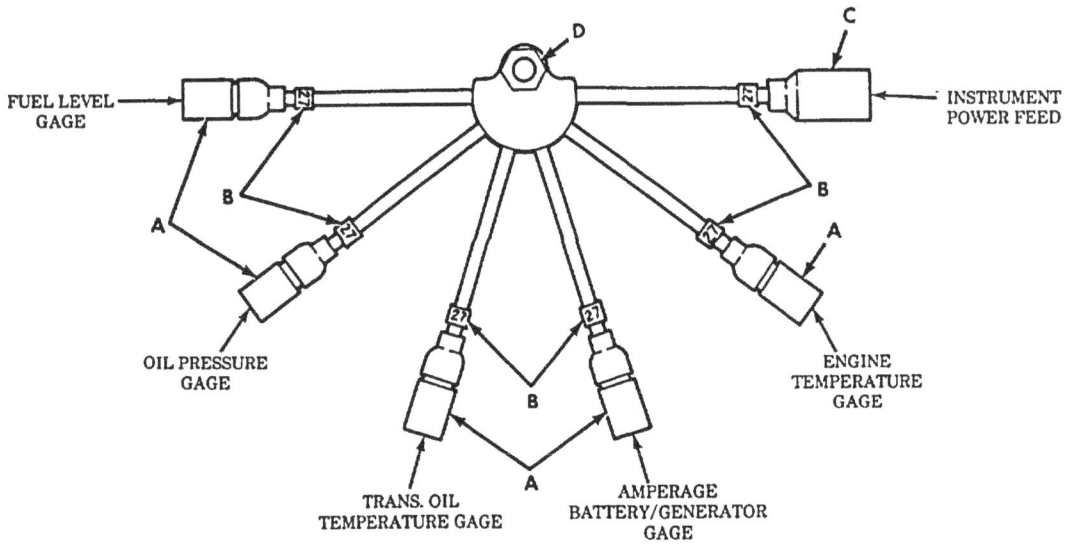

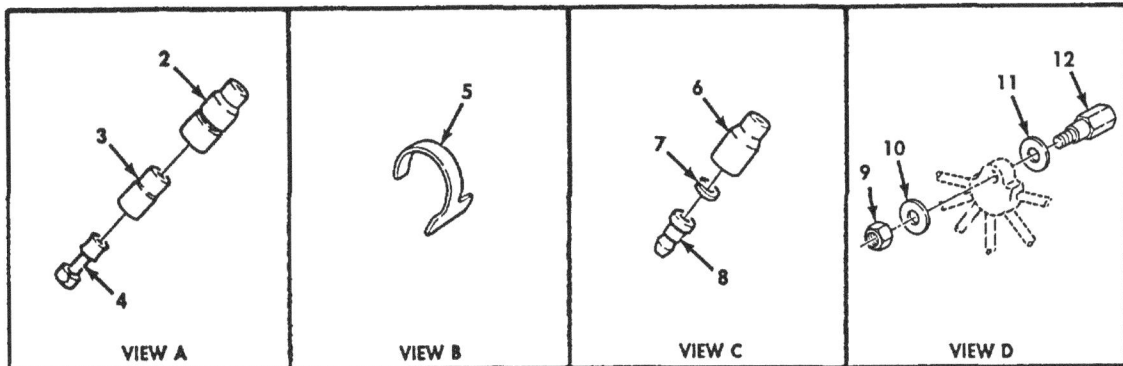

Figure 174. Instrument Cluster Harness.

(1) ITEM NO	(2) SMR CODE	(3) NSN	(4) CAGEC	(5) PART NUMBER	(6) DESCRIPTION AND USABLE ON CODES (UOC)	(7) QTY
					GROUP 0613 CHASSIS WIRING HARNESS	
					FIG. 174 INSTRUMENT CLUSTER HARNESS	
1	PFOZZ	6150010831161	19207	12256007	CABLE ASSEMBLY SET, ...	1
2	PAOZZ	5935008338561	19207	8338561	.SHELL,ELECTRICAL CO.............................	5
3	PAOZZ	5970008338562	19207	8338562	.INSULATOR,BUSHING	5
4	PAOZZ	1015007982997	19207	7982997	.TERMINAL,SOLDERED F CABLE ASSEMBLY	5
5	PAOZZ	9905007524649	81349	M43436/1-1	.BAND,MARKER CABLE ASSEMBLY	6
6	PAOZZ	5935005729180	19207	8338566	.SHELL,ELECTRICAL CO.............................	1
7	PAOZZ	5310008338567	19207	8338567	.WASHER,SLOTTED CABLE ASSEMBLY	1
8	PAOZA	5999000572929	19204	572929	.CONTACT,ELECTRICAL CABLE ASSEMBLY..........	1
9	PAOZZ	5310009349751	96906	MS35650-302	NUT,PLAIN,HEXAGON	1
10	PAOZZ	5310007526593	19207	7526593	WASHER,FLAT...	1
11	PAOZZ		19207	7528592	WASHER,FLAT...	1
12	PAOZZ	5307007526594	19207	7526594	STANDOFF,THREADED S	1

END OF FIGURE

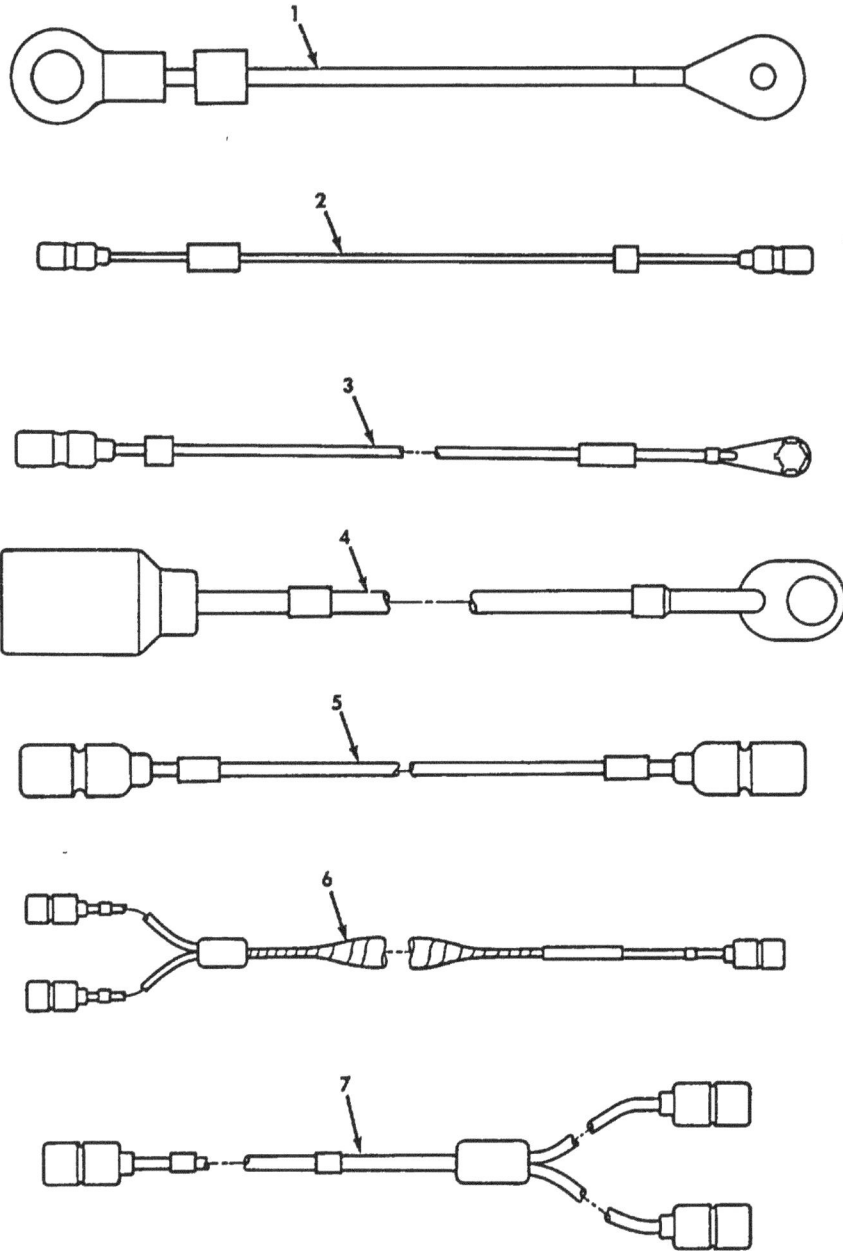

Figure 175. Ground and Jumper Cable Assemblies.

(1) ITEM NO	(2) SMR CODE	(3) NSN	(4) CAGEC	(5) PART NUMBER	(6) DESCRIPTION AND USABLE ON CODES (UOC)	(7) QTY
					GROUP 0613 CHASSIS WIRING HARNESS	
					FIG.175 GROUND AND JUMPER CABLE ASSEMBLIES	
1	PFOZZ	6150010831153	19207	12276953-2	CABLE ASSEMBLY,SPEC FUEL TANK................... GROUND CABLE .... UOC:DAE,DAF,DAG,DAH,DAL,V18,V19,V20, V21,V22,ZAE, ZAF, ZAG, ZAH, ZAL	1
1	PFOZZ	6150010829042	19207	12276953-1	LEAD,ELECTRICAL FUEL TANK GROUND............. CABLE .... UOC:DAA,DAB,DAC,DAD,DAJ,DAK,DAW,DAX, V12,V13,V14,V15,V16,V17,V24,V25,ZAA, ZAB,ZAC,ZAD,ZAJ,ZAK	1
2	PAOZZ	6150010829040	19207	12256098	LEAD,ELECTRICAL AIR TANK JUMPER................. CABLE..	1
3	PFOZZ	6150010835479	19207	12256099	LEAD,ELECTRICAL AIR TANK GROUND .... CABLE..	1
4	PFOZZ	6150011016741	19207	10938046-1	CABLE ASSEMBLY,SPEC HEAD LAMP .... GROUND...	1
5	PFOZZ	6150011310141	19207	12256427-3	CABLE ASSEMBLY,SPEC CIRCUIT .... BREAKER TO BATTERY SWITCH	1
6	PFOZZ	6150010823818	19207	7363004-1	CABLE ASSEMBLY,SPEC AIR TANK JUMPER........ UOC:V39	1
7	PFOZZ	6150011308102	19207	7363005-1	CABLE ASSEMBLY,SPEC FLOODLIGHT................ SWITCH LEAD.. UOC:V39	1

END OF FIGURE

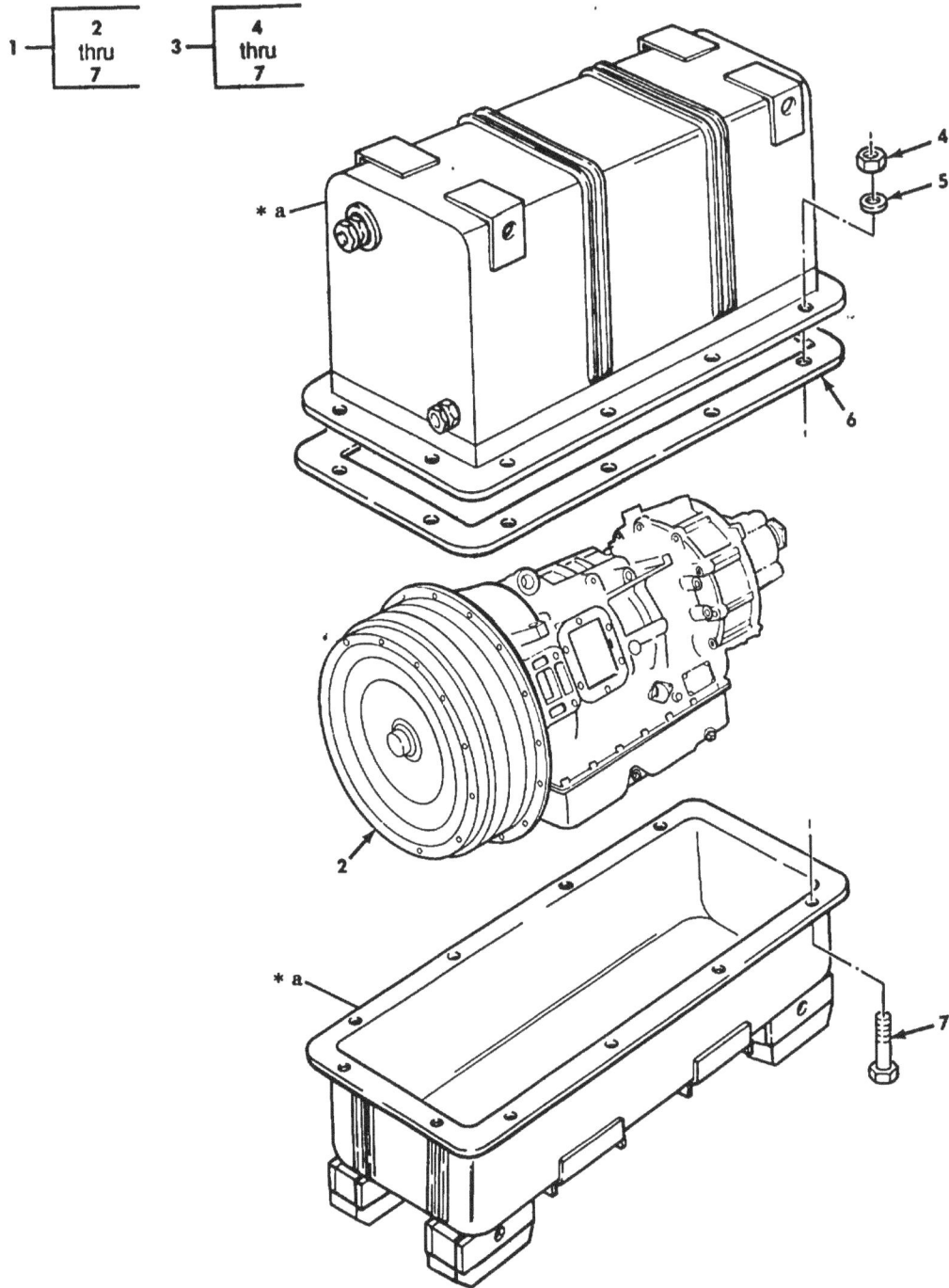

* a PART OF ITEM 3

Figure 176. Transmission Assembly and Shipping Container.

(1) ITEM NO	(2) SMR CODE	(3) NSN	(4) CAGEC	(5) PART NUMBER	(6) DESCRIPTION AND USABLE ON CODES (UOC)	(7) QTY
					GROUP 07 TRANSMISSION 0700 TRANSMISSION ASSEMBLY	
					FIG. 176 TRANSMISSION ASSEMBLY AND SHIPPING CONTAINER	
1	PAFHH	2520011173010	19207	5704512	TRANSMISSION WITH C	1
2	XAFHH		19207	11669085	.TRANSMISSION,MECHAN...................................	1
3	PFFFF	8145011174978	19207	12277382	.SHIPPING AND STORAG....................................	1
4	PAFZZ	5310007680318	96906	MS51967-14	..NUT,PLAIN,HEXAGON.......................................	20
5	PAFZZ	5310008093079	96906	MS27183-19	..WASHER,FLAT...	20
6	PAFZZ	5330011312066	19207	12302613	..GASKET...	1
7	PAFZZ	5305000712069	80204	B1821BH050C150N	..SCREW,CAP,HEXAGON H................................	20

END OF FIGURE

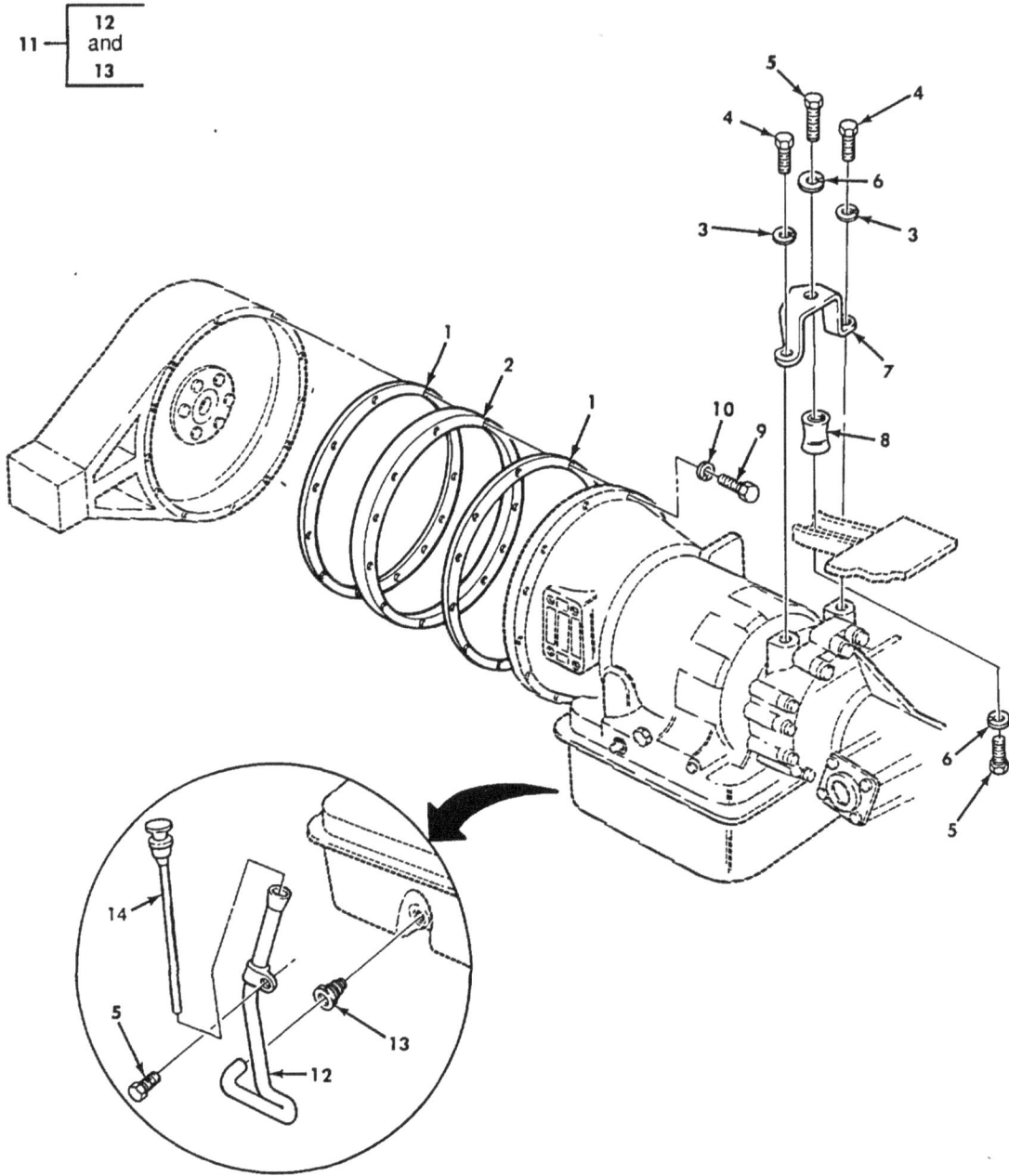

Figure 177. Transmission Mounting Hardware, Fluid Dipstick, and Tube Assembly (M939, M939A1).

(1) ITEM NO	(2) SMR CODE	(3) NSN	(4) CAGEC	(5) PART NUMBER	(6) DESCRIPTION AND USABLE ON CODES (UOC)	(7) QTY
					GROUP 0700 TRANSMISSION ASSEMBLY	
					FIG. 177 TRANSMISSION MOUNTING HARDWARE, FLUID DIPSTICK, AND TUBE ASSEMBLY(M939,M939A1)	
1	PAFZZ	5330007376584	19207	7376584	GASKET... UOC:DAA,DAB,DAC,DAD,DAE,DAF,DAG,DAH, DAJ,DAK,DAL,DAW,DAX,V12,V13,V14,V15, V16,V17,V18,V19,V20,V21,V22,V24,V25, V39	2
2	PAFZZ	5365010667554	19207	12256997	SPACER,RING..................................... UOC:DAA,DAB,DAC,DAD,DAE,DAF,DAG,DAH, DAJ,DAK,DAL,DAW,DAX,V12,V13,V14,V15, V16,V17,V18,V19,V20,V21,V22,V24,V25, V39	1
3	PAFZZ	5310008206653	80045	23MS35338-50	WASHER,LOCK................................... UOC:DAA,DAB,DAC,DAD,DAE,DAF,DAG,DAH, DAJ,DAK,DAL,DAW,DAX,V12,V13,V14,V15, V16,V17,V18 ,V19,V20,V21,V22,V24,V25, V39	2
4	PAFZZ	5305007247221	80204	B1821BH063C175N	SCREW,CAP,HEXAGON H..................... UOC:DAA,DAB,DAC,DAD,DAE,DAF,DAG,DAH, DAJ,DAK,DAL,DAW,DAX,V12 ,V13 ,V14 ,V15, V16,V17,V18,V19,V20,V21,V22,V24,V25, V39	2
5	PAFZZ	5305007320512	80204	B1821BH050C075N	SCREW,CAP,HEXAGON H..................... UOC:DAA,DAB,DAC,DAD,DAE,DAF,DAG,DAH, DAJ,DAK,DAL,DAW,DAX,V12,V13,V14,V15, V16,V17,V18 ,V19,V20,V21,V22,V24,V25, V39	3
6	PAFZZ	5310000034094	01276	210104-8S	WASHER,LOCK................................... UOC:DAA,DAB,DAC,DAD,DAE,DAF,DAG,DAH, DAJ,DAK,DAL,DAW,DAX,V12 ,V13,V14,V15, V16,V17,V18,V19,V20,V21,V22,V24,V25, V39	2
7	PAFZZ	5340010835443	19207	12255736	BRACKET,DOUBLE ANGL...................... UOC:DAA,DAB,DAC,DAD,DAE,DAF,DAG,DAH, DAJ,DAK,DAL,DAW,DAX,V12,V13,V14,V15, V16,V17,V18,V19,V20,V21,V22,V24,V25, V39	1
8	PAFZZ	5342011267910	19207	11669174	MOUNT,RESILIENT.............................. UOC:DAA,DAB,DAC,DAD,DAE,DAF,DAG,DAH, DAJ,DAK,DAL,DAW,DAX,V12,V13,V14,V15, V16,V17,V18,V19,V20,V21,V22,V24,V25, V39	1
9	PAFZZ	5305002693217	96906	MS90725-67	SCREW,CAP,HEXAGON H..................... UOC:DAA,DAB,DAC,DAD,DAE,DAF,DAG,DAH, DAJ,DAK,DAL,DAW,DAX,V12,V13,V14,V15, V16,V17,V18,V19,V20,V21,V22,V24,V25, V39	12
10	PAFZZ	5310000045033	88044	AN935-616	WASHER,LOCK...................................	12

(1) ITEM NO	(2) SMR CODE	(3) NSN	(4) CAGEC	(5) PART NUMBER	(6) DESCRIPTION AND USABLE ON CODES (UOC)	(7) QTY
					UOC:DAA,DAB,DAC,DAD,DAE,DAF,DAG,DAH, DAJ,DAK,DAL,DAW,DAX,V12,V13,V14,V15, V16,V17,V18,V19,V20,V21,V22,V24,V25, V39	
11	PAOZZ	4710011049099	19207	12256018	TUBE ASSEMBLY,METAL.................................... UOC:DAA,DAB,DAC,DAD,DAE,DAF,DAG,DAH, DAJ,DAK,DAL,DAW,DAX,V12,V13,V14,V15, V16,V17,V18,V19,V20,V21,V22,V24,V25, V39	1
12	PAOZZ	4710011147761	19207	12256018-1	.TUBE ASSEMBLY,METAL.................................... UOC:DAA,DAB,DAC,DAD,DAE,DAF,DAG,DAH, DAJ,DAK,DAL,DAW,DAX,V12,V13,V14,V15, V16,V17,V18,V19,V20,V21,V22,V24,V25, V39	1
13	PAOZZ	4730000190236	96906	MS51874-10	.INVERTED NUT,TUBE C.................................... UOC:DAA,DAB,DAC,DAD,DAE,DAF,DAG,DAH, DAJ,DAK,DAL,DAW,DAX,V12,V13,V14,V15, V16,V17,V18,V19,V20,V21,V22,V24,V25, V39	1
14	PAOZZ	6680012962758	19207	11669173-1	GAGE ROD-CAP,LIQUID.................................... UOC:DAA,DAB,DAC,DAD,DAE,DAF,DAG,DAH, DAJ,DAK,DAL,DAW,DAX,V12,V13,V14,V15, V16,V17,V18,V19,V20,V21,V22,V24,V25, V39	1

END OF FIGURE

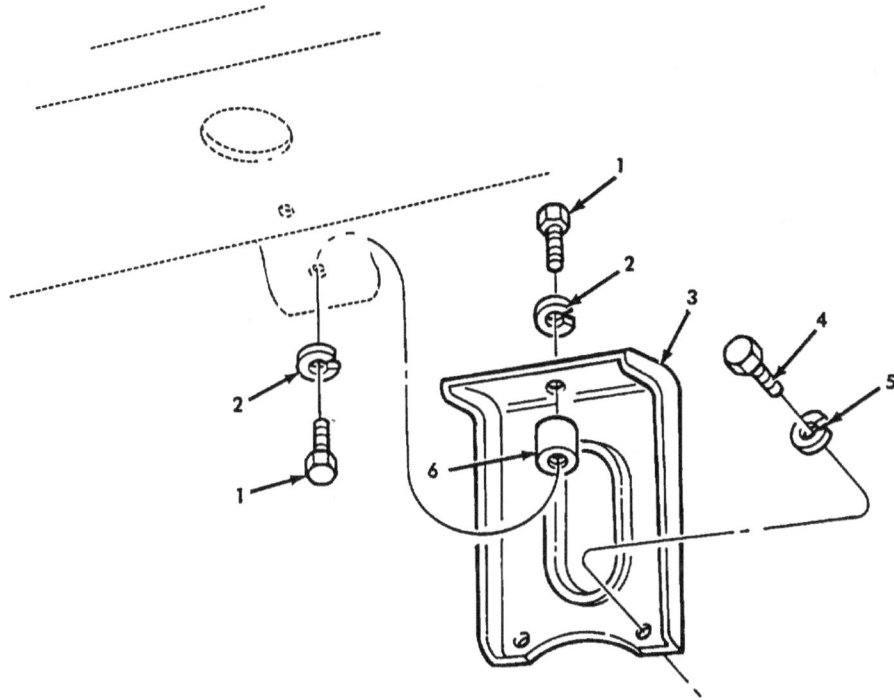

Figure 178. Transmission Mounting Bracket (M939A2).

(1) ITEM NO	(2) SMR CODE	(3) NSN	(4) CAGEC	(5) PART NUMBER	(6) DESCRIPTION AND USABLE ON CODES (UOC)	(7) QTY
					GROUP 0700 TRANSMISSION ASSEMBLY	
					FIG. 178 TRANSMISSION MOUNTING BRACKET (M939A2)	
1	PAOZZ	5305007320512	80204	B1821BH050C075N	SCREW,CAP,HEXAGON H..................................... UOC:ZAA,ZAB,ZAC,ZAD,ZAE,ZAF,ZAG,ZAH, ZAJ,ZAK,ZAL	2
2	PAOZZ	5310000034094	01276	210104-8S	WASHER,LOCK.................................... UOC:ZAA,ZAB,ZAC,ZAD,ZAE,ZAF,ZAG,ZAH, ZAJ,ZAK,ZAL	2
3	PGOZZ	5340012849652	47457	20510302	BRACKET,MOUNTING............................ UOC:ZAA,ZAB,ZAC,ZAD,ZAE,ZAF,ZAG,ZAH, ZAJ,ZAK,ZAL	1
4	PAOZZ	5305007247221	80204	B1821BH063C175N	SCREW,CAP,HEXAGON H..................................... UOC:ZAA,ZAB,ZAC,ZAD,ZAE,ZAF,ZAG,ZAH, ZAJ,ZAK,ZAL	2
5	PAOZZ	5310008206653	96906	MS35338-50	WASHER,LOCK.................................... UOC:ZAA,ZAB,ZAC,ZAD,ZAE,ZAF,ZAG,ZAH, ZAJ,ZAK,ZAL	2
6	PAOZZ	5342011267910	19207	11669174	MOUNT,RESILIENT............................... UOC:ZAA,ZAB,ZAC,ZAD,ZAE,ZAF,ZAG,ZAH, ZAJ,ZAK,ZAL	1

END OF FIGURE

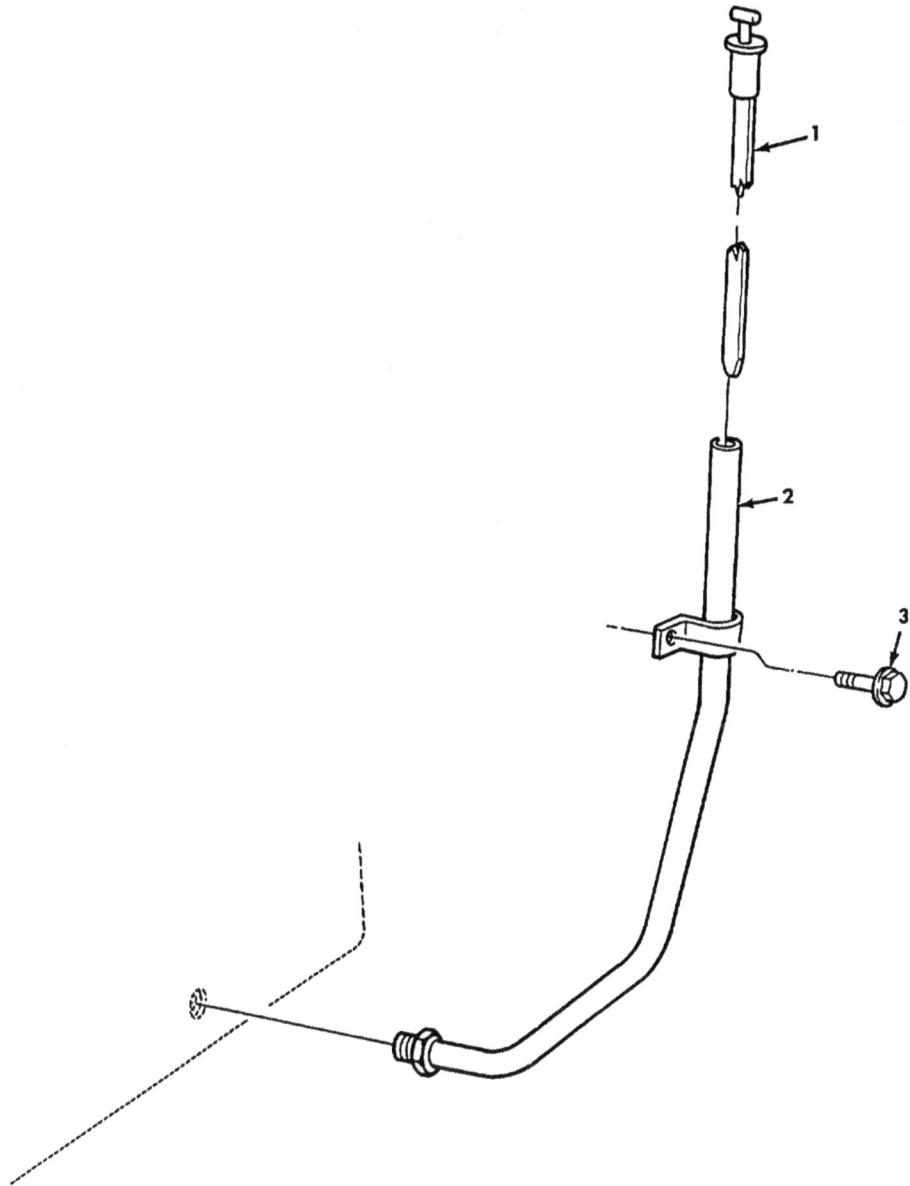

Figure 179. Transmission Dipstick Tube (M939A2).

(1) ITEM NO	(2) SMR CODE	(3) NSN	(4) CAGEC	(5) PART NUMBER	(6) DESCRIPTION AND USABLE ON CODES (UOC)	(7) QTY
					GROUP 0700 TRANSMISSION ASSEMBLY	
					FIG. 179 TRANSMISSION DIPSTICK TUBE (M939A2)	
1	PAOZZ	6680012872153	0U276	42124	GAGE ROD-BREATHER, L .... UOC:ZAA, ZAB, ZAC, ZAD, ZAE, ZAF, ZAG, ZAH, ZAJ, ZAK, ZAL	1
2	PAOZZ	2590012856261	4F744	20510396	FILLER NECK .... UOC:ZAA, ZAB, ZAC, ZAD, ZAE, ZAF, ZAG, ZAH, ZAJ, ZAK, ZAL	1
3	PAOZZ	5305011922036	15434	3912072	SCREW .... UOC:ZAA, ZAB, ZAC, ZAD, ZAE, ZAF, ZAG, ZAH, ZAJ, ZAK, ZAL	1

END OF FIGURE

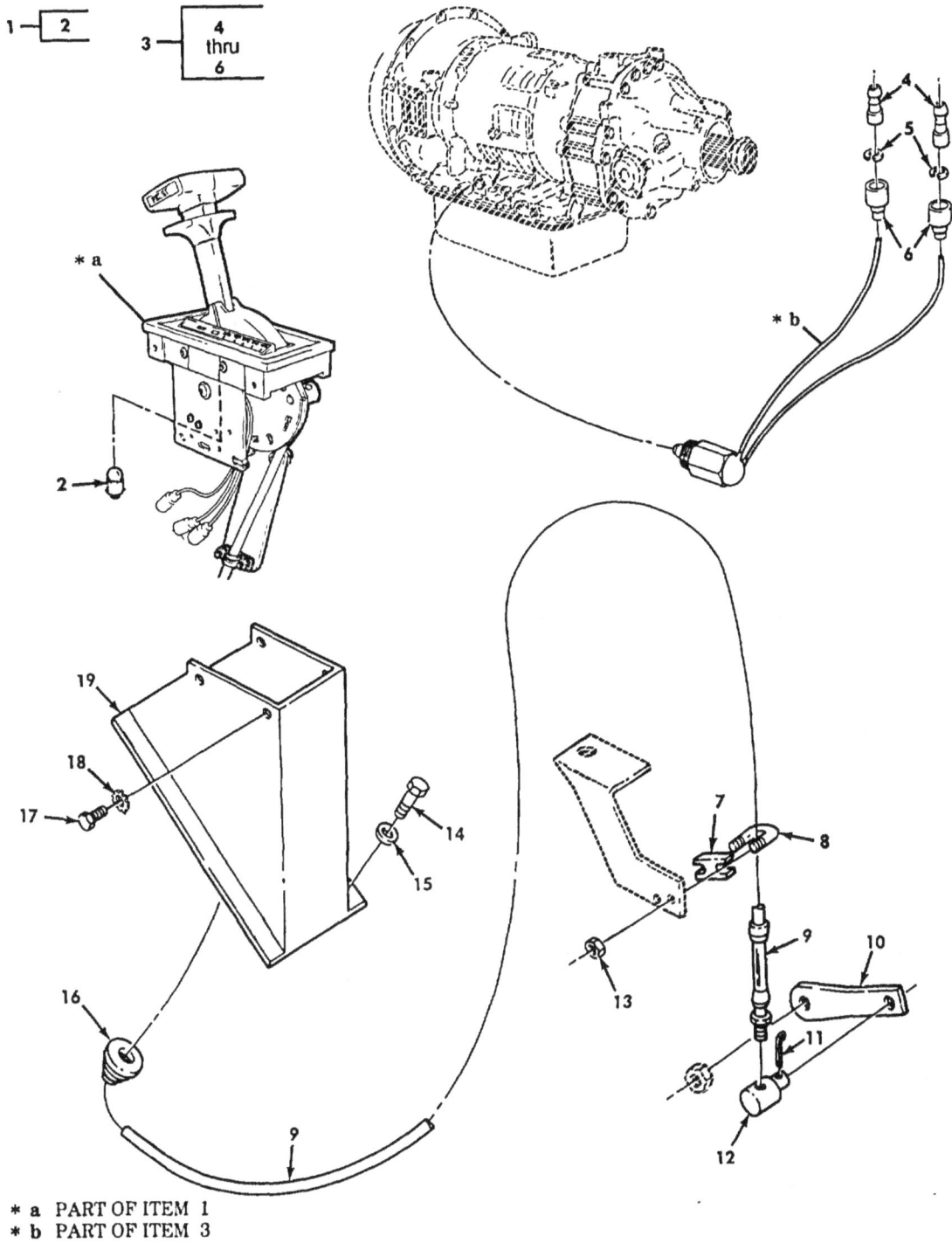

* a   PART OF ITEM 1
* b   PART OF ITEM 3

Figure 180. Transmission Shift Quadrant Control Cable and Neutral Safety Switch (M939, M939A1).

(1) ITEM NO	(2) SMR CODE	(3) NSN	(4) CAGEC	(5) PART NUMBER	(6) DESCRIPTION AND USABLE ON CODES (UOC)	(7) QTY
					GROUP 0705 GEAR SHIFT AND CONTROLS	
					FIG. 180 TRANSMISSION SHIFT QUADRANT CONTROL CABLE AND NEUTRAL SAFETY SWITCH(M939,M939A1)	
1	PBFZZ	2520010907633	60602	55303-2342	TRANSMISSION CONTRO .... UOC:DAA, DAB, DAC, DAD, DAE, DAF, DAG, DAH, DAJ, DAK, DAL, DAW, DAX, V12, V13, V14, V15, V16, V17, V18, V19, V20, V21, V22, V24, V25, V39	1
2	PAOZZ	6240001557866	96906	MS15573-3	.LAMP, INCANDESCENT .... UOC:DAA, DAB, DAC, DAD, DAE, DAF, DAG, DAH, DAJ, DAK, DAL, DAW, DAX, V12, V13, V14, V15, V16, V17, V18, V19, V20, V21, V22, V24, V25, V39	1
3	PAOOO	2920011147538	13445	92113-06	SWITCH, SAFETY, NEUTR TRANS CONTROL .... UOC:DAA, DAB, DAC, DAD, DAE, DAF, DAG, DAH, DAJ, DAK, DAL, DAW, DAX, V12, V13, V14, V15, V16, V17, V18, V19, V20, V21, V22, V24, V25, V39	1
4	PAOZA	5999009263144	96906	MS27148-3	.CONTACT, ELECTRICAL .... UOC:DAA, DAB, DAC, DAD, DAE, DAF, DAG, DAH, DAJ, DAK, DAL, DAW, DAX, V12, V13, V14, V15, V16, V17, V18, V19, V20, V21, V22, V24, V25, V39	2
5	PAOZZ	5310006560067	19207	8724497	.WASHER, SLOTTED .... UOC:DAA, DAB, DAC, DAD, DAE, DAF, DAG, DAH, DAJ, DAK, DAL, DAW, DAX, V12, V13, V14, V15, V16, V17, V18, V19, VV20, V21, V22, V24, V25, V39	2
6	PAOZZ	5935006915591	19207	8724495	.SHELL, ELECTRICAL CO .... UOC:DAA, DAB, DAC, DAD, DAE, DAF, DAG, DAH, DAJ, DAK, DAL, DAW, DAX, V12, V13, V14, V15, V16, V17, V18, V19, V20, V21, V22, V24, V25, V39	2
7	PAFZZ	5340010909331	19207	12256023	STRAP, RETAINING..................... UOC:DAA, DAB, DAC, DAD, DAE, DAF, DAG, DAH, DAJ, DAK, DAL, DAW, DAX, V12, V13, V14, V15, V16, V17, V18, V19, V20, V21, V22, V24, V25, V39	1
8	PAFZZ	5306011126560	19207	12255984	BOLT, U .... UOC:DAA, DAB, DAC, DAD, DAE, DAF, DAG, DAH, DAJ, DAK, DAL, DAW, DAX, V12, V13, V14, V15, V16, V17, V18, V19, V20, V21, , V22, V24, V25, V39	1
9	PAFZZ	3040012085897	60602	4333-60	CONTROL ASSEMBLY, PU .... UOC:DAA, DAB, DAC, DAD, DAE, DAF, DAG, DAH, DAJ, DAK, DAL, DAW, DAX, V12, V13, V14, V15, V16, V17, V18, V19, V20, V21, V22, V24, V25, V39	1
10	PAFZZ	3040011049151	19207	11669346	LEVER, REMOTE CONTRO	1

(1) ITEM NO	(2) SMR CODE	(3) NSN	(4) CAGEC	(5) PART NUMBER	(6) DESCRIPTION AND USABLE ON CODES (UOC)	(7) QTY
					UOC:DAA, DAB, DAC, DAD, DAE, DAF, DAG, DAH, DAJ, DAK, DAL, DAW, DAX, V12, V13, V14, V15, V16, V17, V18, V19, V20, V21, V22, V24, V25, V39	
11	PAFZZ	5315002341863	96906	MS24665-300	PIN, COTTER	1
					UOC:DAA, DAB, DAC, DAD, DAE, DAF, DAG, DAH, DAJ, DAK, DAL, DAW, DAX, V12, V13, V14, V15, V16, V17, V18, V19, V20, V21, V22, V24, V25, V39	
12	PAFZZ	5315011448675	19207	11669345	PIN, SHOULDER, HEADLE	1
					UOC:DAA, DAB, DAC, DAD, DAE, DAF, DAG, DAH, DAJ, DAK, DAL, DAW, DAX, V12, V13, V14, V15, V16, V17, V18, V19, V20, V21, V22, V24, V25, V39	
13	PAFZZ	5310008775797	96906	MS21044-N3	NUT, SELF-LOCKING, HE	2
					UOC:DAA, DAB, DAC, DAD, DAE, DAF, DAG, DAH, DAJ, DAK, DAL, DAW, DAX, V12, V13, V14, V15, V16, V17, V18, V19, V20, V21, V22, V24, V25, V39	
14	PAFZZ	5306011666895	21450	425736	BOLT, ASSEMBLED WASH	4
					UOC:DAA, DAB, DAC, DAD, DAE, DAF, DAG, DAH, DAJ, DAK, DAL, DAW, DAX, V12, V13, V14, V15, V16, V17, V18, V19, V20, V21, V22, V24, V25, V39	
15	PAFZZ	5310000806004	96906	MS27183-14	WASHER, FLAT..	4
					UOC:DAA, DAB, DAC, DAD, DAE, DAF, DAG, DAH, DAJ, DAK, DAL, DAW, DAX, V12, V13, V13, V14, V15, V16, V17, V18, V19, V20, V21, V22, V24, V25, V39	
16	PAFZZ	5325011064125	19207	12256578-5	GROMMET, NONMETALLIC BRACKET.................. ASSEMBLY .... UOC:DAA, DAB, DAC, DAD, DAE, DAF, DAG, DAH, DAJ, DAK, DAL, DAW, DAX, V12, V13, V14, V15, V16, V17, V18, V19, V20, V21, V22, V24, V25, V39	1
17	PAFZZ	5305000680500	96906	MS90725-3	SCREW, CAP, HEXAGON H	4
					UOC:DAA, DAB, DAC, DAD, DAE, DAF, DAG, DAH, DAJ, DAK, DAL, DAW, DAX, V12, V13, V14, V15, V16, V17, V18, V19, V20, V21, V22, V24, V25, V39	
18	PAFZZ	5310002090786	96906	MS35335-33	WASHER, LOCK	4
					UOC:DAA, DAB, DAC, DAD, DAE, DAF, DAG, DAH, DAJ, DAK, DAL, DAW, DAX, V12, V13, V14, V15, V16, V17, V18, V19, V20, V21, V22, V24, V25, V39	
19	PFFZZ	5340011037587	19207	12255944	BRACKET, MOUNTING...............................	1
					UOC:DAA, DAB, DAC, DAD, DAE, DAF, DAG, DAH, DAJ, DAK, DAL, DAW, DAX, V12, V13, V14, V15, V16, V17, V18, V19, V20, V21, V22, V24, V25, V39	

END OF FIGURE

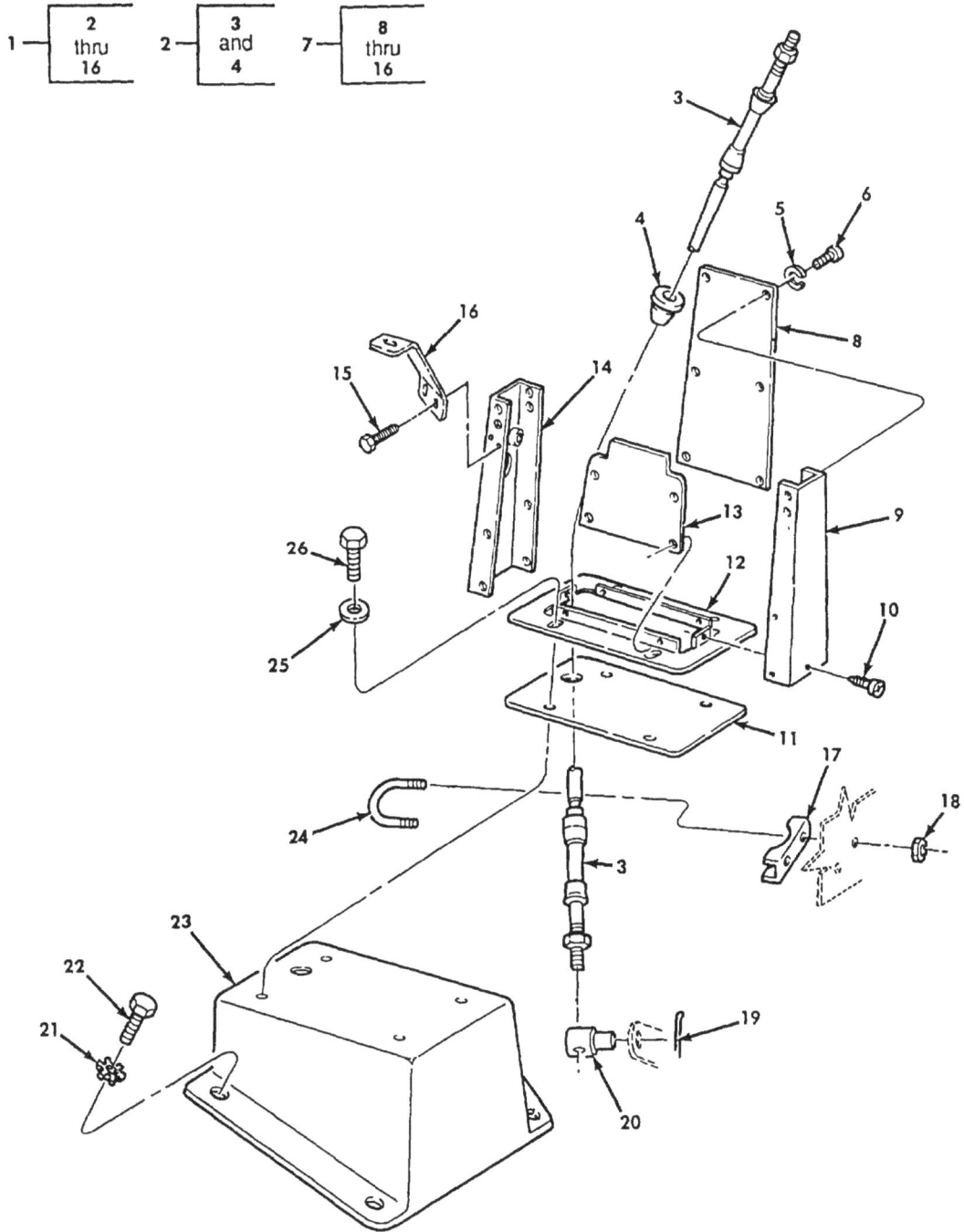

Figure 181. Transmission Shift Control Cable and Related Parts (M939A2).

(1) ITEM NO	(2) SMR CODE	(3) NSN	(4) CAGEC	(5) PART NUMBER	(6) DESCRIPTION AND USABLE ON CODES (UOC)	(7) QTY
					GROUP 0705 GEAR SHIFT AND CONTROLS	
					FIG. 181 TRANSMISSION SHIFT CONTROL CABLE AND RELATED PARTS(M939A2)	
1	PAFZZ	5340012849246	60602	59188	BRACKET, SPECIAL TRANSMISSION SHIFT.......... CONTROLS .... UOC:ZAA, ZAB, ZAC, ZAD, ZAE, ZAF, ZAG, ZAH, ZAJ, ZAK, ZAL	1
2	PAFZZ	2520012865650	60602	45737-40	.CONTROL ASSEMBLY, GA TRANS SHIFT............ CONTROL .... UOC:ZAA, ZAB, ZAC, ZAD, ZAE, ZAF, ZAG, ZAH, ZAJ, ZAK, ZAL	1
3	PAFZZ		19207	12363371-1	..CABLE ASSEMBLY TRANSMISSION................... CONTROL .... UOC:ZAA, ZAB, ZAC, ZAD, ZAE, ZAF, ZAG, ZAH, ZAJ, ZAK, ZAL	1
4	PAFZZ		19207	12432431	..GROMMET, TRANSMISSION .... UOC:ZAA, ZAB, ZAC, ZAD, ZAE, ZAF, ZAG, ZAH, ZAJ, ZAK, ZAL	1
5	PAFZZ	5310005825965	96906	MS35338-44	.WASHER, LOCK .... UOC:ZAA, ZAB, ZAC, ZAD, ZAE, ZAF, ZAG, ZAH, ZAJ, ZAK, ZAL	4
6	PAFZZ	5305009881723	96906	MS35206-279	.SCREW, MACHINE .... UOC:ZAA, ZAB, ZAC, ZAD, ZAE, ZAF, ZAG, ZAH, ZAJ, ZAK, ZAL	4
7	PAFZZ	5340012849247	60602	59189	.BRACKET, MOUNTING CONTROL TOWER .... UOC:ZAA, ZAB, ZAC, ZAD, ZAE, ZAF, ZAG, ZAH, ZAJ, ZAK, ZAL	1
8	PAFZZ	5340012820452	60602	52443	..COVER, ACCESS.. UOC:ZAA, ZAB, ZAC, ZAD, ZAE, ZAF, ZAG, ZAH, ZAJ, ZAK, ZAL	1
9	PAFZZ	5340012849655	60602	52479	..BRACKET, MOUNTING TOWER .... UOC:ZAA, ZAB, ZAC, ZAD, ZAE, ZAF, ZAG, ZAH, ZAJ, ZAK, ZAL	1
10	PAFZZ	5305012718337	24617	9406129	..SCREW, TAPPING, THREA .... UOC:ZAA, ZAB, ZAC, ZAD, ZAE, ZAF, ZAG, ZAH, ZAJ, ZAK, ZAL	12
11	PAFZZ	5330012851601	60602	50161-2	..GASKET.. UOC:ZAA, ZAB, ZAC, ZAD, ZAE, ZAF, ZAG, ZAH, ZAJ, ZAK, ZAL	1
12	PAFZZ	5445012846173	60602	52442	..PLATE, BASE, TOWER SU .... UOC:ZAA, ZAB, ZAC, ZAD, ZAE, ZAF, ZAG, ZAH, ZAJ, ZAK, ZAL	1
13	PAFZZ	5340012843789	60602	52478	..COVER, ACCESS.. UOC:ZAA, ZAB, ZAC, ZAD, ZAE, ZAF, ZAG, ZAH, ZAJ, ZAK, ZAL	1
14	PAFZZ	5340012849654	60602	52456	..BRACKET, MOUNTING TOWER .... UOC:ZAA, ZAB, ZAC, ZAD, ZAE, ZAF, ZAG, ZAH, ZAJ, ZAK, ZAL	1
15	PAFZZ		24617	9427724	..SCREW.. UOC:ZAA, ZAB, ZAC, ZAD, ZAE, ZAF, ZAG, ZAH, ZAJ, ZAK, ZAL	2

(1) ITEM NO	(2) SMR CODE	(3) NSN	(4) CAGEC	(5) PART NUMBER	(6) DESCRIPTION AND USABLE ON CODES (UOC)	(7) QTY
16	PAFZZ		19207	12363442	..BRACKET, SHIFT TOWER SUPPORT .... UOC:ZAA, ZAB, ZAC, ZAD, ZAE, ZAF, ZAG, ZAH, ZAJ, ZAK, ZAL	1
17	PAFZZ	5340012850575	47547	12256023	STRAP, RETAINING .... UOC:ZAA, ZAB, ZAC, ZAD, ZAE, ZAF, ZAG, ZAH, ZAJ, ZAK, ZAL	1
18	PAFZZ	5310000617326	96906	MS21045-3	NUT, SELF-LOCKING, HE.................................... UOC:ZAA, ZAB, ZAC, ZAD, ZAE, ZAF, ZAG, ZAH, ZAJ, ZAK, ZAL	2
19	PAFZZ	5315002341863	96906	MS24665-300	PIN, COTTER .... UOC:ZAA, ZAB, ZAC, ZAD, ZAE, ZAF, ZAG, ZAH, ZAJ, ZAK, ZAL	1
20	PAFZZ	5315011448675	19207	11669345	PIN, SHOULDER, HEADLE .... UOC:ZAA, ZAB, ZAC, ZAD, ZAE, ZAF, ZAG, ZAH, ZAJ, ZAK, ZAL	1
21	PAFZZ	5310002090786	96906	MS35335-33	WASHER, LOCK .... UOC:ZAA, ZAB, ZAC, ZAD, ZAE, ZAF, ZAG, ZAH, ZAJ, ZAK, ZAL	11
22	PAFZZ	5305002678954	80204	B1821BH025F125N	SCREW, CAP, HEXAGON H.................................. UOC: ZAA, ZAB, ZAC, ZAD, ZAE, ZAF, ZAG, ZAH, ZAJ, ZAK, ZAL	7
23	PAFZZ	5340013222906	7A964	20510403	BRACKET, MOUNTING... UOC:ZAA, ZAB, ZAC, ZAD, ZAE, ZAF, ZAG, ZAH, ZAJ, ZAK, ZAL	1
24	PAFZZ	5306011136560	19207	12255984	BOLT, U .... UOC:ZAA, ZAB, ZAC, ZAD, ZAE, ZAF, ZAG, ZAH, ZAJ, ZAK, ZAL	1
25	PAFZZ	5310008094058	96906	MS27183-10	WASHER, FLAT .... UOC:ZAA, ZAB, ZAC, ZAD, ZAE, ZAF, ZAG, ZAH, ZAJ, ZAK, ZAL	7
26	PAFZZ	5305002678953	80204	B1821BH025F063N	SCREW, CAP, HEXAGON H .... UOC:ZAA, ZAB, ZAC, ZAD, ZAE, ZAF, ZAG, ZAH, ZAJ, ZAK, ZAL	4

END OF FIGURE

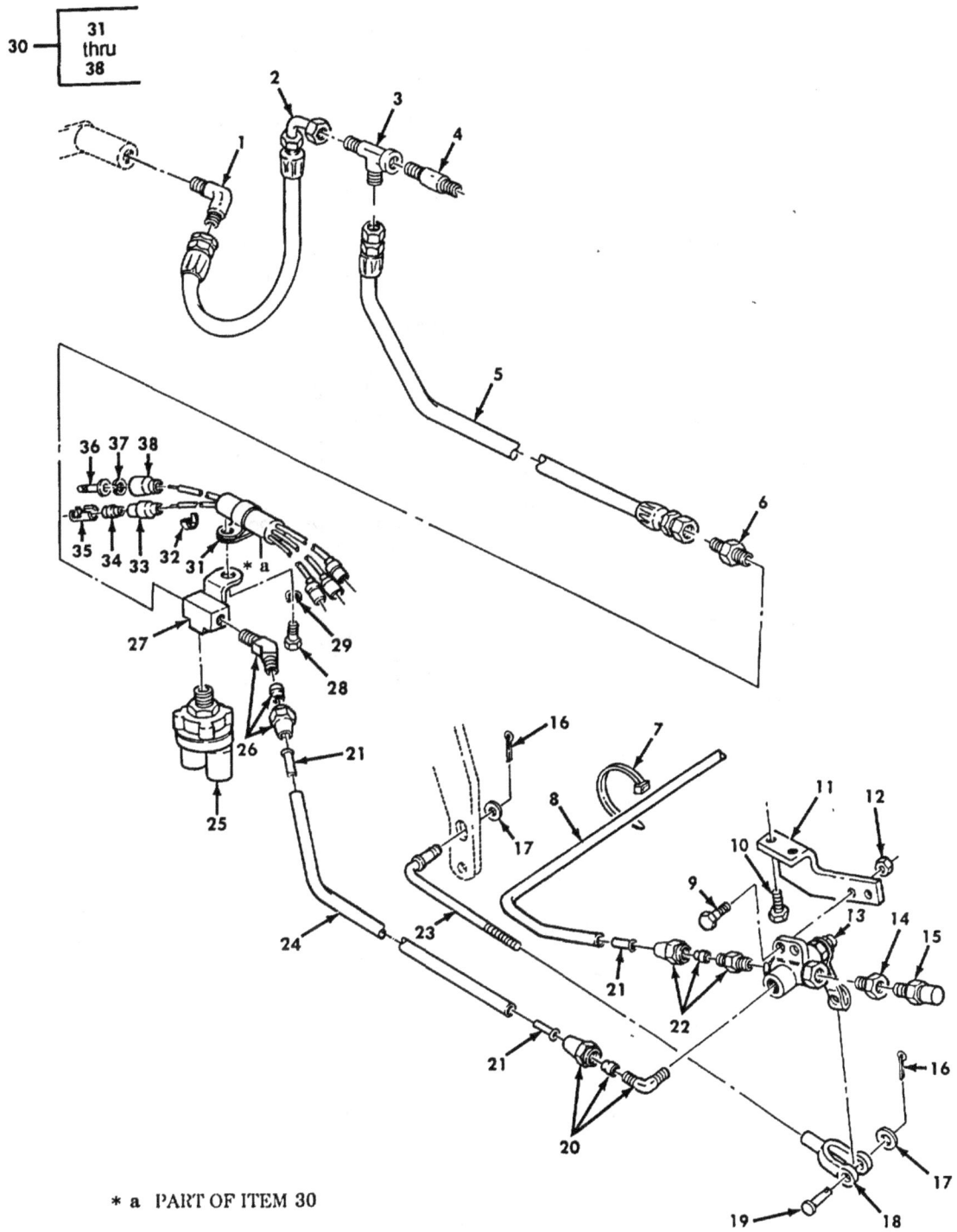

Figure 182. Wrecker Governor Piping and Capacitor Assembly.

* a PART OF ITEM 30

(1) ITEM NO	(2) SMR CODE	(3) NSN	(4) CAGEC	(5) PART NUMBER	(6) DESCRIPTION AND USABLE ON CODES (UOC)	(7) QTY
					GROUP 0705 GEAR SHIFT AND CONTROLS	
					FIG. 182 WRECKER GOVERNOR PIPING AND CAPACITOR ASSEMBLY	
1	PAOZZ	4730008035765	96906	MS51504-A6-2	ELBOW, PIPE TO TUBE .... UOC:DAL, V18, ZAL	1
2	PAOZZ	4720013040688	19207	12302866	HOSE ASSEMBLY, NONME .... UOC:DAL, V18, ZAL	1
3	PAOZZ	4730011464113	96906	MS51514A6	TEE, PIPE TO TUBE .... UOC:DAL, V18, ZAL	1
4	PAOZZ	4730010278943	96906	MS51873-32B	NIPPLE, PIPE .... UOC:DAL, V18, ZAL	1
5	PBOZZ	4720012761252	19207	12302867	HOSE ASSEMBLY, NONME .... UOC:DAL, V18, ZAL	1
6	PAOZZ	4730003228457	96906	MS51500-A6-6	ADAPTER, STRAIGHT, PI .... UOC:DAL, V18, ZAL	1
7	PAOZZ	5975009846582	96906	MS3367-1-0	STRAP, TIEDOWN, ELECT .... UOC:DAL, V18, ZAL	3
8	MOOZZ		19207	CPR104420-2-60	TUBING AIR TANK TO VALVE, MAKE FROM.......... TUBING P/N CPR104420-2 UOC:DAL, V18, ZAL	1
9	PAOZZ	5305000680515	80204	B1821BH025F100N	SCREW, CAP, HEXAGON H .... UOC:DAL, V18, ZAL	2
10	PAOZZ	5305004423217	96906	MS24629-66	SCREW, TAPPING .... UOC:DAL, V18, ZAL	2
11	PAOZZ	5340011675535	19207	12302646	BRACKET, ANGLE..................... UOC:DAL, V18, ZAL	1
12	PAOZZ	5310009359022	96906	MS51943-32	NUT, SELF-LOCKING, HE .... UOC:DAL, V18, ZAL	2
13	PAOZZ	4820008086905	70411	SP2346CM	VALVE, PLUG .... UOC:DAL, V18, ZAL	1
14	PAOZZ	4730001960930	89346	112877	BUSHING, PIPE .... UOC:DAL, V18, ZAL	1
15	PAOZZ	4820007264719	19207	5196397	VALVE, VENT .... UOC:DAL, V18, ZAL	1
16	PAOZZ	5315008392325	96906	MS24665-132	PIN, COTTER .... UOC:DAL, V18, ZAL	2
17	PAOZZ	5310008094058	96906	MS27183-10	WASHER, FLAT .... UOC:DAL, V18, ZAL	2
18	PAOZZ	5340008659496	96906	MS35812-2	CLEVIS, ROD END .... UOC:DAL, V18, ZAL	1
19	PAOZZ	5315000817874	96906	MS20392-3C23	PIN, STRAIGHT, HEADED .... UOC:DAL, V18, ZAL	1
20	PAOZZ	4730000691187	81343	6-4 100202BA	ELBOW, PIPE TO TUBE .... UOC:DAL, V18, ZAL	1
21	PAOZZ	4730010798821	19207	CPR102321-1	INSERT, TUBE FITTING .... UOC:DAL, V18, ZAL	6
22	PAOZZ	4730000691186	81343	6-4 120102BA	ADAPTER, STRAIGHT, PI .... UOC:DAL, V18, ZAL	1
23	PFOZZ	5315011730397	19207	12302647	PIN, STRAIGHT, HEADED .... UOC:DAL, V18, ZAL	1

(1) ITEM NO	(2) SMR CODE	(3) NSN	(4) CAGEC	(5) PART NUMBER	(6) DESCRIPTION AND USABLE ON CODES (UOC)	(7) QTY
24	MOOZZ		19207	CPR104420-2-22	TUBING .... UOC:DAL, V18, ZAL	1
25	PAOZZ	2540007896192	19207	11602160	SWITCH, PRESSURE .... UOC:DAL, V18, ZAL	1
26	PAOZZ	4730005417500	81343	6-6 120302BA(LON	ELBOW, PIPE TO TUBE .... G NUT) UOC:DAL, V18, ZAL	1
27	PAOZZ	4730011147541	19207	11669081	TEE, PIPE .... UOC:DAL, V18, ZAL	1
28	PAOZZ	5305009900695	80205	B1821BH050F088N	SCREW, CAP, HEXAGON H .... UOC:DAL, V18, ZAL	1
29	PAOZZ	5310000034094	01276	210104-8S	WASHER, LOCK .... UOC:DAL, V18, ZAL	1
30	PAOOO	5910011655255	19207	12302648	CAPACITOR, FIXED, MET .... UOC:DAL, V18, ZAL	1
31	PAOZZ	5340011705007	19207	8352678	.CLAMP, LOOP .... UOC:DAL, V18, ZAL	1
32	PAOZZ	9905010138723	81349	M43436/3-1	.BAND, MARKER .... UOC:DAL, V18, ZAL	5
33	PAOZZ	5975006605962	19207	8724494	.CABLE NIPPLE, ELECTR .... UOC:DAL, V18, ZAL	3
34	PAOZZ	5970008338562	19207	8338562	.INSULATOR, BUSHING .... UOC:DAL, V18, ZAL	3
35	PAOZZ	5940003996676	19207	8338564	.TERMINAL ASSEMBLY .... UOC:DAL, V18, ZAL	3
36	PAOZA	5999009263144	96906	MS27148-3	.CONTACT, ELECTRICAL .... UOC:DAL, V18, ZAL	2
37	PAOZZ	5310006560067	19207	8724497	.WASHER, SLOTTE D .... UOC:DAL, V18, ZAL	2
38	PAOZZ	5935006915591	19207	8724495	.SHELL, ELECTRICAL CO .... UOC:DAL, V18, ZAL	2

END OF FIGURE

7 ─ 8

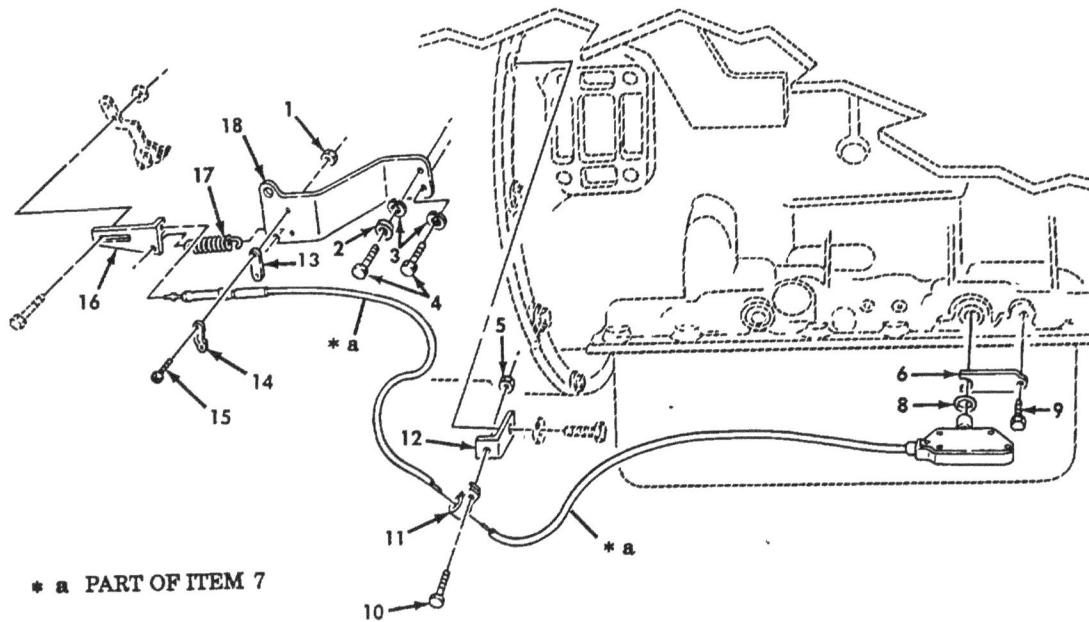

Figure 183. Transmission Modulator and Cable Assembly (M939, M939A1).

* a   PART OF ITEM 7

(1) ITEM NO	(2) SMR CODE	(3) NSN	(4) CAGEC	(5) PART NUMBER	(6) DESCRIPTION AND USABLE ON CODES (UOC)	(7) QTY
					GROUP 0705 GEAR SHIFT AND CONTROLS FIG. 183 TRANSMISSION MODULATOR AND CABLE ASSEMBLY(M939, M939A1)	
1	PAOZZ	5310008775797	96906	MS21044-N3	NUT, SELF-LOCKING, HE .... UOC:DAA, DAB, DAC, DAD, DAE, DAF, DAG, DAH, DAJ, DAK, DAL, DAW, DAX, V12, V13, V14, V15, V16, V17, V18, V19, V20, V21, V22, V24, V25, V39	2
2	PAOZZ	5310000685285	96906	MS27183-20	WASHER, FLAT .... UOC:DAA, DAB, DAC, DAD, DAE, DAF, DAG, DAH, DAJ, DAK, DAL, DAW, DAX, V12, V13, V14, V15, V16, V17, V18, V19, V20, V21, V22, V24, V25, V39	1
3	PAOZZ	5310001670680	96906	MS35338-49	WASHER, LOCK .... UOC:DAA, DAB, DAC, DAD, DAE, DAF, DAG, DAH, DAJ, DAK, DAL, DAW, DAX, V12, V13, V14, V15, V16, V17, V18, V19, V20, V21, V22, V24, V25, V39	2
4	PAOZZ	5305009029338	96906	MS90726-134	SCREW, CAP, HEXAGON H .... UOC:DAA, DAB, DAC, DAD, DAE, DAF, DAG, DAH, DAJ, DAK, DAL, DAW, DAX, V12, V13, V14, V15, V16, V17, V18, V19, V20, V21, V22, V24, V25, V39	2
5	PAOZZ	5310000874652	96906	MS51922-17	NUT, SELF-LOCKING, HE .... UOC:DAA, DAB, DAC, DAD, DAE, DAF, DAG, DAH, DAJ, DAK, DAL, DAW, DAX, V12, V13, V14, V15, V16, V17, V18, V19, V20, V21, V22, V24, V25, V39	1
6	PAOZZ	2520005576619	73342	8627650	RETAINER, MODULATOR .... UOC:DAA, DAB, DAC, DAD, DAE, DAF, DAG, DAH, DAJ, DAK, DAL, DAW, DAX, V12, V13, V14, V15, V16, V17, V18, V19, V20, V21, V22, V24, V25, V39	1
7	PAOOZ	2590010791506	41625	310777-000-0060.0	MODULATOR AND CABLE .... UOC:DAA, DAB, DAC, DAD, DAE, DAF, DAG, DAH, DAJ, DAK, DAL, DAW, DAX, V12, V13, V14, V15, V16, V17, V18, V19, V20, V21, V22, V24, V25, V39	1
8	PAOZZ	5331002483847	96906	MS29513-115	.O-RING .... UOC:DAA, DAB, DAC, DAD, DAE, DAF, DAG, DAH, DAJ, DAK, DAL, DAW, DAX, V12, V13, V14, V15, V16, V17, V18, V19, V20, V21, V22, V24, V25, V39	1
9	PAOZZ	5306002258496	96906	MS90725-31	BOLT, MACHINE .... UOC:DAA, DAB, DAC, DAD, DAE, DAF, DAG, DAH, DAJ, DAK, DAL, DAW, DAX, V12, V13, V14, V15, V16, V17, V18, V19, V20, V21, V22, V24, V25, V39	1
10	PAOZZ	5305002692803	96906	MS90726-60	SCREW, CAP, HEXAGON H .... UOC:DAA, DAB, DAC, DAD, DAE, DAF, DAG, DAH,	1

(1) ITEM NO	(2) SMR CODE	(3) NSN	(4) CAGEC	(5) PART NUMBER	(6) DESCRIPTION AND USABLE ON CODES (UOC)	(7) QTY
					DAJ, DAK, DAL, DAW, DAX, V12, V13, V14, V15, V16, V17, V18, V19, V20, V21, V22, V24, V25, V39	
11	PAOZZ	5340005235999	19207	7373218	CLAMP, LOOP .... UOC:DAA, DAB, DAC, DAD, DAE, DAF, DAG, DAH, DAJ, DAK, DAL, DAW, DAX, V12, V13, V14, V15, V16, V17, V18, V19, V20, V21, V22, V24, V25, V39	1
12	PFOZZ	5340013121136	19207	12256026-1	BRACKET, ANGLE .... UOC:DAA, DAB, DAC, DAD, DAE, DAF, DAG, DAH, DAJ, DAX, DAL, DAW, DAX, V12, V13, V14, V15, V16, V17, V18, V19, V20, V21, V22, V24, V25, V39	1
13	PAOZZ	5365010911630	19207	12256024	SPACER, PLATE .... UOC:DAA, DAB, DAC, DAD, DAE, DAF, DAG, DAH, DAJ, DAK, DAL, DAW, DAX, V12, V13, V14, V15, V16, V17, V18, V19, V20, V21, V22, V24, V25, V39	1
14	PAOZZ	5340010909331	19207	12256023	STRAP, RETAINING .... UOC:DAA, DAB, DAC, DAD, DAE, DAF, DAG, DAH, DAJ, DAK, DAL, DAW, DAX, V12, V13, V14, V15, V16, V17, V18, V19, V20, V21, V22, V24, V25, V39	1
15	PAOZZ	5305009931848	96906	MS35207-265	SCREW, MACHINE .... UOC:DAA, DAB, DAC, DAD, DAE, DAF, DAG, DAH, DAJ, DAK, DAL, DAW, DAX, V12, V13, V14, V15, V16, V17, V18, V19, V20, V21, V22, V24, V25, V39	2
16	PAOZZ	5340010904484	19207	12256025	HOLDER, SPRING .... UOC:DAA, DAB, DAC, DAD, DAE, DAF, DAG, DAH, DAJ, DAX, DAL, DAW, DAX, V12, V13, V14, V15, V16, V17, V18, V19, V20, V21, V22, V24, V25, V39	1
17	PAOZZ	5360010909335	66131	PS-566	SPRING, HELICAL, EXTE .... UOC:DAA, DAB, DAC, DAD, DAE, DAF, DAG, DAH, DAJ, DAK, DAL, DAW, DAX, V12, V13, V14, V15, V16, V17, V18, V19, V20, V21, V22, V24, V25, V39	1
18	PFOZZ	5342010909334	19207	12256094	BRACKET ASSEMBLY, MO .... UOC:DAA, DAB, DAC, DAD, DAE, DAF, DAG, DAH, DAJ, DAK, DAL, DAW, DAX, V12, V13, V14, V15, V16, V17, V18, V19, V20, V21, V22, V24, V25, V39	1

END OF FIGURE

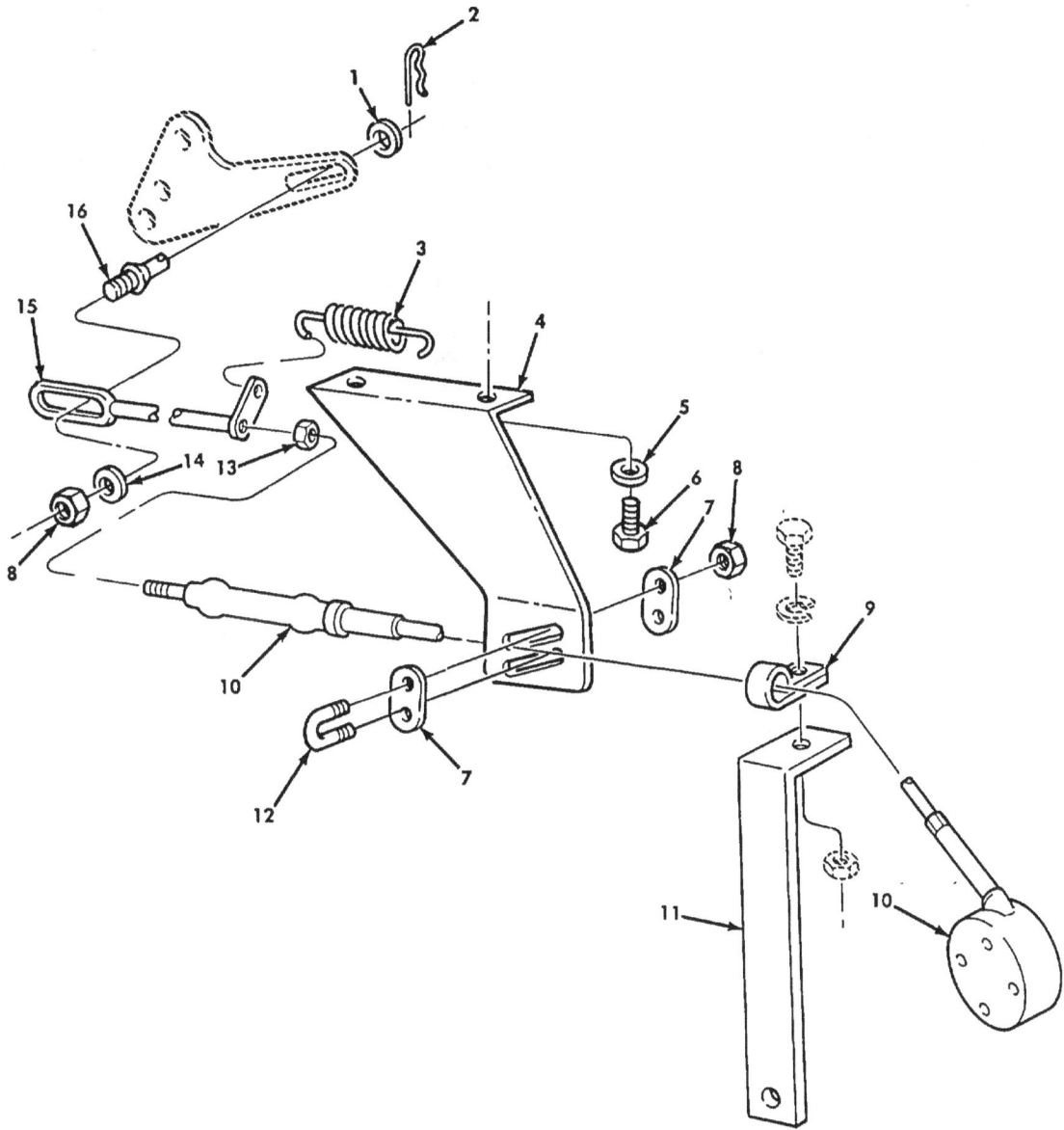

Figure 184. Transmission Modulator and Cable Assembly (M939A2).

(1) ITEM NO	(2) SMR CODE	(3) NSN	(4) CAGEC	(5) PART NUMBER	(6) DESCRIPTION AND USABLE ON CODES (UOC)	(7) QTY
					GROUP 0705 GEAR SHIFT AND CONTROLS	
					FIG. 184 TRANSMISSION MODULATOR AND CABLE ASSEMBLY(M939A2)	
1	PAOZZ	5310013327265	24617	9421861	WASHER, FLAT .... UOC:ZAA, ZAB, ZAC, ZAD, ZAE, ZAF, ZAG, ZAH, ZAJ, ZAK, ZAL	1
2	PAOZZ	5315014469007	58536	A-A-55487-8	PIN, LOCK FOR M939A2 VEHICLES WITH .... MODIFICATION KIT 57K0281 INSTALLED.............. UOC:ZAA, ZAB, ZAC, ZAD, ZAE, ZAF, ZAG, ZAH, ZAJ, ZAK, ZAL	1
2	PAOZZ	5315012849812	60602	10166	PIN, LOCK .... UOC:ZAA, ZAB, ZAC, ZAD, ZAE, ZAF, ZAG, ZAH, ZAJ, ZAK, ZAL	1
3	PAOZZ	5360010909335	66131	PS-566	SPRING, HELICAL, EXTE .... UOC:ZAA, ZAB, ZAC, ZAD, ZAE, ZAF, ZAG, ZAH, ZAJ, ZAK, ZAL	1
4	PAOZZ	5340012307097	47457	20510476	BRACKET, MOUNTING CABLE .... UOC:ZAA, ZAB, ZAC, ZAD, ZAE, ZAF, ZAG, ZAH, ZAJ, ZAK, ZAL	1
5	PAOZZ	5310012865452	24617	11511514	WASHER, FLAT .... UOC:ZAA, ZAB, ZAC, ZAD, ZAE, ZAF, ZAG, ZAH, ZAJ, ZAK, ZAL	2
6	PAOZZ	5306012899197	24617	11500713	BOLT, MACHINE .... UOC:ZAA, ZAB, ZAC, ZAD, ZAE, ZAF, ZAG, ZAH, ZAJ, ZAK, ZAL	2
7	PAOZZ	5365012848138	60602	39048	SPACER, PLATE .... UOC:ZAA, ZAB, ZAC, ZAD, ZAE, ZAF, ZAG, ZAH, ZAJ, ZAKX, ZAL	2
8	PAOZZ	5310008775796	96906	MS21044-N4	NUT, SELF-LOCKING, HE .... UOC:ZAA, ZAB, ZAC, ZAD, ZAE, ZAF, ZAG, ZAH, ZAJ, ZAK, ZAL	3
9	PAOZZ	5340009546014	96906	MS21333-121	CLAMP, LOOP .... UOC:ZAA, ZAB, ZAC, ZAD, ZAE, ZAF, ZAG, ZAH, ZAJ, ZAK, ZAL	1
10	PAOZZ	2590012844547	60602	2242-54	CONTROL ASSEMBLY, PU .... UOC:ZAA, ZAB, ZAC, ZAD, ZAE, ZAF, ZAG, ZAH, ZAJ, ZAK, ZAL	1
11	PAOZZ	5340012806874	47457	20511127	BRACKET, ANGLE .... UOC:ZAA, ZAB, ZAC, ZAD, ZAE, ZAF, ZAG, ZAH, ZAJ, ZAK, ZAL	1
12	PAOZZ	5306012851703	60602	50662	BOLT, U .... UOC:ZAA, ZAB, ZAC, ZAD, ZAE, ZAF, ZAG, ZAH, ZAJ, ZAK, ZAL	1
13	PAOZZ	5310007680319	96906	MS51968-2	NUT, PLAIN, HEXAGON .... UOC:ZAA, ZAB, ZAC, ZAD, ZAE, ZAF, ZAG, ZAH, ZAJ, ZAK, ZAL	1
14	PAOZZ	5310002090786	96906	MS35335-33	WASHER, LOCK .... UOC:ZAA, ZAB, ZAC, ZAD, ZAE, ZAF, ZAG, ZAH, ZAJ, ZAK, ZAL	1
15	PAOZZ	3040012846232	60602	50452-2	CONNECTING LINK, RIG	1

(1) ITEM NO	(2) SMR CODE	(3) NSN	(4) CAGEC	(5) PART NUMBER	(6) DESCRIPTION AND USABLE ON CODES (UOC)	(7) QTY
					UOC:ZAA, ZAB, ZAC, ZAD, ZAE, ZAF, ZAG, ZAH, ZAJ, ZAK, ZAL	
16	PAOZZ	5315012855562	60602	50451	PIN, SHOULDER, HEADLE	1
					UOC:ZAA, ZAB, ZAC, ZAD, ZAE, ZAF, ZAG, ZAH, ZAJ, ZAK, ZAL	

END OF FIGURE

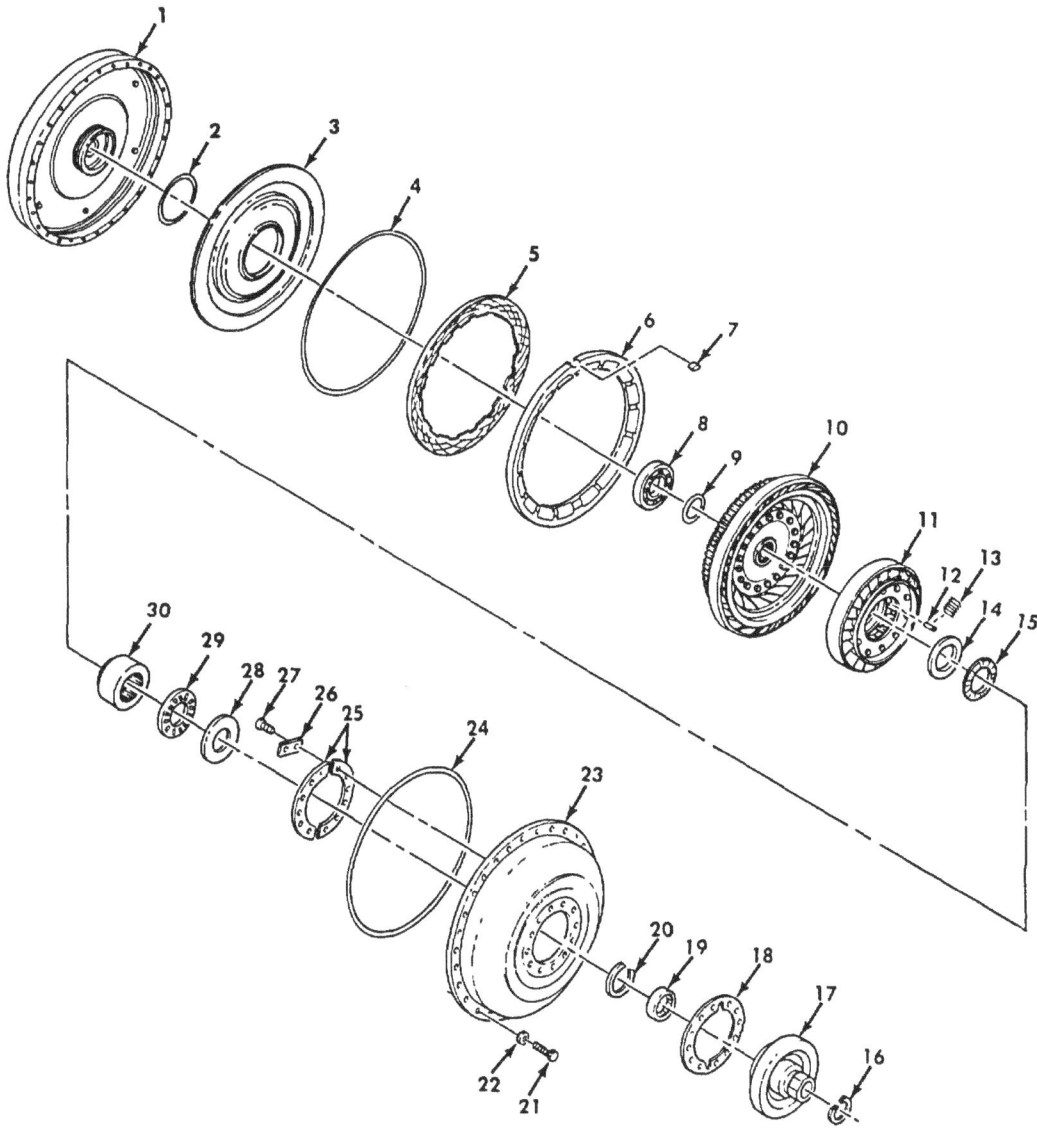

Figure 185. Figure Converter and Lockup Clutch.

(1) ITEM NO	(2) SMR CODE	(3) NSN	(4) CAGEC	(5) PART NUMBER	(6) DESCRIPTION AND USABLE ON CODES (UOC)	(7) QTY
					GROUP 0708 TORQUE CONVERTER	
					FIG. 185 TORQUE CONVERTER AND LOCKUP CLUTCH	
1	PBFHH	2520011340898	73342	23016995	FLYWHEEL, TORQUE CON TRANSMISSION	1
2	PAHZZ	5330009993760	73342	6770492	PACKING, PREFORMED	1
3	PAHZZ	3040012554406	73342	29503808	PISTON, LINEAR ACTUA	1
4	PAHZZ	5330005820456	73342	6758740	GASKET	1
5	PAHZZ	2520013178288	73342	23048511	DISK, CLUTCH, VEHICUL	1
6	PAHZZ	2520004062303	73342	23016957	PLATE, LOCKUP CLUTCH	1
7	PAHZZ	5315004020421	73342	6772552	KEY, LOCKUP BACKING	1
8	PAHZZ	3110001982848	73342	954528	BEARING, BALL, ANNULA	1
9	PAHZZ	5365010800482	73342	6839976	SPACER BEARING, GOLD .015 V	
9	PAHZZ	5365010800483	73342	6839977	SPACER BEARING, SILVER .030 V	
9	PAHZZ	5365010800484	73342	6839978	SPACER BEARING, PLAIN .042 V	
9	PAHZZ	5365010800485	73342	6839979	SPACER BEARING, BLACK .060 V	
9	PAHZZ	5365010800486	73342	6839980	SPACER BEARING, COPPER .072 V	
10	PAHZZ	2520010985117	73342	6839975	TURBINE	1
11	PAHZZ	2520010944751	73342	6837167	STATOR ASSEMBLY	1
12	PAHZZ	3110009413830	73342	6837976	ROLLER, BEARING	10
13	PAHZZ	5360011579921	63005	23012925	SPRING, FLAT	10
14	PAHZZ	3110009169273	73342	9417722	SEAT, BEARING	1
15	PAHZZ	3110008308802	73342	457118	RETAINER AND ROLLER	1
16	PAHZZ	2520004051842	73342	6830187	RING, PISTON	1
17	PAHZZ	2520010985108	73342	23017446	HUB, CONVERTER PUMP USE WITH TRANS .... P/N 6885292	1
18	PAHZZ	5330009119411	73342	6771366	GASKET	1
19	PAHZZ	3110011791161	73342	23048030	BEARING, ROLLER, CYLI	1
20	PAHZZ	5325010796526	73342	6750020	RING, RETAINING	1
21	PAHZZ	5305002924595	73342	9409037	SCREW, CAP, HEXAGON H	30
22	PAHZZ	5310007767670	73342	6769636	WASHER, FLAT	30
23	PAHZZ	4320004035223	73342	23010620	HOUSING, LIQUID PUMP	1
24	PAHZZ	5330011419579	73342	6880389	GASKET	1
25	PAHZZ	3110009164286	73342	6771070	PLATE, RETAINING, BEA	2
26	PAHZZ	5340010560037	73342	6880899	STRIP RETAINER	6
27	PAHZZ	5305002693238	80204	B1821BH038F125N	SCREW, CAP, HEXAGON H	12
28	PAHZZ	3110011467144	73342	9436973	RACE, BEARING..	1
29	PAHZZ	3110010797049	73342	9436972	RETAINER AND ROLLER	1
30	PAHZZ	2520010800448	73342	6839985	RACE	1

END OF FIGURE

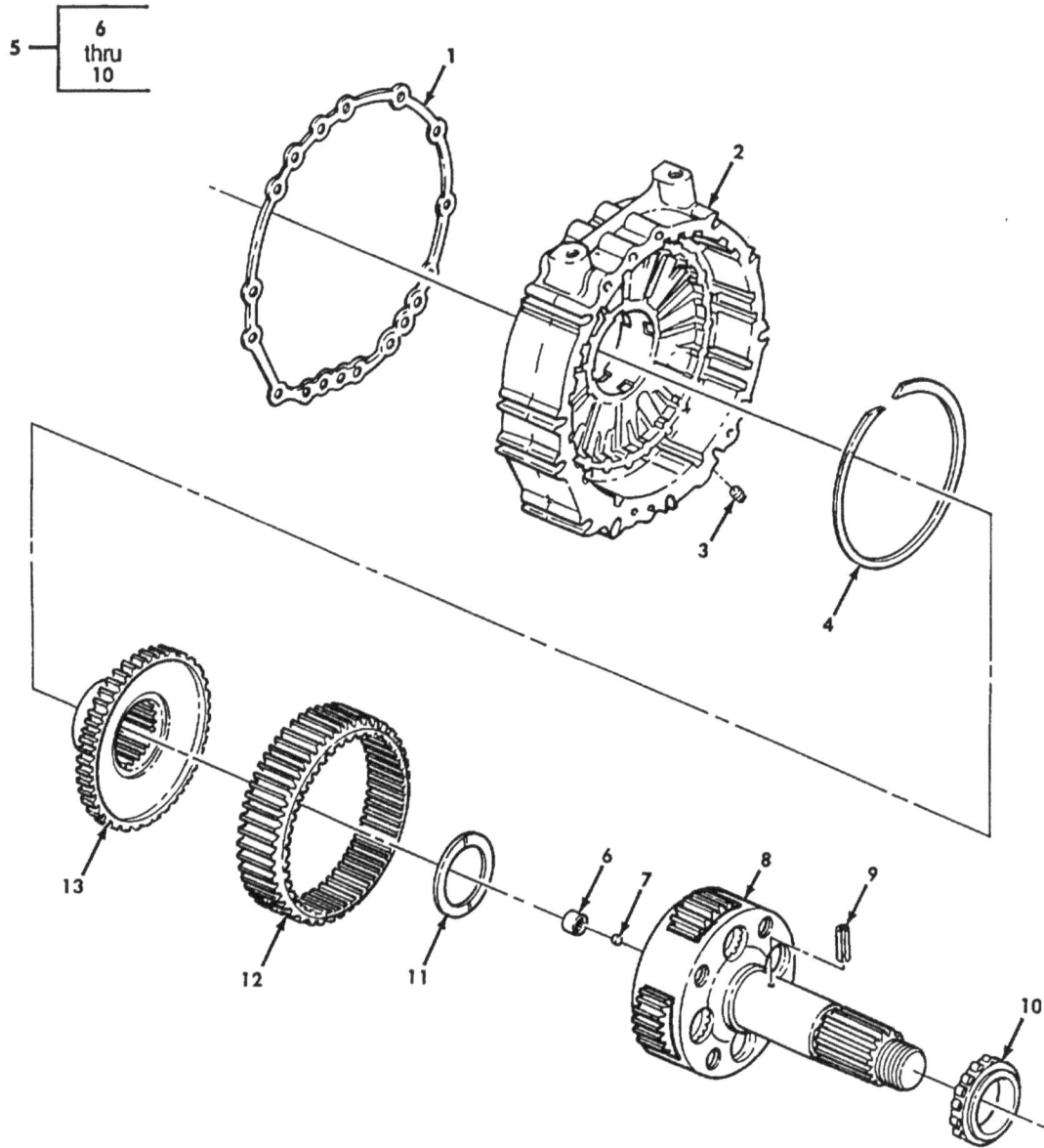

Figure 186. Adapter Housing and Low Planetary Carrier Assembly.

(1) ITEM NO	(2) SMR CODE	(3) NSN	(4) CAGEC	(5) PART NUMBER	(6) DESCRIPTION AND USABLE ON CODES (UOC)	(7) QTY
					GROUP 0710 TRANSMISSION ASSEMBLY	
					FIG. 186 ADAPTER HOUSING AND LOW PLANETARY CARRIER ASSEMBLY	
1	PAHZZ	5330012192555	73342	23045099	GASKET	1
2	PAHZZ	3040011513568	73342	23045053	HOUSING, MECHANICAL	1
3	PAHZZ	4730012066162	81349	WW-P-471ACABCD	PLUG, PIPE	1
4	PAHZZ	5325011334679	73342	6881581	RING, RETAINING	1
5	PAHHH	2520011231648	73342	23016862	CARRIER ASSEMBLY, TR	1
6	PAHZZ	3110010072609	73342	9418483	.BEARING, ROLLER, NEED	1
7	PAHZZ	4730010070802	73342	8623484	.PLUG, CUP	1
8	PAHZZ	2520011225169	73342	23045372	.CARRIER ASSEMBLY, LO	1
9	PAHZZ	5315010583487	80205	NAS561P6-12	.PIN, SPRING	1
10	PFHZZ	3110011476681	73342	23048028	.RING, BEARING, OUTER	1
11	PAHZZ	3120011261097	73342	6880024	BEARING, WASHER, THRU	1
12	PAHZZ	3020011491239	73342	6883999	GEAR, SPUR	1
13	PAHZZ	3020011225906	73342	6881580	GEAR, SPUR	1

END OF FIGURE

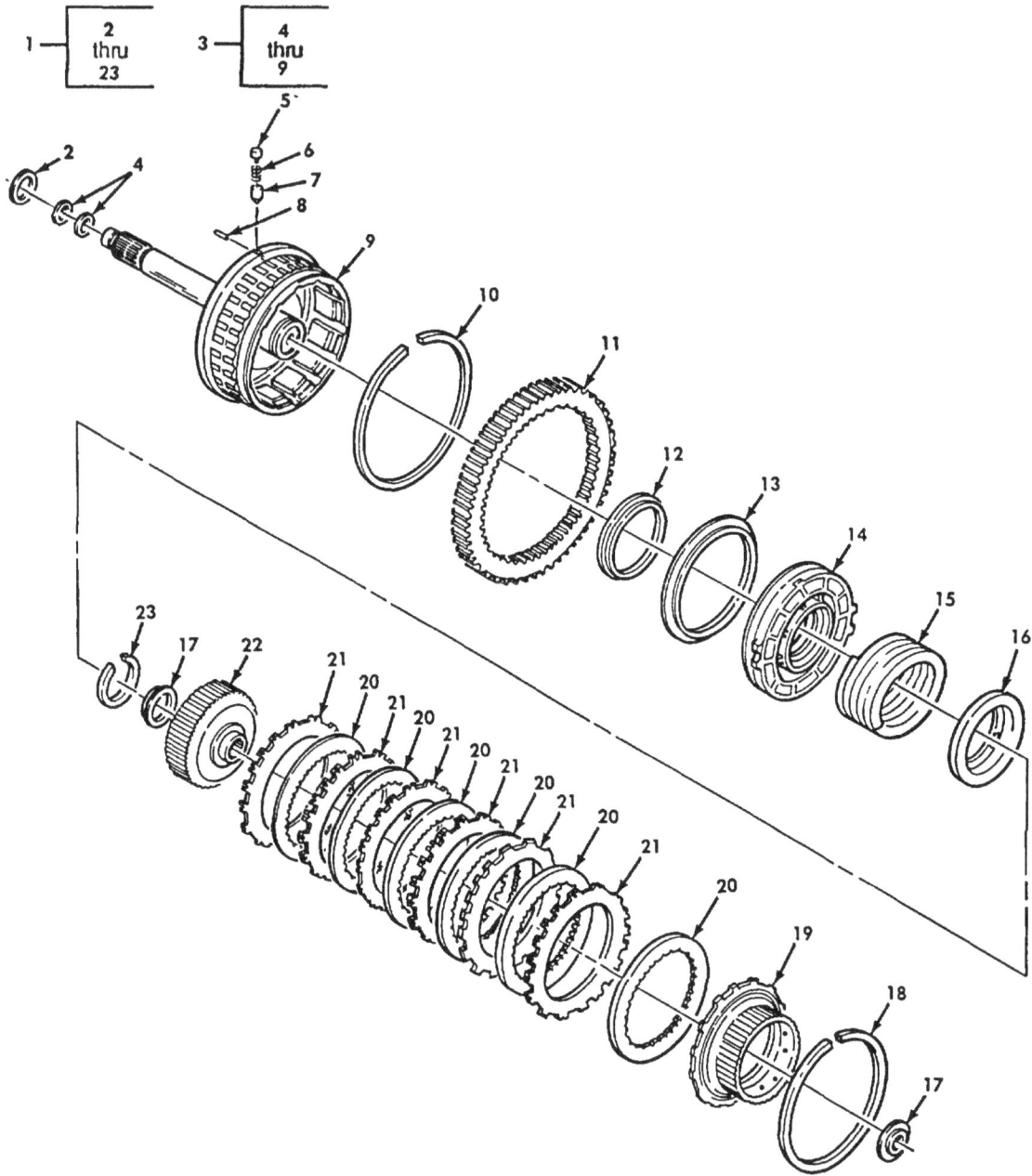

Figure 187. Housing Assembly and Forward Clutch.

(1) ITEM NO	(2) SMR CODE	(3) NSN	(4) CAGEC	(5) PART NUMBER	(6) DESCRIPTION AND USABLE ON CODES (UOC)	(7) QTY
					GROUP 0710 TRANSMISSION ASSEMBLY	
					FIG. 187 HOUSING ASSEMBLY AND FORWARD CLUTCH	
1	PAHHH	2520011491830	73342	23017701	CLUTCH ASSEMBLY, FRI	1
2	PAHZZ	5330003744873	73342	6839163	.SEAL RING, METAL	1
3	PAHHH	2520011305770	73342	23017696	.HOUSING ASSEMBLY, CL	1
4	PAHZZ	5365010109688	73342	6833999	..RING, HOOK TYPE SEAL	2
5	PAHZZ	4820011489278	73342	23012446	..PLUG, ROTARY, VALVE	1
6	PAHZZ	5360011509693	73342	23013747	..SPRING, HELICAL, COMP	1
7	PAHZZ	2520011491785	73342	23013746	..VALVE, CENTRIFUGAL	1
8	PAHZZ	5315008103701	96906	MS16562-36	..PIN, SPRING	1
9	XAHZZ		73342	23017172	..HOUSING, MECHANICAL	1
10	PAHZZ	5365010109687	73342	6838364	.RING, EXTERNAL SNAP	1
11	PAHZZ	3020010650183	73342	6885146	.GEAR, SPUR	1
12	PAHZZ	5330011466053	72582	23015880	.SEAL, PLAIN	1
13	PAXZZ	5365010109689	73342	6833981	.RING, LIP TYPE SEAL PART OF KIT P/N 6884259	1
14	PAHZZ	2520011500900	73342	6885154	.PISTON, FORWARD CLUT 807-.817	1
14	PAHZZ	2520011493808	73342	6885153	.PISTON, FORWARD CLUT 837-.847	1
14	PAHZZ	2520011493438	73342	6885155	.PISTON, FORWARD CLUT 777-.787	1
15	PAHZZ	5360010793096	73342	6836773	.SPRING, CLUTCH	1
16	PAHZZ	2520010650841	73342	6834369	.RETAINER, CLUTCH SPR	1
17	PAHZZ	3110005576163	73342	23015799	.RING, BEARING, INNER	2
18	PAHZZ	5325012192580	73342	23018867	.RING, RETAINING	1
19	PAHZZ	2520011489279	73342	6883090	.HUB, DRIVING, CLUTCH	1
20	PAHZZ	2520010067116	79370	WPC-4946	.DISK, CLUTCH, VEHICUL	6
21	PAHZZ	2520010650077	73342	23016610	.DISK, CLUTCH	6
22	PAHZZ	3020011493836	73342	6883020	.GEAR, SPUR	1
23	PAHZZ	5325011456921	73342	6885156	.RING, RETAINING	1

END OF FIGURE

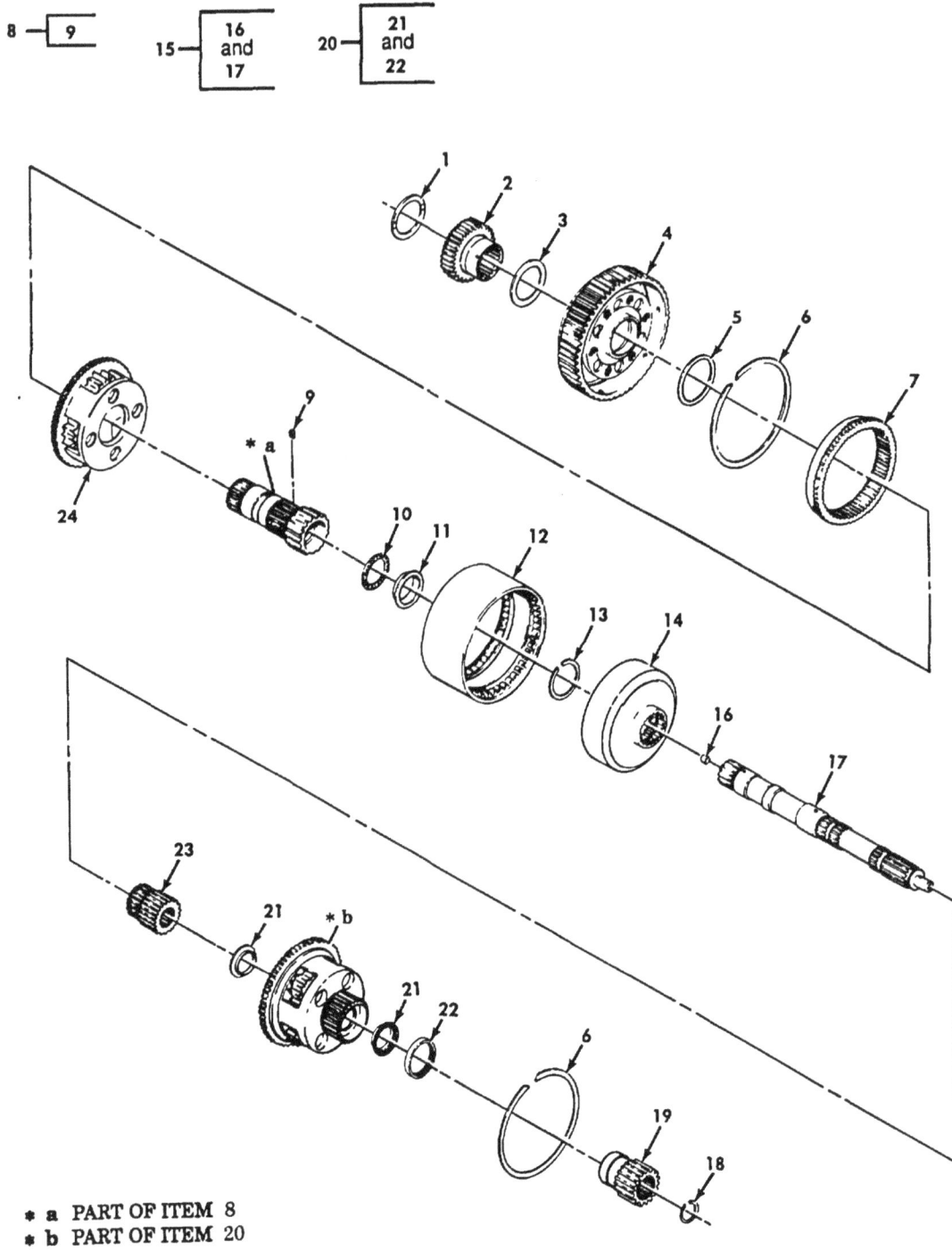

Figure 188. Main Shaft and Carrier Assembly.

* a PART OF ITEM 8
* b PART OF ITEM 20

(1) ITEM NO	(2) SMR CODE	(3) NSN	(4) CAGEC	(5) PART NUMBER	(6) DESCRIPTION AND USABLE ON CODES (UOC)	(7) QTY
					GROUP 0710 TRANSMISSION ASSEMBLY	
					FIG. 188 MAIN SHAFT AND CARRIER ASSEMBLY	
1	PAHZZ	5310005576568	73342	6839364	WASHER, THRUST	1
2	PAHZZ	3020010650871	61849	444969C1	GEAR, SPUR	1
3	PAHZZ	5310005575943	73342	6835386	WASHER, THRUST	1
4	PAHZZ	2520005575974	73342	6882565	CARRIER ASSEMBLY, FR	1
5	PAHZZ	5310005686118	73342	6834389	WASHER, THRUST	1
6	PAHZZ	5325005575835	73342	6834512	RING, RETAINING...	2
7	PAHZZ	3020010082769	73342	23045281	GEAR, RING...	1
8	PAHHZ	2520005576549	7Y635	21606	SHAFT ASSY, SUN GEAR	1
9	PAHZZ	5315010109777	73342	6834940	.PIN, SPRING	2
10	PAHZZ	3110005576708	60380	JF45495	RING, BEARING, OUTER.....................................	1
11	PAHZZ	3110005576666	60380	JF45496	RING, BEARING, INNER.....................................	1
12	PAHZZ	2520010652530	61849	444958C1	DRUM, PLANETARY CARR	1
13	PAHZZ	5325000891262	73342	6769319	RING, RETAINING...	1
14	PAHZZ	3020010082770	73342	6835561	GEAR, INTERNAL...	1
15	PAHHH	2520011312694	73342	23017699	SHAFT ASSEMBLY, TRAN	1
16	PAHZZ	4730011276900	73342	6883707	.PLUG, PIPE	1
17	XAHZZ		73342	23017700	.SHAFT	1
18	PAHZZ	5325005666577	73342	6834583	RING, RETAINING	1
19	PAHZZ	3020011243421	73342	6885151	GEAR, LOW SUN BRASS	1
20	PAHHH	2520011229928	73342	6880008	CARRIER ASSEMBLY, RE	1
21	PAHZZ	3110010652469	24617	9434057	.BEARING	2
22	PAHZZ	3110012391253	73342	9436116	.BALL, BEARING	1
23	PAHZZ	3020011528895	73342	6883901	GEAR CLUSTER, SPUR	1
24	PAHHH	2520005575980	61849	444965C1	CARRIER ASSY, CENTER	1

END OF FIGURE

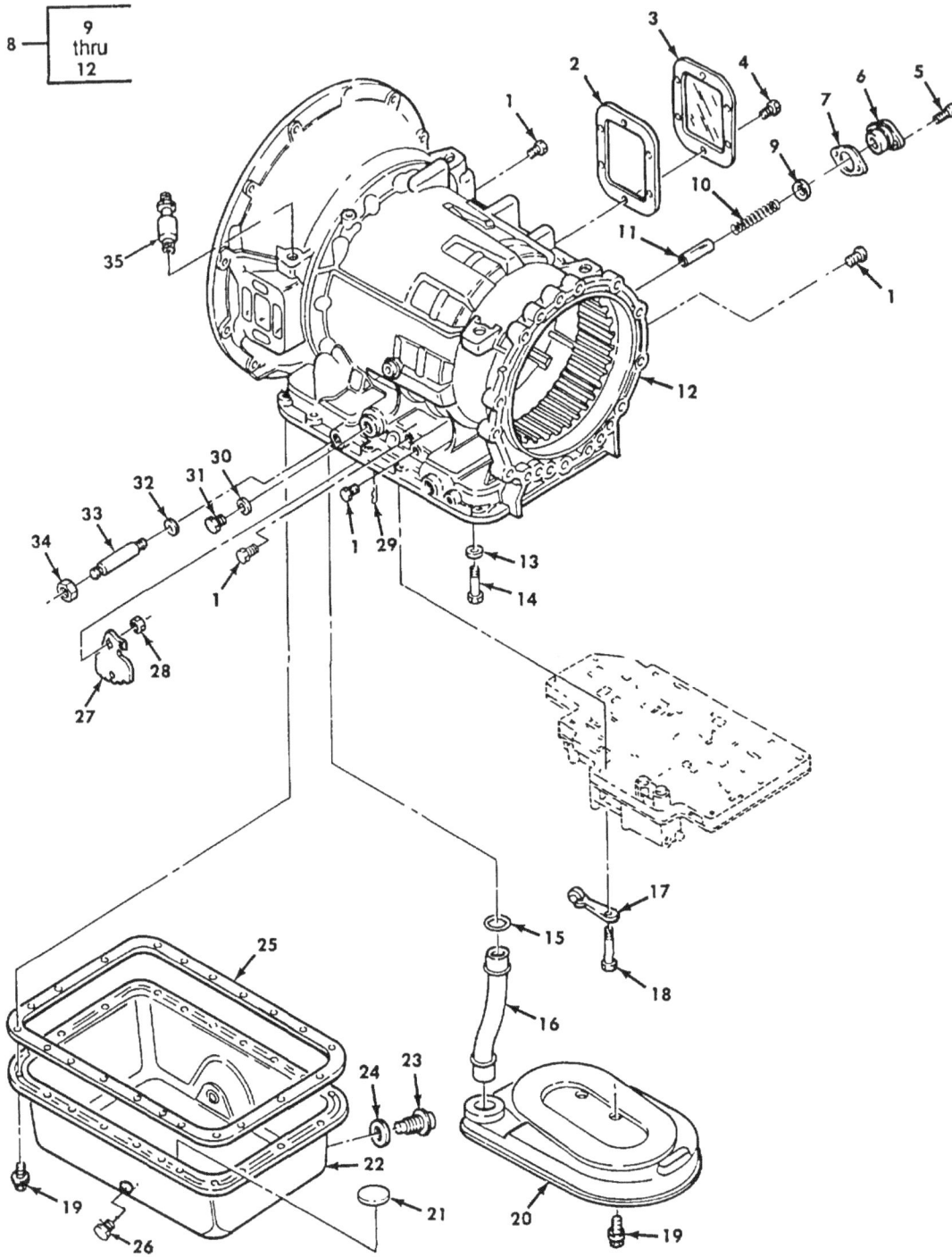

Figure 189. Transmission Housing, Oil Filter, Oil Pan, and Mounting Hardware.

(1) ITEM	(2) SMR	(3)	(4)	(5) PART	(6)	(7)

SECTION II

GROUP 0710 TRANSMISSION ASSEMBLY

FIG. 189 TRANSMISSION HOUSING, OIL FILTER, OIL PAN, AND MOUNTING HARDWARE

NO	CODE	NSN	CAGEC	NUMBER	DESCRIPTION AND USABLE ON CODES (UOC)	QTY
1	PAHZZ	4730011461075	73342	23012036	PLUG,PIPE	4
2	PAHZZ	5330007817774	19207	12288013	GASKET PART OF KIT P/N 6884259	1
3	PAHZZ	2520009197240	73342	6774322	COVER	1
4	PAHZZ	5305005434372	80204	B1821BH038C075N	SCREW,CAP,HEXAGON H	6
5	PAHZZ	5305011245779	73342	6882586	SCREW,CAP,SOCKET HE	2
6	PAHZZ	2520011060826	73342	6881227	ADAPTER,LUBE VALVE	1
7	PAHZZ	5330011119291	73342	6884872	GASKET PART OF KIT P/N 23012502	1
8	PFHZZ	2520011465254	73342	23011821	HOUSING ASSEMBLY,TR	1
9	PADZZ	2520005575900	73342	6834624	.VALVE,LUBE	1
10	PADZZ	5360011235483	73342	6836928	.SPRING,HELICAL,COMP	1
11	PADZZ	4710005575885	73342	23045085	.TUBE,VALVE GUIDE	1
12	XAHZZ		73342	6880967	.HOUSING	1
13	PAHZZ	5310011456923	73342	23014094	WASHER,FLAT PART OF KIT P/N 6884259.	1
14	PAHZZ	5306005708942	73342	23013398	BOLT,MACHINE PART OF KIT P/N 6884259.	1
15	PAOZZ	5331010109693	73342	6762127	O-RING PART OF KIT P/N 23019201 PART OF KIT P/N 6884259	1
16	PAOZZ	4710010788748	73342	6883046	TUBE,BENT,METALLIC	1
17	PAHZZ	2520000087361	73342	8625431	ROLLER AND SPRING	1
18	PAHZZ	5305000712513	80204	B1821BH025C250N	SCREW,CAP,HEXAGON H	1
19	PAOZZ	5306000246580	73342	3829139	BOLT,MACHINE	22
20	PAOZZ	2520011246469	73342	6883044	FILTER ELEMENT,FLUI PART OF KIT P/N 23019201 PART OF KIT P/N 6884259.	1
21	PAOZZ	2920011431263	73342	6775703	MAGNET,OIL PAN PART OF KIT P/N 23012502.	1
22	PBOZZ	2520012069575	73342	23013668	OIL PAN.	1
23	PAOZZ	5365010893573	73342	3921988	PLUG,MACHINE THREAD PART OF KIT P/N 23012502.	1
24	PAOZZ	5330001073925	11862	14079550	GASKET PART OF KIT P/N 23012502	1
25	PAOZZ	5330011208090	73342	29501160	GASKET PART OF KIT P/N 23019201 PART OF KIT P/N 6884259.	1
26	PAOZZ	4730013158280	24617	9436642	PLUG,TUBE FITTING,T	1
27	PAHZZ	3040011225201	73342	6838278	LEVER,REMOTE CONTRO	1
28	PAHZZ	5310005686077	24617	117212	NUT,PLAIN,HEXAGON	1
29	PAHZZ	5315001081112	73342	6831774	PIN,RETAINING	1
30	PAHZZ	5310010841768	73342	6839761	WASHER .	1
31	PAHZZ	5340000957146	73342	445090	PLUG,SHIPPING	1
32	PAHZZ	2840011419503	73342	23010610	SEAL,AIR,GAS TURBIN SELECTOR SHAFT PART OF KIT P/N 6884259.	1
33	KFHZZ	2520012459803	73342	23017610 KIT	SHAFT,MANUAL PART OF KIT P/N 6885213	1
34	PAHZZ	5310011430512	24617	11501033	NUT,SELECTOR SHAFT PART OF KIT P/N 6885213	1
35	PAOZZ	4730013101137	19207	12368370	ADAPTER,STRAIGHT,PI	1

END OF FIGURE

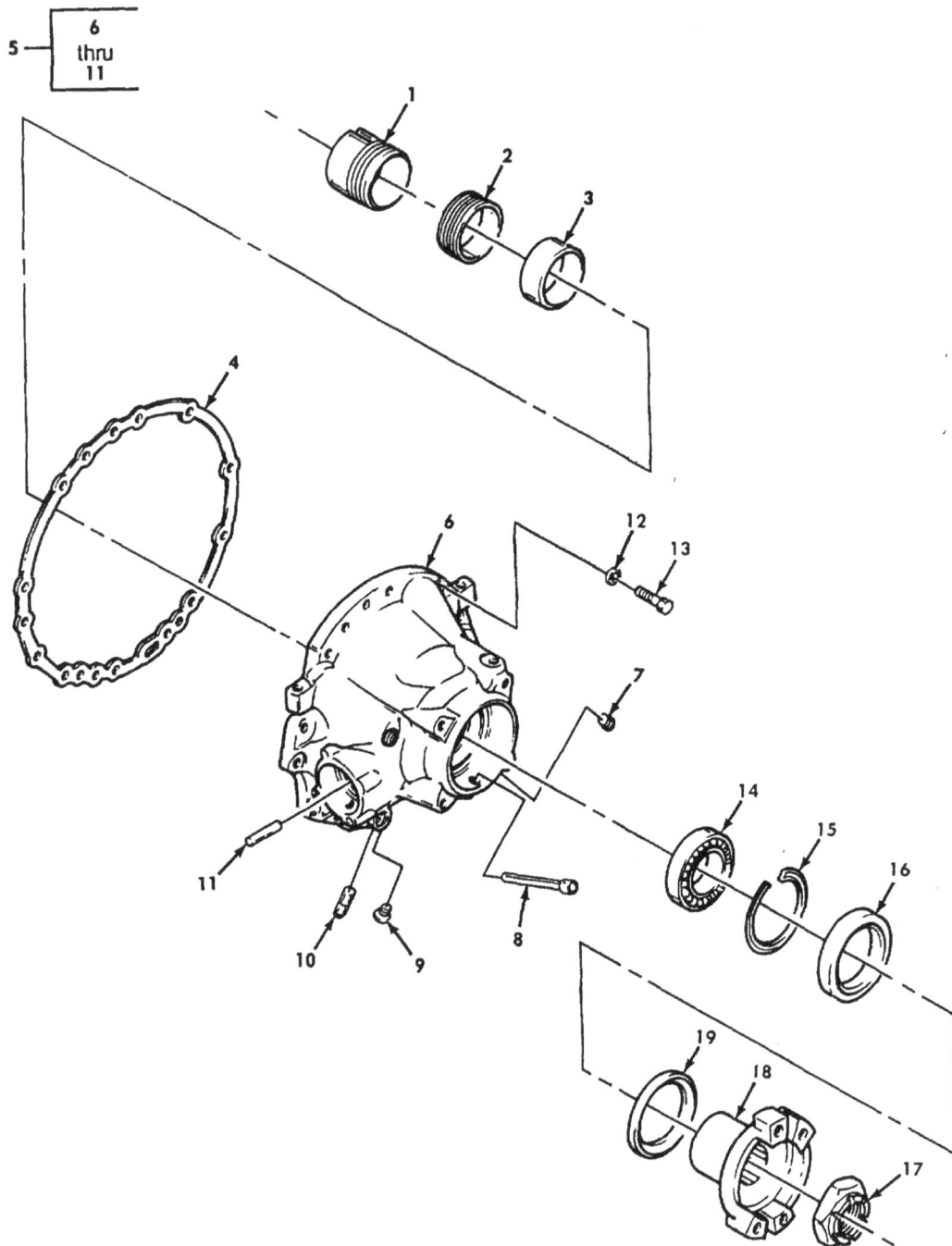

Figure 190. Rear Cover Assembly and Related Parts.

(1) ITEM NO	(2) SMR CODE	(3) NSN	(4) CAGEC	(5) PART NUMBER	(6) DESCRIPTION AND USABLE ON CODES (UOC)	(7) QTY
					GROUP 0710 TRANSMISSION ASSEMBLY	
					FIG. 190 REAR COVER ASSEMBLY AND RELATED PARTS	
1	PAHZZ	3020011797515	73342	29502387	GEAR,BEVEL	1
2	PAHZZ	3020011341830	73342	29503795	GEAR,SPEEDOMETER DR	1
3	PAHZZ	5365011242831	73342	6834556	SPACER,SLEEVE	1
4	PAHZZ	5330012192555	73342	23045099	GASKET PART OF KIT P/N 6884259	1
5	PAHHH	2520011500899	73342	23016018	COVER ASSEMBLY,REAR	1
6	XAHZZ		73342	23016019	.COVER,REAR..	1
7	PAHZZ	4730000575555	29930	444697	.PLUG,PIPE	1
8	PAHZZ	4710011312729	73342	6883974	.TUBE,DRAIN	1
9	PAHZZ	4730011461075	73342	23012036	.PLUG,PIPE	1
10	PAHZZ	4820011265379	73342	6882811	.VALVE,CHECK	1
11	PAHZZ	5315002417523	24617	141231	.PIN,STRAIGHT,H EADLE...................................	1
12	PAHZZ	5310010842362	63005	9411417	WASHER	14
13	PAHZZ	5305011262619	73342	9422961	SCREW,CAP,HEXAGON H	14
14	PAHZZ	3110010560031	43334	3212BAXR1A	BEARING,BALL,ANNULA	1
15	PAHZZ	5325005575897	73342	6834567	RING,RETAINING	1
16	PAFZZ	5330009993752	73342	6773311	SEAL,PLAIN ENCASED PART OF KIT P/N 6884259	1
17	PAFZZ	5310004780548	73342	23040890	NUT,SELF-LOCKING,HE	1
18	PAFZZ	2520011143691	19207	11669428	YOKE,UNIVERSAL JOIN	1
19	PAFZZ	5330009231409	73342	6757563	RETAINER,PACKING REAR COVER...................... PART OF KIT P/N 6884259 USE PRIOR TO S/N 2410227062	1
19	PAFZZ	5330013025092	73342	23016017	GASKET USE AFTER S/N 2410227062	1

END OF FIGURE

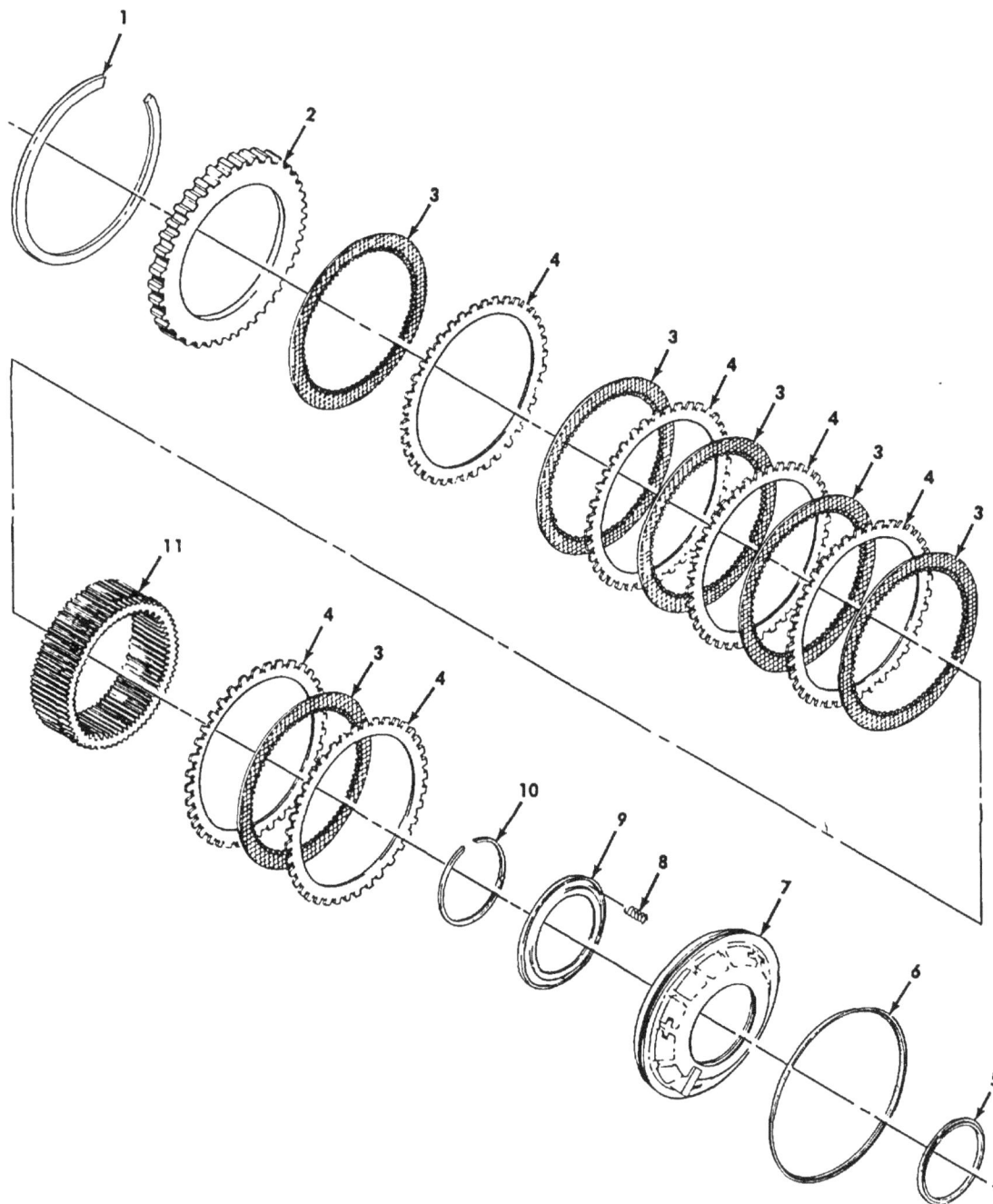

Figure 191. First Clutch.

(1) ITEM NO	(2) SMR CODE	(3) NSN	(4) CAGEC	(5) PART NUMBER	(6) DESCRIPTION AND USABLE ON CODES (UOC)	(7) QTY
					GROUP 0713 INTERMEDIATE CLUTCH	
					FIG. 191 FIRST CLUTCH	
1	PAHZZ	5325000072969	73342	6884275	RING,RETAINING……………………………………………	1
2	PAHZZ	2520005575981	73342	23046720	DISK,CLUTCH 0.702-0.712 ………………………………	1
2	PAHZZ	2520010648847	73342	23046721	DISK,CLUTCH 0.671-0.681 ………………………………	1
2	PAHZZ	2520005636041	73342	23046722	DISK,CLUTCH 0.640-0.650 ………………………………	1
3	PAHZZ	2520001630713	13475	BD809990	DISK,CLUTCH,VEHICUL …………………………………	6
4	PAHZZ	2520005575807	73342	23016606	DISK,CLUTCH ……………………………………………	6
5	KFHZZ	5330010833065	73342	6883031	SEAL,PLAIN PART OF KIT P/N 6884259………………	1
6	KFHZZ	2840010796700	73342	6883033	SEAL,AIR,GAS TURBIN PART OF KIT P/N…………… 6884259 ……………………………………………………	1
7	PAHZZ	2520011461034	73342	23011665	PISTON,CLUTCH TRANS …………………………………	1
8	PAHZZ	5360010793097	73342	6880251	SPRING,PISTON RELEA …………………………………	26
9	PAHZZ	2520010648849	73342	6834339	RETAINER,PISTON SPR …………………………………	1
10	PAHZZ	5325005575794	73342	6833993	RING,RETAINING …………………………………………	1
11	PAHZZ	3020011108251	73342	6835568	GEAR,CLUTCH……………………………………………	1

END OF FIGURE

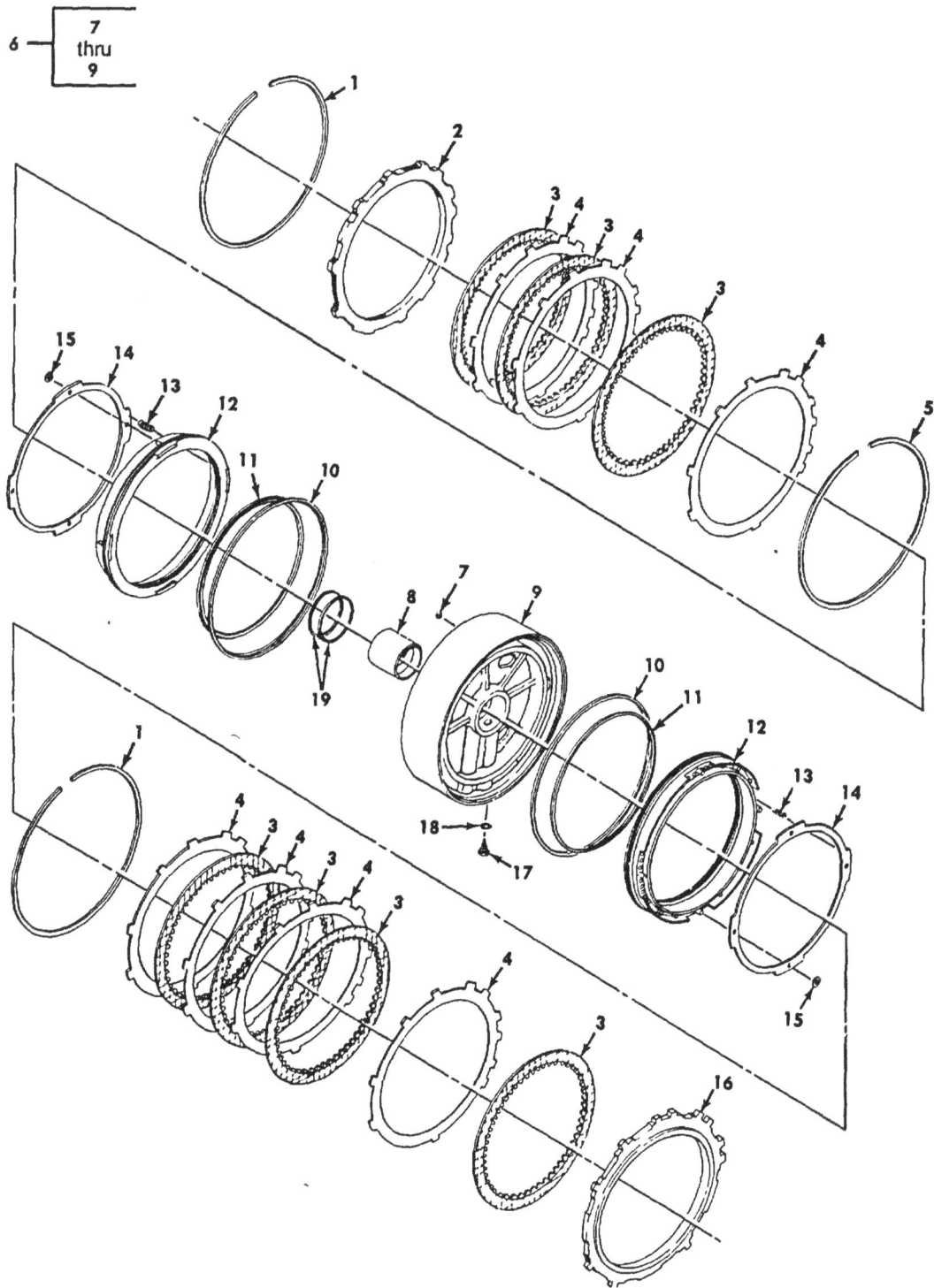

Figure 192. Second, Third Clutch and Center Support

(1) ITEM NO	(2) SMR CODE	(3) NSN	(4) CAGEC	(5) PART NUMBER	(6) DESCRIPTION AND USABLE ON CODES (UOC)	(7) QTY
					GROUP 0713 INTERMEDIATE CLUTCH	
					FIG. 192 SECOND, THIRD CLUTCH ASSEMBLY AND CENTER SUPPORT (TRANSMISSION P/N 6885292)	
1	PAHZZ	5325005576207	73342	6836267	RING,RETAINING	2
2	PAHZZ	2520005575799	73342	23046745	DISK,CLUTCH 0.476-0.486	1
2	PAHZZ	2520005575924	73342	23046746	DISK,CLUTCH 0.450-0.460	1
2	PAHZZ	5365012779878	73342	23046718	SPACER,PLATE 0.463-0.473	1
3	PAHZZ	2520005576090	73342	6835720	DISK,CLUTCH	7
4	PAHZZ	2520005576076	73342	23016608	DISK,CLUTCH	7
5	PAHZZ	5325005576164	73342	6836265	RING,RETAINING 0.148-0.150 WHITE	1
5	PAHZZ	5325005576183	73342	6836266	RING,RETAINING 0.152-0.154 YELLOW	1
5	PAHZZ	5325005576207	73342	6836267	RING,RETAINING 0.155-0.157 GREEN	1
5	PAHZZ	5325005576210	73342	6836268	RING,RETAINING 0.158-0.160 RED	1
6	PBHZZ	2520011681632	73342	23018593	SUPPORT,CENTER CLUT	1
7	PFHZZ	3110011174893	96906	MS19061-10007	.BALL,BEARING...	1
8	PFHZZ	4730010782732	73342	23011924	.BUSHING,PIPE...	1
9	XAHZZ		73342	23017361	.CENTER SUPPORT	1
10	KFHZZ	5330011048934	73342	6883035	SEAL,RING PART OF KIT P/N 6884259..................	2
11	KFHZZ	4310010064952	73342	23011471	RING,PISTON PART OF KIT P/N 6884259.	2
12	PAHZZ	2520010067120	73342	6834230	PISTON,TRANSMISSION	2
13	PAHZZ	5360002119547	73342	6831656	SPRING,HELICAL,COMP	40
14	PAHZZ	2520010650078	73342	6834354	RETAINER,CLUTCH SPR	2
15	PAHZZ	5310011430542	1862	3909063	PUSH ON NUT	8
16	PAHZZ	2520010785564	73342	23046748	DISK,CLUTCH	1
16	PAHZZ	2520010786123	73342	6882321	DISK,CLUTCH 0.234-0.244	1
16	PAHZZ	5365012805643	73342	23046747	SPACER,PLATE 0.221-0.231	1
17	PAHZZ	2940011408227	73342	23010654	FILTER ELEMENT,FLUI	1
18	PAHZZ	5330005798108	96906	MS28775-111	O-RING	1
19	PAHZZ	2520005576211	73342	23014632	RING,SEAL	2

END OF FIGURE

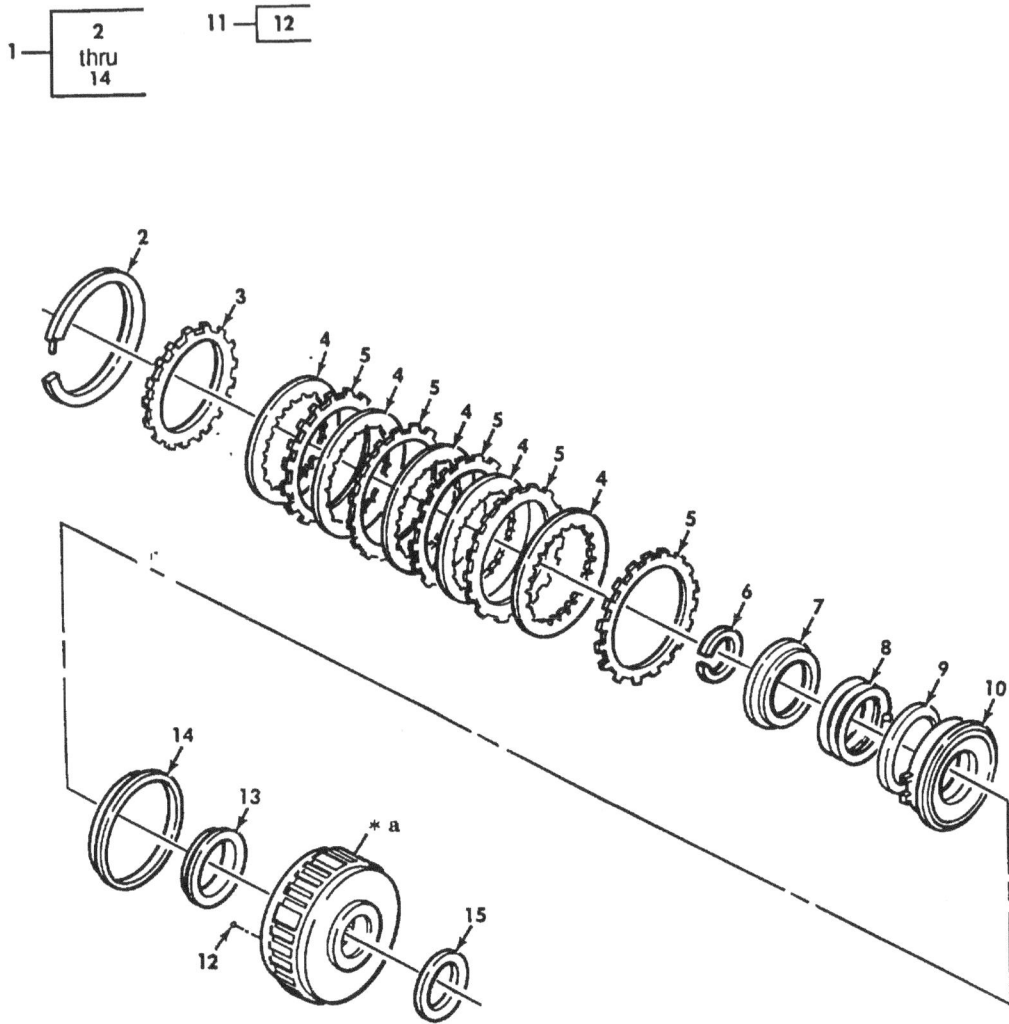

* a   PART OF ITEM 11

Figure 193. Fourth Clutch Assembly

(1) ITEM NO	(2) SMR CODE	(3) NSN	(4) CAGEC	(5) PART NUMBER	(6) DESCRIPTION AND USABLE ON CODES (UOC)	(7) QTY
					GROUP 0713 INTERMEDIATE CLUTCH	
					FIG. 193 FOURTH CLUTCH ASSEMBLY	
1	PAHHH	2520011500901	73342	23013789	FOURTH CLUTCH ASSEMBLY,VEH	1
2	PAHZZ	5325012192580	73342	23018867	.RING,RETAINING	1
3	PAHZZ	2520010067118	73342	29516333	.DISK,CLUTCH	1
4	PAHZZ	2520010067116	79370	WPC-4946	.DISK,CLUTCH...	5
5	PAHZZ	2520010650077	73342	6834374	.DISK,CLUTCH...	5
6	PAHZZ	5325011456921	73342	6885156	.RING,RETAINING	1
7	PAHZZ	2520010650841	73342	6834369	.RETAINER,CLUTCH SPR	1
8	PAHZZ	5360010793096	73342	6836773	.SPRING,CLUTCH	1
9	PAHZZ	5365011974545	73342	6884653	.SPACER,RING	1
10	PAHZZ	2520011493438	73342	6885155	.PISTON,FOURTH CLUTCH	1
					0.777-0.787 THK PRIOR TO;'S/N	1
					242002626502,AND PISTON WITH	
					LETTER "A" AS AN IDENTIFIER	
10	PAHZZ	2520011500900	73342	6885154	.PISTON,FOURTH CLUTCH	1
					0.807-0.817 THK USE PRIOR TO S/N...................	
					2420026502 AND PISTON WITH	
					LETTER "B" AS AN IDENTIFIER	
10	PAHZZ	2520011493808	73342	6885153	.PISTON,FOURTH CLUTCH	1
					0.837-0.847 THK PRIOR TO S/N.	
					2420026502 AND PISTON WITH	
					LETTER "C" AS AN IDENTIFIER	
10	PAHZZ	3040013368217	73342	23017260	PISTON,LINEAR ACTUA FOURTH CLUTCH	1
					0.970-0.980.THK USE AFTER S/N	
					2400256502	
10	PAHZZ	3040013374168	73342	23017261	.PISTON, LINEAR ACTU FOURTH CLUTCH............	1
					0.995-1.005 THK USE AFTER S/N	
					2420062502	
10	PAHZZ	3040013374167	73342	23017262	.PISTON,LINEAR ACTUA............ FOURTH CLUTCH	1
					1.020-1.030 THK USE AFTER S/N	
					2420026502	
11	PAHHH	2520011305772	73342	23011825	.HOUSING ASSEMBLY TR USE PRIOR TO	1
					S/N 2420026502..	
11	PAHZZ	3040013374155	73342	23017257	.HOUSING,MECHANICAL USE AFTER	1
					S/N 2420026502..	
12	XADZZ	2520000087306	73342	8622757	..BALL	
13	PAHZZ	5330011466053	72582	23015880	.SEAL,PLAIN FOURTH CLUTCH PART OF 1	
					KIT P/N 6884259	
14	PAHZZ	5365010109689	73342	6833981	.RING,LIP TYPE SEAL PART OF KIT P/N...............	1
					6884259	
15	PAHZZ	3110010655842	60380	FG42642	BEARING	1

END OF FIGURE

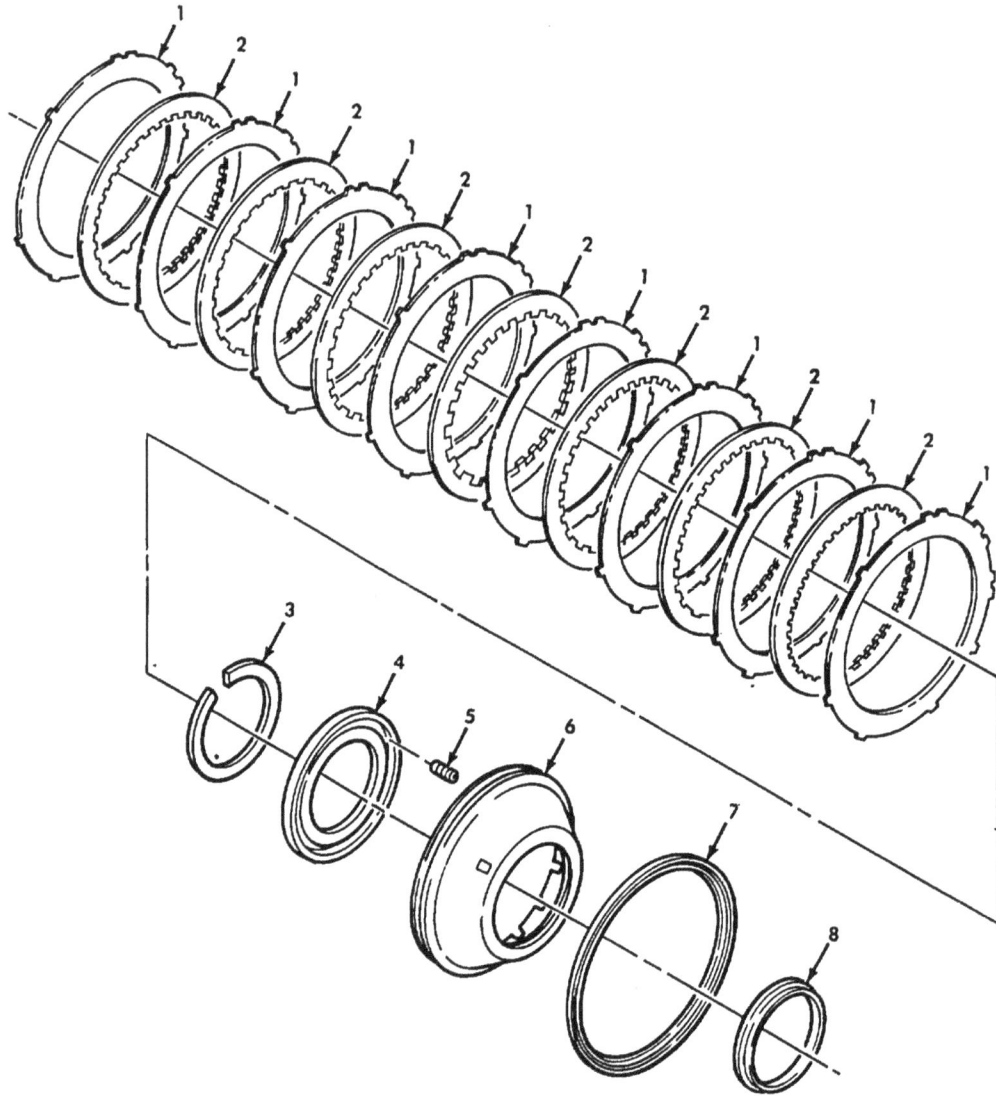

Figure 194. Low Clutch.

(1) ITEM NO	(2) SMR CODE	(3) NSN	(4) CAGEC	(5) PART NUMBER	(6) DESCRIPTION AND USABLE ON CODES (UOC)	(7) QTY
					GROUP 0713 INTERMEDIATE CLUTCH	
					FIG. 194 LOW CLUTCH	
1	PAHZZ	2520012263399	73342	23045054	PLATE,CLUTCH	8
2	PAHZZ	2590012717861	73342	23042210	DISK,CLUTCH	7
3	PAHZZ	5325005575794	73342	6833993	RING,RETAINING	1
4	PAHZZ	2520010648849	73342	6834339	RETAINER,PISTON SPR	1
5	PAHZZ	5360010793097	73342	6880251	SPRING,PISTON RELEA	26
6	PAHZZ	2520011512628	73342	23045496	PISTON,LOW CLUTCH 0.858-0.868	1
6	PAHZZ	2520011503690	73342	23045497	PISTON,LOW CLUTCH 0.829-0.839	1
6	PAHZZ	2520011493439	73342	23045498	PISTON,LOW CLUTCH 0.800-0.810	1
7	KFHZZ	5330011048934	73342	6883035	SEAL,RING PART OF KIT P/N 6884259.................	1
8	XFHZZ	5330010833065	73342	6883031	SEAL,PLAIN PART OF KIT P/N 6884259................	1

END OF FIGURE

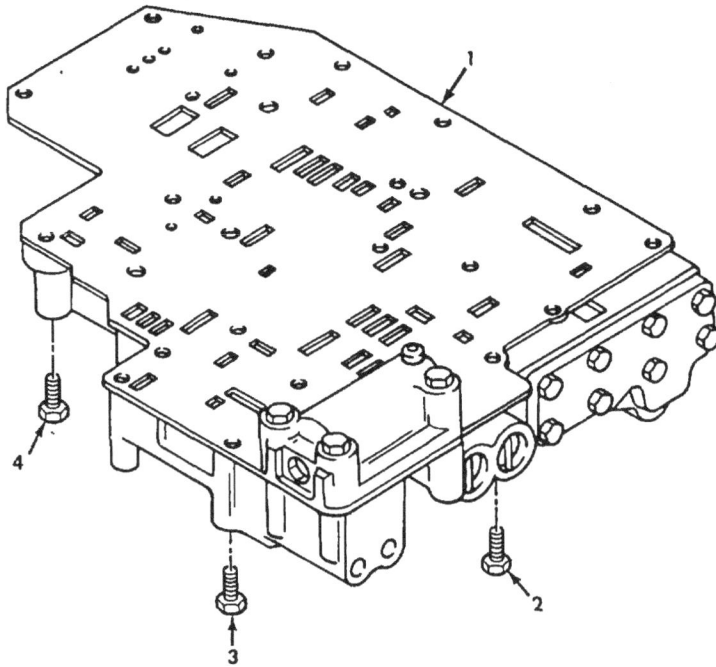

Figure 195. Control Valve Assembly and Mounting Hardware.

(1) ITEM NO	(2) SMR CODE	(3) NSN	(4) CAGEC	(5) PART NUMBER	(6) DESCRIPTION AND USABLE ON CODES (UOC)	(7) QTY
					GROUP 0714 SERVO UNIT	
					FIG. 195 CONTROL VALVE ASSEMBLY AND MOUNTING HARDWARE	
1	KFHDD		73342	23019443	CONTROL VALVE PART OF KIT P/N.................. 230195 96..	1
2	PAHZZ	5305010574265	63005	454815	SCREW,CAP,HEXA GON H..................................	1
3	PAHZZ	5306004005541	73342	273340	BOLT,MACHINE...	3
4	PAHZZ	5305000712515	80204	B1821BH025C300N	SCREW,CAP,HEXAGON H..................................	15

END OF FIGURE

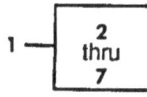

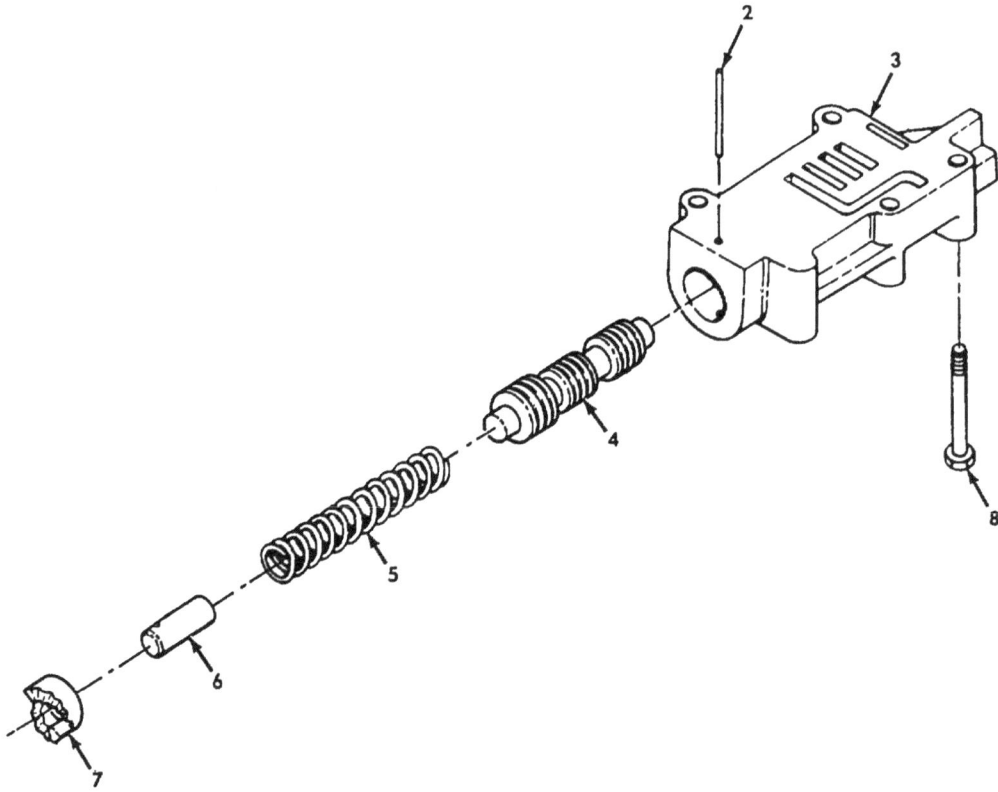

Figure 196. Modulated Lockup Valve and Mounting Hardware.

(1) ITEM NO	(2) SMR CODE	(3) NSN	(4) CAGEC	(5) PART NUMBER	(6) DESCRIPTION AND USABLE ON CODES (UOC)	(7) QTY
					GROUP 0714 SERVO UNIT	
					FIG. 196 MODULATED LOCKUP VALVE AND MOUNTING HARDWARE	
1	PAHDD	4820012215415	73342	23016406	LOCKUP,VALVE PART OF KIT P/N.......................... 23019596...	1
2	PBHZZ	3110010109779	73342	9428493	.BEARING,ROLLER,NEED	1
3	PAHZZ	4810012201185	73342	6881007	.BODY,VALVE	1
4	PBHZZ	3040012212092	73342	6883577	.VALVE,LOCKUP,TRANSM	1
5	PBHZZ	5360012249218	73342	23016366	.SPRING,HELICAL,COMP	1
6	PBHZZ	4820010070962	73342	6833896	.STOP,VALVE	1
7	PBHZZ	5365000389592	19207	12267583	.RING,SPRING...	1
8	PAHZZ	5306005708940	73342	454817	BOLT,MACHINE...	3

END OF FIGURE

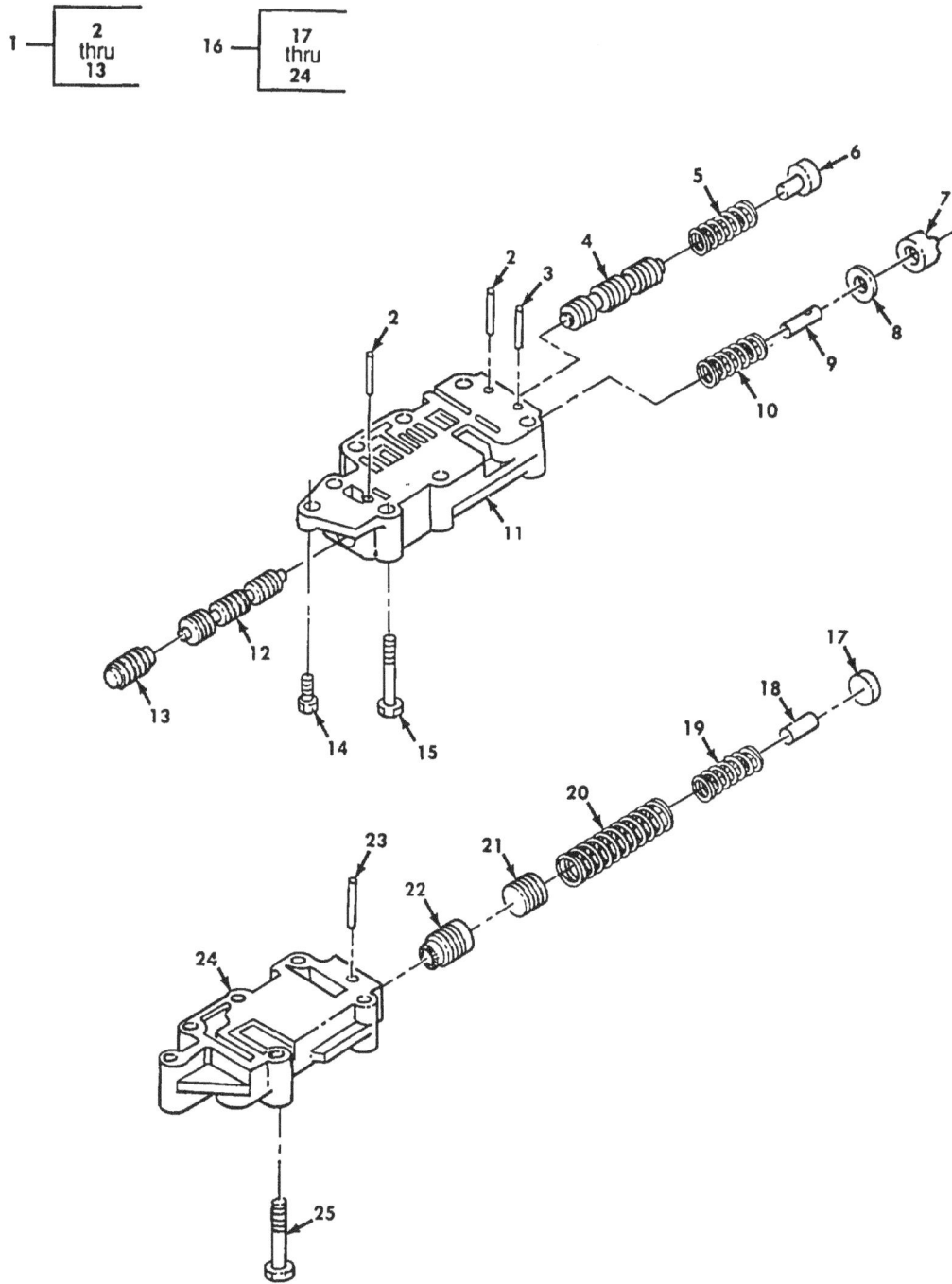

Figure 197. Low Shift Signal Value, Low Trimmer Valve, and Mounting Hardware.

(1) ITEM NO	(2) SMR CODE	(3) NSN	(4) CAGEC	(5) PART NUMBER	(6) DESCRIPTION AND USABLE ON CODES (UOC)	(7) QTY
					GROUP 0714 SERVO UNIT	
					FIG. 197 LOW SHIFT SIGNAL VALVE, LOW TRIMMER VALVE AND MOUNTING HARDWARE	
1	KFHDZ	2520011271675	73342	6883579	VALVE ASSEMBLY,LOW PART OF KIT P/N............ 23019596..	1
2	PBHZZ	5315010836352	73342	9420965	.PIN	2
3	PBHZZ	3110007994903	73342	6700213	.ROLLER,BEARING	
4	PBHZZ	4730011331461	73342	6885188	.PLUG,PIPE	1
5	PBHZZ	5360010748305	73342	6839214	.SPRING,HELICAL,COMP	1
6	PBHZZ	4820010832127	73342	6881138	.STOP,RELAY VALVE	1
7	PBHZZ	5365000389592	19207	12267583	.RING,SPRING..	1
8	PBHZZ	5310000070260	73342	6833949	.WASHER,FLAT	1
9	PBHZZ	5315010044836	19207	12267602	.PIN,STRAIGHT,HEADLE	1
10	PBHZZ	5360004500346	73342	6778156	.SPRING,CUTOFF	1
11	XAHZZ		73342	6881009	.BODY,VALVE	1
12	PBHZZ	2520011305821	73342	6883581	.VALVE,LOW SHIFT SIG	1
13	PBHZZ	4730011326849	73342	6880152	.PLUG,PIPE	1
14	PAHZZ	5305000680509	80204	B1821BH025C125N	SCREW,CAP,HEXAGON H	1
15	PAHZZ	5306005708940	73342	454817	BOLT,MACHINE..	2
16	PAHDD	2520011275005	73342	6881386	VALVE ASSEMBLY,TRAN	1
17	PBHZZ	4730010839925	73342	6838750	.PLUG,PIPE	
18	PBHZZ	5315010044835	73342	6839122	.PIN,STRAIGHT,HEADLE	1
19	PBHZZ	5360011139615	73342	6885166	.SPRING	1
20	PBHZZ	5360010044863	73342	6839271	.SPRING,HELICAL,COMP	1
21	PBHZZ	4730010069629	73342	6835921	.PLUG,TRIMMER	1
22	PBHZZ	4820010070350	73342	23014097	.VALVE,TRIMMER..	1
23	PBHZZ	5315010836351	73342	9423346	.PIN	1
24	XAHZZ	2530013587616	73342	6881387	.BODY,VALVE	1
25	PAHZZ	5305010845370	63005	9424899	SCREW,CAP,HEXAGON H	6

END OF FIGURE

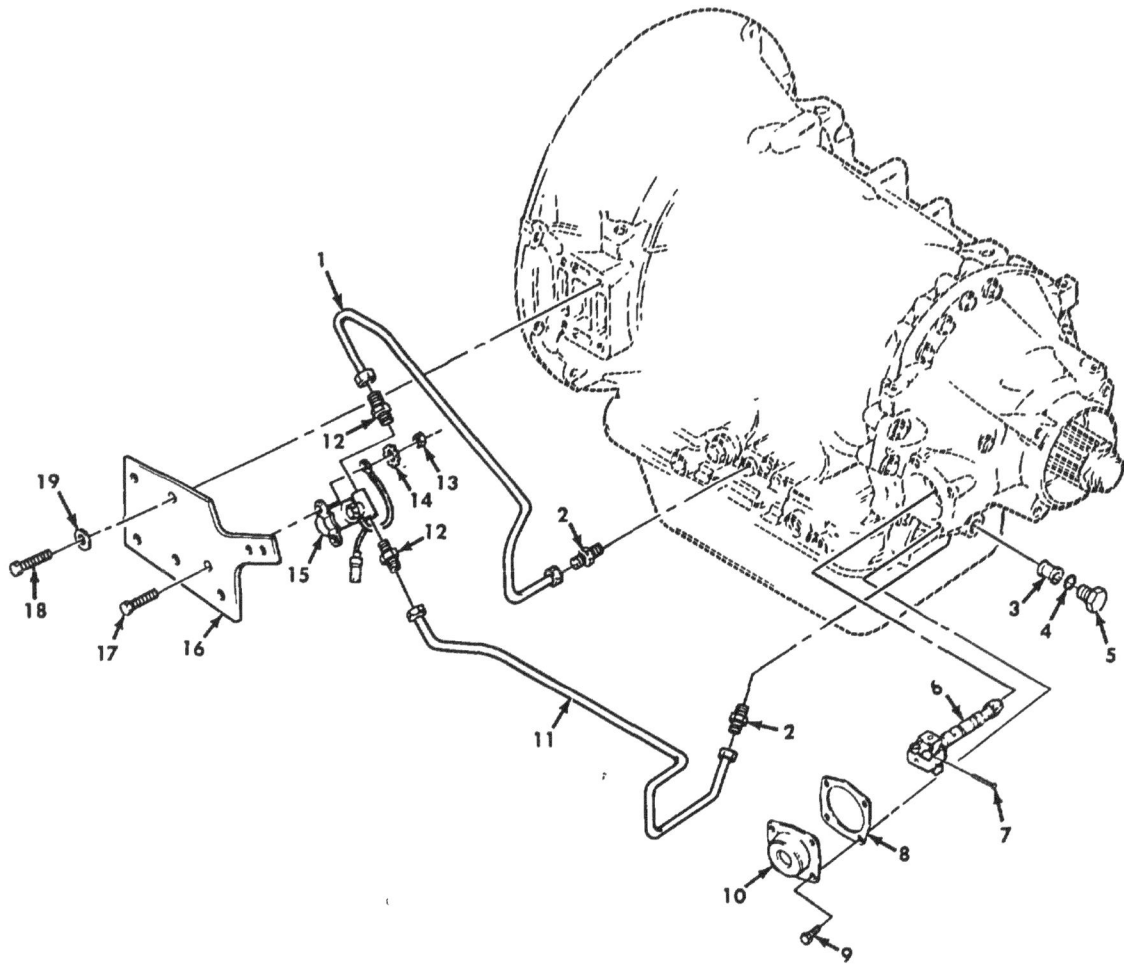

Figure 198. Transmission Tubing, Governor, and Filter Assembly.

(1) ITEM NO	(2) SMR CODE	(3) NSN	(4) CAGEC	(5) PART NUMBER	(6) DESCRIPTION AND USABLE ON CODES (UOC)	(7) QTY
					GROUP 0714 SERVO UNIT	
					FIG. 198 TRANSMISSION TUBING, GOVERNOR, AND FILTER ASSEMBLY	
1	PAHZZ	4710011957644	19207	12302660	TUBE ASSEMBLY,METAL	1
2	PAHZZ	4730006188497	81343	8-2 070102CA	ADAPTER,STRAIGHT,PI	2
3	PAOZZ	4330010749642	73342	6882687	PARTS KIT,FLUID PRE PART OF KIT P/N.............. 6884749	1
4	PAOZZ	5330010803254	73342	6882689	O-RING PART OF KIT P/N 6884749	1
5	PAOZZ	5365011348660	24617	9410360	PLUG	1
6	PAHHH	2910010519444	73342	6885571	GOVERNOR,TRANSMISSI	1
7	KFHZZ		73342	8623232	PIN,GOV,ASSY PART OF KIT P/N 6880353	2
8	KFHZZ	5330000011984	73342	23011670	GASKET PART OF KIT P/N 6880353	1
9	PAHZZ	5306002258496	96906	MS90725-31	BOLT,MACHINE	4
10	PAHZZ	2520011273969	11862	8623262	COVER,GOVERNOR,TRAN	1
11	PAHZZ	4710011957645	19207	12302661	TUBE ASSEMBLY,METAL	1
12	PAHZZ	4730006206904	96906	MS51500-AS-4	ADAPTER,STRAIGHT,PI	2
13	PAFZZ	5310000617326	96906	MS21045-3	NUT,SELF-LOCKING,HE	2
14	PAFZZ	5310005967691	96906	MS35335-32	WASHER,LOCK	1
15	PAFZZ	4810011408221	19207	11669826	VALVE,SOLENOID	1
16	PFHZZ	5945011408242	19207	12302662	BRACKET,SOLENOID	1
17	PAHZZ	5305000187838	24617	187838	SCREW,CAP,HEXAGON H	2
18	PAHZZ	5306011184889	80204	B1821BH050C075L	BOLT,MACHINE	4
19	PAHZZ	5310008095998	96906	MS27183-18	WASHER,FLAT..	4

END OF FIGURE

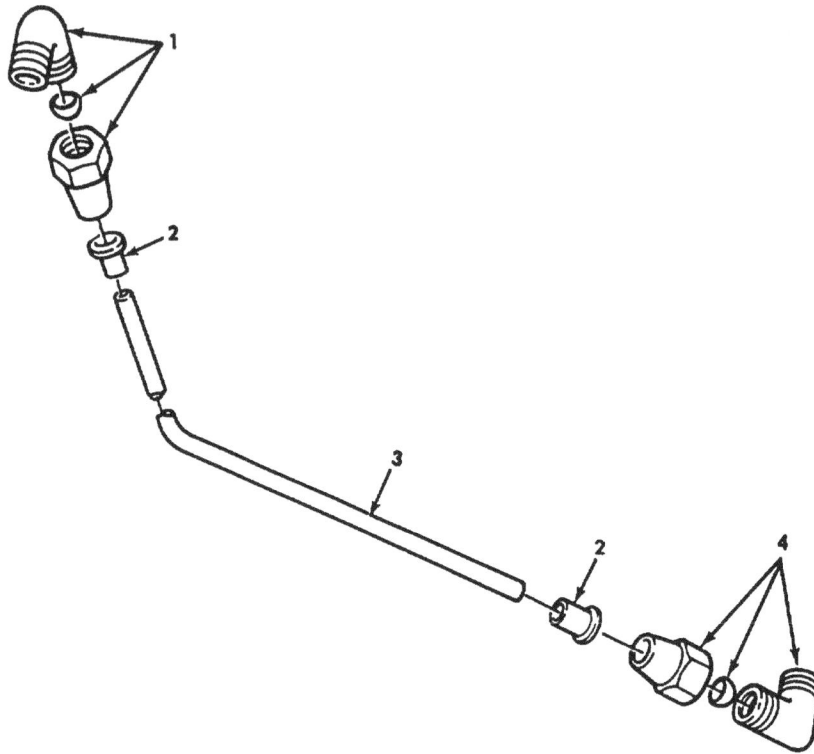

Figure 199. Transmission Vent Line.

(1) ITEM NO	(2) SMR CODE	(3) NSN	(4) CAGEC	(5) PART NUMBER	(6) DESCRIPTION AND USABLE ON CODES (UOC)	(7) QTY
					GROUP 0721 COOLERS, PUMPS, MOTORS	
					FIG. 199 TRANSMISSION VENT LINE	
1	PAOZZ	4730002871604	81343	6-2 120202BA	ELBOW,PIPE TO TUBE	1
2	PAOZZ	4730010798821	19207	CPR102321-1	INSERT,TUBE FITTING	2
3	MOOZZ		19207	CPR104420-2-24	HOSE,NONMETALLIC MAKE FROM TUBE,P/......... N CPR104420-2,24 INCHES LONG........................	1
4	PAOZZ	4730000691187	81343	6-4 100202BA	ELBOW,PIPE TO TUBE	1

END OF FIGURE

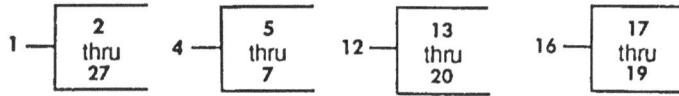

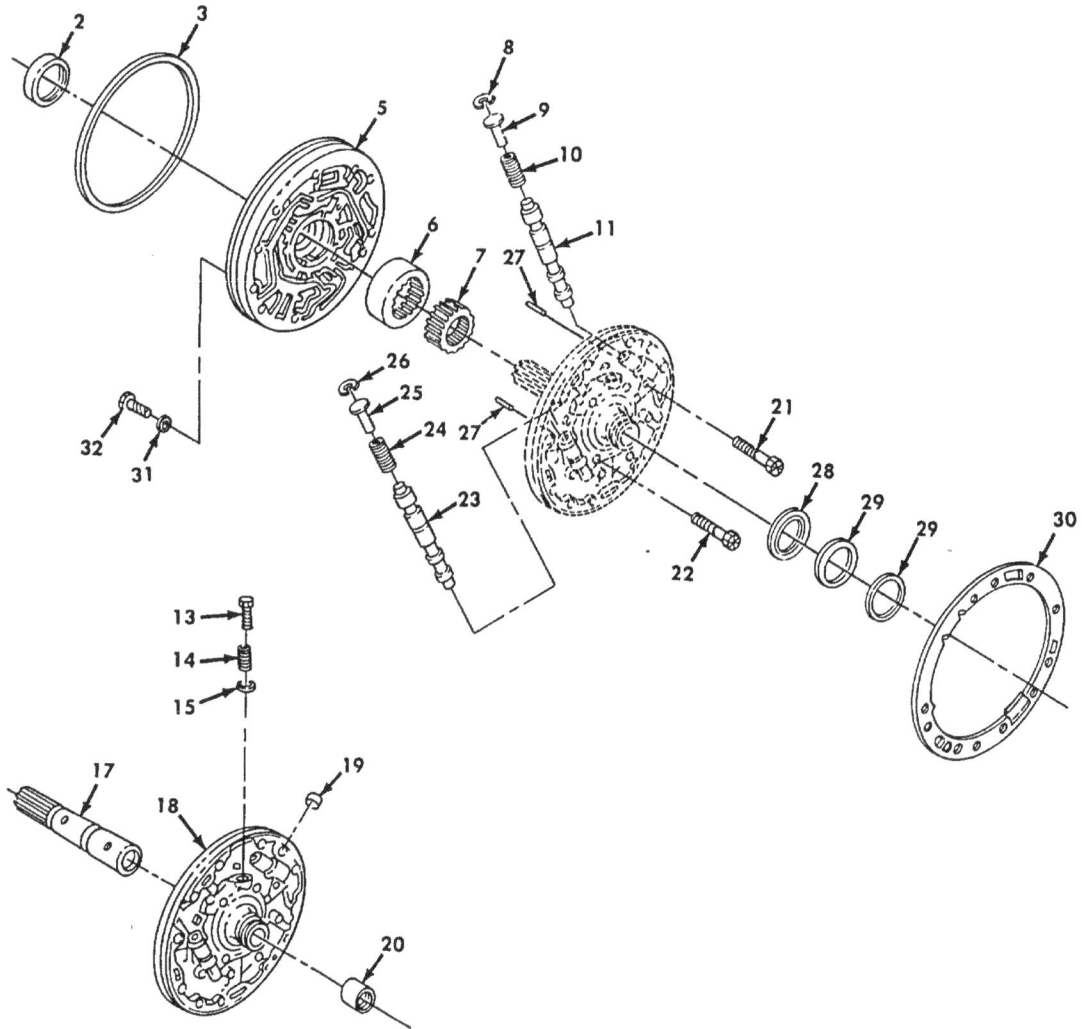

Figure 200. Oil Pump and Front Support Assembly (Transmission P/N 6885292).

(1) ITEM NO	(2) SMR CODE	(3) NSN	(4) CAGEC	(5) PART NUMBER	(6) DESCRIPTION AND USABLE ON CODES (UOC)	(7) QTY
					GROUP 0721 COOLERS, PUMPS, MOTORS	
					FIG. 200 OIL PUMP AND FRONT SUPPORT ASSEMBLY(TRANSMISSION P/N 6885292)	
1	PAHHH	2520011463039	73342	23015870	PUMP ASSEMBLY,TRANS	1
2	PAHZZ	5330012192375	73342	23016643	.SEAL PART OF KIT P/N 6884259........................	1
3	PAHZZ	2840010681713	73342	23016347	.SEAL,AIR,GAS TURBIN PART OF KIT P/N 6884259	1
4	PAHHH	3020010986742	73342	29527509	.BODY,GEAR ASSEMBLY USE WITH..................... TRANS. P/N 6885292	1
5	XAHZZ		73342	6881645	..BODY..	1
6	PAHZZ	3020012774635	73342	23016107	..GEAR,INTERNAL 819-.820	1
6	PAHZZ	3020010780627	73342	23016106	..GEAR,DRIVE MOTION	1
7	PAHZZ	3020012774636	73342	2957492	..GEAR,SPUR 0.8185-0.8190	1
7	PAHZZ	3020012774637	73342	23015398	..GEAR,SPUR 0.8190-0.8195	1
7	PAHZZ	3020011043803	73342	23015400	..GEAR,DRIVE 8200-.8205	1
7	PAHZZ	3020012774638	73342	23015399	..GEAR,SPUR 0.8195-0.8200	1
8	PAHZZ	5325008044997	96906	MS16625-1077	.RING,RETAINING	1
9	PAHZZ	2520010070345	73342	6834413	.STOP,VALVE	1
10	PAHZZ	5360010796704	73342	6836277	.SPRING,HELICAL,COMP	1
11	PAHZZ	2520010070346	73342	6834414	.VALVE,LOCKUP	1
12	PAHHH	2520011346534	73342	23015872	.SUPPORT,ASSEMBLY,TR USE WITH .... TRANS. P/N 6885292	1
13	PAHZZ	2520010111068	73342	6834410	..GUIDE,VALVE	11
14	PAHZZ	5360010796702	73342	6881072	..SPRING	1
15	PAHZZ	4820010111069	73342	6836202	..SEAT,VALVE	1
16	XAHHH		73342	23015874	..SUPPORT,ASSY	1
17	XAHZZ		73342	23013437	...SLEEVE,CONVERTER GR	1
18	XAHZZ		73342	23015876	...SUPPORT.	1
19	PAHZZ	5365010069622	73342	6762187	...PLUG,MACHINE THREAD.................................	1
20	PAHZZ	3110011490876	24617	9438017	..BEARING,ROLLER,NEED	1
21	PAHZZ	5306010522402	96906	MS35764-1297	.BOLT,SELF-LOCKING	2
22	PAHZZ	5305006382362	24617	9409225	.SCREW,CAP,HEXAGON H.................................	12
23	PAHZZ	2520010791615	73342	6881380	.VALVE,MAIN PRESSURE	1
24	PAHZZ	5360010796703	73342	6881381	.SPRING..	1
25	PAHZZ	2520010111067	73342	6834412	.STOP,VALVE	1
26	PAHZZ	5365003498518	73342	274613	.RING,RETAINING.	1
27	PAHZZ	5315011557302	24617	9418910	.PIN,STRAIGHT,HEADLE USE STARTING.............. WITH S/N2420021628..	2
28	PAHZZ	3110010655842	60380	FG42642	BEAARING	1
29	PAHZZ	2520005576211	73342	23014632	RING,SEAL	2
30	PAHZZ	5330005576518	73342	23014221	GASKET PART OF XIT P/N 6884259....................	1
31	PAHZZ	5310005575942	73342	6834908	WASHER,SPECIAL PART OF KIT P/N.................... 688425 9	12
32	PAHZZ	5305005591022	73342	179397	SCREW,CAP,HEXAGON H.................................	12

END OF FIGURE

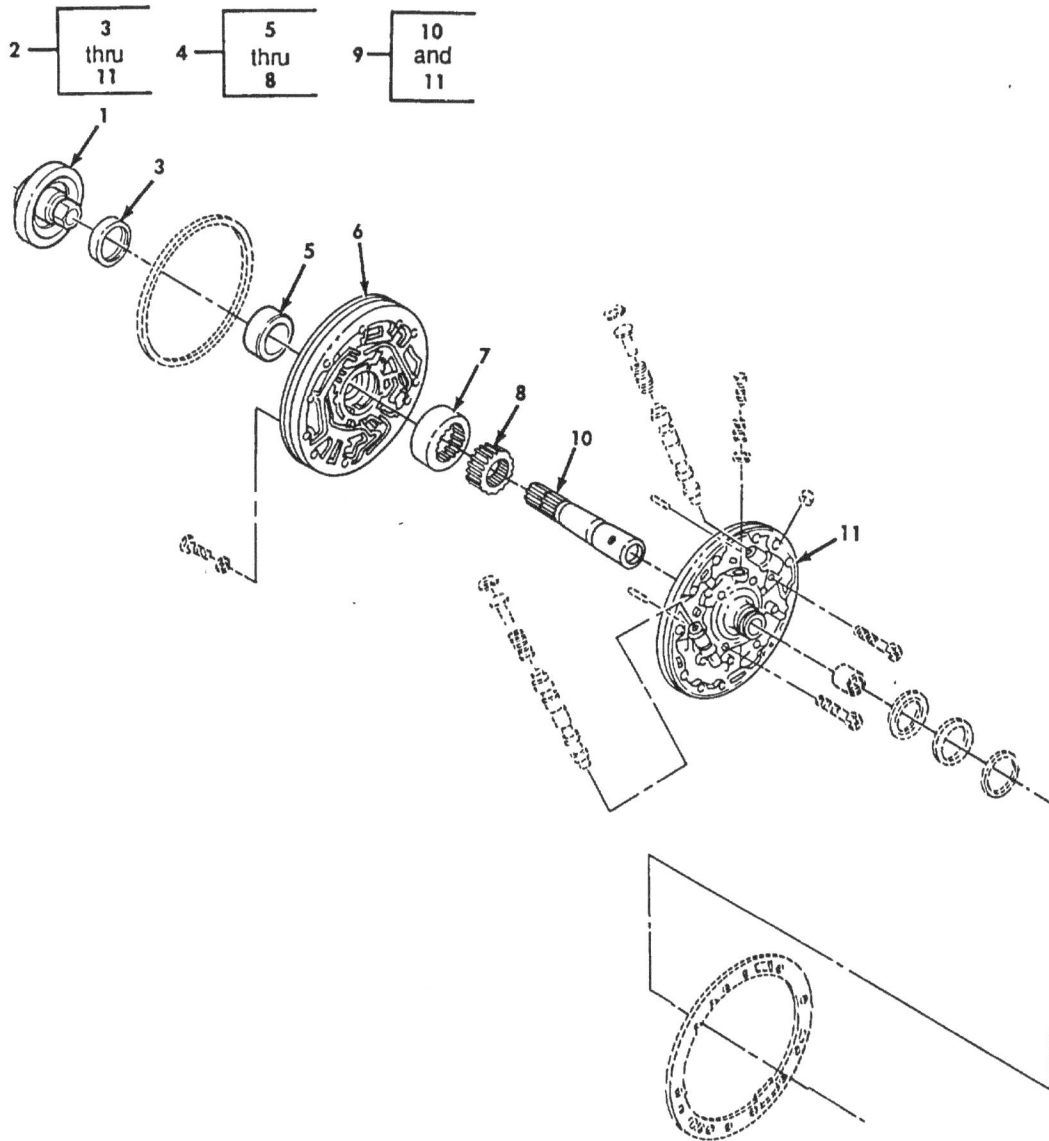

Figure 201. Oil Pump and Front Support Assembly (Transmission PIN 23040127).

(1) ITEM NO	(2) SMR CODE	(3) NSN	(4) CAGEC	(5) PART NUMBER	(6) DESCRIPTION AND USABLE ON CODES (UOC)	(7) QTY
					GROUP 0721 COOLERS,PUMPS,MOTORS	
					FIG. 201 OIL PUMP AND FRONT SUPPORT ASSEMBLY(TRANSMISSION P/N 23040127)	
1	PAHZZ	3040013374379	73342	23016723	HUB,BODY USE WITH TRANSMISSION P/N 23040127 PART OF KIT P/N 23043043	1
2	PAHHH	2815013400377	73342	23017115	OIL PUMP ASSEMBLY,E	1
3	PAHZZ	5330013390708	73342	23045556	.SEAL,PLAIN	1
4	PAHHH	4320013370608	73342	23017114	.HOUSING,LIQUID PUMP USE WITH TRANSMISSION P/N 23040127 PART OF KIT P/N 23043043	1
5	PAHZZ	3120013416519	73342	23046152	..BUSHING,SLEEVE	1
6	XAHZH	2815013400377	73342	23017115	..OIL PUMP ASSEMBLY,E	1
7	PAHZZ	3020013400403	73342	23017112	..GEAR,INTERNAL 0.998-0.999	1
7	PAHZZ	3020013400402	73342	23017113	..GEAR,HELICAL 0.999-1.000	1
8	PAHZZ	3020013400407	73342	23017311	..GEAR,SPUR 0.9985-0.9990	1
8	PAHZZ	3020013472855	73342	23017312	..GEAR,SPUR 0.9990-0.9950	1
8	PAHZZ	3020013400408	73342	29527485	..GEAR,SPUR 0.9995-1.0000	1
8	PAHZZ	3020013400404	73342	23017314	..GEAR,SPUR 1.0000-1.0005	1
9	PAHHH	3040013374161	73342	23016763	.HOUSING,MECHANICAL USE WITH TRANSMISSION P/N 23040127 PART OF KIT P/N 23043043	1
10	XAHZZ		73342	23042126	..SLEEVE,CONNECTOR	1
11	XAHZH	3040013374161	73342	23016763	..HOUSING,MECHANICAL USE WITH TRANSMISSION P/N 23040127 PART OF KIT P/N 23043043	1

END OF FIGURE

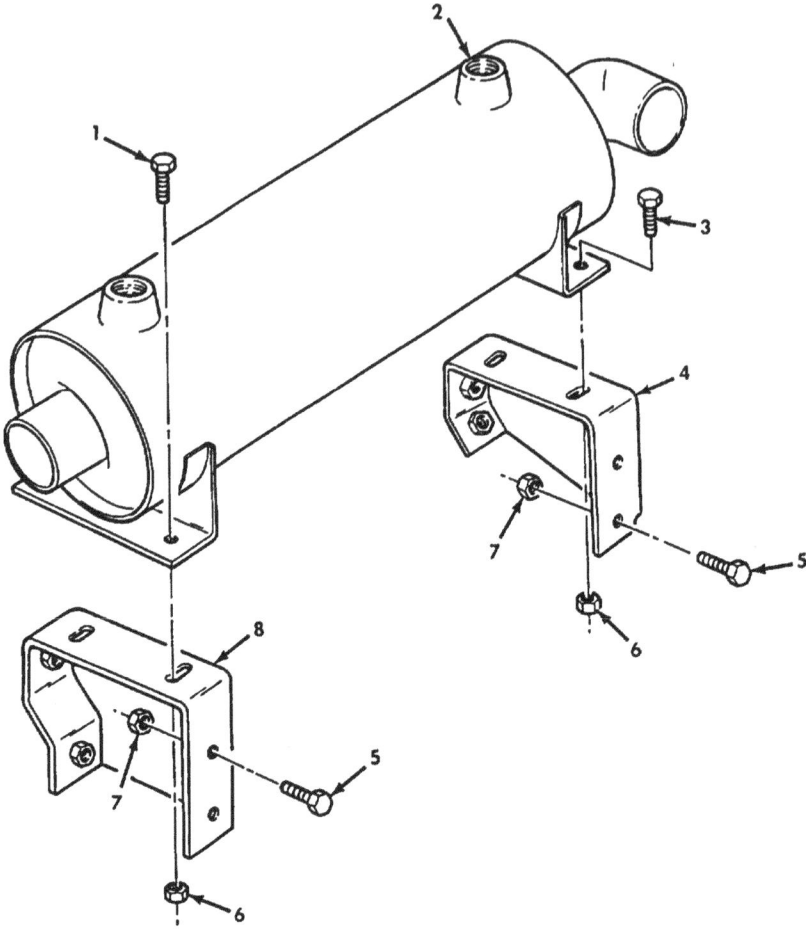

Figure 202. Transmission Oil Cooler and Mounting Hardware.

(1) ITEM NO	(2) SMR CODE	(3) NSN	(4) CAGEC	(5) PART NUMBER	(6) DESCRIPTION AND USABLE ON CODES (UOC)	(7) QTY
					GROUP 0721 COOLERS,PUMPS,MOTORS	
					FIG. 202 TRANSMISSION OIL COOLER AND MOUNTING HARDWARE	
1	PAFZZ	5305002693238	80204	B1821BH038F125N	SCREW,CAP,HEXAGON H	2
2	PAFZZ	2520010985124	19207	11669164	COOLER,FLUID,TRANSM	1
3	PAFZZ	5305009125113	96906	MS51096-359	SCREW,CAP,HEXAGON H	2
4	PFFZZ	5340012319151	19207	12302629	BRACKET,ENGINE ACCE	1
5	PAFZZ	5305007250154	96906	MS90727-112	SCREW,CAP,HEXAGON H	4
6	PAFZZ	5310008140672	96906	MS51943-36	NUT,SELF-LOCKING,HE	4
7	PAFZZ	5310008775795	96906	MS21044-N8	NUT,SELF-LOCKING,HE	4
8	PFFZZ	5340011066015	19207	12256165	BRACKET,SPECIAL	1

END OF FIGURE

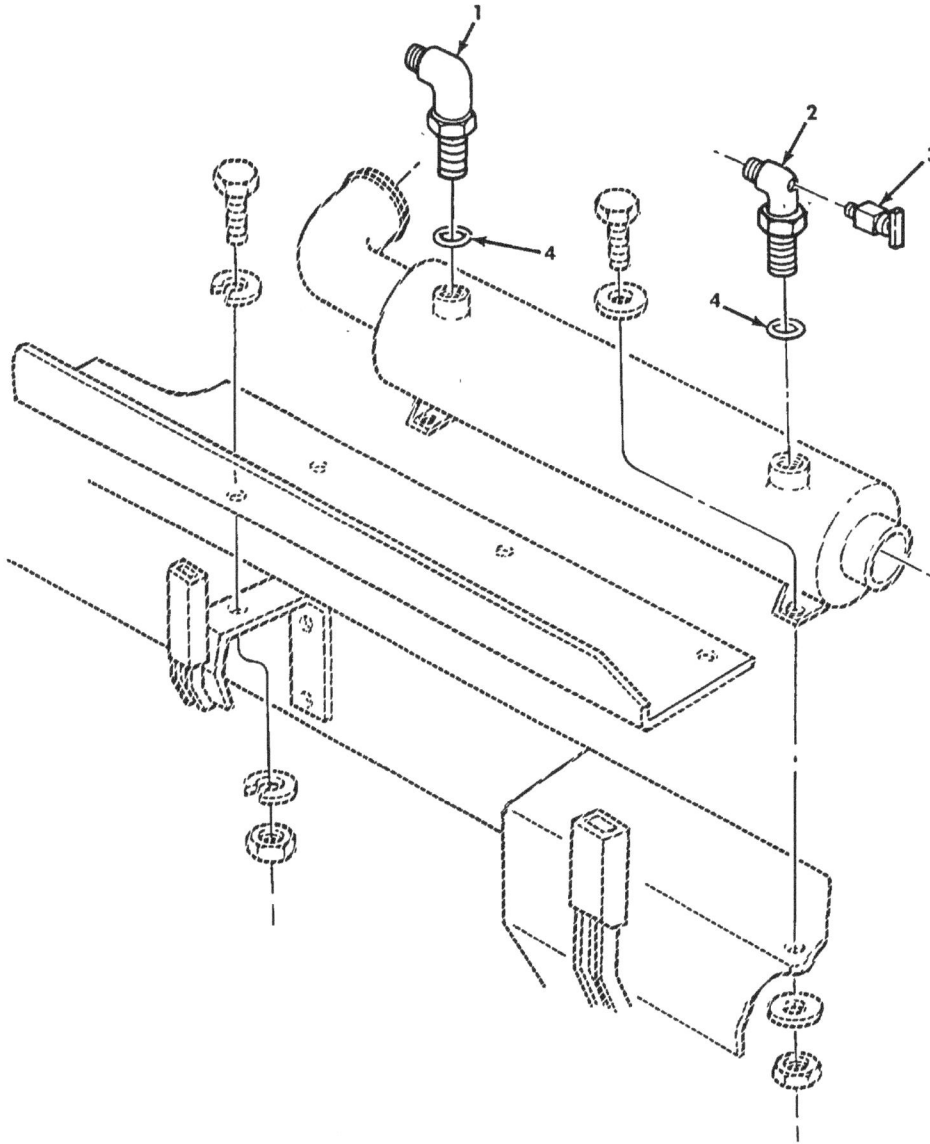

Figure 203. Transmission Oil Cooler Fittings (A1939A2)

(1) ITEM NO	(2) SMR CODE	(3) NSN	(4) CAGEC	(5) PART NUMBER	(6) DESCRIPTION AND USABLE ON CODES (UOC)	(7) QTY
					GROUP 0721 COOLERS,PUMPS,MOTORS	
					FIG. 203 TRANSMISSION OIL COOLER FITTINGS (M939A2)	
1	PAOZZ	4730011104059	19207	12256456-4	ELBOW,TUBE TO BOSS .... UOC: ZAA, ZAB, ZAC, ZAD, ZAE,ZAF, ZAG, ZAR, ZAJ,ZAK,ZAL	1
2	PAOZZ	4730011555449	19207	12256456-5	ELBOW,TUBE TO BOSS .... UOC:ZAA,ZAB,ZAC,ZAD,ZAE,ZAF,ZAG,ZAH, ZAJ,ZAX, ZAL	1
3	PAOZZ	4820008451096	96906	MS35783-2	COCK,DRAIN .... UOC:ZAA,ZAB,ZAC,ZAD,ZAE,ZAF,ZAG,ZAH, ZAJ,ZAK,ZAL	1
4	PAOZZ	5330002859842	96906	MS28778-10	O-RING .... UOC:ZAA,ZAB,ZAC,ZAD,ZAE,ZAF,ZAG,ZAH, ZAJ,ZAK,ZAL	2

END OF FIGURE

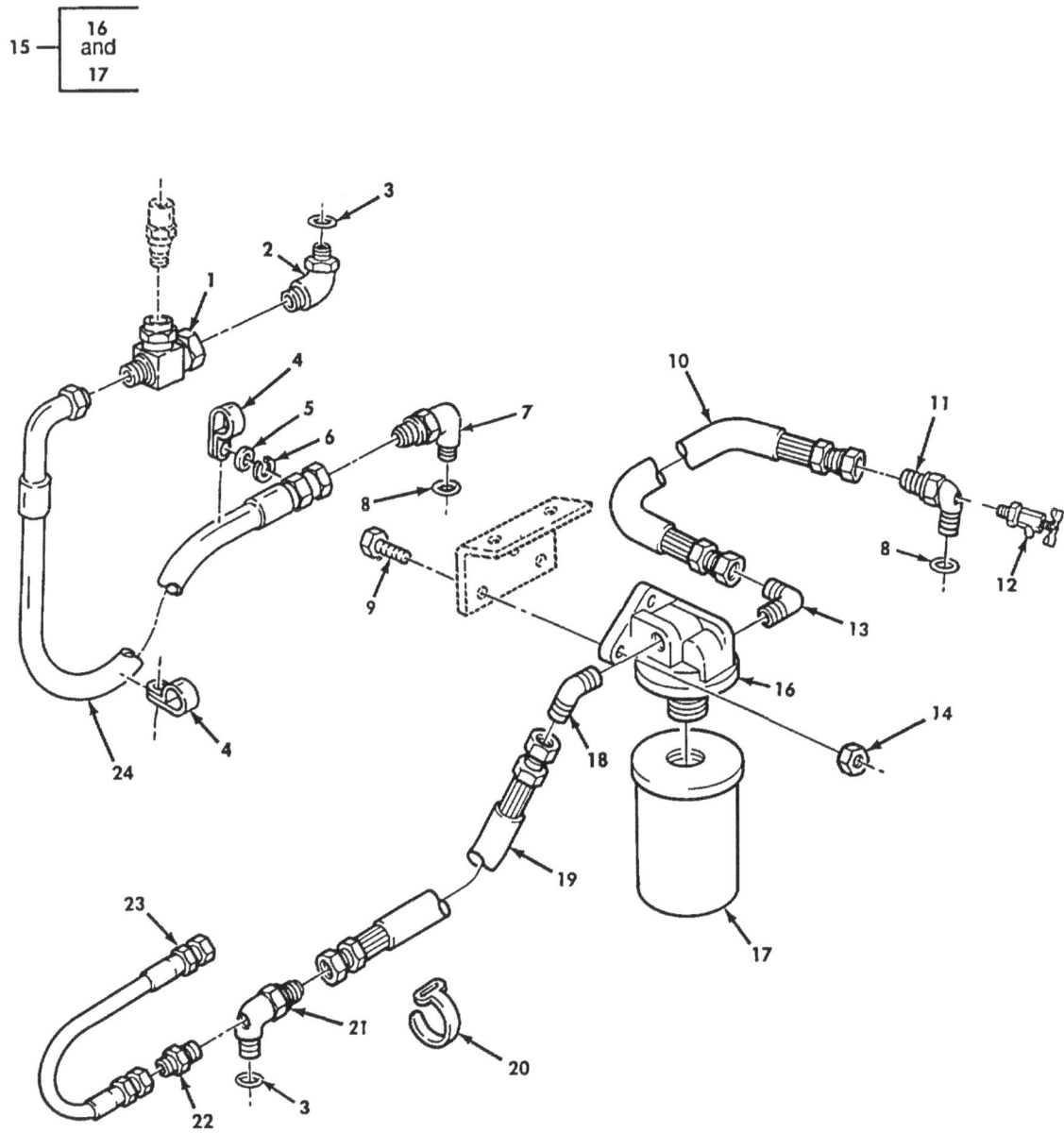

Figure 204. Oil Hoses, Filter, and Mounting Hardware (M939, M939A1).

(1) ITEM NO	(2) SMR CODE	(3) NSN	(4) CAGEC	(5) PART NUMBER	(6) DESCRIPTION AND USABLE ON CODES (UOC)	(7) QTY
					GROUP 0721 COOLERS, PUMPS, MOTORS	
					FIG. 204 OIL HOSES, FILTER, AND MOUNTING HARDWARE(M939,M939A1)	
1	PAFZZ	4730010910266	19207	12256531	TEE, PIPE TO TUBE .... UOC:DAA, DAB, DAC, DAD, DAE, DAF, DAG, DAH, DAJ, DAX, DAL, DAW, DAX, V12, V13, V14, V15, V16, V17, V18, V19, V20, V21, V22, V24, V25, V39	1
2	PAFZZ	4730010893071	19207	12256456-2	ELBOW ASSEMBLY, PIPE .... UOC:DAA, DAB, DAC, DAD, DAE, DAF, DAG, DAH, DAJ, DAK, DAL, DAW, DAX, V12, V13, V14, V15, V16, V17, V18, V19, V20, V21, V22, V24, V25, V39	1
3	PAFFF	5331007762830	81343	AS3551-12	O-RING .... UOC:DAA, DAB, DAC, DAD, DAE, DAF, DAG, DAH, DAJ, DAK, DAL, DAW, DAX, V12, V13, V14, V15, V16, V17, V18, V19, V20, V21, V22, V24, V25, V39	2
4	PAFZZ	5340011068346	19207	7529133-5	CLAMP, LOOP .... UOC:DAA, DAB, DAC, DAD, DAE , DAF, DAG, DAH, DAJ, DAK, DAL, DAW, DAX, V12, V13, V14, V15, V16, V17, V18, V19, V20, V21, V22, V24, V25, V39	2
5	PAFZZ	5310000806004	96906	MS27183-14	WASHER, FLAT .... UOC:DAA, DAB, DAC, DAD, DAE, DAF, DAG, DAH, DAJ, DAK, DAL, DAW, DAX, V12 , V13 , V14 , V15, V16, V17, V18, V19, V20, V21, V22, V24, V25, V39	1
6	PAFZZ	5310000034094	01276	210104-8S	WASHER, LOCK .... UOC:DAA, DAB, DAC, DAD, DAE, DAF, DAG, DAH, DAJ, DAK, DAL, DAW, DAX, V12, V13, V14, V15, V16, V17, V18, V19, V20, V21, V22, V24, V25, V39	1
7	PAFFF	4730011104059	19207	12256456-4	ELBOW, TUBE TO BOSS .... UOC:DAA, DAB, DAC, DAD, DAE, DAF, DAG, DAH, DAJ, DAK, DAL, DAW, DAX, V12, V13, V14, V15, V16, V17 , V18, V19 , V20, V21, V22, V24, V25, V39	1
8	PAFZZ	5331002859842	96906	MS28778-10	O-RING .... UOC:DAA, DAB, DAC, DAD, DAE, DAF, DAG, DAH, DAJ, DAK, DAL, DAW, DAX, V12 , V13 , V14 , V15, V16, V17, V18, V19, V20, V21, V22, V24, V25, V39	2
9	PAFZZ	5306000514077	80204	B1821BH031F113N	BOLT, MACHINE .... UOC:DAA, DAB, DAC, DAD, DAE, DAF, DAG, DAH, DAJ, DAK, DAL, DAW, DAX, V12, V13, V14, V15, V16, V17, V18, V19, V20, V21, V22, V24, V25, V39	2
10	PAFZZ	4720011364454	19207	12277398-1	HOSE ASSEMBLY, NONME .... UOC:DAA, DAB, DAC, DAD, DAE, DAF, DAG, DAH,	1

(1) ITEM NO	(2) SMR CODE	(3) NSN	(4) CAGEC	(5) PART NUMBER	(6) DESCRIPTION AND USABLE ON CODES (UOC)	(7) QTY
					DAJ, DAK, DAL, DAW, DAX, V12, V13, V14, V15, V16, V17, V17, V119, V20, V21, V22, V24, V25, V39	
11	PAFZZ	4730011555449	19207	12256456-5	ELBOW, TUBE TO BOSS .... UOC:DAA, DAB, DAC, DAD, DAE, DAF, DAG, DAH, DAJ, DAK, DAL, DAW, DAX, V2, V13, V14, V155, V16, V17, V18, V19, V20, V21, V22, V24, V25, V39	1
12	PAOZZ	4820008451096	96906	MS35783-2	COCK, DRAIN .... UOC:DAA, DAB, DAC, DAD, DAE, DAF, DAG, DAH, DAJ, DAK, DAL, DAW, DAX, V12, V13, V14, V15, V16, V17, V18, V19, V20, V2, V2, V22, V24, V25, V39	1
13	PAFZZ	4730002894912	96906	MS51504A12	ELBOW, PIPE TO TUBE COVER .... UOC:DAA, DAB, DAC, DAD, DAE, DAF, DAG, DAH, DAJ, DAK, DAL, DAW, DAX, V12, V13, V14, V15, V16, V17, V18 , V19, V20, V21, V22, V24, V25, V39	1
14	PAFZZ	5310002416658	96906	MS51943-34	NUT, SELF-LOCKING, HE .... UOC:DAA, DAB, DAC, DAD, DAE, DAF, DAG, DAH, DAJ, DAK, DAL, DAW, DAX, V12, V13, V14, V15, V16, V17, V18, V19, V20, V21, V2, V22, V24, V25, V39	2
15	PAFFF	2520011232649	70040	25011159	FILTER, FLUID, PRESSU TRANSMISSION .... UOC:DAA, DAB, DAC, DAD, DAE, DAF, DAG, DAH, DAJ, DAK, DAL, DAW, DAX, V12, V13, V14, V15, V16, V17, V18, V19, V20, V21, V22, V24, V25, V39	1
16	XAFZZ	2940011645804	70040	25011151	.COVER, OIL FILTER .... UOC:DAA, DAB, DAC, DAD, DAE, DAF, DAG, DAH, DAJ, DAK, DAL, DAW, DAX, V12, V13 , V14, V15, V16, V17, V18, V19, V20, V21, V22, V24, V25, V39	
17	PAOZZ	2940011102489	73370	PH3519	.FILTER ELEMENT, FLUI .... UOC:DAA, DAB, DAC, DAD, DAE, DAF, DAG, DAH, DAJ, DAK, DAL, DAW, DAX, V12, V13, V14, V15, V16, V17, V18, V19, V20, V21, V22, V24, V25, V39	1
18	PAFZZ	4730010288147	96906	MS51508-B12	ELBOW, PIPE TO TUBE .... UOC:DAA, DAB, DAC, DAD, DAE, DAF, DAG, DAH, DAJ, DAK, DAL, DAW, DAX, V12, V13, V14, V15, V16, V17, V18, V19, V20, V21, V22, V24, V25, V39	1
19	PAFZZ	4720011371453	19207	12277398-2	HOSE ASSEMBLY, NONME .... UOC:DAA, DAB, DAC, DAD, DAE, DAF, DAG, DAH, DAJ, DAK, DAL, DAW, DAX, V12, V13, V14, V15, V16, V17, V18, V19, V20, V21, V22, V24, V25, V39	1
20	PAFZZ	5975009856630	96906	MS3367-3-0	STRAP, TIEDOWN, ELECT .... UOC:DAA, DAB, DAC, DAD, DAE, DAF, DAG, DAH, DAJ, DAK, DAL, DAW, DAX, V12, V13, V14, V15, V16, V17, V18, V19, V20, V21, V22, V24, V25,	2

(1) ITEM NO	(2) SMR CODE	(3) NSN	(4) CAGEC	(5) PART NUMBER	(6) DESCRIPTION AND USABLE ON CODES (UOC)	(7) QTY
					V39	
21	PAFZZ	4730010893072	19207	122'56456-3	ELBOW ASSEMBLY, PIPE .... UOC:DAA, DAB, DAC, DAD, DAE, DAF, DAG, DAH, DAJ, DAK, DAL, DAW, DAX, V12, V13 , V14, V15, V16, V17, V18, V19, V20, V21, V22, V24, V25, V39	1
22	PAFZZ	4730002890383	96906	MS51500-A4	ADAPTER, STRAIGHT, PI .... UOC:DAA, DAB, DAC, DAD, DAE, DAF, DAG, DAH, DAJ, DAK, DAL, DAW, DAX, V12 , V13 , V14, V15, V16, V17, V18, V19, V20, V21, V22, V24, V25, V39	2
23	PAFZZ	4720010893074	19207	12256489	HOSE ASSEMBLY, NONME .... UOC:DAA, DAB, DAC, DAD, DAE, DAF , DAG , DAH, DAJ, DAK, DAL, DAW, DAX, V12 , V13, V14 , V15, V16, V17, V18, V19, V20, V21, V22, V24, V25, V39	1
24	PAFZZ	4720011152274	19207	12277179	HOSE ASSEMBLY, NONME .... UOC:DAA, DAB, DAC, DAD, DAE, DAF, DAG, DAH, DAJ, DAK, DAL, DAW, DAX, V12, V13, V14, V15, V16, V17, V18, V19, V20, V21, V22, V24, V25, V39	1

END OF FIGURE

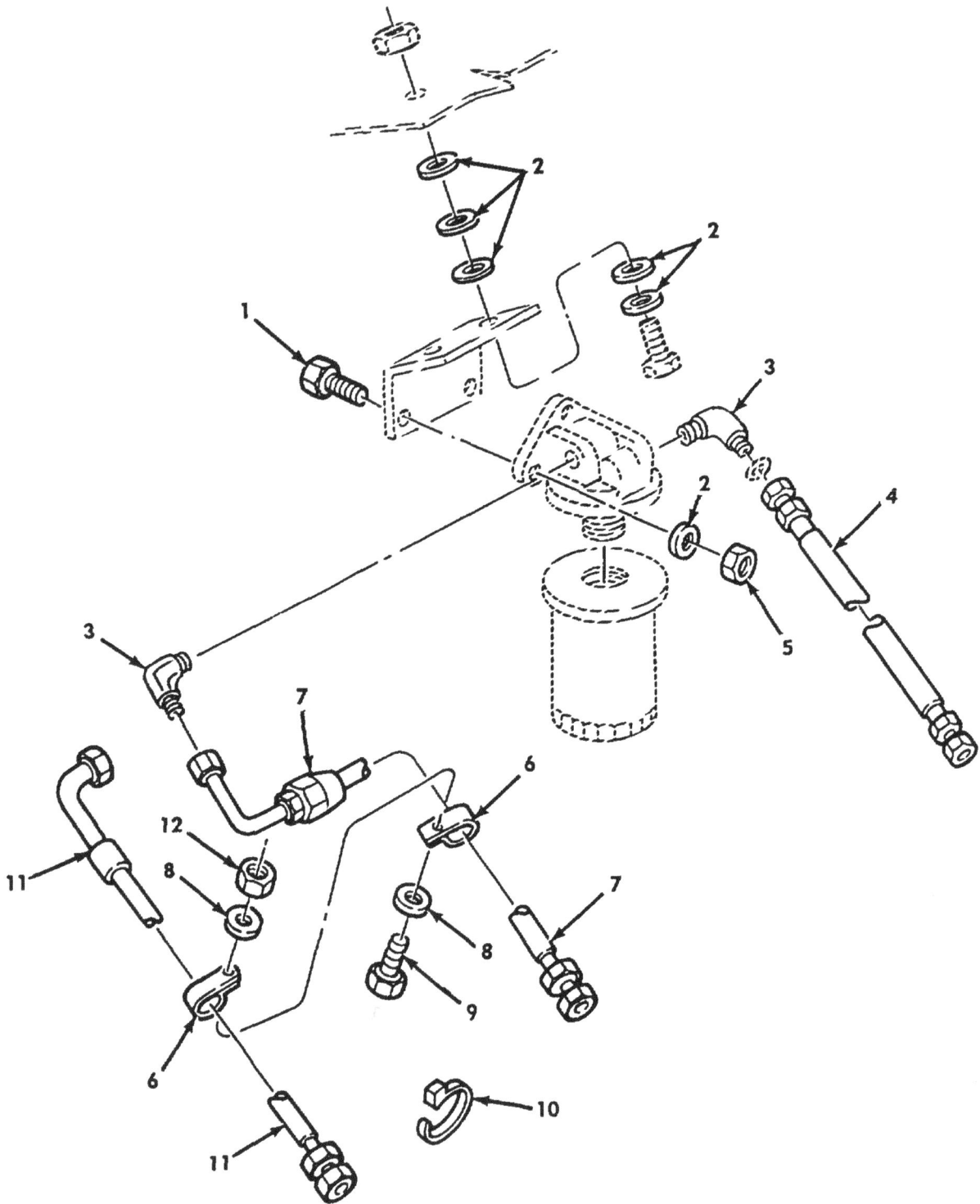

Figure 205. Oil Hoses (M939A2).

(1) ITEM NO	(2) SMR CODE	(3) NSN	(4) CAGEC	(5) PART NUMBER	(6) DESCRIPTION AND USABLE ON CODES (UOC)	(7) QTY
					GROUP 0721 COOLERS, PUMPS, MOTORS	
					FIG. 205 OIL HOSES(M939A2)	
1	PAOZZ	5305002259091	96906	MS90726-36	SCREW, CAP, HEXAGON H .... UOC: ZAA, ZAB, ZAC, ZAD, ZAE, ZAF, ZAG, ZAH, ZAJ, ZAK, ZAL	3
2	PAOZZ	5310000814219	96906	MS27183-12	WASHER, FLAT .... UOC:ZAA, ZAB, ZAC, ZAD, ZAE, ZAF, ZAG, ZAH, ZAJ, ZAK, ZAL	11
3	PAOZZ	4730006400264	81343	12-12 070202BA	ELBOW, PIPE TO TUBE .... UOC: ZAA, ZAB, ZAC, ZAD, ZAE, ZAF, ZAG, ZAH, ZAJ, ZAK, ZAL	2
4	PAOZZ	4720012720504	98441	A3917-1	HOSE ASSEMBLY, NONME .... UOC:ZAA, ZAB, ZAC, ZAD, ZAE, ZAF, ZAG, ZAH, ZAJ, ZAK, ZAL	1
5	PAOZZ	5310002416658	96906	MS51943-34	NUT, SELF-LOCKING, HE .... UOC:ZAA, ZAB, ZAC, ZAD, ZAE, ZAF, ZAG, ZAH, ZAJ, ZAK, ZAL	3
6	PAOZZ	5340011068346	19207	7529133-5	CLAMP, LOOP .... UOC:ZAA, ZAB, ZAC, ZAD, ZAE, ZAF, ZAG, ZAH, ZAJ, ZAK, ZAL	2
7	PAOZZ	4720012717997	98441	A3915-3	HOSE ASSEMBLY, NONME .... UOC:ZAA, ZAB, ZAC, ZAD, ZAE, ZAF, ZAG, ZAH, ZAJ, ZAK, ZAL	1
8	PAOZZ	5310010842362	63005	9411417	WASHER .... UOC:ZAA, ZAB, ZAC, ZAD, ZAE, ZAF, ZAG, ZAH, ZAJ, ZAK, ZAL	2
9	PAOZZ	5305007215492	80204	B1821BH038C063N	SCREW, CAP, HEXAGON H .... UOC: ZAA, ZAB, ZAC, ZAD, ZAE, ZAF, ZAG, ZAH, ZAJ, ZAK, ZAL	1
10	PAOZZ	5975004515001	96906	MS3367-3-9	STRAP, TIEDOWN, ELECT .... UOC:ZAA, ZAB, ZAC, ZAD, ZAE, ZAF, ZAG, ZAH, ZAJ, ZAK, ZAL	2
11	PAOZZ	4720012717996	9844	A3916-2	HOSE ASSEMBLY, NONME .... UOC:ZAA, ZAB, ZAC, ZAD, ZAE, ZAF, ZAG, ZAH, ZAJ, ZAK, ZAL	1
12	PAOZZ	5310009359021	96906	MS51943-35	NUT, SELF-LOCKING, HE .... UOC:ZAA, ZAB, ZAC, ZAD, ZAE, ZAF, ZAG, ZAH, ZAJ, ZAK, ZAL	1

END OF FIGURE

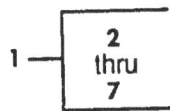

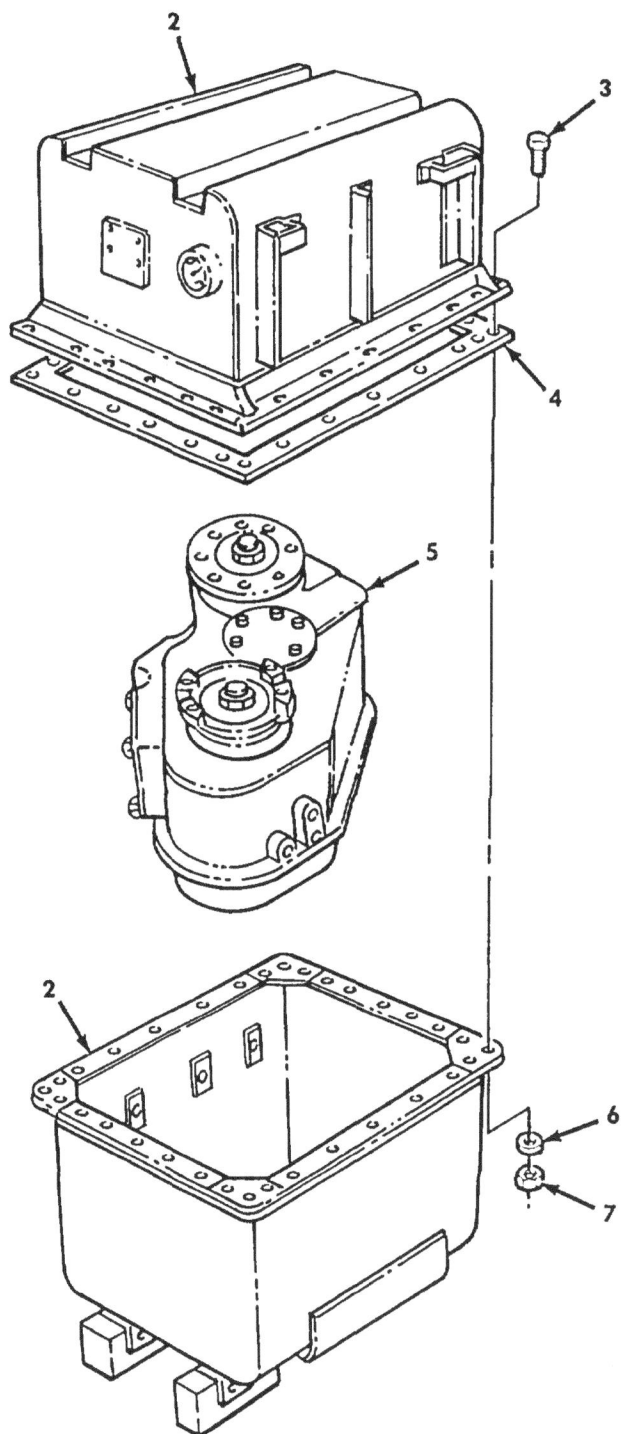

Figure 206. Transfer Case and Shipping Container.

(1) ITEM NO	(2) SMR CODE	(3) NSN	(4) CAGEC	(5) PART NUMBER	(6) DESCRIPTION AND USABLE ON CODES (UOC)	(7) QTY
					GROUP 08 TRANSFER CASE 0800 TRANSFER CASE	
					FIG. 206 TRANSFER CASE AND SHIPPING CONTAINER	
1	PAFHH	2520011441528	19207	5704517	TRANSFER AND CONTAI	1
2	XDFFH	8145009070695	19207	10932088	.SHIPPING AND STORAG	1
3	PAFZZ	5305007247222	80204	B1821BH063C200N	.SCREW, CAP, HEXAGON H	30
4	PAFZZ	5330001712776	19207	10932085	.SEAL, NONMETALLIC RO	1
5	PRFFH		19207	11669112	.TRANSFER, TRANSMISSI	1
6	PAFZZ	5310008206653	96906	MS35338-50	.WASHER, LOCK	30
7	PAFZZ	5310007638920	96906	MS51967-20	.NUT, PLAIN, HEXAGON	30

END OF FIGURE

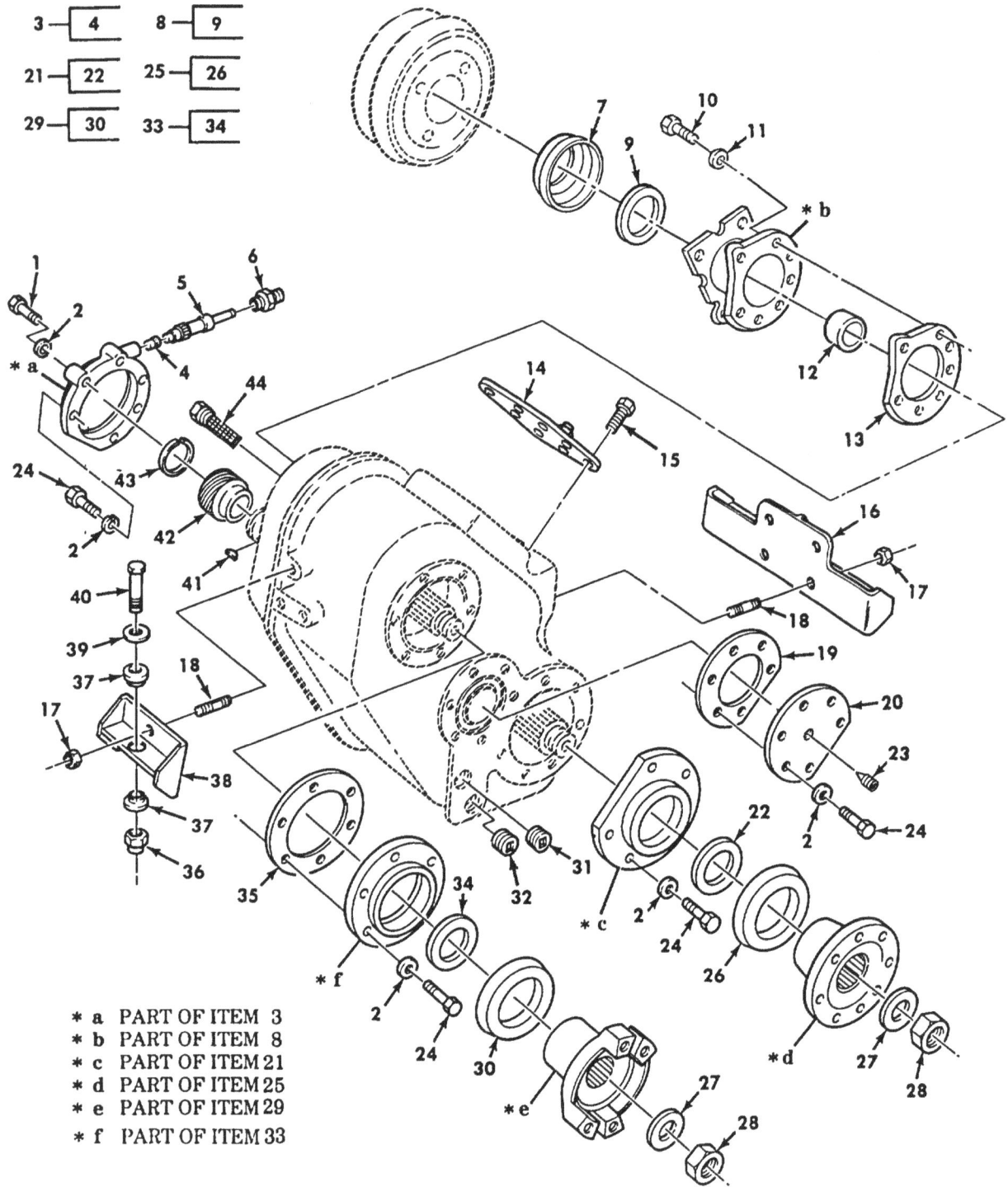

Figure 207. Transfer Case Mounting Hardware, Housing, Flanges, and Speedometer Drive.

* a   PART OF ITEM 3
* b   PART OF ITEM 8
* c   PART OF ITEM 21
* d   PART OF ITEM 25
* e   PART OF ITEM 29
* f   PART OF ITEM 33

(1) ITEM NO	(2) SMR CODE	(3) NSN	(4) CAGEC	(5) PART NUMBER	(6) DESCRIPTION AND USABLE ON CODES (UOC)	(7) QTY
					GROUP 0801 POWER TRANSFER AND FINAL DRIVE ASSEMBLIES	
					FIG. 207 TRANSFER CASE MOUNTING HARDWARE, HOUSING, FLANGES, AND SPEEDOMETER DRIVE	
1	PAFZZ	5305000712058	80204	B1821BH044C225N	SCREW, CAP, HEXAGON H	2
2	PAFZZ	5310008933381	19207	7748744	WASHER, FLAT	24
3	PAFZZ	2520011463447	78500	A3266-W-621	COVER AND SEAL ASSE	1
4	PAFZZ	3120005167516	78500	1199-M-1105	.BEARING, SLEEVE	1
5	PAFZZ	2520011272624	78500	2297-S-3815	SHAFT, SPEEDOMETER, G	1
6	PAFZZ	4730002029539	78500	1199-N-1106	REDUCER, TUBE	1
7	PAOZZ	2520011374843	78500	3264-W-101	SHIELD AND DEFLECTO	1
8	PAFZZ	2530011340900	78500	A3268-R-1084	SPIDER, BRAKE	1
9	PAOZZ	5330010230269	19207	12470105	.SEAL, PLAIN ENCASED	1
10	PAFZZ	5305011353536	70960	S-2914-C-1	SCREW, CAP, HEXAGON H	7
11	PAFZZ	5310007251983	81205	BACW10P53S	WASHER, FLAT	7
12	PAFZZ	5365011296590	78500	1244-D-1590	SPACER	1
13	PAFZZ	5365011352829	78500	2203-V-7848	SHIM .003	1
13	PAFZZ	5365011352830	78500	2203-W-7849	SHIM .005	1
13	PAFZZ	5365011352831	78500	2203-X-7850	SHIM .010	1
14	PAFZZ	2520011394233	78500	A-3289-G-189	COVER, INSPECTION	1
15	PAFZZ	5305011303267	78500	S-265-1-C	SCREW, CAP, HEXAGON H	8
16	PAFZZ	5340010892989	19207	12256188	BRACKET, MOUNTING	1
17	PAFZZ	5310002416664	96906	MS51943-44	NUT, SELF-LOCKING, HE	7
18	PAFZZ	5307011219883	19207	8331791-1	STUD, PLAIN	7
19	PAFZZ	5365011285061	78500	2203-L-7786	SHIM .003	1
19	PAFZZ	5365011333749	78500	2203-M-7787	SHIM .005	1
19	PAFZZ	5365011333750	78500	2203-N-7788	SHIM .010	1
20	PAFZZ	2520011393124	78500	3266-F-864	COVER, SHAFT	1
21	PAFZZ	2520011269364	78500	A3266-A-625	COVER AND SEAL ASSE	1
22	PAOZZ	5330010230269	19207	12470105	.SEAL, PLAIN ENCASED	1
23	PAFZZ	4730000103875	81348	WW-P-471ACBBUE	PLUG, PIPE	1
24	PAFZZ	5305003385162	78500	S-2710-1-C	SCREW, SPECIAL	22
25	PAFZZ	2520011269363	78500	A-3280-U-7535	FLANGE, DEFLECTOR	1
26	PAFZZ	2520002555149	97286	2297-C-627	.SLINGER	1
27	PAOZZ	5310011294373	78500	1229-T-1450	WASHER, FLAT	2
28	PAFZZ	5310010990397	14262	1227-K-1051	NUT	2
29	PAFFF	2520011273595	78500	A-3260-A-79	YOKE AND DEFLECTOR	1
30	PAFZZ	2520002555149	97286	2297-C-627	.SLINGER	1
31	PAFZZ	4730001874210	96906	MS51884-9	PLUG, PIPE	1
32	PAFZZ	4730005954827	19207	7521743	PLUG, PIPE, MAGNETIC	1
33	PAFZZ	2520011343665	78500	A-3266-Y-623	COVER, TRANSFER CASE	1
34	PAOZZ	5330010230269	19207	12470105	.SEAL, PLAIN ENCASED	1
35	PAFZZ	5365011285062	78500	2203-P-7790	SHIM .003	1
35	PAFZZ	5365011285063	78500	2203-Q-7791	SHIM .005	1
35	PAFZZ	5365011285064	78500	2203-R-7792	SHIM .010	1
36	PAFZZ	5310002416664	96906	MS51943-44	NUT, SELF-LOCKING, HE	3
37	PAFZZ	5340007411068	19207	7411068	MOUNT, RESILIENT	6
38	PFFZZ	5340010892988	19207	12256083	BRACKET, DOUBLE ANGL	1
39	PAFZZ	5310005319120	19207	7418962	WASHER, FLAT	3

(1) ITEM NO	(2) SMR CODE	(3) NSN	(4) CAGEC	(5) PART NUMBER	(6) DESCRIPTION AND USABLE ON CODES (UOC)	(7) QTY
40	PAFZZ	5305007262556	96906	MS90727-169	SCREW, CAP, HEXAGON H	3
41	PAFZZ	5315006165514	96906	MS35756-6	KEY, WOODRUFF	1
42	PAFZZ	3020011268444	78500	2297-T-3816	GEAR, SPEED DRIVE	1
43	PAFZZ	5325011296849	78500	1229-U-2829	RING, RETAINING	1
44	PAFZZ	4730011326848	78500	A-2206-D-56	COUPLING, PIPE	1

END OF FIGURE

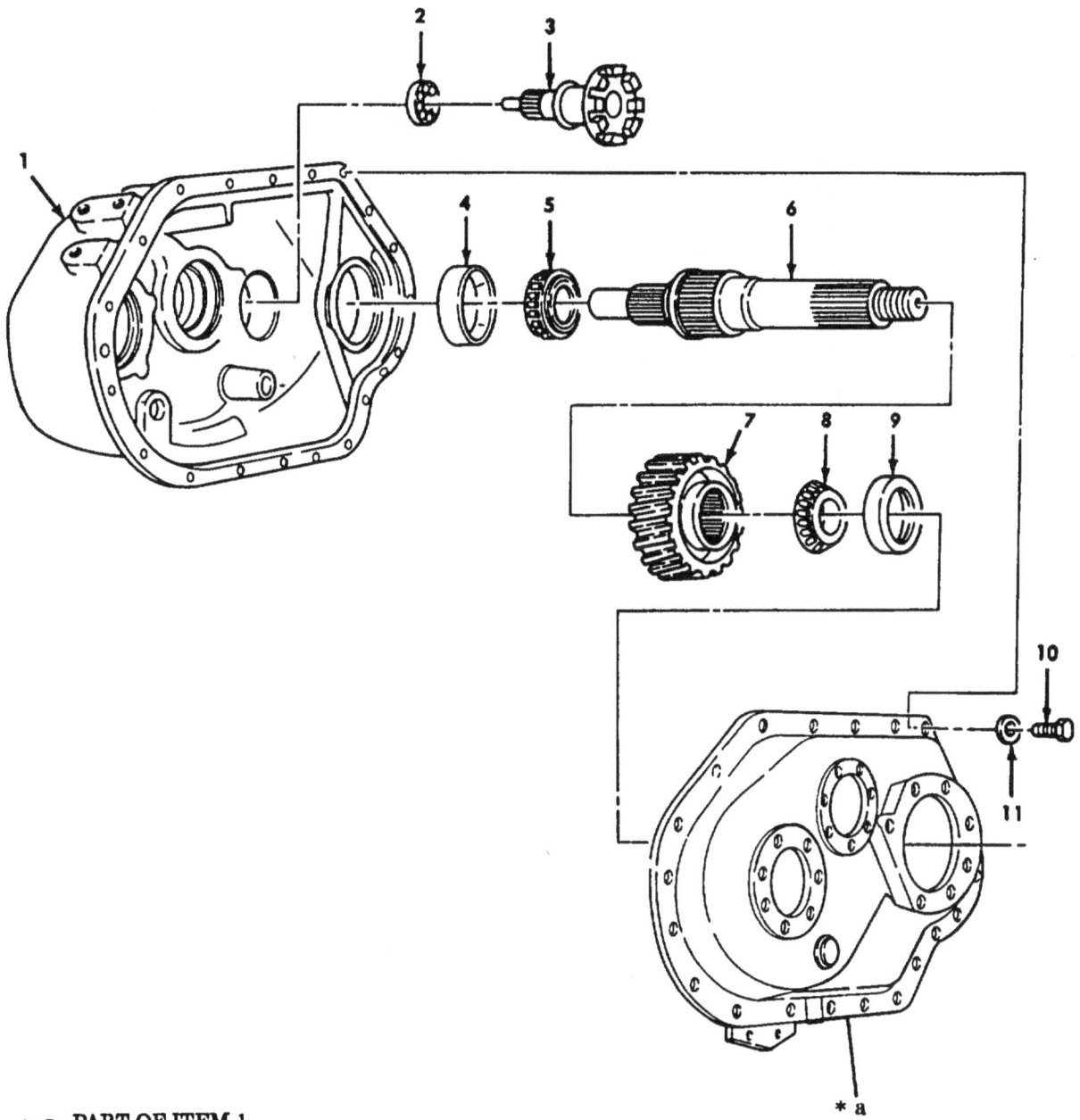

* a   PART OF ITEM 1

Figure 208.   Transfer Case Housing, Output Shaft, and Cover.

(1) ITEM NO	(2) SMR CODE	(3) NSN	(4) CAGEC	(5) PART NUMBER	(6) DESCRIPTION AND USABLE ON CODES (UOC)	(7) QTY

GROUP 0801 POWER TRANSFER AND FINAL DRIVE ASSEMBLIES

FIG. 208 TRANSFER CASE HOUSING, OUTPUT SHAFT, AND COVER

(1) ITEM NO	(2) SMR CODE	(3) NSN	(4) CAGEC	(5) PART NUMBER	(6) DESCRIPTION AND USABLE ON CODES (UOC)	(7) QTY
1	PAHZZ	3040012483995	78500	A3290Y233	HOUSING, MECHANICAL	1
2	PAHZZ	3110002938439	78500	1828-S-149	BEARING, BALL, ANNULA	1
3	PAHZZ	2520011271676	78500	A-3297-K-63	SHAFT ASSEMBLY, TRAN FRONT OUTPUT	1
4	PAHZZ	3110001000305	78500	394A	CUP, TAPERED ROLLER	1
5	PAHZZ	3110001000231	24617	441294	CONE AND ROLLERS, TA	1
6	PAHZZ	2520011272625	78500	3297-L-64	SHAFT, TRANSFER TRAN	1
7	PAHZZ	3020011428090	78500	3892-A-4473	GEAR, HELICAL	1
8	PAHZZ	3110011439242	78500	39578	CONE AND ROLLERS, TA	1
9	PAHZZ	3110001437586	78500	39520	CUP, TAPERED ROLLER	1
10	PAHZZ	5306010845389	78500	S-2712-1	BOLT, MACHINE	19
11	PAFZZ	5310008933381	19207	7748744	WASHER, FLAT	19

END OF FIGURE

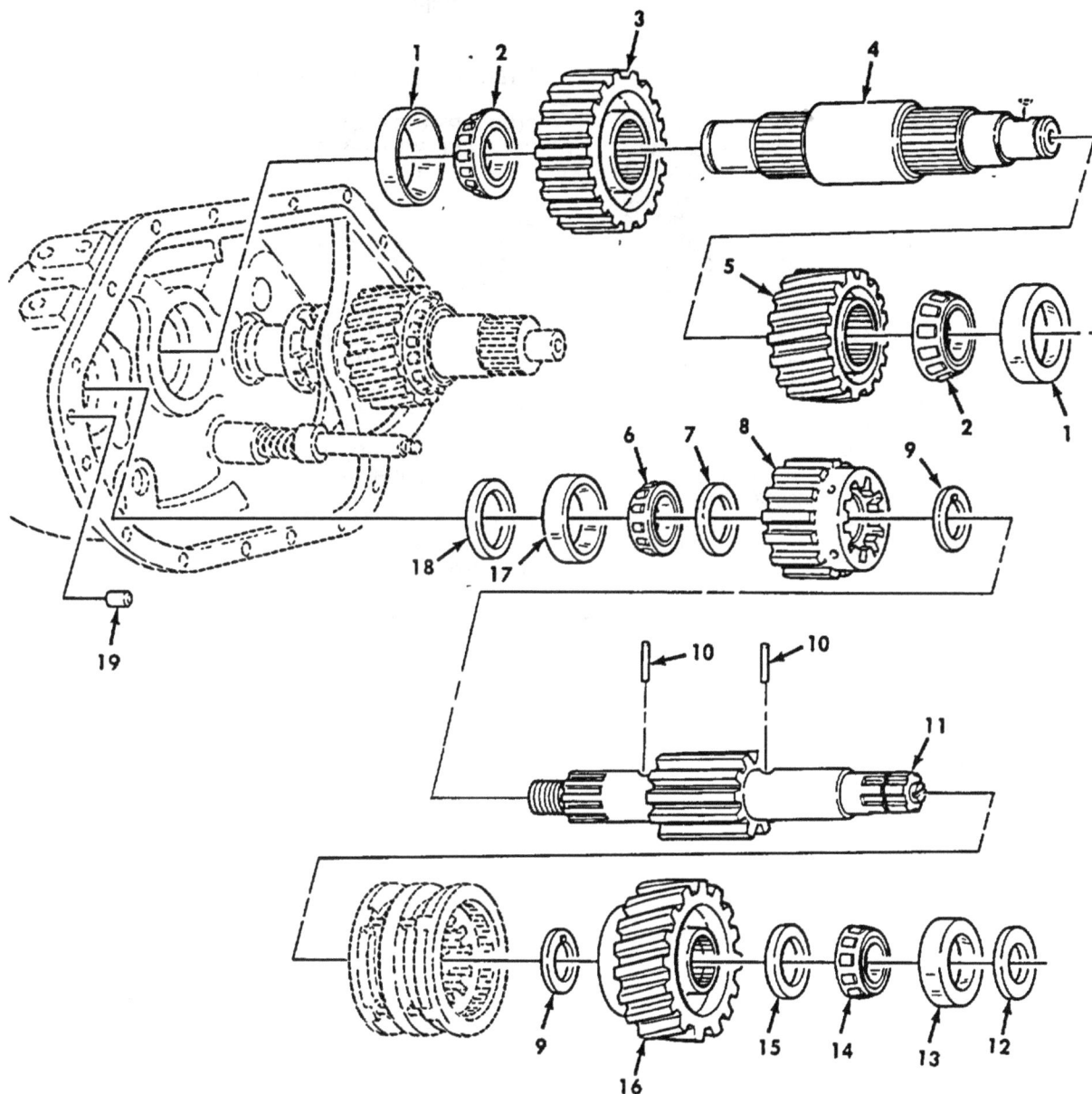

Figure 209. Transfer Housing Input and Intermediate Driveshafts.

(1) ITEM NO	(2) SMR CODE	(3) NSN	(4) CAGEC	(5) PART NUMBER	(6) DESCRIPTION AND USABLE ON CODES (UOC)	(7) QTY
					GROUP 0801 POWER TRANSFER AND FINAL DRIVE ASSEMBLIES	
					FIG. 209 TRANSFER HOUSING INPUT AND INTERMEDIATE DRIVESHAFTS	
1	PAHZZ	3110002274667	60038	HM807010	CUP, TAPERED ROLLER	2
2	PAHZZ	3110008426572	60038	HM807049	CONE AND ROLLERS, TA	2
3	PAHZZ	3020011411554	78500	3892-Y-4471	GEAR, SPUR	1
4	PBHZZ	2520011272336	78500	3280-V-7536	SHAFT, GEAR TRANSFER	1
5	PFHZZ	3020011464198	78500	3892-B-4474	GEAR, SPUR	1
6	PFHZZ	3110005197833	78500	JM-207049	CONE AND ROLLERS, TA	1
7	PAHZZ	5310011453991	78500	1229-U-2803	WASHER, FLAT	1
8	PAHZZ	3020011341015	78500	A-3892-E-4555	GEAR, SPUR	1
9	PAHZZ	5310011333614	78500	1229-N-2796	WASHER	2
10	PFHZZ	5315012779765	78500	1246-Q-615	PIN, STRAIGHT, HEADLE	2
11	PAHZZ	2520011272623	78500	A-3297-T-72	SHAFT, TRANSFER TRAN	1
12	PAHZZ	5310011418464	78500	2221-B-2	WASHER, FLAT	1
13	PAHZZ	3110001000316	78500	454	CUP, TAPERED ROLLER	1
14	PAHZZ	3110001010836	78500	455S	CONE AND ROLLERS, TA	1
15	PAHZZ	3120011439249	78500	1229-R-2800	BEARING, WASHER, THRU	1
16	PAHZZ	2520011275766	78500	A-3892-D-4554	GEAR AND BUSHING AS	1
17	PAHZZ	3110001600338	60038	JM207010	CUP, TAPERED ROLLER	1
18	PAHZZ	5365011330069	78500	1244-Y-1585	SPACER, RING	1
19	PAHZZ	5315011434220	78500	1846-M-351	PIN, STRAIGHT, HEADLE TRANSFER TRANSMISSION HOUSING	2

END OF FIGURE

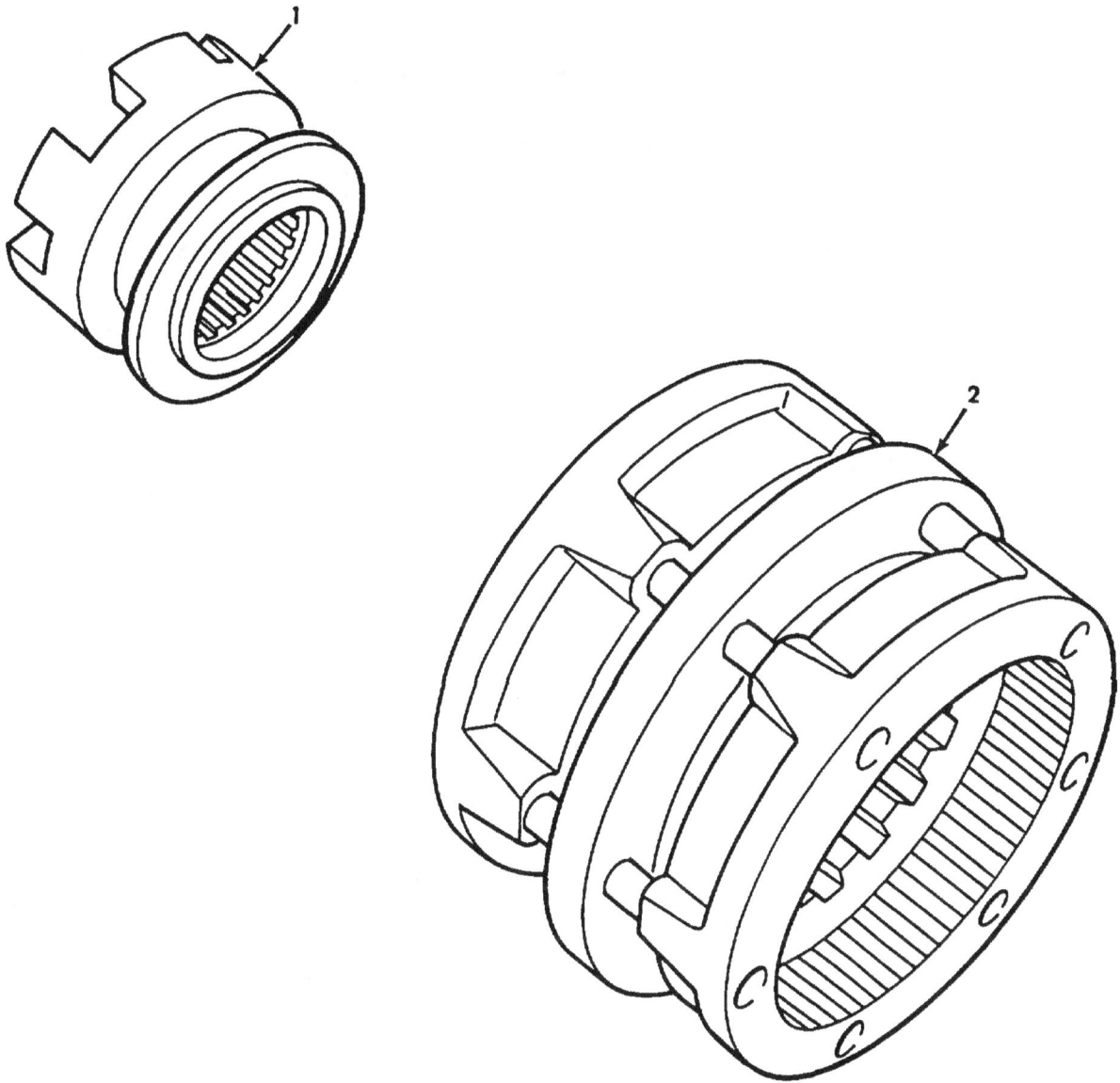

Figure 210. Transfer Clutch Assembly and Synchronizer Assembly.

(1) ITEM NO	(2) SMR CODE	(3) NSN	(4) CAGEC	(5) PART NUMBER	(6) DESCRIPTION AND USABLE ON CODES (UOC)	(7) QTY
					GROUP 0802 CLUTCH AND CLUTCH CONTROLS	
					FIG. 210 TRANSFER CLUTCH ASSEMBLY AND SYNCHRONIZER ASSEMBLY	
1	PAHZZ	2520011276901	78500	3107-E-31	CLUTCH ASSEMBLY, SL	1
2	PAHZZ	2520011350085	78500	A-3107-M-39	SYNCHRONIZER, HIGH A CLUTCH ASSEMBLY	1
					END OF FIGURE	

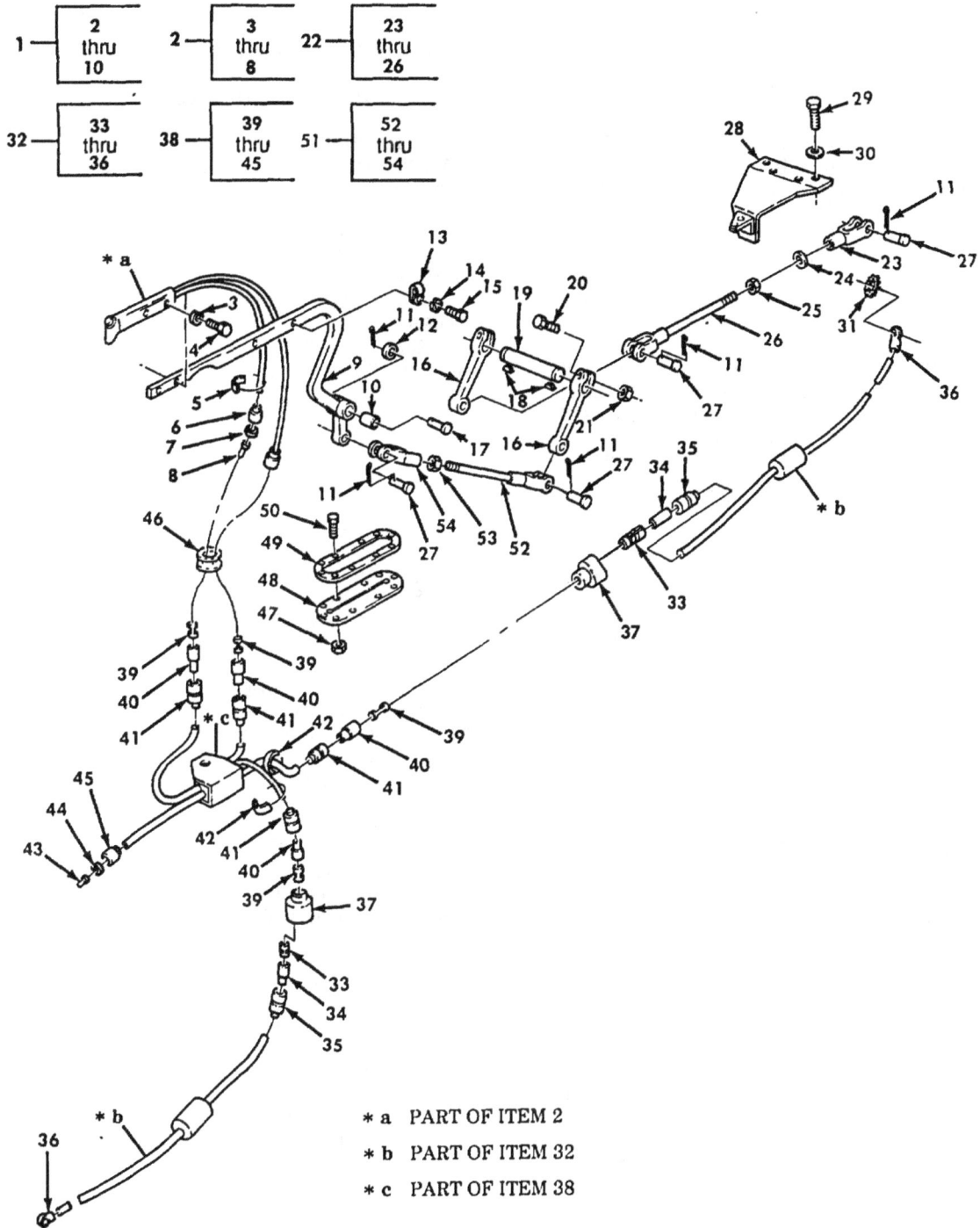

Figure 211. Transfer Controls.

* a   PART OF ITEM 2
* b   PART OF ITEM 32
* c   PART OF ITEM 38

SECTION II

(1) ITEM NO	(2) SMR CODE	(3) NSN	(4) CAGEC	(5) PART NUMBER	(6) DESCRIPTION AND USABLE ON CODES (UOC)	(7) QTY

GROUP 0803 GEAR SHIFT CONTROLS

FIG. 211 TRANSFER CONTROLS

(1) ITEM NO	(2) SMR CODE	(3) NSN	(4) CAGEC	(5) PART NUMBER	(6) DESCRIPTION AND USABLE ON CODES (UOC)	(7) QTY
1	PAFFF	3040011239681	19207	12277363	LEVER, MANUAL CONTRO	1
2	PAFFF	5930011418414	81901	P9838	.SWITCH, PUSH	1
3	PAFZZ	5310005500284	96906	MS35336-9	..WASHER, LOCK	4
4	PAFZZ	5305009592739	96906	MS35191-237	..SCREW, MACHINE	4
5	PAFZZ	9905007524649	81349	M43436/1-1	..BAND, MARKER	2
6	PAFZZ	5935006915591	19207	8724495	..SHELL, ELECTRICAL CO	2
7	PAFZZ	5310006560067	19207	8724497	..WASHER, SLOTTED	2
8	PAFFA	5999009263144	96906	MS27148-3	..CONTACT, ELECTRICAL	2
9	XAFZZ		19207	12277362	.LEVER	1
10	PAFZZ	3120006926153	89346	63942HA	.BEARING, SLEEVE	1
11	PAFZZ	5315000120123	96906	MS24665-355	PIN, COTTER	5
12	PAFZZ	5310008098533	96906	MS27183-23	WASHER, FLAT	1
13	PAFZZ	5340000797837	96906	MS21333-67	CLAMP, LOOP	2
14	PAFZZ	5310000453296	96906	MS35338-43	WASHER, LOCK	2
15	PAFZZ	5305008550957	96906	MS24629-46	SCREW, TAPPING	2
16	PAFZZ	3040011217744	19207	12277094	LEVER, REMOTE CONTRO	2
17	PAFZZ	5315010087084	96906	MS20392-10C91	PIN, STRAIGHT, HEADED	1
18	PAFZZ	5315000124553	96906	MS35756-17	KEY, WOODRUFF	2
19	PAFZZ	3040010889401	19207	12256551	SHAFT, STRAIGHT	1
20	PAFZZ	5305002692811	96906	MS90726-67	SCREW, CAP, HEXAGON H	2
21	PAFZZ	5310008140672	96906	MS51943-36	NUT, SELF-LOCKING, HE	2
22	PAFFF	3040010892991	19207	12256547-2	CONNECTING LINK, RIG	1
23	PAFZZ	5340004664948	71843	2708-6A	.CLEVIS, ROD END	1
24	PAFZZ	5310012249142	19207	12302875	.WASHER, FLAT	1
25	PAFZZ	5310011925760	19207	12302874	.NUT, SELF-LOCKING, HE	1
26	PAFZZ	3040011217741	19207	12256548-2	.CONNECTING LINK, RIG	1
27	PAFZZ	5315001401938	96906	MS35810-6	PIN, STRAIGHT, HEADED	4
28	PFFZZ	5340010911608	19207	12276939	BRACKET, ANGLE	1
29	PAFZZ	5305000680511	80204	B1821BH038C125N	SCREW, CAP, HEXAGON H	4
30	PAFZZ	5310000806004	96906	MS27183-14	WASHER, FLAT	4
31	PAFZZ	5310008892527	96906	MS45904-72	WASHER, LOCK	1
32	PFFFF	5961011805634	19207	12302643	SEMICONDUCTOR DEVIC	2
33	PAFZZ	5940003996676	19207	8338564	.TERMINAL ASSEMBLY	1
34	PAFZZ	5970008338562	19207	8338562	.INSULATOR, BUSHING	1
35	PAFZZ	5935008338561	19207	8338561	.SHELL, ELECTRICAL CO	1
36	PAFZZ	5940008923151	96906	MS35436-9	.TERMINAL, LUG	1
37	PAFZA	5935009006281	96906	MS27147-1	ADAPTER, CONNECTOR	2
38	PAFFF	5910011217603	19207	12277240	CAPACITOR ASSEMBLY	1
39	PAFZZ	5940003996676	19207	8338564	.TERMINAL ASSEMBLY	4
40	PAFZZ	5970008338562	19207	8338562	.INSULATOR, BUSHING	4
41	PAFZZ	5975006605962	19207	8724494	.CABLE NIPPLE, ELECTR	4
42	PAFZZ	9905007524649	81349	M43436/1-1	.BAND, MARKER	5
43	PAFFA	5999009263144	96906	MS27148-3	.CONTACT, ELECTRICAL	1
44	PAFZZ	5310006560067	19207	8724497	.WASHER, SLOTTED	1
45	PAFZZ	5935006915591	19207	8724495	.SHELL, ELECTRICAL CO	1
46	PAFZZ	5325002636651	59875	TD97203	GROMMET, NONMETALLIC	1
47	PAFZZ	5310002081918	88044	AN365-1024A	NUT, SELF-LOCKING, HE	10

(1) ITEM NO	(2) SMR CODE	(3) NSN	(4) CAGEC	(5) PART NUMBER	(6) DESCRIPTION AND USABLE ON CODES (UOC)	(7) QTY
48	PAFZZ	5330010892992	19207	12256552	SEAL, NONMETALLIC SP	1
49	PAFZZ	5330010893046	19207	12256555	GASKET	1
50	PAFZZ	5306000440502	21450	440502	BOLT, MACHINE	10
51	PAFFF	5340010892990	19207	12256547-1	CLEVIS ASSEMBLY COM	1
52	PAFZZ	5340004664948	71843	2708-6A	.CLEVIS, ROD END	1
53	PAFZZ	5310008911733	96906	MS35691-38	.NUT, PLAIN, HEXAGON	1
54	PAFZZ	3040011217742	19207	12256548-1	.CONNECTING LINK, RIG	1

END OF FIGURE

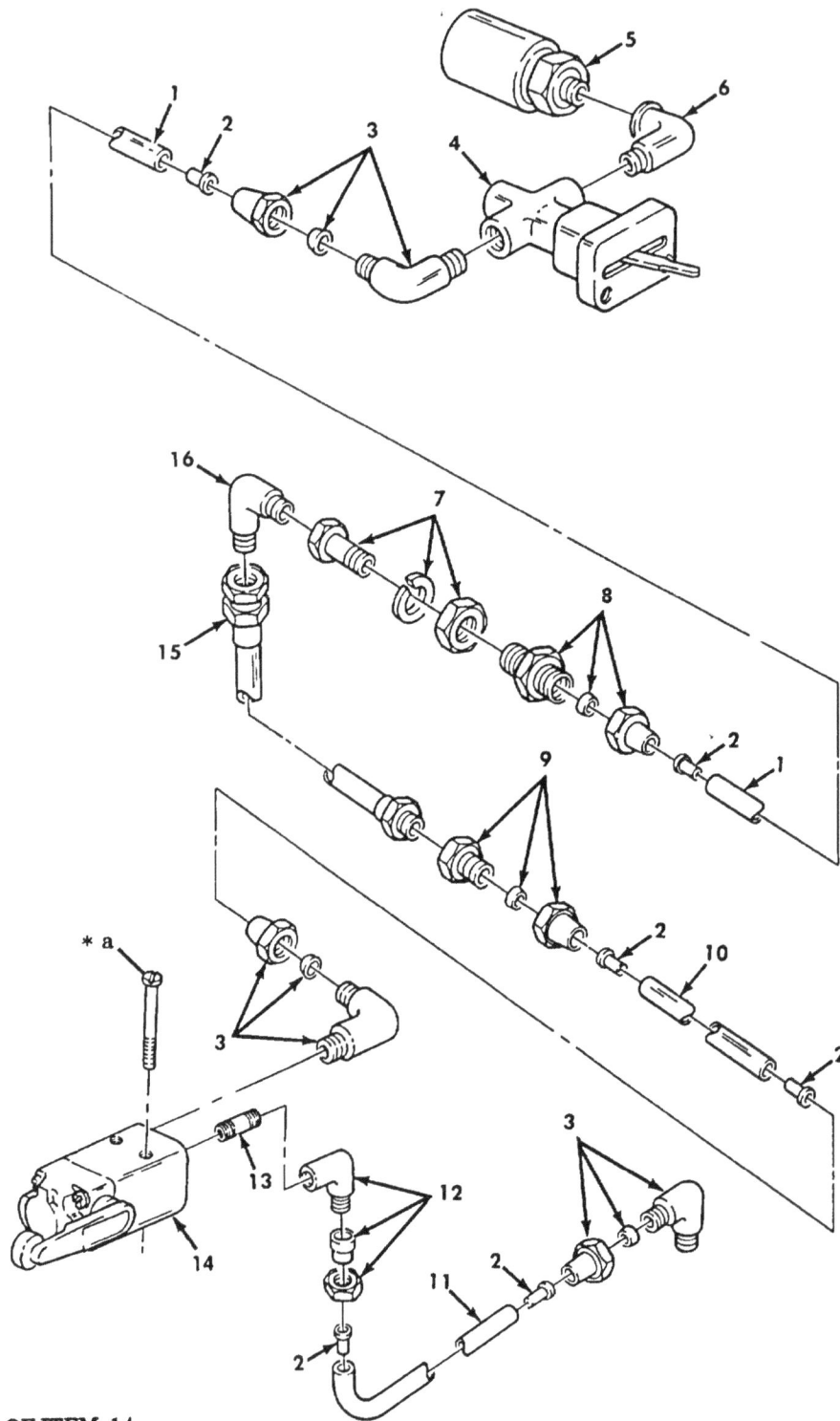

* a PART OF ITEM 14

Figure 212. Transfer Air Control Lines.

(1) ITEM NO	(2) SMR CODE	(3) NSN	(4) CAGEC	(5) PART NUMBER	(6) DESCRIPTION AND USABLE ON CODES (UOC)	(7) QTY
					GROUP 0803 GEAR SHIFT CONTROLS	
					FIG. 212 TRANSFER AIR CONTROL LINES	
1	MOOZZ		19207	CPR104420-2-18	HOSE NONMETALLIC MAKE FROM HOSE, ... P/N CPR104420-2, 18 INCHES LONG	1
2	PAOZZ	4730010798821	19207	CPR102321-1	INSERT, TUBE FITTING	6
3	PAOZZ	4730002871604	81343	6-2 120202BA	ELBOW, PIPE TO TUBE	3
4	PAOZZ	4820008328077	19207	10924753	VALVE, LINEAR, DIRECT	1
5	PAOZA	5930004539367	19207	11609301	SWITCH, PRESSURE	1
6	PAOZZ	4730000530266	96906	MS51952-1	ELBOW, PIPE	1
7	PAOZZ	4730002786318	19207	8328782	COUPLING, PIPE.	1
8	PAOZZ	4730000691186	81343	6-4 120102BA	ADAPTER, STRAIGHT, PI	1
9	PAOZZ	4730002778770	81343	6-6 120103BA	ADAPTER, STRAIGHT, PI	1
10	MOOZZ		19207	CPR104420-2-45	HOSE NONMETALLIC MAKE FROM HOSE, ... P/N CPR104420-2, 45 INCHES LONG	1
11	MOOZZ		19207	CPR104420-2-26	HOSE NONMETALLIC MAKE FROM HOSE, ... P/N CPR104420-2, 26 INCHES LONG....................	1
12	PAOZZ	4730002894052	81343	6-2 120203BA	ELBOW, PIPE TO TUBE	1
13	PAOZZ	4730001961966	96906	MS51846-13	NIPPLE, PIPE	1
14	PAOZZ	4820013293245	97902	LSC-3270-S26022-1	VALVE, LINEAR, DIRECT	1
15	PAOZZ	4720011079939	19207	12256272	HOSE ASSEMBLY, NONME	1
16	PAOZZ	4730008127999	96906	MS51504A6	ELBOW, PIPE TO TUBE	1

END OF FIGURE

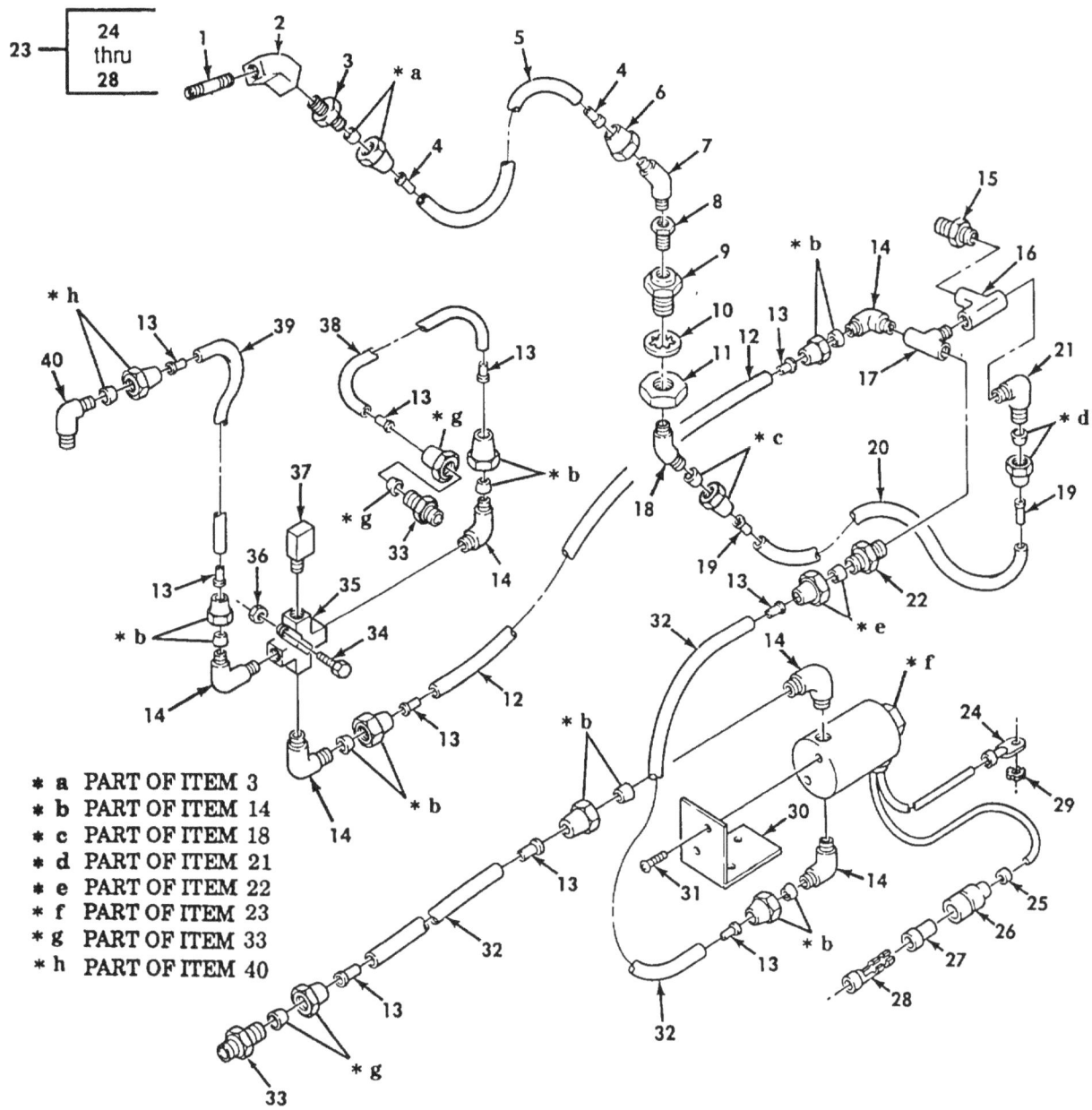

* a PART OF ITEM 3
* b PART OF ITEM 14
* c PART OF ITEM 18
* d PART OF ITEM 21
* e PART OF ITEM 22
* f PART OF ITEM 23
* g PART OF ITEM 33
* h PART OF ITEM 40

Figure 213. Transfer Air-Actuated Controls and Interlock Value.

(1) ITEM NO	(2) SMR CODE	(3) NSN	(4) CAGEC	(5) PART NUMBER	(6) DESCRIPTION AND USABLE ON CODES (UOC)	(7) QTY
					GROUP 0803 GEAR SHIFT CONTROLS	
					FIG. 213 TRANSFER AIR-ACTUATED CONTROLS AND INTERLOCK VALVE	
1	PAOZZ	4730001961966	96906	MS51846-13	NIPPLE, PIPE	1
2	PAOZZ	4730011937684	24617	9409919	ELBOW, PIPE	1
3	PAOZZ	4730003529793	81343	4-2 120103BA	ADAPTER, STRAIGHT, PI	1
4	PAOZZ	4730001324588	79146	H0-159-4	INSERT, TUBE FITTING	2
5	MOOZZ		19207	CPR104420-1-6	HOSE, NONMETALLIC MAKE FROM HOSE, ... P/N CPR104420-1, 6 INCHES LONG....................	1
6	PAOZZ	4730000116452	79470	1110X4	NUT, TUBE COUPLING	1
7	PAOZZ	4730012023351	19207	7336402-1	ELBOW, PIPE TO TUBE	1
8	PAOZZ	4730001805038	24617	444026	BUSHING, PIPE	1
9	PAOZZ	4730004213924	19207	7339982	ADAPTER, STRAIGHT, PI	1
10	PAOZZ	5310005826714	96906	MS35333-49	WASHER, LOCK	1
11	PAOZZ	5310002410157	19207	5331179	NUT, PLAIN, HEXAGON	1
12	MOOZZ		19207	CPR104420-2-55	HOSE, NONMETALLIC MAKE FROM HOSE, ... P/N CPR104420-2, 55 INCHES LONG....................	1
13	PAOZZ	4730010798821	19207	CPR102321-1	INSERT, TUBE FITTING	10
14	PAOZZ	4730000691187	81343	6-4 100202BA	ELBOW, PIPE TO TUBE	6
15	PAOZZ	4730008022560	21450	443987	REDUCER, PIPE	1
16	PAOZZ	4730010926442	21450	444120	TEE, PIPE	1
17	PAOZZ	4730000888666	19207	444134	TEE, PIPE	1
18	PAOZZ	4730002890051	81343	8-6 120202BA	ELBOW, PIPE TO TUBE	1
19	PAOZZ	4730010326038	19207	CPR102321-4	INSERT, TUBE FITTING	2
20	MOOZZ		19207	CPR104420-3-48	HOSE, NONMETALLIC MAKE FROM HOSE, ... P/N CPR104420-3, 48 INCHES LONG....................	1
21	PAOZZ	4730004097854	81343	8-4 120202BA	ELBOW, PIPE TO TUBE	1
22	PAOZZ	4730000691186	81343	6-4 120102BA	ADAPTER, STRAIGHT, PI	1
23	PAOZZ	5945011367640	97902	ESM-3302-S-26021	SOLENOID, ELECTRICAL	1
24	PAOZZ	5940008923151	96906	MS35436-9	.TERMINAL, LUG	1
25	PAOZZ	9905010138723	81349	M43436/3-1	.BAND, MARKER	2
26	PAOZZ	5975006605962	19207	8724494	.CABLE NIPPLE, ELECTR	1
27	PAOZZ	5970008338562	19207	8338562	.INSULATOR, BUSHING	1
28	PAOZZ	5940003996676	19207	8338564	.TERMINAL ASSEMBLY	1
29	PAOZZ	5310000611258	96906	MS45904-76	WASHER, LOCK	1
30	PAOZZ	5340011231333	19207	12277361	BRACKET, ANGLE	1
31	PAOZZ	5305000889044	96906	MS35207-260	SCREW, MACHINE	2
32	MOOZZ		19207	CPR104420-2-15	HOSE, NONMETALLIC MAKE FROM TUBING, ... P/N CPR104420-2, 15 INCHES LONG	2
33	PAOZZ	4730008371177	81343	6-8 120102BA	ADAPTER, STRAIGHT, PI	2
34	PAOZZ	5305002692804	96906	MS90726-61	SCREW, CAP, HEXAGON H .... UOC:DAA, DAB, DAC, DAD, DAE, DAF, DAG, DAH, DAJ, DAK, DAW, V12, V13, V14, V15, V16, V17, V19, V20, V21, V22, V24, V25, ZAA, ZAB, ZAC, ZAD, ZAE, ZAF, ZAG, ZAH, ZAJ, ZAK	1
34	PAOZZ	5305002693238	80204	B1821BH038F125N	SCREW, CAP, HEXAGON H .... UOC:DAL, V18, V39, ZAL	1
35	PAOZZ	4730011126561	19207	11669082	CROSS, PIPE	1
36	PAOZZ	5310008140672	96906	MS51943-36	NUT, SELF-LOCKING, HE	1
37	PAOZZ	4730011079692	19207	12255985	ELBOW, PIPE	1

(1) ITEM NO	(2) SMR CODE	(3) NSN	(4) CAGEC	(5) PART NUMBER	(6) DESCRIPTION AND USABLE ON CODES (UOC)	(7) QTY
38	MOOZZ		19207	CPR104420-2-17	HOSE, NONMETALLIC MAKE FROM HOSE, ... P/N CPR104420-2, 17 INCHES LONG.....................	1
39	MOOZZ		19207	CPR104420-2-24	HOSE, NONMETALLIC MAKE FROM HOSE, ... P/N CPR104420-2, 24 INCHES LONG....................	1
40	PAOZZ	4730002871604	81343	6-2 120202BA	ELBOW, PIPE TO TUBE	1

END OF FIGURE

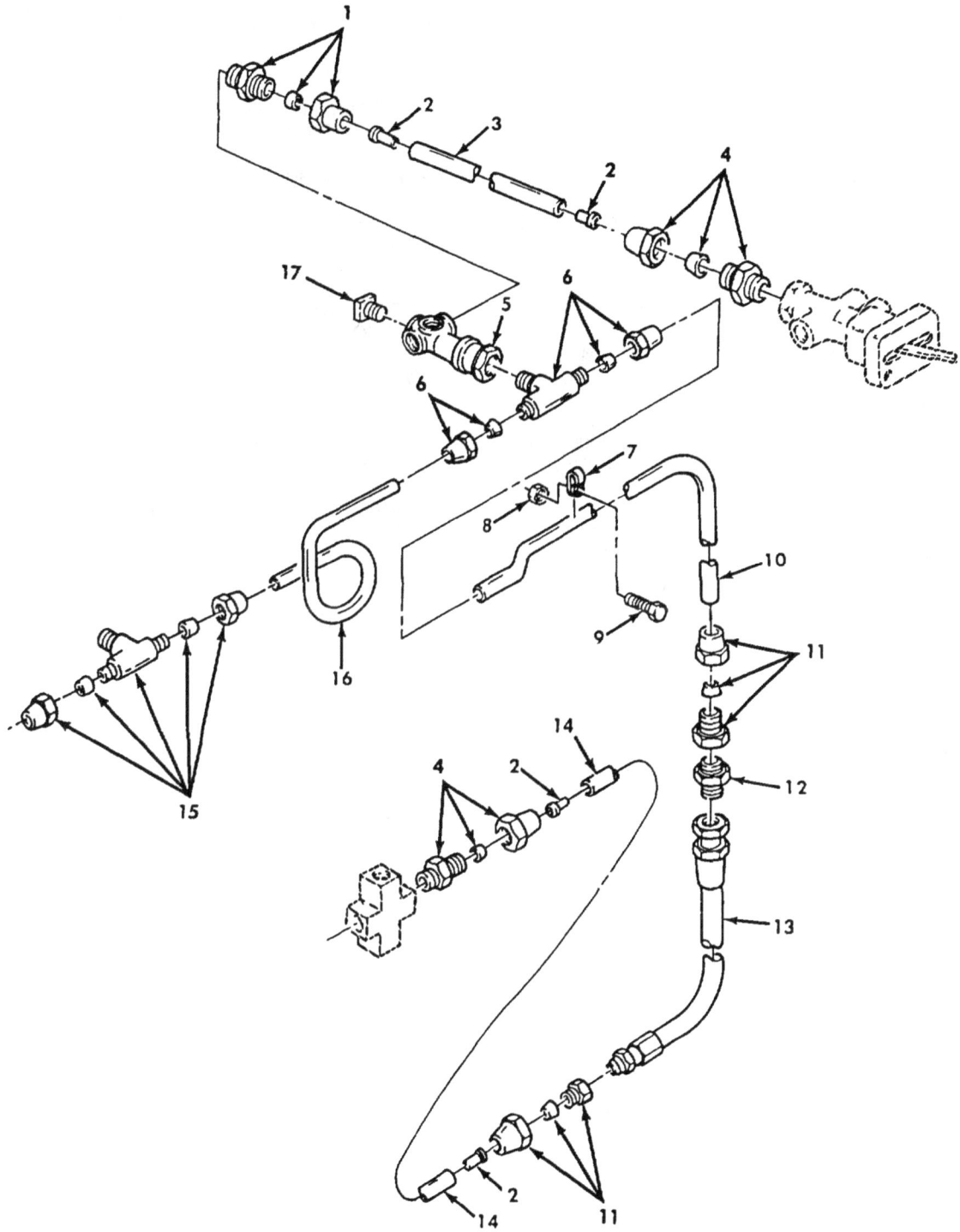

Figure 214. Transfer Air-Actuated Controls and Manifold Fittings.

(1) ITEM NO	(2) SMR CODE	(3) NSN	(4) CAGEC	(5) PART NUMBER	(6) DESCRIPTION AND USABLE ON CODES (UOC)	(7) QTY
					GROUP 0803 GEAR SHIFT CONTROLS	
					FIG. 214 TRANSFER AIR-ACTUATED CONTROLS AND MANIFOLD FITTINGS	
1	PAOZZ	4730000691186	81343	6-4 120102BA	ADAPTER, STRAIGHT, PI	1
2	PAOZZ	4730010798821	19207	CPR102321-1	INSERT, TUBE FITTING	4
3	MOOZZ		19207	CPR104420-2-30	HOSE NONMETALLIC MAKE FROM HOSE, ... P/N CPR104420-2, 30 INCHES LONG	1
4	PAOZZ	4730001423075	81343	6-2 120102BA	ADAPTER, STRAIGHT, PI	2
5	PAOZZ	4730007539300	19207	7539300	CONNECTOR, MULTIPLE, ...	1
6	PAOZZ	4730004946580	81343	6-6-4 120425BA	TEE, PIPE TO TUBE	1
7	PAOZZ	5340008091492	56232	310-8	CLAMP, LOOP	1
8	PAOZZ	5310008775796	96906	MS21044-N4	NUT, SELF-LOCKING, HE	1
9	PAOZZ	5305002678953	80204	B1821BH025F063N	SCREW, CAP, HEXAGON H	1
10	PAOZZ	4710011234576	19207	12256261	TUBE, BENT, METALLIC	1
11	PAOZZ	4730002778770	81343	6-6 120103BA	ADAPTER, STRAIGHT, PI	2
12	PAOZZ	4730003228457	96906	MS51500A6-6	ADAPTER, STRAIGHT, PI	1
13	PAOZZ	4720011079939	19207	12256272	HOSE ASSEMBLY, NONME	1
14	MOOZZ		19207	CPR104420-2-64	HOSE NONMETALLIC MAKE FROM HOSE, ... P/N CPR104420-2, 64 INCHES LONG	1
15	PAOZZ	4730007825461	81343	6-6-2 120425BA	TEE, PIPE TO TUBE	1
16	PAOZZ	4710011253608	19207	12256262	TUBE, BENT, METALLIC	1
17	PAOZZ	4730009541281	81348	WW-P-471ACABCB	PLUG, PIPE	2

END OF FIGURE

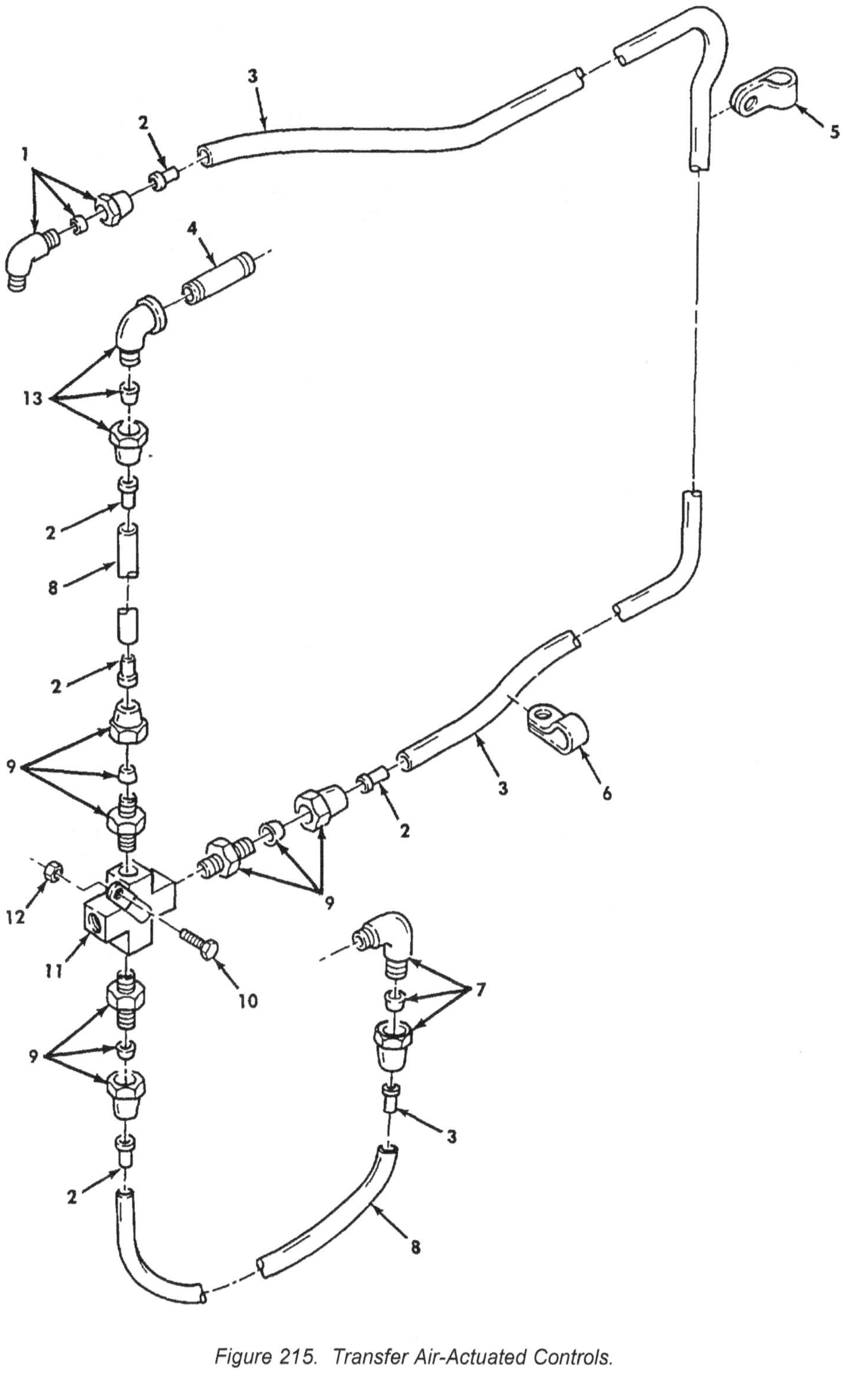

Figure 215. Transfer Air-Actuated Controls.

(1) ITEM NO	(2) SMR CODE	(3) NSN	(4) CAGEC	(5) PART NUMBER	(6) DESCRIPTION AND USABLE ON CODES (UOC)	(7) QTY
					GROUP 0803 GEAR SHIFT CONTROLS	
					FIG. 215 TRANSFER AIR-ACTUATED CONTROLS	
1	PAOZZ	4730000691187	81343	6-4 100202BA	ELBOW, PIPE TO TUBE .... UOC:DAA, DAB, DAE, DAF, DAG, DAH, DAW, DAX, V12, V13, V14, V15, V19, V20, V21, V22, ZAA, ZAB, ZAE, ZAF, ZAG, ZAH	1
1	PAOZZ	4730007017677	96906	MS39185-1	ELBOW, PIPE TO TUBE .... UOC:DAC, DAD, DAJ, DAK, V16, V17, V24, V25, V39, ZAC, ZAD, ZAJ, ZAK	1
2	PAOZZ	4730010798821	19207	CPR102321-1	INSERT, TUBE FITTING....................................	6
3	MOOZZ		19207	CPR104420-2-172	HOSE, NONMETALLIC MAKE FROM HOSE, ... P/N CPR104420-2, 172 INCHES LONG .... UOC:DAL, V18, ZAL	1
3	MOOZZ		19207	CPR104420-2-174	HOSE, NONMETALLIC MAKE FROM HOSE, ... P/N CPR104402-2, 174 INCHES LONG .... UOC:DAC, DAD, DAJ, DAK, V16, V17, V24, V25, V39, ZAC, ZAD, ZAJ, ZAK	1
3	MOOZZ		19207	CPR104420-2-146	HOSE NONMETALLIC MAKE FROM HOSE, ... P/N CPR104420-2, 146 INCHES LONG .... UOC:DAA, DAB, DAW, DAX, V12, V13, V14, V15, ZAA, ZAB	1
3	MOOZZ		19207	CPR104420-2-159	HOSE, NONMETALLIC MAKE FROM HOSE, ... P/N CPR104420-2, 159 INCHES LONG .... UOC:DAE, DAF, DAG, DAH, V19, V20, V21, V22, ZAE, ZAF, ZAG, ZAH	1
4	PAOZZ	4730001961964	24617	142851	NIPPLE, PIPE	1
5	PAOZZ	5340001519651	96906	MS21333-129	CLAMP, LOOP	2
6	PAOZZ	5340011977597	19207	7529133-7	CLAMP, LOOP	1
7	PAOZZ	4730000691187	79146	HD-169-6X4	ELBOW, PIPE TO TUBE	1
8	MOOZZ		19207	CPR104420-2-24	HOSE NONMETALLIC MAKE FROM HOSE, ... P/N CPR104420-2, 24 INCHES LONG....................	2
9	PAOZZ	4730000691186	81343	6-4 120102BA	ADAPTER, STRAIGHT, PI	2
10	PAOZZ	5305002692804	96906	MS90726-61	SCREW, CAP, HEXAGON H	1
11	PAOZZ	4730011126561	19207	11669082	CROSS, PIPE	1
12	PAOZZ	5310008140672	96906	MS51943-36	NUT, SELF-LOCKING, HE	1
13	PAOZZ	4730002894052	81343	6-2 120203BA	ELBOW, PIPE TO TUBE	1

END OF FIGURE

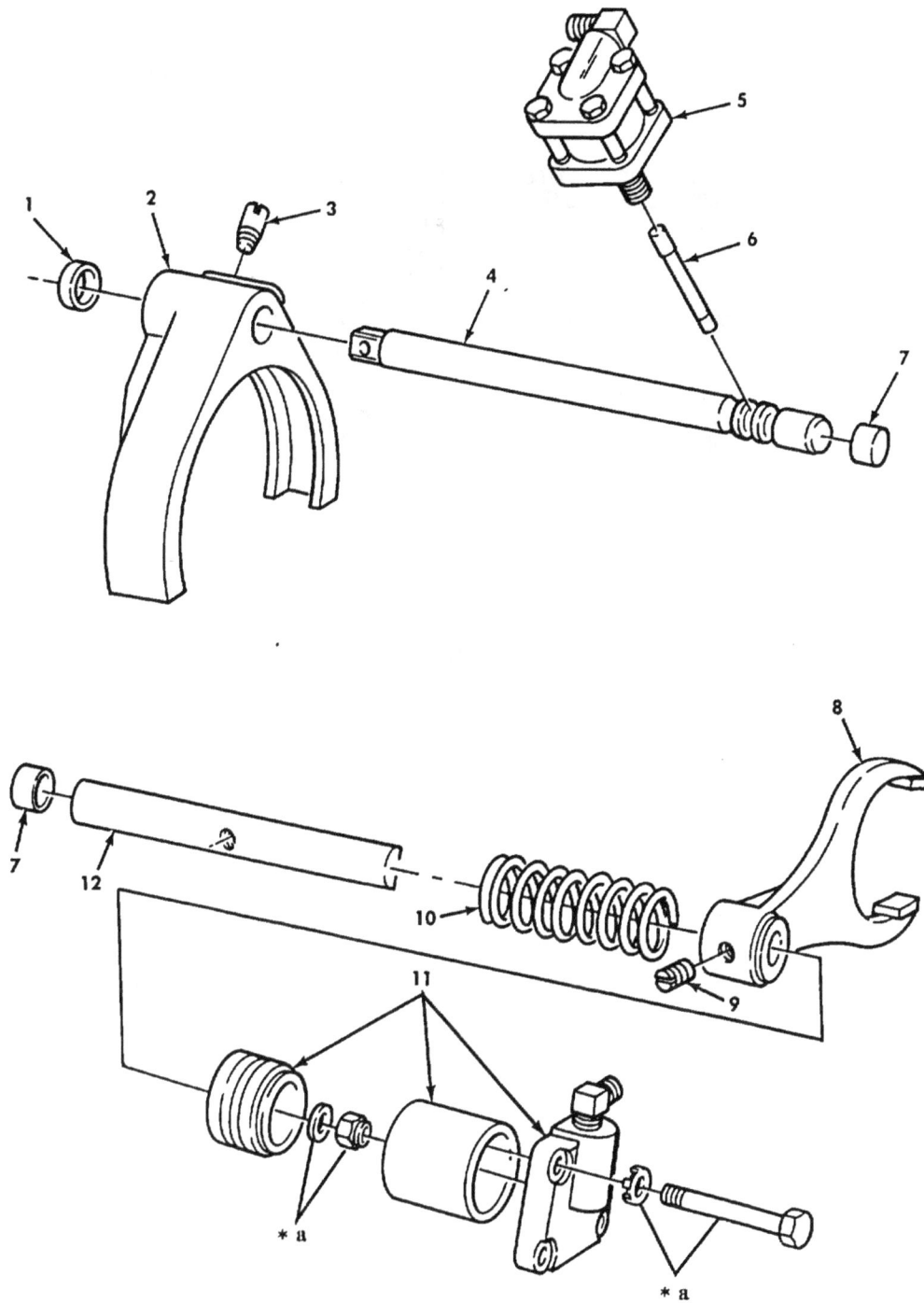

* a PART OF ITEM 11

Figure 216. Transfer Shift Mechanisms.

(1) ITEM NO	(2) SMR CODE	(3) NSN	(4) CAGEC	(5) PART NUMBER	(6) DESCRIPTION AND USABLE ON CODES (UOC)	(7) QTY
					GROUP 0803 GEAR SHIFT CONTROLS	
					FIG. 216 TRANSFER SHIFT MECHANISMS	
1	PAFZZ	5330011328346	78500	A-1205-P-1758	SEAL, PLAIN ENCASED	1
2	PAHZZ	2520011368717	78500	3296-C-107	SHIFTER FORK	1
3	PAHZZ	5305011422793	78500	26X-235	SETSCREW	1
4	PAHZZ	3040011341318	78500	2244-Z-26	SHAFT, STRAIGHT.......................................	1
5	PAFZZ	2520011340899	78500	A1-3261-D-290	INTERLOCK CYLINDER, ...	1
6	PAHZZ	2520011347657	78500	2244-D-30	LOCK, PUSHROD	1
7	PAHZZ	5340007521372	78500	1850-Z-78	PLUG, EXPANSION	2
8	PAHZZ	2520011368718	78500	3296-V-74	FORK, FRONT DECLUTCH	1
9	PAHZZ	5305011422794	78500	26X-230	SETSCREW	1
10	PAHZZ	5360011474787	78500	2258-P-640	SPRING	1
11	PAFZZ	3040011491111	78500	A-3261-S-253	CYLINDER ASSEMBLY, A	1
12	PAHZZ	2520011368719	78500	2244-Y-25	SHAFT, DECLUTCH	1

END OF FIGURE

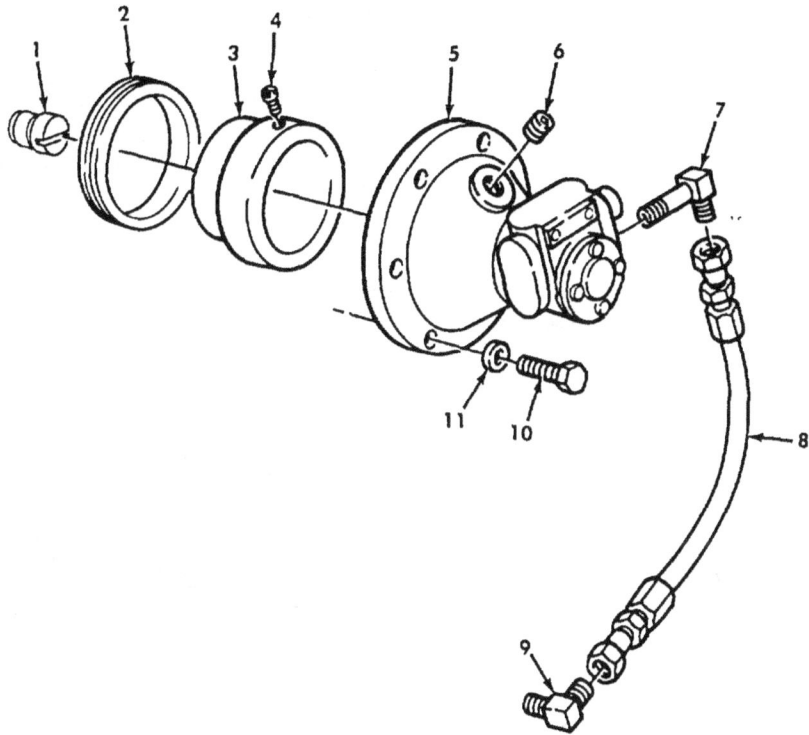

Figure 217. Transfer Oil Pump and Related Parts.

(1) ITEM NO	(2) SMR CODE	(3) NSN	(4) CAGEC	(5) PART NUMBER	(6) DESCRIPTION AND USABLE ON CODES (UOC)	(7) QTY
					GROUP 0804 LUBRICATION,COOLING,OR HYDRAULIC COMPONENTS	
					FIG. 217 TRANSFER OIL PUMP AND RELATED PARTS	
1	PAFZZ	2520011421979	78500	2297-J-3832	BUTTON, PUMP DRIVE .... UOC:DAA, DAB, DAC, DAD, DAE, DAF, DAG, DAH, DAJ, DAK, DAW, DAX, V12 , V13, V14, V15, V16, V17, V20, V21, V22, V24, V25, V39, ZAA, ZAB, ZAC, ZAD, ZAE, ZAF, ZAG, ZAH, ZAJ, ZAK	1
2	PAFZZ	5330011374799	78500	1205-Y-1663	SEAL, PACKING .... UOC:DAA, DAB, DAC, DAD, DAE, DAPF, DAG, DAH, DAJ, DAK, DAW, DAX, V12, V13, V14, V15, V16, V20, V21, V22, V24, V25, V39, ZAA, ZAB, ZAC, ZAD, ZAE, ZAF, ZAG, ZAH, ZAJ	1
3	PAFZZ	3040011343605	78500	2207-K-11	COLLAR, SHAFT .... UOC:DAA, DAB, DAC, DAD, DAE, DAF, DAG, DAH, DAJ, DAK, DAW, DAX, V12, V13, V14, V15, V16, V20, V21, V22, V24, V25, V39, ZAA, ZAB, ZAC, ZAD, ZAE, ZAF, ZAG, ZAH, ZAJ, ZAK	1
4	PAFZZ	5305011422792	78500	S-853	SETSCREW .... UOC:DAA, DAB, DAC, DAD, DAE, DAF, DAG, DAH, DAJ, DAK, DAW, DAX, V12, V13, V14, V15, V16, V17, V20, V21, V22, V24, V25, V39, ZAA, ZAB, ZAC, ZAD, ZAE, ZAF, ZAG, ZAH, ZAJ, ZAK	1
5	PAFZZ	2815011368720	78500	A-3303-J-10	OIL PUMP ASSEMBLY TRANSFER .... TRANSMISSION UOC:DAA, DAB, DAC, DAD, DAE, DAF, DAG, DAH, DAJ, DAX, DAW, DAX, V12 , V13 , V14 , V15 , V16, V17, V19, V20, V21, V22, V24, V25, V39, ZAA, ZAB, ZAC, ZAD, ZAE, ZAF, ZAG, ZAH, ZAJ, ZAK	1
6	PAFZZ	5365011296777	78500	P-26-C	PLUG, MACHINE THREAD .... UOC:DAA, DAB, DAC, DAD, DAE, DAF, DAG, DAH, DAJ, DAK, DAW, DAX, V12, V13, V14, V15, V16, V17, V19, V20, V21, V22, V24, V25, V39, ZAA, ZAB, ZAC, ZAD, ZAE, ZAF, ZAG, ZAH, ZAJ, ZAK	1
7	PAFZZ	4730011277346	78500	2206-J-88	ELBOW, PIPE TO HOSE .... UOC:DAA, DAB, DAC, DAD, DAE, DAF, DAG, DAH, DAJ, DAK, DAW, DAX, V12 , V13 , V14 , V15 , V16, V17, V20, V21, V22, V24, V25, V39, ZAA, ZAB, ZAC, ZAD, ZAE, ZAF, ZAG, ZAH, ZAJ, ZAK	1
8	PAFZZ	4720011269555	78500	A-2296-C-81	HOSE ASSEMBLY, NONME .... UOC:DAA, DAB, DAC, DAD, DAE, DAF, DAG, DAH, DAJ, DAK, DAW, DAX, V12, V13, V14, V15, V16, V17, V20, V21, V22, V24, V25, V39, ZAA, ZAB, ZAC, ZAD, ZAE, ZAF, ZAG, ZAH, ZAJ, ZAK	1
9	PAFZZ	4730011346988	78500	2206-F-58	ELBOW, PIPE TO TUBE .... UOC:DAA, DAB, DAC, DAD, DAE, DAF, DAG, DAH, DAJ, DAK, DAW, DAX, V12, V13, V14, V15, V16, V17, V19, V20, V21, V22, V24, V25, V39, ZAA, ZAB, ZAC, ZAD, ZAE, ZAF, ZAG, ZAH, ZAJ, ZAK	1
10	PAFZZ	5305003385162	78500	S-2710-1-C	SCREW, SPECIAL .... UOC:DAA, DAB, DAC, DAD, DAE, DAF, DAG, DAR, DAJ, DAK, DAW, DAX, V12, V13, V14, V15, V16, V17, V20, V21, V22, V24, V25, V39, ZAA, ZAB,	6

(1) ITEM NO	(2) SMR CODE	(3) NSN	(4) CAGEC	(5) PART NUMBER	(6) DESCRIPTION AND USABLE ON CODES (UOC)	(7) QTY
11	PAFZZ	5310008933381	19207	7748744	ZAC, ZAD, ZAE, ZAF, ZAG, ZAH, ZAJ, ZAK WASHER, FLAT .... UOC:DAA, DAB, DAC, DAD, DAE, DAF, DAG, DAH, DAJ, DAK, DAW, DAX, V12, V13 , V14, V15, V16, V17, V20, V21, V22, V24, V25, V39, ZAA, ZAB, ZAC, ZAD, ZAE, ZAF, ZAG, ZAH, ZAJ, ZAK	6

END OF FIGURE

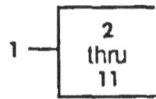

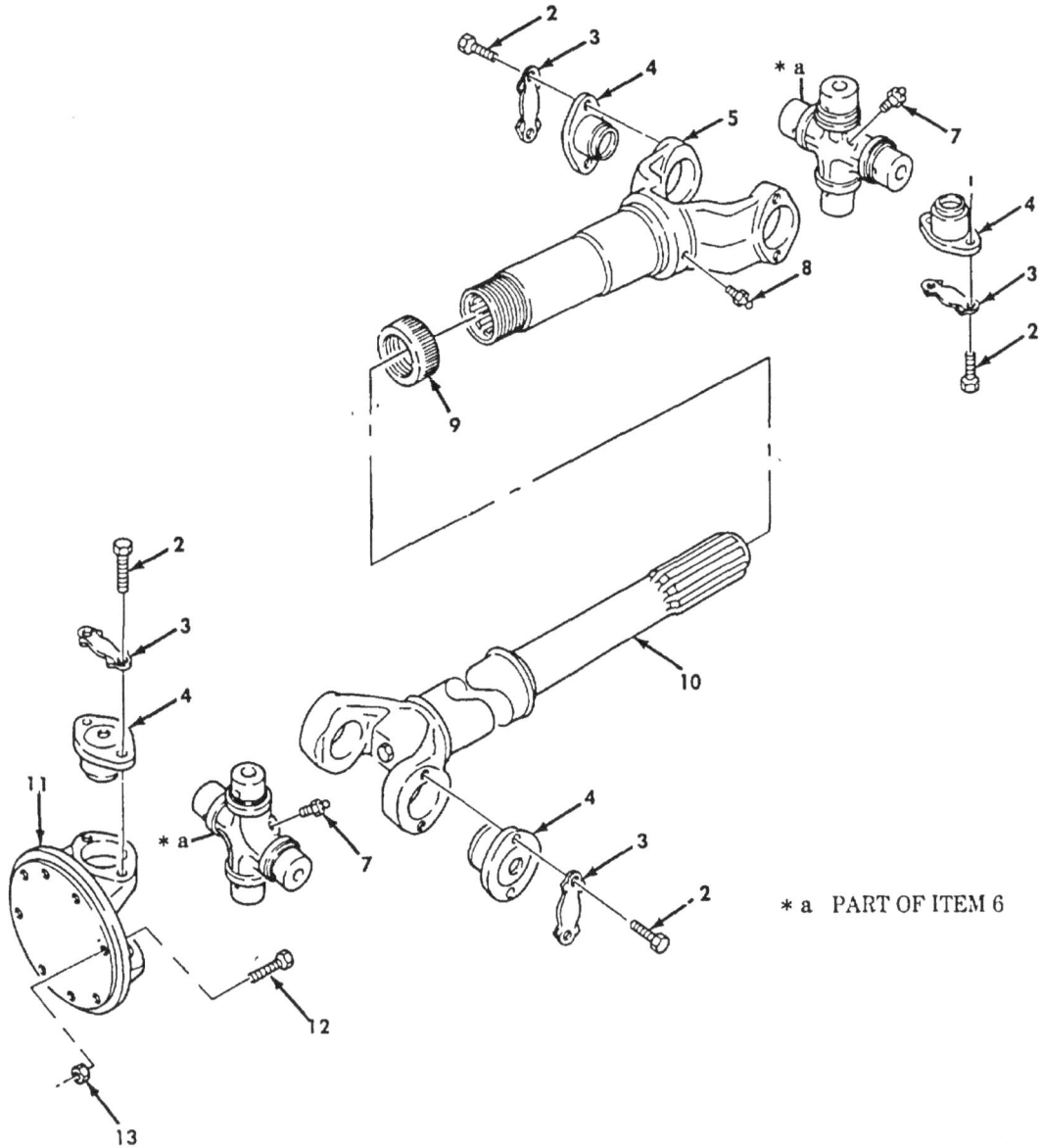

Figure 218. Propeller Shaft Assembly, Front Axle to Transfer Shaft (Dana) (M939, M939A1).

(1) ITEM NO	(2) SMR CODE	(3) NSN	(4) CAGEC	(5) PART NUMBER	(6) DESCRIPTION AND USABLE ON CODES (UOC)	(7) QTY
					GROUP 09 PROPELLER AND PROPELLER SHAFTS	
					0900 PROPELLER SHAFTS	
					FIG. 218 PROPELLER SHAFT ASSEMBLY, FRONT AXLE TO TRANSFER SHAFT (DANA) (M939, M939A1)	
1	PAOOO	2520011128366	95019	912257-0416	DRIVE SHAFT ASSEMBL .... UOC:DAA, DAB, DAC, DAD, DAE, DAF, DAG, DAH, DAJ, DAK, DAL, DAW, DAX, V12, V13, V14, V15, V16, V17, V18, V19, V20, V21, V22, V24, V25, V39	1
2	KFOZZ	5306006798343	61822	920095-15	.BOLT, MACHINE PART OF KIT P/N 5-279X............ UOC:DAA, DAB, DAC, DAD, DAE, DAF, DAG, DAH, DAJ, DAK, DAL, DAW, DAX, V12, V13, V14, V15, V16, V17, V18, V19, V20, V21, V22, V24, V25, V39	16
3	KFOZZ	5340005214479	95019	98-741	.LOCKING PLATE, NUT A PART OF KIT P/N............ 5-729X.. UOC:DAA, DAB, DAC, DAD, DAE, DAF, DAG, DAH, DAJ, DAK, DAL, DAW, DAX, V12, V13, V14, V15, V16, V17, V18, V19, V20, V21, V22, V24, V25, V39	8
4	KFOZZ		95019	5-6-238X	.BEARING ASSEMBLY PART OF KIT P/N .... 5-279X.. UOC:DAA, DAB, DAC, DAD, DAE, DAF, DAG, DAH, DAJ, DAK, DAL, DAW, DAX, V12, V13, V14, V15, V16, V17, V18, V19, V20, V21, V22, V24, V25, V39	8
5	XAOZZ		95019	5-3-2751X	.YOKE, UNIVERSAL JOIN .... UOC:DAA, DAB, DAC, DAD, DAE, DAF, DAG, DAH, DAJ, DAK, DAL, DAW, DAX, V12, V13, V14, V15, V16, V17, V18, V19, V20, V21, V22, V24, V25, V39	1
6	KFOZZ		95019	5-278X	.CROSS ASSEMBLY PART OF KIT P/N.................. 5-279X.. UOC:DAA, DAB, DAC, DAD, DAE, DAF, DAG, DAH, DAJ, DAK, DAL, DAW, DAX, V12, V13, V14, V15, V16, V17, V18, V19, V20, V21, V22, V24, V25, V39	2
7	PAOZZ	4730000504203	96906	MS15001-1	..FITTING, LUBRICATION................................ UOC:DAA, DAB, DAC, DAD, DAE, DAF, DAG, DAH, DAJ, DAK, DAL, DAW, DAX, V12, V13, V14, V15, V16, V17, V18, V19, V20, V21, V22, V24, V25, V39	1
8	PAOZZ	4730009060982	95019	500168-2	.FITTING, LUBRICATION PROPSHAFT YOKE......... UOC:DAA, DAB, DAC, DAD, DAE, DAF, DAG, DAH, DAJ, DAK, DAL, DAW, DAX, V12, V13, V14, V15, V16, V17, V18, V19, V20, V21, V22, V24, V25, V39	1

(1) ITEM NO	(2) SMR CODE	(3) NSN	(4) CAGEC	(5) PART NUMBER	(6) DESCRIPTION AND USABLE ON CODES (UOC)	(7) QTY
9	PAOZZ	2520011423210	95019	5-86-68	.PARTS OIT, DUST BOOT .... UOC:DAA, DAB, DAC, DAD, DAE, DAF, DAG, DAH, DAJ, DAK, DAL, DAW, DAX, V12, V13, V14, V15, V16, V17, V18, V19, V20, V21, V22, V24, V25, V39	1
10	XAOZZ		95019	5-60-285	.SHAFT, PROP, ASSEMBLY .... UOC:DAA, DAB, DAC, DAD, DAE, DAF, DAG, DAH, DAJ, DAK, DAL, DAW, DAX, V12, V13, V14, V15, V16, V17, V18, V19, V20, V21, V22, V24, V25, V39	1
11	PAOZZ	2520010906673	95019	5-2-629	.YOKE, UNIVERSAL JOIN .... UOC:DAA, DAB, DAC, DAD, DAE, DAF, DAG, DAH, DAJ, DAK, DAL, DAW, DAX, V12, V13, V14, V15, V16, V17, V18, V19, V20, V21, V22, V24, V25, V39	1
12	PAOZZ	5306007208747	96906	MS20073-06-12	BOLT, MACHINE .... UOC:DAA, DAB, DAC, DAD, DAE, DAF, DAG, DAH, DAJ, DAK, DAL, DAW, DAX, V12, V13, V14, V15, V16, V17, V18, V19, V20, V21, V22, V24, V25, V39	8
13	PAOZZ	5310008140672	96906	MS51943-36	NUT, SELF-LOCKING, HE .... UOC:DAA, DAB, DA C, DAD, DAE, DAF, DAG, DAH, DAJ, DAK, DAL, DAW, DAX, V12, V13, V14, V15, V16, V17, V18, V19, V20, V21, V22, V24, V25, V39	8

END OF FIGURE

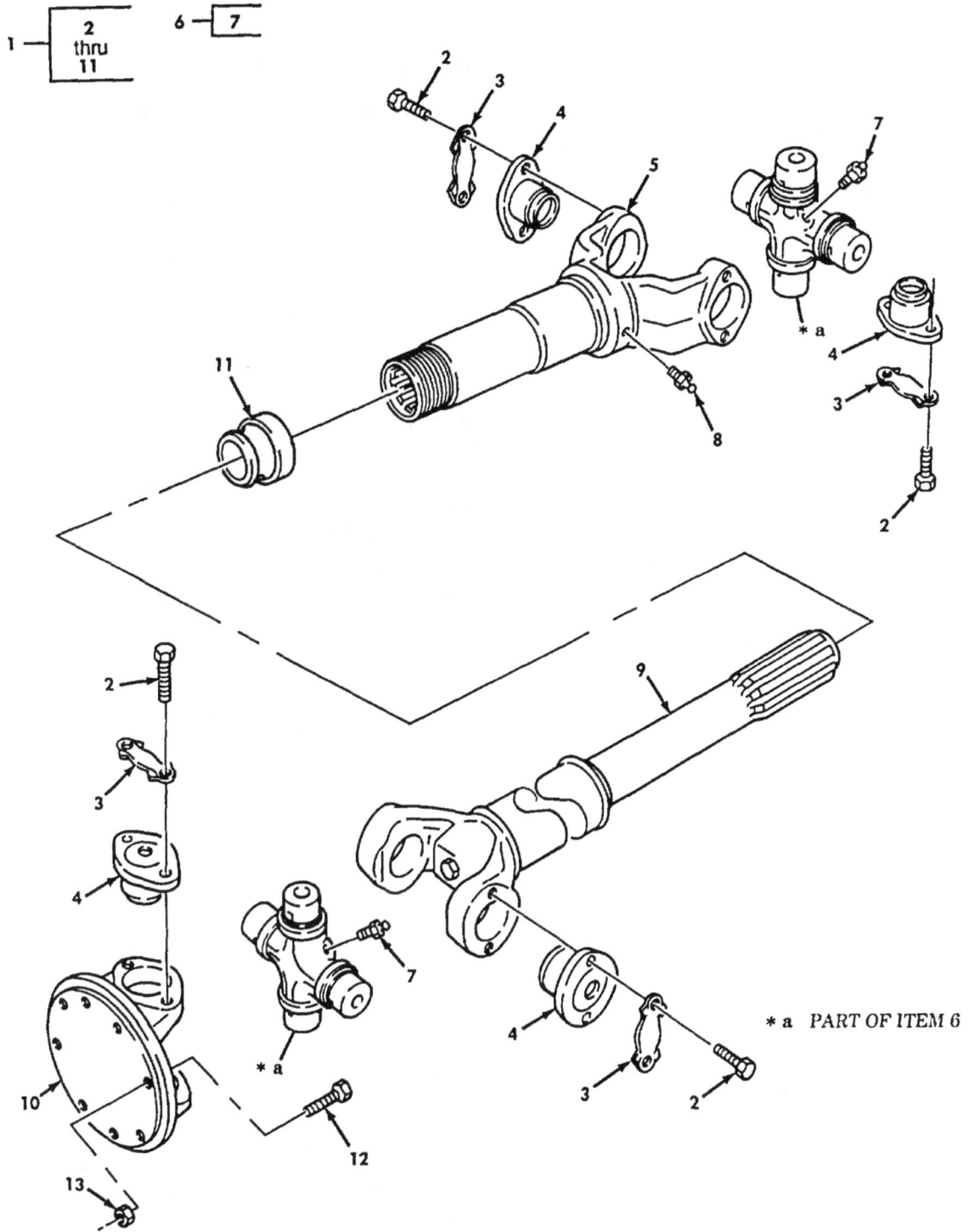

Figure 219. Propeller Shaft Assembly, Front Axle to Transfer Shaft (Rockwell) (A1939, M939A1).

* a   PART OF ITEM 6

(1) ITEM NO	(2) SMR CODE	(3) NSN	(4) CAGEC	(5) PART NUMBER	(6) DESCRIPTION AND USABLE ON CODES (UOC)	(7) QTY
					GROUP 0900 PROPELLER SHAFTS	
					FIG. 219 PROPELLER SHAFT ASSEMBLY, FRONT AXLE TO TRANSFER SHAFT (ROCKWELL) (M939, M939A1)	
1	PAOOO	2520011276950	78500	9830444760	PROPELLER SHAFT .... UOC:DAA, DAB, DAC, DAD, DAE, DAF, DAG, DAH, DAJ, DAK, DAL, DAW, DAX, V12, V13, V14, V15, V16, V17, V18, V19, V20, V21, V22, V24, V25, V39	1
2	KFOZZ		70960	CS-H5-24-15	.SCREW PART OF KIT P/N CP16NS .... UOC:DAA, DAB, DAC, DAD, DAE, DAF, DAG, DAH, DAJ, DAK, DAL, DAW, DAX, V12, V13, V14, V15, V16, V17, V18, V19, V20, V21, V22, V24, V25, V39	16
3	KFOZZ		70960	LPHC37-1	.LOCKPLATE PART OF RKIT P/N CP16NS .... UOC:DAA, DAB, DAC, DAD, DAE, DAF, DAG, DAH, DAJ, DAK, DAL, DAW, DAX, V12, V13, V14, V15, V16, V17, V18 , V19, V20, V21, V22, V24, V25, V39	8
4	KFOZZ		70960	16N4-3A	.BEARING ASSEMBLY PART OF KIT P/N............... CP16NS .... DAA, DAB, DAC, DAD, DABE, DAF, DAG, DAH, DAJ, DAK, DAL, DAW, DAX, V12, V13, V14, V15, V16, V17, V18, V19, V20, V21, V22, V24, V25, V39	8
5	XAOZZ		70960	16NLS32-2	.YOKE UOC:DAA, DAB, DAC, DAD, DAE, DAF, DAG, DAH, DAJ, DAK, DAL, DAW, DAX, V12, V13, V14, V15, V16, V17, V18, V19, V20, V21, V22, V24, V25, V39	
6	KFOZZ		70960	16N1-1A	.CROSS ASSEMBLY PART OF KIT P/N.................. CP16NS .... UOC:DAA, DAB, DAC, DAD, DAE, DAF, DAG, DAH, DAJ, DAK, DAL, DAW, DAX, V12, V13, V14, V15, V16, V17, V18, V19, V20, V21, V22, V24, V25, V39	2
7	PAOZZ	4730001720031	96906	MS15003-1	..FITTING, LUBRICATION .... UOC:DAA, DAB, DAC, DAD, DAE, DAF, DAG, DAH, DAJ, DAK, DAL, DAW, DAX, V12, V13, V14, V15, V16, V17, V18, V19, V20, V21, V22, V24, V25, V39	1
8	PAOZZ	4730001720031	96906	MS15003-1	.FITTING, LUBRICATION .... UOC:DAA, DAB, DAC, DAD, DAE, DAF, DAG, DAH, DAJ, DAK, DAL, DAW, DAX, V12, V13, V14, V15, V16, V17, V18, V19, V20, V21, V22, V24, V25, V39	1
9	XAOZZ		70960	983-55-44760	.SHAFT ASSEMBLY, PROP .... UOC:DAA, DAB, DAC, DAD, DAE, DAF, DAG, DAH, DAJ, DAK, DAL, DAW, DAX, V12, V13, V14, V15, V16, V17, V18, V19, V20, V21, V22, V24, V25, V39	1

(1) ITEM NO	(2) SMR CODE	(3) NSN	(4) CAGEC	(5) PART NUMBER	(6) DESCRIPTION AND USABLE ON CODES (UOC)	(7) QTY
10	XAOZZ		70960	16NF3	.YOKE SHAFT FLANGE .... UOC:DAA, DAB, DAC, DAD, DAE, DAF, DAG, DAH, DAJ, DAK, DAL, DAW, DAX, V12, V13, V14, V15, V16, V17, V18, V19, V20, V21, V22, V24, V25, V39	1
11	PAOZZ	5330011353376	78500	SE-RUR25-2	.SEAL, SHAFT... UOC:DAA, DAB, DAC, DAD, DAE, DAF, DAG, DAH, DAJ, DAK, DAL, DAW, DAX, V12, V13, V14, V15, V16, V17, V18, V19, V20, V21, V22, V24, V25, V39	1
12	PAOZZ	5306007208747	96906	MS20073-06-12	BOLT, MACHINE. .... UOC:DAA, DAB, DAC, DAD, DAE, DAF, DAG, DAH, DAJ, DAK, DAL, DAW, DAX, V12, V13, V14, V15, V16, V17, V18, V19, V20, V21, V22, V24, V25, V39	8
13	PAOZZ	5310008140672	96906	MS51943-36	NUT, SELF-LOCKING. .... UOC:DAA, DAB, DAC, DAD, DAE, DAF, DAG, DAH, DAJ, DAK, DAL, DAW, DAX, V12, V13, V14, V15, V16, V17, V18, V19, V20, V21, V22, V24, V25, V39	8

END OF FIGURE

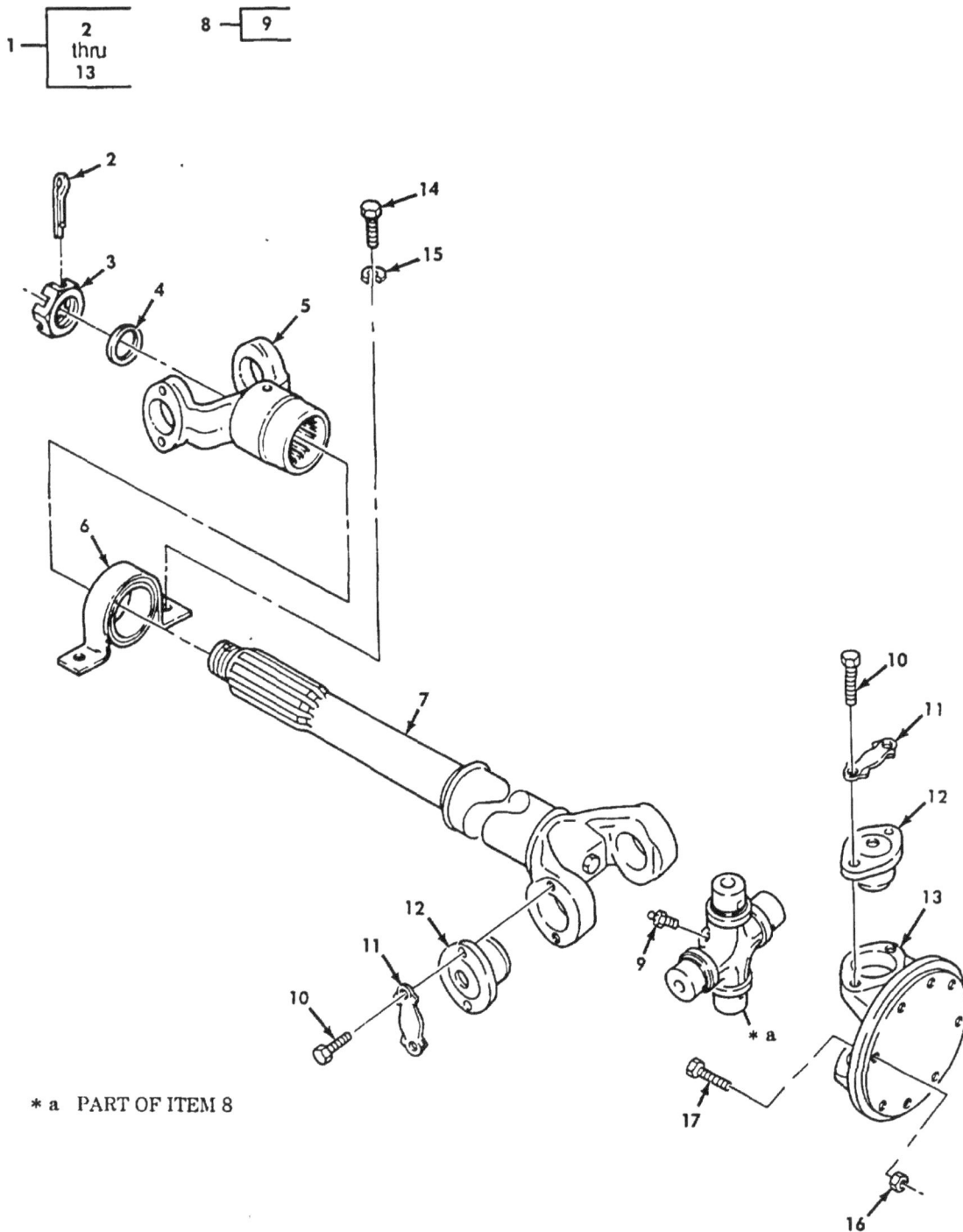

*a PART OF ITEM 8

Figure 220. Propeller Shaft Assembly, Transfer to Front Axle Shaft (Dana) (M939, M939A1).

(1) ITEM NO	(2) SMR CODE	(3) NSN	(4) CAGEC	(5) PART NUMBER	(6) DESCRIPTION AND USABLE ON CODES (UOC)	(7) QTY
					GROUP 0900 PROPELLER SHAFTS	
					FIG. 220 PROPELLER SHAFT ASSEMBLY, TRANSFER TO FRONT AXLE SHAFT (DANA) (M939, M939A1)	
1	PAOOO	2520011122157	72447	908930-3716	PROPELLER SHAFT ASS USE WITH SHAFT ASSEMBLY P/N912257-0416 UOC:DAA,DAB,DAC,DAD,DAE,DAF,DAG,DAH, DAJ,DAK,DAL,DAW,DAX,V12,V13,V14,V15, V16,V17,V18 ,V19,V20,V21,V22,V24,V25, V39	1
2	PAOZZ	5315002981498	95097	70000-362	.PIN,COTTER UOC:DAA,DAB,DAC,DAD,DAE,DAF,DAG,DAH, DAJ,DAK,DAL,DAW,DAX,V12,V13,V14,V15, V16,V17,V18,V19,V20,V21,V22,V24,V25, V39	1
3	PAOZZ	5310011356699	95019	230160-1	.NUT UOC:DAA,DAB,DAC,DAD,DAE,DAF,DAG,DAH, DAJ,DAK,DAL,DAW,DAX,V12,V13,V14,V15, V16,V17,V18,V19,V20,V21,V22,V24,V25, V39	1
4	PAOZZ	5310011356700	95019	230123-8	.WASHER,FLAT UOC:DAA,DAB,DAC,DAD,DAE,DAF,DAG,DAH, DAJ,DAK,DAL,DAW,DAX,V12,V13,V14,V15, V16,V17,V18,V19,V20,V21,V22,V24,V25, V39	1
5	XAOZZ		95019	5-4-1721	.END YOKE PROPELLOR SHAFT UOC:DAA,DAB,DAC,DAD,DAE,DAF,DAG,DAH, DAJ,DAX,DAL,DAW,DAX,V12,V13,V14,V15, V16,V17,V18,V19,V20,V21,V22,V24,V25, V39	1
6	PAOZZ	3130009088589	95019	210084-3X	.HOUSING,BEARING UNI UOC:DAA,DAB,DAC,DAD,DAE,DAF,DAG,DAH, DAJ,DAK,DAL,DAW,DAX,V12,V13,V14,V15, V16,V17,V18,V19,V20,V21,V22,V24,V25, V39	1
7	XAOZZ		95019	5-60-249	.SHAFT,PROP,AXLE UOC:DAA,DAB,DAC,DAD,DAE,DAF,DAG,DAH, DAJ,DAK,DAL,DAW,DAX,V12,V13,V14,V15, V16,V17,V18,V19,V20,V21,V22,V24,V25, V39	1
8	KFOZZ		95019	5-5-278X	.CROSS ASSEMBLY PART OF KIT P/N 5-279X UOC:DAA,DAB,DAC,DAD,DAE,DAF,DAG,DAH, DAJ,DAK,DAL,DAW,DAX,V12,V13,V14,V15, V16,V17,V18,V19,V20,V21,V22,V24,V25, V39	1
9	PAOZZ	4730000504203	96906	MS15001-1	..FITTING,LUBRICATION UOC:DAA,DAB,DAC,DAD,DAE,DAF,DAG,DAH, DAJ,DAK,DAL,DAW,DAX,V12,V13,V14 ,V15,	1

(1) ITEM NO	(2) SMR CODE	(3) NSN	(4) CAGEC	(5) PART NUMBER	(6) DESCRIPTION AND USABLE ON CODES (UOC)	(7) QTY
					V16, V17, V18, V19, V20, V21, V22, V24, V25, ... V39	
10	PAOZZ	5306002707387	19207	5214483	.BOLT, MACHINE PART OF KIT P/N .... 5-279X UOC:DAA, DAB, DAC, DAD, DAE, DAF, DAG, DAH, DAJ, DAK, DAL, DAW, DAX, V12, V13, V14, V15, V16, V17, V18, V19, V20, V21, V22, V24, V25, V39	8
11	PAOZZ	5340005214479	19207	5214479	.LOCKING PLATE, NUT A PART OF KIT .... P/N 5-279X .... UOC:DAA, DAB, DAC, DAD, DAE, DAF, DAG, DAH, DAJ, DAK, DAL, DAW, DAX, V12, V13, V14, V15, V16, V17, V18, V19, V20, V21, V22, V24, V25, V39	4
12	KFOZZ		19207	7066101	.BEARING ASSEMBLY PART OF KIT P/N .... 5-279X .... UOC:DAA, DAB, DAC, DAD, DAE, DAF, DAG, DAH, DAJ, DAK, DAL, DAW, DAX, V12, V13, V14, V15, V16, V17, V18, V19, V20, V21, V22, V24, V25, V39	4
13	PAOZZ	2520010906673	95019	5-2-629	.YOKE, UNIVERSAL JOIN .... UOC:DAA, DAB, DAC, DAD, DAE, DAF, DAG, DAH, DAJ, DAK, DAL, DAW, DAX, V12, V13, V14 , V15, V16, V17, V18, V19, V20, V21, V22, V24, V25, V39	1
14	PAOZZ	5305002692804	96906	MS90726-61	SCREW, CAP, HEXAGON H .... UOC:DAA, DAB, DAC, DAD, DAE, DAF, DAG, DAH, DAJ, DAK, DAL, DAW, DAX, V12, V13, V14, V15, V16, V17, V18, V19, V20, V21, V22, V24, V25, V39	2
15	PAOZZ	5310002090965	96906	MS35338-47	WASHER, LOCK .... UOC:DAA, DAB, DAC, DAD, DAE, DAF, DAG, DAH, DAJ, DAK, DAL, DAW, DAX, V12, V13, V14, V15, V16, V17, V18, V19, V20, V21, V22, V24, V25, V39	2
16	PAOZZ	5310008140672	96906	MS51943-36	NUT, SELF-LOCKING, HE .... UOC:DAA, DAB, DAC, DAD, DAE, DAF, DAG, DAH, DAJ, DAK, DAL, DAW, DAX, V12, V13, V14, V15, V16, V17, V18, V19, V20, V21, V22, V24, V25, V39	8
17	PAOZZ	5306007208747	96906	MS20073-06-12	BOLT, MACHINE.. UOC:DAA, DAB, DAC, DAD, DAE, DAF, DAG, DAH, DAJ, DAK, DAL, DAW, DAX, V12, V13, V14, V15, V16, V17, V18, V19, V20, V21, V22, V24, V25, V39	8

END OF FIGURE

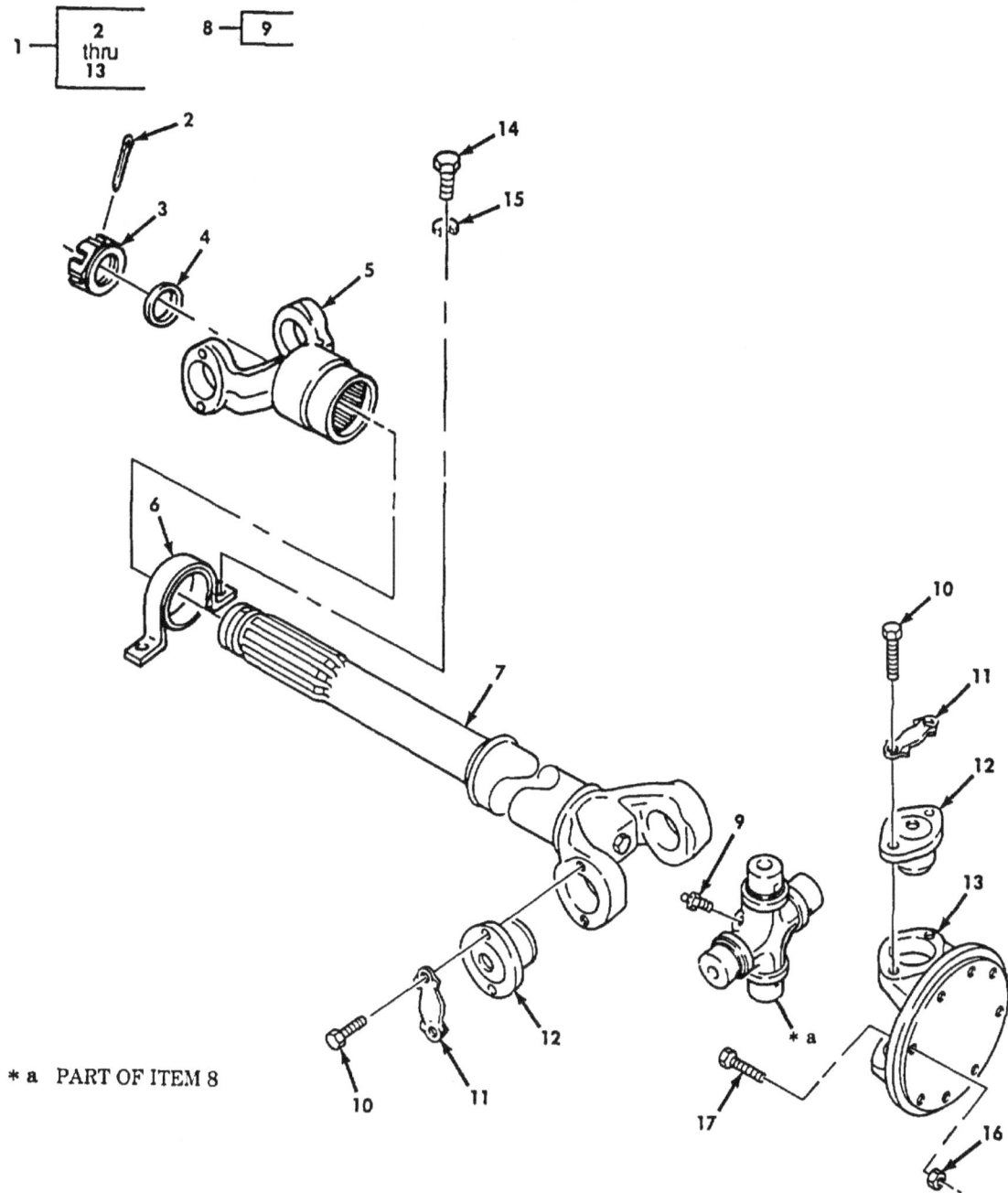

Figure 221. Propeller Shaft Assembly, Transfer to Front Axle Shaft (Rockwell) (M939, M939A1).

* a   PART OF ITEM 8

(1) ITEM NO	(2) SMR CODE	(3) NSN	(4) CAGEC	(5) PART NUMBER	(6) DESCRIPTION AND USABLE ON CODES (UOC)	(7) QTY
					GROUP 0900 PROPELLER SHAFT ASSEMBLY	
					FIG. 221 PROPELLER SHAFT ASSEMBLY, TRANSFER TO FRONT AXLE SHAFT (ROCKWELL) (M939, M939A1)	
1	PAOOO	2520011276951	78500	9836444761	PROPELLER SHAFT WIT USE WITH SHAFT........... ASSY P/N9830444760 .... UOC:DAA, DAB, DAC, DAD, DAE, DAF, DAG, DAH, DAJ, DAK, DAL, DAW, DAX, V12, V13, V14, V15, V16, V17, V18, V19, V20, V21, V22, V24, V25, V39	1
2	PAOZZ	5315011361659	78500	C02-40-4	.PIN, COTTER YOKE, SLOTTED NUT .... UOC:DAA, DAB, DAC, DAD, DAE, DAF, DAG, DAH, DAJ, DAK, DAL, DAW, DAX, V12, V13, V14, V15, V16, V17, V18, V19, V20, V21, V22, V24, V25, V39	1
3	PAOZZ	5310011359506	78500	NU-PT20-18	.NUT, PLAIN, SLOTTED, H .... UOC:DAA, DAB, DAC, DAD, DAE, DAF, DAG, DAH, DAJ, DAK, DAL, DAW, DAX, V12, V13, V14, V15, V16, V17, V18, V19, V20, V21, V22, V24, V25, V39	1
4	PAOZZ	5310011356752	78500	WAR-21-3	.WASHER, FLAT .... UOC:DAA, DAB, DAC, DAD, DAE, DAF, DAG, DAH, DAJ, DAK, DAL, DAW, DAX, V12, V13, V14, V15, V16, V17, V18, V19, V20, V21, V22, V24, V25, V39	1
5	XAOZZ		70960	16NYS28-21	.YOKE .... UOC:DAA, DAB, DAC, DAD, DAE, DAF, DAG, DAH, DAJ, DAK, DAL, DAW, DAX, V12, V13, V14, V15, V16, V17, V18, V1 9, V20, V21, V22, V24, V25, V39	1
6	PAOZZ	2520011337876	78500	UCB-104	.BEARING ASSEMBLY, YO .... UOC:DAA, DAB, DAC, DAD, DAE, DAF, DAG, DAH, DAJ, DAK, DAL, DAW, DAX, V12, V13, V14, V15, V16, V17, V17, V1, V, V20, V21, V22, V24, V25, V39	1
7	XAOZZ		70960	983-65-44-761	.SHAFT .... UOC:DAA, DAB, DAC, DAD, DAE, DAF, DAG, DAH, DAJ, DAK, DAL, DAW, DAX, V12, V13, V13, V, V15, V16, V17, V17, V119, V20, V21, V22, V24, V25, V39	1
8	KFOZZ		70960	16N1-1A	.CROSS ASSEMBLY PART OF KIT P/N................ CP16NS .... UOC:DAA, DAB, DAC, DAD, DAE, DAF, DAG, DAH, DAJ, DAK, DAL, DAW, DAX, V12, V13, V14, V15, V16, V17, V18, V19, V20, V21, V22, V24, V25, V39	1
9	PAOZZ	4730001720031	96906	MS15003-5	..FITTING, LUBRICATION .... UOC:DAA, DAB, DAC, DAD, DAE, DAF, DAG, DAH, DAJ, DAK, DAL, DAW, DAX, V12, V13, V14, V15, V16, V17, V18, V19, V20, V21, V22, V24, V25, V39	1

(1) ITEM NO	(2) SMR CODE	(3) NSN	(4) CAGEC	(5) PART NUMBER	(6) DESCRIPTION AND USABLE ON CODES (UOC)	(7) QTY
10	KFOZZ		70960	CS-H5-24-15	.SCREW PART OF KIT P/N CP16NS .... UOC:DAA, DAB, DAC, DAD, DAE, DAF, DAG, DAH, DAJ, DAK, DAL, DAW, DAX, V12, V13, V14, V15, V16, V17, V18, V19, V20, V21, V22, V24, V25, V39	8
11	KFOZZ		70960	LPHC37-1	.LOCKPLATE PART OF KIT P/N CP16NS .... UOC:DAA, DAB, DAC, DAD, DAE, DAF, DAG, DAH, DAJ, DAK, DAL, DAW, DAX, V12, V13, V14, V15, V16, V17, V18, V19, V20, V21, V22, V24, V25, V39	4
12	KFOZZ		70960	16N4-3A	.BEARING ASSEMBLY PART OF KIT P/N.............. CP16NS .... UOC:DAA, DAB, DAC, DAD, DAE, DAF, DAG, DAH, DAJ, DAK, DAL, DAW, DAX, V12, V13, V14, V15, V16, V17, V18, V19, V20, V21, V22, V24, V25, V39	4
13	XAOZZ		70960	16NF3	.YOKE UNIVERSAL JOIN .... UOC:DAA, DAB, DAC, DAD, DAE, DAF, DAG, DAH, DAJ, DAK, DAL, DAW, DAX, V12, V13, V14, V15, V16, V17, V18, V19, V20, V21, V22, V24, V25, V39	1
14	PAOZZ	5305002692804	96906	MS90726-61	SCREW, CAP, HEXAGON H.............. UOC:DAA, DAB, DAC, DAD, DAE, DAF, DAG, DAH, DAJ, DAK, DAL, DAW, DAX, V12, V13, V14, V15, V16, V17, V18, V19, V20, V21, V22, V24, V25, V39	2
15	PAOZZ	5310002090965	96906	MS35338-47	WASHER, LOCK .... UOC:DAA, DAB, DAC, DAD, DAE, DAF, DAG, DAH, DAJ, DAK, DAL, DAW, DAX, V12, V13, V14, V15, V16, V17, V18, V19, V20, V21, V22, V24, V25, V39	2
16	PAOZZ	5310008140672	96906	MS51943-36	NUT, SELF-LOCKING, HE.............. UOC:DAA, DAB, DAC, DAD, DAE, DAF, DAG, DAH, DAJ, DAK, DAL, DAW, DAX, V12, V13, V14, V15, V16, V17, V18, V19, V20, V21, V22, V24, V25, V39	8
17	PAOZZ	5306007208747	96906	MS20073-06-12	BOLT, MACHINE .... UOC:DAA, DAB, DAC, DAD, DAE, DAF, DAG, DAH, DAJ, DAK, DAL, DAW, DAX, V12, V13, V14, V15, V16, V17, V18, V19, V20, V21, V22, V24, V25, V39	8

END OF FIGURE

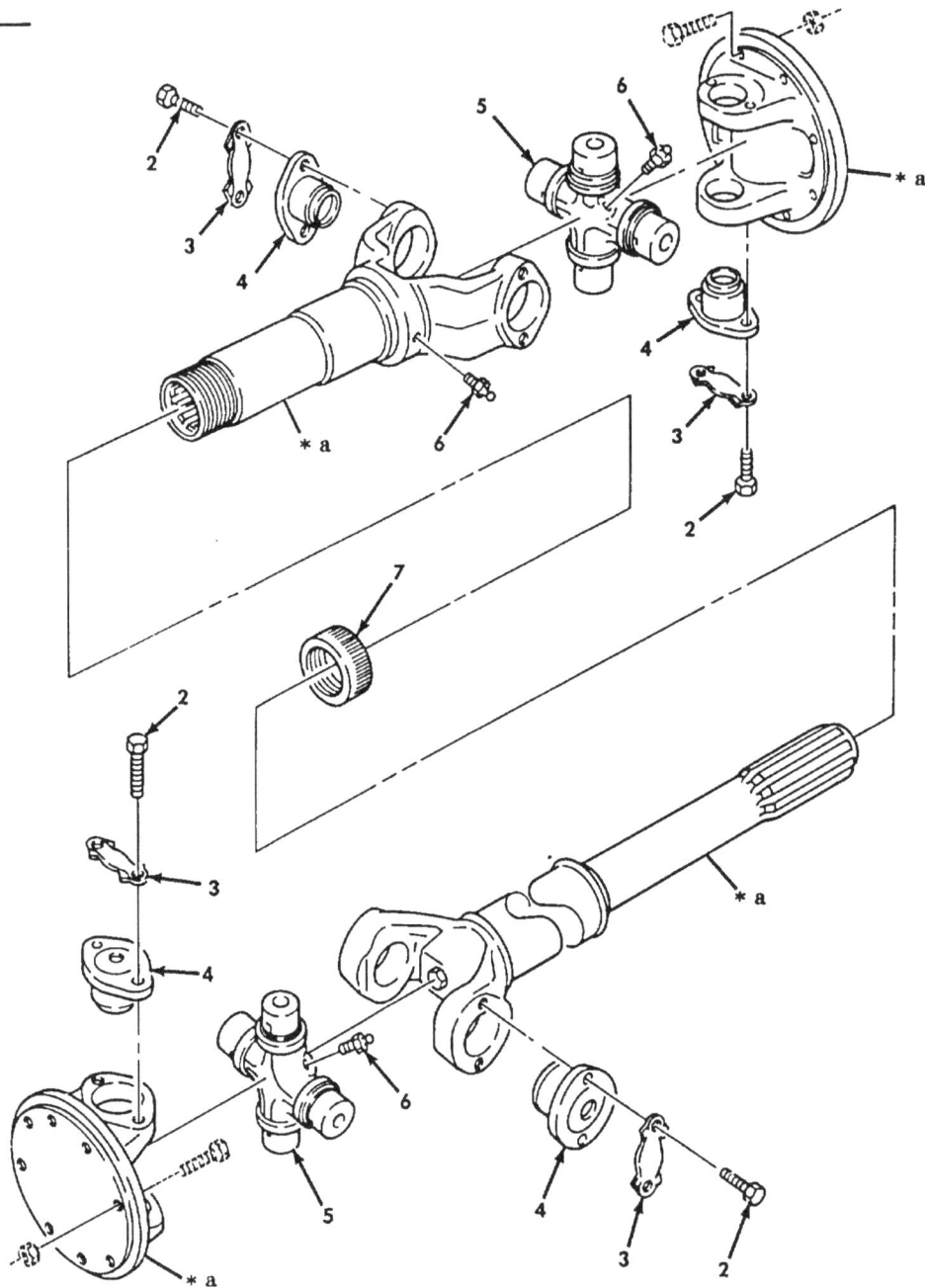

* a PART OF ITEM 1

Figure 222. Propeller Shaft Assembly, Transfer to Front Axle Shaft (M939A2).

(1) ITEM NO	(2) SMR CODE	(3) NSN	(4) CAGEC	(5) PART NUMBER	(6) DESCRIPTION AND USABLE ON CODES (UOC)	(7) QTY
					GROUP 0900 PROPELLER SHAFTS	
					FIG. 222 PROPELLER SHAFT ASSEMBLY, TRANSFER TO FRONT AXLE SHAFT (M939A2)	
1	PAOOO	2520013391648	78500	983-04-48-798	PROPELLER SHAFT .... UOC:ZAA, ZAB, ZAC, ZAD, ZAE, ZAF, ZAG, ZAH, ZAJ, ZAK, ZAL	1
2	KFOZZ		70960	CS-H5-24-15	.SCREW PART OF KIT P/N CP16NS .... UOC:ZAA, ZAB, ZAC, ZAD, ZAE, ZAF, ZAG, ZAH, ZAJ, ZAK, ZAL	16
3	KFOZZ		70960	LPHC37-1	.LOCKPLATE PART OF KIT P/N CP16NS .... UOC:ZAA, ZAB, ZAC, ZAD, ZAE, ZAF, ZAG, ZAH, ZAJ, ZAK, ZAL	8
4	KFOZZ		70960	16N4-3A	.BEARING ASSEMBLY PART OF KIT P/N............... CP16NS .... UOC:ZAA, ZAB, ZAC, ZAD, ZAE, ZAF, ZAG, ZAH, ZAJ, ZAK, ZAL	8
5	KFOZZ		70960	16N1-1A	.CROSS ASSEMBLY PART OF KIT P/N.................. CP16NS .... UOC:ZAA, ZAB, ZAC, ZAD, ZAE, ZAF, ZAG, ZAH, ZAJ, ZAK, ZAL	2
6	PAOZZ	4730001720031	96906	MS15003-5	.FITTING, LUBRICATION .... UOC:ZAA, ZAB, ZAC, ZAD, ZAE, ZAF, ZAG, ZAH, ZAJ, ZAK, ZAL	2
7	PAOZZ	5330011353376	78500	SE-RUR25-2	.SEAL, SHAFT .... UOC:ZAA, ZAB, ZAC, ZAD, ZAE, ZAF, ZAG, ZAH, ZAJ, ZAK, ZAL	1

END OF FIGURE

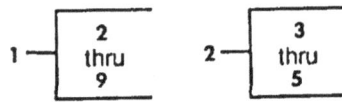

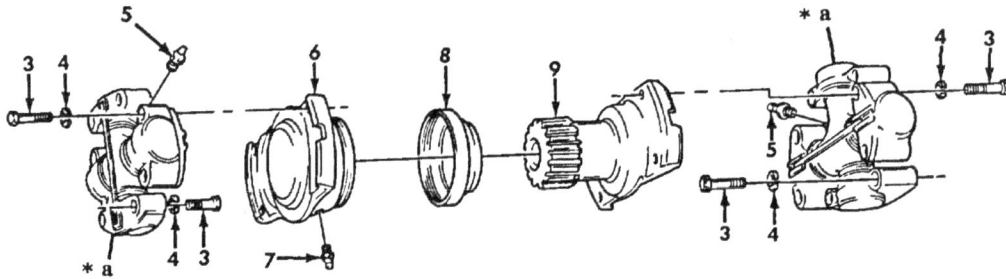

*a   PART OF ITEM 2

Figure 223. Propeller Shaft Assembly, Transmission to Transfer (M939, M939A1).

(1) ITEM NO	(2) SMR CODE	(3) NSN	(4) CAGEC	(5) PART NUMBER	(6) DESCRIPTION AND USABLE ON CODES (UOC)	(7) QTY
					GROUP 0900 PROPELLER SHAFTS	
					FIG. 223 PROPELLER SHAFT ASSEMBLY, TRANSMISSION TO TRANSFER (M939, M939A1)	
1	PAOOO	2520011122156	19207	11669426-1	PROPELLER SHAFT .... UOC:DAA, DAB, DAC, DAD, DAE, DAF, DAG, DAH, DAJ, DAK, DAL, DAW, DAX, V12, V13, V14, V15, V16, V17, V18, V19, V20, V21, V22, V24, V25, V39	1
2	PAOZZ	2520012804129	78500	CP85WB62	.PARTS KIT, UNIVERSAL .... UOC:DAA, DAB, DAC, DAD, DAE, DAF, DAG, DAH, DAJ, DAK, DAL, DAW, DAX, V12, V13, V14, V15, V16, V17, V18, V19, V20, V21, V22, V24, V25, V39	2
3	PAOZZ	5305001006791	78500	CS-HB-20-16	..SCREW, CAP, HEXAGON H............................ UOC:DAA, DAB, DAC, DAD, DAE, DAF, DAG, DAH, DAJ, DAK, DAL, DAW, DAX, V12, V13, V14, V15, V16, V17, V18, V19, V20, V21, V22, V24, V25, V39	4
4	PAOZZ	5310000312673	70960	WA-LH8-3	..WASHER, LOCK .... UOC:DAA, DAB, DAC, DAD, DAE, DAF, DAG, DAH, DAJ, DAK, DAL, DAW, DAX, V12, V13, V14, V15, V16, V17, V18, V19, V20, V21, V22, V24, V25, V39	4
5	PAOZZ	4730001720031	96906	MS15003-5	..FITTING, LUBRICATION.............................. UOC:DAA, DAB, DAC, DAD, DAE, DAF, DAG, DAH, DAJ, DAK, DAL, DAW, DAX, V12, V13, V14, V15, V16, V17B, 18, V19, V20, V21, V22, V24, V25, V39	1
6	XAOZZ		70960	85WBL548-19	.YOKE, REAR, UNIVERSAL.............................. UOC:DAA, DAB, DAC, DAD, DAE, DAF, DAG, DAH, DAJ, DAK, DAL, DAW, DAX, V12, V13, V14, V15, V16, V17, V18, V19, V20, V21, V22, V24, V25, V39	1
7	PAOZZ	4730000504208	96906	MS15003-1	.FITTING, LUBRICATION .... UOC:DAA, DAB, DAC, DAD, DAE, DAF, DAG, DAH, DAJ, DAK, DAL, DAW, DAX, V12, V13, V14, V15, V16, V17, V18, V19, V20, V21, V22, V24, V25, V39	1
8	PAOZZ	2520010834404	78500	SERUR40-3	.CAP, DUST, PROPELLER .... UOC:DAA, DAB, DAC, DAD, DAE, DAF, DAG, DAH, DAJ, DAK, DAL, DAW, DAX, V12, V13, V14, V15, V16, V17, V18, V19, V20, V21, V22, V24, V25, V39	1
9	XAOZZ		70960	85WBYSM48-68	.YOKE, PROPELLOR SHAF .... UOC:DAA, DAB, DAC, DAD, DAE, DAF, DAG, DAH, DAJ, DAK, DAL, DAW, DAX, V12, V13, V14, V15, V16, V17, V18, V19, V20, V21, V22, V24, V25, V39	1

END OF FIGURE

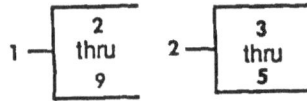

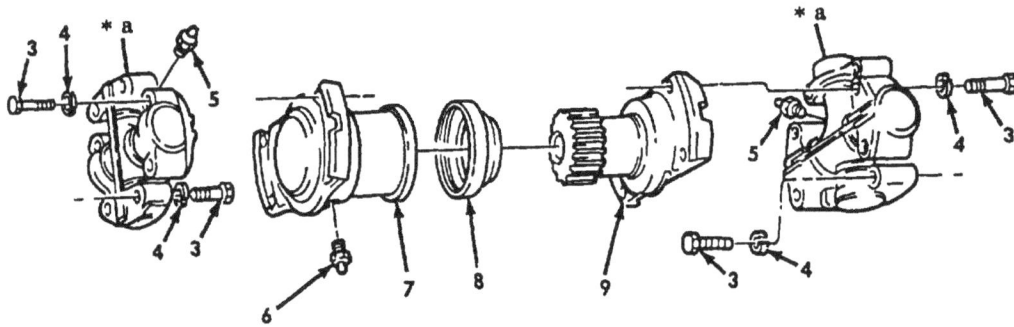

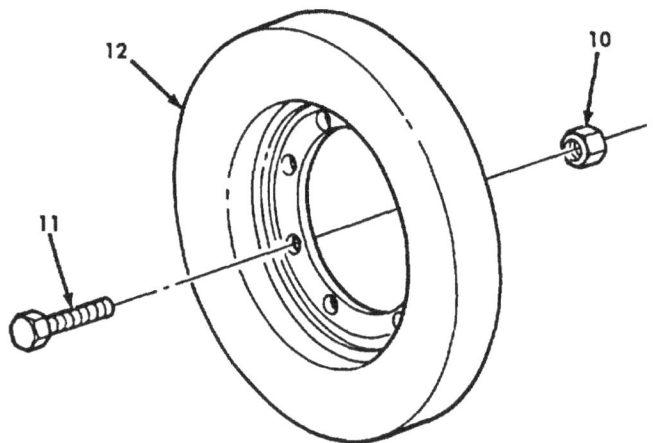

* a PART OF ITEM 2

Figure 224. Propeller Shaft Assembly, Transmission to Transfer (M939A2).

(1) ITEM NO	(2) SMR CODE	(3) NSN	(4) CAGEC	(5) PART NUMBER	(6) DESCRIPTION AND USABLE ON CODES (UOC)	(7) QTY
					GROUP 0900 PROPELLER SHAFTS	
					FIG. 224 PROPELLER SHAFT ASSEMBLY, TRANSMISSION TO TRANSFER(M939A2)	
1	PAOZZ	2520012856283	70960	963-92-48-799	PROPELLER SHAFT ASSEMBLY .... UOC:ZAA, ZAB, ZAC, ZAD, ZAE, ZAF, ZAG, ZAH, ZAJ, ZAK, ZAL	1
2	PAOZZ	2520012804129	78500	CP85WB62	.PARTS KIT, UNIVERSAL...................................... UOC:ZAA, ZAB, ZAC, ZAD, ZAE, ZAF, ZAG, ZAH, ZAJ, ZAK, ZAL	2
3	PAOZZ	5305001006791	78500	CS-HB-20-16	..SCREW, CAP, HEXAGON H .... UOC:ZAA, ZAB, ZAC, ZAD, ZAE, ZAF, ZAG, ZAH, ZAJ, ZAK, ZAL	4
4	PAOZZ	5310000312673	70960	WA-LH8-3	..WASHER, LOCK .... UOC:ZAA, ZAB, ZAC, ZAD, ZAE, ZAF, ZAG, ZAH, ZAJ, ZAK, ZAL	4
5	PAOZZ	4730001720031	96906	MS15003-5	..FITTING, LUBRICATION................................... UOC:ZAA, ZAB, ZAC, ZAD, ZAE, ZAF, ZAG, ZAH, ZAJ, ZAK, ZAL	1
6	PAOZZ	4730000504208	96906	MS15003-1	..FITTING, LUBRICATION................................... UOC:ZAA, ZAB, ZAC, ZAD, ZAE, ZAF, ZAG, ZAH, ZAJ, ZAK, ZAL	1
7	XAOZZ		70960	85WBLS40-34	.YOKE, SLIP .... UOC:ZAA, ZAB, ZAC, ZAD, ZAE, ZAF, ZAG, ZAH, ZAJ, ZAK, ZAL	1
8	PAOZZ	5330010821818	78500	SERUR33-4	.SEAL .... UOC:ZAA, ZAB, ZAC, ZAD, ZAE, ZAF, ZAG, ZAH, ZAJ, ZAK, ZAL	1
9	PAOZZ	4320012856262	70960	PS40-16-34	.YOKE, HYDRAULIC MOTO .... UOC:ZAA, ZAB, ZAC, ZAD, ZAE, ZAF, ZAG, ZAH, ZAJ, ZAK, ZAL	1
10	PAOZZ	5310008140672	96906	MS51943-36	NUT, SELF-LOCKING, HE .... UOC:ZAA, ZAB, ZAC, ZAD, ZAE, ZAF, ZAG, ZAH, ZAJ, ZAK, ZAL	8
11	PAOZZ	5305012718383	24617	9415757	SCREW, CAP, HEXAGON H .... UOC:ZAA, ZAB, ZAC, ZAD, ZAE, ZAF, ZAG, ZAH, ZAJ, ZAK, ZAL	8
12	PAOZZ	2815012842284	74480	9711085-002	DAMPER, DRIVELINE TO .... UOC:ZAA, ZAB, ZAC, ZAD, ZAE, ZAF, ZAG, ZAH, ZAJ, ZAK, ZAL	1

END OF FIGURE

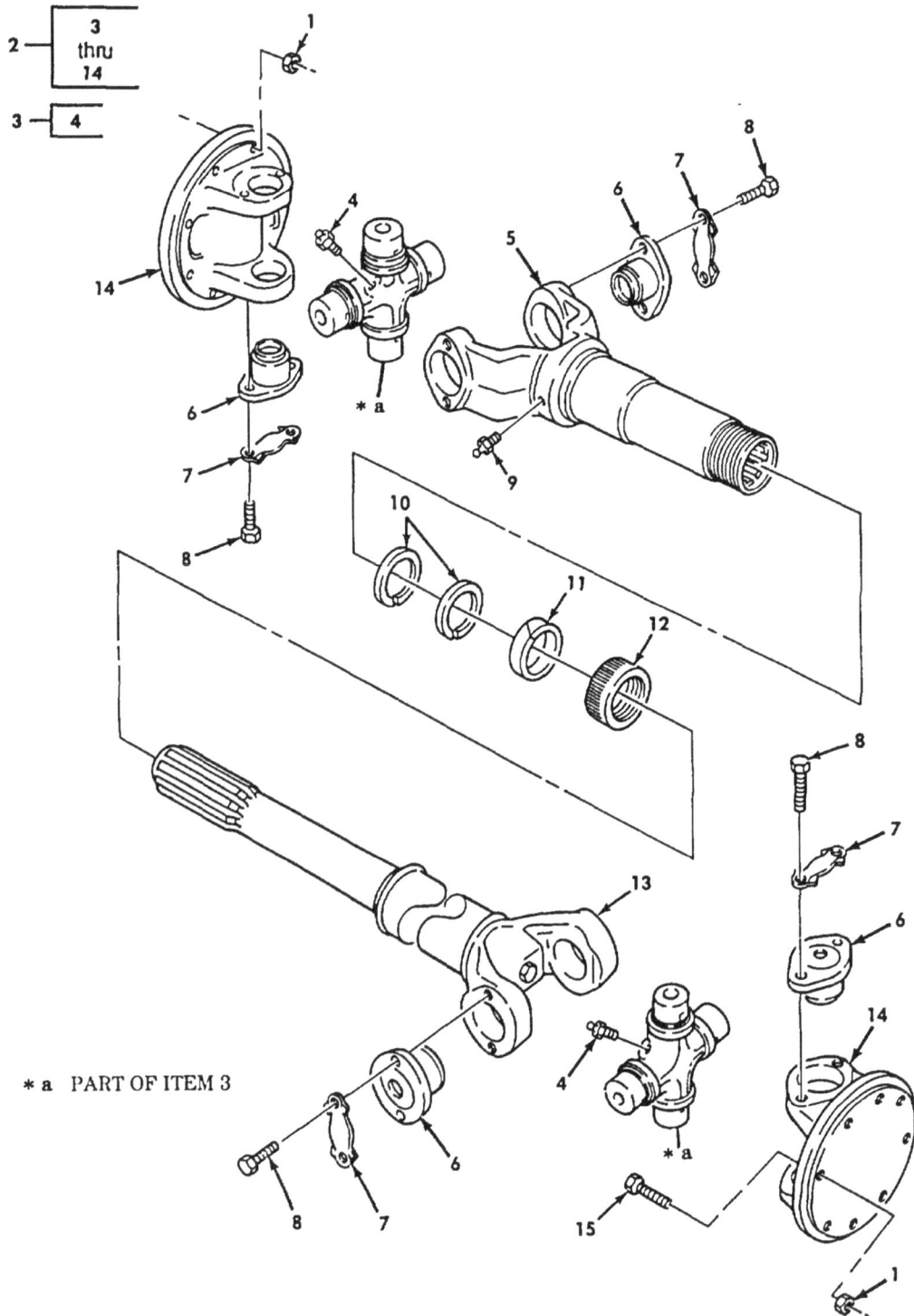

*a   PART OF ITEM 3

Figure 225. Propeller Shaft Assembly, Transfer to Forward Rear Axle (Long Wheelbase).

(1) ITEM NO	(2) SMR CODE	(3) NSN	(4) CAGEC	(5) PART NUMBER	(6) DESCRIPTION AND USABLE ON CODES (UOC)	(7) QTY
					GROUP 0900 PROPELLER SHAFTS	
					FIG. 225 PROPELLER SHAFT ASSEMBLY, TRANSFER TO FORWARD REAR AXLE (LONG WHEELBASE)	
1	PAOZZ	5310008140672	96906	MS51943-36	NUT, SELF-LOCKING, HE .... UOC: DAA, DAB, DAL, DAW, DAX, V12, V13, V14, V15, V18, ZAA, ZAB, ZAL	16
2	PAOOO	2520011908485	78500	984-04-44758	PROPELLER SHAFT WIT (ROCKWELL) .... UOC: DAA, DAB, DAL, DAW, DAX, V12, V13, V14, V15, V18, ZAA, ZAB, ZAL	1
2	PAOOO	2520011712360	72447	204581-3	PROPELLER SHAFT (DANA).............................. UOC: DAA, DAB, DAL, DAW, DAX, V12, V13, V14, V15, V18, ZAA, ZAB, ZAL	1
3	KFOZZ		19207	8764802	.CROSS, ASSEMBLY, UNIV PART OF KIT............. P/N 5-280X .... UOC: DAA, DAB, DAL, DAW, DAX, V12, V13, V14, V15, V18, ZAA, ZAB, ZAL	2
4	PAOZZ	4730000504203	96906	MS15001-1	..FITTING, LUBRICATION .... UOC: DAA, DAB, DAL, DAW, DAX, V12, V13, V14, V15, V18, ZAA, ZAB, ZAL	1
5	XAOZZ		19207	7335052	.YOKE.. UOC: DAA, DAB, DAL, DAW, DAX, V12, V13, V14, V15, V18, ZAA, ZAB, ZAL	1
6	KFOZZ		19207	7335051	.BEARING, ASSY, JOINT PART OF KIT P/N........... 5-280X... UOC: DAA, DAB, DAL, DAW, DAX, V12, V13, V14, V15, V18, ZAA, ZAB, ZAL	8
7	PAOZZ	5340011195682	19207	7335053	.LOCKING PLATE, NUT A.................................... UOC: DAA, DAB, DAL, DAW, DAX, V12, V13, V14, V15, V18, ZAA, ZAB, ZAL	8
8	PAOZZ	5306011195681	19207	7335054	.BOLT, HEX HEAD PART OF KIT P/N.................... 5-280X... UOC: DAA, DAB, DAL, DAW, DAX, V12, V13, V14, V15, V18, ZAA, ZAB, ZAL	16
9	PAOZZ	4730000504208	96906	MS15003-1	.FITTING, LUBRICATION...................................... UOC: DAA, DAB, DAL, DAW, DAX, V12, V13, V14, V15, V18, ZAA, ZAB, ZAL	1
10	PAOZZ	5310006798346	1CW61	920095-4	.WASHER, SPLIT .... UOC: DAA, DAB, DAL, DAW, DAX, V12, V13, V14, V15, V18, ZAA, ZAB, ZAL	2
11	PAOZZ	5330007412636	19207	7412636	.PACKING, PREFORMED .... UOC: DAA, DAB, DAL, DAW, DAX, V12, V13, V14, V15, V18, ZAA, ZAB, ZAL	1
12	PAOZZ	2520007348101	19207	7348101	.CAP, DUST, PROPELLER.................................... UOC: DAA, DAB, DAL, DAW, DAX, V12, V13, V14, V15, V18, ZAA, ZAB, ZAL	1
13	XAOZZ		19207	10883109	.SHAFT ASSEMBLY, PROP.................................. UOC: DAA, DAB, DAL, DAW, DAX, V12, V13, V14, V15, V18, ZAA, ZAB, ZAL	1

(1) ITEM NO	(2) SMR CODE	(3) NSN	(4) CAGEC	(5) PART NUMBER	(6) DESCRIPTION AND USABLE ON CODES (UOC)	(7) QTY
14	XAOZZ		19207	7348215	.FLANGE, COMPANION, VE............................ UOC: DAA, DAB, DAL, DAW, DAX, V12, V13, V14, V15, V18, ZAA, ZAB, ZAL	2
15	PAOZZ	5306007208747	96906	MS20073-06-12	BOLT, MACHINE .... UOC: DAA, DAB, DAL, DAW, DAX, V12, V13, V14, V15, V18, ZAA, ZAB, ZAL	8

END OF FIGURE

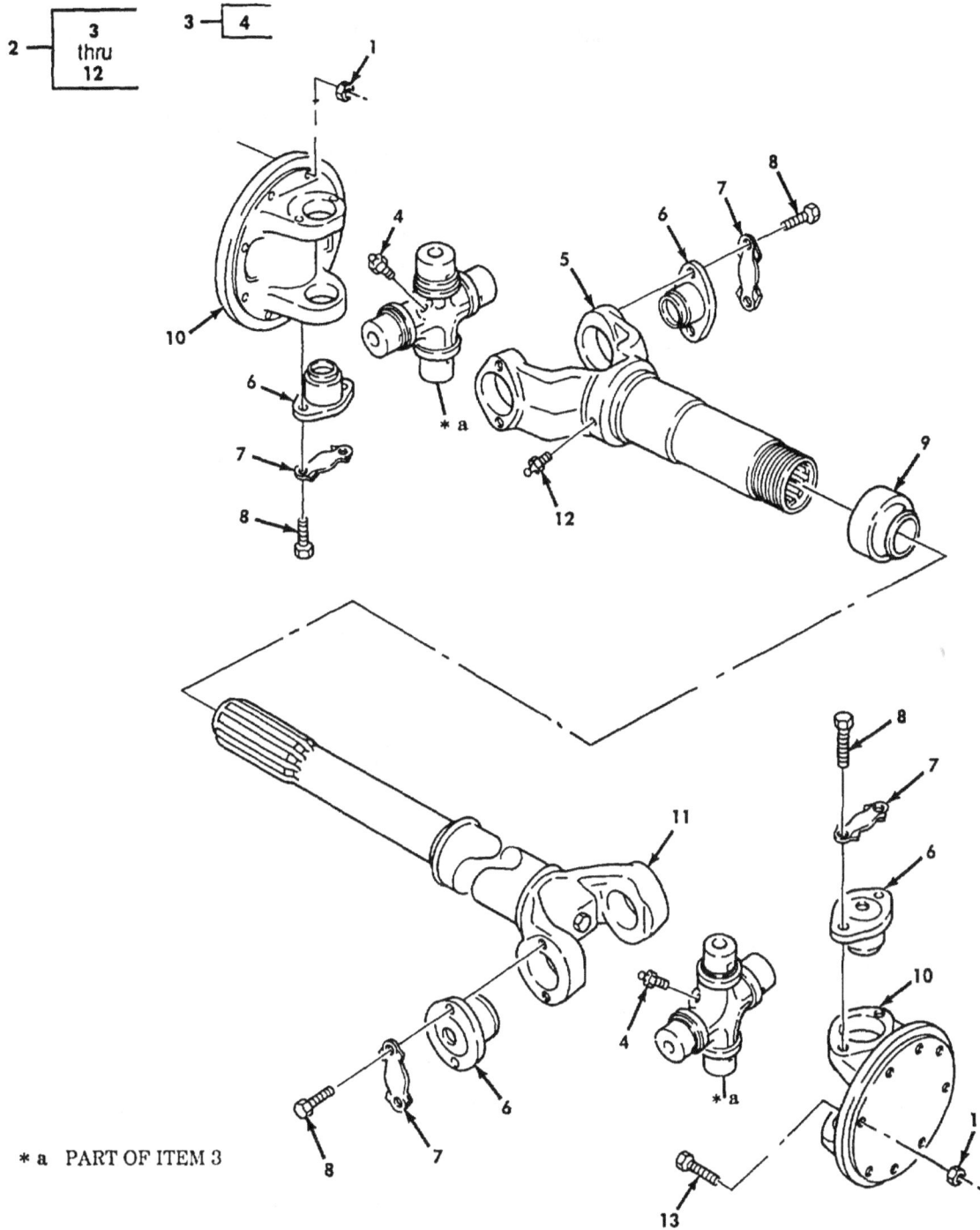

Figure 226. Propeller Shaft Assembly, Transfer to Forward Rear Axle (Short Wheelbase).

* a   PART OF ITEM 3

(1) ITEM NO	(2) SMR CODE	(3) NSN	(4) CAGEC	(5) PART NUMBER	(6) DESCRIPTION AND USABLE ON CODES (UOC)	(7) QTY
					GROUP 0900 PROPELLER SHAFTS	
					FIG. 226 PROPELLER SHAFT ASSEMBLY, TRANSFER TO FORWARD REAR AXLE (SHORT WHEELBASE)	
1	PAOZZ	5310008140672	96906	MS51943-36	NUT, SELF-LOCKING, HE .... UOC: DAE, DAF, DAG, DAH, V19, V20, V21, V22, ZAE, ZAG, ZAH	16
2	PAOOO	2520011147690	19207	11669144	PROPELLER SHAFT WIT (DANA) .... UOC: DAE, DAF, DAG, DAH, V19, V20, V21, V22, ZAE, ZAF, ZAG, ZAH (ROCKWELL)....................	1
2	PAOOO	2520013236342	72447	916210-4823	PROPELLER SHAFT WIT .... UOC: DAE, DAF, DAG, DAH, V19, V20, V21, V22, ZAE, ZAF, ZAG , ZAH	1
3	KFOZZ		70960	17N1-19A	.CROSS ASSEMBLY PART OF KIT P/N 5-.............. 280X UOC: DAE, DAF, DAG, DAH, V19, V20, V21, V22, ZAE, ZAF, ZAG, ZAH...................................	2
4	PAOZZ	4730000504203	96906	MS15001-1	..FITTING, LUBRICATION................................. UOC: DAE, DAF, DAG, DAH, V19, V20, V21, V22, ZAE, ZAF, ZAG, ZAH.	1
5	XAOZZ		70960	17NLS40-24	.YOKE.. UOC: DAE, DAF, DAG, DAH, V19, V20, V21, V22, ZAE, ZAF, ZAG, ZAH	1
6	KFOZZ		70960	17N4-15A	.BEARING ASSEMBLY PART OF KIT P/N 5-........... 0280. UOC: DAE, DAF, DAG, DAH, V19, V20, V21, V22, ZAE, ZAF, ZAG, ZAH.	8
7	KFOZZ		70960	LPHC39-3	.LOCKPLATE PART OF KIT P/N 5-0280.................. UOC: DAE, DAF, DAG, DAH, V19, V20, V21, V22, ZAE, ZAF, ZAG, ZAH.	8
8	KFOZZ		70960	CS-H6-24-49	.SCREW PART OF KIT P/N 5-0280 .... UOC: DAE, DAF, DAG, DAH, V19, V20, V21, V22, ZAE, ZAF, ZAG, ZAH.	16
9	PAOZZ	5330010821818	78500	SE-RUR33-4	.SEAL .... UOC: DAE, DAF, DAG, DAH, V19, V20, V21, V22, ZAE, ZAF, ZAG, ZAH.	1
10	XAOZZ		70960	17NF1	.YOKE SHAFT FLANGE .... UOC: DAE, DAF, DAG, DAH, V19, V20, V21, V22, ZAE, ZAF, ZAG, ZAH.	2
11	XAOZZ		70960	984-55-44759	.SHAFT ASSEMBLY, PROP .... UOC: DAE, DAF, DAG, DAH, V19, V20, V21, V22, ZAE, ZAF, ZAG, ZAH....................................	1
12	PAOZZ	4730000504208	96906	MS15003-1	.FITTING, LUBRICATION .... UOC: DAE, DAF, DAG, DAH, V19, V20, V21, V22, ZAE, ZAF, ZAG, ZAH....................................	1
13	PAOZZ	5306007208747	96906	MS20073-06-12	BOLT, MACHINE.. UOC: DAE, DAF, DAG, DAH, V19, V20, V21, V22, ZAE, ZAF, ZAG, ZAH.	8

END OF FIGURE

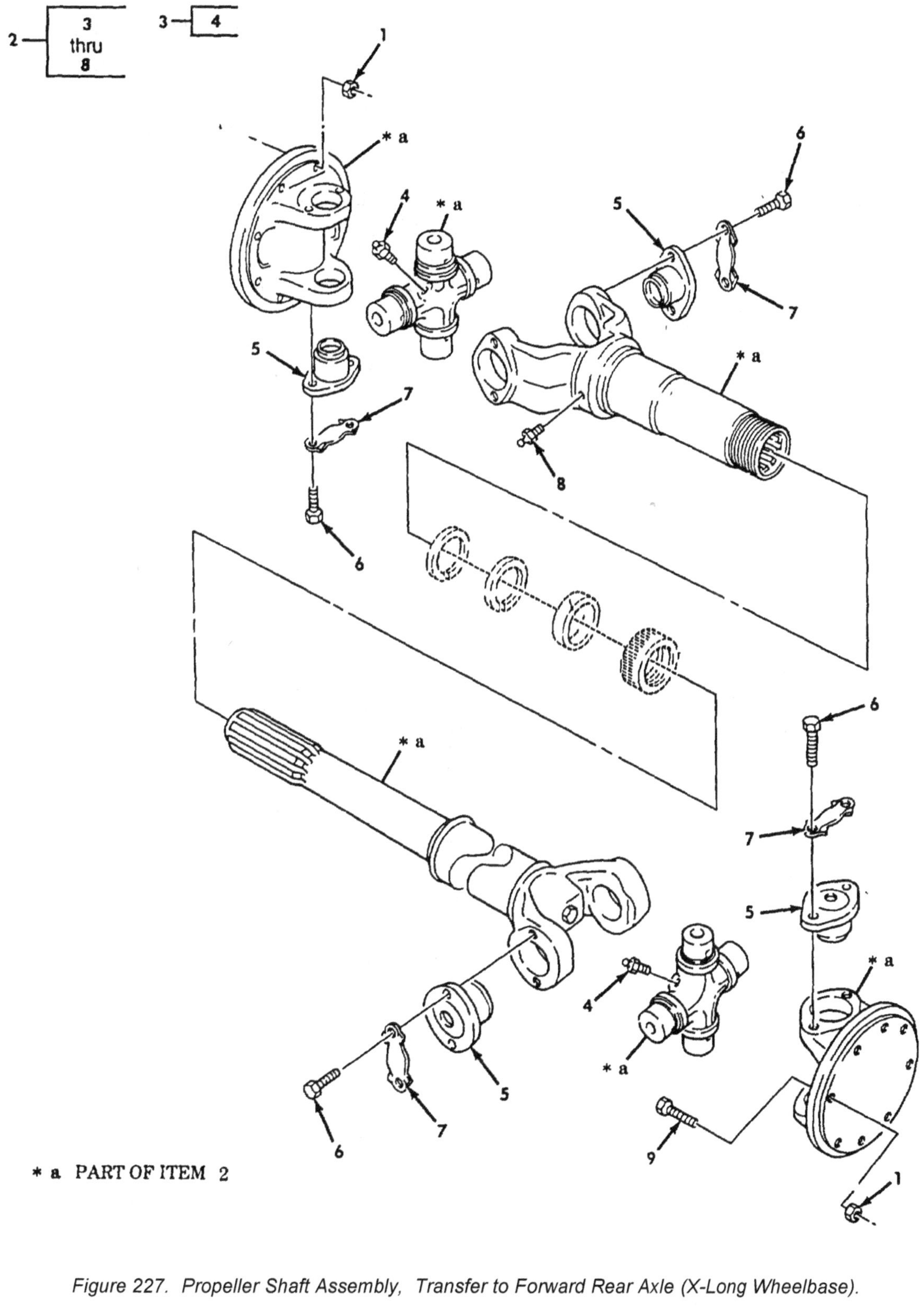

Figure 227. Propeller Shaft Assembly, Transfer to Forward Rear Axle (X-Long Wheelbase).

* a PART OF ITEM 2

SECTION II

(1) ITEM NO	(2) SMR CODE	(3) NSN	(4) CAGEC	(5) PART NUMBER	(6) DESCRIPTION AND USABLE ON CODES (UOC)	(7) QTY
					GROUP 0900 PROPELLER SHAFTS	
					FIG. 227 PROPELLER SHAFT ASSEMBLY, TRANSFER TO FORWARD REAR AXLE (X-LONG WHEELBASE)	
1	PAOZZ	5310008140672	96906	M551943-36	NUT, SELF-LOCKING, HE .... UOC: DAC, DAD, DAJ, DAK, V16, V16, V17, V18, V24, V25, V39, ZAC, ZAD, ZAJ, ZAK	16
2	PAOOO		72447	916210-4823	PROPELLER SHAFT WIT (DANA)........................... UOC: DAC, DAD, DAJ, DAK, V16, V16, V17, V18, V24, V25, V39, ZAC, ZAD, ZAJ, ZAK	1
2	PAOOO		78500	984-04-48-325	PROPELLER SHAFT WIT (ROCKWELL)................. UOC: DAC, DAD, DAJ, DAK, V16, V16, V17, V18, V24, V25, V39, ZAC, ZAD, ZAJ, ZAK	1
3	KFOZZ		19207	8764802	.CROSS ASSEMBLY PART OF KIT P/N 5-............... 280 .... UOC: DAC, DAD, DAJ, DAK, V16, V16, V17, V18, V24, V25, V39, ZAC, ZAD, ZAJ, ZAK	2
4	PAOZZ	4730000504203	96906	MS15001-1	.FITTING, LUBRICATION .... UOC: DAC, DAD, DAJ, DAK, V16, V16, V17, V18, V24, V25, V39, ZAC, ZAD, ZAJ, ZAK	1
5	KFOZZ		19207	7335051	.BEARING ASSEMBLY PART OF KIT P/N 5-............ 280X .... UOC: DAC, DAD, DAJ, DAK, V16, V16, V17, V18, V24, V25, V39, ZAC, ZAD, ZAJ, ZAK 280X	2
6	PAOZZ	5306011195681	19207	7335054	.BOLT, HEX HEAD, PART OF KIT P/N 5-................ 280 .... UOC: DAC, DAD, DAJ, DAK, V16, V16, V17, V18, V24, V25, V39, ZAC, ZAD, ZAJ, ZAK	4
7	PAOZZ	5340011195682	19207	7335053	.LOCKING PLATE, NUT A.PATR OF KIT .... PART OF KIT P/N 5-280X.................................... UOC: DAC, DAD, DAJ, DAK, V16, V16, V17, V18, V24, V25, V39, ZAC, ZAD, ZAJ, ZAK	4
8	PAOZZ	5306007208747	96906	MS15003-1	.FITTING, LUBRICATION .... UOC: DAC, DAD, DAJ, DAK, V16, V16, V17, V18, V24, V25, V39, ZAC, ZAD, ZAJ, ZAK	1
9	PAOZZ	5306007208747	96906	MS20073-06-12	BOLT, MACHINE .... UOC: DAC, DAD, DAJ, DAK, V16, V16, V17, V18, V24, V25, V39, ZAC, ZAD, ZAJ, ZAK	8

END OF FIGURE

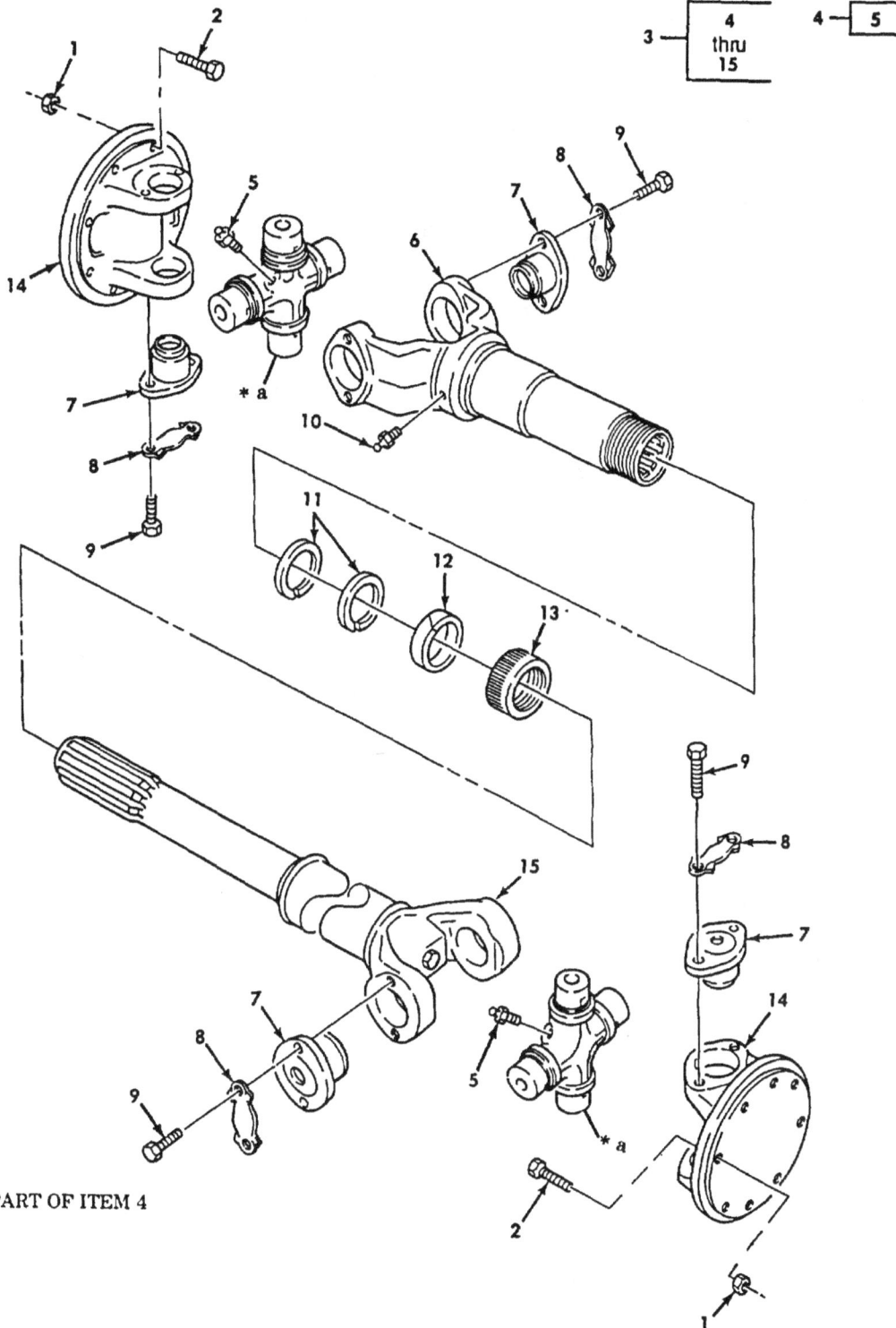

* a   PART OF ITEM 4

Figure 228. Propeller Shaft Assembly, Forward Rear Axle to Rear Rear Axle (Dana).

(1) ITEM NO	(2) SMR CODE	(3) NSN	(4) CAGEC	(5) PART NUMBER	(6) DESCRIPTION AND USABLE ON CODES (UOC)	(7) QTY
					GROUP 0900 PROPELLER SHAFTS	
					FIG. 228 PROPELLER SHAFT ASSEMBLY, FORWARD REAR AXLE TO REAR REAR AXLE (DANA)	
1	PAOZZ	5310008140672	96906	MS51943-36	NUT, SELF-LOCKING, HE	16
2	PAOZZ	5306007208747	96906	MS20073-06-12	BOLT, MACHINE	16
3	PAOOO	2520010934274	19207	8332248	PROPELLER SHAFT WIT	1
					UOC: DAA, DAB, DAC, DAD, DAE, DAF, DAG, DAH, DAJ, DAK, DAL, DAW, DAX, V12, V13, V14, V15, V16, V17, V18, V19, V20, V21, V22, V24, V25, ZAA, ZAB, ZAC, ZAD, ZAE, ZAF, ZAG, ZAH, ZAJ, ZAK, ZAL	
3	PAOOO	2520003180983	72447	200825-3	PROPELLER SHAFT WIT	1
					UOC: V39 V39	
4	KFOZZ		19207	7066097	.CROSS, ASSEMBLY, UNIV.PART OF KIT P/N....... 5-279X..	2
5	PAOZZ	4730000504203	96906	MS15001-1	..FITTING, LUBRICATION....................................	2
6	XAOZZ		19207	7066093	.YOKE	1
7	KFOZZ		19207	7066101	.BEARING, ASSY, NEEDLE PART OF KIT P/N....... 5-279X..	8
8	KFOZZ	5340005214479	19207	5214479	.LOCKING PLATE, NUT A.PART OF KIT P/N.......... 5-279X..	8
9	KFOZZ	5306002707387	19207	5214483	.BOLT, MACHINE PART OF KIT P/N 5-279X............	16
10	PAOZZ	4730000504208	96906	MS15003-1	.FITTING, LUBRICATION	1
11	PAOZZ	5310002763012	19207	7994440	.WASHER, SPLIT	2
12	PAOZZ	5330010671740	34623	7066104	.SEAL, PROPELLER SHAF	1
13	PAOZZ	2520004576660	72447	4-14-19	.CAP, DUST, PROPELLER	1
14	XAOZZ	2520000095304	19207	7066092	.YOKE, UNIVERSAL JOIN	2
15	XAOZZ		19207	10883167	.PROPELLER, SHAFT	1
					UOC: DAA, DAB, DAC, DAD, DAE, DAF, DAG, DAH, DAJ, DAK, DAL, DAW, DAX, V12, V13, V14, V15, V16, V17, V18, V19, V20, V21, V22, V24, V25, ZAA, ZAB, ZAC, ZAD, ZAE, ZAF, ZAG, ZAH, ZAJ, ZAK, ZAL..	
15	XAOZZ		19207	10883168	.PROPELLER SHAFT .... UOC: V39	1

END OF FIGURE

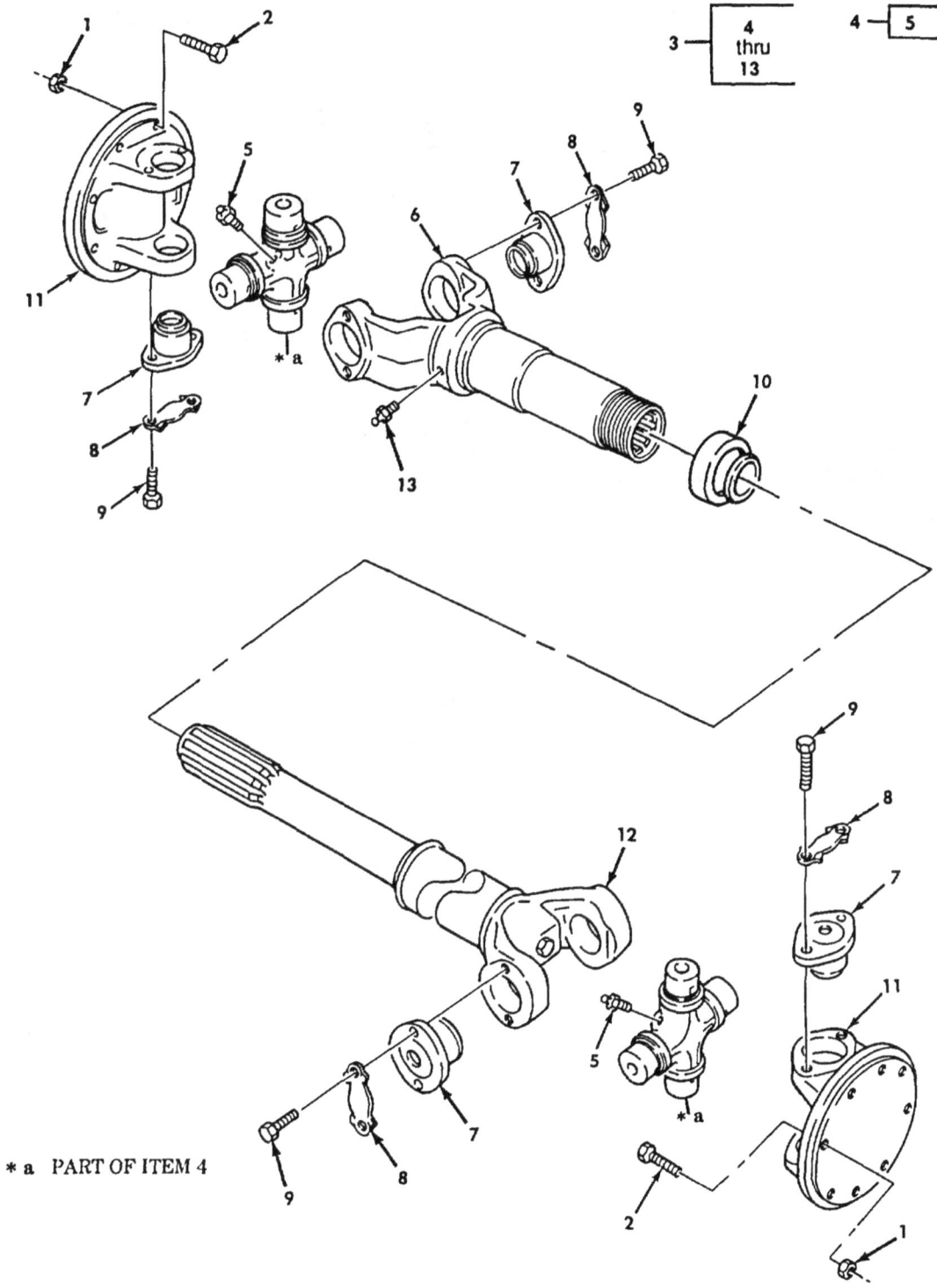

Figure 229. Propeller Shaft Assembly, Forward Rear Axle to Rear Rear Axle (Rockwell).

(1) ITEM NO	(2) SMR CODE	(3) NSN	(4) CAGEC	(5) PART NUMBER	(6) DESCRIPTION AND USABLE ON CODES (UOC)	(7) QTY
					GROUP 0900 PROPELLER SHAFTS	
					FIG. 229 PROPELLER SHAFT ASSEMBLY, FORWARD REAR AXLE TO REAR REAR AXLE (ROCKWELL)	
1	PAOZZ	5310008140672	96906	MS51943-36	NUT, SELF-LOCKING, HE	16
2	PAOZZ	5306007208747	96906	MS20073-06-12	BOLT, MACHINE	16
3	PAOOO	2520011276953	78500	983-04-44757	PROPELLER SHAFT WIT .... UOC: DAA, DAB, DAC, DAD, DAE, DAF, DAG, DAH, DAJ, DAK, DAL, DAW, DAX, V12, V13, V14, V15, V16, V17, V18, V19, V20, V21, V22, V24, V25, ZAA, ZAB, ZAC, ZAD, ZAE, ZAF, ZAG, ZAH, ZAJ, ZAK, ZAL -	1
3	PAOOO	2520011276954	78500	9830444478	PROPELLER SHAFT WIT .... UOC: V39	1
4	KFOZZ		70960	16N1-1A	.CROSS ASSEMBLY PART OF KIT P/N.................. CP16NS	2
5	PAOZZ	4730001720031	96906	MS15003-5	..FITTING, LUBRICATION..............................	1
6	XAOZZ		70960	16NLS32-2	.YOKE	1
7	KFOZZ		70960	16N4-3A	.BEARING ASSEMBLY PART OF KIT P/N.............. CP16NS	8
8	KFOZZ		70960	LPHC37-1	.LOCKPLATE PART OF KIT P/N CP16NS	8
9	KFOZZ		70960	CS-H5-24-15	.SCREW PART OF KIT P/N CP16NS	16
10	PAOZZ	5330011353376	78500	SE-RUR25-2	.SEAL, SHAFT	1
11	XAOZZ		70960	16NF3	.FLANGE	2
12	XAOZZ		70960	983-55-44757	.SHAFT ASSEMBLY, PROP .... UOC: DAA, DAB, DAC, DAD, DAE, DAF, DAG, DAH, DAJ, DAK, DAL, DAW, DAX, V12, V13, V14, V15, V16, V17, V18, V19, V20, V21, V22, V24, V25, ZAA, ZAB, ZAC, ZAD, ZAE, ZAF, ZAG, ZAH, ZAJ, ZAK, ZAL	1
12	XAOZZ		70960	983-55-44478	.SHAFT ASSEMBLY, PROP .... UOC: V39	1
13	PAOZZ	4730000504208	96906	MS15003-1	.FITTING, LUBRICATION	1

END OF FIGURE

Figure 230. Front Axle and Housing Assembly (M939, M939A1).

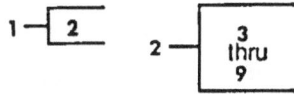

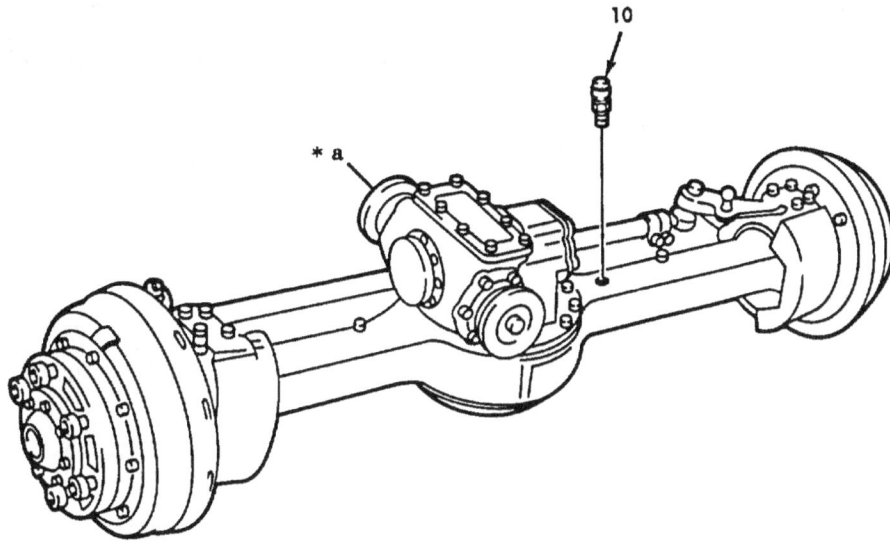

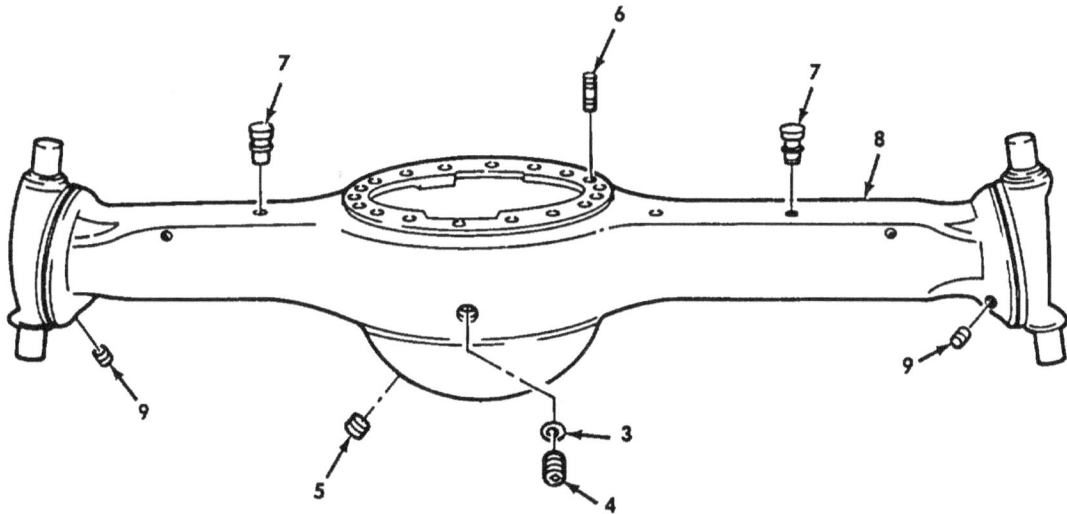

* a   PART OF ITEM 1

(1) ITEM NO	(2) SMR CODE	(3) NSN	(4) CAGEC	(5) PART NUMBER	(6) DESCRIPTION AND USABLE ON CODES (UOC)	(7) QTY
					GROUP 10 FRONT AXLE 1000 FRONT AXLE ASSEMBLY	
					FIG. 230 FRONT AXLE AND HOUSING ASSEMBLY(M939, M939A1)	
1	PGFHH	2520010911659	19207	12256540	AXLE ASSEMBLY, AUTOM .... UOC: DAA, DAB, DAC, DAD, DAE, DAF, DAG, DAH, DAJ, DAK, DAL, DAW, DAX, V12, V13, V14, V15, V16, V17, V18, V19, V20, V21, V22, V24, V25	1
1	PGFHH	2520010911660	19207	12256540-1	AXLE ASSEMBLY, AUTOM .... UOC: V39	1
2	PBFZZ	3040009307864	19207	8758345	.HOUSING, MECHANICAL .... UOC: DAA, DAB, DAC, DAD, DAE, DAF, DAG, DAH, DAJ, DAK, DAL, DAW, DAX, V12, V13, V14, V15, V16, V17, V18, V19, V20, V21, V22, V24, V25, V39	1
3	PAOZZ	5330005143289	96906	MS35769-21	..GASKET	1
4	PAOZZ	5365007326126	19207	7326126	..PLUG, MACHINE THREAD...............................	1
5	PAOZZ	4730009686129	96906	MS49006-10	..PLUG, PIPE, MAGNETIC	1
6	PAFZZ	5307004341783	78500	SN1020-1	..STUD, PLAIN	18
7	PAFZZ	5315007521651	19207	7521651	..PIN, SHOULDER, HEADLE	2
8	XAFZZ		19207	8758343	..HOUSING	1
9	PAFZZ	4730009541281	81348	WW-P-471ACABCB	..PLUG, PIPE	2
10	PAOZZ	4820007264719	57733	5196397	VALVE, VENT..	1

END OF FIGURE

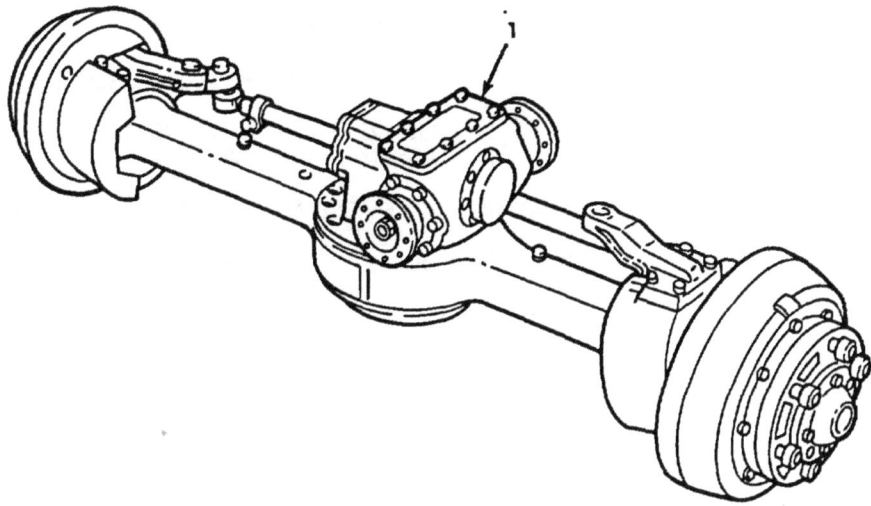

Figure 231. Front Axle and Housing Assembly (M939A2).

(1) ITEM NO	(2) SMR CODE	(3) NSN	(4) CAGEC	(5) PART NUMBER	(6) DESCRIPTION AND USABLE ON CODES (UOC)	(7) QTY
					GROUP 1000 FRONT AXLE ASSEMBLY	
					FIG. 231 FRONT AXLE AND HOUSING ASSEMBLY (M939A2)	
1	PGFHH	2520012919992	78500	M-1240-RDAX-14-644	AXLE ASSEMBLY, AUTOM FRONT, WITHOUT....... BRACKETS .... UOC: ZAA, ZAB, ZAC, ZAD, ZAE, ZAF, ZAG, ZAH, ZAJ, ZAK, ZAL	1
					END OF FIGURE	

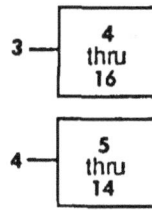

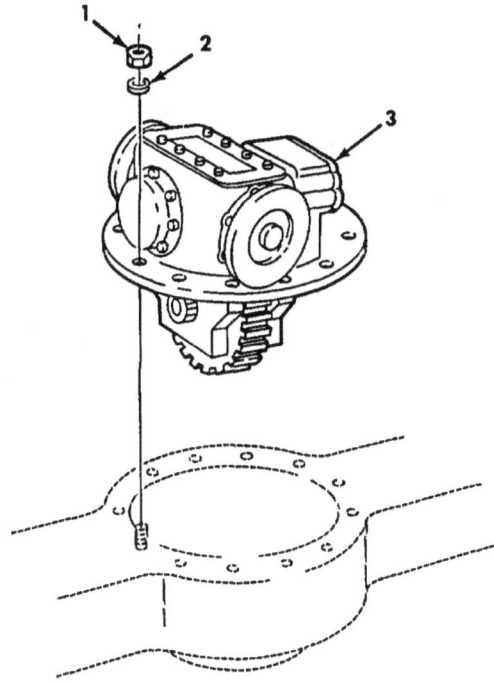

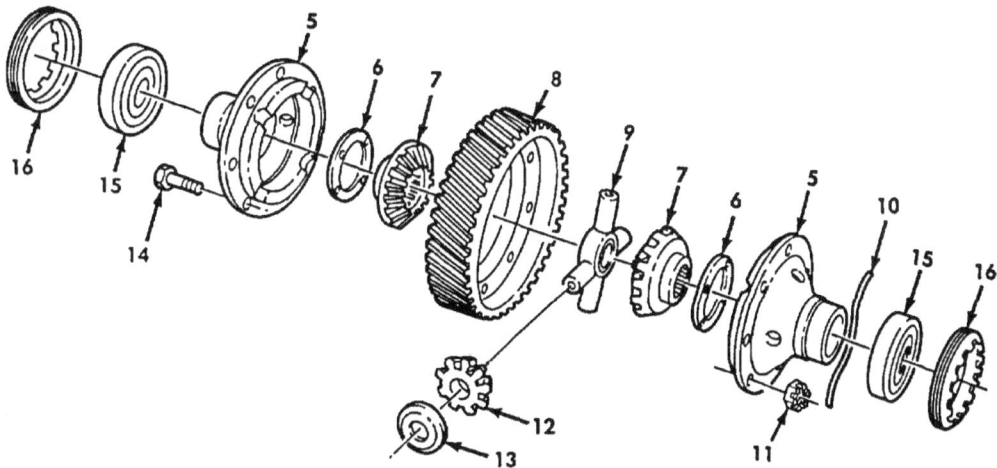

Figure 232. Front Axle Differential Assembly.

(1) ITEM NO	(2) SMR CODE	(3) NSN	(4) CAGEC	(5) PART NUMBER	(6) DESCRIPTION AND USABLE ON CODES (UOC)	(7) QTY
					GROUP 1002 DIFFERENTIAL	
					FIG. 232 FRONT AXLE DIFFERENTIAL ASSEMBLY	
1	PAFZZ	5310002747721	19207	7346813	NUT, PLAIN, HEXAGON	18
2	PAFZZ	5310004832266	19207	10900409	WASHER, FLAT	18
3	PAFHH	2520007346970	78500	A3800E473	DIFFERENTIAL, DRIVIN	1
4	PAHHH	2520004199422	19207	8758285	.DIFFERENTIAL GEAR U	1
5	XAHZZ		19207	8758252	..CARRIER, DIFFERENTIA	2
6	KFHZZ	3120003034977	78500	1229-H-1022	..BEARING, WASHER, THRU PART OF KIT .... P/N 7346959...	2
7	KFHZZ		19207	8758203	..GEAR, BEVEL PART OF KIT P/N 7346959............	2
8	KFHZZ		19207	7346824	..GEAR, HELICAL PART OF KIT P/N..................... 10899232..	1
9	KFHZZ		19207	8758206	..SPIDER PART OF KIT P/N 7346959	1
10	MHHZZ		80244	22W1642125X36	..WIRE, NONELECTRICAL MAKE FROM WIRE, ... P/N 22-W-1642-125, 36 INCHES LONG	1
11	PAHZZ	5310007207627	19207	8761279	..NUT, PLAIN, SLOTTED, H.................................	8
12	KFHZZ		78500	2233-W-101	..GEAR, BEVEL PART OF KIT P/N 7346959............	4
13	KFHZZ		7850	1229-Q-1031	..WASHER, THRUST PART OF KIT P/N................. 7346959	4
14	PAHZZ	5306007346822	19207	7346822	..BOLT, MACHINE	8
15	PAHZZ	3110001004216	21450	712699	.BEARING, ROLLER, TAPE	2
16	PAHZZ	5365007346818	19207	7346818	.RING, EXTERNALLY THR	2

END OF FIGURE

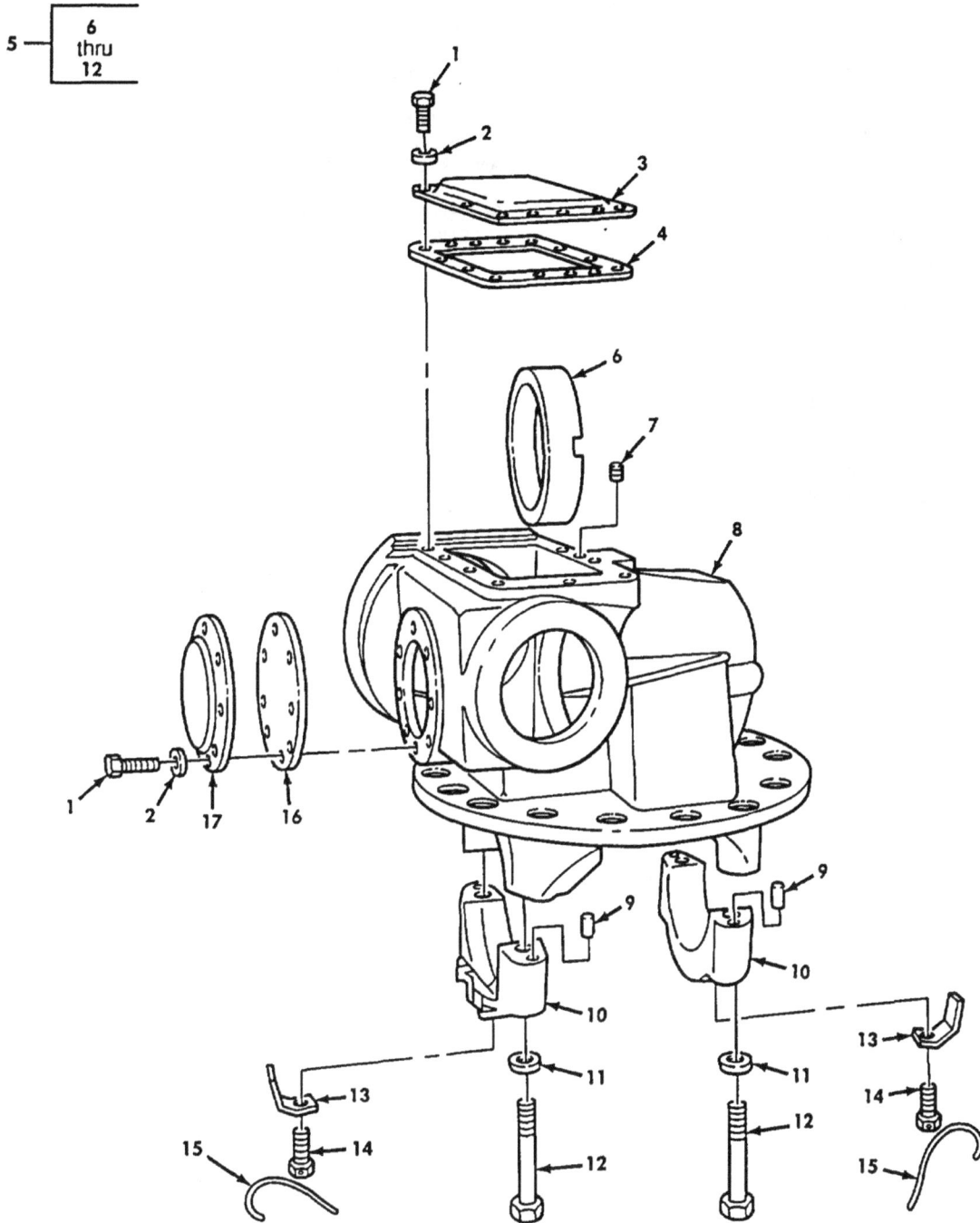

Figure 233. Front Axle Differential Carrier Assembly, Access Covers, and Related Parts.

(1) ITEM NO	(2) SMR CODE	(3) NSN	(4) CAGEC	(5) PART NUMBER	(6) DESCRIPTION AND USABLE ON CODES (UOC)	(7) QTY
					GROUP 1002 DIFFERENTIAL	
					FIG. 233 FRONT AXLE DIFFERENTIAL CARRIER ASSEMBLY, ACCESS COVERS, AND RELATED PARTS	
1	PAOZZ	5305005434372	80204	B1821BH038C075N	SCREW, CAP, HEXAGON H	18
2	PAOZZ	5310001770892	19207	7748743	WASHER, FLAT	18
3	PFOZZ	5340007346820	78500	3866-K-557	COVER, ACCESS	1
4	PAFZZ	533000138838	19207	7535079	GASKET PART OF KIT P/N 7346807	1
5	PAFHH	2520007347548	78500	A2-3800E473	CARRIER AND CAP ASS	1
6	PAHZZ	3120007346883	19207	7346883	.BEARING, SLE EVE	1
7	PAHZZ	5305007521718	19207	7521718	.SETSCREW	1
8	XAHZZ		19207	8758260	.CARRIER, DIFF, DR, AXLE	1
9	PAHZZ	5315005972723	19207	7534717	.PIN, SHOULDER, HEADLE	4
10	XBHZZ		19207	8758249	.CAP, DIFFERENTIAL SU	2
11	PAHZZ	5310003555453	19207	10900408	.WASHER, FLAT	4
12	PAHZZ	5306007346817	19207	7346817	.BOLT, MACHINE	4
13	PAHZZ	2530007346819	19207	7346819	LOCK	2
14	PAHZZ	5305009037794	96906	MS51095-410	SCREW, CAP, HEXAGON H	2
15	MHHZZ		80244	22W1642125X12	IN WIRE, NONELECTRICAL MAKE WIRE, P/N....... 22-W-1642-125, 12 INCHES LONG	1
16	PAOZZ	5330007346899	19207	7346899	GASKET..	1
17	PFOZZ	5340004454555	19207	8758248	COVER, ACCESS CARRIER	1

END OF FIGURE

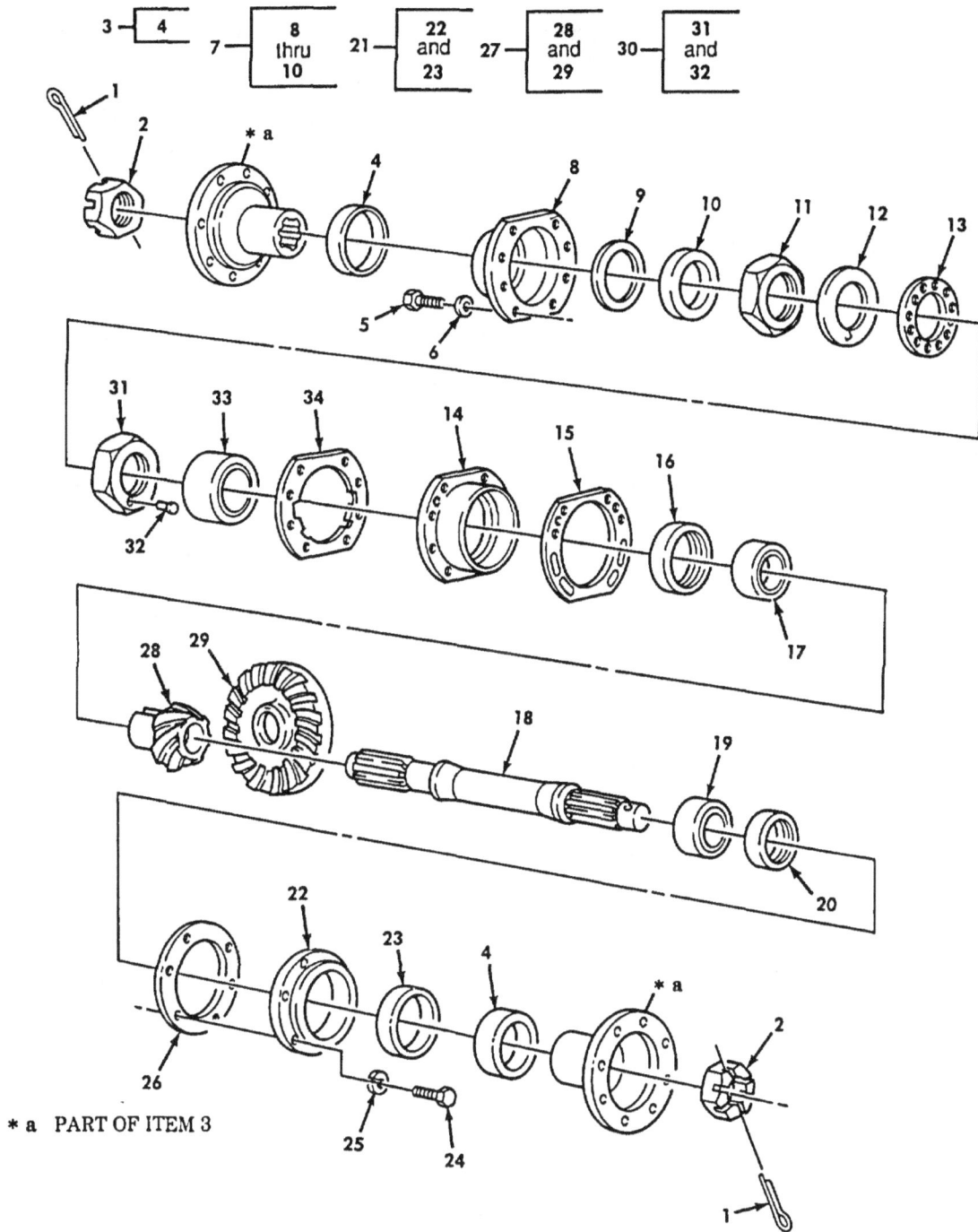

Figure 234. Front Axle Differential Through Shaft, Gears, and Flange Assemblies.

(1) ITEM NO	(2) SMR CODE	(3) NSN	(4) CAGEC	(5) PART NUMBER	(6) DESCRIPTION AND USABLE ON CODES (UOC)	(7) QTY
					GROUP 1002 DIFFERENTIAL	
					FIG. 234 FRONT AXLE DIFFERENTIAL THROUGH SHAFT, GEARS, AND FLANGE ASSEMBLIES	
1	PAFZZ	5315000590184	96906	MS24665-361	PIN, COTTER	2
2	PAFZZ	5310002825661	19207	7346893	NUT, PLAIN, SLOTTED, H	2
3	PAFZZ	2520007346802	78500	A3280G1957	FLANGE, COMPANION, UN	2
4	PAFZZ	2520005638309	19207	7535048	.DEFLECTOR, DIRT AND	1
5	PAFZZ	5305000180178	24617	180178	SCREW, CAP, HEXAGON H	8
6	PAFZZ	5310005159627	38597	50-4-18-17-7	WASHER, FLAT	8
7	PAFZZ	2520007346897	19207	7346897	COVER ASSEMBLY	1
8	XAFZZ		19207	8758251	.COVER, BEARING RETAI	1
9	PAFZZ	5330004195872	19207	10900396	.GASKET	1
10	PAFZZ	5330010230269	19207	12470105	.SEAL, PLAIN ENCASED	1
11	PAHZZ	5310003532427	19207	7979263	NUT, PLAIN, OCTAGON	1
12	PAHZZ	5310001473274	19207	8758258	WASHER, KEY	1
13	PAHZZ	5310007007089	19207	5139123	WASHER, KEY	1
14	PAHZZ	3110004199471	19207	8758202	PLATE, RETAINING, BEA	1
15	PAHZZ	5365010778564	19207	7346888	SHIM .003 PART OF KIT P/N 7346807	1
15	KFHZZ	5365007346889	19207	7346889	SHIM .005	1
15	KFHZZ	5365007346890	19207	7346890	SHIM .010	1
16	PAHZZ	5365006926119	19207	8758209	SPACER, BEVEL PINION 367 PART OF .... KIT P/N 5704278	1
16	KFHZZ		19207	8758211	COLLAR 369 PART OF KIT P/N 5704278	1
16	KFHZZ		19207	8758213	COLLAR 371 PART OF KIT P/N 5704278	1
16	KFHZZ		19207	8758215	COLLAR, BEARING 373 PART OF KIT P/N 5704278	1
16	KFHZZ	3040006926121	19207	8758217	COLLAR, SHAFT 375 PART OF KIT P/N 5704278	1
16	KFHZZ		19207	8758219	COLLAR, BEARING 377 PART OF KIT P/N 5704278	1
16	KFHZZ		19207	8758221	COLLAR, BEARING 379 PART OF KIT P/N 5704278	1
16	KFHZZ		19207	8758223	COLLAR 381 PART OF KIT P/N 5704278	1
16	KFHZZ	3040006926123	19207	8758225	COLLAR, SHAFT 383 PART OF KIT P/N 5704278	1
16	KFHZZ		19207	8758227	COLLAR, BEARING 385 PART OF KIT P/N 5704278	1
16	KFHZZ	3120006926124	19207	8758229	BEARING, WASHER, THRU 387 PART OF .... KIT P/N 5704278	1
16	KFHZZ		19207	8758231	COLLAR, BEARING 389 PART OF KIT P/N 5704278	1
16	KFHZZ		19207	8758233	COLLAR, BEARING 391 PART OF KIT P/N 5704278	1
16	KFHZZ		19207	8758235	COLLAR, BEARING 393 PART OF KIT P/N 5704278	1
16	KFHZZ		19207	8758237	COLLAR, BEARING 395 PART OF KIT P/N 5704278	1
16	KFHZZ		19207	8758239	COLLAR, BEARING 397 PART OF KIT P/N 5704278	1

(1) ITEM NO	(2) SMR CODE	(3) NSN	(4) CAGEC	(5) PART NUMBER	(6) DESCRIPTION AND USABLE ON CODES (UOC)	(7) QTY
16	KFHZZ		19207	8758241	COLLAR, BEARING 399 PART OF KIT P/N............ 5704278	1
16	KFHZZ		19207	8758243	COLLAR, BEARING 401 PART OF KIT P/N............ 5704278	1
16	KFHZZ		19207	8758245	COLLAR, BEARING 403 PART OF KIT P/N............ 5704278	1
17	PAHZZ	3110001005355	05840	323W231-J	BEARING, ROLLER, TAPE	1
18	PAHZZ	3040007346892	19207	7346892	SHAFT, SHOULDERED.	1
19	PAHZZ	3110001950460	78500	712148	BEARING, ROLLER, CYLI	1
20	PAHZZ	3120006623370	19207	7346894	BEARING, WASHER, THRU.	1
21	PAFZZ	3110007346895	19207	7346895	PLATE, RETAINING, BEA	1
22	XAHZZ		19207	8758207	.CAP	1
23	PAOZZ	5330010230269	19207	12470105	.SEAL, PLAIN ENCASED	1
24	PAFZZ	5305005432419	80204	B1821BH038C113N	SCREW, CAP, HEXAGON H...............................	6
25	PAFZZ	5310006379541	96906	MS35338-46	WASHER, LOCK	6
26	PAHZZ	5330006412466	19207	7346896	GASKET CAP	1
27	PAHZZ	3020007346881	78500	MPS516	GEAR SET, BEVEL, MATC	1
28	XAHZZ		19207	8758204	.PINION, BEVEL.	1
29	XAHZZ		19207	8758205	.GEAR, BEVEL.	1
30	PAHZZ	5310003740836	19207	7979308	NUT, PLAIN, OCTAGON.	1
31	XAHZZ	5310011350049	19207	7979309	.NUT, PLAIN, OCTAGON...................................	1
32	PAHZZ	5315010587268	19207	7979310	.PIN, ADJUSTMENT	1
33	PAHZZ	3110001004177	21450	703189	BEARING, ROLLER, TAPE.................................	1
34	PAHZZ	5330007346886	19207	7346886	GASKET..	1

END OF FIGURE

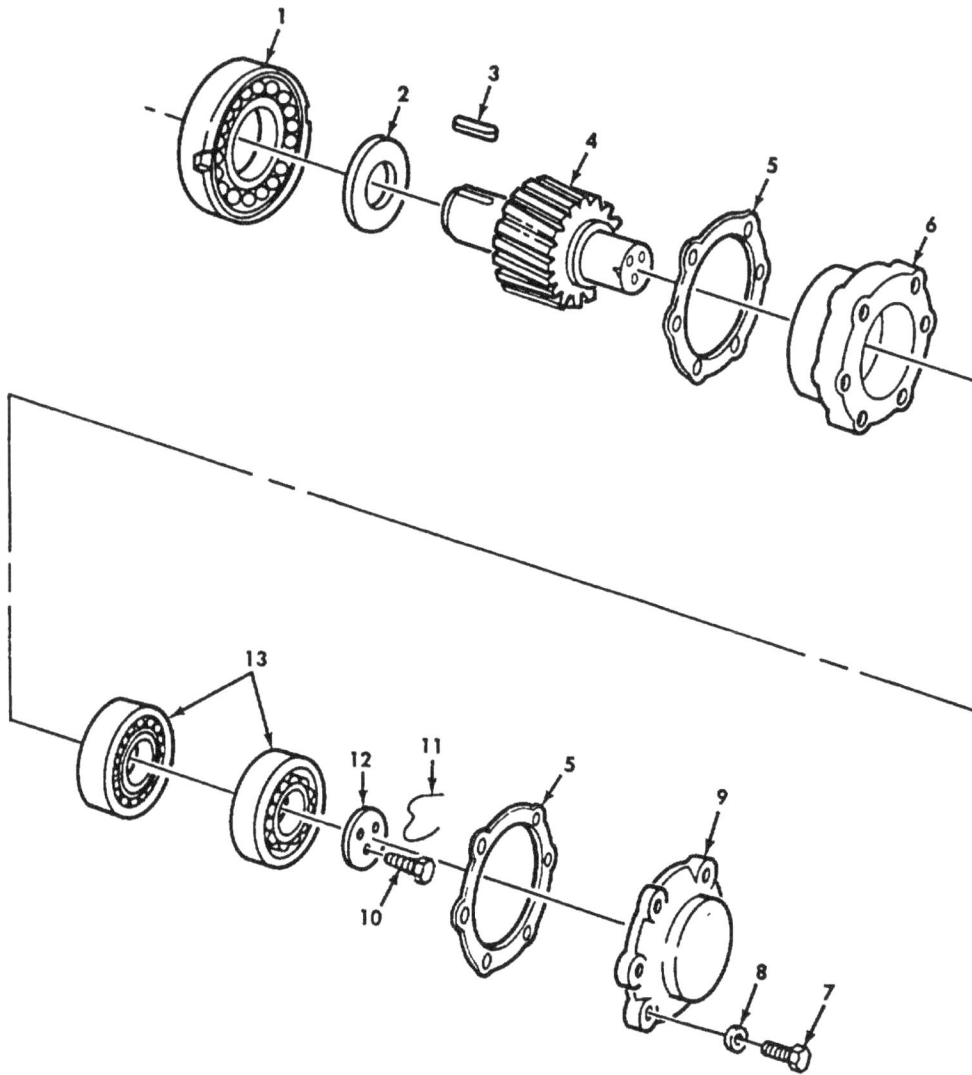

Figure 235. Front Axle Differential Helical Drive Pinion and Related Parts.

(1) ITEM NO	(2) SMR CODE	(3) NSN	(4) CAGEC	(5) PART NUMBER	(6) DESCRIPTION AND USABLE ON CODES (UOC)	(7) QTY
					GROUP 1002 DIFFERENTIAL	
					FIG. 235 FRONT AXLE DIFFERENTIAL HELICAL DRIVE PINION AND RELATED PARTS	
1	PAHZZ	3110001586011	43334	U1218TAM	BEARING, ROLLER, CYLI	1
2	PAHZZ	5310000336012	19207	7346882	WASHER, RECESSED	1
3	PAHZZ	5315002765385	19207	7534781	KEY, MACHINE	1
4	PAHZZ	3040007346825	19207	7346825	GEARSHAFT, HELICAL PART OF KIT P/N............. 10899232..	1
5	PAHZZ	5365007346880	19207	7346880	SHIM 010 PART OF KIT P/N 7346807...................V	
5	PAHZZ	5365007346879	19207	7346879	SHIM 005 PART OF KIT P/N 7346807...................V	
5	PAHZZ	5330007346878	19207	7346878	GASKET 010 PART OF KIT P/N 7346807 V	
6	PAHZZ	2520004946586	19207	8758208	CAP, BEARING, DIFFERE	1
7	PAEZZ	5305000712071	80204	B1821BH050C200N	SCREW, CAP, HEXAGON H	6
8	PAHZZ	5310000034094	01276	210104-8S	WASHER, LOCK	6
9	PAHZZ	3110007346877	19207	7346877	PLATE, RETAINING, BEA	1
10	PAHZZ	5305009419460	96906	MS51095-386	SCREW, CAP, HEXAGON H	3
11	MHHZZ		80244	22W1642125X12	WIRE, NONELECTRICAL MAKE FROM WIRE, ...V P/N 22-W-1642-125, 12 INCHES LONG	
12	PAHZZ	3110003505407	78500	1829S149	PLATE, RETAINING, BEA	1
13	PAHZZ	3110001005825	21450	712627	BEARING, ROLLER, TAPE	2

END OF FIGURE

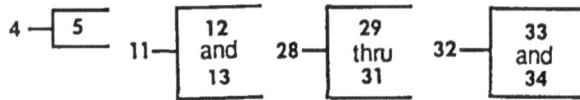

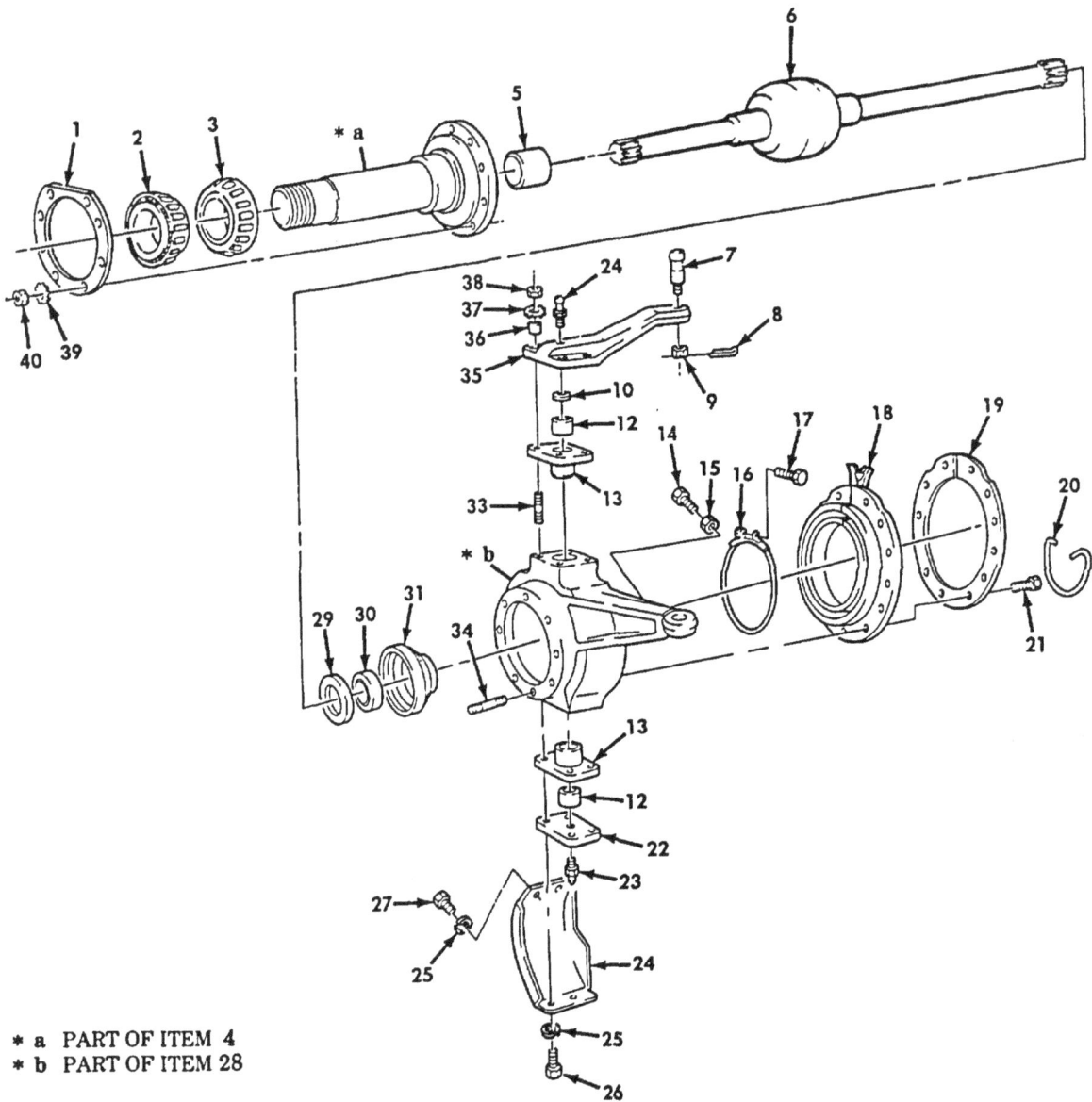

Figure 236. Spindle Shaft Assembly, Universal Shaft, and Knuckle Assembly.

* a PART OF ITEM 4
* b PART OF ITEM 28

(1) ITEM NO	(2) SMR CODE	(3) NSN	(4) CAGEC	(5) PART NUMBER	(6) DESCRIPTION AND USABLE ON CODES (UOC)	(7) QTY
					GROUP 1004 STEERING	
					FIG. 236 SPINDLE SHAFT ASSEMBLY, UNIVERSAL SHAFT, AND KNUCKLE ASSEMBLY	
1	PAOZZ	5330012715151	78500	3286-N-1054	SEAL, PLAIN	2
2	PAOZZ	3110001000650	60038	598	CONE AND ROLLERS, TA .... UOC: ZAA, ZAB, ZAC, ZAD, ZAE, ZAF, ZAG, ZAH, ZAJ, ZAK, ZAL	2
3	PAOZZ	3110009509700	60038	594A	CONE AND ROLLERS, TA .... UOC: ZAA, ZAB, ZAC, ZAD, ZAE, ZAF, ZAG, ZAH, ZAJ, ZAK, ZAL	2
4	PAOZZ	2530012720542	78500	A-3213X1558	SPINDLE, WHEEL, DRIVI .... UOC: ZAA, ZAB, ZAC, ZAD, ZAE, ZAF, ZAG, ZAH, ZAJ, ZAK, ZAL	2
4	PAOFF	2530007346982	19207	7346982	SPINDLE, WHEEL, DRIVI .... UOC: DAA, DAB, DAC, DAD, DAE, DAF, DAG, DAH, DAJ, DAK, DAL, DAW, DAX, V12, V13, V14, V15, V16, V17, V18, V19, V20, V21, V22, V24, V25, V39	2
5	PAOZZ	3120005370614	19207	7346983	.BEARING, SLEEVE	1
6	PAOZF	2520007346985	02413	811E1139	SHAFT, AXLE, AUTOMOTI LEFT HAND	1
6	PAOZZ	2520007346984	13038	J950112-6	SHAFT, AXLE, AUTOMOTI RIGHT HAND	1
7	PAFZZ	5307004831105	19207	11664599	STUD, BALL	1
8	PAFZZ	5315002981481	96906	MS24665-357	PIN, COTTER	2
9	PAFZZ	5310008352140	96906	MS35692-69	NUT, PLAIN, SLOTTED, H	2
10	PAFZZ	5310011281592	78500	1224-D-551	WASHER, FLAT 145 V	
10	PAFZZ	5310011340209	19207	8758282	WASHER, FLAT 155 V	
10	PAFZZ	5310010542568	19207	8758280	WASHER, FLAT 135 V	
10	PAFZZ	5310011345791	19207	8758283	WASHER, FLAT 165 V	
10	PAFZZ	5365010558769	78500	1244-G-553	SHIM 175 V	
11	PAFFF	2530007346971	19207	7346971	SLEEVE ASY	4
12	PAHZZ	3120002881889	19207	7346972	.BEARING, SLEEVE	1
13	XAFZZ		19207	8758253	.SLEEVE, STEERING KNU	1
14	PAFZZ	5306002811651	19207	5225875	BOLT, MACHINE	2
15	PAFZZ	5310007638919	96906	MS51967-18	NUT, PLAIN, HEXAGON	2
16	PAOZZ	4730011185972	19207	12300575	CLAMP, HOSE PART OF KIT P/N 5704510............ UOC: DAA, DAB, DAC, DAD, DAE, DAF, DAG, DAH, DAJ, DAK, DAL, DAW, DAX, V12, V13, V14, V15, V16, V17, V18, V19, V20, V21, V22, V24, V25, V39	2
17	PAOZZ	5305001433266	96906	MS27039-0825	SCREW, MACHINE PART OF KIT P/N.................. 5704510	2
17	PAOZZ	5305007521693	19207	7071098	SCREW, MACHINE. .... UOC: V15	2
18	PAOZZ	2530008327123	92679	98883R91	BOOT, VEHICULAR COM PART OF KIT P/N.......... 5704510	2
19	PAFZZ	2520007346975	19207	7346975	RING, DIFFERENTIAL C	2
20	MOOZZ		10001	22-W-1642-100-40	WIRE, NONELECTRICAL PART OF KIT P/N.......... 5704510	2
21	PAOZZ	5305005515097	96906	MS35265-94	SCREW, MACHINE	24

(1) ITEM NO	(2) SMR CODE	(3) NSN	(4) CAGEC	(5) PART NUMBER	(6) DESCRIPTION AND USABLE ON CODES (UOC)	(7) QTY
22	PAFZZ	5340007346974	19207	7346974	COVER, ACCESS	2
23	PAOZZ	4730000504208	96906	MS15003-1	FITTING, LUBRICATION	4
24	PAFZZ	2530007346977	19207	7346977	GUARD, BRUSH KNUCKLE AND STUD................ ASSEMBLY, LEFT SIDE....................	1
24	PAFZZ	2530007346978	19207	7346978	GUARD, BRUSH KNUCKLE AND STUD................ ASSEMBLY, RIGHT SIDE....................	1
25	PAFZZ	5310008206653	80045	23MS35338-50	WASHER, LOCK	12
26	PAFZZ	5305007247223	80204	B1821BH063C225N	SCREW, CAP, HEXAGON H	8
27	PAFZZ	5305007247219	80204	B1821BH063C125N	SCREW, CAP, HEXAGON H	4
28	PAOOO	4320007346951	78500	A1244J556	SEAL ASSEMBLY, SHAFT	2
29	PAOZZ	5310000336007	19207	7346986	.WASHER, FLAT	1
30	PAOZZ	5330001458355	78500	A1805H60	.SEAL, PLAIN ENCASED	1
31	PAOZZ	5330010986668	19207	8758268	.RETAINER, OIL SEAL	1
32	PAFFF	2530012722926	78500	A2-3111-D-2604	SPINDLE, WHEEL, DRIVI RIGHT HAND .... UOC: ZAA, ZAB, ZAC, ZAD, ZAE, ZAF, ZAG, ZAH, ZAJ, ZAK, ZAL	1
32	PAFFF	2530012719824	78500	A2-3111-E-2605	SPINDLE, WHEEL, DRIVI LEFT HAND .... UOC: ZAA, ZAB, ZAC, ZAD, ZAE, ZAF, ZAG, ZAH, ZAJ, ZAK, ZAL	1
32	PAFZZ	2530007346815	19207	7346815	SPINDLE, WHEEL, DRIVI LEFT HAND .... UOC: DAA, DAB, DAC, DAD, DAE, DAF, DAG, DAH, DAJ, DAK, DAL, DAW, DAX, V12, V13, V14, V15, V16, V17, V18, V19, V20, V21, V22, V24, V25, V39	1
32	PAFZZ	2530000067469	19207	7346816-1	SPINDLE, WHEEL, DRIVI RIGHT HAND .... UOC: DAA, DAB, DAC, DAD, DAE, DAF, DAG, DAH, DAJ, DAK, DAL, DAW, DAX, V12, V13, V14, V15, V16, V17, V18, V19, V20, V21, V22, V24, V25, V39	1
33	PAFZZ	5307007346900	19207	7346900	.STUD, PLAIN	4
34	PAFZZ	5307012817219	78500	SN-925-1C	.STUD, PLAIN .... UOC: ZAA, ZAB, ZAC, ZAD, ZAE, ZAF, ZAG, ZAH, ZAJ, ZAK, ZAL	9
34	PAFZZ	5307012775053	78500	SN-921-1-C	.STUD, PLAIN . ... UOC: DAA, DAB, DAC, DAD, DAE, DAF, DAG, DAH, DAJ, DAK, DAL, DAW, DAX, V12, V13, V14, V15, V16, V17, V18, V19, V20, V21, V22, V24, V25, V39	9
35	PAFZZ	2530010889357	19207	11664597-1	ARM, STEERING GEAR LEFT HAND	1
35	PAFZZ	2530002310178	19207	11664598	ARM, STEERING RIGHT HAND	1
36	PAOZZ	5365004272282	19207	7346812	BUSHING, TAPERED	8
37	PAFZZ	5310008000695	96906	MS35335-39	WASHER, LOCK	8
38	PAFZZ	5310002747721	19207	7346813	NUT, PLAIN, HEXAGON	8
39	PAOZZ	5310006166354	96906	MS35335-38	WASHER, LOCK	20
40	PAOZZ	5310007638911	96906	MS51968-17	NUT, PLAIN, HEXAGON	20

END OF FIGURE

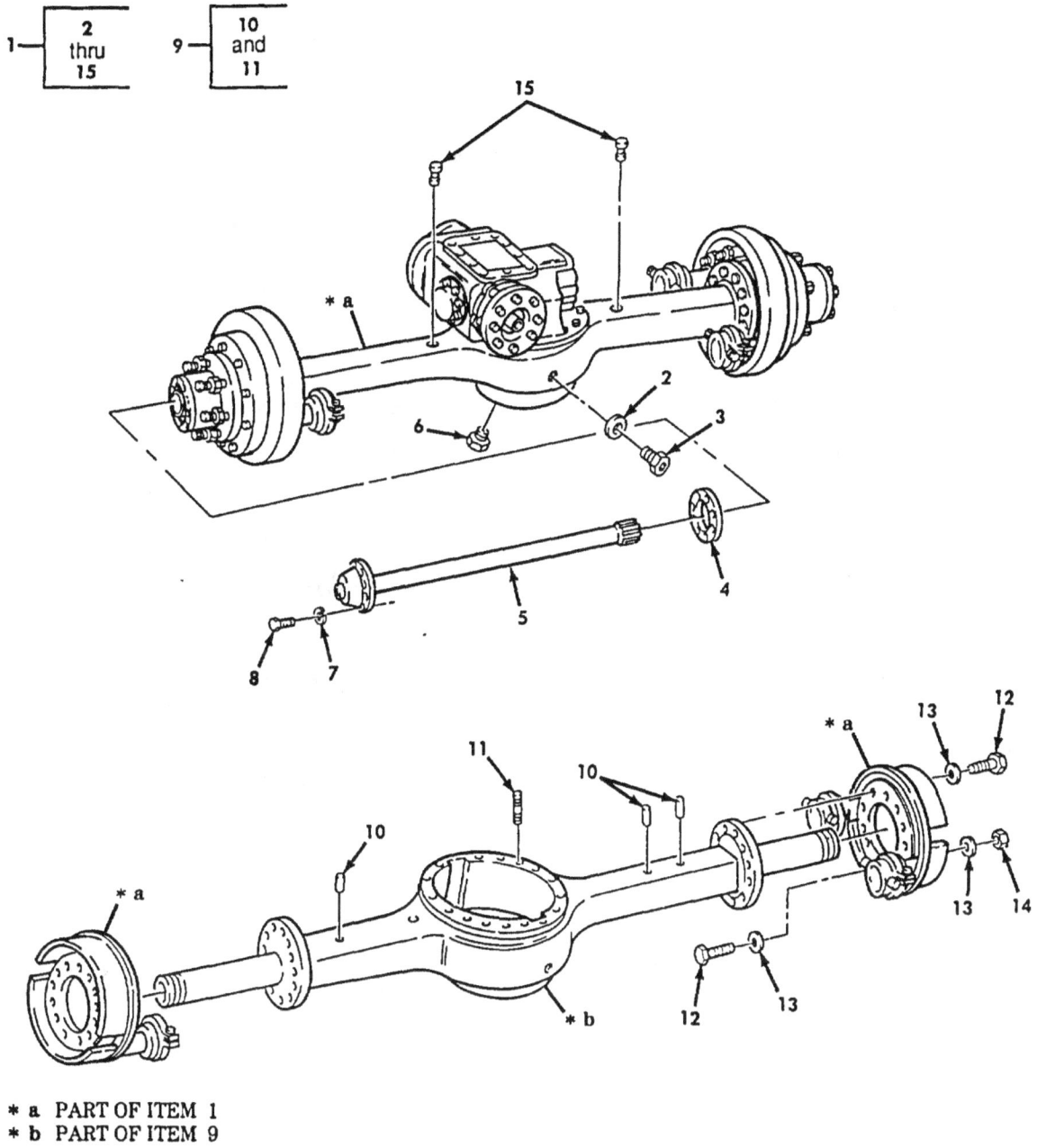

*a PART OF ITEM 1
*b PART OF ITEM 9

Figure 237. Forward Rear and Rear Rear Axle Assemblies (M939, M939A1).

(1) ITEM NO	(2) SMR CODE	(3) NSN	(4) CAGEC	(5) PART NUMBER	(6) DESCRIPTION AND USABLE ON CODES (UOC)	(7) QTY
					GROUP 11 REAR AXLE 1100 REAR AXLE ASSEMBLY	
					FIG. 237 FORWARD REAR AND REAR REAR AXLE ASSEMBLIES(M939, M939A1)	
1	PAFHH	2520011176592	19207	12256156-1	AXLE ASSEMBLY, AUTOM REAR REAR................ UOC:DAA, DAB, DAC, DAD, DAE, DAF, DAG, DAH, DAJ, DAK, DAL, DAW, DAX, V12, V13, V14, V15, V16, V17, V18, V19, V20, V21, V22, V24, V25	1
1	PAFHH	2520011173014	19207	12256156-2	AXLE ASSEMBLY, AUTOM FORWARD REAR......... UOC:DAA, DAB, DAC, DAD, DAE, DAF, DAG, DAH, DAJ, DAK, DAL, DAW, DAX, V12, V13, V14, V15, V16, V17, V18, V19, V20, V21, V22, V24, V25	1
1	PAFHH	2520011174933	19207	12256156-3	AXLE ASSEMBLY, AUTOM DUAL REAR, REAR. UOC:V39	1
1	PAFHH	2520011173015	19207	12256156-4	AXLE ASSEMBLY, AUTOM DUAL WHEEL, ... REAR AXLE UOC:V39	1
2	PAOZZ	5330005143289	96906	MS35769-21	.GASKET .... UOC:DAA, DAB, DAC, DAD, DAE, DAF, DAG, DAH, DAJ, DAK, DAL, DAW, DAX, V12, V13, V14, V15, V16, V17, V18, V19, V20, V21, V22, V24, V25, V39	1
3	PAOZZ	5365007326126	19207	7326126	.PLUG, MACHINE THREAD.................... UOC:DAA, DAB, DAC, DAD, DAE, DAF, DAG, DAH, DAJ, DAK, DAL, DAW, DAX, V12, V13, V14, V15, V16, V17, V18, V19, V20, V2 1, V22, V24, V25, V39	1
4	PAOZZ	5330013779775	14153	02764	.GASKET NON-ASBESTOS.................................. UOC:DAA, DAB, DAC, DAD, DAE, DAF, DAG, DAH, DAJ, DAK, DAL, DAW, DAX, V12, V13, V14, V15, V16, V17, V18, V19, V20, V21, V22, V24, V25, V39	2
5	PAOZZ	2520007346960	78500	3202J2610	.SHAFT, AXLE, AUTOMOTI .... UOC:DAA, DAB, DAC, DAD, DAE, DAF, DAG, DAH, DAJ, DAK, DAL, DAW, DAX, V12, V13, V14, V15, V16, V17, V18, V19, V20, V21, V22, V24, V25	2
5	PAOZZ	2520007411142	19207	7411142	.SHAFT, AXLE, AUTOMOTI REAR REAR .... HOUSING ASSEMBLY UOC:V39	2
6	PAOZZ	4730005954827	78500	1250Z182	.PLUG, PIPE, MAGNETIC..................................... UOC:DAA, DAB, DAC, DAD, DAE, DAF, DAG, DAH, DAJ, DAK, DAL, DAW, DAX, V12, V13, V14, V15, V16, V17, V18, V19, V20, V21, V22, V24, V25, V39	1
7	PAOZZ	5310000034094	01276	210104-8S	.WASHER, LOCK. .... UOC:DAA, DAB, DAC, DAD, DAE, DAF, DAG, DAH, DAJ, DAK, DAL, DAW, DAX, V12, V13, V14, V15, V16, V17, V18, V19, V20, V21, V22, V24, V25, V39	20
8	PAOZZ	5306001829267	19207	10896726	.BOLT, MACHINE..	20

(1) ITEM NO	(2) SMR CODE	(3) NSN	(4) CAGEC	(5) PART NUMBER	(6) DESCRIPTION AND USABLE ON CODES (UOC)	(7) QTY
					UOC:DAA, DAB, DAC, DAD, DAE, DAF, DAG, DAH, DAJ, DAK, DAL, DAW, DAX, V12, V13, V14, V15, V16, V17, V18, V19, V20, V21, V22, V24, V25, V39	
9	XAFFD		19207	8758344	.HOUSING WITH STUDS	1
					UOC:DAA, DAB, DAC, DAD, DAE, DAF, DAG, DAH, DAJ, DAK, DAL, DAW, DAX, V12, V13, V14, V15, V16, V17, V18, V19, V20, V21, V22, V24, V25	
9	XAFFD		19207	8758322	.HOUSING WITH STUDS	1
					UOC:V39	
10	PAFZZ	5315007521651	19207	7521651	..PIN, SHOULDER, HEADLE	3
					UOC:DAA, DAB, DAC, DAD, DAE, DAF, DAG, DAH, DAJ, DAK, DAL, DAW, DAX, V12, V13, V14, V15, V16, V17, V18, V19, V20, V21, V22, V24, V25, V39	
11	PAFZZ	5307004341783	78500	SN1020-1	..STUD, PLAIN	18
					UOC:DAA, DAB, DAC, DAD, DAE, DAF, DAG, DAH, DAJ, DAK, DAL, DAW, DAX, V12, V13, V14, V15, V16, V17, V18, V19, V20, V21, V22, V24, V25, V39	
12	PAFZZ	5305007167454	80204	B1821BH056C175N	.SCREW, CAP, HEXAGON H..............................	20
					UOC:DAA, DAB, DAC, DAD, DAE, DAF, DAG, DAH, DAJ, DAK, DAL, DAW, DAX, V12, V13, V14, V15, V16, V17, V18, V19, V20, V21, V22, V24, V25, V39	
13	PAFZZ	5310007251983	81205	BACW10P53S	.WASHER, FLAT	36
					UOC:DAA, DAB, DAC, DAD, DAE, DAF, DAG, DAH, DAJ, DAK, DAL, DAW, DAX, V12, V13, V14, V15, V16, V17, V18, V19 , V20, V21, V22, V24, V25, V39	
14	PAFZZ	5310012052830	78500	1229-Y-1507	WASHER, FLAT...	4
14	XDFZZ		56152	401569	NUT, PLAIN, HEXAGON	16
					UOC:DAA, DAB, DAC, DAD, DAE, DAF, DAG, DAH, DAJ, DAK, DAL, DAW, DAX, V12, V13, V14, V15, V16, V17, V18, V19, V20, V21, V22, V24, V25, V39	
15	PAOZZ	4820007264719	57733	5196397	.VALVE, VENT..	2
					UOC:DAA, DAB, DAC, DAD, DAE, DAF, DAG, DAH, DAJ, DAX, DAL, DAW, DAX, V12, V13, V14, V15, V17, V18, V19, V20, V21, V22, V24, V25, V39	

END OF FIGURE

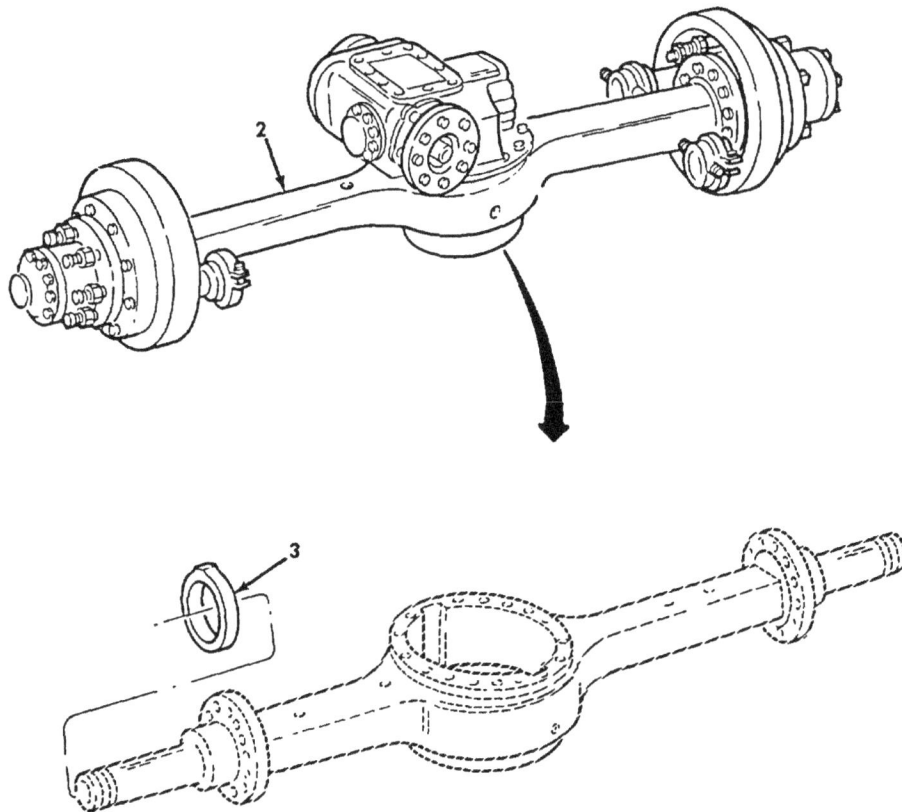

Figure 238. Forward Rear and Rear Rear Axle Assemblies (M939A2).

(1) ITEM NO	(2) SMR CODE	(3) NSN	(4) CAGEC	(5) PART NUMBER	(6) DESCRIPTION AND USABLE ON CODES (UOC)	(7) QTY
					GROUP 1100 REAR AXLE ASSEMBLY	
					FIG. 238 FORWARD REAR AND REAR REAR AXLE ASSEMBLIES(M939A2)	
1	PAFHH	2520012919993	78500	M-1240-RDAX-29-644	AXLE ASSEMBLY, AUTOM REAR REAR, WITHOUT BRACKETS UOC:ZAA, ZAB, ZAC, ZAD, ZAE, ZAF, ZAG, ZAH, ZAJ, ZAK, ZAL	1
1	PAFHH	2520012919994	78500	M-1240-RDAX-30-644	AXLE ASSEMBLY, AUTOM FORWARD REAR, WITHOUT BRACKETS UOC:ZAA, ZAB, ZAC, ZAD, ZAE, ZAF, ZAG, ZAH, ZAJ, ZAK, ZAL	1
2	XAFZZ		78500	C3-3201-W-7589	.HOUSING, AXLE UOC:ZAA, ZAB, ZAC, ZAD, ZAE, ZAF, ZAG, ZAH, ZAJ, ZAK, ZAL	1
3	PAFZZ	2520012717007	78500	3268-L-1338	.ADAPTER, FIN UOC:ZAA, ZAB, ZAC, ZAD, ZAE, ZAF, ZAG, ZAH, ZAJ, ZAK, ZAL	1

END OF FIGURE

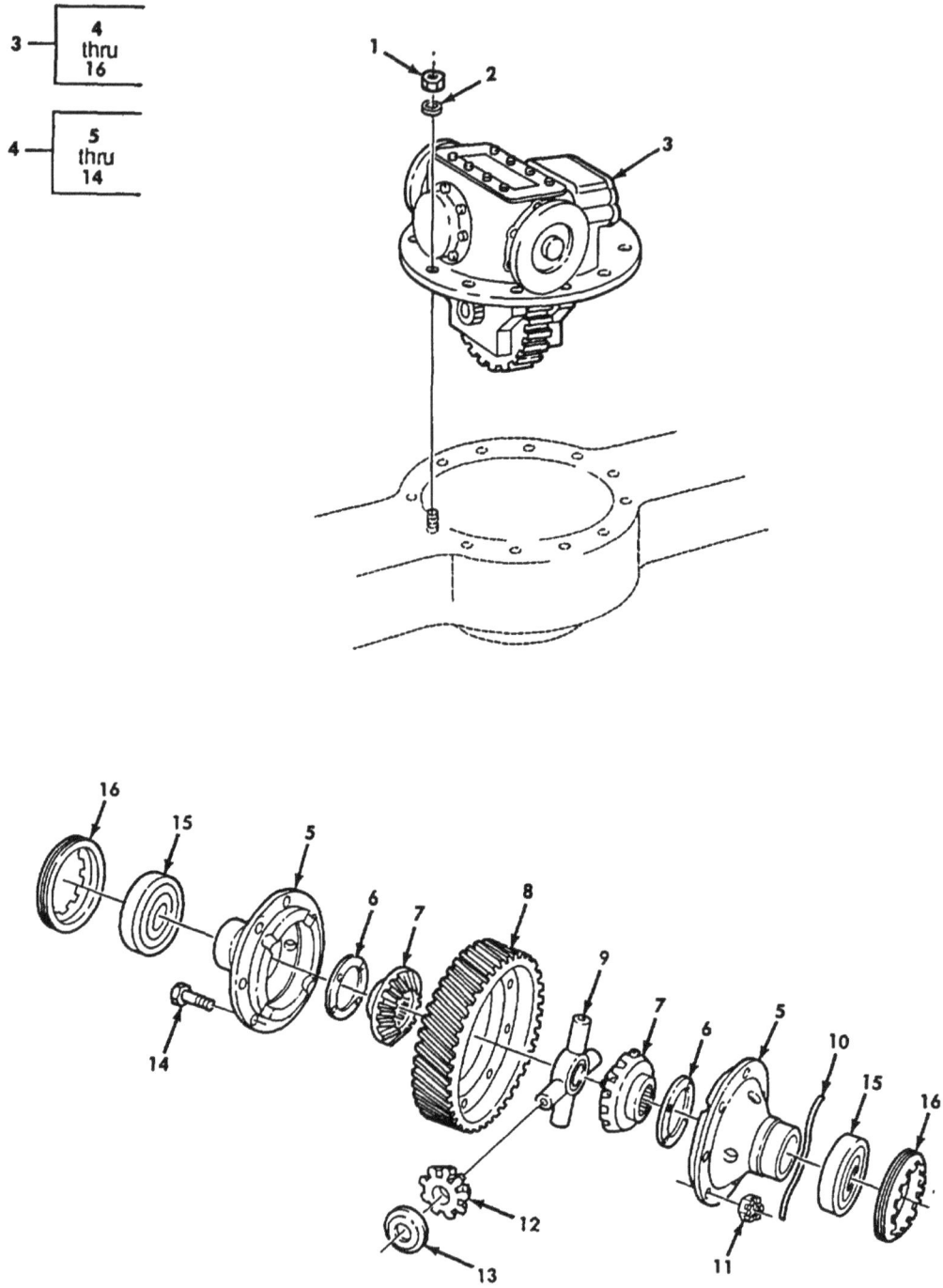

Figure 239. Rear Axle Differential Assembly and Mounting Hardware.

(1) ITEM NO	(2) SMR CODE	(3) NSN	(4) CAGEC	(5) PART NUMBER	(6) DESCRIPTION AND USABLE ON CODES (UOC)	(7) QTY
					GROUP 1102 DIFFERENTIAL	
					FIG. 239 REAR AXLE DIFFERENTIAL ASSEMBLY AND MOUNTING HARDWARE	
1	PAFZZ	5310002747721	19207	7346813	NUT, PLAIN, HEXAGON	18
2	PAFZZ	5310004832266	19207	10900409	WASHER, FLAT	18
3	PAFHH	2520007346970	78500	A3800E473	DIFFERENTIAL, DRIVIN	2
4	PAHHH	2520004199422	19207	8758285	.DIFFERENTIAL GEAR U	1
5	XDHZZ		19207	8758252	..CARRIER, DIFFERENTIA	2
6	KFHZZ		78500	1229-H-1022	..BEARING, WASHER, THRU PART OF KIT .... P/N 7346959..	2
7	KFHZZ		19207	8758203	..GEAR, BEVEL PART OF KIT P/N 7346959............	2
8	KFHZZ		19207	7346824	..GEAR, HELICAL PART OF KIT P/N...................... 10899232...	1
9	KFHZZ	2520011143729	19207	8758206-KF	..SPIDER, DIFFERENTIAL PART OF KIT .... P/N 7346959..	1
10	MHHZZ		80244	22W1642125X36	..WIRE, NONELECTRICAL MAKE FROM WIRE .... P/N 22-W-1642-124, 36 INCHES LONG	1
11	PAHZZ	5310007207627	19207	8761279	..NUT, PLAIN, SLOTTED, H................................	8
12	KFHZZ		78500	2233-W-101	..GEAR, BEVEL PART OF KIT P/N 7346959............	4
13	KFHZZ		78500	1229-Q-1031	..WASHER, THRUST PART OF KIT P/N................. 7346959...	4
14	PAHZZ	5306007346822	19207	7346822	.BOLT, MACHINE	8
15	PAHZZ	3110001004216	21450	712699	.BEARING, ROLLER, TAPE	2
16	PAHZZ	5365007346818	19207	7346818	.RING, EXTERNALLY THR	2

END OF FIGURE

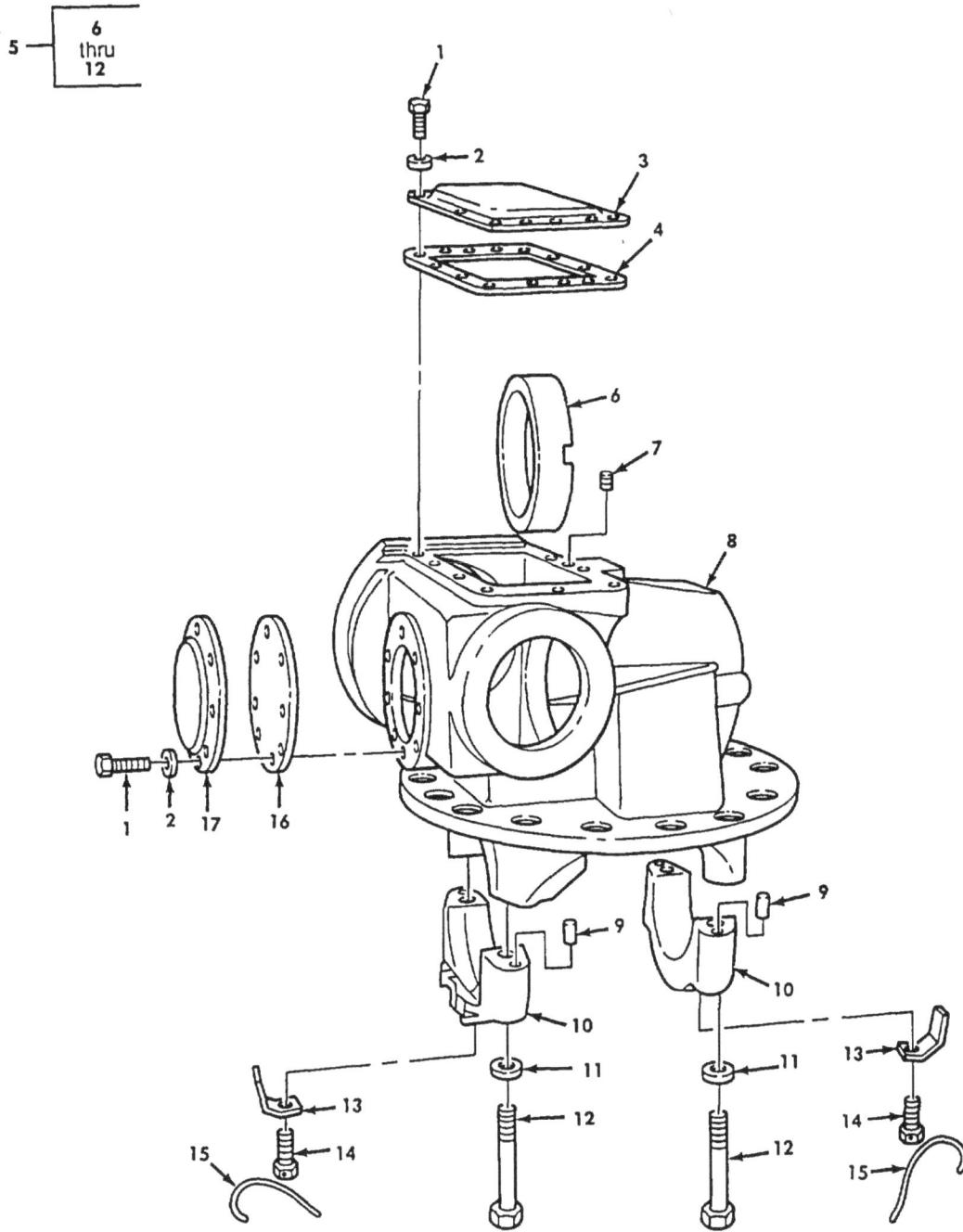

Figure 240. Rear Axle Differential Carrier Assembly, Access Cover, and Related Parts.

(1) ITEM NO	(2) SMR CODE	(3) NSN	(4) CAGEC	(5) PART NUMBER	(6) DESCRIPTION AND USABLE ON CODES (UOC)	(7) QTY
					GROUP 1102 DIFFERENTIAL	
					FIG. 240 REAR AXLE DIFFERENTIAL CARRIER ASSEMBLY, ACCESS COVER, AND RELATED PARTS	
1	PAFZZ	5305001159526	80204	B1821BH038C075D	SCREW, CAP, HEXAGON H	18
2	PAOZZ	5310001770892	19207	7748743	WASHER, FLAT	18
3	PFOZZ	5340007346820	78500	3866-X-557	COVER, ACCESS	1
4	PAOZZ	5330001388388	19207	7535079	GASKET PART OF KIT P/N 7346807	1
5	PAHHH	2520007347548	78500	A2-3800E473	CARRIER AND CAP ASS	2
6	PAHZZ	3120007346883	19207	7346883	.BEARING, SLEEVE	1
7	PAHZZ	5305007521718	19207	7521718	.SETSCREW..	1
8	XAHZZ		19207	8758260	.CARRIER, HOUSING	1
9	PAHZZ	5315005972723	19207	7534717	.PIN, SHOULDER, HEADLE	4
10	XBHZZ		19207	8758249	.CAP, DIFFERENTIAL SU	2
11	PAHZZ	5310003555453	19207	10900408	.WASHER, FLAT	4
12	PAHZZ	5306007346817	19207	7346817	.BOLT, MACHINE	4
13	PAHZZ	2530007346819	19207	7346819	.LOCK	2
14	PAHZZ	5305009037794	96906	MS51095-410	SCREW, CAP, HEXAGON H	2
15	MHHZZ		80244	22W1642125X12 IN	WIRE, NONELECTRICAL	2
16	PAOZZ	5330007346899	19207	7346899	GASKET..	1
17	PFOZZ	5340004454555	19207	8758248	COVER, ACCESS	1

END OF FIGURE

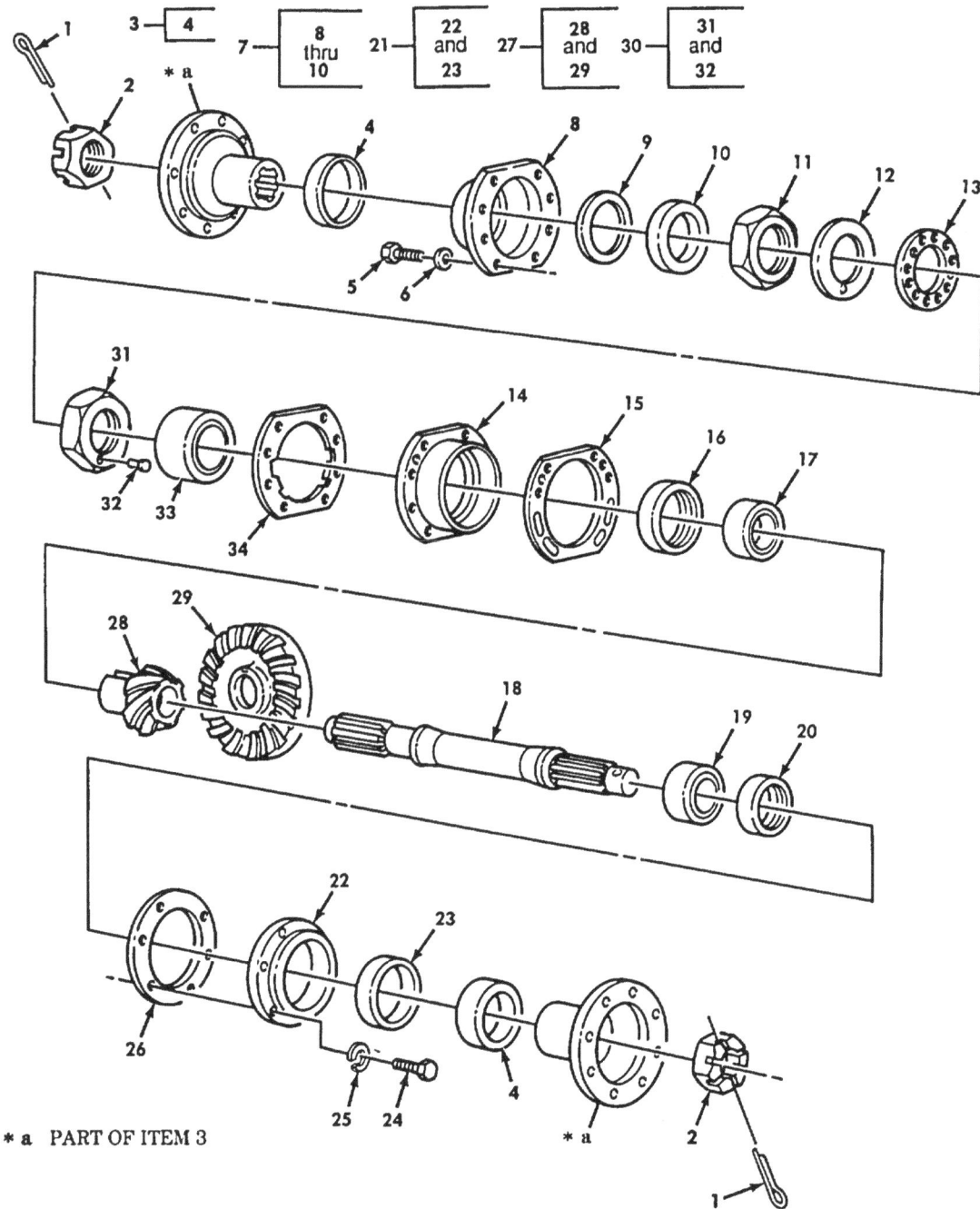

* a PART OF ITEM 3

Figure 241. Rear Axle Differential Through Shaft, Gears, and Flange Assemblies.

(1) ITEM NO	(2) SMR CODE	(3) NSN	(4) CAGEC	(5) PART NUMBER	(6) DESCRIPTION AND USABLE ON CODES (UOC)	(7) QTY
					GROUP 1102 DIFFERENTIAL	
					FIG. 241 REAR AXLE DIFFERENTIAL THROUGH SHAFT, GEARS, AND FLANGE ASSEMBLIES	
1	PAFZZ	5315000590184	96906	MS24665-361	PIN, COTTER	2
2	PAFZZ	5310002825661	19207	7346893	NUT, PLAIN, SLOTTED, H	2
3	PAFZZ	2520007346802	78500	A3280G1957	FLANGE, COMPANION, UN	2
4	PAFZZ	2520005638309	19207	7535048	.DEFLECTOR, DIRT AND	1
5	PAFZZ	5305000180178	24617	180178	SCREW, CAP, HEXAGON H	8
6	PAFZZ	5310005159627	38597	50-4-18-17-7	WASHER, FLAT	8
7	PAFZZ	2520007346897	19207	7346897	COVER ASSEMBLY	1
8	XAFZZ		19207	8758251	.COVER, BRG RETAINER	1
9	PAFZZ	5330004195872	19207	10900396	.GASKET	1
10	PAOZZ		19207	12470105	.SEAL, PLAIN ENCASED	1
11	PAHZZ	5310003532427	19207	7979263	NUT, PLAIN, OCTAGON	1
12	PAHZZ	5310001473274	19207	8758258	WASHER, KEY	1
13	PAHZZ	5310007007089	19207	5139123	WASHER, KEY	1
14	PAHZZ	3110004199471	19207	8758202	PLATE, RETAINING, BEA	1
15	KFHZZ	5365010778564	19207	7346888	SHIM .003 PART OF KIT P/N 7346807.........	1
15	KFHZZ	5365007346889	19207	7346889	SHIM .005 PART OF KIT P/N 7346807.........	5
15	KFHZZ	5365007346890	19207	7346890	SHIM .010 PART OF KIT P/N 7346807.........	1
16	PAHZZ	5365006926119	19207	8758209	SPACER, BEVEL PINION 367 PART OF .... KIT P/N 5704278	1
16	KFHZZ		19207	8758211	COLLAR .369 PART OF KIT P/N 5704278.......	1
16	KFHZZ		19207	8758213	COLLAR .371 PART OF KIT P/N 5704278.......	1
16	KFHZZ		19207	8758215	COLLAR .373 PART OF KIT P/N 5704278.......	1
16	PAHZZ	3040006926121	19207	8758217	COLLAR, SHAFT .375 PART OF KIT P/N........ 5704278	1
16	KFHZZ		19207	8758219	COLLAR .377 PART OF KIT P/N 5704278.......	1
16	PAHZZ	3110010777134	19207	8758221	COLLAR, BEARING .379 PART OF KIT P/N...... 5704278	1
16	KFHZZ		19207	8758223	COLLAR .381 PART OF KIT P/N 5704278.......	1
16	PAHZZ	3040006926123	19207	8758225	COLLAR, SHAFT .383 PART OF KIT P/N........ 5704278	1
16	KFHZZ		19207	8758227	COLLAR .385 PART OF KIT P/N 5704278.......	1
16	KFHZZ	3120006926124	19207	8758229	BEARING, WASHER, THRU .387	1
16	KFHZZ		19207	8758231	COLLAR .389 PART OF KIT P/N 5704278.......	1
16	XFHZZ		19207	8758233	COLLAR .391 PART OF KIT P/N 5704278.......	1
16	KFHZZ		19207	8758235	COLLAR .393 PART OF KIT P/N 5704278.......	1
16	KFHZZ		19207	8758237	SPACER, BEVEL PINION .395 PART OF .... KIT P/N 5704278	1
16	KFHZZ		19207	8758239	COLLAR .397 PART OF KIT P/N 5704278.......	1
16	KFHZZ		19207	8758241	COLLAR .399 PART OF KIT P/N 5704278.......	1
16	KFHZZ		19207	8758243	COLLAR .401 PART OF KIT P/N 5704278.......	1
16	KFHZZ		19207	8758245	COLLAR .403 PART OF KIT P/N 5704278.......	1
17	PAHZZ	3110001005355	05840	323W231-J	BEARING, ROLLER, TAPE	1
18	PAHZZ	3040007346892	78500	3880M533	SHAFT, SHOULDERED	1
19	PAHZZ	3110001950460	78500	712148	BEARING, ROLLER, CYLI	1
20	PAHZZ	3120006623370	19207	7346894	BEARING, WASHER, THRU	1

(1) ITEM NO	(2) SMR CODE	(3) NSN	(4) CAGEC	(5) PART NUMBER	(6) DESCRIPTION AND USABLE ON CODES (UOC)	(7) QTY
21	PAFZZ	3110007346895	19207	7346895	PLATE, RETAINING, BEA	1
22	XAFZZ		19207	8758207	.CAP	1
23	PAFZZ	5330010230269	78500	A-1205-U-1737	.SEAL, PLAIN ENCASED	1
24	PAFZZ	5305005432419	80204	B1821BH038C113N	SCREW, CAP, HEXAGON H	6
25	PAFZZ	5310006379541	96906	MS35338-46	WASHER, LOCK	6
26	PAHZZ	5330006412466	19207	7346896	GASKET PART OF KIT P/N 7346807	1
27	PAHZZ	3020007346881	78500	MPS516	GEAR SET, BEVEL, MATC	1
28	XAHZZ		19207	8758204	.PINION, BEVEL	1
29	XAHZZ		19207	8758205	.GEAR, BEVEL	1
30	PAHZZ	5310003740836	19207	7979308	NUT, PLAIN, OCTAGON	1
31	XAHZZ	5310011350049	19207	7979309	.NUT, PLAIN, OCTAGON	1
32	PAHZZ	5315010587268	19207	7979310	.PIN, ADJUSTMENT	1
33	PAHZZ	3110001004177	21450	703189	BEARING, ROLLER, TAPE	1
34	PAHZZ	5330007346886	19207	7346886	GASKET COVER PART OF KIT P/N...................... 7346807	1

END OF FIGURE

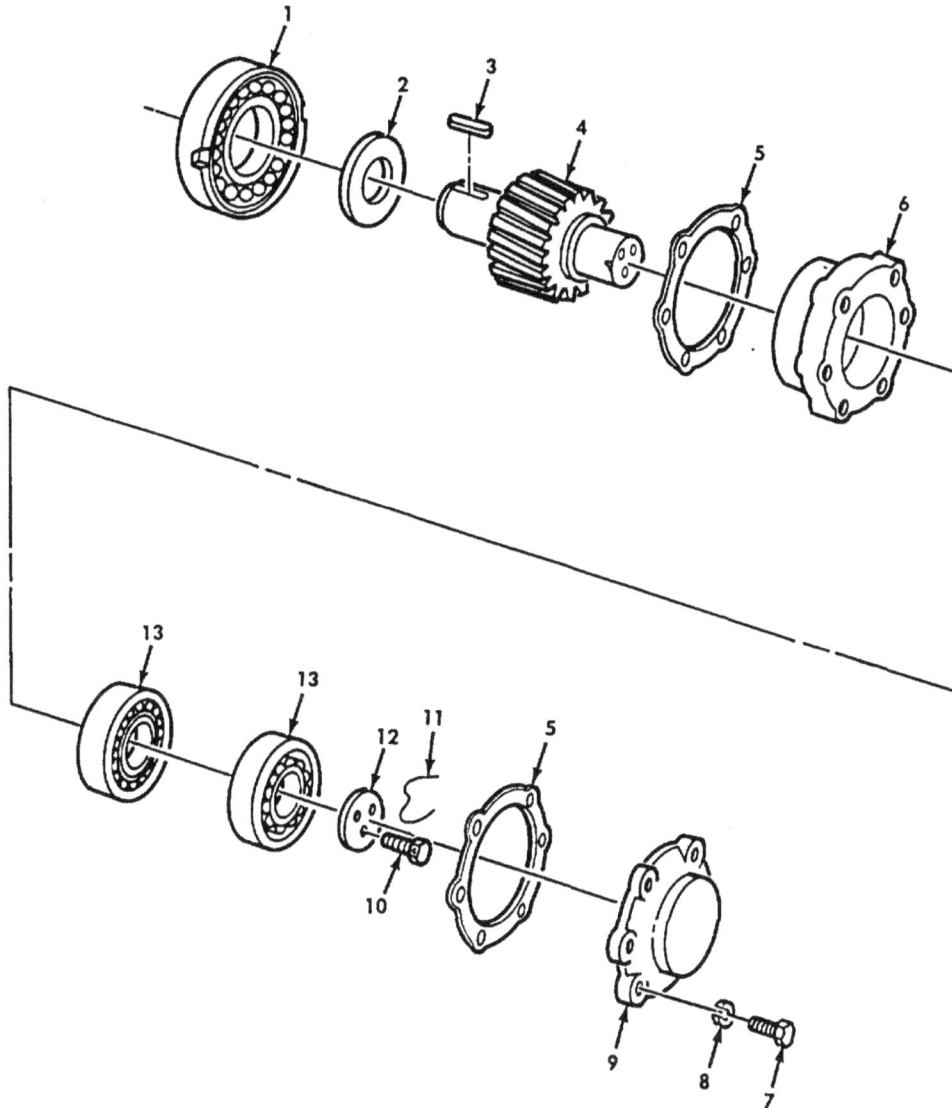

Figure 242. Rear Axle Differential Helical Drive Pinion and Related Parts.

(1) ITEM NO	(2) SMR CODE	(3) NSN	(4) CAGEC	(5) PART NUMBER	(6) DESCRIPTION AND USABLE ON CODES (UOC)	(7) QTY
					1102 DIFFERENTIAL	
					FIG. 242 REAR AXLE DIFFERENTIAL HELICAL DRIVE PINION AND RELATED PARTS	
1	PAHZZ	3110001586011	43334	U1218TAM	BEARING, ROLLER, CYLI	1
2	PAHZZ	5310000336012	19207	7346882	WASHER, RECESSED	1
3	PAHZZ	5315002765385	19207	7534781	KEY, MACHINE	1
4	KFHZZ	3040007346825	19207	7346825	GEARSHAFT, HELICAL PART OF KIT P/N.............. 10899232................................	1
5	KFHZZ	5330007346878	19207	7346878	SHIM .010 PART OF KIT P/N 7346807.................V	
5	PAHZZ	5365007346879	19207	7346879	SHIM .005 PART OF KIT P/N 7346807.................V	
5	KFHZZ	5365007346880	19207	7346880	SHIM .003 PART OF KIT P/N 7346807.................V	
6	PAHZZ	2520004946586	19207	8758208	CAP, BEARING, DIFFERE	1
7	PAHZZ	5305000712071	80204	B1821BH050C200N	SCREW, CAP, HEXAGON H	6
8	PAHZZ	5310000034094	01276	210104-8S	WASHER, LOCK	6
9	PAHZZ	3110007346877	19207	7346877	PLATE, RETAINING, BEA	1
10	PAHZZ	5305009419460	96906	MS51095-386	SCREW, CAP, HEXAGON H	3
11	MHHZZ		80244	22W1642125X12	WIRE, NONELECTRICAL RETAINING PLATE, ... MAKE FROM P/N 22-W-1642-125, 12 INCHES LONG................................	1
12	PAHZZ	3110003505407	78500	1829S149	PLATE, RETAINING, BEA................................	1
13	PAHZZ	3110001005825	21450	712627	BEARING, ROLLER, TAPE	2

END OF FIGURE

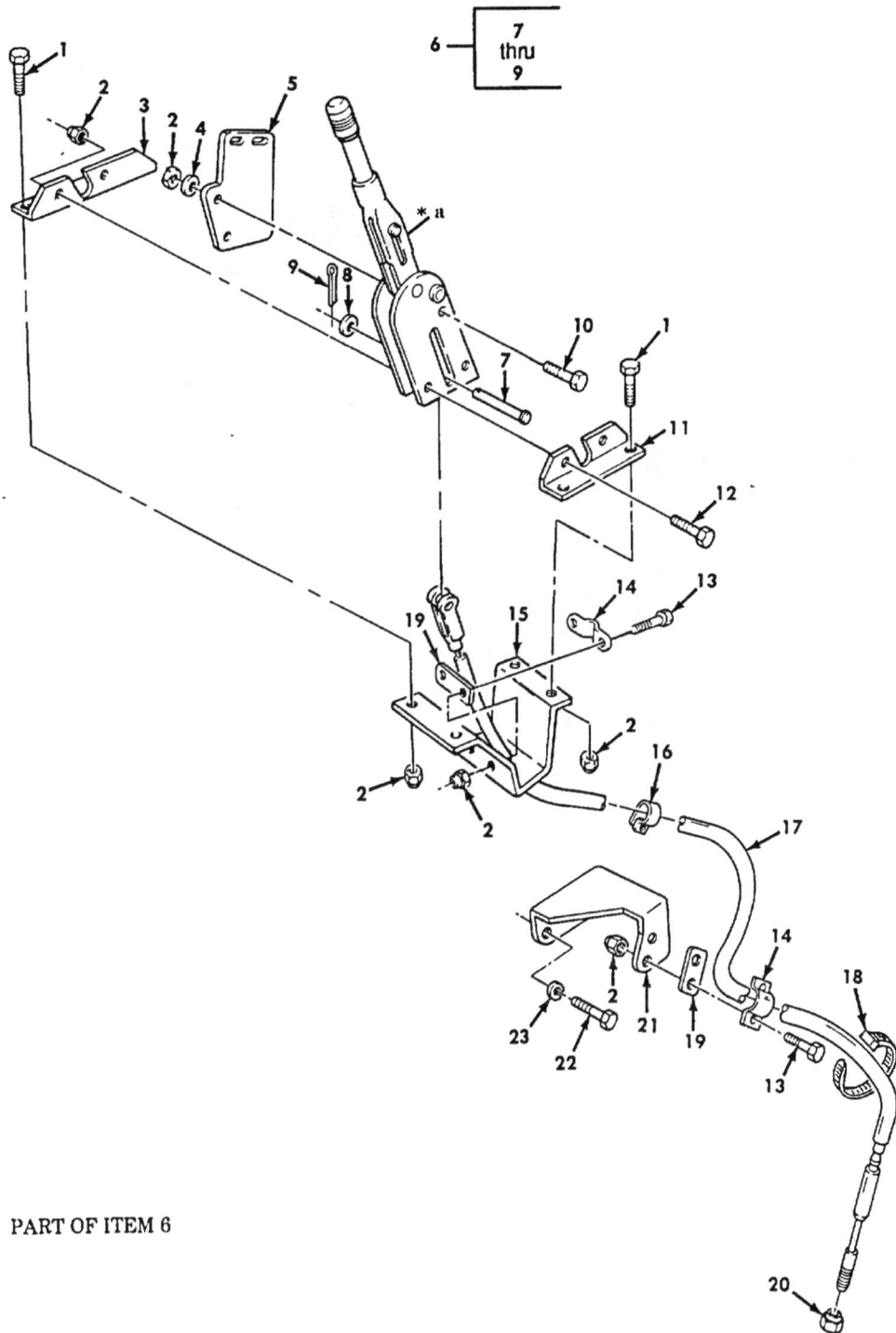

* a   PART OF ITEM 6

Figure 243. Parking Brake Lever Assembly, Cable, and Mounting Hardware.

(1) ITEM NO	(2) SMR CODE	(3) NSN	(4) CAGEC	(5) PART NUMBER	(6) DESCRIPTION AND USABLE ON CODES (UOC)	(7) QTY
					GROUP 12 BRAKES 1201 HAND BRAKES	
					FIG. 243 PARKING BRAKE LEVER ASSEMBLY, CABLE, AND MOUNTING HARDWARE	
1	PAOZZ	5305002692803	96906	MS90726-60	SCREW, CAP, HEXAGON H	4
2	PAOZZ	5310008140672	96906	MS51943-36	NUT, SELF-LOCKING, HE	14
3	PFOZZ	5340004807608	19207	11608772	BRACKET, ANGLE	1
4	PAOZZ	5310000806004	96906	MS27183-14	WASHER, FLAT	1
5	PFOZZ	5340013338147	19207	12375362	BRACKET, ANGLE PART OF KIT P/N .... 12375360..........	1
6	PAOZZ	3040000402401	19207	8365663	LEVER, MANUAL CONTRO	1
7	PAOZZ	5315007415746	19207	7415746	.PIN, STRAIGHT, HEADED	1
8	PAOZZ	5310006255756	96906	MS15795-812	.WASHER, FLAT	1
9	PAOZZ	5315008423044	96906	MS24665-283	.PIN, COTTER PART OF KIT P/N 12375360............	1
10	PAOZZ	5305002692811	96906	MS90726-67	SCREW, CAP, HEXAGON H	2
11	PFOZZ	5340007409361	19207	7409361	BRACKET, ANGLE	1
12	PAOZZ	5305002693242	80204	B1821BH038F200N	SCREW, CAP, HEXAGON H	1
13	PAOZZ	5305002693238	80204	B1821BH038F125N	SCREW, CAP, HEXAGON H	2
14	PAOZZ	5340007409366	19207	7409366	STRAP, RETAINING	2
15	PFOZZ	5340009998591	19207	10883130	BRACKET, CLAMP	2
16	PAOZZ	5340000538994	96906	MS21333-126	CLAMP, LOOP	1
17	PAOZZ	2590011368721	19207	12255644-1	CONTROL ASSEMBLY, PU	1
18	PAOZZ	5975001563253	96906	MS3367-2-9	STRAP, TIEDOWN, ELECT	1
19	PAOZZ	3040007409372	19207	7409372	CONNECTING LINK, RIG	2
20	PAOZZ	5310002697044	19207	7373244	NUT, SELF-LOCKING, SI	1
21	PFOZZ	2590011122167	19207	12302657	BRACKET, VEHICULAR C	1
22	PAOZZ	5305007098516	96906	MS90727-86	SCREW, CAP, HEXAGON H	2
23	PAOZZ	5310002090965	96906	MS35338-47	WASHER, LOCK	2

END OF FIGURE

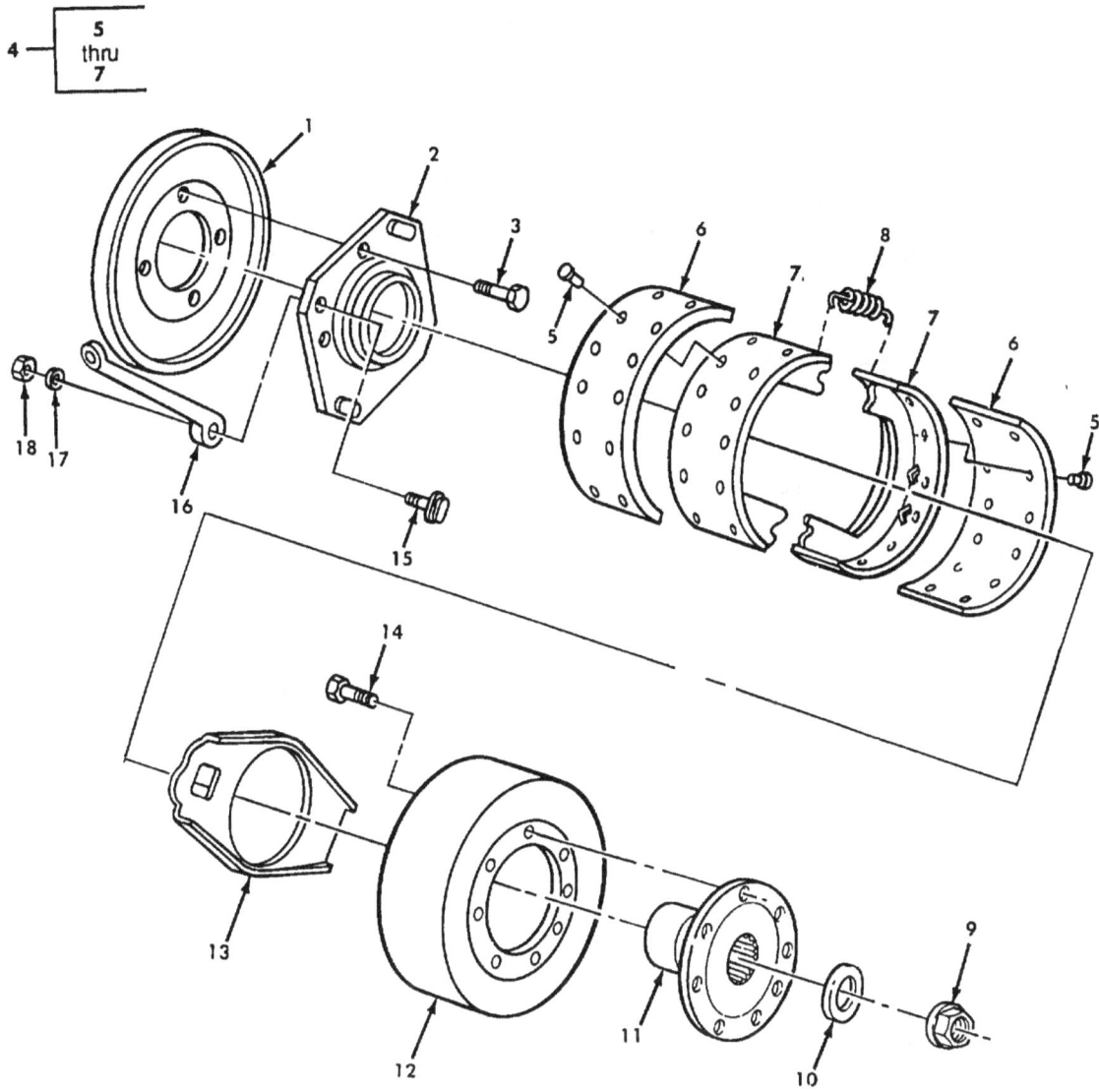

Figure 244. Parking Brake Shoe Assembly, Drum, Dust Shield, and Related Parts.

(1) ITEM NO	(2) SMR CODE	(3) NSN	(4) CAGEC	(5) PART NUMBER	(6) DESCRIPTION AND USABLE ON CODES (UOC)	(7) QTY
					GROUP 1201 HAND BRAKE	
					FIG. 244 PARKING BRAKE SHOE ASSEMBLY, DRUM, DUST SHIELD, AND RELATED PARTS	
1	PAOZZ	2530011382016	78500	3264-V-100	DUST SHIELD, BRAKE	1
2	PAOZZ	2530011346626	78500	A3736-K-375	PLATE, BACKING, BRAKE	1
3	PAOZZ	5306011328271	78500	10X-1250	BOLT, MACHINE	4
4	PAOZZ	2530011400107	70960	A13722-D-420	BRAKE SHOE	2
5	PFFZZ	5320011469582	78500	RV-876	.RIVET	24
6	XAFZZ		70960	A3722-D-420	.SHOE BRAKE BACKING PLATE	2
7	PAFZZ	2530013266127	78500	2000-P-1446	.LINING, FRICTION	2
8	PAOZZ	5360007586456	78500	2758-A-53	SPRING, HELICAL, EXTE	2
9	PAOZZ	5310010990397	14262	1227-K-1051	NUT	1
10	PAOZZ	6310011294373	78500	1229-T-1450	WASHER, FLAT	1
11	PAOZZ	2520011269363	78500	A-3280-U-7535	FLANGE, DEFLECTOR	1
12	PAOFF	2530011382015	78500	3219-H-4064	BRAKE DRUM	1
13	PAOZZ	2530011341332	78500	A-3787-N-14	PLATE, BACKING, BRAKE	1
14	PAOZZ	5306011290327	78500	10X-507-C	BOLT, MACHINE BRAKE DRUM	8
15	PAOZZ	2530011328272	78500	2710-U-151	STUD	1
16	PAOZZ	3040011388578	78500	302-Z-676	LEVER, MANUAL CONTRO	1
17	PAOZZ	5310000685285	96906	MS27183-20	WASHER, FLAT	1
18	PAOZZ	5310009496280	78500	1779-Z-260	NUT, LOCK	1

END OF FIGURE

Figure 245. Front Axle Brake and Spider Assembly.

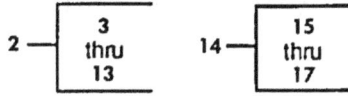

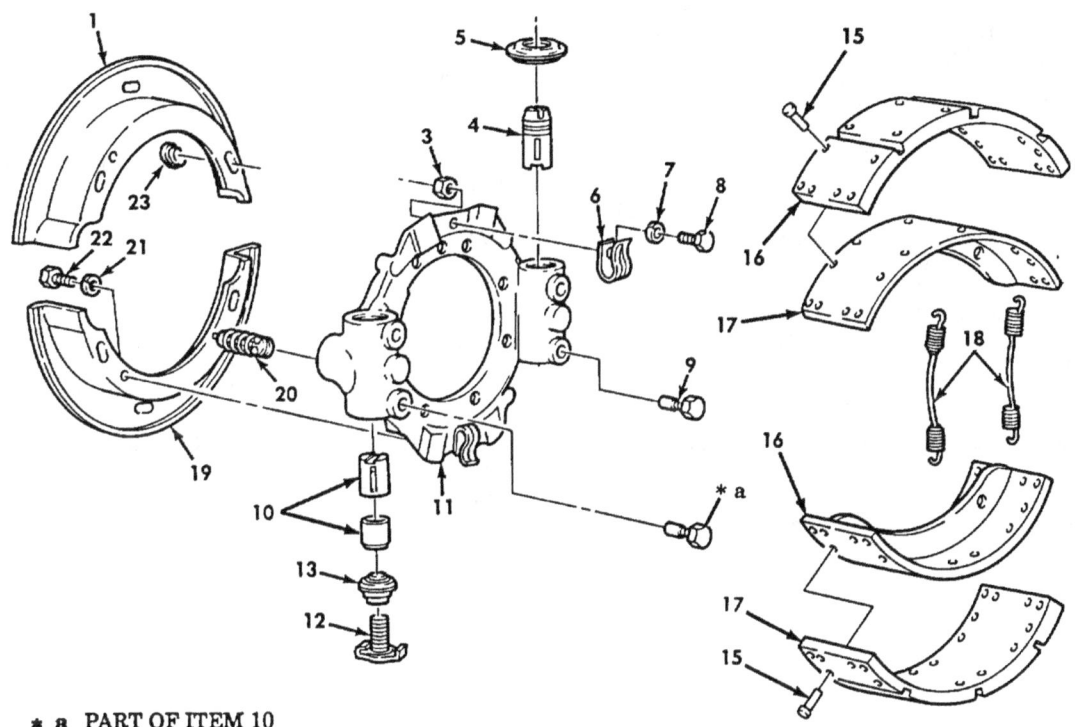

* a PART OF ITEM 10

(1) ITEM NO	(2) SMR CODE	(3) NSN	(4) CAGEC	(5) PART NUMBER	(6) DESCRIPTION AND USABLE ON CODES (UOC)	(7) QTY
					GROUP 1202 HAND BRAKES	
					FIG. 245 FRONT AXLE BRAKE AND SPIDER ASSEMBLY	
1	PAOZZ	2530011226015	78500	3236-J-2012	DUST SHIELD, BRAKE .... UOC:DAA, DAB, DAC, DAD, DAE, DAF, DAG, DAH, DAJ, DAK, DAL, DAW, DAX, V12, V13, V14, V15, V16, V18, V19, V20, V21, V22, V24, V25, V39	2
2	PAOFF	2530011340901	78500	A-3211-H-2868	SPIDER, BRAKE L.H	1
2	PAOFF	2530011341333	78500	A-3211-L-2872	SPIDER, BRAKE R.H	1
3	PAFZZ	5310007680319	96906	MS51968-2	.NUT, PLAIN, HEXAGON	2
4	PAFZZ	2530011273596	78500	2297-K-3885	.PISTON, HYDRAULIC BR L.H. PART OF .... KIT P/N.1254	2
4	PAFZZ	2530011317741	78500	2297-L-3886	.PLUNGER, ANCHOR, SPID R.H. PART OF .... KIT P/N.1254	2
5	PAFZZ	5330012929573	78500	1205-B-2004	.SEAL, PLAIN ENCASED PART OF KIT P/N........... 1254.	2
6	PAFZZ	5340006137784	78500	1718-D-134	.CLIP	2
7	PAFZZ	5310010632299	78500	WA-14-C	.WASHER, LOCK	2
8	PAFZZ	5306010623148	78500	S-146-C	.BOLT, MACHINE	2
9	KFFZZ	2530012717006	78500	A-3280-Z-8190	.GUIDE, ANCHOR PART OF KIT P/N.1254...........	2
10	PAFZZ	3040042804156	78500	A-3280-V-8186	.PAWL PART OF KIT P/N 1218	2
11	XAFZZ		78500	3211-H-2868	.SPIDER L.H	1
11	XAFZZ		78500	3211-L-2872	.SPIDER R.H	1
12	KFFZZ		78500	A-2297-V-5326	.ADJUSTING SCREW ASS PART OF KIT P/N.1218	2
13	PAFZZ	2530012719347	78500	1205-C-2005	.SEAL, PLAIN ENCASED PART OF KIT P/N........... 1218.	2
14	PAOZZ	2530012863257	19207	5705696	PARTS KIT, BRAKE SHO QTY-4 PER SET	1
15	PAOZZ	5320012906360	96906	MS16536-243	.RIVET, TUBULAR PART OF KIT P/N................... 5705695	96
16	KFOZZ		19207	12356896	.LINING, BRAKE PART OF KIT P/N 5705695	8
17	XAOZZ		19207	12356897	.SHOE, BRAKE	4
18	PAOZZ	5360004824422	78500	2758-W-127	SPRING PART OF KIT 1218	2
19	PAOZZ	2530011226016	78500	3236-K-2013	DUST, SHIELD, BRAKE .... UOC:DAA, DAB, DAC, DAD, DAE, DAF, DAG, DAH, DAJ, DAK, DAL, DAW, DAX, V12, V13, V14, V15, V16, V18, V19, V20, V21, V22, V24, V25, V39	2
20	PAOZZ	2530001179144	78500	A-2747-H-112	WEDGE ASSEMBLY	4
21	PAOZZ	5310010623384	78500	1229-S-513-C	WASHER, LOCK	8
22	PAOZZ	5305013571656	78500	S255Z	SCREW, CAP, HEXAGON H	8
23	PAOZZ	5340001811546	78500	1707-C-3	CAP-PLUG, PROTECTIVE	2

END OF FIGURE

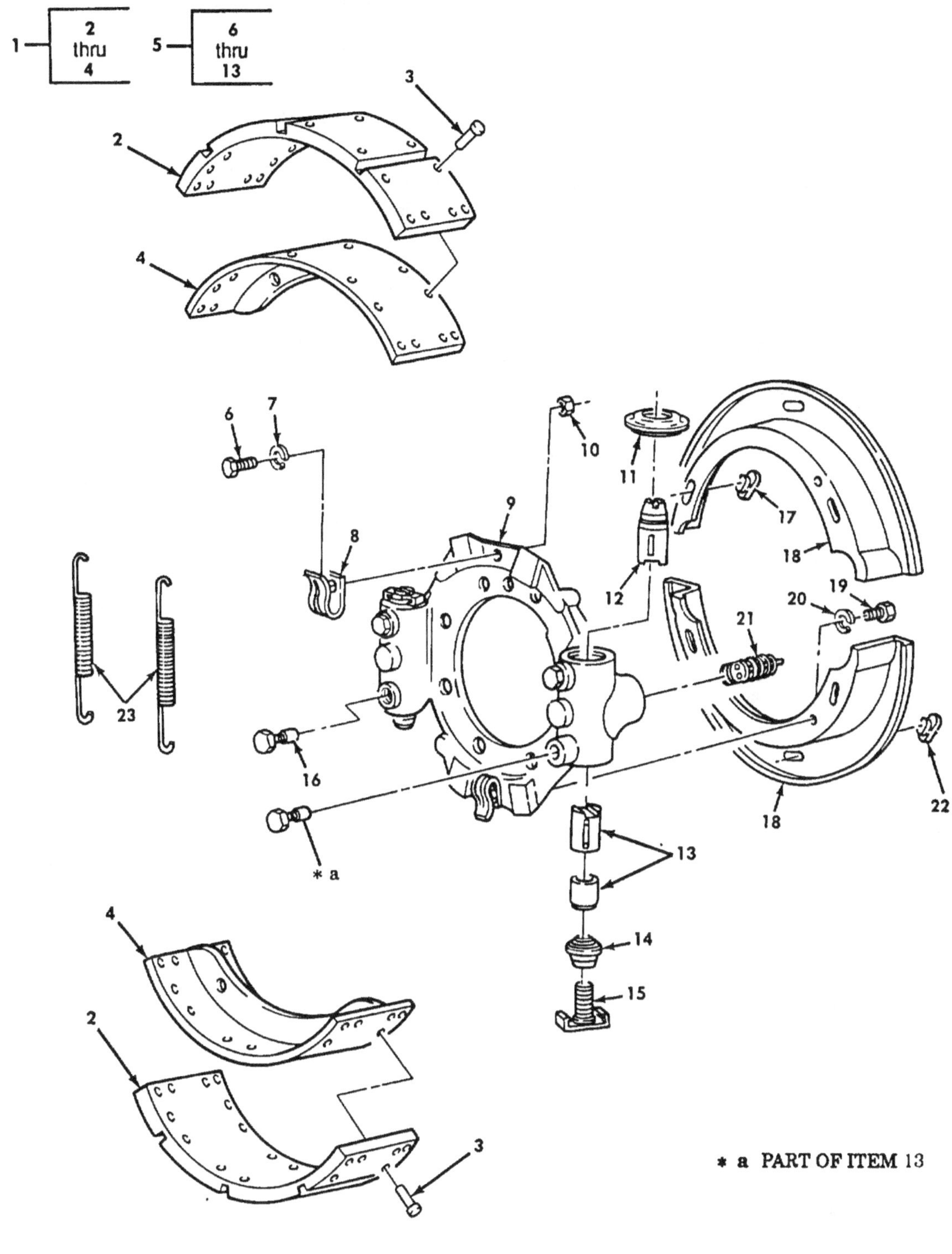

Figure 246. Rear Axle Brake and Spider Assembly.

* a PART OF ITEM 13

(1) ITEM NO	(2) SMR CODE	(3) NSN	(4) CAGEC	(5) PART NUMBER	(6) DESCRIPTION AND USABLE ON CODES (UOC)	(7) QTY
					GROUP 1202 HAND BRAKES	
					FIG. 246 REAR AXLE BRAKE AND SPIDER ASSEMBLY	
1	PAOZZ	2530012863257	19207	5705696	PARTS KIT, BRAKE SHO QTY-4 PER SET	1
2	KFOZZ		19207	12356896	.LINING, BRAKE PART OF KIT P/N...................... 5705695..	8
3	PAOZZ	5320012906360	96906	MS16536-243	.RIVET, TUBULAR PART OF KIT P/N..................... 5705695..	96
4	KFOZZ		19207	12356897	.SHOE, BRAKE	4
5	PAOFF	2530011246530	78500	A-3211-K-2871	SPIDER ASSEMBLY FWD REAR, L.H. AND........... REAR REAR, R.H	2
5	PAOFF	2530011254280	78500	A-3211-D-2994	SPIDER ASSEMBLY FWD REAR, R.H. AND .... REAR REAR, L.H	1
6	PAFZZ	5306010623148	78500	S-146-C	.BOLT	2
7	PAFZZ	5310010632299	78500	WA-14-C	.WASHER, LOCK	2
8	PAFZZ	5340006137784	78500	1718-D-134	.CLIP	2
9	XAFZZ	2530011346618	78500	A1-3211-K-2871	.SPIDER, BRAKE FWD REAR, L.H. AND .... REAR REAR, R.H	1
9	XAFZZ	2530011231229	78500	A1-3211-D-2994	.SPIDER, BRAKE FWD REAR, R.H. AND.............. REAR REAR, L.H	1
10	PAFZZ	5310012372615	78500	N-14-C	.NUT, HEXAGON	2
11	PAFZZ	5330012929573	78500	1205-B-2004	.SEAL, ANCHOR PART OF KIT P/N 1173, ... 1174..	2
12	PAFZZ	5315012708268	78500	2297-Y-5329	.PLUNGER, ANCHOR LEFT, PART OF KIT .... P/N 1173...	2
12	PAFZZ	5315012708269	78500 2	297-Z-5330	.PLUNGER, ANCHOR RIGHT, PART OF KIT........... P/N 1174...	2
13	PAFZZ	3040012804156	78500	A-3280-V-8186	.PLUNGER ASSEMBLY, ADJUSTING PART........... OF KIT P/N 1164..	2
14	PAFZZ	5330012719347	78500	1205-C-2005	.SEAL ADJUSTER PART OF KIT P/N..................... 1164..	2
15	KFFZZ		78500	A-2297-V-5326	.BOLT ASSEMBLY, ADJUSTER PART OF KIT.......... P/N 1164...	2
16	KFFZZ		78500	A-3280-Z-8190	.GUIDE, PLUNGER PART OF KIT P/N 1173, ... 1174..	2
17	PAOZZ	5340001811546	78500	1707-C-3	CAP-PLUG, PROTECTIVE	8
18	PAOZZ	2530010769567	78500	3236-L-2014	SHIELD, BRAKE DISK .... UOC:DAA, DAB, DAC, DAD, DAE, DAF, DAG, DAH, DAJ, DAK, DAL, DAW, DAX, V12, V13, V14, V15, V16, V18, V19, V20, V21, V22, V24, V25, V39	8
19	PAOZZ	5305013571656	78500	S255Z	SCREW, CAP, HEXAGON H	16
20	PAOZZ	5310010623384	78500	1229-S-513-C	WASHER, LOCK	16
21	PAOZZ	2530001179144	78500	A-2747-H-112	WEDGE ASSEMBLY	8
22	PAOZZ	2530004066785	78500	1707B	COVER, DUST	8
23	PAOZZ	5360012750545	78500	2258-S-1033	SPRING, SHOE RETURN PART OF KIT .... P/N 1164...	2

END OF FIGURE

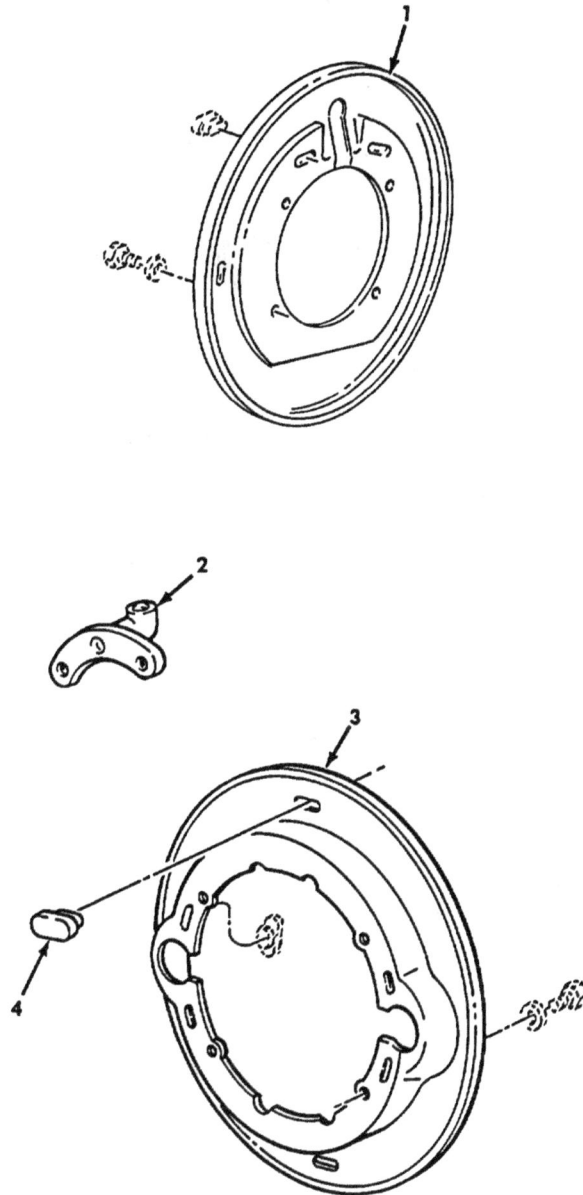

Figure 247. Service Brake Mounting Plate (M939A2).

(1) ITEM NO	(2) SMR CODE	(3) NSN	(4) CAGEC	(5) PART NUMBER	(6) DESCRIPTION AND USABLE ON CODES (UOC)	(7) QTY
					GROUP 1202 HAND BRAKES	
					FIG. 247 SERVICE BRAKE MOUNTING PLATE (M939A2)	
1	PAFZZ	2530012717074	78500	3264-B-1068	SHIELD, BRAKE DISK .... UOC:ZAA, ZAB, ZAC, ZAD, ZAE, ZAF, ZAG, ZAH, ZAJ, ZAK, ZAL	2
2	PAOZZ	4730012717187	78500	3280-W-8551	MANIFOLD, AIR LINE .... UOC:ZAA, ZAB, ZAC, ZAD, ZAE, ZAF, ZAG, ZAH, ZAJ, ZAK, ZAL	2
3	PAFZZ	2530012717075	78500	3264-A-1067	SHIELD, BRAKE DISK .... UOC:ZAA, ZAB, ZAC, ZAD, ZAE, ZAF, ZAG, ZAH, ZAJ, ZAK, ZAL	2
4	PAFZZ	2530004066785	78500	1707B2	COVER, DUST .... UOC:ZAA, ZAB, ZAC, ZAD, ZAE, ZAF, ZAG, ZAH, ZAJ, ZAK, ZAL	4

END OF FIGURE

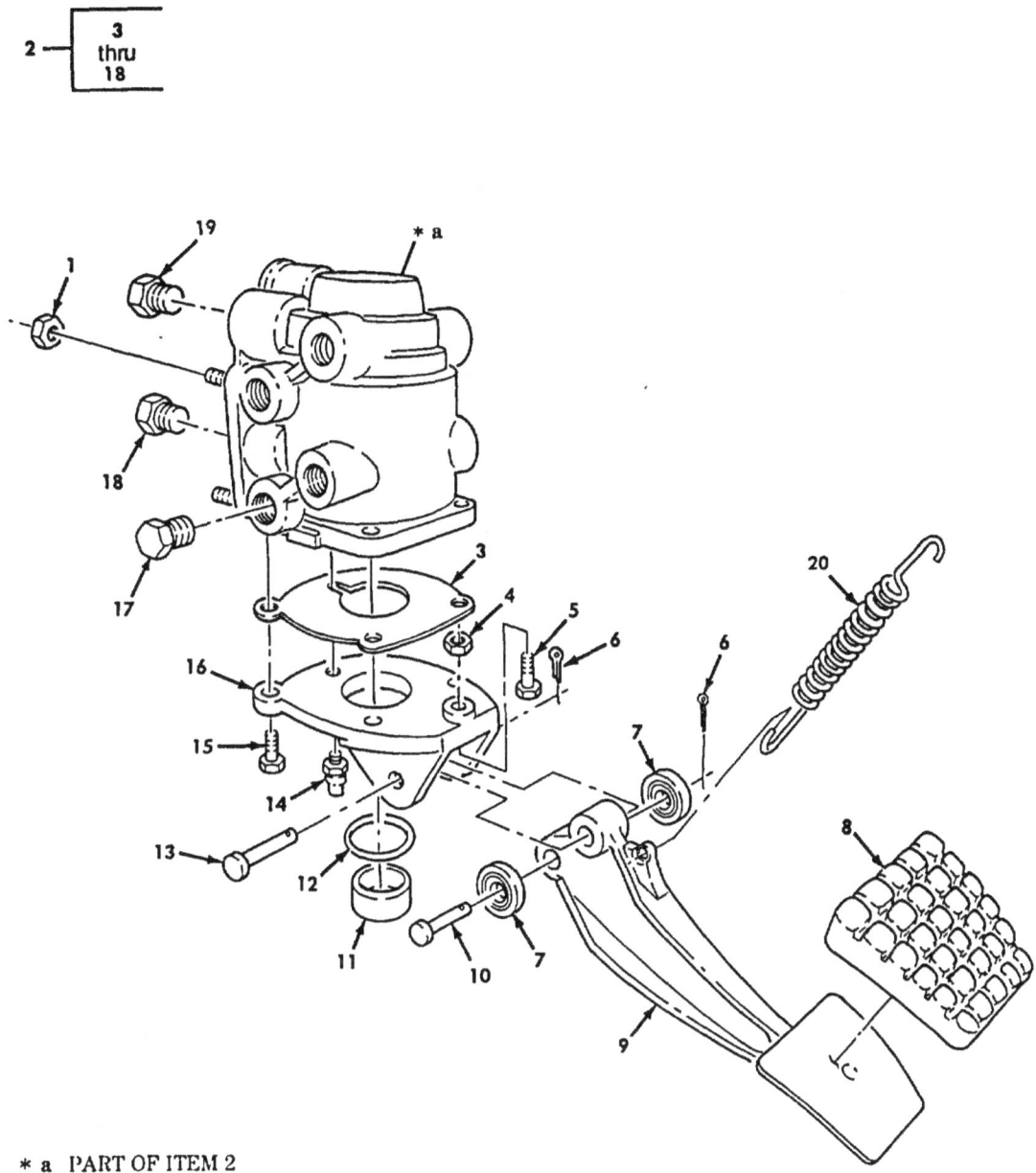

* a PART OF ITEM 2

Figure 248. Treadle Value Assembly.

(1) ITEM NO	(2) SMR CODE	(3) NSN	(4) CAGEC	(5) PART NUMBER	(6) DESCRIPTION AND USABLE ON CODES (UOC)	(7) QTY
					GROUP 1206 MECHANICAL BRAKE SYSTEM	
					FIG. 248 TREADLE VALVE ASSEMBLY	
1	PAOZZ		96906	MS51922-21	NUT, SELF-LOCKING	3
2	PAOHH	2530011126435	06853	102352	VALVE, BRAKE PNEUMAT	1
3	PAFZZ	5330011236409	06853	291882	.GASKET	1
4	PAFZZ	5310011337216	06853	230250	.NUT, PLAIN, HEXAGON	1
5	PAFZZ	5305011337193	06853	243890	.THUMBSCREW	1
6	PAOZZ	5315012464339	06853	210492	.PIN, COTTER	1
7	PAOZZ	3120011325579	06853	245118	.ROLLER, BRAKE PEDAL	2
8	PAOZZ	2540011039128	06853	244682	.PAD, PEDAL	1
9	PFOZZ	2540011319639	06853	292532	.PEDAL, BRAKE VALVE	1
10	PAOZZ	5315011059475	06853	200981	.PIN, STRAIGHT, HEADED	1
11	PFHZZ	4820011049313	06853	244680	.SLEEVE, DIRECTIONAL	1
12	PFHZZ	5330000742692	06853	233955	.O-RING	1
13	PAFZZ	5315011124507	06853	240445	.PIN, STRAIGHT, HEADED	1
14	PAFZZ	4820007264719	57733	5196397	.VALVE, VENT BRAKE VALVE	1
15	PAFZZ	5306011241225	06853	293337	.BOLT, ASSEMBLED WASH	3
16	PAFZZ	2530011233105	06853	291883	.PLATE, MOUNTING, BRAK	1
17	PAFZZ	4730010304950	24617	272977	.PLUG, PIPE..	1
18	PAHZZ	4730000127951	06853	230111	.PLUG, PIPE	6
19	PAOZZ	4730010304950	96906	MS49005-2C	PLUG, PIPE	1
20	PAOZZ	5360011126546	19207	12255975	SPRING, HELICAL, EXTE...	1

END OF FIGURE

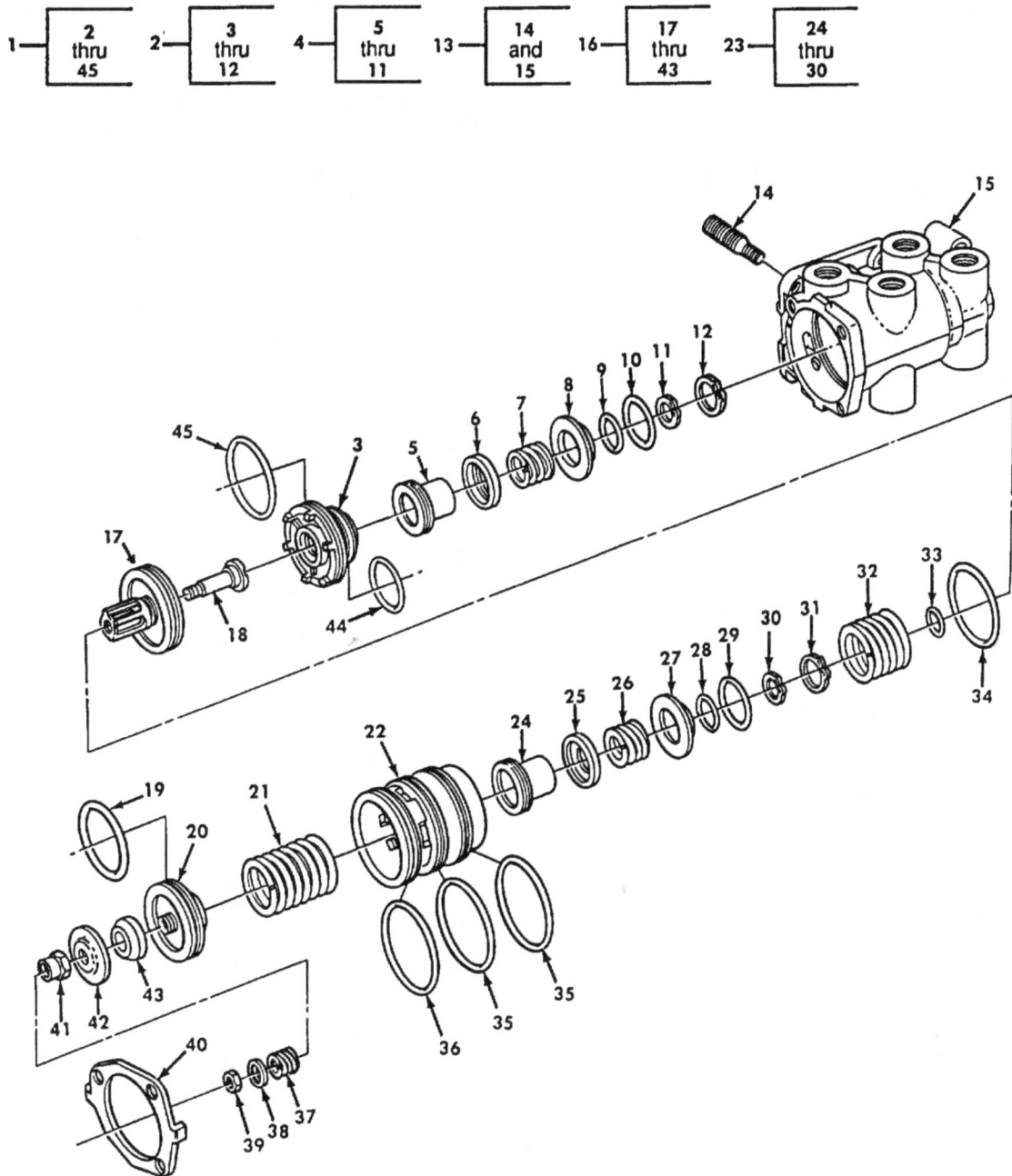

Figure 249. Treadle Valve Assembly, Basic.

(1) ITEM NO	(2) SMR CODE	(3) NSN	(4) CAGEC	(5) PART NUMBER	(6) DESCRIPTION AND USABLE ON CODES (UOC)	(7) QTY
					GROUP 1206 MECHANICAL BRAKE SYSTEM	
					FIG. 249 TREADLE VALVE ASSEMBLY, BASIC	
1	PAFFF	2530011271677	06853	7022-27	VALVE, BRAKE PNEUMAT	1
2	PAFZZ	2530011272337	06853	7022-23	.PISTON, VALVE...	1
3	XDFZZ	4810011609604	06853	247213	..PISTON, VALVE	
4	PAFFF	2530011341836	06853	7022-24	..INLET, EXHAUST VALVE	1
5	PAFZZ	2530011341838	06853	292896	...VALVE, INLET AND EXH PART OF KIT.............. P/N 289352...	1
6	PAFZZ	2530011273971	06853	247216	...RETAINER, HELICAL CO	1
7	PAFZZ	5360011246811	06853	247217	...SPRING, HELICAL, COMP	1
8	PAFZZ	5330011256280	06853	292898	...RETAINER, PACKING	1
9	PAFZZ	5330011237038	06853	292899	..O-RING	1
10	PAFZZ	5330004540364	06853	239029	..O-RING	1
11	PAFZZ	5325008362131	06853	292897	..RING, RETAINING	1
12	PAFZZ	5365005597574	96906	MS16627-1137	..RING, RETAINING	1
13	XBFZZ		06853	289022	.BODY, BRAKE VALVE	1
14	PAFZZ	5307002155399	06853	202998	..STUD, PLAIN	3
15	XAFZZ		06853	248114	.BODY	1
16	PAFZZ	4820011316123	06853	7022-25	.PISTON, VALVE	1
17	PAFZZ	2530011229933	06853	290202	.PISTON, VALVE	1
18	PAFZZ	5305011250929	06853	290186	.SCREW, SHOULDER.	1
19	PAFZZ	5330000853494	06853	239643	..O-RING PART OF KIT P/N 289352.....................	1
20	PAFZZ	2530011609652	06853	290184	..PISTON, HYDRAULIC BR	1
21	PAFZZ	5360011251671	06853	247231	..SPRING, HELICAL, COMP PART OF KIT.............. P/N 289352...	1
22	XAFZZ	2530012102751	06853	247222	..PISTON, BRAKE VALVE	1
23	PAFZZ	2805011341837	06853	7022-26	..VALVE, POPPET, ENGINE	1
24	PAFZZ	2530010838102	06853	112442	...VALVE, INLET AND EXH PART OF KIT.............. P/N 289352...	1
25	PAFZZ	5340010836420	06853	244435	...SEAT, HELICAL COMPRE	1
26	PAFZZ	5360010839766	06853	244437	...SPRING, HELICAL	1
27	PAFZZ	5330011203733	06853	292894	...RETAINER, PACKING	1
28	PAFZZ	5330003119234	06853	234045	...PACKING, PREFORMED	1
29	PAFZZ	5330004656453	06853	239219	...O-RING	1
30	PAFZZ	5365008042025	96906	MS16624-1087	...RING, RETAINING	1
31	PAFZZ	5340007387552	06853	240345	..CLIP, RETAINING.	1
32	PAFZZ	5360011241402	06853	247224	..SPRING, HELICAL, COMP PART OF KIT.............. P/N 289352...	1
33	PAFZZ	5330008832799	06853	239136	..O-RING PART OF KIT P/N 289352.....................	1
34	PAFZZ	5330011261233	06853	247226	..O-RING PART OF KIT P/N 289352.....................	1
35	PAFZZ	5330011324734	06853	247233	..O-RING PART OF KIT P/N 289352 PART.............. OF KIT P/N 289353..	2
36	PAFZZ	5330011234536	06853	247235	..O-RING PART OF KIT P/N 289352 PART.............. OF KIT P/N 289353..	1
37	PAFZZ	5360011251670	06853	290185	..SPRING, HELICAL, COMP PART OF KIT.............. P/N 289352...	1
38	PAFZZ	5340011354292	06853	290189	..RETAINER, HELICAL CO	1
39	PAFZZ	5310008775797	96906	MS21044N3	..NUT, SELF-LOCKING, HE PART OF KIT.............. P/N 289352...	1
40	PAFZZ	2530001273970	06853	247236	..RETAINER PART OF KIT P/N 289352, ... 289353...	1
41	PAFZZ	5310011268834	06853	290188	..NUT, SLEEVE PART OF KIT P/N 289352..............	1

(1) ITEM NO	(2) SMR CODE	(3) NSN	(4) CAGEC	(5) PART NUMBER	(6) DESCRIPTION AND USABLE ON CODES (UOC)	(7) QTY
					289352	
42	PAFZZ	4820011320606	06853	290187	..SEAT, VALVE, PISTON	1
43	PAFZZ	5340009314527	06853	241559	..MOUNT, RESILIENT PART OF KIT P/N................	1
44	PAFZZ	5330011245720	06853	247221	.O-RING PART OF KIT P/N 289352 PART...............	1
					OF KIT P/N 289353	
45	PAFZZ	5330011246405	06853	247234	.PACKING, PREFORMED PART OF KIT P/N...........	1
					289352 PART OF KIT P/N 289353..........................	

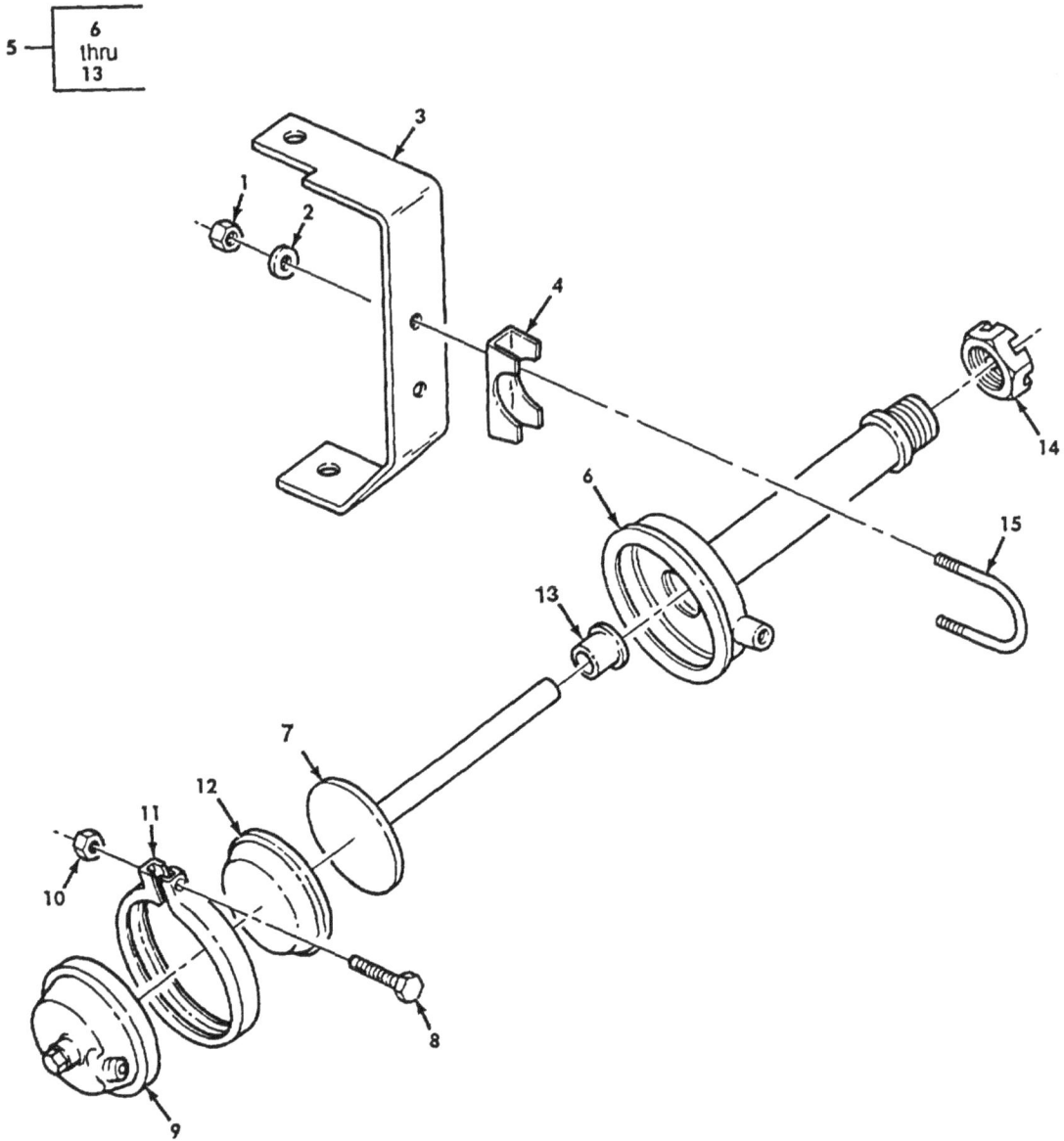

Figure 250. Air Chambers, Front and Rear Brakes.

(1) ITEM NO	(2) SMR CODE	(3) NSN	(4) CAGEC	(5) PART NUMBER	(6) DESCRIPTION AND USABLE ON CODES (UOC)	(7) QTY
					GROUP 1208 AIR BRAKES	
					FIG. 250 AIR CHAMBERS, FRONT AND REAR BRAKES	
1	PAOZZ	5310011260566	78500	NL-25-1-C	NUT	8
2	PAOZZ	5310012009879	78500	1229-E-1669-C	WASHER, FLAT	8
3	PAOZZ	5340012908884	78500	3299-M-5369	BRACKET, MOUNTING RIGHT FRONT- .... REAR AXLE AND LEFT REAR-REAR AXLE	2
3	PAOZZ	5340012908883	78500	3299-N-5370	BRACKET, MOUNTING LEFT FRONT- .... REAR AXLE AND RIGHT REAR- REAR AXLE	2
4	PAOZZ	5340012895030	78500	2255-H-86	BRACKET, DOUBLE ANGL	4
5	PAOFF	2530011267869	78500	E1-3276-L-12	BRAKE CHAMBER ASSY FRONT AXLE	2
6	XAFZZ		78500	A5-3280-F-5232	.HSG ASSY, NON-PRESS FRONT AXLE .... AIR BRAKE CHAMBER	1
7	PAOZZ	2530011341336	78500	A71-1779-V-230	.PUSH ROD, CHAMBER BRAKE HOUSING............ FRONT AXLE	1
5	PAOFF	2530010917814	78500	E-3276-L-12	CHAMBER, AIR BRAKE FRONT-REAR AND............ REAR-REAR AXLE .... UOC:DAA,DAB,DAC,DAD,DAE,DAF,DAG,DAH, DAJ,DAK,DAL,DAW,DAX,V12,V13,V14,V15, V16,V17,V18,V19,V20,V21,V22,V24,V25, ZAA,ZAB,ZAC,ZAD,ZAE,ZAF,ZAG,ZAH,ZAJ, ZAK,ZAL	4
6	XAFZZ		78500	A4-3280-F-5232	.HSG, ASSY, NON-PRESS FRONT-REAR AND........ REAR-REAR AXLE BRAKE CHAMBER .... UOC:DAA,DAB,DAC,DAD,DAE,DAF,DAG,DAH, DAJ,DAK,DAL,DAW,DAX,V12,V13,V14,V15, V16,V17,V18,V19,V20,V21,V22,V24,V25, ZAA,ZAB,ZAC,ZAD,ZAE,ZAF,ZAG,ZAH,ZAJ, ZAK,ZAL	1
7	PAFZZ	2530010642630	78500	A45-1779-V-230	.PUSH ROD, CHAMBER BRAKE HOUSING............. FRONT-REAR AND REAR-REAR AXLE .... UOC:DAA,DAB,DAC,DAD,DAE,DAF,DAG,DAH, DAJ,DAK,DAL,DAW,DAX,V12,V13,V14,V15, V16,V17,V18,V19,V20,V21,V22,V24,V25, ZAA,ZAB,ZAC,ZAD,ZAE,ZAF,ZAG,ZAH,ZAJ, ZAK,ZAL	1
5	PAOZZ	2530014259520	78500	E5-3276L-12	CHAMBER, AIR BRAKE. FRONT-REAR AND........... REAR-REAR AXLE UOC:V39	4
6	XAOFF		78500	A9-3280-F-5232	.HSG, ASSY, NON-PRESS FRONT-REAR AND........ REAR-REAR AXLE BRAKE CHAMBER UOC:V39	1
7	PAOZZ	2530014171645	78500	A59-1779-V-230	.PUSH ROD, CHAMBER BRAKE HOUSING............ FRONT-REAR AND REAR-REAR AXLE UOC:V39	1
8	PAFZZ	5306004987209	79780	15X725	.BOLT, SPECIAL RIM CLENCHING CLAMP	1
9	PAFZZ	2530012898359	78500	A2-3780-J-62	.HOUSING, AIR BRAKE C....................................	1

(1) ITEM NO	(2) SMR CODE	(3) NSN	(4) CAGEC	(5) PART NUMBER	(6) DESCRIPTION AND USABLE ON CODES (UOC)	(7) QTY
10	PAFZZ	5310001771258	78500	N-35-P	.NUT, PLAIN, HEXAGON PRESSURE .... HOUSING CLAMP	1
11	PAFZZ	5340000157560	78500	2797-V-100	.CLAMP, SYNCHRO PRESSURE HOUSING............	1
12	PAFZZ	2530004302392	16662	AD17544	.DIAPHRAGM, CHAMBER, B.................................	1
13	PAFZZ	2530000870163	78500	1779-Q-433	.GUIDE, BRAKE ASSEMBL HOUSING.................... ASSEMBLY	1
14	PAOZZ	5310001232572	78500	1727-N-40	NUT, PLAIN, ROUND HOUSING ASSEMBLY...........	4
15	PAOZZ	5306011966632	78500	2297-Y-4341	BOLT, U..	4

END OF FIGURE

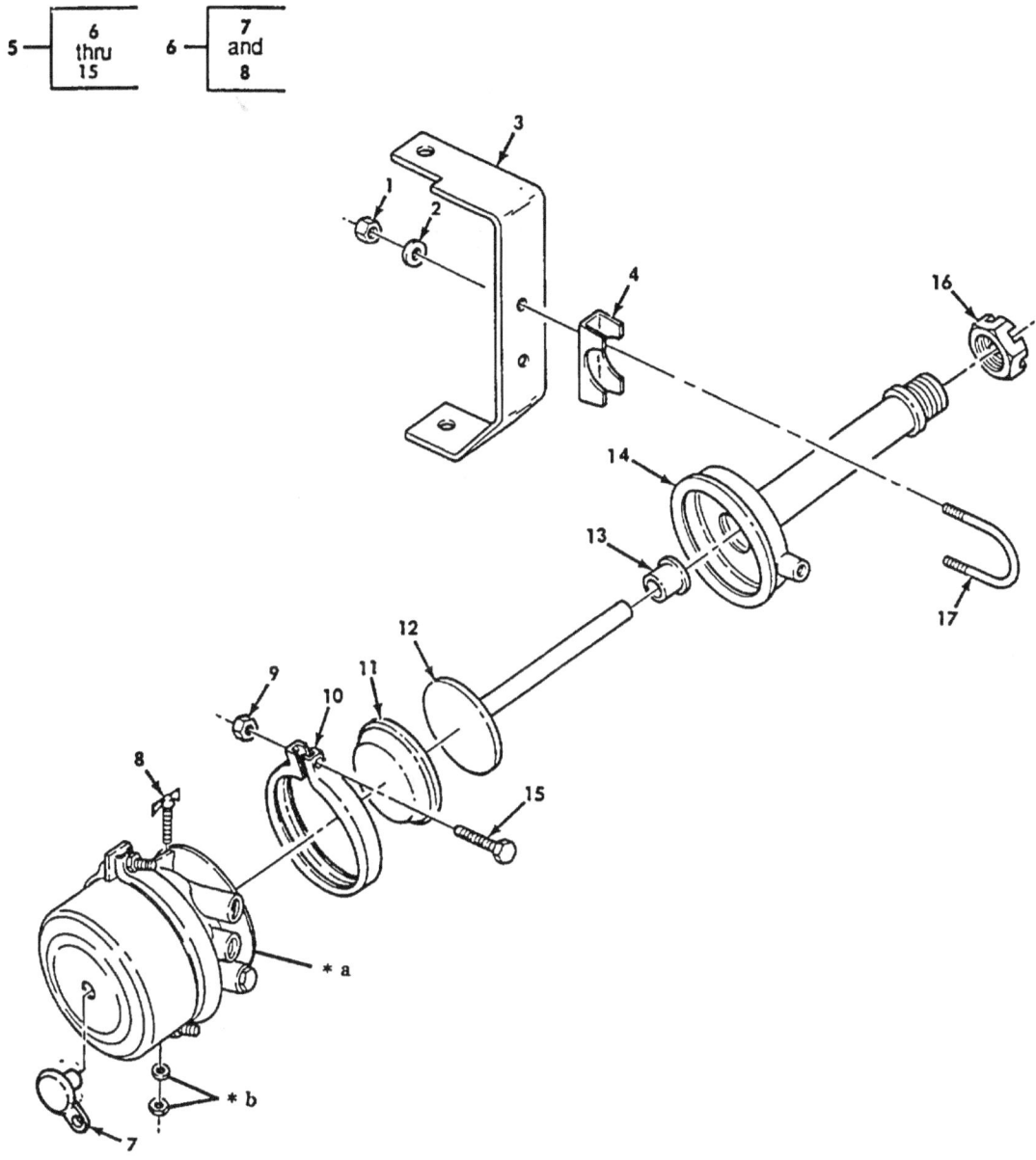

* a PART OF ITEM 6
* b PART OF ITEM 8

Figure 251. Rear Axle Fail-safe Brake Chamber Assembly.

(1) ITEM NO	(2) SMR CODE	(3) NSN	(4) CAGEC	(5) PART NUMBER	(6) DESCRIPTION AND USABLE ON CODES (UOC)	(7) QTY
					GROUP 1208 AIR BRAKES	
					FIG. 251 REAR AXLE FAIL-SAFE BRAKE CHAMBER ASSEMBLY	
1	PAOZZ	5310011260566	78500	NL-25-1-C	NUT	8
2	PAOZZ	5310012009879	78500	1229-E-1669-C	WASHER, FLAT	8
3	PAOZZ	5340012908882	78500	3299-F-5362	BRACKET, MOUNTING LEFT FRONT-.................... REAR AXLE AND RIGHT REAR-REAR AXLE	2
3	PAOZZ	5342012974454	78500	3299-G-5363	BRACKET, MOUNTING RIGHT FRONT- .... REAR. AXLE AND LEFT REAR-REAR AXLE	2
4	PAOZZ	5340012895030	78500	2255-H-86	CLAMP, U-BOLT	4
5	PAOFF	2530011256076	78500	X76-3276-L-12	BRAKE CHAMBER, FAIL REAR	4
6	XAFZZ		78500	A-3280-E-6817	.HOUSING ASSEMBL, BRA BRAKE CHAMBER.	1
7	PAOZZ	2530010846975	50153	11M012	.PLUG, CHAMBER TOP FAIL SAFE HOUSING........	1
8	PAOZZ	2530010953561	50153	11M011	..STUD ASSEMBLY, RELEA HOUSING	1
9	PAFZZ	5310001771258	78500	N-35-P	..NUT, PLAIN, HEXAGON..................................	1
10	PAFZZ	5340000157560	78500	2797-V-100	.CLAMP, SYNCHRO	1
11	PAFZZ	2530004302392	14892	234226	.DIAPHRAGM, CHAMBER, B HOUSING	1
12	PAFZZ	2530010642630	78500	A45-1779-V-230	.ROD, CHAMBER, AIR BRA HOUSING	1
13	PAFZZ	2530000870163	78500	1779-Q-433	.GUIDE, BRAKE ASSEMBL PUSH ROD	1
14	XAFZZ		78500	A4-3280-F-5232	.HOUSING, NON-PRESS BRAKE CHAMBER...........	1
15	PAFZZ	5306004987209	79780	15X-725	.BOLT, SPECIAL RIM CLENCHING CLAMP	1
16	PAOZZ	5310001232572	78500	1727-N-40	NUT, PLAIN, ROUND	1
17	PAOZZ	5306011966632	78500	2297-Y-4341	BOLT, U..	4

END OF FIGURE

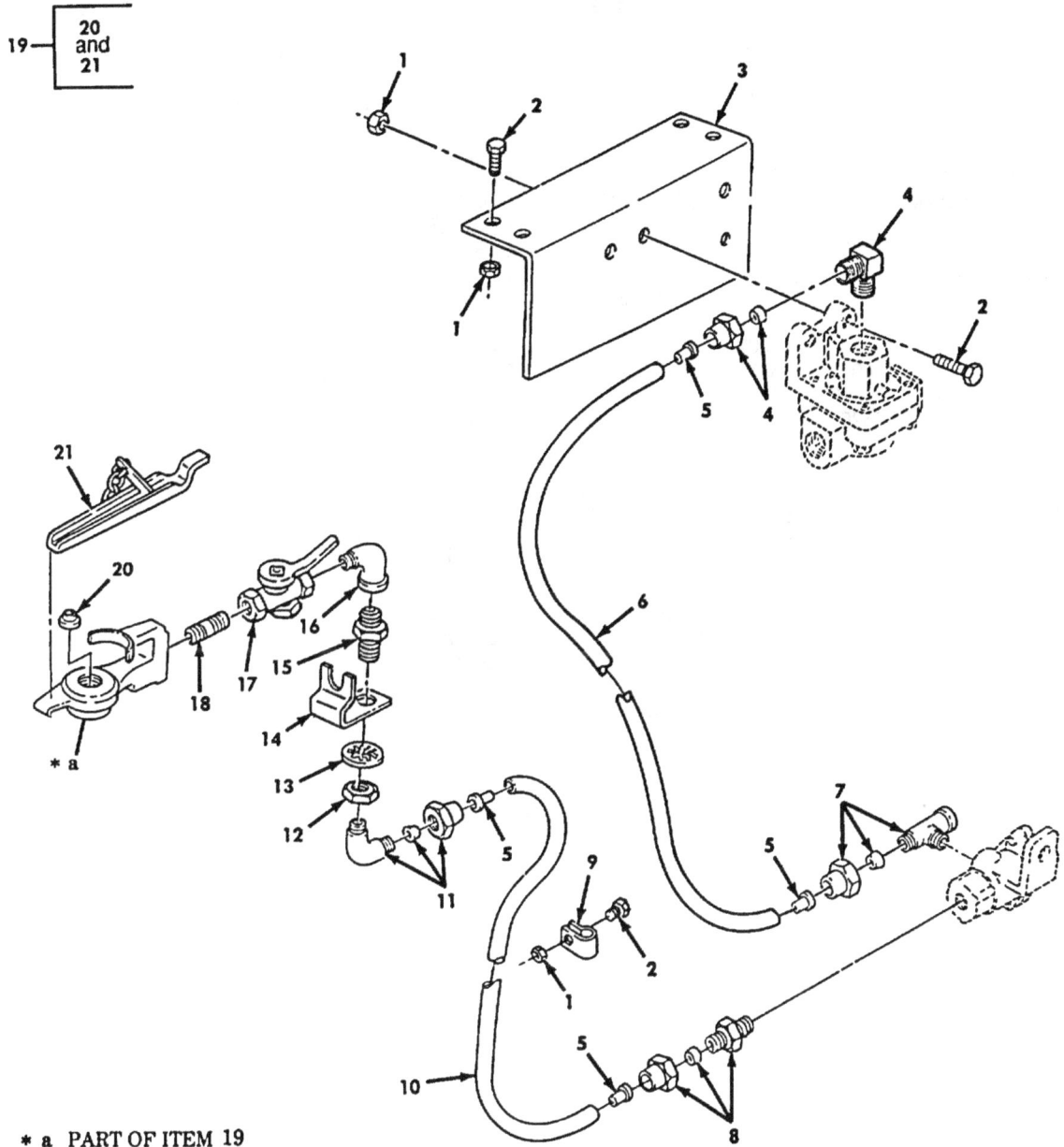

* a PART OF ITEM 19

Figure 252. Front Service Half Coupling, Air Brake Valve, and Related Parts.

(1) ITEM NO	(2) SMR CODE	(3) NSN	(4) CAGEC	(5) PART NUMBER	(6) DESCRIPTION AND USABLE ON CODES (UOC)	(7) QTY
					GROUP 1208 AIR BRAKES	
					FIG. 252 FRONT SERVICE HALF COUPLING, AIR BRAKE VALVE, AND RELATED PARTS	
1	PFOZZ	5310002416658	96906	MS51943-34	NUT, SELF-LOCKING, HE	6
2	PAOZZ	5306002259088	96906	MS90726-33	BOLT, MACHINE	6
3	PAOZZ		19207	12255970-1	BRACKET	1
4	PAOZZ	4730002890051	81343	8-6 120202BA	ELBOW, PIPE TO TUBE	1
5	PAOZZ	4730010326038	19207	CPR102321-4	INSERT, TUBE FITTING	4
6	MOOZZ		19207	CPR104420-3-25	HOSE, NONMETALLIC MAKE FROM HOSE, P/ .... N CPR104420-3	1
6	MOOZZ		19207	CPR104420-3-23	HOSE, NONMETALLIC	1
7	PAOZZ	4730002788902	96906	MS39189-2	TEE, PIPE TO TUBE	1
8	XDOZZ	4730001423076	81343	8-6 120102BA	ADAPTER, STRAIGHT, PI................................	1
9	PAOZZ	5340007255267	96906	MS21333-115	CLAMP, LOOP	2
10	MOOZZ		19207	CPR104420-3-45	HOSE, NONMETALLIC MAKE FROM HOSE P/ .... N CPR104420-3	1
10	MOOZZ		19207	CPR104420-3-46	HOSE MAKE FROM HOSE P/N C608	1
11	PAOZZ	4730004097854	81343	8-4 120202BA	ELBOW, PIPE TO TUBE	1
12	PAOZZ	5310008913428	96906	MS35691-77	NUT, PLAIN, HEXAGON	1
13	PAOZZ	5310005826714	96906	MS35333-49	WASHER, LOCK	1
14	PAOZZ	5340011079929	19207	12256265	BRACKET, ANGLE	1
15	PAOZZ	4730004199424	19207	8376311	REDUCER, PIPE	1
16	PAOZZ	4730002775555	96906	MS51952-4	ELBOW, PIPE	1
17	PAOZZ		19207	CPR109458	COCK, CUT OFF	1
18	PAOZZ		96906	MS51953-75	NIPPLE	1
19	PAOZZ	4730005950083	58536	A52484-1	COUPLING HALF, QUICK	1
20	PAOZZ	5330000902128	06853	213630	.PACKING, PREFORMED	1
21	PAOZZ	2530007409445	19207	7409445	.DUMMY COUPLING, AUTO	1

END OF FIGURE

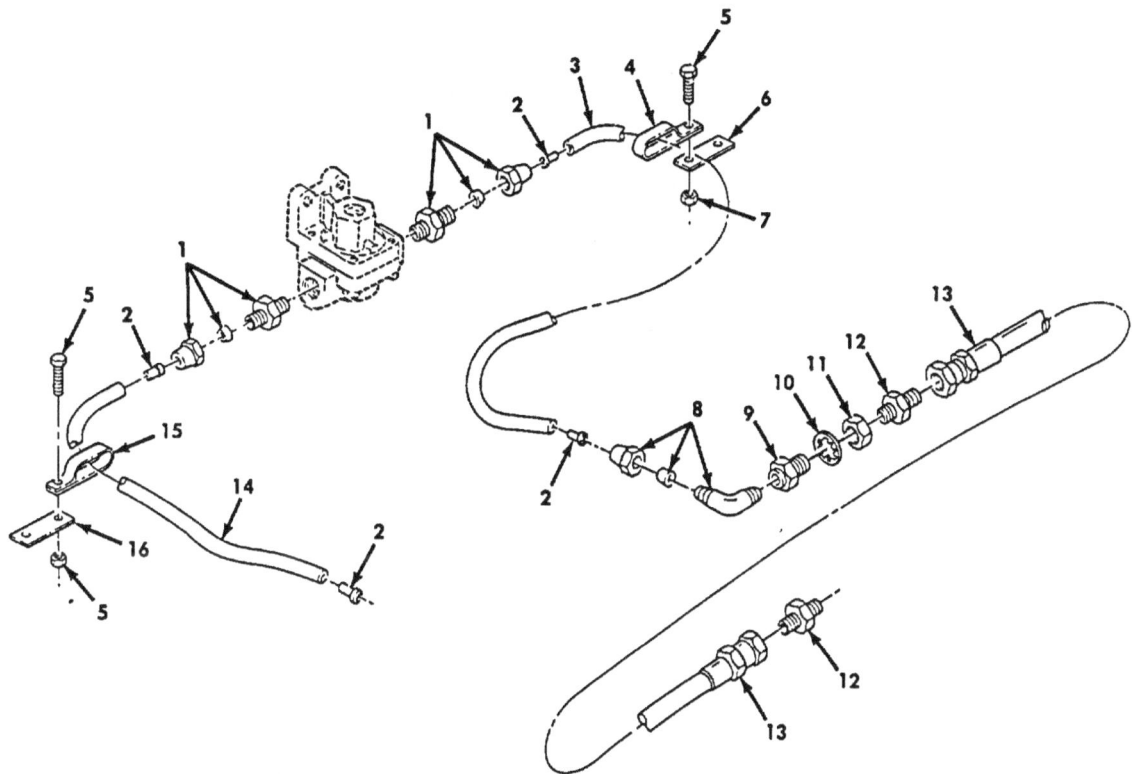

Figure 253. Air Control Lines, Front Service Check Valve to Brake Chambers.

(1) ITEM NO	(2) SMR CODE	(3) NSN	(4) CAGEC	(5) PART NUMBER	(6) DESCRIPTION AND USABLE ON CODES (UOC)	(7) QTY
					GROUP 1208 AIR BRAKES	
					FIG. 253 AIR CONTROL LINES, FRONT SERVICE CHECK VALVE TO BRAKE CHAMBERS	
1	PAOZZ		81343	8-6 100102BA	ADAPTER	2
2	PAOZZ	4730010326038	19207	CPR102321-4	INSERT, TUBE FITTING..................................	4
3	MOOZZ		19207	CPR104420-3-37	HOSE NONMETALLIC	1
4	PAOZZ	5340009891771	96906	MS21333-123	CLAMP, LOOP	1
5	PAOZZ	5305002692803	96906	MS90726-60	SCREW, CAP, HEXAGON H.............................	4
6	PAOZZ	5340011979300	19207	12302694	PLATE, MENDING..	2
7	PAOZZ	5310008140672	96906	MS51943-36	NUT, SELF-LOCKING, HE...............................	4
8	PAOZZ	4730002890051	81343	8-6 120202BA	ELBOW, PIPE TO TUBE	2
9	PAOZZ	4730004213924	19207	7339982	ADAPTER, STRAIGHT, PI FRONT LEFT AND.......... RIGHT HAND BRAKE CHAMBER ELBOW	2
10	PAOZZ	5310005826714	96906	MS35333-49	WASHER, LOCK	2
11	PAOZZ	5310002410157	19207	5331179	NUT, PLAIN, HEXAGON..................................	2
12	PAOZZ	4730008139611	96906	MS51500-B8	ADAPTER, STRAIGHT, PI FRONT LEFT AND.......... RIGHT HAND BRAKE CHAMBER HOSE ASSEMBLY	4
13	PAOZZ	4720011450371	19207	12256270-1	HOSE ASSEMBLY, NONME	2
14	MOOZZ		19207	CPR104420-3-38	HOSE NONMETALLIC	1
15	PAOZZ	5340011179876	19207	7397785	CLAMP, LOOP..	1

END OF FIGURE

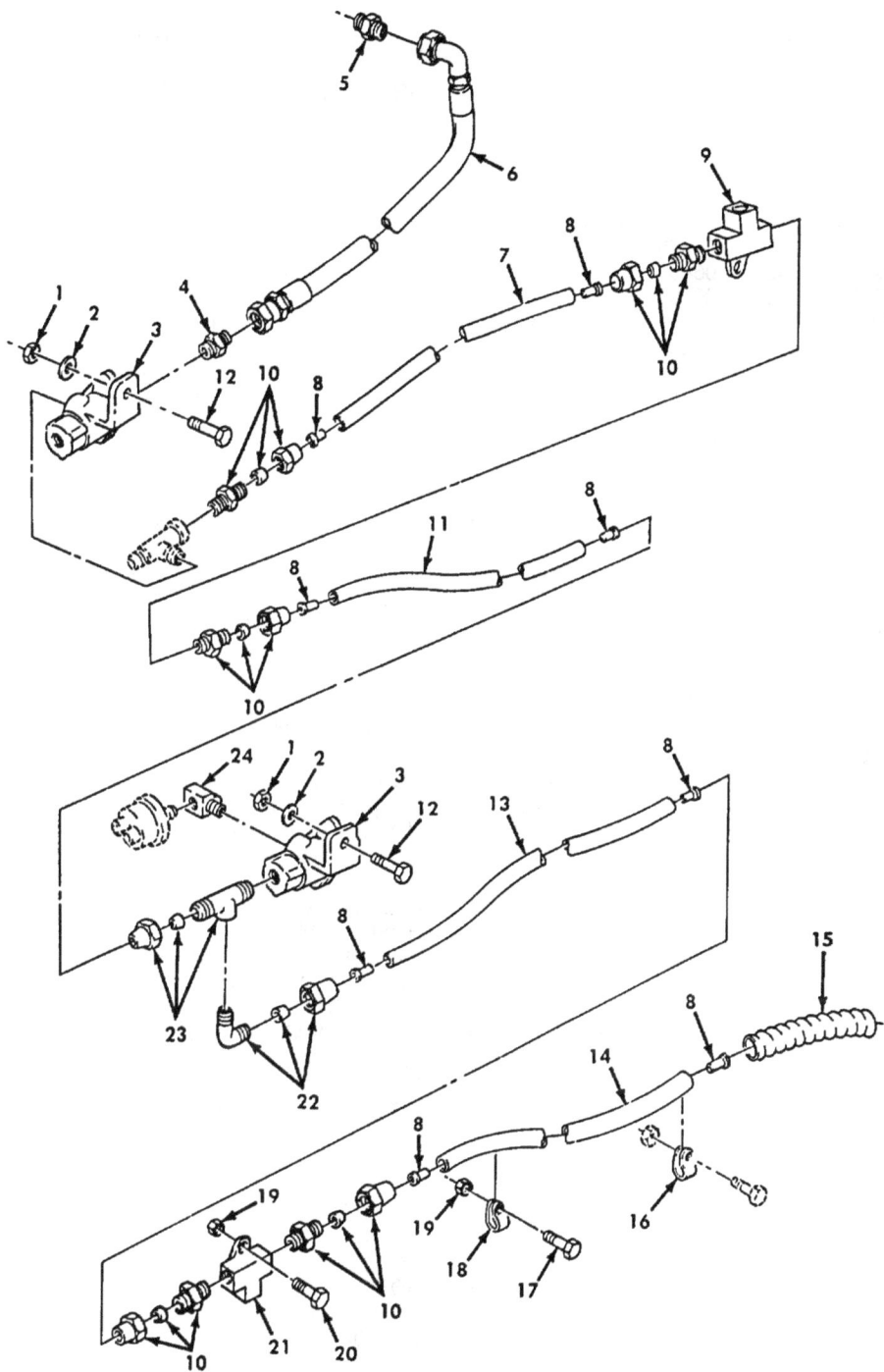

Figure 254. Double-check Valves and Related Service Air Lines.

(1) ITEM NO	(2) SMR CODE	(3) NSN	(4) CAGEC	(5) PART NUMBER	(6) DESCRIPTION AND USABLE ON CODES (UOC)	(7) QTY
					GROUP 1208 AIR BRAKES	
					FIG. 254 DOUBLE-CHECK VALVES AND RELATED SERVICE AIR LINES	
1	PAOZZ	5310002416658	96906	MS51943-34	NUT, SELF-LOCKING, HE......................................	2
2	PAOZZ	5310000814219	96906	MS27183-12	WASHER, FLAT	2
3	PAOZZ	4820007287467	06853	278614	VALVE, LINEAR, DIRECT	1
4	PAOZZ	4730008139611	96906	MS51500A8	ADAPTER, STRAIGHT, PI	1
5	PAOZZ	4730008099427	96906	MS51500A8-8	ADAPTER, STRAIGHT PIPE	1
6	PAOZZ	4720011122329	19207	12256269-3	ROSE ASSEMBLY, NONME .... UOC:DAA, DAB, DAC, DAD, DAE, DAF, DAG, DAH, DAJ, DAK, DAW, DAX, V12, V13, V14, V15, V16, V21, V22, V24, V25, ZAA, ZAB, ZAC, ZAD, ZAE, ZAF, ZAG, ZAH, ZAJ, ZAK	1
6	PAOZZ	4720012902510	19207	12302893	HOSE ASSEMBLY, NONME .... UOC:DAL, V18, ZAL	1
7	MOOZZ		19207	CPR104420-2-80	HOSE, NONMETALLIC MAKE FROM HOSE, ... P/N CPR104420-2, 80 INCHES LONG .... UOC:DAA, DAB, DAW, DAX, V12, V13, V14, V15, ZAA, ZAB	1
8	PAOZZ	4730010798821	19207	CPR102321-1	INSERT, TUBE FITTING	8
9	PAOZZ	4730011085103	19207	10937774-1	TEE, PIPE	1
10	PAOZZ	4730000691186	81343	6-4 120102BA	ADAPTER, STRAIGHT, PI	5
11	MOOZZ		19207	CPR104420-2-43	HOSE, NONMETALLIC MAKE FROM HOSE, ... P/N CPR104420-2, 43 INCHES LONG	1
12	PAOZZ	5305002259093	96906	MS90726-38	SCREW, CAP, HEXAGON H	2
13	MOOZZ		19207	CPR104420-2-54	HOSE, NONMETALLIC MAKE FROM HOSE, ... P/N CPR104420-2, 54 INCHES LONG .... UOC:DAG, DAH, V21, V22, ZAG, ZAH	1
13	MOOZZ		19207	CPR104420-2-77	HOSE, NONMETALLIC MAKE FROM HOSE, ... P/N CPR104420-2, 77 INCHES LONG .... UOC:DAL, V18, ZAL	1
13	MOOZZ		19207	CPR104420-2-83	HOSE, NONMETALLIC MAKE FROM HOSE, ... P/N CPR104420-2, 83 INCHES LONG .... UOC:DAA, DAB, DAW, DAX, V12, V13, V14, V15, ZAA, ZAB	1
13	MOOZZ		19207	CPR104420-2-70	HOSE, NONMETALLIC MAKE FROM HOSE, ... P/N CPR104420-2, 70 INCHES LONG .... UOC:DAE, DAF, V19, V20	1
13	MOOZZ		19207	CPR104420-2-113	HOSE, NONMETALLIC MAKE FROM HOSE, ... P/N CPR104420-2, 113 INCHES LONG .... UOC:DAC, DAD, DAJ, DAK, V16, V17, V24, V25, V39, ZAC, ZAD, ZAJ, ZAK	1
14	MOOZZ		19207	CPR104420-2-28	HOSE, NONMETALLIC MAKE FROM HOSE, ... P/N CPR104420-2, 28 INCHES LONG .... UOC:DAL, V18, ZAL	1
14	MOOZZ		19207	CPR104420-2-13	HOSE, NONMETALLIC MAKE FROM HOSE, ... P/N CPR104420-2, 13 INCHES LONG .... UOC:DAE, DAF, V19, V20, ZAE, ZAF	1

(1) ITEM NO	(2) SMR CODE	(3) NSN	(4) CAGEC	(5) PART NUMBER	(6) DESCRIPTION AND USABLE ON CODES (UOC)	(7) QTY
14	MOOZZ		19207	CPR104420-2-41	HOSE, NONMETALLIC MAKE FROM HOSE, ... P/N CPR104420-2, 41 INCHES LONG .... UOC:DAB, DAE, DAF, DAW, DAX, V12, V13, V14, V15, V19, V20, ZAA, ZAB, ZAE, ZAF	1
14	MOOZZ		19207	CPR104420-2-94	HOSE, NONMETALLIC MAKE FROM HOSE, ... P/N CPR104420-2, 94 INCHES LONG .... UOC:DAC, DAD, V16, V17, ZAC, ZAD	1
14	MOOZZ		19207	CPR104420-2-73	HOSE, NONMETALLIC MAKE FROM HOSE, ... P/N CPR104420-2, 73 INCHES LONG UOC:DAJ, DAK, V24, V25, ZAJ, ZAK	1
14	MOOZZ		19207	CPR104420-2-80	HOSE, NONMETALLIC MAKE FROM HOSE, ... CPR104420-2, 80 INCHES LONG UOC:V39	1
15	MOOZZ		19207	12302690-6.	TUBING, PLASTIC, SPIR MAKE FROM .... TUBING, P/N 12302690, LENGTH AS REQUIRED..	1
16	PAOZZ	5340002221653	19207	7415752	CLAMP, LOOP... UOC:DAL, V18, ZAL	2
16	PAOZZ	5340002221653	19207	7415752	CLAMP, LOOP .... UOC:DAA, DAB, DAC, DAD, DAE, DAF, DAJ, DAK, DAW, DAX, V12, V13, V14, V15, V16, V17, V19, V20, V22, V24, V25, V39, ZAA, ZAB, ZAC, ZAD, ZAE, ZA, , ZAF, ZAJ, ZAK	1
17	PAOZZ	5305002692804	96906	MS90726-61	SCREW, CAP, HEXAGON H .... UOC:DAA, DAB, DAW, DAX, V12, V13, V14, V15, ZAA, ZAB	1
17	PAOZZ	5305002692804	96906	MS90726-61	SCREW, CAP, HEXAGON H .... UOC:DAC, DAD, DAJ, DAK, V16, V17, V24, V25, V39, ZAC, ZAD, ZAJ, ZAK	2
18	PAOZZ	5340009546014	96906	MS21333-121	CLAMP, LOOP .... UOC:DAA, DAB, DAW, DAX, V12, V13, V14, V15,	1
18	PAOZZ	5340009546014	96906	MS21333-121	CLAMP, LOOP .... UOC:DAC, DAD, DAJ, DAK, V16, V17, V24, V25, V39, ZAC, ZAD, ZAJ, ZAK	2
19	PAOZZ	5310008140672	96906	MS51943-36	NUT, SELF-LOCKING, HE .... UOC:DAA, DAB, DAW, DAX, V12, V13, V14, V15, ZAA, ZAB	2
19	PAOZZ	5310008140672	96906	MS51943-36	NUT, SELF-LOCKING, HE .... UOC:DAC, DAD, DAJ, DAK, V16, V17, V24, V25, V39, ZAC, ZAD, ZAJ, ZAK	3
20	PAOZZ	5305002693238	80204	B1821BH038F125N	SCREW, CAP, HEXAGON H	1
21	PAOZZ	4730011079690	19207	10937774	TEE, PIPE	1
22	PAOZZ	4730000691187	81343	6-4 100202BA	ELBOW, PIPE TO TUBE	1
23	PAOZZ	4730011950095	19207	12302759	TEE, PIPE TO TUBE.......................................	1
24	PAOZZ	4730002784822	72582	444042	ELBOW, PIPE... UOC:DAA, DAB, DAC, DAD, DAE, DAF, DAJ, DAK, DAL, DAW, DAX, V12, V13, V14, V15, V16, V17, V18, V19, V20, V22, V24, V25, V39, ZAA, ZAB, ZAC, ZAD, ZAE, ZAF, ZAJ, ZAK, ZAL	1

END OF FIGURE

13 — 14

* a   PART OF ITEM 13

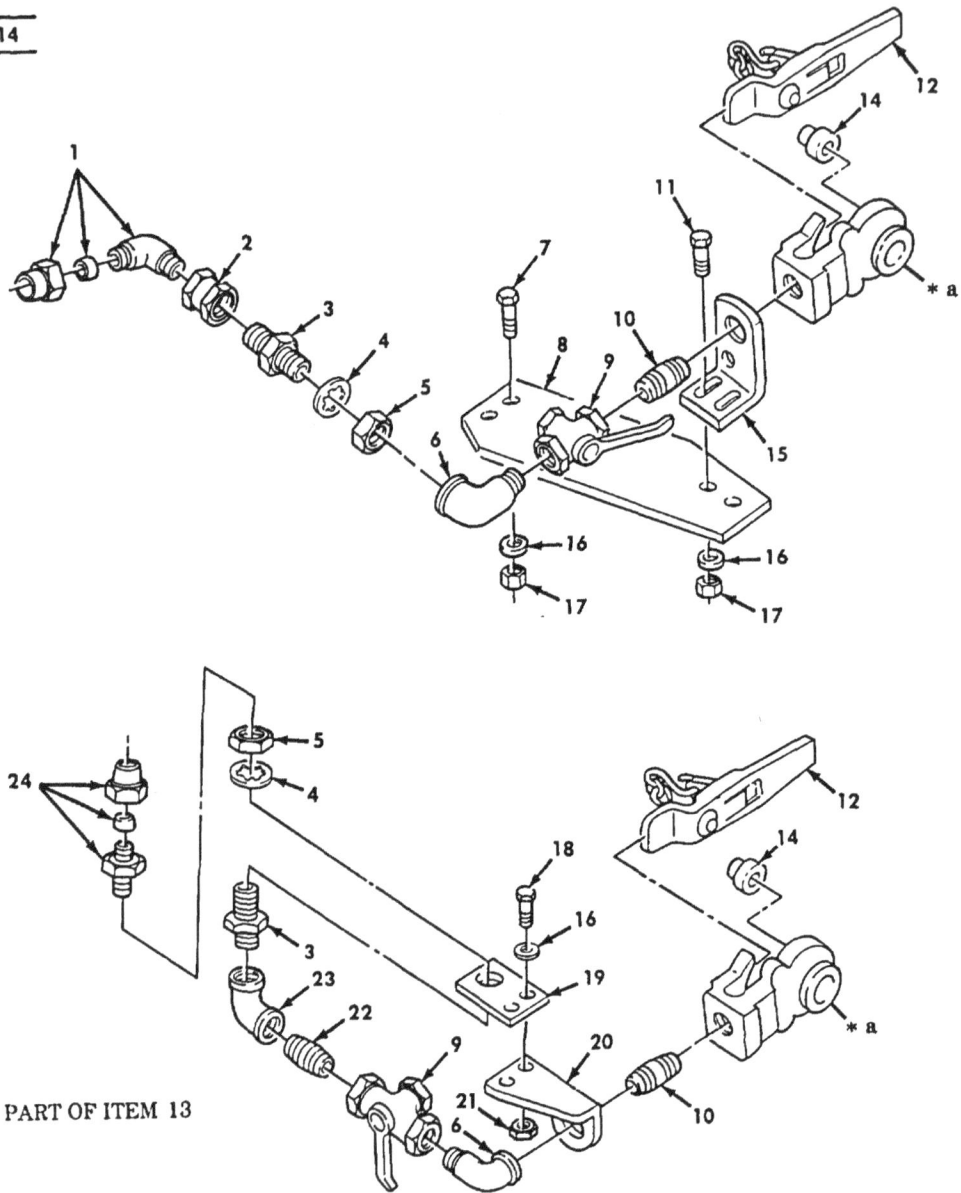

Figure 255. Rear Service Half Coupling and Mounting Hardware.

(1) ITEM NO	(2) SMR CODE	(3) NSN	(4) CAGEC	(5) PART NUMBER	(6) DESCRIPTION AND USABLE ON CODES (UOC)	(7) QTY
					FIG. 1208 AIR BRAKES	
					FIG. 255 REAR SERVICE HALF COUPLING AND MOUNTING HARDWARE	
1	PAOZZ	4730001439282	81343	6-8 120202BA	ELBOW .... UOC:DAA, DAB, DAW, DAX, V12, V13, V14, V15, ZAA, ZAB	1
2	PAOZZ	4730005417790	21450	144077	COUPLING, PIPE .... UOC:DAA, DAB, DAW, DAX, V12, V13, V14, V15, ZAA, ZAB	1
3	PAOZZ	4730004199424	19207	8376311	REDUCER, PIPE...	1
4	PAOZZ	5310005826714	96906	MS35333-49	WASHER, LOCK	1
5	PAOZZ	5310000219760	21450	219760	NUT, PLAIN, HEXAGON	1
6	PAOZZ	4730002775555	96906	MS51952-4	ELBOW, PIPE... UOC:DAA, DAB	1
7	PAOZZ	5306000680514	80204	B1821BH025F088N	BOLT, MACHINE . UOC:DAA, DAB, DAW, DAX, V12, V13, V14, V15, ZAA, ZAB	2
8	PFOZZ	2530004785865	19207	11593371	PLATE, MOUNTING.. UOC:DAA, DAB, DAW, DAX, V12, V13, V14, V15, ZAA, ZAB	1
9	PAOOZ	4820004205499	06853	285172	VALVE, BALL .... UOC:DAA, DAB, DAC, DAD, DAE, DAF, DAG, DAH, DAJ, DAK, DAL, DAW, DAX, V12, V13, V14, V15, V16, ZAC, ZAD, ZAE, ZAF, ZAG, ZAH, ZAJ, ZAK, ZAL	1
10	PAOZZ	4730001961504	24617	192075	NIPPLE, PIPE	1
11	PAOZZ	5305002693239	80204	B1821BH038F138N	SCREW, CAP, HEXAGON H .... UOC:DAA, DAB, DAW, DAX, V12, V13, V14, V15, ZAA, ZAB	2
12	PAOZZ	2530002703878	19207	7014965	DUMMY COUPLING, AUTO	1
13	PAOOZ	4730005950083	58536	A52484-1	COUPLING HALF, QUICK	1
14	PAOZZ	5330000902128	06853	213630	.PACKING, PREFORMED	1
15	PFOZZ	5340004199487	19207	10883438	BRACKET, ANGLE .... UOC:DAA, DAB, DAW, DAX, V12, V13, V14, V15, ZAA, ZAB	1
16	PAOZZ	5310008094058	96906	MS27183-10	WASHER, FLAT .... UOC:DAA, DAB, DAW, DAX, V12, V13, V14, V15, ZAA, ZAB	4
16	PAOZZ	5310008094058	96906	MS27183-10	WASHER, FLAT .... UOC:DAC, DAD, DAE, DAF, DAG, DAH, DAJ, DAK, DAL, V16, V17, V18, V19, V20, V21, V22, V24, V25, V39, ZAC, ZAD, ZAE, ZAF, ZAG, ZAH, ZAJ, ZAK, ZAL	2
17	PAOZZ	5310009359022	96906	MS51943-32	NUT, SELF-LOCKING, HE .... UOC:DAA, DAB, DAW, DAX, V12, V13, V14, V15, ZAA, ZAB	4
18	PAOZZ	5305002678954	80204	B1821BH025F125N	SCREW, CAP, HEXAGON H .... UOC:DAC, DAD, DAE, DAF, DAG, DAH, DAJ, DAK, DAL, V16, V17, V18, V19, V20, V21, V22, V24, V25, V39, ZAC, ZAD, ZAE, ZAF, ZAG, ZAH, ZAJ, ZAK, ZAL	2

(1) ITEM NO	(2) SMR CODE	(3) NSN	(4) CAGEC	(5) PART NUMBER	(6) DESCRIPTION AND USABLE ON CODES (UOC)	(7) QTY
19	PFOZZ	5340001577938	19207	10883331	PLATE, MENDING .... UOC:DAC, DAD, DAE, DAF, DAG, DAH, DAJ, DAK, DAL, V16, V17, V18, V19, V20, V21, V22, V24, V25, V39, ZAC, ZAD, ZAE, ZAF, ZAG, ZAH, ZAJ, ZAK, ZAL	1
20	PFOZZ	5340001797123	19207	10883328	BRACKET, COUPLING AI .... UOC:DAC, DAD, DAE, DAF, DAG, DAH, DAJ, DAK, DAL, V16, V17, V18, V19, V20, V21, V22, V24, V25, V39, ZAC, ZAD, ZAE, ZAF, ZAG, ZAH, ZAJ, ZAK, ZAL	1
21	PAOZZ	5310009359022	96906	MS51943-32	NUT, SELF-LOCKING, HE .... UOC:DAC, DAD, DAE, DAF, DAG, DAH, DAJ, DAK, DAL, V16, V17, V18, V19, V20, V21, V22, V24, V25, V39, ZAC, ZAD, ZAE, ZAF, ZAG, ZAH, ZAJ, ZAK, ZAL	2
22	PAOZZ	4730001961993	96906	MS51846-67	NIPPLE, PIPE .... UOC:DAL, V18, ZAL	1
23	PAOZZ	4730002493885	96906	MS51845-4	ELBOW, PIPE... UOC:DAC, DAD, DAE, DAF, DAG, DAH, DAJ, DAK, DAL, V16, V17, V18, V19, V20, V21, V22, V24, V25, V39, ZAC, ZAD, ZAE, ZAF, ZAG, ZAH, ZAJ, ZAK, ZAL	1
24	PAOZZ	4730000691186	81343	6-4 120102BA	ADAPTER, STRAIGHT, PI .... UOC:DAC, DAD, DAE, DAF, DAG, DAH, DAJ, DAK, DAL, V16, V17, V18, V19, V20, V21, V22, V24, V25, V39, ZAC, ZAD, ZAE, ZAF, ZAG, ZAH, ZAJ, ZAK, ZAL	1

END OF FIGURE

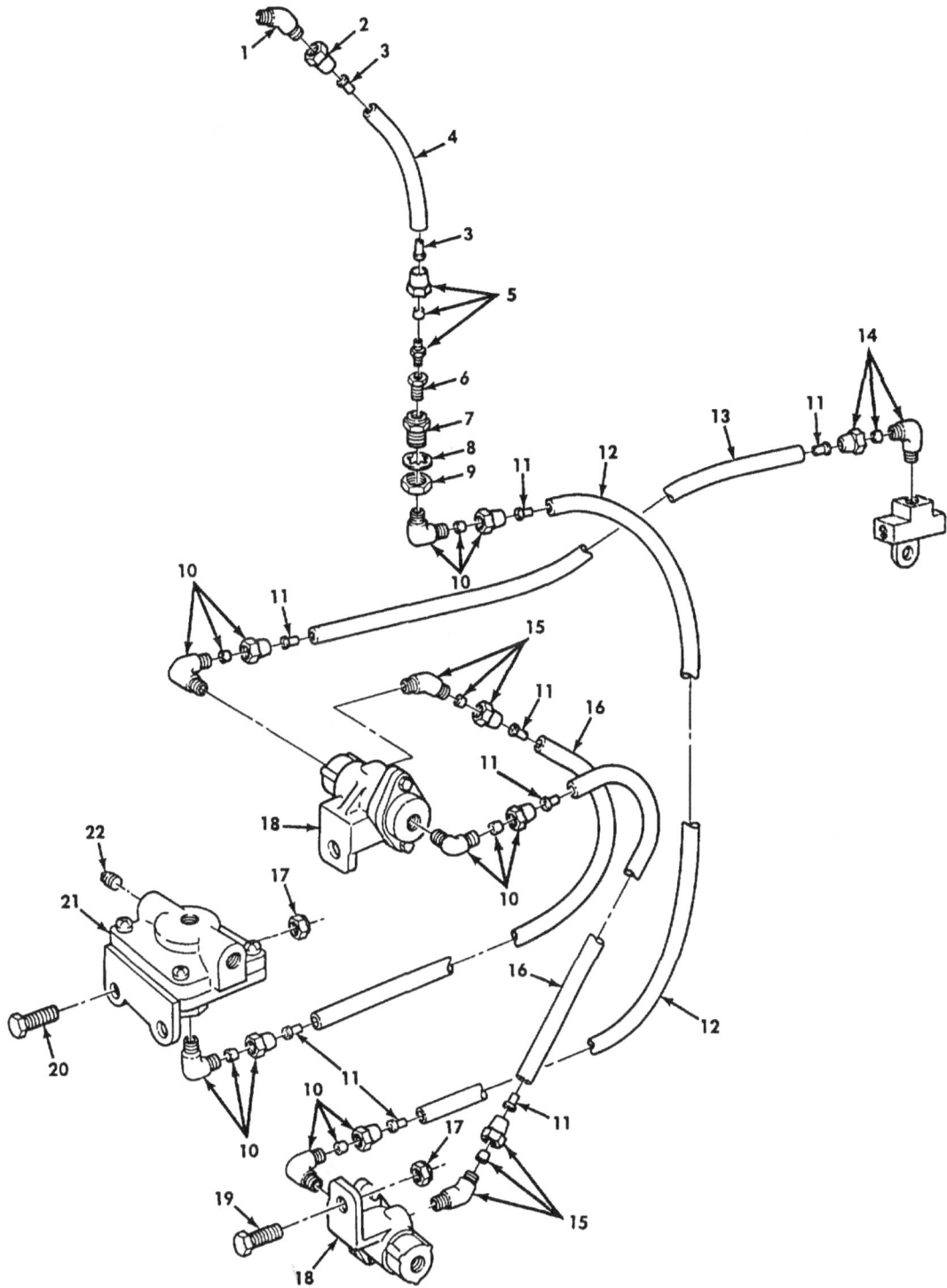

Figure 256. Quick-release Valve, Air Brake Valves, and Related Lines.

(1) ITEM NO	(2) SMR CODE	(3) NSN	(4) CAGEC	(5) PART NUMBER	(6) DESCRIPTION AND USABLE ON CODES (UOC)	(7) QTY
					GROUP 1208 AIR BRAKES	
					FIG. 256 QUICK-RELEASE VALVE, AIR BRAKE VALVES, AND RELATED LINES	
1	PAOZZ	4730012023351	96906	7336402-1	ELBOW	1
2	PAOZZ	4730000116452	96906	MS39166-3	NUT, TUBE COUPLING	1
3	PAOZZ	4730001324588	79146	HO-159-4	INSERT, TUBE FITTING	2
4	MOOZZ		19207	CPR104420-2-6	HOSE, NONMETALLIC MAKE FROM HOSE, P/ .... N CPR104420-1, 6 INCHES LONG........................	1
5	PAOZZ	4730002778750	81343	4-2 120102BA	ADAPTER, STRAIGHT, PI	1
6	PAOZZ	4730001805038	24617	444026	BUSHING, PIPE	1
7	PAOZZ	4730004213924	19207	7339982	ADAPTER, STRAIGHT, PI	1
8	PAOZZ	5310005826714	96906	MS35333-49	WASHER, LOCK	1
9	PAOZZ	5310002410157	19207	5331179	NUT, PLAIN, HEXAGON	1
10	PAOZZ	4730002890051	81343	8-6 120202BA	ELBOW, PIPE TO TUBE	2
11	PAOZZ	4730010326038	19207	CPR102321-4	INSERT, TUBE FITTING	8
12	MOOZZ		19207	CPR104420-3-83	HOSE, NONMETALLIC	1
13	MOOZZ		19207	CPR104420-3-74	HOSE, NONMETALLIC MAKE FROM HOSE, P/ .... N CPR104420-3, FROM NSN 4720-01-003- 6706, APPROX 74 INCHES LONG .... UOC:DAA, DAB, DAW, DAX, V12, V13, V14, V15, ZAA, ZAB	1
13	MOOZZ		19207	CPR104420-3-69	HOSE, NONMETALLIC .... UOC:DAC, DAD, DAE, DAF, DAG, DAH, DAJ, DAK, DAL, V16, V17, V18, V19, V20, V21, V22, V24, V25, V39, ZAC, ZAD, ZAE, ZAF, ZAG, ZAH, ZAJ, ZAK, ZAL	1
14	PAOZZ	4730004097854	81343	8-4 120202BA	ELBOW, PIPE TO TUBE	1
15	PAOZZ	4730011171614	81343	8-6 120302BA	ELBOW, PIPE TO TUBE	2
16	MOOZZ		19207	CPR104420-3-9	HOSE, NONMETALLIC MAKE FROM HOSE P/ .... N CPR104420-3	2
17	PAOZZ	5310002416658	96906	MS51943-34	NUT, SELF-LOCKING, HE	4
18	PAOZZ	4820007287467	06853	278614	VALVE, LINEAR, DIRECT	2
19	PAOZZ	5305002259093	96906	MS90726-34	SCREW, CAP, HEXAGON H	2
20	PAOZZ	5306000501238	96906	MS90727-32	BOLT, MACHINE	2
21	PAOZZ	4820008572737	19207	11602155	VALVE, LINEAR, DIRECT	1
22	PAOZZ	4730012244152	81348	WW-P-471ACABCC	PLUG, PIPE	2

END OF FIGURE

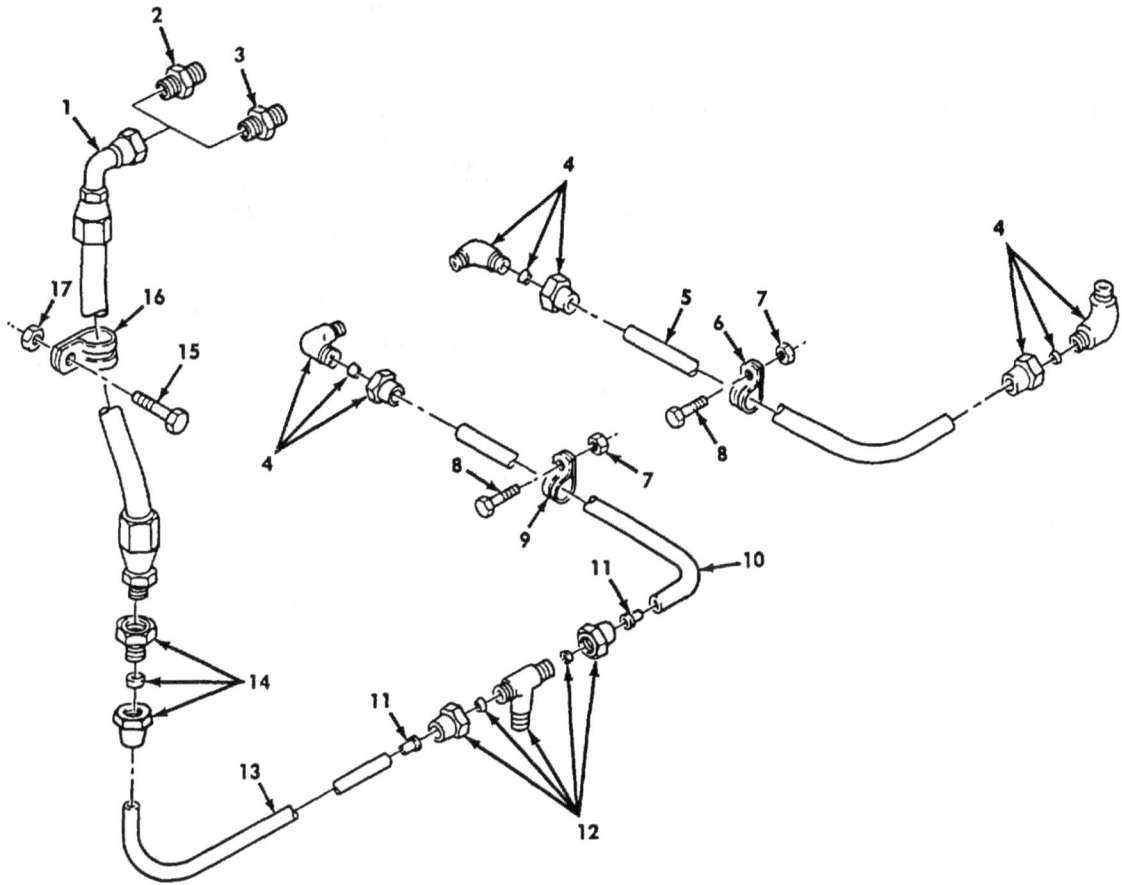

Figure 257. Treadle Valve to Secondary Relay Valve and Primary Relay Valve to Half Coupling Lines.

(1) ITEM NO	(2) SMR CODE	(3) NSN	(4) CAGEC	(5) PART NUMBER	(6) DESCRIPTION AND USABLE ON CODES (UOC)	(7) QTY
					GROUP 1208 AIR BRAKES	
					FIG. 257 TREADLE VALVE TO SECONDARY RELAY VALVE AND PRIMARY RELAY VALVE TO HALF COUPLING LINES	
1	PAOZZ	4720011122325	19207	12256263	HOSE ASSEMBLY, NONME	1
2	PAOZZ	4730008099427	96906	MS51500-AS-8	ADAPTER, STRAIGHT, PI .... UOC:DAA, DAB, DAC, DAD, DAE, DAF, DAG, DAH, DAJ, DAK, DAL, DAW, DAX, V12, V13, V14, V15, V16, V21, V22, V24, V25, ZAA, ZAB, ZAC, ZAD, ZAE, ZAF, ZAG, ZAH, ZAJ, ZAK	1
3	PAOZZ		81343	6-4 100202BA	ADAPTER, STRAIGHT PIPE	1
4	PAOZZ	4730000691187	81343	6-4 100202BA	ELBOW, PIPE TO TUBE	3
5	MOOZZ		19207	CPR104420-2-35	HOSE, NONMETALLIC MAKE FROM HOSE P/ .... N CPR104420-2, 35 INCHES LONG...................... UOC:DAA, DAB, DAW, DAX, V12, V13, V14, V15, ZAA, ZAB	1
5	MOOZZ		19207	CPR104420-2-37	HOSE, NONMETALLIC MAKE FROM HOSE P/ .... N CPR104420-2, 37 INCHES LONG...................... UOC:DAC, DAD, DAE, DAF, DAG, DAH, DAJ, DAK, DAL, V16, V17, V18, V19, V20, V21, V22, V24, V25, V39, ZAC, ZAD, ZAE, ZAF, ZAG, ZAH, ZAJ, ZAK, ZAL	1
6	PAOZZ	5340009546014	96906	MS21333-121	CLAMP, LOOP	1
7	PAOZZ	5310008140672	96906	MS51943-36	NUT, SELF-LOCKING, HE	2
8	PAOZZ	5305002692804	96906	MS90726-61	SCREW, CAP, HEXAGON H	2
9	PAOZZ		96906	MS21333-121	CLAMP, LOOP	1
10	MOOZZ		19207	CPR104420-2-50	HOSE, NONMETALLIC MAKE FROM HOSE P/ .... N CPR104420-2, 50 INCHES LONG...................... UOC:DAG, DAH, V21, V22, ZAG, ZAH	1
10	MOOZZ		19207	CPR104420-2-78	HOSE, NONMETALLIC CHECK VALVE TO.............. FRONT RELAY VALVE, MAKE FROM HOSE, P/ N CPR104420-2, 78 INCHES LONG...................... UOC:DAC, DAD, DAE, DAF, DAJ, DAK, DAL, V16, V17, V18, V19, V20, V24, V25, V39	1
11	PAOZZ	4730010798821	19207	CPR102321-1	INSERT, TUBE FITTING.....................................	6
12	PAOZZ	4730011950095	19207	12302759	TEE, PIPE TO TUBE	1
13	MOOZZ		19207	CPR104420-2-56	HOSE MAKE FROM HOSE CPR104420-2, ... 56 INCHES LONG UOC:DAA, DAB, DAW, DAX, V12, V13, V14, V15, ZAA, ZAB	1
13	MOOZZ		19207	CPR104420-2-66	HOSE, NONMETALLIC .... MAKE FROM HOSE, P/ N CPR104420-2, 66 INCHES LONG...................... UOC:DAC, DAD, DAE, DAF, DAJ, DAK, DAL, DAW, V16, V17, V18, V19, V20, V24, V25, V39	1
14	PAOZZ	4730002778770	81343	6-6 120103BA	ADAPTER, STRAIGHT, PI	1
15	PAOZZ	5305008573367	96906	MS9316-13	SCREW, MACHINE	1
16	PAOZZ	5340011059143	96906	MS21334-42	CLAMP, LOOP	1
17	PAOZZ	5310000617326	96906	MS21045-3	NUT, SELF-LOCKING, HE	1

END OF FIGURE

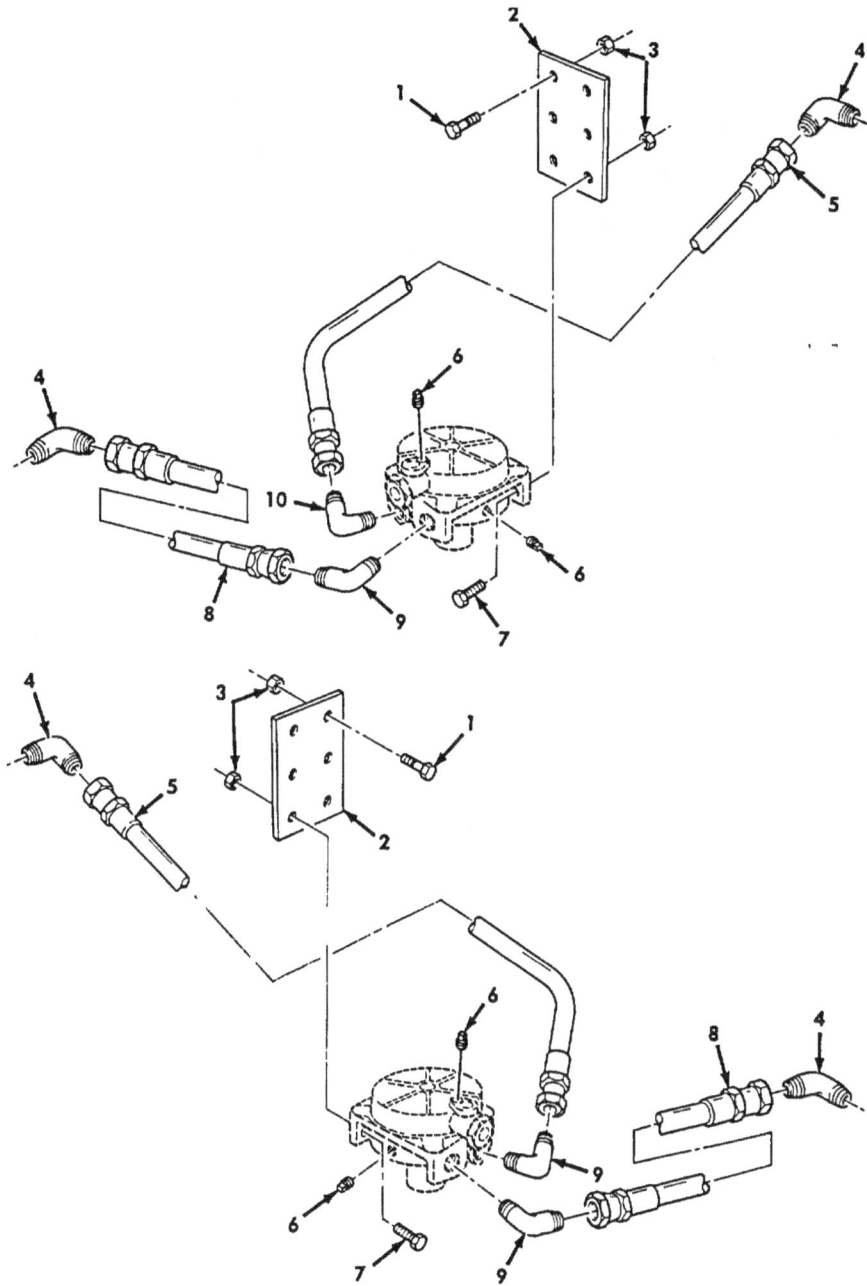

Figure 258. Primary and Secondary Relay Valve, Mounting Hardware, and Related Lines.

(1) ITEM NO	(2) SMR CODE	(3) NSN	(4) CAGEC	(5) PART NUMBER	(6) DESCRIPTION AND USABLE ON CODES (UOC)	(7) QTY
					GROUP 1208 AIR BRAKES	
					FIG. 258 PRIMARY AND SECONDARY RELAY VALVE, MOUNTING HARDWARE, AND RELATED LINES	
1	PAOZZ	5305002692804	96906	MS90726-61	SCREW, CAP, HEXAGON H	8
2	PFOZZ	5340011122164	19207	12256288	BRACKET, MOUNTING	2
3	PAOZZ	5310008140672	96906	MS51943-36	NUT, SELF-LOCKING, HE	12
4	PAOZZ	4730005551764	96906	MS51504A8	ELBOW, PIPE TO TUBE	4
5	PAOZZ	4720011885139	19207	12256271-6	HOSE ASSEMBLY, NONME	2
6	PAOZZ	4730003593872	81348	WW-P-471ACABCE	PLUG, PIPE	4
7	PAOZZ	5305002693239	80204	B1821BH038F138N	SCREW, CAP, HEXAGON H	4
8	PAOZZ	4720011122333	19207	12256271-1	HOSE ASSEMBLY, NONME	2
9	PAOZZ	4730000523420	96906	MS51508A8-8	ELBOW, PIPE TO TUBE	3
10	PAOZZ	4730002736686	81343	8-8 070202CA(CAD)	ELBOW, PIPE TO TUBE	1

END OF FIGURE

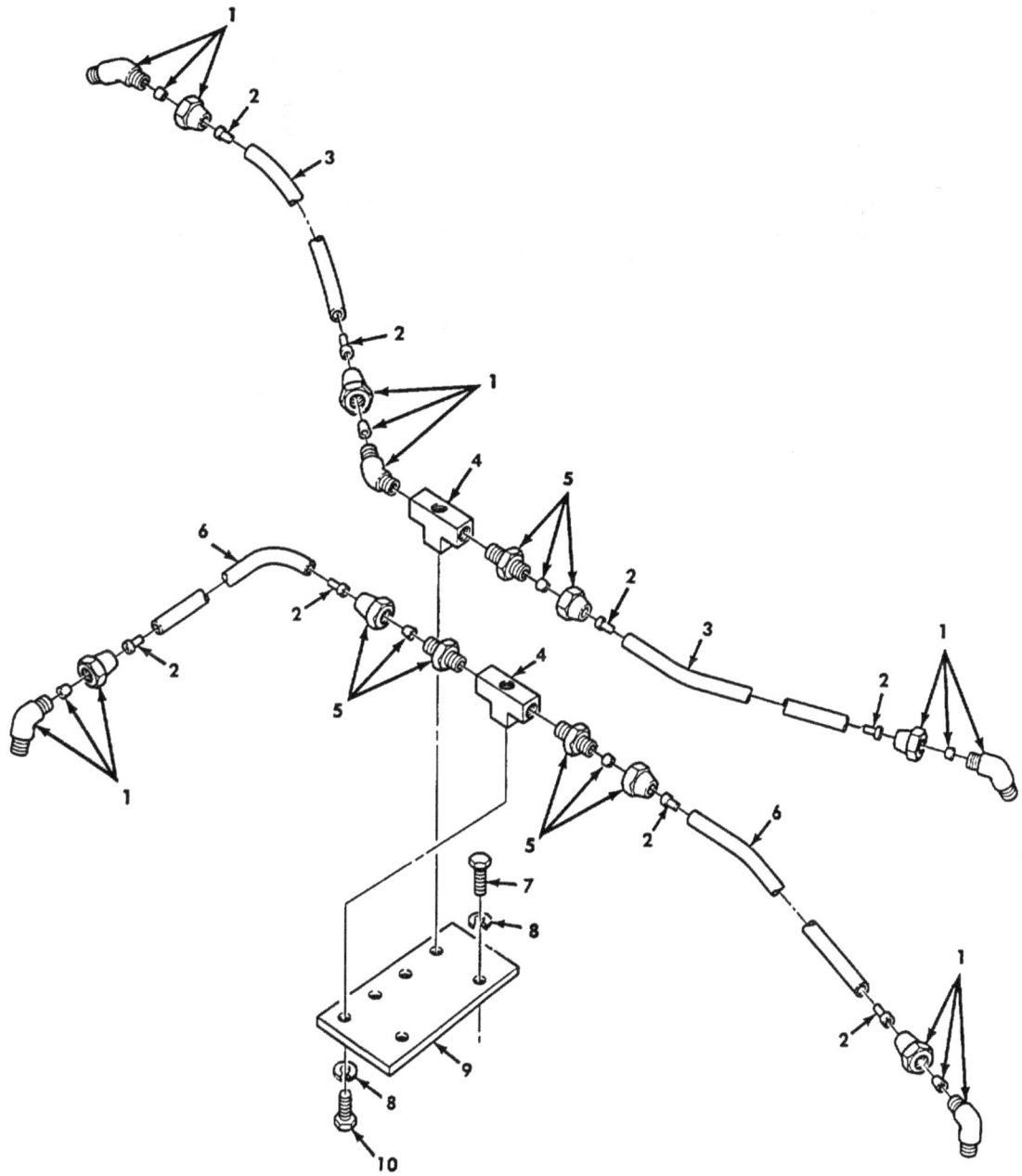

Figure 259. Forward Rear Axle Air Brake Chamber Control Lines and Mounting Hardware.

(1) ITEM NO	(2) SMR CODE	(3) NSN	(4) CAGEC	(5) PART NUMBER	(6) DESCRIPTION AND USABLE ON CODES (UOC)	(7) QTY
					GROUP 1208 AIR BRAKES	
					FIG. 259 FORWARD REAR AXLE AIR BRAKE CHAMBER CONTROL LINES AND MOUNTING HARDWARE	
1	PAOZZ	4730011171614	81343	8-6 120302BA	ELBOW, PIPE TO TUBE …………………………………	5
2	PAOZZ	4730010326038	19207	CPR102321-4	INSERT, TUBE FITTING …………………………………	8
3	MOOZZ		19207	CPR104420-3-21	HOSE, NONMETALLIC MAKE FROM HOSE, P/ ……. N CPR104420-3, 21 INCHES LONG………………………	2
4	PAOZZ	4730011951884	19207	12302675	TEE, PIPE ………………………………………………	4
5	PAOZZ		81343	8-6 120102BA	ADAPTER, STRAIGHT, PI ………………………………	3
6	MOOZZ		19207	CPR104420-3-23	HOSE NONMETALLIC MAKE FROM HOSE, P/ ……… N CPR104420-3 ……………………………………………	2
7	PAOZZ	5305009422196	80204	B1821BH038C100D	SCREW, CAP, HEXAGON H ……………………………	2
8	PAOZZ	5310006379541	96906	MS35338-46	WASHER, LOCK …………………………………………	6
9	PAOZZ	5340011885086	19207	12302691	BRACKET, MOUNTING …………………………………	1
10	PAOZZ	5305002692803	96906	MS90726-60	SCREW, CAP, HEXAGON H ……………………………	4

END OF FIGURE

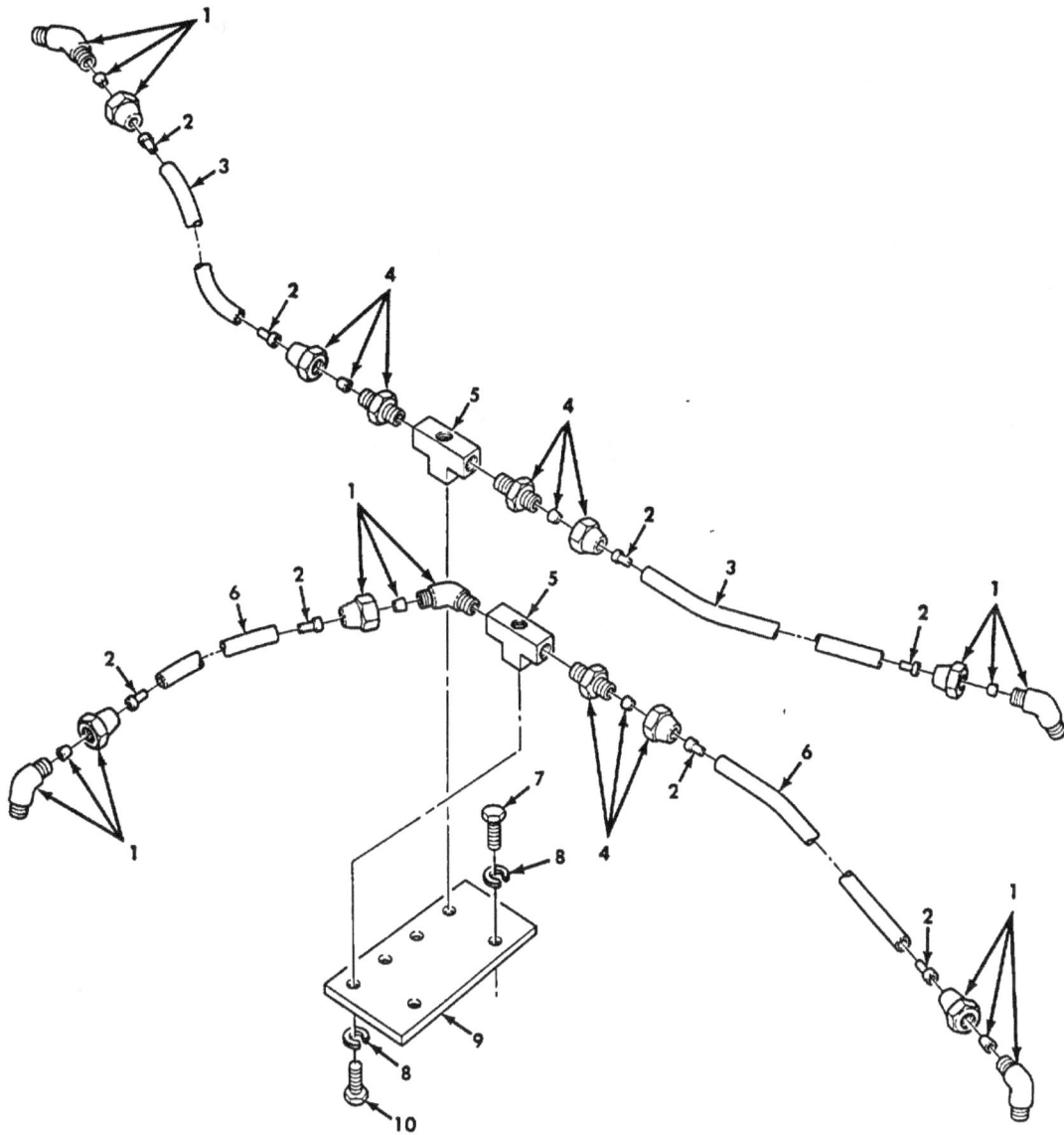

Figure 260. Rear Axle Air Brake Chamber Control Lines and Mounting Hardware.

(1) ITEM NO	(2) SMR CODE	(3) NSN	(4) CAGEC	(5) PART NUMBER	(6) DESCRIPTION AND USABLE ON CODES (UOC)	(7) QTY
					GROUP 1208 AIR BRAKES	
					FIG. 260 REAR AXLE AIR BRAKE CHAMBER CONTROL LINES AND MOUNTING HARDWARE	
1	PAOZZ	4730011171614	81343	8-6 120302BA	ELBOW, PIPE TO TUBE .... UOC:DAA, DAB, DAC, DAD, DAE, DAF, DAG, DAH, DAJ, DAK, DAL, DAW, DAX, V12, V13, V14, V15, V16, V17, V18, V19, V20, V21, V22, V24, V25, V39	5
2	PAOZZ	4730010326038	19207	CPR102321-4	INSERT, TUBE FITTING................................	8
3	MOOZZ		19207	CPR104420-3-23	HOSE, NONMETALLIC MAKE FROM HOSE............ CPR104420-3, 23 INCHES LONG	2
4	PAOZZ		81343	8-6 120102BA	ADAPTER, STRAIGHT, PI	3
5	PAOZZ	4730011951884	19207	12302675	TEE, PIPE	4
6	MOOZZ		19207	CPR104420-3-19	HOSE, NONMETALLIC MAKE FROM HOSE, ... CPR104420-3, 19 INCHES LONG	2
7	PAOZZ	5305009422196	80204	B1821BH038C100D	SCREW, CAP, HEXAGON H	2
8	PAOZZ	5310006379541	96906	MS35338-46	WASHER, LOCK	6
9	PAOZZ	5340011885086	19207	12302691	BRACKET, MOUNTING	1
10	PAOZZ	5305002692803	96906	MS90726-60	SCREW, CAP, HEXAGON H	4

END OF FIGURE

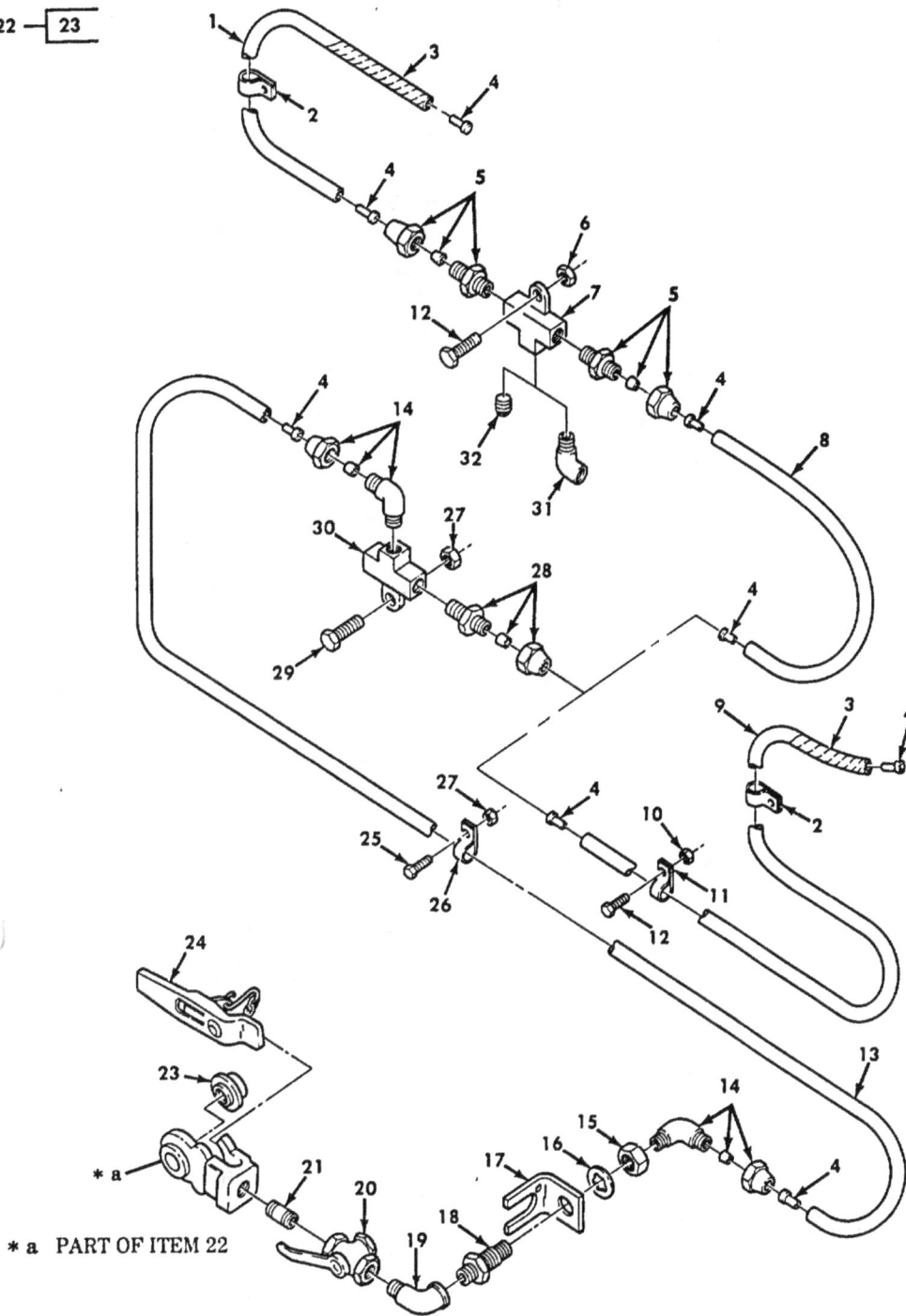

Figure 261. Front Emergency Half Coupling and Related Lines.

* a   PART OF ITEM 22

(1) ITEM NO	(2) SMR CODE	(3) NSN	(4) CAGEC	(5) PART NUMBER	(6) DESCRIPTION AND USABLE ON CODES (UOC)	(7) QTY
					GROUP 1208 AIR BRAKES	
					FIG. 261 FRONT EMERGENCY HALF COUPLING AND RELATED LINES	
1	MOOZZ		19207	CPR104420-2-38	HOSE,NONMETALLIC MAKE FROM HOSE P/N....... CPR104420-2, 38 INCHES LONG UOC:DAL,V18	1
2	PAOZZ	5340002221653	19207	7415752	CLAMP,LOOP	1
3	MOOZZ		19207	12302690-6	TUBING,PLASTIC,SPIR MAKE FROM P/N.............. 12302690, 6 INCHES LONG	1
4	PAOZZ	4730010798821	19207	CPR102321-1	INSERT,TUBE FITTING.....................................	8
5	PAOZZ	4730000691186	81343	6-4 120102BA	ADAPTER,STRAIGHT,PI..................................... UOC:DAL,V18,ZAL	3
6	PAOZZ	5310008140672	96906	MS51943-36	NUT,SELF-LOCKING,HE UOC:DAL,V18,ZAL	1
7	PAOZZ	4730011079690	19207	10937774	TEE,PIPE. UOC:DAL,V18,ZAL	2
8	MOOZZ		19207	CPR104420-2-66	HOSE,NONMETALLIC MAKE FROM HOSE,P/........ N CPR104420-2, 66 INCHES LONG UOC:DAL,V18,ZAL	1
9	MOOZZ		19207	CPR104420-2-101	HOSE,NONMETALLIC MAKE FROM HOSE,P/........ N CPR104420-2, 101 INCHES LONG .... UOC:DAA,DAB,DAW,DAX,V12,V13,V14,V15, ZAA,ZAB	1
9	MOOZZ		19207	CPR104420-2-77	HOSE,NONMETALLIC MAKE FROM HOSE,P/........ N CPR104420-2, 77 INCHES LONG .... UOC:DAE,DAF,DAG,DAH,V21,V22, ZAG,ZAH	1
9	MOOZZ		19207	CPR104420-2-148	HOSE,NONMETALLIC AMKE FROM HOSE,P/........ N CPR104420-2, 148 INCHES LONG UOC:DAC,DAD,V16,V17,V39,ZAC,ZAD	1
9	MOOZZ		19207	CPR104420-2-169	HOSE,NONMETALLIC MAKE FROM HOSE,P/........ N CPR104420-2, 169 INCHES LONG UOC:DAJ,DAK,V24,V25,ZAJ,ZAK	1
10	PAOZZ	5310008140672	96906	MS51943-36	NUT,SELF-LOCKING .... UOC:DAA,DAB,DAC,DAD,DAE,DAF,DAG,DAH, DAJ,DAK,DAL,DAW,DAX,V12,V13,V14,V15, V16,V17,V18,V19,V20,V21,V22,V24,V25, V39,ZAA,ZAB,ZAE,ZAF,ZAG,ZAH,ZAJ,ZAK, ZAL.	4
11	PAOZZ	5340009546014	96906	MS21333-121	CLAMP,LOOP .... UOC:DAA,DAB,DAG,DAH,DAW,DAX,V12,V13, V14,V15,V21,V22,ZAA,ZAB,ZAG,ZAH.	1
11	PAOZZ	5340009546014	96906	MS21333-121	CLAMP,LOOP UOC:DAC,DAD,DAJ,DAK,V16,V17,V24,V25, V39,ZAC,ZAD,ZAJ,ZAK	3
12	PAOZZ	5305002693238	80204	B1821BH038F125N	SCREW,CAP,HEXAGON H........................... UOC:DAA,DAB,DAW,DAX,V12,V13,V14,V15, ZAA,ZAB	2
12	PAOZZ	5305002693238	80204	BIB21BH038F125N	SCREW,CAP,HEXAGON H UOC:DAC,DAD,DAJ,DAK,V16,V17,V24,V25, V39,ZAC,ZAD,ZAJ,ZAK	3

(1) ITEM NO	(2) SMR CODE	(3) NSN	(4) CAGEC	(5) PART NUMBER	(6) DESCRIPTION AND USABLE ON CODES (UOC)	(7) QTY
13	MOOZZ		19207	CPR104420-2-178	HOSE,NONMETALLIC MAKE FROM HOSE,P/N....... CPR104420-2-178, 178 INCHES LONG UOC:DAE,DAF,DAG,DAH,V19,V20,V21,V22 ZAE,ZAF,ZAG,ZAH	1
13	MOOZZ		19207	CPR104420-2-203	HOSE,NONMETALLIC MAKE FROM HOSE,P/......... N CPR104420-2, 203 INCHES LONG .... UOC:DAA,DAB,DAW,DAX,V12,V13,V14,V15, ZAA,ZAB	1
13	MOOZZ		19207	CPR104420-2-191	HOSE,NONMETALLIC MAKE FROM HOSE,P/......... N CPR104420-2, 191 INCHES LONG UOC:DAL,V18,ZAL	1
13	MOOZZ		19207	CPR104420-2-229	HOSE,NONMETALLIC MAKE FROM HOSE,P/......... N CPR104420-2, 229 INCHES LONG UOC:DAC,DAD,DAJ,DAK,V16,V17,V24,V25, V39,ZAC,ZAD,ZAJ,ZAK	1
14	PAOZZ	4730000691187	81343	6-4 100202BA	ELBOW,PIPE TO TUBE	2
15	PAOZZ	5310008913428	96906	MS35691-77	NUT,PLAIN,HEXAGON	1
16	PAOZZ	5310005826714	96906	MS35333-49	WASHER,LOCK	1
17	PAOZZ	5340011079929	19207	12256265	BRACKET,ANGLE	1
18	PAOZZ	4730004199424	19207	8376311	REDUCER,PIP E.	1
19	PAOZZ	4730002775555	96906	MS51952-4	ELBOW,PIPE	1
20	PAOOZ	4820004205499	06853	285172	VALVE,BALL	1
21	PAOZZ	4730001961504	24617	192075	NIPPLE,PIPE	1
22	PAOZZ	4730005950083	58536	A52484-1	COUPLING HALF,QUICK	1
23	PAOZZ	5330000902128	06853	213630	.PACKING,PREFORMED	1
24	PAOZZ	2530002703878	19207	7014965	DUMMY COUPLING,AUTO	1
25	PAOZZ	5305002692804	96906	MS90726-61	SCREW,CAP,HEXAGON H	1
26	PAOZZ	5340000538994	96906	MS21333-126	CLAMP,LOOP	1
27	PAOZZ	5310008140672	96906	MS51943-36	NUT,SELF-LOCKING,HE	2
28	PAOZZ	4730000691186	81343	6-4 120102BA	ADAPTER,STRAIGHT ,PI.	1
29	PAOZZ	5305002693240	80204	B1821BH038F150N	SCREW,CAP,HEXAGON H .... UOC:DAA,DAB,DAC,DAD,DAE,DAF,DAG,DAH, DAJ,DAK,DAL,DAW,DAX,V12,V13,V14,V15, V16,V17,V18,V19,V20,V21,V22,V24,V25, V39	1
30	PAOZZ	4730011079690	19207	10937774	TEE,PIPE	1
31	PAOZZ	4730002315598	88044	AN914-4	ELBOW,PIPE UOC:DAL,V18,ZAL	1
32	PAOZZ	4730002212137	96906	MS20913-2S	PLUG,PIPE UOC:DAL,V18,ZAL	

END OF FIGURE

5 —| 6 |

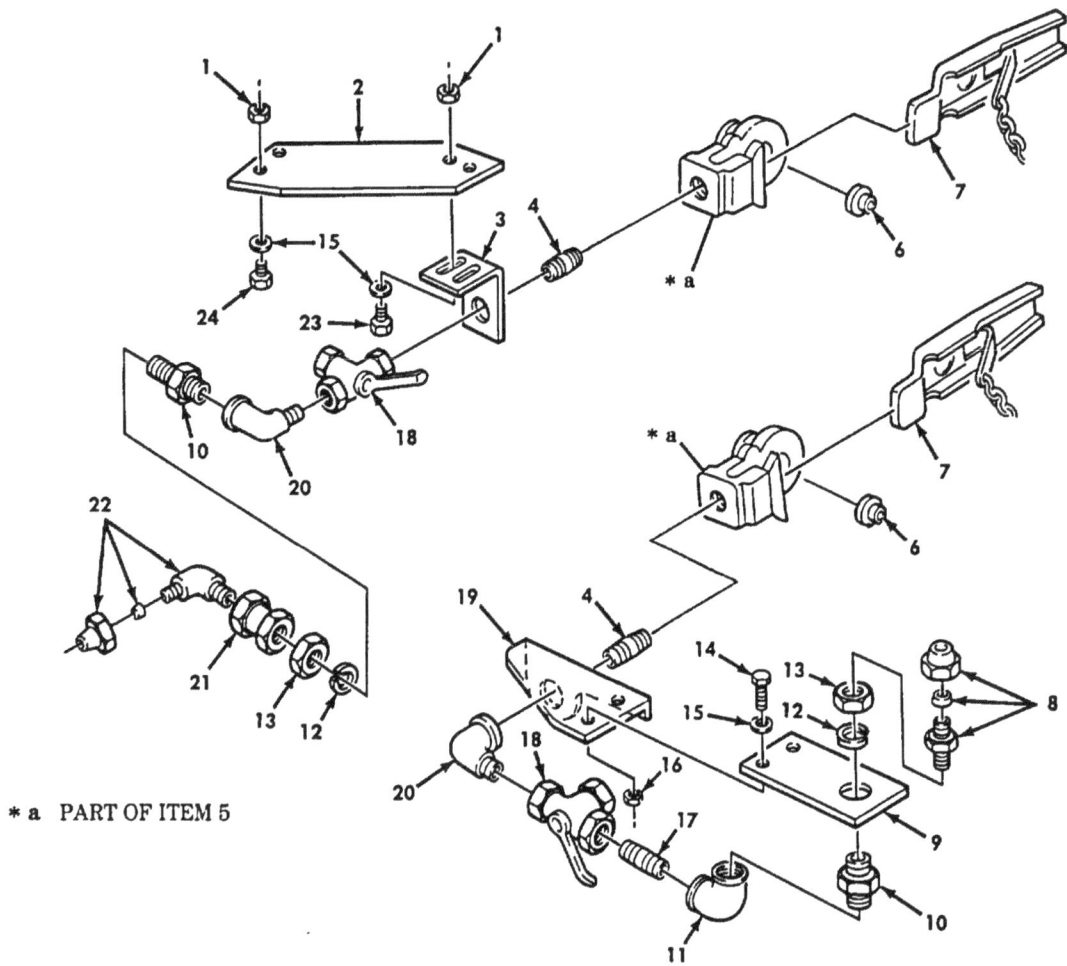

*a   PART OF ITEM 5

Figure 262. Rear Emergency Half Couplings and Mounting Hardware.

(1) ITEM NO	(2) SMR CODE	(3) NSN	(4) CAGEC	(5) PART NUMBER	(6) DESCRIPTION AND USABLE ON CODES (UOC)	(7) QTY
					GROUP 1208 AIR BRAKES	
					FIG. 262 REAR EMERGENCY HALF COUPLINGS AND MOUNTING HARDWARE	
1	PAOZZ	5310009359022	96906	MS51943-32	NUT, SELF-LOCKING, HE .... UOC:DAA, DAB, DAW, DAX, V12, V13, V14, V15, ZAA, ZAB	4
2	PFOZZ	2510001274768	19207	10883437	PLATE, REAR, SUPPORT .... UOC:DAA, DAB, DAW, DAX, V12, V13, V14, V15, ZAA, ZAB	1
3	PFOZZ	5340004199487	19207	10883438	BRACKET, ANGLE .... UOC:DAA, DAB, DAW, DAX, V12, V13, V14, V15, ZAA, ZAB	1
4	PAOZZ	4730001961504	24617	192075	NIPPLE, PIPE	1
5	PAOZZ	4730005950083	58536	A52484-1	COUPLING HALF, QUICK	2
6	PAOZZ	5330000902128	06853	213630	.PACKING, PREFORMED	1
7	PAOZZ	2530002703878	19207	7014965	DUMMY COUPLING, AUTO	1
8	PAOZZ	4730000691186	81343	6-4 120102BA	ADAPTER, STRAIGHT, PI	1
9	PFOZZ	5340001577938	19207	10883331	PLATE, MENDING	1
10	PAOZZ	4730004199424	19207	8376311	REDUCER, PIPE	2
11	PAOZZ	4730002493885	96906	MS51845-4	ELBOW, PIPE	1
12	PAOZZ	5310005826714	96906	MS35333-49	WASHER, LOCK	2
13	PAOZZ	5310000219760	21450	219760	NUT, PLAIN, HEXAGON	2
14	PAOZZ	5305002678954	80204	B1821BH025F125N	SCREW, CAP, HEXAGON H	2
15	PAOZZ	5310008094058	96906	MS27183-10	WASHER, FLAT	4
16	PAOZZ	5310009359022	96906	MS51943-32	NUT, SELF-LOCKING, HE	2
17	PAOZZ	4730001961991	96906	MS51846-64	NIPPLE, PIPE	1
18	PAOZZ	4820004205499	06853	285172	VALVE, BALL	1
19	PFOZZ	5340004484073	19207	10883329	BRACKET, ANGLE	1
20	PAOZZ	4730002775555	96906	MS51952-4	ELBOW, PIPE	2
21	PAOZZ	4730005417790	21450	144077	COUPLING, PIPE .... UOC:DAA, DAB, DAW, DAX, V12, V13, V14, V15, ZAA, ZAB	1
22	PAOZZ	4730001439282	02570	MS39182-8	ELBOW, PIPE TO TUBE .... UOC:DAA, DAB, DAW, DAX, V12, V13, V14, V15, ZAA, ZAB	1
23	PAOZZ	5306000680514	80204	B1821BH025F088N	BOLT, MACHINE .... UOC:DAA, DAB, DAW, DAX, V12, V13, V14, V15, ZAA, ZAB	2
24	PAOZZ	5305002693239	80204	B1821BH038F138N	SCREW, CAP, HEXAGON H .... UOC:DAA, DAB, DAW, DAX, V12, V13, V14, V15, ZAA, ZAB	2

END OF FIGURE

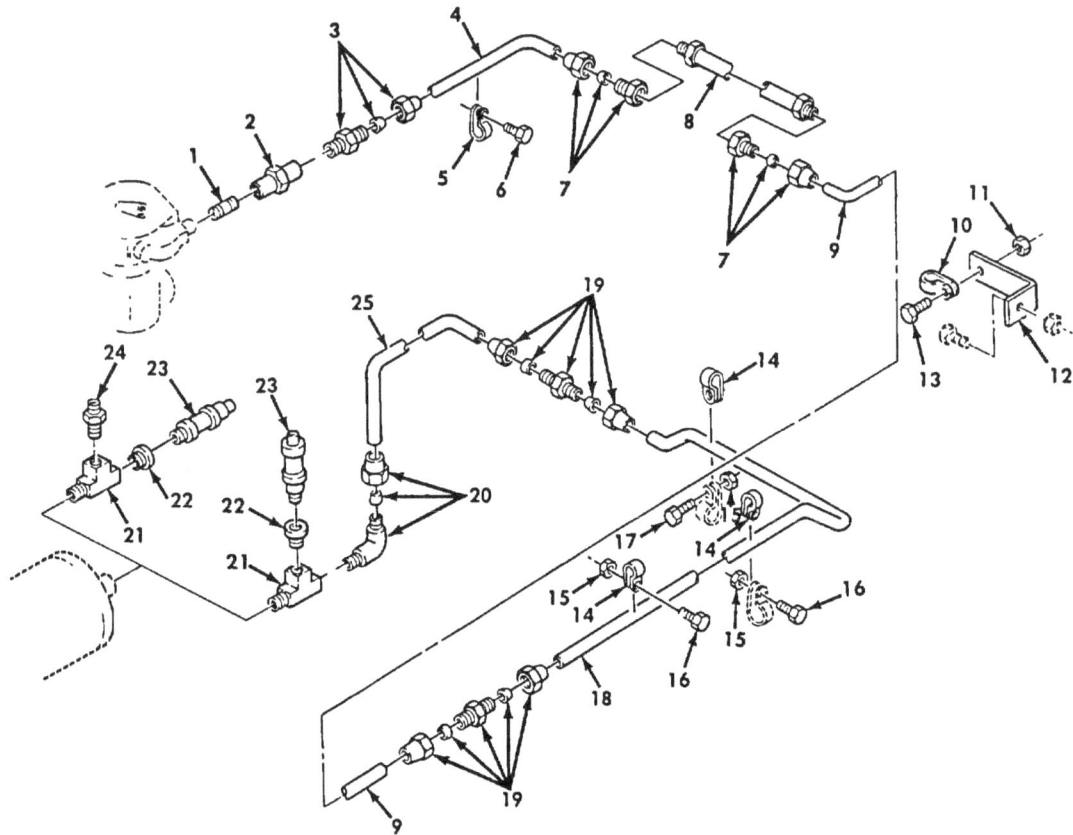

Figure 263. Air Compressor to Wet Reservoir Air Supply Lines.

(1) ITEM NO	(2) SMR CODE	(3) NSN	(4) CAGEC	(5) PART NUMBER	(6) DESCRIPTION AND USABLE ON CODES (UOC)	(7) QTY
					GROUP 1208 AIR BRAKES	
					FIG. 263 AIR COMPRESSOR TO WET RESERVOIR AIR SUPPLY LINES	
1	PAOZZ	4730001961495	96906	MS51953-80	NIPPLE, PIPE	1
2	PAOZZ	4730005425598	79470	C3309X8	COUPLING, PIPE	1
3	PAOZZ	4730002029036	81343	8-8 120102BA	ADAPTER, STRAIGHT, PI	1
4	MOOZZ		24617	12302615	TUBE, COPPER MAKE FROM TUBE, P/N .... 10-00-2, AS REQUIRED...................................	1
5	PAOZZ	5340011128909	19207	11648730	CLAMP, LOOP	1
6	PAOZZ	5305012770461	96906	MS90727-130	SCREW, CAP, HEXAGON H	1
7	PAOZZ	4730001731867	81343	8-6 120103BA	ADAPTER, STRAIGHT, PI...................................	2
8	PAOZZ	4720000969630	19207	11648644	HOSE ASSEMBLY, NONME	1
9	MOOZZ		19207	12255962	TUBE, COPPER MAKE FROM TUBE, P/N .... 10-00-2, AS REQUIRED...................................	1
10	PAOZZ	5340000573043	96906	MS21333-112	CLAMP, LOOP	1
11	PAOZZ	5310008140673	96906	MS51943-33	NUT, SELF-LOCKING, HE	1
12	PAOZZ	5340011476759	19207	12256266	BRACKET, ANGLE	1
13	PAOZZ	5306000501238	96906	MS90727-32	BOLT, MACHINE	1
14	PAOZZ	5340009891771	96906	MS21333-123	CLAMP, LOOP	3
15	PAOZZ	5310008140672	96906	MS51943-36	NUT, SELF-LOCKING, HE .... UOC:DAL, V18, ZAL	3
15	PAOZZ	5310008140672	96906	MS51943-36	NUT, SELF-LOCKING .... UOC:DAA, DAB, DAC, DAD, DAE, DAF, DAG, DAH, DAJ, DAK, DAW, DAX, V12, V13, V14, V15, V16, V17, V19, V20, V21, V22, V24, V25, V39, ZAA, ZAB, ZAC, ZAD, ZAE, ZAF, ZAG, ZAH, ZAJ, ZAK	3
16	PAOZZ	5305002693238	96906	MS90727-62	SCREW, CAP, HEXAGON .... UOC:DAA, DAB, DAC, DAD, DAE, DAF, DAG, DAH, DAJ, DAK, DAW, DAX, V12, V13, V14, V15, V16, V17, V19, V20, V21, V22, V24, V25, V39, ZAA, ZAB, ZAC, ZAD, ZAE, ZAF, ZAG, ZAH, ZAJ, ZAK	2
16	PZOZZ	5305002693238	96906	MS90727-62	SCREW, CAP, HEXAGON .... UOC:DAL, V18, ZAL	1
17	PAOZZ	5305002692804	96906	MS90726-61	SCREW, CAP, HEXAGON H	3
18	MOOZZ		19207	12255968-1	TUBE, COPPER AIR COMPRESSOR TO WET........ RESERVOIR, MAKE FROM TUBE, P/N 8689210 UOC:DAA, DAB, DAL, DAW, DAX, V12, V13, V14, V15, V18, ZAA, ZAB, ZAL	1
18	MOOZZ		19207	12255968-2	TUBE, COPPER AIR COMPRESSOR TO WET........ RESERVOIR, MAKE FROM TUBE, P/N 8689210 UOC:DAE, DAF, V19, V20, V21, V22, ZAE, ZAF, ZAG, ZAH	1
18	MOOZZ		19207	12255968-3	TUBE, COPPER AIR SUPPLY MAKE FROM............ TUBE, P/N 8689210 .... UOC:DAC, DAD, DAG, DAH, DAJ, DAK, V16, V17, V24, V25, V39, ZAC, ZAD, ZAJ, ZAK	1
19	PAOZZ	4730002783220	81343	8-8 120101BA	NIPPLE, TUBE	2
20	PAOZZ	4730005952572	06853	216310	ELBOW, PIPE TO TUBE WET RESERVOIR............. TEE .... UOC:DAA, DAB, DAC, DAD, DAE, DAF, DAG, DAH, DAJ, DAK, DAW, DAX, V12, V13, V14, V15, V16,	1

(1) ITEM NO	(2) SMR CODE	(3) NSN	(4) CAGEC	(5) PART NUMBER	(6) DESCRIPTION AND USABLE ON CODES (UOC)	(7) QTY
					V17,V19,V20,V21,V22,V24,V25,V39,ZAA, ZAB,ZAC,ZAD,ZAE,ZAF,ZAG,ZAH,ZAJ,ZAK	
21	PAOZZ	4730005402745	24617	444152	TEE,PIPE	2
22	PAOZZ	4730011924729	81348	WW-P-471BDQBCDB	BUSHING,PIPE	2
23	PAOZZ	4820005952761	19207	7066008	VALVE,SAFETY RELIEF.....................................	2
24	PAOZZ	4730002029036	81343	8-8 120102BA	ADAPTER,STRAIGH T ,PIPE	1
					UOC:DAL,V18,ZAL	
25	MOOZZ		19207	12256101	TUBE,COPPER MAKE FROM TUBING,P/N .... 8989210 UOC:DAE,DAF,DAG,DAH,V19,V20,V21,V22, ZAE,ZAF,ZAG,ZAH	1
25	MOOZZ		19207	12255967-1	TUBE,COPPER AIR COMPRESSOR TO WET.......... RESERVOIR,MAKE FROM TUBING P/N 8689210 .... UOC:DAA,DAB,DAE,.DAF,DAL,DAW,DAX,V12, V13,V14,V15,V18,ZAA,ZAE,ZAF,ZAB,ZAL	1
25	MOOZZ		19207	12255967-3	TUBE,COPPER AIR COMPRESSOR TO WET.......... RESERVOIR,MAKE FROM TUBE,P/N 8689210 UOC:DAC,DAD,DAJ,DAK,V24,V25,V39,ZAJ, V16,V17,ZAC,ZAD,ZAK	1

END OF FIGURE

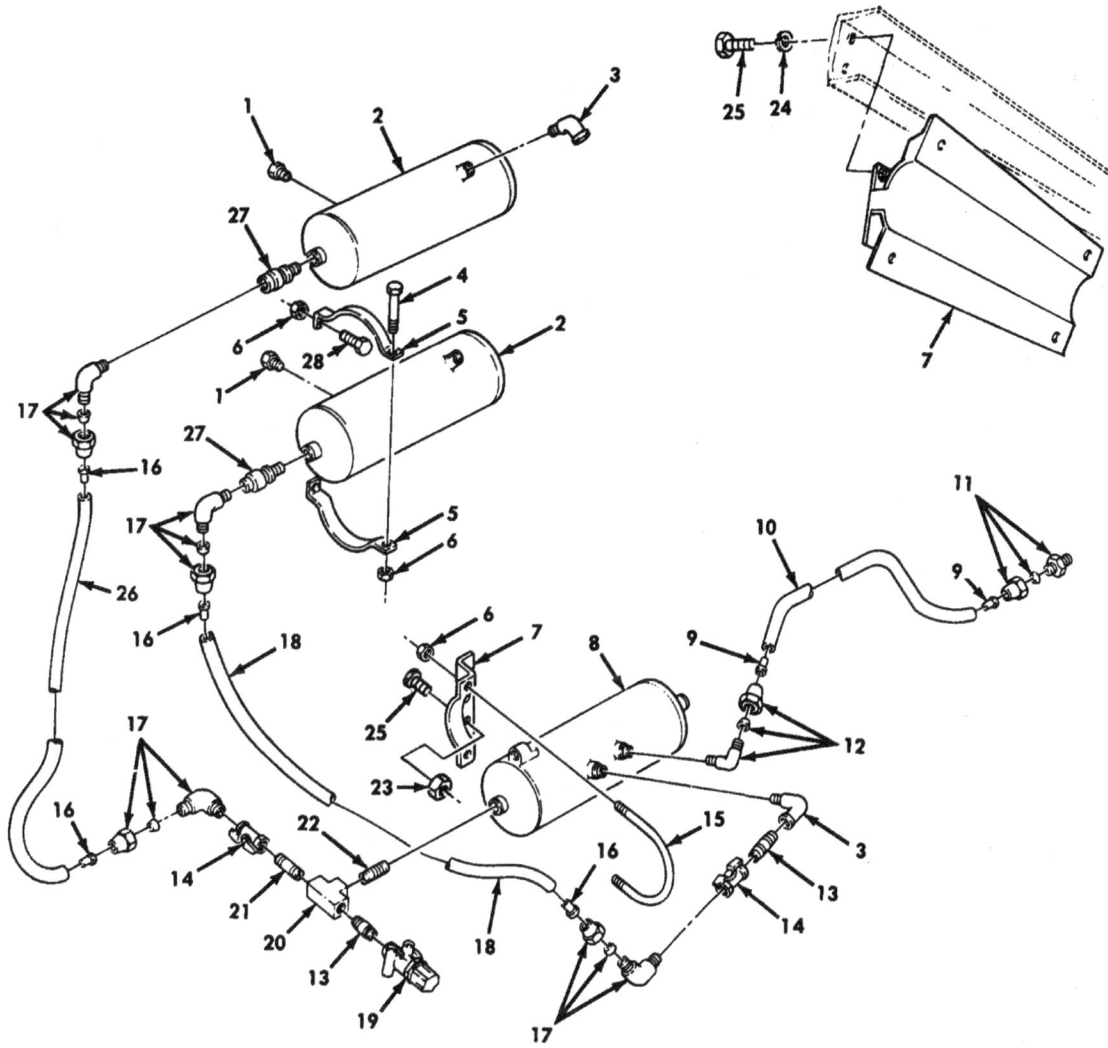

Figure 264. Primary, Secondary, Wet Reservoirs, and Related Lines (Sheet 1 of 2).

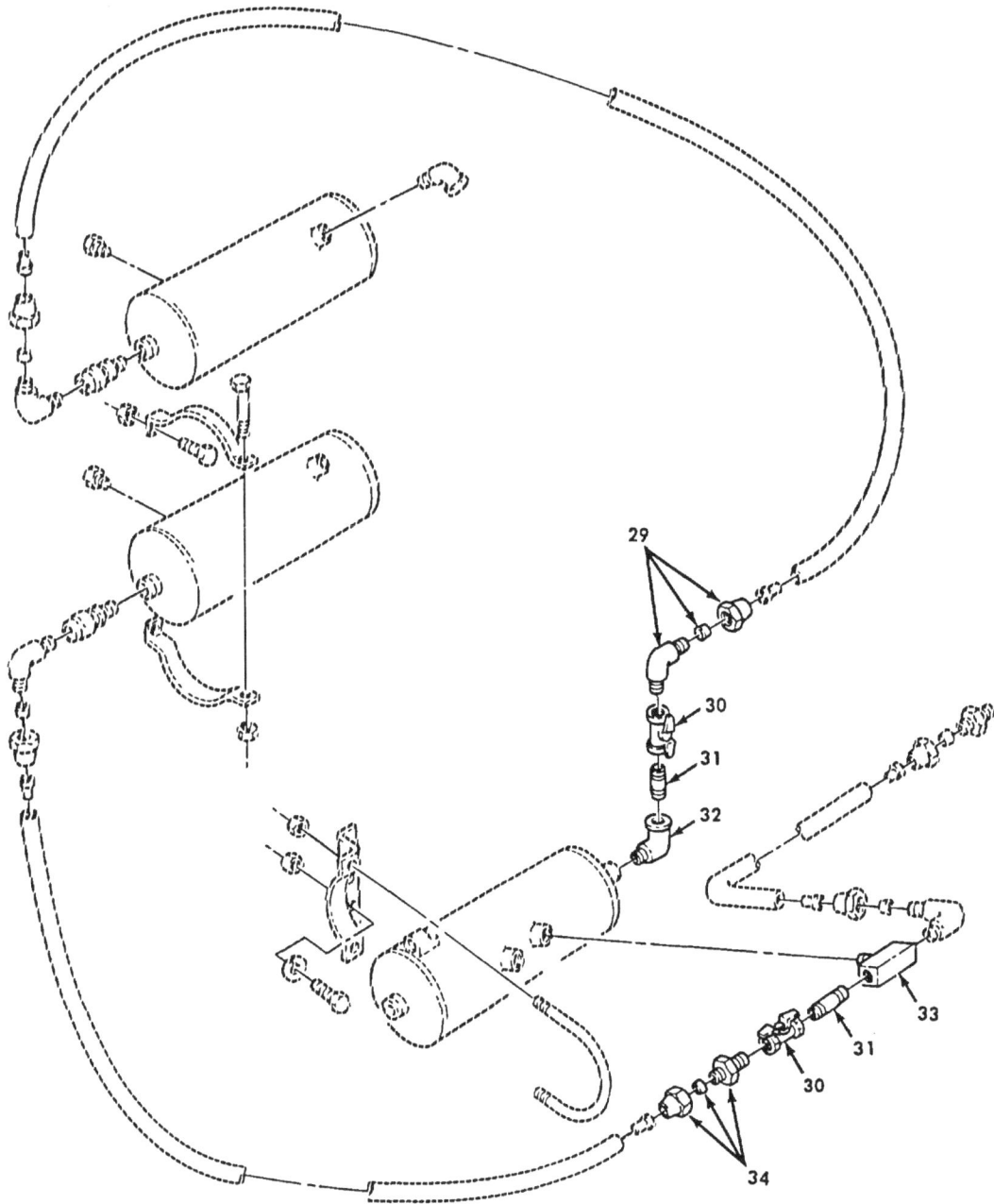

Figure 264. Primary, Secondary, Wet Reservoirs, and Related Lines (Sheet 2 of 2).

(1) ITEM NO	(2) SMR CODE	(3) NSN	(4) CAGEC	(5) PART NUMBER	(6) DESCRIPTION AND USABLE ON CODES (UOC)	(7) QTY

GROUP 1208 AIR BRAKES   )

FIG. 264 PRIMARY, SECONDARY, WET RESERVOIRS, AND RELATED LINES

(1)	(2)	(3)	(4)	(5)	(6)	(7)
1	PAOZZ	4730012066162	81349	WW-P-471ACABCD	PLUG,PIPE	2
2	PAOZZ	2530011112260	19207	12255982	RESERVOIR,AIR	2
3	PAOZZ	4730002775555	96906	MS51952-4	ELBOW,PIPE	3
					UOC:DAA,DAB,DAC,DAD,DAE,DAF,DAG,DAH, DAJ,DAK,DAW,DAX,V12,V13,V14,V15,V16, V17,V19,V20,V21,V22,V24,V25,V39,ZAA, ZAB,ZAC,ZAD,ZAE,ZAF,ZAG,ZAH,ZAJ,ZAK	
4	PAOZZ	5305011337201	24617	147601	SCREW,CAP,HEXAGON H	4
5	PAOZZ	5340011126554	19207	12255969-1	STRAP,RETAINING	8
6	PAOZZ	5310008140672	96906	MS51943-36	NUT,SELF-LOCKING,HE	14
					UOC:DAA,DAB,DAC,DAD,DAE,DAF,DAG,DAH, DAJ,DAK,DAL,DAW,DAX,V21,V22,V24,V25, V39,ZAA,ZAB,ZAC,ZAD,ZAE,ZAF,ZAG,ZAH, ZAJ,ZAK,ZAL	
7	PAOZZ	5340011126436	19207	12255974	BRACKET,MOUNTING	2
					UOC:DAA,DAB,DAE,DAF,DAG,DAH,DAW,DAX, V12,V13,V14,V15,V19,V20,V21,V22	
7	PAOZZ	5340011885088	19207	12302755	BRACKET,MOUNTING	1
					UOC:DAC,DAD,DAJ,DAK,DAL,V16,V17,V18, V24,V25,V39,ZAC,ZAD,ZAJ,ZAK,ZAL	
8	PAOZZ	2530011112261	19207	12255983	RESERVOIR,AIR	1
9	PAOZZ	4730010798821	19207	CPR102321-1	INSERT,TUBE FITTING....................................	2
10	MOOZZ		19207	CPR104420-2-49	HOSE,NONMETALLIC WET RESERVOIR, MAKE FROM HOSE,P/N CPR104420-2, 49 INCHES LONG..	1
					UOC:DAA,DAB,DAL,DAW,DAX,V12,V13,V14, V15,V18,ZAA,ZAB,ZAL	
10	MOOZZ		19207	CPR104420-2-39	HOSE,NONMETALLIC WET RESERVOIR, MAKE FROM HOSE,P/N CPR104420-2, 39 INCHES LONG..	1
					UOC:DAA,DAB,DAL,DAW,DAX,V12,V13,V14, V15,V18,V19,V20,V21,V22,ZAA,ZAB,ZAE, ZAF,ZAH,ZAJ,ZAL	
10	MOOZZ		19207	CPR104420-2-89	HOSE,NONMETALLIC WET RESERVOIR, MAKE FROM HOSE,P/N CPR104420-2, 89 INCHES LONG..	1
					UOC:DAC,DAD,DAJ,DAK,V16,V17,V24,V25, V39,ZAC,ZAD,ZAJ,ZAK,ZAL	
11	PAOZZ	4730000691186	81343	6-4 120102BA	ADAPTER,STRAIGHT,PI	1
12	PAOZZ	4730001439282	02570	MS39182-8	ELBOW,PIPE TO TUBE	1
					UOC:DAA,DAB,DAC,DAD,DAE,DAF,DAG,DAH, DAJ,DAK,DAW,DAX,V12,V13,V14,V15,V16, V17,V19,V20,V21,V22,V24,V25,V39,ZAA, ZAB,ZAC,ZAD,ZAE,ZAF,ZAG,ZAH,ZAJ,ZAK	
13	PAOZZ	4730002221839	96906	MS51846-58	NIPPLE,PIPE	2
					UOC:DAA,DAB,DAC,DAD,DAE,DAF,DAG,DAH, DAJ,DAK,DAW,DAX,V12,V13,V14,V15,V16,	

(1) ITEM NO	(2) SMR CODE	(3) NSN	(4) CAGEC	(5) PART NUMBER	(6) DESCRIPTION AND USABLE ON CODES (UOC)	(7) QTY
14	PAOOZ	4820004205499	06853	285172	V17,V19,V20,V21,V22,V24,V25,V39,ZAA, ZAB,ZAC,ZAD,ZAE,ZAF,ZAG,ZAH,ZAJ,ZAK VALVE,BALL .... UOC:DAA,DAB,DAC,DAD,DAE,DAF,DAG,DAH, DAJ,DAK,DAW,DAX,V12,V13,V14,V15,V16, V17,V19,V20,V21,V22,V24,V25,V39,ZAA, ZAB,ZAC,ZAD,ZAE,ZAF,ZAG,ZAH,ZAJ,ZAK	2
15	PAOZZ	5306002074932	19207	8327473	BOLT,U	1
16	PAOZZ	4730010326038		CPR102321-4	INSERT,TUBE	4
17	PAOZZ	4730005952572	06853	216310	ELBOW,PIPE TO TUBE .... UOC:DAA,DAB,DAC,DAD,DAE,DAF,DAG,DAH, DAJ,DAK,DAW,DAX,V12,V13,V14,V15,V16, V17,V19,V20,V21,V22,V24,V25,V39,ZAA, ZAB,ZAC,ZAD,ZAE,ZAF,ZAG,ZAH,ZAJ,ZAK	4
18	MOOZZ		19207	CPR104420-3-68	HOSE,NONMETALLIC MAKE FROM HOSE, ... P/N CPR104420-2, 68 INCHES LONG .... UOC:DAC,DAD,DAE,DAF,DAG,DAH,DAJ,DAK, DAL,V16,V17,V18,V19,V,V20,V21,V22,V24, V25,V39,ZAC,ZAD,ZAE,ZAF,ZAG,ZAH,ZAJ, ZAK,ZAL	1
18	MOOZZ		19207	CPR104420-3-64	HOSE,NONMETALLIC MAKE FROM HOSE,P/N....... CPR104420-3, 64 INCHES LONG .... UOC:DAA,DAB,DAW,DAX,V12,V13,V14,V15, ZAA,ZAB	1
19	PAOZZ	4820011147543	19207	11669104	VALVE,SAFETY RELIEF .... UOC:DAA,DAB,DAC,DAD,DAE,DAF,DAG,DAH, DAJ,DAK,DAW,DAX,V12,V13,V14,V15,V16, V17,V19,V20,V21,V22,V24,V25,V39,ZAA, ZAB,ZAC,ZAD,ZAE,ZAF,ZAG,ZAH,ZAJ,ZAK, ZAL	1
20	PAOZZ	4730005402745	24617	444152	TEE,PIPE .... UOC:DAA,DAB,DAC,DAD,DAE,DAF,DAG,DAH, DAJ,DAK,DAW,DAX,V12,V13,V14,V15,V16, V17,V19,V20,V21,V22,V24,V25,V39,ZAA, ZAB,ZAC,ZAD,ZAE,ZAF,ZAG,ZAH,ZAJ,ZAK	1
21	PAOZZ	4730001932709	96906	MS51846-24	NIPPLE,PIPE .... UOC:DAA,DAB,DAC,DAD,DAE,DAF,DAG,DAH, DAJ,DAK,DAW,DAX,V12,V13,V14,V15,V16, V17,V19,V20,V21,V22,V24,V25,V39,ZAA, ZAB,ZAC,ZAD,ZAE,ZAF,ZAG,ZAH,ZAJ,ZAK	1
22	PAOZZ	4730011924729	81348	WW-P-471BDQBCDB	BUSHING,PIPE .... UOC:DAA,DAB,DAC,DAD,DAE,DAF,DAG,DAH, DAJ,DAK,DAW,DAX,V12,V13,V14,V15,V16, V17,V19,V20,V21,V22,V24,V25,V39,ZAA, ZAB,ZAC,ZAD,ZAE,ZAF,ZAG,ZAH,ZAJ,ZAK	1
23	PAOZZ	5310008140672	96906	MS51943-36	NUT,SELF-LOCKING,HE	4
24	PAOZZ	5310006379541	96906	MS35338-46	WASHER,LOCK UOC:DAC,DAD,DAJ,DAK,DAL,V16,V1,VV18, V24,V25,V39,ZAC,ZAD,ZAJ,ZAK,ZAL	4
25	PAOZZ	5305002692803	96906	MS90726-60	SCREW,CAP,HEXAGON H UOC:DAC,DAD,DAJ,DAK,DAL,V16,V17,V18, V24,V25,V39,ZAC,ZAD,ZAJ,ZAK,ZAL	4

(1) ITEM NO	(2) SMR CODE	(3) NSN	(4) CAGEC	(5) PART NUMBER	(6) DESCRIPTION AND USABLE ON CODES (UOC)	(7) QTY
26	MOOZZ		19207	CPR104420-3-45	HOSE,NONMETALLIC MAKE FROM HOSE, ... P/N CPR104420-3, 45 INCHES LONG .... UOC:DAA,DAB,DAW,DAX,V12,V13,V14,V15, ZAA,ZAB	1
26	MOOZZ		19207	CPR104420-3-54	HOSE,NONMETALLIC MAKE FROM HOSE, ... P/N CPR104420-3, 54 INCHES LONG .... UOC:DAC,DAD,DAE,DAF,DAG,DAB,DAJ,DAK, DAL,V16,V17,V18,V19,V20,V21,V22,V24, V25,V39,ZAC,ZAD,ZAE,ZAF,ZAG,ZAH,ZAJ, ZAK,ZAL	1
27	PAOZZ	4820006333523	06721	N13526H	VALVE,CHECK.	2
28	PAOZZ	5305002692803	96906	MS90726-60	SCREW,CAP,HEXAGON B	8
29	PAOZZ	4730005952572	06853	216310	ELBOW,PIPE TO TUBE UOC:DAL,V18,ZAL	1
30	PAOOZ	4820004205499	06853	285172	VALVE,BALL UOC:DAL,V18,ZAL	2
31	PAOZZ	4730002221839	96906	MS51846-58	NIPPLE,PIPE. UOC:DAL,V18,ZAL	2
32	PAOZZ	4730002775555	96906	MS51952-4	ELBOW,PIPE UOC:DAL,V18,ZAL	1
33	PAOZZ	4730005402745	24617	444152	TEE,PIPE UOC:DAL,V18,ZAL	1
34	PAOZZ	4730002029036	81343	8-8 120102BA	ADAPTER,STRAIGHT,PI UOC:DAL,V18,ZAL	1

END OF FIGURE

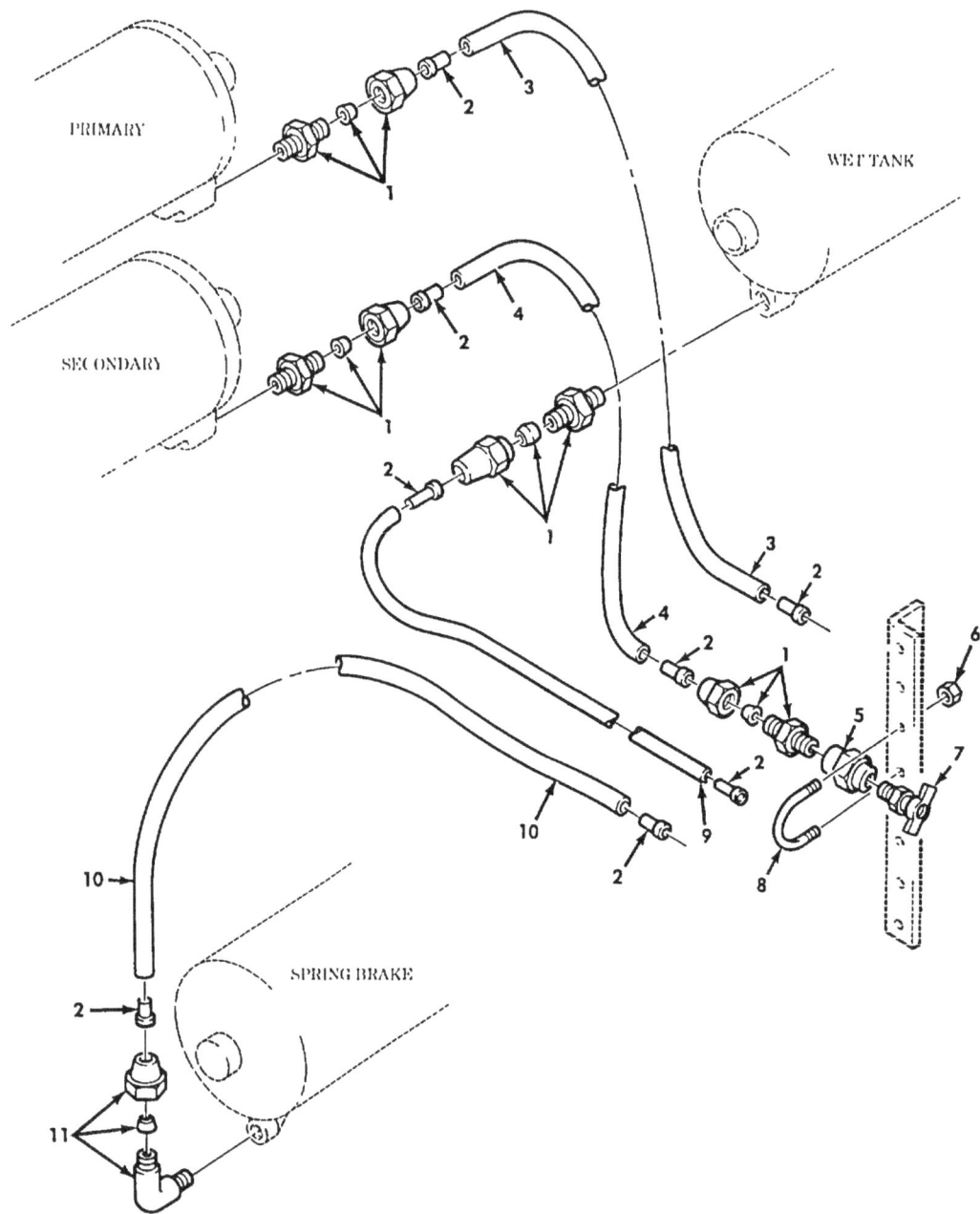

Figure 265. Air Tank Reservoirs, Draincocks, and Related Lines.

(1) ITEM NO	(2) SMR CODE	(3) NSN	(4) CAGEC	(5) PART NUMBER	(6) DESCRIPTION AND USABLE ON CODES (UOC)	(7) QTY
					GROUP 1208 AIR BRAKES	
					FIG. 265 AIR TANK RESERVOIRS, DRAINCOCKS, AND RELATED LINES	
1	PAOZZ	4730000691186	81343	6-4 120102BA	ADAPTER,STRAIGHT,PI	6
2	PAOZZ	4730010798821	19207	CPR102321-1	INSERT,TUBE FITTING....................................	8
3	MOOZZ		19207	CPR104420-2-11	HOSE,NONMETALLIC PRIMARY RESERVOIR, ... MAKE FROM HOSE,P/N CPR104420-2,11 INCHES LONG..	1
4	MOOZZ		19207	CPR104420-2-16	HOSE,NONMETALLIC SECONDARY .... RESERVOIR,MAKE FROM HOSE,P/N CPR104420-2,16 INCHES LONG	1
5	PAOZZ	4730005807408	04164	272M0075P00	COUPLING,PIPE	4
6	PAOZZ	5310000617326	96906	MS21045-3	NUT,SELF-LOCKING,HE	8
7	PAOZZ	4820008491220	96906	MS35782-5	COCK,DRAIN...	4
8	PAOZZ	5306011126560	19207	12255984	BOLT,U...	4
9	MOOZZ		19207	CPR104420-2-26	HOSE NONMETALLIC MAKE FROM TUBING, ... P/N CPR104420-2,26 INCHES LONG	1
10	MOOZZ		19207	CPR104420-2-115	HOSE,NONMETALLIC SPRING BRAKE .... RESERVOIR,MAKE FROM HOSE,P/N CPR104420-2,115 INCHES LONG UOC:DAC,DAD,DAJ,DAK,V16,V17,V24,V25, V39,ZAC,ZAD,ZAJ,ZAK	1
10	MOOZZ		19207	CPR104420-2-110	HOSE,NONMETALLIC .... SPRING BRAKE RESERVOIR,MAKE FROM HOSE,P/N CPR104420-2,110 INCHES LONG .... UOC:DAA,DAB,DAE,DAF,DAG,DAH,DAL,DAW, DAX,V12,V13,V14,V15,V18,V19,V20,V21, V22,ZAA,ZAB,ZAE,ZAF,ZAG,ZAH,ZAL	1
11	PAOZZ	4730000691187	81343	6-4 100202BA	ELBOW,PIPE TO TUBE	2

END OF FIGURE

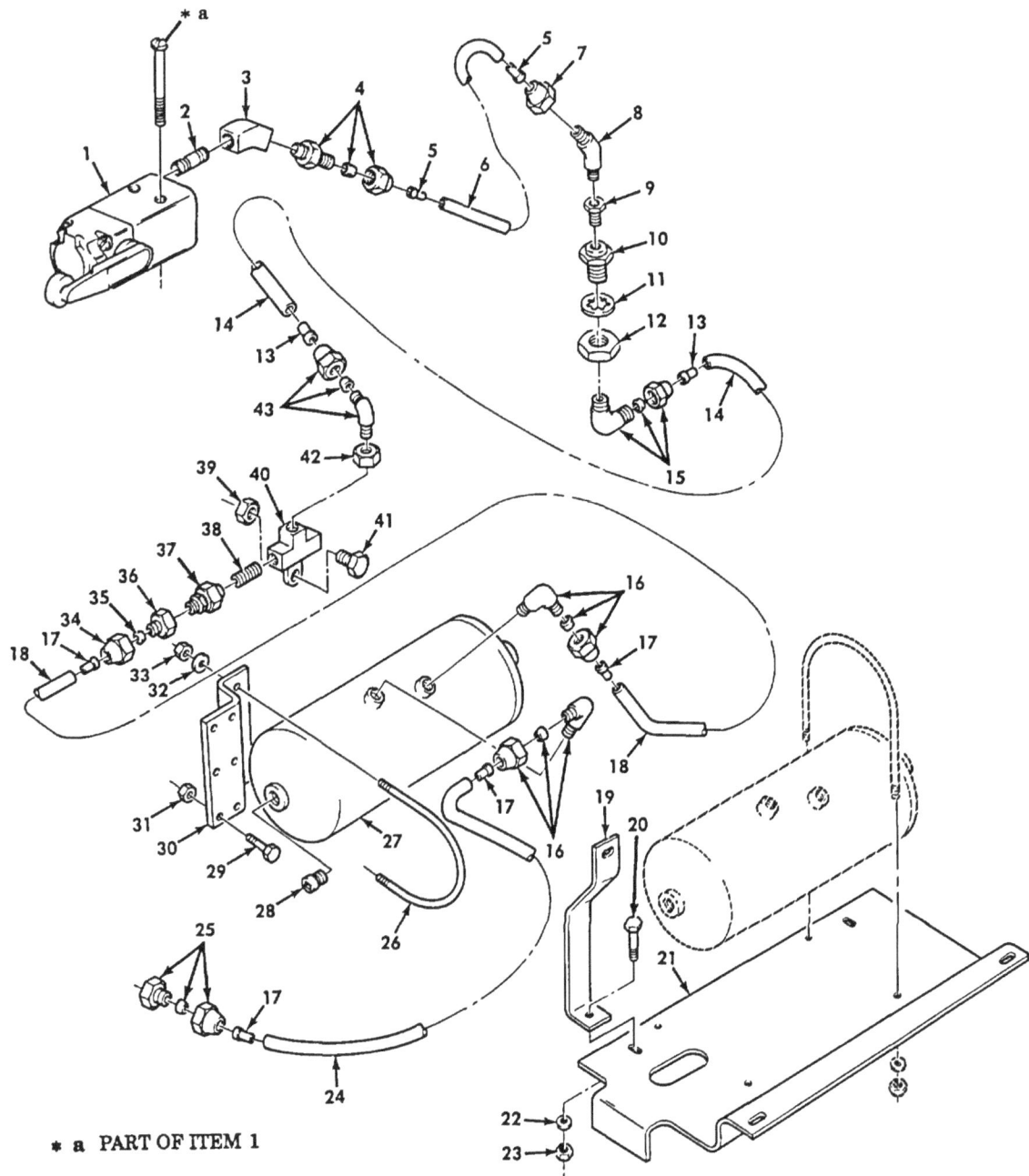

Figure 266. Spring Brake Reservoir, Mounting Hardware, Spring Brake Release Valve, and Related Air Supply Lines.

(1)	(2)	(3)	(4)	(5)	(6)	(7)
ITEM NO	SMR CODE	NSN	CAGEC	PART NUMBER	DESCRIPTION AND USABLE ON CODES (UOC)	QTY

GROUP 1208 AIR BRAKES

FIG. 266 SPRING BRAKE RESERVOIR, MOUNTING HARDWARE, SPRING BRAKE RELEASE VALVE, AND RELATED AIR SUPPLY LINES

(1)	(2)	(3)	(4)	(5)	(6)	(7)
1	PAOZZ	4820013293245	97902	LSC-3270-S26022-1	VALVE,LINEAR,DIRECT	1
2	PAOZZ	4730001961966	96906	MS51846-13	NIPPLE,PIPE	1
3	PAOZZ	4730001257979	29510	444136	TEE,PIPE	1
4	PAOZZ	4730002704580	96906	MS39179-2	ADAPTER,STRAIGHT,PI.	1
5	PAOZZ	4730001324588	79470	2030X4A	INSERT,TUBE FITTING..	2
6	MOOZZ		19207	CPR104420-1-6	HOSE,NONMETALLIC MAKE FROM HOSE, ... P/N CPR104420-1, 6 INCHES LONG....................	1
7	PAOZZ	4730000116452	81343	4-010110B	NUT,TUBE COUPLING	1
8	PAOZZ	4730012023351	19207	7336402-1	ELBOW,PIPE TO TUBE	1
9	PAOZZ	4730001805038	24617	444026	BUSHING,PIPE	1
10	PAOZZ	4730004213924	19207	7339982	ADAPTER,STRAIGHT ,PI.	1
11	PAOZZ	5310005826714	96906	MS35333-49	WASHER,LOCK	1
12	PAOZZ	5310002410157	19207	5331179	NUT,PLAIN,HEXAG ON	1
13	PAOZZ	4730010326038	19207	CPR102321-4	INSERT,TUBE FITTING..	2
14	MOOZZ		19207	CPR104420-3-25	HOSE,NONMETALLIC MAKE FROM HOSE, ... P/N CPR104420-3, 25 INCHES LONG....................	1
15	PAOZZ	4730002890051	81343	8-6 120202BA	ELBOW,PIPE TO TUBE	1
16	PAOZZ	4730005952572	06853	216310	ELBOW,PIPE TO TUBE	2
17	PAOZZ	4730001324588	79146	H0-159-4	INSERT,TUBE FITTING..	4
18	MOOZZ		19207	CPR104420-3-7	HOSE,NONMETALLIC MAKE FROM HOSE, ... P/N CPR104420-3, 7 INCHES LONG......................	1
19	PFOZZ	5340012298365	19207	12302870	BRACKET,MOUNTING. UOC:DAD,DAJ,DAK,DAL,V16,V17,V18,V24, V25,V39,ZAC,ZAD,ZAJ,ZAK,ZAL	2
20	PAOZZ	5305002692804	96906	MS90726-61	SCREW,CAP,HEXAGON H UOC:DAC,DAD,DAJ,DAK,DAL,V16,V17,V18, V24,V25,V39,ZAC,ZAD,ZAJ,ZAK,ZAL	6
21	PFOZZ	5340011982415	19207	12302756	BRACKET,DOUBLE ANGL.................................... UOC:DAC,DAD,DAJ,DAK,DAL,V16,V17,V18, V24,V25,V39,ZAC,ZAD,ZAJ,ZAK, ZAL	1
22	PAOZZ	5310000806004	96906	MS27183-14	WASHER,FLAT RESERVOIR BRACKET................ UOC:DAC,DAD,DAJ,DAK,DAL,V16,V17,V18, V24,V25,V39,ZAC,ZAD,ZAJ,ZAK,ZAL	6
23	PAOZZ	5310009591488	96906	MS51922-21	NUT,SELF-LOCKING... UOC:DAC,DAD,DAJ,DAK,DAL,V16,V17,V18, V24,V25,V39,ZAC,ZAD,ZAJ,ZAK,ZAL	6
24	MOOZZ		19207	CPR104420-3-20	HOSE,NONMETALLIC MAKE FROM HOSE, ... P/N CPR104420-3, 20 INCHES LONG....................	1
25	PAOZZ	4730001731867	81343	8-6 120103BA	ADAPTER,STRAIGH T,PI.....................................	1
26	PAOZZ	5306007539242	19207	7539242	BOLT,U..	2
27	PAOZZ	2530011112261	19207	12255983	RESERVOIR,AIR..	1
28	PAOZZ	4730012066162	81349	WW-P-471ACABCD	PLUG,PIPE	2

(1) ITEM NO	(2) SMR CODE	(3) NSN	(4) CAGEC	(5) PART NUMBER	(6) DESCRIPTION AND USABLE ON CODES (UOC)	(7) QTY
29	PAOZZ	5305002692804	96906	MS90726-61	SCREW,CAP,HEXAGON H .... UOC:DAA,DAB,DAE,DAF,DAG,DAH,DAW,DAX, V12,V13,V14,V15,V19,V20,V21,V22,ZAA, ZAB,ZAE,ZAF,ZAG,ZAH	8
30	PFOZZ	2510011147542	19207	12276929	SUPPORT,AIR TANK .... UOC:DAA,DAB,DAE,DAF,DAG,DAH,DAW,DAX, V12,V13,V14,V15,V19,V20,V21,V22,ZAA, ZAB,ZAE,ZAF,ZAG,ZAH	2
31	PAOZZ	5310008140672	96906	MS51943-36	NUT,SELF-LOCKING,HE .... UOC:DAA,DAB,DAE,DAF,DAG,DAH,DAW,DAX, V12,V13,V14,V15,V19,V20,V21,V22,ZAA, ZAB,ZAE,ZAF,ZAG,ZAH	8
32	PAOZZ	5310008094058	96906	MS27183-10	WASHER,FLAT	4
33	PAOZZ	5310008775796	96906	MS21044N4	NUT,SELF-LOCKING,HE	4
34	PAOZZ	4730000542572	81343	8 120111B	NUT,TUBE COUPLING	1
35	PAOZZ	4730000542571	81343	8 120115B	SLEEVE	1
36	PAOZZ	4730011167886	19207	12276950	ADAPTER,STRAIGHT,PI	1
37	PAOZZ	4820006333523	06721	N13526H	VALVE,CHECK	1
38	PAOZZ	4730001961504	24617	192075	NIPPLE,PIPE	1
39	PAOZZ	5310002416658	96906	MS51943-34	NUT,SELF-LOCKING,HE	1
40	PAOZZ	4730011122168	19207	12256594	TEE ASSEMBLY,PIPE....................................	1
41	PAOZZ	5306002259089	96906	MS90726-34	BOLT,MACHINE	1
42	PAOZZ	4730011000109	96906	MS51887-7C	BUSHING,PIPE	1
43	PAOZZ	4730011171614	81343	8-6 120302BA	ELBOW,PIPE TO TUBE.................................	1

END OF FIGURE

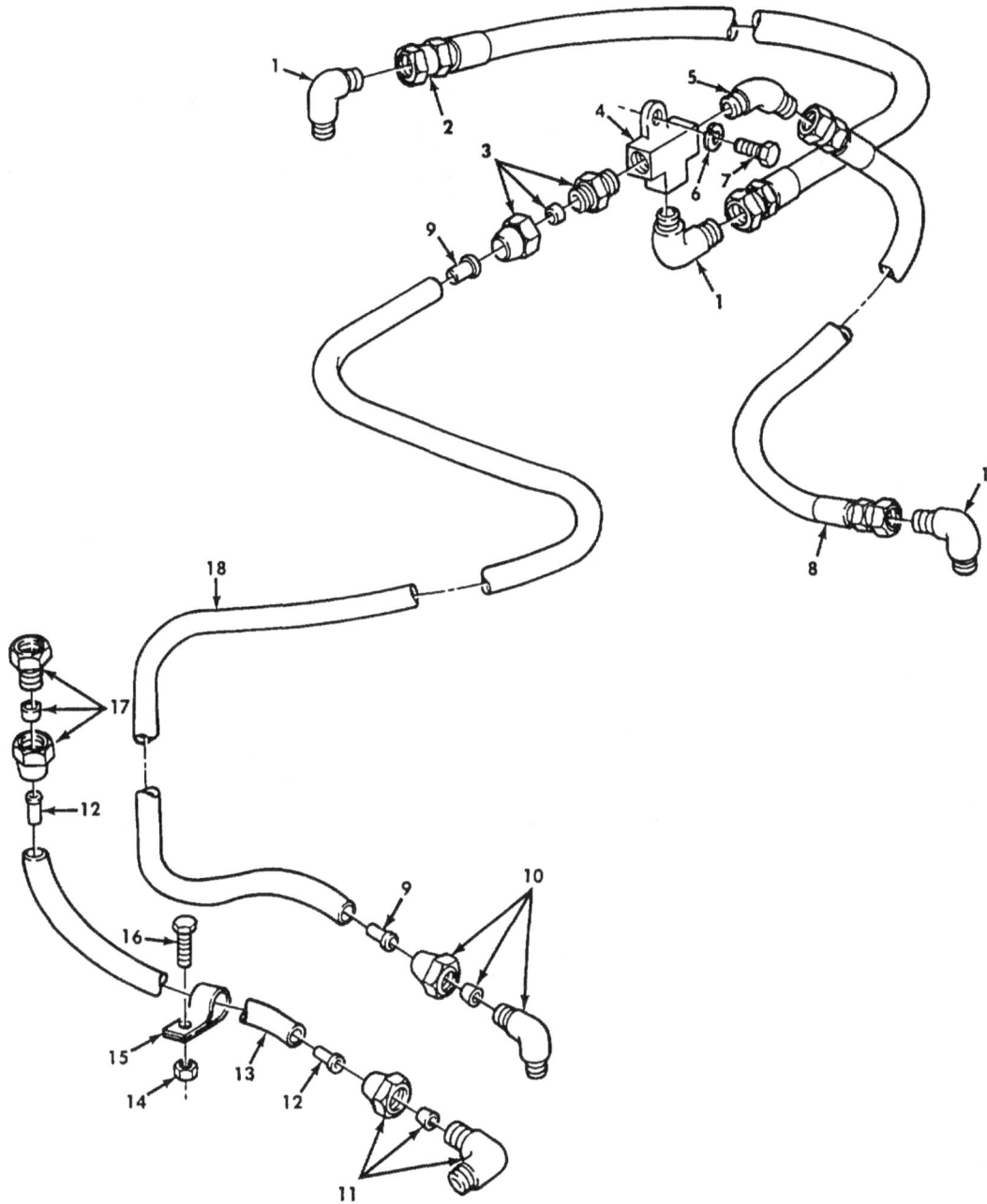

Figure 267. Air Supply Lines, Double-check Valve to Coupling to Push-pull Valve, and Supply Line Quick-release Valve to Tee to Fail-safe Tee.

(1) ITEM NO	(2) SMR CODE	(3) NSN	(4) CAGEC	(5) PART NUMBER	(6) DESCRIPTION AND USABLE ON CODES (UOC)	(7) QTY
					GROUP 1208 AIR BRAKES	
					FIG. 267 AIR SUPPLY LINES, DOUBLE-CHECK VALVE TO COUPLING TO PUSH-PULL VALVE, AND SUPPLY LINE QUICK-RELEASE VALVE TO TEE TO FAIL-SAFE TEE	
1	PAOZZ	4730005551764	96906	MS51504A8	ELBOW,PIPE TO TUBE FRONT REAR................... BRAKE CHAMBER..	3
2	PAOZZ	4720011126577	19207	12256271-5	HOSE ASSEMBLY,NONME FRONT REAR	1
BRAKE CHAMBER						
3	PAOZZ	4730002704616	81343	6-6 060102B	ADAPTER	1
4	PAOZZ	4730011951884	19207	12302675	TEE,PIPE.	1
5	PAOZZ	4730013245071	96906	MS51508A8Z	ELBOW,PIPE TO TUBE FRONT REAR...................	1
BRAKE CHAMBER						
6	PAOZZ	5310006379541	96906	MS35338-46	WASHER,LOCK.	1
7	PAOZZ	5305002693239	80204	B1821BH038F138N	SCREW,CAP,HEXAGON H	1
8	PAOZZ	4720011126576	19207	12256271-2	HOSE ASSEMBLY	1
9	PAOZZ	4730010798821	19207	CPR102321-1	INSERT,TUBE FITTING......................................	2
10	PAOZZ	4730002890155	81343	6-6 120202BA	ELBOW,PIPE TO TUBE	1
11	PAOZZ	4730002890051	81343	8-6 120202BA	ELBOW,PIPE TO TUBE	1
12	PAOZZ	4730010326038	19207	CPR102321-4	INSERT,TUBE FITTING......................................	2
13	MOOZZ		19207	CPR104420-3-44	HOSE,NONMETALLIC MAKE FROM HOSE, ... P/N CPR104420-3, 44 INCHES LONG...................	1
14	PAOZZ	5310008140672	96906	MS51943-36	NUT,SELF-LOCKING,HE	2
15	PAOZZ	5340012315357	19207	7529133-9	CLAMP,LOOP	2
16	PAOZZ	5305009125113	96906	MS51096-359	SCREW,CAP,HEXAGON H	2
17	PAOZZ	4730001731867	81343	8-6 120103BA	ADAPTER,STRAIGHT,PI.	1
18	MOOZZ		19207	CPR104420-2-206	HOSE,NONMETALLIC MAKE FROM HOSE, ... P/N CPR104420-2, 206 INCHES LONG .... UOC:DAA,DAB,DAW,DAX,V12,V13,V14,V15, ZAA,ZAB	1
18	MOOZZ		19207	CPR104420-2-191	HOSE,NONMETALLIC MAKE FROM HOSE, ... P/N CPR104420-2, 191 INCHES LONG UOC:DAL,V18,ZAL	1
18	MOOZZ		19207	CPR104420-2-178	HOSE,NONMETALLIC MAKE FROM HOSE, ... P/N CPR104420-2, 178 INCHES LONG UOC:DAE,DAF,DAG,DAH,V19,V20,V21,V22	1
18	MOOZZ		19207	CPR104420-2-229	HOSE,NONMETALLIC MAKE FORM HOSE, ... P/N CPR104420-2, 229 INCHES LONG UOC:DAC,DAD,DAJ,DAK,V16,V17,V24,V25, V39,ZAC,ZAD,ZAJ,ZAK	1

END OF FIGURE

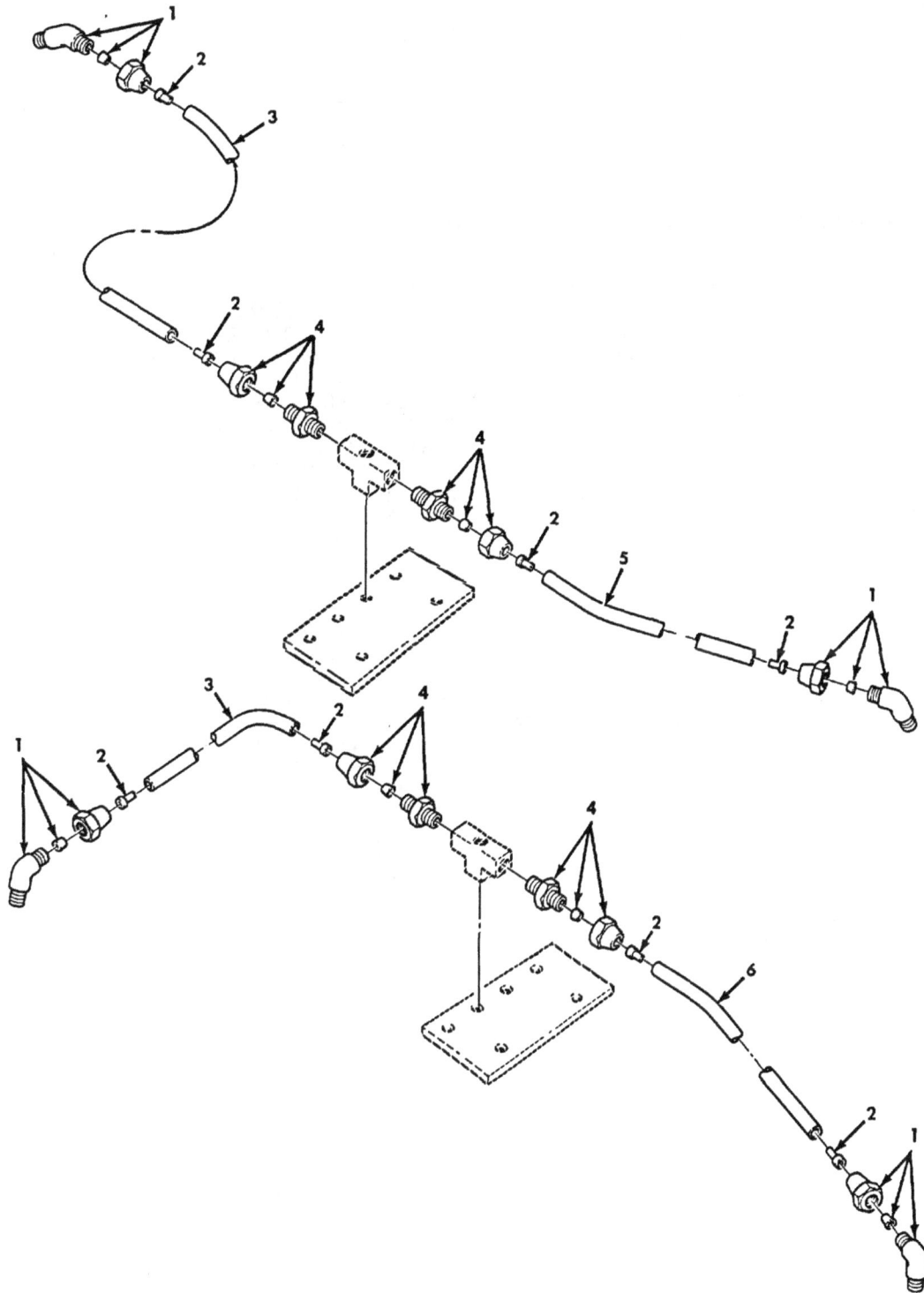

Figure 268. Fail-safe Air Brake Chamber Lines.

(1) ITEM NO	(2) SMR CODE	(3) NSN	(4) CAGEC	(5) PART NUMBER	(6) DESCRIPTION AND USABLE ON CODES (UOC)	(7) QTY
					GROUP 1208	
					FIG. 268 FAIL-SAFE AIR BRAKE CHAMBER LINES	
1	PAOZZ	4730005417500	81343	6-6 120302BA(LON G NUT)	ELBOW,PIPE TO TUBE	4
2	PAOZZ	4730010798821	19207	CPR102321-1	INSERT,TUBE FITTIN..	8
3	MOOZZ		19207	CPR104420-2-14	HOSE,NONMETALLIC MAKE FROM TUBING, ... P/NCPR104420-2,18 INCHES LONG.................... UOC:DAE,DAF,V19,V20,ZAE,ZAF	2
3	MOOZZ		19207	CPR104420-2-18	HOSE,NONMETALLIC MAKE FROM HOSE,P/......... N CPR104420-2,18 INCHES LONG................. UOC:DAA,DAB,DAC,DAD,DAG,DAH,DAJ,DAK, DAL,DAW,DAX,V12,V13,V14,V15,V16,V17, V18,V21,V22,V24,V25,V39,ZAA,ZAB,ZAC, ZAD,ZAG,ZAH,ZAJ,ZAK,ZAL	1
4	PAOZZ	4730002704616	81343	6-6060102B	ADAPTER,STRAIGHT,PI...	4
5	MOOZZ		19207	CPR104420-2-20	HOSE,NONMETALLIC MAKE FROM TUBING, ... P/N CPR104420-2,20 INCHES LONG.................... UOC:DAA,DAB,DAC,DAD,DAG,DAH,DAJ,DAK, DAL,DAW,DAX,V12,V13,V14,V15,V16,V17, V18,V21,V22,V24,V25,V39,ZAA,ZAB,ZAC, ZAD,ZAG,ZAH,ZAJ,ZAK,ZAL	1
5	MOOZZ		19207	CPR104420-2-16	HOSE,NONMETALLIC MAKE FROM HOSE,P/......... N CPR104420-2, 16 INCHES LONG UOC:DAE,DAF,V19,V20	1
6	MOOZZ		19207	CPR104420-2-20	HOSE,NONMETALLICMAKE FROM HOSE,P/.......... N CPR104420-3, 19 INCHES LONG.................... UOC:DAA,DAB,DAC,DAD,DAG,DAH,DAJ,DAK, DAL,DAW,DAX,V12,V13,V14,V15,V16,V17, V18,V21,V22,V24,V25,V39	1
6	MOOZZ		19207	CPR104420-2-19	HOSE,NONMETALLIC MAKE FROM TUBING, ... P/N CPR104420-3,20 INCHES LONG.................... UOC:DAE,DAF,V19,V20,ZAE,ZAF	1

END OF FIGURE

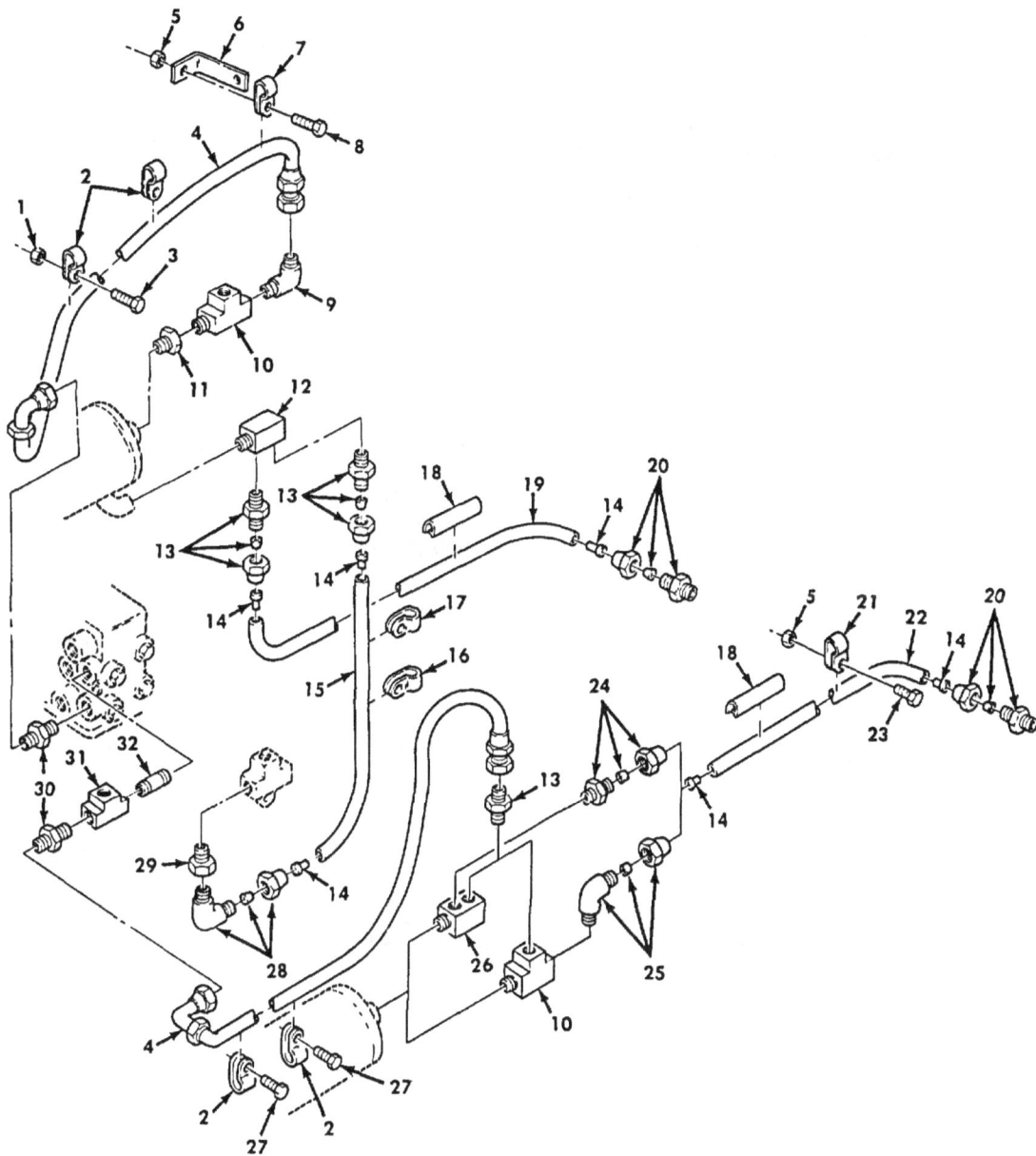

Figure 269. Air Supply Lines, Treadle Value to Primary and Secondary Air Tanks, and Relay Values.

(1) ITEM NO	(2) SMR CODE	(3) NSN	(4) CAGEC	(5) PART NUMBER	(6) DESCRIPTION AND USABLE ON CODES (UOC)	(7) QTY
					GROUP 1208 AIR BRAKES	
					FIG. 269 AIR SUPPLY LINES, TREADLE VALVE TO PRIMARY AND SECONDARY AIR TANKS, AND RELAY VALVES	
1	PAOZZ	5310009359022	96906	MS51943-32	NUT,SELF-LOCKING,HE	1
2	PAOZZ	5340008091494	96906	MS21333-105	CLAMP,LOOP	6
3	PAOZZ	5305002693239	80204	B1821BH038F138N	SCREW,CAP,HEXAGON H	1
4	PAOZZ	4720011122328	19207	12256269-1	HOSE ASSEMBLY,NONME	2
5	PAOZZ	5310008140672	96906	MS51943-36	NUT,SELF-LOCKING,HE	2
6	PAOZZ	5340011978214	19207	12302693	BRACKET,ANGLE	1
7	PAOZZ	5340000538994	96906	MS21333-126	CLAMP,LOOP	2
8	PAOZZ	5305002692803	96906	MS90726-60	SCREW,CAP,HEXAGON H	1
9	PAOZZ	4730002736686	96906	MS510504A8-8	ELBOW,PIPE TO TUBE	1
10	PAOZZ	4730005402745	24617	444152	TEE,PIPE UOC:DAL,V18,V39,ZAL	2
10	PAOZZ	4730005402745	24617	444152	TEE,PIPE .... UOC:DAA,DAB,DAC,DAD,DAE,DAF,DAG,DAH, DAJ,DAK,DAW,DAX,V12,V13,V14,V15,V16, V17,V19,V21,V22,V24,V25,ZAA,ZAB,ZAC, ZAD,ZAE,ZAF,ZAG,ZAH,ZAJ,ZAK	1
11	PAOZZ	4730011924729	81348	WW-P-471BDQBCDB	BUSHING,PIPE	1
12	PAOZZ	4730011150433	19207	12255986	CONNECTOR,MULTIPLE	1
13	PAOZZ	4730002029036	81343	8-8 120102BA	ADAPTER,STRAIGHT,PI.................................	2
14	PAOZZ	4730010326038	19207	CPR102321-4	INSERT,TUBE FITTING	6
15	MOOZZ		19207	CPR104420-3-143	HOSE,NONMETALLIC PRIMARY TO..................... SECONDARY RESERVOIR,MAKE FROM HOSE, P/N CPR104420-3, 143 INCHES LONG .... UOC:DAE,DAF,DAG,DAH,V19,V20,V21,V22, ZAE,ZAF,ZAG,ZAH	1
15	MOOZZ		19207	CPR104420-3-166	HOSE,NONMETALLIC MAKE FROM HOSE, ... P/N CPR104420-3, 166 INCHES LONG UOC:DAA,DAB,DAL,DAW,DAX,V12,V13,V14, V15,V18,ZAA,ZAB,ZAL	1
15	MOOZZ		19207	CPR104420-3-192	HOSE,NONMETALLIC PRIMARY TO..................... SECONDARY RESERVOIR,MAKE FROM HOSE, P/N CPR104420-3, 192 INCHES LONG .... UOC:DAC,DAD,DAJ,DAK,V16,V17,V24,V25, ZAC,ZAD,ZAJ,ZAK	1
16	PAOZZ	5340011979350	19207	7529133-8	CLAMP,LOOP .... UOC:DAA,DAB,DAE,DAF,DAG,DAH,DAL,DAW, DAX,V12,V13,V14,V15,V18,V19,V20,V21, V22,ZAA,ZAB,ZAE,ZAF,ZAG,ZAH,ZAL	1
16	PAOZZ	5340011979350	19207	7529133-8	CLAMP,LOOP UOC:DAC,DAD,DAJ,DAK,V16,V17,V24,V25, V39,ZAC,ZAD,ZAJ,ZAK	2
17	PAOZZ	5340003379619	19207	7373381	CLAMP,LOOP	1
18	PAOZZ	5340011885087	19207	12302768	CLIP,RETAINING	4

(1) ITEM NO	(2) SMR CODE	(3) NSN	(4) CAGEC	(5) PART NUMBER	(6) DESCRIPTION AND USABLE ON CODES (UOC)	(7) QTY
19	MOOZZ		19207	CPR104420-3-126	HOSE,NONMETALLIC PRIMARY RESERVOIR .... TO REAR RELAY VALVE,MAKE FROM HOSE, P/N CPR104420-3,126 INCHES LONG UOC:DAA,DAB,DAL,DAW,DAX,V12,V13,V14, V15,V18,ZAA,ZAB,ZAL	1
19	MOOZZ		19207	CPR104420-3-95	HOSE,NONMETALLIC PRIMARY RESERVOIR .... TO REAR RELAY VALVE,MAKE FROM HOSE, P/N CPR104420-3, 95 INCHES LONG .... UOC:DAE,DAF,DAG,DAH,V19,V20,V21,V22, ZAE,ZAF,ZAG,ZAH	1
19	MOOZZ		19207	CPR104420-3-187	HOSE,NONMETALLIC MAKE FROM HOSE, ... P/N CPR104420-3, 187 INCHES LONG UOC:DAL,V18,ZAL	1
19	MOOZZ		19207	CPR104420-3-107	HOSE,NONMETALLIC MAKE FROM HOSE, ... P/N CPR104420-3, 107 INCHES LONG UOC:DAL,V18,ZAL	1
19	MOOZZ		19207	CPR104420-3-145	HOSE,NONMETALLIC PRIMARY RESERVOIR .... TO REAR RELAY VALVE,MAKE FROM HOSE, P/N CPR104420-3, 145 INCHES LONG .... UOC:DAB,DAC,DAJ,DAK,V16,V17,V24,V25, ZAC,ZAJ,ZAK	1
20	PAOZZ	4730000504358	06853	212193	ADAPTER,STRAIGHT,PI.....................................	2
21	PAOZZ	5340001519651	96906	MS21333-129	CLAMP,LOOP .... UOC:DAE,DAF,DAG,DAH,V19,V20,V21,V22, ZAE,ZAF,ZAG,ZAH	1
22	MOOZZ		19207	CPR104420-3-92	HOSE,NONMETALLIC SECONDARY .... RESERVOIR TO FRONT RELAY VALVE,MAKE FROM HOSE,P/N CPR104420-3,92 INCHES LONG.. UOC:DAA,DAB,DAL,DAW,DAX,V12,V13,V14, V15,V18,ZAA,ZAB,ZAL	1
22	MOOZZ		19207	CPR104420-3-80	HOSE,NONMETALLIC SECONDARY .... RESERVOIR TO FRONT RELAY VALVE,MAKE FROM HOSE,P/N CPR104420-3,80 INCHES LONG.. UOC:DAE,DAF,DAG,DAH,V19,V20,V21,V22, ZAE,ZAF,ZAG,ZAH	1
22	MOOZZ		19207	CPR104420-3-130	HOSE,NONMETALLIC SECONDARY .... RESERVOIR TO FRONT RELAY VALVE,MAKE FROM HOSE,P/N CPR104420-3,130 INCHES LONG.. UOC:DAC,DAD,DAJ,DAK,V16,V17,V24,V25, ZAC,ZAD,ZAJ,ZAK	1
23	PAOZZ	5305002693238	80204	B1821BH038F125N	SCREW,CAP,HEXAGON H UOC:DAC,DAD,DAJ,DAK,V13,V14,V15,V16, V17,V18,V24,V25,V39	3
24	PAOZZ	4730002029036	81343	8-8 120102BA	ADAPTER,STRAIGHT,PI..................................... UOC:DAL,V18,V21,V22,V24,V25,V39,ZAA, ZAB,ZAC,ZAD,ZAE,ZAF,ZAG,ZAH,ZAJ,ZAK	1

(1) ITEM NO	(2) SMR CODE	(3) NSN	(4) CAGEC	(5) PART NUMBER	(6) DESCRIPTION AND USABLE ON CODES (UOC)	(7) QTY
25	PAOZZ	4730002736686	81343	8-8 070202CA(CAD)	ELBOW,PIPE TO TUBE .... UOC:DAA,DAB,DAC,DAD,DAE,DAF,DAG,DAH, DAJ,DAK,DAL,DAW,DAX,V12 ,V13 ,V14 ,V15, V16,V17,V18,V19,V20,V21,V22,V24,V25, V39	1
26	PAOZZ	4730011150433	19207	12255986	CONNECTOR,MULTIPLE .... UOC:DAA,DAB,DAC,DAD,DAE,DAF,DAG,DAH, DAJ,DAK,DAW,DAX,V12,V13,V14,V15,V16, V17,V19,V21,V22,V24,V25,ZAA,ZAB,ZAC, ZAD,ZAE,ZAF,ZAG,ZAH,ZAJ,ZAK	1
27	PAOZZ	5306000680513	60285	6893-2	BOLT,MACHIN E..	2
28	PAOZZ	4730011171614	81343	8-6 120302BA	ELBOW,PIPE TO TUBE	1
29	PAOZZ	4730011924729	81348	WW-P-471BDQBCDB	BUSHING,PIPE	1
30	PAOZZ	4730008099427	96906	MS51500A8-8	ADAPTER,STRAIGHT,PI	2
31	PAOZZ	4730007659102	21450	144089	TEE,PIPE .... UOC:DAA,DAB,DAC,DAD,DAE,DAF,DAG,DAH, DAJ,DAK,DAL,DAW,DAX,V12,V13,V14,V15, V16,V17,V21,V22,V24,V25,ZAA,ZAB,ZAC, ZAD,ZAE,ZAF,ZAG,ZAH,ZAJ,ZAK	1
32	PAOZZ	4730001961495	96906	MS51953-80	NIPPLE,PIPE .... UOC:DAA,DAB,DAC,DAD,DAE,DAF,DAG,DAH, DAJ,DAK,DAL,DAW,DAX,V12,V13,V14 ,V15, V16,V17,V21,V22,V24,V25,ZAA,ZAB,ZAC, ZAD, ZAE,ZAF, ZAG, ZAH,ZAJ,ZAK	1

END OF FIGURE

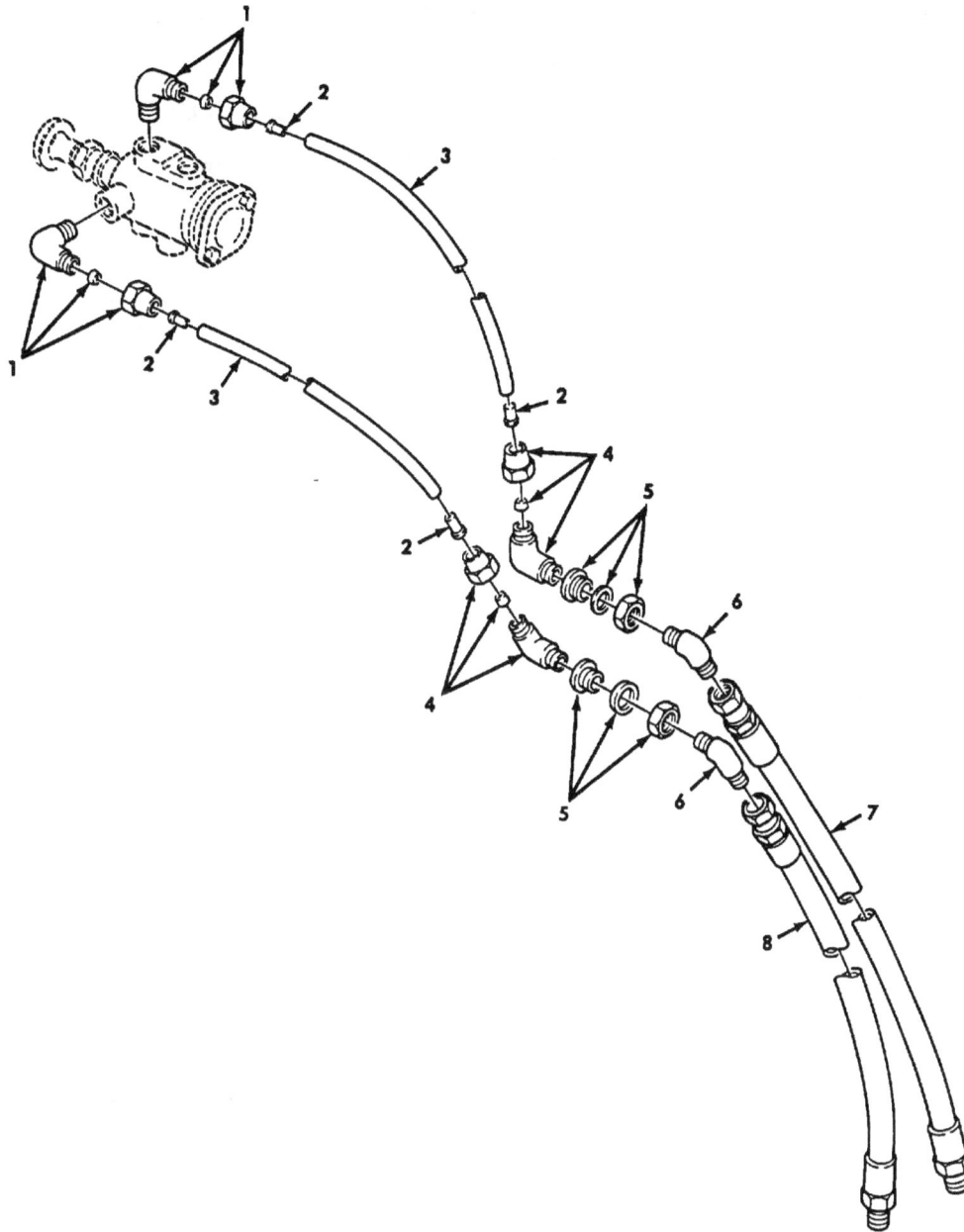

Figure 270. Dash Control Air Supply Lines, Spring Brake Control Valve to Primary and Secondary Air Reservoirs.

(1) ITEM NO	(2) SMR CODE	(3) NSN	(4) CAGEC	(5) PART NUMBER	(6) DESCRIPTION AND USABLE ON CODES (UOC)	(7) QTY
					GROUP 1208 AIR BRAKES	
					FIG. 270 DASH CONTROL AIR SUPPLY LINES, SPRING BRAKE CONTROL VALVE TO PRIMARY AND SECONDARY AIR RESERVOIRS	
1	PAOZZ	4730002871604	81343	6-2 120202BA	ELBOW, PIPE TO TUBE	2
2	PAOZZ	4730010798821	19207	CPR102321-1	INSERT, TUBE FITTING	4
3	MOOZZ		19207	CPR104420-2-29	HOSE, NONMETALLIC MAKE FROM HOSE, P/ .... N CPR104420-2, 29 INCHES LONG......................	2
4	PAOZZ	4730000691187	81343	6-4 100202BA	ELBOW, PIPE TO TUBE	2
5	PAOZZ	4730002786318	19207	8328782	COUPLING, PIPE	2
6	PAOZZ	4730009300982	96906	MS20823-6B	ELBOW, PIPE TO TUBE	2
7	PAOZZ	4720011195844	19207	12256272-2	HOSE ASSEMBLY, NONME	1
8	PAOZZ	4720011195843	19207	12256272-1	HOSE ASSEMBLY, NONME	1

END OF FIGURE

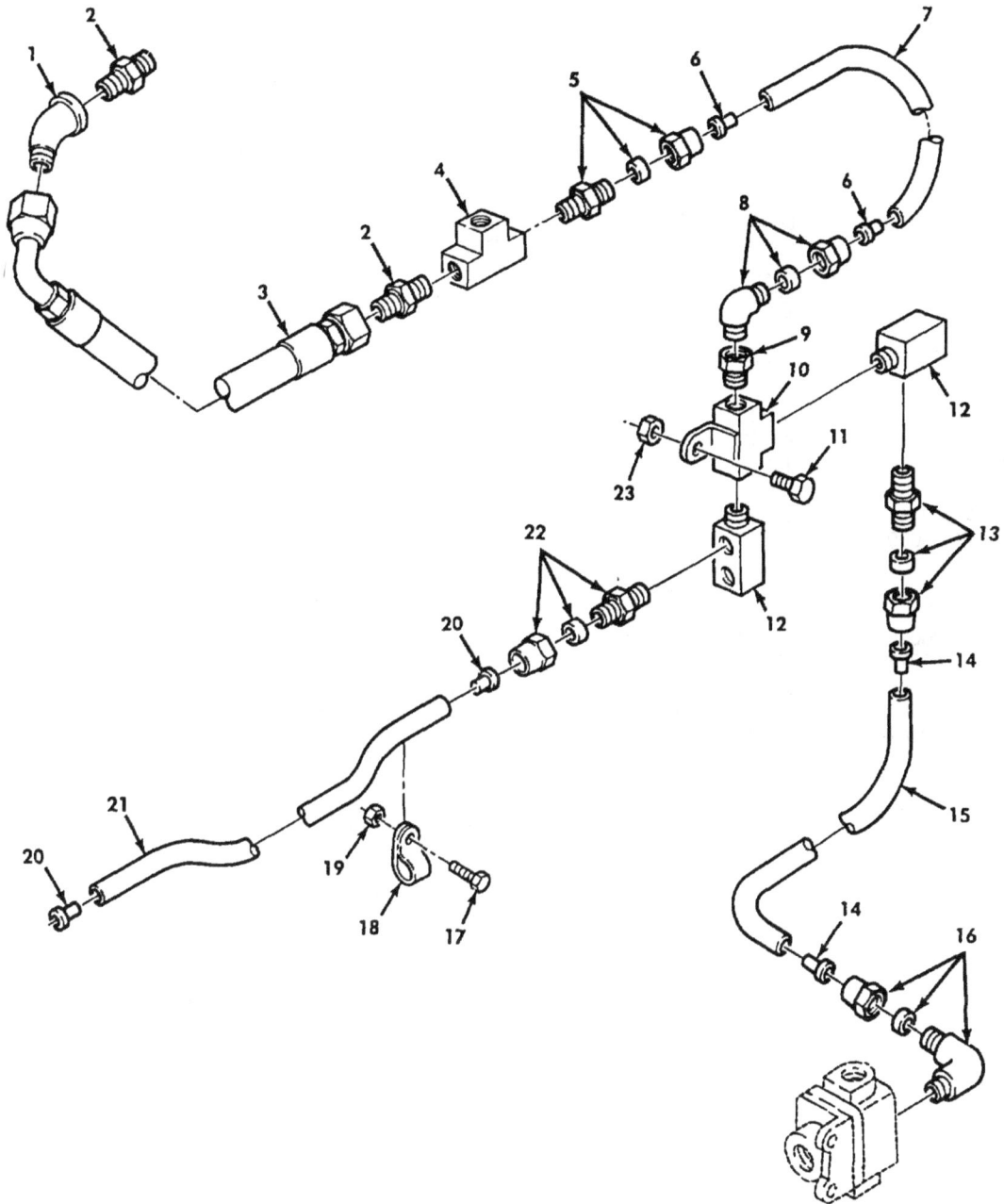

Figure 271. Treadle Valve to Quick-release Valve and Air Intake Tubing.

(1) ITEM NO	(2) SMR CODE	(3) NSN	(4) CAGEC	(5) PART NUMBER	(6) DESCRIPTION AND USABLE ON CODES (UOC)	(7) QTY
					GROUP 1208 AIR BRAKES	
					FIG. 271 TREADLE VALVE TO QUICK- RELEASE VALVE AND AIR INTAKE TUBING	
1	PAOZZ	4730002493935	96906	MS39231-4	ELBOW, PIPE	1
2	PAOZZ	4730008659251	96906	MS51500-A12-8	ADAPTER, STRAIGHT, PI	2
3	PAOZZ	4720011959518	19207	12302873	HOSE ASSEMBLY, NONME	1
4	PAOZZ	4730005402745	24617	444152	TEE, PIPE	1
5	PAOZZ	4730002029035	81343	12-8 120102BA	ADAPTER, STRAIGHT, PI	1
6	PAOZZ	4730001807031	19207	CPR102321-2	INSERT, TUBE FITTING	2
7	MOOZZ		19207	CPR104420-5-15	HOSE, NONMETALLIC MAKE FROM HOSE, P/ .... N CPR104420-5, 17 INCHES LONG	1
8	PAOZZ	4730002890155	81343	6-6 120202BA	ELBOW, PIPE TO TUBE	1
9	PAOZZ	4730004576295	24617	15666151	REDUCER, PIPE	1
10	PAOZZ	4730011122168	19207	12256594	TEE ASSEMBLY, PIPE	1
11	PAOZZ	5306000514077	80204	B1821BH031F113N	BOLT, MACHINE	1
12	PAOZZ	4730011150433	19207	12255986	CONNECTOR, MULTIPLE	2
13	PAOZZ	4730002029036	81343	8-8 120102BA	ADAPTER, STRAIGHT, PI	1
14	PAOZZ	4730010326038	19207	CPR102321-4	INSERT, TUBE FITTING	2
15	MOOZZ		19207	CPR104420-3-22	HOSE, NONMETALLIC MAKE FROM HOSE, P/ .... N CPR104420-3, 22 INCHES LONG	1
16	PAOZZ	4730002890051	81343	8-6 120202BA	ELBOW, PIPE TO TUBE	1
17	PAOZZ	5305002692804	96906	MS90726-61	SCREW, CAP, HEXAGON H	1
18	PAOZZ	5340000538994	96906	M321333-126	CLAMP, LOOP	1
19	PAOZZ	5310008140672	96906	MS51943-36	NUT, SELF-LOCKING, HE	1
20	PAOZZ	4730010798821	19207	CPR102321-1	INSERT, TUBE FITTING	2
21	MOOZZ		19207	CPR104420-2-45	HOSE, NONMETALLIC MAKE FROM HOSE, P/ .... N CPR104420-2, 45 INCHES LONG	1
22	PAOZZ	4730008371177	81343	6-8 120102BA	ADAPTER, STRAIGHT, PI	1
23	PAOZZ	4730010798821	19207	CPR102321-1	INSERT, TUBE FITTING	2

END OF FIGURE

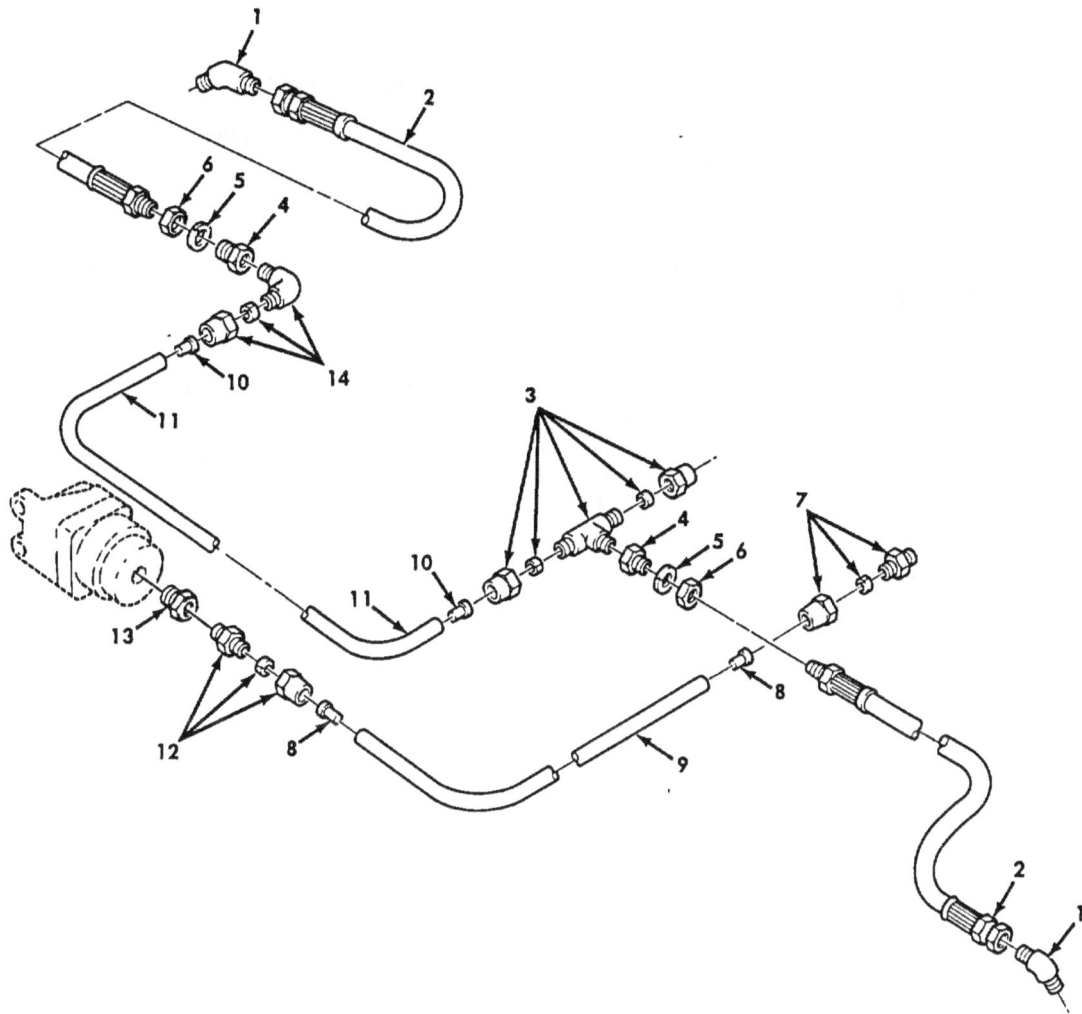

Figure 272. Front Brake Chamber Vent Lines and Limiting Valve.

(1) ITEM NO	(2) SMR CODE	(3) NSN	(4) CAGEC	(5) PART NUMBER	(6) DESCRIPTION AND USABLE ON CODES (UOC)	(7) QTY
					GROUP 1208 AIR BRAKES	
					FIG. 272 FRONT BRAKE CHAMBER VENT LINES AND LIMITING VALVE	
1	PAOZZ	4730010050623	96906	MS51508A6-6S	ELBOW, PIPE TO TUBE	2
2	PAOZZ	4720011538240	19207	12276944-4	HOSE ASSEMBLY, NONME	2
3	PAOZZ	4730010955833	79470	1372X6X6X6	TEE, PIPE TO TUBE	1
4	PAOZZ	4730004213924	19207	7339982	ADAPTER, STRAIGHT, PI	2
5	PAOZZ	5310005826714	96906	MS35333-49	WASHER, LOCK	2
6	PAOZZ	5310002410157	19207	5331179	NUT, PLAIN, HEXAGON	2
7	PAOZZ	4730002029036	81343	8-8 120102BA	ADAPTER, STRAIGHT, PI	1
8	PAOZZ	4730010326038	19207	CPR102321-4	INSERT, TUBE FITTING	2
9	MOOZZ		19207	CPR104420-3-80	HOSE, NONMETALLIC MAKE FROM HOSE, P/ .... N CPR104420-3, 80 INCHES LONG	1
10	PAOZZ	4730010798821	19207	CPR102321-1	INSERT, TUBE FITTING......................................	2
11	MOOZZ		19207	CPR104420-2-100	HOSE, NONMETALLIC MAKE FROM HOSE, P/N..... CPR104420-2, 100 INCHES LONG........................	1
12	PAOZZ	4730005260284	06853	217690	ADAPTER, STRAIGHT, PI	1
13	PAOZZ	4730005291487	24617	444012	REDUCER, PIPE	1
14	PAOZZ	4730002890155	81343	6-6 120202BA	ELBOW, PIPE TO TUBE	1

END OF FIGURE

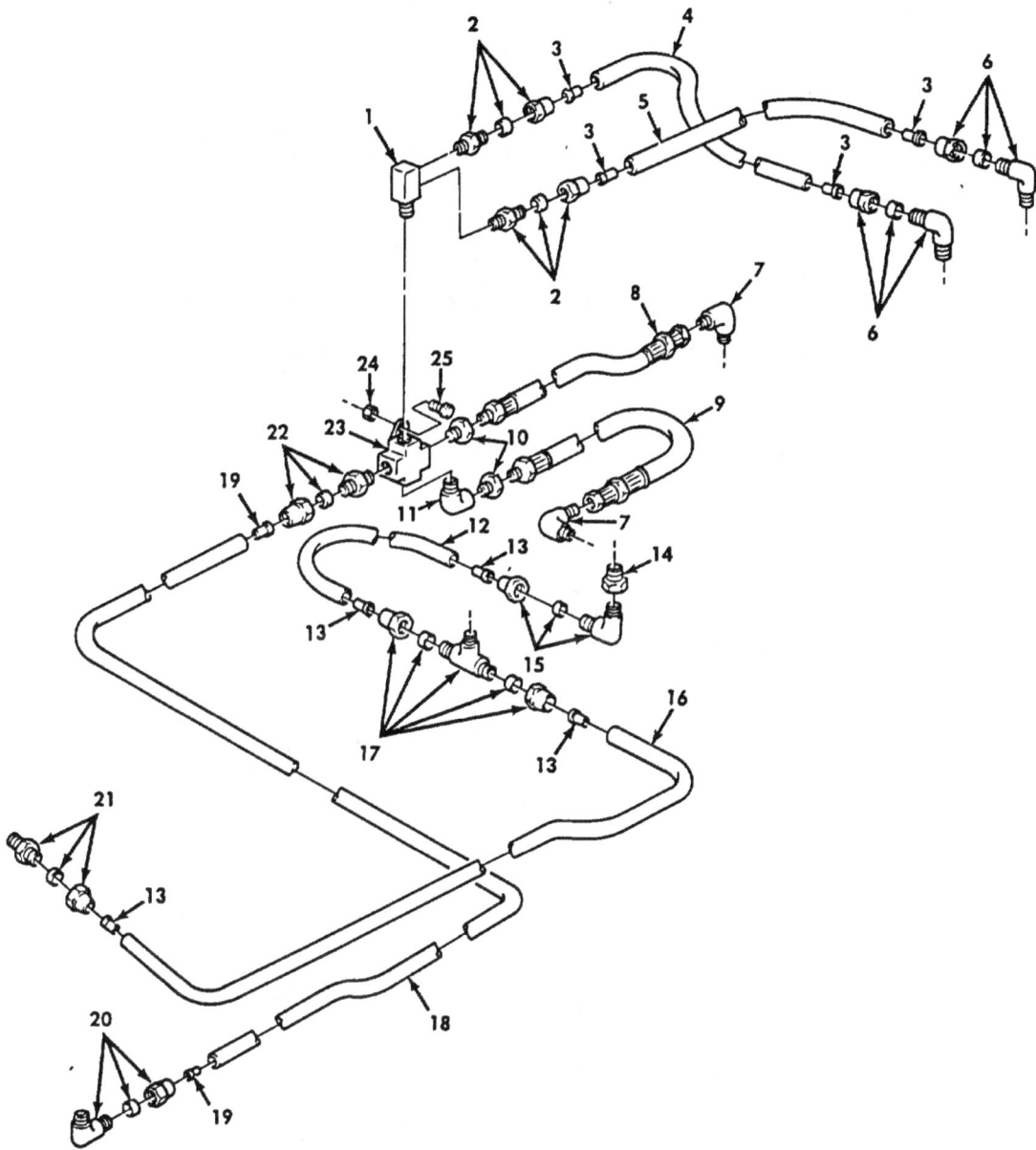

Figure 273. Vent Lines, Air Intake Manifold to Primary and Secondary Relay Values, and Rear Axle Vent Lines.

(1) ITEM NO	(2) SMR CODE	(3) NSN	(4) CAGEC	(5) PART NUMBER	(6) DESCRIPTION AND USABLE ON CODES (UOC)	(7) QTY
					GROUP 1208 AIR BRAKES	
					FIG. 273 VENT LINES, AIR INTAKE MANIFOLD TO PRIMARY AND SECONDARY RELAY VALVES, AND REAR AXLE VENT LINES	
1	PAOZZ	4730011079692	19207	12255985	ELBOW, PIPE	1
2	PAOZZ	4730000691186	81343	6-4 120102BA	ADAPTER, STRAIGHT, PI	2
3	PAOZZ	4730010798821	19207	CPR102321-1	INSERT, TUBE FITTING	4
4	MOOZZ		19207	CPR104420-2-14	HOSE, NONMETALLIC MAKE FRON HOSE, ... P/N CPR104420-2, 14 INCHES LONG	1
5	MOOZZ		19207	CPR104420-2-33	HOSE, NONMETALLIC MAKE FROM HOSE, ... P/N CPR 104420-2, 33 INCHES LONG	1
6	PAOZZ	4730002871604	81343	6-2 120202BA	ELBOW, PIPE TO TUBE	2
7	PAOZZ	4730008353003	96906	MS51504A6-6	ELBOW, PIPE TO TUBE	2
8	PAOZZ	4720011885140	19207	12276944-5	HOSE ASSEMBLY, NONME	1
9	PAOZZ	4720011122326	19207	12276944-1	HOSE ASSEMBLY, NONME .... UOC:DAC, DAD, DAG, DAH, DAJ, DAK, V16, V17, V21, V22, V24, V25	1
10	PAOZZ	4730002000257	24617	444014	REDUCER, PIPE	2
11	PAOZZ	4730002775553	24617	444040	ELBOW, PIPE	1
12	MOOZZ		19207	CPR104420-5-37	HOSE, NONMETALLIC MAKE FROM HOSE, P/ .... N CPR104420-5, 37 INCHES LONG	1
13	PAOZZ	4730001807031	19207	CPR102321-2	INSERT, TUBE FITTING	4
14	PAOZZ	4730004576295	24617	15666151	REDUCER, PIPE	1
15	PAOZZ	4730011238824	81343	12-12 120202BA	ELBOW, PIPE TO TUBE	1
16	MOOZZ		19207	CPR104420-5-144	HOSE, NONMETALLIC AIR INTAKE VALVE............ TO FRONT RELAY VALVE, MAKE FROM HOSE, P/N CPR104420-15, 144 INCHES LONG................. UOC:DAA, DAB, DAW, DAX, V12, V13, V14, V15, ZAA, ZAB	1
16	MOOZZ		19207	CPR104420-5-151	HOSE, NONMETALLIC AIR INTAKE TO .... FRONT RELAY VALVE, MAKE FROM HOSE, P/N CPR104420-5, 151 INCHES LONG .... UOC:DAE, DAF, DAG, DAL, V18, V19, V20, V22, ZAD, ZAF, ZAG, ZAL	1
16	MOOZZ		19207	CPR104420-5-140	HOSE, NONMETALLIC AIR INTAKE TO .... FRONT RELAY VALVE, MAKE FROM HOSE, P/N CPR104420-5, 140 INCHES LONG .... UOC:DAG, DAH, V21, V22, ZAG, ZAH	1
16	MOOZZ		19207	CPR104420-5-200	HOSE, NONMETALLIC AIR INTAKE TO .... FRONT RELAY VALVE, MAKE FROM HOSE, P/ N CPR104420-5, 200 INCHES LONG .... UOC:DAC, DAD, DAJ, DAK, V16, V17, V24, V25, V39, ZAC, ZAD, ZAJ, ZAK	1
17	PAOZZ	4730014426120	24617	144387	TEE, PIPE	1
18	MOOZZ		19207	CPR104420-3-124	HOSE, NONMETALLIC DOUBLE PIPE ELBOW........ TO REAR BRAKE CHAMBER TEE, MAKE FROM HOSE, P/N CPR104420-3, 124 INCHES LONG UOC:DAG, DAH, V21, V22, ZAG, ZAH	1
18	MOOZZ		19207	CPR104420-3-120	HOSE, NONMETALLIC AIR COMPRESSOR TO	1

(1) ITEM NO	(2) SMR CODE	(3) NSN	(4) CAGEC	(5) PART NUMBER	(6) DESCRIPTION AND USABLE ON CODES (UOC)	(7) QTY
					WET RESERVOIR, MAKE FROM TUBING, P/N CPR104420-3, 120 INCHES LONG............................ UOC:DAE, DAF, V19, V20, ZAE, ZAF	
18	MOOZZ		19207	CPR104420-3-136	TUBE, NONMETALLIC ELBOW, PIPE TO .... REAR BRAKE CHAMBER TEE, MAKE FROM HOSE, P/N CPR104420-3, 136 INCHES LONG UOC:DAA, DAB, DAL, DAW, DAX, V12, V13, V14, V15, V18, ZAA, ZAB, ZAL	1
18	MOOZZ		19207	CPR104420-3-180	HOSE, NONMETALLIC ELBOW, PIPE TO............... REAR BRAKE CHAMBER TEE, MAKE FROM HOSE, P/N CPR104420-3, 180 INCHES LONG UOC:DAC, DAD, DAJ, DAK, V16, V17, V24, V25, V39, ZAC, ZAD, ZAJ, ZAK	1
19	PAOZZ	4730010326038	19207	CPR102321-4	INSERT, TUBE FITTING...............................	2
20	PAOZZ	4730005952572	06853	216310	ELBOW, PIPE TO TUBE	1
21	PAOZZ	4730002029035	81343	12-8 120102BA	ADAPTER, STRAIGHT, PI...........................	1
22	PAOZZ	4730005260284	06853	217690	ADAPTER, STRAIGHT, PI...........................	1
23	PAOZZ	4730011126561	19207	11669082	CROSS, PIPE	1
24	PAOZZ	5310008140672	96906	MS51943-36	NUT, SELF-LOCKING, HE.........................	1
25	PAOZZ	5305002693238	80204	B1821BH038F125N	SCREW, CAP, HEXAGON H.........................	1

END OF FIGURE

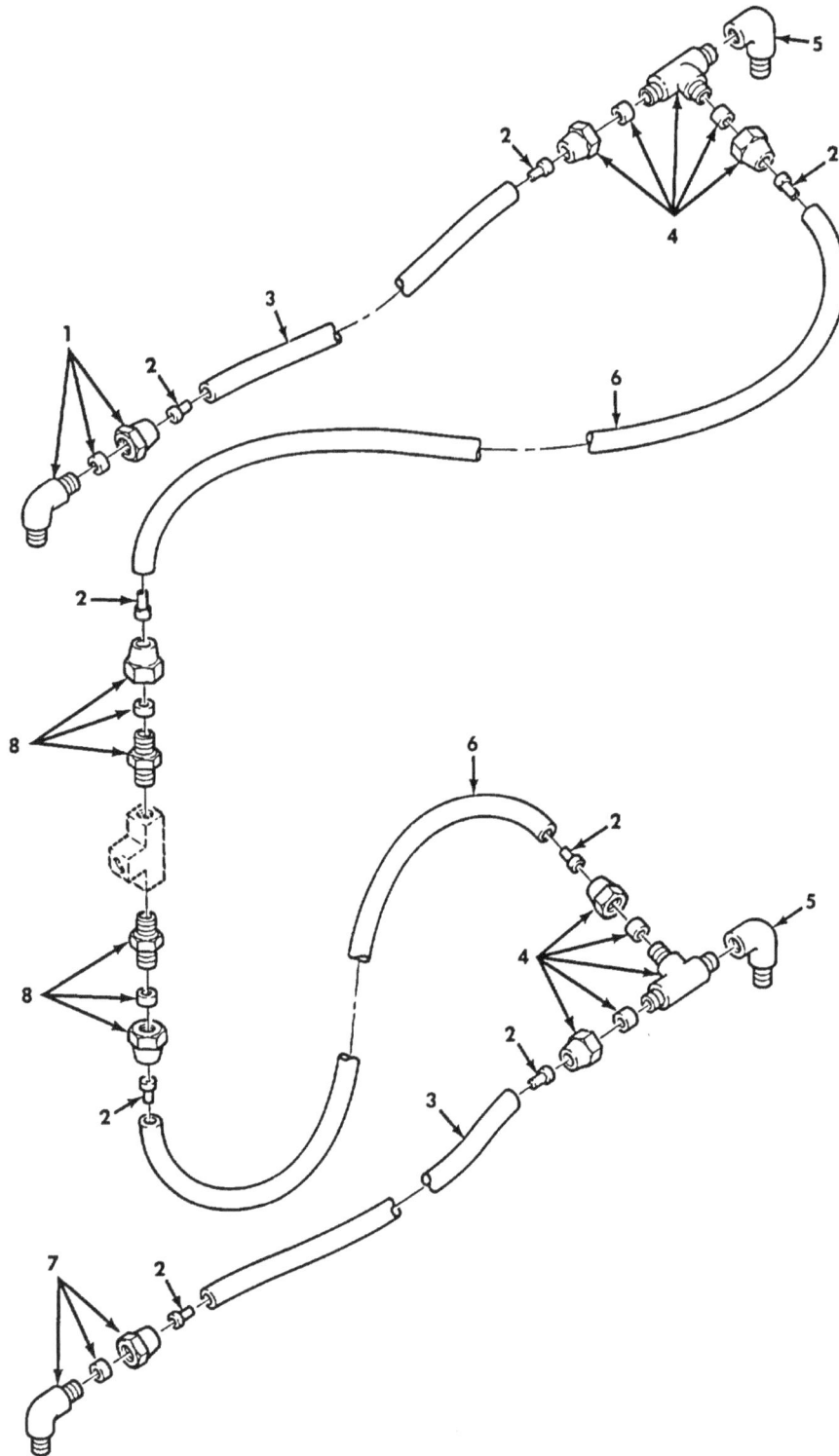

Figure 274. Secondary Vent Lines, Rear Axle Brake Chambers to Fail-safe Brake Chamber.

(1) ITEM NO	(2) SMR CODE	(3) NSN	(4) CAGEC	(5) PART NUMBER	(6) DESCRIPTION AND USABLE ON CODES (UOC)	(7) QTY
					GROUP 1208 AIR BRAKES	
					FIG. 274 SECONDARY VENT LINES, REAR AXLE BRAKE CHAMBERS TO FAIL-SAFE BRAKE CHAMBER	
1	PAOZZ		81343	6-6 120302BA	ELBOW	1
2	PAOZZ	4730010798821	19207	CPR102321-1	INSERT, TUBE FITTING...................................	8
3	MOOZZ		19207	CPR104420-2-17	HOSE MAKE FROM TUBING, P/N 3-4006..............	2
4	PAOZZ	4730009306354	96906	MS39391-3	TEE	2
5	PAOZZ	4730002784822	19207	7364214	ELBOW	2
6	MOOZZ		19207	CPR104420-2-23	HOSE MAKE FROM TUBING, P/N 3-4006..............	2
7	PAOZZ	4730002890155	6-6	120202BA	ELBOW	1
8	PAOZZ	4730002704616	6-6	060102B	ADAPTER, STRAIGHT, PIPE................................	2

END OF FIGURE

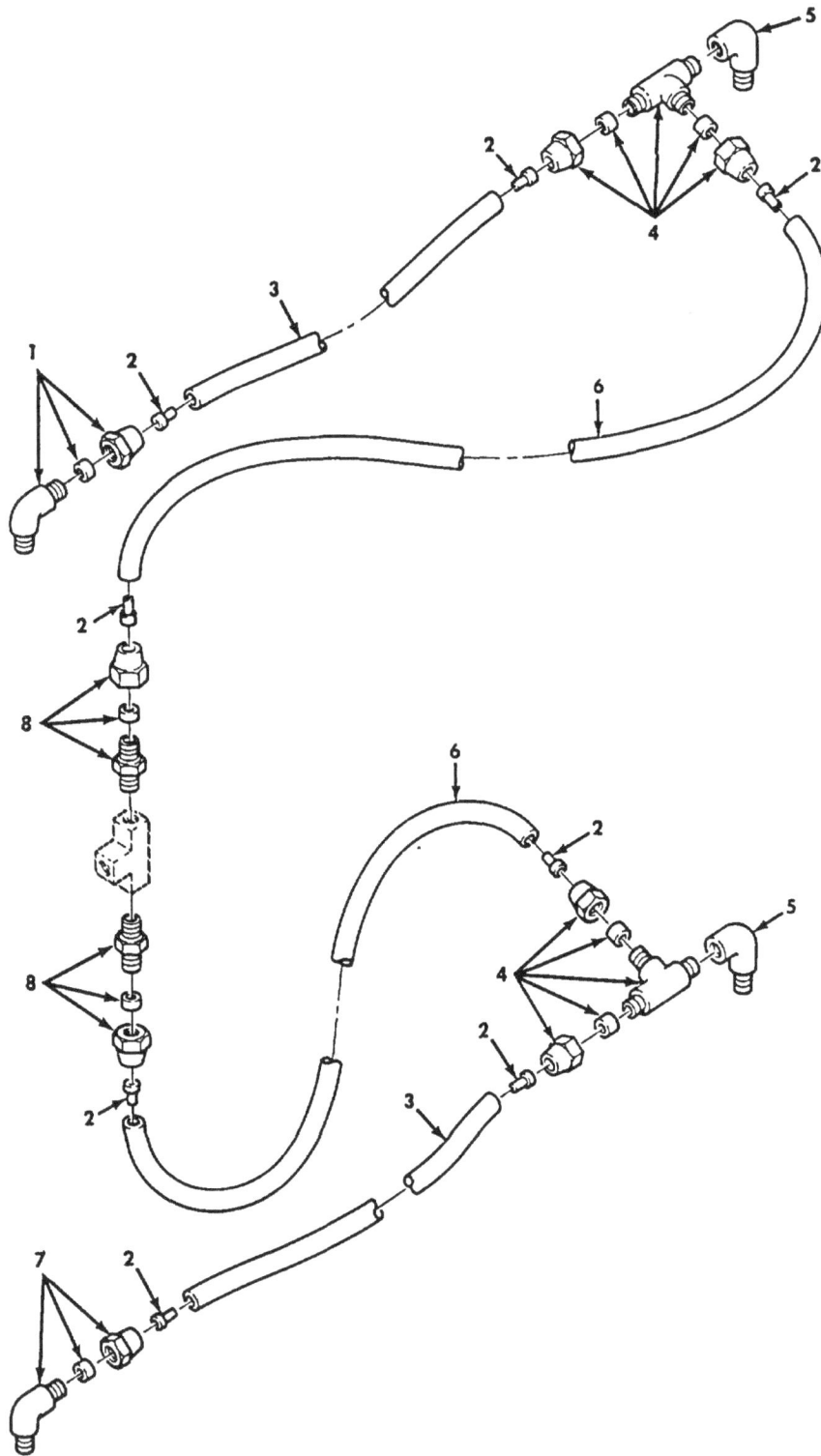

Figure 275. Primary Vent Lines, Rear Axle Air Brake.

(1) ITEM NO	(2) SMR CODE	(3) NSN	(4) CAGEC	(5) PART NUMBER	(6) DESCRIPTION AND USABLE ON CODES (UOC)	(7) QTY

GROUP 1208 AIR BRAKES

FIG. 275 PRIMARY VENT LINES, REAR
AXLE AIR BRAKE

(1) ITEM NO	(2) SMR CODE	(3) NSN	(4) CAGEC	(5) PART NUMBER	(6) DESCRIPTION AND USABLE ON CODES (UOC)	(7) QTY
1	PAOZZ	4730005417500	81343	6-6 120302BA(LON G NUT)	ELBOW, PIPE TO TUBE	1
1	PAOZZ	4730005417500	81343	6-6 120302BA(LON G NUT)	ELBOW, PIPE TO TUBE	1
2	PAOZZ	4730010798821	19207	CPR102321-1	INSERT, TUBE FITTING	8
2	PAOZZ	4730010798821	19207	CPR102321-1	INSERT, TUBE FITTING...	8
3	MOOZZ		19207	CPR104420-2-17	HOSE, NONMETALLIC MAKE FROM TUBING, ... P/N CPR104420-2, 17 INCHES LONG....................	2
3	MOOZZ		19207	CPR104420-2-17	HOSE, NONMETALLIC MAKE FROM TUBING, ... P/N CPR104420-2, 17 INCHES LONG....................	2
4	PAOZZ	4730009306354	96906	MS39191-3	TEE, PIPE TO TUBE..	2
4	PAOZZ	4730009306354	96906	MS39191-3	TEE, PIPE TO TUBE..	2
5	PAOZZ	4730002784822	19207	7364214	ELBOW, PIPE..	2
5	PAOZZ	4730002784822	19207	7364214	ELBOW, PIPE..	2
6	MOOZZ		19207	CPR104420-2-21	HOSE, NONMETALLIC MAKE FROM TUBING, ... P/N CPR 104420-2, 23 INCHES LONG....................	2
6	MOOZZ		19207	CPR104420-2-23	HOSE, NONMETALLIC MAKE FROM TUBING, ... P/N CPR104420-2, 23 INCHES LONG....................	2
7	PAOZZ	4730002890155	81343	6-6 120202BA	ELBOW, PIPE TO TUBE	1
7	PAOZZ	4730002890155	81343	6-6 120202BA	ELBOW, PIPE TO TUBE	1
8	PAOZZ	4730002704616	81343	6-6060102B	ADAPTER, STRAIGHT, PI......................................	2
8	PAOZZ	4730002704616	81343	6-6060102B	ADAPTER, STRAIGHT, PI......................................	2

END OF FIGURE

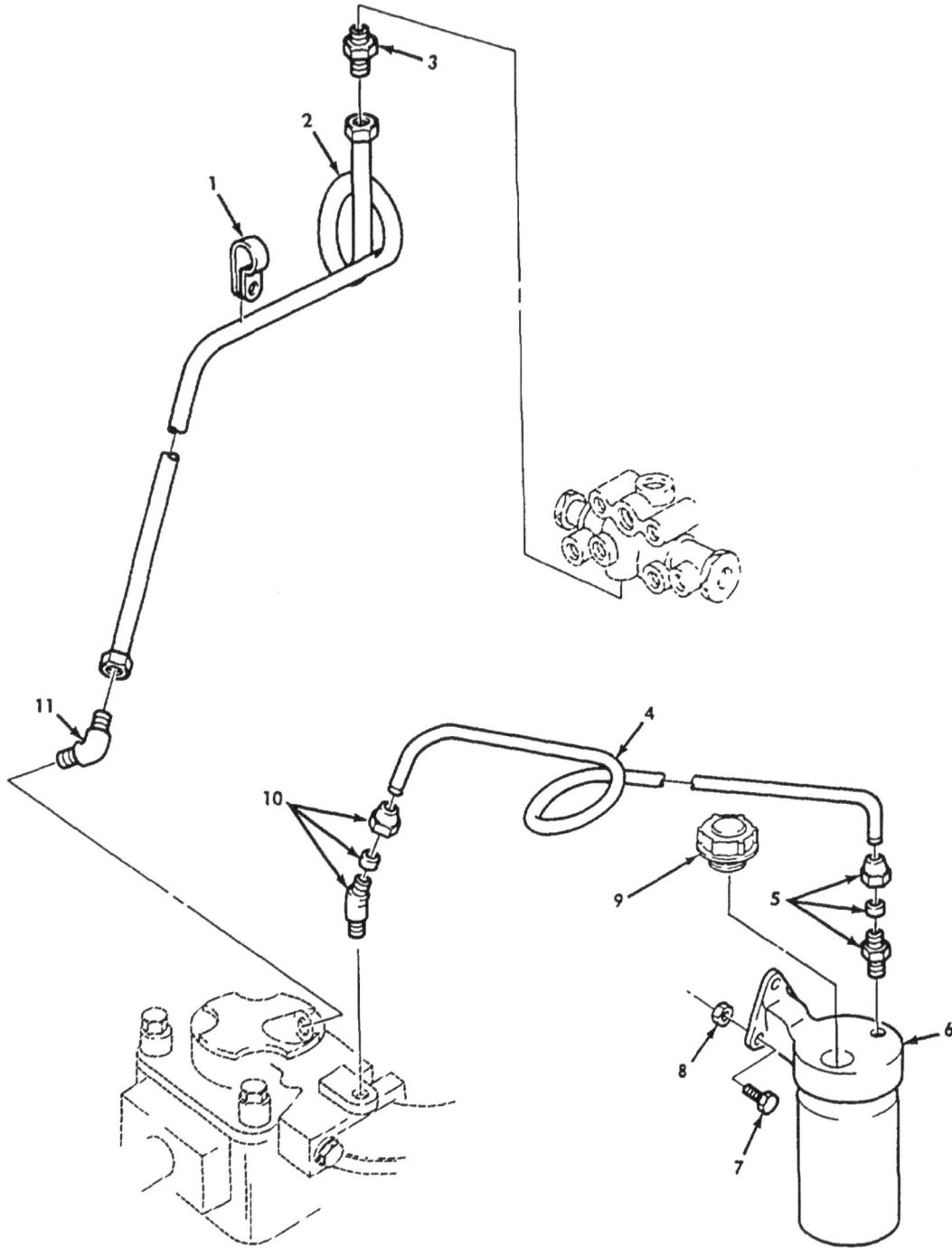

Figure 276. Alcohol Evaporator to Air Compressor to Air Governor Lines.

(1) ITEM NO	(2) SMR CODE	(3) NSN	(4) CAGEC	(5) PART NUMBER	(6) DESCRIPTION AND USABLE ON CODES (UOC)	(7) QTY

GROUP 1208 AIR BRAKES

FIG. 276 ALCOHOL EVAPORATOR TO AIR
COMPRESSOR TO AIR GOVERNOR LINES

(1)	(2)	(3)	(4)	(5)	(6)	(7)
1	PAOZZ	5340000509077	96906	MS21333-119	CLAMP, LOOP	1
2	PAOZZ	4710011208538	19207	12256264	TUBE ASSEMBLY, METAL .... UOC:DAA, DAB, DAC, DAD, DAE, DAF, DAG, DAH, DAJ, DAK, DAL, DAW, DAX, V12, V1 , V134, V14, V15, V16, V17, V18, V19, V20, V21, V22, V24, V25, V39	1
3	PAOZZ	4730002890383	96906	MS51500-A4	ADAPTER, STRAIGHT, PI	1
4	MOOZZ		19207	8689206-34	TUBE, METALLIC MAKE FROM TUBE, P/N............ 8689206 .... UOC:DAA, DAB, DAC, DAD, DAE, DAF, DAG, DAH, DAJ, DAK, DAL, DAW, DAX, V12, V13, V14, V15, V16, V17, V18, V19, V20, V21, V22, V24, V25, V39	1
5	PAOZZ	4730002778750	81343	4-2 120102BA	ADAPTER, STRAIGHT, PI................................ UOC:DAA, DAB, DAC, DAD, DAE, DAF, DAG, DAH, DAJ, DAK, DAL, DAW, DAX, V12, V13, V14, V15, V16, V17, V18, V19, V20, V21, V22, V24, V25, V39	1
6	PAOZZ	2530008597335	96906	MS53067-1	EVAPORATOR, AIR BRAK .... UOC:DAA, DAB, DAC, DAD, DAE, DAF, DAG, DAH, DAJ, DAK, DAL, DAW, DAX, V12, V13, V14, V15, V16, V17, V18, V19, V20, V21, V22, V24, V25, V39	1
7	PAOZZ	5305002693239	80204	B1821BH038F138N	SCREW, CAP, HEXAGON H .... UOC:DAA, DAB, DAC, DAD, DAE, DAF, DAG, DAH, DAJ, DAK, DAL, DAW, DAX, V12, V13, V14, V15, V16, V17, V18, V19, V20, V21, V22, V24, V25, V39	3
8	PAOZZ	5310009359022	96906	MS51943-32	NUT, SELF-LOCKING, HE.................................. UOC:DAA, DAB, DAC, DAD, DAE, DAF, DAG, DAH, DAJ, DAK, DAL, DAW, DAX, V12, V13, V14, V15, V16, V17, V18, V19, V20, V21, V22, V24, V25, V39	3
9	PAOZZ	5340011609589	06853	248824	CAP, EVAPORATOR .... UOC:DAA, DAB, DAC, DAD, DAE, DAF, DAG, DAH, DAJ, DAK, DAL, DAW, DAX, V12, V13, V14, V15, V16, V17, V18, V19, V20, V21, V22, V24, V25, V39	1
10	PAOZZ	4730002784651	96906	MS51817-3	ELBOW, PIPE TO TUBE .... UOC:DAA, DAB, DAC, DAD, DAE, DAF, DAG, DAH, DAJ, DAK, DAL, DAW, DAX, V12, V13, V14, V15, V16, V17, V18, V19, V20, V21, V22, V24, V25, V39	1
11	PAOZZ	4730006473207	96906	MS51504A4	ELBOW, PIPE TO TUBE .... UOC:DAA, DAB, DAC, DAD, DAE, DAF, DAG, DAH, DAJ, DAK, DAL, DAW, DAX, V12, V13, V14, V15, V16, V17, V18, V19, V20, V21, V22, V24, V25, V39	1

END OF FIGURE

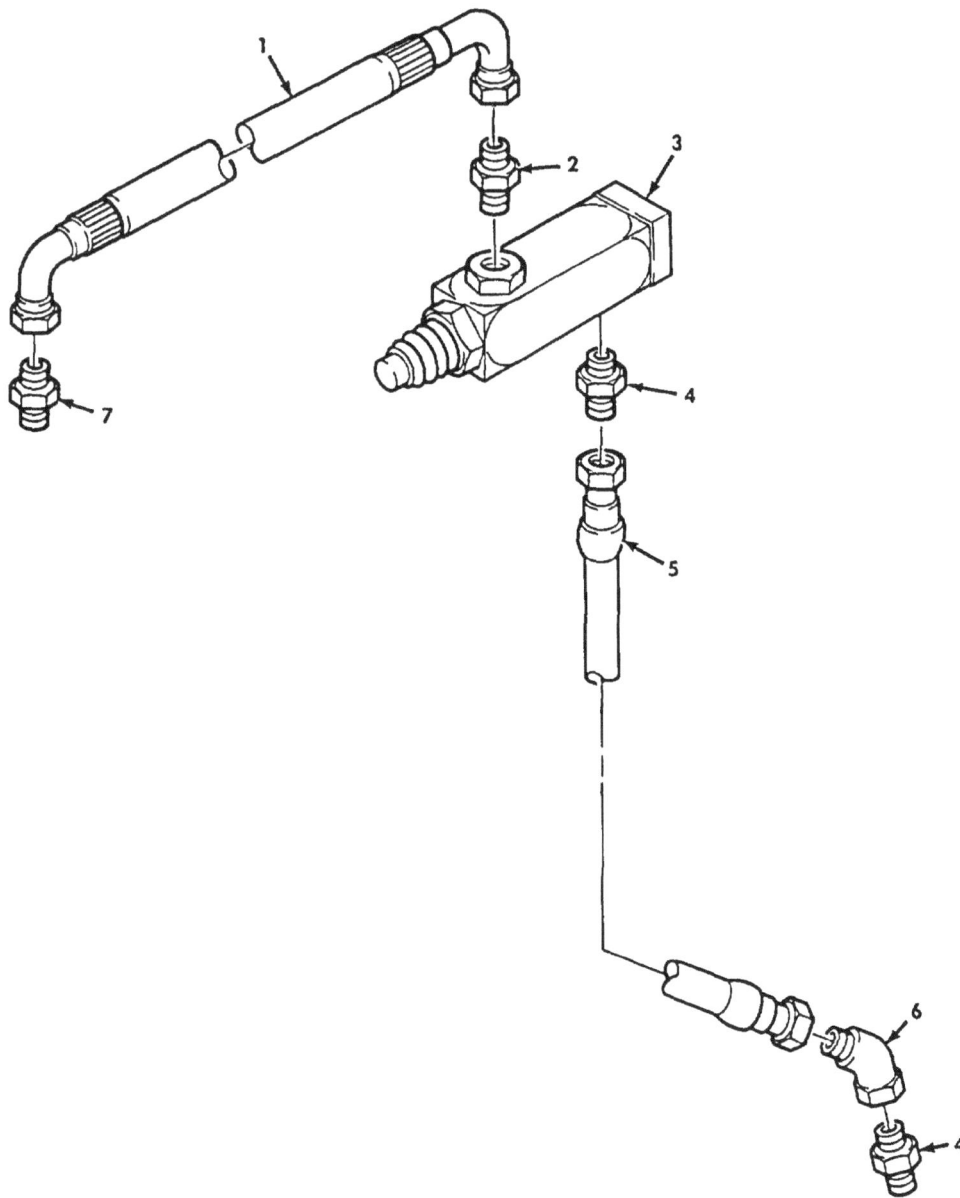

Figure 277. Clutch Fan Actuator and Related Air Lines.

(1) ITEM NO	(2) SMR CODE	(3) NSN	(4) CAGEC	(5) PART NUMBER	(6) DESCRIPTION AND USABLE ON CODES (UOC)	(7) QTY
					GROUP 1208 AIR BRAKES	
					FIG. 277 CLUTCH FAN ACTUATOR AND RELATED AIR LINES	
1	PAOZZ	4720011343847	19207	12277227	HOSE ASSEMBLY, NONME	1
2	PAOZZ	4730008035776	96906	MS51500A6-2	ADAPTER, STRAIGHT, PI	1
3	PAOZZ	4820011491317	21102	CA200046	VALVE, REGULATING, FL	1
4	PAOZZ	4730006188497	01276	2021-2-53S	ADAPTER, STRAIGHT, PI	2
5	PAOZZ	4720011308086	19207	12256281-3	HOSE ASSEMBLY, NONME	1
6	PAOZZ	4730011394231	96906	MS51522-A5	ELBOW, TUBE	1
7	PAOZZ	4730002890382	96906	MS51500A8-8	ADAPTER, STRAIGHT, PIPE	1

END OF FIGURE

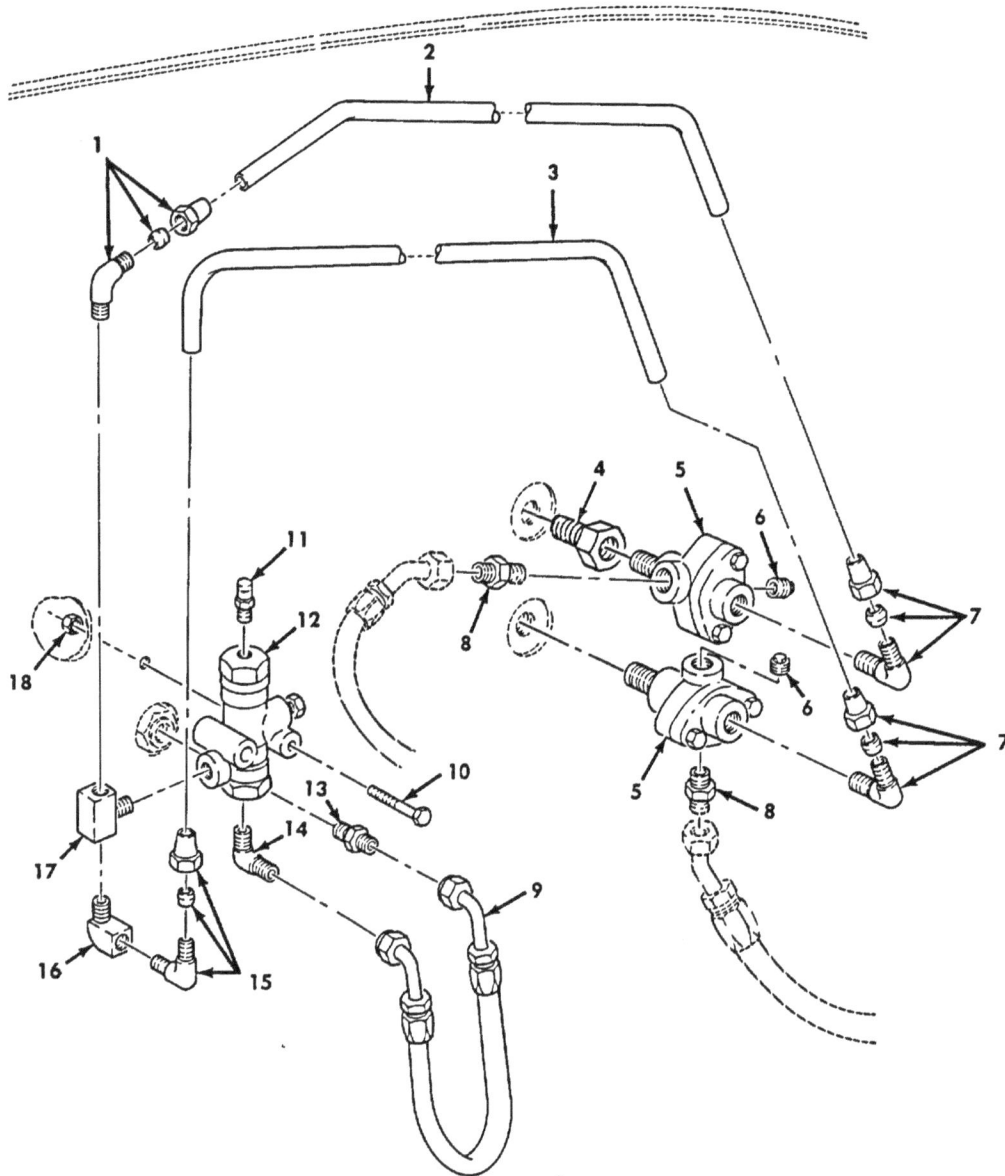

Figure 278. Pressure Regulator, Double-check Valves, and Air Lines.

(1) ITEM NO	(2) SMR CODE	(3) NSN	(4) CAGEC	(5) PART NUMBER	(6) DESCRIPTION AND USABLE ON CODES (UOC)	(7) QTY
					GROUP 1208 AIR BRAKES	
					FIG. 278 PRESSURE REGULATOR, DOUBLE-CHECK VALVES, AND AIR LINES	
1	PAOZZ	4730010669484	81343	6-4 120302BA	ELBOW, PIPE TO TUBE .... UOC:DAL, V18, ZAL	1
2	MOOZZ		19207	12302864	TUBE MAKE FROM TUBING, P/N 8689208............. UOC:DAL, V18, ZAL	1
3	MOOZZ		19207	12302865	TUBE MAKE FROM TUBING, P/N 8689208............. UOC:DAL, V18, ZAL	1
4	PAOZZ	4730004755168	24617	444004	COUPLING, PIPE.. UOC:DAL, V18, ZAL	1
5	PAOZZ	2530009818736	06853	278599	VALVE, BRAKE PNEUMAT .... UOC:DAL, V18, ZAL	2
6	PAOZZ	4730001874202	81348	WW-P-471AASBCC	PLUG, PIPE .... UOC:DAL, V18, ZAL	2
7	PAOZZ	4730002890155	81343	6-6 120202BA	ELBOW... UOC:DAL, V18, ZAL	2
8	PAOZZ	4730008139611	96906	MS51500-B8	ADAPTER .... UOC:DAL, V18, ZAL	2
9	PFOZZ	4720012761253	19207	12302894	HOSE ASSEMBLY, NONME .... UOC:DAL, V18, ZAL	1
10	PAOZZ	5305002678958	80204	B1821BH025F200N	SCREW, CAP, HEXA GON H.............................. UOC:DAL, V18, ZAL	2
11	PFOZZ	4820007264719	19207	5196397	VALVE, VENT.. UOC:DAL, V18, ZAL	1
12	PAOZZ	4820012511699	06853	103998	VALVE, CALIBRATED FL................................ UOC:ZAL	1
12	PAOZZ	2530012228795	06721	N30190DA	GOVERNOR ASSEMBLY, A .... UOC:DAL, V18	1
13	PAOZZ	4730000888908	96906	MS51500-A6	ADAPTER .... UOC:DAL, V18, ZAL	1
14	PAOZZ	4730008127999	96906	MS51504-A6	ELBOW, PIPE TO TUBE .... UOC:DAL, V18, ZAL	1
15	PAOZZ	4730000691187	79146	HD-169-6X4	ELBOW, PIPE TO TUBE .... UOC:DAL, V18, ZAL	1
16	PAOZZ	4730002775553	96906	MS51952-2	ELBOW, PIPE... UOC:DAL, V18, ZAL	1
17	PAOZZ	4730000888666	19207	444134	TEE, PIPE .... UOC:DAL, V18, ZAL	1
18	PAOZZ	5310009359022	96906	MS51943-32	NUT, SELF-LOCKING, HE.................................. UOC:DAL, V18, ZAL	2

END OF FIGURE

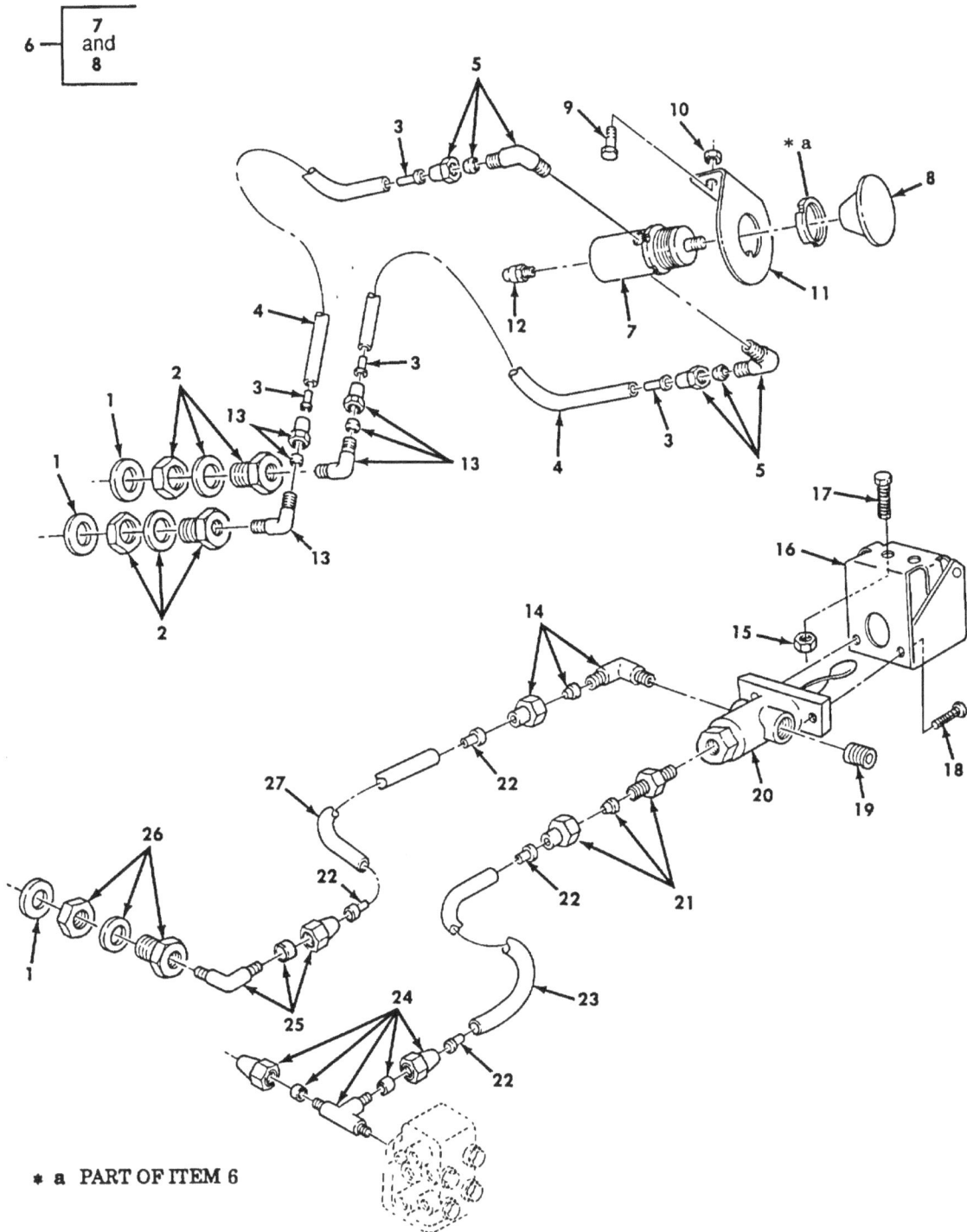

Figure 279. Brake Lock Valve and Air Lines.

* a PART OF ITEM 6

(1) ITEM NO	(2) SMR CODE	(3) NSN	(4) CAGEC	(5) PART NUMBER	(6) DESCRIPTION AND USABLE ON CODES (UOC)	(7) QTY
					GROUP 1208 AIR BRAKES	
					FIG. 279 BRAKE LOCK VALVE AND AIR LINES	
1	PAOZZ	5310008098533	96906	MS27183-23	WASHER, FLAT .... UOC:V18, V39, ZAL	3
2	PFOZZ	4730002786318	19207	8328782	COUPLING, PIPE .... UOC:DAL, V18, ZAL	2
3	PAOZZ	4730010798821	19207	CPR102321-1	INSERT, TUBE FITTING .... UOC:DAL, V18, ZAL	4
4	MOOZZ		19207	CPR104420-2-18	HOSE NONMETALLIC MAKE FROM TUBING, ... P/N CPR104420-2, 18 INCHES LONG .... UOC:DAL, V18, ZAL	2
5	PAOZZ	4730002871604	81343	6-2 120202BA	ELBOW, PIPE TO TUBE .... UOC:DAL, V18, ZAL	2
6	PFOZZ	4820012771486	97902	12302904-2	VALVE, LINEAR, DIRECT .... UOC:DAL, V18, ZAL	1
7	PFOZZ	4820012771486	97902	12302904-2	.VALVE, LINEAR, DIRECT .... UOC:DAL, V18, ZAL	1
8	PFOZZ	5355013182811	19207	12302904-1	.KNOB .... UOC:DAL, V18, ZAL	1
9	PAOZZ	5305000680505	96906	MS90727-5	SCREW, CAP, HEXAGON . ... UOC:DAL, V18, ZAL	2
10	PAOZZ	5310009359022	96906	MS51943-32	NUT, SELF-LOCKING, HE .... UOC:DAL, V18, ZAL	2
11	PFOZZ	5340012773023	19207	12302898	BRACKET, ANGLE .... UOC:DAL, V18, ZAL	1
12	PFOZZ	4820007264719	19207	5196397	VALVE, VENT .... UOC:DAL, V18, ZAL	1
13	PAOZZ	4730000691187	79146	HD-169-6X4	ELBOW, PIPE TO TUBE .... UOC:DAL, V18, ZAL	2
14	PAOZZ	4730002871604	81343	6-2 120202BA	ELBOW, PIPE TO TUBE UOC:V39	1
15	PAOZZ	5310009359022	96906	MS51943-32	NUT, SELF-LOCKING, HE UOC:V39	2
16	PFOZZ		19207	12302896	COVER ASSEMBLY UOC:V39	1
17	PAOZZ	5305002678953	80204	B1821BH025F063N	SCREW, CAP, HEXAGON H UOC:V39	2
18	PAOZZ	5305009846208	96906	MS35206-261	SCREW, MACHINE UOC:V39	2
19	PAOZZ	4730010304950	24617	272977	PLUG, PIPE UOC:V39	1
20	PAOZZ	4820008328077	19207	10924753	VALVE, LINEAR, DIRECT UOC:V39	1
21	PAOZZ	4730001423075	81343	6-2 120102BA	ADAPTER, STRAIGHT, PI UOC:V39	1
22	PAOZZ	4730010798821	19207	CPR102321-1	INSERT, TUBE FITTING UOC:V39	4
23	MOOZZ		19207	CPR104420-2-14	HOSE, NONMETALLIC MAKE FROM TUBING, ... P.N CPR104420-2, 14 INCHES LONG UOC:V39	1

(1) ITEM NO	(2) SMR CODE	(3) NSN	(4) CAGEC	(5) PART NUMBER	(6) DESCRIPTION AND USABLE ON CODES (UOC)	(7) QTY
24	PAOZZ	4730008137811	81343	6-4-6 120424BA	TEE, PIPE TO TUBE UOC:V39	1
25	PAOZZ	4730000691187	79146	HD-169-6X4	ELBOW, PIPE TO TUBE UOC:V39	1
26	PFOZZ	4730002786318	19207	8328782	COUPLING, PIPE UOC:V39	2
27	MOOZZ		19207	CPR104420-2-9	HOSE, NONMETALLIC MAKE FROM TUBING, ... P/N CPR104420-2, 9 INCHES LONG...................... UOC:V39	1

END OF FIGURE

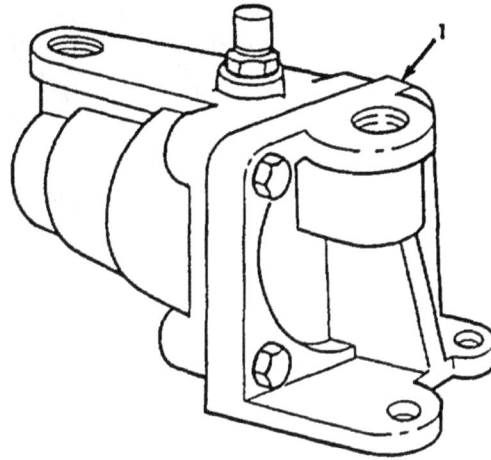

Figure 280. Front Air Brake Chamber Valve Assembly.

(1) ITEM NO	(2) SMR CODE	(3) NSN	(4) CAGEC	(5) PART NUMBER	(6) DESCRIPTION AND USABLE ON CODES (UOC)	(7) QTY
					GROUP 1208 AIR BRAKES	
					FIG. 280 FRONT AIR BRAKE CHAMBER LIMITING VALVE ASSEMBLY	
1	PAOZZ	2530011285552	06853	289849	LIMITING VALVE ASSEMBLY, AIR .... UOC:DAL, V18, ZAL	1
1	PFOZZ	4820008572737	06853	288251	LIMITING VALVE ASSEMBLY, AIR .... UOC:DAA, DAB, DAC, DAD, DAE, DAF, DAG, DAH, DAJ, DAK, DAW, DAX, V12, V13, V14, V15, V16, V17, V19, V20, V21, V22, V24, V25, V39, ZAA, ZAB, ZAC, ZAD, ZAE, ZAF, ZAG, ZAH, ZAJ, ZAK	1
					END OF FIGURE	

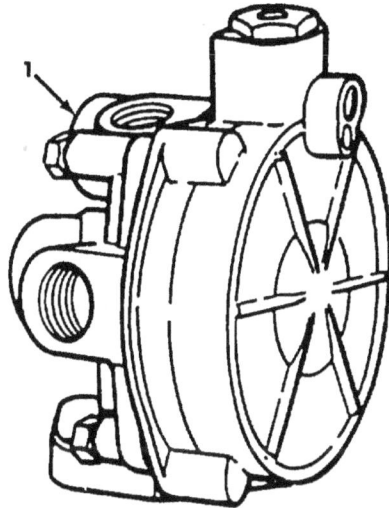

Figure 281. Rear Air Brake Relay Value Assembly.

(1) ITEM NO	(2) SMR CODE	(3) NSN	(4) CAGEC	(5) PART NUMBER	(6) DESCRIPTION AND USABLE ON CODES (UOC)	(7) QTY
					GROUP 1208 AIR BRAKES	
					FIG. 281 REAR AIR BRAKE RELAY VALVE ASSEMBLY (PRIMARY AND SECONDARY PRESSURE RELIEF)	
1	PAOZZ	4820011079695	19207	11669105	VALVE, PRESSURE RELIEF, PRIMARY .... AND SECONDARY	2
					END OF FIGURE	

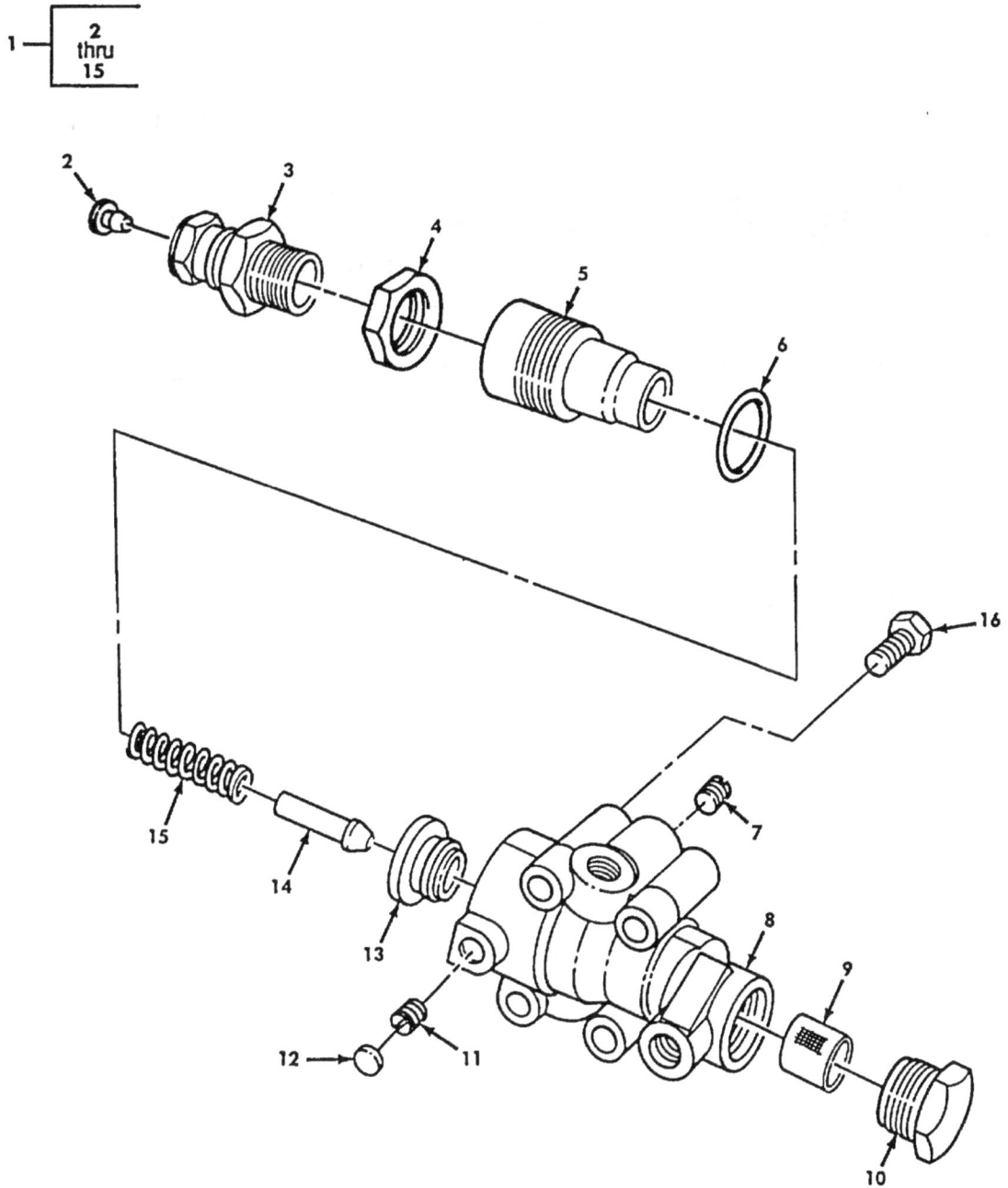

Figure 282. Air Brake Governor Assembly and Mounting Hardware.

(1) ITEM NO	(2) SMR CODE	(3) NSN	(4) CAGEC	(5) PART NUMBER	(6) DESCRIPTION AND USABLE ON CODES (UOC)	(7) QTY
					GROUP 1208 AIR BRAKES	
					FIG. 282 AIR BRAKE GOVERNOR ASSEMBLY AND MOUNTING HARDWARE	
1	PAOZZ	2530012228795	06721	N30190DA	GOVERNOR ASSEMBLY, A OLD VEHICLES	1
1	PAOZZ	2530008544457	55207	0624A0428	GOVERNOR ASSEMBLY, A NEW VEHICLES	1
2	PAFZZ	4820011343788	06721	100880	VALVE, CHECK	1
3	PAFZZ	2530011340902	06721	N-13807-C	ADAPTER, BRAKE GOVER	1
4	PAFZZ	5310001996581	40342	2-X-39	NUT, PLAIN, HEXAGON	1
4	PAOZZ	4730000691186	81343	6-4 120102BA	ADAPTER, STRAIGHT, PI .... UOC:DAL, V18, V39	2
5	PAFZZ	2530011343667	06721	N-13556-E	HOUSING, AIR BRAKE G	1
6	KFFZZ		40342	N-11728-BY	O-RING PART OF KIT P/N RN-32-G	1
7	PAFZZ	5365011387102	40342	7-X-106	PLUG, MACHINE THREAD	3
8	XAFZZ		40342	200743-A	BODY	1
9	KFFZZ		40342	601068	FILTER	1
10	PAFZZ	5365011352077	40342	N-14370	PLUG, MACHINE THREAD	1
11	PAFZZ	5305011355447	06721	3-X-285	SETSCREW	1
12	PAFZZ	5340011361665	06721	8-X-19	PLUG, EXPANSION	1
13	KFFZZ		40342	101374	VALVE ASSEMBLY PART OF KIT P/N RN-.............. 32-G2	1
14	PAFZZ	2530011340903	06721	N-13808	GUIDE, SPRING, GOVERN	1
15	KFFZZ		40342	N-13484-C	SPRING	1
16	PAFZZ	5305006965291	21450	455172	SCREW, ASSEMBLED WAS	2

END OF FIGURE

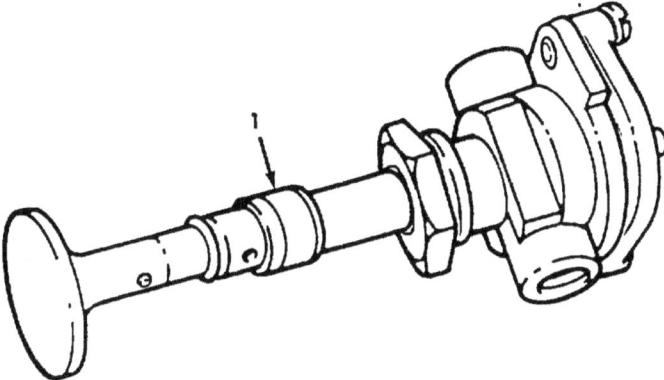

Figure 283. Spring Brake Release Control Valve Assembly.

(1) ITEM NO	(2) SMR CODE	(3) NSN	(4) CAGEC	(5) PART NUMBER	(6) DESCRIPTION AND USABLE ON CODES (UOC)	(7) QTY
					GROUP 1208 AIR BRAKES	
					FIG. 283 SPRING BRAKE RELEASE CONTROL VALVE ASSEMBLY	
1	PAOZZ	4820011147539	06853	104689	VALVE	1
					END OF FIGURE	

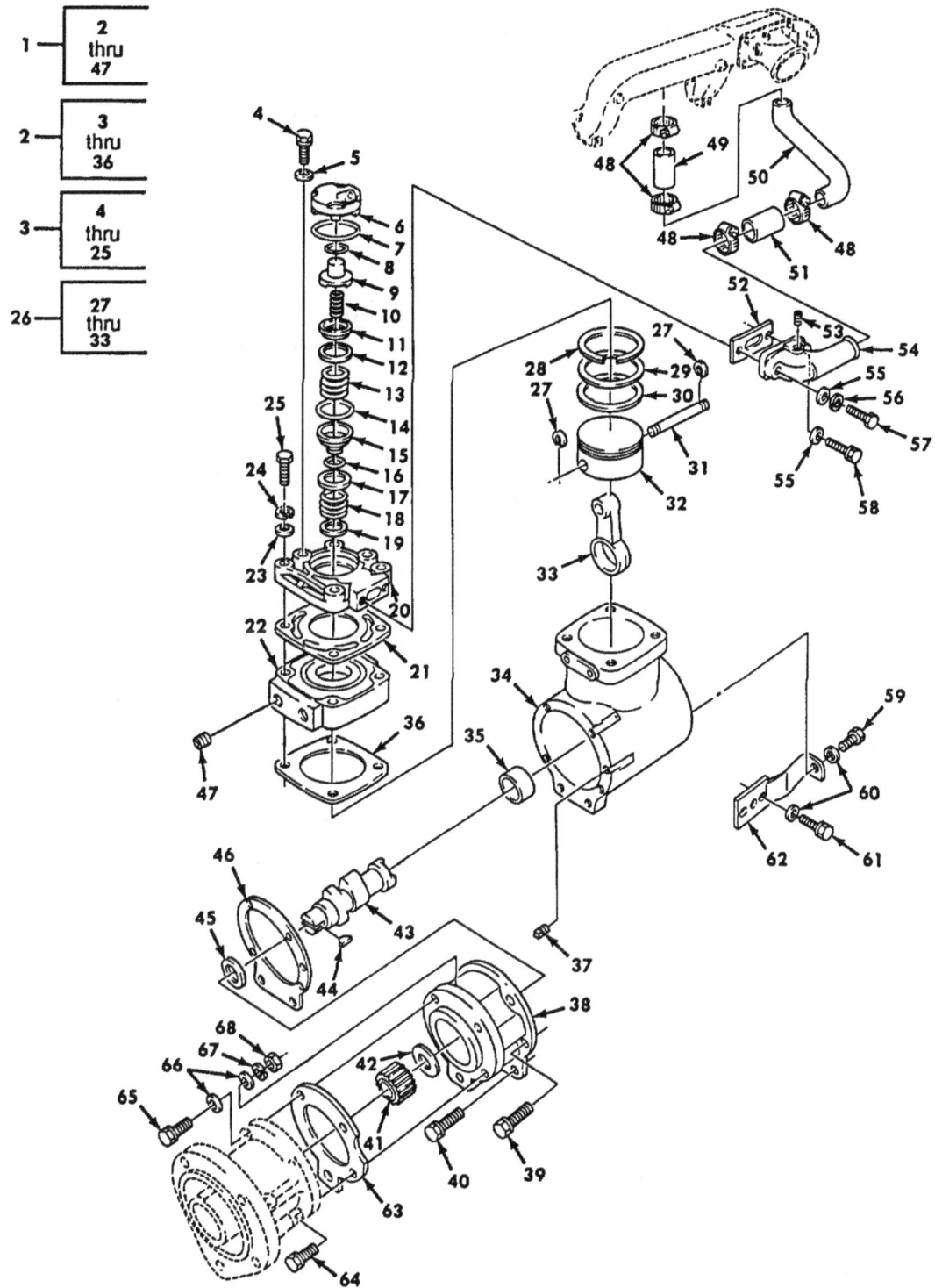

Figure 284. Air Compressor Assembly (M939, M939A1).

(1) ITEM NO	(2) SMR CODE	(3) NSN	(4) CAGEC	(5) PART NUMBER	(6) DESCRIPTION AND USABLE ON CODES (UOC)	(7) QTY
					GROUP 1209 AIR COMPRESSOR	
					FIG. 284 AIR COMPRESSOR ASSEMBLY (M939,M939A1)	
1	PAFFF	2990010800533	15434	3018527	COMPRESSOR, AIR .... UOC:DAA, DAB, DAC, DAD, DAE, DAF, DAG, DAH, DAJ, DAK, DAL, DAW, DAX, V12, V13, V14, V15, V16, V17, V18, V19, V20, V21, V22, V24, V25, V39	1
2	PAFFF	2530011830651	15434	3018491	.HOUSING, AIR COMPRES .... UOC:DAA, DAB, DAC, DAD, DAE, DAF, DAG, DAH, DAJ, DAK, DAL, DAW, DAX, V12, V13, V14, V15, V16, V17, V18, V19, V20, V21, V22, V24, V25, V39	1
3	PAFFF	2530006031506	15434	3029803	..HEAD ASSEMBLY, AIR V .... UOC:DAA, DAB, DAC, DAD, DAE, DAF, DAG, DAH, DAJ, DAK, DAL, DAW, DAX, V12, V13, V14, V15, V16, V17, V18, V19, V20, V21, V22, V24, V25, V39	1
4	PAFZZ	5305011446233	15434	3021470	...SCREW, CAP, HEXAGON H .... UOC:DAA, DAB, DAC, DAD, DAE, DAF, DAG, DAH, DAJ, DAK, DAL, DAW, DAX, V12, V13, V14, V15, V16, V17, V18, V19, V20, V21, V22, V24, V25, V39	2
5	PAFZZ	5310001867403	15434	109557	...WASHER, FLAT .... UOC:DAA, DAB, DAC, DAD, DAE, DAF, DAG, DAH, DAJ, DAK, DAL, DAW, DAX, V12, V13, V14, V15, V16, V17, V18, V19, V20, V21, V22, V24, V25, V39	2
6	PAFZZ	2530011427939	15434	153966	...BODY, UNLOADER VALVE .... UOC:DAA, DAB, DAC, DAD, DAE, DAF, DAG, DAH, DAJ, DAK, DAL, DAW, DAX, V12, V13, V14, V15, V16, V17, V18, V19, V20, V21, V22, V24, V25, V39	1
7	PAFZZ	5330004410145	15434	128086	...O-RING PART OF KIT P/N 3011472 .... UOC:DAA, DAB, DAC, DAD, DAE, DAF, DAG, DAH, DAJ, DAK, DAL, DAW, DAX, V12, V13, V14, V15, V16, V17, V18, V19, V20, V21, V22, V24, V25, V39	1
8	PAFZZ	5330009413762	15434	127936	...O-RING PART OF KIT P/N 3011472 .... UOC:DAA, DAB, DAC, DAD, DAE, DAF, DAG, DAH, DAJ, DAK, DAL, DAW, DAX, V12, V13, V14, V15, V16, V17, V18, V19, V20, V21, V22, V24, V25, V39	1
9	PAFZZ	4310010847148	8X479	191037	...CAP, UNLOADER .... UOC:DAA, DAB, DAC, DAD, DAE, DAF, DAG, DAH, DAJ, DAK, DAL, DAW, DAX, V12, V13, V14, V15, V16, V17, V18, V19, V20, V21, V22, V24, V25, V39	1
10	PAFZZ	5360010863480	15434	3023101	...SPRING, HELICAL, COMP .... UOC:DAA, DAB, DAC, DAD, DAE, DAF, DAG, DAH,	1

(1) ITEM NO	(2) SMR CODE	(3) NSN	(4) CAGEC	(5) PART NUMBER	(6) DESCRIPTION AND USABLE ON CODES (UOC)	(7) QTY
					DAJ, DAK, DAL, DAW, DAX, V12, V13, V14, V15, V16, V17, V18, V19, V20, V21, V22, V24, V25, V39	
11	PAFZZ	3805009555320	15434	144948	...VALVE INTAKE COMPRE .... UOC:DAA, DAB, DAC, DAD, DAE, DAF, DAG, DAH, DAJ, DAK, DAL, DAW, DAX, V12, V13, V14, V15, V16, V17, V18, V19, V20, V21, V22, V24, V25, V39	1
12	PAFZZ	4820009094175	15434	145028	...SEAT, VALVE .... UOC:DAA, DAB, DAC, DAD, DAE, DAF, DAG, DAH, DAJ, DAK, DAL, DAW, DAX, V12, V13, V14, V15, V16, V17, V18, V19, V20, V21, V22, V24, V25, V39	1
13	PAFZZ	5360001299415	15434	190334	...SPRING, HELICAL, COMP .... UOC:DAA, DAB, DAC, DAD-, DAE, DAF, DAG, DAH, DAJ, DAK, DAL, DAW, DAX, V12, V13, V14, V15, V16, V17, V18, V19, V20, V21, V22, V24, V25, V39	1
14	PAFZZ	5330010609061	15434	211315	...PACKING, PREFORMED PART OF KIT P/N .... 3011472 .... UOC:DAA, DAB, DAC, DAD, DAE, DAF, DAG, DAH, DAJ, DAK, DAL, DAW, DAX, V12, V13, V14, V15, V16, V17, V18, V19, V20, V21, V22, V24, V25, V39	1
15	PAFZZ	4820012787385	15434	3043947	...SEAT, VALVE PART OF KIT P/N .... 11602159 .... UOC:DAA, DAB, DAC, DAD, DAE, DAF, DAG, DAH, DAJ, DAK, DAL, DAW, DAX, V12, V13, V14, V15, V16, V17, V18, V19, V20, V21, V22, V24, V25, V39	1
16	PAFZZ	5330009052679	15434	128085	...O-RING PART OF KIT P/N 3011472 .... UOC:DAA, DAB, DAC, DAD, DAE, DAF, DAG, DAH, DAJ, DAK, DAL, DAW, DAX, V12, V13, V14, V15, V16, V17, V18, V19, V20, V21, V22, V24, V25, V39	1
17	PAFZZ	4820004450610	15434	127940	...DISK, VALVE .... UOC:DAA, DAB, DAC, DA D, DAE, DAF, DAG, DAH, DAJ, DAK, DAL, DAW, DAX, V12, V13, V14, V15, V16, V17, V18, V19, V20, V21, V22, V24, V25, V39	1
18	PAFZZ	5360008953216	15434	128080	...SPRING, HELICAL, COMP .... UOC:DAA, DAB, DAC, DAD, DAE, DAF, DAG, DAH, DAJ, DAK, DAL, DAW, DAX, V12, V13, V14, V15, V16, V17, V18, V19, V20, V21, V22, V24, V25, V39	1
19	PAFZZ	5365003694729	15434	183429	...SHIM .... UOC:DAA, DAB, DAC, DAD, DAE, DAF, DAG, DAH, DAJ, DAK, DAL, DAW, DAX, V12, V13, V14, V15, V16, V17, V18, V19, V20, V21, V22, V24, V25, V39	1
20	PFFZZ	4310011465921	15434	153964	...COVER, AIR COMPRESSO .... UOC:DAA, DAB, DAC, DAD, DAE, DAF, DAG, DAH,	1

(1) ITEM NO	(2) SMR CODE	(3) NSN	(4) CAGEC	(5) PART NUMBER	(6) DESCRIPTION AND USABLE ON CODES (UOC)	(7) QTY
					DAJ, DAK, DAL, DAW, DAX, V12, V13, V14, V15, V16, V17, V18, V19, V20, V21, V22, V24, V25, V39	
21	PAFZZ	5330001317072	15434	3047159	...GASKET PART OF KIT P/N 3011472 .... UOC:DAA, DAB, DAC, DAD, DAE, DAF, DAG, DAH, DAJ, DAK, DAL, DAW, DAX, V12, V13, V14, V15, V16, V17, V18, V19, V20, V21, V22, V24, V25, V39	1
22	XAFZZ	4310011461097	15434	218793	...CYLINDER HEAD, COMPR .... UOC:DAA, DAB, DAC, DAD, DAE, DAF, DAG, DAH, DAJ, DAK, DAL, DAW, DAX, V12, V13, V14, V15, V16, V17, V18, V19, V20, V21, V22, V24, V25, V39	1
23	PAFZZ	5310005626558	15434	S626	...WASHER, FLAT .... UOC:DAA, DAB, DAC, DAD, DAE, DAF, DAG, DAH, DAJ, DAK, DAL, DAW, DAX, V12, V13, V14, V15, V16, V17, V18, V19, V20, V21, V22, V24, V25, V39	4
24	PAFZZ	5310004079566	19207	7410218	...WASHER, LOCK .... UOC:DAA, DAB, DAC, DAD, DAE, DAF, DAG, DAH, DAJ, DAK, DAL, DAW, DAX, V12, V13, V14, V15, V16, V17, V18, V19, V20, V21, V22, V24, V25, V39	4
25	PAFZZ	5305002258507	96906	MS90725-43	...BOLT, MACHINE .... UOC:DAA, DAB, DAC, DAD, DAE, DAF, DAG, DAH, DAJ, DAK, DAL, DAW, DAX, V12, V13, V14, V15, V16, V17, V18, V19, V20, V21, V22, V24, V25, V39	4
26	PAFFF	4310006031510	15434	BM-77410	...PISTON, COMPRESSOR .... UOC:DAA, DAB, DAC, DAD, DAE, DAF, DAG, DAH, DAJ, DAK, DAL, DAW, DAX, V12, V13, V14, V15, V16, V17, V18, V19, V20, V21, V22, V24, V25, V39	1
27	PAFZZ	5325009229101	15434	119859	...RING, RETAINING .... UOC:DAA, DAB, DAC, DAD, DAE, DAF, DAG, DAH, DAJ, DAK, DAL, DAW, DAX, V12, V13, V14, V15, V16, V17, V18, V19, V20, V21, V22, V24, V25, V39	2
28	KFFZZ	4310010795245	15434	650330	...RING, PISTON PART OF KIT P/N .... AR73350 .... UOC:DAA, DAB, DAC, DAD, DAE, DAF, DAG, DAH, DAJ, DAK, DAL, DAW, DAX, V12, V13, V14, V15, V16, V17, V18, V19, V20, V21, V22, V24, V25, V39	1
29	KFFZZ	4310011971882	15434	187350	...RING, PISTON PART OF KIT P/N .... AR73350 .... UOC:DAA, DAB, DAC, DAD, DAE, DAF, DAG, DAH, DAJ, DAK, DAL, DAW, DAX, V12, V13, V14, V15, V16, V17, V18, V19, V20, V21, V22, V24, V25, V39	1

(1) ITEM NO	(2) SMR CODE	(3) NSN	(4) CAGEC	(5) PART NUMBER	(6) DESCRIPTION AND USABLE ON CODES (UOC)	(7) QTY
30	KFFZZ	2815010793290	15434	180810	...RING, PISTON PART OF KIT P/N .... AR73350 .... UOC:DAA, DAB, DAC, DAD, DAE, DAF, DAG, DAH, DAJ, DAK, DAL, DAW, DAX, V12, V13, V14, V15, V16, V17, V18, V19, V20, V21, V22, V24, V25, V39	1
31	PAFZZ	4310009037174	15434	119810	...PIN, PISTON .... UOC:DAA, DAB, DAC, DAD, DAE, DAF, DAG, DAH, DAJ, DAK, DAL, DAW, DAX, V12, V13, V14, V15, V16, V17, V18, V19, V20, V21, V22, V24, V25, V39	1
32	XAFZZ	4310010796938	15434	165430	...PISTON, COMPRESSOR .... UOC:DAA, DAB, DAC, DAD, DAE, DAF, DAG, DAH, DAJ, DAK, DAL, DAW, DAX, V12, V13, V14, V15, V16, V17, V18, V19, V20, V21, V22, V24, V25, V39	1
33	PAFZZ	2815003697846	15434	3558655	...CONNECTING ROD, PIST .... UOC:DAA, DAB, DAC, DAD, DAE, DAF, DAG, DAH, DAJ, DAK, DAL, DAW, DAX, V12, V13, V14, V15, V16, V17, V18, V19, V20, V21, V22, V24, V25, V39	1
34	PAFZZ	4310010793319	15434	3558653	..HOUSING, AIR COMPRES .... UOC:DAA, DAB, DAC, DAD, DAE, DAF, DAG, DAH, DAJ, DAK, DAL, DAW, DAX, V12, V13, V14, V15, V16, V17, V18, V19, V20, V21, V22, V24, V25, V39	1
35	XAFZZ	3120010164883	15434	147610	..BEARING, SLEEVE .... UOC:DAA, DAB, DAC, DAD, DAE, DAF, DAG, DAH, DAJ, DAK, DAL, DAW, DAX, V12, V13, V14, V15, V16, V17, V18, V19, V20, V21, V22, V24, V25, V39	1
36	PAFZZ	5330008527347	15434	154018	..GASKET .... UOC:DAA, DAB, DAC, DAD, DAE, DAF, DAG, DAH, DAJ, DAK, DAL, DAW, DAX, V12, V13, V14, V15, V16, V17, V18, V19, V20, V21, V22, V24, V25, V39	1
37	PAFZZ	5365004042934	15434	S965E	.PLUG, MACHINE THREAD .... UOC:DAA, DAB, DAC, DAD, DAE, DAF, DAG, DAH, DAJ, DAK, DAL, DAW, DAX, V12, V13, V14, V15, V16, V17, V18, V19, V20, V21, V22, V24, V25, V39	1
38	PAFZZ	2530011302339	15434	3005152	.SUPPORT, AIR COMPRES .... UOC:DAA, DAB, DAC, DAD, DAE, DAF, DAG, DAH, DAJ, DAK, DAL, DAW, DAX, V12, V13, V14, V15, V16, V17, V18, V19, V20, V21, V22, V24, V25, V39	1
39	PAFZZ	5305011458359	15434	3019573	.SCREW, ASSEMBLED WAS .... UOC:DAA, DAB, DAC, DAD, DAE, DAF, DAG, DAH, DAJ, DAK, DAL, DAW, DAX, V12, V13, V14, V15, V16, V17, V18, V19, V20, V21, V22, V24, V25, V39	1

(1) ITEM NO	(2) SMR CODE	(3) NSN	(4) CAGEC	(5) PART NUMBER	(6) DESCRIPTION AND USABLE ON CODES (UOC)	(7) QTY
40	PAFZZ	5305010886019	15434	3010596	.SCREW, ASSEMBLED WAS .... UOC:DAA, DAB, DAC, DAD, DAE, DAF, DAG, DAH, DAJ, DAK, DAL, DAW, DAX, V12, V13, V14, V15, V16, V17, V18, V19, V20, V21, V22, V24, V25, V39	2
41	PAFZZ	3010010852732	15434	3000174	.COUPLING HALF, SHAFT .... UOC:DAA, DAB, DAC, DAD, DAE, DAF, DAG, DAH, DAJ, DAK, DAL, DAW, DAX, V12, V13, V14, V15, V16, V17, V18, V19, V20, V21, V22, V24, V25, V39	1
42	PAFZZ	3120011297659	15434	211662	.BEARING, WASHER, THRU .... UOC:DAA, DAB, DAC, DAD, DAE, DAF, DAG, DAH, DAJ, DAK, DAL, DAW, DAX, V12, V13, V14, V15, V16, V17, V18, V19, V20, V21, V22, V24, V25, V39	1
43	PAFZZ	4310010793383	15434	AR10922	.CRANKSHAFT, COMPRESS .... UOC:DAA, DAB, DAC, DAD, DAE, DAF, DAG, DAH, DAJ, DAK, DAL, DAW, DAX, V12, V13, V14, V15, V16, V17, V18, V19, V20, V21, V22, V24, V25, V39	1
44	PAFZZ	5315008861432	15434	S-315	.KEY, WOODRUFF .... UOC:DAA, DAB, DAC, DAD, DAE, DAF, DAG, DAH, DAJ, DAK, DAL, DAW, DAX, V12, V13, V14, V15, V16, V17, V18, V19, V20, V21, V22, V24, V25, V39	1
45	PAFZZ	3120001299200	15434	188040	.BEARING, WASHER, THRU STANDARD .... UOC:DAA, DAB, DAC, DAD, DAE, DAF, DAG, DAH, DAJ, DAK, DAL, DAW, DAX, V12, V13, V14, V15, V16, V17, V18, V19, V20, V21, V22, V24, V25, V39	1
45	PAFZZ	3120001299206	15434	188042	.BEARING, WASHER, THRU .004 OVERSIZE .... UOC:DAA, DAB, DAC, DAD, DAE, DAF, DAG, DAH, DAJ, DAK, DAL, DAW, DAX, V12, V13, V14, V15, V16, V17, V18, V19, V20, V21, V22, V24, V25, V39	1
45	PAFZZ	3120001299210	15434	188044	.BEARING, WASHER, THRU .002 OVERSIZE .... UOC:DAA, DAB, DAC, DAD, DAE, DAF, DAG, DAH, DAJ, DAK, DAL, DAW, DAX, V12, V13, V14, V15, V16, V17, V18, V19, V20, V21, V22, V24, V25, V39	1
46	PAFZZ	5330001299389	15434	176027	.GASKET PART OF KIT P/N 3011472 .... UOC:DAA, DAB, DAC, DAD, DAE, DAF, DAG, DAH, DAJ, DAK, DAL, DAW, DAX, V12, V13, V14, V15, V16, V17, V18, V19, V20, V21, V22, V24, V25, V39	1
47	PAFZZ	4730011472223	15434	3008468	.PLUG, PIPE .... UOC:DAA, DAB, DAC, DAD, DAE, DAF, DAG, DAH, DAJ, DAK, DAL, DAW, DAX, V12, V13, V14, V15, V16, V17, V18, V19, V20, V21, V22, V24, V25, V39	1

(1) ITEM NO	(2) SMR CODE	(3) NSN	(4) CAGEC	(5) PART NUMBER	(6) DESCRIPTION AND USABLE ON CODES (UOC)	(7) QTY
48	PAOZZ	4730005558263	70403	A11	CLAMP, HOSE .... UOC:DAA, DAB, DAC, DAD, DAE, DAF, DAG, DAH, DAJ, DAK, DAL, DAW, DAX, V12, V13, V14, V15, V16, V17, V18, V19, V20, V21, V22, V24, V25, V39	4
49	PAOZZ	4720011356694	15434	200517	HOSE, NONMETALLIC .... UOC:DAA, DAB, DAC, DAD, DAE, DAF, DAG, DAH, DAJ, DAK, DAL, DAW, DAX, V12, V13, V14, V15, V16, V17, V18, V19, V20, V21, V22, V24, V25, V39	1
50	PAOZZ	4710011331538	15434	204904	TUBE, BENT, METALLIC .... UOC:DAA, DAB, DAC, DAD, DAE, DAF, DAG, DAH, DAJ, DAK, DAL, DAW, DAX, V12, V13, V14, V15, V16, V17, V18, V19, V20, V21, V22, V24, V25, V39	1
51	PAOZZ	4720009189634	15434	61554	HOSE, PREFORMED .... UOC:DAA, DAB, DAC, DAD, DAE, DAF, DAG, DAH, DAJ, DAK, DAL, DAW, DAX, V12, V13, V14, V15, V16, V17, V18, V19, V20, V21, V22, V24, V25, V39	1
52	PAOZZ	5330011810630	15434	3201386	GASKET PART OF KIT P/N 3011472 .... UOC:DAA, DAB, DAC, DAD, DAE, DAF, DAG, DAH, DAJ, DAK, DAL, DAW, DAX, V12, V13, V14, V15, V16, V17, V18, V19, V20, V21, V22, V24, V25, V39	1
53	PAOZZ	4730000113175	15434	70295	PLUG, PIPE PART OF KIT P/N 3011472 .... UOC:DAA, DAB, DAC, DAD, DAE, DAF, DAG, DAH, DAJ, DAK, DAL, DAW, DAX, V12, V13, V14, V15, V16, V17, V18, V19, V20, V21, V22, V24, V25, V39	1
54	PAOZZ	4730010854156	15434	196282	ELBOW, FLANGE TO HOS .... UOC:DAA, DAB, DAC, DAD, DAE, DAF, DAG, DAH, DAJ, DAK, DAL, DAW, DAX, V12, V13, V14, V15, V16, V17, V18, V19, V20, V21, V22, V24, V25, V39	1
55	PAOZZ	5310005626558	15434	S626	WASHER, FLAT .... UOC:DAA, DAB, DAC, DAD, DAE, DAF, DAG, DAH, DAJ, DAK, DAL, DAW, DAX, V12, V13, V14, V15, V16, V17, V18, V19, V20, V21, V22, V24, V25, V39	2
56	PAOZZ	5310004079566	96906	MS35338-45	WASHER, LOCK .... UOC:DAA, DAB, DAC, DAD, DAE, DAF, DAG, DAH, DAJ, DAK, DAL, DAW, DAX, V12, V13, V14, V15, V16, V17, V18, V19, V20, V21, V22, V24, V25, V39	1
57	PAOZZ	5305002264831	80204	B1821BH031C150N	SCREW, CAP, HEXAGON H .... UOC:DAA, DAB, DAC, DAD, DAE, DAF, DAG, DAH, DAJ, DAK, DAL, DAW, DAX, V12, V13, V14, V15, V16, V17, V18, V19, V20, V21, V22, V24, V25, V39	1
58	PAOZZ	5305011294214	15434	3022590	SCREW, CAPTIVE .... UOC:DAA, DAB, DAC, DAD, DAE, DAF, DAG, DAH, DAJ, DAK, DAL, DAW, DAX, V12, V13, V14, V15, V16, V17, V18, V19, V20, V21, V22, V24, V25, V39	1

(1) ITEM NO	(2) SMR CODE	(3) NSN	(4) CAGEC	(5) PART NUMBER	(6) DESCRIPTION AND USABLE ON CODES (UOC)	(7) QTY
59	PAFZZ	5306011194271	15434	3000173	BOLT, INTERNALLY REL .... UOC:DAA, DAB, DAC, DAD, DAE, DAF, DAG, DAH, DAJ, DAK, DAL, DAW, DAX, V12, V13, V14, V15, V16, V17, V18, V19, V20, V21, V22, V24, V25, V39	1
60	PAFZZ	5310004862505	15434	108330	WASHER, FLAT .... UOC:DAA, DAB, DAC, DAD, DAE, DAF, DAG, DAH, DAJ, DAK, DAL, DAW, DAX, V12, V13, V14, V15, V16, V17, V18, V19, V20, V21, V22, V24, V25, V39	4
61	PAFZZ	5305011768018	15434	3010592	SCREW, ASSEMBLED WAS .... UOC:DAA, DAB, DAC, DAD, DAE, DAF, DAG, DAH, DAJ, DAK, DAL, DAW, DAX, V12, V13, V14, V15, V16, V17, V18, V19, V20, V21, V22, V24, V25, V39	3
62	PFFZZ	5340011777580	15434	202356	BRACKET, ANGLE .... UOC:DAA, DAB, DAC, DAD, DAE, DAF, DAG, DAH, DAJ, DAK, DAL, DAW, DAX, V12, V13, V14, V15, V16, V17, V18, V19, V20, V21, V22, V24, V25, V39	1
63	PAFZZ	5330011810631	15434	320-1850	GASKET .... UOC:DAA, DAB, DAC, DAD, DAE, DAF, DAG, DAH, DAJ, DAK, DAL, DAW, DAX, V12, V13, V14, V15, V16, V17, V18, V19, V20, V21, V22, V24, V25, V39	1
64	PAFZZ	5305010288869	15434	S155	SCREW, CAP, HEXAGON H .... UOC:DAA, DAB, DAC, DAD, DAE, DAF, DAG, DAH, DAJ, DAK, DAL, DAW, DAX, V12 , V13, V14, V15, V16, V17, V18, V19, V20, V21, V22, V24, V25, V39	3
65	PAFZZ	5305000712056	80204	B1821BH044C175N	SCREW, CAP, HEXAGON H .... UOC:DAA, DAB, DA C, DAD, DAE, DAF, DAG, DAH, DAJ, DAK, DAL, DAW, DAX, V12, V13, V14, V15, V16, V17, V18, V19, V20, V21, V22, V24, V25, V39	2
66	PAFZZ	5310011124307	15434	69324	WASHER, FLAT .... UOC:DAA, DAB, DAC, DAD, DAE, DAF, DAG, DAH, DAJ, DAK, DAL, DAW, DAX, V12, V13, V14, V15, V16, V17, V18, V19, V20, V21, V22, V24, V25, V39	4
67	PAFZZ	5310002090965	96906	MS35338-47	WASHER, LOCK .... UOC:DAA, DAB, DAC, DAD, DAE, DAF, DAG, DAH, DAJ, DAK, DAL, DAW, DAX, V12, V13, V14, V15, V16, V17, V18, V19, V20, V21, V22, V24, V25, V39	2
68	PAFZZ	5310006500187	15434	S217	NUT, PLAIN, HEXAGON .... UOC:DAA, DAB, DAC, DAD, DAE, DAF, DAG, DAH, DAJ, DAK, DAL, DAW, DAX, V12, V13, V14, V15, V16, V17, V18, V19, V20, V21, V22, V24, V25, V39	2

END OF FIGURE

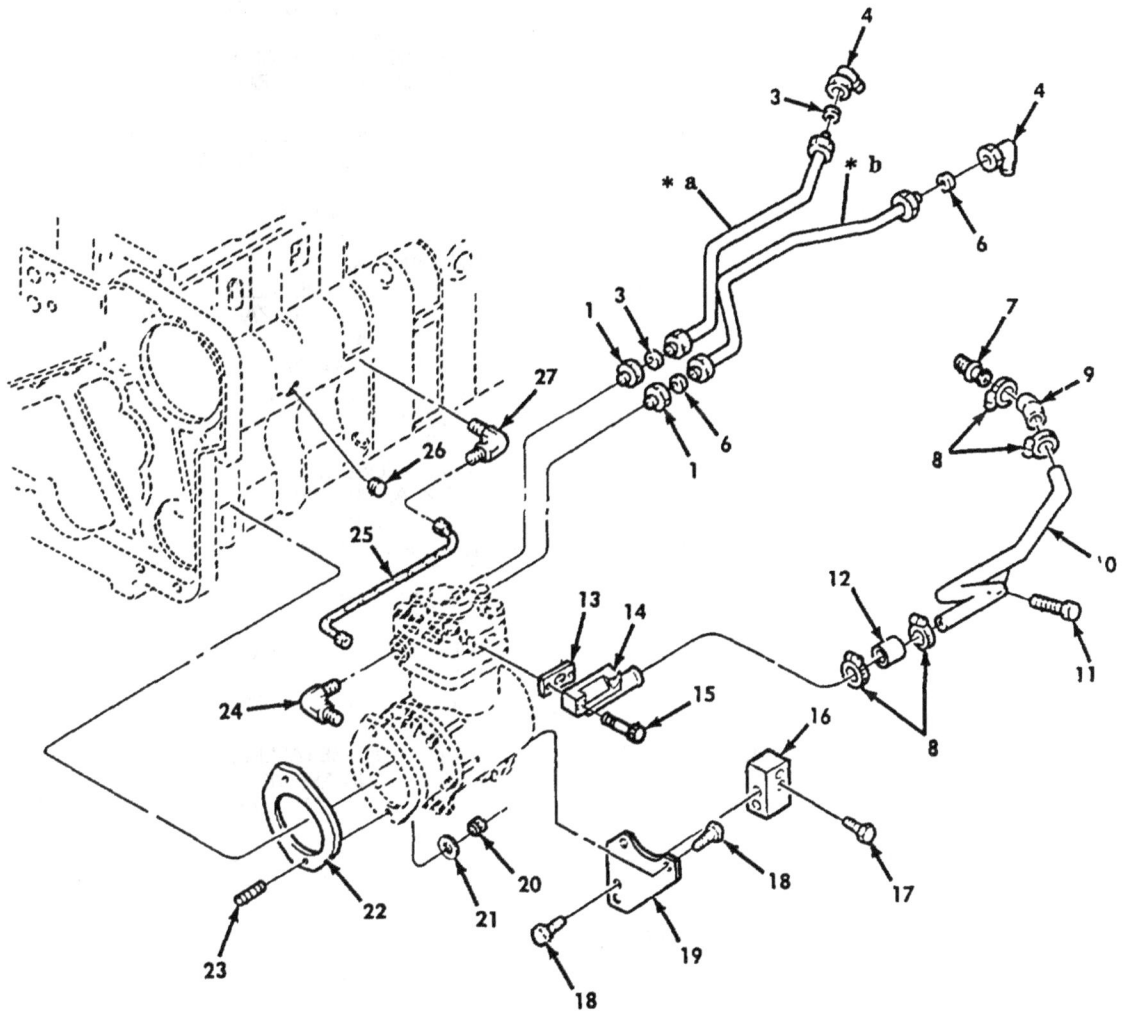

*a PART OF ITEM 2
*b PART OF ITEM 5

Figure 285. Air Compressor Mounting and Lines (M939A2).

(1) ITEM NO	(2) SMR CODE	(3) NSN	(4) CAGEC	(5) PART NUMBER	(6) DESCRIPTION AND USABLE ON CODES (UOC)	(7) QTY
					GROUP 1209 AIR COMPRESSOR	
					FIG. 285 AIR COMPRESSOR MOUNTING AND LINES (M939A2)	
1	PAOZZ	4730003652690	15434	S-1002-A	ADAPTER, STRAIGHT, TU .... UOC:ZAA, ZAB, ZAC, ZAD, ZAE, ZAF, ZAG, ZAH, ZAJ, ZAK, ZAL	2
2	PAOZZ	4710012718031	15434	3911937	TUBE ASSEMBLY, METAL .... UOC:ZAA, ZAB, ZAC, ZAD, ZAE, ZAF, ZAG, ZAH, ZAJ, ZAK, ZAL	1
3	PAOZZ	5365005985255	15434	S-1003-A	.BUSHING, NONMETALLIC .... UOC:ZAA, ZAB, ZAC, ZAD, ZAB, ZAF, ZAG, ZAH, ZAJ, ZAK, ZAL	2
4	PAOZZ	4730012717956	15434	204995	ELBOW, PIPE TO TUBE .... UOC:ZAA, ZAB, ZAC, ZAD, ZAE, ZAF, ZAG, ZAH, ZAJ, ZAK, ZAL	2
5	PAOZZ	4710012718032	15434	3911456	TUBE ASSEMBLY, METAL .... UOC:ZAA, ZAB, ZAC, ZAD, ZAE, ZAF, ZAG, ZAH, ZAJ, ZAK, ZAL	1
6	PAOZZ	5365005985255	15434	S-1003-A	.BUSHING, NONMETALLIC .... UOC:ZAA, ZAB, ZAC, ZAD, ZAE, ZAF, ZAG, ZAH, ZAJ, ZAK, ZAL	2
7	PAOZZ	4730012717969	15434	3907167	ADAPTER, STRAIGHT, TU ENGINE S/N .... 44487829 AND BELOW UOC:ZAA, ZAB, ZAC, ZAD, ZAE, ZAF, ZAG, ZAH, ZAJ, ZAK, ZAL	1
8	PAOZZ	4730005558263	70403	A11	CLAMP, HOSE .... UOC:ZAA, ZAB, ZAC, ZAD, ZAE, ZAF, ZAG, ZAH, ZAJ, ZAK, ZAL	4
9	PAOZZ	4720012722841	15434	3911935	HOSE, PREFORMED ENGING S/N 44487829 .... AND BELOW .... UOC:ZAA, ZAB, ZAC, ZAD, ZAE, ZAF, ZAG, ZAH, ZAJ, ZAK, ZAL	1
10	PAOZZ	4710012717942	15434	3911934	TUBE, BENT, METALLIC ENGINE S/N .... 44478729 AND BELOW .... UOC:ZAA, ZAB, ZAC, ZAD, ZAE, ZAF, ZAG, ZAH, ZAJ, ZAK, ZAL	1
10	PAOZZ	4710013694814	15434	3912627	TUBE, BENT, METALLIC ENGINE S/N .... 44487830 AND ABOVE .... UOC:ZAA, ZAB, ZAC, ZAD, ZAE, ZAF, ZAG, ZAH, ZAJ, ZAK, ZAL	1
11	PAOZZ	5305012716449	15434	3909582	SCREW, CAP, HEXAGON H .... UOC:ZAA, ZAB, ZAC, ZAD, ZAE, ZAF, ZAG, ZAH, ZAJ, ZAK, ZAL	1
12	PAOZZ	4720012718000	15434	3911936	HOSE, NONMETALLIC ENGINE S/N .... 44487829 AND BELOW .... UOC:ZAA, ZAB, ZAC, ZAD, ZAE, ZAF, ZAG, ZAH, ZAJ, ZAK, ZAL	1
12	PAOZZ	4720012718000	15434	3911936	HOSE, NONMETALLIC ENGINE S/N .... 44487830 AND ABOVE	1

(1) ITEM NO	(2) SMR CODE	(3) NSN	(4) CAGEC	(5) PART NUMBER	(6) DESCRIPTION AND USABLE ON CODES (UOC)	(7) QTY
					UOC:ZAA, ZAB, ZAC, ZAD, ZAE, ZAF, ZAG, ZAH, ZAJ, ZAK, ZAL	
13	PAFZZ	5330011810630	15434	3201386	GASKET	1
					UOC:ZAA, ZAB, ZAC, ZAD, ZAE, ZAF, ZAG, ZAH, ZAJ, ZAK, ZAL	
14	PAFZZ	5340012712346	15434	3911533	CONNECTION, AIR INLET	1
					UOC:ZAA, ZAB, ZAC, ZAD, ZAE, ZAF, ZAG, ZAH, ZAJ, ZAK, ZAL	
15	PAFZZ	5305002264831	80204	B1821BH031C150N	SCREW, CAP, HEXAGON H	2
					UOC:ZAA, ZAB, ZAC, ZAD, ZAE, ZAF, ZAG, ZAH, ZAJ, ZAK, ZAL	
16	PAFZZ	5365012875762	7U263	3914723	SPACER, PLATE	1
					UOC:ZAA, ZAB, ZAC, ZAD, ZAE, ZAF, ZAG, ZAH, ZAJ, ZAK, ZAL	
17	PAFZZ	5306012881011	7U263	3903112	BOLT, MACHINE	4
					UOC:ZAA, ZAB, ZAC, ZAD, ZAE, ZAF, ZAG, ZAH, ZAJ, ZAK, ZAL	
18	PAOZZ	5305012881012	7U263	3903464	SCREW, CAP, HEXAGON H	2
					UOC:ZAA, ZAB, ZAC, ZAD, ZAE, ZAF, ZAG, ZAH, ZAJ, ZAK, ZAL	
19	PAOZZ	5340012875699	7U263	3914722	BRACKET, MOUNTING	1
					UOC:ZAA, ZAB, ZAC, ZAD, ZAE, ZAF, ZAG, ZAH, ZAJ, ZAK, ZAL	
20	PAOZZ	5310012875742	7U263	3913371	NUT, SELF-LOCKING, HE	2
					UOC:ZAA, ZAB, ZAC, ZAD, ZAE, ZAF, ZAG, ZAH, ZAJ, ZAK, ZAL	
21	PAOZZ	5310013035531	15434	3035806	WASHER, FLAT	2
					UOC:ZAA, ZAB, ZAC, ZAD, ZAE, ZAF, ZAG, ZAH, ZAJ, ZAK, ZAL	
22	PAOZZ	5330012718306	15434	3916042	GASKET PART OF KIT P/N 3802389 PART .... OF KIT P/N 3802389	1
					UOC:ZAA, ZAB, ZAC, ZAD, ZAE, ZAF, ZAG, ZAH, ZAJ, ZAK, ZAL	
23	PAFZZ	5307012870855	7U263	3913370	STUD, PLAIN	1
					UOC:ZAA, ZAB, ZAC, ZAD, ZAE, ZAF, ZAG, ZAH, ZAJ, ZAK, ZAL	
24	PAOZZ	4730012023351	19207	7336402-1	ELBOW, PIPE TO TUBE	2
					UOC:ZAA, ZAB, ZAC, ZAD, ZAE, ZAF, ZAG, ZAH, ZAJ, ZAK, ZAL	
25	PAOZZ		15434	3906747	HOSE ASSEMBLY, NONME	1
					UOC:ZAA, ZAB, ZAC, ZAD, ZAE, ZAF, ZAG, ZAH, ZAJ, ZAK, ZAL	
26	PAFZZ	4730000189566	24617	G1251	PLUG, PIPE	1
					UOC:ZAA, ZAB, ZAC, ZAD, ZAE, ZAF, ZAG, ZAH, ZAJ, ZAK, ZAL	
27	PAOZZ	4730011428524	15434	68139	ELBOW, PIPE TO TUBE	1
					UOC:ZAA, ZAB, ZAC, ZAD, ZAE, ZAF, ZAG, ZAH, ZAJ, ZAK, ZAL	

END OF FIGURE

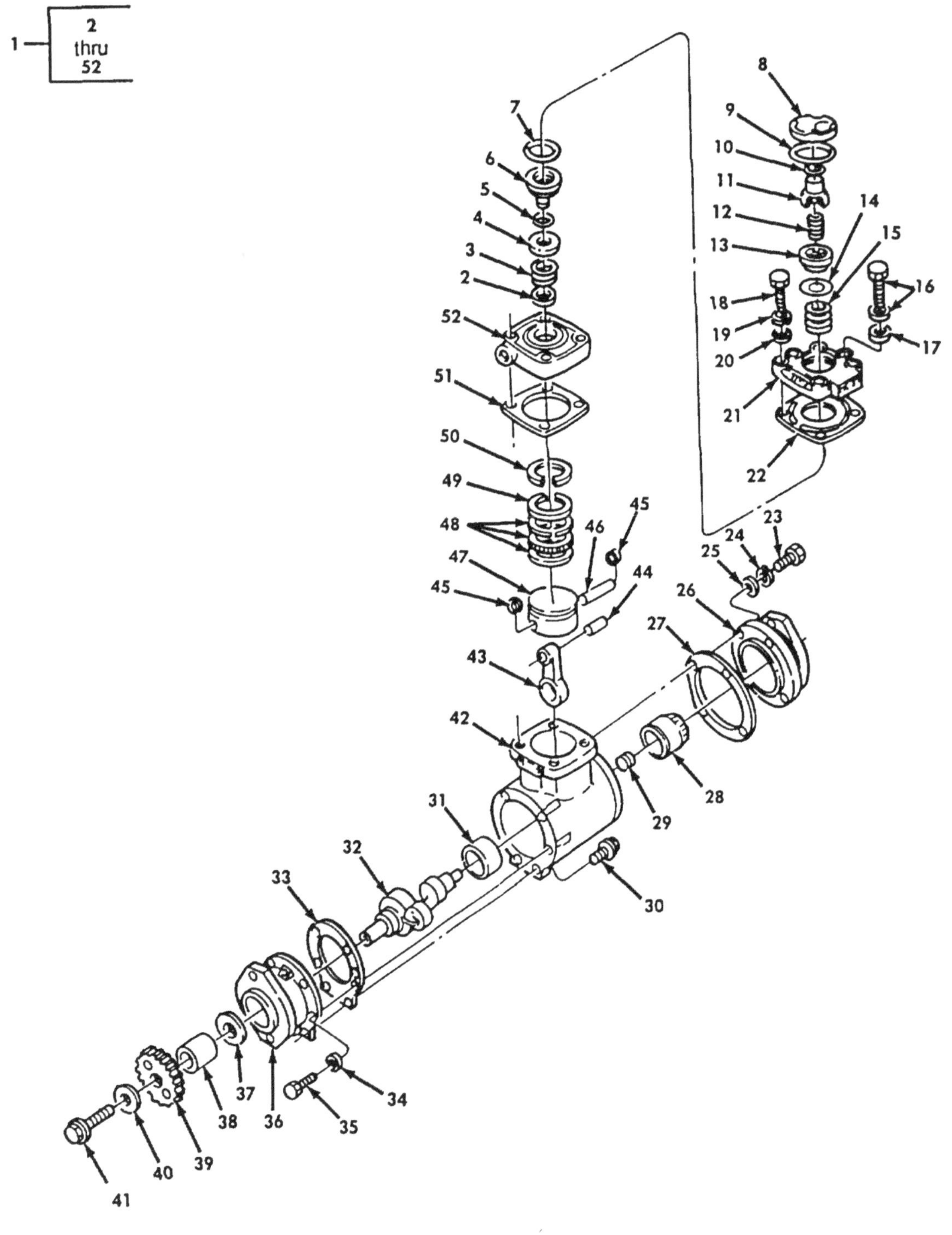

Figure 286. Air Compressor Assembly (M939A2).

(1) ITEM NO	(2) SMR CODE	(3) NSN	(4) CAGEC	(5) PART NUMBER	(6) DESCRIPTION AND USABLE ON CODES (UOC)	(7) QTY
					GROUP 1209 AIR COMPRESSOR	
					FIG. 286 AIR COMPRESSOR ASSEMBLY (M939A2)	
1	PAFFF	2530012688740	15434	3558006	COMPRESSOR, RECIPROC .... UOC:ZAA, ZAB, ZAC, ZAD, ZAE, ZAF, ZAG, ZAH, ZAJ, ZAK, ZAL	1
2	PAFZZ	5365003694729	15434	183429	.SHIM PART OF KIT P/N 3801808 .... UOC:ZAA, ZAB, ZAC, ZAD, ZAE, ZAF, ZAG, ZAH, ZAJ, ZAK, ZAL	1
3	PAFZZ	5360008953216	15434	128080	.SPRING, HELICAL, COMP PART OF KIT P/N .... 3801808 .... UOC:ZAA, ZAB, ZAC, ZAD, ZAE, ZAF, ZAG, ZAH, ZAJ, ZAK, ZAL	1
4	PAFZZ	4820004450610	15434	127940	.DISK, VALVE PART OF KIT P/N 3801808. .... UOC:ZAA, ZAB, ZAC, ZAD, ZAE, ZAF, ZAG, ZAH, ZAJ, ZAK, ZAL	1
5	PAFZZ	5330009052679	15434	128085	.O-RING PART OF KIT P/N 3801808 .... UOC:ZAA, ZAB, ZAC, ZAD, ZAE, ZAF, ZAG, ZAH, ZAJ, ZAK, ZAL	1
6	PAFZZ	4820009094174	15434	144714	.SEAT, VALVE PART OF KIT P/N 3801808. .... UOC:ZAA, ZAB, ZAC, ZAD, ZAE, ZAF, ZAG, ZAH, ZAJ, ZAK, ZAL	1
7	PAFZZ	5330004410145	15434	128086	.O-RING PART OF KIT P/N 3801808 .... UOC:ZAA, ZAB, ZAC, ZAD, ZAE, ZAF, ZAG, ZAH, ZAJ, ZAK, ZAL	1
8	PAFZZ	2530011427939	15434	153966	.BODY, UNLOADER VALVE PART OF KIT P/N .... 3801808 .... UOC:ZAA, ZAB, ZAC, ZAD, ZAE, ZAF, ZAG, ZAH, ZAJ, ZAK, ZAL	1
9	PAFZZ	5330012718289	15434	3043995	.O-RING PART OF KIT P/N 3801808 .... UOC:ZAA, ZAB, ZAC, ZAD, ZAE, ZAF, ZAG, ZAH, ZAJ, ZAK, ZAL	1
10	PAFZZ	5330009413762	15434	127936	.O-RING PART OF KIT P/N 3801808 .... UOC:ZAA, ZAB, ZAC, ZAD, ZAE, ZAF, ZAG, ZAH, ZAJ, ZAK, ZAL	1
11	PAFZZ	4310010847148	8X479	191037	.CAP, UNLOADER PART OF KIT P/N .... 3801808 .... UOC:ZAA, ZAB, ZAC, ZAD, ZAE, ZAF, ZAG, ZAH, ZAJ, ZAK, ZAL	1
12	PAFZZ	5360010863480	15434	3023101	.SPRING, HELICAL, COMP PART OF KIT P/N .... 3801808 .... UOC:ZAA, ZAB, ZAC, ZAD, ZAE, ZAF, ZAG, ZAH, ZAJ, ZAK, ZAL	1
13	PAFZZ	4820009094175	15434	145028	.SEAT, VALVE .... UOC:ZAA, ZAB, ZAC, ZAD, ZAE, ZAF, ZAG, ZAH, ZAJ, ZAK, ZAL	1
14	PAFZZ	3805009555320	15434	144948	.VALVE INTAKE COMPRE PART OF KIT P/N .... 3801808 .... UOC:ZAA, ZAB, ZAC, ZAD, ZAE, ZAF, ZAG, ZAH, ZAJ, ZAK, ZAL	1
15	PAFZZ	5360001299415	15434	190334	.SPRING, HELICAL, COMP PART OF KIT P/N 3801808	1

(1) ITEM NO	(2) SMR CODE	(3) NSN	(4) CAGEC	(5) PART NUMBER	(6) DESCRIPTION AND USABLE ON CODES (UOC)	(7) QTY
					UOC:ZAA, ZAB, ZAC, ZAD, ZAE, ZAF, ZAG, ZAH, ZAJ, ZAK, ZAL	
16	PAFZZ	5305011446233	15434	3021470	.SCREW, CAP, HEXAGON H	2
					UOC:ZAA, ZAB, ZAC, ZAD, ZAE, ZAF, ZAG, ZAH, ZAJ, ZAK, ZAL	
17	PAFZZ	5310001867403	15434	109557	.WASHER, FLAT	1
					UOC:ZAA, ZAB, ZAC, ZAD, ZAE, ZAF, ZAG, ZAH, ZAJ, ZAK, ZAL	
18	PAFZZ	5305002258507	96906	MS90725-43	.SCREW, CAP, HEXAGON H	4
					UOC:ZAA, ZAB, ZAC, ZAD, ZAE, ZAF, ZAG, ZAH, ZAJ, ZAK, ZAL	
19	PAFZZ	5310004079566	96906	MS35338-45	.WASHER, LOCK	4
					UOC:ZAA, ZAB, ZAC, ZAD, ZAE, ZAF, ZAG, ZAH, ZAJ, ZAK, ZAL	
20	PAFZZ	5310005626558	15434	S-626	.WASHER, FLAT	4
					UOC:ZAA, ZAB, ZAC, ZAD, ZAE, ZAF, ZAG, ZAH, ZAJ, ZAK, ZAL	
21	PAFZZ	4310011465921	15434	153964	.COVER, AIR COMPRESSO	1
					UOC:ZAA, ZAB, ZAC, ZAD, ZAE, ZAF, ZAG, ZAH, ZAJ, ZAK, ZAL	
22	PAFZZ	5330001317072	15434	3047159	.GASKET	1
					UOC:ZAA, ZAB, ZAC, ZAD, ZAE, ZAF, ZAG, ZAH, ZAJ, ZAK, ZAL	
23	PAFZZ	5305000711788	80204	B1821BH044C125N	.SCREW, CAP, HEXAGON H	4
					UOC:ZAA, ZAB, ZAC, ZAD, ZAE, ZAF, ZAG, ZAH, ZAJ, ZAK, ZAL	
24	PAFZZ	5310002090965	96906	MS35338-47	.WASHER, LOCK	4
					UOC:ZAA, ZAB, ZAC, ZAD, ZAE, ZAF, ZAG, ZAH, ZAJ, ZAK, ZAL	
25	PAFZZ	5310005626557	15434	S-622	.WASHER, FLAT	4
					UOC:ZAA, ZAB, ZAC, ZAD, ZAE, ZAF, ZAG, ZAH, ZAJ, ZAK, ZAL	
26	PAFZZ	4730012879084	7U263	3558516	.ADAPTER, STRAIGHT, TU	1
					UOC:ZAA, ZAB, ZAC, ZAD, ZAE, ZAF, ZAG, ZAH, ZAJ, ZAK, ZAL	
27	PAFZZ	2330013384829	15434	3069103	.GASKET	1
					UOC:ZAA, ZAB, ZAC, ZAD, ZAE, ZAF, ZAG, ZAH, ZAJ, ZAK, ZAL	
28	PAFZZ	5340012712348	15434	3050926	.ADAPTER, SPLINE	1
					UOC:ZAA, ZAB, ZAC, ZAD, ZAE, ZAF, ZAG, ZAH, ZAJ, ZAK, ZAL	
29	PAFZZ	5365004042934	15434	S-965-E	.PLUG, MACHINE THREAD	1
					UOC:ZAA, ZAB, ZAC, ZAD, ZAE, ZAF, ZAG, ZAH, ZAJ, ZAK, ZAL	
30	PAFZZ	5305011124312	15434	3012472	.SCREW, CAP, HEXAGON H	1
					UOC:ZAA, ZAB, ZAC, ZAD, ZAE, ZAF, ZAG, ZAH, ZAJ, ZAK, ZAL	
31	PAFZZ	3120010164883	15434	147610	.BEARING, SLEEVE	1
					UOC:ZAA, ZAB, ZAC, ZAD, ZAE, ZAF, ZAG, ZAH, ZAJ, ZAK, ZAL	
32	PAFZZ	4310012713807	15434	3558749	.CRANKSHAFT, COMPRESS	1
					UOC:ZAA, ZAB, ZAC, ZAD, ZAE, ZAF, ZAG, ZAH, ZAJ, ZAK, ZAL	

(1) ITEM NO	(2) SMR CODE	(3) NSN	(4) CAGEC	(5) PART NUMBER	(6) DESCRIPTION AND USABLE ON CODES (UOC)	(7) QTY
33	PAFZZ	5330001299389	15434	176027	.GASKET .... UOC:ZAA, ZAB, ZAC, ZAD, ZAE, ZAF, ZAG, ZAH, ZAJ, ZAK, ZAL	1
34	PAFZZ	5310010842362	63005	9411417	.WASHER .... UOC:ZAA, ZAB, ZAC, ZAD, ZAE, ZAF, ZAG, ZAH, ZAJ, ZAK, ZAL	4
35	PAFZZ	5305010886019	15434	3010596	.SCREW, ASSEMBLED WAS .... UOC:ZAA, ZAB, ZAC, ZAD, ZAE, ZAF, ZAG, ZAH, ZAJ, ZAK, ZAL	1
36	PAFZZ	5340012712347	15434	3050367	.SUPPORT, AIR COMPRES .... UOC:ZAA, ZAB, ZAC, ZAD, ZAE, ZAF, ZAG, ZAH, ZAJ, ZAK, ZAL	1
37	PAFZZ	3120012723271	15434	3050924	.BEARING, SLEEVE .... UOC:ZAA, ZAB, ZAC, ZAD, ZAE, ZAF, ZAG, ZAH, ZAJ, ZAK, ZAL	1
38	PAFZZ	5365012708376	15434	3050927	.SPACER, SLEEVE .... UOC:ZAA, ZAB, ZAC, ZAD, ZAE, ZAF, ZAG, ZAH, ZAJ, ZAK, ZAL	1
39	PAFZZ	3020012355055	15434	3902595	.GEAR, SPUR .... UOC:ZAA, ZAB, ZAC, ZAD, ZAE, ZAF, ZAG, ZAH, ZAJ, ZAK, ZAL	1
40	PAFZZ	5310012708387	15434	3053093	.WASHER, FLAT .... UOC:ZAA, ZAB, ZAC, ZAD, ZAE, ZAF, ZAG, ZAH, ZAJ, ZAK, ZAL	1
41	PAFZZ	5306010544485	15434	554316	.BOLT, MACHINE .... UOC:ZAA, ZAB, ZAC, ZAD, ZAE, ZAF, ZAG, ZAH, ZAJ, ZAK, ZAL	1
42	PAFZZ	4310010793319	15434	BM-98685	.HOUSING, AIR COMPRES .... UOC:ZAA, ZAB, ZAC, ZAD, ZAE, ZAF, ZAG, ZAH, ZAJ, ZAK, ZAL	1
43	PAFZZ	2815003697846	15434	3558655	.CONNECTING ROD, PIST .... UOC:ZAA, ZAB, ZAC, ZAD, ZAE, ZAF, ZAG, ZAH, ZAJ, ZAK, ZAL	1
44	PAFZZ	3120011467196	15434	3018153	.BUSHING, SLEEVE .... UOC:ZAA, ZAB, ZAC, ZAD, ZAE, ZAF, ZAG, ZAH, ZAJ, ZAK, ZAL	1
45	PAFZZ	5325009229101	15434	119859	.RING, RETAINING .... UOC:ZAA, ZAB, ZAC, ZAD, ZAE, ZAF, ZAG, ZAH, ZAJ, ZAK, ZAL	2
46	PAFZZ	4310009037174	15434	119810	.PIN, PISTON .... UOC:ZAA, ZAB, ZAC, ZAD, ZAE, ZAF, ZAG, ZAH, ZAJ, ZAK, ZAL	1
47	PAFZZ	4310012715103	15434	3045670	.PISTON, COMPRESSOR .... UOC:ZAA, ZAB, ZAC, ZAD, ZAE, ZAF, ZAG, ZAH, ZAJ, ZAK, ZAL	1
48	PAFZZ	2815010793290	15434	180810	.RING, PISTON .... UOC:ZAA, ZAB, ZAC, ZAD, ZAE, ZAF, ZAG, ZAH, ZAJ, ZAK, ZAL	1
49	PAFZZ	4310010795245	15434	650330	.RING, PISTON .... UOC:ZAA, ZAB, ZAC, ZAD, ZAE, ZAF, ZAG, ZAH, ZAJ, ZAK, ZAL	1
50	PAFZZ	4310011971882	15434	187350	.RING, PISTON	1

(1) ITEM NO	(2) SMR CODE	(3) NSN	(4) CAGEC	(5) PART NUMBER	(6) DESCRIPTION AND USABLE ON CODES (UOC)	(7) QTY
					UOC:ZAA, ZAB, ZAC, ZAD, ZAE, ZAF, ZAG, ZAH, ZAJ, ZAK, ZAL	
51	PAFZZ	5330008527347	15434	154018	.GASKET	1
					UOC:ZAA, ZAB, ZAC, ZAD, ZAE, ZAF, ZAG, ZAH, ZAJ, ZAK, ZAL	
52	PAFZZ	4310012713808	15434	3558762	.CYLINDER HEAD, COMPR	1
					UOC:ZAA, ZAB, ZAC, ZAD, ZAE, ZAF, ZAG, ZAH, ZAJ, ZAK, ZAL	

END OF FIGURE

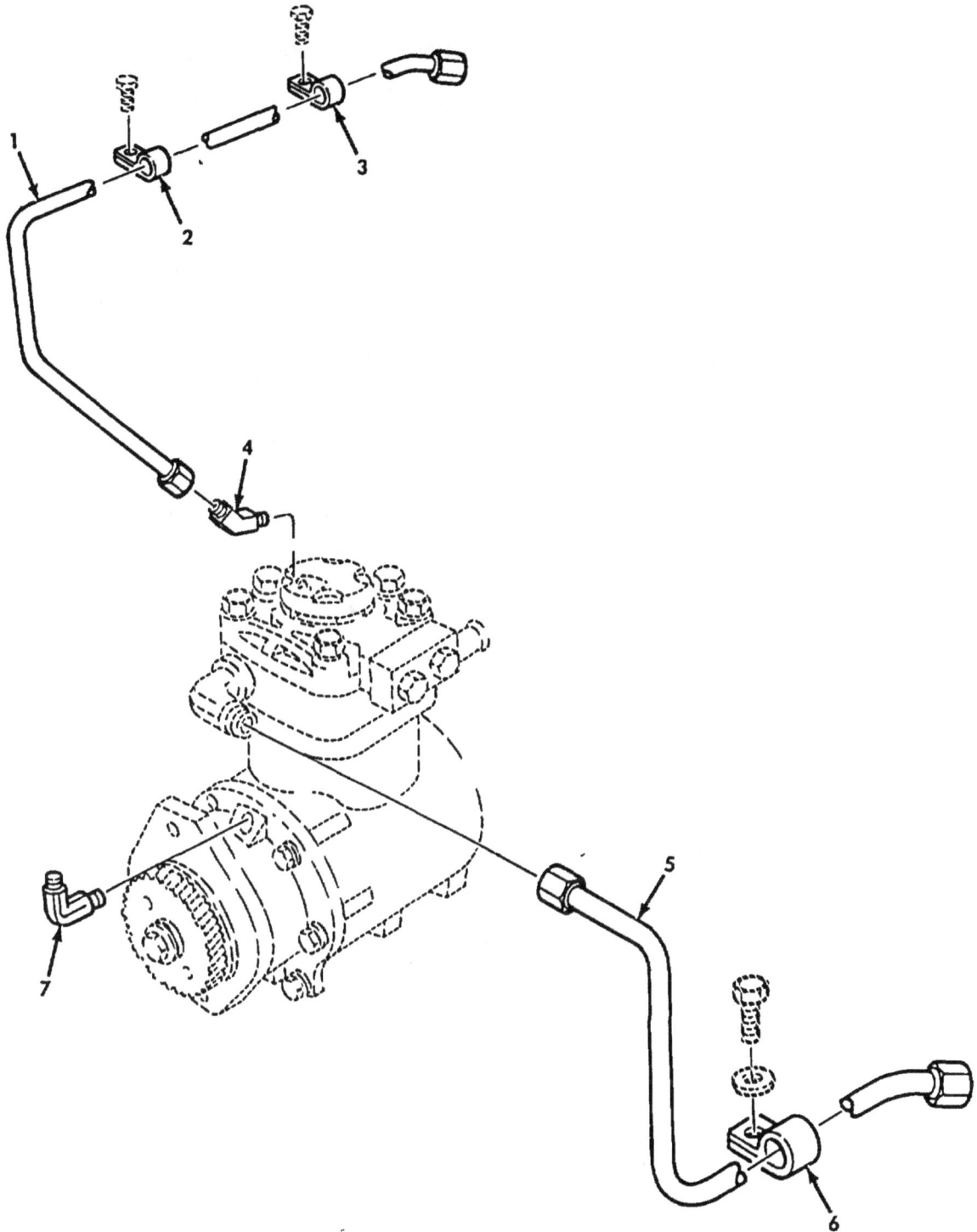

Figure 287. Air Compressor Lines (M939A2).

(1) ITEM NO	(2) SMR CODE	(3) NSN	(4) CAGEC	(5) PART NUMBER	(6) DESCRIPTION AND USABLE ON CODES (UOC)	(7) QTY
					GROUP 1209 AIR COMPRESSOR	
					FIG. 287 AIR COMPRESSOR LINES (M939A2)	
1	PAOZZ	4710012793162	34805	20510404	TUBE ASSEMBLY, METAL .... UOC:ZAA, ZAB, ZAC, ZAD, ZAE, ZAF, ZAG, ZAH, ZAJ, ZAK, ZAL	1
2	PAOZZ	5340009546014	96906	MS21333-121	CLAMP, LOOP .... UOC:ZAA, ZAB, ZAC, ZAD, ZAE, ZAF, ZAG, ZAH, ZAJ, ZAK, ZAL	1
3	PAOZZ	5340000509077	96906	MS21333-119	CLAMP, LOOP .... UOC:ZAA, ZAB, ZAC, ZAD, ZAE, ZAF, ZAG, ZAH, ZAJ, ZAK, ZAL	1
4	PAOZZ	4730012838150	93061	159F-4-4	ELBOW, PIPE TO TUBE .... UOC:ZAA, ZAB, ZAC, ZAD, ZAE, ZAF, ZAG, ZAH, ZAJ, ZAK, ZAL	1
5	PAOZZ	4710012793158	4F744	20511201	TUBE ASSEMBLY, METAL OUTLET .... UOC:ZAA, ZAB, ZAC, ZAD, ZAE, ZAF, ZAG, ZAH, ZAJ, ZAK, ZAL	1
6	PAOZZ	5340002827509	96906	MS21333-62	CLAMP, LOOP .... UOC:ZAA, ZAB, ZAC, ZAD, ZAE, ZAF, ZAG, ZAH, ZAJ, ZAK, ZAL	1
7	PAOZZ	4730012023351	19207	7336402-1	ELBOW, PIPE TO TUBE .... UOC:ZAA, ZAB, ZAC, ZAD, ZAE, ZAF, ZAG, ZAH, ZAJ, ZAK, ZAL	1

END OF FIGURE

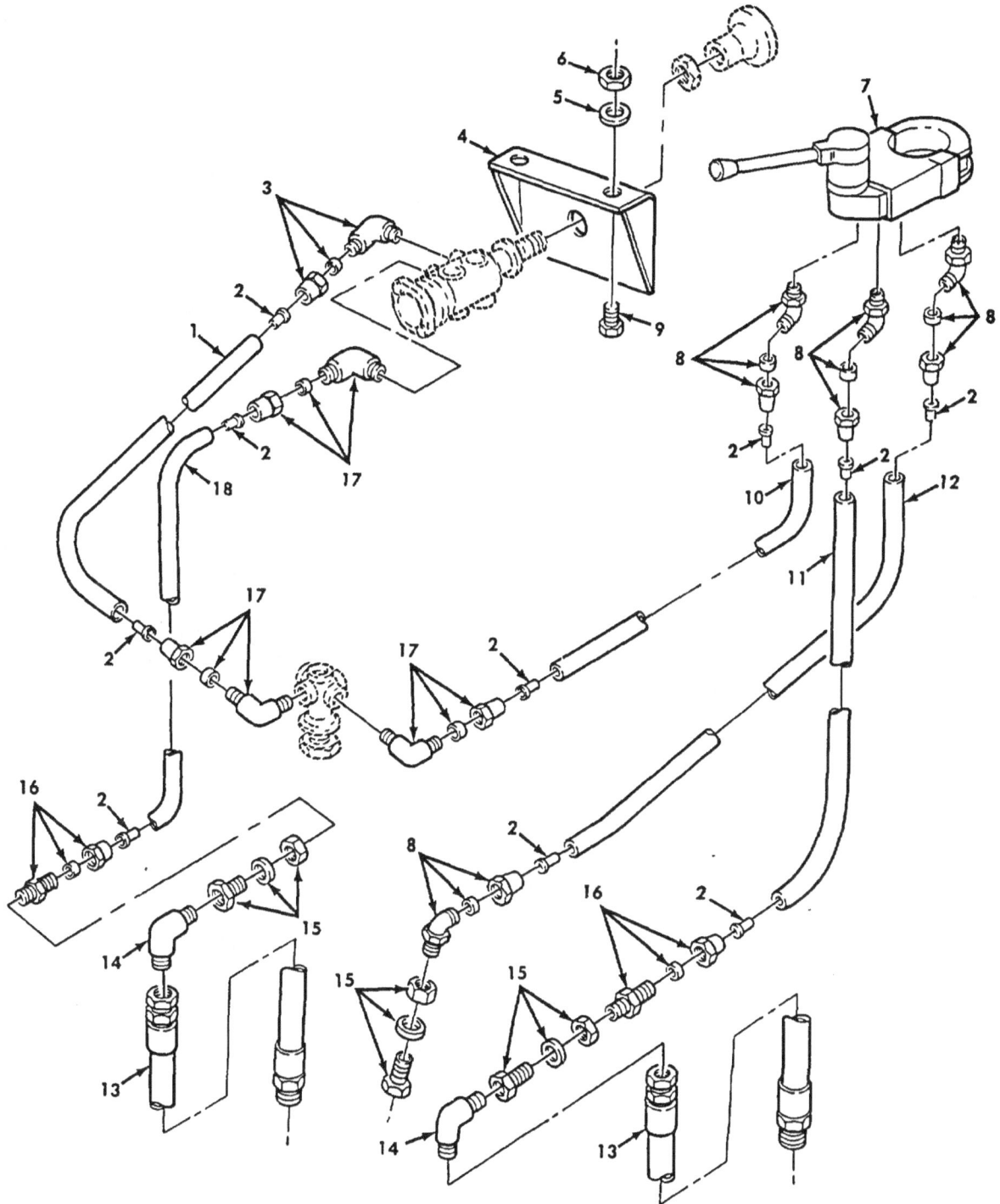

Figure 288. Trailer Brake Control Valve, Hand Valve, and Related Lines.

(1) ITEM NO	(2) SMR CODE	(3) NSN	(4) CAGEC	(5) PART NUMBER	(6) DESCRIPTION AND USABLE ON CODES (UOC)	(7) QTY
					GROUP 1211 TRAILER BRAKE CONNECTIONS AND CONTROLS	
					FIG.288 TRAILER BRAKE CONTROL VALVE, HAND VALVE, AND RELATED LINES	
1	MOOZZ		19207	CPR104420-2-26	HOSE, NONMETALLIC MAKE FROM HOSE, P/N .... CPR104420-2, 26 INCHES LONG .... UOC:DAG, DAH, V21, V22, ZAG, ZAH	1
2	PAOZZ	4730010798821	19207	CPR102321-1	INSERT, TUBE FITTING .... UOC:DAG, DAH, V21, V22, ZAG, ZAH	10
3	PAOZZ	4730002871604	81343	6-2 120202BA	ELBOW, PIPE TO TUBE .... UOC:DAG, DAH, V21, V22, ZAG, ZAH	1
4	PFOZZ	5340011079693	19207	12255848	BRACKET .... UOC:DAG, DAH, V21, V22, ZAG, ZAH	1
5	PAOZZ	5310006379541	96906	MS35338-46	WASHER, LOCK .... UOC:DAG, DAH, V21, V22, ZAG, ZAH	2
6	PAOZZ	5310000581626	96906	MS35650-3382	NUT, PLAIN, HEXAGON .... UOC:DAG, DAH, V21, V22, ZAG, ZAH	2
7	PAOZZ	4820000629719	19207	11602159	VALVE, ROTARY, DIRECT .... UOC:DAG, DAH, V21, V22, ZAG, ZAH	1
8	PAOZZ	4730007017677	96906	MS39185-1	ELBOW, PIPE TO TUBE .... UOC:DAG, DAH, V21, V22, ZAG, ZAH	4
9	PAOZZ	5305002693234	96906	MS90727-58	SCREW, CAP, HEXAGON H .... UOC:DAG, DAH, V21, V22, ZAG, ZAH	2
10	MOOZZ		19207	CPR104420-2-34	HOSE, NONMETALLIC MAKE FROM HOSE, P/N .... CPR104420-2, 34 INCHES LONG .... UOC:DAG, DAH, V21, V22, ZAG, ZAH	1
11	MOOZZ		19207	CPR104420-2-22	HOSE, NONMETALLIC MAKE FROM HOSE, P/N .... CPR104420-2, 22 INCHES LONG .... UOC:DAG, DAH, V21, V22, ZAG, ZAH	1
12	MOOZZ		19207	CPR104420-2-21	HOSE, NONMETALLIC MAKE FROM HOSE, P/N .... CPR104420-2, 21 INCHES LONG .... UOC:DAG, DAH, V21, V22, ZAG, ZAH	1
13	PAOZZ	4720011079939	19207	12256272	HOSE ASSEMBLY, NONME .... UOC:DAG, DAH, V21, V22, ZAG, ZAH	2
14	PAOZZ	4730008127999	96906	MS51504-A6	ELBOW, PIPE TO TUBE .... UOC:DAG, DAH, V21, V22, ZAG, ZAH	2
15	PAOZZ	4730002786318	19207	8328782	COUPLING, PIPE .... UOC:DAG, DAH, V21, V22, ZAG, ZAH	3
16	PAOZZ	4730000691186	81343	6-4 120102BA	ADAPTER, STRAIGHT, PI .... UOC:DAG, DAH, V21, V22, ZAG, ZAH	2
17	PAOZZ	4730000691187	81343	6-4 100202BA	ELBOW, PIPE TO TUBE .... UOC:DAG, DAH, V21, V22, ZAG, ZAH	3
18	MOOZZ		19207	CPR104420-2-9	HOSE, NONMETALLIC MAKE FROM TUBING, ... P/N CPR104420-2, 9 INCHES LONG .... UOC:DAG, DAH, V21, V22, ZAG, ZAH	1

END OF FIGURE

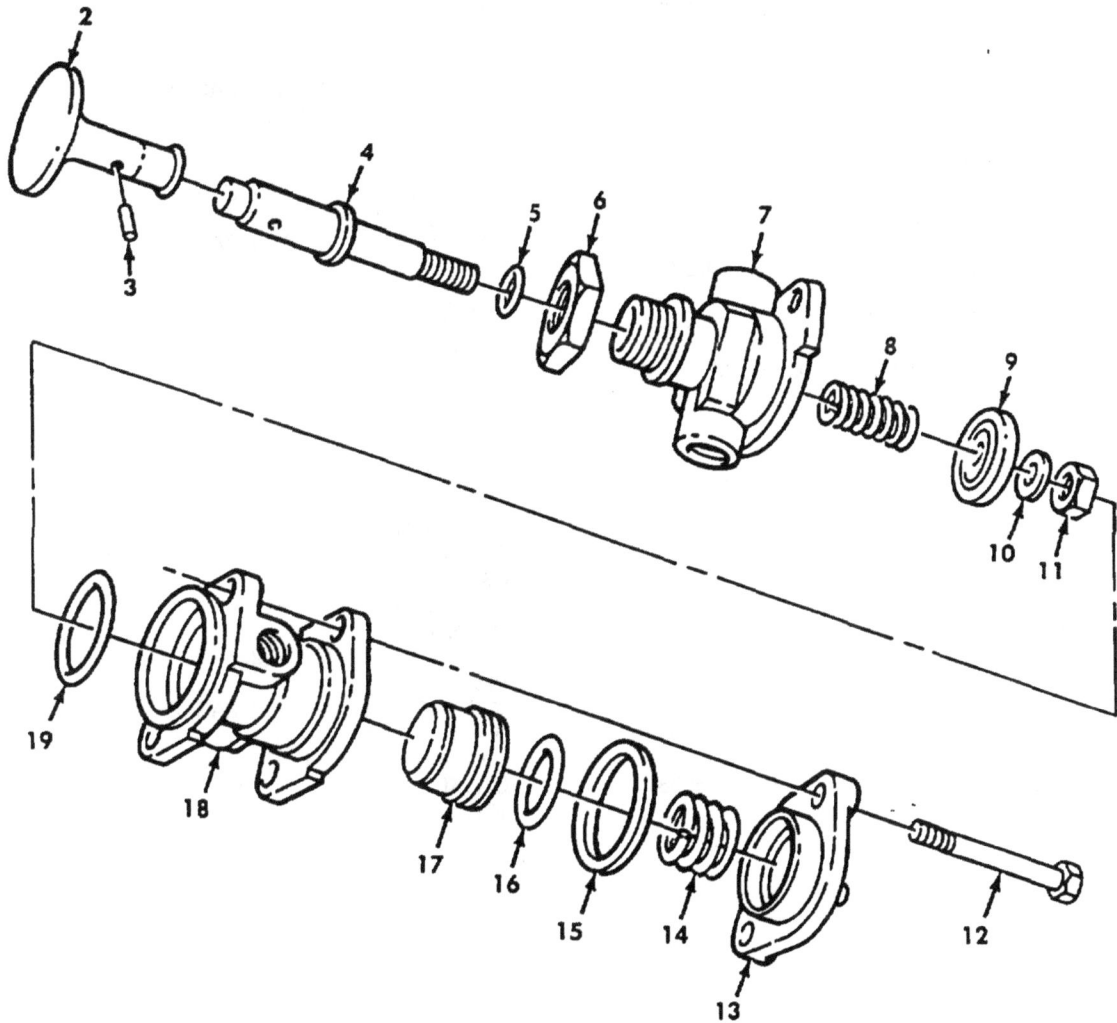

Figure 289. Trailer Air Supply Valve Assembly.

(1) ITEM NO	(2) SMR CODE	(3) NSN	(4) CAGEC	(5) PART NUMBER	(6) DESCRIPTION AND USABLE ON CODES (UOC)	(7) QTY
					GROUP 1211 TRAILER BRAKE CONNECTIONS AND CONTROLS	
					FIG. 289 TRAILER AIR SUPPLY VALVE ASSEMBLY	
1	PAOZZ	4810011079694	19207	11669066	VALVE ASSEMBLY, WITH .... UOC:DAG, DAH, V21, V22, ZAG, ZAH	1
2	PAOZZ	5355011085184	06853	244282	.KNOB .... UOC:DAG, DAH, V21, V22, ZAG, ZAH	1
3	PAOZZ	5315011294621	96906	MS51923-361	.PIN, SPRING .... UOC:DAG, DAH, V21, V22, ZAG, ZAH	1
4	PAFZZ	4820011049992	06853	240273	.DISK, VALVE .... UOC:DAG, DAH, V21, V22, ZAG, ZAH	1
5	PAFZZ	5330010678567	19207	8693035-1	.O-RING .... UOC:DAG, DAH, V21, V22, ZAG, ZAH	1
6	PAFZZ	5310011057227	06853	239357	.NUT, PLAIN, HEXAGON .... UOC:DAG, DAH, V21, V22, ZAG, ZAH	1
7	XAFZZ	4820011573556	06853	293514	.BODY, VALVE .... UOC:DAG, DAH, V21, V22, ZAG, ZAH	1
8	PAFZZ	5360011270953	06853	239330	.SPRING, HELICAL, COMP .... UOC:DAG, DAH, V21, V22, ZAG, ZAH	1
9	PAFZZ	4820004843539	06853	240277	.DISK, VALVE .... UOC:DAG, DAH, V21, V22, ZAG, ZAH	1
10	PAFZZ	5310001670835	19207	7402359	.WASHER, FLAT .... UOC:DAG, DAH, V21, V22, ZAG, ZAH	1
11	PAFZZ	5310009359022	96906	MS51943-32	.NUT, SELF-LOCKING, HE .... UOC:DAG, DAH, V21, V22, ZAG, ZAH	1
12	PAFZZ	5306002830407	24617	9409106	.BOLT, ASSEMBLED WASH .... UOC:DAG, DAH, V21, V22, ZAG, ZAH	2
13	PAFZZ	4820011609597	06853	243637	.CAP, VALVE .... UOC:DAG, DAH, V21, V22, ZAG, ZAH	1
14	PAFZZ	5360010207063	06853	243635	.SPRING, HELICAL, COMP .... UOC:DAG, DAH, V21, V22, ZAG, ZAH	1
15	PAFZZ	5330011277134	19207	7060081-3	.GASKET .... UOC:DAG, DAH, V21, V22, ZAG, ZAH	1
16	PAFZZ	5330011278550	19207	8693035-2	.O-RING .... UOC:DAG, DAH, V21, V22, ZAG, ZAH	1
17	PAFZZ	4810011085107	19207	11669076	.PISTON, VALVE .... UOC:DAG, DAH, V21, V22, ZAG, ZAH	1
18	XAFZZ		06853	243633	.BODY, VALVE .... UOC:DAG, DAH, V21, V22, ZAG, ZAH	1
19	PAFZZ	5330011273803	19207	7060081-2	.GASKET .... UOC:DAG, DAH, V21, V22, ZAG, ZAH	1

END OF FIGURE

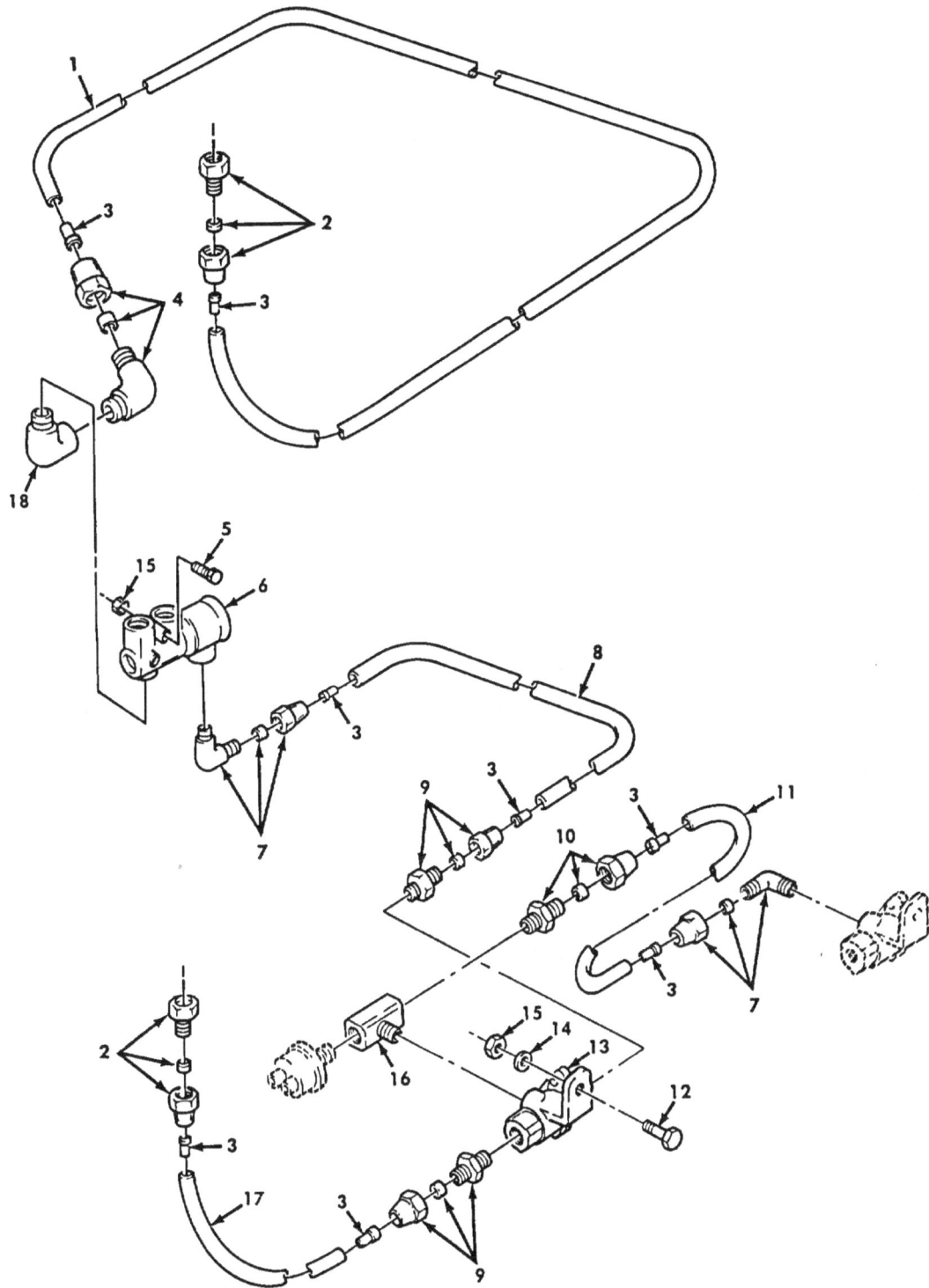

Figure 290. Tractor Protection Valve, Double-check Valve, and Trailer Service Air Lines.

(1) ITEM NO	(2) SMR CODE	(3) NSN	(4) CAGEC	(5) PART NUMBER	(6) DESCRIPTION AND USABLE ON CODES (UOC)	(7) QTY
					GROUP 1211 TRAILER BRAKE CONNECTIONS AND CONTROLS	
					FIG. 290 TRACTOR PROTECTION VALVE, DOUBLE-CHECK VALVE, AND TRAILER SERVICE AIR LINES	
1	MOOZZ		19207	CPR104420-2-130	HOSE NONMETALLIC MAKE FROM HOSE, P/N CPR104420-2, 130 INCHES LONG UOC:DAG, DAH, V21, V22, ZAG, ZAH	1
2	PAOZZ	4730002778770	81343	6-6 120103BA	ADAPTER, STRAIGHT, PI UOC:DAG, DAH, V21, V22, ZAG, ZAH	2
3	PAOZZ	4730010798821	19207	CPR102321-1	INSERT, TUBE FITTING UOC:DAG, DAH, V21, V22, ZAG, ZAH	8
4	PAOZZ	4730007017677	96906	MS39185-1	ELBOW, PIPE TO TUBE UOC:DAG, DAH, V21, V22, ZAG, ZAH	1
5	PAOZZ	5305002259099	96906	MS90726-44	SCREW, CAP, HEXAGON H UOC:DAG, DAH, V21, V22, ZAG, ZAH	2
6	PAOZZ	4820004363033	06853	279000	VALVE, FLOW CONTROL UOC:DAG, DAH, V21, V22, ZAG, ZAH	1
7	PAOZZ	4730002890155	81343	6-6 120202BA	ELBOW, PIPE TO TUBE UOC:DAG, DAH, V21, V22, ZAG, ZAH	2
8	MOOZZ		19207	CPR104420-2-49	HOSE, NONMETALLIC MAKE FROM TUBING, P/N CPR104420-2, 49 INCHE LONG UOC:DAG, DAH, V21, V22, ZAG, ZAH	1
9	PAOZZ	4730002890155	81343	6-6 120202BA	ELBOW, PIPE TO TUBE UOC:DAG, DAH, V21, V22, ZAG, ZAH	2
10	PAOZZ	4730000691186	81343	6-4 120102BA	ADAPTER, STRAIGHT, PI UOC:DAG, DAH, V21, V22, ZAG, ZAH	1
11	MOOZZ		19207	CPR104420-2-14	HOSE, NONMETALLIC MAKE FROM HOSE, P/N CPR104420-2, 14 INCHES LONG UOC:DAG, DAH, V21, V22, ZAG, ZAH	1
12	PAOZZ	5306000514077	80204	B1821BH031F113N	BOLT, MACHINE UOC:DAG, DAH, V21, V22, ZAG, ZAH	1
13	PAOZZ	4820007287467	06853	278614	VALVE, LINEAR, DIRECT UOC:DAG, DAH, V21, V22, ZAG, ZAH	1
14	PAOZZ	5310000814219	96906	MS27183-12	WASHER, FLAT UOC:DAG, DAH, V21, V22, ZAG, ZAH	1
15	PAOZZ	5310002416658	96906	MS51943-34	NUT, SELF-LOCKING, HE UOC:DAG, DAH, V21, V22, ZAG, ZAH	1
16	PAOZZ	4730001257979	81343	6-6-6-130425B	TEE, PIPE UOC:DAG, DAH, V21, V22, ZAG, ZAH	1
17	MOOZZ		19207	CPR104420-2-70	HOSE, NONMETALLIC MAKE FROM HOSE, P/N CPR104420-2, 70 INCHES LONG UOC:DAG, DAH, V21, V22, ZAG, ZAH	1
18	PAOZZ	4730002775553	96906	MS51952-2	ELBOW, PIPE UOC:DAG, DAH, V21, V22, ZAG, ZAH	1

END OF FIGURE

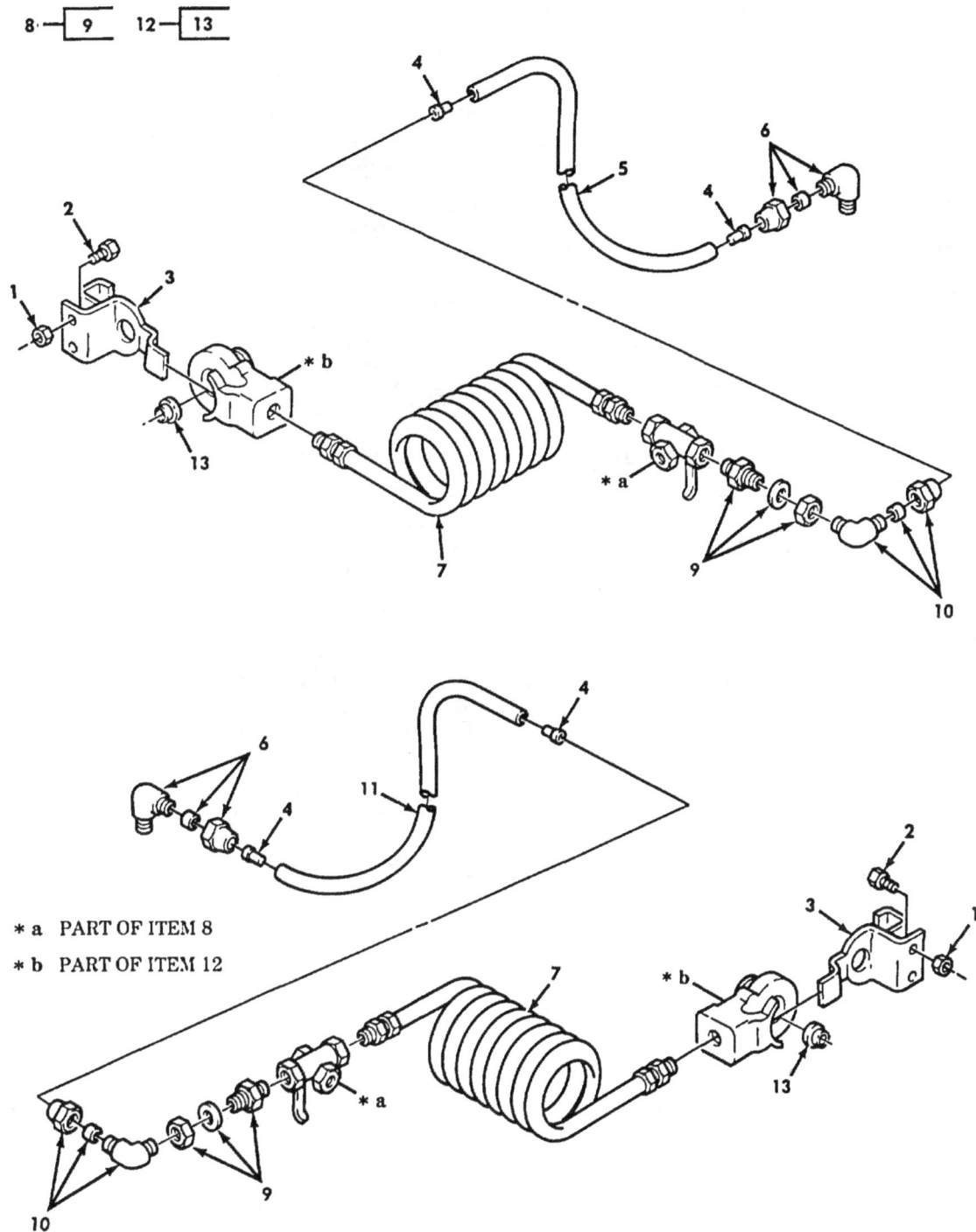

Figure 291. Trailer Brake Lines, Emergency and Service Couplings, and Related Lines.

(1) ITEM NO	(2) SMR CODE	(3) NSN	(4) CAGEC	(5) PART NUMBER	(6) DESCRIPTION AND USABLE ON CODES (UOC)	(7) QTY
					GROUP 1211 TRAILER BRAKE CONNECTIONS AND CONTROLS	
					FIG. 291 TRAILER BRAKE LINES, EMERGENCY AND SERVICE COUPLINGS, AND RELATED LINES	
1	PAOZZ	5310002416658	96906	MS51943-34	NUT, SELF-LOCKING, HE .... UOC:DAG, DAH, V21, V22, ZAG, ZAH	2
2	PAOZZ	5306000501238	96906	MS90727-32	BOLT, MACHINE .... UOC:DAG, DAH, V21, V22, ZAG, ZAH	2
3	PAOZZ	2530005455406	06583	212227	DUMMY COUPLING, AUTO .... UOC:DAG, DAH, V21, V22, ZAG, ZAH	2
4	PAOZZ	4730010798821	19207	CPR102321-1	INSERT, TUBE FITTING .... UOC:DAG, DAH, V21, V22, ZAG, ZAH	4
5	MOOZZ		19207	CPR104420-2-19	HOSE, NONMETALLIC MAKE FROM HOSE, P/ .... N CPR104420-2, 19 INCHES LONG .... UOC:DAG, DAH, V21, V22, ZAG, ZAH	1
6	PAOZZ	4730001439282	02570	MS39182-8	ELBOW, PIPE TO TUBE .... UOC:DAG, DAH, V21, V22, ZAG, ZAH	2
7	PAOZZ	4720011600733	85757	326C-78002	HOSE ASSEMBLY, NONME .... UOC:DAG, DAH, V21, V22, ZAG, ZAH	2
8	PAOZZ	4820005953669	06853	285696	COCK, PLUG .... UOC:DAG, DAH, V21, V22, ZAG, ZAH	2
9	PAOZZ	4730002449848	40670	11682888	.NIPPLE, TANK .... UOC:DAG, DAH, V21, V22, ZAG, ZAH	1
10	PAOZZ	4730000691187	81343	6-4 100202BA	ELBOW, PIPE .... UOC:DAG, DAH, V21, V22, ZAG, ZAH	2
11	MOOZZ		19207	CPR104420-2-23	HOSE, NONMETALLIC MAKE FROM HOSE, P/ .... N CPR104420-2, 23 INCHES LONG .... UOC:DAG, DAH, V21, V22, ZAG, ZAH	1
12	PAOZZ	4730005950083	58536	A52484-1	COUPLING HALF, QUICK .... UOC:DAG, DAH, V21, V22, ZAG, ZAH	2
13	PAOZZ	5330000902128	06853	213630	PACKING, PREFORMED .... UOC:DAG, DAH, V21, V22, ZAG, ZAH	2

END OF FIGURE

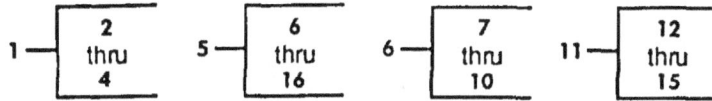

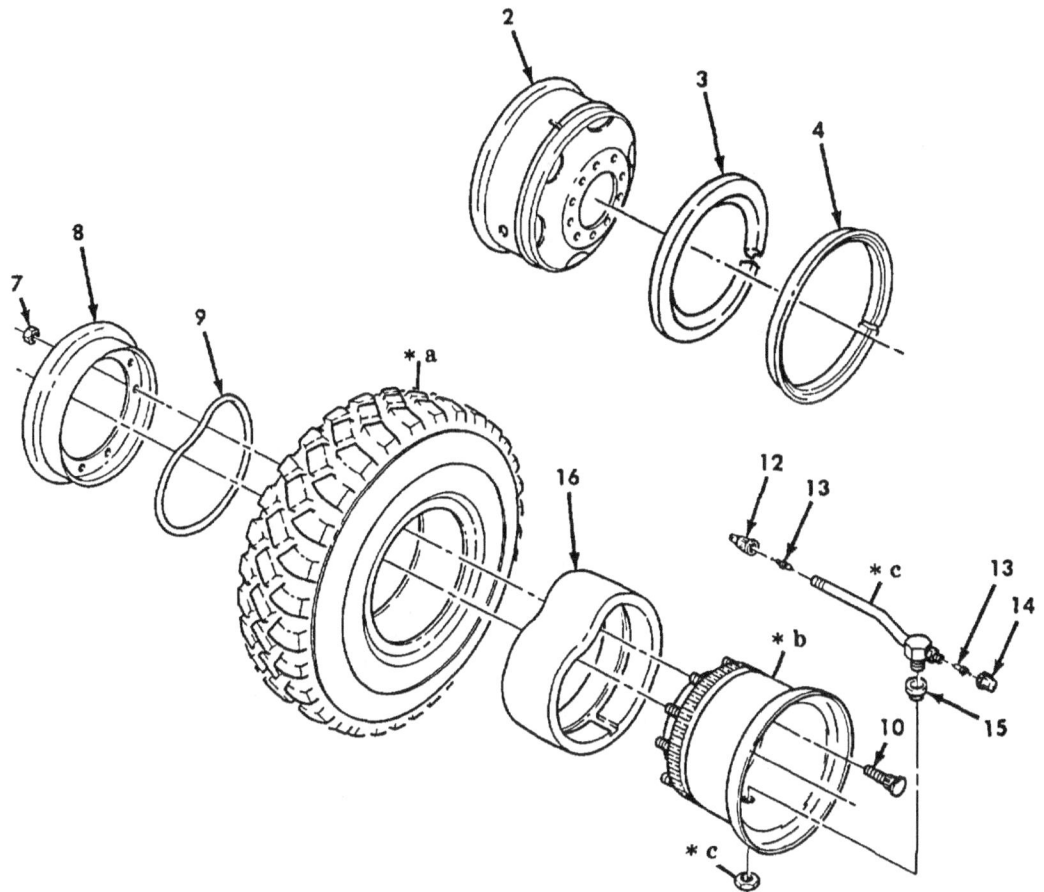

* a PART OF ITEM 5
* b PART OF ITEM 6
* c PART OF ITEM 11

Figure 292. Wheel Assembly.

(1) ITEM NO	(2) SMR CODE	(3) NSN	(4) CAGEC	(5) PART NUMBER	(6) DESCRIPTION AND USABLE ON CODES (UOC)	(7) QTY
					GROUP 13 WHEELS 1311 WHEEL ASSEMBLY	
					FIG. 292 WHEEL ASSEMBLY	
1	PAOZZ	2530012893962	19207	12356821	WHEEL, PNEUMATIC TIR UOC:V39	11
1	PAOZZ	2530006035768	19207	7388820	WHEEL, PNEUMATIC TIR .... UOC:V12, V13, V14, V15, V16, V17 , V18, V9, V20, V21, V22, V24, V25	11
2	XDOZZ	2530007389493	19207	7389493	.RIM, WHEEL, PNEUMATIC UOC:V12, V13, V14, V15, V16, V16, VV1, V18, V9, V20, V21, V22, V24, V25	1
3	PAOZZ	2530012885877	19207	12356823	.RING UOC:V12, V13, V14, V15, V16, V17, V18, V19 V20, V21, V22, V24, V25	1
4	PAOZZ	2530007389061	19207	7389061	.RING, SIDE, AUTOMOTIV ....WHEEL ASSEMBLY UOC:V12, V13, V14, V15, V16, V17, V1 , V1 9, V20, V21, V22, V24, V25	1
5	AOOOO		19207	12301128-1	WHEEL & BEADLOCK....................................... UOC:DAA, DAB, DAC, DAD, DAE, DAF, DAG, DAH, DAJ, DAK, DAL, DAW, DAX, ZAA, ZAB, ZAC, ZAD, ZAE, ZAF, ZAG, ZAH, ZAJ, ZAK, ZAL	7
6	PAOOO	2530013030801	19207	12363603	.WHEEL, PNEUMATIC TIR .... UOC:DAA, DAB, DAC, DAD, DAE, DAF, DAG, DAH, DAJ, DAK, DAL, DAW, DAX, ZAA, ZAB, ZAC, ZAD, ZAE, ZAF, ZAG, ZAH, ZAJ, ZAK, ZAL	1
7	PAOZZ	5310011022711	19207	12257242	..NUT, SELF-LOCKING, EX................................ UOC:DAA, DAB, DAC, DAD, DAE, DAF, DAG, DAH, DAJ, DAK, DAL, DAW, DAX, ZAA, ZAB, ZAC, ZAD, ZAE, ZAF, ZAG, ZAH, ZAJ, ZAK, ZAL	10
8	XAOZZ		19207	12363607	..CLAMP, RIM CLENCHING................................ UOC:DAA, DAB, DAC, DAD, DAE, DAF, DAG, DAH, DAJ, DAK, DAL, DAW, DAX, ZAA, ZAB, ZAC, ZAD, ZAE, ZAF, ZAG, ZAH, ZAJ, ZAK, ZAL	1
9	PAOZZ	5331013147598	19207	12363606	..O-RING .... UOC:DAA, DAB, DAC, DAD, DAE, DAF, DAG, DAH, DAJ, DAK, DAL, DAW, DAX, ZAA, ZAB, ZAC, ZAD, ZAE, ZAF, ZAG, ZAH, ZAJ, ZAK, ZAL	1
10	PAOZZ	5306013146742	19207	12363604	..BOLT, FINNED NECK .... UOC:DAA, DAB, DAC, DAD, DAE, DAF, DAG, DAH, DAJ, DAK, DAL, DAW, DAX, ZAA, ZAB, ZAC, ZAD, ZAE, ZAF, ZAG, ZAH, ZAJ, ZAK, ZAL	10
11	PAOOO	4820012108821	17875	VS-1072	.VALVE, ANGLE .... UOC:DAA, DAB, DAC, DAD, DAE, DAF, DAG, DAH, DAJ, DAK, DAL, DAW, DAX, ZAA, ZAB, ZAC, ZAD, ZAE, ZAF, ZAG, ZAH, ZAJ, ZAK, ZAL	1
12	PAOZZ	2640000603550	81348	ZZ-V-25/TYPEIV/ CLASS1/TR-VC-2	..CAP, PNEUMATIC VALVE............................... UOC:DAA, DAB, DAC, DAD, DAE, DAF, DAG, DAH, DAJ, DAK, DAL, DAW, DAX, ZAA, ZAB, ZAC, ZAD, ZAE, ZAF, ZAG, ZAH, ZAJ, ZAK, ZAL	1

(1) ITEM NO	(2) SMR CODE	(3) NSN	(4) CAGEC	(5) PART NUMBER	(6) DESCRIPTION AND USABLE ON CODES (UOC)	(7) QTY
13	PAOZZ	2640011115467	81348	TYV/CL2/TR C1/STANDARD LENGTH	..VALVE CORE .... UOC:DAA, DAB, DAC, DAD, DAE, DAF, DAG, DAH, DAJ, DAK, DAL, DAW, DAX, ZAA, ZAB, ZAC, ZAD, ZAE, ZAF, ZAG, ZAH, ZAJ, ZAK, ZAL	2
14	PAOZZ	2640000501235	81348	TY IV/CL1/TR VC3	..CAP, PNEUMATIC VALVE................................. UOC:DAA, DAB, DAC, DAD, DAE, DAF, DAG, DAH, DAJ, DAK, DAL, DAW, DAX, ZAA, ZAB, ZAC, ZAD, ZAE, ZAF, ZAG, ZAH, ZAJ, ZAK, ZAL	1
15	PAOZZ	5325012317594	19207	12303016	..GROMMET, NONMETALLIC .... UOC:DAA, DAB, DAC, DAD, DAE, DAF, DAG, DAH, DAJ, DAK, DAW, DAX, ZAA, ZAB, ZAC, ZAD, ZAE, ZAF, ZAG, ZAH, ZAJ, ZAK, ZAL	1
16	PAOZZ	2530012118405	19207	12301119	.BEADLOCK, PNEUMATIC .... UOC:DAA, DAB, DAC, DAD, DAE, DAF, DAG, DAH, DAJ, DAK, DAL , DAW, DAX, ZAA, ZAB, ZAC, ZAD, ZAE, ZAF, ZAG, ZAH, ZAJ, ZAK, ZAL	1

END OF FIGURE

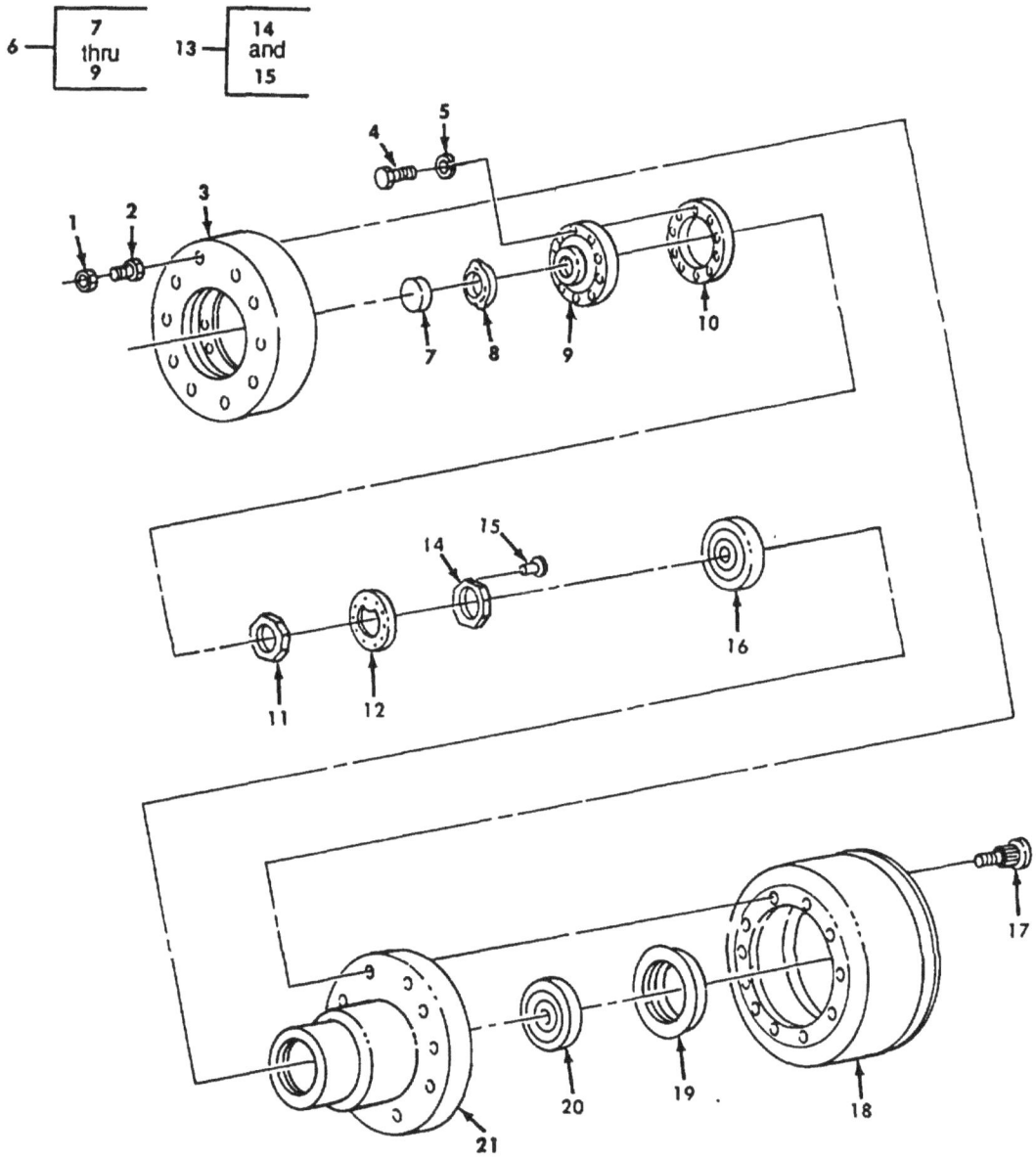

Figure 293. Front Wheel Hub and Drum Assembly (M939, M939A1).

(1) ITEM NO	(2) SMR CODE	(3) NSN	(4) CAGEC	(5) PART NUMBER	(6) DESCRIPTION AND USABLE ON CODES (UOC)	(7) QTY
					GROUP 1311 WHEEL ASSEMBLY	
					FIG. 293 FRONT WHEEL HUB AND DRUM ASSEMBLY (M939 , M939A1)	
1	PAOZZ	5310013753107	24614	3A7313	NUT, PLAIN, SINGLE BA WHEEL STUD, R.H.......... UOC:DAA, DAB, DAC, DAD, DAE, DAF, DAG, DAH, DAJ, DAK, DAL, DAW, DAX, V12 , V13, V14, V15, V16, V17, V18, V19, V20, V21 , V22, V24, V25	10
1	PAOZZ	5310002737771	78500	1199G111	NUT, PLAIN, SINGLE BA WHEEL STUD, L.H.......... UOC: DAA, DAB, DAC, DAD, DAE, DAF, DAG, DAH, DAJ, DAK, DAW, DAX, V12, V13 , V14, V15, V16, V17, V18, V19, V20, V21, V22, V24, V25	10
1	PAOZZ	5310012298029	96906	MS51983-8	NUT, PLAIN, SINGLE BA WHEEL STUD, R.H. 10 UOC :V39	
1	PAOZZ	5310012705463	58536	A-A-52427	NUT, PLAIN, SINGLE BA WHEEL STUD, L.H.......... UOC :V39	10
2	PAOZZ	2530003591162	78500	1199J114C	NUT, CAP, DUAL WHEEL, R.H UOC:V39	10
2	PAOZZ	2530006931029	52304	10709	NUT, CAP, DUAL WHEEL, L.M UOC:V39	10
3	PAOZZ	2530007411105	78500	3280-B-1978	ADAPTER, BRAKE DRUM .... UOC:V39..	2
4	PAOZZ	5306001829267	19207	10896726	BOLT, MACHINE .... UOC:DAA, DAB, DAC, DAD, DAE, DAF, DAG, DAH, DAJ, DAK, DAL, DAW, DAX, V12, V13 , V14, V15, V16, V17, V18, V19, V20, V21, V22, V24, V25, V39	20
5	PAOZZ	5310000034094	01276	210104-8S	WASHER, LOCK .... UOC:DAA, DAB, DAC, DAD, DAE, DAF, DAG, DAH, DAJ, DAK, DAL, DAW, DAX, V12, V13, V14, V15, V16, V17, V18, V19, V20, V21, V22, V24, V25, V39	20
6	PAOZZ	2520007346991	19207	7346991	FLANGE, COMPANION, VE .... UOC:DAA, DAB, DAC, DAD, DAE, DAF, DAG, DAH, DAJ, DAK, DAL, DAW, DAX, V12 , V13 , V14, V15, V16, V17, V18, V19, V20, V21, V24, V25, V39	2
7	PAOZZ	5340007346992	19207	7346992	.PLUG, EXPANSION.. UOC: DAA, DAB, DAC, DAD, DAE, DAF, DAG, DAH, DAJ, DAK, DAL, DAW, DAX, V12 , V13 , V14 , V15, V16, V17, V18, V19, V20, V21, V22, V24, V25, V39	1
8	PAOZZ	5330002913273	96906	MS29513-229	.PACKING, PREFORMED .... UOC: DAA, DAB , DAC, DAD , DAE, DA, DAG , DAH, DAJ, DAK, DAL, DAW, DAX, V12, V13, V14, V15, V16, V17, V18, V19, V20, V21, V22, V24, V25, V39	1
9	XAOZZ		19207	8758278	.FLANGE ASSEMBLY, COMPANION, UN NHA......... P/N 7346991 19207 .... UOC: DAA, DAB, DAC, DAD, DAE, DAF, DAG, DAH, DAJ, DAK, DAL, DAW, DAX, V12 , V13, V14 , V15, V16, V17, V18, V19, V20, V21, V22, V24, V25, V39	1
10	PAOZZ	5330013779775	14153	02764	GASKET, NON-ASBESTOS	2

(1) ITEM NO	(2) SMR CODE	(3) NSN	(4) CAGEC	(5) PART NUMBER	(6) DESCRIPTION AND USABLE ON CODES (UOC)	(7) QTY
					UOC:DAA, DAB, DAC, DAD, DAE, DAF, DAG, DAH, DAJ, DAK, DAL, DAW, DAX, V12, V13, V14, V15, V16, V17, V18, V19, V20, V21, V22, V24, V25, V39	
11	PAOZZ	5310003532427	19207	7979263	NUT, PLAIN, OCTAGON .... UOC: DAA, DAB, DAC, DAD, DAB, DAF, DAG, DAH , DAJ, DAK, DAL, DAW, DAX, V12, V13, V14, V15, V16, V17, V18, V19, V20, V21, V22, V24, V25, V39	2
12	PAOZZ	5310007007089	19207	5139123	WASHER, KEY, WHEEL BEARING LOCK .... UOC: DAA, DAB, DAC, DAD, DAE , DAF, DAG, DAH, DAJ, DAK, DAL, DAW, DAX, V12 , V13, V14 , V15, V16, V17, V18, V19, V20, V21, V22, V24, V25, V39	2
13	PAOZZ	5310003740836	19207	7979308	NUT, PLAIN, OCTAGON .... UOC: DAA, DAB, DAC, DAD, DAE, DAF, DAG, DAH, DAJ, DAK, DAL, DAW, DAX, V12, V13, V14, V15, V16, V17, V18, V19, V20, V21, V22, V24, V25, V39	2
14	XAOZZ	5310011350049	19207	7979309	.NUT, PLAIN, OCTAGON .... UOC: DAA, DAB, DAC, DAD, DAE, DAF, DAG, DAH, DAJ, DAK, DAL, DAW, DAX, V12, V13, V14, V15, V16, V17, V18, V19, V20, V21, V22, V24, V25, V39	1
15	PAOZZ	5315010587268	19207	7979310	.PIN, ADJUSTMENT, OUTER WHEEL.................... BEARING NUT .... UOC:DAA, DAB, DAC, DAD, DAE, DAF, DAG , DAH, DAJ, DAK, DAL, DAW, DAX, V12 , V13 , V14 , V15, V16, V17, V18, V19, , 20, V21, V22, V24, V25, V39	1
16	PAOZZ	3110001004223	96906	MS19081-181	BEARING, ROLLER, TAPE .... UOC: DAA, DAB, DAC, DAD, DAE, DAF, DAG, DAH, DAJ, DAK, DAL, DAW, DAX, V12 , V13 , V14 , V15, V16, V17, V18, V19, , 20, V21, V22, V24, V25, V39	2
17	PAOZZ	5307011424783	78500	20X-1529	STUD, LEFT FRONT WHEEL UOC:V39	10
17	PAOZZ	5307011424784	78500	20X-1528	STUD, RIGHT FRONT WHEEL UOC:V39	10
17	PAOZZ	5306012715842	78500	20X-1815-Z	STUD, LEFT FRONT WHEEL........................ UOC: DAA, DAB, DAC, DAD, DAE, DAF, DAG, DAH, DAJ, DAK, DAL, DAW, DAX, V12 , V13 , V14 , V15, V16, V17, V18, V19, V20, V221, V22, V2 2, V2 5	10
17	PAOZZ	5306012715843	78500	20X-1816-Z	BOLT, RIBBED, SHOULDE, RIGHT FRONT............ WHEEL .... UOC: DAA, DAB, DAC , DAD, DAE, DAF, DAG, DAH, DAJ, DAK, DAL, DAW, DAX, V12 , V13 , V14 , V15, V16, V17, V18, V19 , V20, V21, V22, V24, V25	10
18	PAOZZ	2530011346570	78500	3219-C-4345	BRAKE DRUM .... UOC: DAA, DAB, DAC, DAD, DAE, DAF, DAG, DAH, DAJ, DAK, DAL, DAW, DAX, V12 , V13 , V14 , V15, V16, V17, V18, V9 , V20, V21, V22, V24, V25, V39	2

(1) ITEM NO	(2) SMR CODE	(3) NSN	(4) CAGEC	(5) PART NUMBER	(6) DESCRIPTION AND USABLE ON CODES (UOC)	(7) QTY
19	PAOZZ	533001448350	19207	12375801	SEAL, PLAIN, ENCASED .... UOC:DAA, DAB, DAC, DAD, DAE, DAF, DAG, DAH, DAJ, DAK, DAL, DAW, DAX, V12, V13, V14, V15, V16, V17, V18, V19, V20, V21, V22, V24, V25, V39..	2
20	PAOZZ	3110006898250	96906	MS19081-182	BEARING, ROLLER, TAPE, INNER .... UOC:DAA, DAB, DAC, DAD, DAE, DAF, DAG, DAH, DAJ, DAK, DAL, DAW, DAX, V12, V13, V14, V15, V16, V17, V18, V19, V20, V21, V22, V24, V25, V39	2
21	PAOZZ	3040011391611	78500	A333-T-2334	HUB, FRONT WHEEL .... UOC:DAA, DAB, DAC, DAD, DAE, DAF, DAG, DAH, DAJ, DAK, DAL, DAW, DAX, V12 , V13 , V14 , V15, V16, V17, V18, V19, V20, V21, V22, V24, V25, V39	2

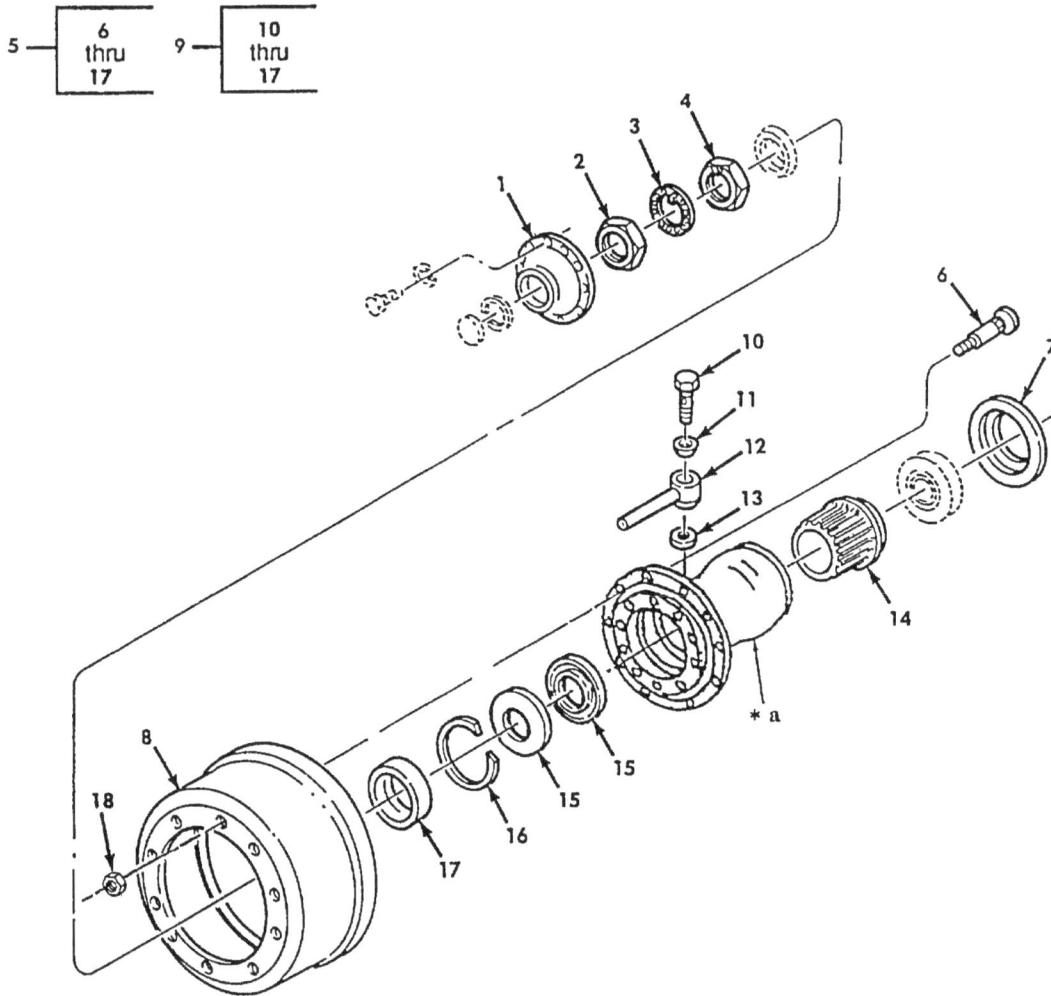

* **a** PART OF ITEM 9

Figure 294. Front Wheel Hub and Drum Assembly (M939A2).

(1) ITEM NO	(2) SMR CODE	(3) NSN	(4) CAGEC	(5) PART NUMBER	(6) DESCRIPTION AND USABLE ON CODES (UOC)	(7) QTY
					GROUP 1311 WHEEL ASSEMBLY	
					FIG. 294 FRONT WHEEL HUB AND DRUM ASSEMBLY(M939A2)	
1	PAOZZ	2530012719809	78500	A-3270-R-1032	FLANGE, DRIVE AXLE .... UOC:ZAA, ZAB, ZAC, ZAD, ZAE, ZAF, ZAG, ZAH, ZAJ, ZAK, ZAL	2
2	PAOZZ	5310012708425	78500	1227-E-1357	NUT, PLAIN, OCTAGON....................................... UOC:ZAA, ZAB, ZAC, ZAD, ZAE, ZAF, ZAG, ZAH, ZAJ, ZAK, ZAL	2
3	PAOZZ	5310003219974	78500	1229-U-1009	WASHER, KEY .... UOC:ZAA, ZAB, ZAC, ZAD, ZAE, ZAF, ZAG, ZAH, ZAJ, ZAK, ZAL	2
4	PAOZZ	5310012712467	78500	1227-D-1356	NUT, PLAIN, OCTAGON....................................... UOC:ZAA, ZAB, ZAC, ZAD, ZAE, ZAF, ZAG, ZAH, ZAJ, ZAK, ZAL	2
5	PAOZZ	2530012828619	78500	A2-333-X-3534	BRAKE DRUM, ASSEMBLY, RIGHT HAND .... UOC:ZAA, ZAB, ZAC, ZAD, ZAE, ZAF, ZAG, ZAH, ZAJ, ZAK, ZAL	1
5	PAOZZ	2530012842310	78500	A3-333-X-3534	HUB, WHEEL, VEHICULAR, ASSEMBLY, LEFT........ HAND .... UOC:ZAA, ZAB, ZAC, ZAD, ZAE, ZAF, ZAG, ZAH, ZAJ, ZAK, ZAL	1
6	PAOZZ	5306012715842	78500	20X-1815-Z	.BOLT, RIBBED SHOULDE, LEFT SIDE .... UOC:ZAA, ZAB, ZAC, ZAD, ZAE, ZAF, ZAG, ZAH, ZAJ, ZAK, ZAL	10
6	PAOZZ	5306012715843	78500	20X-1816-Z	.BOLT, RIBBED SHOULDE, RIGHT SIDE............... UOC:ZAA, ZAB, ZAC, ZAD, ZAE, ZAF, ZAG, ZAH, ZAJ, ZAK, ZAL	10
7	PAOZZ	5330012719410	01212	4591SCR	.SEAL, PLAIN ENCASED .... UOC:ZAA, ZAB, ZAC, ZAD, ZAE, ZAF, ZAG, ZAH, ZAJ, ZAK, ZAL	2
8	PAOFF	2530012718023	78500	3219-E-4815	.BRAKE DRUM .... UOC:ZAA, ZAB, ZAC, ZAD, ZAE, ZAF, ZAG, ZAH, ZAJ, ZAK, ZAL	1
9	PAOZZ	2530012844399	78500	A-333-X-3534	.HUB, WHEEL, VEHICULAR .... UOC:ZAA, ZAB, ZAC, ZAD, ZAE, ZAF, ZAG, ZAH, ZAJ, ZAK, ZAL	1
10	PAOZZ	5305012716450	78500	2206-W-1011	..SCREW, CAP, HEXAGON H .... UOC:ZAA, ZAB, ZAC, ZAD, ZAE, ZAF, ZAG, ZAH, ZAJ, ZAK, ZAL	1
11	PAOZZ	4730012720582	78500	2297-N-5630	..SEAL, CONICAL, FLARED .... UOC:ZAA, ZAB, ZAC, ZAD, ZAE, ZAF, ZAG, ZAH, ZAJ, ZAK, ZAL	1
12	PAOZZ	4730012717978	78500	A-2206-V-1010	..TEE, TUBE .... UOC:ZAA, ZAB, ZAC, ZAD, ZAE, ZAF, ZAG, ZAH, ZAJ, ZAK, ZAL	1
13	PAOZZ	5330012721148	78500	2208-S-1033	..GASKET... UOC:ZAA, ZAB, ZAC, ZAD, ZAE, ZAF, ZAG, ZAH, ZAJ, ZAK, ZAL	1
14	PAOZZ	3120012718353	78500	2245-H-1022	..BUSHING, SLEEVE	1

(1) ITEM NO	(2) SMR CODE	(3) NSN	(4) CAGEC	(5) PART NUMBER	(6) DESCRIPTION AND USABLE ON CODES (UOC)	(7) QTY
					UOC:ZAA, ZAB, ZAC, ZAD, ZAE, ZAF, ZAG, ZAH, ZAJ, ZAK, ZAL	
15	PAOZZ	5330012721147	78500	A-1205-N-2120	..RETAINER, OIL SEAL …………………………	2
					UOC:ZAA, ZAB, ZAC, ZAD, ZAE, ZAF, ZAG, ZAH, ZAJ, ZAK, ZAL	
16	PAOZZ	5365012711837	78500	1229-A-3095	..RING, RETAINING …………………………	1
					UOC:ZAA, ZAB, ZAC, ZAD, ZAE, ZAF, ZAG, ZAH, ZAJ, ZAK, ZAL	
17	PAOZZ	3110001424390	78500	592A	..CUP, TAPERED ROLLER …………………………	2
					UOC:ZAA, ZAB, ZAC, ZAD, ZAE, ZAF, ZAG, ZAH, ZAJ, ZAK, ZAL	
18	PAOZZ	5310013753107	2Y614	3A7313	NUT, PLAIN, HEXAGON, RIGHT HAND…………………	10
					UOC:ZAA, ZAB, ZAC, ZAD, ZAE, ZAF, ZAG, ZAH, ZAJ, ZAK, ZAL	
18	PAOZZ	5310002737771	78500	1199G111	NUT, PLAIN, SINGLE BA LEFT HAND…………………	10
					UOC:ZAA, ZAB, ZAC, ZAD, ZAE, ZAF, ZAG, ZAH, ZAJ, ZAK, ZAL	

END OF FIGURE

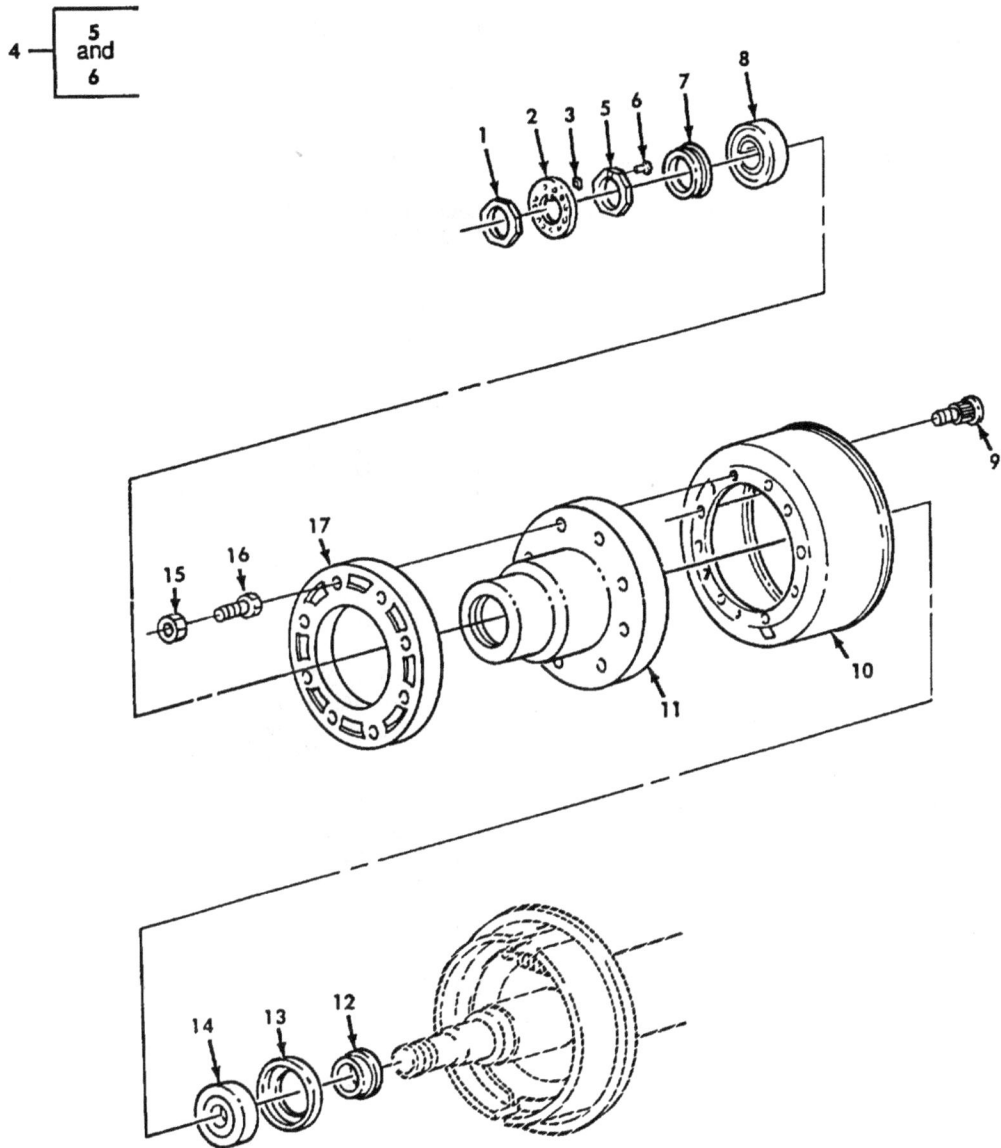

Figure 295. Rear Wheel Hub and Drum Assembly (M939, M939A1).

(1) ITEM NO	(2) SMR CODE	(3) NSN	(4) CAGEC	(5) PART NUMBER	(6) DESCRIPTION AND USABLE ON CODES (UOC)	(7) QTY
					GROUP 1311 WHEEL ASSEMBLY	
					FIG. 295 REAR WHEEL HUB AND DRUM ASSEMBLY(M939, M939A1)	
1	PAOZZ	5310003532427	19207	7979263	NUT, PLAIN, OCTAGON .... UOC:DAA, DAB, DAC, DAD, DAE, DAF, DAG, DAH, DAJ, DAK, DAL, DAW, DAX, V12, V13, V14, V15, V16, V17, V18, V19, V20, V21, V22, V24, V25, V39	2
2	PAOZZ	5310007007089	19207	5139123	WASHER, KEY .... UOC:DAA, DAB, DAC, DAD, DAE, DAF, DAG, DAH, DAJ, DAK, DAL, DAW, DAX, V12, V13, V14, V15, V16, V17, V18, V19, V20, V21, V22, V24, V25, V39	2
3	PAOZZ	5330011337262	78500	5-X-633	GASKET .... UOC:DAA, DAB, DAC, DAD, DAE, DAF, DAG, DAR, DAJ, DAK, DAL, DAW, DAX, V12, V13, V14, V15, V16, V17, V18, V19, V20, V21, V22, V24, V25, V39	2
4	PAOZZ	5310003740836	19207	7979308	NUT, PLAIN, OCTAGON .... UOC:DAA, DAB, DAC, DAD, DAE, DAF, DAG, DAH, DAJ, DAK, DAL, DAW, DAX, V12, V13, V14, V15, V16, V17, V18, V19, V20, V21, V22, V24, V25, V39	2
5	PAOZZ	5310011350049	19207	7979309	.NUT, PLAIN, OCTAGON .... UOC:DAA, DAB, DAC, DAD, DAE, DAF, DAG, DAH, DAJ, DAK, DAL, DAW, DAX, V12, V13, V14, V15, V16, V17, V18, V19, V20, V21, V22, V24, V25, V39	1
6	PAOZZ	5315010587268	19207	7979310	.PIN, ADJUSTMENT..................................... UOC:DAA, DAB, DAC, DAD, DAE, DAF, DAG, DAH, DAJ, DAK, DAL, DAW, DAX, V12, V13, V14, V15, V16, V17, V18, V19, V20, V21, V22, V24, V25, V39	1
7	PAOZZ	5330009613596	19207	7413447	SEAL, PLAIN .... UOC:DAA, DAB, DAC, DAD, DAE, DAF, DAG, DAH, DAJ, DAK, DAL, DAW, DAX, V12, V13, V14, V15, V16, V17, V18, V19, V20, V21, V22, V24, V25, V39	2
8	PAOZZ	3110001004223	96906	MS19081-181	BEARING, ROLLER, TAPE .... UOC:DAA, DAB, DAC, DAD, DAE, DAF, DAG, DAH, DAJ, DAK, DAL, DAW, DAX, V12, V13, V14, V15, V16, V17, V18, V19, V20, V21, V22, V24, V25, V39	2
9	PAOZZ	5307011328273	78500	20X-1337Z	STUD, LEFT HAND THREAD REAR WHEEL........... HUB.. UOC:DAA, DAB, DAC, DAD, DAE, DAF, DAG, DAH, DAJ, DAK, DAL, DAW, DAX, V12, V13, V14, V15, V16, V17, V18, V19, V20, V21, V22, V24, V25	20
9	PAOZZ	5306011328274	78500	20X-1336-Z	BOLT, RIBBED SHOULDE RIGHT HAND................ THREAD REAR WHEEL HUB	20

(1) ITEM NO	(2) SMR CODE	(3) NSN	(4) CAGEC	(5) PART NUMBER	(6) DESCRIPTION AND USABLE ON CODES (UOC)	(7) QTY
					UOC:DAA, DAB, DAC, DAD, DAE, DAF, DAG, DAH, DAJ, DAK, DAL, DAW, DAX, V12, V13, V14, V15, V16, V17, V18, V19, V20, V21, V22, V24, V25	
9	PAOZZ	5307011359290	78500	20X-1527	STUD	20
					UOC:V39	
9	PAOZZ	5307011359291	78500	20X-1526	STUD	20
					UOC:V39	
10	PAOZZ	2530011346570	78500	3219-C-4345	BRAKE DRUM	2
					UOC:DAA, DAB, DAC, DAD, DAE, DAF, DAG, DAH, DAJ, DAK, DAL, DAW, DAX, V12, V13 , V14 , V15, V16, V17, V18, V19, V20, V21, V22, V24, V25, V39	
11	PAOZZ	3040011341184	78500	A-333-U-801	HUB, BODY	2
					UOC: DAA, DAB, DAC, DAD, DAE, DAF, DAG, DAH, DAJ, DAK, DAL, DAW, DAX, V12, V13, V14, V15, V16, V17, V18, V19, V20, V21, V22, V24, V25, V39	
12	PAOZZ	2590007409553	19207	7409553	RING, WIPER	2
					UOC: DAA, DAB, DAC, DAD, DAE, DAF, DAG, DAH, DAJ, DAK, DAL, DAW, DAX, V12, V13, V14, V15, V16, V17, V18, V19, V20, V21, V22, V24, V25, V39	
13	PAOZZ	5330014448350	19207	12375801	SEAL, PLAIN ENCASED, INNER WHEEL .... BEARING	2
					UOC:DAA, DAB, DAC, DAD, DAE, DAF, DAG, DAH, DAJ, DAK, DAL, DAW, DAX, V12 , V13, V14, V15, V16, V17, V18, V19, V20, V21, V22, V24, V25, V39	
14	PAOZZ	3110006898250	96906	MS19081-182	BEARING, ROLLER, TAPE	2
					UOC: DAA, DAB, DAC, DAD, DAE, DAF, DAG, DAH, DAJ, DAK, DAL, DAW, DAX, V12, V13, V14, V15, V16, V17, V18, V19, V20, V21, V22, V24, V25, V39	
15	PAOZZ	5310002737771	78500	1199G111	NUT, PLAIN, SINGLE BA, WHEEL STUD, ... LEFT HAND	20
					UOC: DAA, DAB, DAC, DAD, DAE, DAF, DAG, DAH, DAJ, DAK, DAL, DAW, DAX, V12, V13, V14, V15, V16, V17, V18, V19, V20, V21, V22, V24, V25,	
15	PAOZZ	5310012705463	58536	A-A-52427	NUT, PLAIN, SINGLE BA, WHEEL STUD, ... LEFT HAND	20
					V39	
15	PAOZZ	5310013753107	24614	3A7313	NUT, PLAIN, SINGLE BA, WHEEL STUD, ... RIGHT HAND	20
					UOC:DAA, DAB, DAC, DAD, DAE, DAF, DAG, DAH, DAJ, DAK, DAL, DAW, DAX, V12, V13, V14, V15, V16, V17, V18, V19, V20, V21, V22, V24, V25,	
15	PAOZZ	5310012298029	96906	MS51983-8	NUT, PLAIN, SINGLE BA, WHEEL STUD, ... RIGHT HAND	20
					V39	
16	PAOZZ	2530006931029	52304	10709	NUT, CAP, DUAL WHEEL, LEFT HAND	20
					UOC: V39	
16	PAOZZ	2530003591162	78500	1199J114C	NUT, CAP, DUAL WHEEL, RIGHT HAND	20
					UOC:V39	

(1) ITEM NO	(2) SMR CODE	(3) NSN	(4) CAGEC	(5) PART NUMBER	(6) DESCRIPTION AND USABLE ON CODES (UOC)	(7) QTY
17	PFOZZ	5365011322015	78500	3280-A-1977	SPACER UOC:V39	2

END OF FIGURE

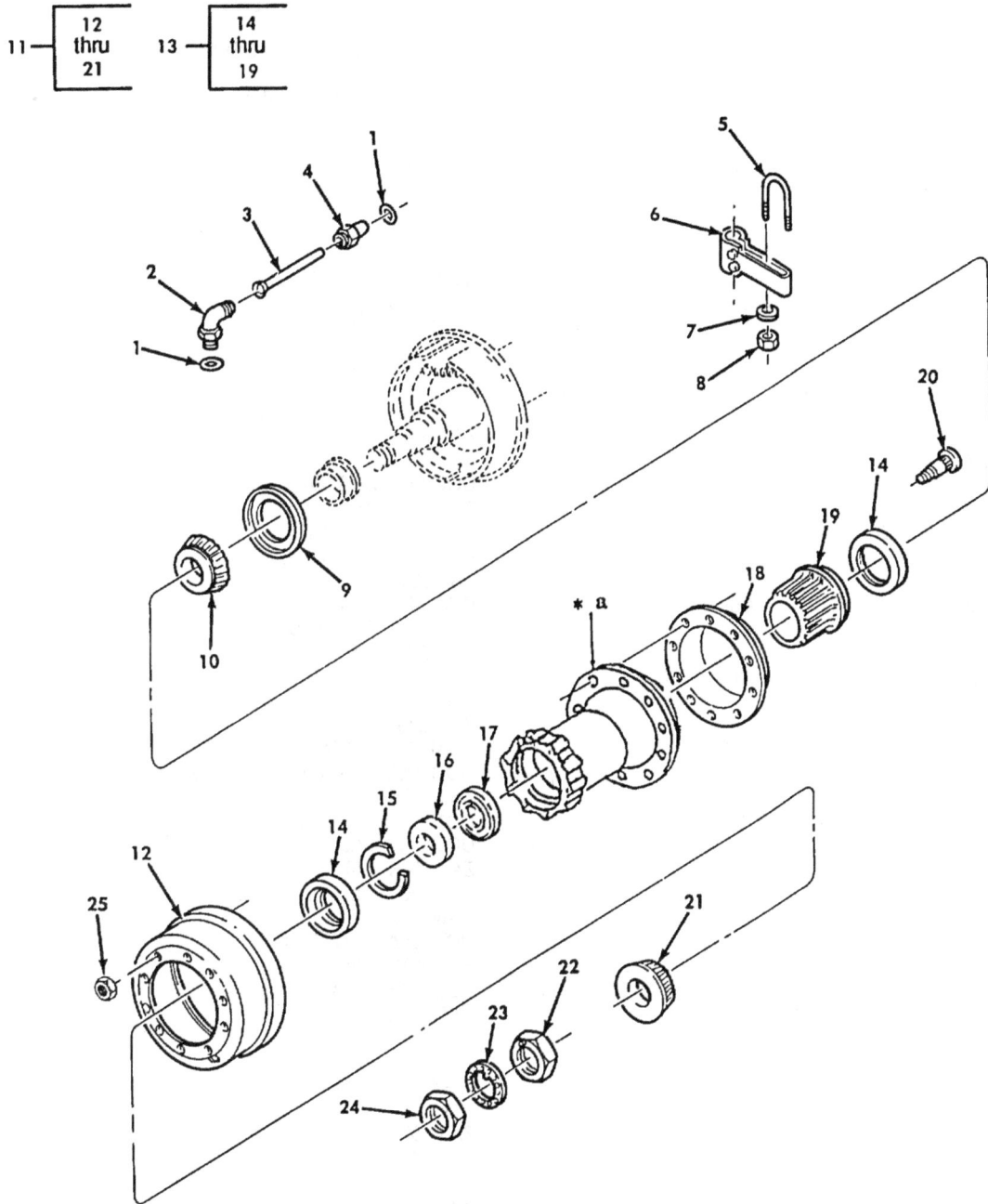

* a PART OF ITEM 13

Figure 296. Rear Wheel Hub and Drum Assembly (M939A2).

(1) ITEM NO	(2) SMR CODE	(3) NSN	(4) CAGEC	(5) PART NUMBER	(6) DESCRIPTION AND USABLE ON CODES (UOC)	(7) QTY
					GROUP 1311 WHEEL ASSEMBLY	
					FIG. 296 REAR WHEEL HUB AND DRUM ASSEMBLY (M939A2)	
1	PAOZZ	5330012929575	78500	3780-Q-381	O-RING .... UOC:ZAA, ZAB, ZAC, ZAD, ZAE, ZAF, ZAG, ZAH, ZAJ, ZAK, ZAL	4
2	PAOZZ	4730013182821	78500	A-2206-Y-1013	ELBOW, HOSE TO BOSS .... UOC:ZAA, ZAB, ZAC, ZAD, ZAE, ZAF, ZAG, ZAH, ZAJ, ZAK, ZAL	2
3	PAOZZ	4710012717939	78500	2296-N-1002	TUBE, METALLIC .... UOC:ZAA, ZAB, ZAC, ZAD, ZAE, ZAF, ZAG, ZAH, ZAJ, ZAK, ZAL	2
4	PAOZZ	5310012715872	78500	1227-X-1350	NUT, SLEEVE .... UOC:ZAA, ZAB, ZAC, ZAD, ZAE, ZAF, ZAG, ZAH, ZAJ, ZAK, ZAL	2
5	PAOZZ	5306012812333	78500	1199-A-3927	BOLT, U .... UOC:ZAA, ZAB, ZAC, ZAD, ZAE, ZAF, ZAG, ZAH, ZAJ, ZAK, ZAL	2
6	PAOZZ	5340013192349	78500	A-3299-C-6139	BRACKET, MOUNTING, CHAMBER .... UOC:ZAA, ZAB, ZAC, ZAD, ZAE, ZAF, ZAG, ZAH, ZAJ, ZAK, ZAL	8
7	PFOZZ	5310011353554	06848	2666715-060	WASHER, FLAT .... UOC:ZAA, ZAB, ZAC, ZAD, ZAE, ZAF, ZAG, ZAH, ZAJ, ZAK, ZAL	4
8	PAOZZ	5310011260566	78500	NL-25-1-C	NUT .... UOC:ZAA, ZAB, ZAC, ZAD, ZAE, ZAF, ZAG, ZAH, ZAJ, ZAK, ZAL	4
9	PAOZZ	5330012719410	01212	4591SCR	SEAL, PLAIN ENCASED .... UOC:ZAA, ZAB, ZAC, ZAD, ZAE, ZAF, ZAG, ZAH, ZAJ, ZAK, ZAL	2
10	PAOZZ	3110009509700	60038	594A	CONE AND ROLLERS, TA .... UOC:ZAA, ZAB, ZAC, ZAD, ZAE, ZAF, ZAG, ZAH, ZAJ, ZAK, ZAL	2
11	PAOZZ	2530012842311	78500	A2-333-Z-3536	HUB, WHEEL, VEHICULAR, ASSEMBLY, RIGHT...... HAND .... UOC:ZAA, ZAB, ZAC, ZAD, ZAE, ZAF, ZAG, ZAH, ZAJ, ZAK, ZAL	1
11	PAOZZ	2530012847446	78500	A3-333-Z-3536	HUB, WHEEL, VEHICULAR, ASSEMBLY, LEFT....... HAND .... UOC:ZAA, ZAB, ZAC, ZAD, ZAE, ZAF, ZAG, ZAH, ZAJ, ZAK, ZAL	1
12	PAOFF	2530012718023	78500	3219-E-4815	.BRAKE DRUM .... UOC:ZAA, ZAB, ZAC, ZAD, ZAE, ZAF, ZAG, ZAH, ZAJ, ZAK, ZAL	1
13	PAOOO	2530012839694	78500	A1-333-Z-3536	.HUB, WHEEL, VEHICULAR .... UOC:ZAA, ZAB, ZAC, ZAD, ZAE, ZAF, ZAG, ZAH, ZAJ, ZAK, ZAL	1
14	PAOZZ	3110001424390	78500	592A	..CUP, TAPERED ROLLER .... UOC:ZAA, ZAB, ZAC, ZAD, ZAE, ZAF, ZAG, ZAH, ZAJ, ZAK, ZAL	2

(1) ITEM NO	(2) SMR CODE	(3) NSN	(4) CAGEC	(5) PART NUMBER	(6) DESCRIPTION AND USABLE ON CODES (UOC)	(7) QTY
15	PAOZZ	5365012711837	78500	1229-A-3095	..RING, RETAINING............................... UOC:ZAA, ZAB, ZAC, ZAD, ZAE, ZAF, ZAG, ZAH, ZAJ, ZAK, ZAL	1
16	PAOZZ	5330013080175	78500	A-1205-D-2162	..SEAL, PLAIN ENCASED..................................... UOC:ZAA, ZAB, ZAC, ZAD, ZAE, ZAF, ZAG, ZAH, ZAJ, ZAK, ZAL	2
17	PAOZZ	5330012721147	78500	A-1205-N-2120	..RETAINER, OIL SEAL .... UOC:ZAA, ZAB, ZAC, ZAD, ZAE, ZAF, ZAG, ZAH, ZAJ, ZAK, ZAL	2
18	PAOZZ	2530012853563	78500	3286-P-1056	..DEFLECTOR, DIRT AND........................... UOC:ZAA, ZAB, ZAC, ZAD, ZAE, ZAF, ZAG, ZAH, ZAJ, ZAK, ZAL	2
19	PAOZZ	3120012718353	78500	2245-H-1022	..BUSHING, SLEEVE .... UOC:ZAA, ZAB, ZAC, ZAD, ZAE, ZAF, ZAG, ZAH, ZAJ, ZAK, ZAL	1
20	PAOZZ	5306012715842	78500	20X-1815-Z	.BOLT, RIBBED SHOULDE, LEFT HAND .... UOC:ZAA, ZAB, ZAC, ZAD, ZAE, ZAF, ZAG, ZAH, ZAJ, ZAK, ZAL	10
20	PAOZZ	5306012715843	78500	20X-1816-Z	.BOLT, RIBBED SHOULDE, RIGHT HAND.............. UOC:ZAA, ZAB, ZAC, ZAD, ZAE, ZAF, ZAG, ZAH, ZAJ, ZAK, ZAL	10
21	PAOZZ	3110001000650	60038	598	.CONE AND ROLLERS, TA .... UOC:ZAA, ZAB, ZAC, ZAD, ZAE, ZAF, ZAG, ZAH, ZAJ, ZAK, ZAL	2
22	PAOZZ	5310012712467	78500	1227-D-1356	NUT, PLAIN, OCTAGON..................................... UOC:ZAA, ZAB, ZAC, ZAD, ZAE, ZAF, ZAG, ZAH, ZAJ, ZAK, ZAL	2
23	PAOZZ	5310003219974	78500	1229-U-1009	WASHER, KEY .... UOC:ZAA, ZAB, ZAC, ZAD, ZAE, ZAF, ZAG, ZAH, ZAJ, ZAK, ZAL	2
24	PAOZZ	5310012708425	78500	1227-E-1357	NUT, PLAIN, OCTAGON .... UOC:ZAA, ZAB, ZAC, ZAD, ZAE, ZAF, ZAG, ZAH, ZAJ, ZAK, ZAL	2
25	PAOZZ	5310013753107	2Y614	3A7313	NUT, PLAIN, HEXAGON, RIGHT HAND................... UOC:ZAA, ZAB, ZAC, ZAD, ZAE, ZAF, ZAG, ZAH, ZAJ, ZAK, ZAL	10
25	PAOZZ	5310002737771	78500	1199G111	NUT, PLAIN, SINGLE BA, LEFT HAND................... UOC:ZAA, ZAB, ZAC, ZAD, ZAE, ZAF, ZAG, ZAH, ZAJ, ZAK, ZAL	10

END OF FIGURE

9 —[10]

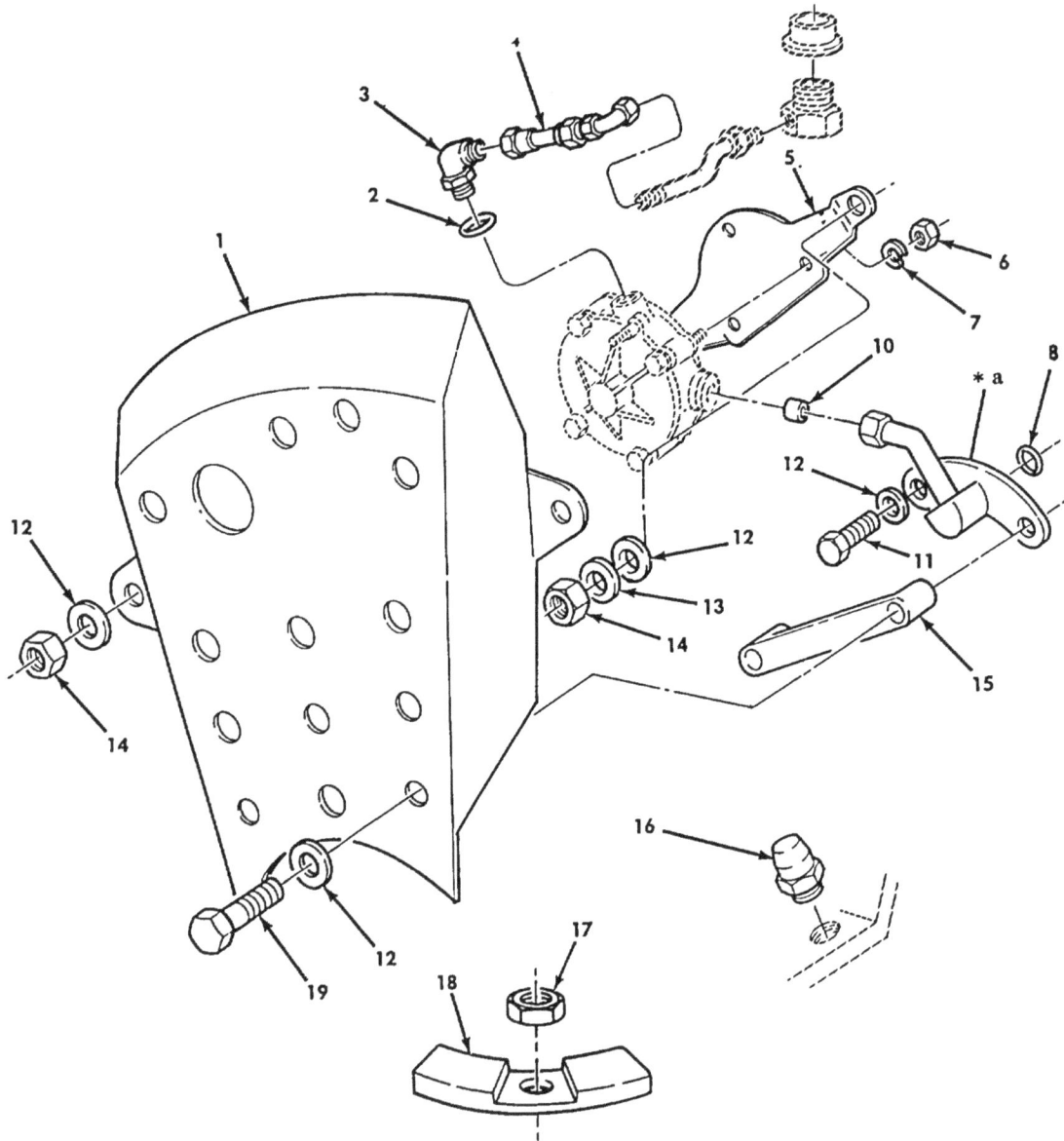

* a PART OF ITEM 9

Figure 297. Front Axle Wheel Valve Installation (M939A2).

(1) ITEM NO	(2) SMR CODE	(3) NSN	(4) CAGEC	(5) PART NUMBER	(6) DESCRIPTION AND USABLE ON CODES (UOC)	(7) QTY
					GROUP 1311 WHEEL ASSEMBLY	
					FIG. 297 FRONT AXLE WHEEL VALVE INSTALLATION(M939A2)	
1	PAOZZ	2530012808989	47457	20510675	SHIELD, WHEEL VALVE .... UOC:ZAA, ZAB, ZAC, ZAD, ZAE, ZAF, ZAG, ZAH, ZAJ, ZAK, ZAL	2
2	PAOZZ	5330008045695	96906	MS28788-6	O-RING .... UOC:ZAA, ZAB, ZAC, ZAD, ZAE, ZAF, ZAG, ZAH, ZAJ, ZAK, ZAL	2
3	PAOZZ	4730012781001	96906	MS51527-A6Z	ELBOW, TUBE TO BOSS.................................... UOC:ZAA, ZAB, ZAC, ZAD, ZAE, ZAF, ZAG, ZAH, ZAJ, ZAK, ZAL	1
4	PAOZZ	4720012793034	98441	A3994-2	HOSE ASSEMBLY, NONME .... UOC:ZAA, ZAB, ZAC, ZAD, ZAE, ZAF, ZAG, ZAH, ZAJ, ZAK, ZAL	2
5	PAOZZ	5340012807099	47457	20510684	BRACKET, MOUNTING . ... UOC:ZAA, ZAB, ZAC, ZAD, ZAE, ZAF, ZAG, ZAH, ZAJ, ZAK, ZAL	2
6	PAOZZ	5310000131245	21450	131245	NUT, SELF-LOCKING, HE .... UOC:ZAA, ZAB, ZAC, ZAD, ZAE, ZAF, ZAG, ZAH, ZAJ, ZAK, ZAL	6
7	PAOZZ	5310005825965	96906	MS35338-44	WASHER, LOCK .... UOC:ZAA, ZAB, ZAC, ZAD, ZAE, ZAF, ZAG, ZAH, ZAJ, ZAK, ZAL	6
8	PAOZZ	5330011335858	96906	MS28775-206	O-RING .... UOC:ZAA, ZAB, ZAC, ZAD, ZAE, ZAF, ZAG, ZAH, ZAJ, ZAK, ZAL	2
9	PAOZZ	4710012793159	19207	12432286	TUBE ASSEMBLY, METAL................................... UOC:ZAA, ZAB, ZAC, ZAD, ZAE, ZAF, ZAG, ZAH, ZAJ, ZAK, ZAL	2
10	PAOZZ	4730013918301	19207	12363523	.SLEEVE, COMPRESSION .... UOC:ZAA, ZAB, ZAC, ZAD, ZAE, ZAF, ZAG, ZAH, ZAJ, ZAK, ZAL	2
11	PAOZZ	5305000712071	80204	B1821BH050C200N	SCREW, CAP, HEXAGON H.................................... UOC:ZAA, ZAB, ZAC, ZAD, ZAE, ZAF, ZAG, ZA-H, ZAJ, ZAK, ZAL	2
12	PAOZZ	5310013029472	47457	20511272	WASHER, FLAT, FOR USE W/TRI-TRAP............... WHEEL... UOC:ZAA, ZAB, ZAC, ZAD, ZAE, ZAF, ZAG, ZAH, ZAJ, ZAK, ZAL	6
12	PAOZZ	5310008238803	96906	MS27183-21	WASHER, FLAT... UOC:ZAA, ZAB, ZAC, ZAD, ZAE, ZAF, ZAG, ZAH, ZAJ, ZAK, ZAL	6
13	PAOZZ	5310008098533	96906	MS27183-23	WASHER, FLAT .... UOC:ZAA, ZAB, ZAC, ZAD, ZAE, ZAF, ZAG, ZAH, ZAJ, ZAK, ZAL	2
14	PAOZZ	5310009353569	96906	MS51943-46	NUT, SELF-LOCKING, HE, FOR USE W/TRI- .... TRAP WHEEL .... UOC:ZAA, ZAB, ZAC, ZAD, ZAE, ZAF, ZAG, ZAH, ZAJ, ZAK, ZAL	6
14	PAOZZ	5310002256408	72962	41NE126	NUT, SELF-LOCKING, HE .... UOC:ZAA, ZAB, ZAC, ZAD, ZAE, ZAF, ZAG, ZAH,	6

(1) ITEM NO	(2) SMR CODE	(3) NSN	(4) CAGEC	(5) PART NUMBER	(6) DESCRIPTION AND USABLE ON CODES (UOC)	(7) QTY
15	PAOZZ	5365012803690	4F744	20510674	ZAJ, ZAK, ZAL SPACER ASSEMBLY .... UOC:ZAA, ZAB, ZAC, ZAD, ZAE, ZAF, ZAG, ZAH, ZAJ, ZAK, ZAL	2
16	PAOZZ	4820007264719	57733	5196397	VALVE, VENT .... UOC:ZAA, ZAB, ZAC, ZAD, ZAE, ZAF, ZAG, ZAH, ZAJ, ZAK, ZAL	2
17	PAOZZ	5310011022711	19207	12257242	NUT, SELF-LOCKING, EX .... UOC:ZAA, ZAB, ZAC, ZAD, ZAE, ZAF, ZAG, ZAH, ZAJ, ZAK, ZAL	2
18	PAOZZ	3040012804382	47457	20511155	WEIGHT, COUNTERBALAN .... UOC:ZAA, ZAB, ZAC, ZAD, ZAE, ZAF, ZAG, ZAH, ZAJ, ZAK, ZAL	2
19	PAOZZ	5305000712077	80204	B1821BH050C350N	SCREW, CAP, HEXAGON H .... UOC:ZAA, ZAB, ZAC, ZAD, ZAE, ZAF, ZAG, ZAH, ZAJ, ZAK, ZAL	4

END OF FIGURE

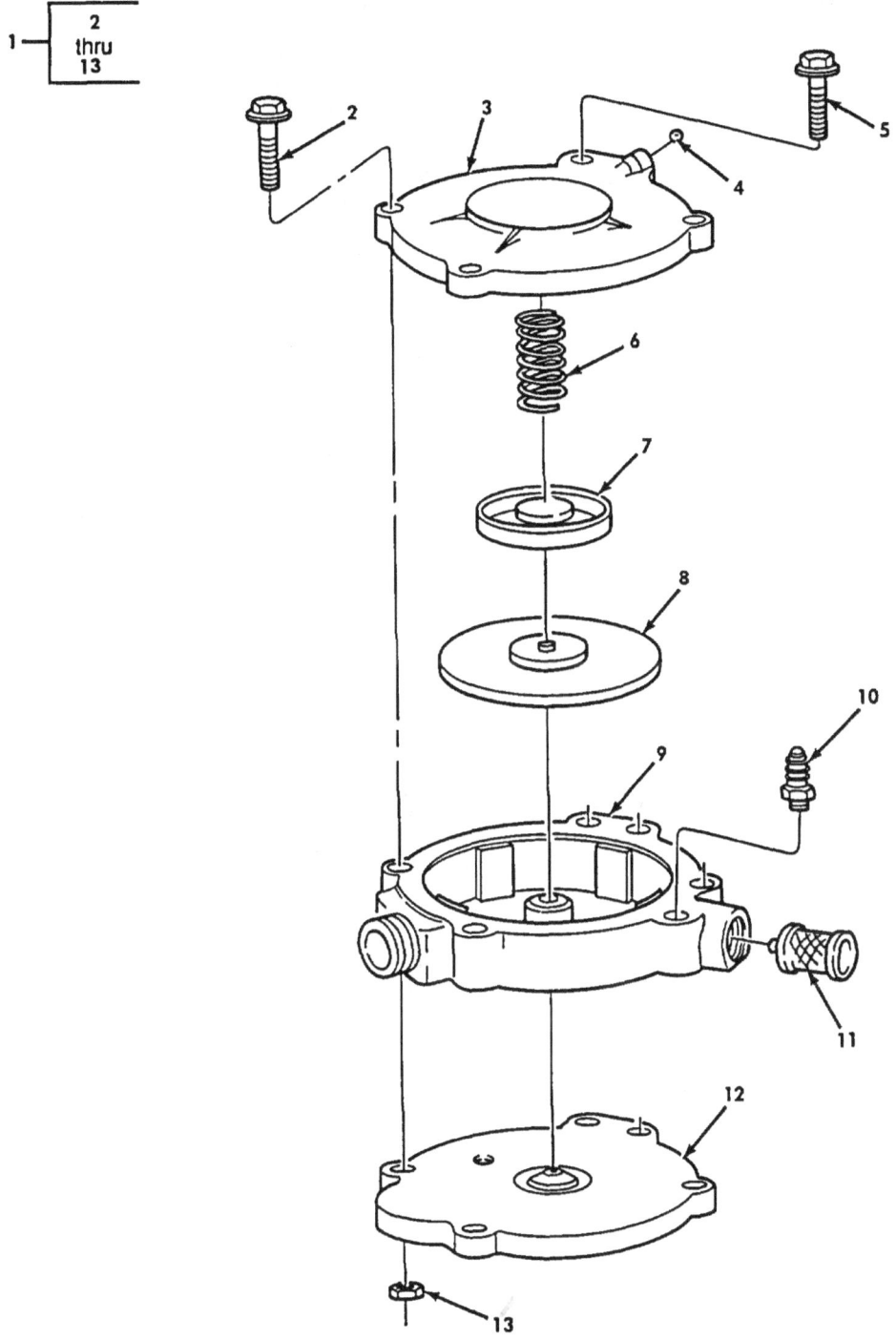

Figure 298. Wheel Valve Assembly (M939A2).

(1) ITEM NO	(2) SMR CODE	(3) NSN	(4) CAGEC	(5) PART NUMBER	(6) DESCRIPTION AND USABLE ON CODES (UOC)	(7) QTY
					GROUP 1311 WHEEL ASSEMBLY	
					FIG. 298 WHEEL VALVE ASSEMBLY (M939A2)	
1	PAOOO	4820012873963	52304	599735	WHEEL VALVE .... UOC:ZAA, ZAB, ZAC, ZAD, ZAE, ZAF, ZAG, ZAH, ZAJ, ZAK, ZAL	6
2	PAOZZ	5305012821529	52304	599811	.SCREW, CAP, HEXAGON H .... UOC:ZAA, ZAB, ZAC, ZAD, ZAE, ZAF, ZAG, ZAH, ZAJ, ZAK, ZAL	1
3	XAOZZ		52304	599618	.COVER .... UOC:ZAA, ZAB, ZAC, ZAD, ZAE, ZAF, ZAG, ZAH, ZAJ, ZAK, ZAL	1
4	KFOZZ		52304	599624	.BALL, PART OF KIT P/N 599913.......................... UOC:ZAA, ZAB, ZAC, ZAD, ZAE, ZAF, ZAG, ZAH, ZAJ, ZAK, ZAL	3
5	PAOZZ	5305012821528	52304	599810	.SCREW, CAP, HEXAGON H PART OF KIT P/N....... 599913 .... UOC:ZAA, ZAB, ZAC, ZAD, ZAE, ZAF, ZAG, ZAH, ZAJ, ZAK, ZAL	3
6	KFOZZ		52304	599812KF	.SPRING PART OF KIT P/N 599913 .... UOC:ZAA, ZAB, ZAC, ZAD, ZAE, ZAF, ZAG, ZAH, ZAJ, ZAK, ZAL	1
7	XAOZZ		52304	599622	.PLUG... UOC:ZAA, ZAB, ZAC, ZAD, ZAE, ZAF, ZAG, ZAH, ZAJ, ZAK, ZAL	1
8	KFOZZ	4820012833929	52304	599805	.DIAPHRAGM, ACTUATOR PART OF KIT P/N......... 599913 .... UOC:ZAA, ZAB, ZAC, ZAD, ZAE, ZAF, ZAG, ZAH, ZAJ, ZAK, ZAL	1
9	XAOZZ		52304	599731	.HOUSING... UOC:ZAA, ZAB, ZAC, ZAD, ZAE, ZAF, ZAG, ZAH, ZAJ, ZAK, ZAL	1
10	XAOZZ		52304	599756	.VALVE, TANK .... UOC:ZAA, ZAB, ZAC, ZAD, ZAE, ZAF, ZAG, ZAH, ZAJ, ZAK, ZAL	1
11	PAOZZ	4460012842344	52304	599791	.FILTER ELEMENT, FLUI .... UOC:ZAA, ZAB, ZAC, ZAD, ZAE, ZAF, ZAG, ZAH, ZAJ, ZAK, ZAL	1
12	XAOZZ		52304	599728	.BASE .... UOC:ZAA, ZAB, ZAC, ZAD, ZAE, ZAF, ZAG, ZAH, ZAJ, ZAK, ZAL	1
13	PAOZZ	5310007616882	96906	MS51967-2	.NUT, PLAIN, HEXAGON .... UOC:ZAA, ZAB, ZAC, ZAD, ZAE, ZAF, ZAG, ZAH, ZAJ, ZAK, ZAL	4

END OF FIGURE

5 — | 6 |

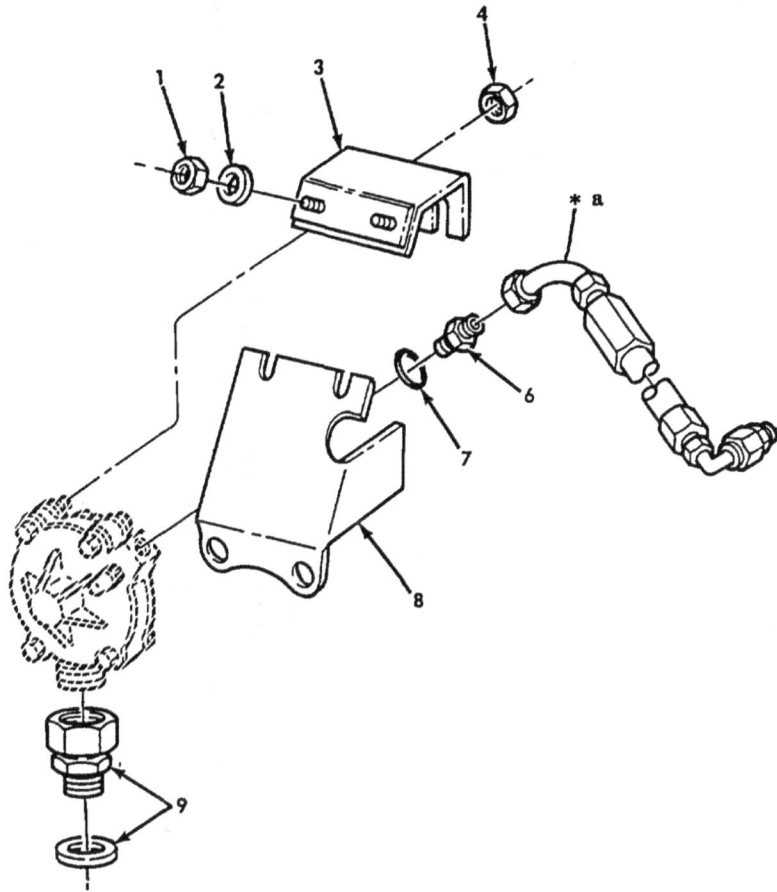

* a PART OF ITEM 5

Figure 299. Rear Axle Wheel Valve Assembly ((M939A2).

(1) ITEM NO	(2) SMR CODE	(3) NSN	(4) CAGEC	(5) PART NUMBER	(6) DESCRIPTION AND USABLE ON CODES (UOC)	(7) QTY
					GROUP 1311 WHEEL ASSEMBLY	
					FIG. 299 REAR AXLE WHEEL VALVE ASSEMBLY (M939A2)	
1	PAOZZ	5310009349758	96906	MS35649-202	NUT, PLAIN, HEXAGON .... UOC:ZAA, ZAB, ZAC, ZAD, ZAE, ZAF, ZAG, ZAH, ZAJ, ZAK, ZAL	2
2	PAOZZ	5310008094058	96906	MS27183-10	WASHER, FLAT .... UOC:ZAA, ZAB, ZAC, ZAD, ZAE, ZAF, ZAG, ZAH, ZAJ, ZAK, ZAL	16
3	PAOZZ	5340012849653	16567	20510669-2	BRACKET, MOUNTING . ... UOC:ZAA; ZAB, ZAC, ZAD, ZAE, ZAF, ZAG, ZAH, ZAJ, ZAK, ZAL	4
4	PAOZZ	5310000131245	21450	131245	NUT, SELF-LOCKING, HE .... UOC:ZAA, ZAB, ZAC, ZAD, ZAE, ZAF, ZAG, ZAH, ZAJ, ZAK, ZAL	16
5	PAOOO	4720013041439	98441	A3995-5	HOSE ASSEMBLY, NONME .... UOC:ZAA, ZAB, ZAC, ZAD, ZAE, ZAF, ZAG, ZAH, ZAJ, ZAK, ZAL	4
6	PAOZZ	4730004919576	96906	MS51525-A6	ADAPTER, STRAIGHT, TU .... UOC:ZAA, ZAB, ZAC, ZAD, ZAE, ZAF, ZAG, ZAH, ZAJ, ZAK, ZAL	1
7	PAOZZ	5331008045695	96906	MS28778-6	O-RING .... UOC:ZAA, ZAB, ZAC, AZD, ZAE, ZAF, ZAG, ZAH, ZAJ, ZAK, ZAL	4
8	PAOZZ	5340012807102	16567	20510669-1	BRACKET, MOUN TING................................. UOC:ZAA, ZAB, ZAC, ZAD, ZAE, ZAF, ZAG, ZAH, ZAJ, ZAK, ZAL	4
9	PAOZZ	4730012791519	47457	20510736	ADAPTER, STRAIGHT, TU .... UOC:ZAA, ZAB, ZAC, ZAD, ZAE, ZAF, ZAG, ZAH, ZAJ, ZAK, ZAL	4

END OF FIGURE

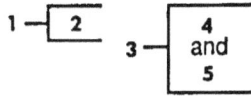

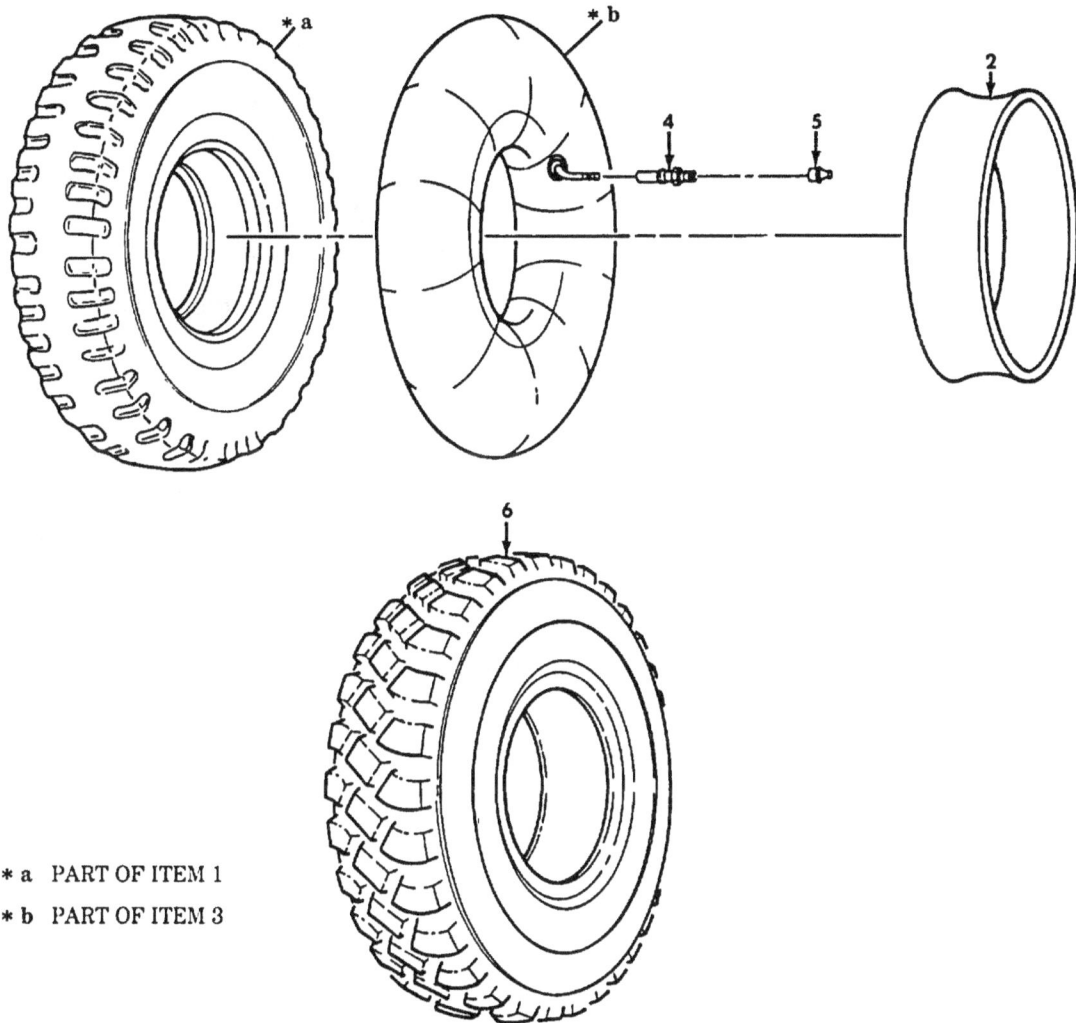

* a PART OF ITEM 1
* b PART OF ITEM 3

Figure 300. Tire and Tube Assembly.

(1) ITEM NO	(2) SMR CODE	(3) NSN	(4) CAGEC	(5) PART NUMBER	(6) DESCRIPTION AND USABLE ON CODES (UOC)	(7) QTY
					GROUP 1313 TIRES	
					FIG. 300 TIRE AND TUBE ASSEMBLY	
1	PCOHH	2610002628653	81348	MIL-T-12459/CLCL /SA/1100-20/F/CC	TIRE, PNEUMATIC SIZE 11.00 X 20 12 .... PLY UOC:V12, V13, V14, V15, V16, V17, V18, V19, V20, V21, V22, V24, V25	11
1	PCOHH	2610013737294	81348	GP3STYLXTYRBCLA/ T/11.00-R20/H/TB	TIRE, PNEUMATIC, VEHI RADIAL, 11 X 20............. UOC:V12, V13, V14, V15, V16, V17, V18, V19, V20, V21, V22, V24, V25	11
1	PCOHH	2610002042545	81348	ZZ-T-381M/GRP3/1 4.00-20/F/TBCC	TIRE, PNEUMATIC 14 X 20 .... UOC:V39	11
2	PAOZZ	2640001585617	19207	11662389-2	.FLAP, INNER TUBE, PNE .... UOC:V12, V13, V14 S, 15, V16, V17, V18, V19, V20, V21, V22, V24, V25, V39	1
3	PAOZZ	2610000519450	81348	11.00-20/TR78A/O N CENTER	INNER TUBE, PNEUMATI UOC:V12, V13, V14, V15, V16, V17, V18, V19, V20, V21, V22, V24, V25	11
3	PAOZZ	2610000519464	81348	ZZ-I-550E/GP2/14 .00-20/TR179A/OC	INNER TUBE, PNEUMATI GROUP 2, 14 X 20........... UOC:V39	11
4	PAFZZ	5365010907769	19207	12255764	.BUSHING, RUBBER.. UOC:V12, V13, V14, V15, V16, V17, V18, , 19, V20, V21, V22, V24, V25, V39	1
5	PAOZZ	2640010982029	17875	TRVC-8	.CAP, PNEUMATIC VALVE UOC:V12, V13, V14, V15, V16, V17, V18, V19, V20, V21, V22, V24, V25, V39	1
6	PCOFF	2610012141344	12195	02021050	TIRE, PNEUMATIC, VEHI .... UOC:DAA, DAB, DAC, DAD, DAE, DAF, DAG, DAH, DAJ, DAK, DAL, DAW, DAX, ZAA, ZAB, ZAC, ZAD, ZAE, ZAF, ZAG, ZAH, ZAJ, ZAK, ZAL	1

END OF FIGURE

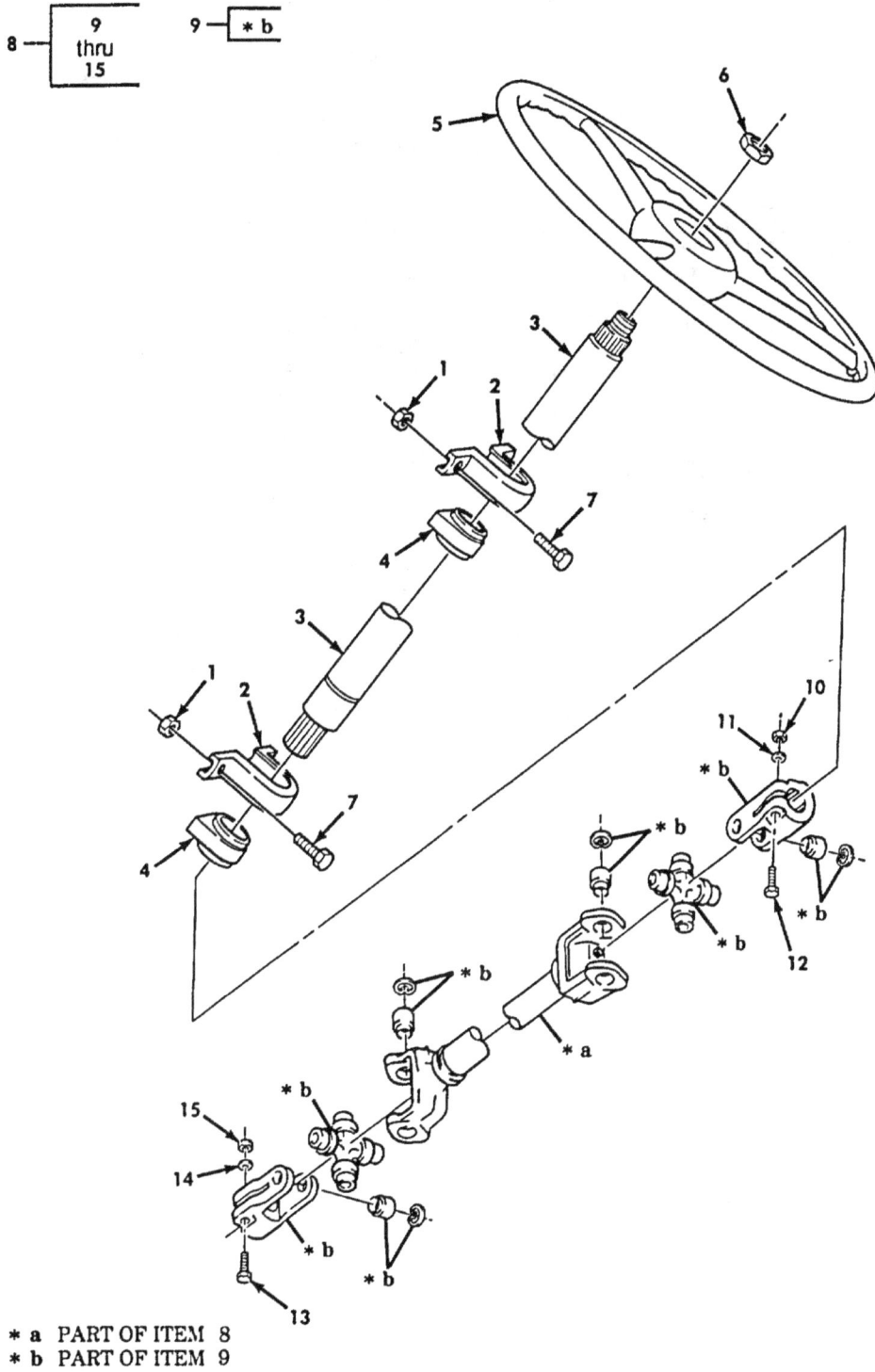

Figure 301. Steering Column, Wheel, and Propeller Shaft.

* a PART OF ITEM 8
* b PART OF ITEM 9

(1) ITEM NO	(2) SMR CODE	(3) NSN	(4) CAGEC	(5) PART NUMBER	(6) DESCRIPTION AND USABLE ON CODES (UOC)	(7) QTY
					GROUP 14 STEERING 1401 MECHANICAL STEERING GEAR	
					FIG. 301 STEERING COLUMN, WHEEL, AND PROPELLER SHAFT	
1	PAOZZ	5310008140672	96906	MS51943-36	NUT, SELF-LOCKING, HE	4
2	PAOZZ	2530010893047	19207	12255752	SUPPORT, STEERING CO	2
3	PAOFF	2530011468941	77640	UC28254	HOUSING, STEERING CO UPPER	1
4	PAOZZ	5365010907769	19207	12255764	BUSHING, NONMETALLIC	2
5	PAOZZ	2530000806572	77915	79210	STEERING WHEEL	1
6	PAOZZ	5310010889298	19207	5167785	NUT, PLAIN, HEXAGON	1
7	PAOZZ	5305002692804	96906	MS90726-61	SCREW, CAP, HEXAGON H	4
8	PAOOO	2520011079932	19207	11664542-1	PROPELLER SHAFT WIT PROP SHAFT .... STEERING(ROCKWELL)...................................	1
8	PAOOO	2520013695335	19207	11664542-2	PROPELLER SHAFT WIT PROP SHAFT .... STEERING (DANA).......................................	1
9	PAOZZ	2520002946752	95019	5-170X	.SPIDER, UNIVERSAL JO DANA	2
9	PAOZZ	2520007557336	78500	CPL6522-P2	.PARTS KIT, UNIVERSAL ROCKWELL	2
10	PAOZZ	5310007320559	96906	MS51968-8	.NUT, PLAIN, HEXAGON	1
11	PAOZZ	5310006379541	96906	MS35338-46	.WASHER, LOCK PART OF KIT P/N 5-170X............ PART OF KIT P/N CPL6522-P2	1
12	PAOZZ	5305002693242	80204	B1821BH038F200N	.SCREW, CAP, HEXAGON H PART OF KIT P/N....... 5-170X PART OF KIT P/N CPL6522-P2...................	1
13	PAOZZ	5305002693242	80204	B1821BH038F200N	.SCREW, CAP, HEXAGON H DANA.....................	1
14	PAOZZ	5310002090965	96906	MS35338-47	.WASHER, LOCK ROCKWELL	1
15	PAOZZ	5310008807745	96906	MS51968-11	.NUT, PLAIN, HEXAGON ROCKWELL	1
15	PAOZZ	5310007320559	96906	MS51968-8	.NUT, PLAIN, HEXAGON DANA	1

END OF FIGURE

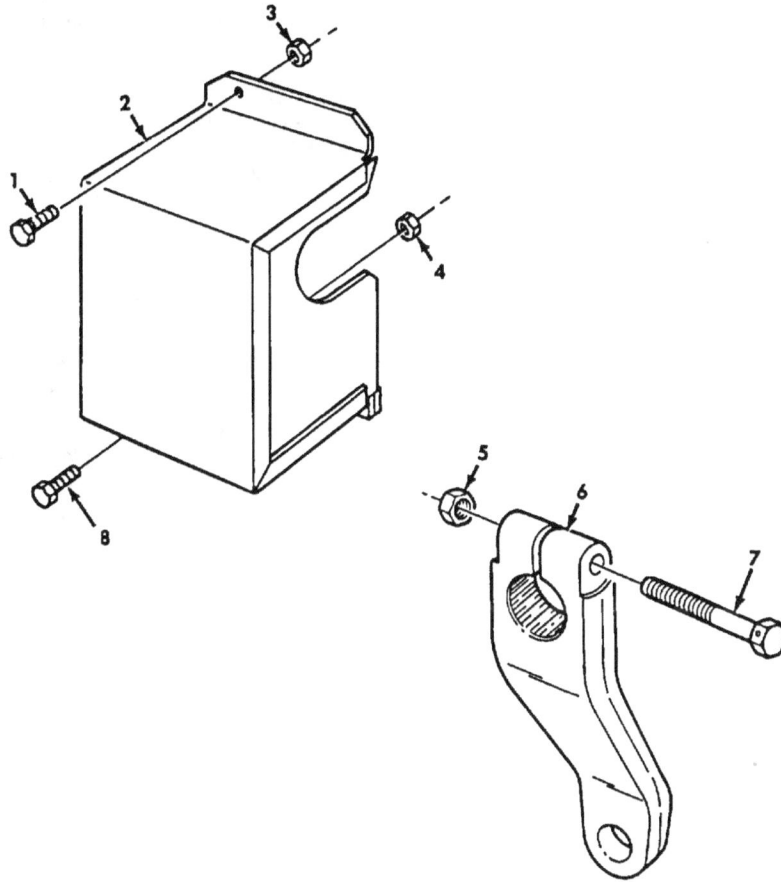

Figure 302. Stone Shield and Pitman Arm.

(1) ITEM NO	(2) SMR CODE	(3) NSN	(4) CAGEC	(5) PART NUMBER	(6) DESCRIPTION AND USABLE ON CODES (UOC)	(7) QTY
					GROUP 1401 MECHANICAL STEERING GEAR	
					FIG. 302 STONE SHIELD AND PITMAN ARM	
1	PAOZZ	5305002693231	80204	B1821BH038F050N	SCREW, CAP, HEXAGON H	2
2	PAOZZ		19207	12363536	SHIELD, STEERING GEA	1
3	PAOZZ	5310002416658	96906	MS51943-34	NUT, SELF-LOCKING, HE	1
4	PAOZZ	5310008140672	96906	MS51943-36	NUT, SELF-LOCKING, HE	1
5	PAOZZ	5310009353569	96906	MS51943-46	NUT, SELF-LOCKING, HE .... UOC:DAA, DAB, DAC, DAD, DAE, DAF, DAG, DAH, DAJ, DAK, DAL, DAW, DAX, V12, V13, V14, V15, V16, V17, V18, V19, V20, V21, V22, V24, V25, V39	1
6	PAOZZ	3040010893081	19207	12255761	CONNECTING LINK, RIG .... UOC:DAA, DAB, DAC, DAD, DAE, DAF, DAG, DAH, DAJ, DAK, DAL, DAW, DAX, V12, V13, V14, V15, V16, V17, V18, V19, V20, V21, V22, V24, V25, V39	1
7	PAOZZ	5305001191620	96906	MS18153-169	SCREW, CAP, HEXAGON .... UOC:DAA, DAB, DAC, DAD, DAE, DAF, DAG, DAH, DAJ, DAK, DAL, DAW, DAX, V12, V13, V14, V15, V16, V17, V18, V19, V20, V21, V22, V24, V25, V39	1
8	PAOZZ	5305002692804	96906	MS90726-61	SCREW, CAP, HEXAGON H	1

END OF FIGURE

Figure 303. Tie Rod Assemblies.

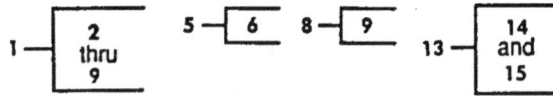

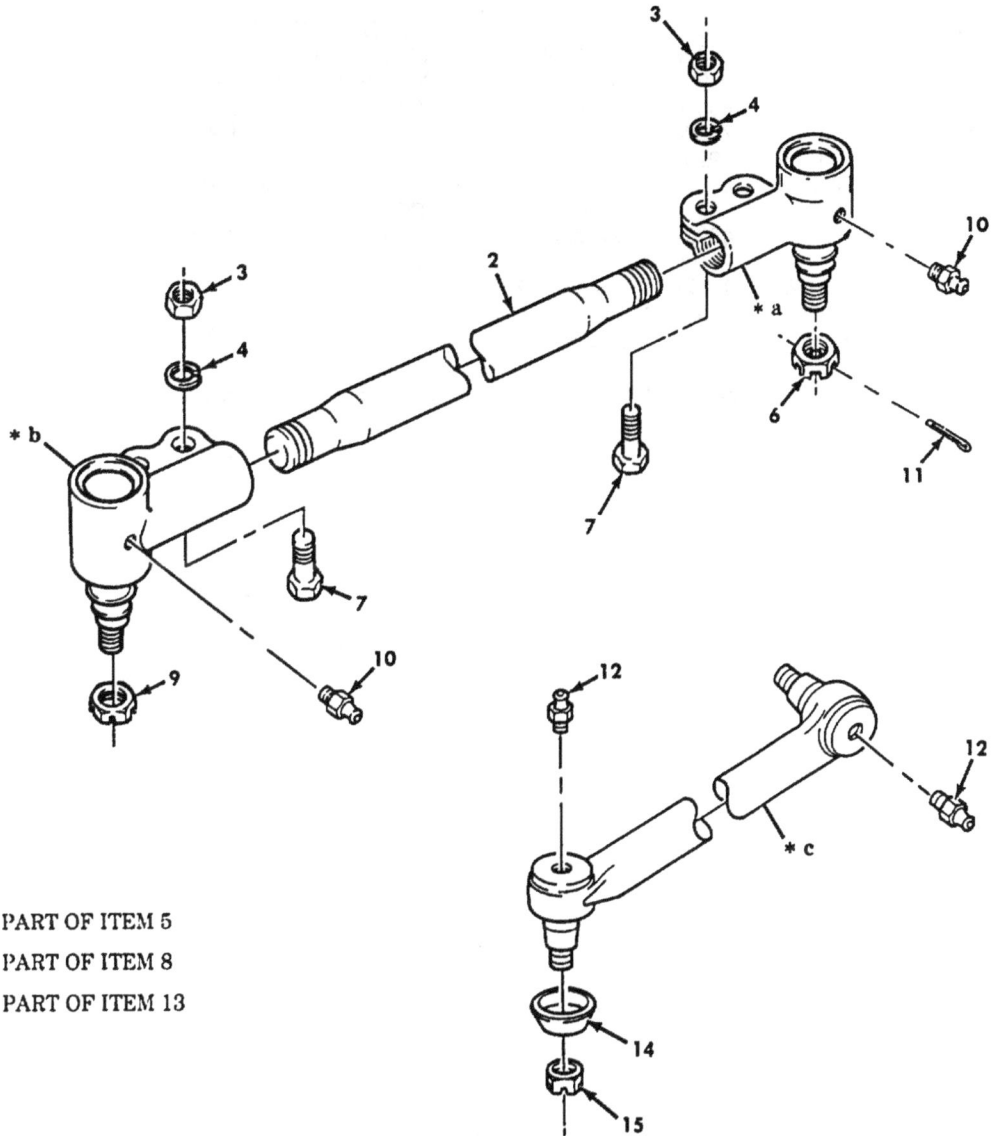

* a PART OF ITEM 5
* b PART OF ITEM 8
* c PART OF ITEM 13

(1) ITEM NO	(2) SMR CODE	(3) NSN	(4) CAGEC	(5) PART NUMBER	(6) DESCRIPTION AND USABLE ON CODES (UOC)	(7) QTY
					GROUP 1401 MECHANICAL STEERING GEAR	
					FIG. 303 TIE ROD ASSEMBLIES	
1	PAOOZ	2530005120032	78500	A3102Y3431	TIE ROD, STEERING	1
2	XAOZZ		19207	12356831-1	.SHAFT, STEERING	1
3	PAOZZ	5310000614651	96906	MS51943-43	.NUT, SELF-LOCKING, HE	4
4	PAOZZ	5310008206653	80045	23MS35338-50	.WASHER, LOCK	4
5	PAOOZ	2530013093803	78500	A3144V438	.TIE ROD END, STEERIN	1
6	PAOZZ	5310002639488	19207	10938304-2	..NUT, PLAIN, SLOTTED, H..................................	2
7	PAOZZ	5305007262552	80204	B1821BH063F225N	.SCREW, CAP, HEXAGON H	4
8	PAOOZ	2530013096203	78500	A3144U437	.TIE ROD END, STEERIN	1
9	PAOZZ	5310002639488	19207	10938304-2	..NUT, PLAIN, SLOTTED, H..................................	1
10	PAOZZ	4730000504208	96906	MS15003-1	FITTING, LUBRICATION	2
11	PAOZZ	5315002981481	96906.	MS24665-357	PIN, COTTER	2
12	PAOZZ	4730001720034	96906	MS15003-6	FITTING, LUBRICATION	2
13	PAOOZ	2530011010084	19207	11669131	TIE ROD, STEERING	1
14	PAOZZ	2530011269209	31033	L-28-VC-126	.COVER, DUST	2
15	PAOZZ	5310008352140	96906	MS35692-69.	.NUT, PLAIN, SLOTTED, H	2

END OF FIGURE

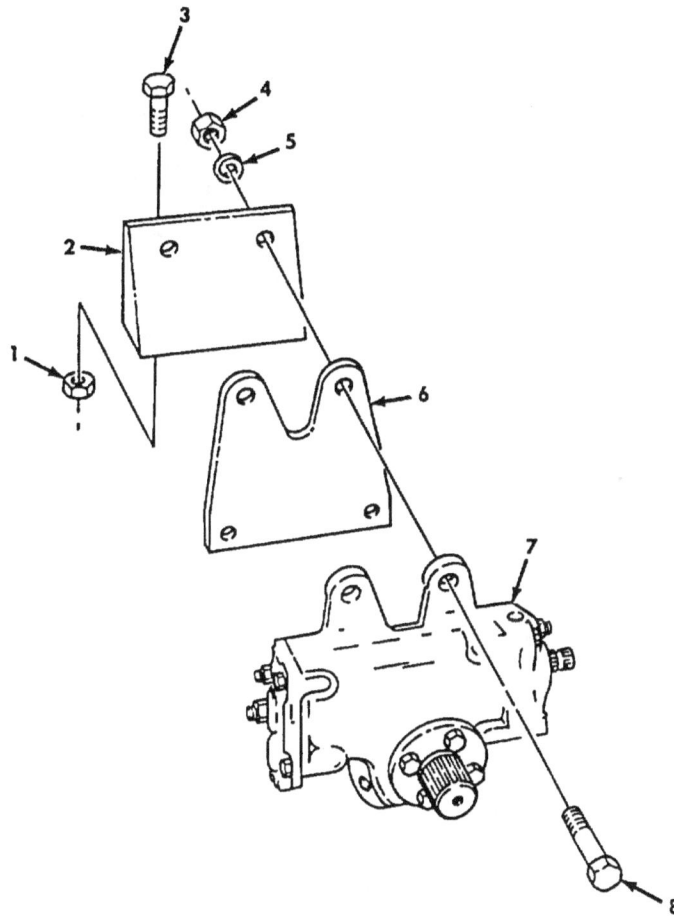

Figure 304. Steering Gear Assembly and Mounting Hardware.

(1) ITEM NO	(2) SMR CODE	(3) NSN	(4) CAGEC	(5) PART NUMBER	(6) DESCRIPTION AND USABLE ON CODES (UOC)	(7) QTY

GROUP 1407 POWER STEERING GEAR

FIG. 304 STEERING GEAR ASSEMBLY
AND MOUNTING HARDWARE

(1) ITEM NO	(2) SMR CODE	(3) NSN	(4) CAGEC	(5) PART NUMBER	(6) DESCRIPTION AND USABLE ON CODES (UOC)	(7) QTY
1	PAFZZ	5310004883888	96906	MS51943-40	NUT, SELF-LOCKING, HE	4
2	PFFZZ	5340010907637	19207	12255758	BRACKET, ANGLE	1
3	PAFZZ	5305007195235	80204	B1821BH050F175N	SCREW, CAP, HEXAGON H	4
4	PAFZZ	5310009825014	96906	MS21045-14	NUT, SELF-LOCKING, HE	4
5	PAFZZ	5310008098540	96906	MS27183-25	WASHER, LOCK	4
6	PFFZZ	5340010907638	19207	12255760	PLATE, MOUNTING	1
7	PAFFF	2530011198710	19207	11668993	STEERING GEAR ASSEMBLY	1
8	PAFZZ	5305002138886	80204	B1821BH088F400N	SCREW, CAP, HEXAGON H	4

END OF FIGURE

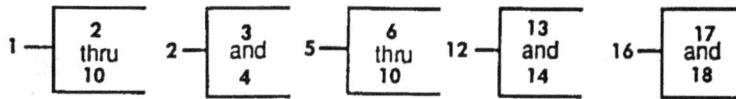

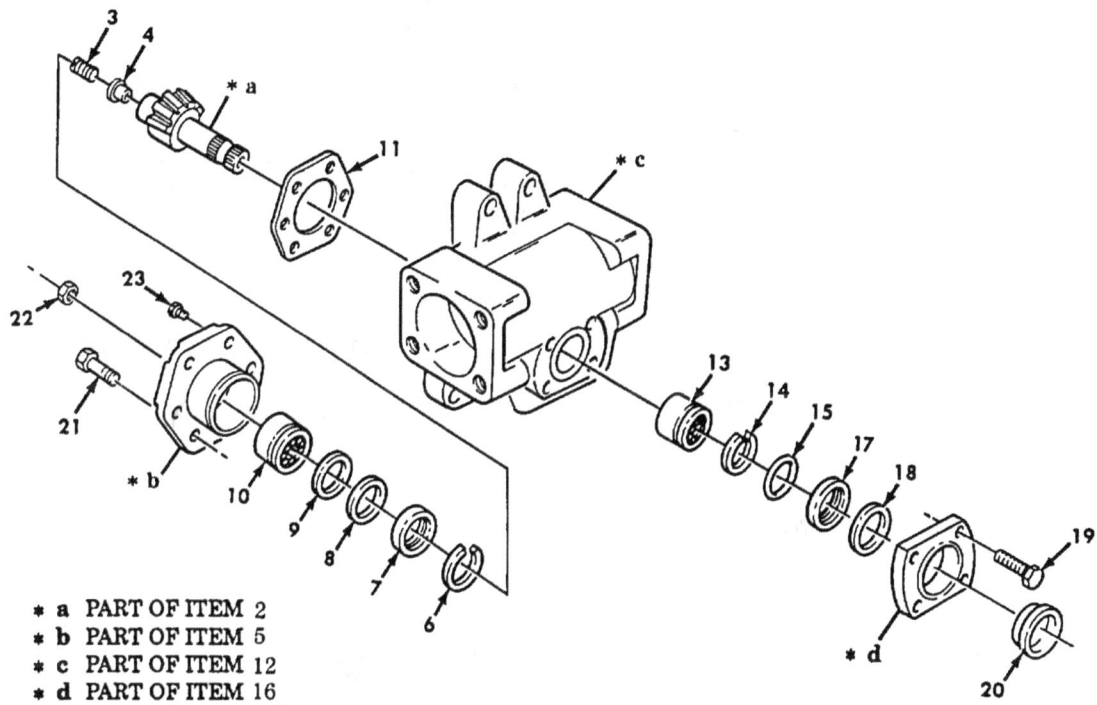

Figure 305. Steering Gear Sector Shaft and Housing Assembly (M939, M939A1).

(1) ITEM NO	(2) SMR CODE	(3) NSN	(4) CAGEC	(5) PART NUMBER	(6) DESCRIPTION AND USABLE ON CODES (UOC)	(7) QTY
					GROUP 1407 POWER STEERING GEAR ASSEMBLY	
					FIG. 305 STEERING GEAR SECTOR SHAFT AND HOUSING ASSEMBLY(M939, M939A1)	
1	PBFFZ	2530011343668	77640	HFB644100-A3-397	SHAFT AND CONTROL A .... UOC:DAA, DAB, DAC, DAD, DAE, DAF, DAG, DAH, DAJ, DAK, DAL, DAW, DAX, V12, V13, V14, V15, V16, V17, V18, V19, V20, V21, V22, V24, V25, V39	1
2	PAFFF	3040011464161	07367	HFB644100-A1-397	.GEAR, SECTOR SHAFT .... UOC:DAA, DAB, DAC, DAD, DAE, DAF, DAG, DAH, DAJ, DAK, DAL, DAW, DAX, V12 , V13 , V14 , V15, V16, V17, V18, V19, V20, V21, V22, V24, V25, V39	1
3	PAFZZ	5305008992049	77640	021200	..SETSCREW .... UOC:DAA, DAB, DAC, DAD, DAE, DAF, DAG, DAH, DAJ, DAK, DAL, DAW, DAX, V12, V13, V14, V15, V16, V17, V18, V19, V20, V21, V22, V24, V25, V39	1
4	PAFZZ	2530002250680	77640	062005	..RETAINER, ADJUSTING .... UOC:DAA, DAB, DAC, DAD, DAE, DAF, DAG, DAH, DAJ, DAK, DAL, DAW, DAX, V12, V13, V14, V15, V16, V17, V18, V19, V20, V21, V22, V24, V25, V39	1
5	PAFFZ	2530011317445	77640	HFB645001-A1	.COVER, STEERING SH................................. UOC:DAA, DAB, DAC, DAD, DAE, DAF, DAG, DAH, DAJ, DAK, DAL, DAW, DAX, V12, V13, V14, V15, V16, V17, V18, V19, V20, V21, V22, V24, V25, V39	1
6	PAFZZ	5325011802448	77640	401445	..RING, RETAINING .... UOC:DAA, DAB, DAC, DAD, DAE, DAF, DAG, DAH, DAJ, DAK, DAL, DAW, DAX, V12 , V13 , V14 , V15, V16, V17, V18, V19, V20, V21, V22, V24, V25, V39	1
7	PAFZZ	5330011354064	77640	032791-A1	..SEAL, PLAIN ENCASED .... UOC:DAA, DAB, DAC, DAD, DAE, DAF, DAG, DAH, DAJ, DAK, DAL, DAW, DAX, V12, V13, V14, V15, V16, V17, V18, V19, V20, V21, V22, V24, V25, V39	1
8	PAFZZ	5310011616131	77640	028435	..WASHER, FLAT .... UOC:DAA, DAB, DAC, DAD, DAE, DAF, DAG, DAH, DAJ, DAK, DAL, DAW, DAX, V12, V13, V14, V15, V16, V17, V18, V19, V20, V21, V22, V24, V25, V39	1
9	PAFZZ	5310011356755	77640	028434	..WASHER, FLAT .... UOC:DAA, DAB, DAC, DAD, DAE, DAF, DAG, DAH, DAJ, DAK, DAL, DAW, DAX, V12, V13, V14, V15, V16, V17, V18, V19, V20, V21, V22, V24, V25, V39	1

(1) ITEM NO	(2) SMR CODE	(3) NSN	(4) CAGEC	(5) PART NUMBER	(6) DESCRIPTION AND USABLE ON CODES (UOC)	(7) QTY
10	XBFZZ		77640	071016	..BEARING, ROLLER, NEED .... UOC:DAA, DAB, DAC, DAD, DAE, DAF, DAG, DAH, DAJ, DAK, DAL, DAW, DAX, V12, V13, V14, V15, V16, V17, V18, V19, V20, V21, V22, V24, V25, V39	1
11	PAFZZ	5330011494415	77640	HFB649000	GASKET .... UOC:DAA, DAB, DAC, DAD, DAE, DAF, DAG, DAH, DAJ, DAK, DAL, DAW, DAX, V12, V13, V14, V15, V16, V17, V18, V19, V20, V21, V22, V24, V25, V39	1
12	XAFFF		07367	HFB642010-A1	HOUSING .... UOC:DAA, DAB, DAC, DAD, DAE, DAF, DAG, DAH, DAJ, DAK, DAL, DAW, DAX, V12, V13, V14, V15, V16, V17, V18, V19, V20, V21, V22, V24, V25, V39	1
13	XDFZZ		77640	071018	.BEARING, ROLLER, NEED .... UOC:DAA, DAB, DAC, DAD, DAE, DAF, DAG, DAH, DAJ, DAK, DAL, DAW, DAX, V12, V13, V14, V15, V16, V17, V18, V19, V20, V21, V22, V24, V25, V39	1
14	PAFZZ	5365006137796	77640	401309	.RING, RETAINING .... UOC:DAA, DAB, DAC, DAD, DAE, DAF, DAG, DAH, DAJ, DAK, DAL, DAW, DAX, V12, V13, V14, V15, V16, V17, V18, V19, V20, V21, V22, V24, V25, V39	1
15	PAFZZ	5330011432780	77640	032586	O-RING .... UOC:DAA, DAB, DAC, DAD, DAE, DAF, DAG, DAH, DAJ, DAK, DAL, DAW, DAX, V12, V13, V14, V15, V16, V17, V18, V19, V20, V21, V22, V24, V25, V39	1
16	PFFFZ 2	30011275006	77640	402368-A1	COVER, STEERING GEAR .... UOC:DAA, DAB, DAC, DAD, DAE, DAF, DAG, DAH, DAJ, DAK, DAL, DAW, DAX, V12, V13, V14, V15, V16, V17, V18, V19, V20, V21, V22, V24, V25, V39	1
17	PAFZZ	5330011350682	77640	032634-A1	.PACKING WITH RETAIN .... UOC:DAA, DAB, DAC, DAD, DAE, DAF, DAG, DAH, DAJ, DAK, DAL, DAW, DAX, V12, V13, V14, V15, V16, V17, V18, V19, V20, V21, V22, V24, V25, V39	1
18	PAFZZ	5310011356754	77640	028433	.WASHER, FLAT .... UOC:DAA, DAB, DAC, DAD, DAE, DAF, DAG, DAH, DAJ, DAK, DAL, DAW, DAX, V12, V13, V14, V15, V16, V17, V18, V19, V20, V21, V22, V24, V25, V39	1
19	PAFZZ	5305011422526	77640	G9429710	SCREW .... UOC:DAA, DAB, DAC, DAD, DAE, DAF, DAG, DAH, DAJ, DAK, DAL, DAW, DAX, V12, V13, V14, V15, V16, V17, V18, V19, V20, V21, V22, V24, V25, V39	4
20	PAFZZ	5330011434185	77640	032591	SEAL .... UOC:DAA, DAB, DAC, DAD, DAE, DAF, DAG, DAH,	1

(1) ITEM NO	(2) SMR CODE	(3) NSN	(4) CAGEC	(5) PART NUMBER	(6) DESCRIPTION AND USABLE ON CODES (UOC)	(7) QTY
					DAJ, DAK, DAL, DAW, DAX, V12, V13, V14, V15, V16, V17, V18, V19, V20, V21, V22, V24, V25, V39	
21	PAFZZ	5305007262525	96906	MS90727-158	SCREW, CAP, HEXAGON H .... UOC:DAA, DAB, DAC, DAD, DAE, DAF, DAG, DAH, DAJ, DAK, DAL, DAW, DAX, V12, V13, V14, V15, V16, V17, V18, V19, V20, V21, V22, V24, V25, V39	6
22	PAFZZ	5310008911733	96906	MS35691-38	NUT, PLAIN, HEXAGON .... UOC:DAA, DAB, DAC, DAD, DAE, DAF, DAG, DAH, DAJ, DAK, DAL, DAW, DAX, V12, V13, V14, V15, V16, V17, V18, V19, V20, V21, V22, V24, V25, V39	1
23	PAFZZ	5340004436132	77640	036141	PLUG, PROTECTIVE, DUS .... UOC:DAA, DAB, DAC, DAD, DAE, DAF, DAG, DAH, DAJ, DAK, DAL, DAW, DAX, V12, V13, V14, V15, V16, V17, V18, V19 , V20, V21, V22, V24 , V25 V39	1

END OF FIGURE

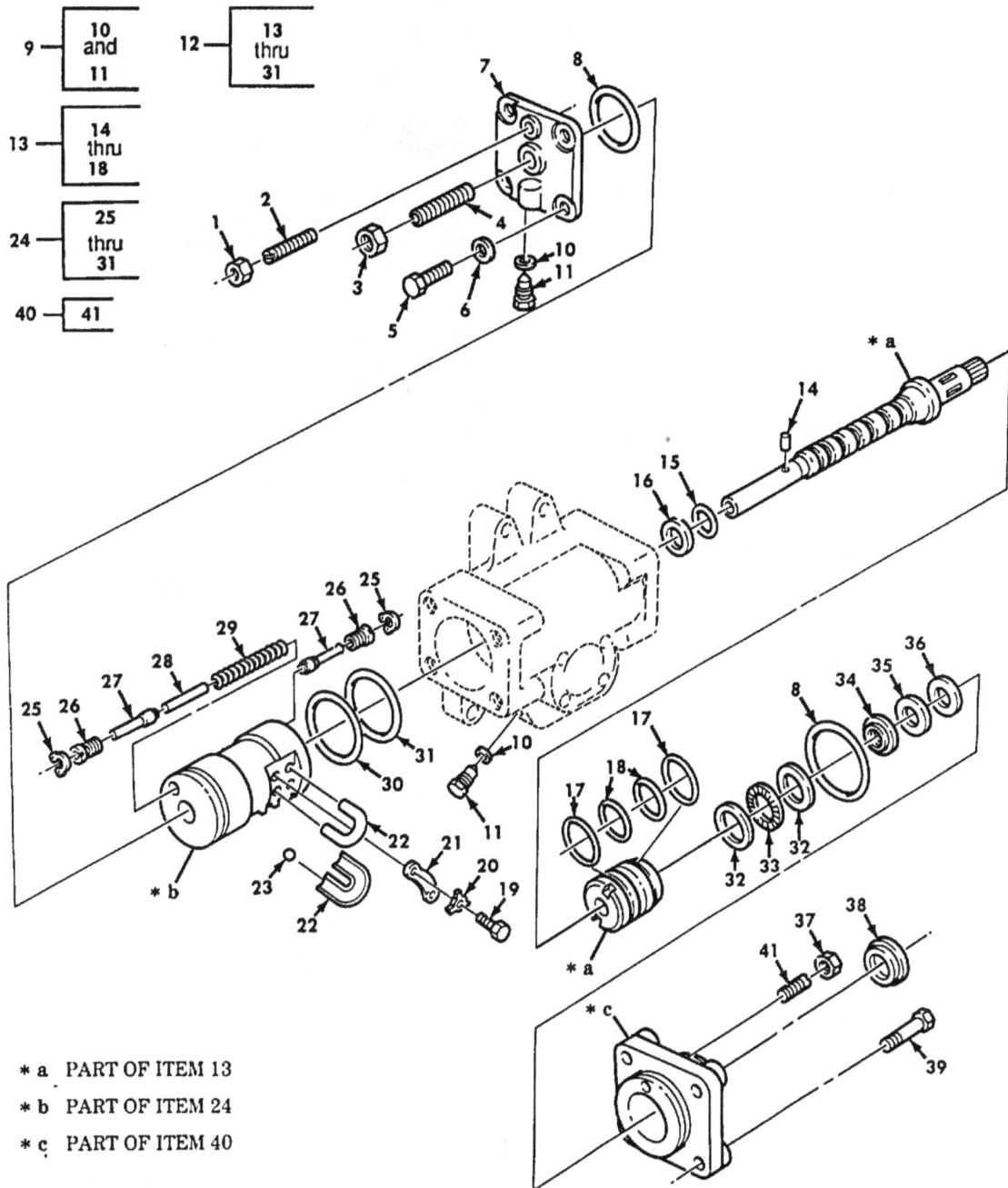

Figure 306. Steering Gear, Worm, and Valve (M939, M939A1).

* a   PART OF ITEM 13
* b   PART OF ITEM 24
* c   PART OF ITEM 40

(1) ITEM NO	(2) SMR CODE	(3) NSN	(4) CAGEC	(5) PART NUMBER	(6) DESCRIPTION AND USABLE ON CODES (UOC)	(7) QTY
					GROUP 1407 POWER STEERING GEAR ASSEMBLY	
					FIG. 306 STEERING GEAR, WORM, AND VALVE(M939, M939A1)	
1	PAFZZ	5310010885851	77640	025121	NUT, PLAIN, HEXAGON .... UOC:DAA, DAB, DAC, DAD, DAE, DAF, DAG, DAH, DAJ, DAK, DAL, DAW, DAX, V12 , V13 , V14 , V15, V16, V17, V18, V19, V20, V21, V22, V24, V25, V39	1
2	PAFZZ	5305011386295	77640	021318	SETSCREW .... UOC:DAA, DAB, DAC, DAD, DAE, DAF, DAG, DAH, DAJ, DAK, DAL, DAW, DAX, V12, V13, V14, V15, V16, V17, V18, V19, V20, V21, V22, V24, V25, V39	1
3	PAFZZ	5310011339186	07367	025122	NUT .... UOC:DAA, DAB, DAC, DAD, DAE, DAF, DAG, DAH, DAJ, DAK, DAL, DAW, DAX, V12, V13, V14, V15, V16, V17, V18, V19, V20, V21, V22, V24, V25, V39	1
4	PAFZZ	5305011296842	77640	021333	SETSCREW .... UOC:DAA, DAB, DAC, DAD, DAE, DAF, DAG, DAH, DAJ, DAK, DAL, DAW, DAX, V12, V13, V14, V15, V16, V17, V18, V19, V20, V21, V22, V24, V25, V39	1
5	PAFZZ	5305011385115	77640	020252	SCREW. .... UOC:DAA, DAB, DAC, DAD, DAE, DAF, DAG, DAH, DAJ, DAIK, DAL, DAW, DAX, V12, V13, V14, V15, V16, V17, V18, V19, V20, V21, V22, V24, V25, V39	4
6	PAFZZ	5310005969763	24446	161A820P1	WASHER, FLAT .... UOC:DAA, DAB, DAC, DAD, DAE, DAF, DAG, DAH, DAJ, DAK, DAL, DAW, DAX, V12, V13, V14, V15, V16, V17, V18, V19, V20, V21, V22, V24, V25, V39	4
7	PAFZZ	2530011268445	07367	402377	COVER, STEERING GEAR .... UOC:DAA, DAB, DAC, DAD, DAE, DAF, DAG, DAH, DAJ, DAK, DAL, DAW, DAX, V12, V13, V14, V15, V16, V17, V18, V19, V20, V21, V22, V24, V25, V39	1
8	PAFZZ	5330011497229	77640	032616	PACKING, PREFORMED .... UOC:DAA, DAB, DAC, DAD, DAE, DAF, DAG, DAH, DAJ, DAK, DAL, DAW, DAX, V12, V13, V14, V15, V16, V17, V18, V19, V20, V21, V22, V24, V25, V39	2
9	PAFFF	5365011340522	77640	415437-A1	PLUG .... UOC:DAA, DAB, DAC, DAD, DAE, DAF, DAG, DAH, DAJ, DAK, DAL, DAW, DAX, V12, V13, V14, V15, V16, V17, V18, V19, V20, V21, V22, V24, V25, V39	2
10	PAFZZ	5330011354807	77640	032229	.O-RING .... UOC:DAA, DAB, DAC, DAD, DAE, DAF, DAG, DAH, DAJ, DAK, DAL, DAW, DAX, V12, V13, V14, V15, V16, V17, V18, V19, V20, V21, V22, V24, V25, V39	1

(1) ITEM NO	(2) SMR CODE	(3) NSN	(4) CAGEC	(5) PART NUMBER	(6) DESCRIPTION AND USABLE ON CODES (UOC)	(7) QTY
11	XAFZZ		77640	G9410358	.PLUG .... UOC:DAA, DAB, DAC, DAD, DAE, DAF, DAG, DAH, DAJ, DAK, DAL, DAW, DAX, V12, V13, V14, V15, V16, V17, V18, V19, V20, V21, V22, V24, V25, V39	1
12	PAFFF	2530012063911	77640	HFB523001-A11	SLEEVE AND WORMGEAR .... UOC:DAA, DAB, DAC, DAD, DAE, DAZ, DAG, DAH, DAJ, DAX, DAL, DAW, DAX, V12, V13, V14, V15, V16, V17, V18, V19, V20, V21, V22, V24, V25, V39	1
13	PBFFF	2520011343418	77640	HFB523001-A1	.WORM AND VALVE ASSE .... UOC:DAA, DAB, DAC, DAD, DAE, DAF, DAG, DAH, DAJ, DAK, DAL, DAW, DAX, V12, V13, V14, V15, V16, V17, V18, V19, V20, V21, V22, V24, V25, V39	1
14	PAFZZ	3110011379725	77640	040127	..ROLLER, BEARING .... UOC:DAA, DAB, DAC, DAD, DAE, DAF, DAG, DAH, DAJ, DAK, DAL, DAW, DAX, V12, V13, V14, V15, V16, V17, V18, V19, V20, V21, V22, V24, V25, V39	1
15	PAFZZ	5330011354066	77640	032536	..RETAINER, PACKING... UOC:DAA, DAB, DAC, DA D, DAE, DAF, DAG, DAH, DAJ, DAK, DAL, DAW, DAX, V12, V13, V14, V15, V16, V17, V18, V19, V20, V21, V22, V24, V25, V39	1
16	PAFZZ	5330011354808	77640	032552	..PACKING, PREFORMED..................................... UOC:DAA, DAB, DAC, DAD, DAE, DAF, DAG, DAH, DAJ, DAK, DAL, DAW, DAX, V12, V13, V14, V15, V16, V17, V18, V19, V20, V21, V22, V24, V25, V39	1
17	PAFZZ	5330011354067	77640	032570	..RETAINER, PACKING .... UOC:DAA, DAB, DAC, DAD, DAE, DAF, DAG, DAH, DAJ, DAK, DAL, DAW, DAX, V12, V13, V14, V15, V16, V17, V18, V19, V20, V21, V22, V24, V25, V39	2
18	PAFZZ	5330011354809	77640	032571	..PACKING, PREFORMED .... UOC:DAA, DAB, DAC, DAD, DAE, DAF, DAG, DAH, DAJ, DAK, DAL, DAW, DAX, V12, V13, V14, V15, V16, V17, V18, V19, V20, V21, V22, V24, V25, V39	2
19	PAFZZ	5305011361618	77640	G179810	.SCREW. .... UOC:DAA, DAB, DAC, DAD, DAE, DAF, DAG, DAH, DAJ, DAK, DAL, DAW, DAX, V12, V13, V14, V15, V16, V17, V18, V19, V20, V21, V22, V24, V25, V39	2
20	PAFZZ	5310011356758	77640	028426	.WASHER, LOCK .... UOC:DAA, DAB, DAC, DAD, DAE, DAF, DAG, DAH, DAJ, DAK, DAL, DAW, DAX, V12, V13, V14, V15, V16, V17, V18, V19, V20, V21, V22, V24, V25, V39	2

(1) ITEM NO	(2) SMR CODE	(3) NSN	(4) CAGEC	(5) PART NUMBER	(6) DESCRIPTION AND USABLE ON CODES (UOC)	(7) QTY
21	PAFZZ	5340011361660	77640	402380	.CLIP, RETAINING .... UOC:DAA, DAB, DAC, DAD, DAE, DAF, DAG, DAH, DAJ, DAK, DAL, DAW, DAX, V12, V13, V14, V15, V16, V17, V18, V19, V20, V21, V22, V24, V25, V39	1
22	PAFZZ	2530011434203	77640	400122-X1	.GUIDE, BALL RETURN .... UOC:DAA, DAB, DAC, DAD, DAE, DAF, DAG, DAH, DAJ, DAK, DAL, DAW, DAX, V12, V13, V14, V15, V16, V17, V18, V19, V20, V21, V22, V24, V25, V39	2
23	PAFZZ	3110011634932	77640	216191-X1	.BALL, BEARING .... UOC:DAA, DAB, DAC, DAD, DAE, DAF, DAG, DAH, DAJ, DAK, DAL, DAW, DAX, V12 , V13 , V14 , V15, V16, V17, V18, V19, V20, V21, V22, V24, V25, V39	27
24	PAFFF	2530011380922	07367	HFB647000-A1	.PISTON ASSEMBLY, STE .... UOC:DAA, DAB, DAC, DAD, DAE, DAF, DAG, DAH, DAJ, DAK, DAL, DAW, DAX, V12, V13, V14 , V15, V16, V17, V18, V19, V20, V21, V22, V24, V25, V39	1
25	PAFZZ	5340011356350	77640	401379	..RETAINER... UOC:DAA, DAB, DAC, DAD, DAE, DAF, DAG, DAH, DAJ, DAK, DAL, DAW, DAX, V12, V13, V14, V15, V16, V17, V18, V19, V20, V21, V22, V24, V25, V39	2
26	PAFZZ	4820011394346	77640	415442	..SEAT, VALVE .... UOC:DAA, DAB, DAC, DAD, DAE, DAF, DAG, DAH, DAJ, DAK, DAL, DAW, DAX, V12, V13, V14, V15, V16, V17, V18, V19, V20, V21, V22, V24, V25, V39	2
27	PAFZZ	4820011346619	77640	040124	..DISK, VALVE .... UOC:DAA, DAB, DAC, DAD, DAE, DAF, DAG, DAH, DAJ, DAK, DAL, DAW, DAX, V12, V13, V14, V15, V16, V17, V18, V19, V20, V21, V22, V24, V25, V39	2
28	PAFZZ	2530011343669	77640	040125	..ROD, ALIGNING, VEHICU .... UOC:DAA, DAB, DAC, DAD, DAE, DAF, DAG, DAH, DAJ, DAK, DAL, DAW, DAX, V12, V13, V14, V15, V16, V17, V18, V19, V20, V21, V22, V24, V25, V39	1
29	PAFZZ	5360011354065	77640	401375	..SPRING, HELICAL, COMP............................. UOC:DAA, DAB, DAC, DAD, DAE, DAF, DAG, DAH, DAJ, DAK, DAL, DAW, DAX, V12, V13, V14, V15, V16, V17, V18, V19, V20, V21, V22, V24, V25, V39	1
30	PAFZZ	5330011354068	77640	032615	..PACKING, PREFORMED................................. UOC:DAA, DAB, DAC, DAD, DAE, DAF, DAG, DAX, DAJ, DAK, DAL, DAW, DAX, V12, V13, V14, V15, V16, V17, V18, V19, V20, V21, V22, V24, V25, V39	1
31	PAFZZ	5330011354069	77640	032590	..O-RING .... UOC:DAA, DAB, DAC, DAD, DAE, DAF, DAG, DAH, DAJ, DAK, DAL, DAW, DAX, V12, V13, V14 , V15, V16, V17, V18, V19, V20, V21, V22, V24, V25, V39	1

(1) ITEM NO	(2) SMR CODE	(3) NSN	(4) CAGEC	(5) PART NUMBER	(6) DESCRIPTION AND USABLE ON CODES (UOC)	(7) QTY
32	PAFZZ	3120011281565	77640	028430	BEARING, WASHER, THRU .... UOC:DAA, DAB, DAC, DAD, DAE, DAF, DAG, DAR, DAJ, DAK, DAL, DAW, DAX, V12, V13, V14, V15, V16, V17, V18, V19, V20, V21, V22, V24, V25, V39	2
33	PAFZZ	3110006499498	77640	067026	RETAINER AND ROLLER .... UOC:DAA, DAB, DAC, DAD, DAE, DAF, DAG, DAH, DAJ, DAK, DAL, DAW, DAX, V12, V13, V14, V15, V16, V17, V18, V19, V20, V21, V22, V24, V25, V39	1
34	PAFZZ	5330011436013	77640	032577-A1	SEAL .... UOC:DAA, DAB, DAC, DAD, DAE, DAF, DAG, DAH, DAJ, DAX, DAL, DAW, DAX, V12, V13, V14, V15, V16, V17, V18, V19, V20, V21, V22, V24, V25, V39	1
35	PAFZZ	5310011418465	77640	028445	WASHER, FLAT .... UOC:DAA, DAB, DAC, DAD, DAE, DAF, DAG, DAH, DAJ, DAK, DAL, DAW, DAX, V12, V13, V14, V15, V16, V17, V18, V19, V20, V21, V22, V24, V25, V39	1
36	PAFZZ	5365004765259	77640	401314	RING, RETAINING . ... UOC:DAA, DAB, DAC, DAD, DAE, DAF, DAG, DAH, DAJ, DAK, DAL, DAW, DAX, V12, V13, V14, V15, V16, V17, V18, V19, V20, V21, V22, V24, V25, V39	1
37	PAFZZ	5310011290193	07367	025124	NUT.. UOC:DAA, DAB, DAC, DAD, DAE, DAF, DAG, DAH, DAJ, DAK, DAL, DAW, DAX, V12, V13, V14, V15, V16, V17, V18, V19, V20, V21, V22, V24, V25, V39	1
38	PAFZZ	5330011434186	77640	032579	SEAL .... UOC:DAA, DAB, DAC, DAD, DAE, DAF, DAG, DAH, DAJ, DAK, DAL, DAW, DAX, V12, V13, V14, V15, V16, V17, V18, V19, V20, V21, V22, V24, V25, V39	1
39	PAFZZ	5305011361619	77640	020251	SCREW.. UOC:DAA, DAB, DAC, DAD, DAE, DAF, DAG, DAH, DAJ, DAK, DAL, DAW, DAX, V12, V13, V14, V15, V16, V17, V18, V19, V20, V21, V22, V24, V25, V39	4
40	PAFFF	3040011317446	77640	HFB646013-A1	PLATE, RETAINING, SHA .... UOC:DAA, DAB, DAC, DAD, DAE, DAF, DAG, DAH, DAJ, DAK, DAL, DAW, DAX, V12, V13, V14, V15, V16, V17, V18, V19, V20, V21, V22, V24, V25, V39	1
41	PAFZZ	5305011361686	77640	021336	..SETSCREW .... UOC:DAA, DAB, DAC, DAD, DAE, DAF, DAG, DAH, DAJ, DAK, DAL, DAW, DAX, V12, V13, V14, V15, V16, V17, V18, V19, V20, V21, V22, V24, V25, V39	1

END OF FIGURE

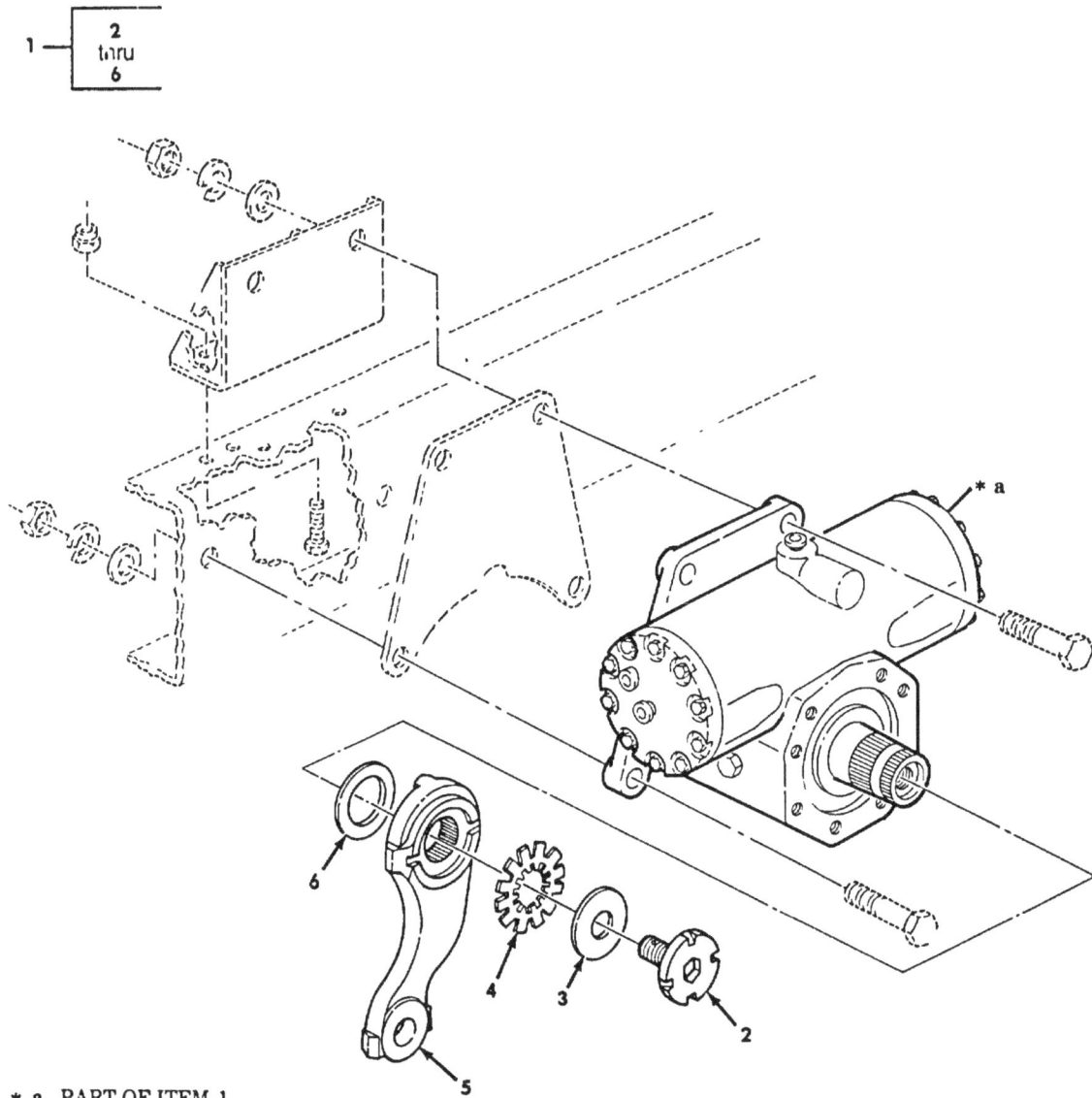

* a PART OF ITEM 1

Figure 307. Sheppard Steering Gear Assembly and Mounting Hardware.

(1) ITEM NO	(2) SMR CODE	(3) NSN	(4) CAGEC	(5) PART NUMBER	(6) DESCRIPTION AND USABLE ON CODES (UOC)	(7) QTY
					GROUP 1407 POWER STEERING GEAR ASSEMBLY	
					FIG. 307 SHEPPARD STEERING GEAR ASSEMBLY AND MOUNTING HARDWARE	
1	PAFFH	2530012844566	78222	292SAF61	STEERING GEAR ASSEMBLY .... UOC:ZAA, ZAB, ZAC, ZAD, ZAE, ZAF, ZAG, ZAH, ZAJ, ZAK, ZAL	1
2	KFHZZ		78222	1790422	.RETAINER PART OF KIT P/N 1790522K. .... UOC:ZAA, ZAB, ZAC, ZAD, ZAE, ZAF, ZAG, ZAH, ZAJ, ZAK, ZAL	1
3	KFHZZ		78222	2262121	.TABWASHER PART OF KIT P/N 1790522K. .... UOC:ZAA, ZAB, ZAC, ZAD, ZAE, ZAF, ZAG, ZAH, ZAJ, ZAK, ZAL	1
4	KFHZZ		78222	2262021	.WASHER, FRICTION PART OF KIT P/N .... 1790522K .... UOC:ZAA, ZAB, ZAC, ZAD, ZAE, ZAF, ZAG, ZAH, ZAJ, ZAK, ZAL	1
5	PAHZZ	2530013398673	78222	2582563	.PITMAN ARM, STEERING .... UOC:ZAA, ZAB, ZAC, ZAD, ZAE, ZAF, ZAG, ZAH, ZAJ, ZAK, ZAL	1
6	KFHZZ		78222	2369001	.SEAL, QUAD-O-DYN SHA PART OF KIT P/N .... 5518441 .... UOC:ZAA, ZAB, ZAC, ZAD, ZAE, ZAF, ZAG, ZAH, ZAJ, ZAK, ZAL	1
	KFHZZ	2530013397913	78222	1790522K	PARTS KIT, STEERING .... RETAINER (1) 307-2 TAB, WASHER (1) 307-3 WASHER, FRICTION (1) 307-4	1

END OF FIGURE

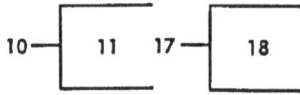

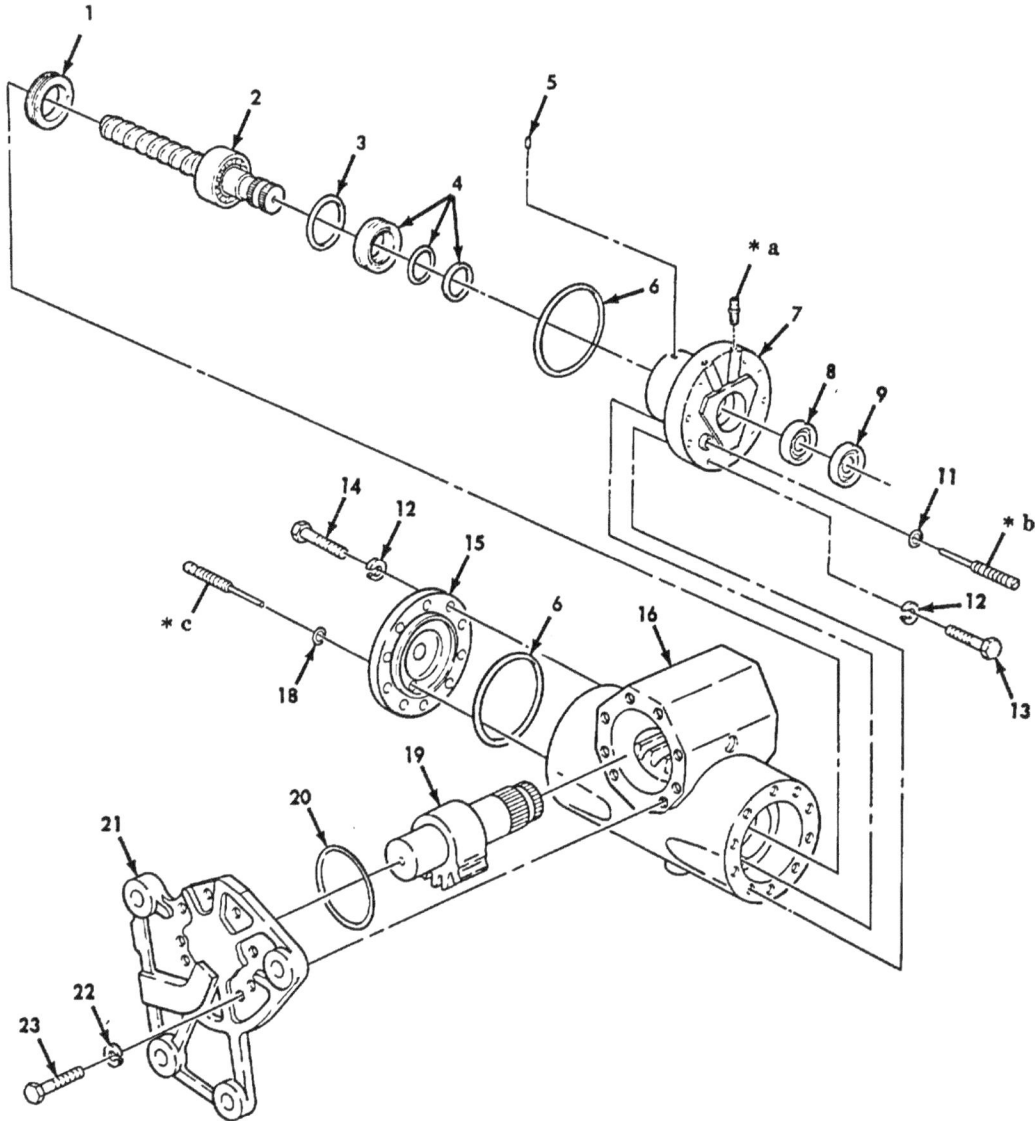

* a PART OF ITEM 7
* b PART OF ITEM 10
* c PART OF ITEM 17

Figure 308. Steering Gear Sector Shaft and Housing Assembly.

(1) ITEM NO	(2) SMR CODE	(3) NSN	(4) CAGEC	(5) PART NUMBER	(6) DESCRIPTION AND USABLE ON CODES (UOC)	(7) QTY
					GROUP 1407 POWER STEERING GEAR ASSEMBLY	
					FIG. 308 STEERING GEAR SECTOR SHAFT AND HOUSING ASSEMBLY	
1	XDHZZ		78222	2411521	NUT, BEARING RETAINI	1
2	KFHZZ		78222	2574418	BEARING SHAFT ACTUA PART OF KIT P/N........... 5523281	1
3	KFHZZ	5330013291879	78222	2360831	O-RING PART OF KIT P/N 5518441	1
4	PFHZZ	5330012886304	78222	2370461K	SEAL KIT PART OF KIT P/N 5518441.....................	1
5	KFHZZ		78222	0856121	PIN, LOCKING PART OF KIT P/N 5518441	1
6	KFHZZ		78222	2370291	SEAL RING, METAL PART OF KIT P/N.................... 5518441	1
7	XAHZZ		78222	2182413K	BEARING CAP SERVICE	1
8	PAHZZ	5330005486114	78222	2350961	SEAL, PLAIN ENCASED PART OF KIT P/N............. 5518441	1
9	PAHZZ	5330013390859	78222	2351081	SEAL, NONMETALLIC SP PART OF KIT P/N........... 5518441	1
10	PFHZZ	5999013412972	78222	1820651K	PARTS KIT, ELECTRONI PART OF KIT P/N............ 5518441	1
11	KFHZZ		78222	2360451 KF	.O-RING, RELIEF PLUNG PART OF KIT P/N........... 5518441	1
12	PAHZZ	5310004079566	96906	MS35338-45	WASHER, LOCK	2
13	PAHZZ	5305002259093	96906	MS90726-38	SCREW, CAP, HEXAGON H 5/16-24 X 1.5..............	1
14	PAHZZ	5306002259089	96906	MS90726-34	BOLT, MACHINE 5/16-24 X 1.0............................	1
15	XDHZZ		78222	0031832K	CYLINDER HEAD SERVI	1
16	XAHZZ		78222	1132375K	STEERING GEAR HOUS	1
17	PFHZZ	2520011522384	45152	35H113	PLUNGER, RELIEF VALV PART OF KIT P/N........... 5518441	1
18	KFHZZ		89619	8QD1038 KF	.O-RING, RELIEF PLUNG PART OF KIT P/N........... 5518441	1
19	XDHZZ		78222	1870213	SHAFT, SECTOR	1
20	KFHZZ		78222	2360651	O-RING, COVER PART OF KIT P/N 5518441	1
21	XDHZZ		78222	0924734K	COVER	1
22	PAHZZ	5310001711735	96906	MS51848-14	WASHER, LOCK 1/2	8
23	PAHZZ	5305007195219	96906	MS90727-111	SCREW, CAP, HEXAGON H 1/2-20 X 1.250.	8
	KFHZZ	5330013416583	78222	5518441	STEERING GASKET AND SEAL SET....................	1

O-RING, HI-PRES SEAL	(	1)	308-3
O-RING, COVER	(	1)	308-20
O-RING, RELIEF PLUNG	(	1)	308-11
O-RING, RELIEF PLUNG	(	1)	308-18
PARTS KIT, SOLENOID	(	1)	308-10
PIN, LOCKING	(	1)	308-5
PLUNGER, RELIEF VALV	(	1)	308-17
SEAL KIT	(	1)	308-4
SEAL RING, METAL	(	1)	308-6
SEAL, NONMETALLIC	(	1)	308-9
SEAL, PLAIN ENCASED	(	1)	308-8

(1) ITEM NO	(2) SMR CODE	(3) NSN	(4) CAGEC	(5) PART NUMBER	(6) DESCRIPTION AND USABLE ON CODES (UOC)	(7) QTY
	KFHZZ	2530013400365	78222	5523281	PARTS KIT, STEERING	1
					BALL,PLUNGER RELIEF (1) 309-11	
					BEARING,SHAFT ACTUA (1) 308-2	
					PIN,LOCKING (3) 308-8	
					PISTON (1) 308-6	
					PISTON PLUG (1) 308-1	
					SEAT,PLUNGER,RELIEF (2) 308-13	
					SET,SCREW (1) 308-7	
					SLIPPER,STRAINER (1) 308-10	
					SPRING,REVERSING (2) 308-3	
					SPRING,RELIEF,PLUNG (1) 308-12	
					TEFLON PISTON RING (1) 308-5	
					VALVE,ADJUSTING NUT (1) 308-2	
					VALVE POSITION PIN (1) 308-9	
					VALVE,STEER GEAR,AC (1) 308-4	

END OF FIGURE

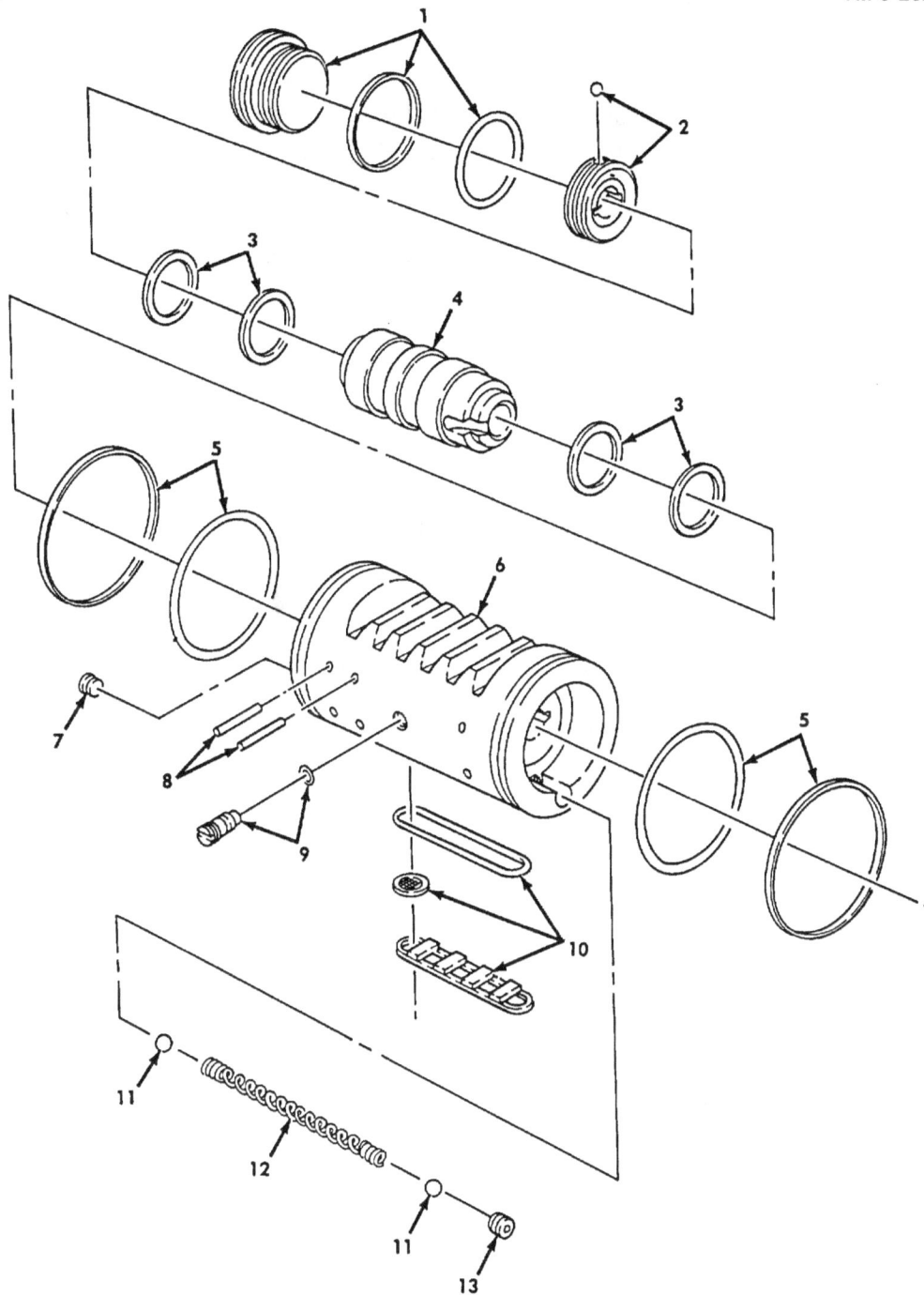

Figure 309. Steering Gear, Worm, and Valve (M939A2).

(1) ITEM NO	(2) SMR CODE	(3) NSN	(4) CAGEC	(5) PART NUMBER	(6) DESCRIPTION AND USABLE ON CODES (UOC)	(7) QTY
					GROUP 1407 POWER STEERING GEAR ASSEMBLY	
					FIG. 309 STEERING GEAR, WORM, AND VALVE(M939A2)	
1	KFHZZ		78222	3331081K	PISTON PLUG PART OF KIT P/N 5523281.	1
2	KFHZZ		78222	2411531K	VALVE ADJUSTING NUT PART OF KIT P/N 5523281	1
3	KFHZZ		78222	0752021	SPRING, REVERSING PART OF KIT P/N 5523281	2
4	KFHZZ		78222	0400511	VALVE, STEER, GEAR, AC PART OF KIT P/N 5523281	1
5	KFHZZ		78222	0130491K	TEFLON PISTON RING PART OF KIT P/N 5523281	2
6	KFHZZ		78222	0857112	PISTON PART OF KIT P/N 5523281	1
7	KFHZZ		78222	0591304	SETSCREW PART OF KIT P/N 5523281	1
8	KFHZZ		78222	0856401	PIN, LOCKING PART OF KIT P/N 5523281.	6
9	KFHZZ		78222	0856311K	VALVE POSITION PIN PART OF KIT P/N 5523281	1
10	KFHZZ		78222	2370392K	SLIPPER, STRAINER PART OF KIT P/N 5523281	1
11	KFHZZ		78222	2040081	BALL, PLUNGER, RELIER PART OF KIT P/N 5523281	2
12	KFHZZ		78222	0751921	SPRING, RELIEF, PLUNG PART OF KIT P/N 5523281	1
13	KFHZZ		78222	0170121	SEAT, PLUNGER, RELIEF PART OF KIT P/N 5523281	1

END OF FIGURE

2 — 3

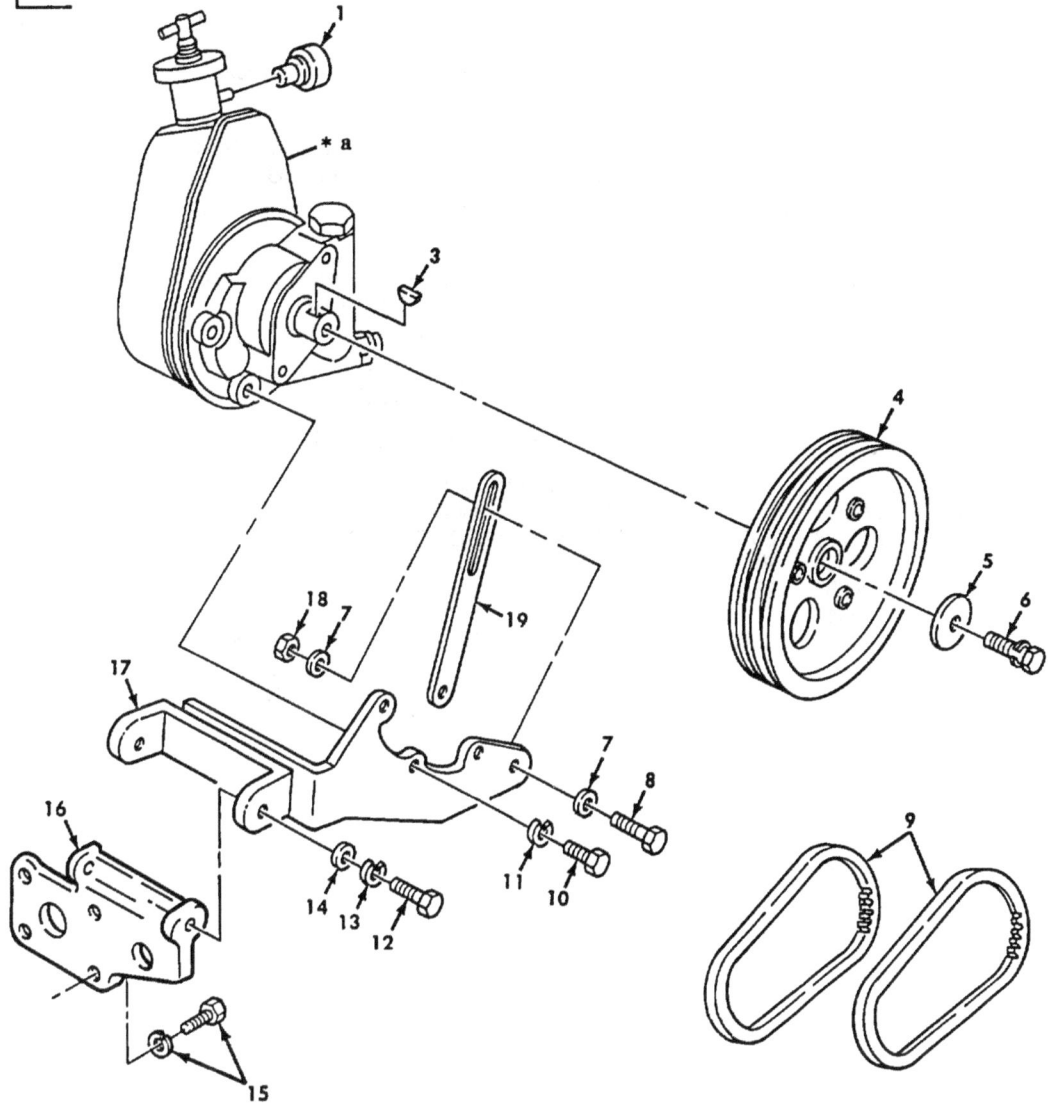

* a PART OF ITEM 2

Figure 310. Power Steering Pump Assembly, Pulley, Mounting Hardware and Belts (M939, M939A1)

(1) ITEM NO	(2) SMR CODE	(3) NSN	(4) CAGEC	(5) PART NUMBER	(6) DESCRIPTION AND USABLE ON CODES (UOC)	(7) QTY
					GROUP 1410 HYDRAULIC PUMP OR FLUID MOTOR ASSEMBLY	
					FIG. 310 POWER STEERING PUMP ASSEMBLY, PULLEY, MOUNTING HARDWARE, AND BELTS (M939, M939A1)	
1	PAOZZ	4310010940791	90005	569020-02	BREATHER	1
2	PAFFF	2530011122161	19954	ER15996-1	PUMP ASSEMBLY, POWER	1
3	PAOZZ	5315006165514	96906	MS35756-6	.KEY, WOODRUFF............................	1
4	PAOZZ	3020001347946	19207	11664244	PULLEY, GROOVE	1
5	PAOZZ	5310002310280	19207	11664241	WASHER, FLAT	1
6	PAOZZ	5306002385661	19207	11664243	BOLT, ASSEMBLED WASH	1
7	PAFZZ	5310000806004	96906	MS27183-14	WASHER, FLAT	2
8	PAFZZ	5305002693239	80204	B1821BH038F138N	SCREW, CAP, HEXAGON H	1
9	PAOZZ	3030008324312	96906	MS51066-48-2	BELTS, V, MATCHED SET	1
10	PAFZZ	5305007252317	80204	B1821BH038C150N	SCREW, CAP, HEXAGON H	3
11	PAFZZ	5310006379541	96906	MS35338-46	WASHER, LOCK	3
12	PAFZZ	5305000711788	80204	B1821BH044C125N	SCREW, CAP, HEXAGON H	2
13	PAFZZ	5310002090965	96906	MS35338-47	WASHER, LOCK	2
14	PAFZZ	5310008094061	96906	MS27183-15	WASHER, FLAT	2
15	PAFZZ	5305010886019	15434	3010596	SCREW, ASSEMBLED WAS	4
16	PAFZZ	5340001743424	15434	127459	BRACKET, ACCESSORY, D	1
17	PFFZZ	2530002303596	19207	11664675	BRACKET ASSEMBLY, PO	1
18	PAFZZ	5310009824908	96906	MS21045-6	NUT, SELF-LOCKING, HE	1
19	PBFZZ	3040002303598	19207	11664549	CONNECTING LINK, RIG	1

END OF FIGURE

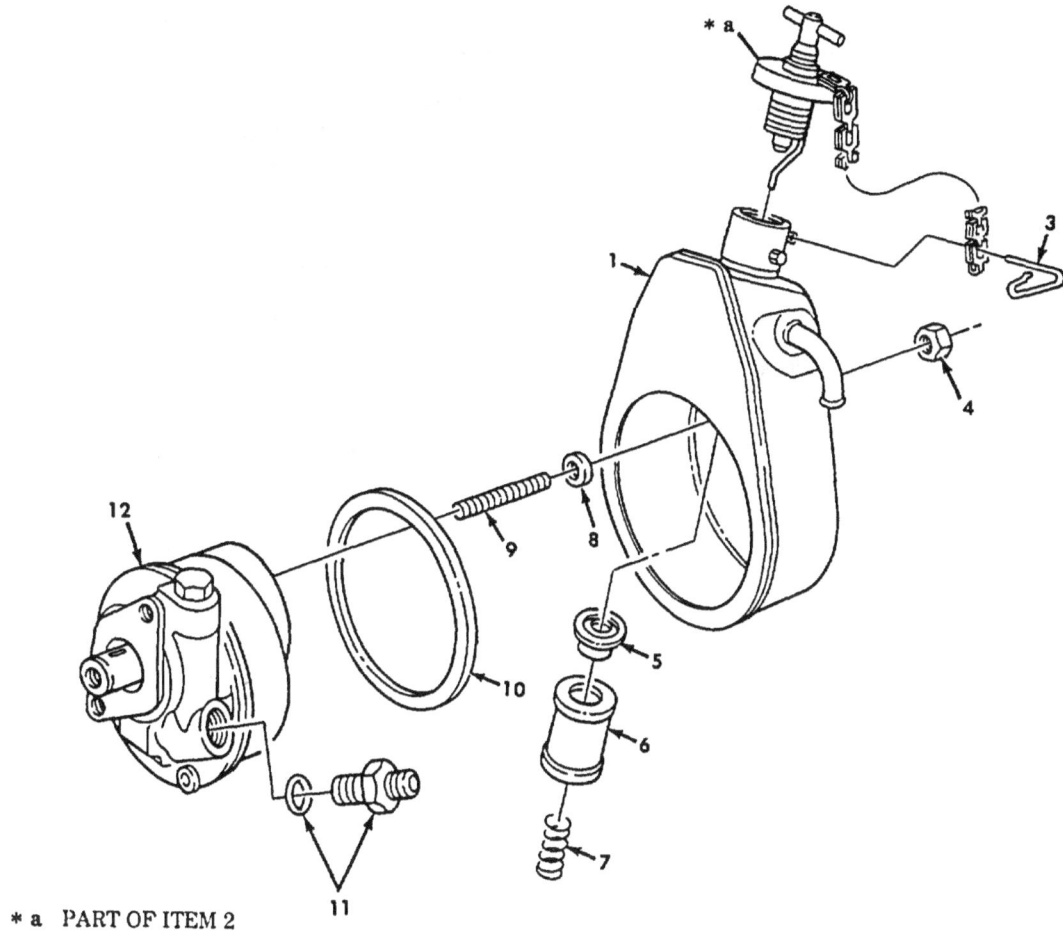

* a   PART OF ITEM 2

Figure 311. Power Steering Pump Assembly (M939, M939A1).

(1) ITEM NO	(2) SMR CODE	(3) NSN	(4) CAGEC	(5) PART NUMBER	(6) DESCRIPTION AND USABLE ON CODES (UOC)	(7) QTY
					GROUP 1410 HYDRAULIC PUMP OR FLUID MOTOR ASSEMBLY	
					FIG. 311 POWER STEERING PUMP ASSEMBLY ,(M939, M939A1)	
1	XAFZZ		19954	ER 27804	RESERVOIR BODY .... UOC:DAA, DAB, DAC, DAD, DAE, DAF, DAG, DAH, DAJ, DAK, DAL, DAW, DAX, V12, V13, V14, V15, V16, V17, V18, V19, V20, V21, V22, V24, V25, V39	1
2	PAFZZ	2530011269303	19954	ER-27512	CAP ASSEMBLY .... UOC:DAA, DAB, DAC, DAD, DAE, DAF, DAG, DAH, DAJ, DAK, DAL, DAW, DAX, V12, V13, V14, V15, V16, V17, V18, V19, V20, V21, V22, V24, V25, V39	1
3	PAFZZ	5315005142660	96906	MS29523-1	.PIN, RETAINING................................ UOC:DAA, DAB, DAC, DAD, DAE, DAF, DAG, DAH, DAJ, DAK, DAL, DAW, DAX, V12, V13, V14, V15, V16, V17, V18, V19, V20, V21, V22, V24, V25, V39	1
4	PAFZZ	5310011294227	19954	ER-99401	NUT PART OF KIT P/N ERS-27785.................. UOC:DAA, DAB, DAC, DAD, DAE, DAF, DAG, DAH, DAJ, DAK, DAL, DAW, DAX, V12, V13, V14, V15, V16, V17, V18, V19, V20, V21, V22, V24, V25, V39	2
5	KFFZZ		19954	ER-93978	WASHER, SEALING PART OF KIT P/N ERS-........... 27785................................ UOC:DAA, DAB, DAC, DAD, DAE, DAF, DAG, DAH, DAJ, DAK, DAL, DAW, DAX, V12, V13, V14, V15, V16, V17, V18, V19, V20, V21, V22, V24, V25, V39	1
6	KFFZZ		19954	ER-99420	ELEMENT, FILTER PART OF KIT P/N ERS-............. 27785................................ UOC:DAA, DAB, DAC, DAD, DAE, DAF, DAG, DAH, DAJ, DAK, DAL, DAW, DAX, V12, V13, V14, V15, V16, V17, V18, V19, V20, V21, V22, V24, V25, V39	1
7	KFFZZ		19954	ER-72269	SPRING, FILTER PART OF KIT P/N ERS-.............. 27785................................ UOC:DAA, DAB, DAC, DAD, DAE, DAF, DAG, DAH, DAJ, DAK, DAL, DAW, DAX, V12, V13, V14, V15, V16, V17, V18, V19, V20, V21, V22, V24, V25, V39	1
8	PAFZZ	5310004832385	19954	ER93765	WASHER, SEAL PART OF KIT P/N ERS-................ 27785................................ UOC:DAA, DAB, DAC, DAD, DAE, DAF, DAG, DAH, DAJ, DAK, DAL, DAW, DAX, V12, V13, V14, V15, V16, V17, V18, V19, V20, V21, V22, V24, V25, V39	2

(1) ITEM NO	(2) SMR CODE	(3) NSN	(4) CAGEC	(5) PART NUMBER	(6) DESCRIPTION AND USABLE ON CODES (UOC)	(7) QTY
9	PAFZZ	5307004786782	19954	ER-97043	STUD, CONTINUOUS THR .... UOC:DAA, DAB, DAC, DAD, DAE, DAF, DAG, DAB, DAJ, DAK, DAL, DAW, DAX, V12, V13, V14, V15, V16, V17, V18, V19, V20, V21, V22, V24, V25, V39	2
10	PAFZZ	5330011290384	19954	ER-82141	GASKET .... UOC:DAA, DAB, DAC, DAD, DAE, DAF, DAG, DAH, DAJ, DAK, DAL, DAW, DAX, V12 , V13, V14, V15, V16, V17, V18, V19, V20, V21, V22, V24, V25, V39	1
11	PAFFZ	4730011276697	19954	ER-22663	ADAPTER, STRAIGHT, TU .... UOC:DAA, DAB, DAC, DA D, DAE, DAF, DAG, DAH, DAJ, DAK, DAL, DAW, DAX, V12, V13, V14, V15, V16, V17, V18, V19, V20, V21, V22, V24, V25, V39	1
12	XAFZZ		19954	ER-15490-1	BASE ASSEMBLY .... UOC:DAA, DAB, DAC, DAD, DAE, DAF, DAG, DAB, DAJ, DAK, DAL, DAW, DAX, V12, V13, V14, V15, V16, V17, V18 , V19, V20, V21, V22, V24, V25, V39	1

END OF FIGURE

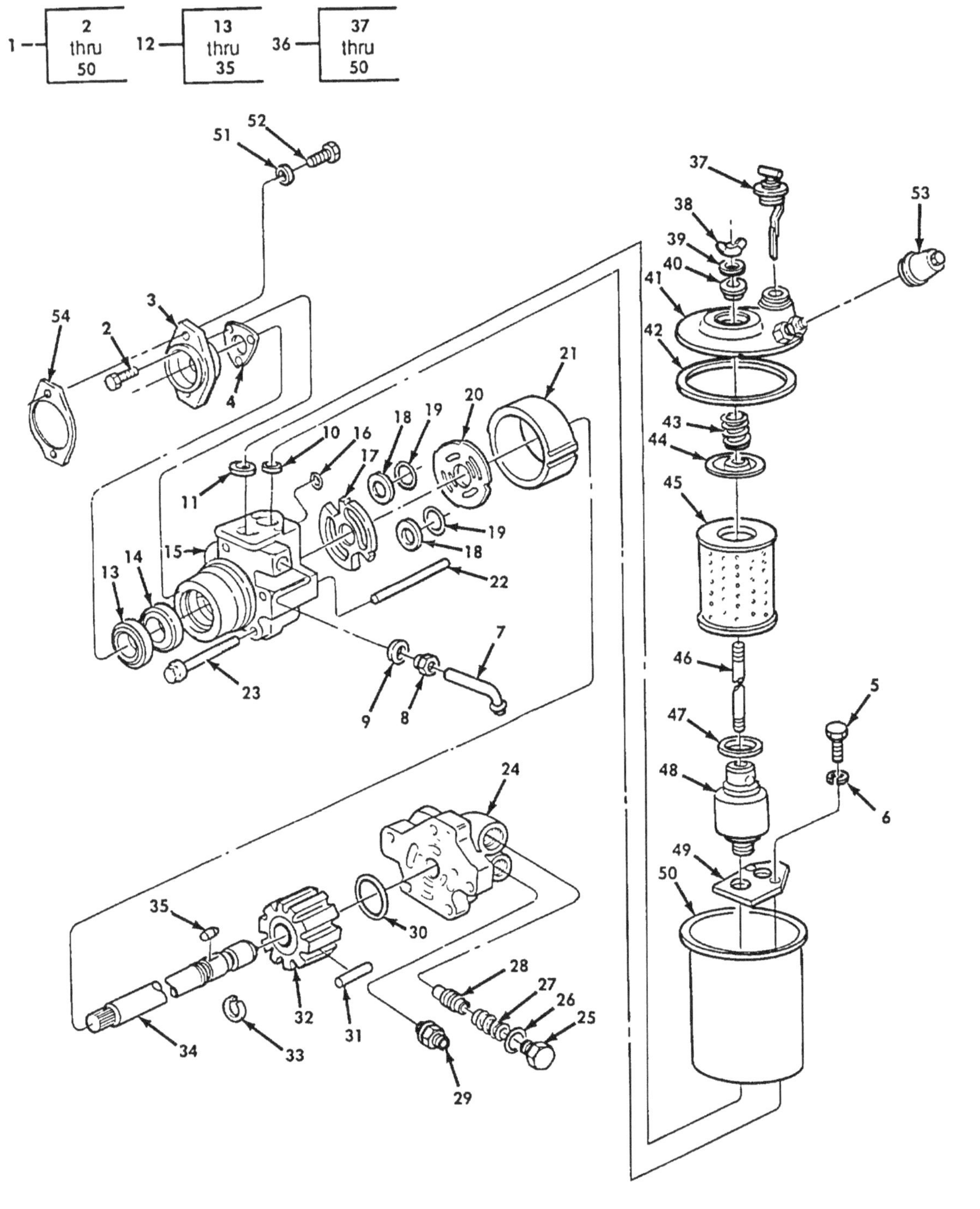

Figure 312. Power Steering Pump Assembly (M939A2).

(1) ITEM NO	(2) SMR CODE	(3) NSN	(4) CAGEC	(5) PART NUMBER	(6) DESCRIPTION AND USABLE ON CODES (UOC)	(7) QTY
					GROUP 1410 HYDRAULIC PUMP OR FLUID MOTOR ASSEMBLY	
					FIG. 312 POWER STEERING PUMP ASSEMBLY (M939A2)	
1	PAOFF	2530012744457	47457	20510093	PUMP ASSEMBLY, POWER .... UOC: ZAA, ZAB, ZAC, ZAD, ZAE, ZAF, ZAG, ZAH, ZAJ, ZAK, ZAL	1
2	PAFZZ	5305012716446	47457	20510093-2Z	.SCREW, CAP, SOCKET HE .... UOC: ZAA, ZAB, ZAC, ZAD, ZAE, ZAF, ZAG, ZAH, ZAJ, ZAK, ZAL	3
3	PAFZZ	5340012715711	47457	20510093-3Z	.FLANGE, SPECIAL................................... UOC: ZAA, ZAB, ZAC, ZAD, ZAE, ZAF, ZAG, ZAH, ZAJ, ZAK, ZAL	1
4	PAFZZ	5330012719407	47457	20510093-4Z	.GASKET.. UOC: ZAA, ZAB, ZAC, ZAD, ZAE, ZAF, ZAG, ZAH, ZAJ, ZAK, ZAL	1
5	PAFZZ	5305012715168	47457	20510093-5Z	.SCREW, CAP, HEXAGON H HEAD, W/............... LOCKWASHER.. UOC: ZAA, ZAB, ZAC, ZAD, ZAE, ZAF, ZAG, ZAH, ZAJ, ZAK, ZAL	2
6	PAFZZ	5310012708406	47457	20510093-6Z	.WASHER, LOCK .... UOC: ZAA, ZAB, ZAC, ZAD, ZAE, ZAF, ZAG, ZAR, ZAJ, ZAK, ZAL	3
7	PAOZZ	4710012720572	47457	20510093-7Z	.TUBE ASSEMBLY, METAL DEGREE .... UOC: ZAA, ZAB, ZAC, ZAD, ZAE, ZAF, ZAG, ZAH, ZAJ, ZAK, ZAL	1
8	XAOZZ		47457	20510093-8Z	.NUT, PLAIN, HEXAGON .... UOC: ZAA, ZAB, ZAC, ZAD, ZAE, ZAF, ZAG, ZAH, ZAJ, ZAK, ZAL	1
9	PAFZZ	5330012719376	47457	20510093-9Z	.O-RING.. UOC: ZAA, ZAB, ZAC, ZAD, ZAE, ZAF, ZAG, ZAH, ZAJ, ZAK, ZAL	1
10	PAOZZ	5330012719408	47457	20510093-10Z	.GASKET PART OF KIT P/N ERS-28001................. UOC: ZAA, ZAB, ZAC, ZAD, ZAE, ZAF, ZAG, ZAH, ZAJ, ZAK, ZAL	2
11	PAOZZ	5330012719409	47457	20510093-11Z	.GASKET PART OF KIT P/N ERS-28001................. UOC: ZAA, ZAB, ZAC, ZAD, ZAE, ZAF, ZAG, ZAH, ZAJ, ZAK, ZAL	2
12	XAFFF		47457	20510093-12Z XA	.PUMP ASSEMBLY, POWER .... UOC: ZAA, ZAB, ZAC, ZAD, ZAE, ZAF, ZAG, ZAH, ZAJ, ZAK, ZAL	1
13	KFFZZ		19954	ER 99658	..SEAL PART OF KIT P/N 20510093-25Z................. UOC: ZAA, ZAB, ZAC, ZAD, ZAE, ZAF, ZAG, ZAH, ZAJ, ZAK, ZAL	1
14	KFFZZ		19954	ER 85349	..SEAL PART OF KIT P/N 20510093-25Z................. UOC: ZAA, ZAB, ZAC, ZAD, ZAE, ZAF, ZAG, ZAH, ZAJ, ZAK, ZAL	1
15	XAFZZ		47457	20510093-14Z	..BODY, SPECIAL..................................... UOC: ZAA, ZAB, ZAC, ZAD, ZAE, ZAF, ZAG, ZAH, ZAJ, ZAK, ZAL	1

(1) ITEM NO	(2) SMR CODE	(3) NSN	(4) CAGEC	(5) PART NUMBER	(6) DESCRIPTION AND USABLE ON CODES (UOC)	(7) QTY
16	KFFZZ		19954	008913-015	..O-RING BY PASS PART OF KIT P/N..................... 20510093-25Z .... UOC: ZAA, ZAB, ZAC, ZAD, ZAE, ZAF, ZAG, ZAH, ZAJ, ZAK, ZAL	1
17	XAFZZ		47457	20510093-15Z	..PLATE, END .... UOC: ZAA, ZAB, ZAC, ZAD, ZAE, ZAF, ZAG, ZAH, ZAJ, ZAK, ZAL	1
18	KFOZZ		19954	19622	..KIDNEY SEAL PART OF KIT P/N........................ 20510093-25Z .... UOC: ZAA, ZAB, ZAC, ZAD, ZAE, ZAF, ZAG, ZAH, ZAJ, ZAK, ZAL	1
19	KFFZZ		19954	ER 82413	..END PLATE PART OF KIT P/N 20510093............... -25Z... UOC: ZAA, ZAB, ZAC, ZAD, ZAE, ZAF, ZAG, ZAH, ZAJ, ZAK, ZAL	1
20	KFFZZ		19954	ER 99562	..PORT PLATE PART OF KIT P/N......................... 20510093-26Z .... UOC: ZAA, ZAB, ZAC, ZAD, ZAE, ZAF, ZAG, ZAH, ZAJ, ZAK, ZAL	1
21	KFFZZ		19954	ER 42505	..CAM PART OF KIT P/N 20510093-26Z.................. UOC: ZAA, ZAB, ZAC, ZAD, ZAE, ZAF, ZAG, ZAH, ZAJ, ZAK, ZAL	1
22	KFFZZ		19954	ER 99635	..LOCATING PIN PART OF KIT P/N....................... 20510093-26Z.. UOC: ZAA, ZAB, ZAC, ZAD, ZAE, ZAF, ZAG, ZAH, ZAJ, ZAK, ZAL	1
23	PAFZZ	5305012744405	47457	20510093-13Z	..SCREW, CAP, HEXAGON H............................. UOC: ZAA, ZAB, ZAC, ZAD, ZAE, ZAF, ZAG, ZAH, ZAJ, ZAK, ZAL	2
24	PAFZZ	5340012715709	47457	20510093-16Z	..COVER, SPECIAL...................................... UOC: ZAA, ZAB, ZAC, ZAD, ZAE, ZAF, ZAG, ZAH, ZAJ, ZAK, ZAL	1
25	PAFZZ	4820002299917	47457	20510093-17Z	..CAP, VALVE.. UOC: ZAA, ZAB, ZAC, ZAD, ZAE, ZAF, ZAG, ZAH, ZAJ, ZAK, ZAL	1
26	KFFZZ		19954	008913-211	..O-RING VALVE CAP PART OF KIT P/N................. 20510093-25Z.. UOC: ZAA, ZAB, ZAC, ZAD, ZAE, ZAF, ZAG, ZAH, ZAJ, ZAK, ZAL	1
27	PAFZZ	5360008111609	81118	ER7614	..SPRING, HELICAL, COMP .... UOC: ZAA, ZAB, ZAC, ZAD, ZAE, ZAF, ZAG, ZAH, ZAJ, ZAK, ZAL	1
28	PAFZZ	4820012716918	19954	ER-27856-150	..VALVE, SAFETY RELIEF .... UOC: ZAA, ZAB, ZAC, ZAD, ZAE, ZAF, ZAG, ZAH, ZAJ, ZAK, ZAL	1
29	XAFZZ		19954	ER-17319 XA	..PLUG .... UOC: ZAA, ZAB, ZAC, ZAD, ZAE, ZAF, ZAG, ZAH, ZAJ, ZAK, ZAL	1
30	KFFZZ		19954	008913-041	..O-RING BOOM COVER PART OF KIT P/N............. 20510093-25Z.. UOC: ZAA, ZAB, ZAC, ZAD, ZAE, ZAF, ZAG, ZAH, ZAJ, ZAK, ZAL	1

(1) ITEM NO	(2) SMR CODE	(3) NSN	(4) CAGEC	(5) PART NUMBER	(6) DESCRIPTION AND USABLE ON CODES (UOC)	(7) QTY
31	KFFZZ		19954	ER 42504	..ROLLERS PART OF KIT P/N 20510093-26Z................ UOC: ZAA, ZAB, ZAC, ZAD, ZAE, ZAF, ZAG, ZAH, ZAJ, ZAK, ZAL	1
32	KFFZZ		19954	ER 42550	..CARRIER PART OF KIT P/N 20510093-26Z............... UOC: ZAA, ZAB, ZAC, ZAD, ZAE, ZAF, ZAG, ZAH, ZAJ, ZAK, ZAL	1
33	PAFZZ	5365012657251	47457	20510093-20Z	..RING, RETAINING................................ UOC: ZAA, ZAB, ZAC, ZAD, ZAE, ZAF, ZAG, ZAH, ZAJ, ZAK, ZAL	2
34	PAFZZ	3040012713649	47457	20510093-21Z	..SHAFT, SHOULDERED............................ UOC: ZAA, ZAB, ZAC, ZAD, ZAE, ZAF, ZAG, ZAH, ZAJ, ZAK, ZAL	1
35	KBFZZ		19954	ER 92739-069	..DRIVE PIN PART OF KIT P/N 20510093-26Z.............. UOC: ZAA, ZAB, ZAC, ZAD, ZAE, ZAF, ZAG, ZAH, ZAJ, ZAK, ZAL	1
36	PAFFF	2530012722911	47457	20510093-22Z	..RESERVOIR, STEERING .... UOC: ZAA, ZAB, ZAC, ZAD, ZAE, ZAF, ZAG, ZAH, ZAJ, ZAK, ZAL	1
37	PAFZZ	5340013198630	47457	20510093-23Z	..CAP, FILLER OPENING .... UOC: ZAA, ZAB, ZAC, ZAD, ZAE, ZAF, ZAG, ZAH, ZAJ, ZAK, ZAL	1
38	XAOZZ		19954	ERS-28103 XA	..WING NUT .... UOC: ZAA, ZAB, ZAC, ZAD, ZAE, ZAF, ZAG, ZAH, ZAJ, ZAK, ZAL	1
39	XAOZZ		19954	ERS-28103 XA	..WASHER .... UOC: ZAA, ZAB, ZAC, ZAD, ZAE, ZAF, ZAG, ZAH, ZAJ, ZAK, ZAL	1
40	KFOZZ		19954	ER 8309	..GASKET PART OF KIT P/N ERS-28001............... PART OF KIT P/N 20511322Z .... UOC: ZAA, ZAB, ZAC, ZAD, ZAE, ZAF, ZAG, ZAH, ZAJ, ZAK, ZAL	1
41	KFOZZ		19954	ER 28105	..LID PART OF KIT P/N 20511322Z .... UOC: ZAA, ZAB, ZAC, ZAD, ZAE, ZAF, ZAG, ZAH, ZAJ, ZAK, ZAL	1
42	KFOZZ		19954	ER 8412	..GASKET PART OF KIT P/N ERS-28001. .... UOC: ZAA, ZAB, ZAC, ZAD, ZAE, ZAF, ZAG, ZAH, ZAJ, ZAK, ZAL	1
43	XAOZZ		19954	ERS-28103 XA	..SPRING...................................... UOC: ZAA, ZAB, ZAC, ZAD, ZAE, ZAF, ZAG, ZAH, ZAJ, ZAK, ZAL	1
44	KFOZZ		19954	ER 99690	..FILTER CAP PART OF KIT P/N ERS-28001............ UOC: ZAA, ZAB, ZAC, ZAD, ZAE, ZAF, ZAG, ZAH, ZAJ, ZAK, ZAL	1
45	KFOZZ		19954	ER 92814	..FILTER ELEMENT PART OF KIT P/N ERS-28001....................... UOC: ZAA, ZAB, ZAC, ZAD, ZAE, ZAF, ZAG, ZAH, ZAJ, ZAK, ZAL	1

(1) ITEM NO	(2) SMR CODE	(3) NSN	(4) CAGEC	(5) PART NUMBER	(6) DESCRIPTION AND USABLE ON CODES (UOC)	(7) QTY
46	XAOZZ		19954	ERS-28103 XA	..STUD .... UOC: ZAA, ZAB, ZAC, ZAD, ZAE, ZAF, ZAG, ZAH, ZAJ, ZAK, ZAL	1
47	KFOZZ		19954	ER 82150	..PACKING PRE. PART OF KIT P/N ERS-............... 28001 .... UOC: ZAA, ZAB, ZAC, ZAD, ZAE, ZAF, ZAG, ZAH, ZAJ, ZAK, ZAL	1
48	XAOZZ		19954	ERS-28103 XA	..VALVE .... UOC: ZAA, ZAB, ZAC, ZAD, ZAE, ZAF, ZAG, ZAH, ZAJ, ZAK, ZAL	1
49	XAFZZ		19954	ERS-28103 XA/A2	..PLATE .... UOC: ZAA, ZAB, ZAC, ZAD, ZAE, ZAF, ZAG, ZAH, ZAJ, ZAK, ZAL	1
50	XAOZZ		19954	ERS-28103 XA	..RESERVOIR .... UOC: ZAA, ZAB, ZAC, ZAD, ZAE, ZAF, ZAG, ZAH, ZAJ, ZAK, ZAL	1
51	PAOZZ	5310002748041	90407	12084P11	WASHER, FLAT .... UOC: ZAA, ZAB, ZAC, ZAD, ZAE, ZAF, ZAG, ZAH, ZAJ, ZAK, ZAL	2
52	PAOZZ	5305000680511	80204	B1821BH038C125N	SCREW, CAP, HEXAGON H .... UOC: ZAA, ZAB, ZAC, ZAD, ZAE, ZAF, ZAG, ZAH, ZAJ, ZAK, ZAL	2
53	PAOZZ	4310010940791	90005	569020-02	BREATHER .... UOC: ZAA, ZAB, ZAC, ZAD, ZAE, ZAG, ZAH, ZAJ, ZAK, ZAL	1
54	PAOZZ	5330010715727	15434	154916	GASKET .... UOC: ZAA, ZAB, ZAC, ZAD, ZAE, ZAF, ZAG, ZAH, ZAJ, ZAK, ZAL	1

END OF FIGURE

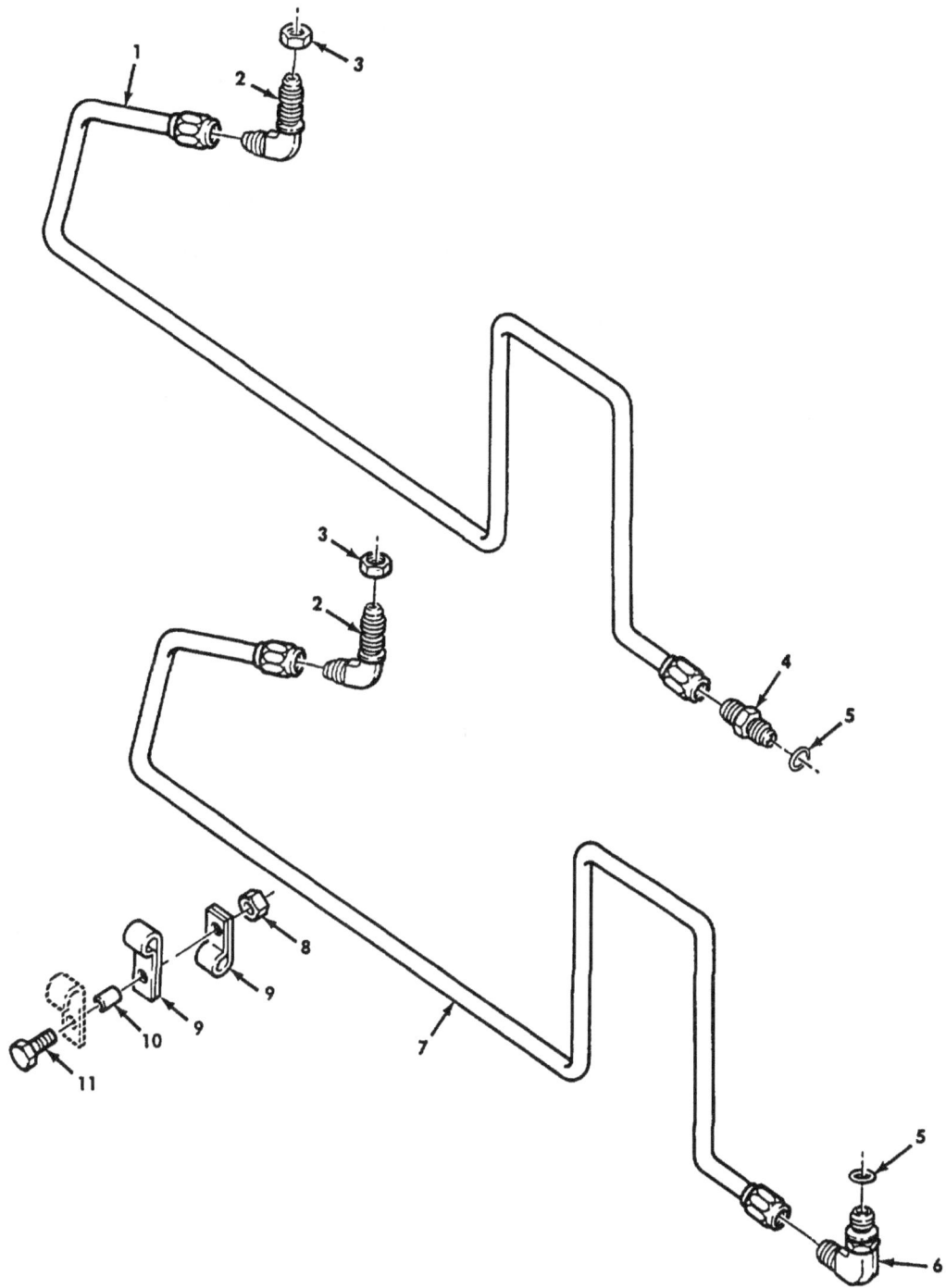

Figure 313. Power Steering Hydraulic Lines and Fittings.

(1) ITEM NO	(2) SMR CODE	(3) NSN	(4) CAGEC	(5) PART NUMBER	(6) DESCRIPTION AND USABLE ON CODES (UOC)	(7) QTY
					GROUP 1411 HOSES, LINES, FITTINGS	
					FIG. 313 POWER STEERING HYDRAULIC LINES AND FITTINGS	
1	PAOZZ	2530011908376	19207	12277387	TUBE AND WORM ASSEM	1
2	PAOZZ	4730008052222	96906	MS24394-6	ELBOW, TUBE	2
3	PAOZZ	5310008392066	96906	MS35691-45	NUT, PLAIN, HEXAGON	2
4	PAOZZ	4730010276590	19207	11648493	ADAPTER, STRAIGHT, TU	1
5	PAOZZ	5330008080794	96906	MS28778-8	O-RING	2
6	PAOZZ	4730001433941	96906	MS51527A6	ELBOW, TUBE TO BOSS	1
7	PAOZZ	4710011908490	19207	12277388	TUBE ASSEMBLY, METAL	1
8	PAOZZ	5310009359022	96906	MS51943-32	NUT, SELF-LOCKING, HE	2
9	PAOZZ	5340008091492	81348	CMDX2-3PT573036	CLAMP, LOOP	4
10	PAOZZ	5365004969706	19207	10883319	SPACER, SLEEVE	2
11	PAOZZ	5305002678956	96906	MS90727-12	SCREW, CAP, HEXAGON H	2

END OF FIGURE

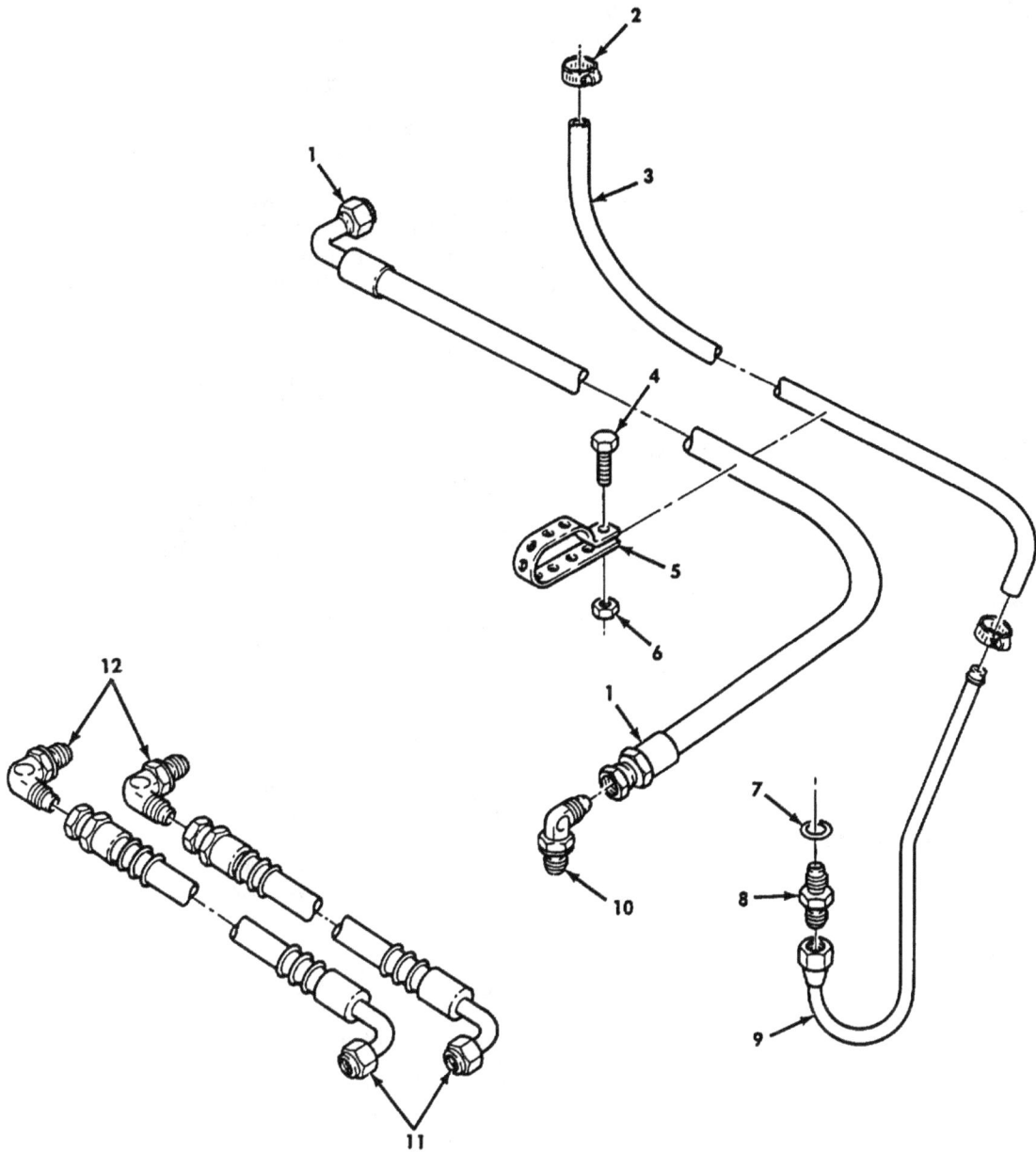

Figure 314. Power Steering Hydraulic Lines and Fittings.

(1) ITEM NO	(2) SMR CODE	(3) NSN	(4) CAGEC	(5) PART NUMBER	(6) DESCRIPTION AND USABLE ON CODES (UOC)	(7) QTY
					GROUP 1411 HOSES, LINES, FITTINGS	
					FIG. 314 POWER STEERING HYDRAULIC LINES AND FITTINGS(CONT'D)	
1	PAOZZ	4720001776162	19207	11664558	HOSE ASSEMBLY, NONME .... UOC: DAA, DAB, DAC, DAD, DAE, DAF, DAG, DAH, DAJ, DAK, DAL, DAW, DAX, V12, V13, V14, V15, V16, V17, V18, V19, V20, V21, V22, V24, V25, V39	1
2	PAOZZ	4730010919370	19207	11608950-2	CLAMP, HOSE.. UOC: DAA, DAB, DAC, DAD, DAE, DAF, DAG, DAH, DAJ, DAK, DAL, DAW, DAX, V12, V13, V14, V15, V16, V17, V18, V19, V20, V21, V22, V24, V25, V39	2
3	MOOZZ		01276	2565-8	BULK HOSE, NONMETALLIC MAKE FROM P/N...... 2565-8, 21 INCHES LONG................................... UOC: DAA, DAB, DAC, DAD, DAE, DAF, DAG, DAH, DAJ, DAK, DAL, DAW, DAX, V12, V13, V14, V15, V16, V17, V18, V19, V20, V21, V22, V24, V25, V39	1
4	PAOZZ	5305009932463	96906	MS35207-279	SCREW, MACHINE .... UOC: DAA, DAB, DAC, DAD, DAE, DAF, DAG, DAH, DAJ, DAK, DAL, DAW, DAX, V12, V13, V14, V15, V16, V17, V18, V19, V20, V21, V22, V24, V25, V39	1
5	PAOZZ	5975003458055	19207	10905840	STRAP, TIEDOWN, ELECT................................. UOC: DAA, DAB, DAC, DAD, DAE, DAF, DAG, DAH, DAJ, DAK, DAL, DAW, DAX, V12, V13, V14, V15, V16, V17, V18 , V19, V20, V21, V22, V24, V25, V39	1
6	PAOZZ	5310007680319	96906	MS51968-2	NUT, PLAIN, HEXAGON .... UOC: DAA, DAB, DAC, DAD, DAE, DAF, DAG, DAH, DAJ, DAK, DAL, DAW, DAX, V12, V13, V14, V15, V16, V17, V18, V19, V20, V21, V22, V24, V25, V39	1
7	PAOZZ	5330008080794	96906	MS28778-8	O-RING .... UOC: DAA, DAB, DAC, DAD, DAE, DAF, DAG, DAH, DAJ, DAK, DAL, DAW, DAX, V12, V13, V14, V15, V16, V17, V18, V19, V20, V21, V22, V24, V25, V39	1
8	PAOZZ	4730000504652	70434	8BCO	ADAPTER, STRAIGHT, TU .... UOC: DAA, DAB, DAC, DAD, DAE, DAF, DAG, DAH, DAJ, DAK, DAL, DAW, DAX, V12, V13, V14, V15, V16, V17, V18, V19, V20, V21, V22, V24, V25, V39	1
9	PAOZZ	471000405196	19207	11648500	TUBE ASSEMBLY, METAL .... UOC: DAA, DAB, DAC, DAD, DAE, DAF, DAG, DAH, DAJ, DAK, DAL, DAW, DAX, V12, V13, V14, V15, V16, V17, V18, V19, V20, V21, V22, V24, V25, V39	1

(1) ITEM NO	(2) SMR CODE	(3) NSN	(4) CAGEC	(5) PART NUMBER	(6) DESCRIPTION AND USABLE ON CODES (UOC)	(7) QTY
10	PAOZZ	4730010260929	19207	11664720	ELBOW, TUBE TO BOSS.. UOC: DAA, DAB, DAC, DAD, DAE, DAF, DAG, DAH, DAJ, DAK, DAL, DAW, DAX, V12, V13, V14, V15, V16, V17, V18, V19, V20, V21, V22, V24, V25, V39	1
11	PAOZZ	4720001776160	19207	11664560	HOSE ASSEMBLY, NONME	2
12	PAOZZ	4730001433941	96906	MS51527A6	ELBOW, TUBE TO BOSS	2

END OF FIGURE

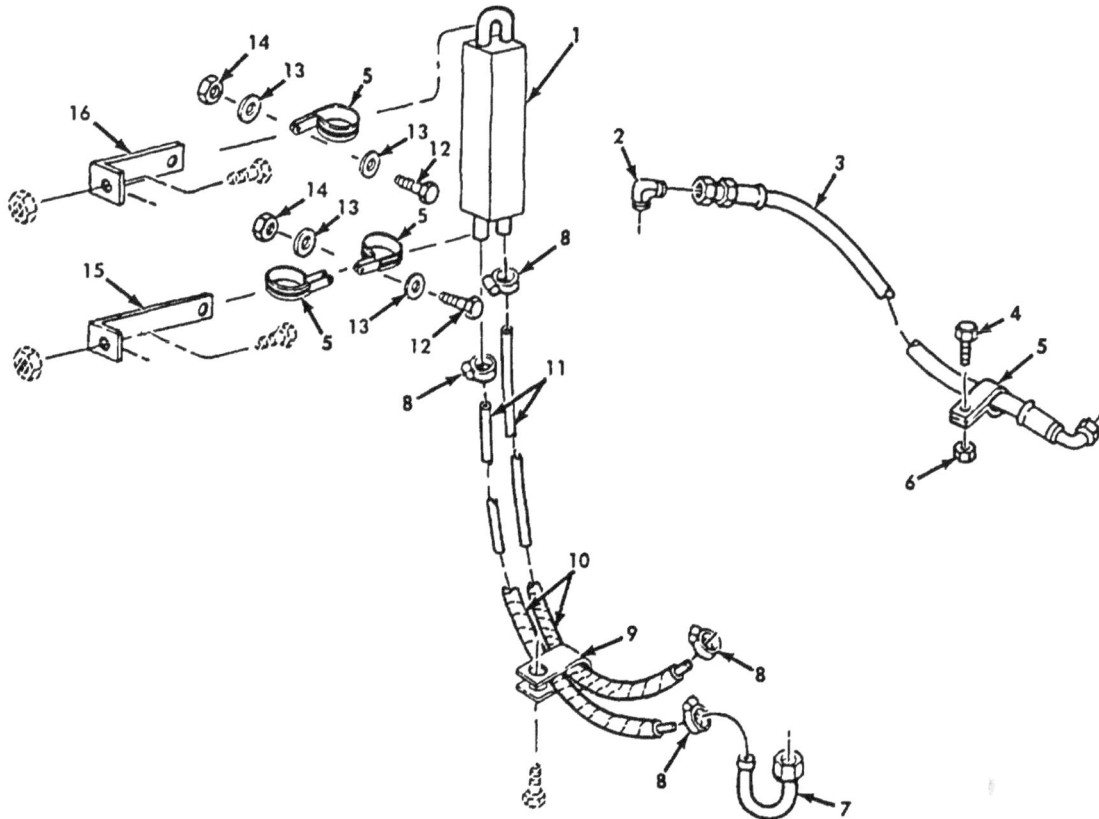

Figure 315. Power Steering Oil Cooler, Hydraulic Lines, and Fittings (M939A2).

(1) ITEM NO	(2) SMR CODE	(3) NSN	(4) CAGEC	(5) PART NUMBER	(6) DESCRIPTION AND USABLE ON CODES (UOC)	(7) QTY
					GROUP 1411 HOSES, LINES, FITTINGS	
					FIG. 315 POWER STEERING OIL COOLER, HYDRAULIC LINES, AND FITTINGS(M939A2)	
1	PAOFF	2520013050457	14205	10217-3736	COOLER, FLUID, TRANSM .... UOC: ZAA, ZAB, ZAC, ZAD, ZAE, ZAF, ZAG, ZAH, ZAJ, ZAK, ZAL	1
2	PAOZZ	4730010260929	19207	11664720	ELBOW, TUBE TO BOSS .... UOC: ZAA, ZAB, ZAC, ZAD, ZAE, ZAF, ZAG, ZAH, ZAJ, ZAK, ZAL	1
3	PAOZZ	4720012793038	47457	20510893	HOSE ASSEMBLY, NONME .... UOC: ZAA, ZAB, ZAC, ZAD, ZAE, ZAF, ZAG, ZAH, ZAJ, ZAK, ZAL	1
4	PAOZZ	5305009125113	96906	MS51096-359	SCREW, CAP, HEXAGON H .... UOC: ZAA, ZAB, ZAC, ZAD, ZAE, ZAF, ZAG, ZAH, ZAJ, ZAK, ZAL	1
5	PAOZZ	534000053B994	96906	MS21333-126	CLAMP, LOOP .... UOC: ZAA, ZAB, ZAC, ZAD, ZAE, ZAF, ZAG, ZAH, ZAJ, ZAK, ZAL	1
5	PAOZZ	5340000573043	96906	MS21333-112	CLAMP, LOOP .... UOC: ZAA, ZAB, ZAC, ZAD, ZAE, ZAF, ZAG, ZAH, ZAJ, ZAK, ZAL	1
6	PAOZZ	5310009591488	96906	MS51922-21	NUT, SELF-LOCKING, HE .... UOC: ZAA, ZAB, ZAC, ZAD, ZAE, ZAF, ZAG, ZAH, ZAJ, ZAK, ZAL	1
7	PAOZZ	4710012793165	7T637	20510895	TUBE ASSEMBLY, METAL .... UOC: ZAA, ZAB, ZAC, ZAD, ZAE, ZAF, ZAG, ZAH, ZAJ, ZAK, ZAL	1
8	PAOZZ	4730010919370	19207	11608950-2	CLAMP, HOSE .... UOC: ZAA, ZAB, ZAC, ZAD, ZAE, ZAF, ZAG, ZAH, ZAJ, ZAK, ZAL	4
9	PAOZZ	5340013142423	99953	DT-1417	CLAMP, LOOP .... UOC: ZAA, ZAB, ZAC, ZAD, ZAE, ZAF, ZAG, ZAH, ZAJ, ZAK, ZAL	1
10	MOOZZ		19207	12302690-6	TUBING, PLASTIC, SPIR MAKE FROM .... TUBING, P/N 2565-8 .... UOC: ZAA, ZAB, ZAC, ZAD, ZAE, ZAF, ZAG, ZAH, ZAJ, ZAK, ZAL	2
11	MOOZZ		47457	20510894 X 44	HOSE, MAKE FROM HOSE, P/N 2565-8, ... 44 INCHES LONG................................. UOC: ZAA, ZAB, ZAC, ZAD, ZAE, ZAF, ZAG, ZAH, ZAJ, ZAK, ZAL	2
12	PAOZZ	5306000680513	60285	6893-2	BOLT, MACHINE .... UOC: ZAA, ZAB, ZAC, ZAD, ZAE, ZAF, ZAG, ZAH, ZAJ, ZAK, ZAL	2
13	PAOZZ	5310008238804	96906	MS27183-9	WASHER, FLAT .... UOC: ZAA, ZAB, ZAC, ZAD, ZAE, ZAF, ZAG, ZAH, ZAJ, ZAK, ZAL	4
14	PAOZZ	5310009359022	96906	MS51943-32	NUT, SELF-LOCKING, HE .... UOC: ZAA, ZAB, ZAC, ZAD, ZAE, ZAF, ZAG, ZAH, ZAJ, ZAK, ZAL	2

(1) ITEM NO	(2) SMR CODE	(3) NSN	(4) CAGEC	(5) PART NUMBER	(6) DESCRIPTION AND USABLE ON CODES (UOC)	(7) QTY
15	PAOZZ	5340013335564	19207	12302693-1	BRACKET, ANGLE .... UOC: ZAL	1
16	PAOZZ	5340011978214	19207	12302693	BRACKET, ANGLE .... UOC: ZAA, ZAB, ZAC, ZAD, ZAE, ZAF, ZAG, ZAH, ZAJ, ZAK	1

END OF FIGURE

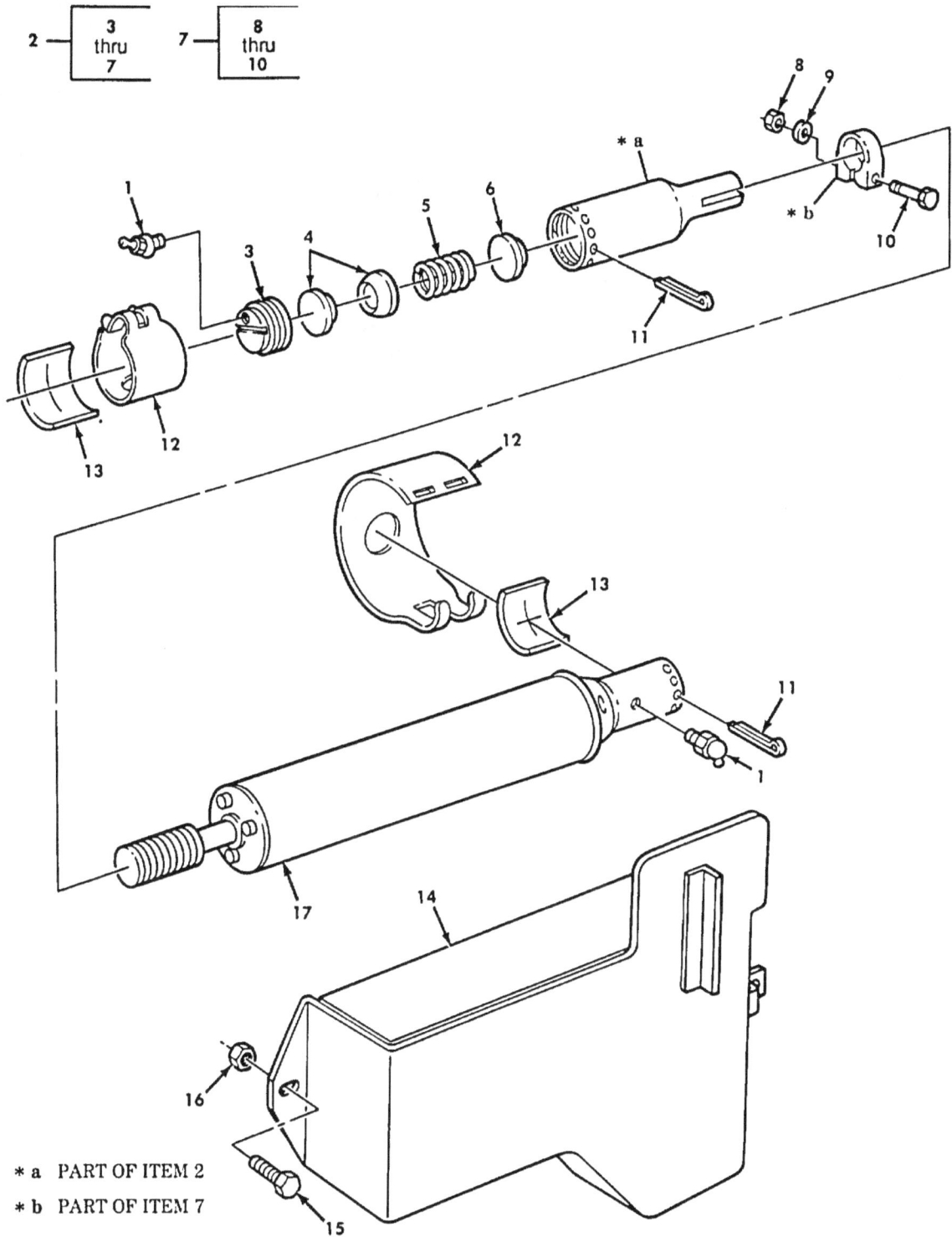

Figure 316. Tie Rod End, Steering Cylinder Assembly, and Booster Shield.

* a   PART OF ITEM 2
* b   PART OF ITEM 7

(1) ITEM NO	(2) SMR CODE	(3) NSN	(4) CAGEC	(5) PART NUMBER	(6) DESCRIPTION AND USABLE ON CODES (UOC)	(7) QTY
					GROUP 1412 HYDRAULIC OR AIR CYLINDERS	
					FIG. 316 TIE ROD END, STEERING CYLINDER ASSEMBLY, AND BOOSTER SHIELD	
1	PAOZZ	4730001720028	96906	MS15003-4	FITTING, LUBRICATION	2
2	PAOOO	2530001344633	19207	11664541	TIE ROD END, STEERING	1
3	PAOZZ	2530011317855	72210	S-13354D	.PLUG, ADJUSTING, BALL	1
4	PFOZZ	4820008484361	72210	S-13227	.SEAT, VALVE	2
5	PFOZZ	5360007036587	19207	7036587	.SPRING, HELICAL, COMP	1
6	XAOZZ		72210	S-13353D	.SEAT, VALVE	1
7	PAOOO	5340011243189	72210	S-15079H	.CLAMP, LOOP	1
8	PAOZZ	5310000103028	96906	MS35690-824	..NUT, PLAIN, HEXAGON	1
9	PAOZZ	5310006374000	77210	S-4281	..WASHER, FLAT...	1
10	PAOZZ	5305011328387	72210	S14955T	..SCREW, CAP, HEXAGON H............................	1
11	PAOZZ	5315002981498	96906	MS24665-362	PIN, COTTER	2
12	PAOZZ	5340004098936	19207	11664287	COVER, ACCESS	2
13	PAOZZ	5330010893073	19207	11664286	FELT, MECHANICAL, PRE	2
14	PBOZZ	2530010911611	19207	12256043	BOOSTER SHIELD ASSE	1
15	PAOZZ	5305002693238	80204	B1821BH038F125N	SCREW, CAP, HEXAGON H	2
16	PAOZZ	5310008140672	96906	MS51943-36	NUT, SELF-LOCKING, HE	2
17	PAOZZ	2530001344619	77640	C36549-1-0900	CYLINDER ASSEMBLY, S	1

END OF FIGURE

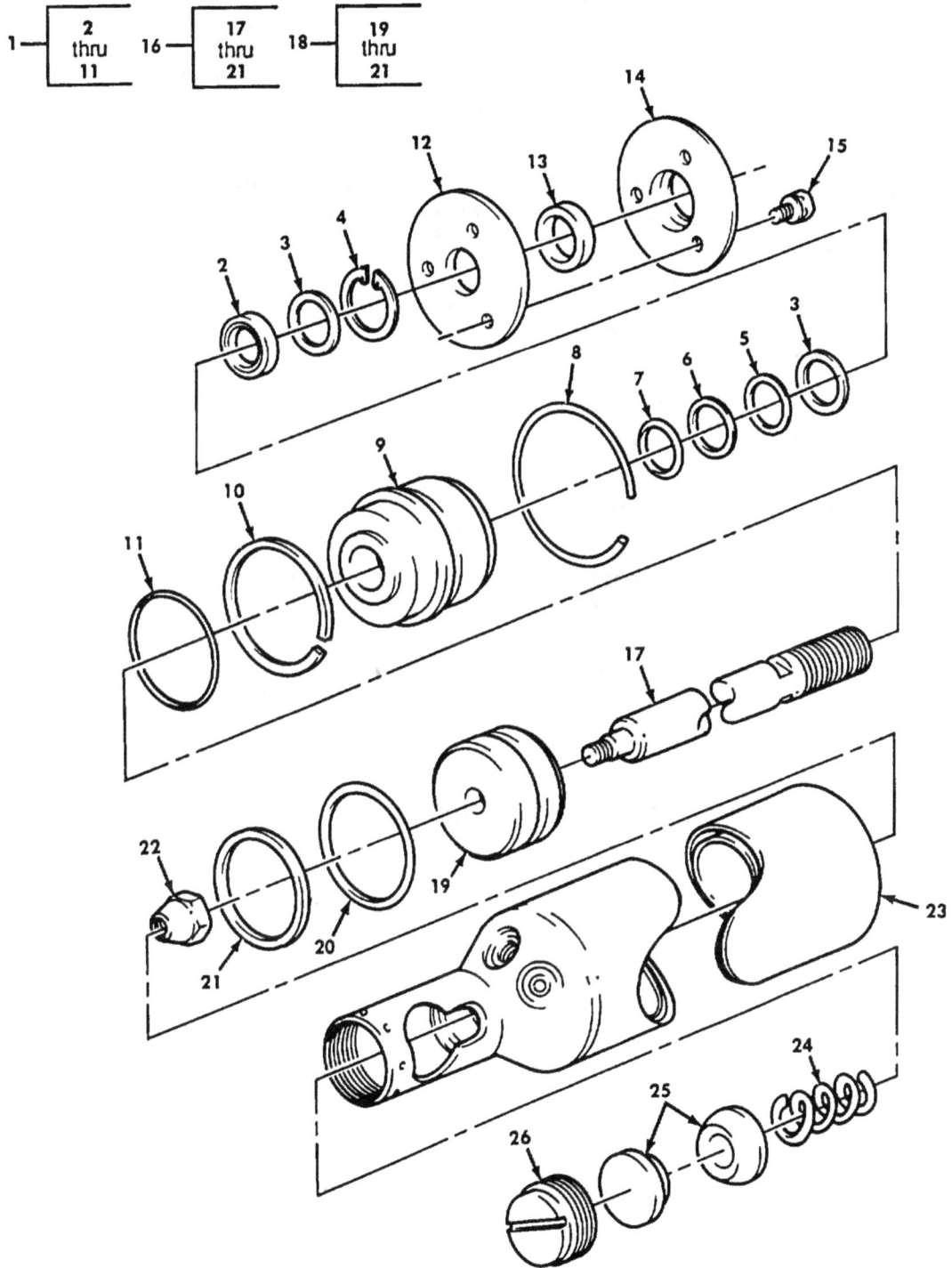

Figure 317. Power Steering Cylinder Assembly.

(1) ITEM NO	(2) SMR CODE	(3) NSN	(4) CAGEC	(5) PART NUMBER	(6) DESCRIPTION AND USABLE ON CODES (UOC)	(7) QTY
					GROUP 1412 HYDRAULIC OR AIR CYLINDERS	
					FIG. 317 POWER STEERING CYLINDER ASSEMBLY	
1	PAFFF	2530011343670	77640	365509-S-1	GLAND ASSEMBLY, POWE	1
2	PAFZZ	5330007339766	77640	032205	.SEAL, PLAIN ENCASED	1
3	PAFZZ	5310008612316	77640	028271	.WASHER, FLAT	2
4	PAFZZ	5325008064105	77640	401233	.RING, RETAINING	1
5	PAFZZ	5365008766862	77640	028272	.SPACER, RING	1
6	PAFZZ	5330007984635	77640	032249	.RING BACKUP	1
7	PAFZZ	5330007339765	77640	032200-17	.O-RING	1
8	PAFZZ	5325008064104	77640	401264	.RING, RETAINING	1
9	XAFZZ		77640	C-365509	.GLAND	1
10	PAFZZ	5330007984637	77640	032272	.GASKET...	1
11	PAFZZ	5330008078993	96906	MS28775-228	.O-RING	1
12	PAFZZ	3040004151373	77640	402230	PLATE, RETAINING, SHA	1
13	PAFZZ	5330011436486	77640	032460	RING, WIPER	1
14	PAFZZ	2590011324654	77640	402341	RETAINER, OIL SEAL	1
15	PAFZZ	5306004195876	77640	020196	BOLT, ASSEMBLED WASH	3
16	PAFFF	2530005451561	77640	C-364007-A-1-1531	ROD, PISTON, VEHICULA	1
17	PAFZZ	2530000413116	73740	C-334101-1531	.ROD, PISTON POWER CY	1
18	PBFFF	3040011342023	77640	C-364007-A-2	.PISTON, LINEAR ACTUA	1
19	XAFZZ		77640	C-364007	..PISTON	1
20	PAFZZ	5330001955757	77640	032361	..O-RING	1
21	PAFZZ	2590011349834	77640	C-366005	..RING, WIPER	1
22	PAFZZ	5310004213991	73740	G-9415992	NUT, POWER CYLINDER	1
23	XAFZZ		77640	C-363833-A-3-1656	HEAD ASSEMBLY	1
24	PAFZZ	5360007956975	77640	401212	SPRING, HELICAL, COMP	1
25	PAFZZ	4820008484361	29510	167961R1	SEAT, VALVE	2
26	PAFZZ	2530011328943	77640	403490	PLUG, ADJUSTING, BALL	1

END OF FIGURE

Figure 318. Frame Assembly, Standard Cargo, and Dropside Cargo.

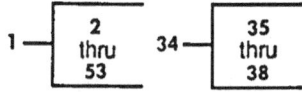

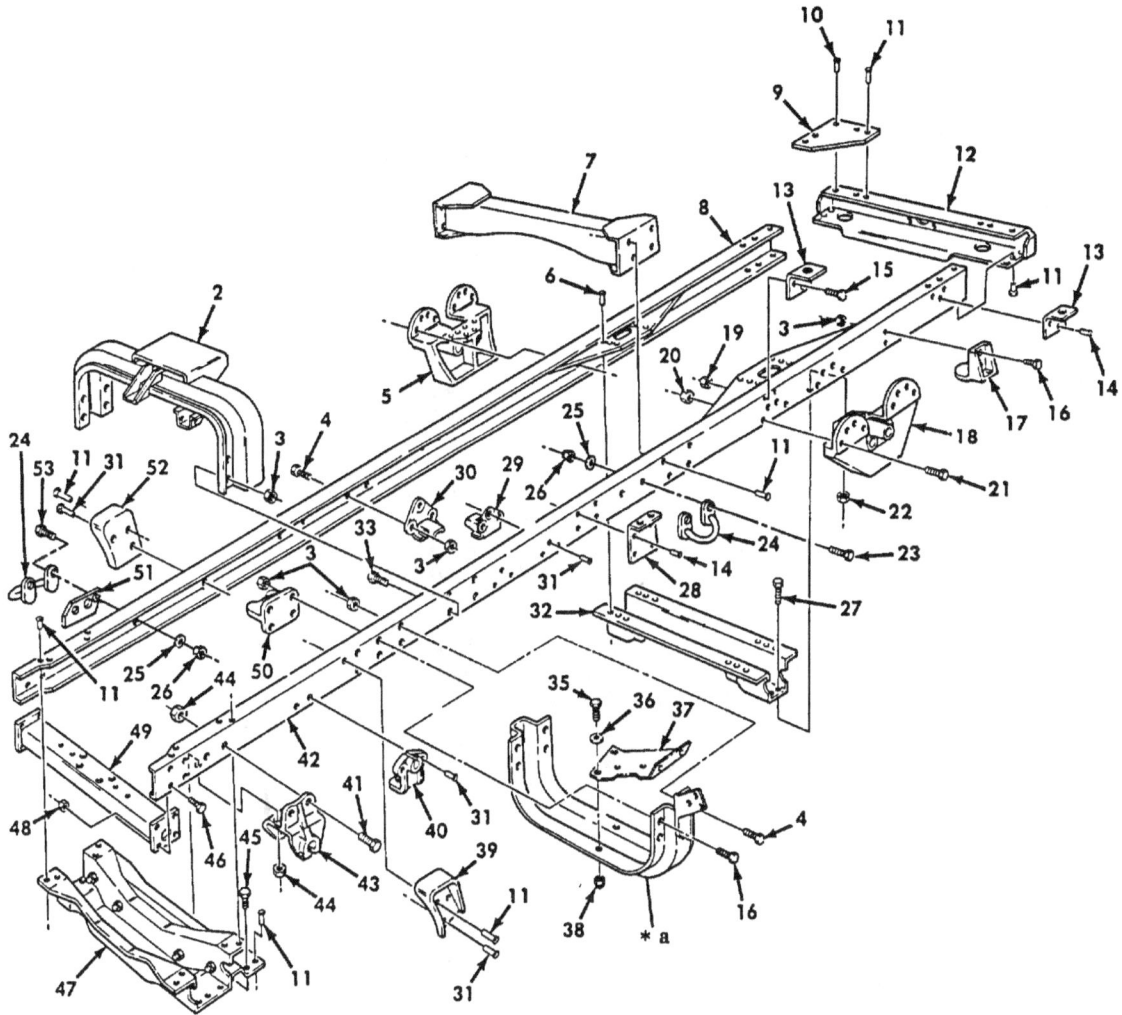

* a PART OF ITEM 34

SECTION II

(1) ITEM NO	(2) SMR CODE	(3) NSN	(4) CAGEC	(5) PART NUMBER	(6) DESCRIPTION AND USABLE ON CODES (UOC)	(7) QTY
					GROUP 15 FRAME, TOWING ATTACHMENTS, AND DRAWBARS 1501 FRAME ASSEMBLY	
					FIG. 318 FRAME ASSEMBLY, STANDARD CARGO, AND DROPSIDE CARGO	
1	XAHHH		19207	12256206-1	FRAME ASSEMBLY .... UOC: V12, V14	1
1	XAHHH		19207	12256206-2	FRAME ASSEMBLY .... UOC: V13, V15	1
1	XAHHH		19207	12256206-3	FRAME ASSEMBLY SEE FIG. 325 FOR A2s .... UOC: DAB, DAX	1
1	XAHHH		19207	12256206-4	FRAME ASSEMBLY SEE FIG. 325 FOR A2s .... UOC: DAA, DAW	1
2	XBFZZ		19207	12256150-1	.CROSSMEMBER .... UOC: DAA, DAB, DAW, DAX, V12, V13, V14, V15, ZAA, ZAB	1
3	PAFZZ	5310004883888	96906	MS51943-40	.NUT, SELF-LOCKING, HE .... UOC: DAA, DAB, DAW, DAX, V12, V13, V14, V15, ZAA, ZAB	16
4	PAFZZ	5305007195221	80204	B1821BH050F150N	.SCREW, CAP, HEXAGON H......................... UOC: DAA, DAB, DAW, DAX, V12, V13, V14, V15, ZAA, ZAB	8
5	PFFZZ	5340004097958	19207	11611648	.BRACKET, MOUNTING .... UOC: DAA, DAB, DAW, DAX, V12, V13, V14, V15, ZAA, ZAB	1
6	PFFZZ	5320011454621	24617	425924	.RIVET, SOLID.. UOC: DAA, DAB, DAW, DAX, V12, V13, V14, V15, ZAA, ZAB	12
7	XBHZZ		19207	12256467	.CROSSMEMBER ASSY .... UOC: DAA, DAB, DAW, DAX, V12, V13, V14, V15, ZAA, ZAB	1
8	XAHZZ		19207	12256108	.RAIL ASSY SEE FIG. 325 FOR A2s..................... UOC: DAA, DAB, DAW, DAX, V12, V13, V14, V15,	1
9	PAHZZ	5340010889153	19207	8758409	.PLATE, MOUNTING.. UOC: DAA, DAB, DAW, DAX, V12, V13, V14, V15, ZAA, ZAB	2
10	PFFZZ	5320000189512	19207	389512	.RIVET, SOLID.. UOC: DAA, DAB, DAW, DAX, V12, V13, V14, V15, ZAA, ZAB	2
11	PFFZZ	5320000104130	21450	104130 XD	.RIVET, SOLID.. UOC: DAA, DAB, DAW, DAX, V12, V13, V14, V15, ZAA, ZAB	38
12	PFHZZ	2510009339577	19207	8331198	.FRAME SECTION, STRUC .... UOC: DAA, DAB, DAW, DAX, V12, V13, V14, V15, ZAA, ZAB	1
13	PFHZZ	5340006649863	19207	7373206	.BRACKET, ANGLE .... UOC: DAA, DAB, DAW, DAX, V12, V13, V14, V15, ZAA, ZAB	2
14	PFFZZ	5320011364495	96906	MS35743-79	.RIVET, SOLID.. UOC: DAA, DAB, DAW, DAX, V12, V13, V14, V15, ZAA, ZAB	6

(1) ITEM NO	(2) SMR CODE	(3) NSN	(4) CAGEC	(5) PART NUMBER	(6) DESCRIPTION AND USABLE ON CODES (UOC)	(7) QTY
15	PAFZZ	5305009146131	96906	MS18153-63	.SCREW, CAP, HEXAGON H .... UOC: DAA, DAB, DAW, DAX, V12, V13, V14, V15, ZAA, ZAB	6
16	PAFZZ	5305007195238	80204	B1821BH050F200N	.SCREW, CAP, HEXAGON H .... UOC: DAA, DAB, DAW, DAX, V12, V13, V14, V15, ZAA, ZAB	8
17	PFHZZ	5340010899130	19207	11682319	.BRACKET, MOUNTING .... UOC: DAA, DAB, DAW, DAX, V12, V13, V14, V15, ZAA, ZAB	4
18	PFHZZ	5340010899128	19207	11611648-1	.BRACKET, BOGIE .... UOC: DAA, DAB, DAW, DAX, V12, V13, V14, V15, ZAA, ZAB	1
19	PAFZZ	5310008140672	96906	MS51943-36	.NUT, SELF-LOCKING, HE .... UOC: DAA, DAB, DAW, DAX, V12, V13, V14, V15, ZAA, ZAB	6
20	PAFZZ	5310009353569	96906	MS51943-46	.NUT, SELF-LOCKING, HE .... UOC: DAA, DAB, DAW, DAX, V12, V13, V14, V15, ZAA, ZAB	16
21	PAFZZ	5305009162345	80204	B1821BH075F200N	.SCREW, CAP, HEXAGON H .... UOC: DAA, DAB, DAW, DAX, V12, V13, V14, V15, ZAA, ZAB	16
22	PAFZZ	5310002416664	96906	MS51943-44	.NUT, SELF-LOCKING, HE .... UOC: DAA, DAB, DAW, DAX, V12, V13, V14, V15, ZAA, ZAB	12
23	PAFZZ	5305007195239	96906	MS90727-116	.SCREW, CAP, HEXAGON H............................ UOC: DAA, DAB, DAW, DAX, V12, V13, V14, V15, ZAA, ZAB	16
24	PAFZZ	5340012919212	19207	12301265	.LOOP, STRAP .... UOC: DAA, DAB, DAW, DAX, V12, V13, V14, V15, ZAA, ZAB	10
25	PAFZZ	5310001808544	19207	10872159	.WASHER, FLAT .... UOC: DAA, DAB, DAW, DAX, V12, V13, V14, V15, ZAA, ZAB	20
26	PAFZZ	5310008775795	96906	MS21044-N8	.NUT, SELF-LOCKING, HE .... UOC: DAA, DAB, DAW, DAX, V12, V13, V14, V15, ZAA, ZAB	2
27	PAFZZ	5305007262550	80204	B1821BH063F175N	.SCREW, CAP, HEXAGON H .... UOC: DAA, DAB, DAW, DAX, V12, V13, V14, V15, ZAA, ZAB	12
28	PFHZZ	5340011307938	19207	12276977	.BRACKET, ANGLE .... UOC: DAA, DAB, DAW, DAX, V12, V13, V14, V15, ZAA, ZAB	2
29	PBHZZ	2510010889157	19207	12256028	.FRAME SECTION, STRUC .... UOC: DAA, DAB, DAW, DAX, V12, V13, V14, V15, ZAA, ZAB	1
30	PBHZZ	2510010889156	19207	12256036	.FRAME SECTION, STRUC .... UOC: DAA, DAB, DAW, DAX, V12, V13, V14, V15, ZAA, ZAB	1
31	PFFZZ	5320011744726	24617	189510	.RIVET, SOLID .... UOC: DAA, DAB, DAW, DAX, V12, V13, V14, V15, ZAA, ZAB	16
32	XBHZZ		19207	12255988	.CROSSMEMBER............................ UOC: DAA, DAB, DAW, DAX, V12, V13, V14, V15,	8

(1) ITEM NO	(2) SMR CODE	(3) NSN	(4) CAGEC	(5) PART NUMBER	(6) DESCRIPTION AND USABLE ON CODES (UOC)	(7) QTY
					ZAA, ZAB	
33	PAFZZ	5305007250154	96906	MS90727-112	.SCREW, CAP, HEXAGON H	8
					UOC: DAA, DAB, DAW, DAX, V12, V13, V14, V15, ZAA, ZAB	
34	PFHZZ	2510010885914	19207	12256033	.FRAME SECTION, STRUC SEE FIG. 325.............	1
					FOR A2s	
					UOC: DAA, DAB, DAW, DAX, V12, V13, V14, V15	
35	PAFZZ	5305007250154	96906	MS90727-112	..SCREW, CAP, HEXAGON H............................	4
					UOC: DAA, DAB, DAW, DAX, V12, V13, V14, V15	
36	PAFZZ	5310008095998	96906	MS27183-18	..WASHER, FLAT	4
					UOC: DAA, DAB, DAW, DAX, V12, V13, V14, V15	
37	PFFZZ	5340011704940	19207	12302672	..BRACKET, MOUNTING	1
					UOC: DAA, DAB, DAW, DAX, V12, V13, V14, V15	
38	PAFZZ	5310004883888	96906	MS51943-40	..NUT, SELF-LOCKING, HE	4
					UOC: DAA, DAB, DAW, DAX, V12, V13, V14, V15	
39	PFHZZ	5342011647597	19207	12277352-1	.BRACKET, SPECIAL FRONT CAB, L.H	1
					UOC: DAA, DAB, DAW, DAX, V12, V13, V14, V15, ZAA, ZAB	
40	PFHZZ	3040010889161	19207	8758403	.BRACKET, EYE, NONROTA	2
					UOC: DAA, DAB, DAW, DAX, V12, V13, V14, V15, ZAA, ZAB	
41	PAOZZ	5305007250140	96906	MS90727-139	.SCREW, CAP, HEXAGON H	14
					UOC: DAA, DAB, DAW, DAX, V12, V13, V14, V15, ZAA, ZAB	
42	XAHZZ		19207	12256104	.RAIL ASSEMBLY SEE FIG. 325 FOR A2s	1
					UOC: DAA, DAB, DAW, DAX, V12, V13, V14, V15, ZAA, ZAB	
43	PFOZZ	2510007409337	19207	7409337	.SHACKLE, LEAF SPRING	4
					UOC: DAA, DAB, DAW, DAX, V12, V13, V14, V15, ZAA, ZAB	
44	PAOZZ	5310002416661	96906	MS51943-42	.NUT, SELF-LOCKING, HE	16
					UOC: DAA, DAB, DAW, DAX, V12, V13, V14, V15, ZAA, ZAB	
45	PAOZZ	5305007250145	80204	B1821BH056F200N	.SCREW, CAP, HEXAGON H	2
					UOC: DAA, DAB, DAW, DAX, V12, V13, V14, V15, ZAA, ZAB	
46	PAFZZ	5305007195235	80204	B1821BH050F175N	.SCREW, CAP, HEXAGON H	4
					UOC: DAA, DAW, V13, V15, ZAA	
47	XBHZZ		19207	11664271-1	.CROSSMEMBER ASSY SEE FIG. 325 FOR	1
					A2s	
					UOC: DAA, DAW, V13, V15, ZAA	
48	PAFZZ	5310004883888	96906	MS51943-40	.NUT, SELF-LOCKING, HE	4
					UOC: DAA, DAW, V13, V15, ZAA	
49	PGFZZ	2510010907639	19207	12302674-1	.FRAME SECTION, STRUC	1
					UOC: DAA, DAB, DAW, DAX, V12, V13, V14, V15, ZAA, ZAB	
50	PFHZZ	2510007409508	19207	7409508	.BRACKET SUPPORT	2
					UOC: DAA, DAB, DAW, DAX, V12, V13, V14, V15, ZAA, ZAB	
51	PAFZZ	5365013049529	19207	12356901	.SPACER, PLATE	2
					UOC: DAA, DAB, DAW, DAX, V12, V13, V14, V15, ZAA, ZAB	
52	PFHZZ	5342011647598	19207	12277352-2	.BRACKET, SPECIAL FRONT CAB, R.H	2
					UOC: DAA, DAB, DAW, DAX, V12, V13, V14, V15, ZAA, ZAB	

(1) ITEM NO	(2) SMR CODE	(3) NSN	(4) CAGEC	(5) PART NUMBER	(6) DESCRIPTION AND USABLE ON CODES (UOC)	(7) QTY
53	PAFZZ	5305007195240	99696	375N300-001-21	.SCREW, CAP, HEXAGON H .... UOC: DAA, DAB, DAW, DAX, V12, V13, V14, V15, ZAA, ZAB	4

END OF FIGURE

Figure 319. Frame Assembly, Dump.

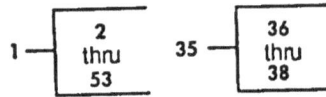

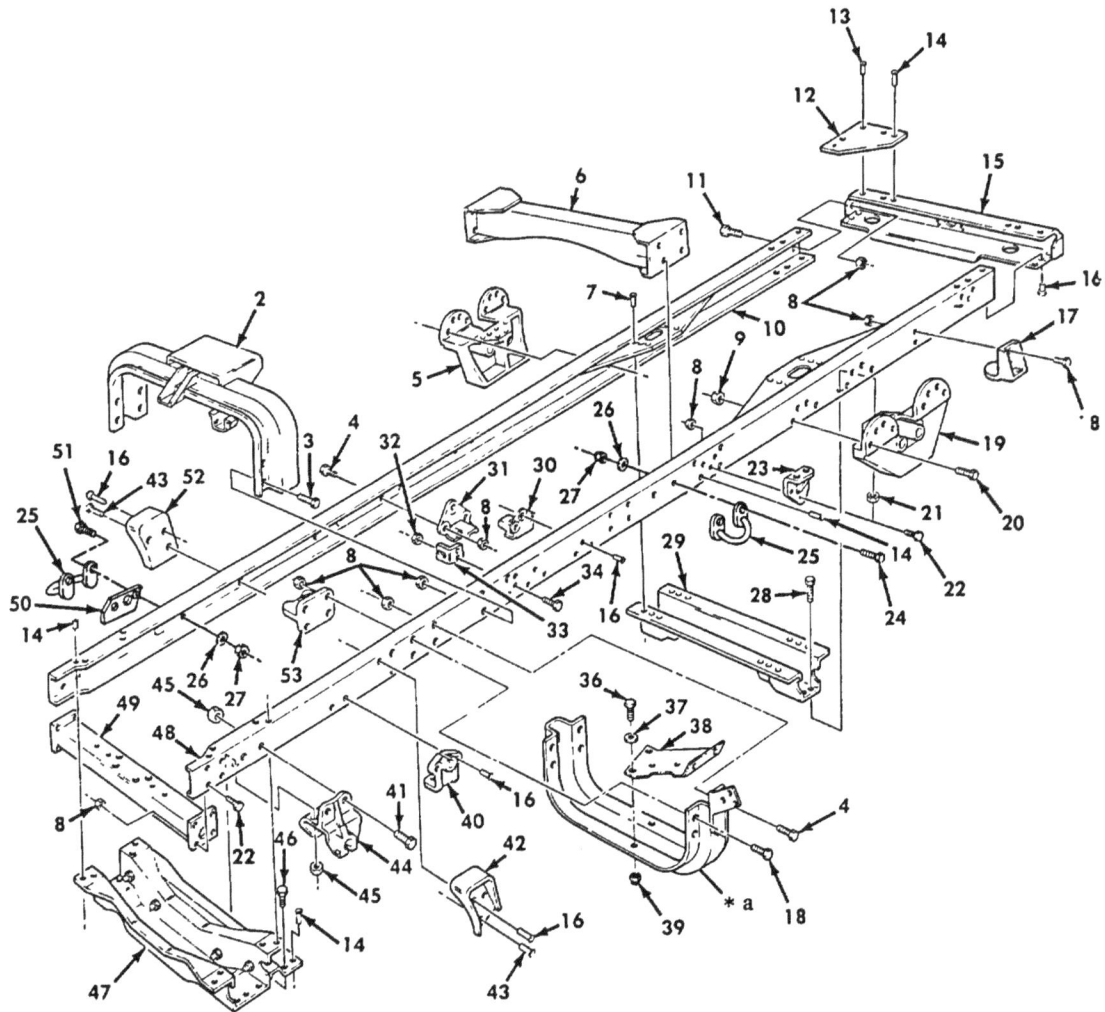

* a PART OF ITEM 35

(1) ITEM NO	(2) SMR CODE	(3) NSN	(4) CAGEC	(5) PART NUMBER	(6) DESCRIPTION AND USABLE ON CODES (UOC)	(7) QTY
					GROUP 1501 FRAME ASSEMBLY	
					FIG. 319 FRAME ASSEMBLY, DUMP	
1	XAHHH		19207	12256413-1	FRAME ASSEMBLY SEE FIG. 325 FOR A2s .... UOC: V19, DAF	1
1	XAHHH		19207	12256413-2	FRAME ASSEMBLY SEE FIG. 325 FOR A2s .... UOC: V20, DAE	1
2	XBFZZ		19207	12256150-1	.CROSSMEMBER .... UOC: DAE, DAF, V19, V20, ZAE, ZAF	1
3	PAFZZ	5305007250154	96906	MS90727-112	.SCREW, CAP, HEXAGON H .... UOC: DAE, DAF, V19, V20, ZAE, ZAF	8
4	PAFZZ	5305007195221	80204	B1821BH050F150N	.SCREW, CAP, HEXAGON H .... UOC: DAE, DAF, V19, V20, ZAE, ZAF	7
5	PFFZZ	5340004097958	19207	11611648	.BRACKET, MOUNTING................................. UOC: DAE, DAF, V19, V20, ZAE, ZAF	1
6	XBHZZ		19207	11611644	.CROSSMEMBER, FRAME .... UOC: DAE, DAF, V19, V20, ZAE, ZAF	1
7	PFFZZ	5320011454621	24617	425924	.RIVET, SOLID .... UOC: DAE, DAF, V19, V20, ZAE, ZAF	12
8	PAFZZ	5310004883888	96906	MS51943-40	.NUT, SELF-LOCKING, HE .... UOC: DAE, DAF, V19, V20, ZAE, ZAF	30
9	PAFZZ	5310009353569	96906	MS51943-46	.NUT, SELF-LOCKING .... UOC: DAE, DAF, V19, V20, ZAE, ZAF	16
10	XAHZZ		19207	12256358	.RAIL ASSEMBLY SEE FIG. 325 FOR A2s .... UOC: DAE, DAF, V19, V20	1
11	PAFZZ	5305007195219	96906	MS90727-111	.SCREW, CAP, HEXAGON H .... UOC: DAE, DAF, V19, V20, ZAE, ZAF	4
12	PFHZZ	2510010889153	19207	8755409	.PLATE, SUPPORT .... UOC: DAE, DAF, V19, V20, ZAE, ZAF	2
13	PFFZZ	5320011453188	24617	451956	.RIVET, SOLID .... UOC: DAE, DAF, V19, V20, ZAE, ZAF	2
14	PFFZZ	5320011453185	24617	142391	.RIVET, SOLID .... UOC: DAE, DAF, V19, V20, ZAE, ZAF	30
15	XBHZZ		19207	8331198	.CROSSMEMBER, FRAME .... UOC: DAE, DAF, V19, V20, ZAE, ZAF	1
16	PFFZZ	5320011453184	24617	142392	.RIVET, SOLID .... UOC: DAE, DAF, V19, V20, ZAE, ZAF	21
17	PFFZZ	5340010899130	19207	11682319	.BRACKET, MOUNTING .... UOC: DAE, DAF, V19, V20, ZAE, ZAF	4
18	PAFZZ	5305007195238	80204	B1821BH050F200N	.SCREW, CAP, HEXAGON H .... UOC: DAE, DAF, V19, V20, ZAE, ZAF	4
19	PFFZZ	5340010899128	19207	11611648-1	.BRACKET, BOGIE .... UOC: DAE, DAF, V19, V20, ZAE, ZAF	1
20	PAFZZ	5305009162345	80204	B1821BH075F200N	.SCREW, CAP, HEXAGON H .... UOC: DAE, DAF, V19, V20, ZAE, ZAF	16
21	PAFZZ	5310002416664	96906	MS51943-44	.NUT, SELF-LOCKING, HE .... UOC: DAE, DAF, V19, V20, ZAE, ZAF	12
22	PAFZZ	5305007195235	80204	B1821BH050F175N	.SCREW, CAP, HEXAGON H .... UOC: DAE, DAF, V19, V20, ZAE, ZAF	5
23	PFFZZ	5340004094008	19207	7351285	.BRACKET, ANGLE .... UOC: DAE, DAF, V19, V20, ZAE, ZAF	2
24	PAFZZ	5305007195239	96906	MS90727-116	.SCREW, CAP, HEXAGON .... UOC: DAE, DAF, V19, V20, ZAE, ZAF	16

(1) ITEM NO	(2) SMR CODE	(3) NSN	(4) CAGEC	(5) PART NUMBER	(6) DESCRIPTION AND USABLE ON CODES (UOC)	(7) QTY
25	PAFZZ	5340012919212	19207	12301265	.LOOP, STRAP .... UOC: DAE, DAF, V19, V20, ZAE, ZAF	10
26	PAFZZ	5310001808544	19207	10872159	.WASHER, FLAT .... UOC: DAE, DAF, V19, V20, ZAE, ZAF	20
27	PAFZZ	5310008775795	96906	MS21044-N8	.NUT, SELF-LOCKING, HE .... UOC: DAE, DAF, V19, V20, ZAE, ZAF	2
28	PAFZZ	5305007262550	80204	B1821BH063F175N	.SCREW, CAP, HEXAGON H .... UOC: DAE, DAF, V19, V20, ZAE, ZAF	12
29	XDHZZ		19207	12255988	.CROSSMEMBER .... UOC: DAE, DAF, V19, V20, ZAE, ZAF	1
30	PBHZZ	2510010889157	19207	12256028	.FRAME SECTION, STRUC .... UOC: DAE, DAF, V19, V20, ZAE, ZAF	2
31	PBFZZ	2510010889156	19207	12256036	.FRAME SECTION, STRUC .... UOC: DAE, DAF, V19, V20, ZAE, ZAF	1
32	PAFZZ	5310008140672	96906	MS51943-36	.NUT, SELF-LOCKING, HE .... UOC: DAE, DAF, V19, V20, ZAE, ZAF	2
33	PFFZZ	5340010893068	19207	12256670	.BRACKET, ANGLE .... UOC: DAE, DAF, V19, V20, ZAE, ZAF	1
34	PAFZZ	5305002693238	80204	B1821BH038F125N	.SCREW, CAP, HEXAGON H .... UOC: DAE, DAF, V19, V20, ZAE, ZAF	2
35	PFHZZ	2510010885914	19207	12256033	.FRAME SECTION, STRUC SEE FIG. 325............. FOR A2s .... UOC: DAE, DAF, V19, V20	1
36	PAFZZ	5305007250154	96906	MS90727-112	..SCREW, CAP, HEXAGON H.................... UOC: DAE, DAF, V19, V20	4
37	PAFZZ	5310008095998	96906	MS27183-18	..WASHER, FLAT.................... UOC: DAE, DAF, V19, V20	4
38	PFFZZ	5340011704940	19207	12302672	..BRACKET, MOUNTING .... UOC: DAE, DAF, V19, V20	1
39	PAFZZ	5310004883888	96906	MS51943-40	.NUT, SELF-LOCKING, HE .... UOC: V20, DAE, DAF, V19, ZAE, ZAF	4
40	PFHZZ	3040010889161	19207	8758403	.BRACKET, EYE, NONROTA .... UOC: DAE, DAF, V19, V20, ZAE, ZAF	2
41	PAOZZ	5305007250140	96906	MS90727-139	.SCREW, CAP, HEXAGON H .... UOC: DAE, DAF, V19, V20, ZAE, ZAF	14
42	PFHZZ	5340011647597	19207	12277352-1	.BRACKET, SPECIAL.................... UOC: DAE, DAF, V19, V20, ZAE, ZAF	1
43	PFFZZ	5320011507745	24617	111274	.RIVET, SOLID.................... UOC: DAE, DAF, V19, V20, ZAE, ZAF	4
44	PFOZZ	2510007409337	19207	7409337	.SHACKLE, LEAF SPRING .... UOC: DAE, DAF, V19, V20, ZAE, ZAF	4
45	PAOZZ	5310002416661	96906	MS51943-42	.NUT, SELF-LOCKING, HE .... UOC: DAE, DAF, V19, V20, ZAE, ZAF	16
46	PAFZZ	5305007250145	80204	B1821BH056F200N	.SCREW, CAP, HEXAGON H.................... UOC: DAE, DAF, V19, V20, ZAE, ZAF	2
47	XBHZZ		19207	11664271-1	.CROSSMEMBER ASSEMBL SEE FIG. 325............. FOR A2s .... UOC: DAE, DAF, V19, V20	1
48	XAHZZ		19207	12256344	.RAIL ASSEMBLY SEE FIG. 325 FOR A2s .... UOC: DAE, DAF, V19, V20	1
49	PGFZZ	2510010907639	19207	12302674-1	.FRAME SECTION, STRUC .... UOC: DAE, V20, ZAE	1
50	PAFZZ	5365013049529	19207	12356901	.SPACER, PLATE .... UOC: DAE, DAF, V19, V20, ZAE, ZAF	2

(1) ITEM NO	(2) SMR CODE	(3) NSN	(4) CAGEC	(5) PART NUMBER	(6) DESCRIPTION AND USABLE ON CODES (UOC)	(7) QTY
51	PAFZZ	5365007195240	99696	375N300-001-21	.SCREW, CAP, HEXAGON H .... UOC:DAE, DAF, V19, V20, ZAE, ZAF	4
52	PFHZZ	5340011647598	19207	12277352-2	.BRACKET, SPECIAL .... UOC:DAE, DAF, V19, V20, ZAE, ZAF	1
53	PFHZZ	2510007409508	19207	7409508	.BRACKET SUPPORT.. UOC:DAE, DAF, V19, V20, ZAE, ZAF	2

END OF FIGURE

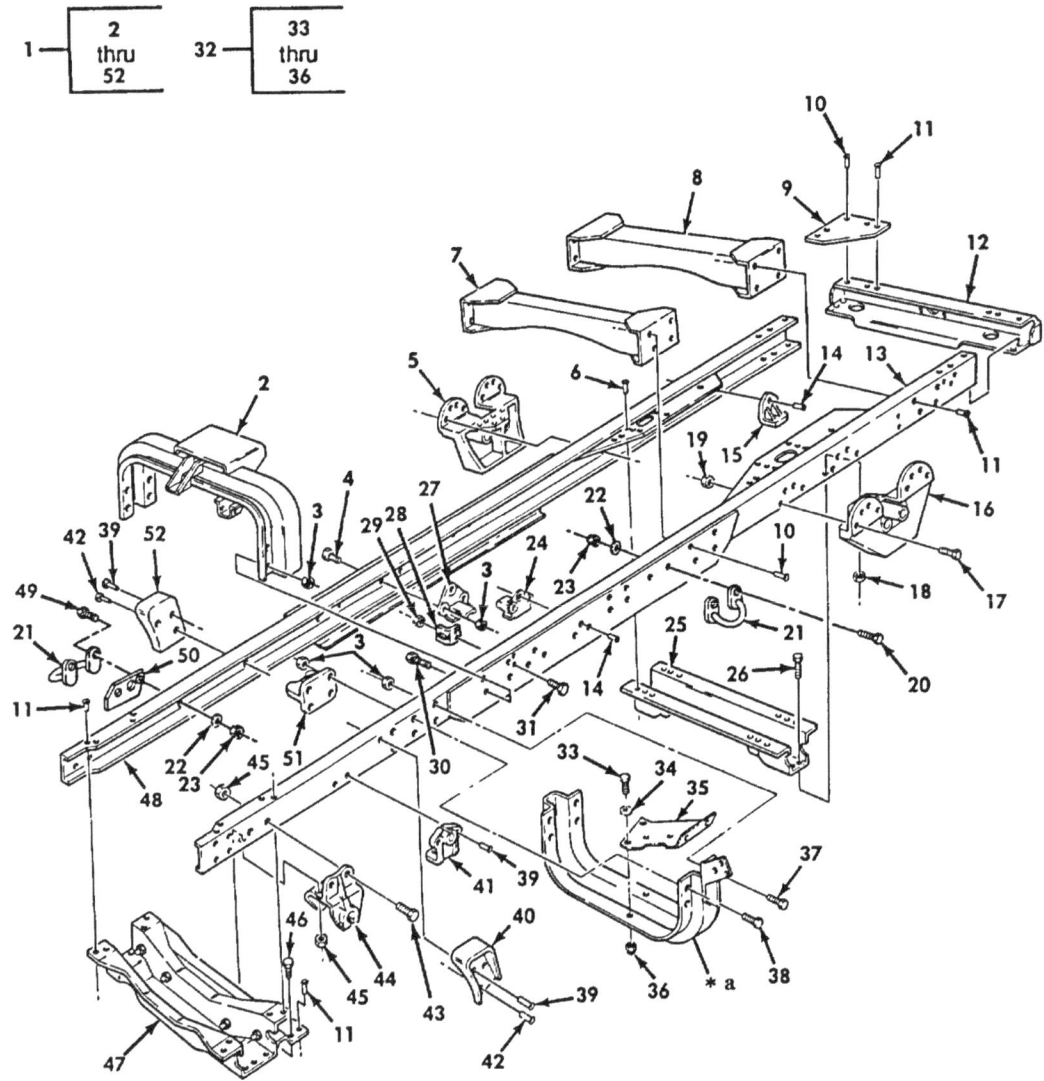

* a PART OF ITEM 32

Figure 320. Frame Assembly, Wrecker.

(1) ITEM NO	(2) SMR CODE	(3) NSN	(4) CAGEC	(5) PART NUMBER	(6) DESCRIPTION AND USABLE ON CODES (UOC)	(7) QTY
					GROUP 1501 FRAME ASSEMBLY	
					FIG. 320 FRAME ASSEMBLY, WRECKER	
1	XAHHH		19207	12256414	FRAME ASSEMBLY SEE FIG. 325 FOR A2s .... UOC:DAL, V18, ZAL	1
2	XBHZZ		19207	12256150-2	.CROSSMEMB ER .... UOC:DAL, V18, ZAL	1
3	PAFZZ	5310004883888	96906	MS51943-40	.NUT, SELF-LOCKING, HE UOC:DAL, V18, ZAL	19
4	PAFZZ	5305007195235	80204	B1821BH050F175N	.SCREW, CAP, HEXAGON H.................... UOC:DAL, V18, ZAL	1
5	PFFZZ	5340004097958	19207	11611648	.BRACKET, MOUN TING. .... UOC:DAL, V18, ZAL	1
6	PFFZZ	5320011454621	24617	425924	.RIVET, SOLID............................. UOC:DAL, V18, ZAL	12
7	XBHZZ		19207	11648420	.CROSSMEMBER ASSY .... UOC:DAL, V18B, ZAL	1
8	XBHZZ		19207	12256727	.CROSSMEMBER ASSY .... UOC:DAL, V18, ZAL	1
9	PAHZZ	5340010889153	19207	8758409	.PLATE, MOUNTING.......................... UOC:DAL, V18, ZAL	2
10	PFFZZ	5320011453188	24617	451956	.RIVET, SOLID .... UOC:DAL, V18, ZAL	10
11	PFFZZ	5320011453185	24617	142391	.RIVET, SOLID .... UOC:DAL, V18, ZAL	30
12	XBHZZ		19207	8331199	.CROSSMEMBER, REAR....................... UOC:DAL, V18, ZAL	1
13	XAHZZ		19207	12256341	.RAIL ASSY, SEE FIG. 325 FOR A2s .... UOC:DAL, V18	1
14	PFFZZ	5320011453186	24617	425923	.RIVET, SOLID .... UOC:DAL, V18, ZAL	15
15	PFHZZ	2510004097960	19207	11608818	.BRACKET, BUMPER .... UOC:DAL, V18, ZAL	4
16	PFFZZ	5340010899128	19207	11611648-1	.BRACKET, BOG.............................. UOC:DAL, V18, ZAL	1
17	PAFZZ	5305009162345	80204	B1821BH075F200N	.SCREW, CAP, HEXAGON H..................... UOC:DAL, V18, ZAL	16
18	PAFZZ	5310002416664	96906	MS51943-44	.NUT, SELF-LOCKING, HE .... UOC:DAL, V18, ZAL	12
19	PAFZZ	5310009353569	96906	MS51943-46	.NUT, SELF-LOCKING, HE .... UOC:DAL, V18, ZAL	16
20	PAFZZ	5305007195239	96906	MS90727-116	.SCREW, CAP, HEXAGON H.................... UOC:DAL, V18, ZAL	16
21	PAFZZ	5340001808544	19207	12301265	.LOOP, STRAP .... UOC:DAL, V18, ZAL	10
22	PAFZZ	5310001808544	19207	10872159	.WASHER, FLAT .... UOC:DAL, V18, ZAL	21
23	PAFZZ	5310008775795	96906	MS21044-N8	.NUT, SELF-LOCKING................................ UOC:DAL, V18, ZAL	1
24	PBHZZ	2510010889157	19207	12256028	.FRAME SECTION, STRUCT................................ UOC:DAL, V18, ZAL	
25	XBHZZ		19207	12255988	.CROSSMEMBER. .... UOC:DAL, V18, ZAL	

(1) ITEM NO	(2) SMR CODE	(3) NSN	(4) CAGEC	(5) PART NUMBER	(6) DESCRIPTION AND USABLE ON CODES (UOC)	(7) QTY
26	PAFZZ	5305007262550	80204	B1821BH063F175N	.SCREW, CAP, HEXAGON H .... UOC:DAL, V18, ZAL	12
27	PBFZZ	2510010889156	19207	12256036	.FRAME SECTION, STRUC .... UOC:DAL, V18, ZAL	1
28	PFFZZ	5340010893068	19207	12256670	.BRACKET, ANGLE .... UOC:DAL, V18, ZAL	1
29	PAFZZ	5310008140672	96906	MS51943-36	.NUT, SELF-LOCKING, HE .... UOC:DAL, V18, ZAL	2
30	PAFZZ	5305007195238	80204	B1821BH050F200N	.SCREW, CAP, HEXAGON H .... UOC:DAL, V18, ZAL	12
31	PAFZZ	5305009146131	96906	MS18153-63	.SCREW, CAP, HEXAGON H .... UOC:DAL, V1B, ZAL	2
32	PFHZZ	2510010885914	19207	12256033	.FRAME SECTION, STRUC, SEE FIG. 325............. FOR A2s ...... UOC:DAL, V18	1
33	PAFZZ	5305007250154	96906	MS90727-112	..SCREW, CAP, HEXAGON H............... UOC:DAL, V18	4
34	PAFZZ	5310008095998	96906	MS27183-18	..WASHER, FLAT .... UOC:DAL, V18	4
35	PFFZZ	5340011704940	19207	12302672	..BRACKET, MOUNTING. .... UOC:DAL, V18	1
36	PAFZZ	5310004883888	96906	MS51943-40	..NUT, SELF-LOCKING, HE................. UOC:DAL, V18	4
37	PAFZZ	5305007195221	80204	B1821BH050F150N	.SCREW, CAP, HEXAGON H .... UOC:DAL, V18, ZAL	4
38	PAFZZ	5305007250154	96906	MS90727-112	.SCREW, CAP, HEXAGON H .... UOC:DAL, V18, ZAL	8
39	PFFZZ	5320011453184	24617	142392	.RIVET, SOLID .... UOC:DAL, V18, ZAL	8
40	PFHZZ	3040010889161	19207	8758403	.BRACKET, EYE, NONROTA............... UOC:DAL, V18, ZAL	2
41	PFHZZ	5340011647597	19207	12277352-1	.BRACKET, SPECIAL .... UOC:DAL, V18, ZAL	1
42	PFFZZ	5320011507745	24617	111274	.RIVET, SOLID .... UOC:DAL, V18, ZAL	4
43	PAOZZ	5305007250140	96906	MS90727-139	.SCREW, CAP, HEXAGON H .... UOC:DAL, V18, ZAL	14
44	PFOZZ	2510007409337	19207	7409337	.SHACKLE, LEAF SPRING .... UOC:DAL, V18, ZAL	4
45	PAOZZ	5310002416661	96906	MS51943-42	.NUT, SELF-LOCKING, HE .... UOC:DAL, V18, ZAL	16
46	PAFZZ	5305007250145	80204	B1821BH056F200N	.SCREW, CAP, HEXAGON H .... UOC:DAL, V18, ZAL	2
47	XBHZZ		19207	11664271-1	.CROSSMEMBER ASSEMBL .... UOC:DAL, V18, ZAL	1
48	XAHZZ		19207	12256359	.RAIL ASSY SEE FIG. 325 FOR A2s ...... UOC:DAL, V18	1
49	PAFZZ	5365007195240	99696	375N300-001-21	.SCREW, CAP, HEXAGON H .... UOC:DAL, V18, ZAL	4
50	PAFZZ	5365013049529	19207	12356901	.SPACER, PLATE .... UOC:DAL, V18, ZAL	2
51	PFHZZ	2510007409508	19207	7409508	.BRACKET SUPPORT.......................... UOC:DAL, V18, ZAL	2
52	PFHZZ	5340011647598	19207	12277352-2	.BRACKET, SPECIAL .... UOC:DAL, V18, ZAL	1

END OF FIGURE

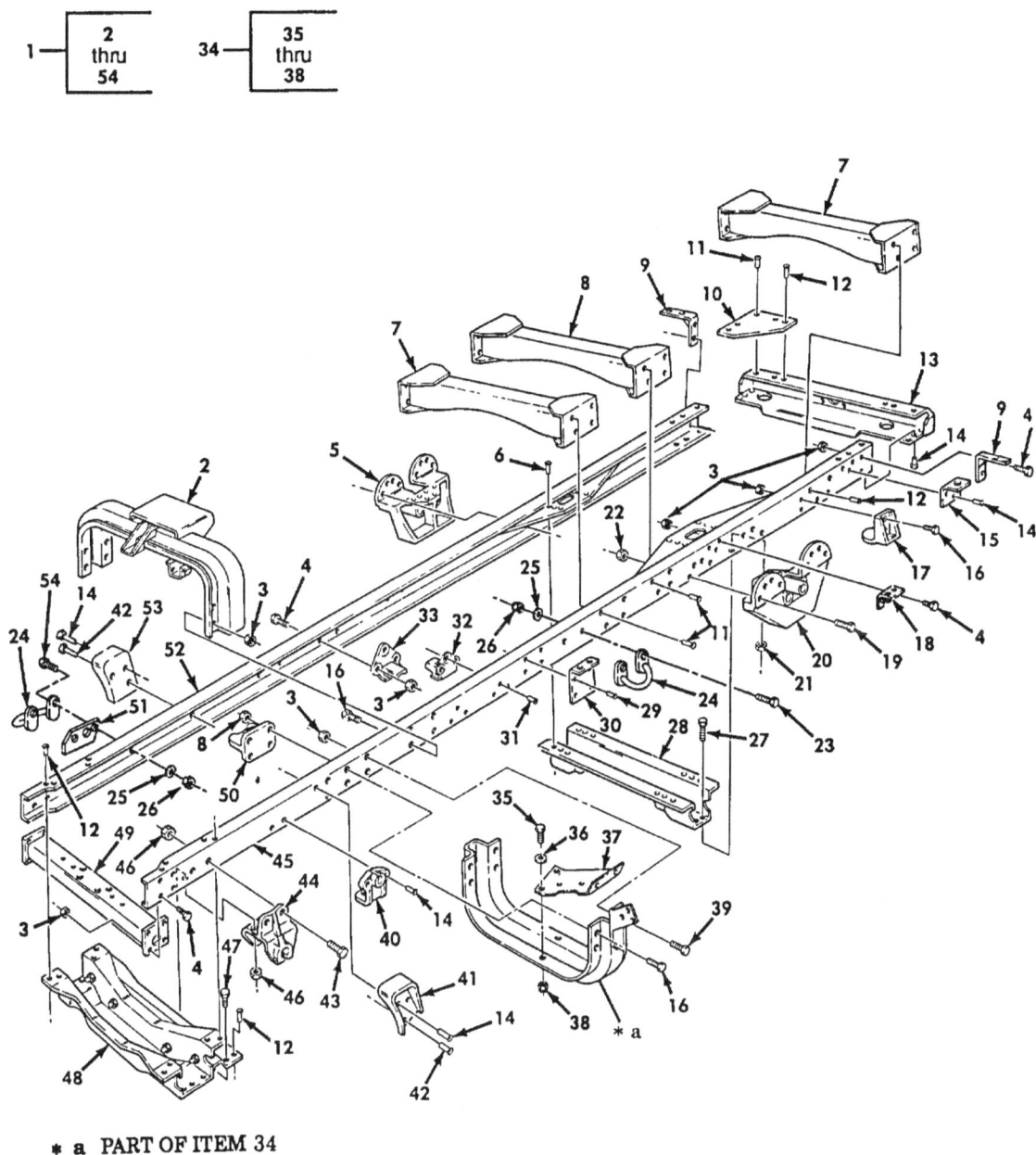

* a PART OF ITEM 34

Figure 321. Frame Assembly, Van.

(1) ITEM NO	(2) SMR CODE	(3) NSN	(4) CAGEC	(5) PART NUMBER	(6) DESCRIPTION AND USABLE ON CODES (UOC)	(7) QTY
					GROUP 1501 FRAME ASSEMBLY	
					FIG. 321 FRAME ASSEMBLY, VAN	
1	XAHHH		19207	12256415	FRAME ASSEMBLY UOC:V24, V25	1
1	XAHHH		19207	12256415-1	FRAME ASSEMBLY SEE FIG. 325 FOR A2s .... UOC:DAJ, DAK	1
2	XBFZZ		19207	12256150-2	.CROSSMEMBER .... UOC:DAJ, DAK, V24, V25, ZAJ, ZAK	1
3	PAFZZ	5310004883888	96906	MS51943-40	.NUT, SELF-LOCKING, HE .... UOC:DAJ, DAK, V24, V25, ZAJ, ZAK	36
4	PAFZZ	5305007195235	80204	B1821BH050F175N	.SCREW, CAP, HEXAGON H............................ UOC:DAJ, DAK, V24, V25, ZAJ, ZAK	15
5	PFFZZ	5340004097958	19207	11611648	.BRACKET, MOUNTING. .... UOC:DAJ, DAK, V24, V25, ZAJ, ZAK	1
6	PFFZZ	5320011454621	24617	425924	.RIVET, SOLID.................................. UOC:DAJ, DAK, V24, V25, ZAJ, ZAK	12
7	XBHZZ		19207	11611789	.CROSSMEMBER ASSY............................. UOC:DAJ, DAK, V24, V25, ZAJ, ZAK	1
8	PAHZZ	2510011571306	19207	11611644	.FRAME SECTION, STRUC .... UOC:DAJ, DAK, V24, V25, ZAJ, ZAK	1
9	PFFZZ	5340004807602	19207	11593218	.BRACKET, ANGLE. .... UOC:DAJ, DAK, V24, V25, ZAJ, ZAK	1
10	PAHZZ	5340010889153	19207	8758409	.PLATE, MOUNTING............................. UOC:DAJ, DAK, V24, V25, ZAJ, ZAK	2
11	PFFZZ	5320011453188	24617	451956	.RIVET, SOLID.................................. UOC:DAJ, DAK, V24, V25, ZAJ, ZAK	18
12	PFFZZ	5320011453185	24617	142391	.RIVET, SOLID.................................. UOC:DAJ, DAK, V24, V25, ZAJ, ZAK	30
13	PFHZZ	2510009339577	19207	8331198	.FRAME SECTION, STRUC .... UOC:DAJ, DAK, V24, V25, ZAJ, ZAK	1
14	PFFZZ	5320011453184	24617	142392	.RIVET, SOLID.................................. UOC:DAJ, DAK, V24, V25, ZAJ, ZAK	20
15	PFHZZ	5340004094018	19207	10871210	.BRACKET, ANGLE. .... UOC:DAJ, V24, V25	2
15	PFHZZ	5340012187159	19207	12303018	.BRACKET, ANGLE. .... UOC:DAJ, DAK, ZAJ, ZAK	1
16	PAFZZ	5305007195238	80204	B1821BH050F200N	.SCREW, CAP, HEXAGON H............................ UOC:DAJ, DAK, V24, V25, ZAJ, ZAK	8
17	PFFZZ	5340010899130	19207	11682319	.BRACKET, MOUNTING. .... UOC:DAJ, DAK, V24, V25, ZAJ, ZAK	4
18	PAFZZ	5340002310278	19207	10871207	.BRACKET, ANGLE. UOC:V24, V25	2
18	PFFZZ	5340012238037	19207	12303021	.BRACKET, ANGLE .... UOC:DAJ, DAK, ZAJ, ZAK	2
19	PAFZZ	5305009162345	80204	B1821BH075F200N	.SCREW, CAP, HEXAGON H............................ UOC:DAJ, DAK, V24, V25, ZAJ, ZAK	16
20	PFFZZ	5340010899128	19207	11611648-1	.BRACKET, BOGIE. .... UOC:DAJ, DAK, V24, V25, ZAJ, ZAK	1
21	PAFZZ	5310002416664	96906	MS51943-44	.NUT, SELF-LOCKING, HE .... UOC:DAJ, DAK, V24, V25, ZAJ, ZAK	12

(1) ITEM NO	(2) SMR CODE	(3) NSN	(4) CAGEC	(5) PART NUMBER	(6) DESCRIPTION AND USABLE ON CODES (UOC)	(7) QTY
22	PAFZZ	5310009353569	96906	MS51943-46	.NUT, SELF-LOCKING, HE .... UOC:DAJ, DAK, V24, V25, ZAJ, ZAK	16
23	PAFZZ	5305007195239	96906	MS90727-116	.SCREW, CAP, HEXAGON H .... UOC:DAB, DAJ, DAK, V24, V25, ZAJ, ZAK	16
24	PAFZZ	5340012919212	19207	12301265	.LOOP, STRAP FASTENER........................ UOC:DAB, DAJ, DAK, V24, V25, ZAJ, ZAK	10
25	PAFZZ	5310001808544	19207	10872159	.WASHER, FLAT .... UOC:DAD, DAJ, DAK, V24, V25, ZAJ, ZAK	20
26	PAFZZ	5310008775795	96906	MS21044-N8	..NUT, SELF-LOCKING, HE........................ UOC:DAJ, DAK, V24, V25, ZAJ, ZAK	2
27	PAFZZ	5305007262550	80204	B1821BH063F175N	.SCREW, CAP, HEXAGON H .... UOC:DAJ, DAK, V24, V25, ZAJ, ZAK	12
28	XBHZZ		19207	12255988	.CROSSMEMBER .... UOC:DAJ, DAK, V24, V25, ZAJ, ZAK	1
29	PFFZZ	5320011453186	24617	425923	.RIVET, SOLID UOC:V24, V25	8
29	PFFZZ	5320011507745	24617	111274	.RIVET, SOLID .... UOC:DAJ, DAX, ZAJ, ZAK	8
30	PFHZZ	5340004094019	19207	10871208	.BRACKET, ANGLE R.H UOC:V24, V25	2
30	PFHZZ	5340012174069	19207	12303020	.BRACKET, ANGLE R.H .... UOC:DAJ, DAK, ZAJ, ZAK	2
30	PFHZZ	5340004094019	19207	10871208-1	.BRACKET, ANGLE L.H UOC:V24, V25	2
30	PFHZZ	5340012174069	19207	12303020-1	.BRACKET, ANGLE L.H .... UOC:DAJ, DAK, ZAJ, ZAK	2
31	PFFZZ	5320011453186	24617	425923	.RIVET, SOLID .... UOC:DAJ, DAK, V24, V25, ZAJ, ZAK	6
32	PBHZZ	2510010889157	19207	12256028	.FRAME SECTION, STRUC .... UOC:DAJ, DAK, V24, V25, ZAJ, ZAK	2
33	PBHZZ	2510010889156	19207	12256036	.FRAME SECTION, STRUC .... UOC:DAJ, DAK, V24, V25, ZAJ, ZAK	1
34	PFHZZ	2510010885914	19207	12256033	.FRAME SECTION, STRUC SEE FIG. 325.............. FOR A2s .... UOC:DAJ, DAK, V24, V25	1
35	PAFZZ	5305007250154	96906	MS90727-112	..SCREW, CAP, HEXAGON H............................. UOC:DAJ, DAK, V24, V25	4
36	PAFZZ	5310008095998	96906	MS27183-18	..WASHER, FLAT........................ UOC:DAJ, DAK, V24, V25	4
37	PFFZZ	5340011704940	19207	12302672	..BRACKET, MOUNTING. .... UOC:DAJ, DAK, V24, V25	1
38	PAFZZ	5310004883888	96906	MS51943-40	..NUT, SELF-LOCKING, HE............................. UOC:DAJ, DAK, V24, V25	4
39	PAFZZ	5305007195221	80204	B1821BH050F150N	.SCREW, CAP, HEXAGON H .... UOC:DAJ, DAK, V24, V25, ZAJ, ZAK	4
40	PFHZZ	3040010889161	19207	8758403	.BRACKET, EYE, NONROTA .... UOC:DAJ, DAK, V24, V25, ZAJ, ZAK	2
41	PFHZZ	5342011647597	19207	12277352-1	.BRACKET, SPECIAL .... UOC:DAJ, DAK, V24, V25, ZAJ, ZAK	1
42	PFHZZ	5320011507745	24617	111274	.RIVET, SOLID .... UOC:DAJ, DAK, V24, V25, ZAJ, ZAK	4
43	PAOZZ	5305007250140	96906	MS90727-139	.SCREW, CAP, HEXAGON H .... UOC:DAJ, DAK, V24, V25, ZAJ, ZAK	14

(1) ITEM NO	(2) SMR CODE	(3) NSN	(4) CAGEC	(5) PART NUMBER	(6) DESCRIPTION AND USABLE ON CODES (UOC)	(7) QTY
44	PFOZZ	2510007409337	19207	7409337	.SHACKLE, LEAF SPRING .... UOC:DAJ, DAK, V24, V25, ZAJ, ZAK	4
45	XAHZZ		19207	12256345	.RAIL ASSY .... UOC:DAJ, DAK, V24, V25, ZAJ, ZAK .... -	1
46	PAOZZ	5310002416661	96906	MS51943-42	.NUT, SELF-LOCKING, HE .... UOC:DAJ, DAK, V24, V25, ZAJ, ZAK	16
47	PAFZZ	5305007250145	80204	B1821BH056F200N	.SCREW, CAP, HEXAGON H .... UOC:DAJ, DAK, V24, V25, ZAJ, ZAK	2
48	XBHZZ		19207	11664271-1	.CROSSMEMBER SEE FIG. 325 FOR A2s .... UOC:DAJ, DAK, V24, V25	1
49	PGHZZ	2510010907639	19207	12302674-1	.FRAME SECTION, STRUC .... UOC:DAJ, DAK, V24, V25, ZAJ, ZAK	1
50	PFHZZ	2510007409508	19207	7409508	.BRACKET SUPPORT .... UOC:DAJ, DAK, V24, V25, ZAJ, ZAK	2
51	PAHZZ	5365013049529	19207	12356901	.SPACER, PLATE .... UOC:DAD, DAJ, DAK, V24, V25, ZAJ, ZAK	2
52	XAHZZ		19207	12256360	.RAIL ASSY SEE FIG. 325 FOR A2s .... UOC:DAJ, DAK, V24, V25, ZAJ, ZAK	1
53	PFHZZ	5342011647598	19207	12277352-2	.BRACKET, SPECIAL .... UOC:DAJ, DAK, V24, V25, ZAJ, ZAK	1
54	PAFZZ	5305007195240	99696	375N300-001-21	.SCREW, CAP, HEXAGON H .... UOC:DAD, DAJ, DAK, V24, V25, ZAJ, ZAK	4

END OF FIGURE

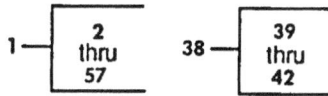

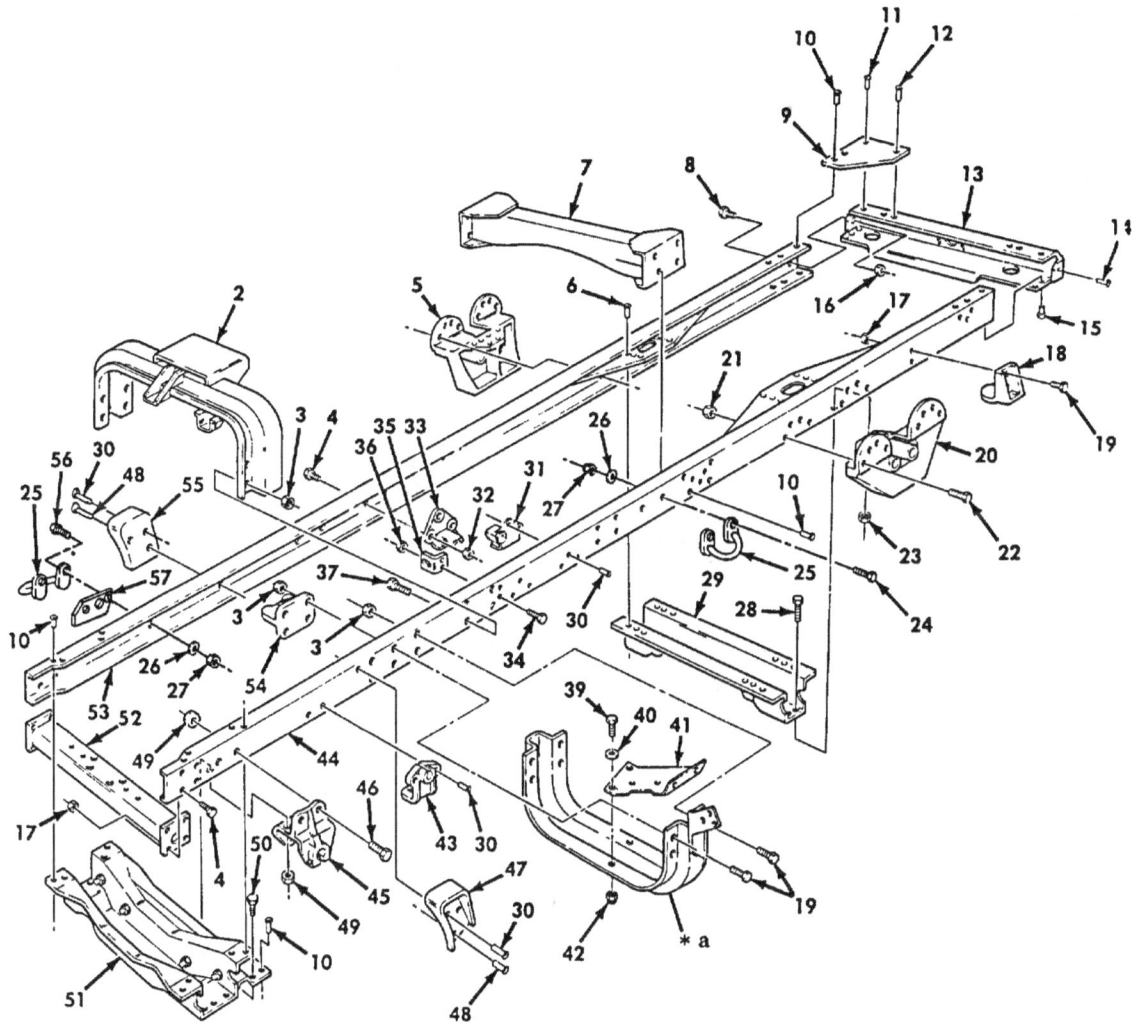

* a PART OF ITEM 38

Figure 322. Frame Assembly, Tractor.

(1) ITEM NO	(2) SMR CODE	(3) NSN	(4) CAGEC	(5) PART NUMBER	(6) DESCRIPTION AND USABLE ON CODES (UOC)	(7) QTY
					GROUP 1501 FRAME ASSEMBLY	
					FIG. 322 FRAME ASSEMBKY, TRACTOR	
1	XAHHH		19207	12256416-2	FRAME ASSEMBLY SEE FIG. 325 FOR A2s	1
					UOC:DAG, V22	
1	XAHHH		19207	12256416-1	FRAME ASSEMBLY SEE FIG. 325 FOR A2s	1
					UOC:DAH, V21	
2	XBFZZ		19207	12256150-1	.CROSSMEMBER	1
					UOC:DAG, DAH, V21, V22, ZAG, ZAH	
3	PAFZZ	5310004883888	96906	MS51943-40	.NUT, SELF-LOCKING, HE	16
					UOC:DAG, DAH, V21, V22, ZAG, ZAH	
4	PAFZZ	5305007195235	80204	B1821BH050F175N	.SCREW, CAP, HEXAGON H............................	12
					UOC:DAG, DAH, V21, V22, ZAG, ZAH	
5	PFFZZ	5340004097958	19207	11611648	.BRACKET, MOUNTING...........................	1
					UOC:DAG, DAH, 'V21, V22, ZAG, ZAH	
6	PFFZZ	5320011454621	24617	425924	.RIVET, SOLID	12
					UOC:DAG, DAH, V21, V22, ZAG, ZAH	
7	XBHZZ		19207	12256467	.CROSSMEMBER ASSY	1
					UOC:DAG, DAH, V21, V22, ZAG, ZAH	
8	PAFZZ	5305007195235	96906	MS90727-114	.SCREW, CAP, HEXAGON H............................	4
					UOC:DAG, DAH, V21, V22, ZAG, ZAH	
9	PFFZZ	5340011202891	19207	10871206	.BRACE, CORNER......................................	2
					UOC:DAG, DAH, V21, V22, ZAG, ZAH	
10	PFHZZ	5320011453185	24617	142391	.RIVET, SOLID	2
					UOC:DAG, DAH, V12, V21, ZAG, ZAH	
11	PFHZZ	5320011453187	24617	142390	.RIVET, SOLID	2
					UOC:DAG, DAH, V21, V22, ZAG, ZAH	
12	PFFZZ	5320011453188	24617	451956	.RIVET, SOLID	2
					UOC:DAG, DAH, V21, V22, ZAG, ZAH	
13	PBHZZ	2510011194094	19207	7413456	.FRAME, STRUCTURAL, VE	1
					UOC:DAG, DAH, V21, V22, ZAG, ZAH	
14	PFHZZ	5320011453184	24617	142392	.RIVET, SOLID	15
					UOC:DAG, DAH, V21, V22, ZAG, ZAH	
15	PFFZZ	5320011453185	24617	142391	.RIVET, SOLID	6
					UOC:DAG, DAH, V21, V22, ZAG, ZAH	
16	PAFZZ	5310000624954	96906	MS21045-8	.NUT, SELF-LOCKING, HE	7
					UOC:DAG, DAH, V21, V22, ZAG, ZAH	
17	PAFZZ	5310004883888	96906	MS51943-40	.NUT, SELF-LOCKING, HE...........................	6
					UOC:DAG, DAH, V21, V22, ZAG, ZAH	
18	PFFZZ	5340010899130	19207	11682319	.BRACKET, MOUN TING.	4
					UOC:DAG, DAH, V21, V22, ZAG, ZAH	
19	PAFZZ	5305007195238	80204	B1821BH050F200N	.SCREW, CAP, HEXAGON H...........................	14
					UOC:DAG, DAH, V12, V21, ZAG, ZAH	
20	PFFZZ	5340010899128	19207	11611648-1	.BRACKET, BOGIE.	1
					UOC:DAG, DAH, V21, V22, ZAG, ZAH	
21	PAFZZ	5310009353569	96906	MS51943-46	.NUT, SELF-LOCKING, HE	16
					UOC:DAG, DAH, V21, V22, ZAG, ZAH	
22	PAFZZ	5305009162345	80204	B1821BH075F200N	.SCREW, CAP, HEXAGON H	16
					UOC:DAG, DAH, V21, V22, ZAG, ZAH	
23	PAFZZ	5310002416664	96906	MS51943-44	.NUT, SELF-LOCKING, HE	12
					UOC:DAG, DAH, V21, V22, ZAG, ZAH	
24	PAFZZ	5305007195239	96906	MS90727-116	.SCREW, CAP, HEXAGON H	16
					UOC:DAG, DAH, V21, V22, ZAG, ZAH	

(1) ITEM NO	(2) SMR CODE	(3) NSN	(4) CAGEC	(5) PART NUMBER	(6) DESCRIPTION AND USABLE ON CODES (UOC)	(7) QTY
25	PAFZZ	5340012919212	19207	12301265	LOOP, STRAP... UOC:DAG, DAH, V21, V22, ZAG, ZAH	10
26	PAFZZ	5310001808544	19207	10872159	.WASHER, FLAT .... UOC:DAG, DAH, V21, V22, ZAG, ZAH	20
27	PAFZZ	5310008775795	96906	MS21044-N8	.NUT, SELF-LOCKING, HE .... UOC:DAG, DAH, V21, V22, ZAG, ZAH	20
28	PAFZZ	5305007262550	80204	B1821BH063F175N	.SCREW, CAP, HEXAGON H.......................... UOC:DAG, DAH, V21, V22, ZAG, ZAH	12
29	XBHZZ		19207	12255988	.CROSSMEMBER .... UOC:DAG, DAH, V21, V22, ZAG, ZAH	1
30	PFFZZ	5320011453184	24617	142392	.RIVET, SOLID... UOC:DAG, DAH, V21, V22, ZAG, ZAH	14
31	PBHZZ	2510010889157	19207	12256028	.FRAME SECTION, STRUC .... UOC:DAG, DAH, V21, V22, ZAG, ZAH	2
32	PAFZZ	5310000624954	96906	MS21045-8	.NUT, SELF-LOCKING, HE .... UOC:DAG, DAH, V21, V22, ZAG, ZAH	3
33	PBHZZ	2510010889156	19207	12256036	.FRAME SECTION, STRUC .... UOC:DAG, DAH, V21, V22, ZAG, ZAH	1
34	PAFZZ	5305009146131	96906	MS18153-63	.SCREW, CAP, HEXAGON H.......................... UOC:DAG, DAH, V21, V22, ZAG, ZAH	2
35	PFFZZ	5340010893068	19207	12256670	.BRACKET, ANGLE. .... UOC:DAG, DAH, V21, V22, ZAG, ZAH	1
36	PAFZZ	5310009824908	96906	MS21045-6	.NUT, SELF-LOCKING, HE .... UOC:DAG, DAH, V21, V22, ZAG, ZAH	2
37	PAFZZ	5305007250154	80204	B1821BH050F138N	.SCREW, CAP, HEXAGON H.......................... UOC:DAG, DAH, V21, V22, ZAG, ZAH	8
38	PFHZZ	2510010885914	19207	12256033	.FRAME SECTION, STRUC SEE FIG. 325............ FOR A2s UOC:DAG, DAH, V21, V22	1
39	PAFZZ	5305007250154	96906	MS90727-112	..SCREW, CAP, HEXAGON H .... UOC:DAG, DAH, V21, V22	4
40	PAFZZ	5310008095998	96906	MS27183-18	..WASHER, FLAT .... UOC:DAG, DAH, V21, V22	4
41	PFFZZ	5340011704940	19207	12302672	..BRACKET, MOUNTING . ... UOC:DAG, DAH, V21, V22	1
42	PAFZZ	5310004883888	96906	MS51943-40	..NUT, SELF-LOCKING, HE .... UOC:DAG, DAH, V21, V22	4
43	PFHZZ	3040010889161	19207	8758403	.BRACKET, EYE, NONROTA.............................. UOC:DAG, DAH, V21, V22, ZAG, ZAH	2
44	XAHZZ		19207	12256299	.RAIL ASSY, SEE FIG. 325 FOR A2s .... UOC:DAG, DAH, V21, V22, ZAG, ZAH	1
45	PFOZZ	2510007409337	19207	7409337	.SHACKLE, LEAF SPRING.................................. UOC:DAG, DAH, V21, V22, ZAG, ZAH	4
46	PAOZZ	5305007250140	96906	MS90727-139	.SCREW, CAP, HEXAGON H.............................. UOC:DAG, DAH, V21, V22, ZAG, ZAH	14
47	PFHZZ	5342011647597	19207	12277352-1	.BRACKET, SPECIAL.. UOC:DAG, DAH, V21, V22, ZAG, ZAH	1
48	PFFZZ	5320011507745	24617	111274	.RIVET, SOLID... UOC:DAG, DAH, V21, V22, ZAG, ZAH	4
49	PAOZZ	5310002416661	96906	MS51943-42	.NUT, SELF-LOCKING, HE .... UOC:DAG, DAH, V21, V22, ZAG, ZAH	16
50	PAFZZ	5305007250145	80204	B1821BH056F200N	.SCREW, CAP, HEXAGON H .... UOC:DAG, DAH, V21, V22, ZAG, ZAH	2
51	XBHZZ		19207	11664271-1	.CROSSMEMBER ASSEMBL SEE FIG. 325............ FOR A2s .... UOC:DAG, DAH, V21, V22	1

(1) ITEM NO	(2) SMR CODE	(3) NSN	(4) CAGEC	(5) PART NUMBER	(6) DESCRIPTION AND USABLE ON CODES (UOC)	(7) QTY
52	PGFZZ	2510010907639	19207	12302674-1	.FRAME SECTION, STRUC .... UOC:DAG, V22, ZAG	1
53	XAHZZ		19207	12256361	.RAIL ASSY SEE FIG. 325 FOR A2s...................... UOC:DAG, DAH, V21, V22	1
54	PFHZZ	2510007409508	19207	7409508	.BRACKET SUPPORT.. UOC:DAG, DAH, V21, V22, ZAG, ZAH	2
55	PFHZZ	5342011647598	19207	12277352-2	.BRACKET , SPECIAL... UOC:DAG, DAH, V21, V22, ZAG, ZAH	1
56	PAFZZ	5305007195240	99696	375N300-001-21	.SCREW, CAP, HEXAGON H............................... UOC:DAG, DAH, V21, V22, ZAG, ZAH	4
57	PAFZZ	5342013049529	19207	12356901	.SPACER, PLATE.. UOC:DAG, DAH, V21, V22, ZAG, ZAH	2

END OF FIGURE

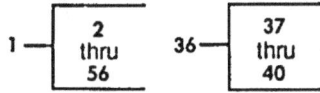

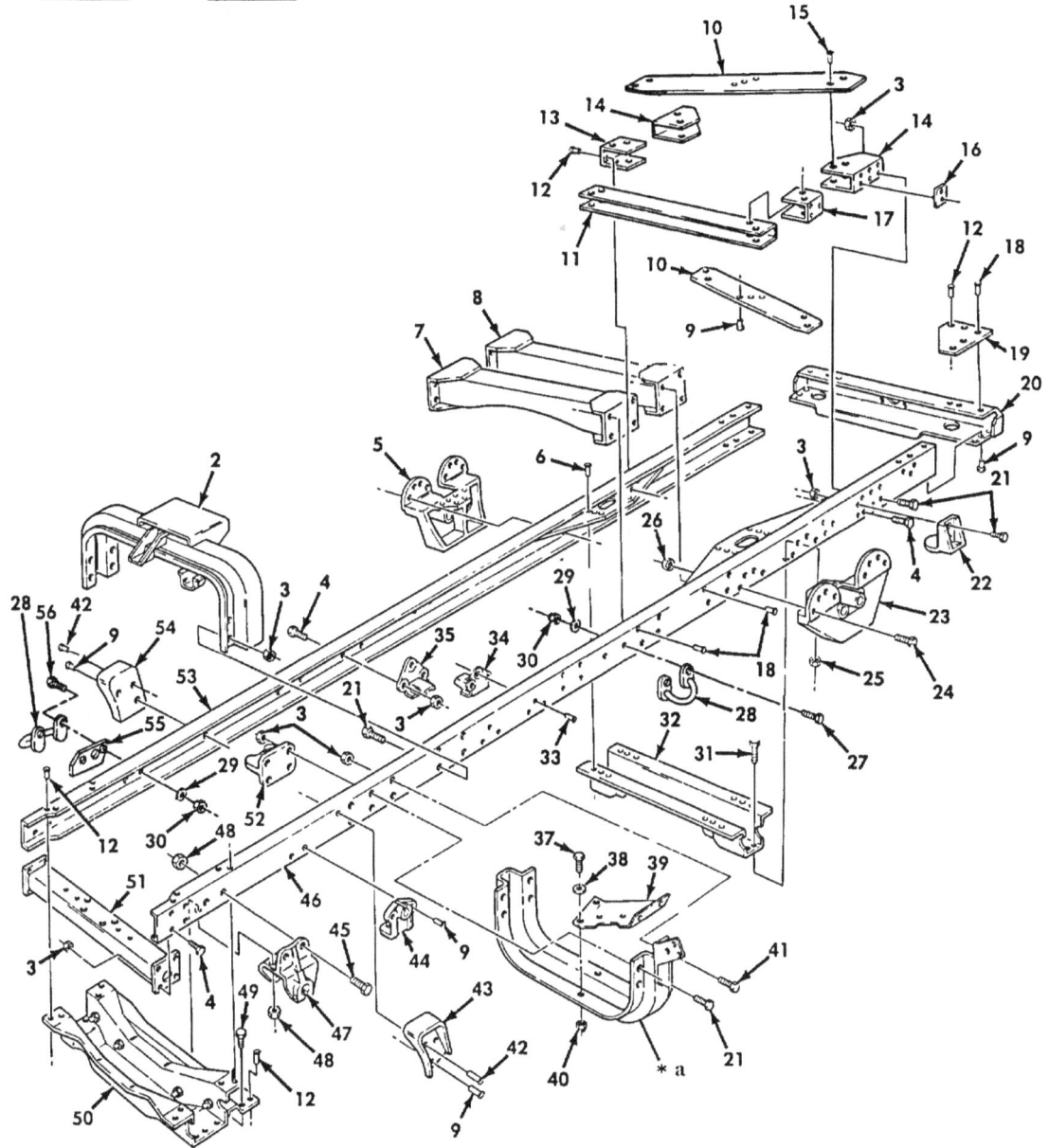

* a PART OF ITEM 36

Figure 323. Frame Assembly, Long Wheel Base, Cargo.

(1) ITEM NO	(2) SMR CODE	(3) NSN	(4) CAGEC	(5) PART NUMBER	(6) DESCRIPTION AND USABLE ON CODES (UOC)	(7) QTY
					GROUP 1501 FRAME ASSEMBLY	
					FIG. 323 FRAME ASSEMBLY, LONG WHEEL BASE, CARGO	
1	XAHHH		19207	12256417-2	FRAME ASSEMBLEY UOC:V17	1
1	XAHHH		19207	12256417-1	FRAME ASSEMBLY UOC:V16	1
1	XAHHH		19207	12256417-3	FRAME ASSEMBLY SEE FIG. 325 FOR A2s .... UOC:DAD	1
1	XAHHH		19207	12256417-4	FRAME ASSEMBLY SEE FIG. 325 FOR A2s .... UOC:DAC	1
2	XBHZZ		19207	12256150-2	.CROSSMEMBER .... UOC:DAC, DAD, V16, V17, ZAC, ZAD	1
3	PAFZZ	5310004883888	96906	MS51943-40	.NUT, SELF-LOCKING, HE UOC:DAC, DAD, V16, V17, ZAC, ZAD	18
4	PAFZZ	5305007195235	80204	B1821BH050F175N	.SCREW, CAP, HEXAGON H.............................. UOC:DAC, DAD, V16, V17, ZAC, ZAD	12
5	PFFZZ	5340004097958	19207	11611648	.BRACKET, MOUNTING. .... UOC:DAC, DAD, V16, V17, ZAC, ZAD	1
6	PFHZZ	5320011454621	24617	425924	.RIVET, SOLID.. UOC:DAC, DAD, V16, V17, ZAC, ZAD	12
7	XBHZZ		19207	11611789	.CROSSMEMBER ASSY .... UOC:DAC, DAD, V16, V17, ZAC, ZAD	1
8	XBHZZ	2510011571306	19207	11611644	.FRAME SECTION, STRUC .... UOC:DAC, DAD, V16, V17, ZAC, ZAD	1
9	PFFZZ	5320011453184	24617	142392	.RIVET, SOLID.. UOC:DAC, DAD, V16, V17, ZAC, ZAD	20
10	PFHZZ	2510011307936	19207	11611806	.REINFORCEMENT, FRAME .... UOC:DAC, DAD, V16, V17, ZAC, ZAD	4
11	PFHZZ	9520011402378	19207	11608831	.CHANNEL, STRUCTURAL.................................. UOC:DAC, DAD, V16, V17, ZAC, ZAD	2
12	PFFZZ	5320011453185	24617	142391	.RIVET, SOLID.. UOC:DAC, DAD, V16, V17, ZAC, ZAD	28
13	PBHZZ	5340011310110	19207	11611804-2	.BRACKET, DOUBLE ANGL............................... UOC:DAC, DAD, V16, V17, ZAC, ZAD	1
14	PFHZZ	9520011307937	19207	11611805-1	.CHANNEL, STRUCTURAL.................................. UOC:DAC, DAD, V16, V17, ZAC, ZAD	2
15	PFFZZ	5320011453189	24617	142393	.RIVET, SOLID .... UOC:DAC, DAD, V16, V17, ZAC, ZAD	16
16	PFFZZ	5365011451310	19207	12256574	.SPACER, PLATE.. UOC:DAC, DAD, V16, V17, ZAC, ZAD	2
17	PFHZZ	5340011310109	19207	11611804-1	.BRACKET, DOUBLE ANGL............................... UOC:DAC, DAD, V16, V17, ZAC, ZAD	1
18	PFFZZ	5320011453188	24617	451956	.RIVET, SOLID .... UOC:DAC, DAD, V16, V17, ZAC, ZAD	18
19	PAHZZ	5340010889153	19207	8758409	.PLATE, MOUNTING. .... UOC:DAC, DAD, V16, V17, ZAC, ZAD	2
20	XBHZZ	2510009339577	19207	8331198	.FRAME SECTION, STRUC .... UOC:DAC, DAD, V16, V17, ZAC, ZAD	1
21	PAFZZ	5305007195238	80204	B1821BH050F200N	.SCREW, CAP, HEXAGON H.............................. UOC:DAC, DAD, V16, V17, ZAC, ZAD	28
22	PFFZZ	5340010899130	19207	11682319	.BRACKET, MOUNTING.	4

(1) ITEM NO	(2) SMR CODE	(3) NSN	(4) CAGEC	(5) PART NUMBER	(6) DESCRIPTION AND USABLE ON CODES (UOC)	(7) QTY
					UOC:DAC, DAD, V16, V17, ZAC, ZAD	
23	PFFZZ	5340010899128	19207	11611648-1	.BRACKET, BOGIE..	1
					UOC:DAC, DAD, V16, V17, ZAC, ZAD	
24	PAFZZ	5305009162345	80204	B1821BH075F200N	.SCREW, CAP, HEXAGON H	16
					UOC:DAC, DAD, V16, V17, ZAC, ZAD	
25	PAFZZ	5310002416664	96906	MS51943-44	.NUT, SELF-LOCKING, HE	12
					UOC:DAC, DAD, V16, V17, ZAC, ZAD	
26	PAFZZ	5310009353569	96906	MS51943-46	.NUT, SELF-LOCKING, HE	16
					UOC:DAC, DAD, V16, V17, ZAC, ZAD	
27	PAFZZ	5305007195239	96906	MS90727-116	.SCREW, CAP, HEXAGON H......................	16
					UOC:DAC, DAD, V16, V17, ZAC, ZAD	
28	PAFZZ	5340012919212	19207	12301265	.LOOP, STRAP	10
					UOC:DAC, DAD, V16, V17, ZAC, ZAD	
29	PAFZZ	5310001808544	19207	10872159	.WASHER, FLAT	20
					UOC:DAC, DAD, V16, V17, ZAC, ZAD	
30	PAFZZ	5310008775795	96906	MS21044-N8	.NUT, SELF-LOCKING, HE	2
					UOC:DAC, DAD, V16, V17, ZAC, ZAD	
31	PAFZZ	5305007262550	80204	B1821BH063F175N	.SCREW, CAP, HEXAGON H......................	12
					UOC:DAC, DAD, V16, V17, ZAC, ZAD	
32	XBHZZ		19207	12255988	.CROSSMSEBER..................................	1
					UOC:DAC, DAD, V16, V17, ZAC, ZAD	
33	PFFZZ	5320011453186	24617	425923	.RIVET, SOLID..................................	9
					UOC:DAC, DAD, V16, V17, ZAC, ZAD	
34	PBHZZ	2510010889157	19207	12256028	.FRAME SECTION, STRUC	2
					UOC:DAC, DAD, V16, V17, ZAC, ZAD	
35	PBHZZ	2510010889156	19207	12256036	.FRAME SECTION, STRUC	1
					UOC:DAC, DAD, V16, V17, ZAC, ZAD	
36	PFHZZ	2510010885914	19207	12256033	.FRAME SECTION, STRUC SEE FIG. 325............... FOR A2s	1
					UOC:DAC, DAD, V16, V17	
37	PAFZZ	5305007250154	96906	MS90727-112	.SCREW, CAP, HEXAGON H	4
					UOC:DAC, V16, V17	
38	PAFZZ	5310008095998	96906	MS27183-18	.WASHER, FLAT	4
					UOC:DAC, V16, V17	
39	PBFZZ	5340011704940	19207	12302672	.BRACKET, MOUNTING........................	1
					UOC:DAC, V16, V17	
40	PAFZZ	5310004883888	96906	MS51943-40	.NUT, SELF-LOCKING, HE	4
					UOC:DAC, V16, V17	
41	PAFZZ	5305007195221	80204	B1821BH050F150N	.SCREW, CAP, HEXAGON H	4
					UOC:DAC, DAD, V16, V17, ZAC, ZAD	
42	PAFZZ	5320011453184	24617	142392	.RIVET, SOLID..................................	14
					UOC:DAC, DAD, V16, V17, ZAC, ZAD	
43	PFHZZ	5342011647597	19207	12277352-1	.BRACKET, SPECIAL..........................	1
					UOC:DAC, DAD, V16, V17, ZAC, ZAD	
44	PFHZZ	3040010889161	19207	8758403	.BRACKET, EYE, NONROTA.....................	2
					UOC:DAC, DAD, V16, V17, ZAC, ZAD	
45	PAOZZ	5305007250140	96906	MS90727-139	.SCREW, CAP, HEXAGON H......................	14
					UOC:DAC, DAD, V16, V17, ZAC, ZAD	
46	XAHZZ		19207	12256340	.RAIL ASSY SEE FIG. 325 FOR A2s......................	1
					UOC:DAC, DAD, V16, V17, ZAC, ZAD	
47	PFOZZ	2510007409337	19207	7409337	.SHACKLE, LEAF SPRING	4
					UOC:DAC, DAD, V16, V17, ZAC, ZAD	
48	PAOZZ	5310002416661	96906	MS51943-42	.NUT, SELF-LOCKING, HE	16
					UOC:DAC, DAD, V16, V17, ZAC, ZAD	

(1) ITEM NO	(2) SMR CODE	(3) NSN	(4) CAGEC	(5) PART NUMBER	(6) DESCRIPTION AND USABLE ON CODES (UOC)	(7) QTY
49	PAFZZ	5305007250145	80204	B1821BH056F200N	.SCREW, CAP, HEXAGON H .... UOC:DAC, DAD, V16, V17, ZAC, ZAD	2
50	XBHZZ		19207	11664271-1	.CROSSMEMBER SEE FIG. 325 FOR A2s .... UOC:DAC, DAD, V16, V17, ZAC, ZAD	1
51	PGFZZ	2510010907639	19207	12302674-1	.FRAME SECTION, STRUC .... UOC:DAC, V17, ZAC	1
52	PFHZZ	2510007409508	19207	7409508	.BRACKET SUPPORT.. UOC:DAC, DAD, V16, V17, ZAC, ZAD	2
53	XAHZZ		19207	12256362	.RAIL, RIGHT SIDE .... UOC:DAC, DAD, V16, V17, ZAC, ZAD	1
54	PFHZZ	5342011647598	19207	12277352-2	.BRACKET, SPECIAL... UOC:DAC, DAD, V16, V17, ZAC, ZAD	1
55	PAFZZ	5365013049529	19207	12356901	.SPACER, PLATE.. UOC:DAC, DAD, V16, V17, ZAC, ZAD	2
56	PAFZZ	5365007195240	99696	375N300-001-21	.SCREW, CAP, HEXAGON H .... UOC:DAC, DAD, V16, V17, ZAC, ZAD	4

END OF FIGURE

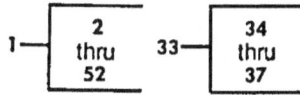

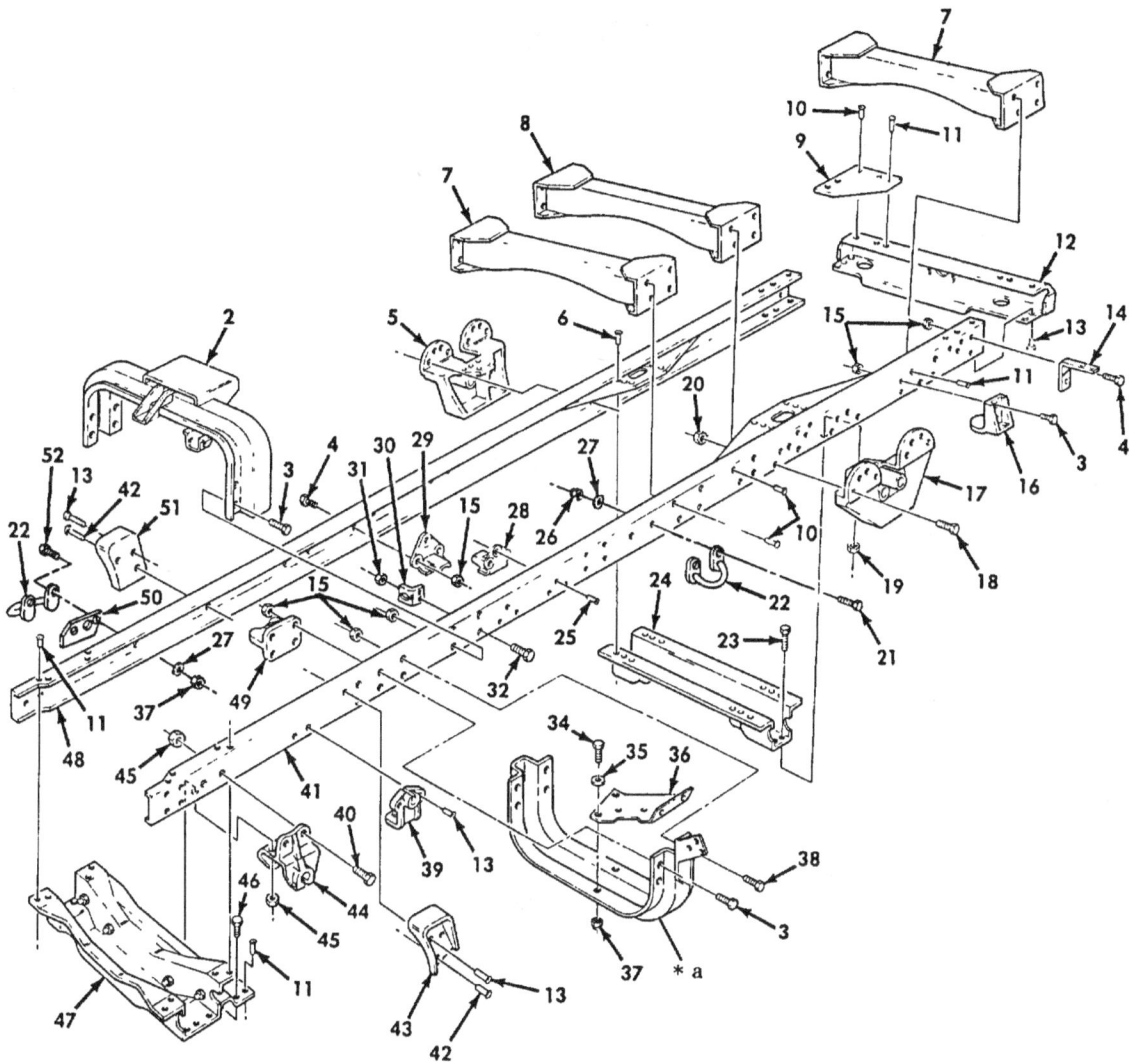

* a PART OF ITEM 33

Figure 324. Frame Assembly, Chassis (M939).

SECTION II

(1) ITEM NO	(2) SMR CODE	(3) NSN	(4) CAGEC	(5) PART NUMBER	(6) DESCRIPTION AND USABLE ON CODES (UOC)	(7) QTY
					GROUP 1501 FRAME ASSEMBLY	
					FIG. 324 FRAME ASSEMBLY, CHASSIS (M939)	
1	XAHHH		19207	12257053	FRAME ASSEMBLY UOC:V39	1
2	XBHZZ		19207	12256150-2	.CROSSMEMBER UOC:V39	1
3	PAFZZ	5305007195238	80204	B1821BH050F200N	.SCREW, CAP, HEXAGON H............................... UOC:V39	20
4	PAFZZ	5305007195235	80204	B1821BH050F175N	.SCREW, CAP, HEXAGON H............................... UOC:V39	7
5	PFFZZ	5340004097958	19207	11611648	.BRACKET, MOUNTING. UOC:V39	1
6	PFFZZ	5320011454621	24617	425924	.RIVET, SOLID.. UOC:V39	12
7	XBFZZ		19207	11611789	.CROSSMEMBER ASSY UOC:V39	1
8	XBHZZ		19207	11608826	.CROSSMEMBER ASSY UOC:V39	1
9	PAHZZ	5340010889153	19207	8758409	.PLATE, MOUNTING.. UOC:V39	2
10	PFFZZ	5320011453188	24617	451956	.RIVET, SOLID UOC:V39	18
11	PFFZZ	5320011453185	24617	142391	.RIVET, SOLID UOC:V39	30
12	XBHZZ	2510009339577	19207	8331198	.FRAME SECTION, STRUC................................... UOC:V39	1
13	PFFZZ	5320011453184	24617	142392	.RIVET, SOLID UOC:V39	14
14	PFFZZ	5340004807602	19207	11593218	.BRACKET, ANGLE UOC:V39	2
15	PAFZZ	5310004883888	96906	MS51943-40	.NUT, SELF-LOCKING, HE UOC:V39	14
16	PFFZZ	5340010899130	19207	11682319	.BRACKET, MOUNTING.. UOC:V39	4
17	PFFZZ	5340010899128	19207	11611648-1	.BRACKET, BOGIE... UOC:V39	1
18	PAFZZ	5305009162345	80204	B1821BH075F200N	.SCREW, CAP, HEXAGON H............................... UOC:V39	16
19	PAFZZ	5310002416664	96906	MS51943-44	.NUT, SELF-LOCKING, HE UOC:V39	12
20	PAFZZ	5310009353569	96906	MS51943-46	.NUT, SELF-LOCKING, HE UOC:V39	16
21	PAFZZ	5305007195239	96906	MS90727-116	.SCREW, CAP, HEXAGON H............................... UOC:V39	16
22	PAFZZ	5340012919212	19207	12301265	.LOOP, STAP UOC:V39	10
23	PAFZZ	5305007262550	80204	B1821BH063F175N	.SCREW, CAP, HEXAGON H............................... UOC:V39	12
24	XBHZZ		19207	12255988	.CROSSMEMBER 	1

(1) ITEM NO	(2) SMR CODE	(3) NSN	(4) CAGEC	(5) PART NUMBER	(6) DESCRIPTION AND USABLE ON CODES (UOC)	(7) QTY
					UOC:V39	
25	PFFZZ	5320011453186	24617	425923	.RIVET, SOLID	7
					UOC:V39	
26	PAFZZ	5310008775795	96906	MS21044-N8	.NUT, SELF-LOCKING, HE	2
					UOC:V39	
27	PAFZZ	5310001808544	19207	10872159	.WASHER, FLAT	20
					UOC:V39	
28	PBHZZ	2510010889157	19207	12256028	.FRAME SECTION, STRUC	2
					UOC:V39	
29	PBHZZ	2510010889156	19207	12256036	.FRAME SECTION, STRUC	1
					UOC:V39	
30	PFFZZ	5340010893068	19207	12256670	.BRACKET, ANGLE	1
					UOC:V39	
31	PAFZZ	5310008140672	96906	MS51943-36	.NUT, SELF-LOCKING, HE	2
					UOC:V39	
32	PAFZZ	5305009146131	96906	MS18153-63	.SCREW, CAP, HEXAGON H	2
					UOC:V39	
33	PFHZZ	2510010885914	19207	12256033	.FRAME SECTION, STRUC	1
					UOC:V39	
34	PAFZZ	5305007250154	96906	MS90727-112	..SCREW, CAP, HEXAGON H...............	4
					UOC:V39	
35	PAFZZ	5310008095998	96906	MS27183-18	..WASHER, FLAT	4
					UOC:V39	
36	PFFZZ	5340011704940	19207	12302672	..BRACKET, MOUNTING.	1
					UOC:V39	
37	PAFZZ	5310004883888	96906	MS51943-40	..NUT, SELF-LOCKING, HE.................	4
					UOC:V39	
38	PAFZZ	5305007195221	80204	B1821BH050F150N	.SCREW, CAP, HEXAGON H	4
					UOC:V39	
39	PFHZZ	3040010889161	19207	8758403	.BRACKET, EYE, NONROTA.................	2
					UOC:V39	
40	PAOZZ	5305007250140	96906	MS90727-139	.SCREW, CAP, HEXAGON H	14
					UOC:V39	
41	XAHZZ		19207	12256346	.RAIL ASSY	1
					UOC:V39	
42	PAFZZ	5320011507745	24617	111274	.RIVET, SOLID	4
					UOC:V39	
43	XBHZZ		19207	12256469-1	.CAB SUPPORT, LEFT	1
					UOC:V39	
44	PFOZZ	2510007409337	19207	7409337	.SHACKLE, LEAF SPRING	4
					UOC:V39	
45	PAOZZ	5310002416661	96906	MS51943-42	.NUT, SELF-LOCKING, HE	16
					UOC:V39	
46	PAFZZ	5305007250145	80204	B1821BH056F200N	.SCREW, CAP, HEXAGON H	2
					UOC:V39	
47	XBHZZ		19207	11664271-1	.CROSSMEMBER..........................	1
					UOC:V39	
48	XAHZZ		19207	12257052	.RAIL ASSY	1
					UOC:V39	
49	PFHZZ	2510007409508	19207	7409508	.BRACKET SUPPORT.........................	2
					UOC:V39	
50	PAFZZ	5342013049529	19207	12356901	.SPACER, PLATE............................	2
					UOC:V39	

(1) ITEM NO	(2) SMR CODE	(3) NSN	(4) CAGEC	(5) PART NUMBER	(6) DESCRIPTION AND USABLE ON CODES (UOC)	(7) QTY
51	PFHZZ	5340010911633	19207	12256469-2	.BRACKET, MOUNTING FRONT CAB, R.H UOC:V39	1
52	PAFZZ	5305007195240	99696	MS375N300-001-21	.SCREW, CAP, HEXAGON H................................ UOC:V39	4

END OF FIGURE

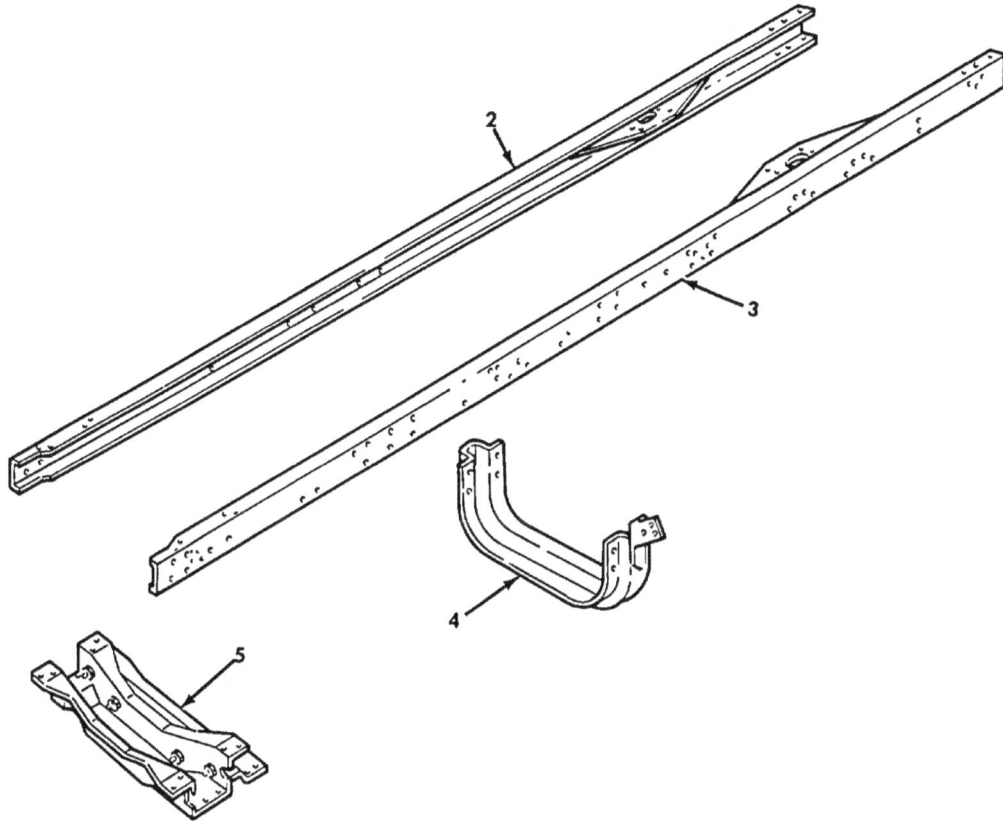

Figure 325. Frame Assembly (M939A2).

(1) ITEM NO	(2) SMR CODE	(3) NSN	(4) CAGEC	(5) PART NUMBER	(6) DESCRIPTION AND USABLE ON CODES (UOC)	(7) QTY
					GROUP 1501 FRAME ASSEMBLY	
					FIG. 325 FRAME ASSEMBLY(M939A2)	
1	XAHHH		19207	12363487-1	FRAME ASSEMBLY, 179 .... UOC:ZAB	1
1	XAHHH		19207	12363487-2	FRAME ASSEMBLY, 179 .... UOC:ZAA	1
1	XAHHH		19207	12363488-1	FRAME ASSEMBLY, 167 .... UOC:ZAF	1
1	XAHHH		19207	12363488-2	FRAME ASSEMBLY, 167 .... UOC:ZAE	1
1	XAHHH		19207	12363489	FRAME ASSEMBLY, 179 .... UOC:ZAL	1
1	XAHHH		19207	12363490	FRAME ASSEMBLY, 215 .... UOC:ZAJ, ZAK	1
1	XAHHH		19207	12363491-1	FRAME ASSEMBLY, 167 .... UOC:ZAH	1
1	XAHHH		19207	12363491-2	FRAME ASSEMBLY, 167 .... UOC:ZAG	1
1	XAHHH		19207	12363492-1	FRAME ASSEMBLY, 215 .... UOC:ZAD	1
1	XAHHH		19207	12363492-2	FRAME ASSEMBLY, 215 .... UOC:ZAC	1
2	XAHZZ		19207	12363414	.RAIL ASSEMBLY, R.H .... UOC:ZAA, ZAB	1
2	XAHZZ		19207	12363420	.RAIL ASSEMBLY, R.H .... UOC:ZAE, ZAF	1
2	XAHZZ		19207	12363421	.RAIL ASSEMBLY, R.H .... UOC:ZAL	1
2	XAHZZ		19207	12363422	.RAIL ASSEMBLY, R.H .... UOC:ZAJ, ZAK	1
2	XAHZZ		19207	12363423	.RAIL ASSEMBLY, R.H .... UOC:ZAC, ZAD	1
2	XAHZZ		19207	12363426	.RAIL ASSEMBLY, R.H .... UOC:ZAG, ZAH	1
3	XAHZZ		19207	12363413	.RAIL ASSEMBLY, L.H .... UOC:ZAA, ZAB	1
3	XAHZZ		19207	12363415	.RAIL ASSEMBLY, L.H .... UOC:ZAC, ZAD	1
3	XAHZZ		19207	12363416	.RAIL ASSEMBLY, L.H .... UOC:ZAL	1
3	XAHZZ		19207	12363417	.RAIL ASSEMBLY, L.H .... UOC:ZAE, ZAF	1
3	XAHZZ		19207	12363418	.RAIL ASSEMBLY, L.H .... UOC:ZAJ, ZAK	1
3	XAHZZ		19207	12363425	.RAIL ASSEMBLY, L.H .... UOC:ZAG, ZAH	1
4	PFHZZ	2510012722923	19207	22363412	.FRAME SECTION, STRUC .... UOC:ZAA, ZAB, ZAC, ZAD, ZAE, ZAF, ZAG, ZAH, ZAJ, ZAK, ZAL	1
5	XBHZZ		19207	12363411	.CROSSMEMBER, FRONT	1

(1) ITEM NO	(2) SMR CODE	(3) NSN	(4) CAGEC	(5) PART NUMBER	(6) DESCRIPTION AND USABLE ON CODES (UOC)	(7) QTY

UOC:ZAA, ZAB, ZAC, ZAD, ZAE, ZAF, ZAG, ZAH, ZAJ, ZAK, ZAL

END OF FIGURE

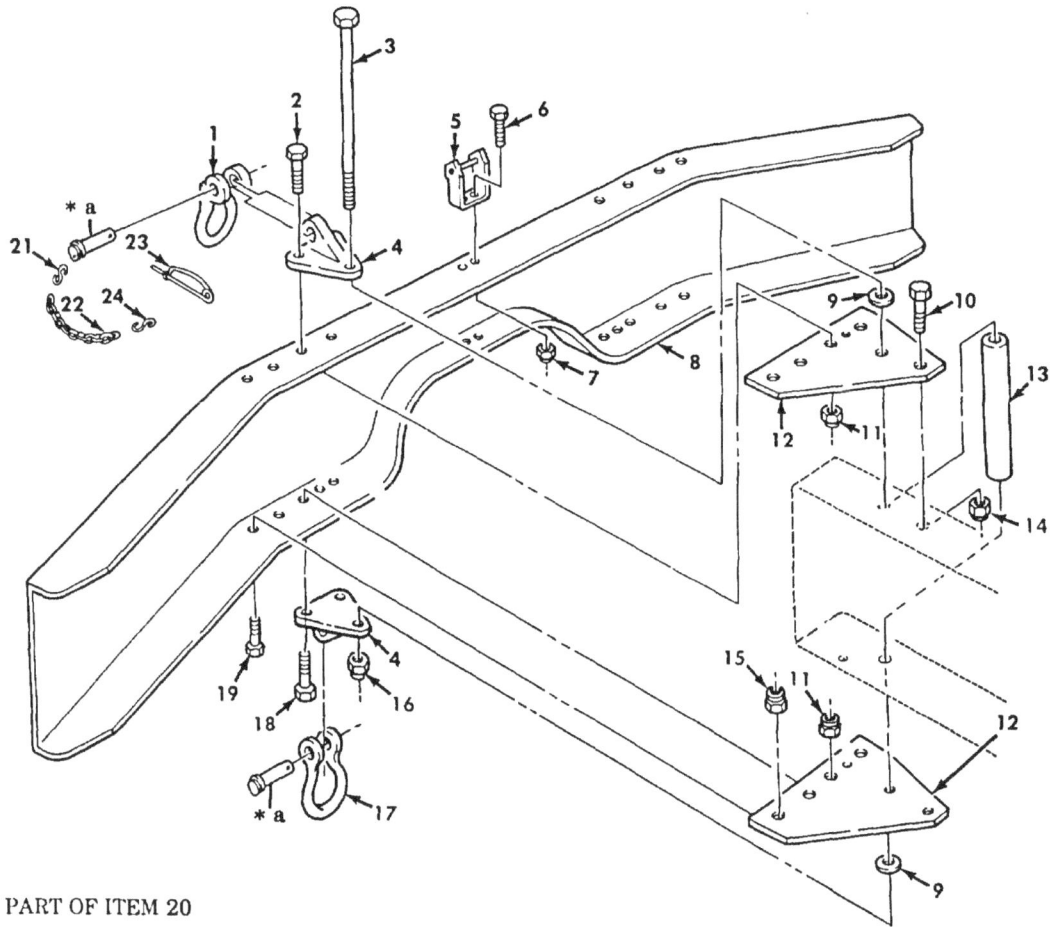

Figure 326. Front Bumper, Lifting Shackle, Hood Latch, and Mounting Hardware.

(1) ITEM NO	(2) SMR CODE	(3) NSN	(4) CAGEC	(5) PART NUMBER	(6) DESCRIPTION AND USABLE ON CODES (UOC)	(7) QTY
					GROUP 1501 FRAME ASSEMBLY	
					FIG. 326 FRONT BUMPER, LIFTING SHACKLE, HOOD LATCH, AND MOUNTING HARDWARE	
1	PAOZZ	4030012226037	19207	12301040	SHACKLE	2
2	PAOZZ	5305007262551	80204	B1821BH063F200N	SCREW, CAP, HEXAGON H	4
3	PAOZZ	5305001404765	19207	10883118	SCREW, CAP, HEXAGON H .... UOC:DAL, V18, ZAL	2
3	PAOZZ	5305007409507	19207	10883119	SCREW, CAP, HEXAGON H .... UOC:DAA, DAB, DAC, DAD, DAE, DAF, DAG, DAH, DAJ, DAK, DAW, DAX, V12, V13, V14, V15, V16, V17, V19, V20, V21, V22, V24, V25, V39, ZAA, ZAB, ZAC, ZAD, ZAE, ZAF, ZAG, ZAH, ZAJ, ZAK	2
4	PFOZZ		19207	12368440	BRACKET, LIFTING................................ UOC:DAA, DAB, DAC, DAD, DAE, DAF, DAG, DAH, DAJ, DAK, DAW, DAX, V12, V13, V14, V15, V16, V17, V19, V20, V21, V22, V24, V25, V39, ZAA, ZAB, ZAC, ZAD, ZAE, ZAF, ZAG, ZAH, ZAJ, ZAK	2
4	PFOZZ		19207	12368440	BRACKET, LIFTING................................ UOC:DAL, V18, ZAL	4
5	PAOZZ	5340011560421	19207	12302678	BRACKET, MOUNTING	1
6	PAOZZ	5305002693239	80204	B1821BH038F138N	SCREW, CAP, HEXAGON H	2
7	PAOZZ	5310009359022	96906	MS51943-32	NUT, SELF-LOCKING, HE	2
8	PFOZZ	2540007376203	19207	7376203	BUMPER, VEHICULAR	1
9	PAOZZ	5310005214534	19207	5214534	WASHER, FLAT	4
10	PAOZZ	5305007254183	96906	MS90726-113	SCREW, CAP, HEXAGON H .... UOC:DAL, V18, ZAL	4
10	PAOZZ	5305007254183	96906	MS90726-113	SCREW, CAP, HEXAGON H .... UOC:DAA, DAB, DAC, DAD, DAE, DA F, DAG, DAH, DAJ, DAK, DAW, DAX, V12, V13, V14, V15, V16, V17, V19, V20, V21, V22, V24, V25, V39, ZAA, ZAB, ZAC, ZAD, ZAE, ZAF, ZAG, ZAH, ZAJ, ZAK	6
11	PAOZZ	5310002416664	96906	MS51943-44	NUT, SELF-LOCKING, HE.................................	8
12	PFOZZ	2510004075084	19207	10883113	PLATE, VEHICLE BUMPE................................. UOC:DAA, DAB, DAC, DAD, DAE, DAF, DAG, DAH, DAJ, DAK, DAW, DAX, V12, V13, V14, V15, V16, V17, V19, V20, V21, V22, V24, V25, V39, ZAA, ZAB, ZAC, ZAD, ZAE, ZAF, ZAG, ZAH, ZAJ, ZAK	2
12	PFOZZ		19207	12375512	PLATE, VEHICLE BUMPE................................. UOC:DAL, V18, ZAL	4
13	PAOZZ	4730007409524	19207	7409524	COUPLING, PIPE	2
14	PAOZZ	5310004883888	96906	MS51943-40	NUT, SELF-LOCKING, HE .... UOC:DAL, V18, ZAL	4
14	PAOZZ	5310004883888	96906	MS51943-40	NUT, SELF-LOCKING, HE................................. UOC:DAA, DAB, DAC, DAD, DAE, DAF, DAG, DAH, DAJ, DAK, DAW, DAX, V12, V13, V14, V15, V16, V17, V19, V20, V21, V22, V24, V25, V39, ZAA, ZAB, ZAC, ZAD, ZAE, ZAF, ZAH, ZAJ, ZAK	6
15	PAOZZ	5310004883888	96906	MS51943-40	NUT, SELF-LOCKING, HE.................................	4
16	PAOZZ	5310002416664	96906	MS51943-44	NUT, SELF-LOCKING, HE.................................	2

(1) ITEM NO	(2) SMR CODE	(3) NSN	(4) CAGEC	(5) PART NUMBER	(6) DESCRIPTION AND USABLE ON CODES (UOC)	(7) QTY
17	PAOZZ	4030013162554	19207	12368441	SHACKLE .... UOC:DAA, DAB, DAC, DAD, DAE, DAF, DAG, DAH, DAJ, DAK, DAW, DAX, V12, V13, V14, V15, V16, V17, V19, V20, V21, V22, V24, V25, V39, ZAA, ZAB, ZAC, ZAD, ZAE, ZAF, ZAF, ZAH, ZAJ, ZAK	2
18	PAOZZ	5305007272283	80204	B1821BH063F150N	SCREW, CAP, HEXAGON H.................................	4
19	PAOZZ	5305007195219	96906	MS90727-111	SCREW, CAP, HEXAGON H.................................	8
20	PAOZZ		19207	12375352	PIN, STRAIGHT, HEADED................................. UOC:DAA, DAB, DAC, DAD, DAE, DAF, DAG, DAH, DAJ, DAK, DAW, DAX, V12, V13, V14, V15, V16, V17, V19, V20, V21, V22, V24, V25, V39, ZAA, ZAB, ZAC, ZAD, ZAE, ZAF, ZAF, ZAH, ZAJ, ZAK	4
20	PAOZZ		19207	12375352	PIN, STRAIGHT, HEADED................................. UOC:DAL, V18, ZAL	2
21	PAOZZ	4030007809350	96906	MS87006-13	.HOOK, CHAIN , S.......................................	1
22	MOOZZ		16003	C43974 X 8	.CHAIN, WELDLESS MAKE FROM CHAIN, P/.......... N C43974 8 INCHES LONG..............................	1
23	PAOZZ	5315007418971	19207	7418971	.PIN, RETAINING.......................................	1
24	PAOZZ	4030009487315	96906	MS87006-33	.HOOK, CHAIN, S.......................................	1

END OF FIGURE

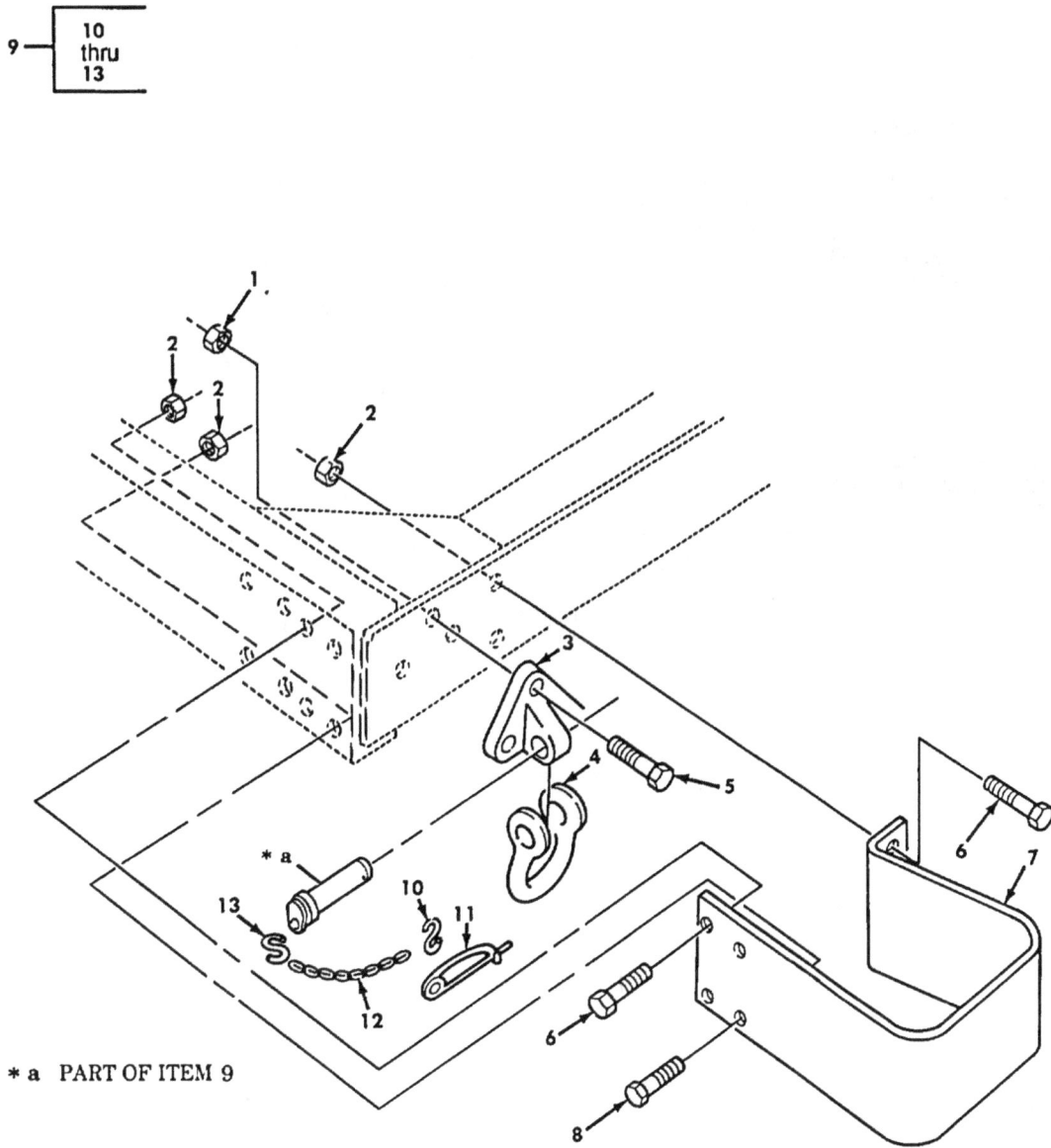

Figure 327. Rear Bumperette and Shackle Assembly.

(1) ITEM NO	(2) SMR CODE	(3) NSN	(4) CAGEC	(5) PART NUMBER	(6) DESCRIPTION AND USABLE ON CODES (UOC)	(7) QTY
					GROUP 1501 FRAME ASSEMBLY	
					FIG. 327 REAR BUMPERETTE AND SHACKLE ASSEMBLY	
1	PAOZZ	5310002416664	96906	MS51943-44	NUT, SELF-LOCKING, HE	6
2	PAOZZ	5310004883888	96906	MS51943-40	NUT, SELF-LOCKING, HE	12
3	PFOZZ	2510007409514	19207	7409514	BRACKET, LIFTING SHACKLE	2
4	PAOZZ	4030007409523	19207	7409523	SHACKLE	2
5	PAOZZ	5305007262550	80204	B1821BH063F175N	SCREW, CAP, HEXAGON H	6
6	PAOZZ	5305007254183	96906	MS90726-113	SCREW, CAP, HEXAGON H	8
7	PAOZZ	2540006930676	89346	99311R1	BUMPER, VEHICULAR .... DAJ, DAK, V12, V13, V14, V15, V16, V17, V25, V39, ZAJ, ZAK	2
7	PAOZZ	2540012106231	19207	12302970	BUMPER, VEHICULAR REAR .... UOC:DAA, DAB, DAC, DAD, DAW, DAX, ZAA, ZAB, ZAC, ZAD	2
8	PAOZZ	5305007254187	96906	MS90726-114	SCREW, CAP, HEXAGON	4
9	PAOOO	5315007412924	19207	7412924	PIN ASSEMBLY	2
10	PAOZZ	4030009487315	96906	MS87006-33	.HOOK, CHAIN, S	1
11	PAOZZ	5315007418971	19207	7418971	.PIN, RETAINING	1
12	MOOZZ		80204	42C15120-205-6	.CHAIN MAKE FROM CHAIN P/N XB-196................	1
13	PAOZZ	4030007809350	96906	MS87006-13	.HOOK, CHAIN, S	1

END OF FIGURE

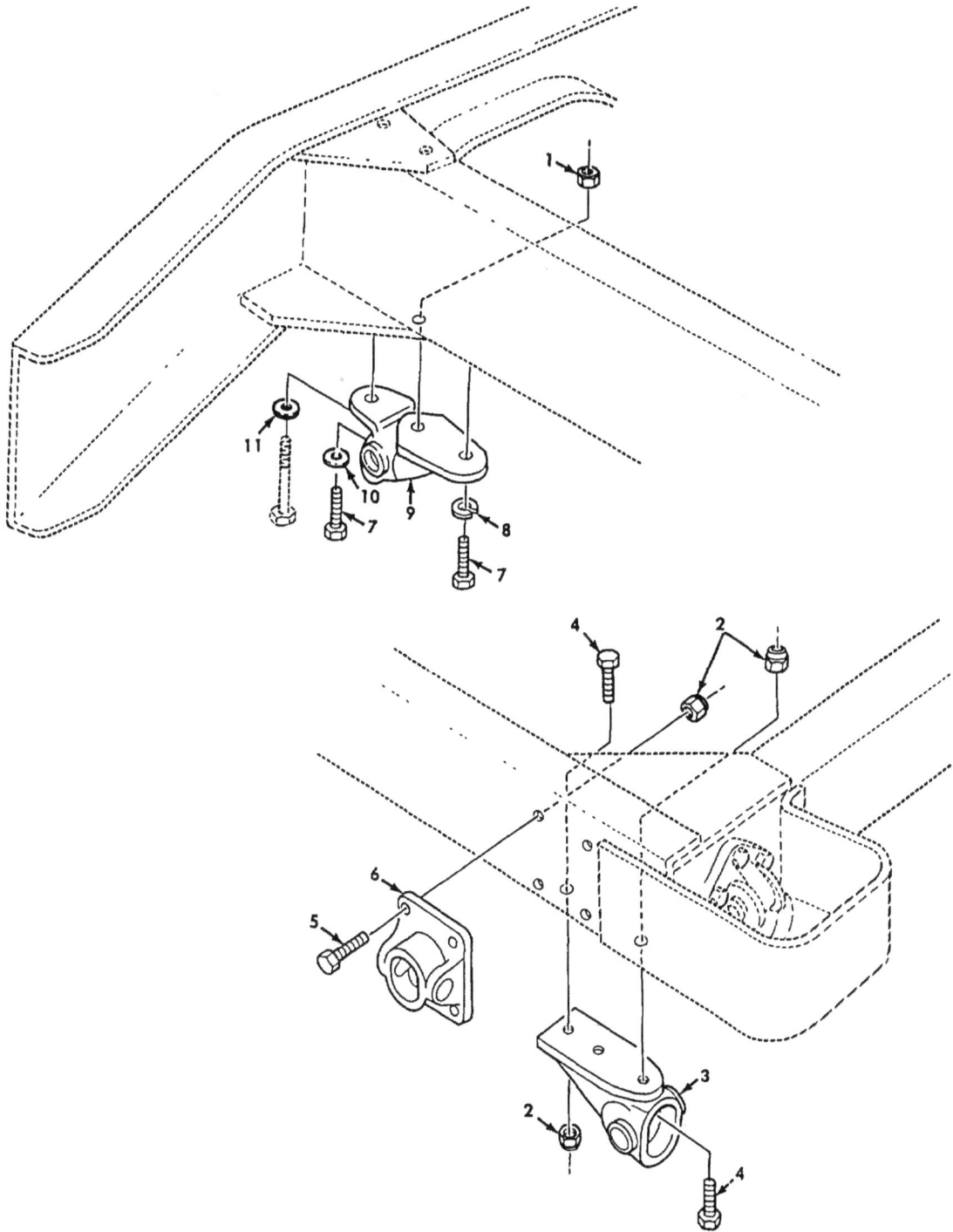

Figure 328. Front and Rear Field Clock Anchors (M939, M939A1).

(1) ITEM NO	(2) SMR CODE	(3) NSN	(4) CAGEC	(5) PART NUMBER	(6) DESCRIPTION AND USABLE ON CODES (UOC)	(7) QTY
					GROUP 1501 FRAME ASSEMBLY	
					FIG. 328 FRONT AND REAR FIELD CHOCK ANCHORS(M939, M939A1)	
1	PAOZZ	5310004883888	96906	MS51943-40	NUT, SELF-LOCKING, HE .... UOC:DAL, V18	2
2	PAOZZ	5310002416664	96906	MS51943-44	NUT, SELF-LOCKING, HE .... UOC:DAL, V18	14
3	PFOZZ	5340011087265	19207	10883114	ANCHOR PLATE .... UOC:DAL, V18	2
4	PAOZZ	5305007262551	80204	B1821BH063F200N	SCREW, CAP, HEXAGON H .... UOC:DAL, V18	6
5	PAOZZ	5305007262550	80204	B1821BH063F175N	SCREW, CAP, HEXAGON H .... UOC:DAL, V18	8
6	PFOZZ	5340011089109	19207	10883115	BRACKET, MOUNTING... UOC:DAL, V18	2
7	PAOZZ	5305007195238	80204	B1821BH050F200N	SCREW, CAP, HEXAGON H .... UOC:DAL, V18	4
8	PAOZZ	5310000034094	01276	210104-8S	WASHER, LOCK .... UOC:DAL, V18	2
9	PFOZZ	5340010822517	19207	10883116-1	BRACKET, MOUNTING... UOC:DAL, V18	2
10	PAOZZ	5310011096755	19207	12256205-1	WASHER , FLAT.. UOC:DAL, V18	2
11	PAOZZ	5310011096756	19207	12256205-2	WASHER, FLAT.. UOC:DAL, V18	2

END OF FIGURE

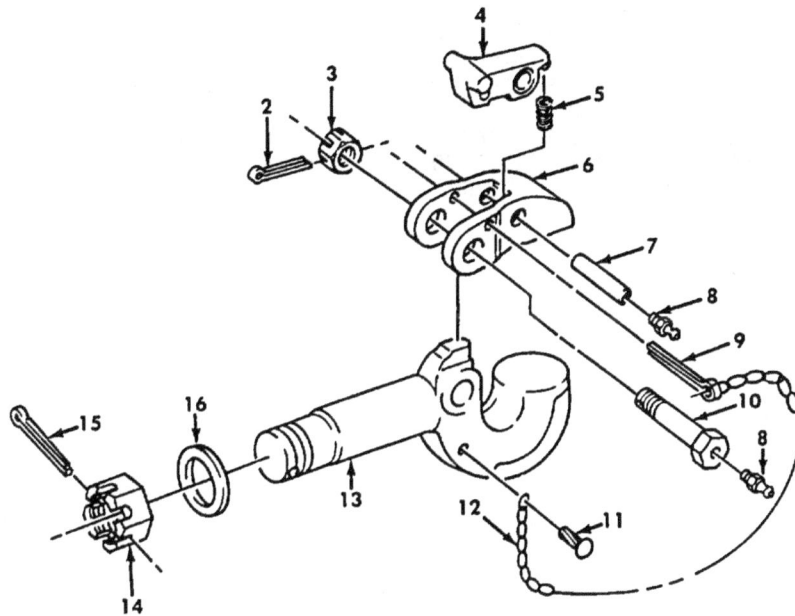

Figure 329. Pintle Hook Assembly.

(1) ITEM NO	(2) SMR CODE	(3) NSN	(4) CAGEC	(5) PART NUMBER	(6) DESCRIPTION AND USABLE ON CODES (UOC)	(7) QTY
					GROUP 1503 PINTLES AND TOWING ATTACHMENTS	
					FIG. 329 PINTLE HOOK ASSEMBLY	
1	PAOOO	2540000473926	19207	8710630	PINTLE ASSEMBLY, TOW	1
2	PAOZZ	5315000137214	96906	MS24665-359	.PIN, COTTER	1
3	PAOZZ	5310000503520	19207	7061871	.NUT, PLAIN, SLOTTED, H	1
4	PFOZZ	2540002373693	19207	7714880	.LATCH, PINTLE HOOK	1
5	PAOZZ	5360007044253	19207	7044253	.SPRING, HELICAL, COMP	1
6	PFOZZ	2540008212277	19207	8710627	.LATCH, PINTLE HOOK	1
7	XDOZZ		19207	8710629	.SHAFT, STRAIGHT	1
8	PAOZZ	4730000504203	96906	MS15001-1	.FITTING, LUBRICATION	2
9	PAOZZ	5315005977399	96906	MS24665-631	.PIN, COTTER	1
10	PAOZZ	4730000473946	19207	8710628	.BOLT, FLUID PASSAGE	1
11	PAOZZ	5305002535626	96906	MS21318-47	.SCREW, DRIVE	1
12	MOOZZ		81348	42C15120-205-6	.CHAIN, MAKE FROM CHAIN, P/N XB-196..............	1
13	PFOZZ	2540009210481	19207	8710626	.HOOK, PINTLE	1
14	PAOZZ	5310007714911	19207	7714911	.NUT, PLAIN, SLOTTED, H PINTLE HOOK.............	1
15	PAOZZ	5315001879591	46717	L6451-101	.PIN, COTTER	1
16	PAOZZ	5310007764909	19207	7764909	WASHER, FLAT PINTLE HOOK	1

END OF FIGURE

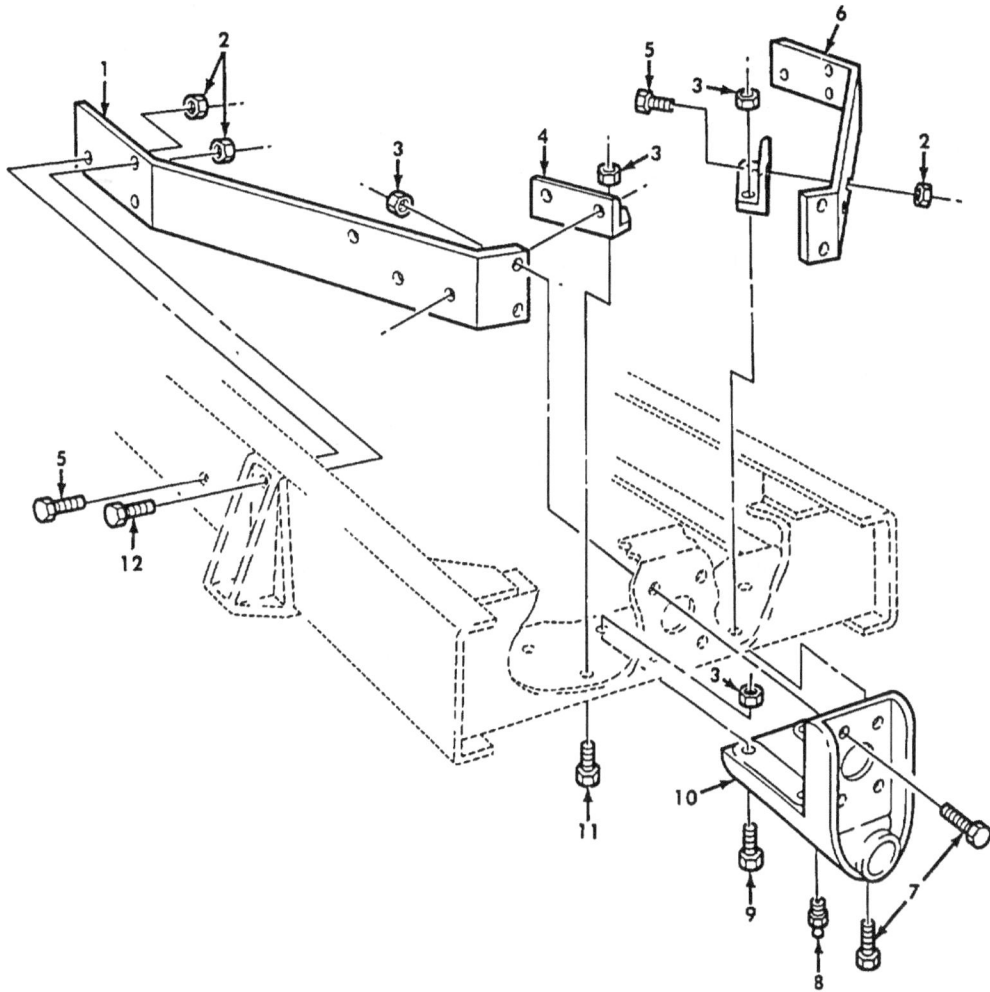

Figure 330. Towing Pintle Mounting Brackets, Tractor.

(1) ITEM NO	(2) SMR CODE	(3) NSN	(4) CAGEC	(5) PART NUMBER	(6) DESCRIPTION AND USABLE ON CODES (UOC)	(7) QTY
					GROUP 1503 PINTLE AND TOWING ATTACHMENTS	
					FIG. 330 TOWING PINTLE MOUNTING BRACKETS, TRACTOR	
1	PFOZZ	2540007409660	19207	7409660	BRACKET, TOW HOOK .... UOC:DAG, DAH, V21, V22, ZAG, ZAH	1
2	PAOZZ	5310004883888	96906	MS51943-40	NUT, SELF-LOCKING, HE .... UOC:DAG, DAH, V21, V22, ZAG, ZAH	10
3	PAOZZ	5310002416664	96906	MS51943-44	NUT, SELF-LOCKING, HE .... UOC:DAG, DAH, V21, V22, ZAG, ZAH	11
4	PFOZZ	5340004193080	19207	10883360	BRACKET, ANGLE .... UOC:DAG, DAH, V21, V22, ZAG, ZAH	2
5	PAOZZ	5305007195235	80204	B1821BH050F175N	SCREW, CAP, HEXAGON H .... UOC:DAG, DAH, V21, V22, ZAG, ZAH	8
6	PFOZZ	2510007409661	19207	7409661	BRACE, PINTLE HOOK .... UOC:DAG, DAH, V21, V22, ZAG, ZAH	1
7	PAOZZ	5305007262552	80204	B1821BH063F225N	SCREW, CAP, HEXAGON H .... UOC:DAG, DAH, V21, V22, ZAG, ZAH	5
8	PAOZZ	4730000504203	96906	MS15001-1	FITTING, LUBRICATION .... UOC:DAG, DAH, V21, V22, ZAG, ZAH	1
9	PAOZZ	5305007262551	80204	B1821BH063F200N	SCREW, CAP, HEXAGON H .... UOC:DAG, DAH, V21, V22, ZAG, ZAH	3
10	PAOZZ	2540008327027	19207	8327027	BRACKET, TOW HOOK .... UOC:DAG, DAH, V21, V22, ZAG, ZAH	1
11	PAOZZ	5305007272283	80204	B1821BH063F150N	SCREW, CAP, HEXAGON H .... UOC:DAG, DAH, V21, V22, ZAG, ZAH	3
12	PAOZZ	5305007195239	96906	MS90727-116	SCREW, CAP, HEXAGON H .... UOC:DAG, DAH, V21, V22, ZAG, ZAH	2

END OF FIGURE

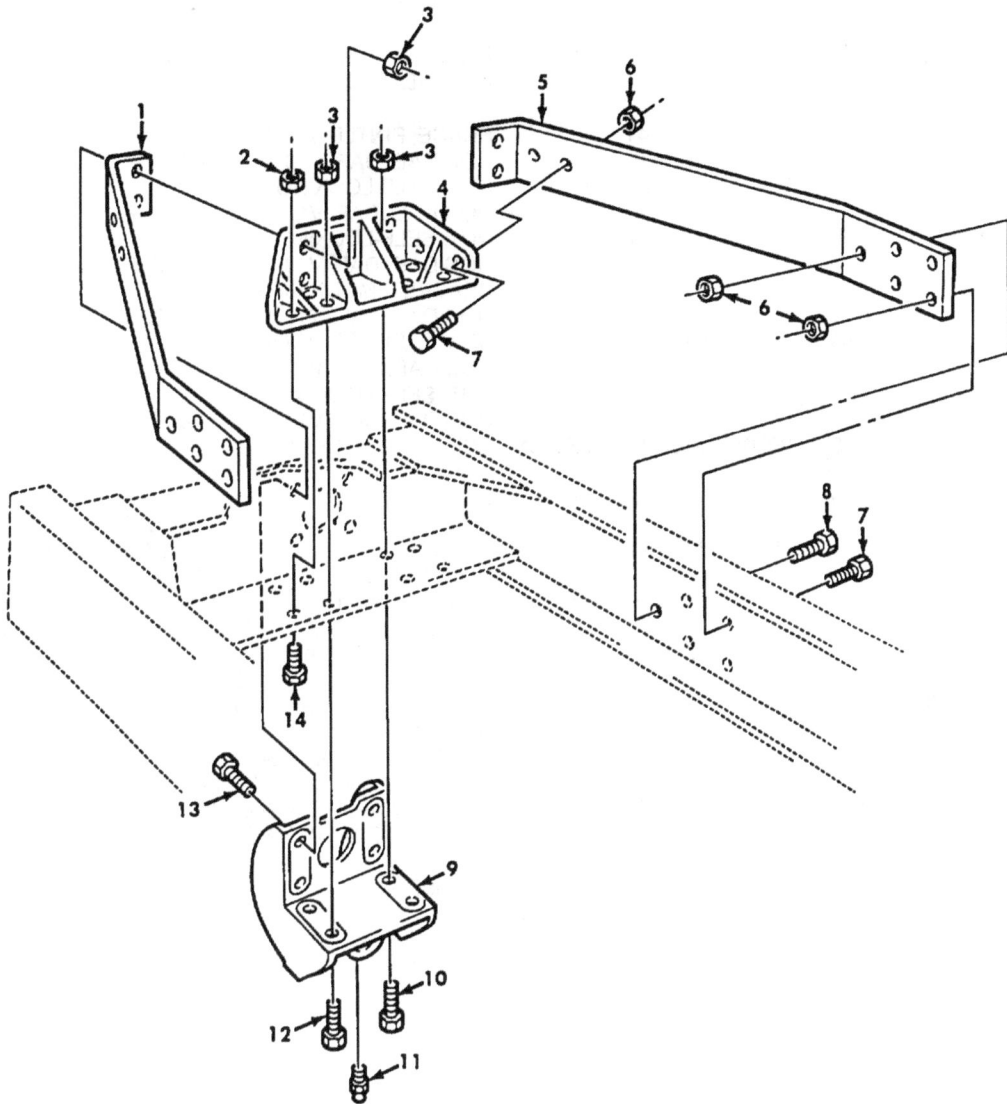

Figure 331. Towing Pintle Mounting Brackets, Dump.

(1) ITEM NO	(2) SMR CODE	(3) NSN	(4) CAGEC	(5) PART NUMBER	(6) DESCRIPTION AND USABLE ON CODES (UOC)	(7) QTY
					GROUP 1503 PINTLES AND TOWING ATTACHMENTS	
					FIG. 331 TOWING PINTLE MOUNTING BRACKETS, DUMP	
1	PFOZZ	2510002317489	19207	8330741	BRACE, PINTLE MOUNTI .... UOC:DAE, DAF, V19, V20, ZAE, ZAF	1
2	PAOZZ	5310002416664	96906	MS51943-44	NUT, SELF-LOCKING, HE .... UOC:DAE, DAF, V19, V20, ZAE, ZAF	4
3	PAOZZ	5310009353569	96906	MS51943-46	NUT, SELF-LOCKING, HE .... UOC:DAE, DAF, V19, V20, ZAE, ZAF	8
4	PBOZZ	5340002317486	19207	8331200	BRACKET, MOUNTING . ... UOC:DAE, DAF, V19, V20, ZAE, ZAF	1
5	PFOZZ	2510002317488	19207	8330737	BRACE, PINTLE HOOK .... UOC:DAE, DAF, V19, V20, ZAE, ZAF	1
6	PAOZZ	5310004883888	96906	MS51943-40	NUT, SELF-LOCKING, HE .... UOC:DAE, DAF, V19, V20, ZAE, ZAF	14
7	PAOZZ	5305007195235	80204	B1821BH050F175N	SCREW, CAP, HEXAGON H .... UOC:DAE, DAF, V19, V20, ZAE, ZAF	12
8	PAOZZ	5305007195239	96906	MS90727-116	SCREW, CAP, HEXAGON H .... UOC:DAE, DAF, V19, V20, ZAE, ZAF	2
9	PBOZZ	5340000402321	19207	8330742	BRACKET, MOUNTING REAR CROSSMEMBER...... UOC:DAE, DAF, V19, V20, ZAE, ZAF	1
10	PAOZZ	5305009261826	80204	B1821BH075F275N	SCREW, CAP, HEXAGON H .... UOC:DAE, DAF, V19, V20, ZAE, ZAF	2
11	PAOZZ	4730000504203	96906	MS15001-1	FITTING, LUBRICATION .... UOC:DAE, DAF, V19, V20, ZAE, ZAF	1
12	PAOZZ	5305007626041	80204	B1821BH075F250N	SCREW, CAP, HEXAGON H .... UOC:DAE, DAF, V19, V20, ZAE, ZAF	2
13	PAOZZ	5305009480803	96906	MS90727-191	SCREW, CAP, HEXAGON H .... UOC:DAE, DAF, V19, V20, ZAE, ZAF	4
14	PAOZZ	5305007262550	80204	B1821BH063F175N	SCREW, CAP, HEXAGON H .... UOC:DAE, DAF, V19, V20, ZAE, ZAF	4

END OF FIGURE

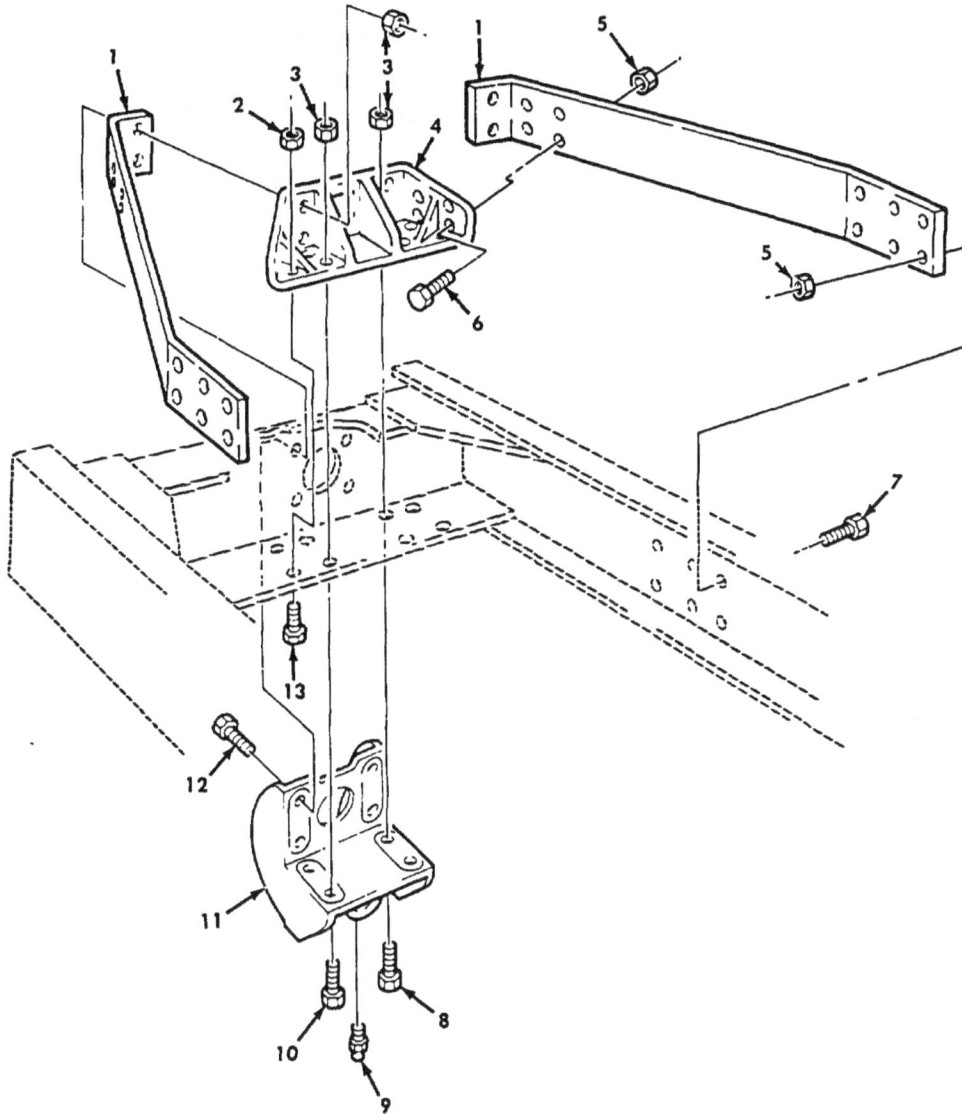

Figure 332. Towing Pintle Mounting Brackets.

(1) ITEM NO	(2) SMR CODE	(3) NSN	(4) CAGEC	(5) PART NUMBER	(6) DESCRIPTION AND USABLE ON CODES (UOC)	(7) QTY
					GROUP 1503 PINTLE AND TOWING ATTACHMENTS	
					FIG. 332 TOWING PINTLE MOUNTING BRACKETS	
1	PFOZZ	5340000402322	19207	8330743	BRACKET, MOUNTING LEFT AND RIGHT............... SIDE FRAME RAILS .... UOC:DAA, DAB, DAJ, DAK, DAL, DAW, DAX, V12, V13, V14, V15, V18, V24, V25, V39, ZAA, ZAB, ZAJ, ZAK, ZAL, ZAW, ZAX	2
1	PFOZZ	5340000969662	19207	8330743-1	BRACKET, SUPPORT, PIN LEFT AND RIGHT.......... SIDE FRAME RAILS .... UOC:DAC, DAD, V16, V17, ZAC, ZAD	2
2	PAOZZ	5310002416664	96906	MS51943-44	NUT, SELF-LOCKING, HE.................................. UOC:DAA, DAB, DAC, DAD, DAJ, DAK, DAL, DAW, DAX, V12, V13, V14, V15, V16, V17, V18, V24, V25, V39, ZAA, ZAB, ZAC, ZAD, ZAJ, ZAK, ZAL	4
3	PAOZZ	5310009353569	96906	MS51943-46	NUT, SELF-LOCKING, HE PINTLE BRACKET.......... TO REAR CROSSMEMBER.................................. UOC:DAA, DAB, DAC, DAD, DAJ, DAK, DAL, DAW, DAX, V12, V13, V14, V15, V16, V17, V18, V24, V25, V29, ZAA, ZAB, ZAC, ZAD, ZAJ, ZAK, ZAL,	6
4	PBOZZ	2540002316486	19207	8331200	BRACKET, PINTLE.. UOC:DAA, DAB, DAC, DAD, DAJ, DAK, DAL, DAW, DAX, V12, V13, V14, V15, V16, V17, V18, V24, V25, V29, ZAA, ZAB, ZAC, ZAD, ZAJ, ZAK, ZAL,	1
5	PAOZZ	5310004883888	96906	MS51943-40	NUT, SELF-LOCKING, HE.................................. UOC:DAA, DAB, DAC, DAD, DAJ, DAK, DAL, DAW, DAX, V12, V13, V14, V15, V16, V17, V18, V24, V25, V39, ZAA, ZAB, ZAC, ZAD, ZAJ, ZAK, ZAL	12
6	PAOZZ	5305007195235	80204	B1821BH050F175N	SCREW, CAP, HEXAGON H................................. UOC:DAA, DAB, DAC, DAD, DAJ, DAK, DAL, DAW, DAX, V12, V13, V14, V15, V16, V17, V18, V24, V25, V39, ZAA, ZAB, ZAC, ZAD, ZAJ, ZAK, ZAL	4
7	PAOZZ	5305007195238	80204	B1821BH050F200N	SCREW, CAP, HEXAGON H................................. UOC:DAC, DAD, DAL, V16, V17, V18, ZAC, ZAD, ZAL	8
7	PAOZZ	5305007254183	96906	MS90726-113	SCREW, CAP, HEXAGON H................................. UOC:DAA, DAB, DAJ, DAK, DAW, DAX, V12, V13, V14, V15, V24, V25, V39, ZAA, ZAB, ZAJ, ZAK	8
8	PAOZZ	5305009261826	80204	B1821BH075F275N	SCREW, CAP, HEXAGON H................................. UOC:DAA, DAB, DAC, DAD, DAJ, DAK, DAL, DAW, DAX, V12, V13, V14, V15, V16, V17, V1S8, V24, V25, V39, ZAA, ZAB, ZAC, ZAD, ZAJ, ZAK, ZAL	2
9	PAOZZ	4730000504203	96906	MS15001-1	FITTING, LUBRICATION .... UOC:DAA, DAB, DAC, DAD, DAJ, DAK, DAL, DAW, DAX, V12, V13, V14, V15, V16, V17, V18, V24, V25, V39, ZAA, ZAB, ZAC, ZAD, ZAJ, ZAK, ZAL	1
10	PAOZZ	5305007626041	80204	B1821BH075F250N	SCREW, CAP, HEXAGON H................................. UOC:DAA, DAB, DAC, DAD, DAJ, DAK, DAL, DAW, DAX, V12, V13, V14, V15, V16, V17, V18, V24, V25, V39, ZAA, ZAB, ZAC, ZAD, ZAJ, ZAK, ZAL	2

(1) ITEM NO	(2) SMR CODE	(3) NSN	(4) CAGEC	(5) PART NUMBER	(6) DESCRIPTION AND USABLE ON CODES (UOC)	(7) QTY
11	PBOZZ	5340000402321	19207	8330742	BRACKET, MOUNTING REAR CROSSMEMBER...... UOC:DAA, DAB, DAC, DAD, DAJ, DAK, DAL, DAW, DAX, V12, V13, V14, V15, V16, V17, V18, V24, V25, V39, ZAA, ZAB, ZAC, ZAD, ZAJ, ZAK, ZAL	1
12	PAOZZ	5305009480803	96906	MS90727-191	SCREW, CAP, HEXAGON H.................................. UOC:DAA, DAB, DAC, DAD, DAJ, DAK, DAL, DAW, DAX, V12, V13, V14, V15, V16, V17, V18, V24, V25, V39, ZAA, ZAB, ZAC, ZAD, ZAJ, ZAK, ZAL	4
13	PAOZZ	5305007262550	80204	B1821BH063F175N	SCREW, CAP, HEXAGON H.................................. UOC:DAA, DAB, DAC, DAD, DAJ, DAK, DAL, DAW, DAX, V12, V13, V14, V15, V16, V17, V18, V24, V25, V39, ZAA, ZAB, ZAC, ZAD, ZAJ, ZAK, ZAL	4

END OF FIGURE

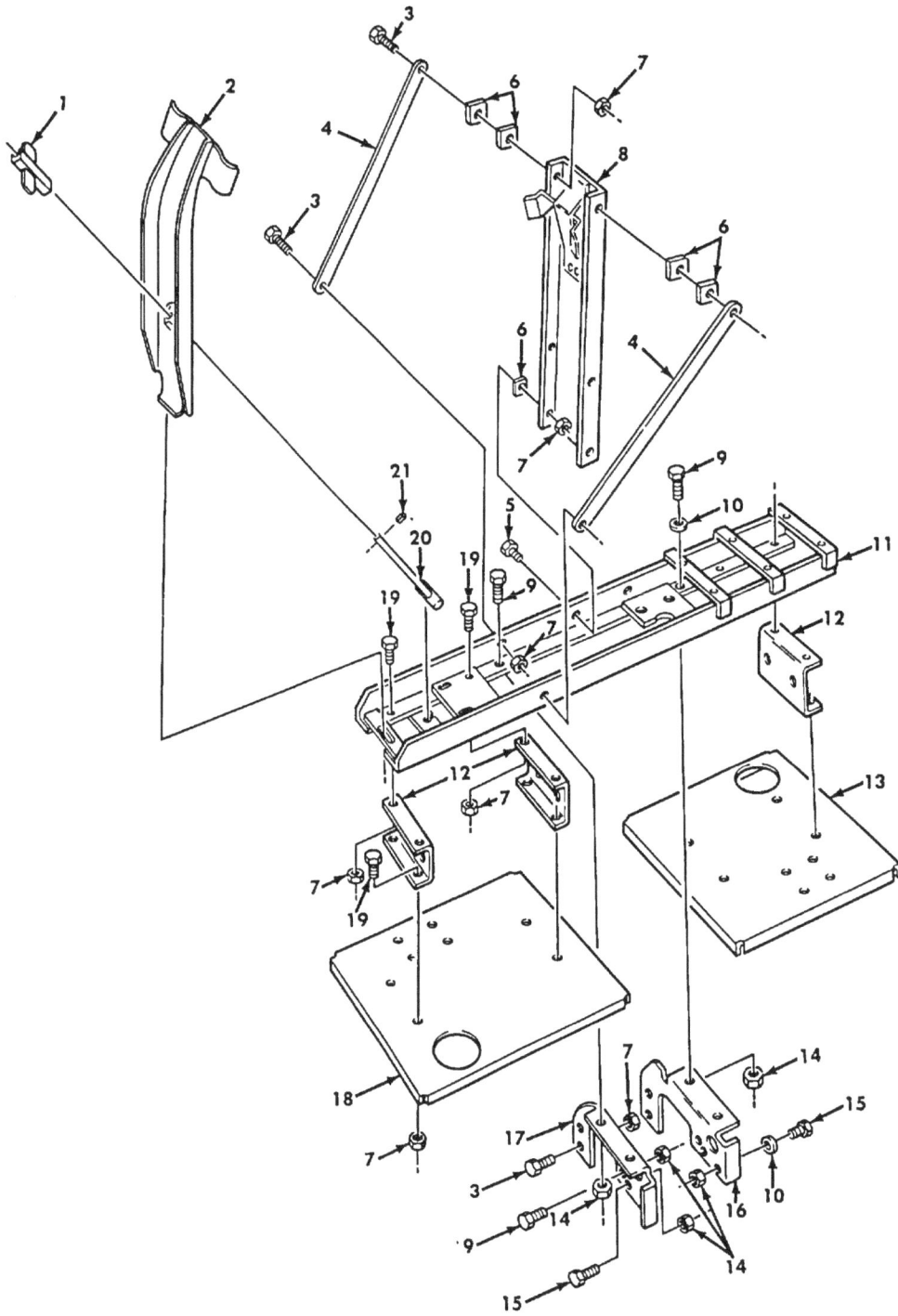

Figure 333. Spare Wheel Carrier, Dump (M939).

(1) ITEM NO	(2) SMR CODE	(3) NSN	(4) CAGEC	(5) PART NUMBER	(6) DESCRIPTION AND USABLE ON CODES (UOC)	(7) QTY
					GROUP 1504 SPARE WHEEL CARRIER AND TIRE LOCK	
					FIG. 333 SPARE WHEEL CARRIER, DUMP (M939)	
1	PAOZZ	5340010893126	19207	12276959	RETAINER, ASSEMBLED UOC:V19, V20	1
2	PFOZZ	2510007413458	19207	7413458	CLAMP, CARRIER, SPARE .... UOC:DAE, DAF, V19, V20	1
3	PAOZZ	5305002693238	80204	B1821BH038F125N	SCREW, CAP, HEXAGON H UOC:V19, V20	6
4	PFOZZ	2510007409534	19207	7409534	BRACE, TIRE CARRIER UOC:V19, V20	2
5	PAOZZ	5305009146131	96906	MS18153-63	SCREW, CAP, HEXAGON H UOC:V19, V20	4
6	PAOZZ	5310004199461	19207	8758194	WASHER, FLAT UOC:V19, V20	2
7	PAOZZ	5310008140672	96906	MS51943-36	NUT, SELF-LOCKING, HE UOC:V19, V20	32
8	PAOZZ	5340011208443	19207	8758195-1	BRACKET, ANGLE UOC:V19, V20	1
9	PAOZZ	5305007254183	96906	MS90726-113	SCREW, CAP, HEXAGON H UOC:V19, V20	6
10	PAOZZ	5310008095998	96906	MS27183-18	WASHER, FLAT... UOC:V19, V20	4
11	XDOZZ		19207	12256386	RACK, TIRE UOC:V19, V20	1
12	PFOZZ	5340010889152	19207	8758442	PLATE, MOUNTING.. UOC:V19, V20	4
13	PAOZZ	9515010911624	19207	12256255-2	PLATE, FLOOR, METAL UOC:V19, V20	1
14	PAOZZ	5310004883888	96906	MS51943-40	NUT, SELF-LOCKING, HE UOC:V19, V20	2
15	PAOZZ	5305007195235	80204	B1821BH050F175N	SCREW, CAP, HEXAGON H UOC:V19, V20	4
16	PFOZZ	2590010911635	19207	12256669	BRACKET, VEHICULAR C UOC:V19, V20	1
17	PFOZZ	2510007409535	19207	7409535	BRACKET, SUPPORT UOC:V19, V20	1
18	PFOZZ	9535011406470	19207	12256255-3	PLATE, FLOOR, METAL UOC:V19, V20	1
19	PAOZZ	5305002692803	96906	MS90726-60	SCREW, CAP, HEXAGON H UOC:V19, V20	22
20	PAOZZ	5306010904544	19207	12276960-1	BOLT, TEE HEAD UOC:V19, V20	1
21	PAOZZ	5315008103701	96906	MS16562-36	PIN, SPRING UOC:V19, V20	1

END OF FIGURE

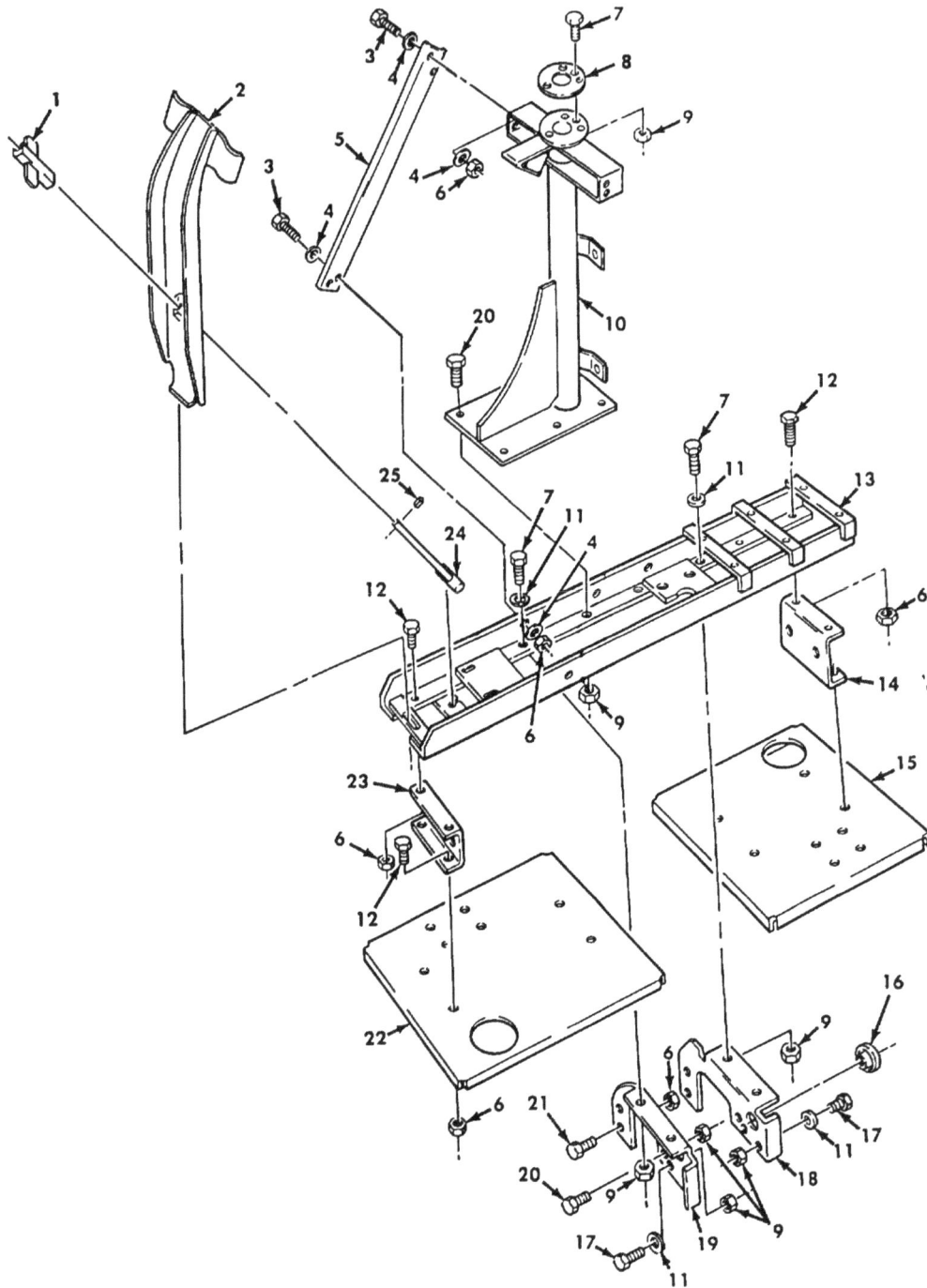

Figure 334. Spare Wheel Carrier, Dump (M939A1, M939A2).

(1) ITEM NO	(2) SMR CODE	(3) NSN	(4) CAGEC	(5) PART NUMBER	(6) DESCRIPTION AND USABLE ON CODES (UOC)	(7) QTY
					GROUP 1504 SPARE WHEEL CARRIER AND TIRE LOCK	
					FIG. 334 SPARE WHEEL CARRIER, DUMP (M939A1, M939A2)	
1	PAOZZ	5340010893126	19207	12276959	RETAINER, ASSEMBLED .... UOC:DAE, DAF, ZAE, ZAF	1
2	PFOZZ	2510012166834	19207	12301101	CLAMP, CARRIER, SPARE .... UOC:DAE, DAF, ZAE, ZAF	1
3	PAOZZ	5305002692804	96906	MS90726-61	SCREW, CAP, HEXAGON H .... UOC:DAE, DAF, ZAE, ZAF	8
4	PAOZZ	5310000806004	96906	MS27183-14	WASHER , FLAT .... UOC:DAE, DAF, ZAE, ZAF	8
5	PAOZZ	9520012102159	19207	12301094	CHANNEL, STRUCTURAL FRONT................ UOC:DAE, DAF, ZAE, ZAF	1
5	PBOZZ	9520012102160	19207	12301095	CHANNEL, STRUCTURAL REAR.................. UOC:DAE, DAF, ZAE, ZAF	1
6	PAOZZ	5310008140672	96906	MS51943-36	NUT, SELF-LOCKING, HE .... UOC:DAE, DAF, ZAE, ZAF	26
7	PAOZZ	5305007195238	80204	B1821BH050F200N	SCREW, CAP, HEXAGON H .... UOC:DAE, DAF, ZAE, ZAF	4
8	PAOZZ	5340012334202	19207	12303064	COVER, ACCESS LOWER DAVIT ASSEMBLY........ UOC:DAE, DAF, ZAE, ZAF	1
9	PAOZZ	5310004883888	96906	MS51943-40	NUT, SELF-LOCKING, HE .... UOC:DAE, DAF, ZAE, ZAF	14
10	XDOZZ		19207	12303028	BRACKET, MOUNTING .... UOC:DAE, DAF, ZAE, ZAF	1
11	PAOZZ	5310008095998	96906	MS27183-1B	WASHER, FLAT .... UOC:DAE, DAF, ZAE, ZAF	5
12	PAOZZ	5305002692803	96906	MS90726-60	SCREW, CAP, HEXAGON H .... UOC:DAE, DAF, ZAE, ZAF	16
13	PFOZZ	2510012116611	19207	12303032	BASE ASSEMBLY, TIRE .... UOC:DAE, DAF, ZAE, ZAF	1
14	PFOZZ	5340010889152	19207	8758442	PLATE, MOUNTING . ... UOC:DAE, DAF, ZAE, ZAF	2
15	PAOZZ	9515010911624	19207	12256255-2	PLATE, FLOOR, METAL FUEL TANK R.H .... UOC:DAE, DAF, ZAE, ZAF	1
16	PAOZZ	5325007397776	19207	7397776	EYELET, METALLIC .... UOC:DAE, DAF, ZAE, ZAF	2
17	PAOZZ	5305007195235	80204	B1821BH050F175N	SCREW, CAP, HEXAGON H .... UOC:DAE, DAF, ZAE, ZAF	4
18	PFOZZ	2590010911635	19207	12256669	BRACKET, VEHICULAR C .... UOC:DAE, DAF, ZAE, ZAF	1
19	PAOZZ	5340012102162	19207	12302919	BRACKET, MOUNTING . ... UOC:DAE, DAF, ZAE, ZAF	1
20	PAOZZ	5305007254183	96906	MS90726-113	SCREW, CAP, HEXAGON H .... UOC:DAE, DAF, ZAE, ZAF	6
21	PAOZZ	5305002693238	80204	B1821BH038F125N	SCREW, CAP, HEXAGON H .... UOC:DAE, DAF, ZAE, ZAF	2
22	PFOZZ	2510012102163	19207	12302966	PANEL, BODY, VEHICULA FUEL TANK L.H............	1

(1) ITEM NO	(2) SMR CODE	(3) NSN	(4) CAGEC	(5) PART NUMBER	(6) DESCRIPTION AND USABLE ON CODES (UOC)	(7) QTY
					UOC:DAE, DAF, ZAE, ZAF	
23	PAOZZ	5340012102164	19207	12302965	BRACKET, DOUBLE ANGL	2
					UOC:DAE, DAF, ZAE, ZAF	
24	PAOZZ	5306012100264	19207	12276960-6	BOLT, TEE HEAD..	1
					UOC:DAE, DAF, ZAE, ZAF	
25	PAOZZ	5315008103701	96906	MS16562-36	PIN, SPRING	1
					UOC:DAE, DAF, ZAE, ZAF	

END OF FIGURE

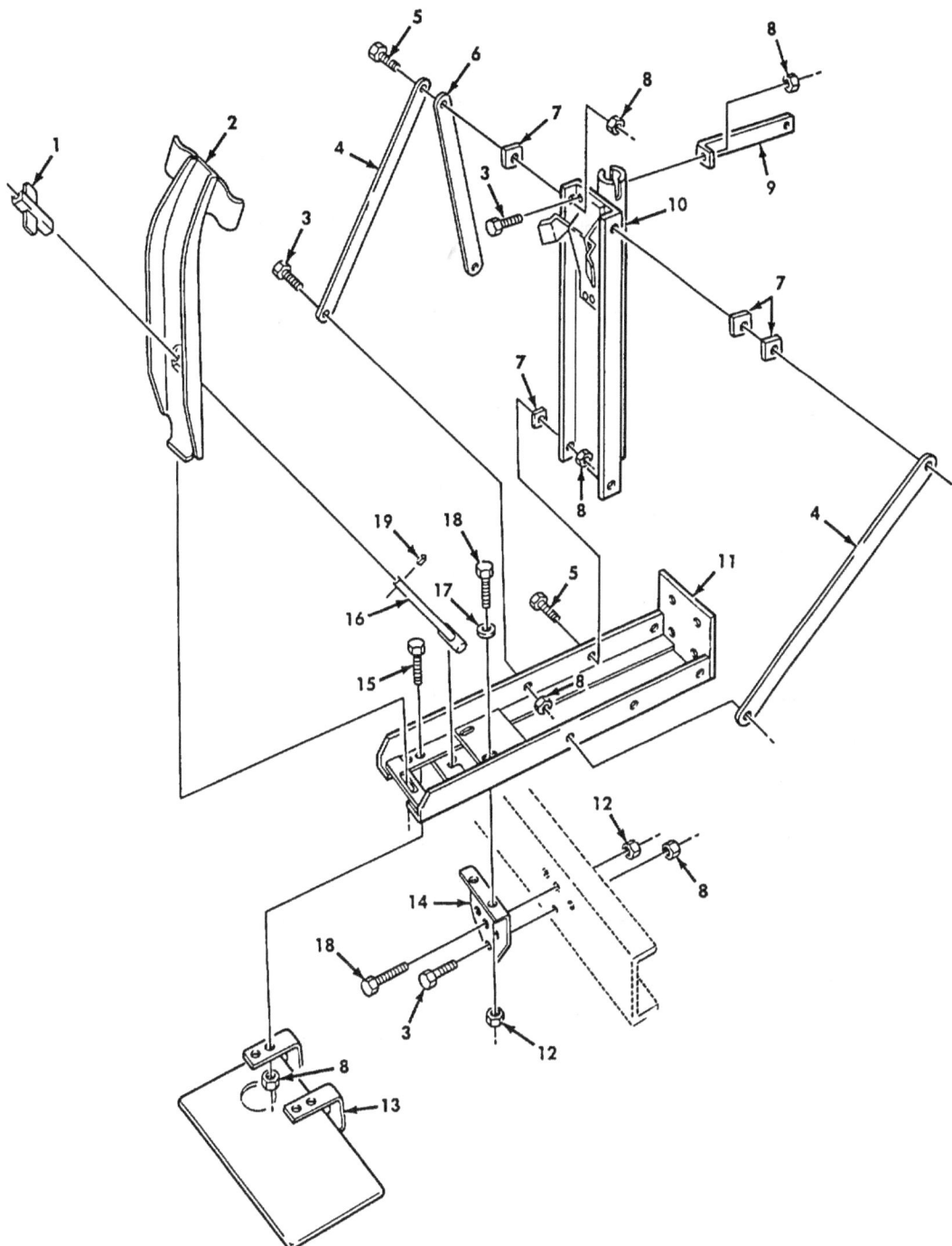

Figure 335. Spare Wheel Carrier, Cargo and Van (M939).

(1) ITEM NO	(2) SMR CODE	(3) NSN	(4) CAGEC	(5) PART NUMBER	(6) DESCRIPTION AND USABLE ON CODES (UOC)	(7) QTY
					GROUP 1504 SPARE WHEEL CARRIER AND TIRE LOCK	
					FIG. 335 SPARE WHEEL CARRIER, CARGO AND VAN (M14939)	
1	PAOZZ	5340010893126	19207	12276959	RETAINER, ASSEMBLED UOC:V12, V13, V14, V15, V16, V17, V24, V25	1
2	PFOZZ	2510007413458	19207	7413458	CLAMP, CARRIER, SPARE CARRIER, SPARE UOC:V12, V13, V14, V15, V16, V17, V24, V2 5	1
3	PAOZZ	5305002693238	80204	B1821BH038F125N	SCREW, CAP, HEXAGON H UOC:V12, V13, V14, V15, V16, V17, V24, V25	7
4	PFOZZ	2510007409534	19207	7409534	BRACE, TIRE CARRIER UOC:V12, V13, V14, V15, V16, V17, V24, V25	2
5	PAOZZ	5305009146131	96906	MS18153-63	SCREW, CAP, HEXAGON H UOC:V12, V13, V14, V15, V16, V17, V24, V25	4
6	PFOZZ	2540010889162	19207	7409534-2	CARRIER, TIRE HOLDER .... UOC:V12, V13, V1, V V15, V16, V17, V 24, V25	1
7	PAOZZ	5310004199461	19207	8758194	WASHER, FLAT UOC:V12, V13, V14, V15, V16, V17, V24, V25	5
8	PAOZZ	5310008140672	96906	MS51943-36	NUT, SELF-LOCKING, HE UOC:V12, V13, V14, V15, V16, V17, V24, V2 5	15
9	PFOZZ	5340010893067	19207	12256668	BRACKET, ANGLE UOC:V12, V13, V14, V15, V16, V17, V24, V25	2
10	PFOZZ	2590010911620	19207	12256249	HOIST UNIT SUPPORT UOC:V12, V13, V14, V15, V16, V17	1
10	PFOZZ	2510004051978	19207	8758195	BRACKET, SPARE WHEEL UOC:V24, V25	1
11	PFOZZ	2540010911622	19207	12256253	CARRIER, TIRE SUPPORT .... UOC:V12, V13, V1, V V15, V16, V17, V 24, V25	1
12	PAOZZ	5310004883888	96906	MS51943-40	NUT, SELF-LOCKING, HE UOC:V12, V13, V14, V15, V16, V17, V24, V25	2
13	PFOZZ	2540010911625	19207	12256256	STEP SUPPORT, TIRE C UOC:V12, V13, V14, V15, V16, V17, V24, V25	1
14	PFOZZ	5340010911621	19207	12256252	BRACKET, ANGLE UOC:V12, V13, V14, V15, V16, V17, V24, V25	1
15	PAOZZ	5305002692804	96906	MS90726-61	SCREW, CAP, HEXAGON H UOC:V12, V13, V14, V15, V16, V17, V24, V25	4
16	PAOZZ	5306010904544	19207	12276960-1	BOLT, TEE HEAD UOC:V12, V13, V14, V15, V16, V17, V24, V2 5	1
17	PAOZZ	5310008095998	96906	MS27183-18	WASHER, FLAT UOC:V12, V13, V14, V15, V16, V17, V24, V25	2
18	PAOZZ	5305007254183	96906	MS90726-113	SCREW, CAP, HEXAGON H .... UOC:V12, V13, V14, V14 , V1516, V17, V24, V25	2
19	PAOZZ	5315008103701	96906	MS16562-36	PIN, SPRING UOC:V12, V13, V14, V15, V16, V17, V24, V25	1

END OF FIGURE

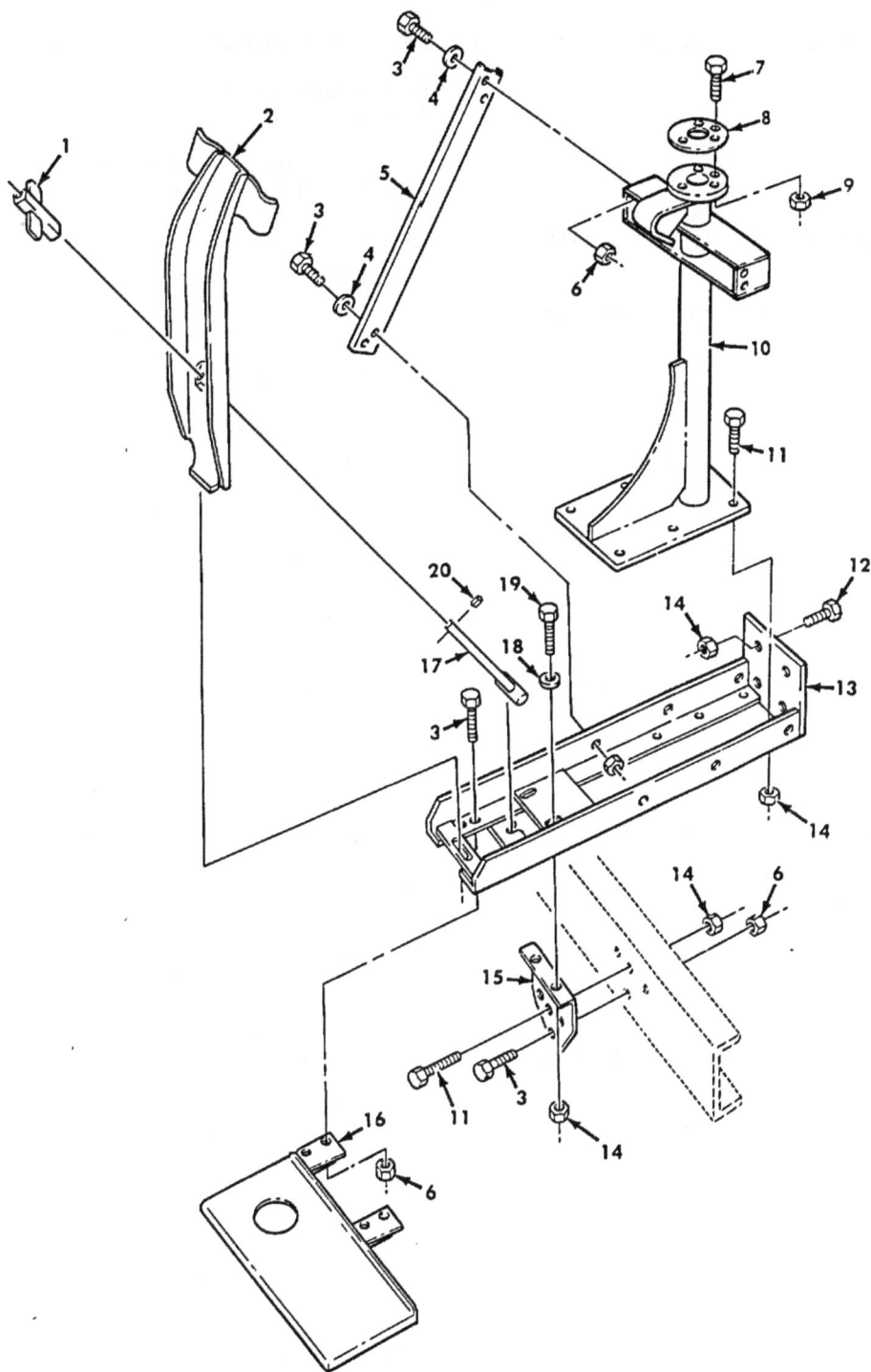

Figure 336. Spare Wheel Carrier, Cargo and Van (M939A1, M939A2).

(1) ITEM NO	(2) SMR CODE	(3) NSN	(4) CAGEC	(5) PART NUMBER	(6) DESCRIPTION AND USABLE ON CODES (UOC)	(7) QTY
					GROUP 1504 SPARE WHEEL CARRIER AND TIRE LOCK	
					FIG. 336 SPARE WHEEL CARRIER, CARGO AND VAN(M939A1, M939A2)	
1	PAOZZ	5340010893126	19207	12276959	RETAINER, ASSEMBLED .... UOC:DAA, DAB, DAC, DAD, DAJ, DAK, DAW, DAX, ZAA, ZAB, ZAC, ZAD, ZAJ, ZAK	1
2	PFOZZ	2510012166834	19207	12301101	CLAMP, CARRIER, SPARE .... UOC:DAA, DAB, DAC, DAD, DAJ, DAK, DAW, DAX, ZAA, ZAB, ZAC, ZAD, ZAJ, ZAK	1
3	PAOZZ	5305002693238	80204	B1821BH038F125N	SCREW, CAP, HEXAGON H .... UOC:DAA, DAB, DA C, DAD, DAJ, DAK, DAW, DAX, ZAA, ZAB, ZAC, ZAD, ZAJ, ZAK	14
4	PAOZZ	5310000806004	96906	MS27183-14	WASHER, FLAT . ... UOC:DAA, DAB, DAC, DAD, DAJ, DAK, DAW, DAX, ZAA, ZAB, ZAC, ZAD, ZAJ, ZAK	8
5	PAOZZ	9520012102159	19207	12301094	CHANNEL, STRUCTURAL FRONT .... UOC:DAA, DAB, DAC, DAD, DAJ, DAK, DAW, DAX, ZAA, ZAB, ZAC, ZAD, ZAJ, ZAK	1
5	PBOZZ	9520012102160	19207	12301095	CHANNEL, STRUCTURAL REAR .... UOC:DAA, DAB, DAC, DAD, DAJ, DAK, DAW, DAX, ZAA, ZAB, ZAC, ZAD, ZAJ, ZAK	1
6	PAOZZ	5310008140672	96906	MS51943-36	NUT, SELF-LOCKING, HE .... UOC:DAA, DAB, DAC, DAD, DAJ, DAK, DAW, DAX, ZAA, ZAB, ZAC, ZAD, ZAJ, ZAK	14
7	PAOZZ	5305007195238	80204	B1821BH050F200N	SCREW, CAP, HEXAGON H .... UOC:DAA, DAB, DAC, DAD, DAJ, DAK, DAW, DAX, ZAA, ZAB, ZAC, ZAD, ZAJ, ZAK	3
8	PAOZZ	5340012334202	19207	12303064	COVER, ACCESS LOWER DAVIT ASSEMBLY......... UOC:DAA, DAB, DAC, DAD, DAJ, DAK, DAW, DAX, ZAA, ZAB, ZAC, ZAD, ZAJ, ZAK	1
9	PAOZZ	5310004883888	96906	MS51943-40	NUT, SELF-LOCKING, HE .... UOC:DAA, DAB, DAC, DAD, DAJ, DAK, DAW, DAX, ZAA, ZAB, ZAC, ZAD, ZAJ, ZAK	3
10	PFOZZ	2590013117213	19207	12375378	BRACKET, VEHICULAR C .... UOC:DAA, DAB, DAC, DAD, DAJ, DAK, DAW, DAX, ZAA, ZAB, ZAC, ZAD, ZAJ, ZAK	1
11	PAOZZ	5305007254183	96906	MS90726-113	SCREW, CAP, HEXAGON H .... UOC:DAA, DAB, DAC, DAD, DAJ, DAK, DAW, DAX, ZAA, ZAB, ZAC, ZAD, ZAJ, ZAK	8
12	PAOZZ	5305007195235	80204	B1821BH050F175N	SCREW, CAP, HEXAGON H .... UOC:DAA, DAB, DAC, DAD, DAJ, DAK, DAW, DAX, ZAA, ZAB, ZAC, ZAD, ZAJ, ZAK	4
13	PFOZZ	2510012104651	19207	12375377	BASE ASSEMBLY , TIRE .... UOC:DAJ, DAK, ZAJ, ZAK	1
13	PAOZZ	2510012102166	19207	12375379	PANEL, BODY, VEHICULA .... UOC:DAA, DAB, DA C, DAD, DAW, DAX, ZAA, ZAB, ZAC, ZAD	1
14	PAOZZ	5310004883888	96906	MS51943-40	NUT, SELF-LOCKING, HE .... UOC:DAA, DAB, DAC, DAD, DAJ, DAK, DAW, DAX, ZAA, ZAB, ZAC, ZAD, ZAJ, ZAK	14

(1) ITEM NO	(2) SMR CODE	(3) NSN	(4) CAGEC	(5) PART NUMBER	(6) DESCRIPTION AND USABLE ON CODES (UOC)	(7) QTY
15	PFOZZ	5340012128476	19207	12301096	BRACKET, ANGLE.. UOC:DAA, DAB, DAC, DAD, DAJ, DAK, DAW, DAX, ZAA, ZAB, ZAC, ZAD, ZAJ, ZAK	1
16	PFOZZ	2510012102167	19207	12301098	STEP ASSEMBLY .... UOC:DAA, DAB, DAC, DAD, DAJ, DAK, DAW, DAX, ZAA, ZAB, ZAC, ZAD, ZAJ, ZAK	1
17	PAOZZ	5306012100264	19207	12276960-6	BOLT, TEE HEAD .... UOC:DAA, DAB, DAC, DAD, DAJ, DAK, DAW, DAX, ZAA, ZAB, ZAC, ZAD, ZAJ, ZAK	1
18	PAOZZ	5310008095998	96906	MS27183-18	WASHER, FLAT.. UOC:DAA, DAB, DAC, DAD, DAJ, DAK, DAW, DAX, ZAA, ZAB, ZAC, ZAD, ZAJ, ZAK	2
19	PAOZZ	5305007168186	96906	MS90726-110	SCREW, CAP, HEXAGON H .... UOC:DAA, DAB, DAC, DAD, DAJ, DAK, DAW, DAX, ZAA, ZAB, ZAC, ZAD, ZAJ, ZAK	2
20	PAOZZ	5315008103701	96906	MS16562-36	PIN, SPRING .... UOC:DAA, DAB, DAC, DAD, DAJ, DAK, DAW, DAX, ZAA, ZAB, ZAC, ZAD, ZAJ, ZAK	1

END OF FIGURE

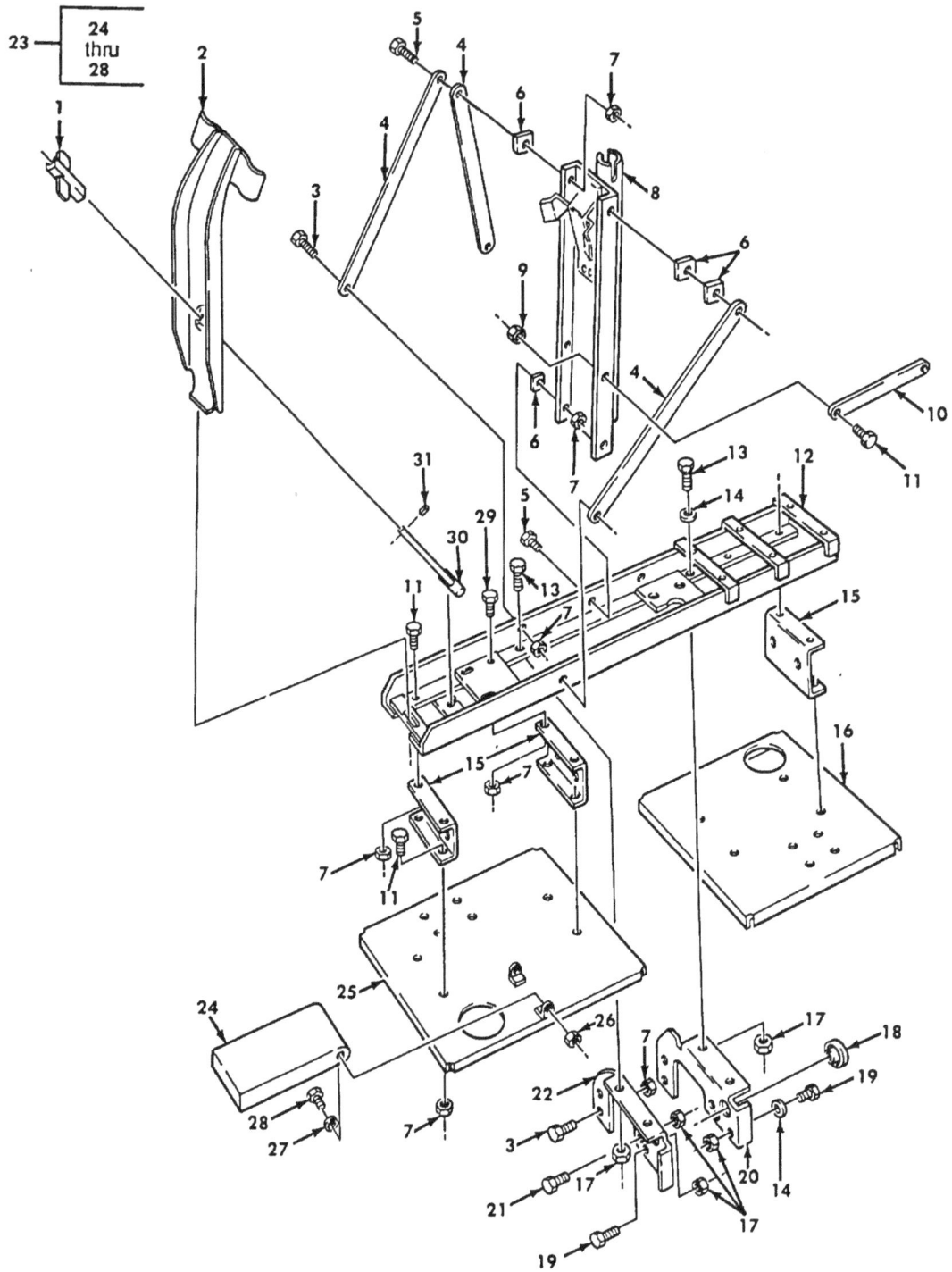

Figure 337. Spare Wheel Carrier, Tractor (M939).

(1) ITEM NO	(2) SMR CODE	(3) NSN	(4) CAGEC	(5) PART NUMBER	(6) DESCRIPTION AND USABLE ON CODES (UOC)	(7) QTY
					GROUP 1504 SPARE WHEEL CARRIER AND TIRE LOCK	
					FIG. 337 SPARE WHEEL CARRIER, TRACTOR(M939)	
1	PAOZZ	5340010893126	19207	12276959	RETAINER, ASSEMBLED UOC:V21, V22	1
2	PFOZZ	2510007413458	19207	7413458	CLAMP, CARRIER, SPARE UOC:V21, V22	1
3	PAOZZ	5305002693238	80204	B1821BH038F125N	SCREW, CAP, HEXAGON H UOC:V21, V22	5
4	PFOZZ	2510007409534	19207	7409534	BRACE, TIRE CARRIER UOC:V21, V22	3
5	PAOZZ	5305009146131	96906	MS18153-63	SCREW, CAP, HEXAGON H UOC:V21, V22	4
6	PAOZZ	5310004199461	19207	8758194	WASHER, FLAT UOC:V21, V22	5
7	PAOZZ	5310008140672	96906	MS51943-36	NUT, SELF-LOCKING, HE UOC:V21, V22	25
8	PFOZZ	2590010911620	19207	12256249	HOIST UNIT SUPPORT UOC:V21, V22	1
9	PAOZZ	5310009591488	96906	MS51922-21	NUT, SELF-LOCKING UOC:V21, V22	2
10	PFOZZ	5340010891355	19207	8758443	PLATE, MENDING UOC:V21, V22	1
11	PAOZZ	5305002692803	96906	MS90726-60	SCREW, CAP, HEXAGON H UOC:V21, V22	16
12	PAOZZ	7125010911619	19207	12256248	RACK, TIRE UOC:V21, V22	1
13	PAOZZ	5305007195238	80204	B1821BH050F200N	SCREW, CAP, HEXAGON H UOC:V21, V22	4
14	PAOZZ	5310008095998	96906	MS27183-18	WASHER , FLAT UOC:V21, V22	4
15	PFOZZ	5340010889152	19207	8758442	PLATE, MOUNTING UOC:V21, V22	4
16	PAOZZ	9515010911624	19207	12256255-2	PLATE, FLOOR, METAL UOC:V21, V22	1
17	PAOZZ	5310004883888	96906	MS51943-40	NUT, SELF-LOCKING, HE UOC:V21, V22	4
18	PAOZZ	5325007397776	19207	7397776	EYELET, METALLIC UOC:V21, V22	1
19	PAOZZ	5305007195235	80204	B1821BH050F175N	SCREW, CAP, HEXAGON H UOC:V21, V22	4
20	PFOZZ	2590010911635	19207	12256669	BRACKET, VEHICULAR C UOC:V21, V22	1
21	PAOZZ	5305007254183	96906	MS90726-113	SCREW, CAP, HEXAGON H UOC:V21, V22	2
22	PFOZZ	2510007409535	19207	7409535	BRACKET, SUPPORT UOC:V21, V22	1
23	PFOZZ	5340011317447	19207	12302654	COVER, ACCESS UOC:V21, V22	1
24	PFOZZ	2510011317449	19207	12302653	DECK, FUEL TANK UOC:V21, V22	1

(1) ITEM NO	(2) SMR CODE	(3) NSN	(4) CAGEC	(5) PART NUMBER	(6) DESCRIPTION AND USABLE ON CODES (UOC)	(7) QTY
25	PFOZZ	2510011317448	19207	12302650	COVER, ACCESS UOC:V21, V22	1
26	PAOZZ	5310008140672	96906	MS51943-36	NUT, SELF-LOCKING, HE UOC:V21, V22	2
27	PAOZZ	5310010531936	81349	M12133/1-6P	WASHER, SPRING TENSI UOC:V21, V22	2
28	PAOZZ	5305009125113	96906	MS51096-359	SCREW, CAP, HEXAGON H UOC:V21, V22	2
29	PAOZZ	5305002692804	96906	MS90726-61	SCREW, CAP, HEXAGON H UOC:V21, V22	2
30	PAOZZ	5306010904544	19207	12276960-1	BOLT, TEE HEAD UOC:V21, V22	1
31	PAOZZ	5315008103701	96906	MS16562-36	PIN, SPRING UOC:V21, V22	1

END OF FIGURE

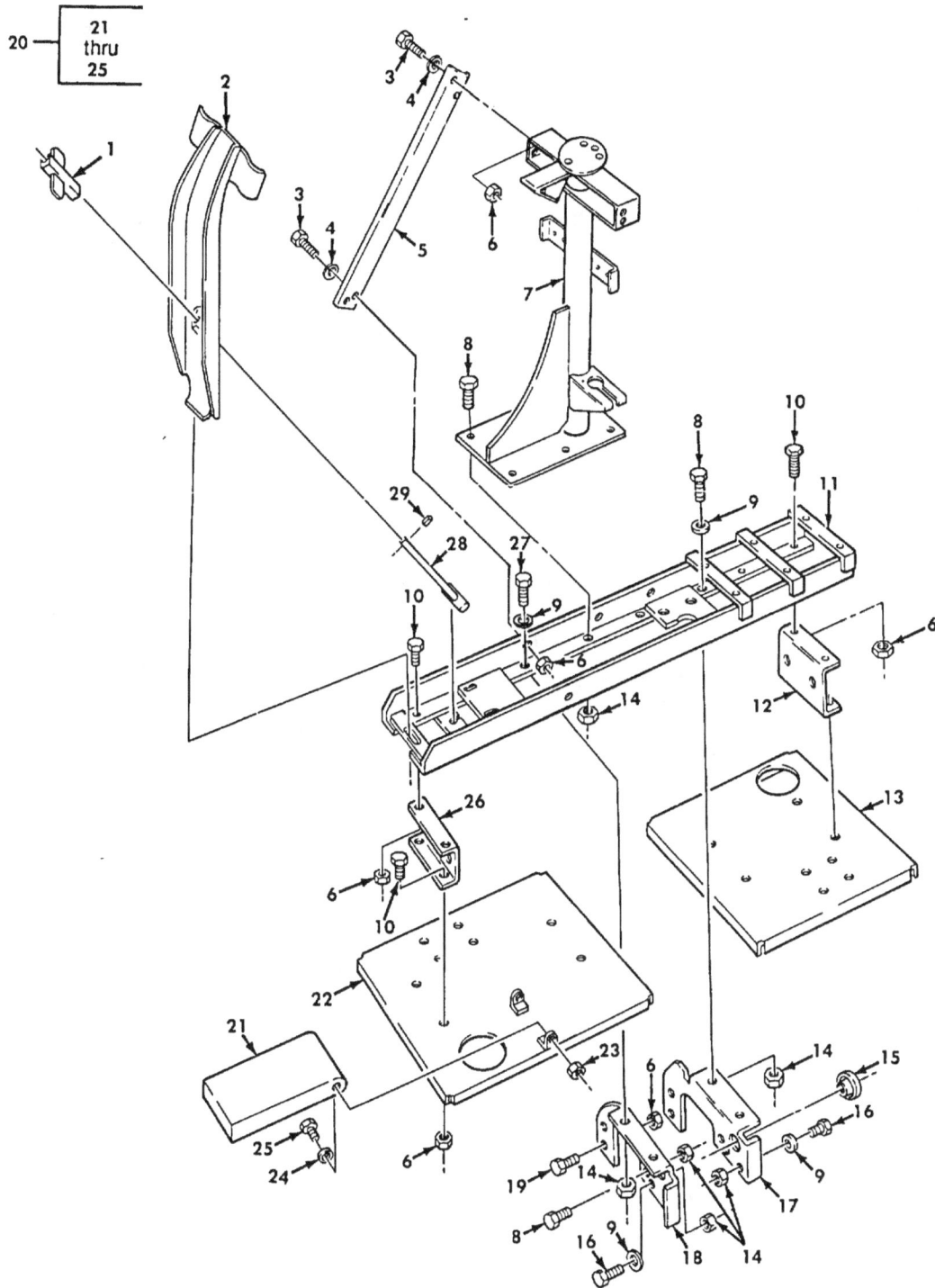

Figure 338. Spare Wheel Carrier, Tractor (M939A1, M939A2).

(1) ITEM NO	(2) SMR CODE	(3) NSN	(4) CAGEC	(5) PART NUMBER	(6) DESCRIPTION AND USABLE ON CODES (UOC)	(7) QTY
					GROUP 1504 SPARE WHEEL CARRIER AND TIRE LOCK	
					FIG. 338 SPARE WHEEL CARRIER, TRACTOR(M939A1, M939A2)	
1	PAOZZ	5340010893126	19207	12276959	RETAINER, ASSEMBLED .... UOC:DAG, DAH, ZAG, ZAH	1
2	PFOZZ	2510012166834	19207	12301101	CLAMP, CARRIER, SPARE .... UOC:DAG, DAH, ZAG, ZAH	1
3	PAOZZ	5305002692804	96906	MS90726-61	SCREW, CAP, HEXAGON H .... UOC:DAG, DAH, ZAG, ZAH	8
4	PAOZZ	5310000806004	96906	MS27183-14	WASHER, FLAT .... UOC:DAG, DAH, ZAG, ZAH	8
5	PAOZZ	9520012102159	19207	12301094	CHANNEL, STRUCTURAL FRONT......................... UOC:DAG, DAH, ZAG, ZAH	1
5	PBOZZ	9520012102160	19207	12301095	CHANNEL, STRUCTURAL REAR........................... UOC:DAG, DAH, ZAG, ZAH	1
6	PAOZZ	5310008140672	96906	MS51943-36	NUT, SELF-LOCKING, HE .... UOC:DAG, DAH, ZAG, ZAH	26
7	PFOZZ	2510012104652	19207	12302918	LOWER DAVIT ASSEMBL .... UOC:DAG, DAH, ZAG, ZAH	1
8	PAOZZ	5305007254183	96906	MS90726-113	SCREW, CAP, HEXAGON H .... UOC:DAG, DAH, ZAG, ZAH	10
9	PAOZZ	5310008095998	96906	MS27183-18	WASHER, FLAT... UOC:DAG, DAH, ZAG, ZAH	5
10	PAOZZ	5305002692803	96906	MS90726-60	SCREW, CAP, HEXAGON H .... UOC:DAG, DAH, ZAG, ZAH	16
11	PFOZZ	2510012104653	19207	12302920	FRAME, STRUCTURAL, VE .... UOC:DAG, DAH, ZAG, ZAH	1
12	PFOZZ	5340010889152	19207	8758442	PLATE, MOUNTING .... UOC:DAG, DAH, ZAG, ZAH	2
13	PAOZZ	9515010911624	19207	12256255-2	PLATE, FLOOR, METAL FUEL TANK R.H .... UOC:DAG, DAH, ZAG, ZAH	1
14	PAOZZ	5310004883888	96906	MS51943-40	NUT, SELF-LOCKING, HE .... UOC:DAG, DAH, ZAG, ZAH	16
15	PAOZZ	5325007397776	19207	7397776	EYELET, METALLIC.. UOC:DAG, DAH, ZAG, ZAH	2
16	PAOZZ	5305007195235	80204	B1821BH050F175N	SCREW, CAP, HEXAGON H .... UOC:DAG, DAH, ZAG, ZAH	4
17	PFOZZ	2590010911635	19207	12256669	BRACKET, VEHICULAR C .... UOC:DAG, DAH, ZAG, ZAH	1
18	PAOZZ	5340012102162	19207	12302919	BRACKET, MOUNTING .... UOC:DAG, DAH, ZAG, ZAH	1
19	PAOZZ	5305002693238	80204	B1821BH038F125N	SCREW, CAP, HEXAGON H .... UOC:DAG, DAH, ZAG, ZAH	2
20	PFOZZ	2510012108799	19207	12302654-1	PANEL, BODY, VEHICULA .... UOC:DAG, DAH, ZAG, ZAH	1
21	PFOZZ	2510011317448	19207	12302650	.COVER, DECK, FUEL TANK .... UOC:DAG, DAH, ZAG, ZAH	1
22	PFOZZ	2510012108800	19207	12302653-1	.PANEL, BODY, VEHICULA .... UOC:DAG, DAH, ZAG, ZAH	1
23	PAOZZ	5310008140672	96906	MS51943-36	.NUT, SELF-LOCKING, HE	2

(1) ITEM NO	(2) SMR CODE	(3) NSN	(4) CAGEC	(5) PART NUMBER	(6) DESCRIPTION AND USABLE ON CODES (UOC)	(7) QTY
					UOC:DAG, DAH, ZAG, ZAH	
24	PAOZZ	5310010531936	81349	M12133/1-6P	.WASHER, SPRING TENSI	2
					UOC:DAG, DAH, ZAG, ZAH	
25	PAOZZ	5305002693235	80204	B1821BH038F088N	.SCREW, CAP, HEXAGON H	2
					UOC:DAG, DAH, ZAG, ZAH	
26	PAOZZ	5340012102164	19207	12302965	BRACKET, DOUBLE ANGL	2
					UOC:DAG, DAH, ZAG, ZAH	
27	PAOZZ	5305007195238	80204	B1821BH050F200N	SCREW, CAP, HEXAGON H	2
					UOC:DAG, DAH, ZAG, ZAH	
28	PAOZZ	5306012100264	19207	12276960-6	BOLT, TEE HEAD...................................	1
					UOC:DAG, DAH, ZAG, ZAH	
29	PAOZZ	5315008103701	96906	MS16562-36	PIN, SPRING	1
					UOC:DAG, DAH, ZAG, ZAH	

END OF FIGURE

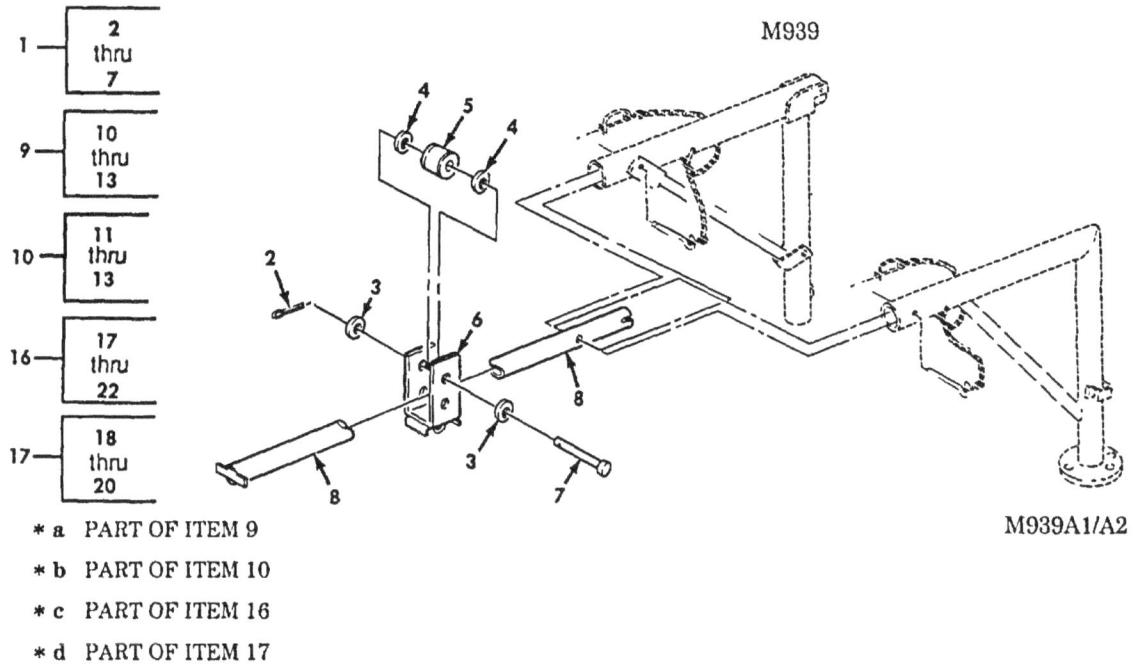

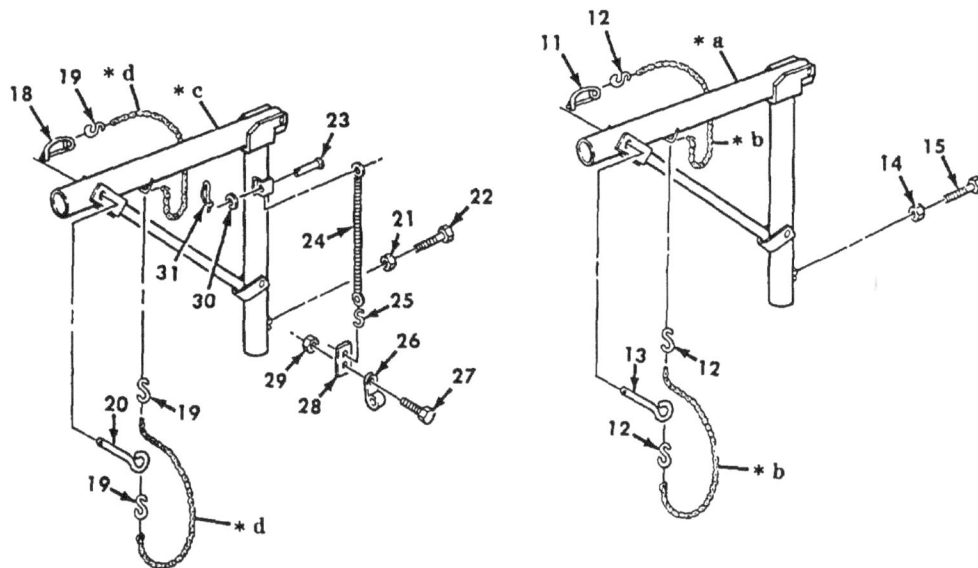

Figure 339. Lifting Device and Davit Assemblies.

(1) ITEM NO	(2) SMR CODE	(3) NSN	(4) CAGEC	(5) PART NUMBER	(6) DESCRIPTION AND USABLE ON CODES (UOC)	(7) QTY
					GROUP 1504 SPARE WHEEL CARRIER AND TIRE LOCK	
					FIG. 339 LIFTING DEVICE AND DAVIT ASSEMBLIES	
1	PAOZZ	3950010941381	19207	12256239	TROLLEY, MONORAIL .... UOC:V12, V13, V14, V15, V16, V17, V21, V22, DAA, DAB, DAC, DAD, DAG, DAH, DAW, DAX, ZAA, ZAB, ZAC, ZAD, ZAG. ZAH.................	1
2	PAOZZ	5315002341864	96906	MS24665-302	.PIN, COTTER................................ UOC:V12, V13, V14, V15, V16, V17, V21, V22, DAA, DAB, DAC, DAD, DAG, DAH, DAW, DAX, ZAA, ZAB, ZAC, ZAD, ZAG.ZAH.................	1
3	PAOZZ	5310008095998	96906	MS27183-18	.WASHER, FLAT .... UOC:V12, V13, V14, V15, V16, V17, V21, V22, DAA, DAB, DAC, DAD, DAG, DAH, DAW, DAX, ZAA, ZAB, ZAC, ZAD, ZAG.ZA	2
4	PAOZZ	5365011208442	19207	12256239-3	.SPACER, RING.................... UOC:V12, V13, V14, V15, V16, V17, V21, V22, DAA, DAB, DAC, DAD, DAG, DAH, DAW, DAX, ZAA, ZAB, ZAC, ZAD, ZAG.ZAH.................	2
5	PAOZZ	3950011281549	19207	12256239-2	.TROLLEY, MONORAIL.................... UOC:V12, V13, V14, V15, V16, V17, V21, V22, DAA, DAB, DAC, DAD, DAG, DAH, DAW, DAX, ZAA, ZAB, ZAC, ZAD, ZAG.ZAH	1
6	PBOZZ	5340011307939	19207	12256239-1	.BRACKET, MOUNTING . ... UOC:V12, V13, V14, V15, V16, V17, V21, V22, DAA, DAB, DAC, DAD, DAG, DAH, DAW, DAX, ZAA, ZAB, ZAC, ZAD, ZAG. ZAH	1
7	PAOZZ	5315009041633	96906	MS20392-7C-117	.PIN, STRAIGHT, HEADED .... UOC:V12, V13, V14, V15, V16, V17, V21, V22, DAA, DAB, DAC, DAD, DAG, DAH, DAW, DAX, ZAA, ZAB, ZAC, ZAD, ZAG.ZAM	1
8	PFOZZ	4710010911618	19207	12256244	HOIST EXTENSION ARM .... UOC:V12, V13, V1, V V15, V16, V17, V21, V22	1
8	PFOZZ	2590012102168	19207	12301100	HOIST EXTENSION . ... UOC:DAA, DAB, DAC, DAD, DAG, DAH, DAW, DAX, ZAA, ZAB , ZAP, ZABG.ZAH ZA	1
9	PFOOO	2510010917628	19207	12256241-1	FRAME SECTION, STRUC UOC:V12, V13, V14, V15, V16, V17	1
10	PAOOO	3940010911617	19207	12256233	.SLING, MULTIPLE LEG UOC:V12, V13, V14, V15, V16, V17	1
11	PAOZZ	5315002901349	19207	7358098	..PIN, RETAINING UOC:V12, V13, V14, V15, V16, V17	1
12	PAOZZ	4030009162141	96906	MS87006-53	..HOOK, CHAIN, S UOC:V12, V13, V14, V15, V16, V17	3
13	PAOZZ	5340011254688	19207	12256251-1	..EYE HOOK UOC:V12, V13, V14, V15, V16, V17	1
14	PAOZZ	5310008348736	96906	MS35691-2	NUT, PLAIN, HEXAGON .... UOC:V12, V13, V1, V V15, V16, V17	1
15	PAOZZ	5305000712509	80204	B1821BH025C150N	SCREW, CAP, HEXAGON H................	1

(1) ITEM NO	(2) SMR CODE	(3) NSN	(4) CAGEC	(5) PART NUMBER	(6) DESCRIPTION AND USABLE ON CODES (UOC)	(7) QTY
					UOC:V12, V13, V14, V15, V16, V17	
16	PBOZZ	2510011147545	19207	12256241-2	DAVIT, TRUCK, BOOM	1
					UOC:V21, V22	
17	PAOZZ	3940010911617	19207	12256233	.SLING, MULTIPLE LEG	1
					UOC:V21, V22	
18	PAOZZ	5315002901349	19207	7358098	..PIN, RETAINING	1
					UOC:V21, V22	
19	PAOZZ	4030009162141	96906	MS87006-53	..HOOK, CHAIN, S.	3
					UOC:V21, V22	
20	PAOZZ	5340011254688	19207	12256251-1	..EYE HOOK	1
					UOC:V21, V22	
21	PAOZZ	5310008348736	96906	MS35691-2	.NUT, PLAIN, HEXAGON	1
					UOC:V21, V22	
22	PAOZZ	5305000712509	80204	B1821BH025C150N	.SCREW, CAP, HEXAGON H	1
					UOC:V21, V22	
23	PAOZZ	5315009579386	96906	MS20392-5C69	PIN, STRAIGHT, HEADED	1
					UOC:V21, V22	
24	PAOZZ	5360010930644	19207	12256245	SPRING, HELICAL, EXTE	1
					UOC:V21, V22	
25	PAOZZ	4030007809350	96906	MS87006-13	HOOK, CHAIN, S	1
					UOC:V21, V22	
26	PAOZZ	5340011208510	19207	7397788	CLAMP, LOOP	1
					UOC:V21, V22	
27	PAOZZ	5305002678951	80204	B1821BH025F044N	SCREW, CAP, HEXAGON H	1
					UOC:V21, V22	
28	PAOZZ	5340011147546	19207	12256234	STRAP, RETAINING	1
					UOC:V21, V22	
29	PAOZZ	5310009359022	96906	MS51943-32	NUT, SELF-LOCKING, HE	1
					UOC:V21, V22	
30	PAOZZ	5310000806004	96906	MS27183-14	WASHER, FLAT	1
					UOC:V21, V22	
31	PAOZZ	5315007418971	19207	7418971	PIN, RETAINING	1
					UOC:V21, V22	

END OF FIGURE

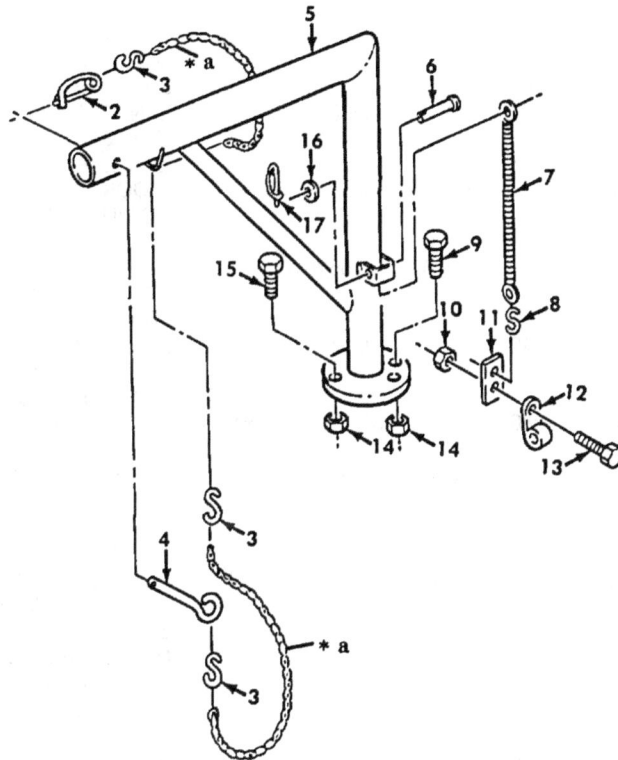

Figure 340. Davit Assembly, Spare Wheel Carrier (M939A1, M939A2).

(1) ITEM NO	(2) SMR CODE	(3) NSN	(4) CAGEC	(5) PART NUMBER	(6) DESCRIPTION AND USABLE ON CODES (UOC)	(7) QTY
					GROUP 1504 SPARE WHEEL CARRIER AND TIRE LOCK	
					FIG. 340 DAVIT ASSEMBLY, SPARE WHEEL CARRIER(M939A1, M939A2)	
1	PAOZZ	3940010911617	19207	12256233	SLING, MULTIPLE LEG .... UOC:DAA, DAB, DAC, DAD, DAG, DAH, DAW, DAX, ZAA, ZAB, ZAC, ZAD, ZAG, ZAH	1
2	PAOZZ	5315002901349	19207	7358098	.PIN, RETAINING .... UOC:DAA, DAB, DAC, DAD, DAG, DAH, DAW, DAX, ZAA, ZAB, ZAC, ZAD, ZAG, ZAH	1
3	PAOZZ	4030009162141	96906	MS87006-53	.HOOK, CHAIN, S .... UOC:DAA, DAB, DAC, DAD, DAG, DAH, DAW, DAX, ZAA, ZAB, ZAC, ZAD, ZAG, ZAH	3
4	PAOZZ	5340011254688	19207	12256251-1	.EYE HOOK... UOC:DAA, DAB, DAC, DAD, DAG, DAH, DAW, DAX, ZAA, ZAB, ZAC, ZAD, ZAG, ZAH	1
5	PAOZZ	2510012116612	19207	12301103	DAVIT, TRUCK, BOOM........................... UOC:DAA, DAB, DAC, DAD, DAG, DAH, DAW, DAX, ZAA, ZAB, ZAC, ZAD, ZAG, ZAH	1
6	PAOZZ	5315009579386	96906	MS20392-5C69	PIN, STRAIGHT, HEADED .... UOC:DAG, DAH, ZAG, ZAH	1
7	PAOZZ	5360010930644	19207	12256245	SPRING, HELICAL, EXTE .... UOC:DAG, DAH, ZAG, ZAH	1
8	PAOZZ	4030007296054	96906	MS87006-63	HOOK, CHAIN, S .... UOC:DAG, DAH, ZAG, ZAH	1
9	PAOZZ	5305007195240	99696	375N300-001-21	SCREW, CAP, HEXAGON H .... UOC:DAA, DAB, DAC, DAD, DAG, DAH, DAW, DAX, ZAA, ZAB, ZAC, ZAD, ZAG, ZAH	3
10	PAOZZ	5310009359022	96906	MS51943-32	NUT, SELF-LOCKING, HE .... UOC:DAA, DAB, DAC, DAD, DAG, DAH, DAW, DAX, ZAA, ZAB, ZAC, ZAD, ZAG, ZAH UOC:DAG, DAH, ZAG, ZAH	1
11	PAOZZ	2510011147586	19207	12256234	STRAP, RETAINING .... UOC:DAG, DAH, ZAG, ZAH	1
12	PAOZZ	5340011208510	19207	7397788	CLAMP, LOOP...................................... UOC:DAG, DAH, ZAG, ZAH	1
13	PAOZZ	5305002678951	80204	B1821BH025F044N	SCREW, CAP, HEXAGON H .... UOC:DAG, DAH, ZAG, ZAH	1
14	PAOZZ	5310004883888	96906	MS51943-40	NUT, SELF-LOCKING, HE .... UOC:DAA, DAB, DAC, DAD, DAG, DAH, DAW, DAX, ZAA, ZAB, ZAC, ZAD, ZAG, ZAH	4
15	PAOZZ	5305007168181	96906	MS90726-118	SCREW, CAP, HEXAGON H .... UOC:DAA, DAB, DAC, DAD, DAG, DAH, DAW, DAX, ZAA, ZAB, ZAC, ZAD, ZAG, ZAH	1
16	PAOZZ	5310000806004	96906	MS27183-14	WASHER, FLAT..................................... UOC:DAG, DAH, ZAG, ZAH	1
17	PAOZZ	5315007418971	19207	7418971	PIN, RETAINING .... UOC:DAG, DAH, ZAG, ZAH	1

END OF FIGURE

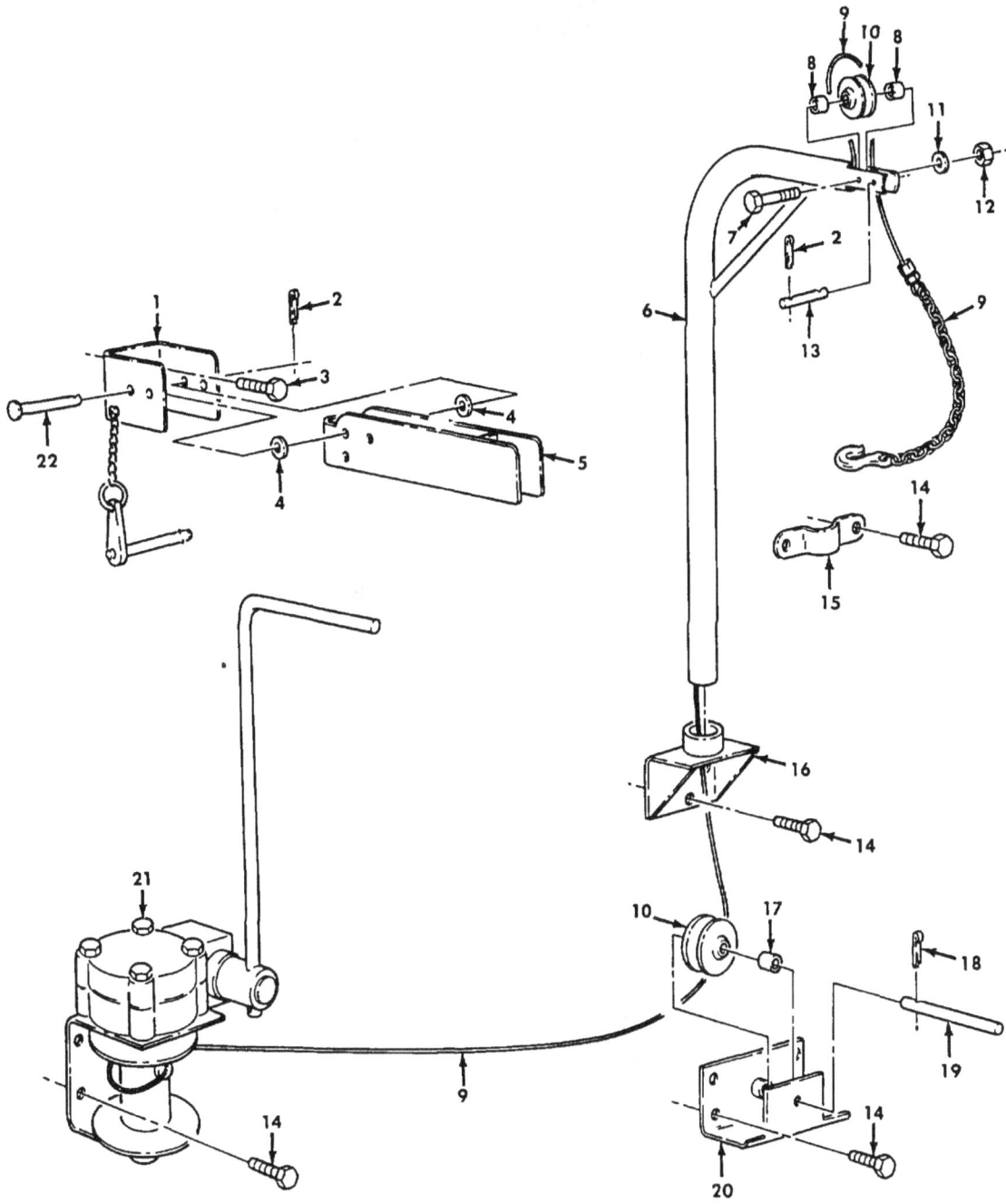

Figure 341. Spare Wheel Davit and Winch Assembly, Van (M939, M939A1).

SECTION II						
(1)	(2)	(3)	(4)	(5)	(6)	(7)
ITEM	SMR			PART		
NO	CODE	NSN	CAGEC	NUMBER	DESCRIPTION AND USABLE ON CODES (UOC)	QTY

TM 9-2320-272-24P-1

GROUP 1504 SPARE WHEEL CARRIER AND TIRE LOCK

FIG. 341 SPARE WHEEL DAVIT AND WINCH ASSEMBLY, VAN(M939A1, M939A2)

(1)	(2)	(3)	(4)	(5)	(6)	(7)
1	PFOZZ	5340012104654	19207	12303051	BRACKET, MOUNTING .... UOC:DAJ, DAK, ZAJ, ZAK	1
2	PAOZZ	5315008395820	96906	MS24665-134	PIN, COTTER .... UOC:DAJ, DAK, ZAJ, ZAK	3
3	PAOZZ	5305001913640	96906	MS51851-85	SCREW, TAPPING .... UOC:DAJ, DAKR, ZAJ, ZAK	3
4	PAOZZ	5365003004379	80205	NAS43HT4-24	SPACER, SLEEVE PULLEY SHAFT .... UOC:DAJ, DAK, ZAJ, ZAK	2
5	PFOZZ	3040012104655	19207	12303056	LEVER, MANUAL CONTRO .... UOC:DAJ, DAK, ZAJ, ZAK	1
6	PFOZZ	2510012104656	19207	12303038	DAVIT ASSEMBLY, SPAR .... UOC:DAJ, DAK, ZAJ, ZAK	1
7	PAOZZ	5306002259098	96906	MS90726-43	BOLT, MACHINE .... UOC:DAJ, DAK, ZAJ, ZAK	1
8	PAOZZ	5365008276452	80205	NAS43HT5-32	SPACER , SLEEVE .... UOC:DAJ, DAK, ZAJ, ZAK	2
9	PAOZZ	4010012106196	19207	12303055	WIRE ROPE ASSEMBLY, ... UOC:DAJ, DAK, ZAJ, ZAK	1
10	PFOZZ	3020005280511	96906	MS20220-A1	PULLEY, GROOVE .... UOC:DAJ, DAK, ZAJ, ZAK	1
11	PAOZZ	5310000814219	96906	MS27183-12	WASHER, FLAT .... UOC:DAJ, DAK, ZAJ, ZAK	1
12	PAOZZ	5310002416658	96906	MS51943-34	NUT, SELF-LOCKING, HE .... UOC:DAJ, DAK, ZAJ, ZAK	1
13	PAOZZ	5315005849809	80205	NAS427W34	PIN, STRAIGHT, HEADLE .... UOC:DAJ, DAK, ZAJ, ZAK	1
14	PAOZZ	5305012195386	96906	MS51851-128	SCREW, TAPPING .... UOC:DAJ, DAK, ZAJ, ZAK	17
15	PFOZZ	5340012190163	19207	12303039	STRAP, RETAINING .... UOC:DAJ, DAK, ZAJ, ZAK	3
16	PFOZZ	5340012104657	19207	12303045	BRACKET, MOUNTING .... UOC:DAJ, DAK, ZAJ, ZAK	1
17	PFOZZ	5365012189917	19207	12303043	SPACER, SLEEVE .... UOC:DAJ, DAK, ZAJ, ZAK	1
18	PAOZZ	5315008161794	89749	1F316	PIN, COTTER .... UOC:DAJ, DAK, ZAJ, ZAK	1
19	PFOZZ	3040012165337	19207	12303040	SHAFT, STRAIGHT .... UOC:DAJ, DAK, ZAJ, ZAK	1
20	PFOZZ	3040012166982	19207	12303050	BRACKET, EYE, ROTATIN .... UOC:DAJ, DAK, ZAJ, ZAK	1
21	PFOZZ	3950012169319	19207	12303057	WINCH, DRUM, HAND OPE .... UOC:DAJ, DAK, ZAJ, ZAK	1
22	PAOZZ	5315012193665	96906	MS9845-34	PIN, STRAIGHT, HEADED .... UOC:DAJ, DAK, ZAJ, ZAK	1

END OF FIGURE

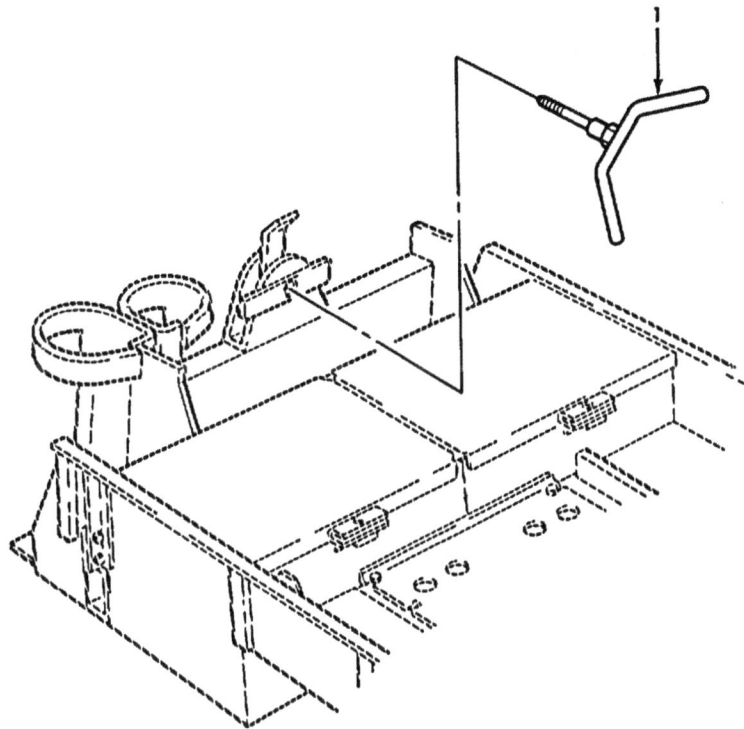

Figure 342. Spare Wheel Carrier, Wrecker (M939).

(1) ITEM NO	(2) SMR CODE	(3) NSN	(4) CAGEC	(5) PART NUMBER	(6) DESCRIPTION AND USABLE ON CODES (UOC)	(7) QTY
					GROUP 1504 SPARE WHEEL CARRIER AND TIRE LOCK	
					FIG. 342 SPARE WHEEL CARRIER, WRECKER (M939)	
1	PAOZZ	2540010853589	19207	10876599	LUG, WING UOC:V18	1
1	PAOZZ		19207	10876599-1	LUG, WING... UOC:DAL, ZAL	1

END OF FIGURE

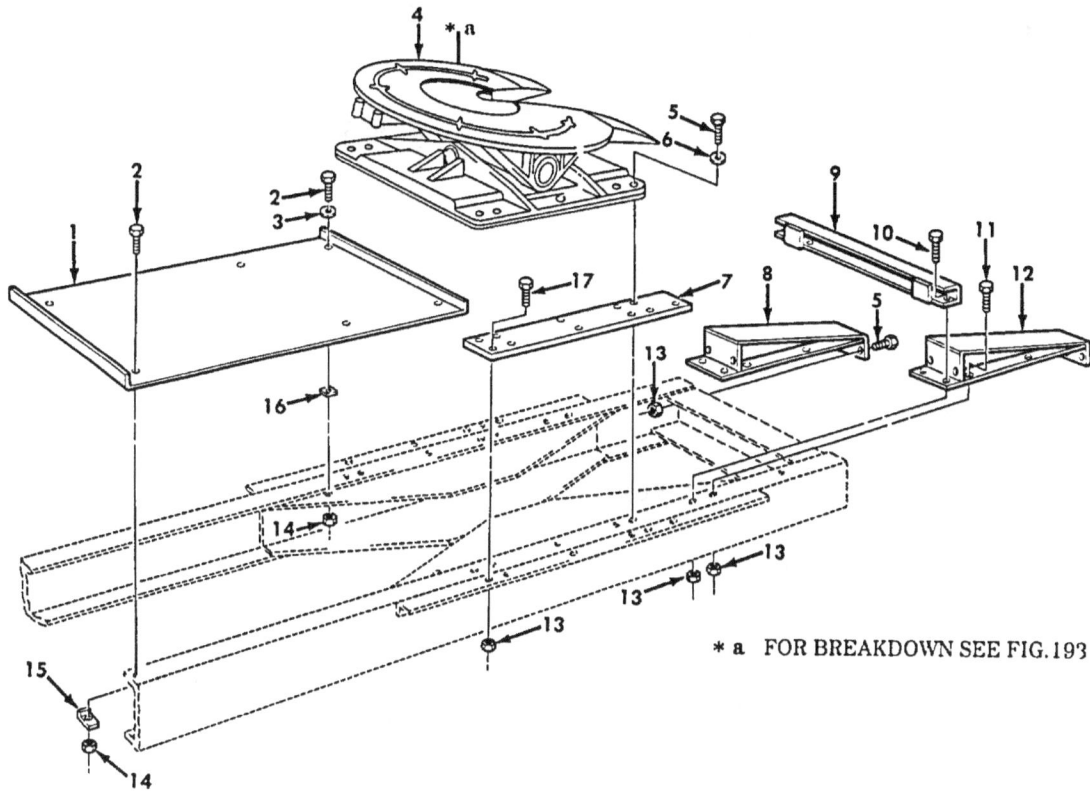

Figure 343. Fifth Wheel Assembly and Mounting Plates

* a   FOR BREAKDOWN SEE FIG. 193

(1) ITEM NO	(2) SMR CODE	(3) NSN	(4) CAGEC	(5) PART NUMBER	(6) DESCRIPTION AND USABLE ON CODES (UOC)	(7) QTY
					GROUP 1506 FIFTH WHEEL	
					FOG. 343 FIFTH WHEEL ASSEMBLY AND MOUNTING PLATES	
1	PFOZZ	2510004084652	19207	11658697	PLATE, MOUNTING, FIFT .... UOC:DAG, DAH, V21, V22, ZAG, ZAH	1
2	PAOZZ	5305002693240	80204	B1821BH038F150N	SCREW, CAP, HEXAGON H .... UOC:DAG, DAH, V21, V22, ZAG, ZAH	6
3	PAOZZ	5310000806004	96906	MS27183-14	WASHER, FLAT .... UOC:DAG, DAH, V21, V22, ZAG, ZAH	2
4	PAOFF	2510010946714	19207	11669413-1	FIFTH WHEEL ASSEMBL .... UOC:DAG, DAH, V21, V22, ZAG, ZAH	1
5	PAOZZ	5305007262550	80204	B1821BH063F175N	SCREW, CAP, HEXAGON H .... UOC:DAG, DAH, V21, V22, ZAG, ZAH	20
6	PAOZZ	5310008000695	96906	MS35335-39	WASHER, LOCK . ... UOC:DAG, DAH, V21, V22, ZAG, ZAH	10
7	PFOZZ	5365004222017	19207	11608882	SPACER, FIFTH WHEEL .... UOC:DAG, DAH, V21, V22, ZAG, ZAH	2
8	PAOZZ	5340007409884	19207	7409884	PLATE, MOUNTING . ... UOC:DAG, DAH, V21, V22, ZAG, ZAH	1
9	PFOZZ	2510010270203	19207	8758365	CHANNEL, REINFORCING .... UOC:DAG, DAH, V21, V22, ZAG, ZAH	1
10	PAOZZ	5305007262552	80204	B1821BH063F225N	SCREW, CAP, HEXAGON H .... UOC:DAG, DAH, V21, V22, ZAG, ZAH	2
11	PAOZZ	5305007262551	80204	B1821BH063F200N	SCREW, CAP, HEXAGON H .... UOC:DAG, DAH, V21, V22, ZAG, ZAH	2
12	PFOZZ	5340007409883	19207	7409883	PLATE, MOUNTING . ... UOC:DAG, DAH, V21, V22, ZAG, ZAH	1
13	PAOZZ	5310002416664	96906	MS51943-44	NUT, SELF-LOCKING, HE .... UOC:DAG, DAH, V21, V22, ZAG, ZAH	26
14	PAOZZ	5310008140672	96906	MS51943-36	NUT, SELF-LOCKING, HE .... UOC:DAG, DAH, V21, V22, ZAG, ZAH	6
15	PAOZZ	2510004084642	19207	10883335	PLATE, RETAINING, FIF .... UOC:DAG, DAH, V21, V22, ZAG, ZAH	4
16	PFOZZ	5365004468770	19207	11658696	SPACER, PLATE .... UOC:DAG, DAH, V21, V22, ZAG, ZAH	2
17	PAOZZ	5305007262553	96906	MS90727-166	SCREW, CAP, HEXAGON H .... UOC:DAG, DAH, V21, V22, ZAG, ZAH	12

END OF FIGURE

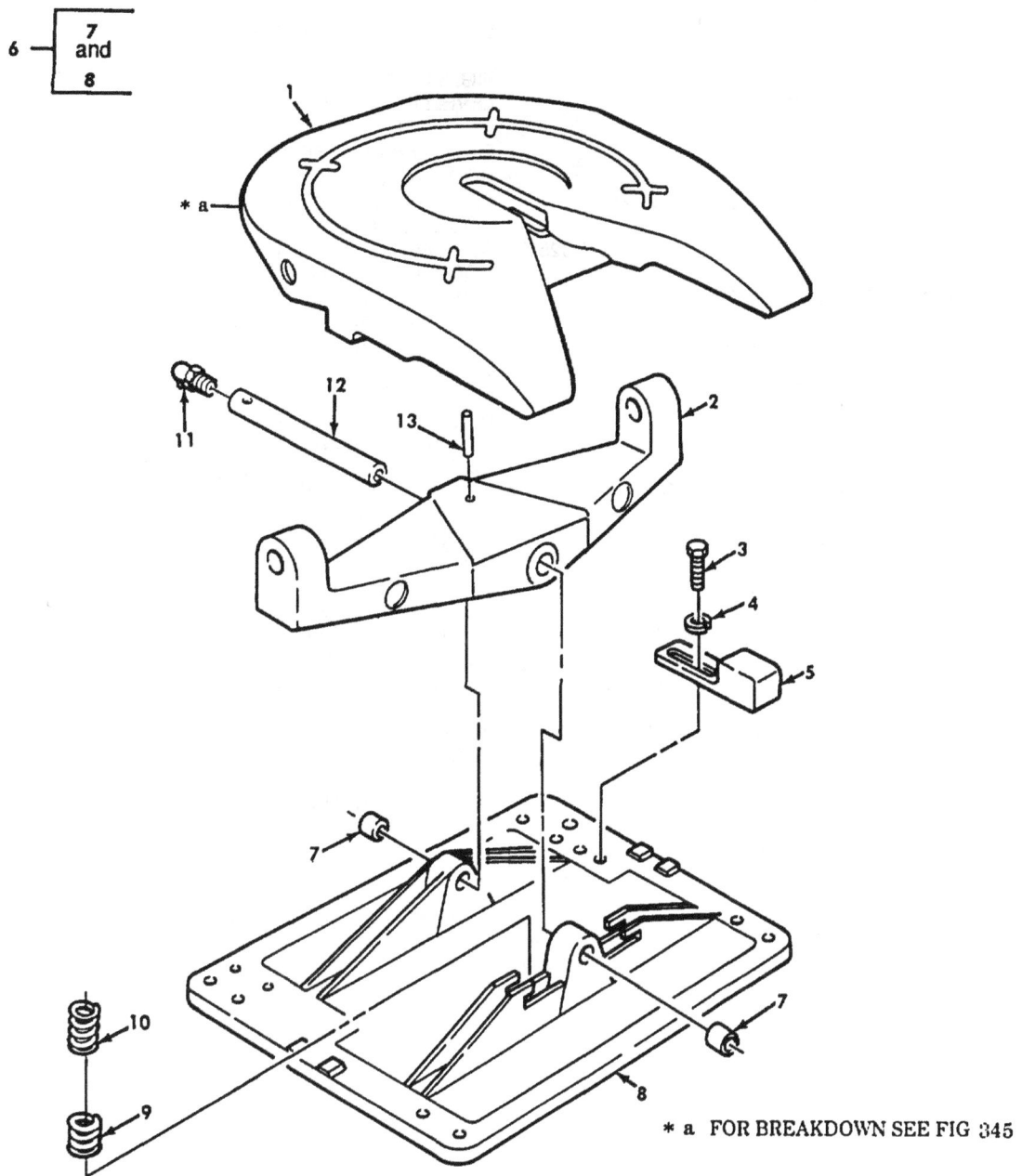

Figure 344. Fifth Wheel Assembly, Basic Leveling Wedge, and Subbase.

(1) ITEM NO	(2) SMR CODE	(3) NSN	(4) CAGEC	(5) PART NUMBER	(6) DESCRIPTION AND USABLE ON CODES (UOC)	(7) QTY
					GROUP 1506 FIFTH WHEEL	
					FIG. 344 FIFTH WHEEL ASSEMBLY, BASIC LEVELLING WEDGE, AND SUBBASE	
1	XAFFF		19207	12300606	FIFTH WHEEL ASSEMBL .... UOC:DAG, DAH, V21, V22, ZAG, ZAH	1
2	XAFFF		19207	12300586	CAST WALKING BEAM .... UOC:DAG, DAH, V21, V22, ZAG, ZAH	1
3	PAFZZ	5305009227994	80204	B1821BH075C250N	SCREW, CAP, HEXAGON H .... UOC:DAG, DAH, V21, V22, ZAG, ZAH	4
4	PAFZZ	5310006805956	88032	300B149X1	WASHER, FLAT........................... UOC:DAG, DAH, V21, V22, ZAG, ZAH	4
5	PFFZZ	2510011580773	19207	12300579	WEDGE, LEVELING .... UOC:DAG, DAH, V21, V22, ZAG, ZAH	2
6	XAFZZ		19207	12300739	FIFTH WHEEL ASSEMBL .... UOC:DAG, DAH, V21, V22, ZAG, ZAH	1
7	XDFZZ		19207	12300738	.BUSHING, SLEEVE .... UOC:DAG, DAH, V21, V22, ZAG, ZAH	4
8	XAFZZ		19207	12300587	.FIFTH WHEEL ASSEMBL .... UOC:DAG, DAH, V21, V22, ZAG, ZAH	1
9	PAFZZ	5360011793105	19207	12300989	SPRING, HELICAL, COMP........................... UOC:DAG, DAH, V21, V22, ZAG, ZAH	2
10	PAFZZ	5360011794599	19207	12300990	SPRING, HELICAL, COMP .... UOC:DAG, DAH, V21, V22, ZAG, ZAH	2
11	PAFZZ	4730001720028	96906	MS15003-4	FITTING, LUBRICATION .... UOC:DAG, DAH, V21, V22, ZAG, ZAH	2
12	PFFZZ	5340011583126	19207	12300582	ROD, STRAIGHT, HEADLE .... UOC:DAG, DAH, V21, V22, ZAG, ZAH	1
13	PBFZZ	5315011192733	19207	12300585	PIN, GROOVED, HEADLES........................... UOC:DAG, DAH, V21, V22, ZAG, ZAH	2

END OF FIGURE

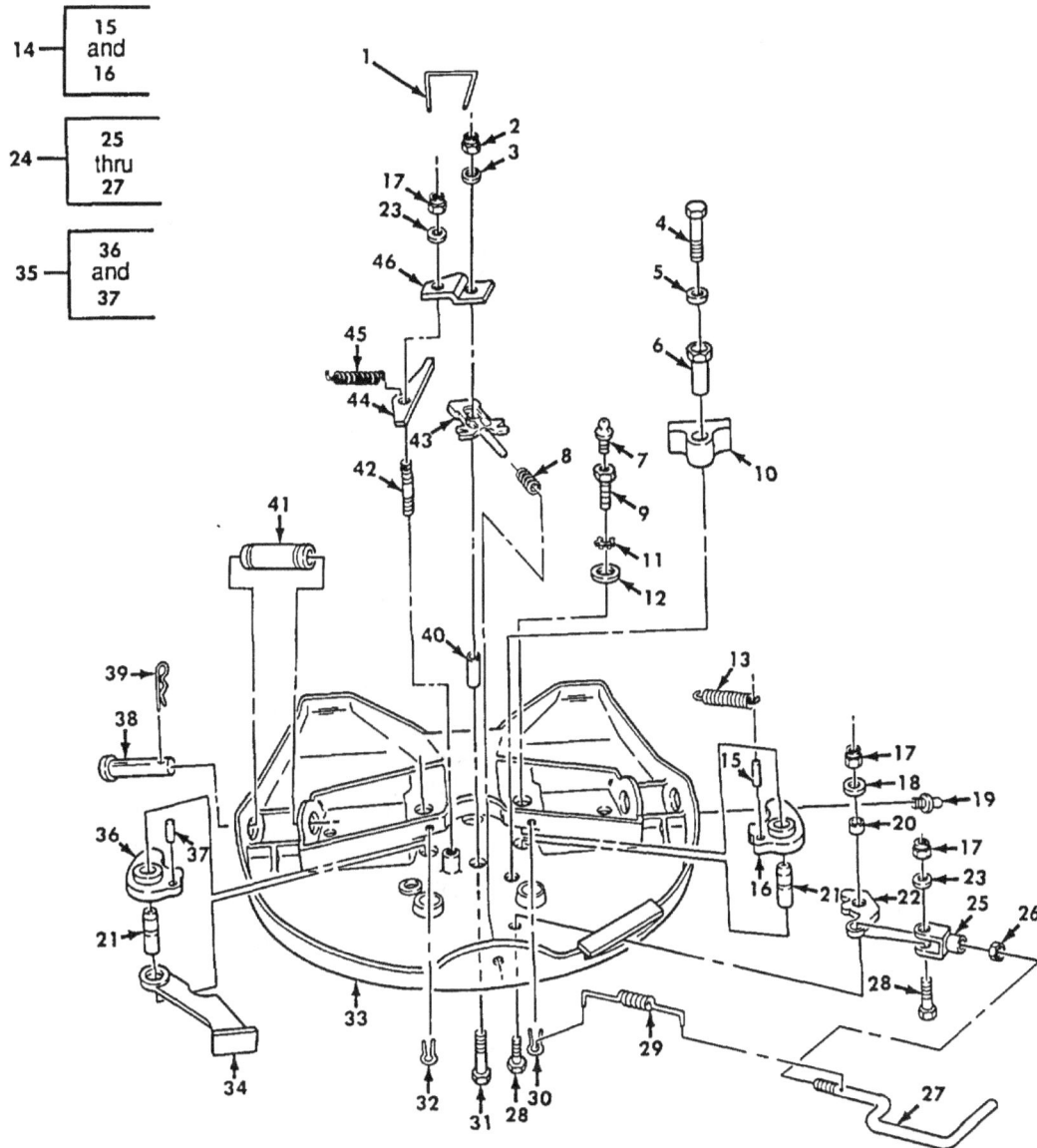

Figure 345. Fifth Wheel Assembly, Basic.

(1) ITEM NO	(2) SMR CODE	(3) NSN	(4) CAGEC	(5) PART NUMBER	(6) DESCRIPTION AND USABLE ON CODES (UOC)	(7) QTY
					GROUP 1506 FIFTH WHEEL	
					FIG. 345 FIFTH WHEEL ASSEMBLY, BASIC	
1	PAFZZ	9505002489842	81346	ASTM A641	WIRE, NONELECTRICAL PART OF KIT P/N............ 59-003128........ UOC:DAG, DAH, V21, V22, ZAG, ZAH	1
2	PAFZZ	5310002256408	96906	MS51922-53	NUT, SELF-LOCKING, HE PART OF KIT P/N........... 59-003128 .... UOC:DAG, DAH, V21, V22, ZAG, ZAH	1
3	PAFZZ	5310010809007	19207	12300609	WASHER, FLAT............ UOC:DAG, DAH, V21, V22, ZAG, ZAH	1
4	PAFZZ	5305007247225	80204	B1821BH063C275N	SCREW, CAP, HEXAGON H............ UOC:DAG, DAH, V21, V22, ZAG, ZAH	1
5	PAFZZ	5310008206653	80045	23MS35338-50	WASHER, LOCK .... UOC:DAG, DAH, V21, V22, ZAG, ZAH	1
6	PBFZZ	2530011192645	19207	12300592	ADJUSTER, SLACK, BRAK .... UOC:DAG, DAH, V21, V22, ZAG, ZAH	1
7	PAOZZ	4730001720034	96906	MS15003-6	FITTING, LUBRICATION .... UOC:DAG, DAH, V21, V22, ZAG, ZAH	2
8	PAFZZ	5360011192736	19207	12300626	SPRING, HELICAL, COMP PART OF KIT P/N .... 59-003128........ UOC:DAG, DAH, V21, V22, ZAG, ZAH	1
9	PAFZZ	4730010800930	19207	12300612	FITTING, LUBRICATION .... UOC:DAG, DAH, V21, V22, ZAG, ZAH	2
10	PBFZZ	5365011192644	19207	12300610	BUSHING, NONMETALLIC .... UOC:DAG, DAH, V21, V22, ZAG, ZAH	1
11	PAFZZ	5310002753683	96906	MS35335-40	WASHER, LOCK .... UOC:DAG, DAH, V21, V22, ZAG, ZAH	2
12	PAFZZ	5310006805956	88032	300B149X1	WASHER, FLAT............ UOC:DAG, DAH, V21, V22, ZAG, ZAH	2
13	PAFZZ	5360010809008	19207	12300590	SPRING PART OF KIT P/N 59-003128.................... UOC:DAG, DAH, V21, V22, ZAG, ZAH	1
14	PAFZZ	2510010806423	19207	12300615-2	JAW ASSEMBLY COUPLE PART OF KIT P/N .... 59-003128........ UOC:DAG, DAH, V21, V22, ZAG, ZAH	1
15	XAFZZ		19207	12300605	.SPRING RTNR PIN........ UOC:DAG, DAH, V21, V22, ZAG, ZAH	1
16	XAFZZ		19207	12300613	.COUPLER JAW .... UOC:DAG, DAH, V21, V22, ZAG, ZAH	1
17	PAFZZ	5310008775795	96906	MS21044-N8	NUT, SELF-LOCKING, HE PART OF KIT P/N........... 59-003128 .... UOC:DAG, DAH, V21, V22, ZAG, ZAH	3
18	PAFZZ	5310010797059	19207	12300608	WASHER .... UOC:DAG, DAH, V21, V22, ZAG, ZAH	1
19	PAOZZ	4730000504208	96906	MS15003-1	FITTING, LUBRICATION .... UOC:DAG, DAH, V21, V22, ZAG, ZAH	2
20	PAFZZ	5365011192739	19207	12300603	SPACER, SLEEVE PART OF KIT P/N 59- .... 003128 .... UOC:DAG, DAH, V21, V22, ZAG, ZAH	1
21	XDFZZ	5315010798060	19207	12300600	PIN PART OF KIT P/N 59-003128..........................	2

(1) ITEM NO	(2) SMR CODE	(3) NSN	(4) CAGEC	(5) PART NUMBER	(6) DESCRIPTION AND USABLE ON CODES (UOC)	(7) QTY
22	PBFZZ	3040011202164	19207	12300596	LEVER, MANUAL CONTRO PART OF KIT P/N......... 59-003128... UOC:DAG, DAH, V21, V22, ZAG, ZAH	1
23	PAFZZ	5310011170610	19207	11640362-1	WASHER, FLAT .... UOC:DAG, DAH, V21, V22, ZAG, ZAH	2
24	PAFZZ	3040011203054	19207	12300584	LEVER, MANUAL CONTRO .... UOC:DAG, DAH, V21, V22, ZAG, ZAH	1
25	PAFZZ	5340011192735	19207	12300583	.CLEVIS, ROD END .... UOC:DAG, DAH, V21, V22, ZAG, ZAH	1
26	PAFZZ	5310007638911	96906	MS51968-17	.NUT, PLAIN, HEXAGON .... UOC:DAG, DAH, V21, V22, ZAG, ZAH	1
27	PAFZZ	5340011192642	19207	12300611	.HANDLE, MANUAL CONTR .... UOC:DAG, DAH, V21, V22, ZAG, ZAH	1
28	PAFZZ	5306011174900	19207	11662860	BOLT, MACHINE .... UOC:DAG, DAH, V21, V22, ZAG, ZAH	2
29	PAFZZ	5360011192737	19207	12300614	SPRING, HELICAL, EXTE .... UOC:DAG, DAH, V21, V22, ZAG, ZAH	1
30	PAFZZ	5315001879414	96906	MS24665-655	PIN, COTTER .... UOC:DAG, DAH, V21, V22, ZAG, ZAH	1
31	PAFZZ	5306010804757	19207	12300595	BOLT PART OF KIT P/N 59-003128 .... UOC:DAG, DAH, V21, V22, ZAG, ZAH	1
32	PAFZZ	5315000187988	96906	MS24665-493	PIN, COTTER .... UOC:DAG, DAH, V21, V22, ZAG, ZAH	1
33	XAFZZ		19207	12300617	BASE .... UOC:DAG, DAH, V21, V22, ZAG, ZAH	1
34	PBFZZ	2510013946119	72540	5900370	PARTS KIT, FIFTH WHE .... UOC:DAG, DAH, V21, V22, ZAG, ZAH	1
35	PAFZZ	2510010806424	19207	12300615-1	JAW ASSEMBLY COUPLE PART OF KIT P/N.......... 59-003128... UOC:DAG, DAH, V21, V22, ZAG, ZAH	1
36	XAFZZ		19207	12300613	.COUPLER JAW... UOC:DAG, DAH, V21, V22, ZAG, ZAH	1
37	XAFZZ		19207	12300605	.SPRING RTNR PIN... UOC:DAG, DAH, V21, V22, ZAG, ZAH	1
38	PBFZZ	5315010814275	19207	12300601	PIN .... UOC:DAG, DAH, V21, V22, ZAG, ZAH	2
39	PAFZZ	5315010798059	19207	12300591	PIN .... UOC:DAG, DAH, V21, V22, ZAG, ZAH	2
40	PAFZZ	5365008096292	19207	11640341	SPACER, SLEEVE PART OF KIT P/N 59- .... 003128... UOC:DAG, DAH, V21, V22, ZAG, ZAH	1
41	PAFZZ	3120010799882	19207	12300588	BUSHING, PIN MOUNT FIFTH WHEEL MOUNT....... UOC:DAG, DAH, V21, V22, ZAG, ZAH	1
42	PBFZZ	5307010800492	19207	12300604	STUD, PLAIN .... UOC:DAG, DAH, V21, V22, ZAG, ZAH	1
43	PBFZZ	5340011192646	19207	12300580	PLUNGER, DETENT PART OF KIT P/N 59-............. 003128... UOC:DAG, DAH, V21, V22, ZAG, ZAH	1
44	PBFZZ	5340011192643	19207	12300594	LEVER, LOCK-RELEASE PART OF KIT P/N............ 59-003128... UOC:DAG, DAH, V21, V22, ZAG, ZAH	1
45	PAFZZ	5360007409903	19207	7409903	SPRING, HELICAL, EXTE .... UOC:DAG, DAH, V21, V22, ZAG, ZAH	1

(1) ITEM NO	(2) SMR CODE	(3) NSN	(4) CAGEC	(5) PART NUMBER	(6) DESCRIPTION AND USABLE ON CODES (UOC)	(7) QTY
46	PFFZZ	5340011346530	19207	12300593	BRACKET, DOUBLE ANGL UOC:DAG,DAH,V21,V22,ZAG,ZAH	1

END OF FIGURE

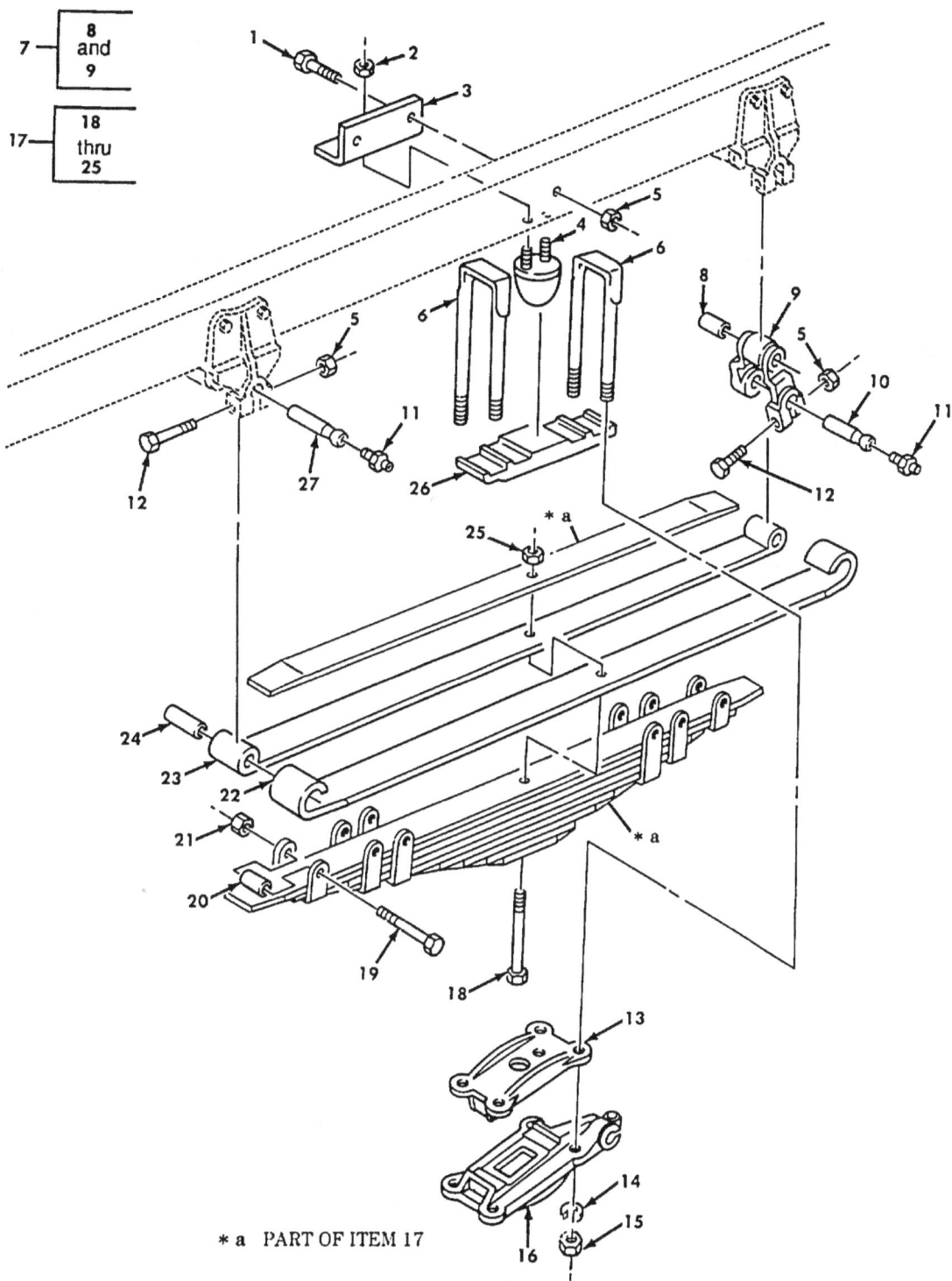

*a PART OF ITEM 17

Figure 346. Front Leaf Spring Assembly and Related Parts.

(1) ITEM NO	(2) SMR CODE	(3) NSN	(4) CAGEC	(5) PART NUMBER	(6) DESCRIPTION AND USABLE ON CODES (UOC)	(7) QTY
					GROUP 16 SPRINGS AND SHOCK ABSORBERS 1601 SPRINGS	
					FIG. 346 FRONT LEAF SPRING ASSEMBLY AND RELATED PARTS	
1	PAOZZ	5305010893070	19207	10883158	SCREW, CAP, HEXAGON H RIGHT SIDE	2
1	PAOZZ	5305007195219	96906	MS90727-111	SCREW, CAP, HEXAGON H LEFT SIDE	2
2	PAOZZ	5310009843806	96906	MS51922-9	NUT, SELF-LOCKING, HE	4
3	PFOZZ	5340004823791	19207	7992548	BRACKET, ANGLE.	2
4	PAOZZ	5340007409335	19207	7409335	BUMPER, NONMETALLIC FRONT AXLE	2
5	PAOZZ	5310000624954	96906	MS21045-8	NUT, SELF-LOCKING, HE	10
6	PAOZZ	5306004457240	19207	11664570	BOLT, U	4
7	PAOZZ	2510007409344	19207	7409344	SHACKLE, TEAF SPRING	2
8	PAOZZ	3120007702941	89346	55602H	.BEARING, SLEEV	2
9	XAOZZ	2510011583243	19207	10871273	.SHACKLE, LEAF SPRING	1
10	PAOZZ	2510007412376	19207	7412376	PIN, VEHICULAR LEAF	5
11	PAOZZ	4730000504208	96906	MS15003-1	FITTING, LUBRICATION	6
12	PAOZZ	5305009409517	96906	MS51106-421	SCREW, CAP, HEXAGON H.	6
13	PAOZZ	2510004172752	19207	11664569	PLATE, WEAR, LEAF SPR	2
14	PAOZZ	5310007542005	96906	MS35338-52	WASHER, LOCK	8
15	PAOZZ	5310002758264	19207	5298551	NUT, PLAIN, HEXAGON	8
16	PAOZZ	5340007409341	19207	7409341	PLATE, MOUNTING RIGHT	1
16	PAOZZ	2510007409340	19207	7409340	PLATE, WEAR, LEAF SPR LEFT	1
17	PAOOO	2510007411110	89346	104842R92	SPRING ASSEMBLY, LEA	2
18	PAOZZ	5305001397074	19207	10883157	.SCREW, MACHINE.	1
19	PAOZZ	5305000695583	96906	MS90725-99	.SCREW, CAP, HEXAGON H	6
20	PAOZZ	5365007001851	60528	3-82-6	.SPACER, SLEEVE	6
21	PAOZZ	5310008911711	96906	MS35691-25	.NUT, PLAIN, HEXAGON	6
22	PAOZZ	2510007411112	19207	7411112	.SPRING, LEAF	1
23	PAOZZ	2510007411111	19207	7411111	.SPRING, LEAF	1
24	PAOZZ	3120007702941	89346	55602H	.BEARING, SLEEVE	2
25	PAOZZ	5310007320560	96906	MS51968-14	.NUT, PLAIN, HEXAGON	1
26	PAOZZ	2540007409343	19207	7409343	SEAT, U-BOLT	2
27	PAOZZ	5315001344632	19207	11664534	PIN, BALL STUD	1

END OF FIGURE

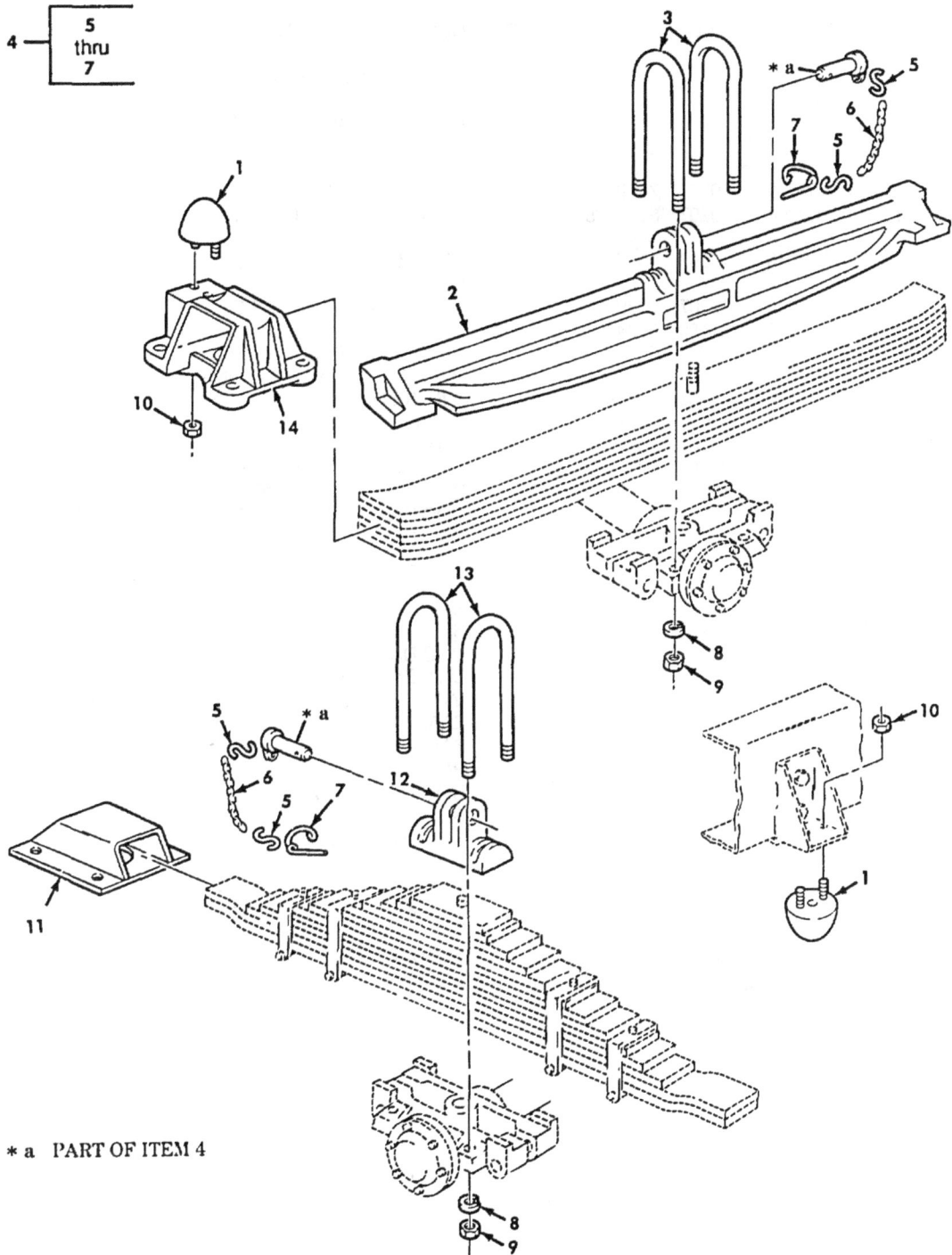

Figure 347. Rear Leaf Spring Mounting Hardware.

(1) ITEM NO	(2) SMR CODE	(3) NSN	(4) CAGEC	(5) PART NUMBER	(6) DESCRIPTION AND USABLE ON CODES (UOC)	(7) QTY
					GROUP 1601 SPRINGS	
					FIG. 347 REAR LEAF SPRING MOUNTING HARDWARE	
1	PAOZZ	5340007409335	19207	7409335	BUMPER, NONMETALLIC	4
2	PAOZZ	2540000402308	19207	8330265	BEAM, SUSPENSION SPR .... UOC:DAL, V18, ZAL	2
3	PAOZZ	5306008696549	19207	10871251	BOLT, U .... UOC:DAL, V18, ZAL	4
4	PAOZZ	5315012062239	19207	7339966-1	PIN, STRAIGHT, HEADED	2
5	PAOZZ	4030005144420	21450	593416	.HOOK, CHAIN, S	2
6	MOOZZ		80244	42C15120-205-6	.CHAIN, WELDLESS MAKE FROM CHAIN, ... P/N XB-196, 6 INCHES LONG	1
7	PAOZZ	5315011087340	19207	7339964	.PIN, RETAINING	1
8	PAOZZ	5310009265885	96906	MS35340-53	WASHER, LOCK	8
9	PAOZZ	5310007007127	19207	7007127	NUT, PLAIN, HEXAGON	8
10	PAOZZ	5310009843806	96906	MS51922-9	NUT, SELF-LOCKING, HE	8
11	PFOZZ	2510007346998	19207	7979363	BRACKET, SPRING MOUN .... UOC:DAA, DAB, DAC, DAD, DAE, DAF, DAG, DAH, DAJ, DAK, DAW, DAX, V12, V13, V14, V15, V16, V17, V19, V20, V21, V22, V23, V24, V25, V39, ZAA, ZAB, ZAC, ZAD, ZAE, ZAF, ZAG, ZAH, ZAJ, ZAK	4
12	PAOZZ	2510001793639	19207	10883205	SADDLE, SUSPENSION .... UOC:DAA, DAB, DAC, DAD, DAE, DAF, DAG, DAH, DAJ, DAK, DAW, DAX, V12, V13, V14, V15, V16, V17, V19, V20, V21, V22, V23, V24, V25, V39, ZAA, ZAB, ZAC, ZAD, ZAE, ZAF, ZAG, ZAH, ZAJ, ZAK	2
13	PAOZZ	5306007411113	19207	10883203	BOLT, U.. UOC:DAL, V18, ZAL	4
13	PAOZZ	5306005755419	19207	8741539	BOLT, U.. UOC:DAA, DAB, DAC, DAD, DAE, DAF, DAG, DAH, DAJ, DAK, DAW, DAX, V12, V13, V14, V15, V16, V17, V19, V20, V21, V22, V23, V24, V25, V39, ZAA, ZAB, ZAC, ZAD, ZAE, ZAF, ZAG, ZAH, ZAJ, ZAK	4
14	PBOZZ	5340004181009	19207	8758275	BRACKET, MOUNTING.. UOC:DAL, V18, ZAL	4

END OF FIGURE

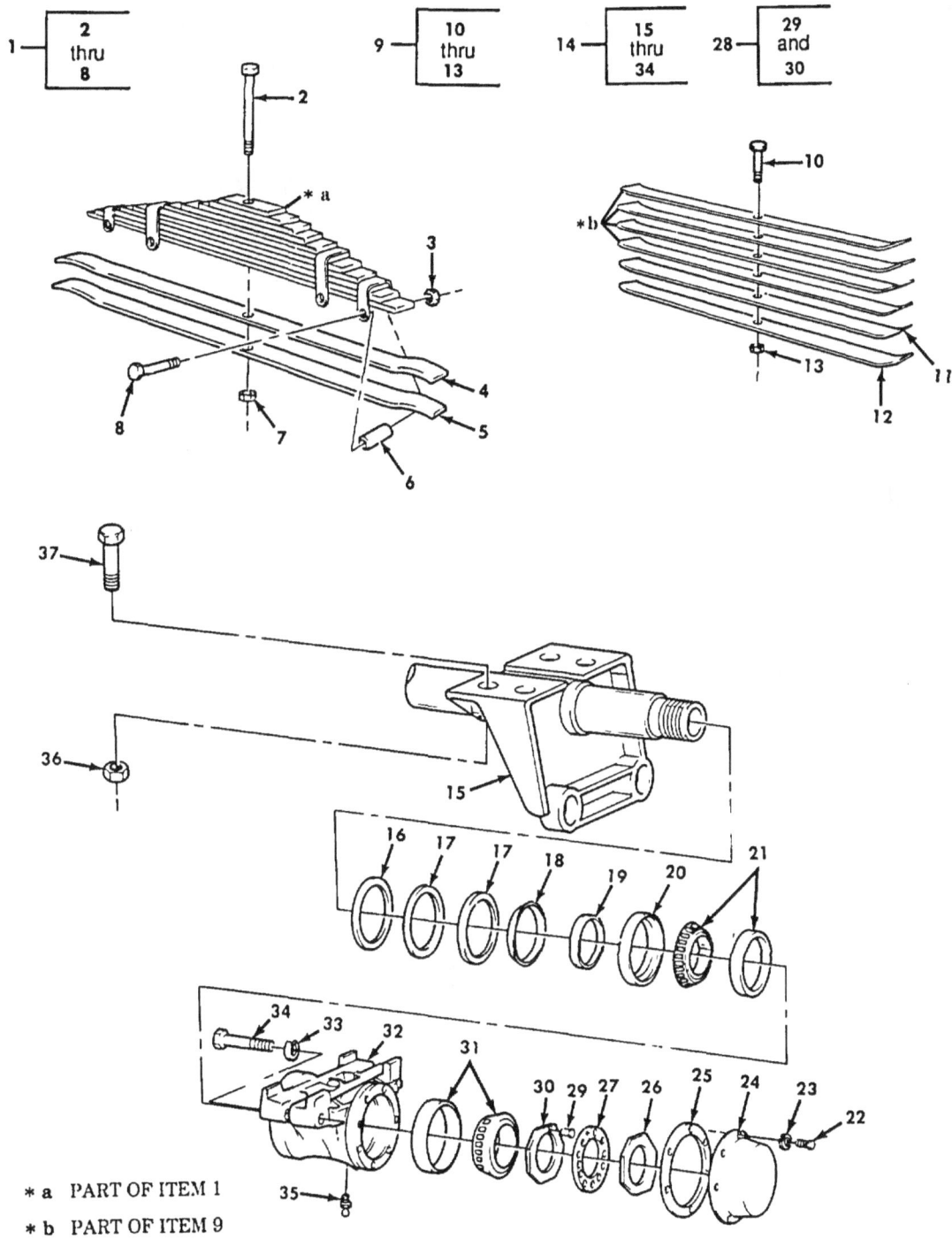

Figure 348. Rear Leaf Spring and Seat Assemblies.

* a   PART OF ITEM 1
* b   PART OF ITEM 9

(1) ITEM NO	(2) SMR CODE	(3) NSN	(4) CAGEC	(5) PART NUMBER	(6) DESCRIPTION AND USABLE ON CODES (UOC)	(7) QTY
					GROUP 1601 SPRINGS	
					FIG. 348 REAR LEAF SPRING AND SEAT ASSEMBLIES	
1	PAOOO	2510007411114	19207	7411114	SPRING ASSEMBLY, LEA UOC:V39	2
1	PAOZZ	2510007409613	19207	7409613	SPRING ASSEMBLY, LEA .... UOC:DAA, DAB, DAC, DAD, DAE, DAF, DAG, DAH, DAJ, DAK, DAW, DAX, V12, V13, V14, V15, V16, V17, V19, V20, V21, V22, V24, V25, ZAA, ZAB, ZAC, ZAD, ZAE, ZAF, ZAG, ZAH, ZAJ, ZAK	2
2	PAOZZ	5305001397075	19207	10883199	.SCREW, MACHINE UOC:V39	1
2	PAOZZ	5305001397072	19207	10883200	.SCREW, MACHINE .... UOC:DAA, DAB, DAC, DAD, DAE, DAF, DAG, DAH, DAJ, DAK, DAW, DAX, V12, V13, V14, V15, V16, V17, V19, V20, V21, V22, V24, V25, ZAA, ZAB, ZAC, ZAD, ZAE, ZAF, ZAG, ZAH, ZAJ, ZAK	1
3	PAOZZ	5310008911711	96906	MS35691-25	.NUT, PLAIN, HEXAGON....................................... UOC:DAA, DAB, DAC, DAD, DAE, DAF, DAG, DAH, DAJ, DAK, DAW, DAX, V12, V13, V1, V V15, V16, V17, V19, V20, V21, V22, V24, V25, V39, ZAA, ZAB, ZAC, ZAD, ZAE, ZAF, ZAG, ZAH, ZAJ, ZAK	4
4	PAOZZ	2510007411116	19207	7411116	.SPRING, LEAF... UOC:V39	1
4	PAOZZ	5360007409602	19207	7409602	.SPRING, LEAF .... UOC:DAA, DAB, DAC, DAD, DAE, DAF, DAG, DAH, DAJ, DAK, DAW, DAX, V12, V13, V14, V15, V16, V17, V19, V20, V21, V22, V24, V25, ZAA, ZAB, ZAC, ZAD, ZAE, ZAF, ZAG, ZAH, ZAJ, ZAK	1
5	PAOZZ	5360007411115	19207	7411115	.SPRING, LEAF UOC:V39	1
5	PAOZZ	2510007409601	19207	7409601	.SPRING, LEAF .... UOC:DAA, DAB, DAC, DAD, DAE, DAF, DAG, DAH, DAJ, DAK, DAW, DAX, V12, V13, V14, V15, V16, V17, V19, V20, V21, V22, V24, V25, ZAA, ZAB, ZAC, ZAD, ZAE, ZAF, ZAG, ZAH, ZAJ, ZAK	1
6	PAOZZ	5365007409612	19207	7409612	.SPACER, BOLT... UOC:DAA, DAB, DAC, DAD, DAE, DAF, DAG, DAH, DAJ, DAK, DAW, DAX, V12, V13, V14, V15, V16, V17, V19, V20, V21, V22, V24, V25, ZAA, ZAB, ZAC, ZAD, ZAE, ZAF, ZAG, ZAK, ZAJ, ZAK	4
7	PAOZZ	5310007320560	96906	MS51968-14	.NUT, PLAIN, HEXAGON....................................... UOC:DAA, DAB, DAC, DAD, DAE, DAF, DAG, DAH, DAJ, DAK, DAW, DAX, V12, V13, V14, V15, V16, V17, V19, V20, V21, V22, V24, V25, V39, ZAA, ZAB, ZAC, ZAD, ZAE, ZAF, ZAG, ZAH, ZAJ, ZAK	1
8	PAOZZ	5305007173999	96906	MS90725-103	.SCREW, CAP, HEXAGON H .... UOC:DAA, DAB, DAC, DAD, DAE, DAF, DAG, DAH, DAJ, DAK, DAW, DAX, V12, V13, V14, V15, V16, V17, V19, V20, V21, V22, V24, V25, V39, ZAA, ZAB, ZAC, ZAD, ZAE, ZAF, ZAG, ZAH, ZAJ, ZAK	4

(1) ITEM NO	(2) SMR CODE	(3) NSN	(4) CAGEC	(5) PART NUMBER	(6) DESCRIPTION AND USABLE ON CODES (UOC)	(7) QTY
9	PAOZZ	2510002779786	19207	8327139	SPRING ASSEMBLY, LEA .... UOC:DAL, V18, ZAL	2
10	PAOZZ	5306004468762	19207	7017002	.BOLT, MACHINE .... UOC:DAL, V18, ZAL	1
11	PAOZZ	2510003180959	19207	8327209	.SPRING, LEAF .... UOC:DAL, V18, ZAL	1
12	PAOZZ	2510003180960	19207	8327210	.SPRING, LEAF .... UOC:DAL, V18, ZAL	1
13	PAOZZ	5310009432141	96906	MS51968-15	.NUT, PLAIN, HEXAGON .... UOC:DAL, V18, ZAL	1
14	PAOZZ	2510001918172	19207	7346962	SEAT ASSEMBLY, SPRIN	1
15	PFOZZ	4710007409607	19207	7409607	.TUBE ASSEMBLY, CROSS	1
16	PAOZZ	5310007409615	19207	7979265	.WASHER, FLAT	2
17	PAOZZ	5330007409312	78500	5X625	.FELT, MECHANICAL, PRE	4
18	PAOZZ	5330007409606	19207	7409606	.RETAINER, PACKING	2
19	PAOZZ	2590007409553	19207	7409553	.RING, WIPER	2
20	PAOZZ	5330014448350	19207	12375801	.SEAL, PLAIN ENCASED	2
21	PAOZZ	3110001004223	96906	MS19081-181	.BEARING, ROLLER, TAPE	2
22	PAOZZ	5306002258496	96906	MS90725-31	.BOLT, MACHINE	12
23	PAOZZ	5310004079566	19207	7410218	.WASHER, LOCK REAR SPRING SEAT, ... ACCESS COVER......................................	12
24	PAOZZ	5340007409391	19207	7979306	.COVER, ACCESS	2
25	PAOZZ	5330007409600	19207	7979274	.GASKET..	2
26	PAOZZ	5310003532427	19207	7979263	.NUT, PLAIN, OCTAGON	2
27	PAOZZ	5310007007089	19207	5139123	.WASHER, KEY	2
28	PAOZZ	5310003740836	19207	7979308	.NUT, PLAIN, OCTAGON	2
29	PAOZZ	5315010587268	19207	7979310	..PIN, ADJUSTMENT	1
30	XAOZZ	5310011350049	19207	7979309	..NUT, PLAIN, OCTAGON	1
31	PAOZZ	3110001004220	96906	MS19081-137	.BEARING, ROLLER, TAPE	2
32	PAOZZ	2510007979305	19207	7979305	.SEAT, SPRING, AXLE	2
33	PAOZZ	5310000609435	96906	MS35338-55	.WASHER , LOCK	4
34	PAOZZ	5306007409608	19207	7979329	.BOLT, MACHINE	4
35	PAOZZ	4730000504208	96906	MS15003-1	FITTING, LUBRICATION	2
36	PAOZZ	5310000577153	96906	MS21045-18	NUT, SELF-LOCKING, HE	8
37	PAOZZ	5306001754967	19207	10871253	BOLT, MACHINE	8

END OF FIGURE

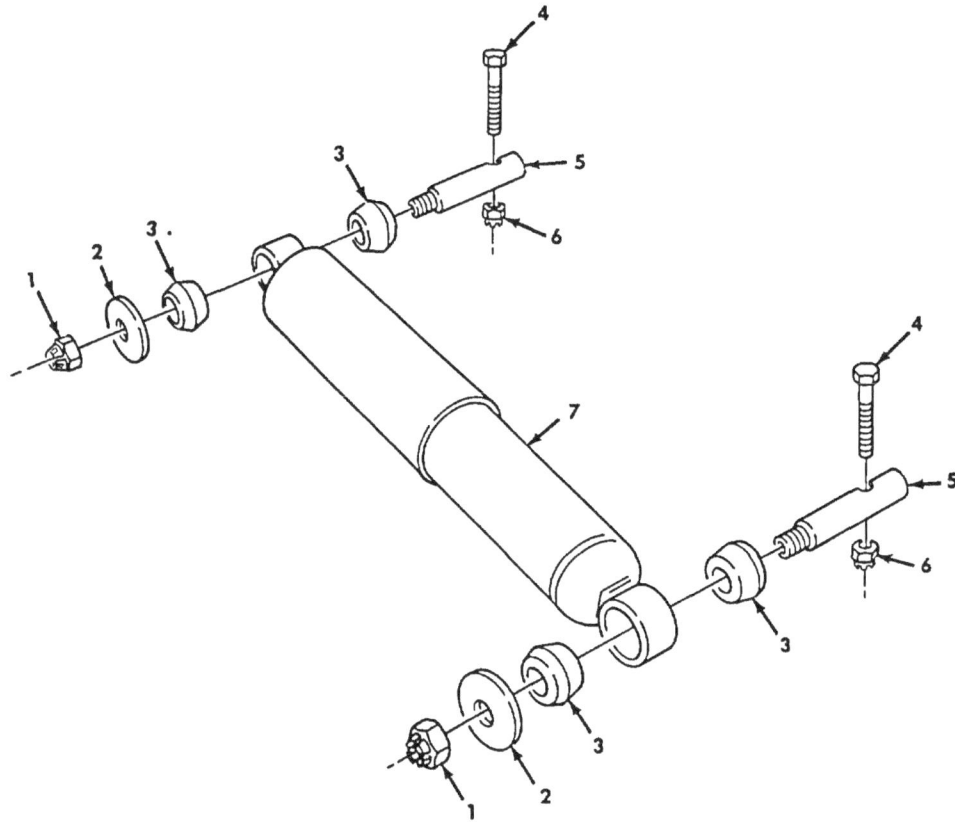

Figure 349. Shock Absorber and Mounting Hardware.

(1) ITEM NO	(2) SMR CODE	(3) NSN	(4) CAGEC	(5) PART NUMBER	(6) DESCRIPTION AND USABLE ON CODES (UOC)	(7) QTY
					GROUP 1604 SHOCK ABSORBER EQUIPMENT	
					FIG. 349 SHOCK ABSORBER AND MOUNTING HARDWARE	
1	PAOZZ	5310008329719	96906	MS51922-61	NUT, SELF-LOCKING, HE	4
2	PAOZZ	5310008098536	96906	MS27183-24	WASHER, FLAT	4
3	PAOZZ	5365007409618	19207	7409618	BUSHING, NONMETALLIC	8
4	PAOZZ	5305007195240	99696	375N300-001-21	SCREW, CAP, HEXAGON H	4
5	PAOZZ	5315007409619	19207	7409619	PIN, SHOULDER, HEADLE	4
6	PAOZZ	5310004883888	96906	MS51943-40	NUT, SELF-LOCKING, HE	4
7	PAOZZ	2510007409617	19207	7409617	SHOCK ABSORBER, DIRE	2

END OF FIGURE

2 —| 3 |

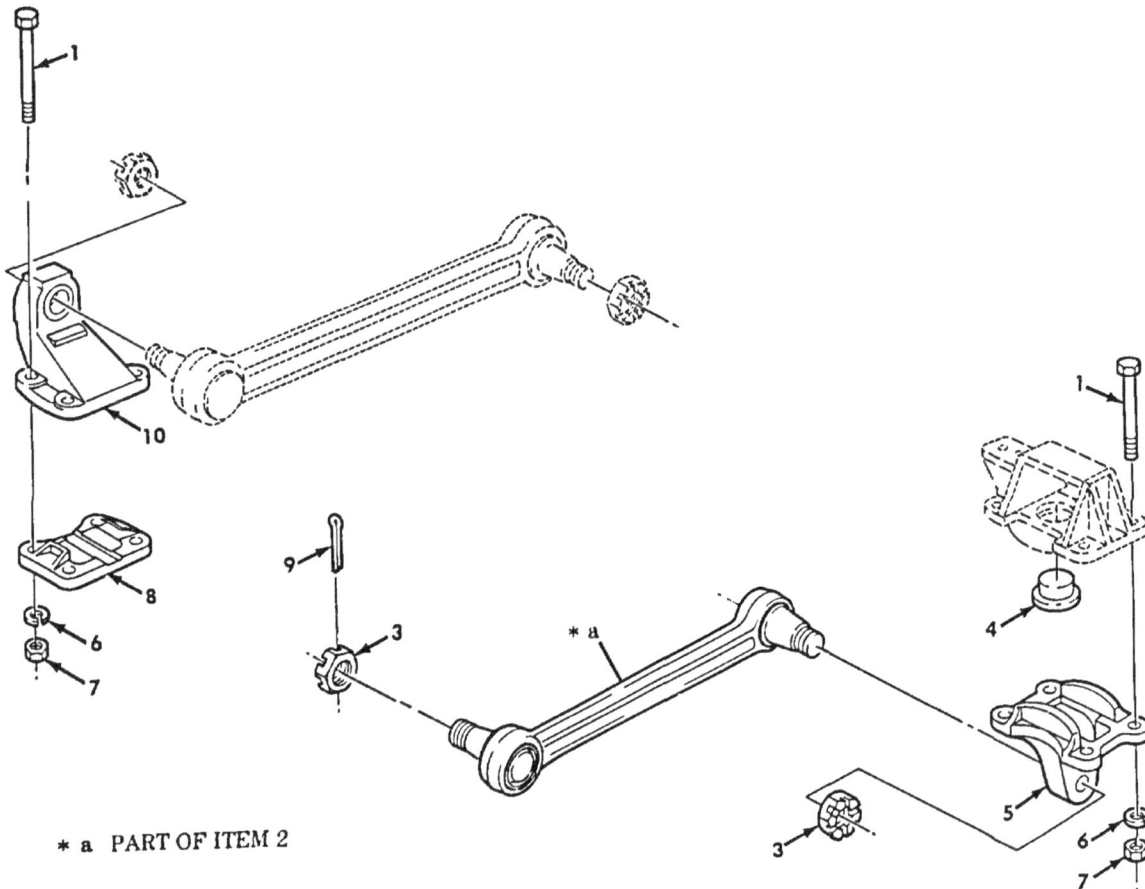

* a   PART OF ITEM 2

Figure 350. Torque Rod and Mounting Hardware.

(1) ITEM NO	(2) SMR CODE	(3) NSN	(4) CAGEC	(5) PART NUMBER	(6) DESCRIPTION AND USABLE ON CODES (UOC)	(7) QTY
					GROUP 1605 TORQUE, RADIUS, AND STABILIZER RODS	
					FIG. 350 TORQUE ROD AND MOUNTING HARDWARE	
1	PAOZZ	5306002723747	19207	7346956	BOLT, MACHINE	24
2	PAOZZ	2530009334941	19207	8728128	ROD, TORQUE .... UOC: V39	6
2	PAOZZ	2530009334941	19207	8728126	ROD, ALIGNING, VEHICU .... UOC: DAA, DAB, DAC, DAD, DAE, DAF, DAG, DAH, DAJ, DAK, DAL, DAW, DAX, V12, V13, V14, V15, V16, V17, V18, V19, V20, V21, V22, V24, V25, ZAA, ZAB, ZAC, ZAD, ZAE, ZAF, ZAG, ZAH, ZAJ, ZAK, ZAL	2
3	PAOZZ	5310007409621	19207	7979183	.NUT, PLAIN, SLOTTED, H	2
4	PAFFF	2510007346952	19207	7979272	PLATE	4
5	PFFZZ	2530007979224	19207	7979224	BRACKET, ROD, TORQUE	4
6	PAOZZ	5310007542005	96906	MS35338-52	WASHER, LOCK	24
7	PAOZZ	5310007346957	19207	7979327	NUT, PLAIN, HEXAGON	24
8	PAFZZ	2530007346954	19207	7346954	CLAMP, BRIDGE	1
9	PAOZZ	5315001879567	96906	MS24665-500	PIN, COTTER	12
10	PAOZZ	3040007346953	19207	7346953	BRACKET, EYE, NONROTA	2

END OF FIGURE

www.ingramcontent.com/pod-product-compliance
Lightning Source LLC
Chambersburg PA
CBHW080412030426
42335CB00020B/2429